Principles of Zoology

WILLIS H. JOHNSON
Wabash College

LOUIS E. DELANNEY
Ithaca College

ELIOT C. WILLIAMS
Wabash College

THOMAS A. COLE
Wabash College

Principles of Zoology

SECOND EDITION

HOLT, RINEHART AND WINSTON
New York Chicago San Francisco Atlanta
Dallas Montreal Toronto London Sydney

Library of Congress Cataloging in Publication Data
Main entry under title:

Principles of zoology.

Includes bibliographies and index.
1. Zoology. I. Johnson, Willis Hugh, 1902-
QL47.2.P74 1977 591 76-50607
ISBN 0-03-012046-2

Printed in the United States of America
890 032 9876543

CREDITS FOR PHOTOGRAPHS

Part Photographs

I: Mitochondria engulfed by lysosomes: Dr. G. E. Palade.

II: Artificial kidney machine: George W. Applegate, M.D.

III: Abnornal chromosome pattern of Down's syndrome: WHO Photo.

IV: *Conolophus subcristatus,* the Galapagos Island land iguana: Dr. E. C. Williams.

V: Portion of coral reef community: Dr. E. C. Williams.

Cover Photo: Dr. E. C. Williams

Cover Design: Louis Scardino

Four-color Section

1, 9, 10, 34, 46: Dr. E. R. Degginger
2, 3, 5, 6, 7, 8, 11, 16, 17, 18, 43: Dr. E. C. Williams
4, 23: Russ Kinne/Photo Researchers
12: R. C. Carpenter/Photo Researchers
13, 15: Ron Church/Photo Researchers
14: Dr. G. Carleton Ray/Photo Researchers
19: J. A. Hancock/Photo Researchers
20, 21: Robert Dunne/Photo Researchers
22: Betty Barford/Photo Researchers
24, 25: Carolina Biological Supply Company
26, 27, 28, 29, 30, 31, 32, 33: Systematics—Ecology Program—MBL
35, 36, 37, 38: Peter M. David
39, 42: Dennis Brokaw
40, 44: Dr. J. C. Cavender
41: Charlie Ott/Photo Researchers
45: J. B. Thurston
47: John Lewis Stage/Photo Researchers

Contents

Preface

The authors firmly believe that all students, in obtaining a liberal education, should have an opportunity to become acquainted with the life sciences. They also believe that such an opportunity can be provided in a single introductory course for students who plan to major in zoology as well as for those who will take no further work in the field. The purpose of this book is to present the basic facts and principles of animal science, to familiarize the student with the vertebrate organism with some emphasis on the human, and to show the great diversity in the animal world by means of a survey of the entire animal kingdom.

The major problem in the preparation of a beginning text in any field is the choice of materials. Since coverage of all phases of zoology is a necessary prerequisite, each phase must receive no more than its just share of the available space. The authors believe that they have struck a reasonable balance, but those who use the text may wish to modify the emphasis by omission of some sections. The organization is such that this can be done without loss of continuity. Those students who wish to delve more deeply into any specific area are guided to additional sources by references at the end of each chapter. A special group of references to *Scientific American* articles is included where appropriate.

Throughout the text, illustrations have been used with two purposes in mind. First, to augment the text and clearly demonstrate the material under discussion. In addition, the figures dealing with details of the anatomy of specific organisms will be quite helpful in laboratory studies.

We are convinced that the cell concept and the concept of organic evolution should be primary, integrating principles throughout the book. In the study of different animal groups, emphasis is placed on their anatomy, physiology, reproduction, and development. Their phylogenetic relationships and their ecological interactions with other organisms are also discussed.

The authors have tried a variety of approaches in the presentation of this material during the past forty years. We recognize the pedagogical advantage of proceeding from the known to the unknown. We have found, however, that starting immediately with the familiar vertebrate animal, without any previous study of the cell, makes the study of structure and function of whole organisms difficult and unsatisfactory. Accordingly, we are convinced that it is necessary to start with a study of cells and their chemistry. In the discussion of the chemistry of cells and in consideration of their function throughout the book, it is essential to make use of certain basic concepts of physics and chemistry. It has been our aim to present these concepts in such a way that beginning students, without any previous study of the physical sciences, can readily understand them and thus appreciate the dependence of the life sciences on the basic physical sciences.

Following the introduction which discusses the nature of the science of zoology (Chapter 1) and reasons for its study (Chapter 2), Part I deals with the Dynamics of Molecules, Cells and Tissues. Chapter 3 is concerned with the nature of matter and in particular with the constituents of living material. This paves the way for the structure of cells and tissues in Chapter 4 followed by a discussion of basic energy sources and the mechanisms whereby living cells store and utilize energy in Chapter 5. In Chapter 6, on bioinformation, the background is given for future discussions of the transfer of information between cells in an individual as well as between generations in the hereditary process. Part I ends with a chapter on cell division, both mitosis and meiosis, along with a discussion of chromosomes, the carriers of the genetic material from cell to daughter cell and from parents to their offspring.

In Part II, Problems that Organisms Face, the major emphasis is on vertebrates with the human as the main example. Comparison with other vertebrates is included where appropriate. Material included in this section involves protection, support and locomotion; nutrition; gaseous exchange and the elimination of metabolic wastes; transport and other functions of the blood including an updated discussion of immune reactions; coordination by hormones and by the nervous system; and a final chapter on learned behavior.

Part III, Continuity of Species, deals with reproduction, development, and genetics. Following chapters on reproduction and on development, classical Mendelian as well as neo-Mendelian concepts are introduced. Then, there is a treatment of the mechanisms of mutation, both chromosomal and gene, the molecular basis of phenotypic change, and the mode of gene action. The last chapter of Part III is concerned with human and applied genetics. It includes a discussion of the roles of heredity and environment, the relation of genes to disease, and the controversial topic of genetic engineering. A discussion of the application of genetic methods to agriculture and the effect of such practices on the world's food supply ends the chapter.

Part IV, Diversity and Adaptation in Animals, is a detailed study of the animal kingdom. Following a chapter on classification, 12 chapters are devoted to the invertebrate phyla. The treatment of each phylum includes a characterization of the phylum, accompanied by a diagrammatic representation of its most distinguishing features. Each class is also characterized and an illustration of a representative member of the class is shown. There is a thorough discussion of each class, using a representative member as an example. The discussion includes anatomy, physiology, reproduction and development, ecology, and phylogenetic relationships. A resumé of the taxonomy of the classes is included down to the order level. Where appropriate, important parasitic species which affect humans are treated in some detail. In connection with the study of the invertebrates, there are over 250 figures, most of them especially drawn for this text, to enable the student to gain some familiarity with these groups—many of which may be entirely new to him.

Five chapters are devoted to the vertebrates, one to the fishes and one to each of the other four classes, amphibians, reptiles, birds, and mammals. The discussion of each group includes their basic diagnostic traits and their relationships to other groups; variations in form and function within the group; structural and behavioral similarities and diversities; and important interactions with other animals including humans. A survey of the classification of each group with appropriate examples is also given. Part IV ends with a chapter on phylogeny including a phylogenetic tree.

Part V, Dynamics of Species, is concerned with some basic and unifying biological principles, utilizing the background obtained from preceding chapters. Throughout the book, evolutionary relationships have been stressed, and the first three chapters of Part IV build upon that base to present the concept of organic evolution. Chapter 40 deals with evidences of evolution from a variety of sources, including important evidence from the now generally accepted concept of continental drift. A survey of current thinking on human evolution is given in Chapter 41, followed by a chapter on the mechanisms of evolution and the origin of life. Chapter 43 contains a thorough discussion of animal behavior, including sections on innate behavior and its modification, genetic aspects of behavior patterns, the evolution of behavior in certain groups, and various types of rhythmic behavior such as circadian rhythms. Three chapters on ecology follow. The first introduces basic species ecology, the second deals with interspecies and ecosystem ecology, and the third with human ecology. The latter takes up many of the pressing environmental problems that beset society today. The final chapter surveys further problems facing humankind in the last quarter of the twentieth century and beyond. Food production, the population explosion, control of disease, and the rapid depletion of our natural resources are all related to our survival.

The study of the life sciences is ever changing and we have tried to indicate this fact wherever appropriate. That many unsolved problems remain is emphasized as is the increasing use of the experimental approach to many problems. Whereas much of the material is based on the cumulative knowledge of the past, an effort has been made to introduce significant new developments that are pertinent to an introductory course. We have also placed some emphasis on the methods used in arriving at many of the important concepts and generalizations.

In this second edition the treatment of the classes of vertebrates has been expanded from a portion of one chapter to five chapters. Behavior is now covered by two chapters instead of one, and instead of a single chapter, ecology now is covered in three chapters. Additional material on ecology has also been added to the discussions of each of the major groups of animals. New illustrations and photographs, many of the latter taken especially for this book by one of the authors, have been added to increase the clarity of presentation. There has also been a general updating of information in all chapters, reflecting the additions to our knowledge since the publication of the first edition.

Although we have tried to be accurate and up-to-date in our statements, we realize the impossibility of completely avoiding errors. Therefore, we will greatly appreciate having our attention called to any errors in fact or any questionable statements. For their helpful reviews of the manuscript, we wish to thank Harold R. Bancroft, Memphis State University; Mark C. Biedeback, California State University, Long Beach; Edward C. Brown, Holyoke Community College; William A. Elmer, Emory University; George L. Harp, Arkansas State University; Grover C. Miller, North Carolina State University; Morgan E. Sisk, Murray State University; James C. Underhill, University of Minnesota; Sandra Winicur, Indiana University at South Bend.

The encouragement, analysis, suggestions, and execution by the following members of the College Department at Holt, Rinehart and Winston—Lyn Peters, Jane Mullen, and Kendall G. Getman—are greatly appreciated.

We are especially grateful for the devoted efforts of Becky Jeffrey in the many tasks involved in preparation of the manuscript. Betty Skeps' skills as a secretary also are much appreciated.

Acknowledgements for permission to reproduce illustrations are given in the legends for the illustrations and, for the four-color plates, on the copyright page. We wish to thank the various publishing companies, the various institutions, business firms, and individuals who kindly gave these permissions.

Crawfordsville, Indiana — *W.H.J.*
*Ithaca, New York** — *L.E.D.**
August, 1976 — *E.C.W.*
T.A.C.

Principles of Zoology

1 What is Zoology

Suppose that you, the reader, were a space traveler from a more distant body approaching the Green Planet, Earth. The view from thousands of miles out in space would be that seen by the astronauts of the 1960s and early 1970s: a montage of blue, green, and white. Only upon close approach to the earth would the first contact with the organisms detailed in this text be made: an occasional high-flying bird of class Aves might be seen. A low-altitude trip around the earth would be necessary to gain some sense of the biotic world. If the space traveler had no greater sensory powers than the average current human inhabitant of the earth, then forests and the world of plants would be more impressive than the world of animals upon the first contact with this planet. Although the traveler might see herds of animals in Africa and the arctic tundra, penguins in Antarctica, and so forth, in addition to high-flying birds, the first impressions, on a purely qualitative basis, would be oriented toward the world of plants which clearly dominate the terrestrial landscape. Additionally, the traveler from outer space undoubtedly would be impressed by the contrast of great cities against the surrounding countryside, lakes, rivers, and oceans. The cities and their elaborate interconnecting sinews of superhighways would tell the visitor that some ingenious form has been able to change locally and selectively the face of the green planet and provide the great contrasts apparent between the cities and their surroundings.

The space traveler might, upon landing, decide to explore the phenomena associated with the world of animals from protozoa to humans. Such a task would involve many avenues and would call upon many resources and talents. Perhaps the facts, principles, and theories set forth in this text would simplify the overall task and provide some guidelines for understanding animals and their interactions with their environments.

LIFE ON EARTH

Although there may be life on other planets, life on the earth is the source of our direct information of life processes. Life on earth is confined to a thin shell on its surface. Life's main abundance is at the interfaces among the sky, sea, and land. At great heights the conditions are unfavorable for life; temperatures are too low, carbon dioxide and oxygen are in low concentrations, and cosmic radiation is too high. The depths of the earth are penetrated by organisms, but no one knows just how far. Eventually, light becomes a limiting factor to life in the deep and dark places of the earth. Organisms that live in these places are dependent on energy sources made available by organisms that do live in light.

The biosphere, this thin shell of sky, sea, and land, depends on the sun, directly or indirectly. Life does exist virtually everywhere in the biosphere, even in hot springs and polar regions. Where does life not exist? Certainly not in the hot ash of a volcano and probably not in the polar ice packs that have been frozen for centuries. The diagrammatic view of the biosphere in Figure 1.1 emphasizes it as a "thin slice of life" relative to the total mass of land, sea, and sky.

Life is not equally distributed over the earth. This applies to all organisms, not only the human population. *Biomass* is the term applied to the living material present in a given area. The total biomass of the earth has never been measured, but for small regions the biomass can be deter-

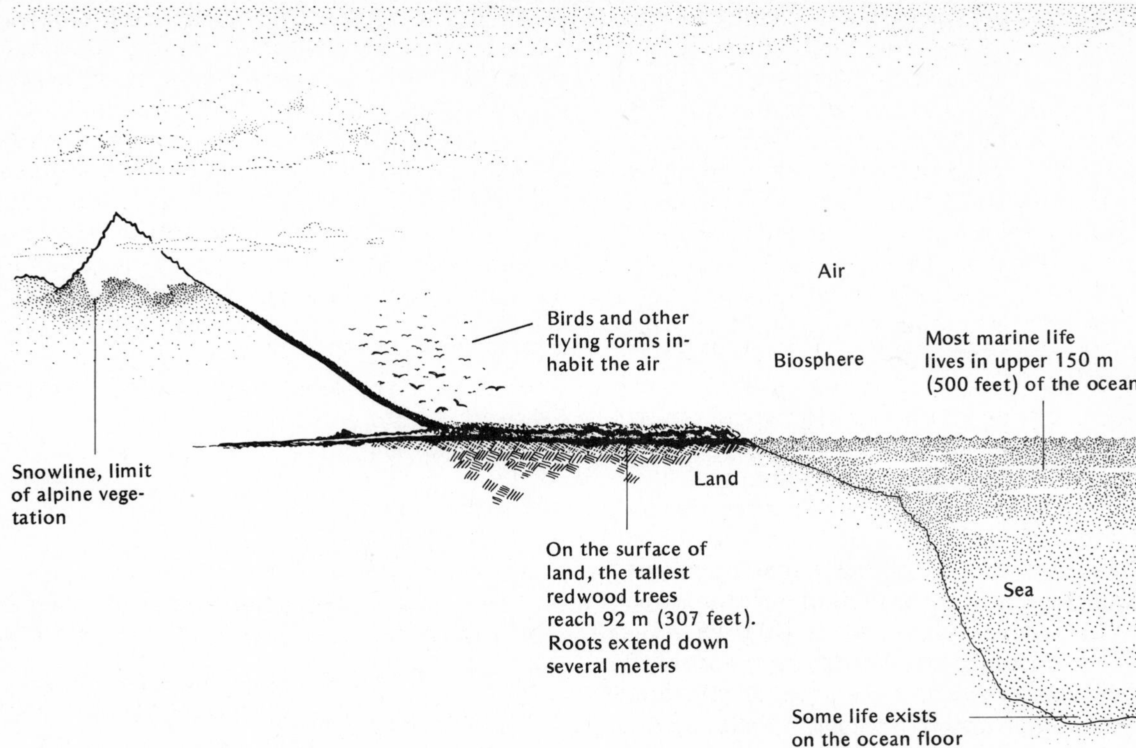

Figure 1.1 Biosphere. Living organisms are present in great numbers and diversity in a small part of the earth and its atmosphere. This has been called "the Thin Slice of Life."

mined or at least estimated. Such measurements and calculations substantiate what one would expect. Some parts of the biosphere contain and produce more biomass than others. In addition to being heterogeneous in quantity, the biomass produced in various regions is also heterogeneous in quality. The availability of light, water, oxygen, carbon dioxide, and minerals in addition to a proper temperature affects the kinds of materials produced by the living organisms of a region.

Biomass production depends fundamentally upon photosynthesis, the mechanism for conversion of light energy into chemical energy. The green plants and other organisms that carry out photosynthesis are called *autotrophs.* The *heterotrophic* organisms, mainly animals, are those that cannot synthesize their own food and must feed upon autotrophs (or other heterotrophs). A fundamental flow of energy and nutrients is thus set up in the life process on the earth. A steady supply of new energy from the sun pushes these cycles. The discipline of ecology studies the interrelationships of organisms and cycles (Chs. 44–46).

SCIENCE AND ZOOLOGY

The word *science* is derived from the Latin verb *scire,* "to know." Originally science meant a state of knowing as opposed to a state of believing. In general, science is related to knowledge which is derived from observation, study, and experimentation. Twentieth-century citizens are constantly exposed to the term science and, in general, science is highly regarded. Any individual in this modern, urban-oriented society would have a difficult time avoiding close contact with science—from the scientific information that influences political discussions of global and supraglobal importance to the "science" involved in a television or newspaper advertisement.

Since science is an inextricable part of our lives, it behooves the twentieth-century citizen to gain some understanding of it. The authors feel that a course *in* science is the best way to learn *about* science. The individual who asserts that he or she is not "going into science" is only preparing him- or herself for a rude awakening. The philosopher, writer, politician, artist, sociologist, and so forth must deal with various aspects of science. Science is creating new solutions to old problems, new methods for implementing old solutions for old problems, and also, new problems. Newspapers, magazines, the visual media, and the study of many areas of science and technology (Ch. 2) illustrate this point. A specific example is the article, "Impact of Space Research on Science and Technology" (see Newell and Jaffe, Suggestions for Further Reading).

Although many people have given definitions of science, none of the definitions has been all-inclusive and entirely satisfactory. Science includes both organized knowledge and the processes which lead to that organized knowledge. Some fields of science deal predominantly with the conservation and transmission of knowledge, while others are involved with new investigations. In the study of science, formal course work in undergraduate curricula is involved mainly with what is already known. At the graduate level, original investigation is heavily emphasized. There is no established time as to when and how the student should be introduced to the investigative aspects of science. Many scientists feel that the investigative aspects are by far the most interesting and they would like to introduce the student to this aspect of science as soon as possible. The first course in a subject necessarily involves a large body of factual information and time is often not available for original experiments. However, in the introductory course the time-honored questions are always appropriate: "How did this come about?," "What is known and not known about this topic?," "How can the unknown become known?," and so forth.

SIZES OF ANIMALS

Clearly, a distinguishing and obvious characteristic of an animal is its size. Animals range in size from about 3 micrometers (μm) (1 μm = 10^{-6} m) in length for the smallest protozoans to about 30 meters (m) (about 100 ft) for the blue whale. Thus the zoologist finds a range in size of about thirty million fold. The blue whale weighs as much as 150,000 kilograms (kg) (330,000 pounds or 165 tons), dwarfing any other animal that lives or has ever lived. A newborn blue whale is larger than a full grown elephant and is reported to consume as much as 455 kg (1000 pounds or ½ ton) of milk per day.

The techniques for studying the smallest protozoan and the largest mammal must vary greatly. Zoologists are specialized with respect to the types of organisms that they study, not only on the basis of the organism's size, but also with respect to its structure or morphology, physiology, ecology, and other attributes or "life styles."

DISTRIBUTION OF ANIMALS

Animals are distributed in various regions on the earth, called biomes. The *biome,* the largest land unit which is recognized conveniently, is characterized by a particular general type of vegetation. Some primary examples of biomes are the coniferous forest, desert, tundra, tropical rain forest, grassland, chaparral, deciduous forest, and savannah (Ch. 45). Additionally, some workers consider the sea as another biome since it is an easily recognizable unit. Although some animal species are cosmopolitan and normally live in more than one biome (protozoans of the same species may be found on every continent), the restricted distribution of most animals to a particular biome emphasizes the interrelationships of the plant and animal worlds, often in highly sophisticated and poorly understood ways.

PROBLEMS THAT ANIMALS FACE

Every animal, whether it be a protozoan or a blue whale, faces certain fundamental problems:

1. Obtaining, storing, and utilizing food
2. Exchanging gases and eliminating wastes
3. Transporting materials in their bodies
4. Coordinating their activities
5. Receiving environmental stimuli
6. Responding to environmental stimuli (behavior)
7. Supporting themselves
8. Moving (locomotion)
9. Protecting and defending themselves
10. Reproducing, developing, and maintaining their species

Parts II and III discuss these problems and some of the major solutions to them. Part IV describes the fantastic diversity found in the animal world; this diversity is a reflection of the diversity by means of which the above problems have been solved.

WHAT IS LIFE?

Zoology is a science that deals with animals and animal life, and the student of zoology should have an understanding of the meaning of the term *life.* Unfortunately, there is no good definition of life. Although one has no difficulty distinguishing between a familiar plant or animal and a rock, other distinctions may not be so obvious. For instance, after a frog has had its brain destroyed, it will die eventually. At what precise point does the frog cease to live? Directly related to this seemingly pedestrian question is the medical–legal question of human death: when is a human legally dead, and therefore, at what point may human organs be taken for transplantation? The processes that are agreed generally to be characteristic of all living organisms are metabolism, growth, reproduction, irritability, and evolution. The consideration of these characteristics illuminates the concept of life more than any single definition can.

All chemical processes that occur in living organisms are grouped under the general term *metabolism.* Metabolism which converts food into a part of the organism is constructive ("build-up") metabolism, or *anabolism.* That action by which food or body constituents are broken down with the release of energy is destructive metabolism, or *catabolism.*

The analogy between a living organism and a machine is not completely accurate, but can be useful. In a relatively complex machine (an automobile or a motorcycle, for example) fuel is burned with the release of energy. This energy is used to move the machine, and waste products are expelled in the exhaust. Fuel is not used in a machine for replacement of parts; worn parts are replaced in a separate, and often expensive, operation. In organisms food is taken in and burned to release energy, and waste products, including heat, are eliminated. However, in addition, food taken into the body is chemically processed and transformed into materials that become part of the organism.

The ratio of anabolism to catabolism in an organism is controlled by several factors: exercise (work), efficiency of the body machinery in carrying out each process, food intake, and nutritional value of the food. If the amount of food that is made a part of the organism is greater than the amount of body structure that is broken down, there is an increase in the mass of the organism. This phenomenon of *growth* of an organism, whether it be localized in one part of the anatomy or experienced by the whole organism, is different from that which occurs in the nonliving world. Growth in nonliving things such as mineral crystals results from additions to the surface, much like additional layers of snow that increase the size of a rolling snowball. On the other hand, growth in an organism involves the addition of new materials to those already present within the body. This is growth from within. Everyone can attest to the phenomenon of growth from within because they have experienced it.

The third characteristic of organisms is the potential for *reproduction.* Each living thing reproduces offspring of its own kind. The new individual may be produced from a part of the old one (asexual reproduction), or it may be produced by a union of two sex cells formed by two individuals normally of the same species (sexual reproduction).

What has been said about reproduction as a criterion of life does not always apply. In the highly organized societies of the social insects some individuals—the workers—are naturally sterile. For example, a honeybee colony consists of a queen, thousands of sterile female workers, and a small number of male drones. A queen mates only once in her lifetime of 10 to 15 years, and she lays both fertilized and unfertilized eggs. The unfertilized eggs develop into drones; the fertilized eggs develop into females. If the female *larvae* are fed a rich diet they develop into fertile queens, but if they are fed a restricted diet they develop into sterile workers. The developmental environment of these workers has negated their potential of reproduction. The same is true for humans who are sterile because of some environmental accident or some inherited malady, for example, castration caused by an accident, or sterility resulting from a hormonal imbalance.

Living things respond to certain changes in the environment; when these changes elicit a response in an organism they are called stimuli. Changes in light intensity, temperature, pressure, and the chemical nature of the air or water are examples of stimuli that elicit responses in living organisms. Some specific examples are insects flying toward a light, animals seeking shade on a hot afternoon, eyes being inflamed in an area of highly polluted air, wild deer fleeing from the hunter's foot sounds, and a dog's ability to follow the trail of a rabbit. This property of organisms to respond to stimuli or changes in the surroundings (environment) is known as *irritability,* and it is usually associated with movement. Types of animal responses are considered in Part II in some detail. Irritability enables living things to *adapt* to changes in the environment. Nonliving things also may be affected by changes in the environment, but such changes and movements are not adaptive.

The final characteristic of living organisms is *evolution,* that is, living organisms change in the characteristics that specify their forms and functions. Organisms undergo changes in their genetic material (mutate); these changes are passed on to the offspring and to succeeding generations. Such changes are the raw materials for evolution. Those mutations which provide an adaptive advantage to an organism will result in the organism contributing a greater number of offspring to the next generation than other members of the population. Thus the property of evolution is a function of populations of organisms and not of individual organisms. (An analysis of evolution is presented in Part V.)

In addition to the five functional characteristics, living organisms are characterized by their structure (body forms), their organization, and their composition. The structure of each organism is no haphazard array of materials. The organism is put together in quite a precise manner, and deviations from the precise order often result in severe incapacitation or death. This text discusses the organization of various organisms at several levels; indeed, there is a system of progressive organization called a *hierarchy of organization.* This hierarchy runs from *atoms* and *molecules* to *cells,* the units of structure and function of organisms. Cells are organized into *tissues,* which are groups of like cells that are structurally modified to perform a particular function. Tissues are combined into *organs,* and a number of organs functioning together constitute an *organ system.*

An organism, then, is composed of a certain number of organ systems that are integrated in the structural and functional organization of the individual. The hierarchy of organization of living things can be extended to the *supraorganismic* level. Organisms affect and are affected by each other; the relationships among organisms depend on both physical and biotic factors. Organisms of the same kind constitute *populations;* several specific populations living in the same environment are called *communities.* The community with its nonliving environment constitutes an *ecosystem.* As mentioned earlier the large, easily recognizable regions of the world (such as grasslands, coniferous forests, and deserts) are called biomes. This text emphasizes this hierarchy of organization from atom to biome.

WHAT IS BIOLOGY? WHAT IS ZOOLOGY?

Biology is the science of living things. The word biology is derived from the Greek *bios,* meaning "life," and *logos,* "the study of." Historically, knowledge about living things was developed somewhat independently by students of plants and by students of animals. As a result, many biologists think of two main subdivisions of biology—*botany,* the study of plants, and *zoology,* the study of animals. Other biologists feel that there are really three types of organisms—plants, animals, and microorganisms—and consider *microbiology* to be a major subdivision of biology. Table 1.1 gives some areas of study according to the taxonomic-based division of biology. This list is not comprehensive, but it gives some idea of the "natural areas" that have arisen according to the classification scheme. Included in Table 1.1 are the origins of these terms. This way of subdividing biology is sometimes called the *vertical method.*

The other major method of subdividing biology into disciplines is based on an operational or functional approach which cuts across taxonomic lines. Some of these are listed in Table 1.2. This type of scheme is sometimes called the *horizontal method* of organization. Since the scope of biology has practically no boundaries, many of the important advances, particularly in the past few years, have been made by workers who defy categorization into a particular branch of biology. How does one classify an individual who studies conditions for the origin of life on the primitive earth—biochemist, evolutionist, microbiologist, philosopher, or jack-of-all-trades?

Table 1.1 The Vertical Method of Subdividing Biology

DISCIPLINE	WORD ORIGIN	MEANING OF PREFIX	THE STUDY OF:	
Anthropology	Gk. *anthrōpos*	Man	Natural history of man	
Mammalogy	L. *mamma*	Breast	Mammals	
Ornithology	Gk. *ornis*	Bird	Birds	
Herpetology	Gk. *herpeton*	Reptile	Reptiles	Zoology
Ichthyology	Gk. *ichthys*	Fish	Fishes	(Gk. *zōon,* animal)
Entomology	Gk. *entomon*	Insect	Insects	
Helminthology	Gk. *helmins*	Worms	Parasitic worms	
Protozoology	Gk. *prōtos*	First	Protozoa	
	zōon	Animal		
Bacteriology	Gk. *baktērion*	Small rod	Bacteria	
Virology	L. *virus*	Poison	Viruses	Microbiology
Phycology	Gk. *phykos*	Seaweed	Algae	
(Algology	L. *alga*	Seaweed	Algae)	Botany
Mycology	Gk. *mykēs*	Fungus	Fungi	(Gk. *botanē,*
Bryology	Gk. *bryon*	Moss	Mosses and liverworts	pasture) or
Pteridology	Gk. *pteris*	Fern	Ferns	Phytology

Table 1.2 The Horizontal Method of Subdividing Biology

DISCIPLINE	WORD ORIGIN	MEANING OF PREFIX	THE STUDY OF:
Morphology	Gk. *morphē*	Form	Form and structure
Anatomy	Gk. *ana*	Up	Organisms as determined by dissection
	tomē	Cutting	
Histology	Gk. *histos*	Tissue	Tissues by microscopy
Cytology	Gk. *kytos*	Hollow vessel	Cells
Physiology	Gk. *physis*	Nature	Functions and activities of cells and organisms
Taxonomy	Gk. *taxis*	Arrangement	Classification
Ecology	Gk. *oikos*	Household	Relationships between organisms and their environment
Genetics	Gk. *genesis*	Descent	Heredity and variation
Embryology	Gk. *embryon*	Embryo	Formation and development of organisms
Paleontology	Gk. *palaios*	Ancient	Fossils and fossil impressions
	on	Being	
Evolution	L. *evolvere*	To unroll	Descent of species
Zoogeography	Gk. *zōon*	Animal	Distribution of animals
	gē	Earth	
	graphein	To write	
Phytogeography	Gk. *phyton*	Plant	Distribution of plants
	gē	Earth	
	graphein	To write	
Biochemistry	Gk. *bios*	Life	Chemistry of organisms
	chemeia	Transmutation	
Biophysics	Gk. *bios*	Life	Biological phenomena in terms of physical principles
	physis	Nature	
Immunology	L. *immunis*	Free	Resistance of organisms to infection
Radiobiology	L. *radius*	Ray	Effects of radioactivity on biological material
	Gk. *bios*	Life	
Endocrinology	Gk. *endon*	Within	Study of hormones and their effects
	krinein	To separate	
Psychobiology	Gk. *psychē*	Soul	Related areas of psychology and biology
Limnology	Gk. *limne*	Lake or marsh	Freshwater lakes, ponds, and streams

Because workers have become specialized in various aspects of biology, many investigations are carried on by *team research* in which zoologists, botanists, microbiologists, chemists, physicists, physicians, and mathematicians collaborate.

With the advent of many new research techniques and a resultant explosion of knowledge, there has been a tendency to divide biology into a new set of horizontal categories consisting of *molecular biology, cellular biology, developmental biology, organismal biology,* and *population* and *community biology.* This method of division will be called *modified horizontal* because the categories are more inclusive than those in the horizontal method of division. Several college biology curricula have been organized along the lines of the modified horizontal method. Molecular biology encompasses biochemistry and biophysics, and presumably can be said to include all aspects of biology which take a molecular approach to the problems and their solution. Cellular biology simply may include all approaches to structure and function of cells. Included are the chemical and physical organization, energetics, transport, mobility and stabilizing mechanisms of cells, growth and division, differentiation, and development. Developmental biology is much broader than the traditional embryology; it includes development from the molecular level to gross structural levels. The phenomena of regeneration, wound repair, and aging are included in developmental biology.

It is obvious that there is considerable overlap among molecular biology, cellular biology, and developmental biology. All of them involve investigations at the molecular level, and all involve the flow of information in biological systems; the term *molecular genetics* is often used to refer to these aspects of biology.

Organismal biology focuses on whole organisms. It is concerned with such matters as functional and developmental anatomy, phylogeny, comparative physiology, psychobiology, and ecology. Population and community biology involve a part of traditional ecology. In this field the primary concern is with the structure and dynamics of populations, communities, ecosystems, and with the ecological aspects of natural selection.

The three methods of dividing biology—vertical, horizontal, and modified horizontal—do not delineate many aspects of applied biology. Biology interacts with the space sciences, earth sciences, mathematics, physical sciences, social sciences, and humanities. Among these applied fields are animal husbandry, plant breeding, medicine, agriculture, pharmacology, conservation, horticulture, bioclimatology, and many others.

CONCEPTS OF ZOOLOGY

This introductory chapter has touched on several areas of zoology as a discipline; the vastness and complexity of the discipline are apparent. It is reasonable for one to wonder at the outset if the concepts of zoology can be stated in a brief outline. These concepts should be fundamental ideas that serve as the skeleton of zoology. The following twelve concepts would be suggested by most professional zoologists.

1. Living systems obey the laws of chemistry and physics.
2. All organisms are composed of cells and cell products (the cell concept).
3. Living cells are able to convert energy from one form to another.
4. Metabolism is catalyzed by enzymes which are under genetic control.
5. Many vitamins are related structurally to coenzymes.
6. Hormones regulate many cellular and organismal activities.
7. As a result of evolution, living systems are organized, structurally and functionally, from atom to biome to carry out a variety of specific functions.
8. Control or regulatory mechanisms exist at all levels of organization.
9. Life comes only from living things by reproduction (biogenesis rather than abiogenesis).
10. DNA is the genetic material (RNA serves this role in certain viruses), and genetic potential is transmitted from generation to generation through genes (gene theory).
11. The primary factor in organic evolution is the number of offspring which survive and become parents of the next generation.
12. Organisms interact with their biological and physical environments.

The remainder of this chapter is devoted to the exploration of each of these concepts in anticipation of the remaining chapters.

Concept 1 The rapid increase in understanding many aspects of zoology has led to the idea that biological systems obey the laws of chemistry and physics without invoking a vitalistic force. Some basic chemistry and physics are discussed in Chapter 3. Other aspects of chemistry and physics are discussed in chapters where they are related particularly to the subject matter.

Concept 2 The cell concept (all organisms are composed of cells and cell products) was formulated by Schleiden and Schwann, a botanist and a zoologist, respectively, in 1838 and 1839. Cells are discussed specifically in Chapter 4 as seen by the light microscope and as revealed by the electron microscope and many other modern techniques. As special functions are taken up in many other chapters, the characteristics of the specialized cells which perform the specialized functions are explored.

Concept 3 Living organisms are able to convert one form of energy to another. Cells which carry out photosynthesis convert light energy into chemical energy. All cells carry out some type of food breakdown. In these processes the chemical energy stored in the food is released with the production of heat, waste products, and energy. The energy is often released in the form of adenosine triphosphate, ATP. ATP can be used to drive most biological processes: cell division, transport across membranes, bioluminescence, synthesis of many compounds by the organism, muscular contraction, and nervous transmission, to name a few. The major concepts involved in biological conversion of energy from one form to another are found in Chapter 5, Bioenergetics. Energy conversion in biological systems is pertinent to many other processes, and the energetic considerations of these processes are taken up in the appropriate chapters.

Concept 4 The metabolic basis of living systems has had an interesting history. The concepts of enzymology, the study of biological catalysts, are

1. All enzymes are proteins or mainly proteins.
2. Enzymes are specific in function.
3. One region of an enzyme, its *active site,* is crucial to the enzyme's activity. The substrate (s) is (are) bound to the three-dimensional active site during the enzyme's activity.
4. The function of an enzyme, like that of all catalysts, is to lower the energy barrier to the reaction.

The fundamentals of enzyme action are set forth in Chapter 3. Enzymes, like other proteins, are under genetic control. This means that each protein is coded by a segment of nucleic acid. A change in a segment of nucleic acid, a gene, results in a corresponding change in the protein product. Such a change in a gene is called a mutation (see Concept 10).

Concept 5 Most vitamins have been shown to be related to coenzymes which are nonprotein but essential parts of certain enzymes. Coenzymes as parts of the biological catalysts, enzymes, are required in extremely small amounts for normal functioning. Thus vitamins are required in small amounts. The role of vitamins in animal metabolism is discussed in Chapter 7.

Concept 6 Hormones are chemical messengers which regulate many cellular activities in plants and animals. The concept of a hormone is somewhat different for plants and animals. In vertebrate animals most hormones travel in the circulatory system and regulate such diverse processes as the sugar level of the blood, constriction of blood vessels, secretions of various sorts, development of sex characteristics, and growth processes. The nature and functions of these chemical messengers and their coordinating activities are taken up in Chapter 12.

Concept 7 As mentioned earlier in the Section, What Is Life?, biological systems are organized along a hierarchy from atom to biome. At each level of this hierarchy of organization there is a close relationship between the form of each structure and the function that each structure performs. This concept is discussed throughout the text.

Concept 8 At each level of organization there are control mechanisms which regulate the functions carried out by biological systems. There are metabolic, genetic, developmental, hormonal, nervous, reproductive, population, and behavioral regulatory mechanisms. These controls are discussed in a variety of places throughout the text.

Concept 9 As mentioned earlier, life itself is elusive of a precise definition, but the general characteristics of living things are understandable (growth, metabolism, reproduction, irritability, and evolution). Life originated about 3.1 to 3.5 billion years ago. Most biologists think that life is not originating spontaneously today. Similarly, many biologists think that there was one spontaneous generation of life on the earth. At one time it was thought that abiogenesis, the spontaneous origin of living things, did occur repeatedly. Around 1680 Francesco Redi showed that maggots could not arise spontaneously from decaying meat. About 200 years later Louis Pasteur proved that microorganisms likewise did not arise spontaneously.

Concept 10 The discipline of molecular genetics has shown clearly that nucleic acid is the genetic material in living systems. The evidence that biological information is stored in the nucleic acids is strong. The functions of the genetic material are

1. Duplication and transmission of the same set of bioinformation to daughter cells in cell division.
2. Control of metabolism and differentiation of cells and organisms by differential expression of units of biological information.
3. The ability to change and transmit these changes of information to future generations. A change of inheritable biological information is called a mutation.

The transmission and expression of genetic information are explored in Chapters 6, 17, 18, as well as other chapters (see Concept 4).

Concept 11 The concept of organic evolution is accepted as a major premise of biology by most biologists. Although many aspects of the general theory of evolution are found in the writings from the Greek philosophers before the Christian period and other writers from the Middle Ages to the 1800s, it was the work of Alfred Wallace and Charles Darwin that welded the data into a cohesive theory. Darwin published his theory of natural selection in 1859, and a controversy over it immediately arose. The theory of evolution by natural selection says that species do not reach their reproductive potential since many offspring fail to survive, but among the offspring, those that have favorable variations will survive in relatively greater numbers and will produce a greater percentage of the next generation. Thus the primary factor in organic evolution is the number of surviving offspring which become the parents of the next generation. Evolution is specifically discussed in Chapters 40–42, but related and supporting material is taken up in many chapters of this text.

Concept 12 The study of the interaction of organisms with their surroundings, physical and biological, is studied in the disciplines of physiology, behavior, and ecology. No citizen in a technological society can be unaware of the principles and concepts of ecology. In this text three chapters, Chapters 44–46, are devoted to specific aspects of basic and applied ecology. Organisms are not randomly distributed about the planet earth, but are associated in communities where they are interdependent as producers, consumers, and decomposers. Like evolution, the discipline of ecology is related to essentially all areas of biology. Physiology is studied in Part II and behavior is taken up in Chapters 14 and 43.

This list of a dozen concepts or fundamental principles of zoology is not an exhaustive list. Most zoologists would agree with the importance of each of these concepts, but would state some in a different manner or add certain others. Each student should keep these twelve concepts in mind and try to formulate his or her own list of concepts and principles during the course of study. The authors feel that this list of twelve concepts constitutes the principles of zoology and have attempted to weave them into the fabric of this text.

SUGGESTIONS FOR FURTHER READING

Bonner, J. T. *The Ideas of Biology.* New York: Harper and Row (1962).

Newell, H. E., and L. Jaffe. "Impact of Space Research on Science and Technology." *Science,* **157,** 29–39 (1967).

Von Bertalanffy, L. *Problems of Life: An Evaluation of Modern Biological and Scientific Thought.* New York: Harper and Row, (1960).

From Scientific American

Hutchinson, G. E. "The Biosphere" (Sept. 1970).

Mausner, B., and J. Mausner. "A Study of the Anti-Scientific Attitude" (Feb. 1955).

2 Why Study Zoology

Another title for this chaper might be, "Zoology and the Future of the Individual." What specific benefits should an individual expect from studying introductory college zoology?

One cannot peruse a newspaper or magazine without seeing several articles that relate to the fundamental principles of zoology—environmental pollution, population problems, organ transplantation, birth control, antibiotics, various kinds of disease, and agricultural problems and progress to mention a few. Therefore it is valid to ask "Why Study Zoology?" Students who use this text may have different interests, different reasons for studying zoology, and different goals in college and postgraduate life. Even though a student's interests and goals may seem entirely unrelated to zoology, the effects of zoological processes are so widely distributed in every phase of life that no student can fail to associate many of the aspects of these processes with himself.

The following are answers to the above question.

Zoology is related directly to one's own existence. A person's body is made of the basic units of structure found in other organisms and it carries out the same basic functions. He or she starts life as an undeveloped form, grows as do other living entities, and arrives at his or her mature morphological form by similar processes. That humans have the same basic units of structure and function as other organisms can be appreciated by studying a variety of animals as well as the human species.

During a study of zoology the student can discover that although the various organisms are alike in their essential processes and functions, they differ considerably in form and nutritional requirements. These differences are the result of adaptations to many kinds of habitats and many ways of obtaining food. No one organism or one kind of organism is independent of other kinds; rather, all kinds of organisms are interdependent in a series of nutritive linkages (predator, prey, parasite). If this were not so, life on this planet would have ceased long ago.

The study of zoology can reveal that the many different kinds of organisms can be placed into several large groups, the members of each group being obviously much alike and distinctly related. However, no one group is isolated entirely from any other. This intergroup relationship marks the first step toward understanding the principle of *organic evolution.* Probably no other biological generalization has had more effect on our thinking than has the idea of evolution, which was so clearly but incompletely presented by Charles Darwin in the nineteenth century. The ability to understand and discuss the validity and implications of this concept requires a rather broad, sound base in zoology.

Humans are social organisms; they live in groups. Zoology is associated closely with two disciplines that study human behavior: psychology and sociology. In attempting to understand the behavior and relationships of humans, psychologists and sociologists gather information from observation and experimentation. Some understanding of the human nervous system, glands, and other organs is necessary for the study of psychology. Some knowledge of the laws of inheritance, for example, is a prerequisite for studies in sociology.

Many other concepts of zoology have direct and practical applications in everyday life. Medicine, dentistry, agriculture, conservation of natural resources, public health, all of which are integral parts of human civilization, are based on zoological knowledge and the methods of science.

Indirectly, zoological concepts must also be treated in other professional activities. For example, consider a writer who is constructing a lengthy novel of social criticism of the twentieth century. Such a work might involve discussions of germ warfare, radioactive fallout and its contamination of the biosphere, sperm banks and artificial insemination, control of the development of individuals, thought control, and psychomimetic drugs. The novel would be dull and would fail to carry its message if the author did not understand the fundamental zoology involved in these problems. Neither could twentieth-century poets, legislators, government employees, artists, financiers, historians, and dramatists be realistic without such knowledge.

THE METHODS OF SCIENCE

The facts and principles discussed in this book have been determined by hundreds of biologists, zoologists, and botanists working through the years. It is customary to say that such facts and principles have been obtained by the scientific method. But what is the scientific method? Actually, there is no single method. At various places in this text accounts are given of the methods used in arriving at some of the great generalizations of biology. There is no single set of procedures that have been used or that can be used in making every scientific discovery. It is, however, possible to indicate a number of procedures and attitudes that are used in establishing scientific facts and principles. One of the primary procedures is careful *observation.*

All observations cannot be direct. One cannot see an enzyme work, but one can measure the results of its action. One cannot see materials transported across a membrane, but one can measure the amounts of compounds after they have passed across it. The records of observations, direct or indirect, are called data. The observations and collection of data lead to a specific statement of the problem. This statement (often involving "creative guesswork") is an *hypothesis.* An hypothesis must then be subjected to further testing. If it holds up under further testing, it may become a *theory.* A theory may become a *law* or a *principle* after it has been tested on a very wide scale.

Observations, experiments, and theories, then, must fit the established patterns of the few principles—or the principles must be revised and restated to accommodate new evidence. The principle of *causality* must always be satisfied. This principle asserts that a phenomenon occurring repeatedly under identical conditions will always have identical results, regardless of where, when, or by whom it is repeated.

The testing referred to here in connection with a theory or a principle is made by many people in all parts of the world. *Repeatability is the essence of science.* No generalization becomes an accepted principle unless it stands the test of a wide variety of investigators. Many of the great biological generalizations have resulted from this procedure of observation, analysis, and conclusion. The cell principle, discussed in the next chapter, was developed in this way. Every student in zoology who examines the tissues of many kinds of organisms in laboratory studies and finds them all composed of cells is engaged in the scientific method and is once again verifying the very important fact of the existence of cells in organisms.

But modern biology makes use of more than the primary procedure of observations. In much of the work in modern biology the *experimental method* is used. Many of the facts to be presented here pertaining to the functions, development, heredity, and other aspects of living organisms have been determined by the experimental method. After an hypothesis has been established, certain predictions may be made from it. Then an experiment may be devised to test a particular hypothesis. Such an experiment should be a *controlled experiment.* In one part of the experiment, the *control,* all conditions are kept constant; in the other part, the *experimental,* all but one of the conditions are kept constant as in the control; this single condition is varied.

The controlled experiment may be illustrated as follows. After many observations and experiments it has been determined that the compound ascorbic acid, commonly called vitamin C, must be present in the human diet to prevent *scurvy,* a disorder characterized by bleeding gums, swollen joints, and general weakness. In a similar way it has been established that the guinea pig has the same requirement. From these facts the hypothesis might be made that ascorbic acid is required by all mammals. From this it would follow that ascorbic acid is required by the rat. An experiment to test this would be set up with litter-mate rats from a pure stock divided into two groups. The two groups would be fed the same complete and balanced diet except that there would be no ascorbic acid in one diet. The group that received ascorbic acid in its diet would be the control group. At the end of a few weeks it would be possible to determine the effects of the lack of ascorbic acid on the experimental group. Actually, in experimenting with the rats, no difference would be found between the experimental and control groups, and therefore the hypothesis that ascorbic acid is required in the diet of *all* mammals would have to be abandoned. The reason for this difference is that the rat, unlike most mammals, synthesizes ascorbic acid from other compounds in its diet.

One trial of an experiment is not enough. The experiment must be repeated a number of times until the experimenter is quite certain that the results warrant the conclusions. (The experiment to test the ascorbic acid requirements of rats has been repeated many times.) *Statistics,* a branch of mathematics used for testing data, is then applied to the large number of trials that have been performed. Statistics, however, never prove an hypothesis; they only indicate whether the results do or do not support the hypothesis.

Scientists usually have an objective point of view. True scientists strive to be honest, not to be influenced by bias or emotion, nor to be swayed by the influence of authority. They base conclusions on facts and must be ready to change their points of view when the data no longer support them.

The researcher's design of an experiment varies greatly with the type of research that is undertaken. Some experiments may be designed to answer one question; a good example is the vitamin C experiment described above. Some "experiments" are really many simple experiments done in parallel, with each set up testing a different variable.

Much has been written on experimental design. One of the most interesting pieces along these lines was written in 1890 by Thomas C. Chamberlin (1853–1928), a geologist who was president of the University of Wisconsin and the American Association for the Advancement of Science in 1908.[1]

Chamberlin suggested that researchers adopt the *method of multiple working hypotheses.* This method involves the making of several hypotheses so that the investigator will not be able to "fasten his affections" on any one hypothesis. Thus, presumably, the investigator is on neutral ground and will test all the hypotheses. This promotes thoroughness in research and exploration of lines of inquiry that would have otherwise been overlooked.

[1] Chamberlin's article, "The Method of Multiple Working Hypotheses," originally appeared in *Science* (old series), **15,** 92 (1890). It was reprinted in *Science,* **148,** 754 (1965). Additionally, it is available from the Institute for Humane Studies, 1175 University Drive, Menlo Park, Calif. 94025.

MEASUREMENT IN SCIENCE AND ZOOLOGY

Scientists go to great pains to obtain accurate measurements of the phenomena they study. Sometimes some measurements are made directly: mass, length, time, volume, and so forth. In other cases even a simple measurement cannot be made directly. What is the volume of a muscle fiber? What is the length of a nerve cell? What is the time between divisions of a human liver cell? Many techniques may be needed to determine a particular quantity. But why measure?

Even those who abhor the tedious task of measurement will realize, upon reflection, that they rely greatly on yardsticks for all sorts of events. Imagine a weather report without a thermometer reading. Tomorrow might be reported as warmer, windier, and more humid than today, but how warm, windy, and humid was today? With no real data, fact gives way to opinion, and the opinion of the persuasive individual may pervade. Just how long was the fish that got away?

If a result can be presented numerically, it can be transmitted to others with a precise meaning. Further, measurement provides a basis for comparison, and therefore it provides a better understanding of the subject. Another objective of measurement is to provide verifications of theories or support for one of a group of theories. A fourth objective of measurement is the evaluation of progress. Music, art, and the understanding of history change, but the subjective evaluations of these topics are assessed as progress by some and as regression by others. There are runners who are able to run a mile faster than ever before—there are data to support that fact. But what evidence is there that wrestlers are better than those of previous generations? The difference is in the measurement.

There is the possibility that the right thing is not being measured. The mere recording of data does not always measure the progress that is being made. Nevertheless, measurement is the best weapon against prejudiced opinion.

WHAT ZOOLOGISTS DO

Not only is it impossible to describe the physical, social, and mental characteristics of a scientist, but it is also impossible to define limits that encompass the job of a zoologist. The trained zoologist may be an experimentalist who works a regular day in the laboratory, alone or assisted by technicians or research personnel. Or a zoologist may be active in research from the design of the experiment, through the actual manipulations involved, to the interpretation of the data. Some zoologists design experiments and interpret the results, leaving the experimental work to technicians or other associates. Thus there is a range in which individuals contribute to the experimental process.

Since the field of zoology is so vast and since its subdivisions have no clearly defined boundaries, biologists are faced with the problem of keeping abreast of recent developments in their "speciality," as well as those in biology in general. New reports flow unceasingly and fill the several hundred journals that publish papers on biology. Some people are very active readers of this current literature; others attend the periodic meetings of their professional societies; some learn most about what is new in biology by listening to their friends or by reading review articles.

The kinds of things that zoologists do are varied. One can make a list of the services and products that he or she uses routinely; how many need the advice of a zoologist during their rendering or production? Put the list at the end of this book and review it after you have read the entire text, and then see how many more items you can add.

INTERACTIONS OF ZOOLOGY

Most persons who have reached college age are aware of the many problems facing the peoples of the world—population, pollution, pesticides, disease, endangered species, famine, poverty, and on and on. Most of the serious problems of the world must be solved in terms of biology. What can a college student contribute to the solutions of these problems? Probably the greatest contribution is to sharpen one's mind for actively solving the problem during the many postgraduate years that lie ahead. But there are things to be done even at the undergraduate level. One can be active in local environmental and conservation clubs. Many colleges have programs in which students may serve the surrounding community. These programs include helping in preschool nursery projects, youth group work, senior citizen problems, and so on.

What does the study of zoology have to do with solving the world's problems? The next 45 chapters represent a partial answer. As another approach to an answer, list specifically the five major problems that you feel most urgently need solutions. Then during the reading of this text determine if there are zoological components in the origins of these problems. Chapters 44–47 will touch upon some of these points in detail.

Some of the "smaller" problems merit some attention. The following is a list of independent student projects that make for worthwhile study.

1. What are the major events in the history of whaling and what are the prospects for the survival of whales?
2. Can society and the grizzly bear coexist?
3. How does the study of zoology relate to the energy crisis?
4. Of what value are fossil beds?
5. What are the ways to alleviate thermal pollution?
6. What are the advantages and disadvantages of vasectomy as a means of birth control.
7. How does the study of zoology relate to an understanding of human genetics?

These and many other topics have a wealth of resource material. Independent study is good training and good practice. Such study has a liberating quality. It permits a student to be one of the architects of his or her own education. Disciplined, analytical thinking is good practice.

In addition, there are many other areas that have direct and practical applications in everyday life. The fields of medicine, dentistry, agriculture, conservation of natural resources, and public health, all of which are integral parts of our civilization, are based on zoological knowledge and the methods of science.

In sum, no matter what the area of specialization, every student will find the study of zoology valuable, not only in its personal application as a member of the animal kingdom, but as a source of understanding in almost any future field of work or study.

THE LANGUAGE OF ZOOLOGY

Introductory courses are often packed full of new material which is immersed in a vocabulary of its own. Essentially, a new language must be learned. The language of zoology depends heavily on Greek and Latin roots. Many of these roots occur several times in the basic zoological vocabulary. It may be of help to make a list of these new words during the reading of each chapter and determine the meaning of the roots. The precise meaning of many elaborate terms may be deduced from a rather modest dictionary of roots. The school library should have some dictionaries of zoology and science that will aid in determining the meaning of those words which are not in the glossary of this text or in a desk dictionary.

Some examples are in order. In Table 1.2 *ecology* is shown to be derived from the Greek word *oikos* which means "house or household." In its simplest statement, ecology is the study of how the household of nature is kept in order. The same Greek word gives rise to the term economics, *-nom-* being derived from a word which means "management." Thus economics is related to the management of a household (state, business, government, community, and so forth). *Oikos* may be seen in other biological words: consider *monoecious* and *dioecious*. These are terms which refer to the sexes of animals. A monoecious animal has both male and female sex organs in one individual; the word is derived directly from the Greek as one house: *monos,* single; *oikos,* house. On the other hand, dioecious animals have separate sexes, two houses.

The language of zoology is extensive because zoology is so diverse and because zoology is related to so many major disciplines—mathematics, physics, chemistry, geology, psychology, and so on. This text has several chapters on topics that were not considered pertinent to introductory zoology courses a few years ago. Chapter 3 on biomolecules involves some physics and chemistry with the biology. Chapter 4 considers information about cells that has been gained mainly with the electron microscope. Chapters 5 and 6 on bioenergetics and bioinformation contain much information that has been gained in recent years with the newer techniques of molecular biology. The emphasis on the human as a biological example at many levels reflects the expanding scientific knowledge about human beings. Chapter 46, Human Ecology, is an addition to this edition of the text.

The problem of the extensive vocabulary introduced in a beginning zoology course has not gone unnoticed by the authors. The text includes an extensive index and a carefully selected glossary. Additionally, new words are defined where they are first used in the text. Nevertheless, the vocabulary of zoology is large, and some students encounter difficulty because of this. A good way to minimize this problem is to use the vocabulary in describing or illustrating the principles of zoology, rather than concentrating on sheer memorization of the vocabulary.

D.E. Green

Garrett Hardin

H. E. Huxley

H. B. D. Kettlewell

Konrad Lorenz

Eugene P. Odum

Figure 2.1 Some contemporary zoologists.

SUGGESTIONS FOR FURTHER READING

Baker, J. J. W., and G. E. Allen. *Hypothesis, Prediction, and Implication in Biology.* Reading, Mass.: Addison-Wesley (1968).

Bonner, J. T. *The Ideas of Biology.* New York: Harper and Row (1962).

Chamberlin, T. C. *Multiple Working Hypotheses.* Menlo Park, Calif.: Inst. for Humane Studies (1965).

Gabriel, M. L., and S. Fogel. *Great Experiments in Biology.* Englewood Cliffs, N.J.: Prentice-Hall (1955).

Glass, B. *The Timely and the Timeless: The Interrelationships of Science, Education, and Society.* New York: Basic Books (1970).

Gray, P. *Student Dictionary of Biology.* New York: Van Nostrand Reinhold (1972).

Grobman, A. B., Ed. *The Social Implications of Biological Education.* Washington, D.C.: Nat. Assoc. of Biology Teachers (1970).

Snow, C. P. *The Two Cultures and the Scientific Revolution.* London: Cambridge Univ. Press (1959).

Snow, C. P. *The Two Cultures: And a Second Look.* London: Cambridge Univ. Press (1964).

From Scientific American

Boehm, G. A. W., H. B. Goodrich, and R. H. Knapp. "The Origins of U.S. Scientists" (July 1951).

Butterfield, H. "The Scientific Revolution" (Sept. 1960).

Eiseley, L. C. "Charles Darwin (Feb. 1956).

Stent, G. S. "Prematurity and Uniqueness in Scientific Discovery" (Dec. 1972).

Terman, L. "Are Scientists Different?" (Jan. 1958).

Dynamics of Molecules, Cells, and Tissues

The first part of this text of zoology begins the hierarchy of organization found in living systems. Atoms, molecules, organelles, cells, and tissues are considered along the lines of the principle of correlation of structure and function. Chapter 3 considers the fundamentals of atomic structure, periodic relationships among the elements, bonding and valence, and biomolecular structure. There are four major classes of organic (compounds of carbon) biomolecules: proteins, carbohydrates, acetogenins, and nucleic acids.

The cell is the fundamental unit of life; it is the unit of structure, function, heredity, and pathology. Chapter 4 discusses how cells are organized into tissues and includes information derived from electron microscopy as well as other techniques used to explore the organization and variety of cells and their arrangement into tissues.

Living systems require energy at all levels of the hierarchy. Chapter 5 on bioenergetics provides the first discussion of energy transformation and storage in living systems. The animal world is dependent on photosynthetic prokaryotes and plants as its ultimate source of energy, the sun. The dynamic relationship of plants and animals in an energetic and nutritional interaction is seen at the levels of organisms, populations, communities, and ecosystems in subsequent parts. Understanding of energy flow at these levels requires an understanding of bioenergetic principles at the cellular level and their relationships to ATP synthesis and utilization.

The proteins and nucleic acids, as informational macromolecules, come to center stage in Chapter 6, Bioinformation. The hereditary material of all cells, DNA, is a major control point in cellular dynamics. DNA is replicated faithfully in the cell cycle which produces new body cells and in meiosis which produces sex cells. It controls cellular events by being differentially expressed: some genes are turned off while others are turned on. In general principle, differential gene activity results in cellular specialization and, as a result, tissue and organ differentiation. The genetic material is also responsible for change by mutation. These genetic changes have profound implications for the well-being of the organism, the continuity of the species, and the makeup of communities, ecosystems, and biomes.

The concluding chapter in this part presents information about chromosomes, their structure, and their movements in mitosis and meiosis. These topics are useful as an integrated theme in which biomolecules, cellular structure, bioenergetics, and bioinformation are shown to contribute to cellular dynamics.

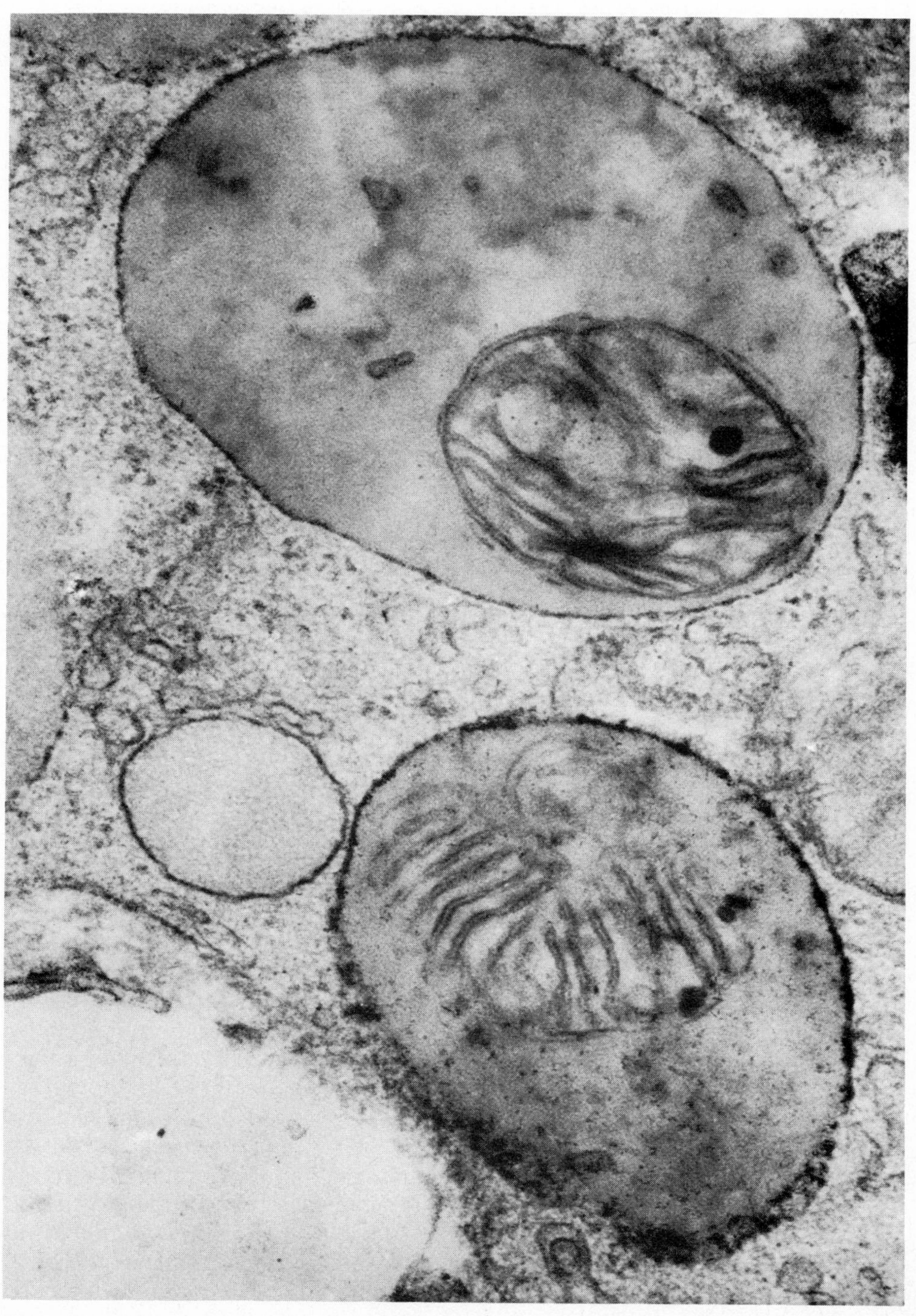

Lysosome containing mitochondrion (Courtesy of Dr. George E. Palade, Yale University Medical School).

3 Biomolecules

Living cells and organisms are made up of many kinds of molecules. These molecules, upon isolation, conform to the same laws of physics and chemistry that apply to lifeless matter. The living organism exhibits properties in addition to those described by the laws of physics and chemistry. Thus these *biomolecules,* the molecules characteristically found in living material, exhibit remarkable properties when they are arranged in complex and highly organized living matter. This highly organized state is maintained by a constant supply of energy which must be provided for it. Each component of a living organism appears to have a specific function; this is true for large organs, such as the liver and brain, down to the individual molecules, such as nucleic acids, proteins, acetogenins, carbohydrates, and water. These molecules interact with each other in what has been called the molecular logic of the living state (see Lehninger, Suggestions for Further Reading). One purpose of Part I is to explore some aspects of this logic, that is, to study the structures and functions of biomolecules and how they interact in forming the physical and chemical basis for living processes.

PHYSICAL AND CHEMICAL PROPERTIES OF ATOMS AND MOLECULES

If living material is composed of intrinsically lifeless molecules, why is living matter so different from lifeless matter? Put another way, why does the living organism appear to be more than the sum of its inanimate parts? To answer these questions are the goals of those biologists who study biomolecules. Some understanding of the physics and chemistry of atoms and molecules is desirable before we explore the biomolecules themselves. The beautiful order and logic of atomic and molecular structure is reflected in the structures of the fantastically complex biomolecules of living organisms.

The Atomic Structure of Matter

All matter presumably consists of small, fundamental entities (atoms) that retain their identities in chemical changes. There are 92 different elements (types of unique atoms) found in nature; these atoms range from hydrogen, the smallest, to uranium, the largest (Fig. 3.1). Additionally, 13 other kinds of atoms have been produced by bombardment in nuclear reactors; these "man-made" atoms are known as the transuranium elements. Atoms are composed of many kinds of smaller particles, but this discussion considers only *protons, electrons,* and *neutrons.* The proton carries a positive unit charge and is about 1850 times as heavy as the electron, which carries a negative unit charge. The neutron is electrically neutral and has about the same mass as the proton. The neutrons and protons are located in the nucleus, the center of the atom; thus most of the mass of an atom is concentrated in the nucleus. The electrons travel around the nucleus in definitive spatial volumes. Since an atom contains the same number of protons and electrons, it is electrically neutral. The nucleus with its orbiting electrons has been compared to the solar system of a central sun and its orbiting planets. Like the solar system, the atom is mostly space.

Two atoms that have the same number of protons (and electrons) but a different number of neutrons are still the same element—only different forms of it. The *atomic number* of an element is the number of protons in the nucleus.

Figure 3.1 Periodic table of the elements. The chemical properties of the elements vary across the periods, while the elements in the same family have similar chemical properties.

Key: Atomic weight 6.939; Atomic number 3; Symbol Li; Name Lithium

IA	IIA	IIIA	IVA	VA	VIA	VIIA	VIII			IB	IIB	IIIB	IVB	VB	VIB	VIIB	Zero
																H 1.00797 1 Hydrogen	He 4.0026 2 Helium
Li 6.939 3 Lithium	Be 9.012 4 Beryllium											B 10.811 5 Boron	C 12.01 6 Carbon	N 14.00 7 Nitrogen	O 15.999 8 Oxygen	F 18.998 9 Fluorine	Ne 20.183 10 Neon
Na 22.99 11 Sodium	Mg 24.31 12 Magnesium											Al 26.98 13 Aluminum	Si 28.08 14 Silicon	P 30.97 15 Phosphorus	S 32.06 16 Sulfur	Cl 35.45 17 Chlorine	Ar 39.95 18 Argon
K 39.10 19 Potassium	Ca 40.08 20 Calcium	Sc 44.96 21 Scandium	Ti 47.91 22 Titanium	V 50.94 23 Vanadium	Cr 52.00 24 Chromium	Mn 54.94 25 Manganese	Fe 55.85 26 Iron	Co 58.93 27 Cobalt	Ni 58.71 28 Nickel	Cu 63.54 29 Copper	Zn 65.37 30 Zinc	Ga 69.72 31 Gallium	Ge 72.59 32 Germanium	As 74.92 32 Arsenic	Se 78.96 34 Selenium	Br 79.91 35 Bromine	Kr 83.80 36 Krypton
Rb 85.47 37 Rubidium	Sr 87.62 38 Strontium	Y 88.91 39 Yttrium	Zr 91.22 40 Zirconium	Nb 92.91 41 Niobium	Mo 95.94 42 Molybde num	Tc 43 Technetium	Ru 101.1 44 Ruthenium	Rh 102.9 45 Rhodium	Pd 106.4 46 Palladium	Ag 107.9 47 Silver	Cd 112.4 48 Cadmium	In 114.8 49 Indium	Sn 118.7 50 Tin	Sb 121.7 51 Antimony	Te 127.6 52 Tellurium	I 126.9 53 Iodine	Xe 131.3 54 Xenon
Cs 132.9 55 Cesium	Ba 137.3 56 Barium	(see below) 57–71	Hf 178.5 72 Hafnium	Ta 180.9 73 Tantalum	W 183.8 74 Tungsten	Re 186.2 75 Rhenium	Os 190.2 76 Osmium	Ir 192.2 77 Iridium	Pt 195.1 78 Platinum	Au 197.0 79 Gold	Hg 200.6 80 Mercury	Tl 204.4 81 Thallium	Pb 20.72 82 Lead	Bi 209.0 83 Bismuth	Po 84 Polonium	At 85 Astatine	Rn 86 Radon
Fr 87 Francium	Ra 88 Radium	(see below) 89–103															

Lanthanides	La 139.9 57 Lanthanum	Ce 140.1 58 Cerium	Pr 140.9 59 Praseody mium	Nd 144.2 60 Neodymium	Pm 61 Promethium	Sm 150.4 62 Samarium	Eu 151.9 63 Europium	Gd 157.3 64 Godolinium	Tb 158.9 65 Terbium	Dy 162.5 66 Dysprosium	Ho 164.9 67 Holmium	Er 167.3 68 Erbium	Tm 168.9 69 Thulium	Yb 173.0 70 Ytterbium	Lu 175.0 71 Lutetium
Actinides	Ac 89 Actinium	Th 232.0 90 Thorium	Pa 91 Protactinium	U 238.0 92 Uranium	Np 93 Neptunium	Pu 94 Plutonium	Am 95 Americium	Cm 96 Curium	Bk 97 Berkelium	Cf 98 Californium	Es 99 Einsteinium	Fm 100 Fermium	Md 101 Mendelev ium	102	Lw 103 Lawrencium

Many of the elements are composed of more than one kind of atom; these atoms differ in the number of neutrons in their nuclei. These forms of the same element (since they contain the same number of protons) are called *isotopes* because they occupy the same place (Gk. *iso,* same; *tope,* place) in the periodic table of elements. Thus isotopes have the same atomic number but have different *mass numbers,* the sum of protons and neutrons. Atomic masses do not influence chemical reactions in a striking way, so isotopes are essentially equivalent in chemical reactions. Certain isotopes are unstable and emit particles or rays at a constant rate. Such isotopes are *radioactive* and have been used in many important experiments in biology. The radioactive isotopes of hydrogen (^{3}H, also called tritium), carbon (^{14}C), phosphorus (^{32}P), sulfur (^{35}S), and sodium (^{22}Na) are among those used often.

The Electronic Structure of the Atom

Hydrogen is the simplest element; it contains one proton and one electron. A shorthand way of describing the hydrogen atom is $^{1}_{1}H$, the subscript denoting the atomic number (protons) and the superscript denoting the mass number (protons and neutrons). The sodium atom is described as $^{23}_{11}Na$, which means that there are 11 protons and electrons and 12 (23 − 11) neutrons in the atom.

One may think of the electrons distributed in space to form diffuse clouds. Electrons travel in orbitals within the atom. An orbital is the volume in which an electron might be found. Orbitals have different shapes and sizes; furthermore, they have different energy levels. The shapes of some orbitals are shown in Figure 3.2.

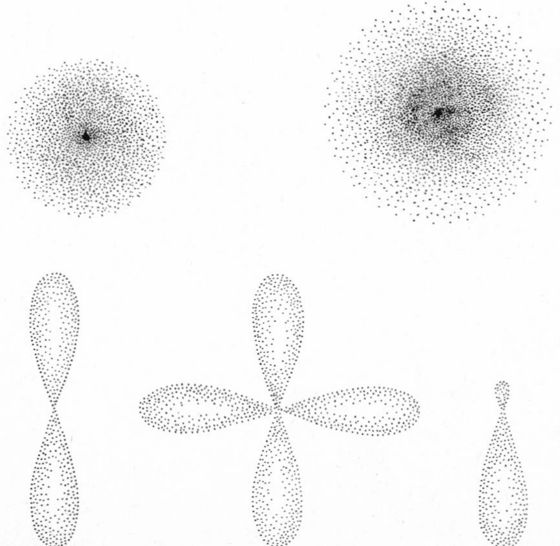

Figure 3.2 Shapes of some atomic orbitals.

Orbitals may be divided simply into two classes, atomic and molecular. As the names suggest, an atomic orbital encompasses one nucleus. Since a molecule is made up of two or more atoms, a molecular orbital encompasses two or more nuclei. The hydrogen atom H and the hydrogen molecule H_2 are used here for illustration. The hydrogen molecule consists of two one-proton nuclei and two electrons that travel around both nuclei. Because orbitals occupy a significant volume, compared to electrons themselves, the different orbitals provide the molecule with its shape or geometry.

Valence, Bonds, and Molecules

The noble gases (He, Ne, Ar, Kr, Xe, Rn) seldom react with other atoms because their outer shells are filled with electrons. Such filled shells, as indicated by their positions at the ends of rows in the periodic table, make them stable elements. When sodium (atomic sodium gas) and chlorine (atomic chlorine gas) are brought together, each sodium atom loses one electron to a chlorine atom. The result of this transfer is that the sodium atom attains the stable electronic configuration of the noble gas neon and is a positive sodium ion. In the process the chlorine atom attains the electron configuration of another noble gas, argon, and becomes the negative chloride ion. Thus an ion is an atom (or a molecule) that has lost or gained electrons and that possesses a charge. Positive and negative ions attract each other. The force of attraction between oppositely charged ions constitutes an *electrovalent,* or *ionic,* bond. The *valence* of an ion is the difference between the number of protons and electrons resulting from the attainment of a stable electronic configuration by loss or addition of electrons. For example, the sodium atom contains 11 protons and electrons, and loss of an electron produces a positive sodium ion with a valence of 1.

As just described, the redistribution of electrons can lead to chemical bonding. The redistribution of electrons does not need to be so extreme as in the formation of ionic bonds (the shift of an electron or electrons from the influence of one nucleus to another). There are shifts that result in the sharing of one or more electrons between two atoms: The force of such an arrangement of shared electrons is another type of bond, the *covalent bond.* As a rule, electrons are shared in pairs, and often covalent bonds are referred to as electron-pair bonds. The concept of valence is pertinent to covalent bonding also.

Here the valence of the atom is the number of electrons it shares with another atom. Carbon shares four of its electrons, generally, and therefore has a valence of 4.

Some atoms, notably fluorine, oxygen, and nitrogen, have electrons in "exposed positions" of outer orbitals that are not bonded to other atoms. Hydrogen atoms attached to other atoms or molecules may interact with such exposed, unshared electrons. This interaction is the *hydrogen bond.* This bond is much weaker than a covalent bond, but when a number of hydrogen bonds exist between two molecules, the energy required to separate them is greater than the bond energies of the individual hydrogen bonds. This phenomenon is called *cooperative hydrogen bonding.* Hydrogen bonding among water molecules is shown in Figure 3.3. Hydrogen bonding between water molecules accounts for the higher boiling point of water than would be predicted by the boiling points of similar compounds. Hydrogen bonding to oxygen and nitrogen is particularly important in the structure of proteins and nucleic acids; these complex organic compounds are discussed later in this chapter.

This discussion of the structure of atoms and molecules should serve as a short review of chemical geometry and bonding and as an indication of the complexity of the elaborate biological molecules, made up of thousands of atoms. The properties of the complex biomolecules rest, ultimately, in the properties of the various atoms and how they are bonded together in the molecules.

THE CONSTITUENTS OF LIVING MATERIAL

Elements Found in Living Material

One characteristic of living material is that it is selective with respect to its environment. The earth's crust is made mainly of oxygen, silicon, aluminum, sodium, calcium, iron, magnesium, and potassium; the remaining elements constitute less than one percent of the entire earth's crust. Living organisms are made mostly of hydrogen, oxygen, carbon, and nitrogen; in fact, 99 percent of many cells is made of these four elements. All of these four readily form covalent bonds by electron sharing. Hydrogen needs one electron, oxygen two, nitrogen three, and carbon four to complete their outer shells and form stable compounds.

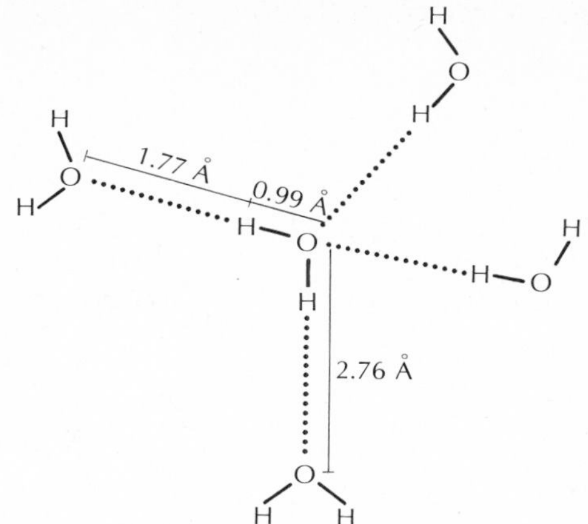

Figure 3.3 Hydrogen bonding around a water molecule in ice.

Considerable amounts of phosphorus (P), potassium (K), sulfur (S), chlorine (Cl), magnesium (Mg), calcium (Ca), sodium (Na), and iron (Fe) are also found in living organisms. The following elements are found in smaller amounts in organisms: copper (Cu), cobalt (Co), manganese (Mn), fluorine (F), vanadium (V), molybdenum (Mo), zinc (Zn), boron (B), silicon (Si), iodine (I), selenium (Se), tin (Sn), chromium (Cr), and aluminum (Al). On the average only 25 of the elements found in the earth's crust are essential components in living organisms.

Compounds Found in Living Material

The chemical elements just referred to are rarely found in cells as the element, but rather they occur in compounds and ions. The compounds are of two kinds, inorganic and organic. Traditionally organic compounds have been called "compounds of carbon." The four common classes of organic compounds found in cells are carbohydrates, acetogenins, proteins, and nucleic acids. Until 1828 most people believed that organic compounds could be formed only by living organisms and could not be synthesized in the laboratory. However, in 1828 the German chemist Friedrich Wöhler succeeded in synthesizing an organic compound, urea [$(NH_2)_2CO$], a waste product of many organisms. Each year many new organic compounds are made in laboratories over the world. Of the several million known chemical compounds, the majority are organic. Of these, many are not made naturally by living organisms but are made only by "test-tube" reactions. Many of the organic compounds are highly complex molecules. Logically, inorganic compounds are all the other "nonorganic" ones. These include water, salts, acids, bases, and gases which are discussed next.

INORGANIC CONSTITUENTS

Water Water is the most abundant compound in the vast majority of cells. The water content of cells of different organisms does vary widely; some cells contain little water while in a dormant state, for example, and others are mainly water. It may seem inconceivable that a living organism could be 96 percent water, but that is the case in some jellyfishes.

All cellular reactions occur in solution or on the surface of membranes that are in contact with an aqueous medium. Thus a cell must always be in direct contact with water in the environment. The more complex organisms may have a water-impervious coat, but their cells have contact with an internal aqueous environment. The importance of water is emphasized by the fact that the lack of water will have disastrous effects on an organism long before the lack of food.

Water is actually an unusual compound and several properties make it an important cellular component and environmental factor for organisms:

WATER AS A SOLVENT Water is the best solvent known; more substances dissolve in water than in any other solvent. Since it is a polar compound, and it attracts the charges on molecules, thereby weakening the attraction between them, they dissolve. The ability of certain compounds to form hydrogen bonds with water is the basis for their solubility in water.

WATER AS A METABOLITE Water itself enters into many chemical reactions. It is used in photosynthesis and is produced in respiration (see Ch. 5). In many reactions a molecule of water is removed from the reactants; this type of reaction is a synthesis. In many other reactions a molecule of water is added in the process of breaking a molecule into smaller ones. This type of reaction is called an *hydrolysis*. These reactions are represented schematically below and are discussed later in this chapter.

$$AH + BOH \underset{\text{hydrolysis}}{\overset{\text{synthesis}}{\rightleftarrows}} AB + HOH$$

WATER AS A TEMPERATURE STABILIZER The high *heat capacity* (number of *calories* required to raise the temperature of 1 gram of substance by 1 degree Celsius) of water is of great protective significance to organisms. The heat capacity of water is 1 calorie per gram (cal/g), while the heat capacities of most organic compounds are around 0.5 cal/g. Even if the air temperature surrounding an organism changes rapidly, the temperature of the organism changes rather slowly because of its water content. The high *heat of vaporization of water* (585 cal are required to vaporize 1 g of water at 20° C; 540 cal/g at 100° C) is an important factor in temperature regulation. The heats of vaporization for most organic compounds are less than 100 cal/g. Water is lost as water vapor by both plants and animals. Large amounts of heat are dissipated by these processes. If this were not the case, a rise in temperature would result. Since water is also a good conductor of heat, the heat absorbed by water is readily distributed, and this tends to equalize the temperature in cells and organisms. The heat of fusion (the amount of heat that must be removed from 1 g of a substance on freezing) is about 80 cal/g for water. Most organic compounds have a heat of fusion about one half of this value. Thus water freezes less readily than most liquids, and this high heat of fusion of water also contributes to temperature stabilization.

SURFACE TENSION OF WATER The strong attraction of water molecules for each other through hydrogen bonding results in water's high surface tension, the highest of all liquids. If water is placed in a capillary tube, it will rise to a point where the strength of the capillary action is held in check by the pull of gravity on the water in the tube. These forces help the blood to complete its flow through the body, even though the heart supplies a pumping force.

DENSITY OF WATER Water has its greatest density at +4° C. Thus as water cools below this temperature it tends to rise to the surface. As a result, lakes and ponds freeze at the top first. Many organisms survive the winter months beneath a layer of ice. If ice were heavier than liquid water, then bodies of water would freeze from the bottom and the bottom-dwelling organisms would be frozen first.

The boiling point of water, 100° C, and the freezing point of water, 0° C, necessarily limit the range of active life. Actually, the upper temperature limit of life is seen in hot springs. Some algae can live at temperatures around 75° C (167° F). The bacteria of hot springs survive at even higher temperatures. At the other end of the temperature scale, metabolism slows down greatly at temperatures below the freezing point, but many organisms tolerate subzero temperatures and resume metabolism when the temperature rises above the freezing point.

Salts, Acids, and Bases Numerous compounds are dissolved in the aqueous matrix of cells. Some compounds dissociate into ions when they go into solution; these compounds are called *electrolytes.* (Similarly, a compound that does not dissociate is a *nonelectrolyte.*) When the electrolyte sodium chloride dissolves, it dissociates into positively charged sodium ions and negatively charged chloride ions. This dissociation may be written as follows:

$$\text{NaCl} \rightleftarrows \underset{\text{sodium ion}}{\text{Na}^+} + \underset{\text{chloride ion}}{\text{Cl}^-}$$

Most inorganic compounds in cells are dissociated into their component ions.

Compounds that produce hydrogen ions (H^+) or protons upon ionization are *acids;* compounds that accept or react with protons are the *bases,* or *alkalis,* for example, the hydroxyl ion (OH^-). Compounds such as NaCl, which upon ionization produce positive and negative ions, none of which are hydrogen or hydroxyl ions, are *salts.* The ionization of various inorganic electrolytes in the aqueous matrix of the cell endows the cell with the important property of electrical conductivity. If pure water is used for the conduction of electricity, very little current will flow. But if an electrolyte such as NaCl is added to the water, good conduction results. The aqueous matrix of the cell is a good conductor of electricity. The property of *irritability* of living cells and organisms is based partly on this electrical conductivity.

The number of the hydrogen and hydroxyl ions in a solution determines its acidity, alkalinity, or neutrality. Water dissociates slightly and is neutral, since equal numbers of hydrogen and hydroxyl ions are produced. A logarithmic scale, the pH scale, has been adopted for the expression of hydrogen-ion concentration. The pH of a solution is defined as the negative logarithm of the hydrogen-ion concentration; thus the pH of pure water is $-\log(10^{-7})$, or 7.0. A pH of 7 indicates a neutral solution in which the numbers of hydrogen and hydroxyl ions are equal. A pH value less than 7 indicates an excess of hydrogen ions, or an acidic solution. A pH value greater than 7 indicates an excess of hydroxyl ions, or an alkaline solution. Note that the *higher* the value of the pH, the *lower* the value of the hydrogen-ion concentration. (In the logarithmic pH scale, a difference of 1 pH unit is a 10-fold difference in hydrogen-ion concentration, a difference of 2 pH units is a 100-fold difference, and so forth.)

A cell is a delicately organized system whose pH is usually near neutrality. Any great deviation from a neutral pH will result in the death of most cells. Chemical systems that function in the maintenance of pH are called *buffers.* Phosphate and carbonate ions are important buffers in cells and organisms. The major extracellular buffer system is the carbonate system. Carbonate ions will react with hydrogen ions in the following manner:

$$H^+ + \underset{\text{carbonate ion}}{CO_3^{--}} \rightleftarrows \underset{\text{bicarbonate ion}}{HCO_3^-}$$

Also, the bicarbonate ion may react with a hydrogen ion and form carbonic acid:

$$H^+ + HCO_3^- \rightleftarrows \underset{\text{carbonic acid}}{H_2CO_3}$$

Thus the removal of excess hydrogen ions tends to keep the pH constant. If hydroxyl ions are introduced, they will be neutralized by hydrogen ions produced by the dissociation of carbonic acid:

$$H_2CO_3 \rightarrow H^+ + HCO_3^-$$
$$H^+ + NaOH \rightarrow H_2O + Na^+$$

Hence,

$$H_2CO_3 + NaOH \rightarrow H_2O + NaHCO_3$$

Thus the carbonate–bicarbonate buffer system provides a mechanism for neutralizing excess hydrogen or hydroxyl ions. The carbonate–bicarbonate system buffers effectively around pH 7.0.

The phosphate system ($H_2PO_4^-$, HPO_4^{--}) has its maximum buffering action around pH 7.2. Body tissues and fluids are buffered by amino acids and proteins as well as other inorganic buffer systems.

Certain kinds and concentrations of ions are necessary for the normal functioning of cells. If a frog's heart is placed in distilled water, it will soon cease beating because water molecules enter and swell the cells. Prevention of such swelling depends on the proper *osmotic pressure* (see Ch. 4) of the bathing solution. If a sugar solution of the proper osmotic pressure is used with the frog's heart, it will beat a longer time. If a solution of sodium chloride of the proper osmotic pressure is used, the heart will beat considerably longer. But if a solution of the proper osmotic pressure and containing the proper proportions of sodium, potassium, calcium, magnesium, and chloride ions is used, the heart may beat for days.

Gases The gases of the air, oxygen (O_2), nitrogen (N_2), and carbon dioxide (CO_2), are soluble in water. Thus they are found in cells. Oxygen is essential for the metabolism of most cells, and its importance is considered in some detail in a later section. Gaseous nitrogen, normally an inert substance, is found in cells, but it is not involved in the metabolism of most cells. Carbon dioxide is found in all cells as a result of metabolism (respiration) and is used by green plants in photosynthesis. Carbon dioxide reacts with water to form carbonic acid,

$$H_2O + CO_2 \rightleftarrows \underset{\text{carbonic acid}}{H_2CO_3}$$

This reaction is a part of the important carbonate–bicarbonate buffer system (see above discussion of buffers).

ORGANIC CONSTITUENTS

As indicated earlier, the organic compounds are compounds of carbon. Since carbon has a valence of 4, it can form four covalent bonds (in most instances) to other atoms. Carbon-containing compounds are numerous, since long chains and rings of carbon atoms are possible. The four major classes of organic compounds of living material are *carbohydrates, acetogenins, proteins* and *nucleic acids.*

Carbohydrates The carbohydrates are compounds of carbon, hydrogen, and oxygen in which the proportion of hydrogen and oxygen is usually the same as in water, 2:1. Sugars, starches, and cellulose are common carbohydrates. Many carbohydrates also contain nitrogen, sulfur, or phosphorus.

Carbohydrates play three major roles in living organisms:

1. As energy sources and storage forms of energy
2. As a source of carbon in the synthesis of other cellular components
3. As structural elements of cells and tissues

A predominant carbohydrate found in cells is glucose. Glucose is a *monosaccharide,* a simple sugar; its formula is $C_6H_{12}O_6$. Two other simple sugars found in cells have the same formula; these are *galactose* and *fructose.* Although the formula shows the same number of each kind of atom in these three sugars, they are different compounds because of the differences in *arrangement* of the atoms. Fructose, galactose, and glucose are isomers (Gk. *iso,* same; *meros,* parts); the differences are shown in the structural formulas in Figure 3.4. The use of structural formulas is important because they give some idea of the spatial relationships of the atoms in a molecule. (Sugars with six carbon atoms, such as glucose, galactose, and fructose, are known as *hexose* sugars; five carbon atoms in a sugar make it a *pentose.*)

Glucose is important as a basic fuel in cells, but it is also important in the synthesis of more complex carbohydrates. When two simple sugars combine, they form a disaccharide. Maltose is produced by the union of two molecules of glucose with the loss of a molecule of water (Fig. 3.5). The reverse reaction, the breakdown of maltose to two molecules of glucose, is a hy-

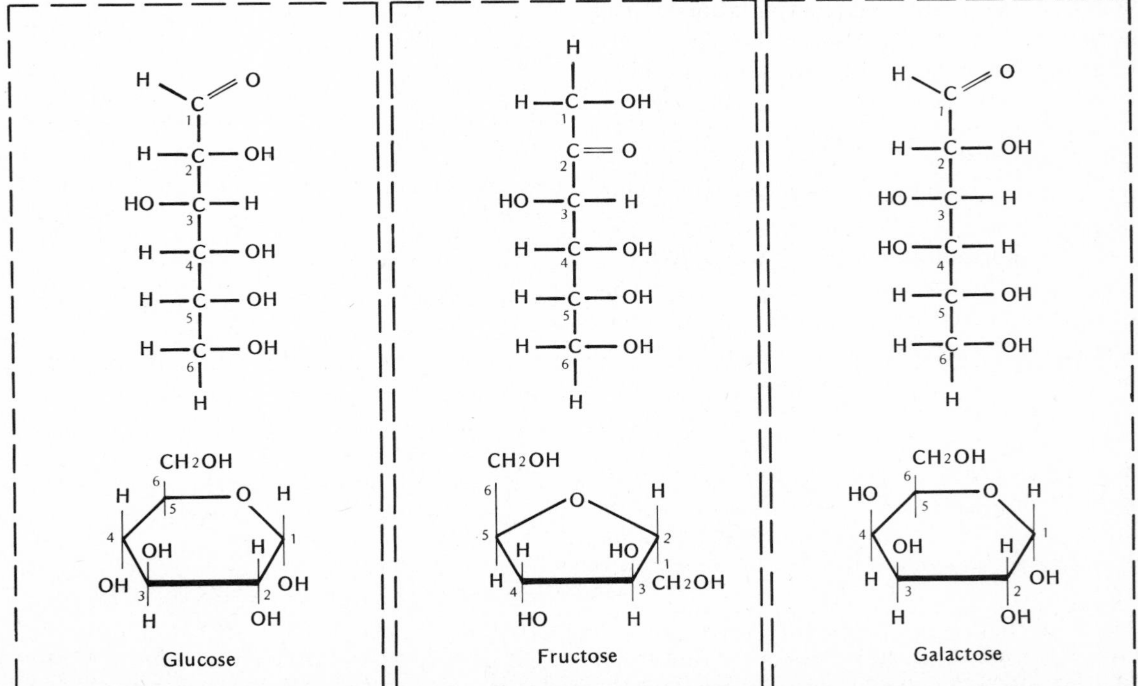

Figure 3.4 Structural formulas of glucose, fructose, and galactose. Both straight-chain and ring structures are shown.

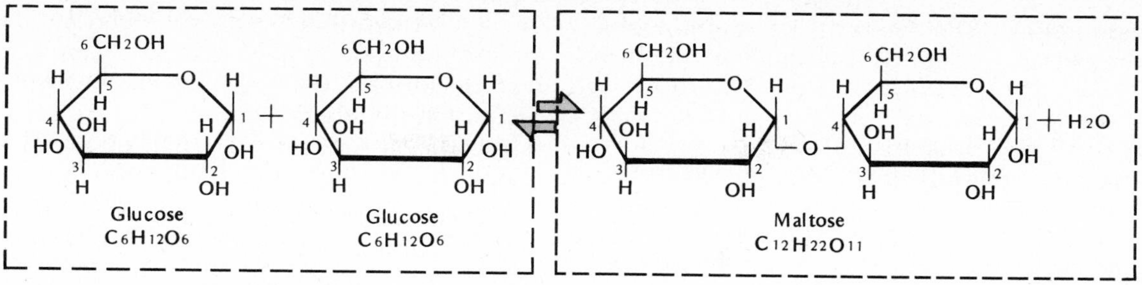

Figure 3.5 Synthesis and hydrolysis of maltose, a disaccharide containing two molecules of glucose.

drolytic reaction, or a hydrolysis. Other disaccharides with the same formula, $C_{12}H_{22}O_{11}$, are sucrose (cane sugar) and lactose (milk sugar). Sucrose is formed by the union of a molecule of glucose and a molecule of fructose; lactose is formed from a molecule of glucose and one of galactose.

The more complex carbohydrates—*starches, glycogen* (animal starch), and *cellulose*—are formed by living cells from glucose. These substances are polysaccharides, and the general formula $(C_6H_{10}O_5)n$ is often used for them; n represents a very large and often unknown number of monosaccharides. The starches are important in living matter because they are energy reserves. They are large and relatively insoluble molecules. Thus unlike the simple sugars, they remain in place until needed by the organism. Most natural starches contain both branched and unbranched chains of glucose units. The linkages between monosaccharides are called *glycosidic bonds.* Glycogen, or animal starch, differs from plant starch in that it is more highly branched and more soluble in water.

Cellulose, which forms much of the support material in plant cell walls, is composed of long chains of glucose units, but the way in which the units are linked together differs from that in starch molecules. A comparison of a short segment of a starch molecule and a cellulose molecule is shown in Figure 3.6. Cellulose is the most abundant organic substance found in nature, and has been utilized by man in many ways. It is used in the manufacture of explosives, photographic film, varnishes, plastics, rayon, and many other products.

Chitin is a polysaccharide found in many organisms including molds, crabs, and insects. Its structure is similar to cellulose, but it is a polymer (a macromolecule made of many similar units) of a nitrogen-containing derivative of glucose (*N*-acetyl glucosamine). Cellulose and chitin, as structural carbohydrates, have the same type of glycosidic linkage which is different from the glycosidic linkage of storage carbohydrates. Thus, it is significant that humans cannot hydrolyze (cleave) the glycosidic linkage of cellulose and chitin.

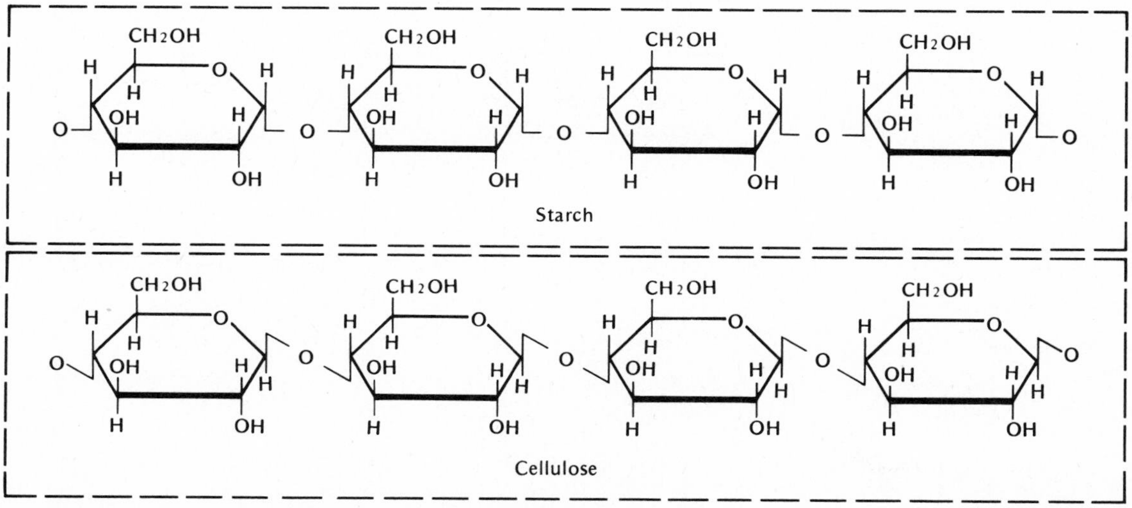

Figure 3.6 Structural representations of starch and cellulose. Notice that the difference between the two polysaccharides is in the way that the glucose units are joined together. Starch is a major foodstuff for humans, but cellulose cannot be hydrolyzed by human enzymes. This difference in the ability to hydrolyze starch but not cellulose resides in the specificity of the human enzymes for the bonds between the glucose units in starch.

Acetogenins The acetogenins comprise a heterogeneous group of substances that are synthesized from the acetyl group $(CH_3{-}\overset{O}{\overset{\|}{C}}{-})$, do not dissolve readily in water, but dissolve in solvents such as ether, chloroform, or alcohol. The most common acetogenins—the neutral lipids, the phospholipids, and the steroids—are discussed here. These substances function in living organisms as structural components (for example, membranes), fuel, and storage forms of metabolic fuel.

The *neutral lipids* are compounds of carbon, hydrogen, and oxygen, but the number of oxygen atoms in fat molecules is small; there is no 2:1 relationship between hydrogen and oxygen atoms as is generally found in carbohydrates. Tristearin is a common compound and is found in beef fat. When tristearin is hydrolyzed, each molecule produces one molecule of glycerol and three molecules of stearic acid, a fatty acid. The simpler fatty acids are hydrocarbon chains with a carboxyl group (—COOH) on one end. Several kinds of fatty acids are found in the neutral fats. A given fat may have all the fatty acids of one kind or may have a mixture of different kinds. The synthesis of fats in a cell involves the union of a molecule of glycerol with the fatty acid molecules. The overall equation for these processes is shown in Figure 3.7.

Some fatty acids have the maximum number of hydrogen atoms. Stearic acid is an example of such a *saturated* (Fig. 3.7) fatty acid. Other fatty acids have hydrogen removed along the carbon chain. Such a fatty acid is *unsaturated*. In general, unsaturated fatty acids have lower melting points than do saturated ones (the unsaturated fatty acids are oils rather than solids at room temperature). Further, unsaturated fatty acids are more easily digested than saturated ones.

In the *phospholipids* one of the fatty acids has been replaced by a phosphate group which may be linked to other compounds. The phospholipids are found commonly in brain, egg yolk, yeast, and liver and are very important constituents of the nervous system and all cellular membranes.

The *steroids,* unlike the neutral lipids and the phospholipids, do not contain fatty acids. All steroids have in common a carbon skeleton composed of four fused rings. The rather complex steroid molecule is synthesized from the simple and common molecule, acetic acid. The predominant steroid in animal tissues is cholesterol (Fig. 3.8). It is the precursor of the sex hormones, the adrenocortical hormones (see Ch. 12), and the bile salts.

Proteins The name "protein" (from the Greek for "preeminent," or "first") was suggested in 1838 for the class of complex organic nitrogenous substances found in large amounts in cells. The name was well chosen. The proteins, as enzymes, are involved in the molecular dynamics of the cell in an intimate way. Furthermore, proteins play an important role in the architecture of cells and tissues as structural

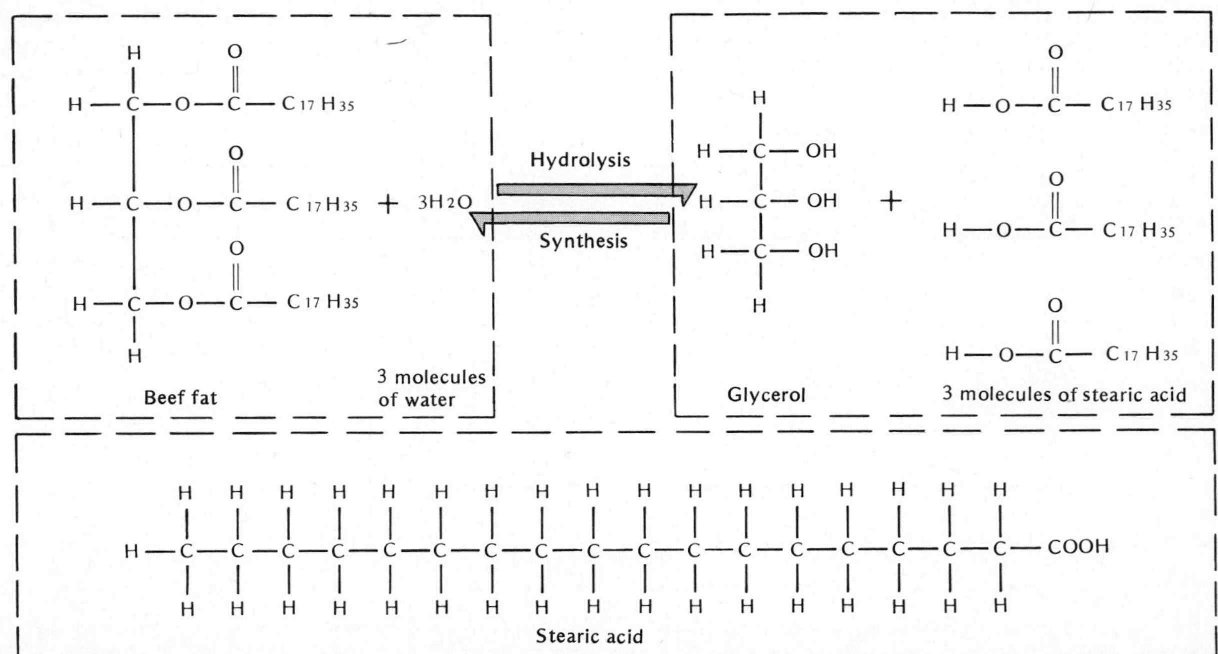

Figure 3.7 Hydrolysis and synthesis of beef fat tristearin. The structural formula for stearic acid is also shown.

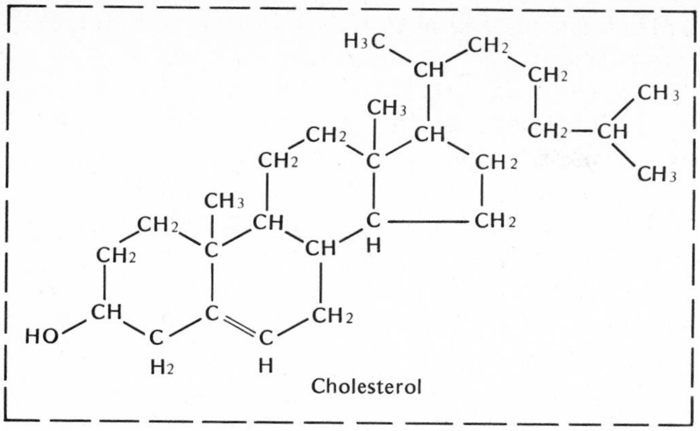

Cholesterol

Figure 3.8 Structural formula of cholesterol.

molecules; hair, wool, and connective tissues contain particularly large amounts of structural proteins. Proteins are the most abundant macromolecules in most cells. Several proteins serve in the transport of oxygen and in muscular contraction. The antibodies that are concerned with immunological defense mechanisms and some of the *hormones,* special chemical messengers within the organism, are proteins. Of course, proteins may also be broken down by catabolism (destructive metabolism, Ch. 1) into waste products and energy.

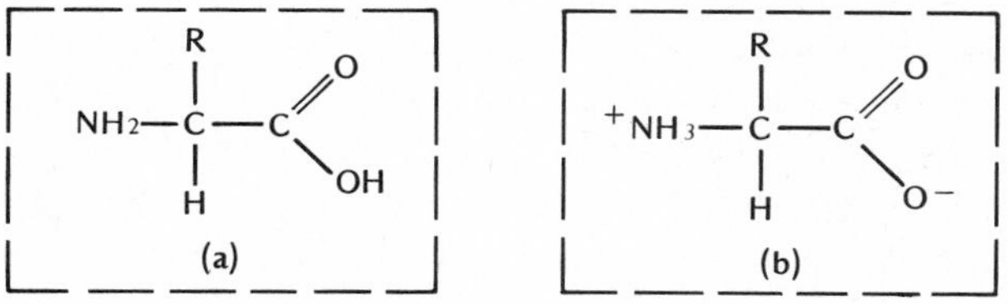

(a) (b)

Figure 3.9 (a) Structural formula of an amino acid. **(b)** Structural formula of an amino acid showing charge relationships.

The amino acids are the building blocks of proteins. The general structural formula of an amino acid is given in Figure 3.9**(a);** this formula shows that all amino acids have common structural features. The *carboxyl* group (—COOH) has acidic properties, whereas the *amino* group ($—NH_2$) has basic ones; under the physiological conditions of the cell, these groups are ionized, and so the amino acid carries both a negative and a positive charge (Fig. 3.9**(b)**). Because amino acids can react with an acid or a base, they are called *amphoteric* (Greek meaning "in both ways") compounds, and they can function as buffers. The R designates the side chain that gives chemical identity to the various amino acids; the side chain may be a simple group or a somewhat complex molecular constellation. Ionically, the side chains may be polar or ionic (interaction with water) or nonpolar (no interaction with water). About 20 amino acids occur generally in proteins. Thus the number of ways these amino acids can be arranged in a protein is theoretically very large. (For a protein composed of 100 amino acids, about 10^{130} different structures could be formed with 20 kinds of amino acids.)

The linear array of the amino acids in a protein molecule is called the primary (I°) structure of the protein. The bonding between the carboxyl group of one amino acid and the amino group of the next is known as a *peptide bond,* or a *peptide linkage.* The formation of the peptide bond is discussed in Chapter 6, but it is pertinent to say here that the net result is the formation of a molecule of water for each peptide linkage formed (Fig. 3.10). Two amino acids linked by a peptide bond constitute a *dipeptide,* three constitute a *tripeptide,* many constitute a *polypeptide.* The terms *protein* and *polypeptide* are used interchangeably in many instances.

It is easy to imagine that a long chain of amino acids may be arranged in several ways. The chain might retain a straight linear form, it might fold up into a ball in more than one way, or it might twist into a spirally wound *helix.* Such arrangements in space constitute the secondary (II°) structure of proteins. The form that many proteins take, at least in part, is the α helix. In the α helix the stability is due to the large number of hydrogen bonds between successive turns of the helix. Figure 3.11 shows a model of the α helix. The fibrous proteins such as silk fibroin are not in an α-helical structure but lie parallel to each other and form sheets. This is called the *pleated sheet arrangement.* The toughness of these proteins is related to this sheet arrangement and

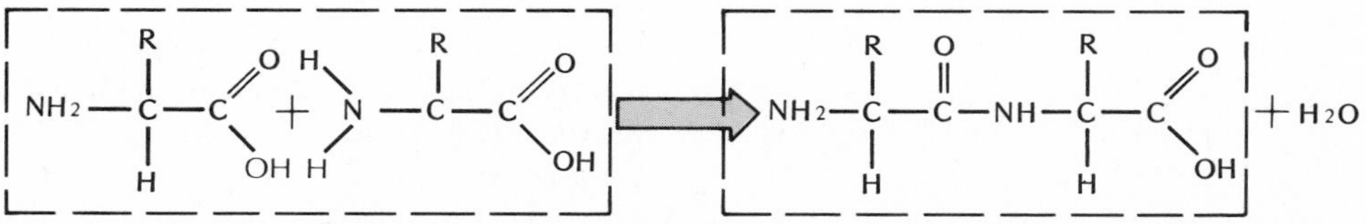

Figure 3.10 Formation of a peptide linkage. Amino acids, the building blocks of proteins, are linked together with the release of a water molecule. The actual mechanism for the formation of a peptide linkage is complex (Ch. 6).

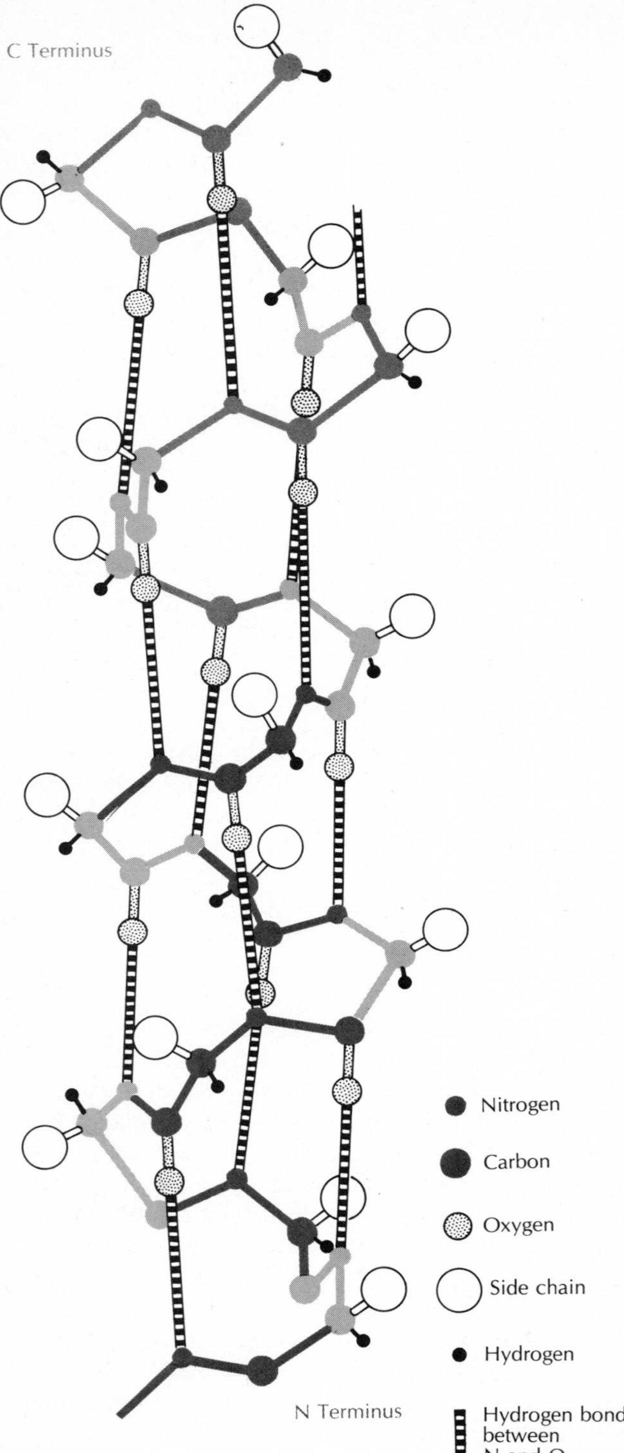

Figure 3.11 The structure of the α helix. Stabilizing forces are hydrogen bonds between nitrogen and oxygen at different levels of the helix (dashed lines). The shading emphasizes the three dimensions of the helix.

the extensive hydrogen bonding between strands and sheets.

The tertiary (III°) structure of a protein is a still higher level of organization. In the globular proteins the helical structure folds upon itself

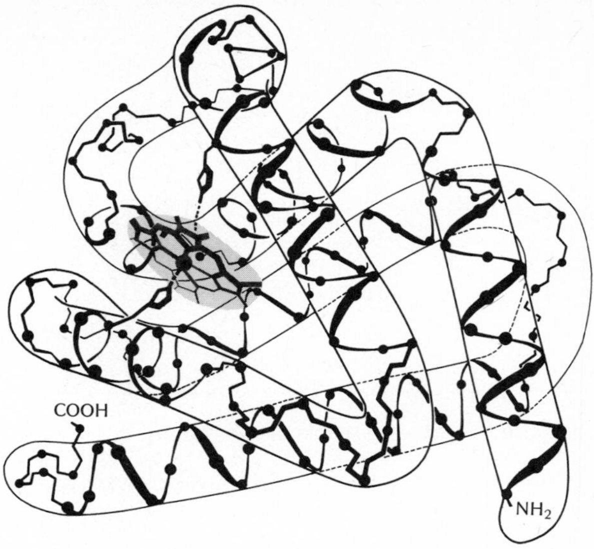

Figure 3.12 The myoglobin molecule. Notice the intricate visceralike folding of the molecule into the tertiary structure. Regions of the α-helix conformation are connected by bends where the helix is disrupted. The iron-containing heme group is represented by the disk. (Redrawn from Lehninger, *Biochemistry,* 2nd ed., Worth.)

and forms a more compact arrangement. The tertiary structure of proteins is very specific; this specific folding of the α helix is stabilized by a number of chemical forces between regions that are close together. Globular proteins are tightly folded into compact shapes. Myoglobin, an example of a folded protein, is shown in Figure 3.12. If there is only one polypeptide chain in a protein, tertiary structure is the highest level attained.

Some proteins do contain more than one polypeptide chain and are said to possess quaternary (IV°) structure. The subunits of a protein with quaternary structure simply do not form an aggregate, but are positioned precisely in relation to each other. Hemoglobin, with four subunits, is a well-studied protein with quaternary structure.

ENZYMES As mentioned earlier the enzymes, or biological catalysts, make up an important group of proteins. Presumably all complex cellular reactions are catalyzed by specific enzymes. Thus the integrity of the enzymes of a cell or organism is paramount to its well-being. Enzymes have properties that reflect their protein nature: enzymes are inactivated by heat, they are very sensitive to changes in pH, and they are sensitive to various ions. Many of the enzymes are *conjugated* proteins; they consist of a protein part (apoenzyme) and a nonprotein prosthetic group (coenzyme):

$$\text{apoenzyme} + \text{coenzyme} \rightarrow \text{(holo-)enzyme}$$

Coenzymes are often called cofactors.

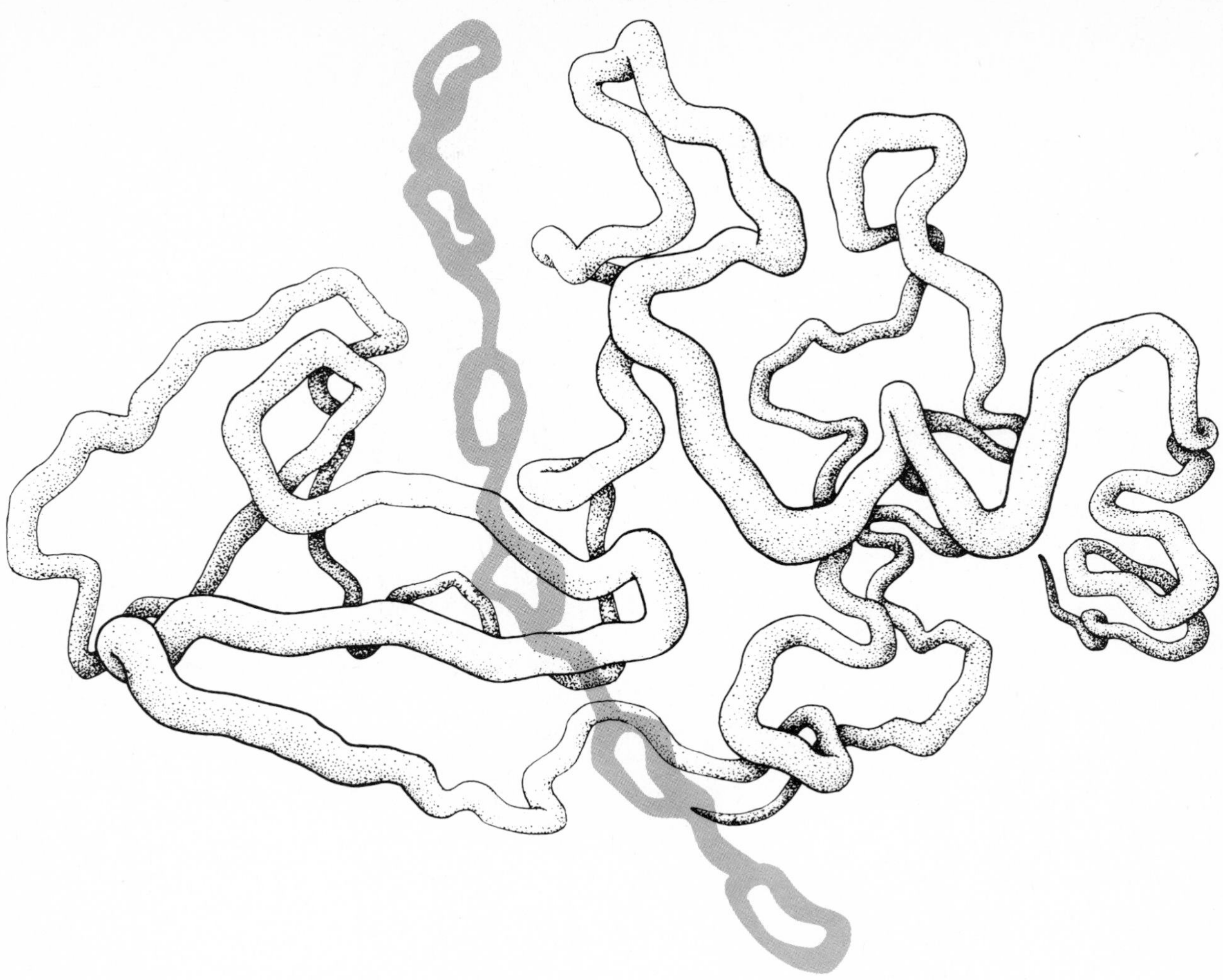

Figure 3.13 A schematic drawing of lysozyme with its active site filled with a portion of a polysaccharide from bacterial cell walls.

The mechanism of enzymatic catalysis is a fundamental problem in biology. During the reaction, the *substrate* (the molecules to be changed by the reaction) binds to a specific region of the enzyme, the *active site* (also called the catalytic site). Studies have shown that amino acids that are widely separated from each other in the linear sequence are involved in the active site (Fig. 3.13). The general conclusion is that the nature of the active site is a three-dimensional function of the complicated folding pattern of the enzyme.

Enzymes are exceptional catalysts. Most enzyme-catalyzed reactions are one hundred million to one hundred trillion (10^8 to 10^{11}) times faster than the corresponding nonenzymatic reactions.

The *turnover number* (number of substrate molecules metabolized per enzyme molecule per minute) for many enzymes is around 1000, but for some enzymes it is greater than one million.

Nucleic Acids As are the proteins and polysaccharides, the nucleic acids are polymers of a repeating unit, a basic building block. The basic building block of nucleic acids is the *nucleotide,* which consists of a nucleobase, a pentose sugar, and a phosphate group. The nucleobases are nitrogen-containing ring compounds of two types: purines and pyrimidines. In nucleic acids five nucleobases occur commonly: adenine (A) and guanine (G), both purines; and cytosine (C), thymine (T), and uracil (U), all pyrimidines. The pentose sugars found in nucleic acids are ribose and 2-deoxyribose. The components of nucleic acids are shown in Figure 3.14. A nucleobase attached to a sugar (without the phosphate) is a *nucleoside* (Fig. 3.15).

On the basis of chemical composition there are two classes of nucleic acids, deoxyribonucleic acid (DNA) and ribonucleic acid (RNA). The names indicate the sugars present. Furthermore, the nucleobases of DNA are A, T, C, G, while RNA contains A, U, C, and G. Both DNA and RNA

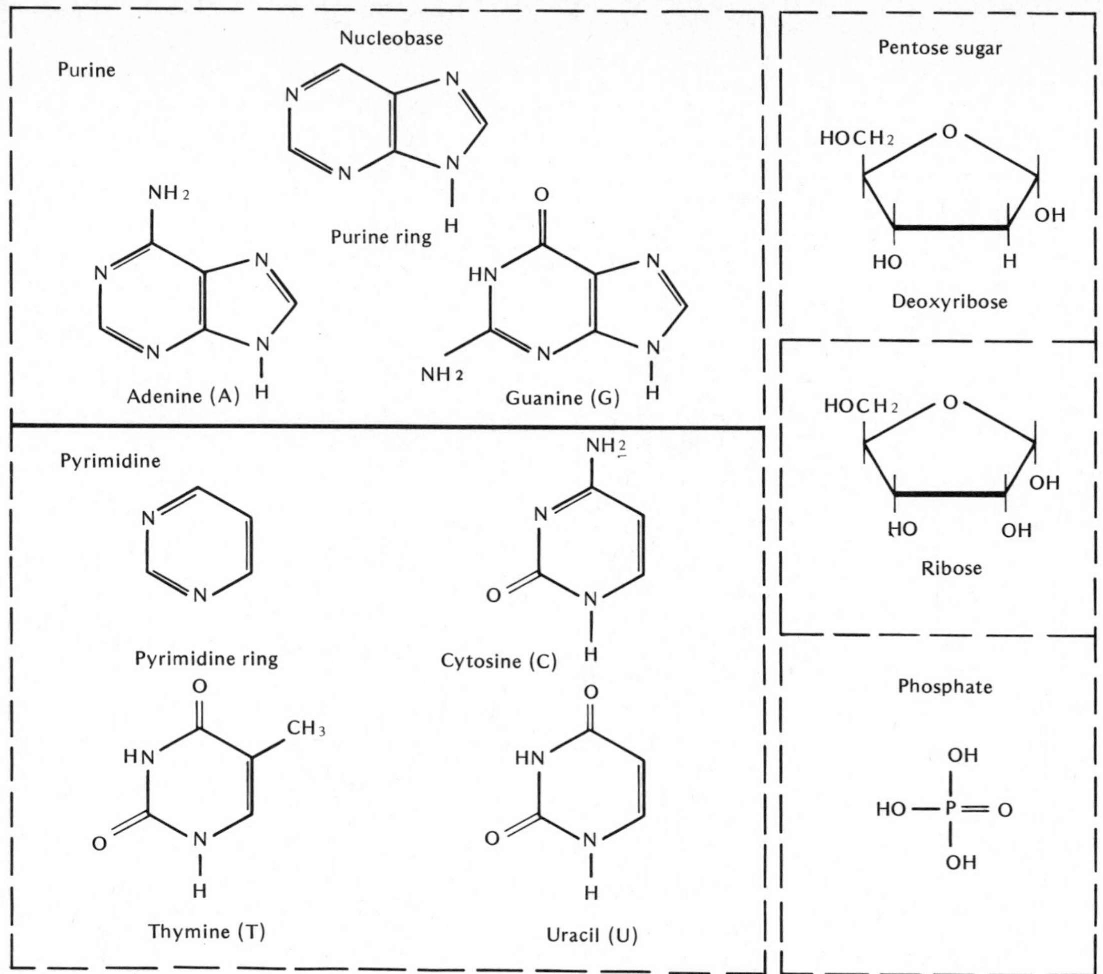

Figure 3.14 Nucleic acid components. The nucleotide, the basic building block of nucleic acids, consists of a nucleobase, a pentose sugar, and a phosphate group. Ribose and deoxyribose differ only in the H— and —OH groups attached to the number 2 carbon atom.

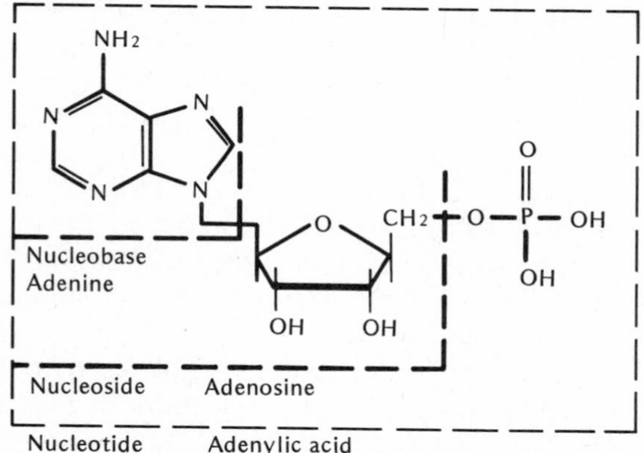

Figure 3.15 Nucleotide, nucleoside, and nucleobase. The relationship of the components of the nucleotide are demonstrated by adenylic acid in this example. If the sugar were deoxyribose, then the nucleoside would be deoxyadenosine. Nucleotides containing other bases are named in similar manner. At physiological pH, the phosphate group is ionized.

are phosphate–sugar chains, with the nucleobases projecting laterally on the sides of the chain. This structural relationship of a polynucleotide chain is shown in Figure 3.16.

The relative amounts of the bases (the base composition) in DNAs obtained from different sources have been determined, and it has been found that the amounts of certain bases are equal. For each adenine there is one thymine, and for each guanine there is one cytosine. James Watson and Francis Crick, using these data and their knowledge derived from studies on the diffraction of X rays by DNA, proposed a model for the structure of DNA. The Watson–Crick model contains two strands wound in a helix, with hydrogen bonding between these pairs. The base-pairing relationships and a schematic diagram of the Watson–Crick model are shown in Figure 3.17.

The function of DNA and the significance of its structure is treated in many chapters throughout the text; here it is sufficient to say that DNA serves as the hereditary material in all true organisms (some viruses contain only RNA).

RNA is divided into three major subgroups on a structural and functional basis; these are *ribosomal RNA, messenger RNA,* and *transfer RNA.* These types of RNA are discussed in Chapter 6.

In addition to the nucleotides found as building blocks of DNA and RNA, two other compounds should be mentioned at this point. The first of these, adenosine triphosphate (ATP), occurs in all cells. The function of ATP is to store and carry energy for many cell activities. The details and use of ATP synthesis are described in

Figure 3.16 Portion of a polynucleotide chain. The sequence of nucleotides in this portion of a polynucleotide strand is ATC. Is this a piece of RNA or DNA?

Figure 3.17 *(below left)* Base-pairing relationships and the Watson-Crick model. *(below right)* The molecular structure of two base pairs. Note that the T═A (or A═T) base pair contains two hydrogen bonds whereas the C≡G (or G≡C) base pair contains three.

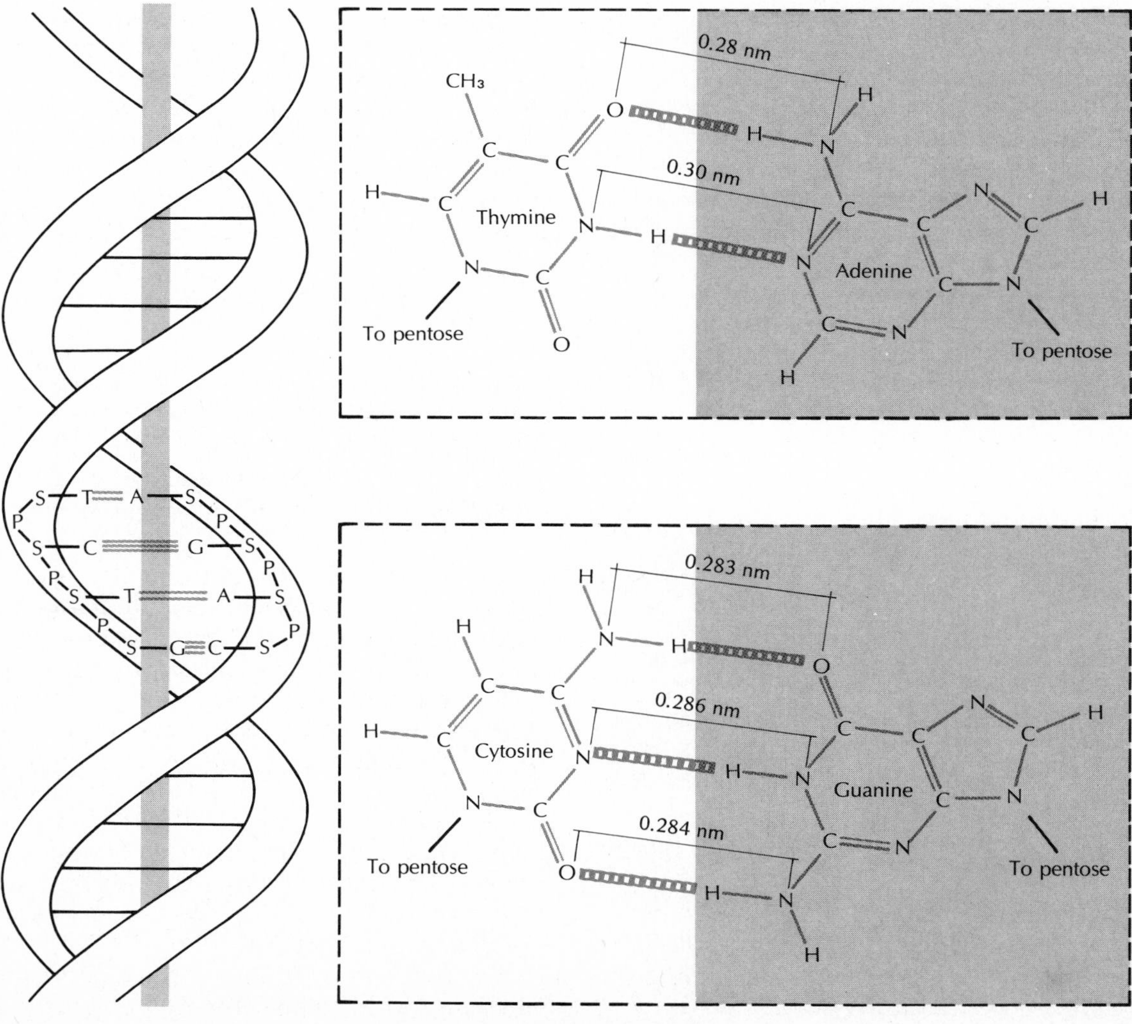

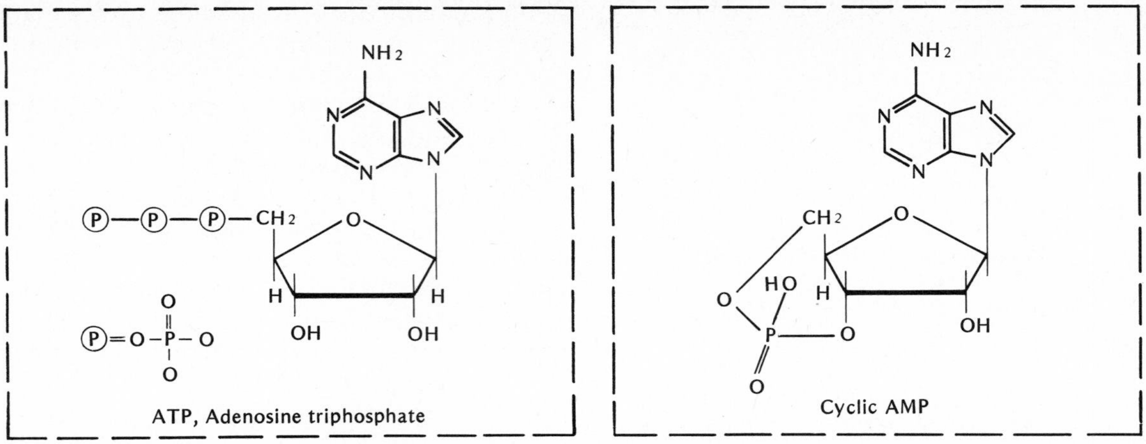

ATP, Adenosine triphosphate

Cyclic AMP

Figure 3.18 Structural formulae for ATP and AMP. ATP consists of adenine, a purine, which is linked to ribose and three phosphates. Cyclic AMP consists of adenine, ribose, and phosphate with the phosphate attached at carbons 3′ and 5′ on the ribose, thereby forming a cyclic structure.

Chapter 5. ATP has three phosphates attached to the ribose sugar (Fig. 3.18).

Cyclic AMP is adenosine monophosphate with the phosphate attached to the ribose at two places (Fig. 3.18). Thus the phosphate and three of the ribose carbons form a ring. Cyclic AMP is involved in the action of hormones in vertebrates (Ch. 12). Cyclic AMP was discovered in cells in 1958 and is involved in many cellular activities throughout the phylogenetic scale of organisms.

INFORMATIONAL MACROMOLECULES

The proteins and nucleic acids are special in that the linear sequence of amino acids or nucleotides in them carries bioinformation (Ch. 6). A comparison of the polysaccharides, which contain many units of one or a few simple sugars (monosaccharides), with the proteins and nucleic acids illustrates the essence of the informational macromolecule: a specific sequence of basic building blocks that carries a message which is readable and interpretable by the cellular machinery. Chapter 6 discusses bioinformation in some detail with an analogy to languages that are also written in a specific linear form.

SUGGESTIONS FOR FURTHER READING

Barker, R. *Organic Chemistry of Biological Compounds.* Englewood Cliffs, N.J.: Prentice-Hall (1971).

Barry, J. M., and E. M. Barry. *An Introduction to the Structure of Biological Molecules.* Englewood Cliffs, N.J.: Prentice-Hall (1969).

Conn, E. E., and P. K. Stumpf. *Outlines of Biochemistry,* 3d ed. New York: Wiley (1972).

Damber, J. G., and A. T. Moore. *Chemistry for the Life Sciences.* New York: McGraw-Hill (1973).

Dickerson, R. E., and I. Geis. *The Structure and Action of Proteins.* New York: Harper and Row (1969).

Edelstein, S. J. *Introductory Biochemistry.* San Francisco: Holden-Day (1973).

Lehninger, A. L. *Short Course in Biochemistry.* New York: Worth Publ. (1973).

Vishniac, R. *Building Blocks of Life.* New York: Charles Scribner's (1972).

White, E. *Chemical Background for the Biological Sciences.* Englewood Cliffs, N.J.: Prentice-Hall (1969).

Wold, F. *Macromolecules: Structure and Function.* Englewood Cliffs, N.J.: Prentice-Hall (1972).

From Scientific American

Crick, F. H. C. "The Structure of the Hereditary Material" (Oct. 1954).

Doty, P. "Proteins" (Sept. 1957).

Feiser, L. F. "Steroids" (Jan. 1955).

Frieden, E. "The Chemical Elements of Life" (July 1972).

Gross, J. "Collagen" (May 1961).

Koshland, D. E., Jr. "Protein Shape and Biological Control" (Oct. 1973).

Merrifield, R. B. "The Automatic Synthesis of Proteins" (Mar. 1968).

Mirsky, A. E. "The Discovery of DNA" (June 1968).

Neurath, H. "Protein-Digesting Enzymes" (Dec. 1964).

Phillips, D. C. "The Three-Dimensional Structure of an Enzyme" (Nov. 1966).

Stein, W. H., and S. Moore. "The Chemical Structure of Proteins" (Feb. 1961).

Stroud, R. M. "A Family of Protein Cutting Proteins" (July 1974).

Stumpf, P. K. "ATP" (Apr. 1953).

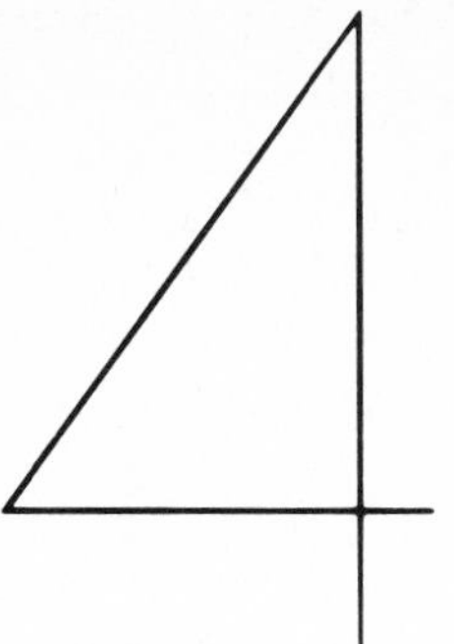

Cells and Tissues

PROBLEMS OF BIOLOGICAL EFFICIENCY AND DIVISION OF LABOR

The biological world is made up of a great variety of shapes and sizes of organisms, and all of them operate according to the principle of molecular economy discussed in Chapter 3. Furthermore, the use of biomolecules is done in a most orderly way. This should immediately suggest that the physical organization of living things also is orderly, and it is. When molecular processes are carried out in small volumes, the surface-to-volume ratio is low and favors an easy exchange between the surface and the interior. Large volumes have relatively small surfaces proportionately unless the volumes are spread out in very thin films. But a man, or a whale, has a seemingly large surface-to-volume ratio. Thus we look to see how life can be organized so that there is orderly biomolecular activity, so that there is an adequately low surface-to-volume ratio, and so that large size can be accounted for. We find that the universal unit which accommodates to these three points is the cell. The cell is the structural and functional unit of life.

The cell concept was stated over 135 years ago (Matthias Schleiden 1838 and Theodor Schwann 1839). It was not possible to determine that all organisms are composed of cells and cell products (Ch. 1) until the development of the microscope because of the small size of most plant and animal cells, which vary in size from 10 to 100 μm. [1 μm = 0.001 millimeter (mm), or approximately 1/25,000 inch.]

THE IMPORTANCE OF THE CELL CONCEPT

In addition to being the unit of structure and function of living matter, the cell is the unit of heredity, pathology, and reproduction. While there is no such thing as a typical cell, there are two basic cell types. One type, the *prokaryotic* cell, is represented by bacteria and blue-green algae; it lacks a well-defined nucleus. The other is the *eukaryotic* cell, which has a well-defined nucleus. The following discussion concentrates on eukaryotic cells found in all animals.

In relating cellular structure and function, four common kinds of functions are discussed to give a broad description of cells.

1. Cells regulate the orderly movement of materials into and out of themselves, and involve the limiting boundary to the cell, the *plasma membrane.*
2. Cells use energy in doing work, synthesizing and degrading molecules, and exporting products. This is accomplished chiefly in the *cytoplasm.* The cytoplasm consists of many parts and, with a few exceptions, makes up most of the mass of the cell.
3. Cells operate under the control of the hereditary material, DNA (deoxyribonucleic acid), which is localized in the *nucleus.* The details of the chemistry of the genetic material are discussed in Chapters 3 and 6.
4. Related to DNA is the function of cell duplication. Usually two cells arise from one cell by cell division. In eukaryotic cells, especially, a complex set of structures is involved in cell division.

PHYSICAL AND FUNCTIONAL PARTS OF A CELL

Historically, the *cell wall,* which is actually a nonliving product of the plant cell, was the first structure to be identified. However, cell walls will be omitted in this discussion for they do not exist with animal cells.

A typical cell has three major parts. The outer, limiting *cell* or *plasma membrane* regulates the traffic of materials into and out of the cell. Inside the plasma membrane is the *cytoplasm,* composing the bulk of most cells and being the site where energy is used and substances are produced. The *nucleus* is usually centrally placed and contains the genetic material (bioinformation; Ch. 6) that is transmitted from one cell generation to the next; it instructs and regulates the activities of the cytoplasm.

The Membranes

PLASMA MEMBRANE

This important membrane bounds the living substance of the cell, although it is invisible with the light microscope. Its capacity to regulate selectively the movement of materials between the cell and the surroundings helps determine whether a cell is living or dead.

There have been numerous models of membrane structure. Presently receiving strong support is the *fluid mosaic* hypothesis (Fig. 4.1). This hypothesis assumes that there are two layers of phospholipids. Attached to, embedded in, or crossing the phospholipids are highly folded globular proteins. This model is especially attractive because of the growing recognition of specific receptor sites on membranes. For example, in animal cells, membranes are known to have specific receptor sites for the pancreatic hormone, insulin (Ch. 12), or the dendritic ends of neurons have specific receptor sites for transmitter substances produced by the axon of an adjacent neuron (Ch. 13). Also, this model permits greater flexibility of form and function of the membrane than some of the other models of membrane structure.

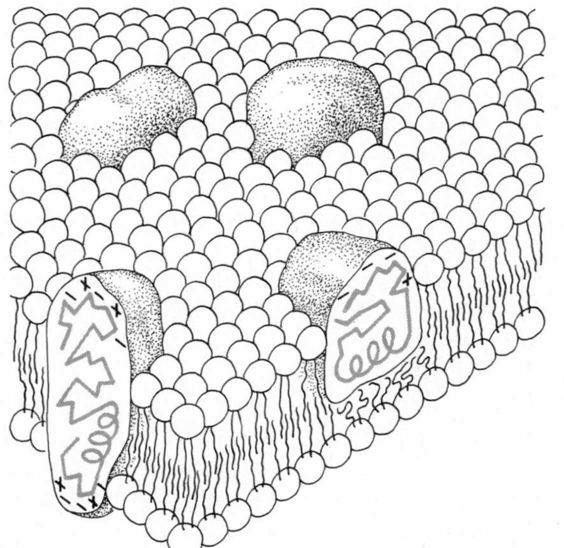

Figure 4.1 Fluid mosaic model of cell membrane. Globular proteins bearing various charges are viewed as distributed throughout the plane of the membrane. They are partially or cross-sectionally embedded in a matrix of two layers of phospholipids.

Some cells secrete materials that lie between cells. These materials are sometimes called cementing substances. However, the precise conditions that lead to cellular adhesion are not fully understood and are actively being investigated.

CELL MEMBRANES AND THE CONTROL OF MOVEMENT OF MATERIALS

Two physical processes involved in the movement of materials are *diffusion* and *osmosis,* which provide methods for examining the role of the plasma membrane. Diffusion is the movement of molecules, atoms, and ions from regions of their high concentration to regions of their low concentration. For instance, the odor of an open container of pungent chemical or perfume will spread through the surroundings. Because of the kinetic energy of the ions and molecules this is accomplished by their tendency to distribute equally in space. Diffusion is an important phenomenon to living cells and organisms because they are composed primarily of water and most cells are surrounded by a watery medium of some proportions. Nearly every student at some time has seen the classical demonstration of diffusion where colored crystals are carefully placed in the bottom of a stationary test tube or glass cylinder and observed over a span of time. The water adjacent to the crystals first becomes colored and the color spreads until it is equally distributed throughout the container.

The principle of diffusion can be related to membranes and their roles. A membrane that allows all molecules to move across it is freely permeable. Living cell membranes are differentially permeable and allow some molecules to diffuse across the membrane but retard or prevent others from moving across. Some nonliving membranes can serve as models of semipermeability. For instance, an artificial system consisting of two solutions of different concentrations separated by a membrane impermeable or semipermeable to the substances in solution,

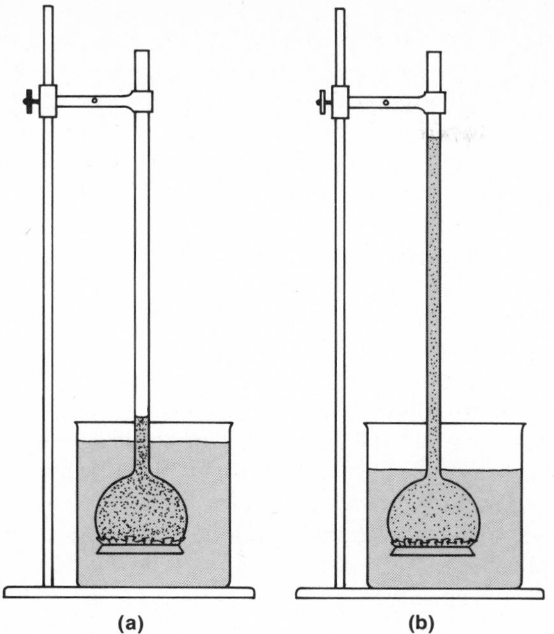

Figure 4.2 Osmometer **(a)** Beginning of the experiment. **(b)** After two hours.

but freely permeable to water, results in a diffusion of water to the solution of lower concentration of water. This net movement of water that tends to equalize the concentrations on each side of the membrane is called osmosis. Figure 4.2 illustrates a reliable kind of osmometer for demonstrating osmosis.

The principles of osmosis are employed in the kitchen in soaking vegetables in water to make them crisp, or in wilting lettuce with vinegar. The following laboratory experimentation illustrates the role of the plasma membrane or identifies its existence.

Red Blood Cells and Osmosis Red blood cells are excellent osmometers. To maintain their normal shape outside the blood stream, they must be kept in an *isotonic* solution, one where the concentration of water and solutes is the same as that of the cytoplasm with water leaving and entering the cell at the same rate (Fig. 4.3). A solution of 0.9 percent NaCl is isotonic for mammalian red blood cells. (An isotonic salt solution is not necessarily a *balanced salt solution;* the latter contains proper proportions of the salts found in the cytoplasm, for instance, sodium, potassium, magnesium, and calcium.) If red blood cells are placed in distilled water, they will tend to swell and burst because distilled water is a *hypotonic* solution (lesser solute concentration) with the concentration of water greater than that within the cell. In contrast to this, if red blood cells are placed, for example, in a 10 percent salt solution, they very

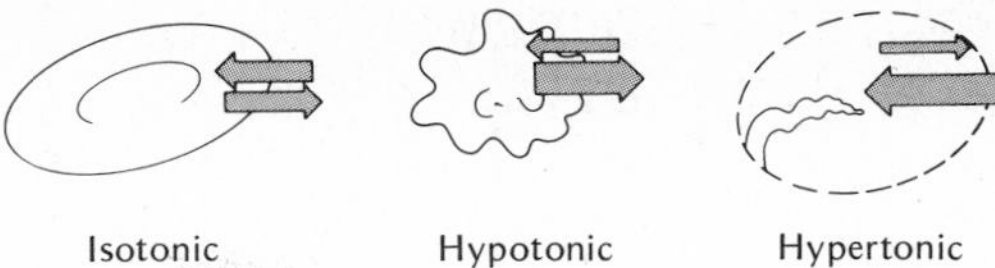

Figure 4.3 Behavior of red blood cells in isotonic, hypertonic, and hypotonic selections.

quickly lose water, shrink, and appear wrinkled. This salt solution is *hypertonic* (greater solute concentration) for the red cells as the concentration of water is less than that in the cytoplasm. The term applied to the shrinking of animal cells is *crenation*. It is obvious from the foregoing statement that cells, such as red blood cells, may be killed by being surrounded by hypotonic or hypertonic solutions. It should also be obvious that in handling living tissues, such as blood in blood banks, during surgery, and so forth, it is important to maintain isotonic conditions.

More dynamic controls on the movement of materials across membranes exist, however, and are referred to as transport mechanisms. Another role of membranes is associated with the second messenger concept (Ch. 12). These are briefly discussed here to relate them to the activities of membranes that will be discussed in numerous places.

Active Transport This is a process acting across membranes such as plasma membranes, mitochondria, and the endoplasmic reticulum. Molecules move across the membrane against a diffusion gradient, from the lower to the higher concentration, and employ the energy of ATP as a "pump." This sometimes concentrates materials to over 1000 times that of its surroundings. It is believed that molecules are moved in active transport by forming loose combinations with proteins that release the molecules at the other surface of the membrane (Fig. 4.4). A number of experiments demonstrate the need for ATP, one being that the lowering of oxygen concentration halts active transport because ATP production is reduced.

CELL MEMBRANES AND SPECIFIC CELLULAR ACTIVITY: THE ROLE OF CYCLIC AMP

Cell membranes have specific receptor sites for hormones on their surfaces. Many hormones function through their action on receptor sites. The major differences in the actions of these hormones is in the presence or absence of receptor sites specific for each specific hormone. When these sites are activated by specific hor-

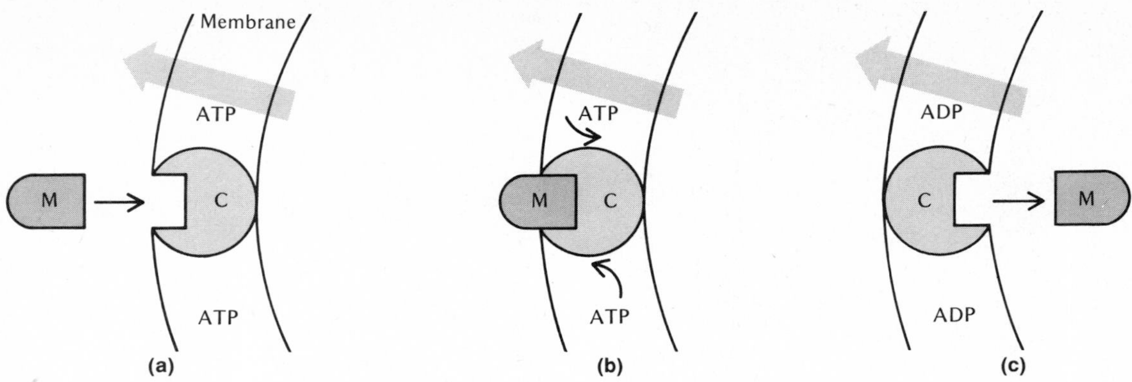

Figure 4.4 Model of active transport. **(a)** Molecule M combines at active site with carrier C. Changes in charge or configuration on C–M combination activate ATP to provide energy to move M to the opposite side of the membrane **(b),** where it dissociates **(c).** The arrow shows the direction of the diffusion gradient that requires no energy source.

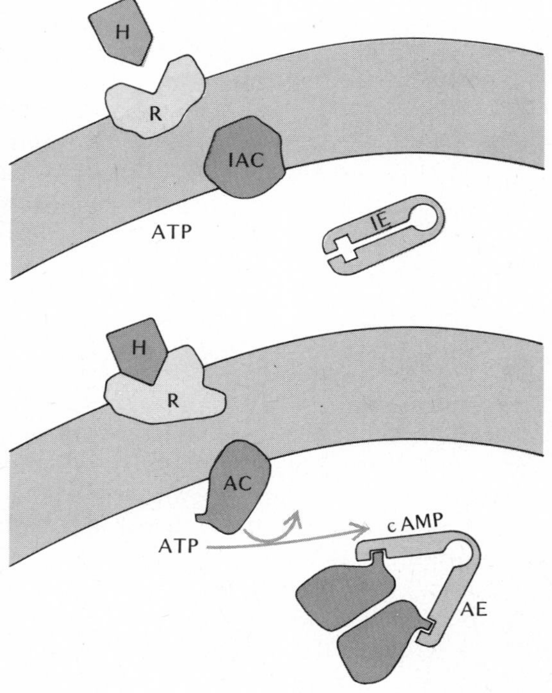

Figure 4.5 Schema of the role of cyclic AMP in activating cellular function. At the cell surface a hormone, H, combines at its receptor site R. The local change activates the inactive adenyl cyclase, IAC, in the membrane converting it to an active form, AC. This catalyzes conversion of ATP to cyclic AMP, cAMP, that acts on inactive enzymes, IE, which become active enzymes, AE, that participate in the physiological responses of the cell.

mones, an inactive enzyme in the membrane, adenylate cyclase, is activated and released toward the interior of the cell. There it catalyzes the hydrolysis of ATP to a monophosphate having the peculiarity that its persisting phosphate is bonded to the pentose sugar at two points, thus cyclic 3′, 5′-adenosine monophosphate (*cyclic AMP* or cAMP) (Fig. 3.18). Cyclic AMP then activates a series of intracellular enzymes that produce specific cellular activities (Fig. 4.5). Thus when epinephrine (adrenalin) reacts with an epinephrine receptor site on the outside of the cell membrane, adenylate cyclase is activated. This causes the ATP to produce cyclic AMP that, in turn, activates a series of enzymatic activities that release glucose from glycogen for delivery into the blood stream. Or, if a thyroid-stimulating hormone is received on the surface of the cells of the thyroid gland, the cAMP activates the secretion of thyroxin.

The Nobel Prize for physiology and medicine was given in 1971 to Earl W. Sutherland, Jr., for the work that led to the generalization that primary chemical messengers, hormones located extracellularly, can activate a second messenger, cAMP, located intracellularly, to produce specific reactions in accordance with the stimulus of the primary messenger. Until this generalization was formulated, most inquiries had labored with the question of how some hormone molecules (Ch. 12) could enter the cell and control the direction of the cell's activities.

Recently, other cyclic monophosphates have been shown to be important at the cell surface. For instance, cyclic guanosine monophosphate (*cGMP*) has been shown to function in an antagonistic way to cAMP. The roles of these second messengers will be discussed elsewhere (see Ch. 12).

CELL MEMBRANES AND CELL SURFACE CONFIGURATIONS

While appearing to be smooth under the light microscope, three aspects of the free surfaces of cells are noteworthy. Cells specialized for absorption such as those lining the intestine and composing kidney tubules show thousands of minute cylindrical processes, *microvilli* (Fig. 4.6). Also, active movement of the surface of the

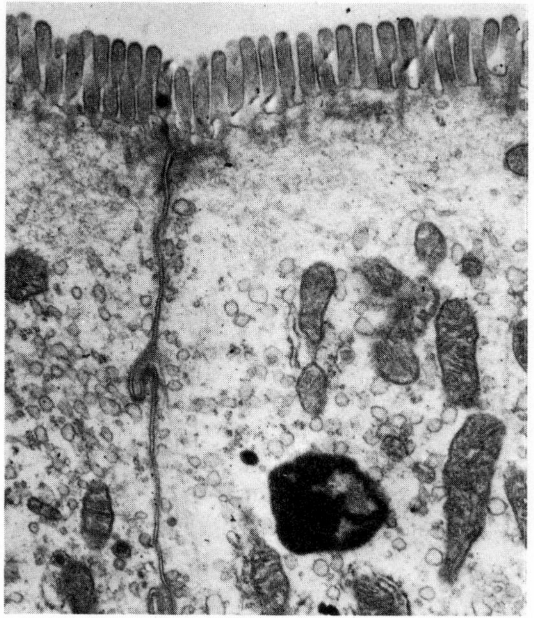

Figure 4.6 Micrograph of microvilli on the surface of intestinal cells (× 15,000). (From Rhodin, *An Atlas of Ultrastructure,* W. B. Saunders, 1963.)

cell, energy-expending processes, enclose substances from the surroundings into vesicles or vacuoles; such enclosing activity is called *endocytosis* and is of two forms. The first, *pinocytosis,* involves the formation of fluid-filled vacuoles at the surface of the cell (Fig. 4.7); the second, *phagocytosis,* as seen in amoebae and in scavenging white blood cells, involves the engulfing of solid materials into vacuoles (Fig. 4.8). In both forms of endocytosis the vacuoles usually move away from the surface to another site in the cell for further processing.

The Cytoplasm

The cytoplasm is not a simple homogeneous material. In addition to water, salts, and biomolecules (Ch. 3), it contains numerous complex organelles. Cytoplasmic organelles are organized structures with particular functions. They partition the cytoplasm into smaller, organized units in which biochemical reactions can take place in a well-ordered way. They function to process food, to extract the chemical

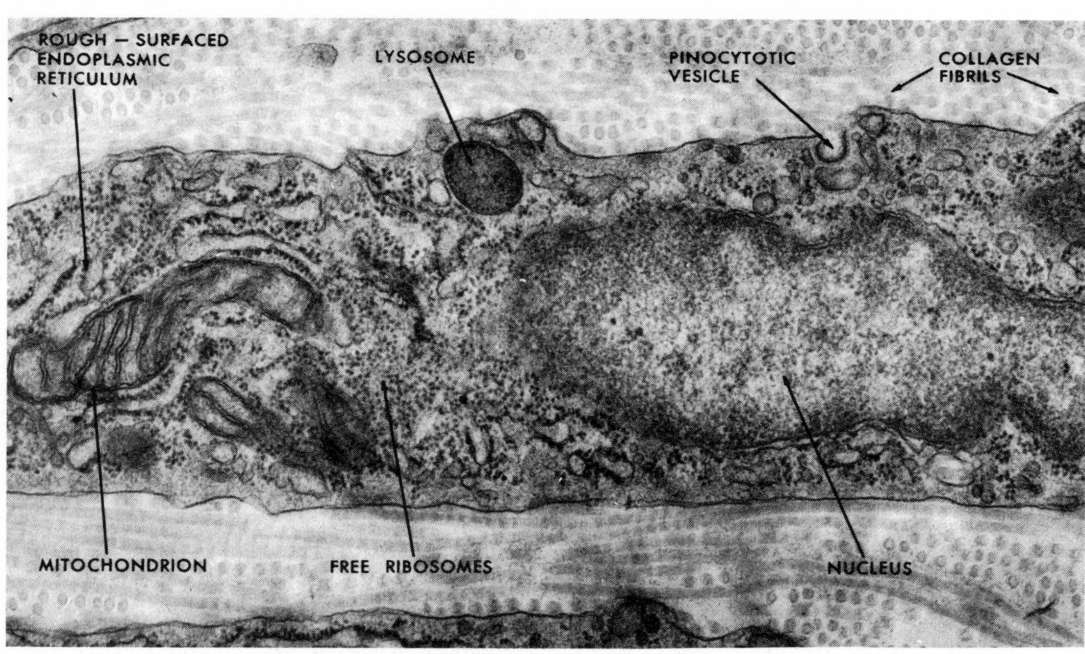

Figure 4.7 Micrograph of fibroblasts in the submucosa of the urinary bladder (× 28,000). (From Rhodin, *An Atlas of Ultrastructure,* W. B. Saunders, 1963.)

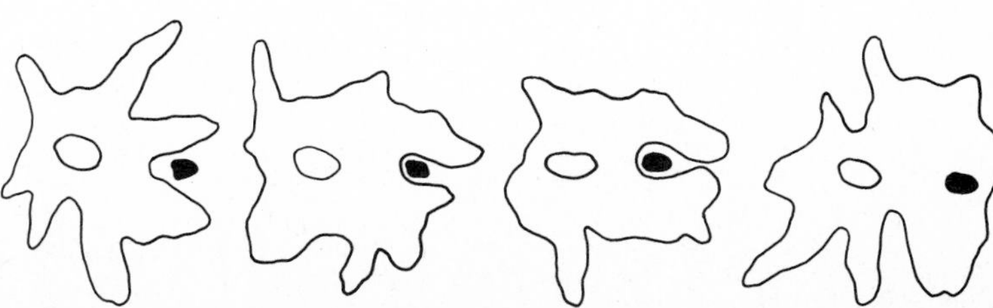

Figure 4.8 Diagram of phagocytosis by an amoeba.

energy from it, to transfer the energy to special molecules in the cell, to assist in building cell parts, and to produce compounds either for use by the cell or for export.

In some cells the total number of organelles clearly distinguishable in the cytoplasm may be small. The cytoplasmic matrix of such cells is called the *ground substance.* On the other hand some cells are so laden with organelles and membranous structures that the ground substance is not observed.

ENERGY-ASSOCIATED ORGANELLES

Living organisms have two major organelles that deal especially with the conversion of light energy to usable chemical energy or to the orderly release of the chemical energy. Conversion of light energy to usable chemical energy is a function of plant organelles, the *chloroplasts.* Interested students will find these organelles discussed in biology or botany textbooks. As they are only rarely found in animal cells and are not produced there, they will not be discussed here. However, it should be noted that they have a highly organized internal membrane structure. The other organelle actively associated with energy utilization is a universal one, found in both plants and animals; this is the *mitochondrion.*

Mitochondria Mitochondria have a double membrane. The outer one is about 7.5 nanometers (nm) thick and conforms to the shape of the organelle (Fig. 4.9). The inner one, 5–5.5 nm thick, is characteristically thrown into shelf-like folds called *cristae,* that can vary among tissues and among species. In some organisms, such as protozoans, or in mammalian kidney tissue, the inner membrane may form tubules. Mitochondria are heavily laden with enzymes, some in the central fluid matrix, others embedded in the inner membrane. When seen under the electron microscope, the inner membrane has particles on 4–5 nm-stalks topped by 7.5–10 nm-spheres (Fig. 4.10). Each mitochondrion may have from 100,000 to 1,000,000 of these elementary particles which possess many enzymes, including those of the hydrogen transport system that synthesize adenosine triphosphate (ATP) (Chs. 3 and 5).

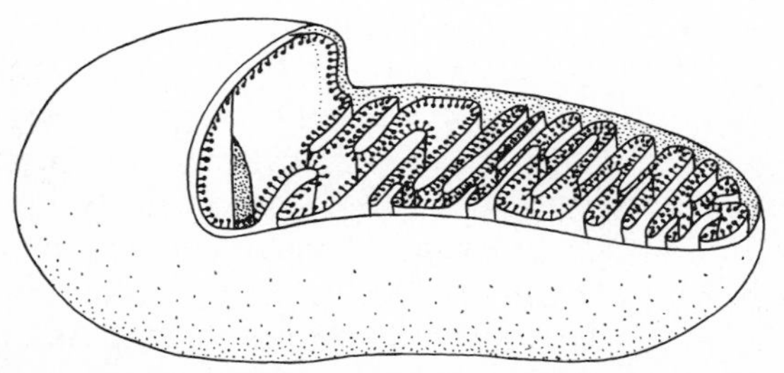

Figure 4.9 Diagram of a mitochondrion showing a few of the elementary particles at the cut edges of the folded inner membrane (*crista*).

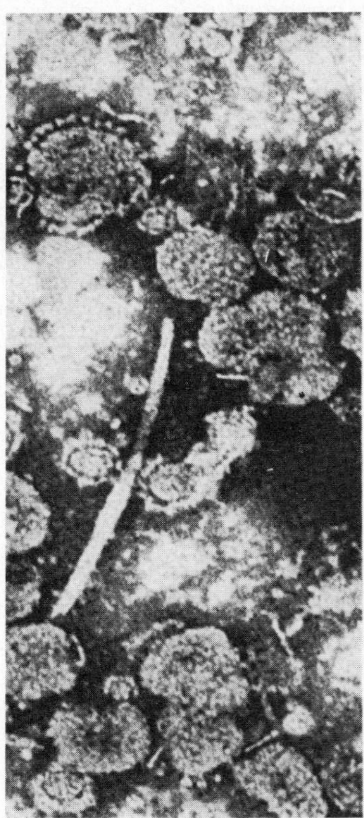

Figure 4.10 Elementary particles attached to the surface of a mitochondrial crista in an adult honey bee (× 230,000). (From B. Chance and D. F. Parsons, *Science,* **142,** 1176–1179, Nov. 29, 1963; © 1963 by the American Association for the Advancement of Science, with permission.)

Because the enzymes of mitochondria control an orderly series of reactions that hydrolyze fats and organic acids while transferring their energy to energy-rich phosphorus-containing molecules, they have been appropriately called the "powerhouses" of the cell. There are four steps in the hydrolysis of molecules whose energy ends up in such molecules as ATP: glycolysis that produces two three-carbon compounds from a six-carbon sugar; a decarboxylation step that converts the three-carbon compound to a two-carbon one; the Krebs citric acid cycle that incorporates the two-carbon compound and proceeds through many steps that liberate hydrogen atoms; and the hydrogen transport system that cascades the electrons associated with the hydrogen atoms in a series of steps that convert adenosine diphosphate (ADP) to high-energy ATP. The details of these steps are discussed in Chapter 5. The point to

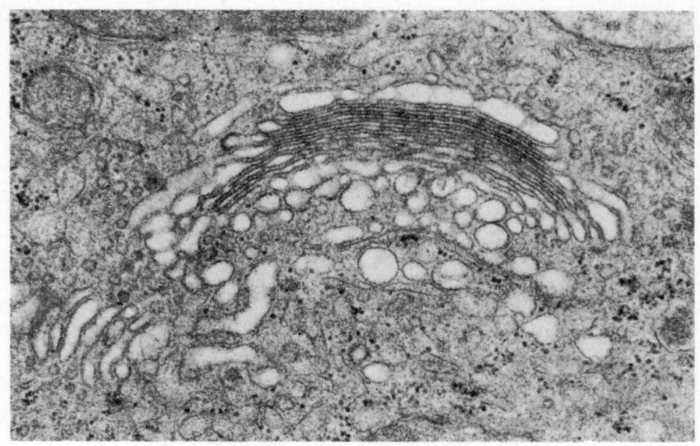

Figure 4.11 Golgi apparatus of a glandular cell from a mouse (× 54,000). (Courtesy of D. Friend; from D. W. Fawcett, *The Cell: An Atlas of Fine Structure,* W. B. Saunders, 1966.)

be made here is that glycolysis and decarboxylation occur outside the mitochondria and the actual production of ATP is small. The last two steps, the Krebs citric acid cycle and hydrogen transport, are closely coupled inside the mitochondrion. These occur in association with the inner membrane and the attached elementary particles and result in production of large amounts of ATP.

In addition to having DNA and RNA that differ from that of the cell's nucleus, both mitochondria and chloroplasts have ribosomes, and those organelles control most, but not all, of their own reproduction.

In most cells mitochondria appear to be randomly placed. However, some cells have specific associations for mitochondria. For instance, mitochondria are located among the contractile elements in muscles and they lie near the junctions of nerve cells. These relationships place them in a position to provide a ready supply of ATP for muscle contraction or nerve impulse transmission. Actively secreting cells have especially numerous mitochondria.

GOLGI APPARATUS (DICTYOSOME)

Another cytoplasmic organelle common to all eukaryote cells is the *Golgi apparatus.* The Golgi apparatus requires special staining for identification with the light microscope. It is composed of flattened smooth vessels whose membranes are tightly packed into layers (Fig. 4.11). Its position in the cell is usually near the nucleus, often as a cap over a pair of centrioles (see below). The Golgi apparatus is prominent in secretory cells. It is polarized, with one edge of the complex usually near the rough endoplasmic reticulum. At this edge it takes up the products of synthesis from the endoplasmic reticulum by accepting these products into small vesicles. It moves and concentrates the products to the opposite face, called the "secretion or maturation face." Here larger vesicles are produced to move the secretions to other parts of the cell or to the surface to be released from the cell. Another important function of the Golgi apparatus is to add carbohydrates to the proteins as they progress through the apparatus.

The Endoplasmic Reticulum and Ribosomes The endoplasmic reticulum may be either smooth or rough (Fig. 4.12), depending on the absence or presence of *ribosomes*. The ribosomes are composed of a small and a large subribosomal particle. Those that are attached to the endoplasmic reticulum synthesize proteins, including enzymes, that are generally exported to other cells, tissues, or the surroundings. The ribosomes that are free in the cytoplasm function to produce enzymes used by the cell.

The ribosomes themselves are synthesized under the control of the RNA messengers (mRNA) that are constructed under the direction of DNA. When ribosomes are linked together by RNA in the process of assembling amino acids into proteins (Ch. 6), they are called polyribosomes (Fig. 4.13). It is interesting to note that the proteins that make up part of ribosome structure are synthesized on cytoplasmic polyribosomes and migrate rapidly into the nucleus to participate in ribosome assembly in the nucleolus. This is an excellent example of the dynamic movements that occur in cells at the molecular level.

Differences in the relationships of ribosomes to the endoplasmic reticulum are illustrated by cells of the pancreas and of striated muscle. The pancreatic cells that export digestive enzymes such as trypsin and lipase have a rough endoplasmic reticulum. Muscle cells, principally composed of contractile proteins (Ch. 8) that are not secreted, have a smooth endoplasmic reticulum called the sarcoplasmic reticulum (Fig. 4.14).

Lysosomes and Secretory Vesicles Lysosomes are membrane-bound packages of enzymes. Usually they are formed on the secreting face of the Golgi apparatus. Sometimes the lysosomes simply store the enzymes. Sometimes they join vacuoles that have formed on the surface of the cell and moved inward; here they can digest ingested material. Sometimes they contribute to the enzymatic breakdown of the cells that produce them. This is an important function in animals undergoing rapid change, such as in amphibian metamorphosis where a tadpole, as it becomes a frog, has its tail rapidly destroyed.

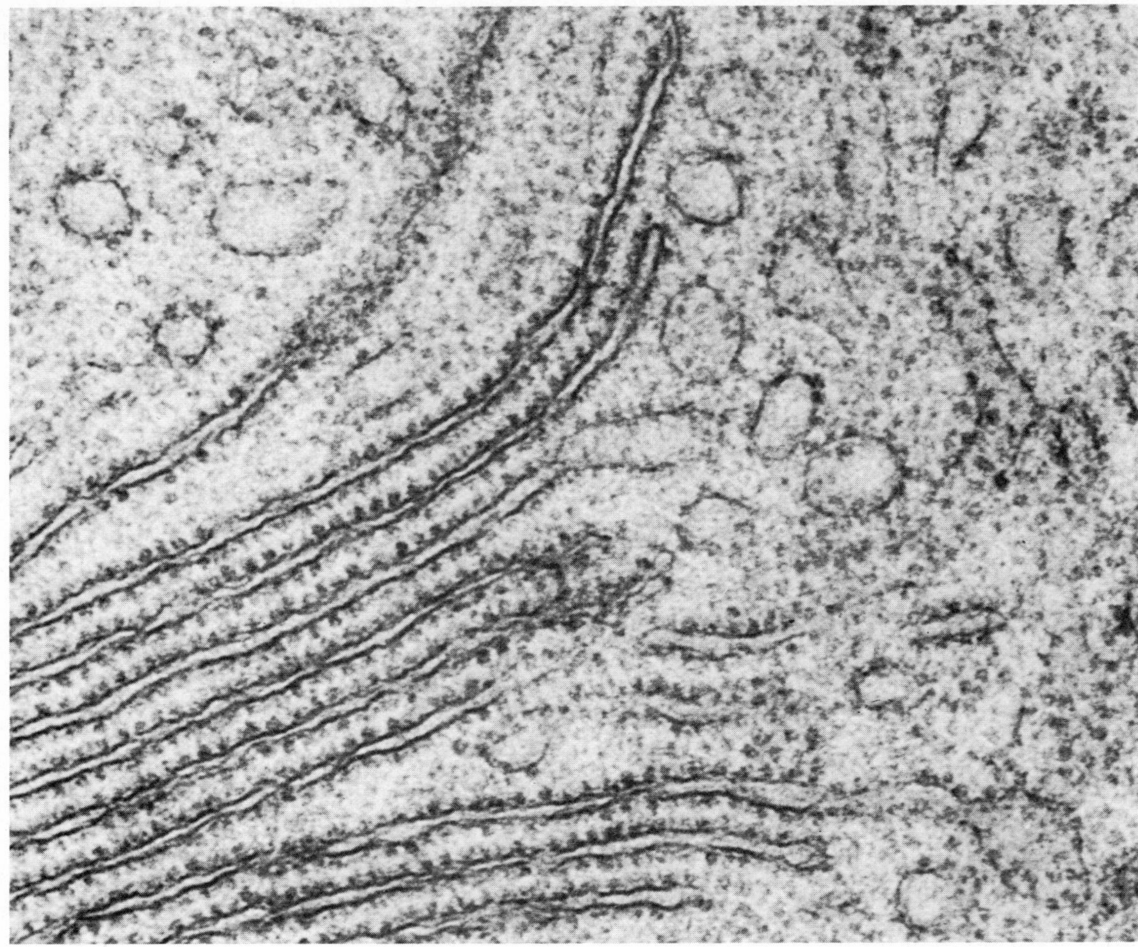

Figure 4.12 Rough (granular) endoplasmic reticulum from an acinar cell of the pancreas (× 90,000). (Courtesy of D. W. Fawcett, Harvard University, Cambridge.)

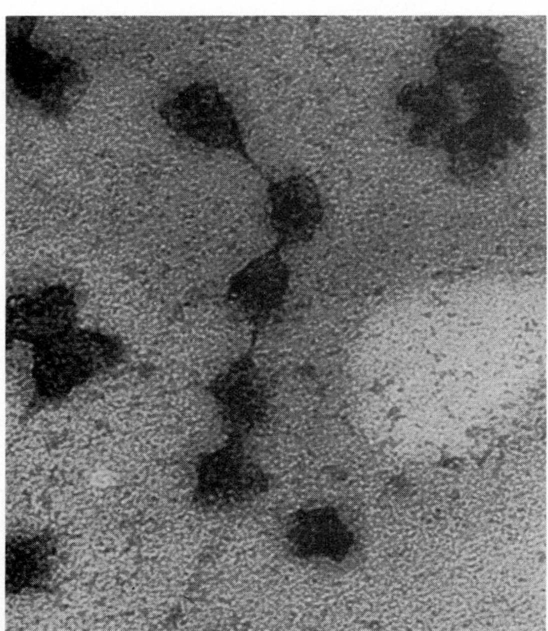

Secretory vesicles also are produced on the secreting face of the Golgi apparatus but move to the surface to discharge their contents and to add to the plasma membrane (Fig. 4.15).

NUCLEAR ENVELOPE AND MEMBRANE SYNTHESIS

A better appreciation of membrane structures will be obtained from viewing the role of the nucleus in controlling membrane synthesis. The nuclear envelope (Fig. 4.15) is composed of two distinct membranes. The inner smooth one, facing the nucleus, is associated with chromatin (nucleic acids and proteins, Ch. 6) and bioinformation is transferred to the outer membrane

Figure 4.13 Electron micrograph of isolated rabbit reticulocyte polyribosomes stained with uranyl nitrate. Especially obvious is the strand of messenger RNA along which the ribosomes are assembled. (Courtesy of A. Rich, Massachusetts Institute of Technology, Cambridge.)

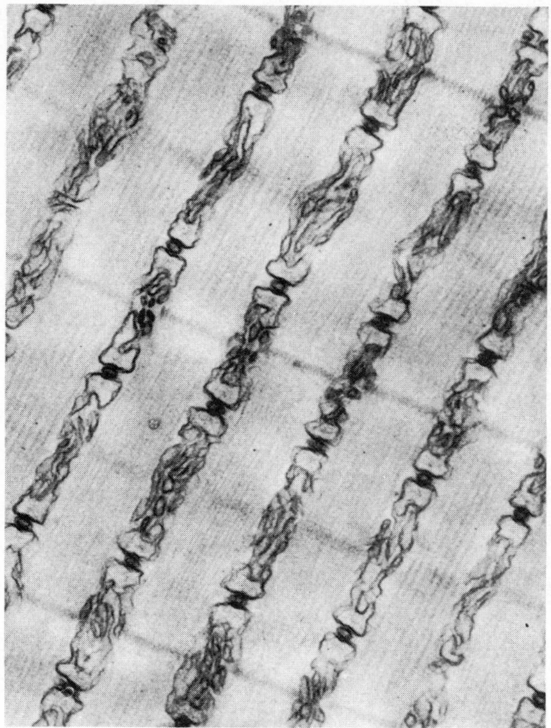

Figure 4.14 Sarcoplasmic reticulum in the muscle of the swim bladder of a fish (× 39,000). (From D. W. Fawcett and J. P. Revel, *J. of Cell Biology* (Suppl.), **10,** 89–109, 1961.)

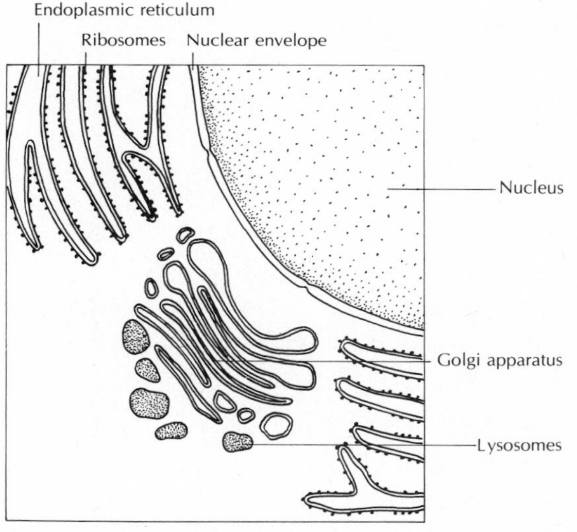

Figure 4.15 Interrelationships among the endoplasmic reticulum, Golgi apparatus, and lysosomes.

that possesses ribosomes. Here a new membrane is formed that moves outward and makes up an *endomembrane* system consisting of endoplasmic reticulum, Golgi apparatus, secretory vesicles, endocytotic vesicles (those that are formed at the edge of the cell) and vacuoles, and the plasma membrane itself. Thus except for the mitochondria and plastids and fibrillar structures in the cytoplasm, all membranous structures are related in their origin near the nucleus, can have continuity, and may be transformed from one membranous structure to another.

OTHER MEMBRANE-BOUND STRUCTURES

The remaining structures of the cytoplasm that can be seen in cells are the *vacuoles.* These are membrane-bound cavities within the cytoplasm filled with water and dissolved salts. In animal cells they are usually small and often numerous.

Vacuoles in some lower animals are utilized to move food through the cytoplasm where it is digested. Some protozoa have contractile vacuoles with associated radiating canals that deliver fluid to the vacuoles; the vacuoles rhythmically discharge the fluid to the exterior. Movement of these organelles is accomplished with the aid of microtubules attached to their membranes (see below). The contractile vacuoles function as a mechanism for maintaining fluid balance in the protozoa. In addition to the principal structures just described, both plants and animals may have nonliving materials, or inclusions, present such as starch grains, crystals, pigment granules, and secretory granules in the cytoplasm, many of which are not membrane bound.

It is important to understand that individual organelles do not operate independently. For instance, mitochondria convert energy and make it more useful to the cell, Golgi bodies package products resulting from the energy provided by the mitochondria, and some vacuoles containing these products may transport them to other parts of the cell or to the exterior.

FILAMENTOUS AND TUBULAR CYTOPLASMIC ORGANELLES

Three major kinds of cellular structures exist in filamentous form. The ones studied longest are the *myofibrils* that form the contractile elements in muscle cells (Ch. 8). They consist of parallel and interdigitating arrays of two kinds of proteins, actin and myosin.

Microtubules form the second kind of filamentous structure. They are formed from a protein, *tubulin.* They are long straight structures that, in cross section, appear as a circle of about 24 nm in outside diameter. The protein in these microtubules is arranged in parallel, slightly pitched subunits that probably have cross-

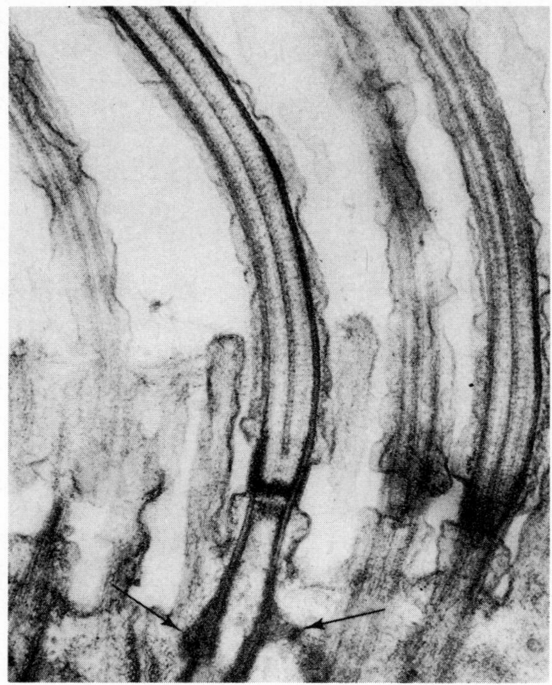

Figure 4.16 Cilia on the gill of the fresh-water mussel as seen in longitudinal section (× 34,000). The shaft of the cilium curves upward from basal bodies (see arrows) in the epithelial cytoplasm. (Courtesy of I. Gibbons, University of Hawaii.)

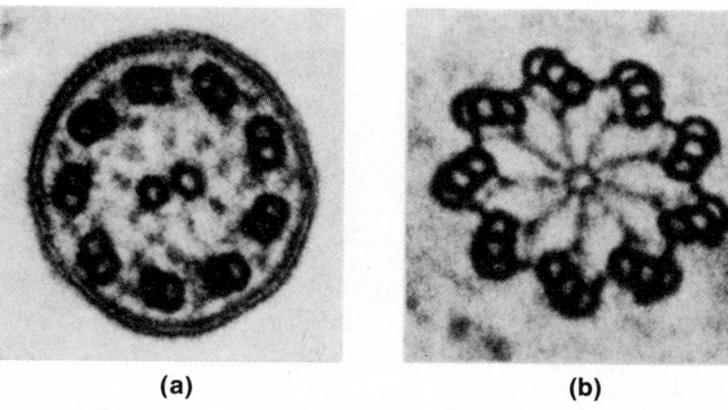

(a) (b)

Figure 4.17 **(a)** Cross section of a flagellum from the protozoan, *Trichonympha* (× 127,500). **(b)** Cross section of a basal body in *Trichonympha* (× 127,500). (Courtesy of I. Gibbons, University of Hawaii.)

bridges between the molecules. Also there are often lateral projections between tubules. Organelles constructed of microtubules are cilia, flagella, basal bodies, sperm tails, centrioles, and spindle fibers. A characteristic organization of such structures is a "9 + 2" microtubular arrangement (Figs. 4.16 and 4.17). Centrioles and spindle fibers (Figs. 4.18 and 4.19) illustrate two forms of microtubular organization. A centriole is a self-replicating organelle whose progeny centriole lies at right angles to the parent organelle; each centriole is approximately 330 nm long and 150 nm in diameter. Each cylinder is composed of 9 sets of triple tubules running in the long axis.

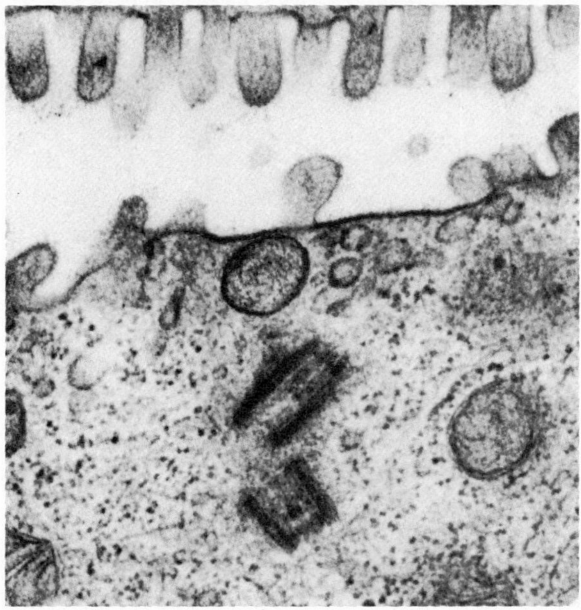

Figure 4.18 Longitudinal section of centrioles from embryonic chick epithelium. (Courtesy of S. Sorokin; from D. W. Fawcett, *The Cell: An Atlas of Fine Structure,* W. B. Saunders, 1966.)

Spindle fibers in animals are synthesized in association with centrioles as the nuclear envelope disappears and the two centrioles move toward opposite poles in karyokinesis (Ch. 7).

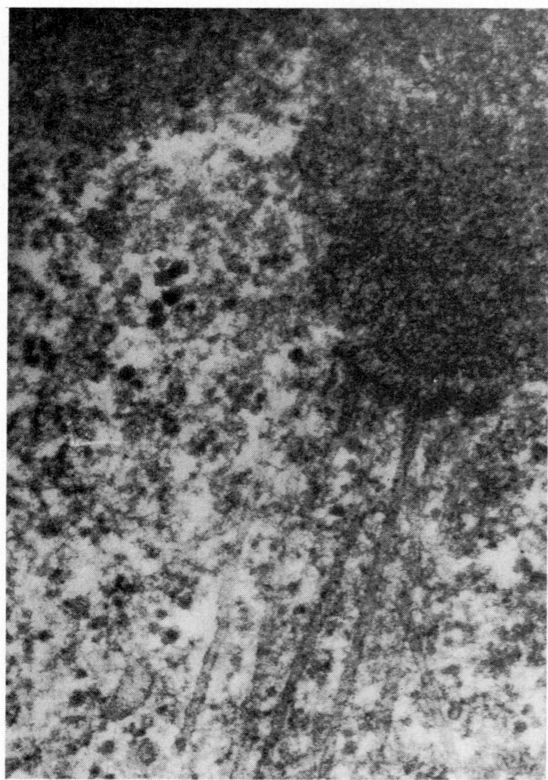

Figure 4.19 Spindle fibers are microtubules. Tubular fibers extend from the centromere of a chromosome seen as the dark area at top right. (From E. Robbins and N. Gonatas, *J. of Cell Biology,* **21,** 429, 1964, with permission.)

The spindle fibers are assembled from shorter preexisting segments rather than by being synthesized on ribosomes at the time. Some spindle microtubules are continuous between the centrioles and some terminate on the chromosomes.

Microfilaments form the third kind of filamentous organelle. As with microtubules, they seem to be universal among eukaryote cells. These structures are not hollow and range between 5 and 8 nm in diameter. Both actin and myosin are found. The microfilaments function in several ways. Noteworthy is their involvement in creating the constriction of the plasma membrane during the cytokinesis process when one cell becomes two. They have been implicated in streaming motions of the cytoplasm of such cells as amoebae, migrating fibroblasts, and glial cells of the nervous system. In some cells it can be seen that the microfilaments are of two kinds or distributions. Some microfilaments lie in parallel distributions in cells while others form a loose network.

There are two inhibitors of the assembly of microtubules and microfilaments, both of which interfere with cell division. One inhibitor, colchicine, specifically prevents the assembly of subunits of microtubules. The other inhibitor, cytochalasin B, interferes with the functioning of microfilaments. These two inhibitors have been selectively used to study the functions of microtubules and microfilaments.

The Nucleus

The nucleus receives expanded treatment in Chapters 6 and 7 and, therefore, is only briefly discussed here. In the interphase stage (Ch. 7), when the nucleus is organized to control cellular functions, electron micrographs show a granular background in which lie a few large, round, dark nucleoli, and areas of dense irregular chromatin masses are associated with the inner membrane of the nuclear envelope. There are no discernible membranes or microtubules within the nucleus, except in protozoa that produce a mitotic spindle within the nuclear membrane.

There are pores in the nuclear envelope (Fig. 4.15) that are not simply holes, but have a complex structure and regulate exchange between the nucleoplasm and the cytoplasm.

JUNCTIONS BETWEEN CELLS

There is usually an identifiable intercellular space of about 20 nm between the plasma membranes of adjacent cells. However, there are three major kinds of junctions between cells. *Tight junctions* are places where the intercellular spaces are absent. They usually exist between unlike kinds of cells or tissue layers. They seem to prevent easy access of materials from one population of cells to the other. *Gap junctions,* on the other hand, are places where membranes of two adjacent cells create passageways between the cytoplasms of the cells, enhancing the molecular traffic between cells. They can be found between smooth muscle cells, for instance. The third junction type, *desmosomes* (Fig. 4.20), can be visualized as localized "spot welds." Locally the intercellular space is filled with a dense, fibrous material, and internally microfilaments, *tonofibrils,* extend into the cytoplasmic matrix. The function of desmosomes seems to be purely mechanical, reenforcing contacts between, for example, epithelial cells where considerable strength is advantageous; a special and large desmosome association exists in the form of intercalated discs of heart muscle, another place where intercellular strength is important.

CELL DIVISION

Although cell division is considered in more detail in Chapter 7, the main features of the process are outlined here because of its importance to the survival of cells.

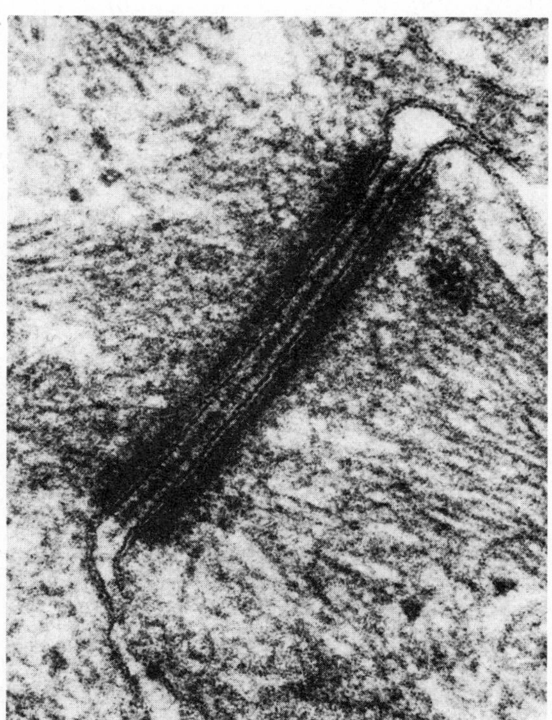

Figure 4.20 Desmosome at the junction of two epithelial cells of a fish capillary (× 140,000). (Courtesy of D. W. Fawcett, Harvard University.)

Cell division involves two processes: nuclear division, called *mitosis;* and a division of the cytoplasm, called *cytokinesis.*

During the intervals between the stages of mitosis, the chromosomes are extended and not easily recognized. A condensation of chromosomes so that they become visible signals the onset of mitosis. By this time, replication has occurred in each of the chromosomes. In most cells, when the chromosomes have stopped contracting, the nuclear envelope disappears. This is accompanied by the formation of a *spindle* composed of microtubules, called spindle fibers, that converge at the two poles. The chromosomes move into position on the equatorial region of the spindle and the replicated, but still joined, chromosomes become attached to the fibers, which converge on centrioles. The two strands of each chromosome then separate and move toward the centrioles at opposite poles where they aggregate, and a new nuclear envelope forms around them. This creates two nuclei, each with identical complements of chromosomes. The process of mitosis is usually followed by cytokinesis.

There are a number of instances in plants and animals where nuclei divide but the cytoplasm does not. In animals, the multinucleate condition is called a *syncytium.* Syncytia can occur in two ways. One is for the nuclei to divide without the cytoplasm dividing. Some embryonic tissues associated with the placenta illustrate this. The other way of creating a syncytium is for originally separate cells to fuse. This occurs in the development of striated muscle in vertebrates.

COMMUNITIES OF CELLS: MULTICELLULAR ORGANIZATION

When cells remain together, forming multicellular organisms, most of the cells become specialized. This provides for a division of functions among cells that usually does not occur in unicellular forms of life such as bacteria, some fungi, protozoa, and some algae. While single-celled organisms are capable of immortality in that their cell division is capable of continuing their kind indefinitely, the multicellular condition is accompanied by a restriction of potential immortality to special *primordial germ cells* (see Ch. 7). These cells are the progenitors of eggs and sperm. The remaining animal cells are called *somatic* cells.

Multicellular Organization of a Complex Animal

The fertilized egg of an animal divides a large number of times (Fig. 4.21). Its cells remain adherent to each other. In time, a stage is reached when adherent groups of cells move as sheets or clusters of cells. Beginning certainly with flatworms and all higher phyla, this produces three kinds of cell layers called the *germ layers.* They

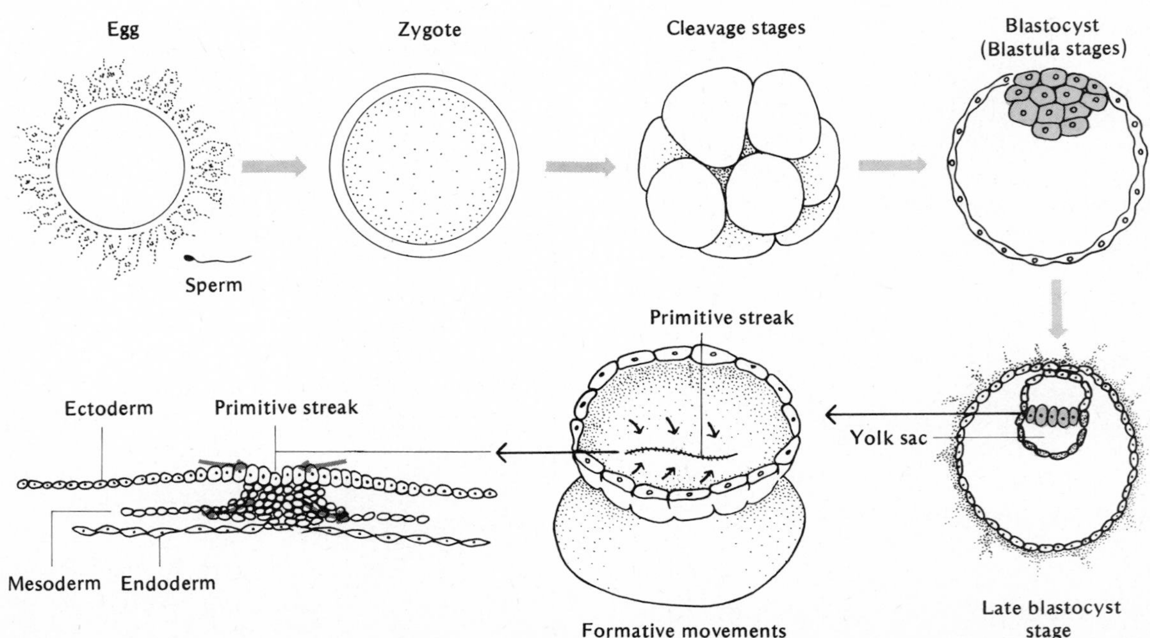

Figure 4.21 Some steps in development from an egg illustrating the appearance of organization and pattern.

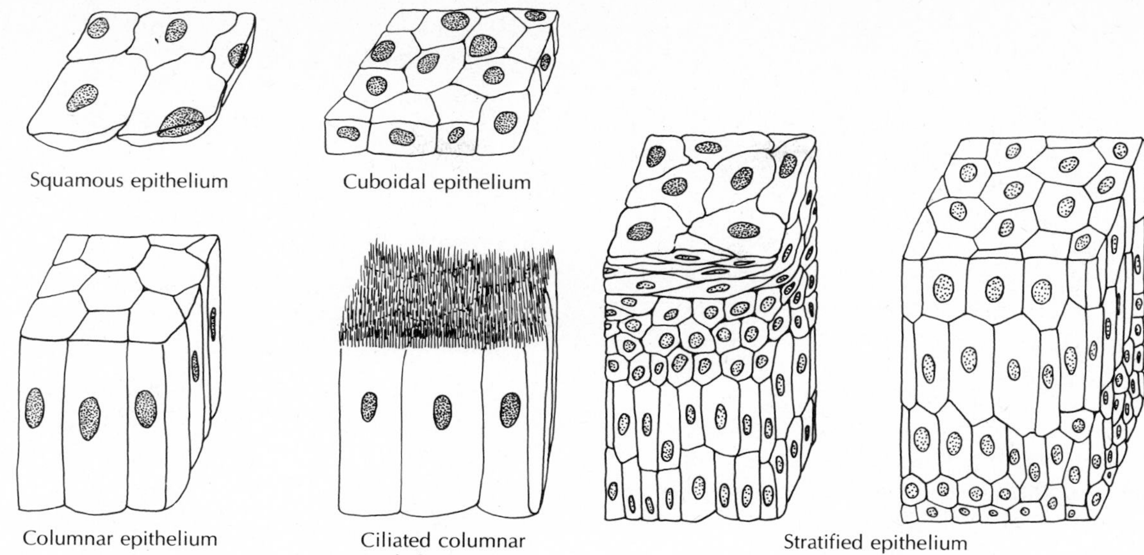

Figure 4.22 Epithelial tissues.

are an outer layer, *ectoderm,* a middle *mesoderm,* and an inner *endoderm.* From these three embryonic layers, cells differentiate and become unlike in structure and function. Those groups of similar cells that remain together constitute *tissues.* In a complex animal such as humans, five basic kinds of tissues are recognized: epithelial, supporting, muscle, nervous, and vascular.

As will be seen in Chapters 22–33, invertebrate developments can be of several kinds. In these animals movement of cells in layers or sheets is not as usual as it is in the vertebrates used here.

EPITHELIAL TISSUE

The cells of epithelial tissues are closely connected, with little intercellular material between them (Fig. 4.22). They cover the body and line the various cavities, ducts, and vessels as continuous layers. The various glands, such as the salivary glands, the pancreas, and the sweat glands, are derived from specialized epithelial cells. The most important general function of epithelial tissue is in connection with the movement of materials. All substances that normally enter and leave the body tissues must pass through an epithelium. The several shapes and arrangements of epithelial cells assist in classifying them. *Squamous* epithelium has thin, flat, tilelike cells. Such an epithelium lines the body cavity, covers many of the internal organs, and composes the mesenteries that form the thin sheets of tissue that support the internal organs of vertebrate animals. *Cuboidal* epithelium may be found in the cells of kidney tubules. *Columnar* epithelium is generally found lining the alimentary tracts of animals, and there the elongated cells are fitted together in sheets, with the long axes of the cells at right angles to the cavity.

Squamous, cuboidal, and columnar are the basic shapes of epithelial cells. Each type may exist as several layers forming a *stratified epithelium.* The outer part of human skin is a *stratified squamous epithelium,* for instance. Also, each type may have cilia, vibratile hairlike extensions of the cytoplasm, on the free surface. The epithelium lining the air passages (the trachea) is a *ciliated epithelium.*

While epithelium can be formed from all three germ layers, it is interesting from a functional standpoint that ectodermally derived epithelia generally serve a protective function, and that one of the common molecules produced by these cells is the protein *keratin* that can be found in skin, nails, claws, hooves, and hair. Epithelia produced by the endoderm tend to be mucous cells providing lubrication and secretions, as well as allowing absorption and transport across the cells. A common product of these cells is *mucin.* Epithelia derived from mesoderm primarily function in the movement of materials in from one side of the cell and out on the other side and do not have prominent special secretions.

SUPPORTING TISSUE

The supporting tissues are made up of *fibrous connective tissue, cartilage,* and *bone.* These tissues support and bind the different parts of the body together. They differ from other tissues

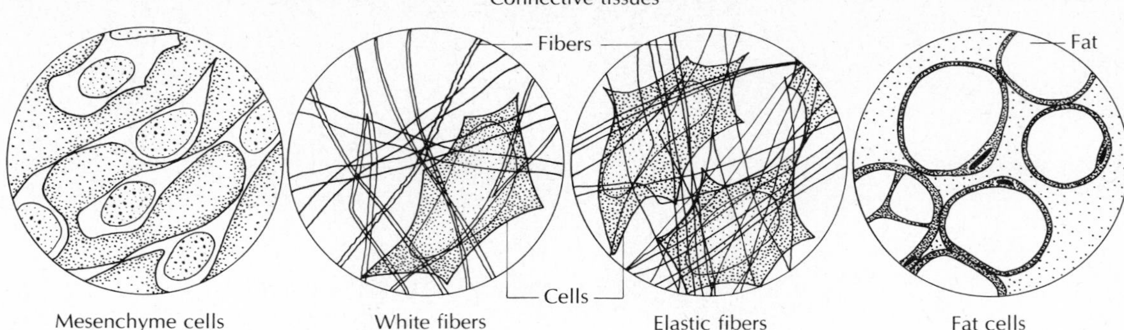

Figure 4.23 Some types of connective tissue.

in that they all produce a very large amount of intercellular material. This intercellular material, the *matrix,* is different in the three types mentioned. The mature cells are distributed in the matrix, which is secreted by the cells during growth and development.

Fibrous connective tissue (Fig. 4.23) is found in many places in the body; it binds the skin to the muscles, attaches muscles to bones, covers the blood vessels, and so on. The matrix of such tissue may contain white fibers, yellow elastic fibers, or both. The white fibers, composed of *collagen* (a source of gelatin), are the most characteristic elements of this type of tissue.

In hyaline cartilage (Fig. 4.24) the cells are scattered through a matrix of uniform appearance. This cartilage is a tough, somewhat elastic material found at the ends of bones, in the nose, and in the ear. In the early embryo most of the skeleton is composed of cartilage, which is later converted into bone. Other types of cartilage have fibers in the matrix.

Bone is the characteristic vertebrate supporting, or skeletal, tissue. Due to the calcium salts deposited in the bone matrix, it forms a very hard, rigid tissue. The bone cells are scattered in this matrix, usually in a definite pattern and with numerous fine processes extending in all directions. As Figure 4.25 shows, the cells are arranged in concentric circles around canals through which the blood vessels and nerves pass. These are called Haversian systems.

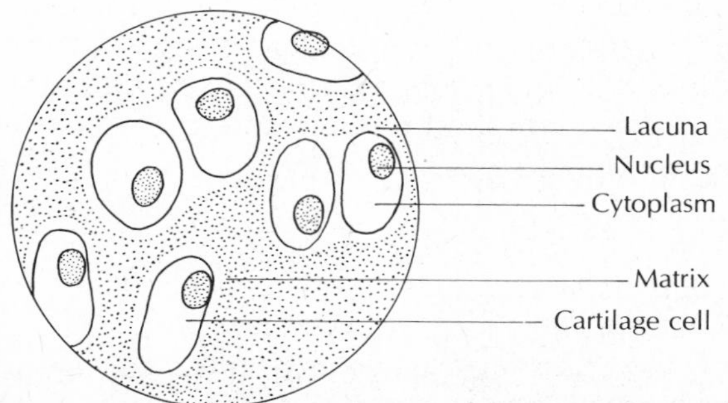

Figure 4.24 Hyaline cartilage showing cartilage cells, lacunae, and matrix.

Adipose tissue (Fig. 4.23), or fat tissue, is considered to be a modified form of fibrous connective tissue. Here the cells have enlarged through the deposition of fat in large vacuoles. Adipose tissue is found in many parts of the body. The degree to which these cells are laden with fat determines the obesity of the person, not the number of fat cells, which is determined early in life. A person who early develops a relatively large number of fat cells will later have greater difficulty in coping with the control of body weight than one who has fewer fat cells.

MUSCLE TISSUE

Muscle tissue functions in contraction. Muscle cells have thin threads, *myofibrils,* in their cytoplasm which are responsible for the contractile qualities of muscle tissues (Ch. 8). Three kinds of muscle tissues are recognized in vertebrates:

1. *Smooth* or *involuntary* muscle found in the various internal organs, the blood vessels, and the ducts of glands
2. *Striated* or *voluntary* muscle associated with the skeleton and constituting the bulk of the body tissues—the "meat" we eat
3. *Cardiac* muscle found in the heart

Smooth muscle cells (Fig. 4.26) are long and spindle shaped. The oval nucleus lies near the widest part of the spindle. With proper staining, the myofibrils in the cytoplasm can be seen. Smooth muscle cells are found in the form of sheets in the various internal organs, especially around fluid-filled sacs. As a rule, these sheets are double, with the axes of the cells in one sheet at right angles to the axes of the cells in the other sheet.

Striated muscle tissue (Fig. 4.27), found attached to the skeleton, is a syncytium composed

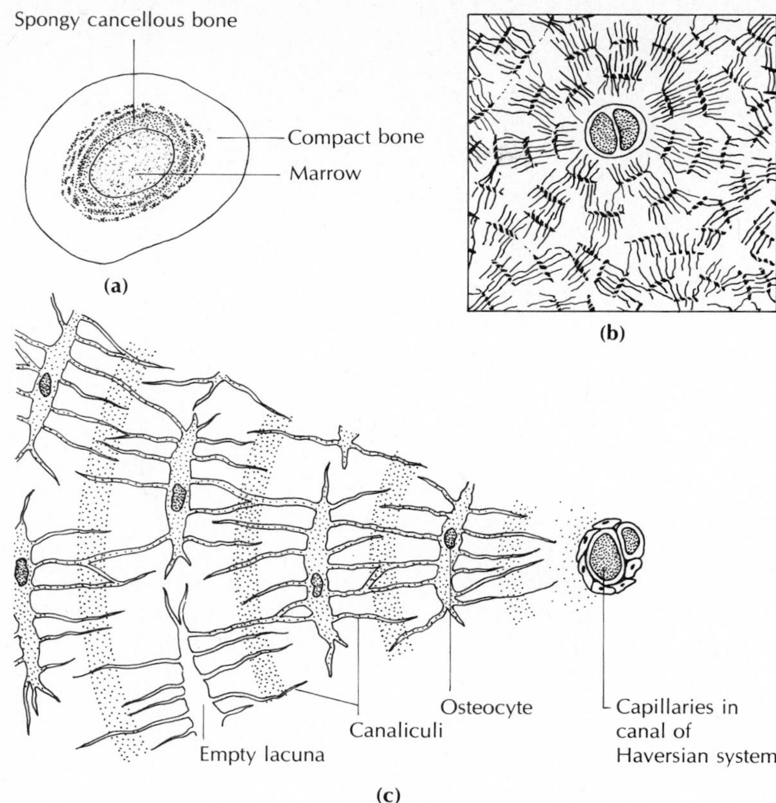

Figure 4.25 Bone. **(a)** Cross section of a long bone. **(b)** A small section of bone. **(c)** Enlargement of **(b)**.

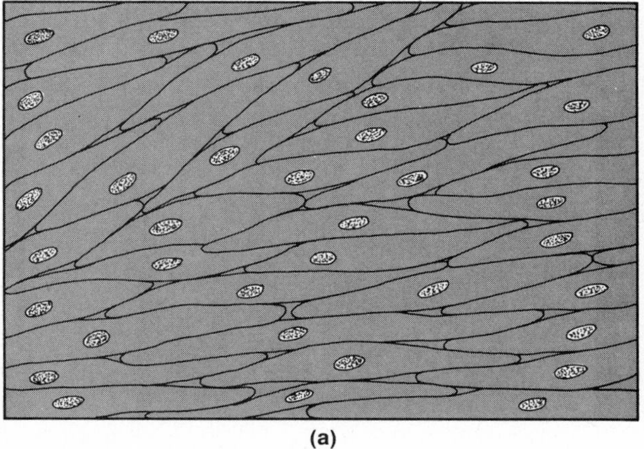

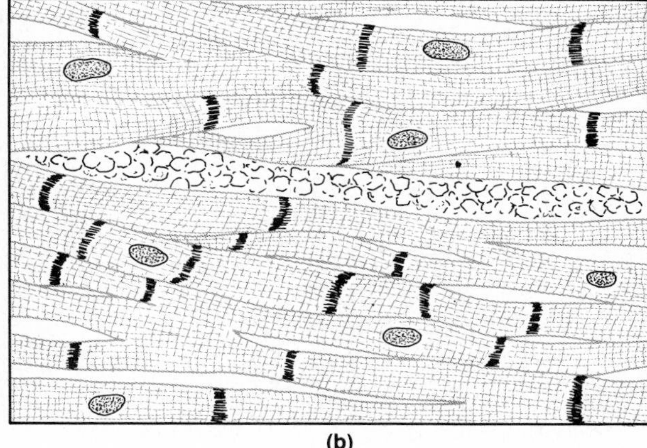

Figure 4.26 Muscle. **(a)** Smooth. **(b)** Cardiac.

of long multinucleate structures called *fibers.* Each fiber is bounded by a thin membrane, the *sarcolemma.* The sarcolemma bounds the cytoplasm, called *sarcoplasm,* and the long *cross-striated myofibrils.* Muscle fibers are organized into bundles that compose whole muscles.

Cardiac muscle (Fig. 4.26) has myofibrils that are cross-striated, but the tissues give the appearance of typical uninucleate, not multinucleate, cells. Each cell, however, is connected with adjacent cells by specializations of the membrane. From a functional standpoint, cardiac muscle differs from both the involuntary and voluntary muscles. It contracts automatically and rhythmically. For instance, the heart muscle of a chick embryo starts contracting during the second day of development, before there are any nerve connections to the heart.

NERVOUS TISSUE

The *nervous tissue* is composed of very specialized cells, called neurons (Fig. 4.28), that are involved in the coordination of the organism and in its conscious experiences. Neurons function in *irritability* and *conductivity* and are of several kinds; however, they all possess some basic

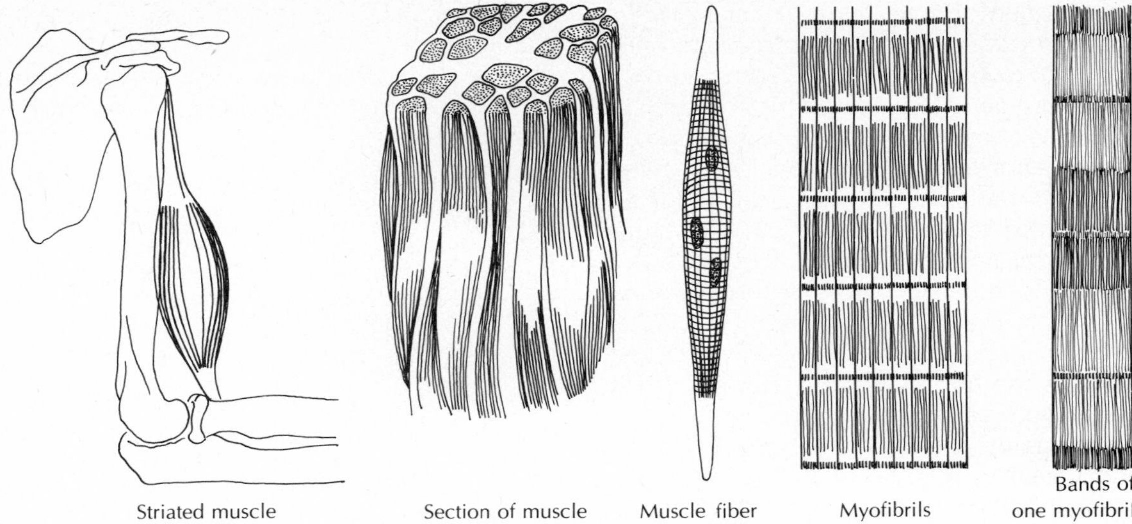

Figure 4.27 (*above*) Striated muscle.

Figure 4.28 Neuron, structural unit of the nervous system. **(a)** Complete motor neuron. **(b)** Organization of cell body. **(c)** Cross section of axon showing myelin sheath. **(d)** Neuromotor junction. **(e)** Variations in the shapes of neurons.

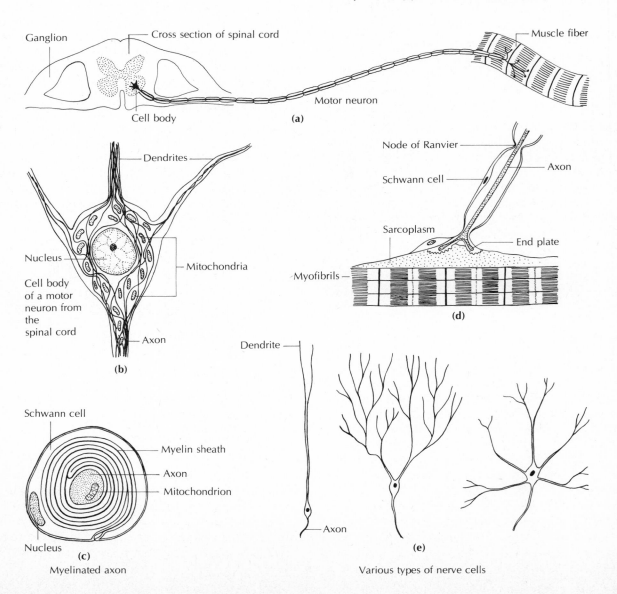

Ganglion
Cross section of spinal cord
Muscle fiber
Motor neuron
Cell body
(a)

Dendrites
Nucleus
Mitochondria
Cell body of a motor neuron from the spinal cord
Axon
(b)

Node of Ranvier
Axon
Schwann cell
Sarcoplasm
End plate
Myofibrils
(d)

Schwann cell
Myelin sheath
Axon
Mitochondrion
Nucleus
(c)
Myelinated axon

Dendrite
Axon
(e)
Various types of nerve cells

features that can be illustrated in a *motor neuron,* one that transmits impulses from the brain or spinal cord to skeletal muscle. Each neuron is composed of a *cell body* containing the nucleus and two kinds of cytoplasmic processes, numerous short *dendrites,* and a single long *axon.* From a functional point of view a dendrite carries an impulse toward the cell body, and the axon carries the impulse away from the cell body. Neurons do not exist singly but in aggregations. What one commonly calls *nerves* are bundles composed of nerve processes, axons and dendrites. When the cell bodies of neurons are in aggregates outside of the brain and spinal cord, these aggregates are called *ganglia* (singular, *ganglion*). Neuron processes may exist as unmyelinated fibers (naked), or they may be surrounded by an insulating sheath, the *myelin* or *medullary sheath,* as shown in Figure 4.27. This sheath is composed of fatty material and tends to give the nerves and other parts of the nervous system, where it is present, a white appearance.

VASCULAR TISSUE

The *vascular tissue* is the blood, a liquid tissue whose matrix is the *blood plasma.* Suspended in the fluid matrix are three general types of cells or cell fragments:

1. Red blood cells, or *erythrocytes,* containing the compound hemoglobin which greatly increases the oxygen-carrying capacities of the blood
2. White blood cells, or *leucocytes,* some of which are amoeboid and capable of engulfing bacteria and foreign substances that may be present in the body while others produce antibodies
3. Blood platelets, or *thrombocytes,* small colorless structures that function in the formation of blood clots

The blood vessels do not qualify as part of the vascular tissue but are organs composed of several tissues, such as epithelial, connective, and muscular. Further details of the blood are presented in Chapter 11.

Organs of an Animal

The different types of tissues described above are combined to form the various organs, such as the heart, lungs, stomach, kidneys, bladder, small intestine, and large intestine. In each of the various organs the kinds of tissues are arranged in a characteristic way, making it possible for the organ to perform its particular functions.

An examination of the small intestine of a frog (Fig. 4.29) shows how tissues are combined into organs. Five rather well-defined regions will be described in order, beginning with the cavity, or *lumen,* of the gut and working out.

The lining layer, or *mucosa,* is composed mainly of columnar epithelium. The cells of the mucosa perform at least three functions. Distributed among the main kinds of columnar cells are *goblet cells,* so-called due to the gobletlike cavity filled with *mucin.* This secretion serves as a lubricant. The majority of the epithelial cells perform the other two major functions: they secrete enzymes into the lumen of the intestine and they absorb digested material from the lumen.

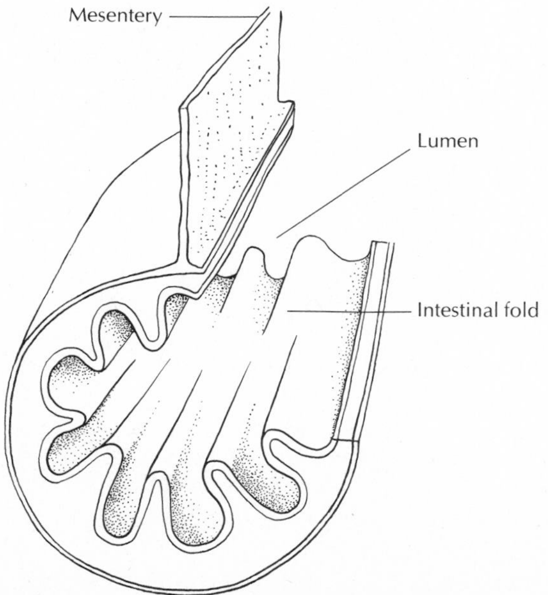

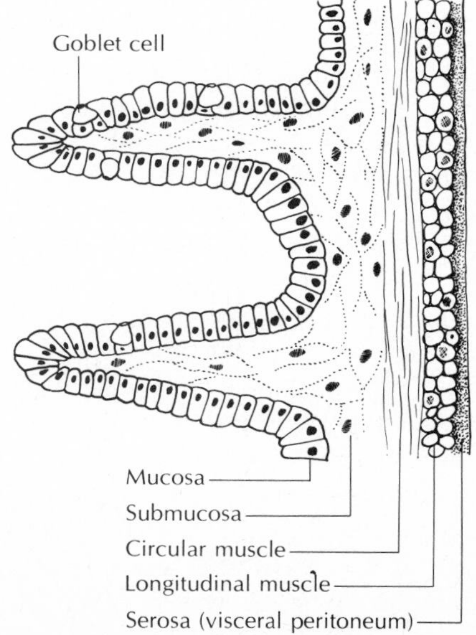

Figure 4.29 Section of a frog's intestine showing the folds and layers.

Next to the mucosa is a layer that is somewhat irregular in appearance. This is the *submucosa,* composed primarily of fibrous connective tissue. Pockets of epithelial cells from the mucosa form *intestinal glands* that project into the submucosa. The submucosa is also richly penetrated by small blood and lymph vessels and by nervous tissue.

Adjoining the submucous layer is the *circular muscle layer* whose smooth muscle cells have their long axes arranged around the lumen.

Just external to the circular muscle layer is the *longitudinal muscle layer* whose smooth muscle cells are at right angles to the circular muscle cells. The contraction of the cells in the longitudinal layer causes the intestine to shorten, while the contraction of the circular muscle layer constricts the lumen and elongates the intestine.

The very thin, outermost layer of the intestine is the *serosa,* or the *visceral peritoneum.* It is composed of a sheet of flat squamous epithelium.

The segment of intestine just described does not function as five separate parts. Instead, the individual cells of each layer operate as members of a community of similar cells, or as a tissue, and the tissues operate in a coordinated fashion as an organ. Cells in the mucosa are primarily involved in the absorption of digested food material. Cells in the submucosa support the blood and lymph vessels into which the absorbed food materials are passed to other parts of the body; the two layers of muscle tissue churn the contents of the lumen and move it along; and the serosa acts to encase and support the other tissues.

Organ Systems

The small intestine is an organ forming a part of a larger *organ system,* the digestive system. In addition to the small intestine, the digestive system includes the mouth, salivary glands, esophagus, stomach, large intestine, pancreas, and liver. Throughout the body, tissues are organized into organs and organs function together as organ systems. There is a definite hierarchy in the organization of most organisms. The *cells* are the basic units of structure and form the different *tissues.* The tissues are combined into *organs,* and a number of organs functioning together constitute an *organ system.* The following organ systems are found in the vertebrates:

1. Digestive system
2. Respiratory system
3. Circulatory system
4. Excretory system
5. Reproductive system
6. Nervous system
7. Muscular system
8. Skeletal system
9. Integumentary system
10. Endocrine system

SUMMARY

Students of the organization of life are impressed by the similarities and differences among living organisms. The flexibility and freedom of animal cells have permitted the development and survival of many cell types. When the animal cells aggregated into communities of cells as multicellular organisms, five kinds of tissues evolved, producing a great diversity in kinds of organs in various animals.

These points are made here to emphasize that organisms that survive and reproduce are successful in the biological sense and that adaptations make them successful. The study of cells, tissues, and organs is thus an introduction to a common biological phenomenon that is repeatedly identified or described throughout this book, namely, variation develops on common themes.

SUGGESTIONS FOR FURTHER READING

Berridge, M. J., and J. L. Oschman. *Transporting Epithelia.* New York: Academic Press (1972).

DuPraw, E. J. *Cellular and Molecular Biology.* New York: Academic Press (1968).

Fawcett, D. W. *An Atlas of Fine Structure.* Philadelphia: W. B. Saunders (1968).

Kimoto, S., and J. C. Russ. "The Characteristics and Applications of the Scanning Electron Microscope." *American Scientist,* **57,** 112–133 (1969).

Lehninger, A. L. *The Mitochondrion.* New York: W. A. Benjamin (1971).

Loewy, A. G., and P. Siekevitz. *Cell Structure and Function,* 2d ed. New York: Holt, Rinehart and Winston (1969).

Palade, G. E. "The Organization of Living Matter." *Proc. Nat. Academy of Science,* (U. S.), **52** (Aug. 1964).

Porter, K. R., and M. A. Bonneville. *An Introduction to the Fine Structure of Cells and Tissues.* Philadelphia: Lea and Febiger (1963).

Swanson, C. P. *The Cell.* Englewood Cliffs, N.J.: Prentice-Hall (1969).

Threadgold, L. T. *The Ultrastructure of the Animal Cell.* New York: Pergamon (1967).

From Scientific American

Allen, R. D. "Amoeboid Movement" (Feb. 1962).

Allison, A. "Lysosomes and Disease" (Nov. 1967).

Basergo, R., and W. E. Kisieleski. "Autobiographies of Cells" (Aug. 1963).

Brachet, J. "The Living Cell" (Sept. 1961).

Capaldi, R. A. "A Dynamic Model of Cell Membranes" (Mar. 1974).

de Duve, C. "The Lysosome" (May 1963).

Fox, C. F. "The Structure of Cell Membranes" (Feb. 1972).

Green, D. C. "The Mitochondrion" (Jan. 1964).

Hokin, L. E., and M. R. Hokin. "The Chemistry of Cell Membranes" (Oct. 1965).

Lehninger, A. L. "How Cells Transform Energy" (Sept. 1961).

Lowenstein, W. R. "Intercellular Communication" (May 1970).

Mazia, D. "How Cells Divide" (Sept. 1961).

Mazia, D. "The Cell Cycle" (Jan. 1974).

Neutra, M., and C. P. Leblond. "The Golgi Apparatus" (Feb. 1969).

Nomura, M. "Ribosomes" (Oct. 1969).

Racker, E. "The Membrane of the Mitochondrion" (Feb. 1968).

Rich, A. "Polyribosomes" (Dec. 1963).

Robertson, J. D. "The Membrane of the Cell" (Apr. 1962).

Satir, P. "Cilia" (Feb. 1961).

Satir, P. "How Cilia Move" (Oct. 1974).

Siekevitz, P. "Powerhouse of the Cell" (July 1957).

Soloman, A. K. "The State of Water in Red Cells" (Feb. 1971).

Stent, G. S. "Cellular Communication" (Sept. 1972).

Wessels, N. K. "How Cells Change Shape" (Oct. 1971).

Zamecnik, P. C. "The Microsome" (Mar. 1958).

5 Bioenergetics

The thousands of compounds found in living organisms are related to each other through *metabolism.* This ability to capture, transform, and store various forms of energy according to the instructions of their genetic material is an essential feature of living organisms. A cell's cytoplasm is not a random suspension of enzymes, substrates, and products, but it possesses an intricate organization that is not completely understood. The specificity of enzymes operates at the molecular level; above the molecular level there is an intricate organization of macromolecular complexes, membranes, particles, fibrils, and organelles. This highly organized state of living material is difficult to maintain from the energetic standpoint; the structural framework of the cell must have a continuous supply of energy provided for it.

Energy is the capacity to do work. Energy may be subdivided into two broad categories: *potential energy* (the energy of position) and *kinetic energy* (the energy of motion). Positional, radiant, electrical, chemical, and atomic energies are forms of potential energy.

MASS AS A FORM OF ENERGY

In 1905 Albert Einstein formulated a statement that mass and energy are interconvertible:

$$E = mc^2$$

where E is the energy, m is the mass, and c is the velocity of electromagnetic radiation (for example, light). On the basis of this statement, the first law of thermodynamics is formulated: *Mass and different forms of energy are interconvertible, or mass-energy may neither be created nor destroyed.*

The conversions between mass and energy occur in nuclear fusion and fission reactions. The helium atom has a mass of 4.0026 and the hydrogen atom has a mass of 1.0080; the relationships are shown in the following equations:

4 hydrogen atoms ⟶ 1 helium atom

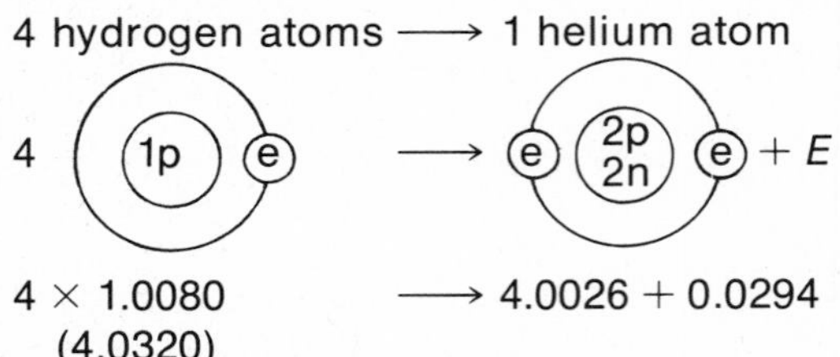

$4 \times 1.0080 \longrightarrow 4.0026 + 0.0294$
(4.0320)

This conversion of four hydrogen atoms to a helium atom thus is accompanied by the disappearance of a small amount of mass; this mass is converted to energy, according to Einstein's equation. This reaction continuously occurs in the sun at a rate of over 100 million tons of mass converted to energy each second.

ENERGY AND REACTIONS

In order for atoms to be joined by specific bonds into molecules, energy is needed to bring the atoms together to form a stable molecule. When the bonds of a molecule are broken, the energy that holds the atoms together may be used to do work. When an organism makes or breaks bonds during metabolism, it is important to know whether energy is consumed or liberated. Consider the schematic reaction of the interconversion between compounds A and B ($A \rightleftarrows B$). The reaction will be *exergonic* if energy is liberated and it will be *endergonic* if energy

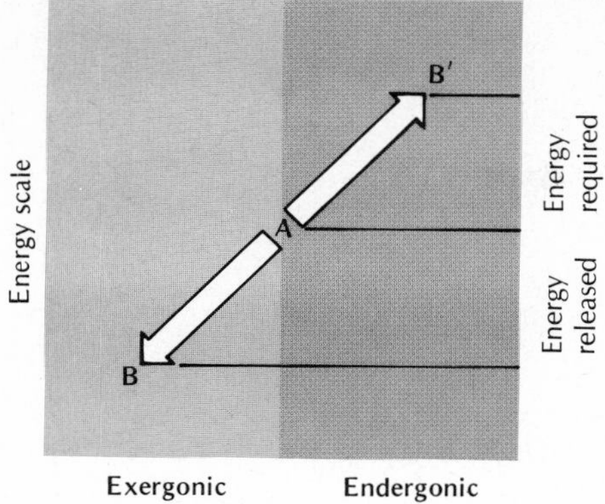

Figure 5.1 Energy relationships of reactions. An exergonic reaction releases energy and an endergonic reaction requires energy.

must be supplied to promote the reaction. These concepts are schematized in Figure 5.1. If energy is given off, not all of this energy is available to do work. That energy actually employed to do work is called free energy.[1]

Although there is no completely satisfactory explanation of chemical reactions, the *collision theory* says that all molecules in a solution do not have the same kinetic energy, for some molecules acquire more energy through collisions than do others. The molecules that are traveling at a fast rate are more likely to react than the ones that are traveling at a slow rate. Thus there is an energy barrier to the reaction of molecules, and this barrier is the *energy of activation.* It is the energy of activation that determines the rate of a reaction; the higher the barrier, the slower the rate of the reaction. The energy of activation of a chemical is shown in Figure 5.2 for the reaction A $\rightleftarrows$ B.

In this section the rudiments of the concepts of energy and chemical reactions have been presented. With respect to biological systems the function of biological catalysts, the enzymes, is to lower the activation energy and therefore speed up the attainment of an equilibrium. A catalyst does not affect the point of equilibrium; the point of equilibrium depends only on the products and the reactants, assuming that all other factors are constant.

[1] In any actual reaction not all the energy is available for useful work. Some energy is transformed into a state called *entropy,* in which it cannot be used to perform work. Entropy is a measure of randomness of a system. All systems, in the absence of a source of free energy, are inclined toward a state of randomness or high entropy. This is a statement of the second law of thermodynamics. Unless useful energy is supplied from other sources (food), an organism, as does any other thermodynamic system, tends toward a disordered and random state of high entropy (death).

ENERGY, MOLECULES, AND OXIDATION

In Chapter 3 the chemical composition of living material was discussed. The lipids, carbohydrates, proteins, and nucleic acids were presented as the four major classes of organic compounds of living systems. The proteins, lipids, and carbohydrates are significant sources of energy for organisms (Ch. 9). The energy content of a molecule depends on the atoms in the molecule and how they are arranged, that is, it depends on the structure of the molecule.

The expression "biological oxidation–reduction" refers to the movements of electrons in living organisms. *Oxidation* means the loss of electrons by an atom or molecule, whereas *reduction* is the gain of electrons. A compound that donates electrons is a *reducing agent,* and

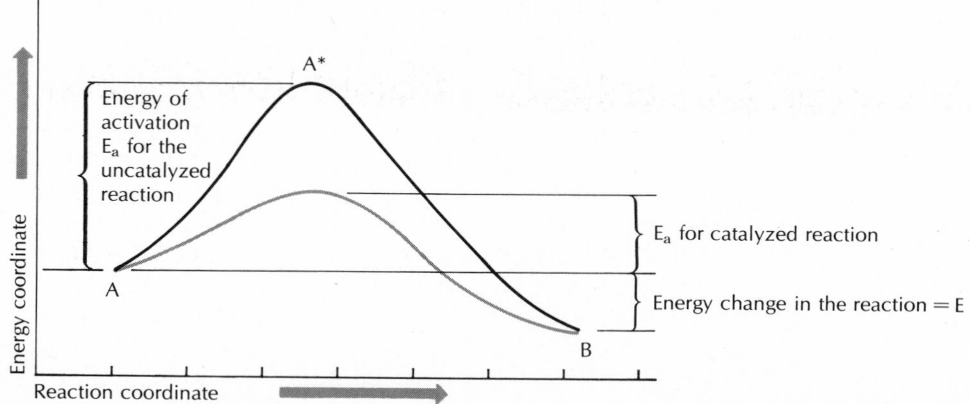

Figure 5.2 Energy diagram for a reaction. The reaction coordinate is the "path" of the reaction from A to B. The energy of activation is the "peak" of the path of the reaction. The function of a catalyst is to lower the energy of activation and speed up the attainment of equilibrium. The energy change is the same whether the reaction is catalyzed or uncatalyzed. The point of equilibrium does not change in a catalyzed reaction; only the reaching of equilibrium is accelerated.

Table 5.1 Free Energies of Hydrolysis of Some Biological Compounds

REACTION	CHANGE IN FREE ENERGY (CAL/MOL)
$ATP + H_2O \rightarrow ADP + H_3PO_4$	7300
$ADP + H_2O \rightarrow AMP + H_3PO_4$	7300
Phosphoenolpyruvic acid + $H_2O \rightarrow$ pyruvic acid + H_3PO_4	14,800
Phosphocreatine + $H_2O \rightarrow$ creatine + H_3PO_4	10,300
Glucose-6-phosphate + $H_2O \rightarrow$ glucose + H_3PO_4	3300
$AMP + H_2O \rightarrow$ adenosine + H_3PO_4	2200

a compound that accepts them is an *oxidizing agent.* Thus in an oxidation–reduction reaction the reducing agent is oxidized and the oxidizing agent is reduced.

For living systems oxidation often refers to conversion of organic compounds to carbon dioxide and water. In oxidation, hydrogen atoms (protons and electrons) are removed from compounds. A molecule that contains a relatively high proportion of hydrogen will release a greater amount of energy than one that contains a relatively low proportion of hydrogen. Thus the *caloric value* of lipids is higher than that of carbohydrates and proteins. The caloric value of a food is the amount of energy given off (expressed in calories) when the food is completely oxidized.

ENERGY-RICH COMPOUNDS

If a cell or an organism does not trap the energy of the activated complex of an exergonic reaction, then the energy is liberated as heat. *Adenosine triphosphate* (ATP) (Fig. 3.18) is a compound that is often used to link exergonic reactions with endergonic reactions, that is, the activated intermediate of a reaction drives the synthesis of ATP, and then ATP drives the synthesis of other compounds.

Energy-rich compounds exhibit a large decrease in free energy when they are hydrolyzed, and they are generally destroyed by acid, alkali, and heat. ATP shows a decrease in free energy of 7300 calories per mole,[2] (cal/mole) when it is hydrolyzed to adenosine diphosphate (ADP); furthermore, ADP also shows a decrease in free energy of 7300 cal/mole when it is hydrolyzed to adenosine monophosphate (AMP). Since these are relatively large changes in free energy, it is often said that ATP contains two high-energy phosphate bonds. Adenosine phosphates are not the only energy-rich compounds found in cells. Table 5.1 shows the free energies of hydrolysis of some other biological compounds in addition to energy-rich phosphates.

[2] A mole is the molecular weight expressed in grams or the weight of 6.02×10^{23} molecules of the substance.

CELLULAR RESPIRATION, ENERGY RELEASE, AND CELL WORK

Consider all the activities that your body has performed today. No matter how long the list grows, it probably will not be complete for many of the body's activities proceed without our consciousness. All the concerted, controlled activities are dependent upon a source of energy. This energy is produced by cellular processes, collectively termed *cellular respiration.* In cellular respiration, carbohydrates, proteins, and lipids are broken down to simpler compounds with the production of usable energy in the form of energy-rich compounds. As just mentioned, ATP is a compound that is often used to drive cellular work performed at various levels of the hierarchy of organization: tissues, organs, organ systems, the organism as a whole, and the various levels of organismic interaction.

Since mitochondria produce the vast majority of ATP, they have been referred to as the "powerhouses of the cell." Muscular contraction has been studied extensively with respect to energy–work relationships, and the tremendous body of data shows that ATP is rapidly synthesized from ADP by transfer of a phosphate group from phosphocreatine, an energy-rich storage molecule. Then ATP participates intimately in the process of muscular contraction. ATP drives nervous transmission as well as active transport, where materials cross membranes at a great rate or against a concentration gradient. The electrical discharge in the electric eel is another type of work produced by living forms.

Secretion is another energy-dependent process carried out by specialized cells. Cellular division also requires the expenditure of energy and is dependent on a supply of ATP. All biochemical syntheses of cells require ATP. In fact, ATP is utilized in the first step in the synthesis of proteins, carbohydrates, fats, and nucleic acids. Even the formation of sucrose, the disaccharide of glucose and fructose, requires ATP. The ubiquitous role of ATP in cellular function should be clear now. Organisms vary greatly in the details of their morphologies; however, most organisms use the same fuels for the production of energy, which is utilized for the various kinds of biological work. Most discussions of cellular respiration center on glucose as one of the most important fuels. The chemical fate of glucose will provide the axial thread for this discussion of respiration. With glucose as the fuel the overall equation for cellular respiration may be written as follows:

$$C_6H_{12}O_6 + 6O_2 \rightarrow 6CO_2 + 6H_2O + \text{energy}$$

Of course this equation is an oversimplification. The breakdown of glucose may be discussed conveniently in four phases:

1. Glycolysis
2. Formation of acetyl coenzyme A from pyruvic acid
3. The Krebs citric acid cycle
4. The hydrogen transport system

Figure 5.3 relates the production of ATP in cellular respiration to several types of cell work. What others can you add?

Glycolysis (phase 1) starts with glucose (or carbohydrates that can be converted to glucose). Glucose, a six-carbon compound, is converted in glycolysis to two molecules of pyruvic acid, a three-carbon compound. In a sense, pyruvic acid is the end product of glycolysis and is the result of breaking glucose apart in its middle between carbon atoms 3 and 4.

Glycolysis is anaerobic, regardless of whether oxygen is present or absent. But in the presence of oxygen, pyruvic acid (phase 2) loses a carbon atom as carbon dioxide, and the remaining two carbons (the acetyl group) are combined with coenzyme A:

$$\underset{\text{pyruvic acid}}{CH_3{-}\overset{\overset{\large O}{\|}}{C}{-}COOH} \xrightarrow{\text{coenzyme A(CoA)}} \underset{\text{acetyl CoA}}{CH_3{-}\overset{\overset{\large O}{\|}}{C}{-}CoA} + \underset{\text{carbon dioxide}}{CO_2}$$

Acetyl CoA then enters the Krebs citric acid cycle (phase 3). The Krebs cycle is a complex pathway, but the net result is that the atoms of the acetyl group are lost as carbon dioxide and hydrogen atoms:

$$\underset{\text{acetyl group}}{CH_3{-}\overset{\overset{\large O}{\|}}{C}{-}} \rightarrow \underset{\text{carbon dioxide}}{CO_2} + \underset{\text{hydrogen atoms}}{xH}$$

First of all, the acetyl (2C) is combined with oxaloacetic acid (4C) to produce citric acid (6C). As the successive reactions occur, a molecule of carbon dioxide is released at two places. Eventually, oxaloacetic acid is reformed and can then combine with another acetyl group of acetyl CoA. Five pairs of hydrogens are removed as each acetyl group is metabolized. One pair is taken out as pyruvic acid and is converted to acetyl CoA (phase 2) and the other four pairs are removed within the cycle itself (phase 3). The five pairs of hydrogens are picked up by cofactors of the enzymes that catalyze the dehydrogeneration reactions. From there they are passed down the hydrogen transport system, phase 4 of cellular respiration under aerobic conditions.

The hydrogen transport system is a complex group of protein molecules (cytochromes and flavoproteins) and they pass the hydrogen atoms (protons plus their electrons) along in a series of oxidation–reduction reactions (see earlier in this chapter). At three places along the transport system ATP is synthesized from ADP and inorganic phosphate (PO_4^{3-}). The energy that drives the synthesis of ATP is derived from the "fall" of electrons from one protein carrier to another. The various protein carriers may be likened to a

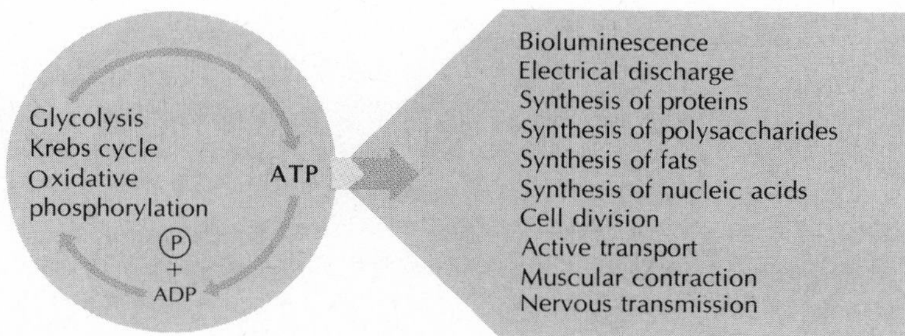

Figure 5.3 Use of energy-rich compounds in cell work. Although ATP is used to represent energy-rich compounds, it must be remembered that several types of energy-rich compounds exist in cells.

series of waterfalls. If the transport of hydrogens and electrons occurred in one step, the delicate biological system would be burned up because too much heat would be produced. Oxygen is the terminal acceptor of the hydrogen transport system. Addition of the hydrogens to oxygen produces water. Returning to the original equation for cellular respiration:

$$C_6H_{12}O_6 + 6O_2 \rightarrow 6CO_2 + 6H_2O + \text{energy (ATP)}$$

it is clear where each component fits into the reaction.

ATP is produced as a result of all four phases of cellular respiration—glycolysis, formation of acetyl CoA, the Krebs cycle, and the hydrogen transport system. Two ATPs are produced in glycolysis. Under anaerobic conditions that is all the ATP formed; under aerobic conditions a total of 36 are produced (Table 5.2).

Table 5.2 ATPs Produced under Aerobic and Anaerobic Conditions

CONDITIONS	END PRODUCT(S)	NUMBER OF ATPs
Anaerobic	Lactic acid or ethanol	2
Aerobic	CO_2 and H_2O	36

THE PARTICIPATION OF OTHER MOLECULES IN RESPIRATION: THE METABOLIC MILL

The discussion of respiration has centered around glucose. Indeed, glucose is a focal point of the respiratory scheme, but the participation of other molecules must be considered also. Any intermediate of the glycolytic–Krebs cycle scheme may be metabolized by the scheme even though produced by some other reaction. For example, pyruvic acid can be produced from the amino acid alanine by a deamination reaction. This molecule of pyruvic acid may be metabolized by the scheme, just as a molecule of pyruvic acid coming from the breakdown of glucose. This emphasizes the fact that like molecules from several different reactions may contribute to the total amount or concentration of that particular molecule in the cell.

The points stressed in the preceding few paragraphs are that (1) the mitochondria serve as energy converters in the cell and (2) the fuel for the mitochondria is prepared by a large number of reactions. Admittedly, these reactions are numerous and complex, but this is necessary for

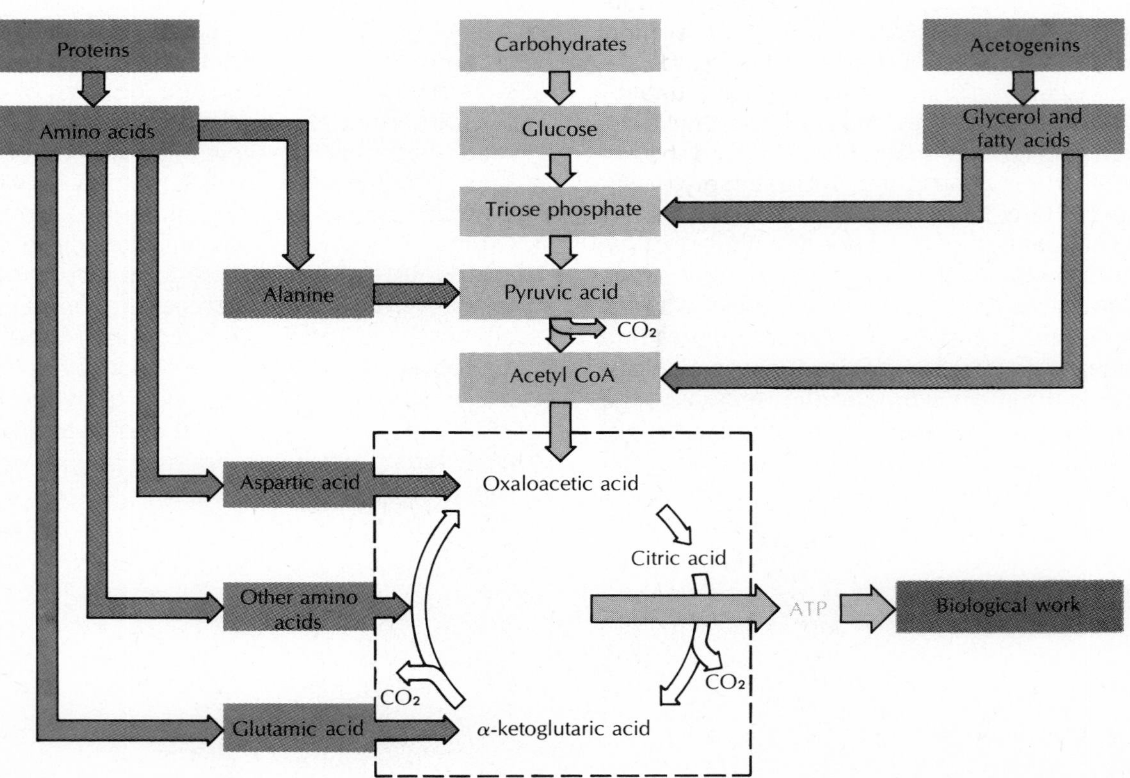

Figure 5.4 Metabolic mill. Diagram is intended to show schematically how the constituents of acetogenins, carbohydrates, and proteins are funneled into the Krebs citric acid cycle with ATP production.

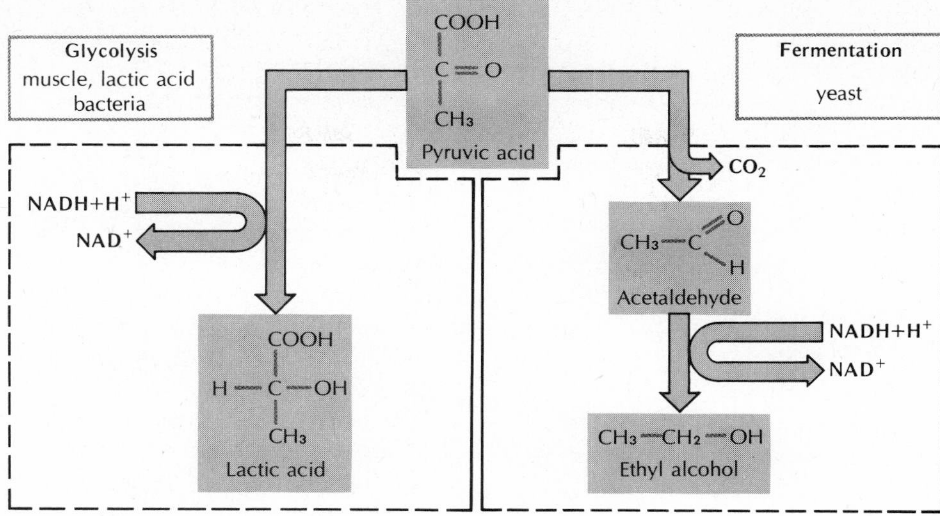

Figure 5.5 Comparison of methods of regeneration of NAD^+ under anaerobic conditions. In each case the NAD^+ (the oxidized form of the nicotinamide cofactor) can be reused in glycolysis.

the transformation of the many kinds of food molecules into the relatively few kinds of molecules that are intermediates in glycolysis and the Krebs citric acid cycle. The schematic diagram in Figure 5.4 is intended to summarize these reactions. It is appropriately labeled "the metabolic mill" and is an example of the molecular logic of living systems.

The formation of pyruvic acid leads to the consideration of the role of oxygen and the terminal anaerobic reactions in fermentation and glycolysis. If oxygen is not present in a cell, usually the pyruvic acid is converted to either ethyl alcohol or lactic acid. NAD^+ is a cofactor for dehydrogenation reactions and is converted in glycolysis to $NADH + H^+$. If the oxidized form of the cofactor (NAD^+) is not regenerated, the entire store of the cofactor will be converted to the reduced form. The lack of oxidized cofactor will bring vital dehydrogenation reactions to a halt, and such a condition results in the death of the cell. Yeast cells regenerate NAD^+ by forming ethyl alcohol; muscle cells form lactic acid. These relationships are shown in Figure 5.5. Fermentation involves a decarboxylation reaction (removal of a carboxyl group in the form of carbon dioxide) that changes pyruvic acid to acetaldehyde; acetaldehyde is reduced to ethanol, and NAD^+ is regenerated. Regeneration of NAD^+ in muscle glycolysis involves the reduction of pyruvic acid to lactic acid.

Phases of Cellular Respiration

The preceding discussion outlined the mechanism of cellular respiration. The four phases now are considered in more detail.

GLYCOLYSIS (PHASE 1)

Glycolysis is catalyzed by the sequential action of several enzymes that have been isolated, crystallized, and thoroughly studied. The details of glycolysis can be seen in Figure 5.6. Notice that ATP is used at points A and B and is produced at points D and E. NAD^+ is the acceptor of hydrogens removed at point C. The following information should be gained from the study of Figures 5.5 and 5.6.

1. A six-carbon compound (glucose) is changed into two three-carbon compounds (two molecules of pyruvic acid).
2. A net synthesis of two molecules of ATP occurs for each glucose molecule that is metabolized.
3. Two NADH molecules are produced for each glucose metabolized (C).
4. Under anaerobic conditions, pyruvic acid is converted to lactic acid or ethanol.

FORMATION OF ACETYL COENZYME A (PHASE 2)

In the presence of oxygen (aerobic conditions), pyruvic acid is metabolized in a more complex way. It is converted to acetyl CoA and carbon dioxide; the acetyl group is funneled into the Krebs citric acid cycle. One pair of hydrogens is removed as pyruvic acid is converted to acetyl CoA (Fig. 5.7, point A).

THE KREBS CITRIC ACID CYCLE (PHASE 3)

A study of Figure 5.7 reveals that for each acetyl fragment entering the cycle four pairs of

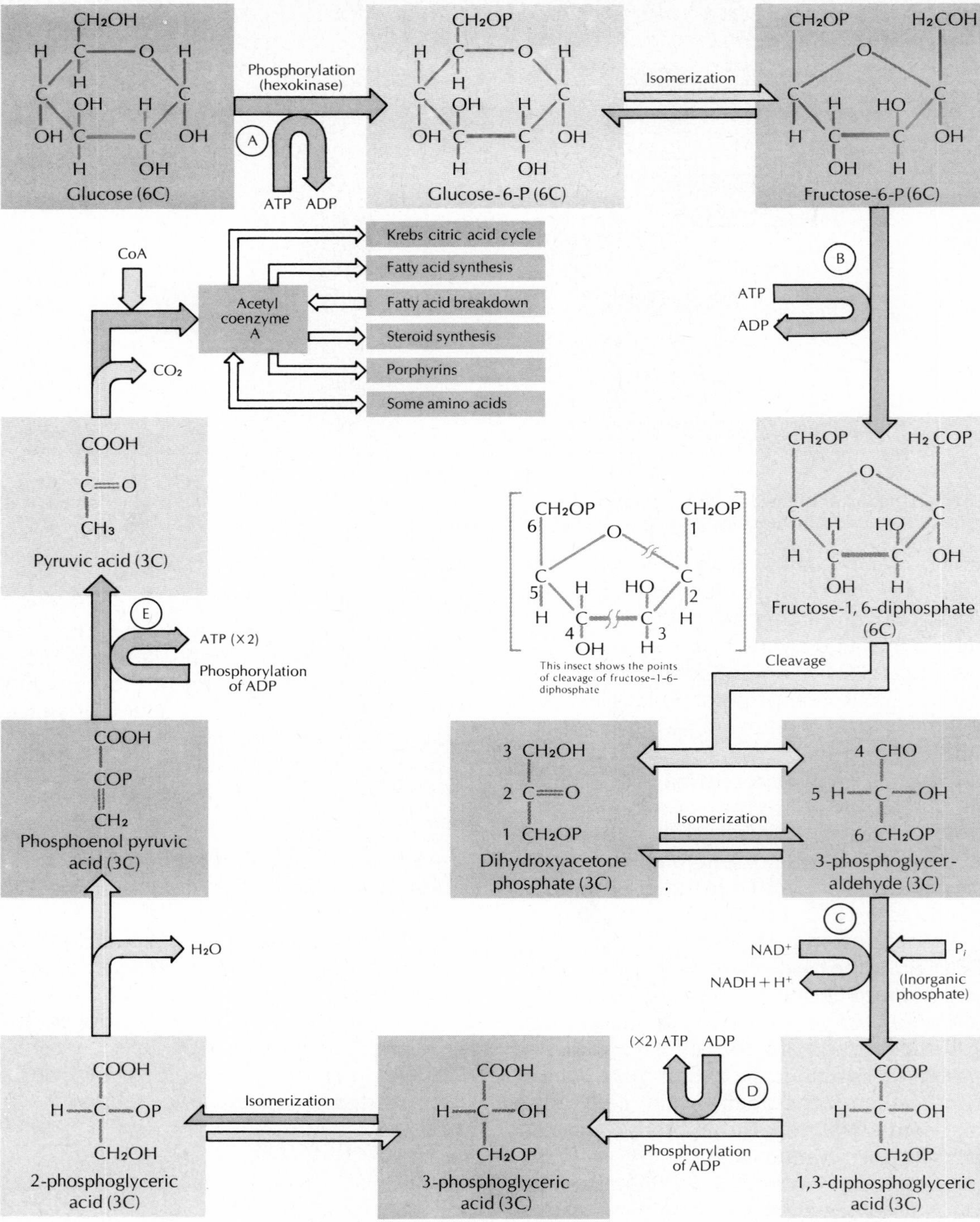

Figure 5.6 Glycolysis, the breakdown of glucose to pyruvic acid. Glycolysis converts glucose (6C) to molecules of pyruvic acid (3C). Note where ATP is used or produced (A, B, D, E) and where hydrogens are removed (C). The net ATP production in glycolysis is two. Under anaerobic conditions, NAD^+ is regenerated by the formation of ethanol or lactic acid (Fig. 5.5). Under aerobic conditions, NAD^+ is regenerated by the hydrogen transport system (Fig. 5.8), and pyruvic acid is converted to carbon dioxide and the acetyl group (2C).

hydrogen atoms are removed from the compounds being metabolized; these points are identified as B, C, E, and F. NAD^+ is the cofactor in three of these reactions, and a flavin compound accepts the hydrogens in the other case (E). ATP is produced at one place; this is designated as D. Since two molecules of acetyl CoA per original molecule of glucose enter the cycle, two molecules of ATP are produced at this point for each initial molecule of glucose. The following points about this phase of glucose metabolism should be emphasized:

1. Five pairs of hydrogens are removed for each acetyl fragment that is metabolized. One pair is removed in phase 2, and four pairs are removed in phase 3. Thus for each original molecule of glucose, ten pairs are removed.
2. ATP is produced directly within the cycle; two molecules are produced for each original molecule of glucose that is metabolized.

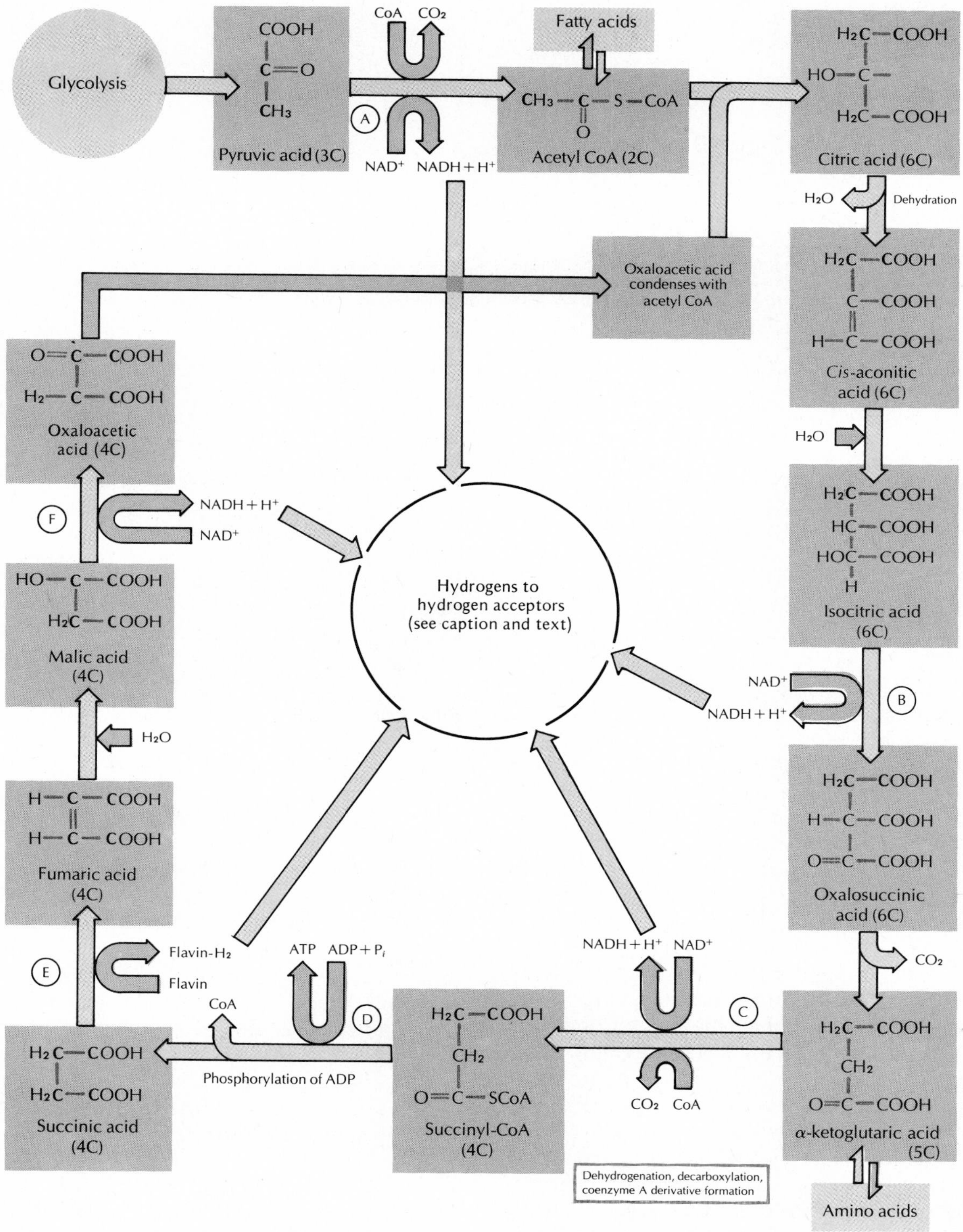

Figure 5.7 Krebs citric acid cycle. At points A, B, C, E, and F hydrogens are removed from the compounds and transferred to hydrogen acceptors in the hydrogen transport system. The hydrogen transport system generates ATP by oxidative phosphorylation. At point D, ATP is generated by a "substrate-level" or "substrate-linked" reaction. Note that two CO_2 molecules are produced as each acetyl group is metabolized by the cycle. Oxaloacetic acid is regenerated in each turn of the complete cycle.

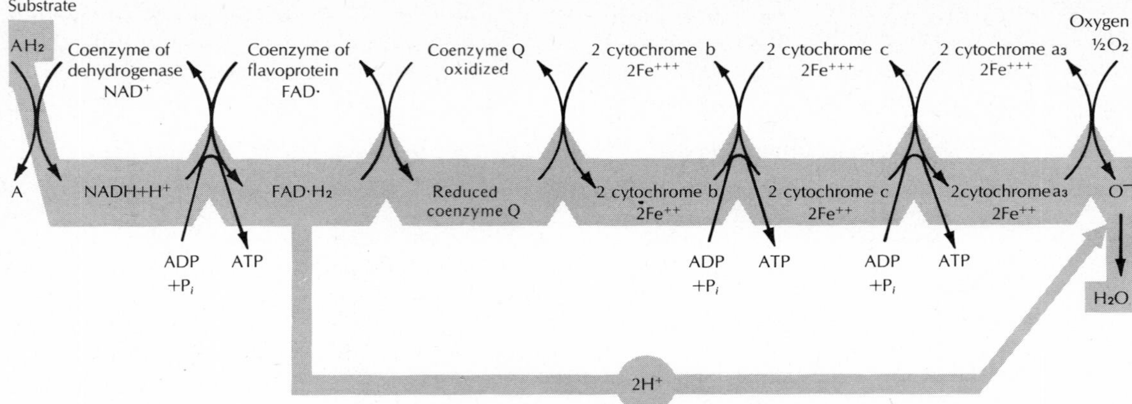

Figure 5.8 Hydrogen transport system, also called the electron transport system and cytochrome system. It is the "main line" of oxidation and reduction in cells that use oxygen as an acceptor of hydrogen atoms which are removed from molecules in metabolism. Each hydrogen atom (proton and electron) is removed from the substrate molecule by a dehydrogenase enzyme. Over 150 dehydrogenase enzymes have been studied. Most dehydrogenases contain tightly bound divalent (+2) metal ions. Hydrogens are removed in pairs and accepted by the NAD^+ coenzyme of the dehydrogenase. The pairs of hydrogens actually are carried by the reduced cofactor (NADH) as a hydride ion (1 proton and 2 electrons, $H{:}^{(-)}$) and a proton (H^+). At the level of the flavoprotein, both protons are left in the mitochondrial matrix while the electrons are passed down the cytochromes. The last cytochrome in the series, cytochrome a_3 or cytochrome oxidase, passes the electrons to oxygen, and the protons are picked up to form water. Note the three places where ATP formation is coupled to this sequential transfer system. Details of the components of the hydrogen transport system are not completely worked out. Nonheme iron proteins are components of some hydrogen transport systems.

3. Two carbon atoms are put into the cycle in the acetyl fragment; two carbon atoms are removed as carbon dioxide.

THE HYDROGEN TRANSPORT SYSTEM (PHASE 4)

In the presence of oxygen the hydrogens are passed through a series of reactions that produce ATP. This is oxidative phosphorylation, since this type of ATP production requires oxygen in the last step of the series. This series of reactions is shown in Figure 5.8 and is frequently called the *main line* of oxidation and reduction in the cell, the *hydrogen transport system,* the *cytochrome system,* or the *electron transport system.*

Each hydrogen atom, consisting of a proton and an electron, is passed to FAD, the prosthetic group (or cofactor) of flavoprotein and then to coenzyme Q. Coenzyme Q liberates the protons (H^+) into the fluid of the mitochondrion while the electrons (e^-) are transferred down the line to successive cytochromes. The cytochromes are proteins that possess the iron-containing heme group. The last cytochrome is cytochrome a_3 (cytochrome oxidase), which transfers the electrons to the oxygen that combines with the protons from the mitochondrial fluid to form water. Thus oxygen is the terminal electron acceptor in the aerobic pathway. Lack of oxygen quickly affects the metabolism of the cell by stopping this energy-release mechanism.

Figure 5.8 shows that ATP is synthesized at three places in the main line. Thus each pair of hydrogens that is passed on by the dehydrogenases (NAD^+ cofactor) results in the synthesis of three ATP molecules. Each pair that is transferred from the substrate directly to the flavoprotein results in the synthesis of two ATPs. Thus for the pairs of hydrogens from phase 2 (formation of acetyl CoA, Fig. 5.7, point A) three ATPs are formed. In the Krebs citric acid cycle three pairs of hydrogens enter at the level of NAD^+ ($3 \times 3 = 9$) and one pair (from succinic acid) enters at the level of flavoprotein ($1 \times 2 = 2$). Thus for each pyruvic acid molecule that is completely metabolized by phases 2 and 3, 14 ATPs are produced in the hydrogen transport system. For each glucose molecule 28 ATPs are produced by the main line of oxidation and reduction.

It is possible to summarize the number of ATPs produced for each original molecule of glucose metabolized to carbon dioxide and water:

Net from glycolysis	2
Hydrogens (2×2) removed in glycolysis and sent through the main line (Fig. 5.6)[3]	4
Hydrogens removed in formation of acetyl CoA and in the citric acid cycle	28
ATPs produced in the Krebs citric acid cycle (Fig. 5.7)	2
	36

[3] The NADH produced in glycolysis does not penetrate the mitochondrial membrane. The hydrogens are transported across the membrane in a complex manner (a transport system called the glycerol phosphate shuttle) and are accepted by FAD. As a result, four rather than six ATPs are formed by oxidative phosphorylation as these hydrogens pass down the cytochrome system.

Thus 36 ATPs are produced for each molecule of glucose metabolized to carbon dioxide and water. In muscle glycolysis, free glucose is not the initial substrate; only glycogen (animal starch) serves as the source of glucose. The first step in the sequence of reactions under these conditions is

$$\underset{\text{glycogen}}{(\text{glucose})_n} + \underset{\substack{\text{inorganic}\\ \text{phosphate}}}{P_i} \rightarrow \text{glucose phosphate} + (\text{glucose})_{n-1}$$

This reaction is repeated many times and supplies glucose phosphate to glycolysis. ATP is not required as the source of phosphate; therefore the net ATP production from glycolysis with glycogen as the source of glucose is 3 and the total for complete glucose oxidation is 37.

The student should not memorize the rather elaborate pathways presented in Figures 5.6, 5.7, and 5.8, but remember the four phases of glucose metabolism and know the significance of each phase. An appreciation of the role of glucose and oxygen in the production of ATP and nicotinamide dinucleotide phosphate (NADPH) should be inherent in the study of these schemes.

This discussion on the utilization of energy in biological systems has emphasized the role of ATP. One must keep in mind that the primary storage of energy is in oxidizable compounds (basic foodstuffs such as glucose) and that there are other storage forms such as reduced NADPH and creatine phosphate (in muscle, Ch. 8).

Localization of Respiratory Enzymes in the Cell

The mitochondrion is found in virtually every type of eukaryotic cell (mammalian erythrocytes are exceptions). Blue-green algae and bacteria, the prokaryotes, do not have highly organized structures comparable to the mitochondria of eukaryotes. In the prokaryotes, electron transport and ATP production is associated with the cell membrane. Since there is no mitochondrial membrane and no glycerol phosphate shuttle (footnote 3, this chapter) the total number of ATPs produced per glucose molecule is 38 in prokaryotes.

In eukaryotic cells the mitochondria are in constant motion and they tend to be clustered strategically near structures that require ATP. High concentrations of mitochondria are located, for example, at the junction of nerve cells where impulses are transmitted across the membranes (Ch. 13), regularly arranged in rows in muscle cells along the ATP-requiring contractile elements (Ch. 8), around actively beating sperm tails (Ch. 15), and near the edges of intestinal cells where absorption is occurring (Ch. 9). Mitochondria also frequently are located near fuel sources such as lipid droplets. The size, shape, structure, and possible evolutionary origin of mitochondria were discussed in Chapter 4.

The carbohydrates, lipids, and proteins that serve as metabolic fuels are broken down by degradative enzymes outside of mitochondria into smaller fragments: pyruvic acid, glycerol, fatty acids, and amino acids (Fig. 5.4). These small molecules can diffuse across the outer membrane. This passage across the inner membrane requires specific transport systems. The enzymes of the citric acid cycle are located in the matrix of the mitochondrion where the breakdown of the small fuel molecules occurs. ATP production is a function of the inner membrane. There is good evidence that the enzymes of the electron transport system are arranged in an orderly pattern on the cristae of the inner membrane.

As previously discussed, the movement of electrons along the transport system is coupled to ATP production. Isolated mitochondria are capable of carrying out these reactions, and furthermore, mitochondrial fragments will carry out oxidative phosphorylation also. These fragmented mitochondria will oxidize NADH under the proper conditions; if ADP is absent, the process is halted and oxygen is not consumed. Thus the flow of electrons is tightly coupled to oxidative phosphorylation. This is a conservation mechanism since the fuel will not be burned unless the energy is trapped in ATP. The inner membrane provides part of this conservational control since the inner membrane allows an ADP molecule to pass into the matrix only if an ATP is transported out. This molecule-for-molecule exchange assures a control of the level of ATP in the cytoplasm. If ATP is being used at a high rate, then the ADP is generated at a high rate also, and this cytoplasmic ADP is exchanged rapidly for ATP across the inner membrane.

PHOTOSYNTHESIS

The overall equation for photosynthesis in green plants is (left to right)

$$CO_2 + H_2O \underset{\text{respiration}}{\overset{\text{photosynthesis}}{\rightleftarrows}} \underset{\text{glucose}}{C_6H_{12}O_6} + 6O_2$$

Inspection of this equation shows that photosynthesis and cellular respiration are reverse reactions of each other. Photosynthetic plants capture the sun's energy to drive the synthesis of glucose (and many other compounds). The animal ultimately depends on the energy trapped by photosynthesis; it consumes plants (or other animals) and derives energy and substance by respiration.

The interlocking of photosynthesis and respiration is fundamental to the understanding of animal nutrition (Ch. 9) and ecology (Chs. 44 and 45).

ORGANISMAL RESPIRATION AND CELLULAR RESPIRATION

The introduction to this chapter mentions the general framework and strategy of metabolism in the biological hierarchy: the organismal and cellular levels. The body of the chapter is concerned with the scheme of cellular respiration, however. Although organismal respiration is not emphasized here, the general relationships of cellular and organismal metabolism are clear. Organismal respiration is dealt with in Chapters 9, 10, and 11.

SUGGESTIONS FOR FURTHER READING

Conn, E. E., and P. K. Stumpf. *Outlines of Biochemistry,* 3rd ed. New York: Wiley (1972).

Dyson, R. *Cell Biology—A Molecular Approach.* Boston: Allyn and Bacon (1974).

Lehninger, A. L. *Bioenergetics.* New York: W. A. Benjamin (1970).

Lehninger, A. L. *Short Course in Biochemistry.* New York: Worth Publ. (1973).

Loewy, A. G., and P. Siekevitz. *Cell Structure and Function.* New York: Holt, Rinehart and Winston (1969).

McElroy, W. D. *Cell Physiology and Biochemistry.* Englewood Cliffs, N.J.: Prentice-Hall (1969).

Wood, W. B., J. H. Wilson, R. M. Benbow, and L. E. Hood. *Biochemistry—A Problems Approach.* Menlo Park, Calif.: W. A. Benjamin (1974).

From Scientific American

Dawkins, J. J. R., and D. Hull. "The Production of Heat by Fat" (Aug. 1965).

Green, D. E. "The Metabolism of Fats" (Jan. 1954).

Green, D. E. "The Mitochondrion" (Jan. 1964).

Lehninger, A. L. "Energy Transformation in the Cell" (May 1960).

Lehninger, A. L. "How Cells Transform Energy" (Sept. 1961).

Siekevitz, P. "Powerhouse of the Cell" (July 1957).

Stumpf, P. K. "ATP" (Apr. 1953).

6 Bioinformation

What is bioinformation? How does it differ from other kinds of information? Strictly defined, bioinformation is any kind of information possessed and processed by living systems. This would include the information needed and used to control cellular activities, the information concerned with the influences of one tissue or organ on another (hormones and development), the processes involved in the nervous system (learning, memory, and behavior), and so forth. Chapter 3 discusses the proteins and nucleic acids as informational macromolecules, that is, they are specific, linear polymers. The pattern in which building blocks (amino acids and nucleotides) are put together constitutes a meaningful code—information. The remainder of this chapter considers bioinformation at the subcellular level, with an emphasis on integrating the storage and "reading" mechanisms of the proteins and nucleic acids.

Analogy of Language and Bioinformation of Macromolecules

Languages, such as English or German, use an alphabet of a few letters which make up sentences, and the sentences, units of coordinated thought, form paragraphs, books, and documents. Bioinformation involves, really, two languages: nucleic acid and protein. The nucleic acid language is written in nucleotide symbols and it has two "dialects," DNA and RNA. In DNA there are four nucleotides, A, T, C, and G. In RNA dialect there are also four letters, A, U, C, and G. (Actually, there are more than four kinds of nucleotides in RNA and DNA, but these other nucleotides are exceptions to the general rule.) The fundamental aspects of the DNA and RNA dialects are similar, but the different functions of the two classes of nucleic acids are related to their different structures—double-strandedness versus single-strandedness, deoxyribose versus ribose, and the nucleotide composition.

After one type of RNA, messenger RNA, has been *transcribed* from DNA, the messengers are *translated* into proteins. Protein language is written with 20 letters (amino acids). Immediately one sees a definite difference between the information stored in the nucleic acids and proteins (4 letters versus 20). Translation, indeed, involves a change from a 4-letter alphabet to a 20-letter alphabet.

THE GENETIC MATERIAL

A convincing body of data indicates that DNA is the genetic material. These data may be divided into two types, direct and indirect. First, certain observations suggest this, but since they are not strong suggestions, in the absence of any other information, they are termed *indirect* or circumstantial evidences. One suggestion arises from the fact that the somatic cells of a species contain about the same amount of DNA, whereas the amounts of RNA and protein may vary greatly, depending on the function and the environment of the cells. Gametes contain about one half as much DNA as do the nuclei of somatic cells; this is to be expected because the gametes have one set of chromosomes while somatic cells have two sets (Ch. 7).

In cells of polyploid organisms (those with more than two sets of chromosomes) and in cells with polytene chromosomes (duplicated repeatedly but not separated, Ch. 7), there is an increase in DNA content that is related directly to the degree of chromosomal duplication.

Ultraviolet light is one of a number of mutagenic agents (Ch. 18). If bacteria or other appropriate organisms (those that have no protective

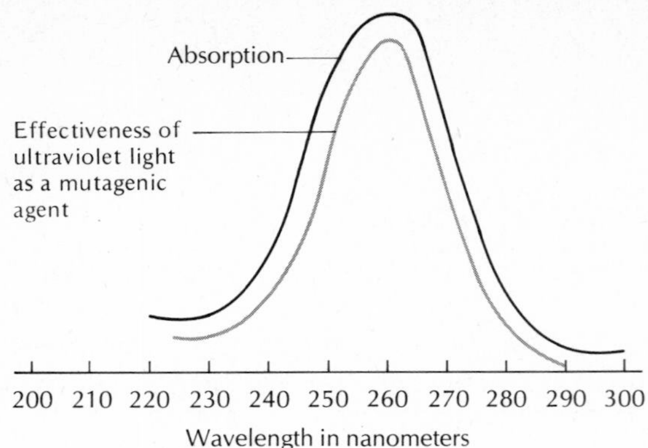

Figure 6.1 Correlation of ultraviolet light absorption by nucleic acids and the action spectrum of induced mutagenesis suggests that the genetic material is nucleic acid.

covering to prevent penetration of the cell nucleus by ultraviolet light) are irradiated with various wavelengths of ultraviolet light, then the effectiveness of each wavelength in producing mutations can be determined. The resulting action spectrum (see Ch. 5) shows that the most effective wavelengths are in the 260 nm (10^{-9} m) region. The action spectrum of ultraviolet-induced mutagenesis is similar to the absorption spectrum of the nucleic acids; this is shown in Figure 6.1. The correlation of these curves suggests that the genetic material is nucleic acid. These ideas support the thought that DNA is the genetic material, a stable but metabolically active control center of the cell.

Other bits of evidence indicate strongly that DNA is the hereditary material; these are *direct* or experimental evidences. The first bit of evidence is the phenomenon of *transformation.* The pneumococcus bacterium *(Diplococcus pneumoniae)* causes pneumonia in humans and mice and exists as several types, each of which can be distinguished immunologically. Injection of the bacteria or bacterial cell walls into an animal such as a rabbit or guinea pig induces the formation of specific antibodies against the type used for the inoculation (see Chapter 11 for a discussion of antibody formation). Some pneumococcal types possess no tough cell walls or coats, and they produce a "rough" colony on a culture plate; the coated types produce "smooth" colonies. The rough strains are avirulent (the coatless bacteria are easily engulfed by phagocytes in the bloodstream), whereas the smooth strains are virulent, or disease causing.

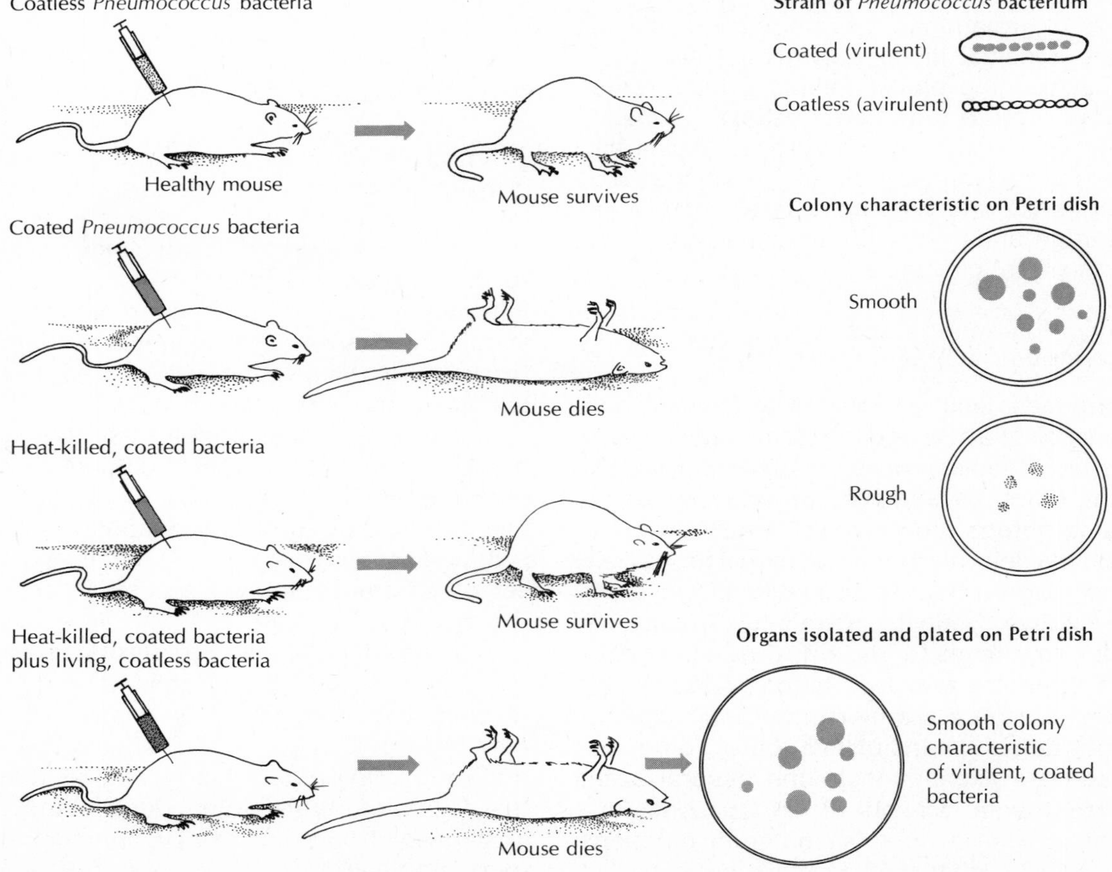

Figure 6.2 Transformation: evidence that DNA is the genetic material.

The first studies on transformation were reported in 1928 by Frederick Griffith, who found that the injection of avirulent bacteria along with heat-killed virulent bacteria caused pneumonia and death in the mouse. Thus some event had changed the avirulent type into the virulent type (Fig. 6.2). Later workers showed that transformation of type specificity could occur in a test-tube culture and that the mouse was not required. After this discovery other workers found that a substance isolated from the virulent type could, when added to a culture of avirulent bacteria, cause the transformation of the avirulent into the virulent type. All physical and chemical tests have indicated that this transforming factor is DNA.

Another bit of direct evidence comes from studies on bacteriophages (bacterial viruses). Viruses possess a protein coat around a core of nucleic acid. Using T_2 bacteriophage (bacterial virus) (see Ch. 7), A. D. Hershey[1] and Martha Chase grew cultures of *Escherichia coli* bacteria on ^{32}P or ^{35}S (radioactive isotopes of phosphorus and sulfur), and after adding T_2 to the cultures, harvested T_2 phages that were labeled with ^{32}P or ^{35}S. (In order to label the bacteriophages with radioactive isotopes, radioactive bacteria must be used because bacteriophages do not metabolize independently.) Then a sample of radioactive phages was added to unlabeled bacteria. After the phages attached to the bacteria, the phage–bacterium complexes were sheared apart in a Waring blendor. By differential centrifugation (sedimentation of the bacteria but not of the bacteriophages) the infected bacteria and the empty phage ghosts were separated. The results of the experiment showed that when ^{35}S was used as the label (sulfur is a constituent of proteins and not of nucleic acids), 80 percent of the radioactivity was found in the ghosts and 20 percent had been transferred to the bacteria. Furthermore, when ^{32}P was the label (phosphorus is a constituent of nucleic acids but is seldom found in proteins), 85 percent of the radioactivity was transferred to the bacteria. The experiment was later refined to show that almost all DNA of the virus entered the bacterium on infection but only 3 percent of the protein was transferred. The interpretation of this experiment (Fig. 6.3) is that the bulk of the phage DNA is transferred to bacteria in the infection process and the bulk of the protein is not.

The Hershey–Chase experiment demonstrated that information required for the synthesis of a virus particle is contained in the DNA that enters the bacterium, rather than in the small amount of transferred protein.

[1] In 1969 Hershey shared the Nobel Prize for Medicine or Physiology for this demonstration that DNA of bacteriophage T_2 and not protein is the genetic material.

THE REPLICATION OF DNA

One of the three properties of the genetic material is that of replication, the capability of exact duplication. (The other two are transcription and mutation; transcription is discussed

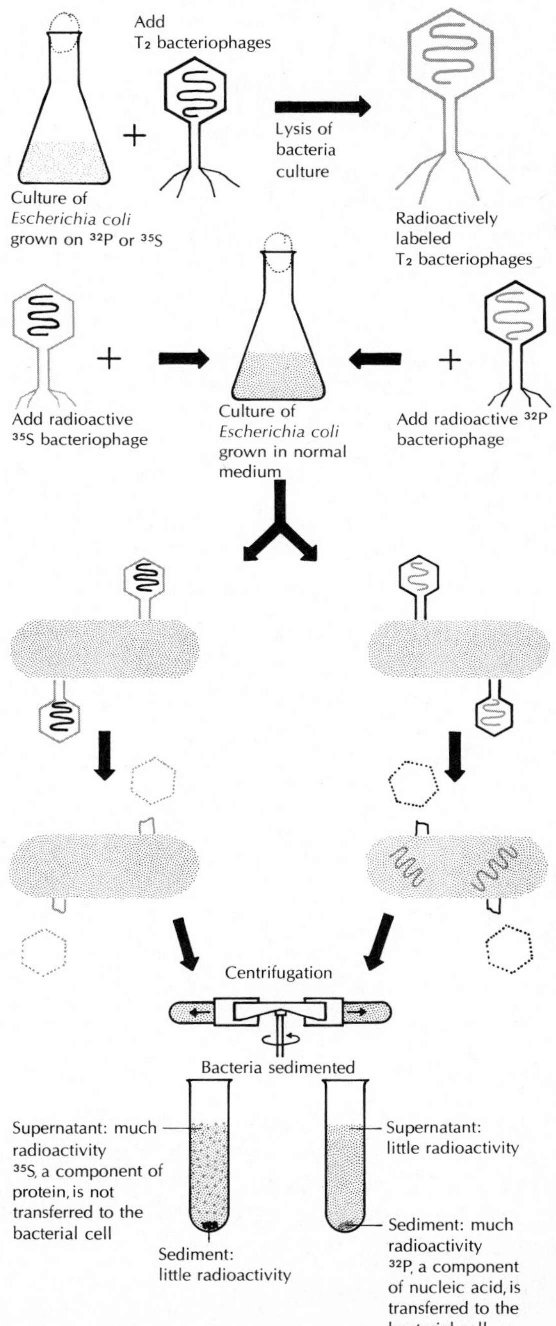

Figure 6.3 Hershey–Chase experiment.

later in this chapter while mutation is discussed in Chapter 19.) All cells of a multicellular organism are derived from the zygote, and presumably all the cells contain the same genetic material; thus the necessity for the exact transmission of the genetic material from parent to progeny cell imposes a replication requirement upon that material. James D. Watson and Francis H. C. Crick were aware of the replication requirement and proposed a mechanism for replication at the same time that they proposed a model for DNA structure. Their proposal for DNA replication was that the two polynucleotide chains of each DNA molecule separate from each other and that each chain serves as a specific surface (a template) for the formation of a complementary chain. The result of this process is two complete DNA double helices. The schematization of DNA replication is shown in Figure 6.4. Notice the role of base pairing in this template-specific process.

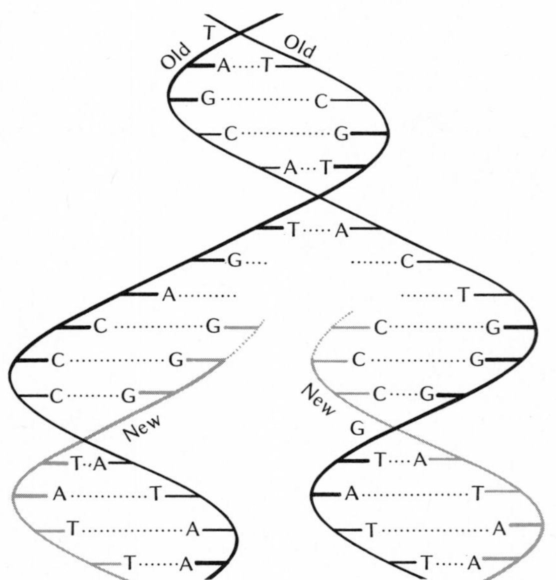

Figure 6.4 Replication of DNA according to Watson and Crick. The two polynucleotide chains of each DNA molecule separate, and each serves as a template for the formation of the complementary chain. The base pairing specificities (A = T, C ≡ G) tend to insure exact duplication. (From Levine, *Genetics,* Holt, Rinehart and Winston, 1968).

The proposal of Watson and Crick for the replication of DNA was tested experimentally in 1958 by Matthew Meselson, Franklin Stahl, and Jerome Vinograd. Bacteria *(Escherichia coli)* were grown for several generations on a medium that contained ^{15}N, the heavy nitrogen isotope, rather than ^{14}N. DNA extracted from such bacteria and containing the heavy nitrogen is more dense than normal DNA and moves faster in a gravitational field produced by an ultracentrifuge. In this type of centrifugation a tube is prepared with a salt solution that increases in concentration from top to bottom—a density gradient. When DNA is applied to the top of and is centrifuged in such a density gradient, it moves to the region of its buoyant density. (Where the density of the fluid equals the density of the DNA, the DNA will stop. Compare density gradient centrifugation with a swim in a salt lake.) Thus heavy DNA will move farther toward the bottom of the tube than will light DNA. The critical experiment was as follows. Cells containing heavy DNA were transferred to light medium (only ^{14}N present) and allowed to continue growing there. Samples of cells in the light medium then were withdrawn at various times. Since the doubling of generation time (the time necessary for doubling of the population) for *E. coli* was known, each of the various sample times could be expressed as some fraction of the generation time. The DNA was isolated from each batch of cells and ultracentrifuged in a density gradient. A schematic comparison of the proposal and the results of the Meselson–Stahl–Vinograd experiment is shown in Figure 6.5. The figure indicates that each of the original heavy DNA double helices was separated into two single chains. Upon each original heavy single chain, a new light chain is produced, and the resulting double helix of DNA is hybrid and intermediate in density. A second round of replication produces two hybrid and two light double helices.

The replication of DNA logically leads to the question: What are the chemical events in this process? Arthur Kornberg and associates have devised experiments that permit the cell-free synthesis of DNA. The enzyme *DNA polymerase* (now called DNA polymerase I) catalyzes the synthesis of macromolecular DNA from nucleotide building blocks. The DNA polymerase reaction requires four nucleotides in the form of triphosphates. (As discussed in Chapter 3, three phosphates are attached to the sugar. This is an energy-rich form.) In addition, DNA polymerase requires some macromolecular (large) DNA as a primer; if the primer is omitted, there is usually no synthesis. Thus it turns out, as might be expected logically of the genetic material, that *the synthesis of DNA is a highly specific process.* DNA polymerases have been found in extracts of all cells—bacterial, plant, and animal—where DNA replication has been measured.

The DNA that is synthesized in the cell-free system is similar to the DNA that is used as a primer; the physical properties, such as the sedimentation rate in an ultracentrifuge and the mo-

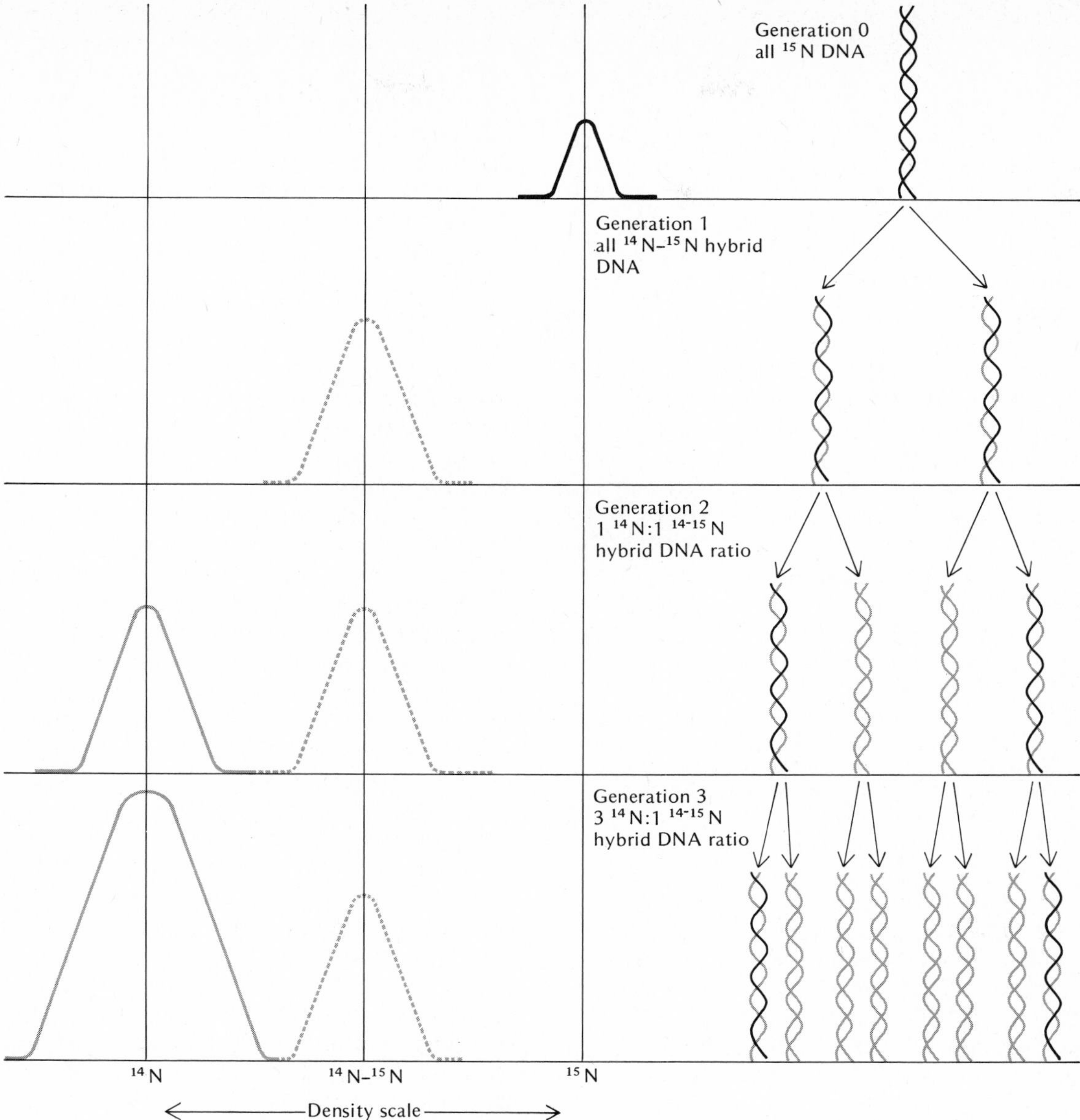

Figure 6.5 Schematic comparison of the Watson–Crick proposal for DNA replication and the Meselson–Stahl–Vinograd experiment.

lecular weight, are similar, while the base composition (the percentage of each kind of base) of the product reflects that of the primer. This ratio is obtained without regard to the extent of synthesis (1–20 times the amount of primer). The interpretation of this observation is that the primer molecules are copied completely and repeatedly.

The DNA synthesized in the cell-free system, as described above, has no biological activity. But in 1967 the cell-free synthesis of a biologically active DNA was accomplished in Kornberg's laboratory. This was done by using the circular single-stranded DNA of a small bacteriophage (ΦX-174) as a template. (A single-stranded DNA molecule is found in a few bacteriophages.) The DNA strand, which is in the bacteriophage, is called the plus strand. It was used to synthesize a double-stranded circle (plus and minus). The minus strand was isolated and was used as a template for the synthesis of plus strands. The plus strands were then assayed by adding them to bacterial protoplasts. The plus strands did enter the protoplasts and produce complete bacteriophages. An outline of this experiment is shown in Figure 6.6.

At this point in the study of DNA replication it was logical to conclude that the basic aspects of the process were well understood. However, a few months later it was shown that mutations could be induced in the gene which specifies DNA polymerase I. These mutations seriously decreased the enzyme's activity. The bacteria were

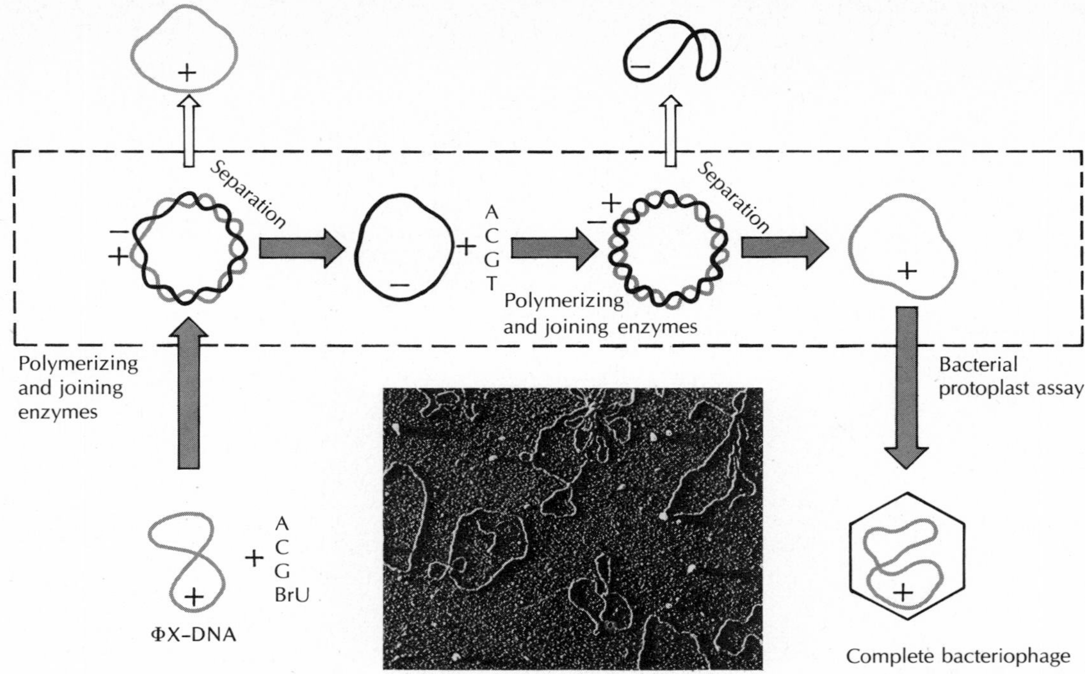

Figure 6.6 Synthesis of infectious (biologically active) DNA in a cell-free system. Virus ϕX-174 has a single circular strand of DNA called the plus (+) strand. (DNA polymerase will replicate the circular strand but will not join the ends; the joining enzyme is able to do this. If ϕX-174 DNA (+) is replicated and joined in the presence of 5-bromouracil (BrU) rather than thymine (T), a heavy (−) strand is formed. The "light" (+) and "heavy" (−) strands can be separated by density gradient ultracentrifugation. The isolated "heavy" (−) strand can then be used as a template for the synthesis of a new light (+) strand in a cell-free system with the two enzymes. The isolated new light (+) strand is then assayed with a bacterial *(E. coli)* protoplast system. Complete bacteriophages are produced showing that the DNA produced in a cell-free system is biologically active. Recent evidence indicates that the DNA polymerase used in these experiments is probably not the true DNA-replicating enzyme in cells, but this DNA polymerase can be "forced" to catalyze the synthesis in the cell-free system. Inset shows electron micrograph of the circular (+) strands of ϕX-174 DNA synthesized in the cell-free system.

not harmed and DNA replication seemed to occur normally in these cells. What could explain these curious observations? Probably DNA polymerase I is not the major enzyme involved in DNA replication, but it is a repair enzyme for broken or imperfect DNA. Under *in vitro* conditions with high concentrations of nucleoside triphosphates the DNA polymerase is able to do more than repair DNA. It is "forced" to synthesize complete DNA molecules.

Which enzyme is the *real* DNA polymerase? Two new enzymes called DNA polymerase II and III, in order of their discovery, have been identified and partially purified. Many details are yet to come, but DNA polymerase II seems to be the fundamental enzyme involved in DNA replication.

DNA polymerases in eukaryotic cells have been studied and they exhibit no unusual properties. In addition to the nuclei, these polymerases are found in mitochondria and chloroplasts, are associated with ribosomes and are found in isolated, smooth endoplasmic reticulum fractions. Since mitochondria and chloroplasts contain unique DNAs and are semiautonomous cellular organelles, the presence of DNA polymerases is logical. The significance of DNA polymerases associated with ribosomes and smooth endoplasmic reticula, which do not contain DNA, is not clear.

In general, DNA polymerases require a template of primer DNA, all four nucleobases in the form of deoxyribonucleoside triphosphates (base–deoxyribose–P–P–P, see Ch. 3), and Mn^{++} or Mg^{++}. The primer DNA directs the specificity of synthesis through the base-pairing relationships, A = T, C ≡ G. Experiments using ^{32}P-radioactive deoxyribonucleoside triphosphates have shown that synthesis always occurs in the 5′ → 3′ direction. Since the double stranded DNA molecule has the two strands arranged in an antiparallel fashion, parallel synthesis on each strand would be impossible. A mechanism of DNA synthesis which accounts for the synthesis in the 5′ → 3′ direction on both strands is shown in Figure 6.7 along with the general reaction and the details of DNA strand extension.

METABOLIC SYSTEMS

Metabolism occurs in orderly pathways. Each step of these metabolic pathways is catalyzed by

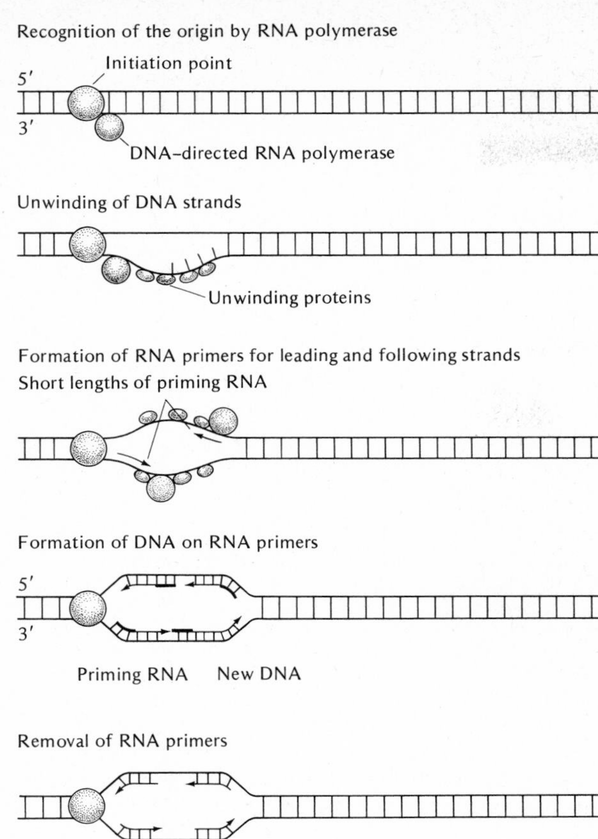

Figure 6.7 General steps of DNA replication. DNA ligase is the joining enzyme. (Redrawn from Lehninger, *Biochemistry*. 2nd ed., Worth.)

an enzyme; a good example is glycolysis (Ch. 5). If one reaction in a pathway is blocked, then the other reactions may be affected significantly. If a blocked pathway is an anabolic one, the compound after the block must be supplied in the diet for normal functioning of the organisms. If the blocked pathway is a catabolic one, then the compound before the block accumulates and is excreted (or accumulated). Figure 6.8 summarizes these relationships.

Some well-known metabolic diseases are directly related to mutant, or altered, enzymes. Such direct influence is called a *primary* block. In the human disease *alcaptonuria* the enzyme that breaks down homogentisic acid into further oxidation products is absent (Fig. 6.9). The homogentisic acid in the urine of alcaptonurics oxidizes spontaneously to a black pigment. Consequently alcaptonuria was one of the first inborn errors of metabolism to be recognized; it is easily detected in infants by the dark stains in their diapers. Alcaptonurics slowly deposit the pigmented material in their connective tissues and often have an unsightly staining of the cartilage in the ears and nose; this also causes a relatively benign arthritic condition. *Phenylketonuria,* a very severe mental disease, is due to a primary block that inactivates the enzyme that converts phenylalanine to tyrosine. In individuals with this disease the phenylalanine concentration builds up in the body fluids and is excreted in the urine. Another block, responsible for *albinism,* is also shown in Figure 6.9. This block prevents the formation of the normal pigments of the skin, hair, and eyes. *Galactosemia* is another disease due to a primary block. Infants which have this disease are unable to metabolize galactose, a monosaccharide contained in lactose (a disaccharide), the principal sugar of milk. As a consequence, the galactose level in the blood is high and galactose is excreted.

The deleterious effects of many metabolic disorders cannot be traced directly to primary blocks; rather, the effects seem to come from other reactions that are influenced by metabolites accumulated at the primary blocks. The accumulation of metabolites above normal levels may result in an inhibition of other reactions; such reactions are secondarily blocked. Examples of secondary blocks can be seen in the diseases just mentioned. In galactosemia, galactose probably acts as an inhibitor of reactions involving glucose. Removal of galactose from the diet of the affected infant permits these reactions to function normally. Adults produce an entirely different enzyme; thus, infant galactosemics outgrow their metabolic disease. In phenylketonuria, phenylalanine is metabolized by other enzymes, and the resulting products serve as inhibitors of the usual phenylalanine pathway.

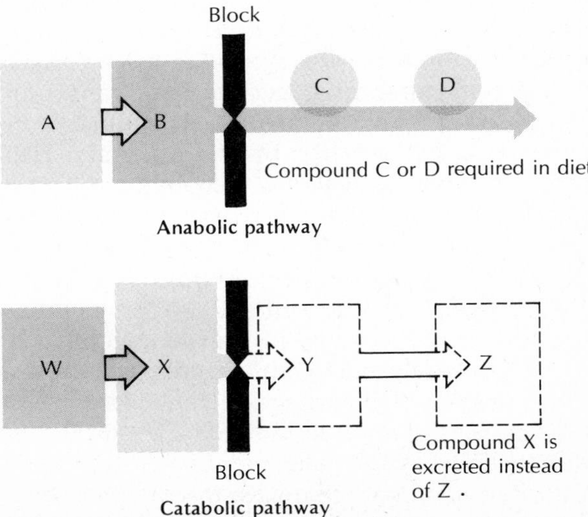

Figure 6.8 Blocks in metabolic pathways.

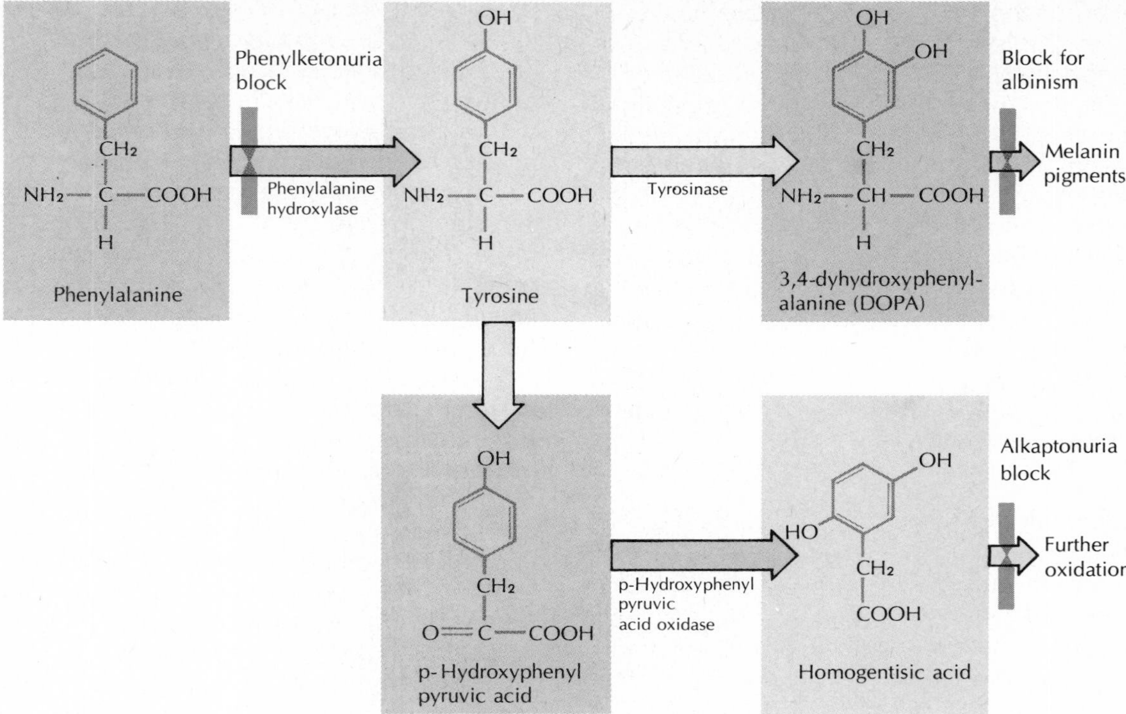

Figure 6.9 Pathways of phenylalanine metabolism showing primary genetic blocks responsible for phenylketonuria, alcaptonuria, and albinism.

Thus, a biochemical reaction is not an isolated entity but affects and is affected by many other reactions occurring in the organism.

TRANSCRIPTION: BIOSYNTHESIS OF RNA

The bioinformation stored in the chromosomes of cells must be "read" before this information can influence the operation of the cell. The reading of bioinformation occurs in two distinct steps: transcription and translation. As mentioned earlier, *transcription* is the synthesis of RNA under the direction of DNA. *Translation* is a complex process of protein synthesis involving messenger RNA (mRNA), transfer RNA (tRNA), ribosomes, several enzymes, and other factors.

The three major kinds of RNA (messenger, ribosomal (rRNA), and transfer) are transcribed from specific regions of the DNA. In eukaryote cells this occurs mainly in the nucleus. Transcription is catalyzed by an enzyme called RNA polymerase, and the reaction requires all four nucleoside triphosphates (ATP, GTP, CTP, and UTP), a divalent ion (Mg^{++} or Mn^{++}), and DNA, which is a primer or template for the reaction:

$$n\text{ATP} + n\text{GTP} + n\text{UTP} + n\text{CTP} + \text{DNA template} \xrightarrow[\text{Mg}^{++}\text{ or Mn}^{++}]{\text{RNA polymerase}} \underbrace{(\text{AMP} - \text{GMP} - \text{UMP} - \text{CMP})n}_{\text{RNA}} + \text{DNA} + 4n\ \text{PP}$$

The first demonstration of a mRNA was reported in 1956. When bacteriophages containing DNA as the genetic material infect bacteria, the infected cells synthesize a new kind of RNA molecule. This newly synthesized RNA has base ratios $(A + T)/(C + G)$ that are different from the DNA of the host bacterial cell. The experiment was performed in the following manner. *Escherichia coli* were grown and then infected with bacteriophage T_2. At the time of infection ^{32}P was added to the medium. A large fraction of RNA in the infected cells was found to contain a large amount of radioactive phosphorus. Hydrolysis of this RNA fraction and subsequent determination of the amount of each kind of nucleobase showed that the base ratio was different from the RNA and the DNA of *E. coli* cells but similar to the DNA of the infecting phage. This was taken to mean that the infecting DNA had been read several times and that the base sequence of T_2 phage DNA is significantly different from the bacterial chromosome. The term "messenger" was coined to indicate that the RNA molecule could convey a message.

Not all of the DNA of a cell is used as a template for the synthesis of mRNA. Some DNA genes specify rRNA and tRNA. Quantitative analysis indicates that, in a certain strain of bacteria, 0.42 percent of the total DNA is complementary to ribosomal RNA and about 0.03 percent is complementary to tRNA. Thus the majority of the remaining DNA is presumably complementary to mRNA. The bulk of the RNA in a cell is ribosomal RNA. This means that a small portion of the DNA is transcribed over and over to produce the rRNA. In higher organisms the regions of DNA that produce rRNA are amplified, and rRNA is abundantly produced. This fact has been used in visualization of the transcription process.

VISUALIZATION OF TRANSCRIPTION

During the early growth of the amphibian egg, the region of the chromosome that contains genes for rRNA synthesis (called the nucleolus organizer region) is multiplied about a thousand times, and these extra copies are released into the nuclear sap. There is evidence that these *extrachromosomal nucleoli* function similarly to familiar nucleoli of mature cells in synthesizing

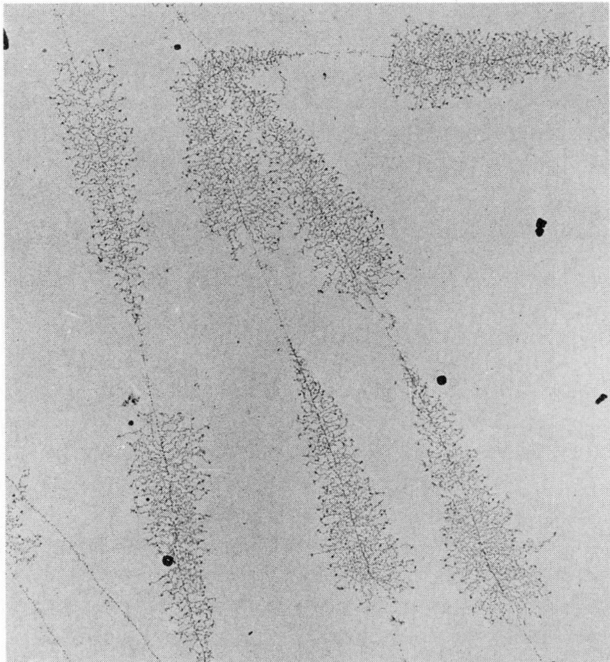

Figure 6.10 Nucelolar gene isolated from an oocyte of the spotted newt, *Triturus viridescens* (× 16,000). Each length of long strand is a DNA molecule, and each fibrillar matrix region is a gene coding for ribosomal RNA molecules. Each fibril in the matrix is an RNA molecule. (Courtesy of O. L. Miller, Jr., and B. R. Beatty, Oak Ridge National Laboratory, *Science*, **164**, 955, 1969; © 1969 by the American Association for the Advancement of Science, with permission.)

ribosomal RNA molecules. Each extrachromosomal nucleolus consists of a compact fibrous core and a granular cortex. If the cells and their nuclei are disrupted in a hypotonic solution, the extrachromosomal nucleolar cores can be isolated and unwound. The core contains a circular molecule of double-stranded DNA which is about 10–30 nm in diameter. It is periodically coated along its length with fibrillar matrix material (Fig. 6.10). The structural arrangement of the fibrils and experiments on RNA synthesis indicate that each matrix-covered DNA region is a gene coding for rRNA molecules. Each fibril in the matrix is an RNA molecule, therefore many RNA molecules are simultaneously being synthesized on each gene for rRNA.

PROTEIN SYNTHESIS: TRANSLATION

It is possible to disrupt cells, isolate some cell components, and bring about protein synthesis in cell-free *(in vitro)* preparations. The components required for protein synthesis are many: amino acids, tRNAs, mRNAs, ribosomes, several specific enzymes, and ATP. The first step in protein synthesis is the enzyme-catalyzed activation of the amino acids. This reaction requires ATP and forms an activated complex, AMP–amino acid, which is bound to the amino-acid-activating enzymes. There is at least one kind of amino-acid-activating enzyme for each kind of amino acid. The next step is the transfer of the amino acids to specific tRNA molecules. The amino acid–tRNA complex then interacts with ribosomes and mRNA.

The ribosomes are the sites of protein synthesis within the cell, but they are not active in protein synthesis unless mRNA is attached to them. In fact the ribosome is made of a large subunit and a small subunit (Ch. 4). The two subunits do not seem to join unless mRNA and the tRNA bearing the first amino acid are first attached to the small subunit. Several ribosomes may be attached to one messenger forming a polyribosome. The actual assembly of the protein molecule involves the polyribosome and the tRNAs charged with amino acids.

A specific base-pairing mechanism between the tRNA and the mRNA is involved in protein synthesis. This pairing is between sequences of three nucleotides. The three nucleotides in the mRNA constitute a *codon,* and three on the tRNA constitute an *anticodon* (Fig. 6.11). As an incoming amino acid is transferred to the growing peptide chain, the messenger moves along the

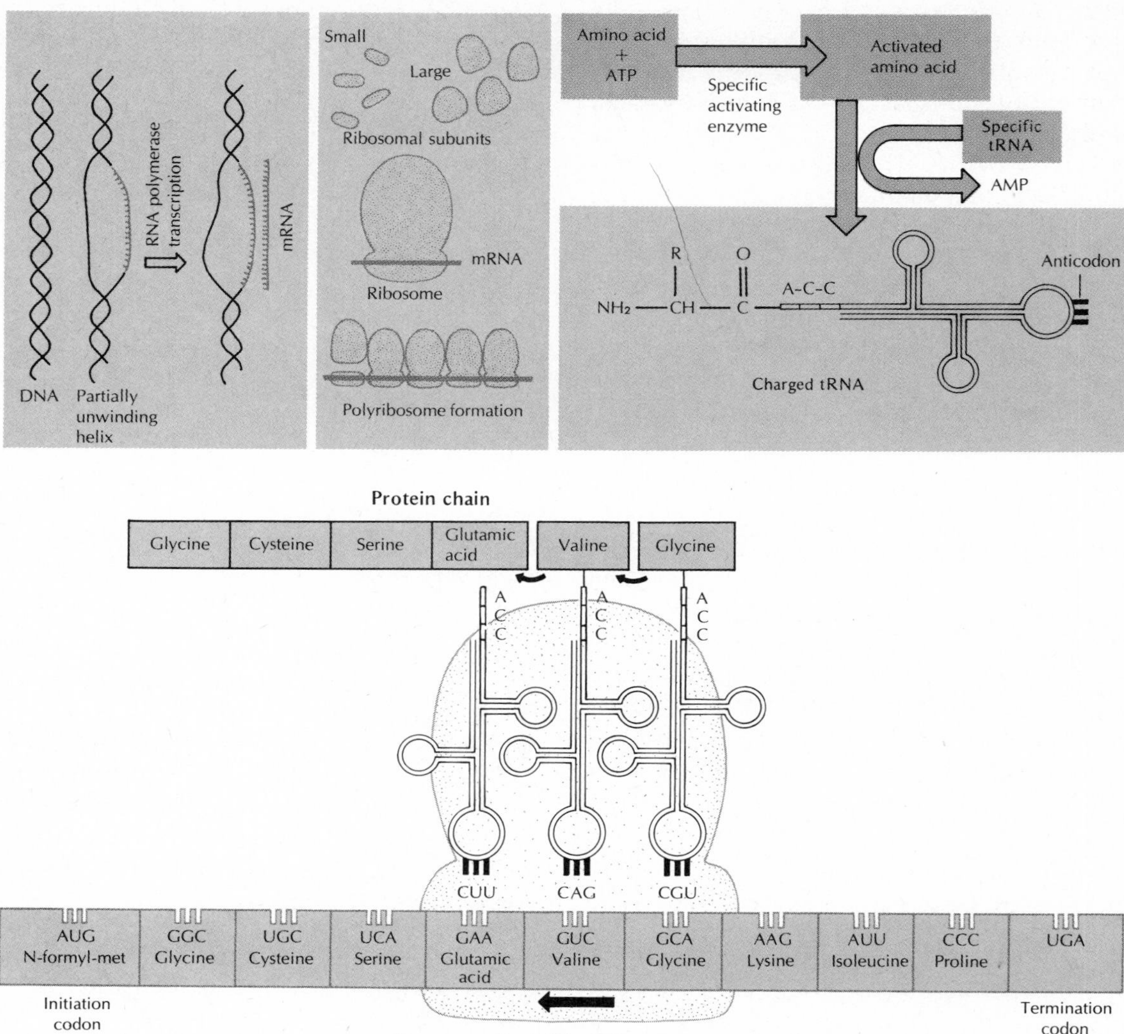

Figure 6.11 A scheme for protein synthesis. Messenger, ribosomal, and transfer RNAs are transcribed from DNA, and all participate in protein synthesis. The actual site of protein synthesis is the polyribosome. Messenger RNA and transfer RNA pair through their codon and anticodon regions and ensure a faithful reading of the message. One cistron (see text) is shown. As the protein is synthesized it folds into its II° and III° structures.

ribosome to bring the next codon for rRNA attachment onto the ribosomes. Thus this picture of the formation of peptide bonds involves the relative movement of the ribosome and the messenger, with the addition of an amino acid at each small movement of the ribosome.

The scheme above suggests that the fate of an amino acid is determined once it is attached to its ribosomal RNA. Indeed, this was shown by an elegant experiment: the tRNA specific for the amino acid cysteine was joined to cysteine, the cysteine was changed to alanine while still on the tRNA, and the behavior of this alanine was studied. The results show that the alanine behaved as cysteine: the alanine was incorporated into protein in the place of cysteine. This is good evidence that the specificity lies in the tRNA and not in the amino acid. The overall scheme for the synthesis of protein specified by the genetic material is diagrammed in Figure 6.11.

In addition to the components of protein synthesis already mentioned there are several other factors involved in the process. These factors, probably proteins, are involved in the initiation of translation (probably three), in elongation of the chain (probably two), and in the release (probably one). These factors and their functions are currently being investigated.

Messenger RNA formation and the attachment of ribosomes to it has been visualized by lysing bacteria and viewing with the electron microscope (Fig. 6.12). The conclusions from these studies are that most of the bacterial chromosome is not genetically active at any one instant, translation is coupled with transcription in prokaryotes, and the genes for the small and large ribosome subunits occur in tandem on the bacterial chromosome.

Some Details of Protein Synthesis

For several years biochemists thought that the mRNA codon for the first or NH_2-terminal amino acid was distinctive and that the ribosome was able to recognize it as the starting point for the synthesis of a polypeptide chain. Analysis of the proteins of *E. coli* and several other bacteria revealed that most, if not all, of them begin with methionine, a sulfur-containing amino acid. Careful studies revealed that the initiating methionine enters as *N*-formyl-methionine attached to a tRNA molecule. *N*-formyl-methionine is the

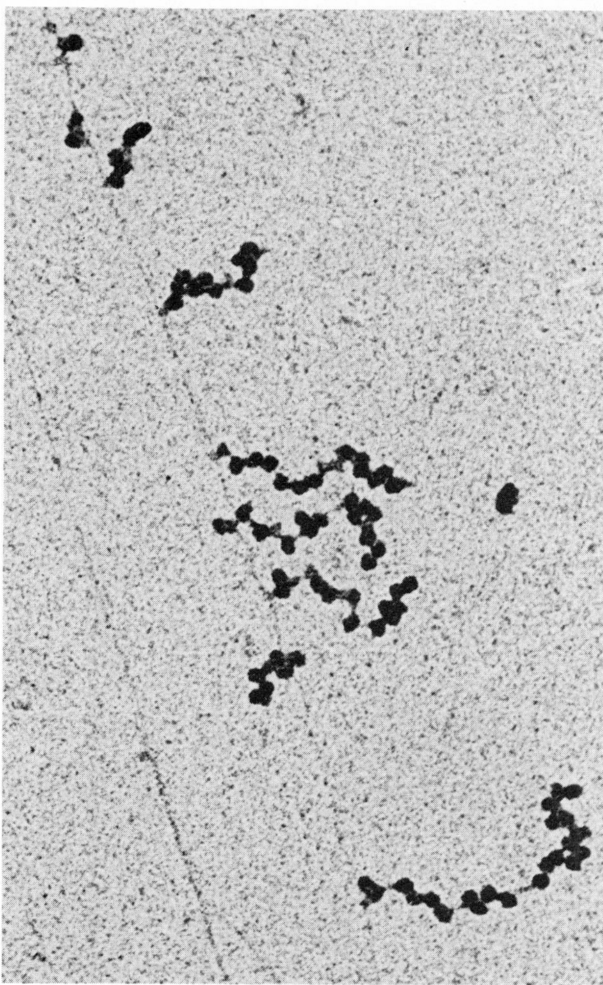

Figure 6.12 Active segment of *Escherichia coli* chromosome showing attached polyribosomes (×85,000). (Courtesy of O. L. Miller, Jr., and B. A. Hamalko, Oak Ridge National Laboratory, *Science,* **169,** 391, 1970, © 1970 by the American Association for the Advancement of Science, with permission.)

result of the addition of a formyl group $\mathrm{H{-}\overset{\overset{\displaystyle O}{\|}}{C}{-}}$ to the amino group of methionine. There are two types of tRNA that can accept methionine, but the methionine attached to only one type of tRNA can be formylated. This tRNA is the initiator tRNA. The blocking of the amino group of methionine by the formyl group prevents the amino group from entering into peptide bond formation and also seems to promote binding of the *N*-formyl-methionine attached to the tRNA to the ribosome. The formyl group does not appear in the finished protein but is removed by enzymatic cleavage.

The binding of the mRNA and the initiator tRNA with its *N*-formyl-methionine is the central event in initiation of the polypeptide chain. Some insight to this binding process resulted from some experiments involving *E. coli* grown on a medium enriched with heavy isotopes of carbon, hydrogen, and nitrogen. The ribosomes of cells grown for several generations on such a medium are "heavy"; they have a greater density than ribosomes isolated from cells grown on normal medium. If *E. coli* cells grown on the heavy medium are transferred to normal medium and allowed to grow further, the ribosomes show two kinds of hybrids; one with a heavy large subunit and a light small subunit, and one with a light large subunit and a heavy small subunit. It was concluded that the ribosome constantly dissociates into subunits and that the subunits reassociate at random. Other workers were able to demonstrate that the small subunit binds to the mRNA and the initiator tRNA to form an *initiation complex,* which then combines with the large subunit. Three specific protein factors are involved with initiation. These initiation factors (F_1, F_2, and F_3) can be extracted from the small subunit of the ribosome so they seem to be a normal ribosomal component. But, interestingly enough, as the initiation complex is formed, the initiation factors apparently are released from the small subunit. This elaborate initiation process is insurance that the ribosomes do not start synthesis in the middle of a messenger.

After initiation, the polypeptide chain is elongated in a cycle of three steps for each amino acid that is added. In the first step the next amino acid attached to its tRNA binds to a site on the ribosome. This site is adjacent to the site at which the initiator tRNA is bound. The binding to the second site requires a protein factor (T, which contains two subunits) and the energy-rich triphosphate, GTP. In the second step the peptide bond is formed between the carboxyl group of the first amino acid, *N*-formyl-methionine, and the amino group of the second amino acid. This reaction breaks the bond between the *N*-formyl-methionine and the initiator tRNA. The reaction is catalyzed by a specific enzyme, which is a part of the large subunit of the ribosome. As a result the growing peptide chain is attached to the tRNA bound to the second site. The initiator tRNA remains bound to the first site. The third step involves a shift of the ribosome and the messenger with respect to each other. As a result, the empty tRNA in the first site is ejected and the tRNA bearing the growing amino acid chain is shifted from the second site to the first. This complex process is thought to be involved with a conformational change in the ribosome driven by energy from GTP hydrolysis. A specific protein factor, called G, is required. The messenger appears to file through the groove between the two ribosomal subunits. The three-step elongation cycle is repeated over and over for each amino acid added to the growing chain. A new G factor–GTP complex is used in the third step of each cycle.

After the last amino acid is added, the completed polypeptide chain is still attached to the last tRNA. There is a specific protein release factor (that is part of the ribosome) that hydrolyses the completed chain from the last tRNA. Little is known about the state in which the completed chain leaves the ribosome in the termination or release step. Probably the chain has folded to a large degree toward its final three-dimensional shape. After the codon for the last amino acid is reached, the next codon is nonsense, that is, it codes for no amino acid and is a terminator codon.

SOME DETAILS OF THE GENETIC CODE

The genetic code is the dictionary used by cells to translate information written in the 4-letter DNA language into the 20-letter protein language. As mentioned above, the group of nucleotides that codes for one amino acid is a *code word,* or *codon.* The simplest possible code would be a *singlet* code in which one nucleotide codes for one amino acid. A singlet code would be inadequate for cells because only four amino acids could be specified. A *doublet* code could specify 16 (4×4) amino acids, whereas a *triplet* code could specify 64 ($4 \times 4 \times 4$) amino acids. Clearly the triplet code is the simplest code that can account for the 20 common amino acids. Some important discoveries about the nature of the genetic code were made using synthetic polynucleotides as messengers in cell-free protein-synthesizing systems. Polyuridylic acid (poly U) which contains only the uridylic acid nucleotide directs the synthesis of a molecule that contains only phenylalanine. This dramatic discovery started a series of productive experiments that have greatly illuminated molecular genetics, most particularly the genetic code. In the triplet code, the RNA code word for phenylalanine would be UUU.

There is much evidence that the genetic code is a triplet one. Crick and coworkers have shown by genetic tests that loss or addition of three nucleotides from the end of a gene did not affect its biological activity. On the other hand, loss or gain of one or two nucleotides (or small multiples thereof) near the beginning or in the middle of a gene seriously affected gene function. This suggested that a three-nucleotide change put

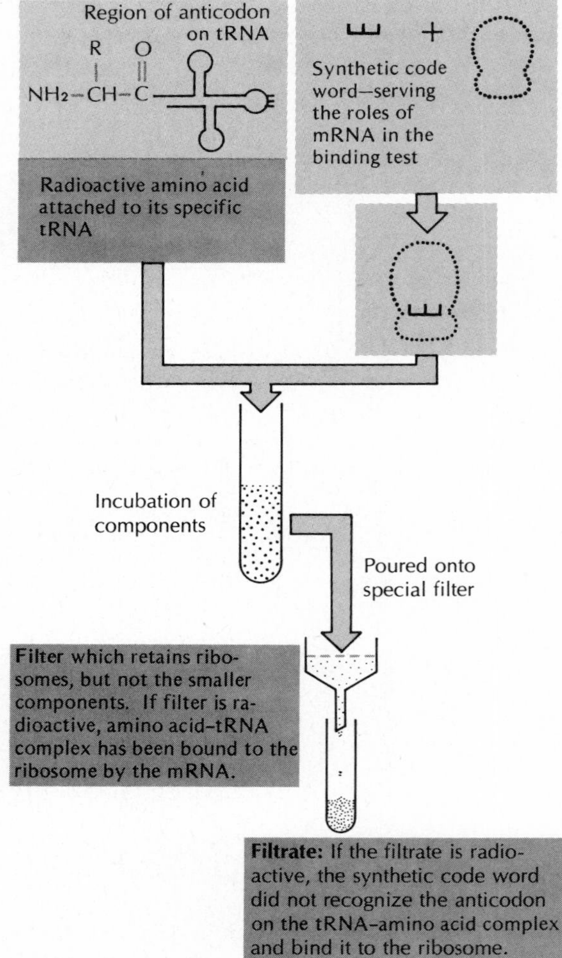

Figure 6.13 Membrane filter technique for testing code word assignments. If a synthetic RNA code word (three letters of known sequence) causes the radioactive amino acid to be retained on the filter by binding to the ribosomes, then the amino acid is coded by that code word.

the reading of the gene back into correct sequence before a critical region of the molecule was reached.

Several techniques were used in the deduction of the dictionary of messenger RNA codons. A particularly important one is based on hydrogen bonding of transfer RNA molecules to the ribosome–messenger complex. Three-letter code words were synthesized and incubated with ribosomes and tRNA that was attached to a radioactive amino acid. If such a single synthetic codon recognized the tRNA molecule, then the radioactive amino acid was bound to the ribosome. The incubation mixture was then poured onto a nitrocellulose filter that retained the ribosomes but not the smaller components. Then if the amino acid was bound to the ribosome by the tRNA, the labeled amino acid stayed on the filter (see Fig. 6.13). If the codon did not recognize the tRNA molecule, the labeled amino acid (and the attached tRNA molecule) went through the filter. Thus the presence or absence of the radioactivity on the filter gave a relatively clear-cut answer to the question: Is this particular amino acid encoded by this particular codon?

Probably the most direct way to confirm the genetic code is to synthesize a mRNA molecule with a defined base sequence and then determine the amino acid sequence of the polypeptide product of that messenger. Some work along these lines has been done, and it supports the other work on the genetic code (Fig. 6.14). H. Gobind Khorana has been a pioneer in the synthesis of nucleic acids. After several years of study on how short chains of nucleotides could by synthesized, Khorana successfully synthesized long chains of repeating dinucleotide, UGUGUGUGUG . . . . . , which contains two triplets, UGU and GUG. The repeating trinucleotide, AGCAGCAGC . . . . , contains three triplets, AGC, GCA, and CAG. Using these long synthetic chains as messengers, polypeptides were synthesized. In the case of the repeating dinucleotide, two amino acids were found in the polypeptide; these were cysteine (UGU) and valine (GUG). For the repeating trinucleotide three polypeptides were found to be formed: polyserine from (AGC)n, polyalanine from (GCA)n and polyglutamine from (CAG)n. With this technique of using synthetic messengers of repeating se-

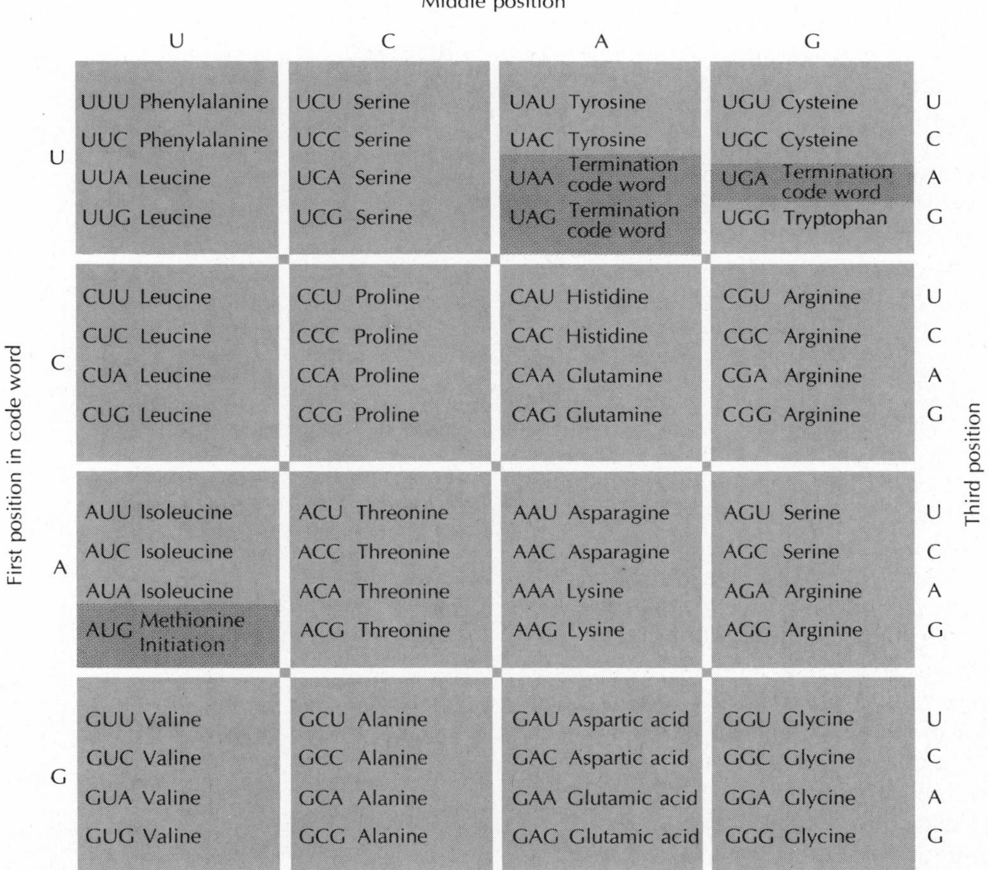

First position in code word	Middle position: U	C	A	G	Third position
U	UUU Phenylalanine	UCU Serine	UAU Tyrosine	UGU Cysteine	U
U	UUC Phenylalanine	UCC Serine	UAC Tyrosine	UGC Cysteine	C
U	UUA Leucine	UCA Serine	UAA Termination code word	UGA Termination code word	A
U	UUG Leucine	UCG Serine	UAG Termination code word	UGG Tryptophan	G
C	CUU Leucine	CCU Proline	CAU Histidine	CGU Arginine	U
C	CUC Leucine	CCC Proline	CAC Histidine	CGC Arginine	C
C	CUA Leucine	CCA Proline	CAA Glutamine	CGA Arginine	A
C	CUG Leucine	CCG Proline	CAG Glutamine	CGG Arginine	G
A	AUU Isoleucine	ACU Threonine	AAU Asparagine	AGU Serine	U
A	AUC Isoleucine	ACC Threonine	AAC Asparagine	AGC Serine	C
A	AUA Isoleucine	ACA Threonine	AAA Lysine	AGA Arginine	A
A	AUG Methionine Initiation	ACG Threonine	AAG Lysine	AGG Arginine	G
G	GUU Valine	GCU Alanine	GAU Aspartic acid	GGU Glycine	U
G	GUC Valine	GCC Alanine	GAC Aspartic acid	GGC Glycine	C
G	GUA Valine	GCA Alanine	GAA Glutamic acid	GGA Glycine	A
G	GUG Valine	GCG Alanine	GAG Glutamic acid	GGG Glycine	G

Figure 6.14 Dictionary of mRNA code words.

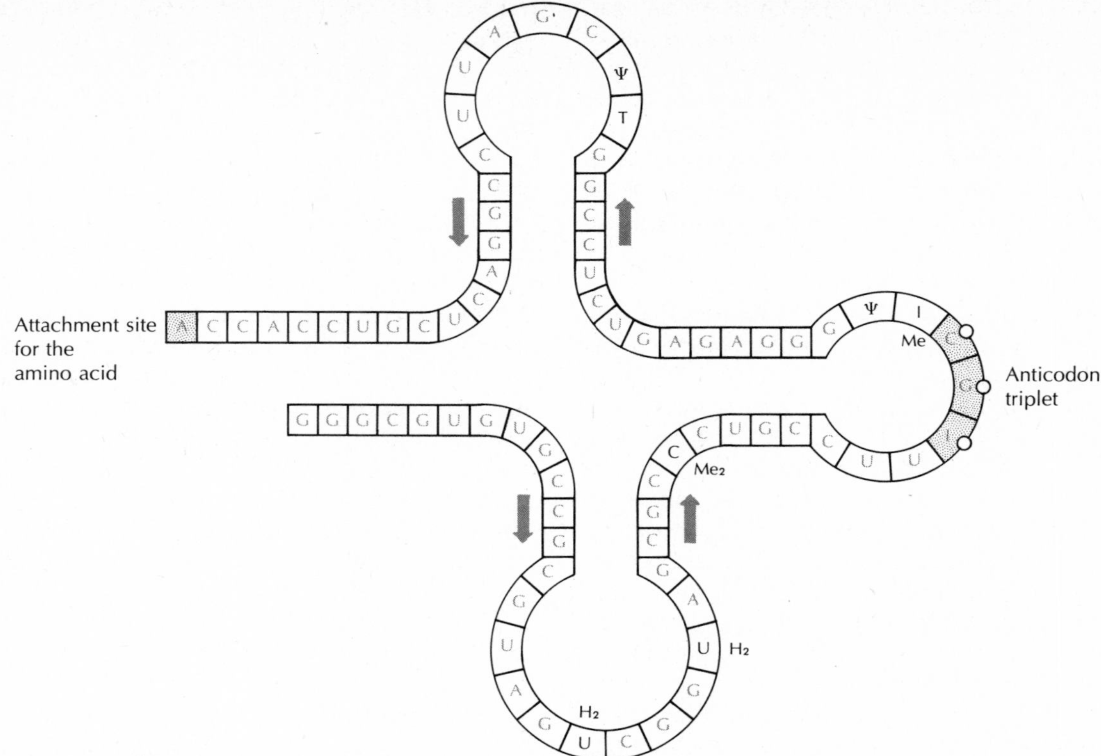

Figure 6.15 Structure of alanyl tRNA—the tRNA that carries alanine, an amino acid, to the site of protein synthesis. A, C, U, and G are symbols for the common RNA nucleobases. Ψ, T, I, I-Me, C-Me_2, and U-H_2 represent "minor" RNA nucleobases. These minor bases are not found to a great extent in RNA, but they apparently play an important role in the structure and function of tRNA molecules.

quence, the base sequences of all the codons for amino acids were determined.

Another method for checking the code is to discover the anticodon in tRNA. The complete determination of the structure of a specific tRNA was finished in 1965 after about seven years of work. The tRNA, obtained from yeast, contains 77 nucleotides and is specific for alanine. The proper sequence of three nucleotides that is complementary to the alanine codon, the anticodon, is located near the middle of the molecule. The two-dimensional structure of alanine tRNA is roughly cloverleaf shaped and is shown in Figure 6.15. The nucleotide sequences of over 60 tRNAs have now been determined. All results show that the anticodon of a given tRNA is complementary to one of the codons that has been assigned to the particular amino acid. The three-dimensional structure of yeast phenylalanine tRNA has been determined by X-ray crystallography. The hydrogen bonding relationships predicted from the nucleotide sequence arranged in a cloverleaf are maintained in this three-dimensional structure (Fig. 6.16).

The elucidation of the nature of the genetic code has been a fascinating story. Its basic nature has been established and it shows how the information of nucleic acids is used to control

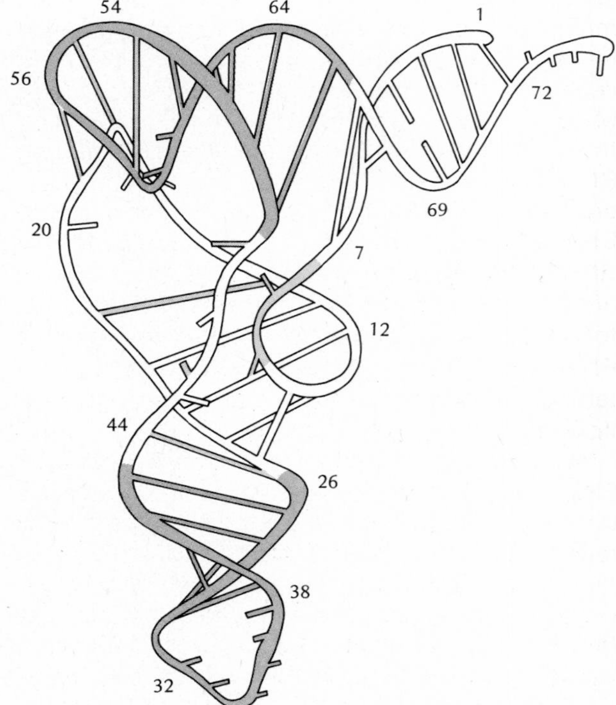

Figure 6.16 Three-dimensional structure of yeast phenylalanine tRNA from X-ray crystallographic studies. The numbers refer to the nucleotide sequence from the 5′-end to the 3′-end. Redrawn from S. H. Kim, et al., *Science*, **185,** 435-439, 1974.

the structure of proteins. The genetic code confirms the theme of molecular genetics. Genetic information can be stored as a one-dimensional message in nucleic acids and is expressed in the linear structure of proteins. A linear segment of nucleotides in DNA from an initiator codon to a termination codon that specifies a mRNA molecule is called a *cistron.* Since the genetic code is most probably a triplet one, the cistron that directs the synthesis of a protein containing 100 amino acids would contain at least 300 nucleotides.

THE FLOW OF GENETIC INFORMATION

This chapter has developed the theme that the flow of genetic information in cells is DNA → RNA → protein. Indeed, this logical and reasonable picture is attractive and has been dubbed "the dogma of molecular biology." With RNA-containing viruses the nature of replication of the RNA to produce more viral RNA has been thought not to involve DNA, that is, RNA serves as a template for RNA synthesis. For some years several workers have thought that tumor-causing RNA viruses replicate inside of cells by synthesizing a DNA intermediate, which then serves as a template for the production of new RNA molecules that are assembled into new viruses. In 1970 Howard Temin, a virologist at the University of Wisconsin, and David Baltimore at the Massachusetts Institute of Technology reported the existence of such DNA in cells infected with RNA viruses. Temin had held this view for several years since he had shown that inhibitors of DNA synthesis would inhibit the replication of RNA tumor viruses in infected cells. Immediately after Temin's report in 1970, other laboratories found similar evidence to support this idea of RNA-directed DNA synthesis. This is an important exception to the idea that bioinformation flows from DNA to RNA to protein.

What is the significance of these findings? First, the fact that a well established concept has not held up in all cases is a good example of how ideas must change in biology and science as new information is gained. Second, some possible applications of this information may be important in medicine. Several tumor-causing viruses contain RNA. If the enzyme that catalyzes the synthesis of DNA from RNA (this enzyme is called *reverse transcriptase*) can be selectively inhibited, it may be possible to inhibit viral production, and therefore, the tumors caused by these viruses, without harming the normal cellular functions. This would be a cure for the cancers caused by these viruses. The situation is quite complex, with many questions to be answered. For example, is the reverse transcriptase an enzyme peculiar to RNA viruses which cause tumors? Does the reverse transcriptase occur in normal cells and has it just not been detected heretofore? There is some evidence that the reverse transcriptase may function in the multiplication of RNA-containing viruses that do not cause tumors. Also, some recent studies indicate that normal cells do contain small amounts of the enzyme. As this story unfolds in the next few years, it will be reported, undoubtedly, in newspapers and news magazines. Thus many people will be able to follow this example of application of basic research to the benefit of human welfare.

REGULATION OF INFORMATION FLOW

If a culture of bacteria is growing in minimal medium (a simple medium containing a few compounds that satisfy the barest nutritional requirements) with glucose as a carbon source, the presence of the enzyme β-galactosidase can be detected only in minute amounts. (β-galactosidase catalyzes the hydrolysis of lactose into galactose and glucose and is also called lactase.) If lactose is then added to the culture medium, β-galactosidase molecules appear very shortly thereafter and can be detected easily. Such an enzyme is called an *inducible* enzyme because its synthesis is induced by a specific small molecule. On the other hand, a culture of bacteria growing on minimal medium synthesizes the amino acid histidine from carbohydrates by a relatively complicated pathway. All enzymes of this pathway can be shown to be synthesized by the cells under these conditions. If histidine is added to the medium, the synthesis of the histidine enzymes is turned off. Such enzymes are called *repressible* enzymes because their synthesis is repressed by a specific small molecule. These are two cases of differential gene activity in which the synthesis of gene products is sensitive to environmental conditions.

Such cellular control mechanisms have been studied, and a model for these phenomena has been proposed. This proposal, called the operon model, involves the following components: regulator genes, operator genes, repressor substances, corepressors, inducers, and structural genes. (The reader may find it helpful to refer to

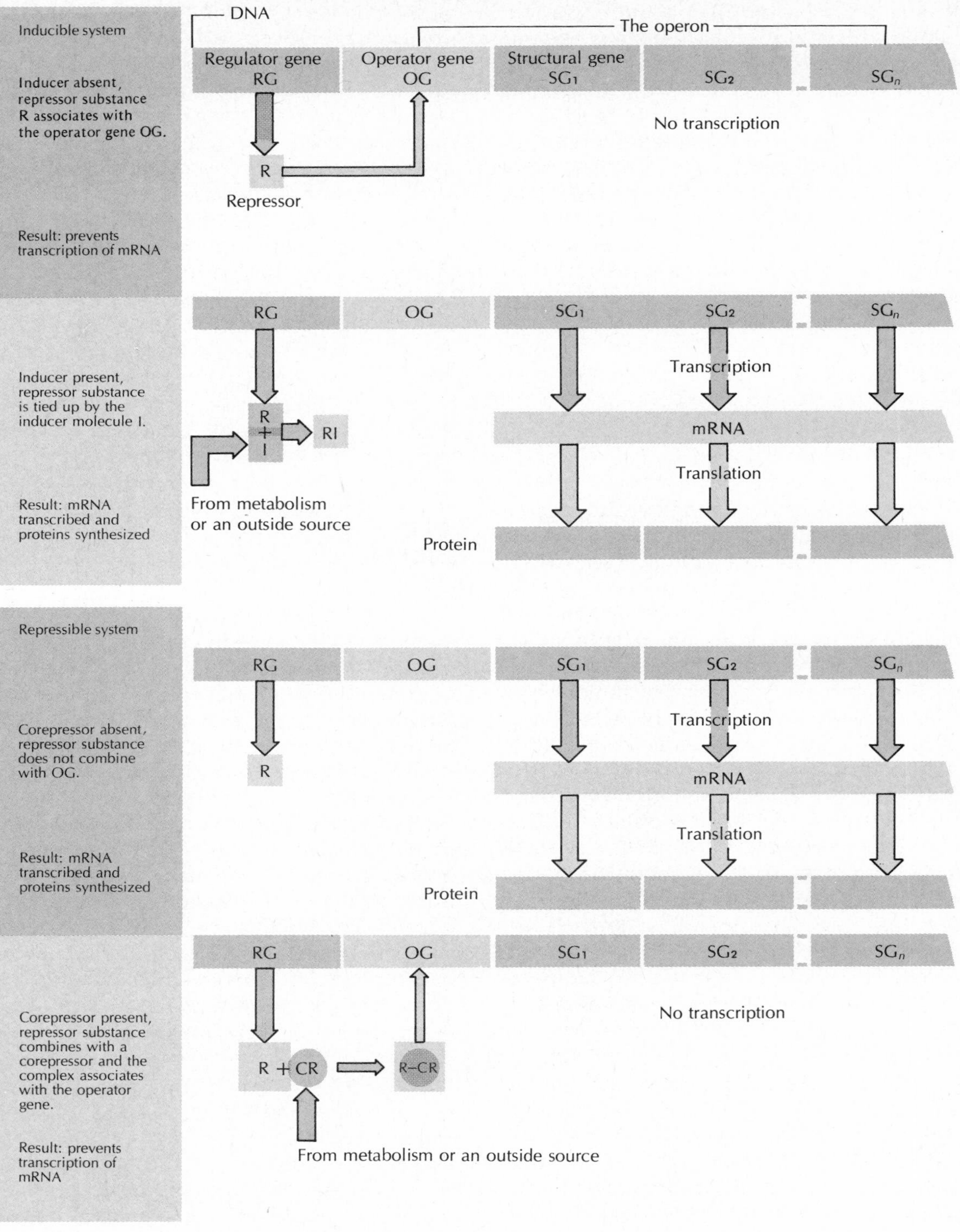

Figure 6.17 Operon model for induction and repression.

Figure 6.17 as he reads the following description.) The regulator gene (RG) produces the repressor substance (R), a protein. The operator gene (OG) turns the transcription of structural genes (SG) on and off in response to the repressor substance. The structural genes (cistrons) function in specifying a protein via mRNA. The inducer or corepressor (for example, lactose or histidine, respectively) combines with the repressor substance to change its ability to combine with the operator gene. Transcription of structural genes occurs when the operator gene is not associated with the repressor substance. The inducible and repressible systems differ in the form of the repressor substance that will associate with the operator gene. In an inducible system the repressor substance normally associates with the operator gene (no transcription); the addition of inducer prevents the association and turns on transcription. In a repressible system the repressor substance normally does not associate with the operator gene (transcription); the addition of the corepressor allows the association of the repressor with the operator gene and turns off transcription. Thus the inducer and corepressor function differently in the two systems. They are small molecules and may be obtained from the outside or from metabolism within the cell. No matter what the source, the concentration of those substances controls transcription within the cell. If a cellular metabolism raises the concentration of a corepressor, this elevation will turn off the production of more corepressors. This mechanism is a sensitive control system, and it helps regulate the concentration of small molecules in the cell.

The mRNA produced by the structural genes that are controlled by one operator gene probably is one long "giant messenger" transcribed from contiguous cistons. The regulator gene is not necessarily adjacent to the operator gene that is regulated. The operator genes and the structural genes under their control constitute an *operon,* which is the unit of transcription.

The operon model for the regulation of gene activity is based on studies done primarily with microorganisms. This model has served as an important stimulus in thinking about ways in which eukaryotic cells might become different from each other as a result of changes in their environment. Recent work indicates that the operon model, although of fundamental importance in understanding regulation of the flow of bioinformation, does not spell out the whole story. Refinements, extensions, and alternatives of this model are even more complex and beyond our scope in this introductory text.

SUMMARY

The understanding of the principles of information flow in cells has been applied to several areas of biology. This will be evident as several succeeding chapters refer to DNA replication, transcription, and translation. Any discussion of genetics is related necessarily to the mechanism and control of gene expression. Physiology and metabolism, the study of dynamics of organisms and cells, are based on the production of specific enzymes and the functions of various cellular organelles. The action of various hormones is implicated strongly in interaction with mechanisms of gene expression. The complex events in development and differentiation likewise are dependent on perfect orchestration of transcription and translations; some genes must be turned on and some must be turned off in these developmental events.

The implications of molecular genetics go beyond the level of the organism. Evolution, behavior, and ecology are interacting disciplines at the intraspecific and interspecific levels. Additionally, some of the major problems facing the human species are being investigated against a background of bioinformation flow and with techniques developed in the field of molecular genetics.

The principles of bioinformation flow and molecular genetics in this chapter may be regarded as a basic statement. Variations and elaborations upon the themes will be made in subsequent chapters, particularly Chapters 7, 12, 15–19, 20, 42 and 47.

SUGGESTIONS FOR FURTHER READING

DuPraw, E. J. *DNA and Chromosomes.* New York: Holt, Rinehart and Winston (1969).

Goodenough, U., and R. P. Levine. *Genetics.* New York: Holt, Rinehart and Winston (1974).

Hood, L. E., J. H. Wilson, and W. B. Wood. *Molecular Biology of Eucaryotic Cells.* Menlo Park, Calif.: W. A. Benjamin (1975).

Ingram, V. M. *The Biosynthesis of Macromolecules,* 2nd ed. New York: W. A. Benjamin (1972).

Lehninger, A. L. *Short Course in Biochemistry.* New York: Worth Publ. (1973).

Loewy, A. G., and P. Siekevitz. *Cell Structure and Function.* New York: Holt, Rinehart and Winston (1969).

Novikoff, A. B., and E. Holtzmann. *Cells and Organelles,* 2nd ed. New York: Holt, Rinehart and Winston (1977).

Watson, J. D. *Molecular Biology of the Gene,* 3rd ed. Menlo Park, Calif.: W. A. Benjamin (1976).

From Scientific American

Brown, D. D. "The Isolation of Genes" (Aug. 1973).

Cairns, J. "The Bacterial Chromosome" (Jan. 1966).

Clark, B. F. C., and R. A. Marcker. "How Proteins Start" (Jan. 1968).

Crick, F. H. C. "The Genetic Code" (Oct. 1962).

Crick, F. H. C. "The Genetic Code: III" (Oct. 1966).

Deering, R. A. "Ultraviolet Radiation and Nucleic Acid" (Dec. 1962).

Dickerson, R. E. "The Structure and History of an Ancient Protein" (Apr. 1972).

Edgar, R. S., and R. H. Epstein. "The Genetics of a Bacterial Virus" (Feb. 1965).

Fraenkel-Conrat, H. "The Genetic Code of a Virus" (Oct. 1964).

Holley, R. W. "The Nucleotide Sequence of Nucleic Acid" (Feb. 1966).

Horowitz, N. "The Gene" (Oct. 1956).

Jacob, F., and E. L. Wollman. "Viruses and Genes" (June 1961).

Kornberg, A. "The Synthesis of DNA" (Oct. 1968).

Maniatis, T., and M. Ptashne. "A DNA Operator-Repressor System" (Jan. 1976).

Miller, O. L., Jr. "The Visualization of Genes in Action" (Mar. 1973).

Nirenberg, M. W. "The Genetic Code: II" (Mar. 1963).

Nomura, M. "Ribosomes" (Oct. 1969).

Ptashne, M., and W. Gilbert. "Genetic Repressors" (June 1970).

Sobell, H. M. "How Actinomycin Binds to DNA" (Aug. 1974).

Stent, G. S., "The Multiplication of Bacterial Viruses" (May 1953).

Temin, H. M. "RNA-Directed Synthesis" (Jan. 1972).

Tomasz, A. "Cellular Factors in Genetic Transformation" (Jan. 1969).

Wood, W. B., and R. S. Edgar. "Building a Bacterial Virus" (July 1967).

7 Chromosomes, Mitosis, and Meiosis

In the discussion of the cell nucleus in Chapter 4 the chromatin material was described as condensing into stainable chromosomes at the time of cell division. The term "chromosome" means colored body (Gk. *chrome,* color; *soma,* body); the name is related to the fact that cells with definite nuclei have chromosomes that stain easily with a variety of biological stains.

CHROMOSOMES

Viral Chromosomes

The viruses do not qualify as cells for several reasons (see Ch. 4). They do not have cytoplasm and therefore do not undergo cell division. Their reproduction occurs only within living cells, where they use materials and energy from the host cell. A virus possesses all the genetic information necessary for the production of new viruses but none or only a little of the machinery necessary for such production. The chromosomes of a few viruses have been well studied. Generally they are naked nucleic acids–DNA or RNA (Fig. 7.1). The nucleic acid of bacterial viruses (bacteriophages) is packed tightly into the capsule of the virus. As present in the mature virus particle, the linear double-stranded DNA helix is about 17μm long in phage lambda and 52 μm long (52,000 nm) in T_2 and T_4 bacteriophages. In T_2 and T_4 the DNA is packed into a head that is 95 nm long. This is a packing ratio of over 500:1. The packing ratio is the degree of shortening because of coiling, supercoiling, and folding of the DNA. Replication of double-stranded DNA viral chromosomes is by a mechanism of separation of the two strands from each other, with synthesis of a new strand on each old separated strand. Some viral chromosomes are

Figure 7.1 Electron micrograph of the chromosome of the lambda virus, a closed circle of DNA (× 19,000). (Courtesy of L. A. MacHattie, University of Toronto.)

single-stranded circular DNA molecules, and replication is somewhat different (Ch. 6). The viral chromosomes in solution, outside of the virus, are influenced by the temperature and the kinds of ions that are used to neutralize the negative charges on the phosphate groups of DNA. The viral chromosomes will coil and supercoil in response to changes in the salt concentration in the solution. The chromosome of the tobacco mosaic virus (TMV) is a single strand of RNA coiled inside a coat of protein subunits. All plant viruses and some bacterial and animal viruses possess RNA rather than DNA. At one time it was thought that viruses could possess only one kind of nucleic acid; recently some RNA tumor viruses have been shown to contain small amounts of DNA in addition to RNA, and one report gives evidence that a bacteriophage contains both DNA and RNA.

The Chromosomes of Prokaryotes

The nucleoid area, or the chromatin of the prokaryotes, may be detected by staining and viewing with the light microscope or by using the electron microscope. There appears to be one bacterial chromosome per nucleoid area (some bacteria have more than one nucleoid area). The chromosome of the bacterium *Hemophilus influenzae* is about 860 μm long and appears to be linear, naked DNA (Fig. 7.2). The chromosome of the colon bacterium, *Escherichia coli,* is about 1000 μm long and is a closed circle. The DNA of *E. coli* is packed into a nucleoid less than 1 μm long at a packing ratio of about 1000:1. Since viral and prokaryotic chromosomes are simple in comparison to eukaryotic chromosomes, some workers have suggested that they be termed *prochromosomes.*

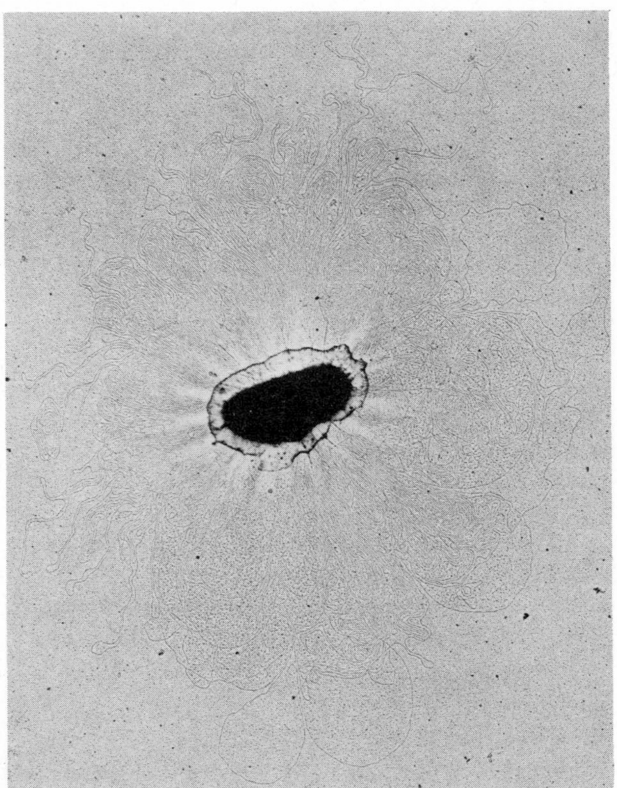

Figure 7.2 The bacterium *Hemophilus influenzae* has been disrupted and is shown with its chromosome of naked DNA spread out around it (× 27,200). (Courtesy of L. A. MacHattie, University of Toronto.)

The Chromosomes of Eukaryotes

The structure of the chromosomes of the eukaryotes is extremely complex. Table 7.1 shows the DNA content of some viruses and prokaryotes in comparison to some eukaryotes. From these data it is clear that a eukaryotic nucleus will contain about 1000 times more DNA than a virus or prokaryotic cell. In addition to a complex chromosome structure, the eukaryotes possess a mitotic apparatus, a nucleus enclosed by a membrane, and a drastic increase in chromosome size over the prochromosomes of viruses and prokaryotes. Furthermore, parts of the eukaryotic chromosome sometimes are metabolically inactive, while at other times they are active—some chemical components are rather constant for long periods and others fluctuate greatly. Chemically, the chromosome contains DNA, RNA, and proteins.

Table 7.1 DNA Content of Some Viruses, Prokaryotic Cells, and Eukaryotic Nuclei

	DALTONS*	NUCLEOTIDES
Viruses		
ΦX-174	1.6×10^6	5500
T_2	130×10^6	400×10^3
T_5	85×10^6	260×10^3
Prokaryotic cells		
Hemophilus influenzae	0.7×10^9	2×10^6
Escherichia coli	2.6×10^9	7.4×10^6
Eukaryotic nuclei		
Drosophila melanogaster (fruit fly)	0.12×10^{12}	0.4×10^9
Frog	28×10^{12}	90×10^9
Carp	2.0×10^{12}	6.6×10^9
Mouse	3.0×10^{12}	8.6×10^9
Human	3.6×10^{12}	11.2×10^9

* The dalton is the unit of atomic weight. The atomic weight of hydrogen is 1 dalton, while the molecular weight of water is 18 daltons, and so on.

There have been several suggestions as to how the DNA is ordered in the eukaryotic chro-

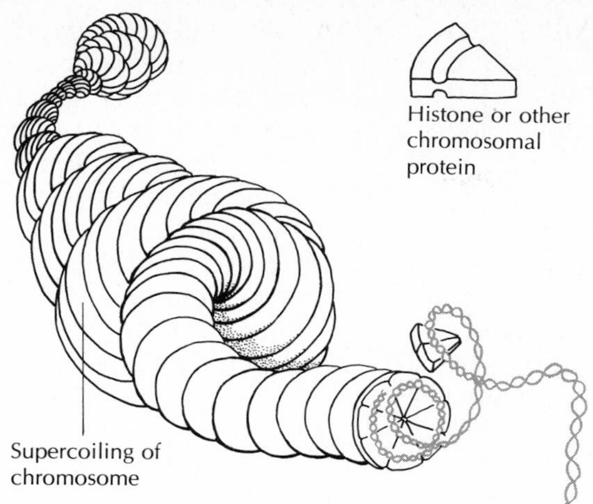

Figure 7.3 Chromosome model. Inside each chromosomal fiber, a single DNA double helix is tightly packed by supercoiling. The DNA molecule is held in place by histones and other proteins which are shown as wedge-shaped molecules. (Redrawn from E. J. DuPraw, *DNA and Chromosomes*, Holt, Rinehart and Winston, 1970.)

mosome (see DuPraw, Suggestions for Further Reading). The suggestion in current favor is that it is a continuous strand from one end to the other with proteins and other materials attached along its length. Figure 7.3 shows a model of the structure of a chromosome that illustrates the coiling and supercoiling to give a packing ratio of 56:1.

Whatever the structure of the eukaryotic chromosome should turn out to be, it must provide for some elaborate and precise movements. These movements are associated with two processes, mitosis and meiosis. As noted in Chapter 4, these complex chromosomal events are means for passing on the same sets of chromosomes to somatic or body cells (mitosis) or reducing by one half the number of chromosomes in a gamete (meiosis).

At certain regions there are constrictions of the chromosome; the primary constriction is the region of the *centromere* (sometimes called the kinetochore), and the other regions are called *secondary constrictions.* The centromere is easily distinguished in stained preparations of cells, since it does not stain or stains only lightly. The centromere has been studied by electron microscopy, and it is thought to be a structureless segment of the chromosome in some species and a set of complex granules or tubules in others. The centromere functions in attaching to the spindle fibers that form in mitosis and meiosis.

Karyotypes

All body or somatic cells (not sperm or eggs) in an organism normally have the same number of chromosomes. This number ranges from two to several hundred in different species. The human chromosome number is 46. Some other chromosome numbers are given in Table 7.2. The chromosomes of body cells occur in pairs. In the formation of the fertilized egg one member of each pair of chromosomes is contributed by the egg, and likewise, one member of each pair is contributed by the sperm. Except for a pair of chromosomes associated with sex in most animals, the members of each pair of chromosomes are normally alike in size and shape. The members of such a pair are called *homologous chromosomes.* In most species the chromosomes of different pairs tend to differ in size and shape. Thus the chromosomes exhibit individuality, and it is possible to study individual chromosomes during the division process. The *karyotype* is the term applied to the description of the chromosome composition of a species as illustrated in Figure 7.4. Usually karyotypes are established by cutting the individual chromosomes from a photograph and arranging them by homologous pairs.

Table 7.2 Chromosome Numbers of Selected Organisms

ORGANISM	DIPLOID (2*n*) NUMBER
Dog	52
Chicken	18
Fruit fly	8
Planaria	14
Hydra	12
Crayfish	200
Frog	26

Each kind of gamete, sperm or egg, contains one set of chromosomes, that is, one of each chromosome in a homologous pair. This chromosome set is a *haploid* set and is abbreviated by the symbol *n.* Somatic cells, on the other hand, are 2*n* or *diploid* since two chromosome sets are present in the nuclei.

Recently methods for studying chromosomes in somatic cells of humans have been improved. This has permitted an accurate and detailed description of the normal karyotype and the discovery of abnormalities responsible for certain congenital malformations (Ch. 19). Several techniques have helped greatly in this regard:

1. Treatment of cultures of cells with the alkaloid drug, *colchicine,* which prevents the formation of the spindle, with the result that many cells arrest at metaphase with definable chromosomes.
2. Treatment of cells with a hypotonic solution that causes nuclei to swell and makes the

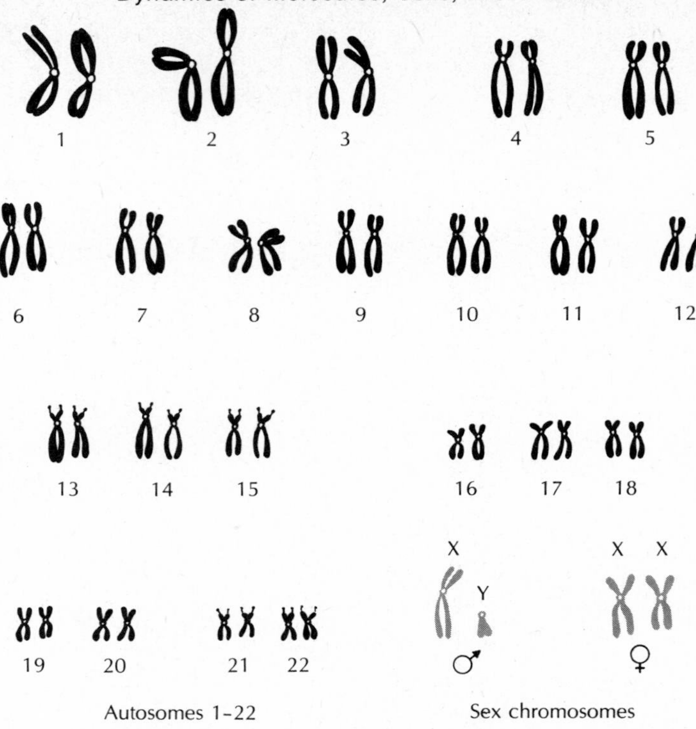

Figure 7.4 Mitotic metaphase chromosomes of the human arranged according to the "Denver classification." Chromosomes are grouped according to the position of the centromere, size, and ratio of arm lengths. Such an arrangement of photographs of chromosomes is the karyotype. The twenty-third pair of human chromosomes is the pair of sex chromosomes, XX in the female and XY in the male.

spreading of the chromosomes much easier.

3. Cells used in karyotype preparation are often obtained from bone marrow, skin, or connective tissue by biopsy. White blood cells are widely used also. The addition of phytohemaglutinin (a substance derived from the broad bean) agglutinates the red blood cells, and also stimulates division of white cells.
4. Tissue culture methods have been important in chromosome analysis.

These procedures have led to a rather rapid advance in human chromosome methodology and in the understanding of the relationship of abnormal chromosome structure to pathology.

The chromosomes are classified according to size: small, medium, and large. Additionally, chromosomes are recognized according to the position of the centromere (the primary constriction) and the resulting relative length of the arms. If the centromere is in the center, then the arms of the chromosomes are the same length. The centromere may be asymmetrically located in the chromosome, or it may be placed at the end. Some chromosomes have secondary constrictions. This makes the chromosomal arm appear as if it contained a stalk with a bulb of chromatin attached. The bulb of chromatin is called a satellite and is useful in identification of chromosomes (Fig. 7.5). The karyotype of the human female has 23 homologous pairs including a pair of X chromosomes; the male has 22 homologous pairs and an X and a Y chromosome. The X and Y chromosomes are called *sex chromosomes* while the others are termed autosomes (see Part III).

Karyotyping of cells by the traditional methods of photographing a stained preparation of mitotic cells and arranging the individual photographs of the chromosomes in order of size and centromere position has allowed cytologists to identify postively only 4 of the 23 pairs of human chromosomes. In the late 1960s karyotyping was advanced by the development of chromosome analysis under a fluorescence microscope. The Y chromosome fluoresces when treated with a class of compounds called quinacrine dyes (normally used in the treatment of malaria). Sperm and cells taken from certain extraembryonic membranes (the amnion and chorion) can be distinguished as to their sexual type. This staining technique can be used for the prenatal determination of the sex of a baby. Cells in the amniotic fluid may be withdrawn through a syringe needle (the human female reproductive system is discussed in Chapter 15). The amniotic fluid cells originally were sloughed from the lower intestinal tract of the baby. This process of withdrawing amniotic fluid cells is called *amniocentesis.* In addition to staining the chromo-

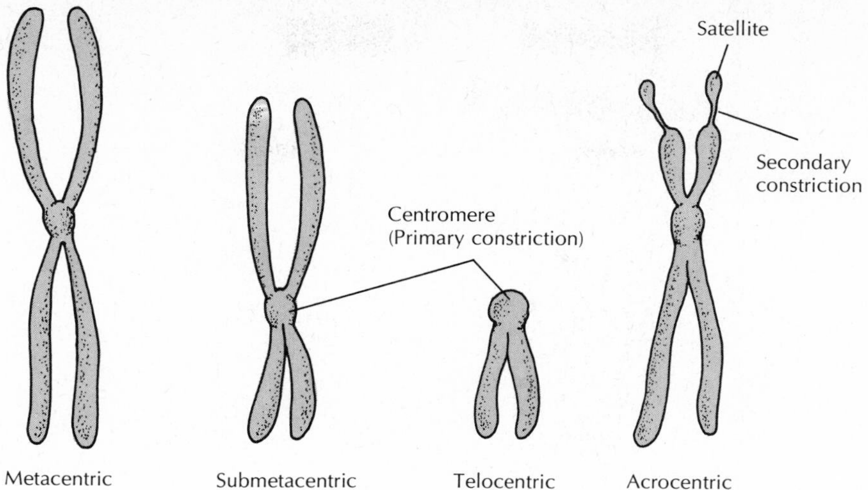

Figure 7.5 Types of metaphase chromosomes.

somes of the cells for determination of the sex of the baby or detection of chromosome abnormalities, the amniotic cells may be cultured and metabolic deficiences of the baby determined (Ch. 19). Since there is a slight chance of miscarriage associated with amniocentesis, it is seldom used for prenatal sex determination exclusively.

In 1970 a truly remarkable discovery in karyotype preparation was made. Treatment of chromosomes with hydrochloric acid to remove basic proteins, with ribonuclease to remove RNA, and with sodium hydroxide (pH 12) to denature the DNA produces a chromosome that shows a specific banding pattern with Giemsa stain (long used for staining blood cells). Precise identification of every human chromosome is possible with this technique (Fig. 7.6). This new technique should allow the identification of chromosome abnormalities involved in birth defects, drug and pollution damage, mental retardation and illness, aging, and cancer.

The quinacrine and Giemsa stain techniques set off an explosion of reports of new information on the karyotypes of many kinds of organisms. Additionally, other staining techniques have been shown to give banding profiles when used on chromosomes that have been "relaxed" by the above or similar techniques.

MITOSIS

Ordinary cell division often is called mitosis. Strictly speaking, the term mitosis refers to nu-

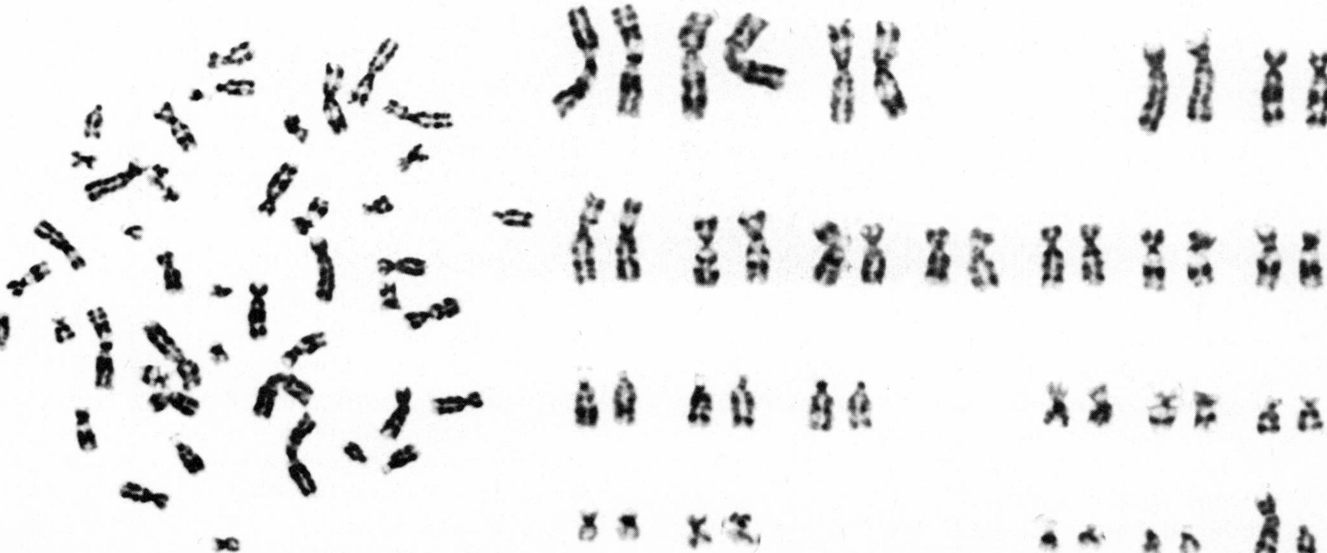

Figure 7.6 Karyotype of a normal human male. The Giemsa stain banding pattern is characteristic for each chromosome. (Courtesy of M. W. Shaw, Medical Genetics Center, University of Texas at Houston.)

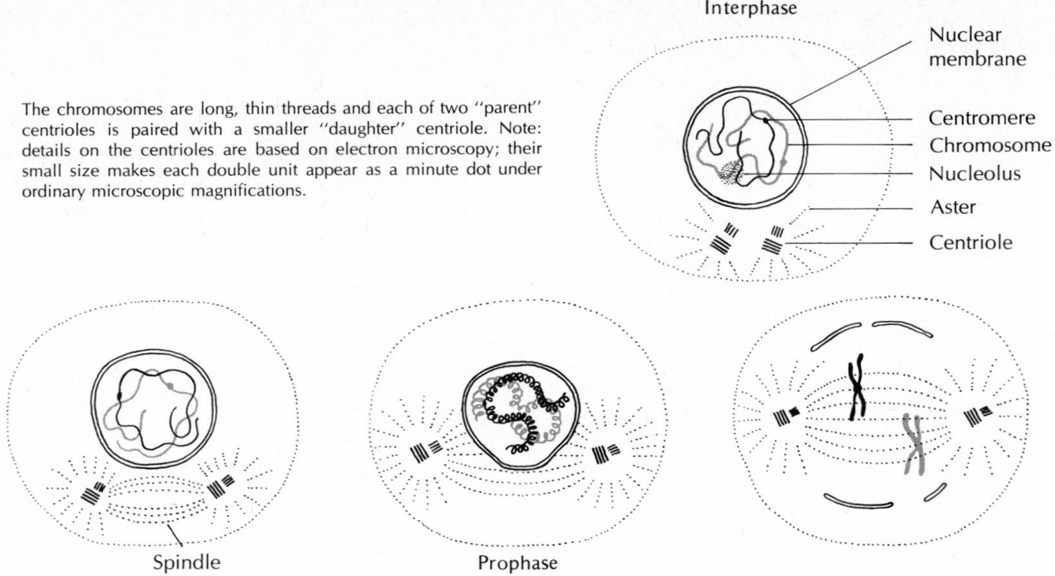

The chromosomes are long, thin threads and each of two "parent" centrioles is paired with a smaller "daughter" centriole. Note: details on the centrioles are based on electron microscopy; their small size makes each double unit appear as a minute dot under ordinary microscopic magnifications.

Astral rays appear and the centrioles separate and begin to move to opposite poles; the spindle fibers begin to form between them. The chromosomes coil and condense; the nuclear membrane and nucleolus break down; connections between the centromeres and the poles become established and the chromosomes move to the "equatorial plate."

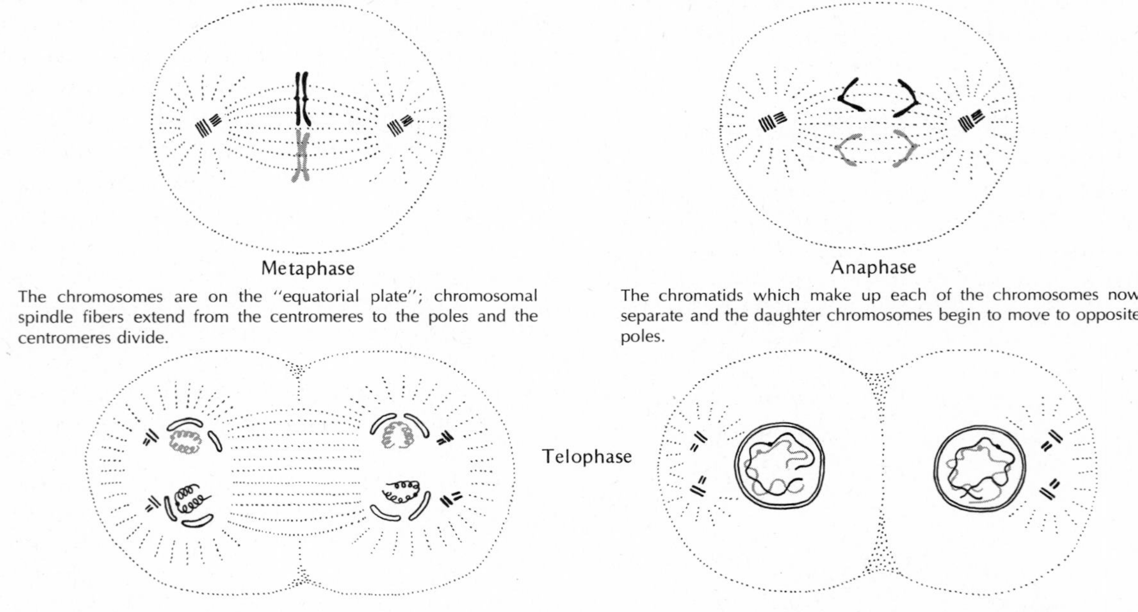

The chromosomes are on the "equatorial plate"; chromosomal spindle fibers extend from the centromeres to the poles and the centromeres divide.

The chromatids which make up each of the chromosomes now separate and the daughter chromosomes begin to move to opposite poles.

Constriction of the cytoplasm, cytokinesis, begins this stage. The chromosomes reach the poles, uncoil, and lengthen; the nucleoli and nuclear membrane re-form; each centriole produces a new one. Constriction completes the separation and two daughter cells now enter interphase.

Figure 7.7 Mitotic division in a cell containing two chromosomes. (By permission from "How Cells Divide" by Daniel Mazia. (Copyright © 1961 by Scientific American, Inc. All rights reserved.)

clear changes during cell division, whereas the division of the cytoplasm more properly is called *cytokinesis.* As mentioned in Chapter 4, mitosis is a mechanism that normally produces two daughter cells, each with the same chromosome complements. This means that the two cells so produced have the same hereditary material, since the chromosomes are the locations of DNA in the nucleus.

Stages of Mitosis

The changes occurring in mitosis are continuous, but for analytic purposes the process is divided into stages. The stage between divisions is known as the *interphase.* The process of division is divided into *prophase, metaphase, anaphase,* and *telophase.* Although the process may be studied in living cells by using a polarizing microscope or a phase microscope, in ordinary studies the cells are killed and stained; thus each cell is analogous to a single frame of a moving picture film. Below are described the stages of mitosis as they occur in a typical animal cell (Fig. 7.7).

INTERPHASE

Although the interphase stage is not a part of the division process, it is a very important stage in the *cell cycle.* In this stage the individual chromosomes cannot be seen as such. Each chromosome is greatly elongated, and the chromatin material (after staining) appears diffuse in the nucleus. There is evidence that the replication of the chromosome occurs during interphase. This means, of course, that the synthesis of more DNA and more protein, the main constituents of chromosomes, occurs in this period. Three steps of interphase are recognized: the G_1 (gap) stage, which is the period preceding DNA synthesis; the S (synthesis) stage, during which DNA synthesis occurs; and the G_2 stage, a period of preparation for actual division. The cell cycle is diagrammed in Figure 7.8.

There also is evidence that the synthesis of the proteins that go into the formation of the spindle (described below) occurs during interphase. With this in mind it can be said that mitosis is a phenomenon that results in the orderly distribution of the already duplicated chromosomes to daughter cells. In addition to the chromatin, the nucleoli are often conspicuous in the stained interphase nucleus. Figure 7.7 gives a schematic view of the mitotic process in a cell containing two chromosomes (one homologous pair).

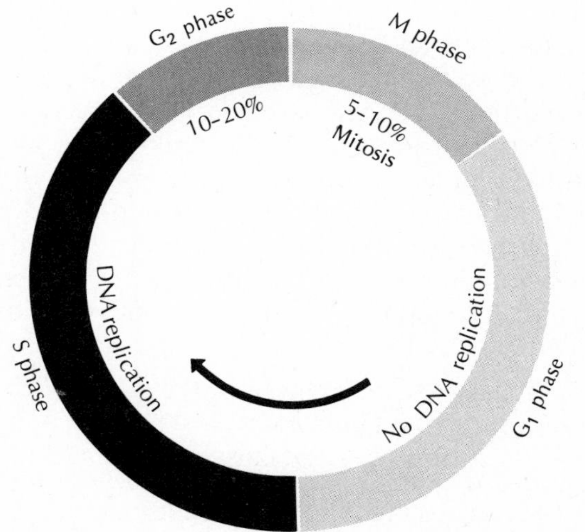

Figure 7.8 Cell cycle. Mitosis and DNA replication (S phase) are generally separated in time by a G_1 and a G_2 phase.

PROPHASE

The first visible change at the beginning of *prophase* is the appearance of the chromosomes as long, thin threads. This is due to the shortening and thickening of the interphase chromatin. These changes result from the loss of water by the chromosomes and by the coiling of the constituent elements.

The chromosomes in prophase are double structures except for the centromere (mentioned already as the primary constriction), which holds them together and which will be the point of attachment to a spindle fiber. The doubling of the chromosomes occurred in interphase (S stage) when they were dispersed as chromatin. The two members of each double chromosome are known as chromatids, the term used for them until division of the centromere occurs at the end of metaphase.

Chromosome movement is associated with the microtubules called *spindle fibers.* One end of a fiber may be attached to the centromere of a chromosome, while the other end may terminate near a small structure at the pole of the spindle; this structure is a *centriole.* Other fibers may run from pole to pole without attaching to chromosomes. The centrioles originate just outside the nucleus. Each centriole (Fig. 4.18) is a cluster of nine groups of triplet microtubules whose structure is identical to basal bodies found at the bases of cilia and flagella.

At the onset of prophase there is a pair of centrioles near the nuclear membrane. Each centriole replicates and each pair of the resulting two pairs moves in opposite direction as the nuclear membrane breaks down. Thus a pair of centrioles is located at each pole. Their dimensions are close to the lower limit of the resolving power of the light microscope. As a result, the centriole appears as a very small dot under a light microscope, and in some cases it is not visible at all.

As the centrioles move toward the poles, fibrillar structures begin to differentiate from each centriole. Each radiating fibrillar structure is called an *aster,* and the fibrils themselves are *astral rays.* Between the two asters the spindle fibers appear. The structure formed by the centrioles, asters, and spindle is called the *mitotic apparatus,* and it is involved intimately in the division process. Figure 7.9 shows asters and spindle fibers in cells of a developing whitefish embryo.

Returning to the events that occur in the nucleus, the chromosomes continue to shorten and thicken, the nucleoli disappear, and as men-

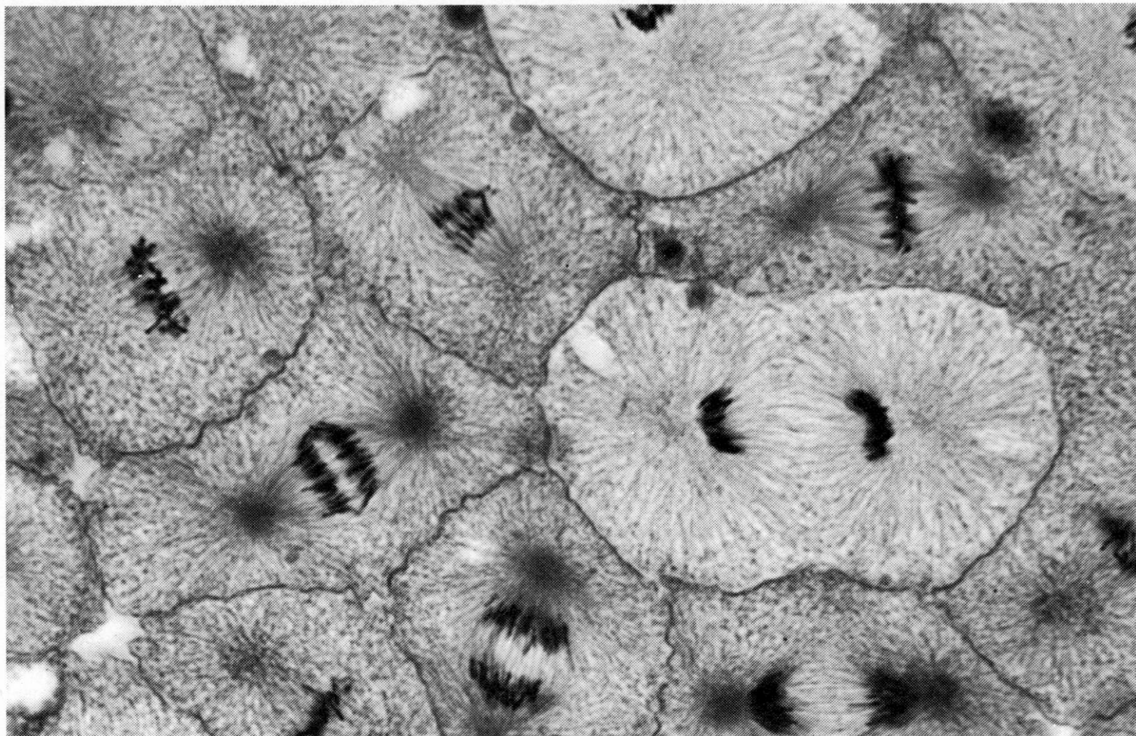

Figure 7.9 Mitosis in whitefish eggs. (Courtesy of Macmillan Science Co., Inc.–Turtox/Cambosco, Chicago.)

tioned before, the nuclear membrane disintegrates during late prophase. Then the chromosomes move to the center of the disintegrating nucleus and begin to arrange themselves in the central plane of the spindle, known as the *equatorial plate.*

METAPHASE

When the nuclear membrane has broken down and the chromosomes have become arranged on the spindle in the equatorial plate, *metaphase* begins. A spindle fiber from each pole appears attached to the centromere of each chromosome. The simultaneous splitting of the centromeres of all chromosomes signals the end of metaphase.

ANAPHASE

Anaphase begins as soon as the centromeres split; the chromatids (now called *daughter chromosomes*) move to opposite poles. The movement of the daughter chromosomes is not really understood. They seem to be pulled to opposite poles by the spindle fibers. Some chromosomes have their centromeres in the middle, some near one end, and occasionally one is found with a terminal centromere. In the movement of the chromosomes they appear as V, J, or as rods. This is taken by many biologists to indicate that the chromosomes are being pulled along by contraction of the spindle fibers. Recent work suggests that small amounts of actin and myosin, contractile proteins characteristically found in muscle cells, are present in spindle fibers.

Although much remains to be learned about this movement of chromosomes, it can be said that both centromere and spindle fibers are essential. A chromosome without a centromere does not move, and one with a centromere but not attached to a spindle fiber does not reach one of the poles.

TELOPHASE

The chromosomes approach the poles and become less compact as the nuclear membrane is reformed and the nucleoli reappear. As the daughter chromosomes approach the poles, a furrow appears at the surface of the cell in the equatorial region, and constriction into two cells begins. No clear understanding of the controlling conditions for the formation of this furrow exists at present, although several hypotheses have been proposed.

Duration of Cell Division

The time required for completion of mitosis varies from cell type to cell type and is related to the length of the G_1 phase. Temperature, within certain limits, also affects the duration of the process. In general the metaphase and anaphase stages are of short duration; most of the time of cell division is spent in prophase and telophase stages. It must be remembered that the period between cell division, interphase, is usually long. For instance, in some mammalian cell cultures 16 to 20 hours between cell divisions is a common occurrence.

Many cells of both plants and animals studied in tissue culture show the following durations in the cell cycle.

	G_1	10–20 hours (usually less than 50 percent of the total)
	S	6–8 hours
	G_2	1–4 or more hours (usually less than 20 percent of the total)
	M	1 hour
Total		18–33 hours

Control and Regulation of Cell Division

Just what initiates the division process has not been determined definitely. Usually it is assumed that the surface–volume relationships of a cell play an important role. As a cell grows in size, its surface increases in proportion to the square of its radius, while its volume increases in proportion to the cube of its radius. Since a living cell must obtain through its surface all materials needed for maintenance and growth, there will come a point at which the surface area is inadequate to supply the large volume. It has been proposed that there is a critical point in the surface–volume relationship at which the division process is initiated. This proposal seems plausible, but it does not explain the initiation of cell division in all cells, since some cells divide at a size at which others continue to enlarge.

Changing the temperature will, within certain limits, speed up or slow down the process. Also spindle formation can be prevented by placing cells in a solution of the alkaloid drug colchicine. In this situation the chromatids of the double chromosomes (not to be confused with double-stranded DNA) are unable to separate because of the disruption of the microtubules composing the spindle fibers.

The Significance of Cell Division

Mitosis is the universal process associated with eukaryotic cell reproduction. Two cells are formed, each usually half the size of the original cell and each containing the same complement of chromosomes. Between cell divisions, cell growth occurs through the assimilation of materials from the outside and their synthesis into new cell parts. Usually, cells must have their normal complement of chromosomes for normal functioning. Addition or loss of chromosomes or parts of chromosomes may lead to malfunction of the cells (Ch. 19). As indicated in Chapters 18 and 42, these changes may also lead to evolution.

In cell replacement, wound healing, and regeneration, new cells are produced by cell division. Although many of the highly specialized cells that result from cell differentiation lose the ability to divide, cell divisions occur throughout life in such regions as the blood-cell-forming tissues, the epithelial tissues of the skin and the intestine, and the gamete-forming tissues of the ovaries and testes. Once differentiated, many cells such as nerve cells may never undergo mitosis.

The process of mitosis is significant in another way. The basic unit of all organisms is the cell, and that the process of cell division in many kinds of organisms is essentially the same is a potent argument for the basic relationship of all living things.

Autoradiographic Studies on Chromosomal Duplication

Autoradiography is a useful technique for cell studies that involve small amounts of materials (Ch. 4). Radioactively labeled compounds are used. After the experiment has been performed, the radioactive atoms are localized by putting a photographic film against the fixed (chemically killed and immobilized) cells. The β rays (electrons) emitted by the radioactive atoms expose the silver grains in the photographic emulsion. Thus by looking at the dark spots on the developed film, one can see the pattern of radioactive compounds in the biological material.

J. Herbert Taylor and his associates studied the transfer of atoms of DNA in the chromosomes of the English broad bean *(Vicia faba)* during mitosis. DNA in the root tips of the broad bean was labeled with tritium (^3H), the radioactive isotope of hydrogen, by exposing the growing tips to a solution containing ^3H-labeled thymidine (the deoxyribonucleoside of thymine, Ch. 3). After about one third of a division cycle the seedlings were transferred to a growth medium without labeled thymidine but containing colchicine. After periods equal to one or two division cycles, the root tips were fixed and pressed against the photographic film. The electrons emitted by the incorporated tritium are of such low energy that they do not penetrate deeply into the film, and they therefore produce images only at their point of entry into the film emulsion. The chromosomes and their autoradiograms can be viewed simultaneously with the light microscope.

The broad bean contains 12 chromosomes ($2n = 12$, $n = 6$). Some nuclei treated in this experiment contained 12 chromosomes, some contained 24, and some contained 48. The chromosomes in cells with 12 metaphase chromosomes have not duplicated following labeling. The chromosomes in cells with 24 and 48 metaphase chromosomes have duplicated once and twice, respectively. The chromosomes in the nuclei containing 12 chromosomes were equally radioactive in the two chromatids. Those chromosomes that had experienced a division and were in the second metaphase after labeling (24 chromosome nuclei) were labeled in one chromatid only. In those cells with 48 chromosomes, two sets of chromosomes were completely unlabeled and the other two sets were labeled like those that had undergone only one division. Figure 7.10 outlines the basic ideas of Taylor's autoradiographic studies on chromosomal duplication. Although the actual arrangement of DNA in chromosomes is yet unknown, the transmission of atoms to daughter chromosomes supports the idea that a DNA double helix runs the entire length of the chromosome. For simplicity of presentation the metaphase chromatids in Figure 7.10 have been represented as containing two linear DNA molecules rather than a double helix.

Unsolved Problems

Although the process of cell division has been studied rather intensively for many years, it has been necessary in the preceding discussion to point out in many places that this or that aspect of the process is not fully understood. In sum, this is true for (1) the initiation of the process; (2) the movements of the chromosomes to the equator of the spindle and their subsequent movements to the poles of the spindle; (3) the formation of the spindle; (4) the constriction of the cytoplasm in animal cells; and (5) the apparent need of centrioles in animal and lower plant cells but not in higher plant cells.

At times, masses of cells become abnormal and divide without normal organization. This results in tumors, both benign and cancerous. What causes this abnormal behavior of cells is still not completely understood. It thus seems that an understanding of the factors that initiate and regulate cell division is essential to the solution of the cancer problem.

As a rule the division of the nucleus and the division of the cytoplasm is synchronized. Yet there are situations when this does not happen. In some of the protozoa, such as the malarial parasite, many nuclear divisions occur before membranes finally are formed around each nucleus. Since normally the division plane for cytokinesis coincides with the equatorial plane of the spindle, how such membranes are formed in the absence of a spindle remains unknown.

MEIOSIS

The discussion of mitosis indicated that the body of cells of each species contain a charac-

The following schematic drawing represents the basic ideas of Taylor's autoradiographic studies on chromosomal duplication.

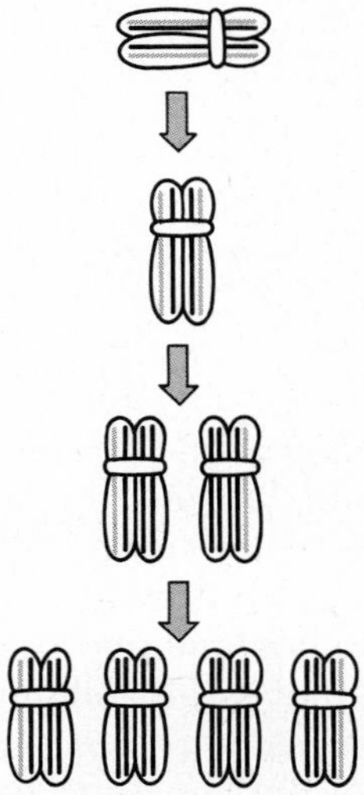

1. Chromosome being labeled by incorporation of ^{3}H-thymidine

Transfer to unlabeled medium with colchicine.

2. First metaphase. Chromosomes have not duplicated following labeling. There are two equally labeled chromatids. [12 chromosome nuclei]

3. Second metaphase. Chromosomes duplicated once. Each chromosome labeled in only one chromatid. [24 chromosome nuclei]

Third metaphase. Chromosomes duplicated twice. Two chromosomes labeled in one chromatid. Two chromosomes not labeled in either chromatid. [48 chromosome nuclei]

Figure 7.10 Taylor's autoradiographic studies on chromosomal duplication.

teristic number of chromosomes. It was indicated also that this characteristic number is the diploid, or $2n$, number, that is, it is made up of homologous pairs of chromosomes. In addition, it was pointed out that this diploid condition is the result of sexual reproduction. This means that the germ cells that unite in fertilization must be haploid and that the process of their formation from diploid cells must involve a mechanism for reducing the number to the haploid, or n, number. As mentioned in Chapter 4 and earlier in this chapter, the mechanism for this reduction is meiosis.

Meiosis actually involves two special cell divisions, and the products of these divisions are four cells, each one of which contains one of each pair of chromosomes that characterizes the diploid cells of the species. Were this not so, the number of chromosomes in sexually reproducing organisms would double with each new generation. Meiosis and subsequent fertilization, however, normally maintain the constancy of the chromosome number of the species. Meiosis in multicellular animals occurs in the testes in the formation of sperm and in ovaries in the formation of eggs.

Meiotic Divisions

FIRST MEIOTIC DIVISION

Both meiotic divisions (Fig. 7.11) may be viewed as a series of stages similar to those in mitosis: prophase, metaphase, anaphase, and telophase. The prophase of the first meiotic division is quite long in duration and differs in several respects from the prophase of mitosis. When the chromosomes become visible in early prophase, they are exceedingly long and thin. Although there is no evidence at this time of the replicated state of the chromosomes, our knowledge of DNA synthesis suggests that replication has occurred by the time this stage is reached.

The pairs of homologous chromosomes move together and unite throughout their lengths in a process known as *synapsis.* Many early studies with the light microscope and recent studies with the electron microscope have revealed that synapsed chromosomes are associated in a very precise way. After pairing, the united chromosomes shorten and thicken, and each fused pair looks like a single entity at this stage. In later prophase the united chromosomes separate slightly, and it is possible to see that each chromosome is now double, composed of two chromatids that are held together at the centromere. Since the fused homologous chromosomes are now made up of four parts (two chromosomes consisting of two chromatids each), each such unit is called a *tetrad.* The number of tetrads is always equal to the haploid number of chromosomes.

As in mitosis, while the nuclear changes described above are occurring, spindle fibers appear, and there may be centrioles and typical asters as well.

The end of prophase of the first meiotic division is marked by the disintegration of the nuclear membrane and the movement of the tetrads to the equatorial plate of the spindle. When the tetrads become aligned on the equatorial plate, the metaphase stage is reached.

In anaphase the homologous chromosomes, which fused during early prophase in synapsis and which are made up of two pairs of chromatids, each held together at the centromere, separate and move to opposite poles. These double structures, each one of which is one half of a tetrad, are called *dyads.*

As in mitosis, the telophase is marked by the beginning of cytokinesis, which occurs by constriction. Following the telophase, each daughter cell may enter an interphase condition similar to that seen in mitosis. However, in many cells the interphase is quite short and in others it is lacking entirely.

SECOND MEIOTIC DIVISION

If the second meiotic division occurs immediately after the first division, as will be seen for *Ascaris* eggs later in this chapter, the dyads soon orient on a second spindle. The division that follows resembles mitosis in that the centromeres divide and the sister chromatids are separated, one of each pair going to each pole as a daughter chromosome.

If an interphase occurs between the first and second meiotic divisions, there will be a prophase, as in mitosis, ending in the breakdown of the nuclear membrane and the completion of the spindle. During anaphase the sister chromatids separate and move to the poles, followed by cytokinesis and the formation of two cells in telophase. Since the complete meiotic process involves two successive divisions, four haploid cells are produced for each cell that undergoes meiosis (see Fig. 7.11).

Comparison of Mitosis and Meiosis

In each mitotic division of diploid cells the number of chromosomes characteristic of the species is maintained; in the two successive divisions of meiosis the diploid number is reduced to the haploid number. The basic differences in the two processes can be seen in a comparison

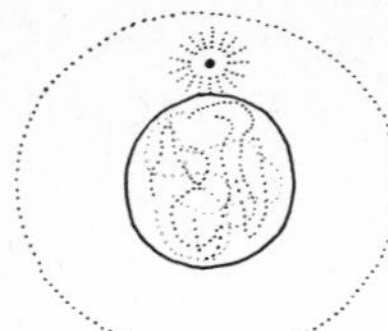

Interphase

Early prophase
The chromosomes shorten and thicken; the centrioles divide and commence to move toward the poles.

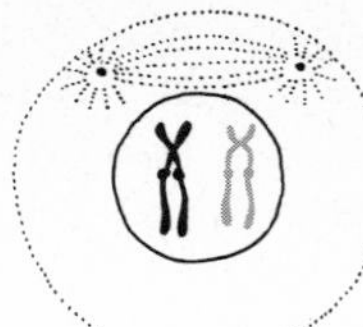

Prophase
Homologous chromosomes pair up side by side in a process called synapsis.

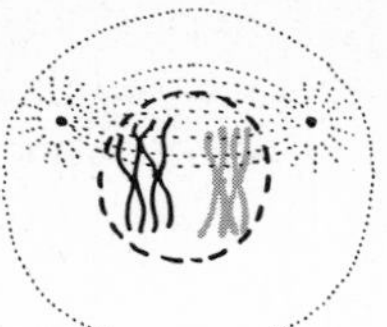

Prophase
Each chromosome separates into two chromatids which are held together at the centromere. Thus, there is one four-part structure, called a tetrad for each pair of homologous chromosomes.

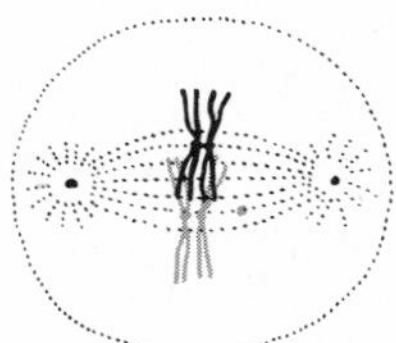

Metaphase
The nuclear membrane disintegrates and the tetrads move to the equatorial plate of the spindle which has formed between the two centrioles. The alignment of the tetrads on the equatorial plate marks the beginning of metaphase.

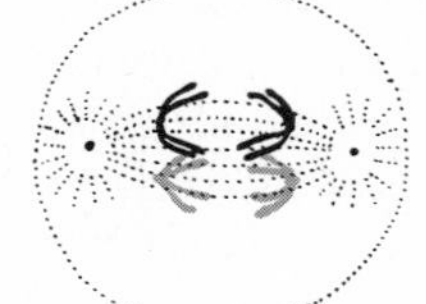

Anaphase
The homologous chromosomes which make up each tetrad separate, producing two dyads for each tetrad which move to opposite poles. Each dyad is composed of two chromatids.

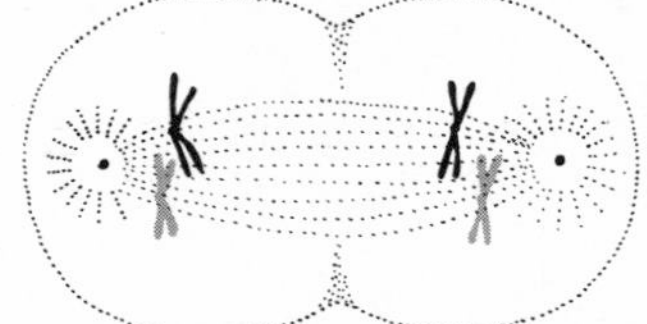

Telophase
As the dyads approach the poles, cytokinesis begins. Note that actual reduction has already occurred since each daughter cell receives one of each kind of chromosome (that is, one member of each homologous pair).

(a)

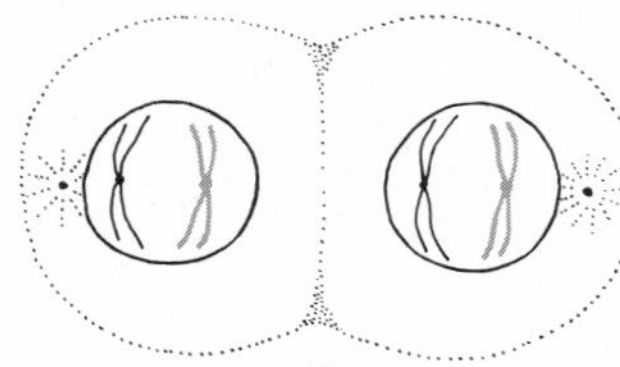

Interphase
Following the first meiotic division, there may be interphase similar to the mitotic interphase. In some cases the interphase lasts a long time, but in others it is quite short while in still other cases there is no interphase at all.

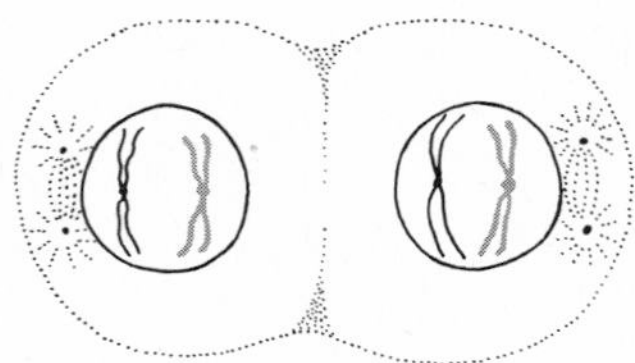

Prophase
Prophase is of short duration during which the chromosomes (dyads) shorten and thicken and move to the equatorial plate. The centrioles divide and move to the poles.

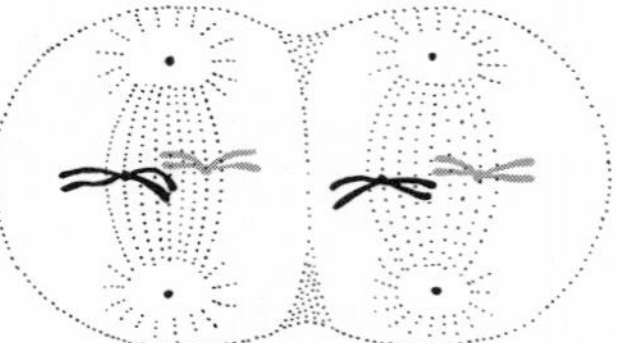

Metaphase
In metaphase the dyads are aligned on the equatorial plate and the centromeres divide.

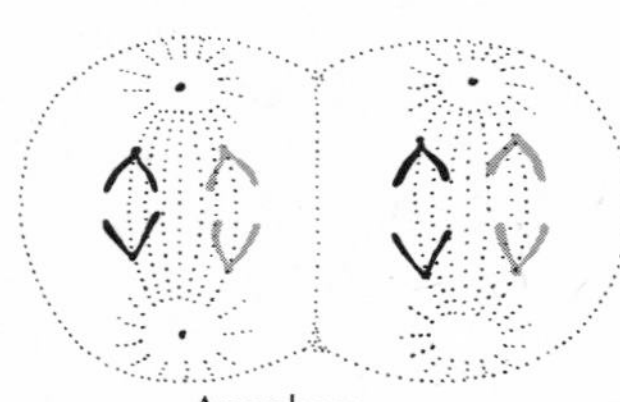

Anaphase
Anaphase is marked by the separation of the chromatids which make up each dyad and the movement of the daughter chromosomes to opposite poles.

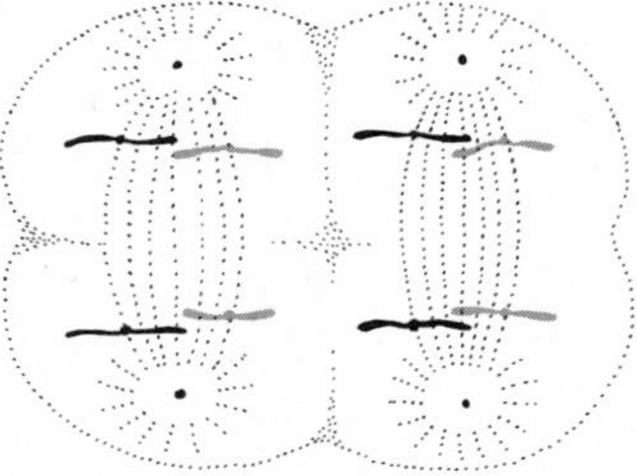

Telophase
With cytokinesis, telophase begins and division is completed with the production of four haploid cells for each cell which undergoes meiosis.

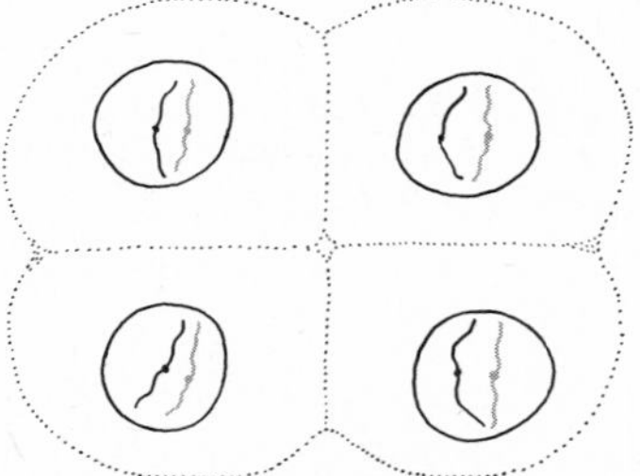

Products of the two meiotic divisions
In the case of sperm formation, four sperm of equal size are produced; while in egg formation there is a single large egg and three small polar bodies. In both cases, however, the nuclear divisions are as outlined above.

(b)

Figure 7.11 Meiosis in a cell containing four chromosomes, that is, two pairs of homologous chromosomes. Homologous chromosomes are indicated by gray or black.

of mitosis and the first meiotic division. In mitosis the replicated chromosomes line up on the spindle, the centromeres divide, and the sister chromatids are distributed to the poles, one of each pair going to each pole. In the prophase of the first meiotic division the chromosomes go through the complicated behavior of synapsis and tetrad formation. When the chromosomes orient on the first meiotic spindle as tetrads, there is no division of the centromeres, and one of each homologous pair moves to each pole as a dyad.

In the second meiotic division the two-stranded chromosomes (dyads) orient on the spindle, their centromeres divide, and the daughter chromosomes move to opposite poles.

Spermatogenesis and Oogenesis

The meiotic process in the formation of sperm (spermatogenesis) differs from that in the formation of eggs (oogenesis) in one important respect. As shown in Figure 7.12, four sperm are formed from each starting cell, whereas the same process in the female produces a single egg and two or three small polar bodies.

SPERM

Each testis contains many seminiferous tubules in which millions of sperm are formed. The walls of these tubules are lined with unspecialized diploid germ cells, the spermatogonia (see Fig. 15.2). During development and until sexual maturity is reached, the spermatogonia increase in numbers by mitotic divisions. Spermatogenesis actually starts with the enlargement of some of the spermatogonia to form primary spermatocytes. These undergo the first meiotic division, each one forming two secondary spermatocytes. These undergo the second meiotic division to form four equal-sized spermatids. The spermatids then metamorphose into motile and functional sperm. In the changes that occur the nucleus of each spermatid shrinks in size and becomes the head of the sperm. Most of the cytoplasm is shed, but a very small part of it goes into the formation of the tail, or flagellum, which is capable of vigorous movement.

Part of the Golgi apparatus of the cytoplasm moves to the anterior end of the sperm and forms a small structure, called the *acrosome,* which aids the sperm in penetrating the egg membrane. Mitochondria present in the cytoplasm of the spermatid collect in the region between the head

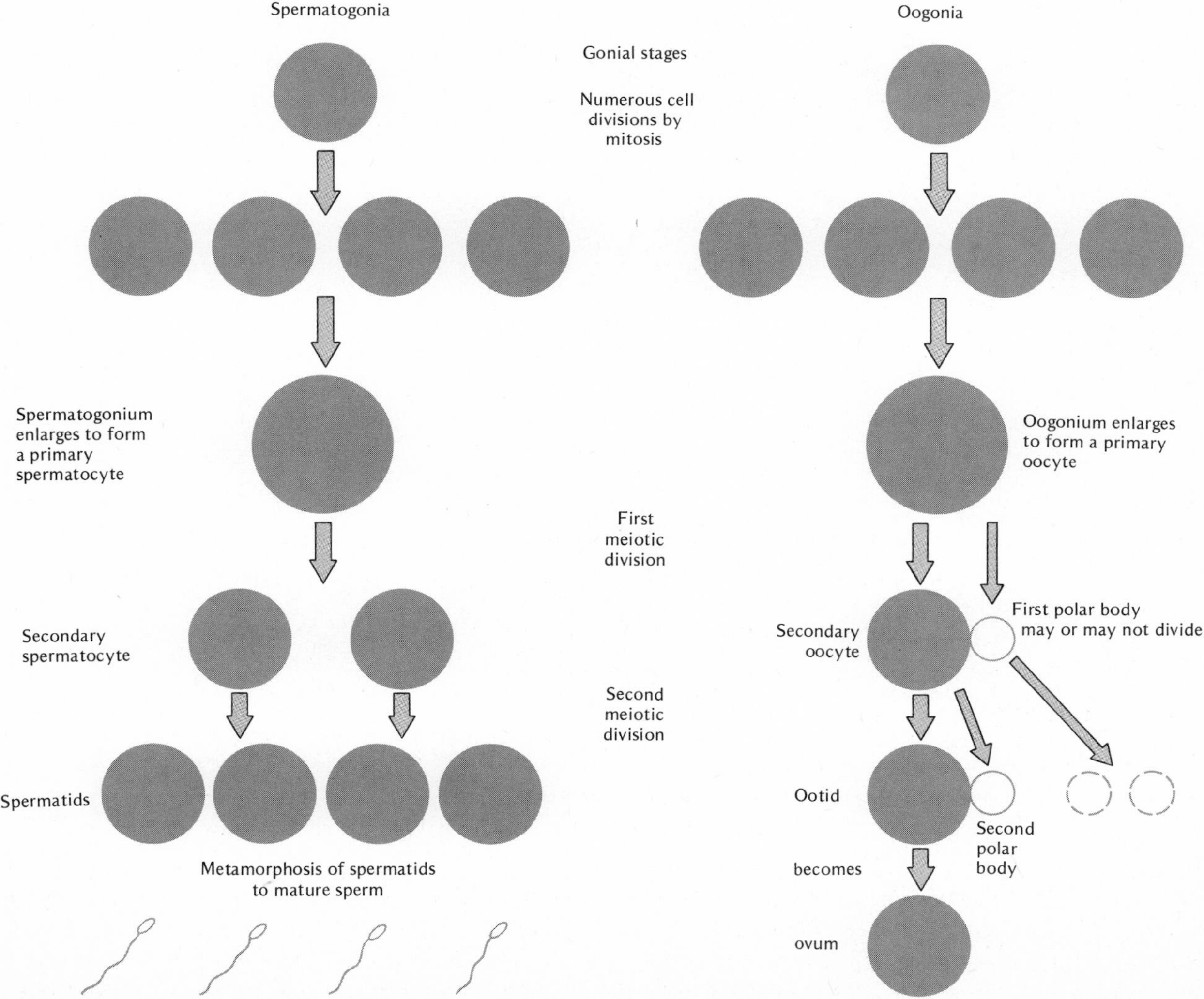

Figure 7.12 Comparison of formation of eggs and sperm.

and the tail of the sperm, forming a small middle piece. These mitochondria provide the energy (in the form of ATP) used in the beating of the tail. All normal sperm are haploid when they are released from the testes. Most sperm have a long tail for locomotion. The *Ascaris* sperm (Fig. 7.13) has no tail; it moves by amoeboid movement as do spermatozoa of other roundworms.

EGGS

The mammalian ovary is covered by a germinal epithelium. These epithelial cells divide mitotically around germ cells that migrated to this region in early development. Clusters of these cells push into the interior of the ovary to form follicles. The beginning of oogenesis is marked by the enlargement of an oogonium in the center of the follicle to form a primary oocyte. These cells become very large comparatively, and their growth takes much longer than the corresponding growth of primary spermatocytes. During this period of growth much stored food or yolk is formed if the egg is of a species lacking uterine implantation; this adaptation ensures that the fertilized egg will have energy reserves during early stages of development. In most mammals, where implantation occurs quickly, the adapta-

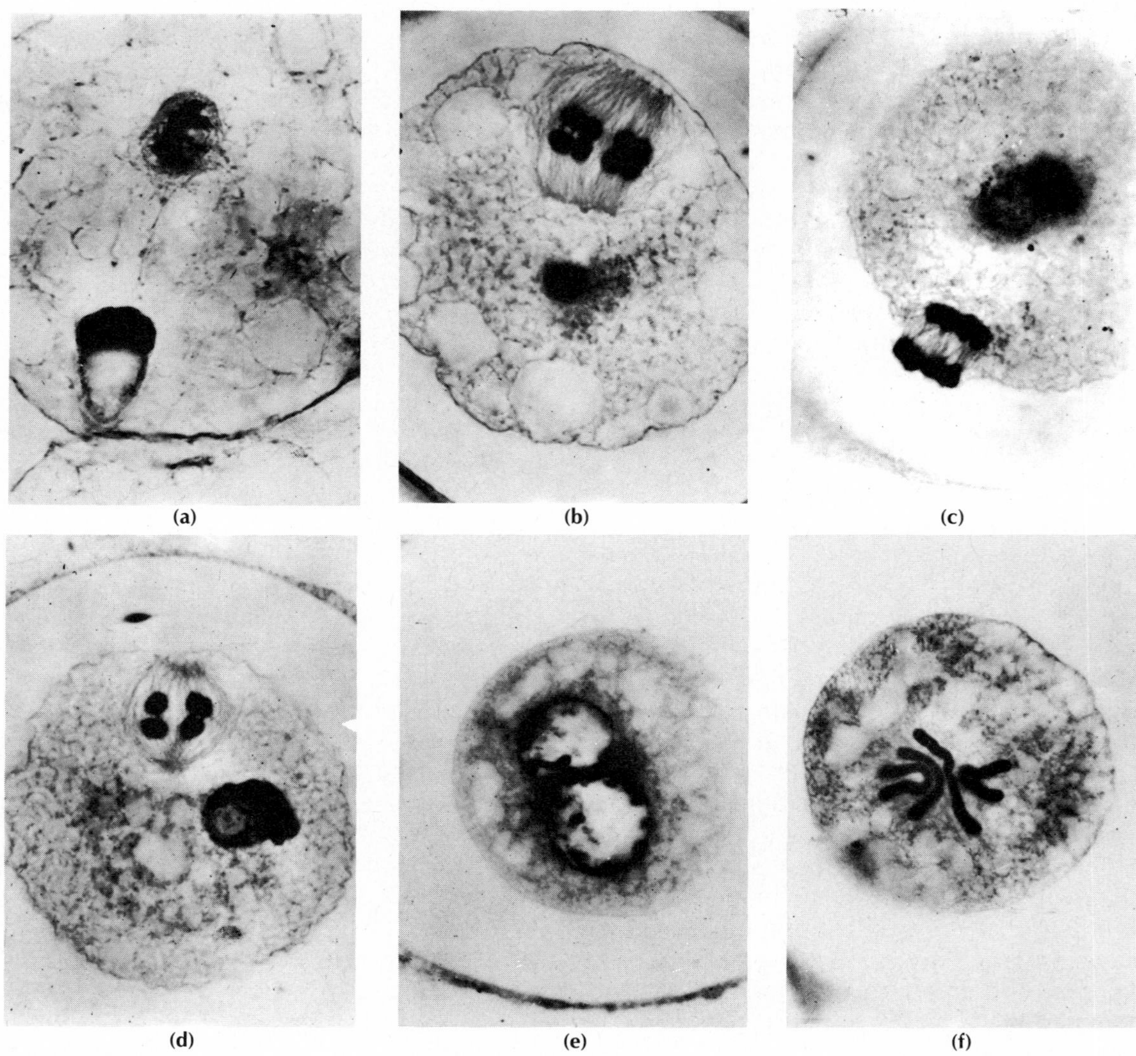

Figure 7.13 Meiosis in the roundworm *Ascaris.* **(a)** Fertilization: the conical structure with the black "cap" is the sperm, the dense area near the top of the photograph is the nucleus of the egg. **(b)** First meiotic division: two tetrads and the maturation spindle are at the upper left edge of the egg, the lower dense mass is nuclear material from the sperm. **(c)** First meiotic division and first polar body formation: being cast off of the surface of the egg at the left are two dyads (one half of each tetrad). **(d)** Second meiotic division: near the top is the small black first polar body lying on the inner surface of the shell, the two remaining dyads are aligned on the spindle, the monad of each dyad lying closest to the surface of the egg will be cast off in the second polar body. **(e)** Fusion of the male and female pronuclei. **(f)** Metaphase plate view of the chromosomes in the first mitotic division after meiosis. (Courtesy of Carolina Biological Supply Company.)

tion for storing yolk in the egg is reduced or absent.

At the end of the growth process the primary oocyte undergoes the first meiotic division. Although the nuclear division, involving the chromosomes, follows the normal sequence described earlier, the cytoplasmic division is unequal and results in one large *secondary oocyte* and a small structure made up almost entirely of nuclear material, called the *first polar body.* The secondary oocyte also divides unequally in the second meiotic division, producing a large ootid and a small *second polar body.* The first polar body may or may not divide, but in any event the polar bodies all soon disintegrate. The ootid becomes the mature egg, or ovum. The single ovum that results from the meiotic divisions of each primary oocyte contains most of the original cytoplasm with its stored food or yolk. Comparison of the strategies involved in sperm and egg maturation show that the egg is provided with as much cytoplasm and cytoplasmic organelles as possible for food, energy, and protein synthesis in the developing zygote and early embryo. On the other hand, the motile sperm has a small amount of cytoplasm; the excess weight of a large amount of cytoplasm would deter the sperm from moving rapidly. These strategies are examples of evolutionary adaptations which illustrate a correlation of structure and function (Concept 7, Ch. 1).

Meiosis in Ascaris Eggs

One of the first careful studies of meiosis was made on egg formation in the parasitic roundworm *Ascaris.* Because this is classic material often used in laboratory studies on meiosis, a brief account of it is given here. The diploid number of chromosomes in this species is four (two homologous pairs), and thus the process is comparatively easy to follow (Fig. 7.13).

In Ascaris, as in all animals, the sperm cell is haploid when it penetrates the egg in fertilization; meiosis has already occurred. The egg cell, at the time that the sperm enter, is actually not a mature ovum but a primary oocyte. When the sperm penetrates the egg, the egg nucleus contains two tetrads. The nuclear membrane now breaks down, a small spindle forms near the periphery of the egg, and the tetrads become arranged in metaphase stage on this first meiotic spindle. At the end of the first meiotic division one half of each tetrad is pinched off in the first polar body, leaving the other half of each tetrad, or two dyads, in the egg.

A second meiotic spindle forms immediately with the two dyads on the spindle. In this division half of each dyad goes into the second polar body and the other half remains in the egg. The chromosome number of the original egg nucleus has now been reduced to the haploid, or *n,* number (in this case, 2). The first polar body may divide to give a total of three polar bodies, but this does not always happen. The haploid egg and the sperm nuclei remain for a period in interphase condition. They are called at this time the male and female pronuclei. Subsequently, a large spindle, the first mitotic spindle, forms; the egg and sperm nuclear membranes break down, and four chromosomes—two from the egg and two from the sperm—orient on the spindle. In the mitotic division that ensues two cells are formed, each containing four chromosomes, one member of each homologous pair having been derived from each parent. Further mitotic divisions result in the formation of a new multicellular animal.

Significance of Meiosis

As indicated in previous discussions, the process of meiosis results in the formation of haploid cells. Through meiosis and subsequent fertilization the diploid number of chromosomes characteristic of each species is maintained. Without meiosis the number of chromosomes in sexually reproducing organisms would double in each generation—clearly an impossible arrangement.

The processes of meiosis and fertilization provide the basis for biparental inheritance, since each parent makes an equal contribution of chromosomes to the new individual. Subsequent discussion of heredity and organic evolution indicate the great significance of biparental inheritance in producing variations.

Meiosis also provides the basis for genetic variation in two other ways. One of these is the random distribution of maternal and paternal chromosomes during the first meiotic division. In a cell with two pairs of chromosomes, for example, all maternal chromosomes may go to one pole and all paternal chromosomes may go to the other; but it is equally probable that one maternal and one paternal chromosome of each pair may go to each pole. Thus in a species with two pairs of chromosomes and with many cells undergoing meiosis, four different kinds of haploid gametes will be produced and in equal numbers.

Genetic variation also results from the phenomenon of *crossing over.* During synapsis in the first meiotic division, when homologous chromosomes are closely intertwined, a break in adjacent chromatids of different chromosomes may occur and segments may be exchanged. This may result in chromatids with new genetic combinations. The phenomenon of crossing over is discussed in some detail in Chapter 18.

Finally, meiosis is significant as further evidence of the basic relatedness of living organisms. The process is remarkably similar wherever it occurs in sexually reproducing, eukaryotic organisms.

SUGGESTIONS FOR FURTHER READING

Brachet, J., and A. E. Mirsky. *The Cell,* vol. III (Chs. by D. Mazia and M. M. Rhoades). New York: Academic Press (1961).

DeRobertis, E. O. P., W. W. Nowinski, and F. A. Saez. *Cell Biology.* Philadelphia: W. B. Saunders (1970).

DuPraw, E. J. *DNA and Chromosomes.* New York: Holt, Rinehart and Winston (1970).

Dyson, R. *Cell Biology: A Molecular Approach.* Boston: Allyn and Bacon (1974).

Dyson, R. *Essentials of Cell Biology.* Boston: Allyn and Bacon (1975).

Gardner, E. J. *Genetics,* 5th ed. New York: Wiley (1975).

Hood, L. E., J. H. Wilson and W. B. Wood. *Molecular Biology of Eukaryotic Cells—A Problems Approach.* Menlo Park, Calif.: W. A. Benjamin (1975).

Kennedy, D. *Cellular and Organismal Biology.* San Francisco: W. H. Freeman (1974).

Stahl, F. W. *The Mechanics of Inheritance.* Englewood Cliffs, N. J.: Prentice-Hall (1969).

From Scientific American

Mazia, D. "How Cells Divide" (Sept. 1961).

Mazia, D. "The Cell Cycle" (Jan. 1974).

McKusick, V. A. "The Mapping of Human Chromosomes" (Apr. 1971).

Taylor, J. H. "The Duplication of Chromosomes" (June 1958).

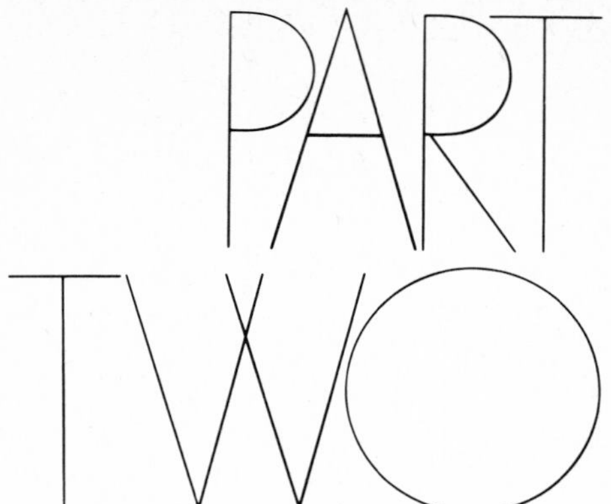

The Problems that Organisms Face

In Part II a jump is made from the cell to the entire organism. For this, the human is the primary focus. Throughout the discussions it will be obvious that an understanding of organismal functions is dependent on some basic knowledge of cell functions.

Using the human as an example, the study is approached by asking the question, "What problems do all organisms have to face in their day-to-day existence?" The following are generally recognized: (1) How do they obtain, store, and utilize food? (2) carry out the exchange of gases and the elimination of wastes? (3) transport materials in their bodies? (4) coordinate their activities? (5) receive environmental stimuli? (6) move? (7) support themselves? (8) behave? (9) defend themselves?

In the course of evolution humans and the other more specialized animals have evolved several organ systems (digestive, respiratory, excretory, circulatory, endocrine, nervous, skeletal, muscular, integumentary, and reproductive which is not discussed here).

After being ground in the mouth food passes through the stomach to the small intestine with enzymes carrying out the extracellular processes of digestion in these different organs. Absorption of the end products of digestion into the blood stream occurs in the small intestine.

For normal activities, gases must be exchanged with the environment and wastes must be eliminated. The important components of the human respiratory system are the lungs and their associated ducts. The primary excretory organs are the kidneys where wastes are eliminated in a series of processes involving filtration, resorption, tubular secretion, and counter-current exchange. Nutrients and wastes, as well as O_2 and CO_2, are transported throughout the body by the beating heart through arteries, capillaries, and veins. Immune mechanisms are considered in the discussion of the circulatory system.

Coordination results from the actions of hormones produced in the endocrine system and the functioning of the nervous system working with the sense organs and the skeletomuscular systems.

Behavior (what an organism does) may be classified as learned or innate. After a brief discussion of innate behavior, the main focus in this section is on learned behavior.

Locomotion in the human is due to powerful striated muscles that are attached to strong supporting internal skeleton of rigid bone. The contraction of muscles at the tissue level is accompanied by the sliding together of two linear proteins, actin and myosin, at the ultrastructural level of organization.

The human defense system is quite elaborate. It involves the skin or integument, mucous membranes, phagocytosis, antibody formation, and a number of behavioral responses.

Artificial kidney (Photo courtesy of Dr. George W. Applegate, Artificial Kidney Unit, Methodist Hospital, Indianapolis, Indiana)

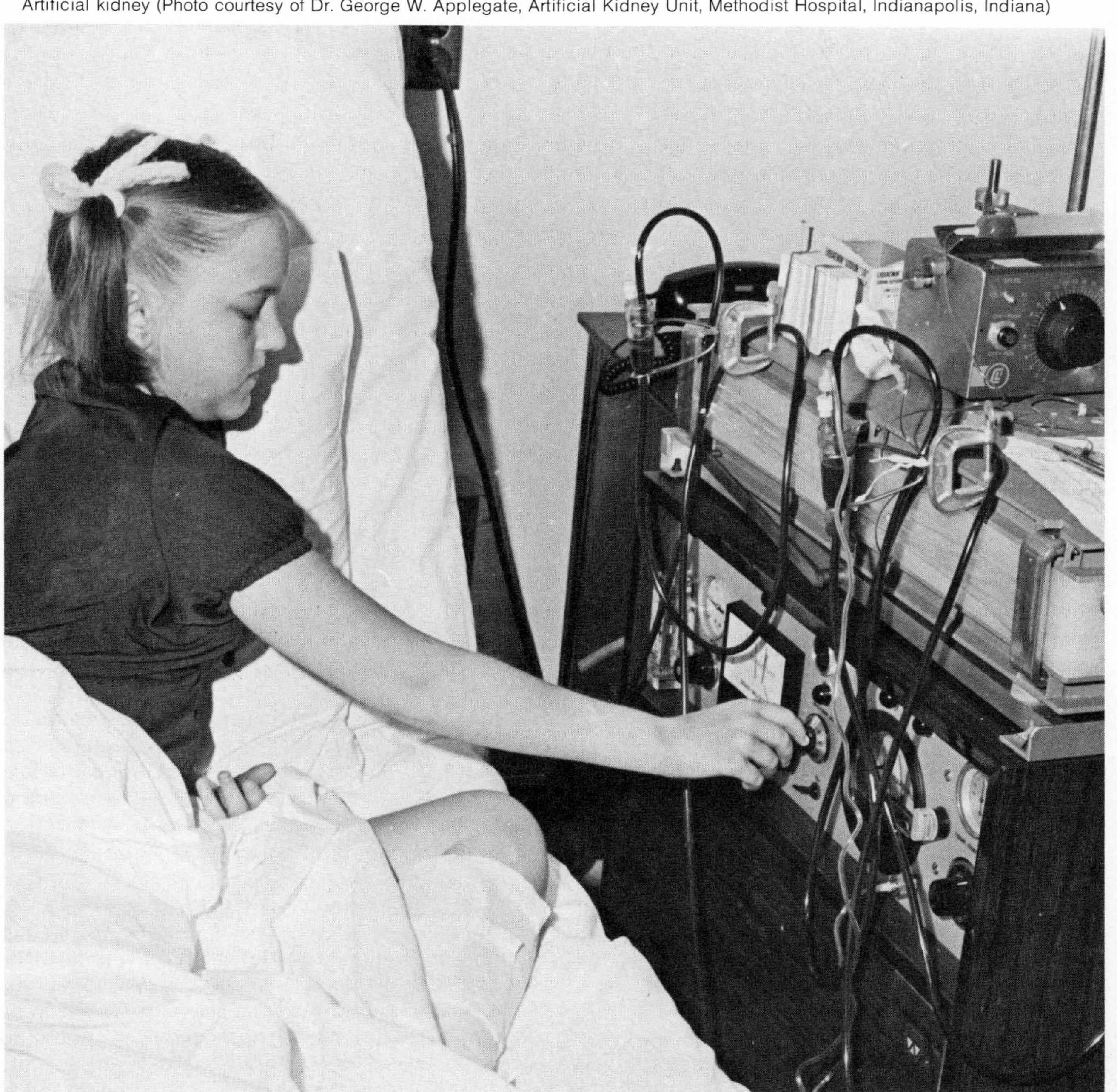

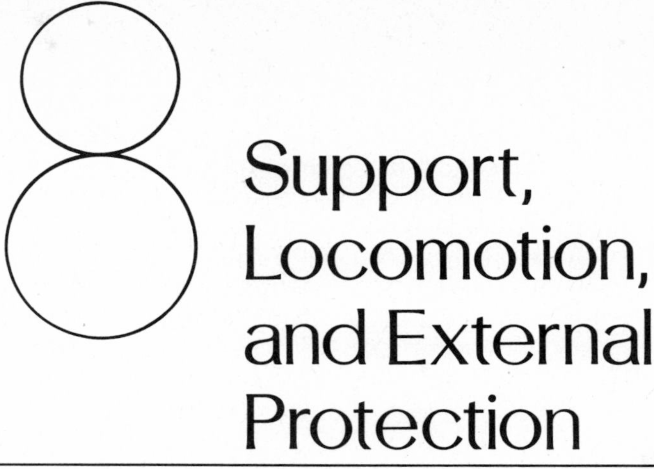

Support, Locomotion, and External Protection

The shape and symmetry of the body and the detailed features that enable one to tell one kind of organism from another and, indeed, to distinguish between individuals, are determined by three sets of structures: the integument or the outer covering of the body, the skeleton, and the muscles.

In the evolution of the different groups of multicellular animals, various supporting or skeletal structures have been developed. In the invertebrate groups these skeletal structures are often on the outside covering the soft parts and are known as *exoskeletons.* Familiar examples of this type of skeleton are found in snails, clams, insects, and crabs. In vertebrates, on the other hand, the skeleton is on the inside and is known as *endoskeleton.* A skeleton, whether inside or outside, serves a number of functions. It supports the soft parts of the body and is responsible in large measure for the shape of the animal. The skeleton also protects the other parts of the body; this is quite obvious in those forms with exoskeletons.

The vertebrate endoskeleton has a similar role as the following discussion will show. In addition to these roles, the skeleton is an integral part of the total effector system of an animal. A large part of the skeleton of either an invertebrate or a vertebrate consists of the levers that move them, and thereby the movements of the animal, in whole or in part, are effected.

THE INTEGUMENT OR SKIN

In invertebrates the skin is often thin and consists of few layers of cells. These often secrete mucus on which the animal can glide; or they may secrete substances that harden as in the production of shells of jointed exoskeletons.

The skin of all vertebrates consists of many layers of cells. It is the largest organ of the human body and it functions in a variety of ways. Montagna has estimated that a human, 1.80 m (6 feet) tall and of average weight, will have about 9.678m^2 (15,000 square inches) of skin area. Before considering its diverse functions, some knowledge of its structure is necessary.

Parts of the Skin

The skin is composed of two principal layers, an outer, thinner *epidermis* of stratified epithelium and an inner, thicker *dermis* (Fig. 8.1). The skin's thickness varies in different parts of the body; it is thickest on the soles of the feet and on the palms of the hands. A close examination of the palms and fingers shows that the epidermis there has numerous ridges, forming whorls and loops, in very specific patterns. These patterns are determined genetically and in humans they serve as a means of personal identification. No two people have identical patterns. They result, in development, very largely from the orientation of the underlying fibers in the dermis.

An examination of the cellular composition of the epidermis shows that cells of the basal layer are columnar or cuboidal in shape. These cells undergo frequent cell divisions, giving rise to layers of cells to the outside. As these layers are pushed to the outside, the cells synthesize large amounts of the fibrous protein, *keratin,* and they also become increasingly compressed or flattened. The outermost layers of the epidermis are

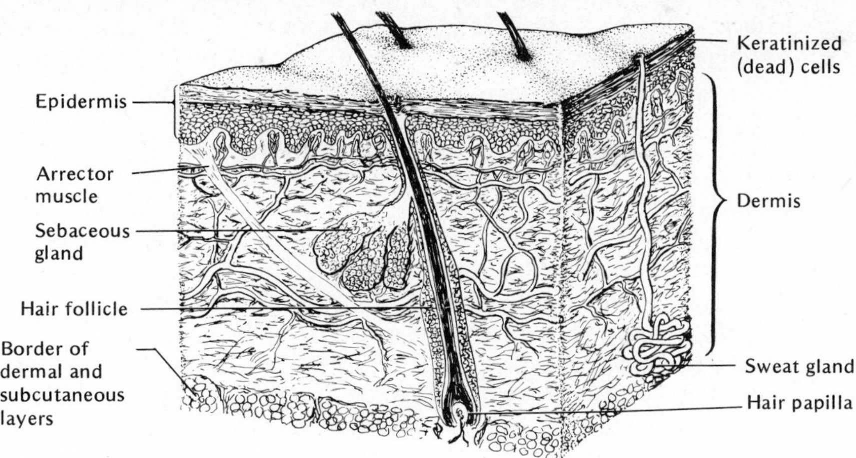

Figure 8.1 Diagrammatic section through mammalian skin to illustrate the layers, hair, sweat glands, and sebaceous glands.

composed, then, of very flat keratinized cells that die and are continually being sloughed off. Together these cells form a horny but flexible layer that protects the underlying parts of the body. These epithelial cells are held together very tightly by numerous submicroscopic *desmosomes* (Ch. 4).

Scattered at the juncture between the deep layers of the epidermis and the dermis are *melanocytes,* cells that produce the dark pigment *melanin.* These melanocytes have numerous long processes that extend for great distances between the epidermal cells. Melanin produced in the melanocytes is incorporated into the epidermal cells where it serves as a protection by absorbing ultraviolet rays. Tanning results from an increase in melanin production as a result of much exposure to the ultraviolet rays of sunlight. It appears that all humans have about the same number of melanocytes in their skin. The difference between light and dark races, however, is due to the fact that melanocytes of dark races produce more melanin.

The juncture of the dermis with the epidermis is uneven. The dermis itself is thicker than the epidermis, and it is composed primarily of connective tissue, both fibrous and elastic. Numerous blood vessels, nerves, and nerve endings, as well as hair follicles, sweat glands, and sebaceous glands, are found in the dermis. Beneath the dermis and attaching the skin to the muscles below is a layer of loose connective tissue. This layer often contains much fat or adipose tissue.

Derivatives of the Skin

The most obvious derivatives of human skin are hairs and nails. Humans are descended from mammalian ancestors that had a covering of hair over most of the body. Hair follicles are found all over the human skin, except oh the palms, soles, and few other regions. Yet long hairs develop only in a few regions of the body. In most mammals the hair coat serves as a protective insulation; in humans, with perhaps the exceptions of the hairs in the nose and those of the eyebrows and eyelashes, hair is largely ornamental. Hairs are formed in inpocketings of the epidermis, the *hair follicles.* At the bottom of each follicle a papilla of connective tissue projects into the follicle. The epithelial cells above this papilla constitute the hair root and by cell divisions form the shaft of the hair, which ultimately extends beyond the surface of the skin. The hair cells of the shaft produce keratin, die, and form a compact mass that is the hair. Hair growth occurs at the bottom of the follicle. From time to time, humans shed hairs. (In many of the mammals the entire coat is replaced in a short period.) When this occurs, the old hair separates from the root region and forms a clublike base. Later a new hair starts growing and pushes the old hair out. Some hairs are straight, some are wavy, and some are kinky. These conditions are determined genetically. Structurally, a straight hair is a round one, a wavy hair is alternately round and oval, while a kinky hair is ribbon-shaped.

Associated with each hair follicle are one or more *sebaceous glands.* Sebaceous glands are also derived from the embryonic epidermis; they secrete the oily *sebum* into the hair follicle and out onto the surface of the skin, making both hair and the surface of the skin more pliable.

Attached to each hair follicle there are small, smooth muscles, the *arrector pili,* which cause the elevation of the hair when they contract. These muscles are involuntary and they contract under the influence of cold or fright. In humans, their contraction produces the familiar "goose flesh."

Scattered through most regions of the skin are numerous sweat glands, which are also derived from the epidermis. Actually, there are two kinds of sweat glands, the *eccrine* glands which are scattered generally over the body surface, and the *apocrine* glands which are associated with hair follicles. The apocrine glands are found primarily in the armpits interspersed with the eccrine type. The apocrine glands are regarded primarily as scent glands. The viscous secretion of these glands is emitted during stress or sexual stimulation, and it becomes malodorous through the decomposing activity of skin bacteria. The watery sweat is produced by the eccrine glands. Whereas humans have these glands scattered over most of the body, in cats, dogs, and rodents these eccrine glands are found only in the foot pads. Mammary glands are also derivatives of the skin and they are somewhat similar to sweat glands in their development and structure.

Nails also develop from inpocketings of epidermis and their growth is somewhat similar to that of hair. Besides the hair and nails found in humans, the feathers, hoofs, claws, scales, and horns found in other vertebrates are also derivatives of the skin. Figure 8.2 shows some of these derivatives.

Functions of the Skin

The horny layers of the skin form a protective shield against blows, friction, and many injurious chemicals. These layers are essentially germproof and, as long as they are not broken, they keep bacteria and other microorganisms from entering the body. The skin is waterproof and thus prevents the loss of body fluids. As mentioned above, the pigment in the outer layers protects the underlying layers from the ultraviolet rays of the sun.

The roles of the skin in the regulation of body temperature, in excretion, and in housing the various receptors for touch, pressure, heat and cold, as well as the nerve endings for pain, will be discussed later. In most mammals the hair follicles have a collar of sensory nerves around them that allows the animal to detect any movement of the hairs. Humans retain some of the sensory function of hairs, and this can be demonstrated by gently touching some of the hairs on one's head. Perhaps the most remarkable prop-

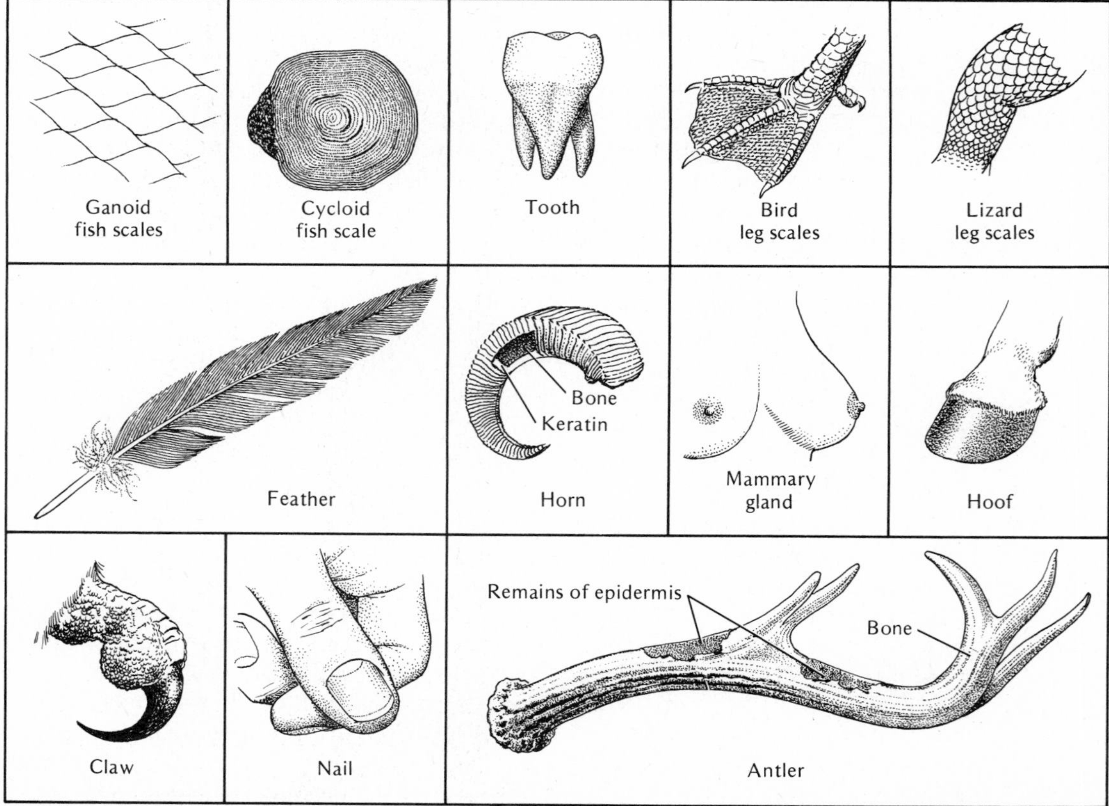

Figure 8.2 Some integumentary derivatives. Some are primarily derived from epidermis, some from dermis, some from both dermis and epidermis.

erty of the skin is its ability to heal or regenerate itself when it is scratched or cut.

THE VERTEBRATE SKELETON

The vertebrate skeleton may be divided into two general parts: the *axial skeleton,* consisting of the *skull,* the *vertebral column,* the *ribs,* and the *sternum;* and the *appendicular skeleton,* consisting of the *appendages* and their *girdles* (Fig. 8.3).

In general the skeletons of all vertebrates are quite similar due to evolutionary relationship.

Parts of the Skeleton

SKULL

The primary function of the vertebrate skull is the protection of the brain. The part of the skull that serves this function is the cranium. The skull also protects the organs of special sense: eyes, ears, nose, and mouth. Furthermore it provides

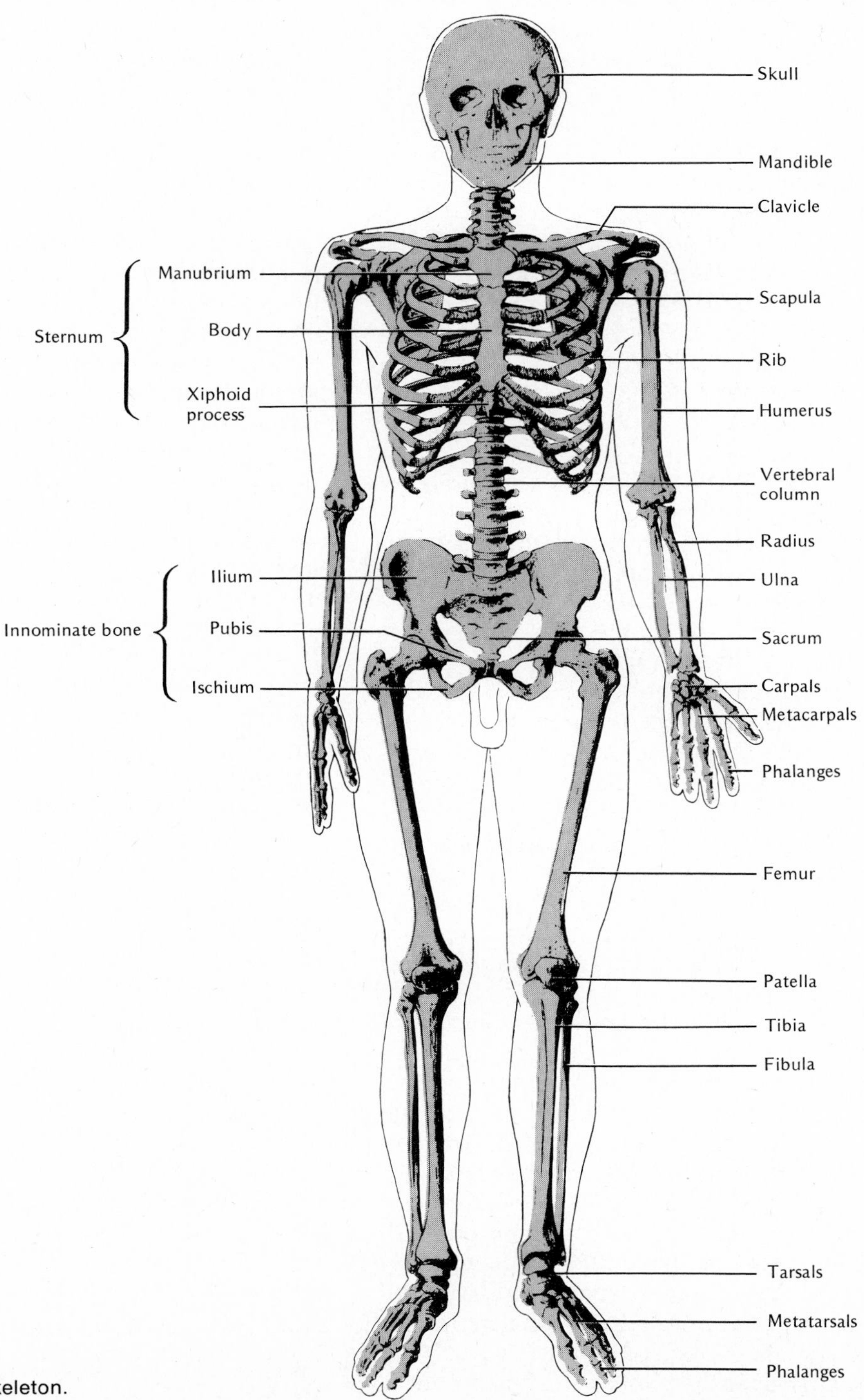

Figure 8.3 The human skeleton.

an attachment and a skeleton for the jaws. The human skull is composed of 22 bones, besides the six tiny ear bones. These bones are joined together by jagged-edged *sutures.* At the time of birth, several of the bones of the cranium are not completely formed, leaving five membranous regions called *fontanelles.* If the bones of the cranium were completely formed and sutured at the time of birth, great difficulty would be experienced in the passage through the birth canal. As it is, the head is somewhat flexible and can undergo changes in shape as it passes through the birth canal.

VERTEBRAL COLUMN

The vertebral column forms a more or less rigid rod extending the length of the trunk (and tail when present). The skull articulates with the vertebral column, and the large openings in the vertebrae form a bony canal in which the spinal cord is housed. This bony rod serves as a support for the various soft parts of the body, and to it are attached the two girdles with their appendages.

Pads of cartilage lie between the centra of the vertebrae, and the whole series of vertebrae are laced together by numerous ligaments. In humans there are 33 or 34 vertebrae in the vertebral column, depending on the number in the coccyx. The human vertebrae are somewhat different in shape and size in the different parts of the body. In the neck region there are seven *cervical* vertebrae; in the thorax, 12 *thoracic* vertebrae; in the lower back region, five *lumbar* vertebrae; in the sacral region, five fused vertebrae form the *sacrum* to which the pelvic girdle is attached; and at the end of the vertebral column is the coccyx or tailbone which consists of four or five small fused vertebrae. The coccyx is the human vestige of a tail. The cervical, thoracic and lumbar vertebrae are articulated so that some movement in those parts of the vertebral column is possible.

A comparison of the human skeleton with that of a typical quadruped, such as the cat, shows some of the changes associated with the transition from a quadrupedal existence to a bipedal existence. In the trunk region of the cat there is one main arch in the vertebral column. This type of arch is commonly used in building for strength. In humans there is an added curve in the lumbar region of the column, giving a reversed S curvature. This increases the strength along the anteroposterior axis. It also makes possible a slight dorsal displacement of the weight and increases the ability to absorb shock.

RIBS

Humans have 12 pairs of ribs (as a rule), one pair articulating with each of the thoracic vertebrae. In front, ten pairs of the ribs are attached to the sternum or breastbone. The sternum has three parts: the upper triangular part, the *manubrium;* the larger middle portion, the *body* or *gladiolus;* and the cartilaginous tip, the *xiphoid process* or *ensiform cartilage.* The lower two pairs of ribs are not attached in front and are known as the "floating" ribs. The ribs form a cage that houses and protects the lungs and the heart. In addition, ribs are moved upward and outward by the intercostal muscles in breathing.

GIRDLES AND APPENDAGES

In the shoulder region the *pectoral girdle* serves for the attachment of the forelimbs, and in the hip region the *pelvic girdle* serves for the attachment of the hindlimbs.

The human pectoral girdle consists of two bones: a broad dorsal scapula, and a slim ventral clavicle that articulates with the upper end of the sternum. Where the scapula and clavicle meet, they form a concavity, the *glenoid fossa,* in which the arm articulates. The scapula is attached to the trunk only by muscles. The freedom of the scapula and the articulation of the clavicle with the sternum offers the human excellent anchorage so that the arm can be used to do work where considerable leverage is required. In some mammals, however, the anchorage of the clavicle would be a hindrance. For instance, running animals, such as horses, lack a clavicle, and in the lithe cats this bone is only a small, sliverlike structure buried in muscle.

The pelvic girdle in humans and other vertebrates is composed of three bones on each side: an *ilium,* an *ischium,* and a *pubis.* In humans these three bones on each side are fused together to form a large *innominate bone* that is attached at the back to the sacrum. The fusion in front is known as the *symphysis pubis.* The sacrum, the two innominate bones, and the coccyx constitute the *pelvis.* The opening in the bony pelvis is the *birth canal.* Fortunately the articulation at the symphysis pubis is of such a nature that some expansion of this joint is possible during childbirth. There is a cavity, where the three bones of the girdle meet on each side, called the *acetabulum,* in which the hindlimb articulates.

The forelimbs of most vertebrates are quite similar. The upper arm bone is the *humerus* and its upper end fits into the glenoid fossa. The forearm contains two bones, the *radius* and the *ulna.* When the human arm is held out to the side at right angles to the body with the palm forward and the thumb up, the radius and ulna are parallel and the radius is uppermost. The ulna has on its end next to the humerus a process often referred to as the "crazy bone." There are eight small *carpal bones* arranged in two rows of four each in the wrist. In the hand there are five *metacarpals,* each corresponding to a digit. There are three *phalanges* in each finger and two in the thumb. While the basic hand or foot pattern has five digits, there is variation among vertebrates. Some salamanders have five hind toes but only four fingers, and the grazing or browsing horses and cattle have the digits reduced to four, two, or one (Ch. 38).

The four limbs of all vertebrates are built on the same general plan, and thus the hindlimbs are similar in construction to the forelimbs. In humans the upper leg bone corresponding to the humerus is the *femur.* It articulates with the pelvic girdle. The lower leg has a *tibia* and a *fibula,* corresponding respectively to the radius and the ulna. The ankle contains seven irregularly shaped bones, the *tarsals,* corresponding to the carpals. The foot proper contains five *metatarsals,* corresponding to the metacarpals of the hand; the bones in the toes are the *phalanges,* two in the big toe and three in the others.

Bone

THE GROWTH OF BONE

The histological nature of bone is indicated in Chapter 4. The living bone cells secrete the hard matrix of calcium phosphate and calcium carbonate. In some vertebrates, such as the sharks and rays, the skeleton is made of cartilage. In many vertebrates most of the skeleton is first laid down as cartilage and the cartilage is replaced by bone. Such bone is known as *cartilage bone.* The bones of the face and the cranium do not pass through a cartilaginous stage and are known as *membrane bone.*

Cartilage bone formation illustrates bone growth and development. Membrane bone formation is similar but ossification proceeds without the steps of cartilage breakdown. Most bones in the body are first formed of cartilage. Each has a shape similar to the bones that will form later in their places. Each bone has a primary center of ossification. The cartilage cells here become swollen and gradually die as calcium salts are deposited in the cartilage matrix. While this occurs, the layer of cells surrounding the cartilage, the *perichondrium,* becomes divided into an outer and inner layer. The outer layer is fibrous but the inner one is cellular, and from this layer blood vessels penetrate to the center of the dying cartilage cells. Special cells, *osteoblasts* and *osteoclasts,* colonize this area. The multinucleate osteoclasts erode and remove the cartilage. In the spaces produced, fibers of collagen are laid down around the salts by the osteoblasts.

In a short bone, such as one of the wrist bones, the cartilage at the edge continues to grow while at the central primary ossification center erosion of cartilage occurs and bone deposition follows. Ossification normally occurs faster than cartilage formation, and the entire structure becomes bone except at surfaces where bones participate in a joint—here a thin layer of cartilage persists.

Growth of long bones is more complicated and involves more than one ossification center (Fig. 8.4). The primary ossification is usually in the center of the shaft. Also in the perichondrium at the level of the primary ossification center the inner layer becomes converted to osteoblasts that add bone to the surface in a collar. This modified perichondrium is now called the *periosteum* and it continues to add to the girth of the bone. Internally, when the shaft of the bone reaches a certain diameter, the osteoclasts destroy the bone and produce a marrow cavity.

Thus along the shaft of a bone it becomes thick due to the deposition of bone by the periosteum, but in cross section it is always a ring of a thickness proportional to its diameter due to the bone-destroying activities of osteoclasts. Furthermore, periosteal deposition proceeds in both directions toward the ends of long bones accompanied internally by cartilage destruction, bone deposition by osteoblasts, and bone destruction by osteoclasts. This extends the marrow cavity lengthwise as bone deposition occurs lengthwise along the shaft.

As ossification proceeds along the length of the shaft, the cartilage at the ends grows to lengthen the bone. Depending on the individual long bone, at various times during the growth process, *secondary ossification* centers appear near the ends of the bones. Some of these appear prenatally but others some years after birth. With the appearance of secondary centers, a long bone then has five identifiable locations along its length. The central shaft area of ossification becomes known as the *diaphysis;* there is a mass of developing bone at each end forming two *epiphyses;* between each epiphysis and the diaphysis a thin plate of growing cartilage persists producing the two *epiphyseal plates.* The two epiphyseal plates, one at each end of a long

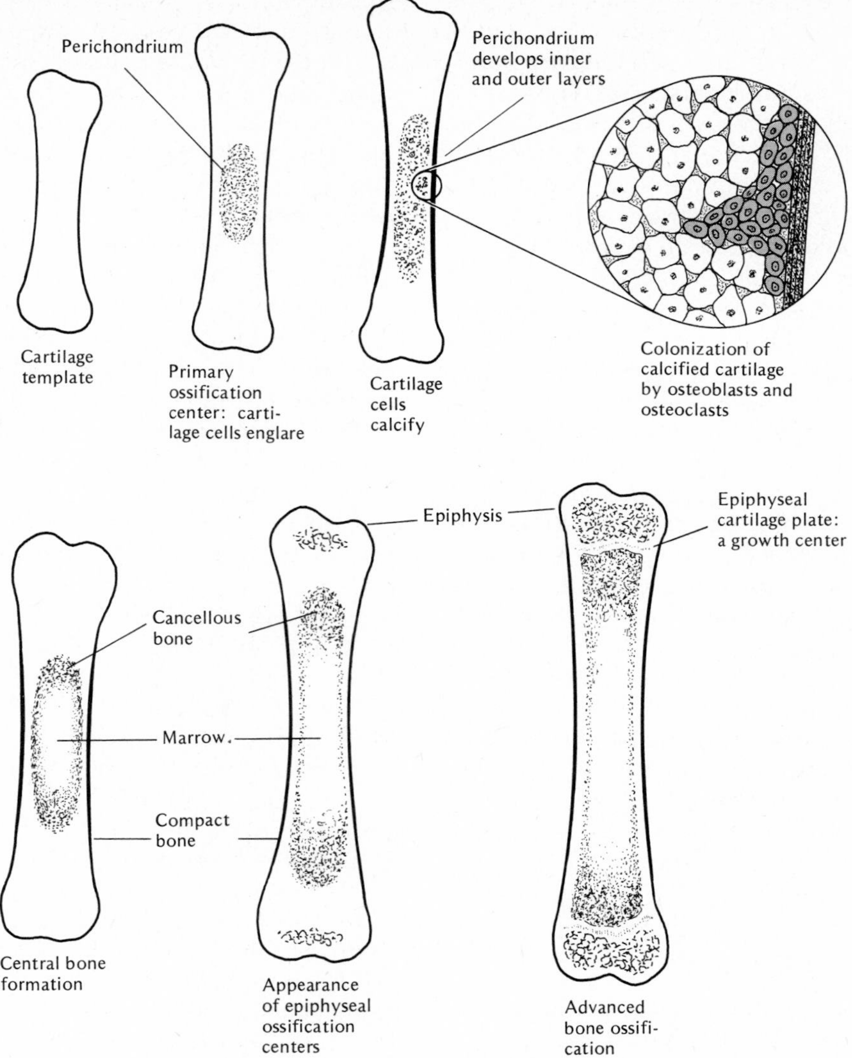

Figure 8.4 Long-bone ossification.

bone, contribute to the increase in length of the bone until maturity for that bone is reached, when the cartilage dies and all that persists is an identifiable epiphyseal line.

Not all bones have the shapes only of short bones or long bones, but the principles of bone growth and development apply to them also. Each bone usually has a primary ossification center, and various processes or prominences may be produced by secondary ossification centers.

An examination of the structure of a long bone, such as a femur, shows several interesting things associated with the functioning of the bone. First, the shaft of a developed bone is strengthened by the *compact bone* formed by the periosteum. Microscopic examination of compact bone shows it to be made up of *Haversian systems.* Each Haversian system consists of a central Haversian canal in which blood capillaries run. Concentrically around each canal the bone has been laid down by osteoblasts that become entrapped in the surrounding bone they produced, and are now called *osteocytes.* The osteocytes have processes which extend throughout the bone and provide a communication route for the exchange of nutrients, gases, and wastes with the capillaries; thus bone, once formed, remains populated with living cells. Second, strength is also achieved internally. Here, where bone replaces cartilage, it forms what is called *spongy bone.* While relatively porous in appearance, the spongy bone is deposited with the spicules of bone organized in girderlike stress lines that provide great strength (Fig. 8.5). Third, spongy bone provides for lightness of weight while providing great strength. Fourth, the marrow cavity provides a

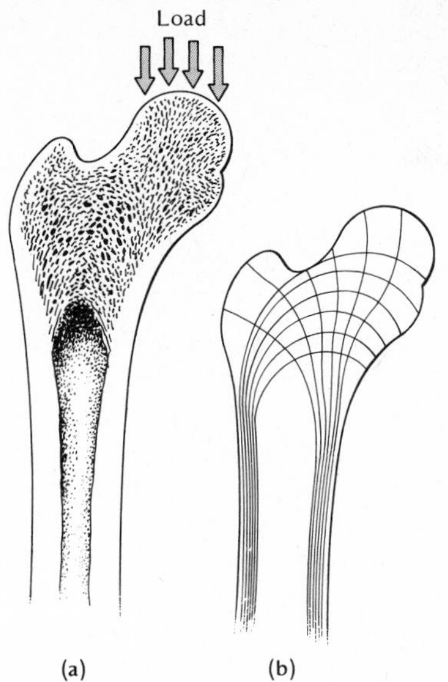

Figure 8.5 **(a)** Sketch of orientation of bony spicules in the upper femur. **(b)** Diagram of computed lines of stress on the upper femur. Thus the inner architecture of the femur is constructed to meet the load requirements of the bone.

site either for blood cell production (red marrow) or as a fat reserve (yellow marrow). Fifth, the skeleton serves as a reservoir for calcium. Under the control of the parathyroid gland calcium ions are released to or withdrawn from the blood. Bone is also a storage site for phosphorus, which is so very important in metabolism.

REPAIR OF FRACTURES

Bone is very strong and much force is needed to break it. When breaking occurs, blood coagulates at the site, producing a *hematoma.* This is invaded in a few days by cells from the inner layer of the periosteum. Some cells produce cartilage and osteoblasts begin their functioning as described for normal bone development. This produces a *provisional callus* that binds the broken ends. This is followed by the production of the *definitive callus* (Fig. 8.6). Depending on how far displaced the broken ends were, the definitive callus may be slim and streamlined across the break or quite massive. Finally, Haversian systems will be built into the repair site and the orientations achieved will reflect the stresses on the reconstructed bone.

Joints

A joint is where two bones meet. Some are only slightly movable and are called *symphyses.* There are two kinds of symphyses. One kind, such as found in the cranium, unites bones firmly and usually permanently. This is called a *fibrous joint,* and when the attachment or bone growth becomes so rigid that no movement occurs, such a joint becomes a *suture.* Another form of symphysis is seen between the pubic bones or between the vertebrae where cartilage lies between the bones. These *secondary cartilaginous joints* usually allow for some movement, as at childbirth.

The *synovial joints* provide the movable joints, as in the knee or the elbow (Fig. 8.7). Here two bones, each with an *articular cartilage* at its end, have stretched between them a fibrous *joint capsule* made up of *ligaments* and lined with a thin vascular membrane, the *synovial membrane.* The function of the synovial membrane is to produce *synovial fluid,* which comes from the bloodstream and lubricates the moving surfaces and nourishes the cartilage. Swollen joints are usually the result of inflammation of the synovial membrane from an injury or misuse. In some joints the synovial membrane is extended through a gap in the capsule to produce a *bursa* between the capsule and surrounding muscles; this is a friction-reducing device. Inflammation of a bursa or deposition of salts in it leads to an infirmity called bursitis.

Joints are identified by the kind of motion they permit. In addition to immovable joints, there are a number of modifications of synovial joints. Some are *hinge* joints, as in the fingers. Between the first and second vertebrae is a *pivotal* joint that allows for rotation in only one direction. *Gliding* joints are found between the short car-

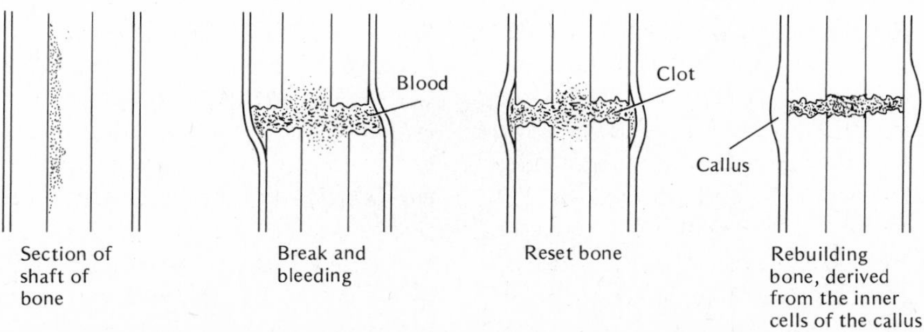

Figure 8.6 Stages in the repair of a fracture of bone.

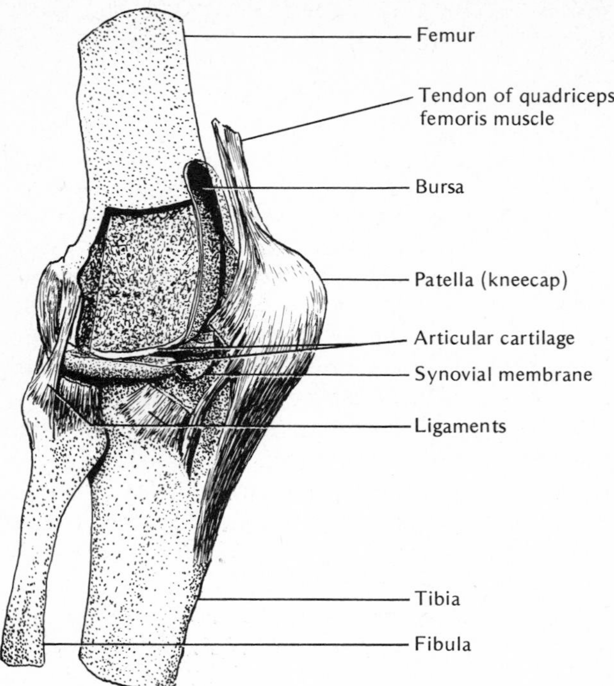

Figure 8.7 Knee joint.

pals and metacarpals of the wrists and ankles. A unique joint in humans is the *saddle* joint between a carpal bone and the metacarpal of the thumb. The organization of this joint provides the unique capacity of the opposable thumb of primates. Also its form seems to be genetically determined as a trait; "hitchhiker's thumb" is inherited. The greatest variety of motion is found in the *spheroidal* or *ball-and-socket* joints, such as are illustrated by the hip and shoulder joints. A modified spheroidal joint is the *condyloid* joint with an eliptical, rather than a spherical, socket. It differs in allowing most motions except rotation, and an example is the joint between the radius and the carpal bones.

MUSCLES AND MUSCLE ACTION

The histological nature of the three kinds of muscle tissue—smooth, cardiac, and striated—was considered in Chapter 4. Here attention will be focused on striated muscles, their relation to the skeleton, and the process of muscle contraction.

The striated muscles, also called voluntary and skeletal, constitute a large part of the bulk and weight of the vertebrate body. What we eat as beefsteak, pork chops, and chicken is primarily skeletal muscle.

Parts and General Functions of Muscles

A typical muscle, such as the *biceps brachii* of the arm shown in Figure 8.8, has three readily recognizable parts: the *belly,* the thick part of the muscle; and the two points of attachment, the *origin* and the *insertion.* The point of attachment that remains relatively fixed when the muscle contracts is the origin, while the point of attachment that moves is the insertion. The two ends of a muscle are usually attached to two bones, although some muscles are directly attached to other muscles. The attachments are made by sheets of fibrous connective tissue or ropelike strands of connective tissue known as *tendons.*

In functioning, muscles always pull; they never push. As a consequence, muscles are always arranged in so-called antagonistic pairs or groups. When the *biceps brachii* (and the underlying *brachialis*) of the arm contracts, the forearm is flexed, while contraction of the *triceps brachii* causes its extension. These two muscles have opposite actions and they represent an antagonistic pair. A muscle that causes the angle of a joint to decrease is called a *flexor,* while one that

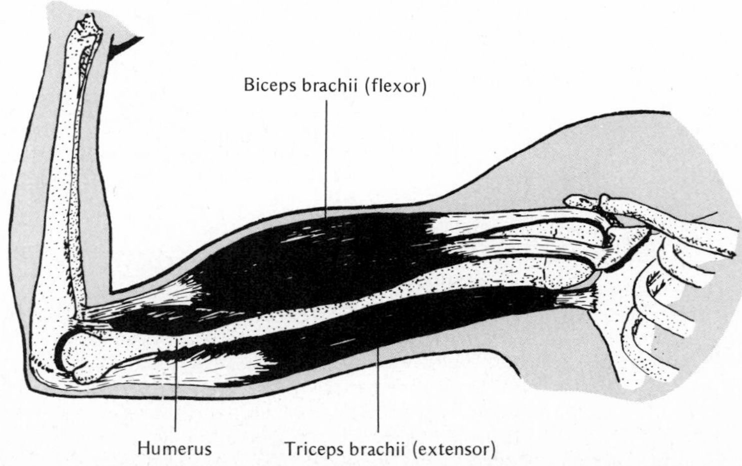

Figure 8.8 Opposing muscles.

causes the angle of a joint to increase is an *extensor.* In a similar way throughout the body, most of the different muscles or muscle groups (often a group of muscles will be involved in a particular movement) are arranged as antagonists. The muscles of the hind leg of the frog afford good material for easy study. The large muscle forming the calf of the leg is the *gastrocnemius* (the same as in humans), and it functions to flex the shank and extend the foot. Opposing this muscle on the front of the shank are the small muscles, the *tibialis anticus longus* and the *peroneus,* that are flexors of the foot and extensors of the shank.

Muscle Contraction

The most commonly used material for the study and observation of muscle contraction is the gastrocnemius muscle of the frog. This large muscle can be easily removed from the body with its large nerve, the *sciatic nerve,* attached. This preparation is called a *nerve–muscle preparation.* When the nerve is pinched or when it is stimulated with an electric current, the muscle contracts, becoming shorter and thicker. In contraction, the whole muscle does not change its volume. This was demonstrated years ago in a simple experiment. If an isolated muscle is placed in a fluid-filled closed chamber with a narrow fluid-filled vertical tube attached, there is no change in the level of the fluid in the upright tube when contraction occurs. From this it must be concluded that the contraction phenomenon involves only changes in the configuration of the molecules that make up muscle fibers.

A method of recording the contractions of a muscle is diagrammed in Figure 8.9. The nerve–muscle preparation is usually kept in a moist chamber to prevent drying, particularly of the delicate nerve. It is attached at one end by the muscle origin to an immovable rod and at the other end the tendon is attached by a thread to a muscle layer. This lever makes contact with the

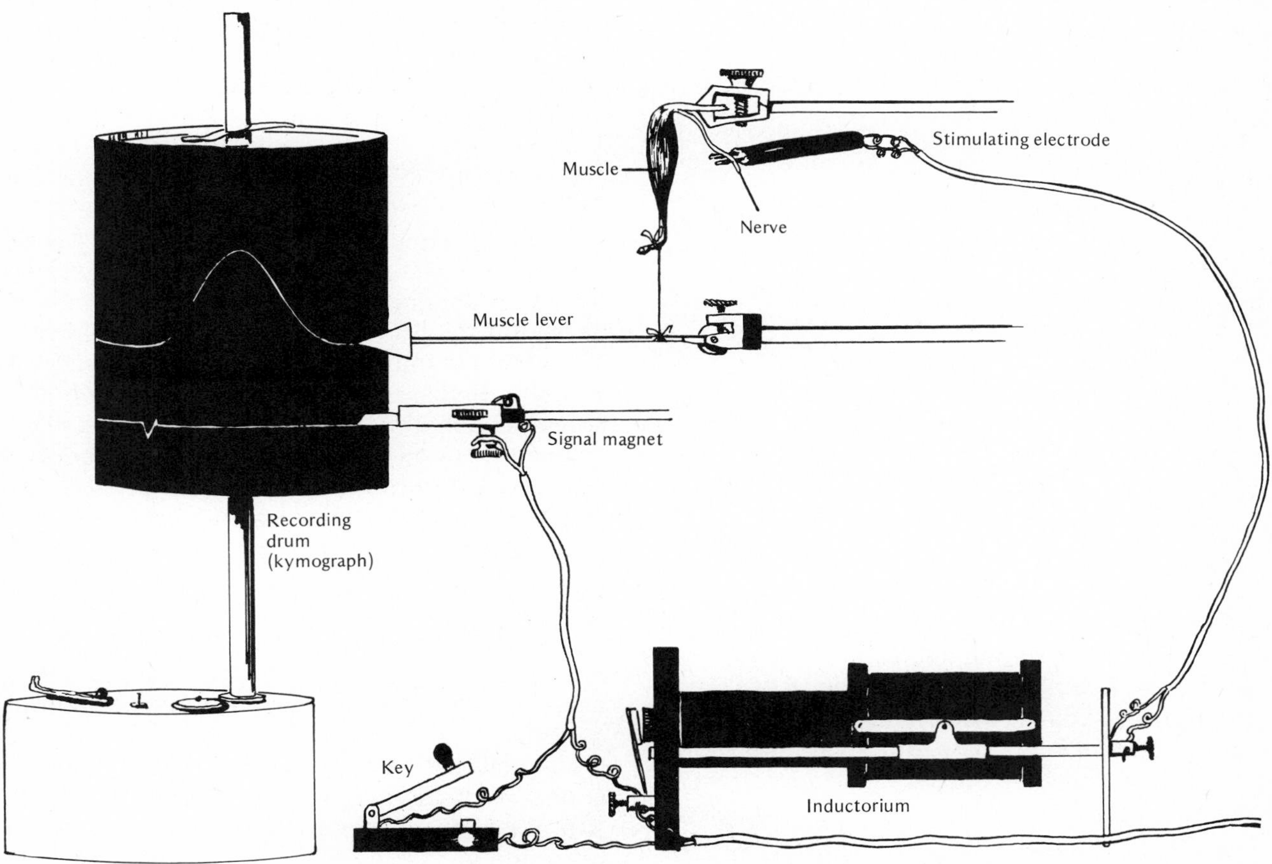

Figure 8.9 A method of recording the contraction of a muscle. The frog nerve–muscle preparation is kept moist and the femur is attached to a clamp. The tendon is attached by means of a thread to a muscle lever which is lifted as the muscle contracts. The electrical stimulus is delivered to the nerve through the stimulating electrode; and in the same circuit is included a signal magnet, shown below the muscle lever, to indicate the instant at which the stimulus is delivered. At the left is shown a kymograph, the large drum of which is turned at a constant desired speed by a clockwork mechanism in the base. Wrapped around the drum is a sheet of recording paper. The writing points trace a line upon the moving drum. The muscle lever and the indicator of the signal magnet carry tips arranged to touch the surface of the moving cylinder in the same vertical line. The record of a simple muscle twitch is shown.

recording paper that covers the surface of a revolving drum. When the muscle contracts, lifting the lever, a mark showing the height and duration of the muscle twitch is made on the recording paper. The recording paper is on the drum of the kymograph, and this drum is revolved at a constant speed by a clockwork mechanism in the base. The instant of stimulation is recorded by an electric signal marker that is included in the stimulating circuit. The muscle may be stimulated directly or through the sciatic nerve.

Under ordinary room conditions, a single twitch of a frog's gastrocnemius muscle lasts for about 0.1 second. A study of the record of such a twitch shows that it is made up of three phases: the *latent period,* the short period (0.01 second or less) between the application of the stimulus and the first signs of contraction; the *contraction period* (0.04 second in duration) when the muscle undergoes shortening; and the *relaxation period* (0.05 second) during which the muscle returns to its original length. During the latent period a change in electric potential occurs in the contracting fibers; a wave of negativity, similar to that to be described in nerve fibers, passes along the muscle fibers.

In the intact animal, muscular activity ordinarily does not involve single twitches but consists of more prolonged contractions. In such contractions, not one impulse but a volley of impulses reaches the muscle and the responses to the different impulses are completely fused. Such a sustained contraction is called *complete tetanus.*

If a single muscle fiber is isolated from a whole muscle, it behaves according to the *all-or-none* law when stimulated; that is, the minimal stimulus that causes contraction causes complete contraction, a stronger stimulus will cause no greater contraction. But the whole muscle does not follow the all-or-none law. The motor nerve that innervates a muscle is composed of many axons, each axon forms many branches, and each branch terminates in one muscle fiber, forming a *myoneural junction* (Fig. 8.10). Thus the impulses passing over one axon would stimulate several muscle fibers, but it would take impulses in all the axons to the muscle to cause all the fibers to contract. It is a common observation that less exertion is required to lift a sheet of paper than an ordinary book. Graded contractions occur in the intact muscle, varying with the amount of stimulation. A muscle is an aggregation of motor units. Impulses may affect a few of these units, or many, with varying degrees of strength of contraction.

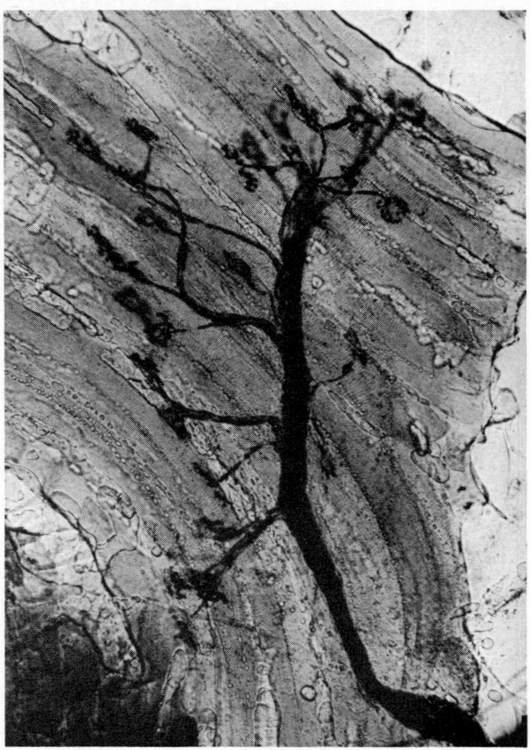

Figure 8.10 Myoneural junctions in striated muscle. (Courtesy of General Biological Supply House, Inc., Chicago.)

The Microscopic Structure of Muscle

Multinucleated cells, called *muscle fibers,* make up a muscle. Each muscle fiber was formed by the fusion, during embryonic life, of hundreds of myoblasts. Surrounding each muscle fiber is the sarcolemma, which is capable of being highly polarized. Periodically tubules of sarcolemma penetrate into the interior of the fiber. These are the so-called T tubules. The interior of a fiber consists of the cytoplasm, called the sarcoplasm, which is largely occupied by parallel arrays of muscle proteins, *myofibrils.* Myofibrils consist of two kinds of filaments, thin filaments and thick filaments. The thin filaments are attached to transverse areas of protein called Z lines. From one Z line to the next along the myofibrils is a sarcomere, the principal unit of contraction of a muscle (Fig. 8.11). In addition to the myofibrils, muscle fibers also contain mitochondria, the source for the production of ATP, and a system of membranes, the sarcoplasmic reticulum (Ch. 4).

The Chemistry of Muscle Contraction

Much research has been devoted to the problem of muscle contraction and the complete story cannot yet be told. What is stated below

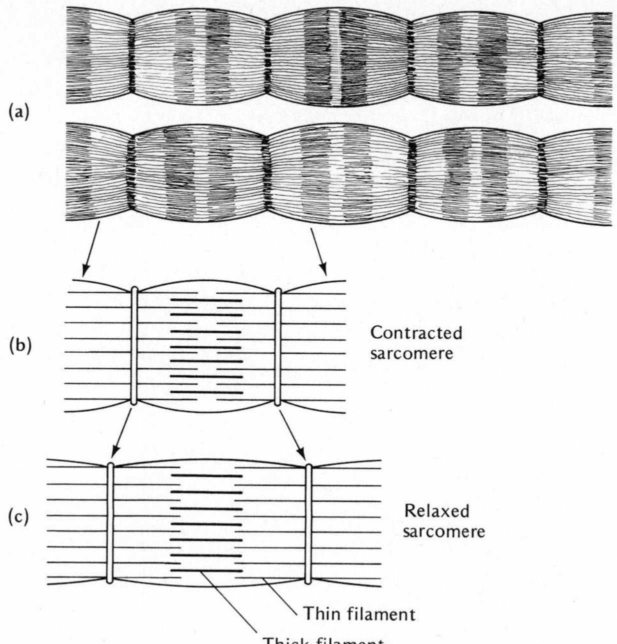

Figure 8.11 **(a)** Sketch of several sarcomeres as seen with the electron microscope. **(b)** Overlapping of thick and thin filaments in contracted muscle. **(c)** Relationship of thick and thin filaments in relaxation. The thick and thin filaments have substructures shown in Figure 8.12. (From "The Mechanism of Muscle Contraction" by H. E. Huxley, Copyright © 1965 by Scientific American, Inc. All rights reserved.)

represents the present views on the matter.

For a long time it has been known that glycogen disappears during muscle contraction, that oxygen is used, that carbon dioxide is formed, and that heat is liberated. These facts suggested that muscular contraction is an oxidative process. But it was possible to show that a muscle can contract a number of times when completely deprived of oxygen. This seemed to establish that, although oxidations are involved in the process, oxidation is not the immediate cause of the contraction.

With the disappearance of glycogen, lactic acid is formed. This breakdown occurs without oxygen and, since in this process energy is liberated rapidly, it was once considered that this reaction was directly responsible for contraction. However, about 1930 it was shown, by poisoning a muscle with iodoacetate which inhibits the breakdown of glycogen into lactic acid, that contractions can still occur. This meant that the energy released in the breakdown of glycogen is not the primary source of energy for contraction.

It is now known that the breakdown of adenosine triphosphate (ATP) to form adenosine diphosphate (ADP) is the immediate source of energy for muscle contraction. The nature of this process was discussed in Chapter 5. In that discussion it was pointed out that each cell has only small amounts of ATP and ADP, and this is true of muscle fibers too. We know that in violent exercise, such as running, much energy is expended. Is the normal rate of formation of ATP in muscle fibers adequate for this? Apparently not, but the muscle cells of all organisms have a reserve of energy-rich phosphate compounds in the form of phosphogens called *phosphocreatine* and *phosphoarginine.* The phosphogen found in vertebrate muscles is phosphocreatine. During periods of intensive muscular activity the phosphocreatine is broken into creatine and an energy-rich phosphate group. This energy-rich phosphate group then unites with ADP to form ATP. According to present evidence the major chemical reactions involved in muscular contraction are as follows.

1. ATP is changed to ADP with a release of energy that causes contraction.
2. Phosphocreatine breaks into creatine and an energy-rich phosphate group that makes possible the resynthesis of ATP.
3. Glycogen is broken into lactic acid in a series of reactions. The energy released in these reactions is available for the resynthesis of phosphocreatine.
4. About one fifth of the lactic acid is oxidized to CO_2 and H_2O. The energy made available in these oxidations may be used to convert the remaining four fifths of the lactic acid to glycogen.

All these reactions are catalyzed by specific enzymes. The first three steps just described do not involve oxygen; they can take place under anaerobic conditions. For the fourth step oxygen is required; if it is not present in adequate amounts, lactic acid accumulates. In severe muscular exercise, a large amount of lactic acid accumulates in muscles and in the blood. If concentrations reach a critical level in the muscle, it is unable to contract. A considerable amount of oxygen must be supplied for the oxidation of the lactic acid after the exercise ceases. This explains the continued hard breathing of persons after they have finished a 100-m dash. The *oxygen debt* that was incurred during the exercise is gradually repaid in the next several minutes following the cessation of exercise.

Mechanism of Contraction

An adequate theory of muscular contraction should explain the mechanism by which the shortening of the contractile elements is produced (Fig. 8.11). Our knowledge of this is not quite complete but large strides have been made in recent years.

Under normal conditions, skeletal muscle contraction is controlled by the peripheral nerves. When an impulse reaches a myoneural junction, it alters the polarization of the sarcolemma of the muscle fibers. The wave of altered polarization passes into the fiber by way of the T tubes where the altered charge causes the release of calcium from the sarcoplasmic reticulum. The released calcium triggers a series of events which cause the thin and thick filaments of the myofibrils to slide past each other and the muscle shortens. To make these events more understandable, it is necessary to examine the thick and thin filaments for their structure and to see how calcium and ATP are related to their structure.

There are four muscle proteins associated with the thick and thin filaments. The thick filaments, which lie midway between Z lines and are unattached to them, consist of bundles of *myosin* molecules (Fig. 8.12). Each myosin molecule is a rod about 0.12 μm in length. At the end of this molecule is an expanded area of a modified myosin, *meromyosin,* which we will simply call the head of the myosin molecule. The heads extend laterally from the thick filament. These heads are active sites where the chemical events of muscle contraction occur.

The thin filaments that extend toward each other from adjacent Z lines consist of three proteins (Fig. 8.12). One protein, the *actin* molecules, is composed of small spherical molecules arrayed to form a twisted double strand resembling beads. A second protein is *tropomyosin;* it is filamentous and long enough to span seven actin myosins in length. It lies near the groove formed by the twisted strand of actins. An elliptical globular protein, *troponin,* is associated with the end of each tropomyosin that lies toward the center of the sarcomere.

In contraction the heads that extend from the sides of the thick filaments come into contact with the thin filaments. Normally, with the presence of calcium and the release of ATP, energy is provided to cause each head to swivel. This pulls the thin filaments toward the center past the thick filaments. The distance between the Z lines is shortened and the muscle fiber is contracted.

To produce contraction, then, the steps are as follows.

1. A nerve impulse alters the polarization of the sarcolemma.
2. This alteration extends into the muscle fiber via the T tubules.
3. Calcium is released from the sarcoplasmic reticulum.
4. Calcium binds to troponin.
5. The information about this binding is conveyed along the tropomyosin, probably shifting its position in the groove between the seven pairs of actin molecules it spans. The thin filament is now in the "on" phase and is ready for the crucial union with the heads bridging across from the thick strands of myosin.
6. To be useful in contraction, however, these heads must be combined with ATP. This produces a charged myosin–ATP intermediate. With the five steps described, it now becomes possible for the myosin–ATP inter-

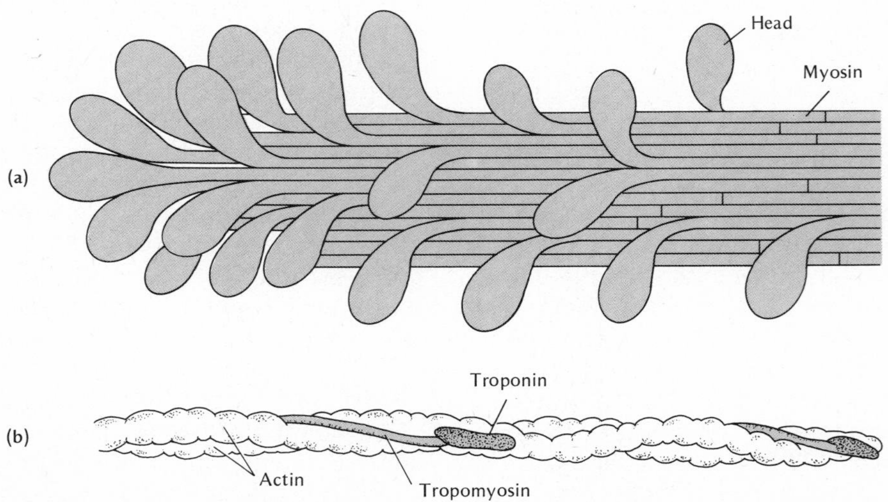

Figure 8.12 Schematic diagrams of the tips of **(a)** a thick filament composed of rodlike myosin molecules tipped at one end with laterally projecting heads, and **(b)** a thin filament composed of two strands of coiled actin in whose grooves lie strands of tropomyosin tipped with troponin. The heads of the myosin rotate on the thin filaments to draw the latter closer together between the Z-lines.

mediate to bind to the actin in the "on" phase.

7. This produces the active complex, namely, the myosin–ATP combined to the actin.
8. With the production of the active complex, hydrolysis occurs and ATP is split into ADP and inorganic phosphate.
9. The energy released from this hydrolysis causes a swiveling of the head that is attached to the thin filament.
10. The thin filaments are drawn toward the centers of the sarcomeres, the muscle shortens, and contraction occurs.
11. With these events, calcium is quickly removed from the troponin on the thin filament and is returned to the sarcoplasmic reticulum by a calcium pump.
12. The withdrawal of calcium permits the myosin head to be uncoupled from the actin, now in the "off" phase, and it can be recharged with another ATP molecule. Of course, the source of the ATP is the mitochondria.

The article by Murray and Weber (Suggestions for Further Reading) will be found especially interesting for its description of the experimentation that led to the concept of muscle contraction and its regulation.

SUGGESTIONS FOR FURTHER READING

Beck, W. S. *Human Design.* New York: Harcourt Brace Jovanovich, (1971).

Bendall, J. R. *Muscles, Molecules, and Movement.* London: Heinemann (1969).

Best, C. H., and N. B. Taylor. *The Human Body: Its Anatomy and Physiology.* New York: Holt, Rinehart and Winston (1963).

Curry, J. *Animal Skeletons.* New York: St. Martin's (1970).

Gergely, J. *Biochemistry of Muscle Contraction.* Boston: Little, Brown (1964).

Griffin, D. R., and A. Novick. *Animal Structure and Function,* 2d ed. New York: Holt, Rinehart and Winston (1970).

Guyton, A. C. *Function of the Human Body,* 3d ed. Philadelphia: W. B. Saunders (1969).

Harrison, R. J., and W. Montagna. *Man.* New York: Appleton-Century-Crofts (1969).

Huxley, H. E. "The Mechanism of Muscle Contraction," *Science,* **164,** 1356–1366 (1969).

Katz, B. *Nerve, Muscle, and Synapse.* New York: McGraw-Hill (1966).

Wilkie, D. R. *Muscle.* London: Edward Arnold (1968).

From Scientific American

Cohen, C. "The Protein Switch of Muscle Contraction" (Nov. 1975).

Hoyle, G. "How Is Muscle Turned on and off?" (Apr. 1970).

Huxley, H. E. "Mechanisms of Muscle Contraction" (Dec. 1965).

Margaria, R. "The Sources of Muscular Energy" (Mar. 1972).

Merton, P. A. "How We Control the Contraction of Our Muscles" (May 1972).

Montagna, W. "The Skin" (Feb. 1965).

Murray, J. M., and A. Weber. "The Cooperative Action of Muscle Proteins" (Feb. 1974).

Porter, K. R., and C. Franzini-Armstrong. "The Sarcoplasmic Reticulum" (Mar. 1965).

Ross, R., and P. Bornstein. "Elastic Fibers in the Body" (June 1971).

Wessells, N. K. "How Living Cells Change Shape" (Oct. 1971).

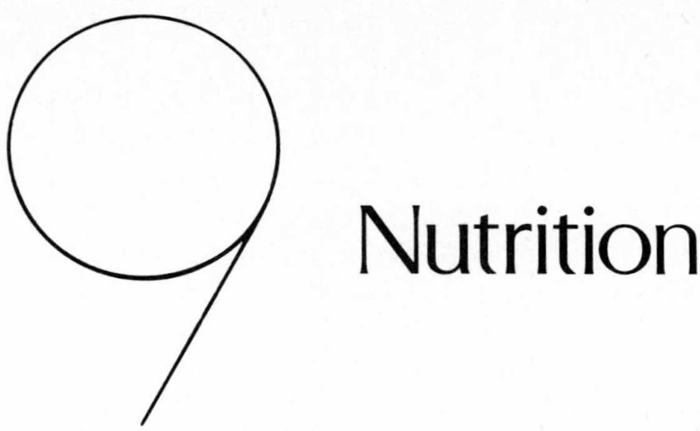

9 Nutrition

All animals must be supplied with materials (food) that serve as energy sources and as building blocks for the synthesis and repair of the living substance.

The total of the processes involved in the absorption and utilization of foods and food accessories is animal nutrition. The basic nutritional requirements of all animals are very similar.

Whereas green plants can synthesize their basic food and structural requirements from water, carbon dioxide, and mineral salts, all animals (unless they are parasites) must ingest solid materials originally made by green plants or other animals, and these materials must undergo digestion.

Digestion is a process in which large complex molecules are converted into smaller simpler molecules that can be utilized in the functioning of living cells. The molecules of most organic food substances are too large to enter the cells which line the intestine. They must be broken down into smaller molecules or else they pass directly through the tract and never really enter the body proper. The processes involved in this breakdown constitute digestion.

In all multicellular animals possessing a mouth, digestion occurs in a specialized region or cavity. In all higher animals, the digestive cavity or tract is a tubular structure differentiated into specialized regions extending from the mouth to the anus. The digestive tracts of all vertebrates are much alike. Similar types of digestive tracts are found in the higher invertebrates, such as the earthworm and the crayfish.

Many different things are utilized as food. Some animals feed only on the flesh of other animals and are called *carnivorous;* some feed only on plants and are said to be *herbivorous;* still others feed on both plants and animals and are *omnivorous.* Many adaptations for food getting are found throughout the animal kingdom. Among these adaptations are the striking modifications found in the beaks and claws of birds and the teeth of mammals. Some animals feed on tiny particles and are called *filter feeders.* They are found in a variety of animal groups, from certain mollusks to whalebone whales. Cows, deer, and other ruminants have several divisions of their stomach (actually three of the four are modifications of the esophagus) which are used for temporary storage of food. They rechew their food after digestion has started in the process known as "chewing their cuds." Birds have no teeth but grind food in a thick-walled gizzard containing small pebbles. Although digestion is brought about primarily by enzymes produced by the animal itself, the activities of bacteria in large intestines, caeca, and rumens of many forms help break down the complex foods into usable forms.

In this chapter we will consider the nature of our digestive system, foods, the process of *digestion, absorption* or the passage of the products of digestion into the blood and lymph vessels, and finally the *assimilation* or use of the products of digestion by the cells.

THE HUMAN DIGESTIVE SYSTEM

The tubular digestive tract or alimentary canal in humans is differentiated into the following regions: *mouth cavity, pharynx, esophagus, stomach, small intestine,* and *large intestine.* In addition, several glands are parts of this system. Three pairs of *salivary glands* secrete saliva into the mouth cavity. The *liver,* the largest gland in the body, and the *pancreas* empty their secretions into the small intestine. Other glands are located in the lining of the digestive tract itself.

A cross section through any higher animal reveals the relationship of the tubular digestive tract to the rest of the body. This tract is in the general *body cavity* or *coelom* and is held in place by *mesenteries.* This characteristic arrangement is called "a tube within a tube" body plan (Fig. 9.1).

In Chapter 4 the cellular detail of the intestine was described. It is composed of five general regions:

1. The lining epithelium or mucosa
2. The submucosa, which is primarily connective tissue with numerous blood and lymph vessels
3. The circular muscle layer
4. The longitudinal muscle layer
5. The outer serosa or peritoneum.

The intestine may be considered as a pattern of structure for the entire alimentary tract. The other regions or organs of the tract have these same layers, but the layers are modified in the different regions. One exception should be made to the last statement. The esophagus, which does not lie in the body cavity, is not surrounded with peritoneum. Some familiarity with the arrangement of the tissues in the alimentary tract is essential if we are to understand how it functions. As we know, some of the materials that are taken in at the mouth are passed through this entire tract and ultimately expelled from the body through the anus. The mechanism for this passage is found in the arrangement of the muscle layers.

Food is moved through the alimentary tract by the alternating contractions of the circular and longitudinal muscles. This type of muscular activity produces a series of waves of contraction and relaxation in the walls of the alimentary tract and is known as *peristalsis.* Thus when food is swallowed, a series of these peristaltic waves forces it through the esophagus into the stomach. The muscle layers of the stomach are very thick. In addition to the circular and longitudinal layers found elsewhere in the digestive tract, there is a layer of *diagonal* fibers. While food material is in the stomach, it is subjected to a *churning* action by the strong peristaltic waves that move from one end of the stomach to the other. As a result of this churning action, the food material in the stomach is reduced to a liquid mass called *chyme.* In the small intestine there are two types of movements, the regular peristaltic waves that move the chyme along and the *segmental movements* that churn the contents of the intestine and thoroughly mix them. These churning movements are produced by alternate contractions and relaxations of successive segments of the intestine. Peristaltic movements are involved in the passage of the material through the large intestine and, ultimately, also in defecation.

The epithelial lining of the entire digestive tract is eroded away continually and it is renewed by cell division about every three days.

The Mouth Cavity

One who has brushed his teeth daily for several years in front of a mirror is quite familiar

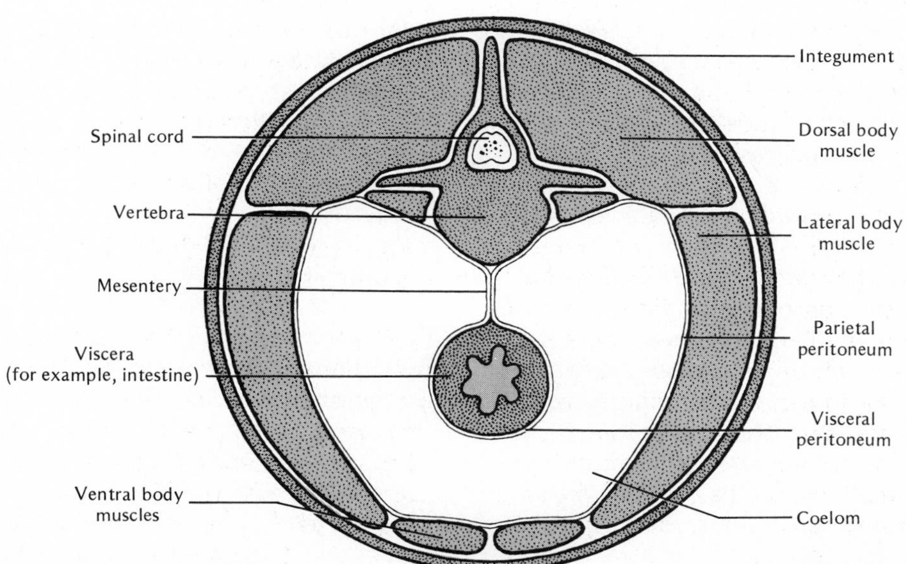

Figure 9.1 Diagrammatic cross section of a vertebrate.

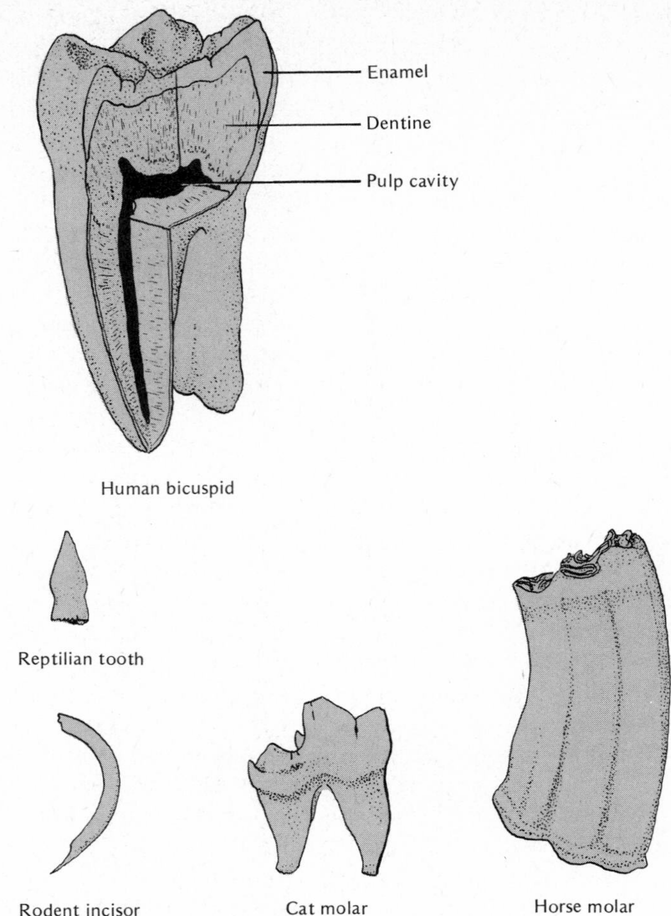

Figure 9.2 Representative tooth types.

with this part of his anatomy. The *mouth opening* is bounded by the *upper* and *lower lips.* Within the lips are the *gums* and the *teeth.* The sides of the cavity are formed by the *cheeks,* the top by the *palate,* and the bottom by the floor beneath the *tongue.* In humans both the tongue and the teeth are important in speech as well as in ingestion and digestion. The muscular tongue moves the food in the mouth during chewing and also shapes the food into a spherical mass or *bolus* for swallowing. In addition, special sensory structures, the *taste buds,* are located on the tongue.

Three pairs of *salivary glands* open into the mouth cavity. The *parotids* are located just below the ear and their ducts open on the inner surface of the cheek opposite the upper second molar tooth. Infection of the parotids causes the disease called *mumps.* The *submaxillary glands* lie below the angle of the jaw, and the *sublinguals* are beneath the tongue. The ducts of these two pairs of glands open on the floor of the mouth beneath the tongue.

An adult human normally has 32 *permanent* teeth, 16 in each jaw. On each side of each jaw there are, from front to back, two *incisors,* one *canine,* two *premolars,* and three *molars.* The third molars are the *wisdom* teeth, and they do not appear in all individuals. The first teeth or *milk teeth* are 20 in number; the eight premolars and the third set of molars are lacking. A typical tooth, as shown in Figure 9.2, consists of a crown, neck, and root. The part projecting above the gum is the *crown,* that surrounded by the gum is the *neck,* and below the neck is the *root* imbedded in the jawbone. The body of a tooth is composed of *dentine,* a substance which is harder than compact bone, although it does resemble bone in its structure, chemical components, and development. Unlike bone, the cells of dentine remain on the surface and send narrow processes into it. The crown is covered with a layer of *enamel* and the root is fastened to the jawbone by a layer of *cement.* The central region of each tooth is the *pulp cavity,* which contains nerves and blood vessels.

The teeth of lower vertebrates—fishes, amphibians, and reptiles—are all relatively simple and unspecialized. The teeth of mammals are not uniform, but are specialized to perform particular functions. Each mammalian type has teeth especially adapted for its particular kind of food. We are all familiar with the large canine teeth of flesh-eating dogs and wolves, and the very large grinding molars of grazing animals like the horse.

The Pharynx

The mouth cavity merges into the pharynx, the cavity behind the soft palate. In addition to the mouth cavity, six other structures connect with the pharynx. These are the two *internal nares* from the nasal cavity, the two *Eustachian tubes* leading to the cavities of the middle ears for equalizing the pressure on the ear drums, the *glottis* opening into the larynx, and the *esophagus* leading to the stomach. When food is swallowed, it is first pushed into the pharynx by movements of the tongue. Reflex mechanisms close all the openings from the pharynx except the one of the esophagus and when the pharynx contracts, the food is forced into the esophagus.

The Esophagus

The esophagus is a muscular tube leading directly to the stomach. It runs through the neck and thorax between the vertebral column and the trachea or windpipe and penetrates the muscular diaphragm before connecting with the stomach. Food is passed rapidly through this tube by the peristaltic movements already described. When food is not being swallowed, the esophagus is a collapsed tube (Fig. 9.3).

The Stomach

The stomach is a large J-shaped sac lying just under the diaphragm on the left side of the abdominal cavity. The rounded upper portion above the cardiac sphincter is called the *fundus,* while the lower part nearest the small intestine is called the *pyloric region.* The central portion between the fundus and the pyloric region is called the *body.* The opening between the stomach and small intestine is controlled by a ring of smooth muscle, a sphincter muscle, called the *pylorus.* The lining of the stomach, or the mucosa, contains many gastric glands that secrete gastric juice containing enzymes and hydrochloric acid.

The Small Intestine

The small intestine in man has a total length of about 6 m (20 feet). The first part of it, about 25 cm (10 inches), is called the *duodenum.* The rest is somewhat arbitrarily divided into the *jejunum* and the *ileum.* Externally these regions cannot be clearly distinguished but internally the jejunum has numerous transverse crescentic folds that disappear in the ileum. About two fifths of the small intestine posterior to the duodenum is jejunum and the remainder is ileum. The duodenum receives the opening of the *common bile duct,* a short tube formed by the union of the *bile duct* and the *pancreatic duct.*

The inner surface of the small intestine is covered with numerous microscopic fingerlike projections, called *villi,* which give it a velvety appearance. These villi greatly increase the area for absorption. The wall of the intestine contains numerous glands that secrete the intestinal juice containing several digestive enzymes.

The Large Intestine

The small intestine empties into the side wall of the large intestine, forming a blind sac, the *caecum,* attached to which is the *vermiform appendix.* In many herbivorous animals, such as the rabbit and guinea pig, this blind end of the large intestine is much longer and is really a functional part of the digestive system.

The large intestine or *colon* is about 1.5 m (5 feet) in length and has the general shape of an inverted U. It is divided into the *ascending, transverse, descending,* and *sigmoid colons.* The terminal 15 cm (6 inches) is a straight tube known as the *rectum.*

Absorption of the products of digestion has occurred by the time the contents of the small intestine enter the large intestine. The primary functions of the large intestine are the absorption of large amounts of water from its contents and the ultimate removal of the remaining material from the body. In addition to the unabsorbed material, large numbers of bacteria, which grow in this material in the large intestine, are present in the feces.

The Liver and Pancreas

The liver and the pancreas are important glands associated with, and embryologically derived from, the digestive tract. Glands are epithelial cells, either single or in groups, that synthesize and secrete special substances. The special substances secreted by glands are synthesized in the cells from materials obtained from the bloodstream, and in the secretion process the cells actually perform work.

The liver is the largest gland in the body; it is located just under the diaphragm, and partly covers the stomach. It is important in digestion because of its secretion, the *bile.* Bile is stored in the *gallbladder,* and when needed, it passes

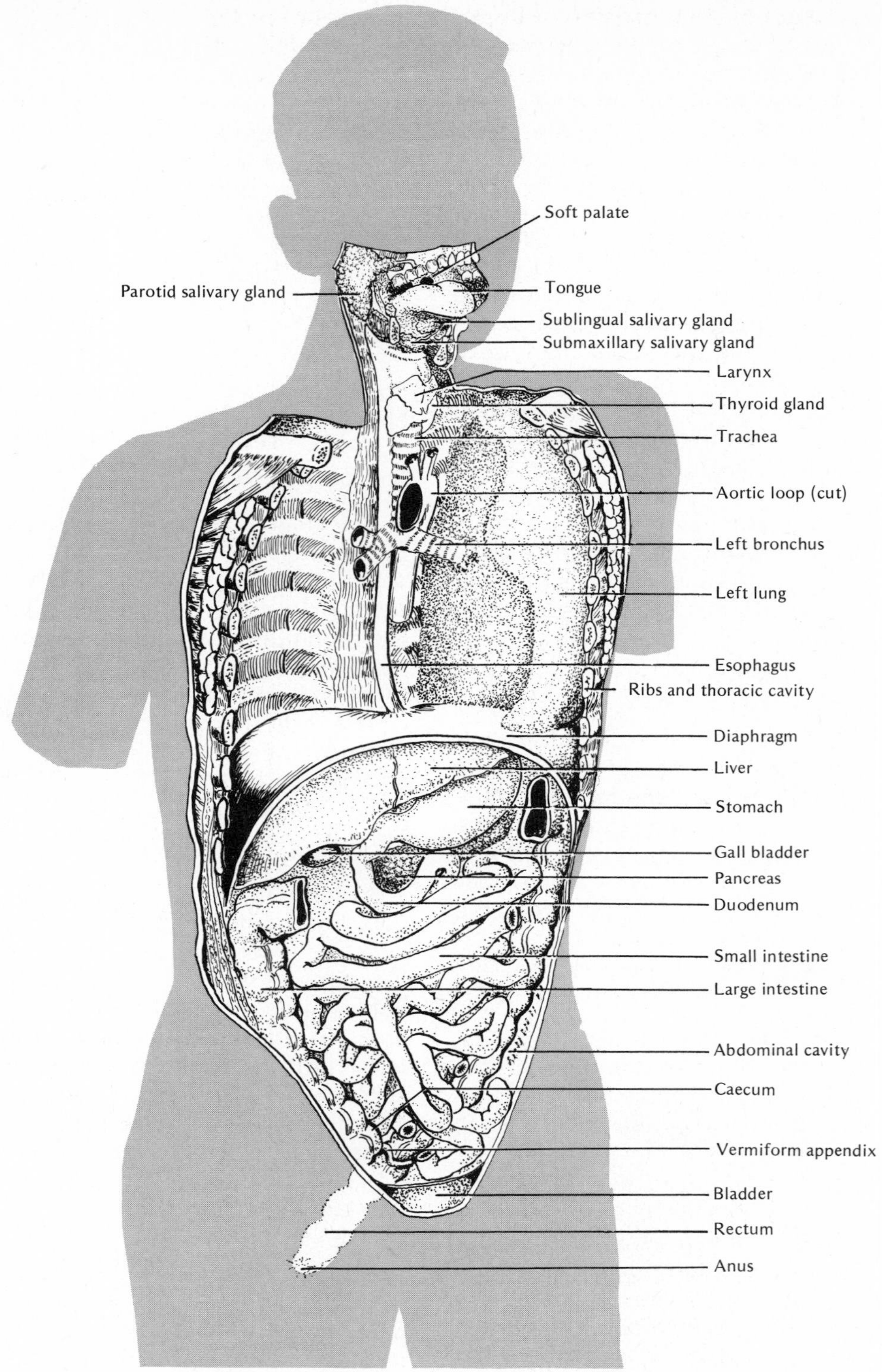

Figure 9.3 Human digestive system.

through the *common bile duct* into the duodenum. There are no digestive enzymes in bile. It acts to emulsify fats in the small intestine. When the fats are broken into tiny droplets by emulsification, there is much more surface area on which the fat-splitting enzyme can work. Bile salts in the bile (primarily the sodium salts of glycocholic and taurocholic acids which are formed in the liver by the conjugation of glycine and taurine with cholic or deoxycholic acid) are responsible for this emulsifying action.

In addition to its role in digestion, this organ is very important in several other functions. Glucose is converted into *glycogen* (animal starch) in the liver and stored. It is constantly being changed back into glucose and released into the bloodstream to maintain a constant level of sugar there. Excess amino acids are deaminated in the liver with the formation of urea and compounds of carbon, hydrogen, and oxygen. The urea is removed by the bloodstream and the compounds of carbon, hydrogen, and oxygen are ultimately used as sources of energy. This gland is a storage organ for vitamins A, D, and B_{12}. In the liver certain toxic substances are changed to nontoxic compounds for elimination. It produces materials involved in blood clotting, and it stores and conserves iron removed from worn-out red blood cells.

All of these activities occurring in the liver are made possible by the variety of special enzymes present.

The pancreas is a much smaller, irregular gland located between the stomach and the duodenum. It is a dual gland. Part of the organ secretes the pancreatic fluid which contains protein-splitting, fat-splitting, and carbohydrate-splitting enzymes. The pancreatic juice is carried to the duodenum by the pancreatic duct and the common bile duct. Other parts of the pancreas, known as the islands of Langerhans, are endocrine in function and will be discussed in Chapter 12.

FOODS

Foods may be defined as substances that yield energy, build tissues, or regulate body functions. The six categories of foods are

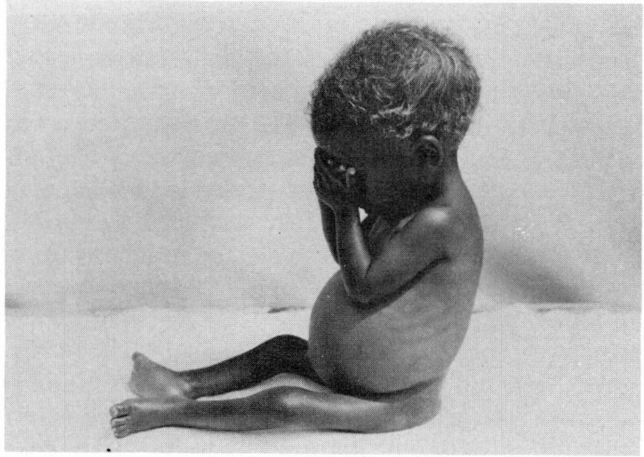

Figure 9.4 Child with kwashiorkor. In many parts of the world, and in Africa in particular, where children are born very close to one another, a child is taken off the mother's milk when quite young. If it is then placed on a protein-poor diet, kwashiorkor results. (Courtesy of FAO of the United Nations.)

1. Carbohydrates
2. Lipids (fats)
3. Proteins
4. Water
5. Inorganic salts
6. Vitamins

In addition to these substances, a proper diet also includes what is commonly called *roughage.* The cellulose in much of the plant material that is eaten serves as roughage. It is relatively indigestible and passes through the digestive tract unabsorbed. The bulk or mass of such material stimulates peristalsis and prevents constipation.

A deficiency in any of these categories will result in abnormal functioning. See Figure 9.4 which shows the effects of a protein-poor diet on a young child.

Carbohydrates

The carbohydrates—sugars and starches—are the chief sources of energy in human food. Potatoes, rice, corn, and other cereals are good sources. Milk is rich in milk sugar or lactose. Cane sugar (sucrose) is used in the preparation of many food items and beverages. Glucose and the other six-carbon sugars can be used in the body without digestion. The complex sugars and starches must be changed in digestion to simple sugars before they can be utilized. The organic acids of many vegetables and fruits serve as sources of energy.

Lipids

Lipids and related compounds are also utilized as sources of energy. They are commonly eaten in the form of butter, fat meat, vegetable oils, and cream. There are some essential fatty acids which are present in unsaturated lipids or vegetable oils. Lipids are changed in digestion into fatty acids and glycerol. These substances may be oxidized by the cells of the body with the release of energy, or they may be synthesized into fats and stored, chiefly in the adipose tissue of the body, as reserve material. Such fatty tissue also serves as insulating and cushioning material. Lipids also carry the lipid-soluble vitamins and are used in the formation of cell membranes.

Proteins

The end products of protein digestion, amino acids, are used in the body primarily for building and repairing cells. The excess amino acids undergo deamination in the liver. The amino groups are removed and the carbon–hydrogen–oxygen parts of the molecules are used in cellular respiration. Important sources of protein in foods are lean meat, milk, cheese, eggs, corn, wheat, peas, and beans. Of the twenty or so amino acids found in the body, eight are said to be *essential* because they cannot be synthesized in the body and must be obtained in the food. Protein foods that contain all the essential amino acids—such as meat, milk, and eggs—are called *complete.* Others—such as corn, peas, and gelatin—that lack one or more of the essential amino acids in their proteins are called *incomplete.* (Recently a strain of corn was studied at Purdue University which has adequate amounts of the essential amino acid *lysine,* which is normally present in inadequate amounts in corn protein.) It is obvious, then, that it is desirable to eat a mixture of different proteins.

Water

Water is an essential part of every living cell. It is the chief solvent in the body, it is the medium for transportation, and it is involved in most of the chemical reactions in the body. Although a person may live for a considerable period of time without other foods, death occurs soon in the absence of water intake.

Inorganic Salts

A dozen elements are known to be essential as mineral salts in the diet. They are sodium, chlorine, potassium, magnesium, calcium, phosphorus, iodine, iron, copper, manganese, zinc, and cobalt. Rarely is the diet deficient in sodium, potassium, magnesium, copper, phosphorus, or chlorine. It is not uncommon, however, to find diets deficient in calcium, iron, iodine, or zinc.

Calcium, magnesium, and phosphorus are essential in the formation of bone and teeth. Milk is an excellent source of calcium and phosphorus. Iron is used in the formation of hemoglobin and of the cytochromes (respiratory enzymes). Iron is best obtained in the following foods: molasses, liver, oatmeal, apricots, eggs, raisins, spinach, and beef.

Iodine is essential for the formation of thyroxin, the thyroid hormone. In certain regions of the world where the soil is deficient in iodine, it is now the practice to furnish the needed iodine by using iodized salt for seasoning.

The body fluids contain about 0.9 percent salt, mostly sodium chloride; this material is essential in maintaining the proper osmotic relations in the body. Because much salt is lost during hot weather through perspiration, especially during strong physical exertion, extra salt in the food or drinking water is necessary to prevent heat exhaustion.

Only traces of some of the elements are necessary. Among these are manganese, zinc, copper, magnesium, molybdenum, fluorine, and cobalt. Some of these are necessary for the formation of some of the enzymes. It has been shown recently that zinc is essential to vitamin A metabolism.

Studies indicate that small amounts of fluorine have a beneficial effect in the maintenance of sound teeth.

Vitamins

Although scientific knowledge of the vitamins is relatively recent, disorders—now called deficiency diseases—resulting from a deficiency of one or more of these substances have been known for centuries. Hippocrates, the Father of Medicine, used ox liver as a cure for night blindness. In 1536 Cartier saved a part of his group of explorers from the ravages of scurvy with a brew of pine needles, which was recommended by some friendly American Indians. After Lind had urged the use of citrus fruits for the prevention of scurvy in the British Navy in the middle of the eighteenth century, the practice was adopted of taking citrus fruit frequently on very long voyages. This practice, which led to the nickname of "limey" for British sailors, successfully prevented scurvy. In 1882 Takaki experimented with the diet of Japanese sailors who often suffered from beriberi, a painful neuritis. He showed that a crew fed with a varied diet rarely suffered from beriberi, whereas crews fed primarily on polished rice often had one third of their members disabled. In 1897 the Dutchman Eijkman, working in Java, showed that chickens and pigeons fed on polished rice alone developed a disorder similar to beriberi, whereas those fed on the unpolished rice never developed such symptoms. Eijkman's observation was important, but he made the wrong interpretation. He assumed that polished rice had some harmful substance that was neutralized by the hulls left on the unpolished rice.

We now define a vitamin as a substance that is required in small quantities for normal body functioning but that cannot be synthesized by the body and must, therefore, be obtained from food. Szent-Györgyi, who first isolated vitamin C, gave this graphic definition in a lecture when he said, "a vitamin is something that makes us sick when we do not eat it."

Although all the vitamins are organic compounds, they are not all related compounds. All of them are needed in only very minute amounts. The daily requirements are usually expressed in terms of micrograms (1 millionth of a gram). As only very small amounts are required, they could not be important as sources of energy in the body. Studies indicate that most, if not all, of the vitamins function as coenzymes (nonprotein parts of enzymes). In the early studies on the vitamins, retarded growth was reported with a deficiency of each of several of the vitamins. Since most of them are now known to function as parts of essential metabolic enzymes, it is not surprising that growth is affected with a deficiency of almost any of them.

Over 20 vitamins have been described for various kinds of organisms but only about a dozen of them have been demonstrated as vitamins for man. What is a vitamin for one organism may not be a vitamin for another. Thus whereas humans and the guinea pig must be furnished with ascorbic acid (vitamin C) in their diets, most of the other animals are able to synthesize this substance in their bodies from raw materials. *Thiamin* (vitamin B_1) is necessary in the food of a number of organisms, including humans. Williams, who has made many major contributions on the nature and role of thiamin in living organisms, has concluded that this substance must be present in all living cells for normal functioning. This means, then, that many organisms are able to synthesize this substance; some, however, must have it furnished. It now seems reasonable to assume that the same concept may be applied to all the other vitamins—they are essential parts of the chemical machinery of living cells. Thus there is an interesting evolutionary aspect to the study of vitamins. All the primitive organisms are able to synthesize their essential compounds from raw materials. In the process of evolution, different organisms have lost this or that power of synthesis, and accordingly a vitamin need originated. It was possible for this to happen in evolution because the needed substance was present in the food of these organisms.

The term vitamin was first used in 1912 by Funk who first isolated from rice polishings a substance, now known as thiamin, that was effective in curing beriberi in birds. In 1916 McCollum repeated the earlier work of Hopkins on the use of a purified diet of fats, proteins, carbohydrates, minerals, and water for rats. Hopkins had found that rats die when fed this diet and that death could be prevented when the diet was supplemented with milk. McCollum found that there were two substances that must be added to the purified diet to permit growth of rats. One of these he called fat-soluble A and the other water-soluble B. Since his first use of letters to designate these essential accessory food substances, it has been common practice to designate the vitamins with letters. We know today that there are several different substances in each of the A and B factors described by McCollum. In most instances today the vitamins are known chemically and may be synthesized in the laboratory. So more and more the practice is developing of referring to each vitamin by its chemical name. Table 9.1 indicates the major vitamins needed by humans.

In addition to these, the following are also needed.

1. Biotin, which is found in liver, milk, eggs, vegetables, and whole-grain cereals. A deficiency results in scaly and itchy skin.
2. Folic acid (folacin) found in liver and leafy vegetables. A deficiency results in anemia and diarrhea.
3. Pantothenic acid found in most foods. A deficiency results in the abnormal functioning of the adrenal glands.

That serious maladjustments may result from a deficiency of any of the established human vitamins is obvious from the preceding discussion. Such disorders do occur. Also, since there are significant individual differences in requirements, people may take in suboptimal amounts of certain vitamins without any marked symptoms. This difference in nutritional requirements among individuals is a good example of *biochemical individuality.* In general the person who eats well-balanced meals is not likely to suffer from lack of vitamins. Evidence has accumulated in recent years which indicates that heavy doses of vitamin C may aid in the prevention of the common cold. However, instead of buying vitamin pills at the suggestion of the common vitamin advertisements, it is much better to consult a physician first about any suspected vitamin deficiency.

Rats are often used to demonstrate vitamin deficiencies. The effects of deficiencies in vitamin A, thiamin, and riboflavin are shown in Figures 9.5, 9.6, and 9.7.

DIGESTION

Digestion is primarily a chemical process, a process of *hydrolysis* in which complex molecules are split into simpler molecules by the action of digestive enzymes known collectively as *hydrolases.* These enzymes, or their precursors, are present in the secretions of the different digestive glands—the salivary glands, the

Table 9.1 Chief Vitamins

VITAMINS	CELL FUNCTION	SOURCES	DEFICIENCY SYMPTOMS IN HUMANS
WATER SOLUBLE			
B_1, thiamin	Coenzyme involved in decarboxylation of pyruvic acid (Ch. 5)	Whole cereals, yeast, lean meat	Beriberi (degeneration of nerves, muscle atrophy)
B_2, riboflavin	Part of cofactor for flavoprotein enzymes which function in hydrogen transport (Ch. 5)	Yeast, meat, cheese, milk, leafy vegetables	Sores on lips, inflammation of eyes
B_6, pyridoxal	Cofactor involved in amino acid synthesis	Cereal grains	Skin lesions and disturbance of central nervous system
B_{12}, cyanocobalamin	Coenzyme involved in nucleic acid metabolism	Liver	Pernicious anemia
Nicotinic acid (nicotinamide, niacin)	Part of NAD^+, cofactor for dehydrogenases	Yeast, meat, wheat	Pellagra (skin inflammation, diarrhea, mental instability)
C, ascorbic acid	Cofactor in hydroxylation reactions, collagen synthesis	Citrus fruits, tomatoes, peppers	Scurvy (bleeding gums, swollen joints)
LIPID SOLUBLE			
A, retinol	Part of rhodopsin, light receptor, pigment in eye	Green and yellow vegetables, liver	Scaly, dry skin night blindness
D, calciferol	Regulation of calcium and phosphorus absorption and utilization*	Fish oils, liver, milk, egg yolk	Rickets in children (abnormal formation of bones and teeth, bowed legs)
E, tocopherol	Prevents abnormal attack of oxygen on lipids	Vegetable oil, wheat germ	Destruction of red blood cells
K, phylloquinone	Prothrombin synthesis	Most higher plants	Faulty blood coagulation

* One form of vitamin D (1, 25-dihydroxycholecalciferol) is now known to be the active form involved in Ca^{++} transport by the intestine and is formed in the kidneys.

gastric glands, the pancreas, and the intestinal glands. Each digestive enzyme works most effectively at a certain pH. The nature of hydrolysis in digestion can be well illustrated by the change that occurs in the disaccharide maltose in the presence of the enzyme maltase found in the intestinal juice. This action is represented as follows:

$$\underset{\text{maltose}}{C_{12}H_{22}O_{11}} + H_2O \xrightarrow{\text{maltase}} \underset{\text{glucose}}{2C_6H_{12}O_6}$$

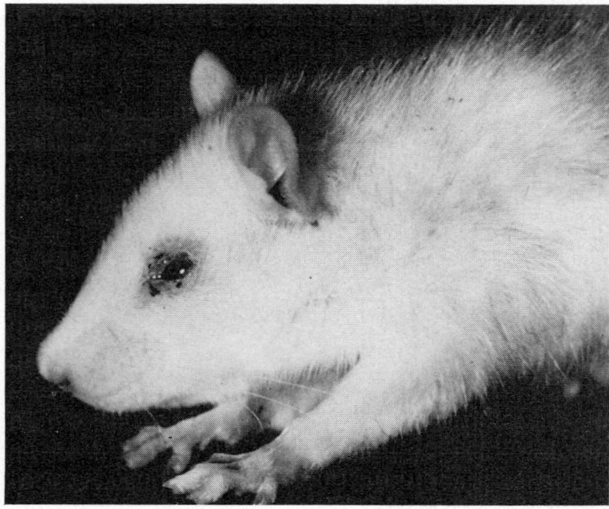

(a)

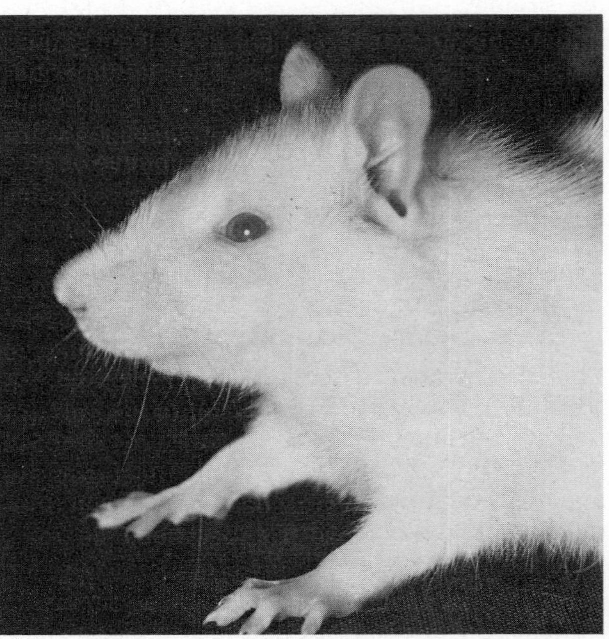

(b)

Figure 9.5 Vitamin A deficiency. **(a)** White rat showing dryness of the cornea and conjunctiva of eye due to lack of adequate vitamin A. **(b)** The same rat after 8 days of vitamin A therapy. (Courtesy of the Upjohn Company, Kalamazoo, Michigan.)

(a)

(b)

Figure 9.6 Thiamin deficiency. **(a)** A thiamin-deficient rat shows the typical arched back and hyperextended hind legs. Such rats have an unsteady gait, turn awkwardly, and lose balance. **(b)** The same rat, 8 hours after receiving thiamin, has normal use of its hind legs and normal balance. (Courtesy of the Upjohn Company, Kalamazoo, Michigan.)

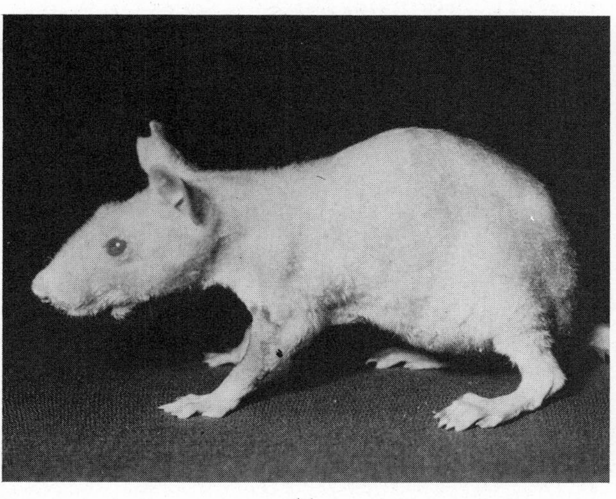

(a)

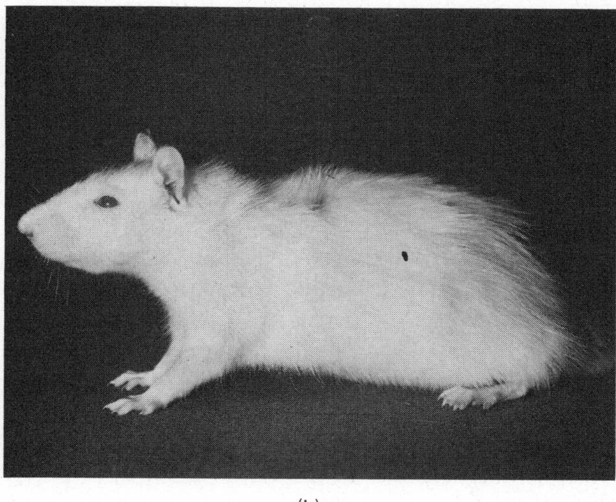

(b)

Figure 9.7 Riboflavin deficiency. **(a)** A riboflavin-deficient rat shows marked changes in the skin and skin lesions. **(b)** After two months of treatment with riboflavin the same rat shows no signs of the original deficiency. (Courtesy of the Upjohn Company, Kalamazoo, Michigan.)

The molecule of maltose is united with a molecule of water to form two simpler molecules of glucose. Glucose is a simple sugar that can be absorbed in the intestine and is an *end product* of carbohydrate digestion. In all digestion, whether it be carbohydrate, fat, or protein digestion, hydrolysis is involved.

Digestion in the Mouth

When food is taken into the mouth, the secretion of *saliva* from the *salivary glands* starts. The saliva contains a large amount of water, the enzyme *salivary amylase* (often called *ptyalin*), and mucus. The mucus helps to hold the food mass together and, as a lubricant, facilitates its passage down the esophagus when swallowing occurs. The salivary secretion is alkaline. Salivary amylase hydrolyzes or splits starches into disaccharides, chiefly maltose. This may be represented by the following overall equation:

$$\underset{\text{starch}}{(C_6H_{10}O_5)_{2n}} + nH_2O \xrightarrow{\text{salivary amylase}} \underset{\text{maltose}}{n(C_{12}H_{22}O_{11})}$$

While in the mouth, digestion of food is initiated by the addition of salivary amylase. The action continues in the food mass as it lies in the stomach until the process is stopped by the action of the stomach secretions, which are acid.

No digestive juices are secreted into the esophagus. The food passes very quickly through this part of the alimentary tract.

Digestion in the Stomach

The gastric glands in the wall of the stomach secrete gastric juice containing two inactive proteins, *prorennin* and *pepsinogen.* Hydrochloric acid, also secreted by the stomach glands, activates or converts these substances to the enzymes *rennin* and *pepsin.* Rennin acts on a protein in milk, *casein,* and, in combination with calcium, forms a curd. As a result of this curdling, the milk proteins remain in the stomach long enough for pepsin to act on them. Pepsin acts only in an acid medium and it starts the process of protein digestion. Pepsin hydrolyzes complex protein molecules. Recent studies have shown that the enzyme is not limited to activity on large molecules but it may act on comparatively simple *peptides,* depending on the chemical structure of the latter. However, pepsin normally hydrolyzes proteins to large fragments, called *polypeptides,* in the stomach. Although pepsin is capable of carrying digestion of proteins to simpler substances, it probably does not do so in the stomach due to the movement of the food mass to the small intestine where the activity of pepsin stops because of the alkaline pH. *Amino acids,* the end products of protein digestion, are not found in large amounts in the stomach. Protein digestion is completed by the action of several enzymes in the small intestine.

Because of the chemical structure of proteins, the initial stages of protein digestion might be described as actions of the enzymes on the "backbone" of the protein molecules. Such enzymes are called *endopeptidases.* Later stages of digestion, to be described in the following paragraphs, would then be a "nibbling" by enzymatic action on the ends rather than on the insides of the molecules. Enzymes involved in this activity are called *exopeptidases.* Enzymes have specific roles and all are not capable of breaking the same bonds.

Why does the pepsin not digest the lining of the stomach, which is primarily composed of proteins? First, because some of the gastric glands keep the lining of the stomach covered with a layer of mucus which is a protection. Second, the hydrochloric acid and pepsinogen are secreted by different gastric glands, and the two substances are mixed in the lumen, away from the lining. Occasionally these protective devices do not work and a portion of the stomach is digested, resulting in a peptic ulcer.

Digestion in the Small Intestine

The digestive process is completed in the small intestine through the action of enzymes in the pancreatic fluid, which is emptied into the duodenum, and through the action of enzymes in the intestinal fluid, which is produced by glands in the wall of the small intestine.

The pancreatic secretion is a complete digestive juice because it contains carbohydrate-splitting, lipid-splitting, nucleic acid-splitting, and protein-splitting enzymes.

The carbohydrate-splitting enzyme from the pancreas is *pancreatic amylase,* and it acts upon starch and glycogen in a similar but even more effective way than salivary amylase, completing the conversion of starch into maltose.

The lipid-splitting enzyme is *pancreatic lipase.* Although a little lipase is found in both the gastric juice and the intestinal juice, almost all fat digestion in humans is mediated by pancreatic lipase. Each lipid molecule acted upon is hydrolyzed into a molecule of *glycerol* and three molecules of free *fatty acid.* These products are ready for absorption. The work of lipase is facilitated, as previously mentioned, by the action of the *bile,* which breaks up the large globules of fat into very small ones, providing much more surface for the enzyme to act upon. This action of the bile salts, called *emulsification,* is also instrumental in facilitating the absorption of fatty acids.

The nucleic acid-splitting enzymes, *ribonuclease* and *deoxyribonuclease,* act upon ribonucleic acids and deoxyribonucleic acids to give nucleotides, which were discussed earlier in Chapter 2.

The protein-splitting enzymes of the pancreatic juice are trypsin and chymotrypsin. These, like pepsin, are endopeptidases; they break peptide bonds within the protein molecules. Trypsin is secreted in an inactive form, *trypsinogen,* which is converted into active trypsin by *enterokinase,* an enzyme present in the intestinal fluid. Chymotrypsin is also secreted in an inactive form, *chymotrypsinogen,* which is converted into the active enzyme by the action of trypsin. Both trypsin and chymotrypsin continue the breakdown of large protein molecules begun by pepsin. Each of these endopeptidases is specific and acts on peptide bonds adjacent to particular amino acids. Trypsin and chymotrypsin and fourteen other enzymes constitute the family of serine proteases. These enzymes are very similar in structure and they apparently evolved from a common ancestral enzyme. The interested student should read the article by Stroud (see Suggestions for Further Reading) in which he discusses their structure and relationship and their mode of action. The enzyme *carboxypeptidase,* also in the pancreatic juice, is an exopeptidase.

This enzyme removes amino acids located at the acid or carboxyl (COOH) end of peptide molecules.

The intestinal fluid contains, in addition to enterokinase just mentioned, several other enzymes that are necessary to complete the digestion of food to simple, absorbable substances. Several of these are involved in the completion of protein digestion. *Aminopeptidase* is another exopeptidase. It acts in a manner similar to carboxypeptidase by removing amino acids from the amino (NH_2) end of peptide molecules instead of the carboxyl end. In addition the intestinal fluid contains *tripeptidases* and *dipeptidases* which hydrolyze tri- and dipeptides into amino acids. These end products of protein digestion are now ready to be absorbed. *Maltase, sucrase,* and *lactase,* which are present in the intestinal fluid, complete carbohydrate digestion by hydrolyzing the disaccharide sugars maltose, sucrose, and lactose to glucose and other simple sugars. Table 9.2 summarizes the process of digestion.

Control of Digestive Secretions

Why is saliva secreted into the mouth when food is eaten? Why do the gastric glands secrete when food enters the stomach? And why does the pancreatic fluid pour into the duodenum when chyme from the stomach enters it? These questions puzzled interested people for a long time; now as a result of much research and observation these and other related questions can be answered.

It has been demonstrated that the flow of saliva is entirely dependent upon a *reflex* mechanism involving the nervous system. This involves the *taste buds* on the tongue as receptors, sensory nerves leading to the brain, motor nerves leading to the salivary glands, and the salivary glands. That this is the case may be demonstrated by cutting the motor nerves to the salivary glands, after which there is no secretion.

Gastric secretion involves both nervous and hormonal control. The gastric glands begin secreting at the start of a meal and continue until after the stomach is emptied. The Russian Pavlov did much to explain the mechanisms of gastric secretion. In one experiment he cut the esophagus of a dog and left the cut end open to the outside. When he fed this dog, the food, of course, never reached the stomach, yet he was able to determine that such a "sham feeding" resulted in the secretion of about one fourth the normal amount of gastric juice. Cutting of the gastric nerves proved that this gastric secretion was under reflex control.

In another experiment Pavlov isolated a small part of the stomach and fastened it to the body wall with an opening to the outside. Through this opening he could introduce food into the stomach pouch without the dog ever seeing, smelling, or tasting the food. When this was done, about three fourths the normal amount of gastric juice was secreted. Cutting the nerves to the stomach did not stop this secretion. The conclusion was that contact of the food with the lining of the stomach causes the cells to secrete into the bloodstream a *hormone* that is carried to the gastric glands to stimulate the secretion. This hormone has been called *gastrin.* Final proof of this mechanism was established in an experiment where the blood supply of two dogs was connected by joining their carotid arteries with rubber tubing. One dog was fed and the other was not. After a short interval, however, the dog that was not fed began to secrete gastric

Table 9.2 Digestion

REGION	SECRETION	ENZYMES	SUBSTRATE ACTED UPON	PRODUCTS
Mouth	Saliva	Salivary amylase	Starches	Maltose
Stomach	Gastric juice	Pepsin	Proteins	Polypeptides
		Rennin	Milk proteins	Curd
Small intestine	Pancreatic juice	Pancreatic amylase	Starches	Maltose
		Lipase	Lipids	Fatty acids and glycerol
		Trypsin	Proteins	Polypeptides
		Chymotrypsin	Proteins	Polypeptides
		Carboxypeptidase	Peptides	Amino acids
		Ribonuclease	Ribonucleic acids	Nucleotides
		Deoxyribonuclease	Deoxyribonucleic acids	Nucleotides
	Intestinal juice	Aminopeptidase	Peptides	Amino acids
		Tripeptidases	Tripeptides	Amino acids
		Maltase	Maltose	Glucose
		Lactase	Lactose	Glucose and galactose
		Sucrase	Sucrose	Glucose and fructose

juice. A *hormone* is a substance that is produced in some part of the body, passes into the bloodstream, and is then carried throughout the body where it produces specific localized effects. This type of experimentation, where the bloodstreams of two organisms are connected, is often used in making tests for suspected hormones.

It has been possible to demonstrate that the flow of both the pancreatic juice and the bile is under hormonal control. When the acidic chyme comes into contact with the cells lining the duodenum, these cells are stimulated to secrete two different hormones. One, *secretin,* is carried by the blood all over the body but functions only to stimulate the pancreas to secrete pancreatic juice. The other, *cholecystokinin–pancreozymin,*[1] stimulates the gall bladder to liberate bile.

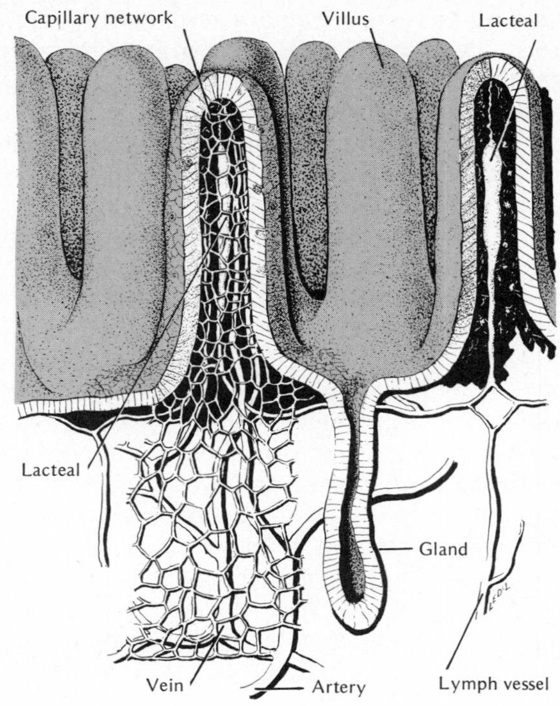

Figure 9.8 Villi.

ABSORPTION AND THE UTILIZATION OF THE END PRODUCTS OF DIGESTION

Absorption

Absorption is the passage of simple sugars, fatty acids, glycerol, amino acids, vitamins, minerals, water, and certain other substances into the circulatory system. This process occurs primarily in the small intestine. The lining of the small intestine, which is the absorbing surface, is a columnar epithelium. It is not smooth but has numerous microscopic fingerlike projections called *villi,* which greatly increase the absorptive surface (see Fig. 9.8). In addition, electron micrographs show that this surface is further increased by numerous *microvilli* (see Fig. 4.6). It has been estimated that as a result of both villi and microvilli the total surface area of the intestinal lining is five times the surface area of the exterior of the body.

Within each villus is a network of blood capillaries and a central lymph vessel, the *lacteal.* In absorption the materials involved pass from the lumen of the intestine through the epithelium and enter either the capillaries or lacteal vessels. Although some substances may pass from the lumen to the circulatory system by simple diffusion, more and more evidence is accumulating that much of the absorption involves work on the part of the epithelial cells; that is, it involves *active transport.* Some substances are absorbed against a diffusion gradient. In addition, many small molecules are absorbed more slowly than larger molecules.

The simple sugars, amino acids, vitamins, minerals, and water enter the blood capillaries in the villi. The absorption of fat and of the end products of fat digestion is different. Only about one third of the ingested fat is hydrolyzed by lipase to glycerol and fatty acids. This hydrolyzed fat, aided by the bile salts and calcium ions, emulsifies the remainder of the fat into tiny globules. There are now indications that these tiny globules of unhydrolyzed fat are taken into the epithelial cells by pinocytosis (Ch. 4). Glycerol, being water soluble, is easily absorbed, but the fatty acids form a complex with bile salts before they are absorbed. It has been demonstrated that the intestinal mucosa contains an enzyme that recombines long-chain fatty acids (those with more than twelve carbon atoms) with glycerol to form lipid. This lipid together with that which enters the mucosal cells without digestion passes into the lacteal vessels. The short-chain fatty acids and some phosphorylated long-chain fatty acids enter the capillaries. Vitamins A, D, and K are absorbed along with fats. The lacteal vessels have very small perforations in their walls, and thus the tiny fat globules are

[1] Cholecystokinin was first described in 1928. In 1947 a second hormone, in addition to secretin, which stimulates the secretion of pancreatic juice, was found in the intestinal mucosa. This was called *pancreozymin.* Recent studies of purified materials indicate that a single polypeptide possesses the actions of both cholecystokinin and pancreozymin.

able to enter them. The fat in the lacteal vessels moves into larger lymph vessels and ultimately into the large *thoracic duct,* which empties into the left subclavian vein. The villi are constantly moving up and down in a kind of pumping action, and this is the initial force in the movement of the absorbed fat in the lymph vessels. Other factors in this movement will be described in the chapter on circulation. After a lipid-rich meal the lacteals have a milky appearance because of the large amount of lipids in them. The fatty acids that enter the capillaries pass into the portal bloodstream (Ch. 11) and into the liver. Although there are two routes taken by the absorbed lipids and fatty acids, they are finally mixed in the bloodstream.

The Utilization of Simple Sugars

In addition to glucose, fructose, and galactose, other simple sugars are absorbed from the intestine. In the liver, fructose and galactose are converted into glucose and most of the glucose is changed into *glycogen,* or animal starch, for storage. The blood normally contains about 0.10 percent glucose. As glucose is taken from the blood by the cells throughout the body, the supply must be maintained. The liver, through hormonal control, is constantly changing glycogen back to glucose for the maintenance of this normal blood-sugar level. As the glucose passes through the body, the cells of certain tissues take it from the blood and synthesize it into fat for storage. But the greatest use and the most important use of glucose in the body is for oxidation. When glucose is oxidized in the cell, carbon dioxide and water are formed and energy is released. This process was described earlier as cellular respiration. The oxidation of glucose is the main source of energy for the vital activities of cells.

Glycogen is also formed and stored in the muscles. Such glycogen, however, is used in the muscles for energy and is not ordinarily returned to the blood.

The Utilization of Lipids

The absorbed lipids are acted on by special cells in the liver and are altered so that they may be stored in the fatty tissues. The stored lipids are available as a source of energy. The liver partially breaks down lipids so that they can be oxidized by other cells in the body. In addition to serving as sources of energy, the stored lipids serve as cushions for some of the internal organs and also as insulating material. Phospholipids, derived from fats, are used in the formation of plasma membranes.

The Utilization of Amino Acids

The primary use of amino acids in the body is in the formation and repair of cells. When the level of amino acids in the blood is greater than that needed to maintain the cells, the excess amino acids are deaminated in the liver with the formation of urea and simple organic acids. These organic acids may be converted into glucose and then oxidized by the cells of the body, or they may be converted into glycogen or lipid for storage.

THE BASAL METABOLIC RATE

The utilization of the various end products of digestion described above constitutes *metabolism.* A part of metabolism is destructive—all of the oxidations are—and this part is called *catabolism.* The building-up processes of metabolism, such as the synthesis of cell parts and of stored materials, is referred to as *anabolism.*

The *basal metabolic rate* is a measure of the amount of energy expended by the body just to maintain a person's vital functions when he is completely at rest and not digesting any food. The large calorie, which is the amount of heat required to raise the temperature of 1 kg of water from 15° to 16° Celsius at a pressure of 1 atmosphere, is used to express the energy requirements of the body. Although the needs of individuals vary, the average person, not engaged in hard labor, requires about 2500 cal daily. Of this, about 1500 to 1800 cal are needed just to maintain the vital functions. Carbohydrates, lipids, and proteins may all serve as sources of the energy. Carbohydrates and proteins, when oxidized in the body, each yield about 4 cal/g, whereas fats yield about 9 cal/g. The average person obtains about two thirds of his energy from the oxidation of carbohydrates.

The determination of basal metabolic rates is common hospital practice today because of its bearing on certain diseases. A person taking the test is kept without food for several hours and, while completely relaxed and resting, his oxygen consumption is determined over a short period of time. The release of energy and the production of heat depend upon the use of oxygen in the oxidation of food materials. Thus the amount of heat produced can be calculated from the amount of oxygen used in a given period.

SUGGESTIONS FOR FURTHER READING

Bayless, W. M., and E. H. Starling. "The Mechanism of Pancreatic Secretion," in M. L. Gabriel and S. Fogel, Eds. *Great Experiments in Biology.* Englewood Cliffs, N.J.: Prentice-Hall (1955).

Best, C. H., and N. G. Taylor. *The Living Body,* 5th ed. New York: Holt, Rinehart and Winston (1968).

Gordon, M. S. *Animal Physiology.* Riverside, N.J.: Macmillan (1972).

Grollman, S. *The Human Body,* 2nd ed. Riverside, N.J.: Macmillan (1969).

Guyton, A. C. *The Function of the Human Body.* Philadelphia: W. B. Saunders (1969).

Wessels, N. K. *Vertebrate Adaptations.* San Francisco: W. H. Freeman (1969).

Williams, R. J. *Biochemical Individuality: The Basis for the Genetrophic Concept.* New York: Wiley (1956).

Wood, D. W. *Principles of Animal Physiology.* New York: American Elsevier (1972).

From Scientific American

Stroud, R. M. "A Family of Protein-Cutting Proteins" (June, 1974).

Young, V. R., and N. S. Scrimshaw. "The Physiology of Starvation" (Oct. 1971).

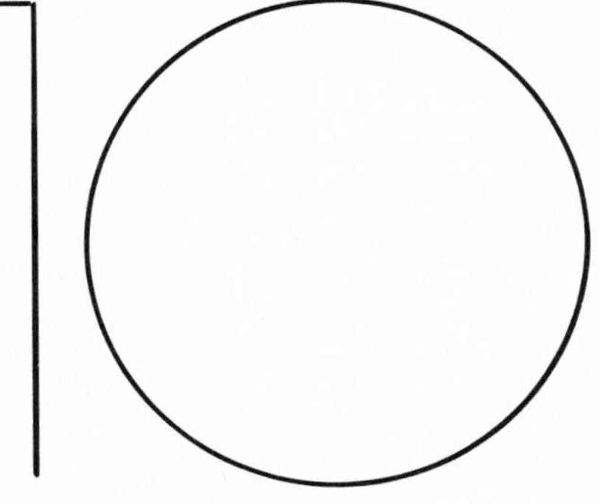

Exchange of Gases and Elimination of Metabolic Wastes

RESPIRATION

The process of *cellular respiration,* which may be anaerobic or aerobic, was discussed in Chapter 5. This process occurs in all living cells. Historically, however, the term respiration has been used to mean different things. Originally it was used to denote breathing, and it is still used that way by many. Breathing, however, is only accessory to the basic process of cellular respiration. The term has also been used to refer to the exchange of gases between the cell and its environment. So in a very general sense the term respiration may be used to include all of the processes by which the cells of organisms receive oxygen and give off carbon dioxide, as well as those processes that take place within the cells resulting in the release of energy and the formation of ATP.

In single-celled animals and simple multicellular animals, the process of the exchange of gases between the cells and their environment is just one of diffusion.

In complex animals, many of the cells are so far away from the source of oxygen that diffusion alone will not suffice and so-called respiratory systems have evolved. In many forms, gills serve in such a capacity; in insects, tracheal tubes are found; in the higher vertebrates, lungs and associated structures are present.

In considering the overall process of respiration in humans, several steps will be recognized:

1. *Breathing,* the movement of air into and out of the lungs
2. *External respiration,* the exchange of oxygen and carbon dioxide between the blood and the air
3. *Transportation,* the carrying of oxygen and carbon dioxide by the blood to and from the cells of the body
4. *Internal respiration,* the exchange of oxygen and carbon dioxide between the blood and the cells of the body

All these steps are just accessory to the basic processes of *cellular respiration.*

THE HUMAN RESPIRATORY SYSTEM

Normally air enters the human system by way of the *nostrils,* but it also enters by way of the mouth. The nostrils lead into the *nasal cavities,* and these chambers are separated from the mouth below by the palate. The surfaces of the nasal cavities are greatly increased by bony folds, called *turbinates.* The cavities are lined with a *mucous membrane* or *epithelium.* As air passes through the cavities, it is warmed and also filtered. The filtering is accomplished by the mucus that is secreted and, also, by long hairs that are present in the anterior parts of the cavities.

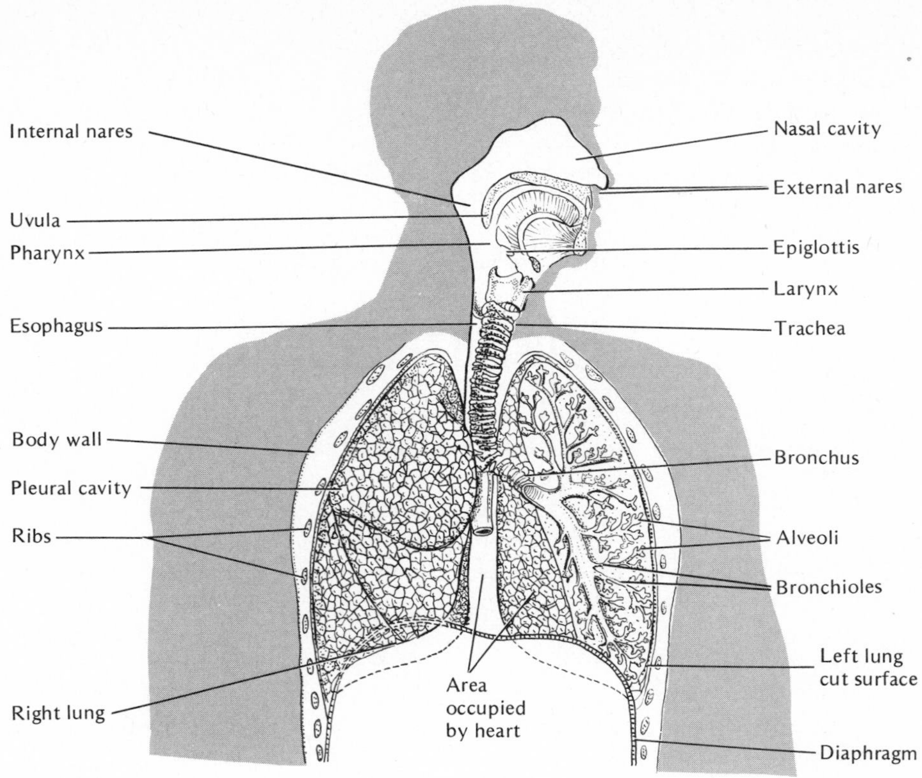

Figure 10.1 Human respiratory system.

Air passes from the nasal cavities into the *pharynx,* through the *glottis,* and into the *larynx.* This structure is often called the "Adam's apple" and is more prominent in men than in women. The *vocal cords* are stretched across the larynx. The glottis, the opening to the larynx, is always open except in swallowing. In that process a flaplike structure, the *epiglottis,* covers the glottis. A long cylindrical tube, the *trachea,* leads from the larynx to the chest region. It is lined with ciliated epithelum. In the middle of the chest region the trachea bifurcates into the *right* and *left bronchi,* leading to the *right* and *left* lungs (Fig. 10.1).

In the lungs each bronchus branches in a treelike manner, forming smaller and smaller tubes, the *bronchioles.* The smaller bronchioles terminate in clusters of cup-shaped cavities, the *alveoli* (Fig. 10.2). It is in the alveoli that the gaseous exchanges occur. The alveoli are lined with a layer of flat epithelial cells and surrounded by networks of blood capillaries. A film of lecithin lines the alveoli of normal lungs; it functions to lower the surface tension and enables the alveoli to remain open. In some newborn babies, and especially in premature ones, this film of lecithin may not be present and as a result the lungs collapse—25,000 die each year from this. This problem is now being attacked by engineers and physicians working together. In one approach the oxygen tension in the arterial blood is monitored until the baby begins to manufacture its own lecithin. In another approach the baby inhales synthetic lecithin aerosol (see Suggestions

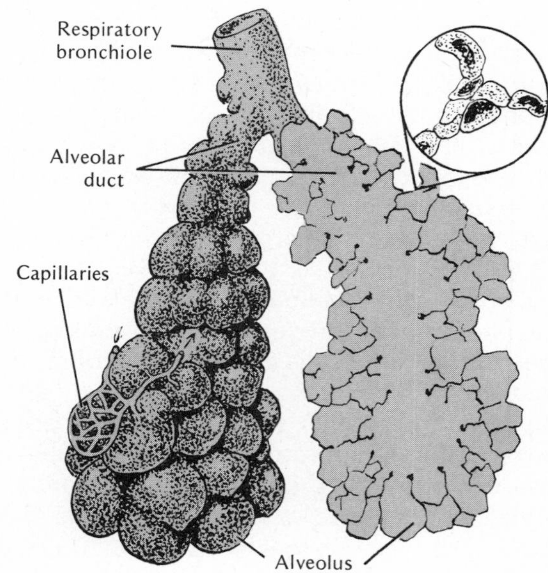

Figure 10.2 Alveoli. **(Left)** Capillaries that are associated with all alveoli are drawn on only one alveolus. **(Right)** Longitudinal section through a cluster of alveoli. **(Magnification)** Some cells of the alveolar wall.

for Further Reading). The parts of the human lung are bound together by connective tissue forming a compact but spongy organ.

The walls of the chest region contain the *ribs* and the rib or *intercostal muscles.* Below, the chest region is separated from the abdominal cavity by the *muscular diaphragm.* Above the diaphragm there are three closed cavities, the two lateral *pleural cavities* that house the lungs, and the median *pericardial cavity* that contains the heart. Externally each lung is covered by a membrane called the *visceral pleura.* The pleura continues, from the point of attachment of each lung, as the lining of each pleural cavity and is called the *parietal pleura.* The two layers of the pleura are normally in contact, and a small amount of lymph is secreted into the cavities to keep the layers of pleura moist and to prevent friction in the breathing movements. When the pleurae are inflamed, with the production of a large amount of fluid in the cavities, the painful condition known as *pleurisy* results.

Breathing

In humans the lungs are spongy and elastic. They are located in airtight cavities, the pleural cavities. The functional parts of the lungs from the standpoint of gaseous exchanges, the alveoli, are connected with the outside air at all times, that is, the nares, nasal chambers, pharynx, glottis, larynx, trachea, bronchi, and bronchioles form a continuous system with the outside and no part of it is ever collapsed. In the trachea there are horseshoe-shaped cartilages in the wall, which keep it open, and in the bronchi and larger bronchioles rings of cartilage function in the same way. So in such a system, when the cavities surrounding the lungs are enlarged, air rushes in. Due to the pressure of the outside air and the elastic nature of the lungs, the lungs will expand to fill the pleural cavities. That the human breathing mechanism is essentially a suction-pump mechanism can be easily demonstrated, as indicated in Figure 10.3. If a rubber balloon[1] is fastened to a glass tube inside a bell jar and a rubber membrane is fastened tightly across the opening of the bell jar, a situation is created simulating a lung in a pleural cavity the bottom of which is the diaphragm. When the rubber membrane is pulled down, increasing the cavity surrounding the balloon, air rushes into the balloon, inflating it. When the rubber membrane is allowed to go back to its original position, air leaves the balloon.

Whereas this illustrates the mechanics of breathing in humans, it does not tell the whole

[1] A lung from a freshly killed sheep or steer may be used very effectively.

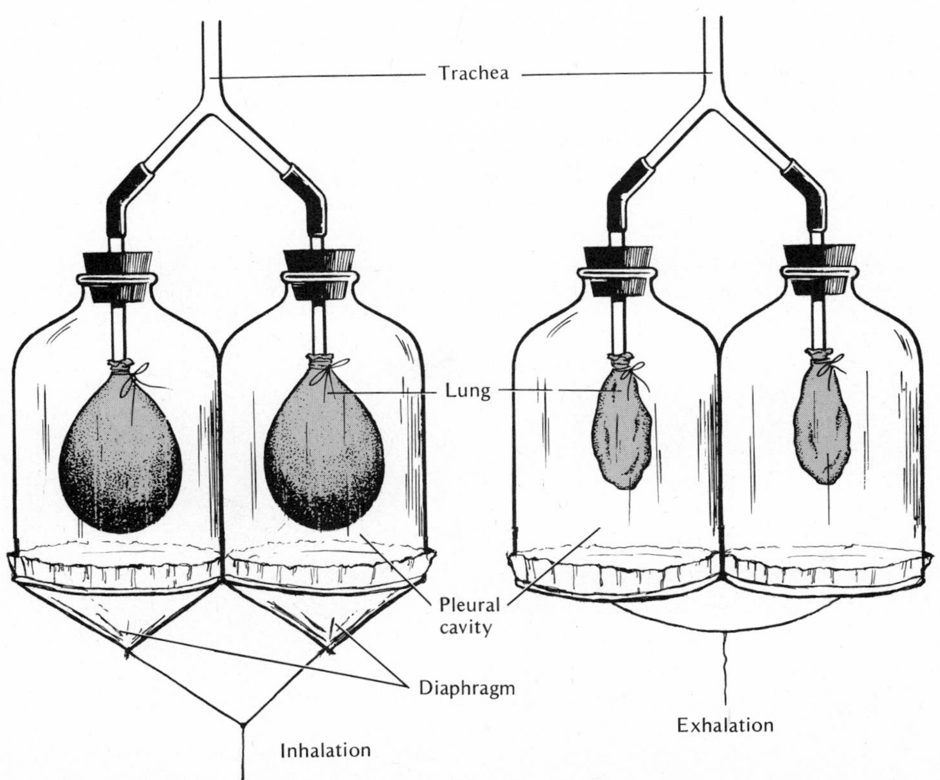

Figure 10.3 Model, showing simulating effects of movements of diaphragm.

story. The rubber membrane illustrated in Figure 10.3 simulates the muscular diaphragm in humans. In inspiration, the muscles of the diaphragm contract, changing it from a dome-shaped structure to a more flattened sheet that presses down on the viscera. In expiration, the muscles relax and the diaphragm assumes its dome-shaped position. These movements of the diaphragm change the vertical dimensions of the pleural cavities. In addition, the pleural cavities also have their front-to-back and side-to-side dimensions changed during breathing. During inspiration, contractions of the intercostal muscles cause the ribs to rise and turn outward. The rise of the ribs causes the front-to-back diameter of the pleural cavities to increase, and the turning out of the ribs causes the side-to-side diameter to increase. In expiration, the ribs return to their former positions, decreasing these diameters of the cavities and allowing the elastic recoil of the lungs to force the air out (Fig. 10.4).

CONTROL OF BREATHING

At rest, breathing occurs about 18 times per minute and we are usually unconscious of the process. Yet we can hold our breath or accelerate the movements in forced breathing, so the process is at least in part under conscious control. But we cannot willfully stop the process for very long; the breathing movements automatically set in again. We know that exercise results in an increase in the breathing movements. What is the nature of the control of this basic process?

If the phrenic nerves that innervate the diaphragm and the intercostal nerves going to the intercostal muscles are cut, all breathing movements cease. It can also be shown that the impulses that pass over these nerves to the breathing muscles originate in a group of nerve cells, in the medulla of the brain, called the *respiratory center.* If this center in the medulla is destroyed, all breathing movements stop. This demonstrates nervous control, but what is responsible for the normal rhythm of breathing and the change in the periodicity of these impulses during exercise?

As mentioned in the preceding discussion, the impulses that pass to the muscles of the diaphragm and the intercostal muscles, resulting in inspiration, originate in the respiratory center. It has been demonstrated that sensory nerves in the lungs are stimulated by the stretching that occurs during inhalation, and as a result, send inhibitory impulses via the vagus nerve into the respiratory center. As a consequence, the motor impulses to the breathing muscles are temporarily halted. When the lungs return to their initial size, the inhibitory impulses cease. The motor impulses can now start again and the cycle is repeated. This explains the normal rhythm of

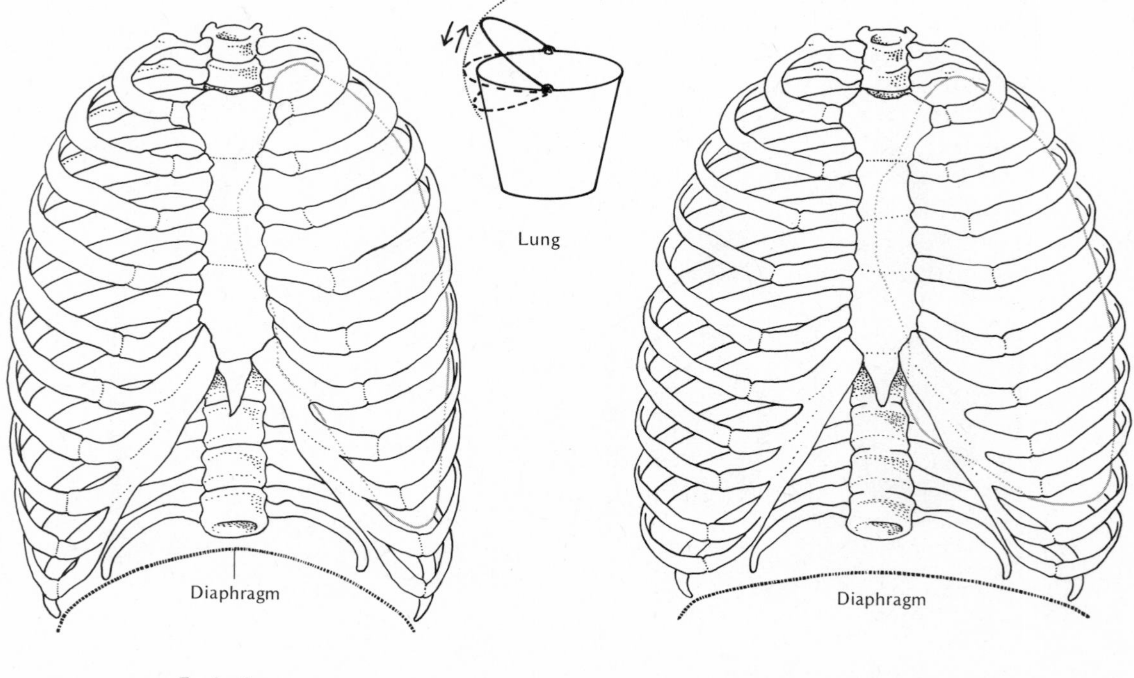

Figure 10.4 Movement of the ribs in breathing. Each rib is attached to the vertebral column in back and to the breastbone in front. Like the handle of a bucket, each rib moves outward as it is pulled up. In addition, the whole breastbone moves up slightly during an inspiration, so that the distance from vertebral column to breastbone is also increased. The downward movement of the diaphragm increases the vertical dimension of the chest cavity.

breathing, but what causes the breathing movements to speed up during exercise?

It has been demonstrated that changes in the carbon dioxide concentration of the blood are responsible for the changes in breathing movements during exercise. During strenuous exercise carbon dioxide is produced in the muscles faster than the lungs can expel it. This increase of carbon dioxide stimulates the respiratory center to send more frequent impulses to the breathing muscles. When the excess carbon dioxide is removed by rapid breathing and there is less carbon dioxide in the blood, the rate of breathing decreases. That this is so may be demonstrated in a variety of ways. If an experimental animal with its head in a closed chamber breathes the same air over and over, breathing movements soon increase. Here the amount of oxygen in the container is decreasing while the carbon dioxide is increasing. If the same animal, with its head in a closed chamber but with an arrangement for absorbing all the carbon dioxide in the air, breathes the same air over and over again, the breathing movements remain the same. In this the amount of oxygen is decreasing while the amount of carbon dioxide remains the same. Again, if the experimental animal is caused to breath an atmosphere containing the normal amount of oxygen but a high amount of carbon dioxide, breathing movements will accelerate immediately. So it is the carbon dioxide, a product of cellular oxidations, that acts on the respiratory center in the normal regulation of breathing movements.

Under certain conditions a lack of oxygen may be the factor responsible for faster breathing movements. An individual flying at a high altitude in a nonpressurized plane will breathe faster than at sea level. This is independent of exercise. As we go up from sea level, the amount of oxygen available in the air decreases. Such lowered oxygen tension indirectly produces increased breathing movements. With an inadequate supply of oxygen, the cells of the body are unable to complete all the normal metabolic oxidations. As a result, there is an accumulation of the products of incomplete oxidations, chiefly lactic acid. The two small *carotid bodies,* one of which is found in the wall of each carotid artery, are quite sensitive to lactic acid. When they are stimulated by the excess lactic acid that accumulates during an oxygen deficiency, they, in turn, stimulate fibers of the ninth cranial nerves, which run from the carotid bodies to the respiratory center. As a result the respiratory center sends out more frequent impulses to the breathing muscles.

THE CAPACITY OF THE LUNGS

In quiet breathing about 500 cubic centimeters (cm^3) of air are taken in and out of the lungs with each inspiration and expiration. This is known as *tidal air.* In the strongest forced inspiration an additional 2500 cm^3 of air may be taken in. This is known as *complemental air.* After a normal expiration, one may forcefully expire another 1000 cm^3 of air known as *supplemental air.* The tidal air, the complemental air, and the supplemental air (about 4000 cm^3) are the *vital capacity.* After a forced expiration there remains in the lungs about 1200 cm^3 of air that cannot be removed. This is the *residual air.* The residual air plus the vital capacity gives the *total lung capacity,* or about 5200 cm^3. Thus there is a great reserve of air in the lungs. When a lung is removed from the body, it does not collapse completely and, when placed in water, it will float. This makes it possible to tell whether a dead infant has been stillborn or born alive. A lung of a stillborn child will sink, whereas a lung from a child born alive will float.

COMPARISON OF INSPIRED AND EXPIRED AIR

The chief constituents of atmospheric air are nitrogen, oxygen, and carbon dioxide. An analysis of atmospheric and expired air shows that there is practically no change in the nitrogen content during breathing, but there is a decrease of oxygen and an increase of carbon dioxide (Table 10.1).

Table 10.1 Comparison of Atmospheric and Expired Air

	OXYGEN	CARBON DIOXIDE	NITROGEN
	(PERCENT BY VOLUME)		
Inspired air (dried)	20.96	0.04	79
Expired air (dried)	16.02	4.38	79+
Change	4.94 (loss)	4.34 (gain)	0+

In addition to the changes in oxygen and carbon dioxide, there is also a change in water content. Regardless of the moisture content of atmospheric air, expired air is always saturated with moisture. Also, expired air is always warm, having approximately body temperature, 37° C (98°–100° F).

EFFECTS OF HIGH ALTITUDES

Mammalian lungs evolved to function at or near sea level, rather than at 7000 or 8000 m (4 or 5 miles) above sea level. At sea level the atmospheric pressure is about 760 mm of mercury and the partial pressure of the oxygen in it is about 150 mm. Under these conditions the blood, in passing through the lungs, becomes practically saturated with oxygen. As one goes up from the surface of the earth, the atmospheric pressure decreases and the partial pressure of oxygen decreases. At 3000 m (9000 feet) the atmospheric pressure is reduced to 550 mm. At 4000 m (14,000 feet) it is about 450 mm and at 5500 m (18,000 feet) elevation the oxygen available in the lungs falls to below half of that at sea level. Under such conditions the blood passing through the lungs is unable to take up enough oxygen to supply the cells of the body, and fainting or impaired function may result. The effects of lowered oxygen tension with increasing elevation begin to be felt between 2500 and 3000 m (8000–10,000 feet) elevation. At 4000 m (14,000 feet) and above the effects are pronounced. It is possible to prevent these effects in mountain climbing and in airplane travel by breathing essentially pure oxygen. In modern planes it is now the practice to use pressurized cabins, where low altitude pressures are maintained. Under these conditions flying at very high altitudes can be comfortable.

It is possible to become acclimated to high altitudes over a period of time. People do live at very high altitudes in the Andes and in other mountainous regions of the world, and outsiders can become acclimated to such places. The physiology of acclimation to high altitudes involves an increase in the number of red blood cells and an increase in the amount of hemoglobin in them.

External Respiration

The figures in Table 10.1 indicate what occurs in humans in external respiration. Oxygen taken in during inspiration diffuses from the alveoli into the blood, and carbon dioxide in the blood diffuses into the alveoli, to be given off in expiration. This process is purely one of diffusion. The concentration of oxygen in the alveoli is greater than the concentration of oxygen in the blood. The thin membranes separating the alveoli and the blood are permeable. Oxygen goes into solution on the moist lining of the alveoli and diffuses from a region of higher to a region of lower concentration. The reverse is true for carbon dioxide and each gas behaves independently.

Internal Respiration

This process takes place throughout the body and is the exchange of gases between the blood and the cells, with oxygen passing from the blood into the cells and carbon dioxide going from the cells into the blood. This is again a process of diffusion, with each gas diffusing from a region of its higher concentration to a region of lower concentration. Actually oxygen does not diffuse directly into the cells nor does carbon dioxide diffuse directly from the cells into the blood. The cells are separated from the capillaries by a film of tissue fluid or *lymph,* and the gases diffuse through this fluid in their passage; this process will be considered further in the chapter on circulation.

Transport of Oxygen and Carbon Dioxide by the Blood

When oxygen diffuses into the blood in external respiration, most of it enters the red blood cells, or erythrocytes, and unites with the hemoglobin in these cells, forming a compound called *oxyhemoglobin.* Hemoglobin is a complex protein ($C_{3032}H_{4816}O_{872}N_{780}S_8Fe_4$) of four chains, 2α and 2β (Ch. 3). The total molecular weight of a normal hemoglobin molecule is about 67,000. The hemoglobin molecule has been studied extensively, and its structure is known at all levels of protein structure—primary through quaternary. The great affinity of hemoglobin for oxygen enables the blood to carry about 50 times more oxygen than the plasma alone can carry. As the blood passes through the alveolar capillaries, the hemoglobin becomes saturated with oxygen. Each 100 milliliters (ml) of blood contains approximately 20 cm³ of oxygen. This reaction may be represented as follows:

$$\underset{\text{hemoglobin}}{Hb} + \underset{\text{oxygen}}{O_2} \rightarrow \underset{\text{oxyhemoglobin}}{HbO_2}$$

Oxyhemoglobin is a very unstable compound, and when the blood reaches the capillaries in the tissues throughout the body where the oxygen tension is low, the compound breaks down into hemoglobin and oxygen, and the oxygen diffuses into the cells. The reaction that occurs in the tissue capillaries is simply the reverse of the above equation:

$$HbO_2 \rightarrow Hb + O_2$$

Oxyhemoglobin is a bright scarlet color, whereas hemoglobin without O_2 is a dull purple. This accounts for the difference in color between most arterial and venous blood.

The property of hemoglobin to combine reversibly with oxygen may be represented by the oxyhemoglobin association–dissociation curve (Fig. 10.5). The curve indicates that hemoglobin is almost completely saturated at an oxygen

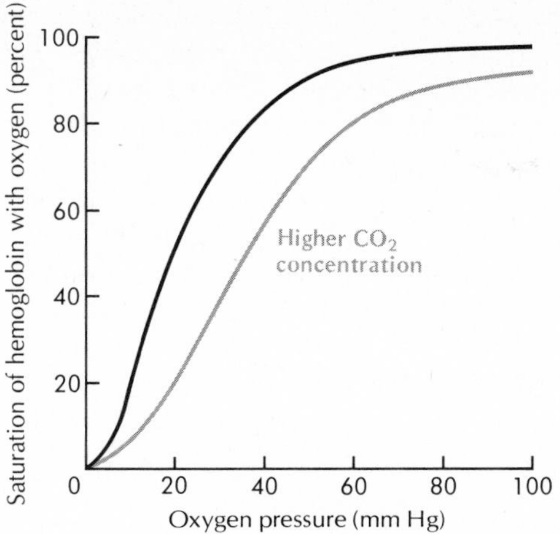

Figure 10.5 Oxygen–hemoglobin dissociation curves. Increase in carbon dioxide concentration shifts the curve to the right, thus favoring liberation of oxygen to the tissues.

pressure of 100 mm Hg. When the oxygen pressure falls below 60 mm Hg the oxygen saturation of hemoglobin falls rapidly. Thus at low oxygen pressures, oxyhemoglobin releases its oxygen very rapidly. Carbon dioxide (or hydrogen ion concentration) affects the ability of hemoglobin to combine with oxygen. This is seen in the shift of the curve in Figure 10.5. This effect of carbon dioxide is due largely to the increased acidity produced as carbon dioxide combines with water in forming carbonic acid. The S-shaped character of the oxyhemoglobin curve illustrates both the high oxygen-carrying capacity of hemoglobin and its ability to unload oxygen very rapidly where the oxygen is needed.

More CO_2 will dissolve in water or blood plasma than O_2, but the blood carries more CO_2 than could be carried in the plasma alone. When CO_2 enters the blood from the tissues, it first combines with water, forming carbonic acid (H_2CO_3). This then reacts with potassium (K) and sodium (Na) ions to form bicarbonates. Most of the CO_2 is transported in the plasma in the form of bicarbonates, whereas some of it is carried in the red blood cells in combination with amino groups of the hemoglobin molecule as *carbamino hemoglobin.* In the alveolar capillaries, due to low CO_2 tension in the alveoli, the bicarbonates and carbamino hemoglobin liberate CO_2. Thus it can be seen that the blood is an important accessory in the completion of respiration. This is true in most higher animals, where a special respiratory system has evolved.

Hemoglobin combines with gases other than oxygen and carbon dioxide. Hydrogen sulfide is a poison as a result of its affinity for hemoglobin. Carbon monoxide CO, formed by incomplete combustion of carbon, combines with hemoglobin to give a stable compound called carboxy-hemoglobin. Carbon monoxide diminishes the amount of oxygen that can be carried by the blood, and therefore victims of carbon monoxide poisoning suffer from a lack of oxygen. Carbon monoxide is one of the most important urban air pollutants. For the urban population in the United States, cigarette smoking is probably the most important source, followed by motor vehicle exhaust. Recent evidence indicates that a person's sense of timing is altered by high carbon monoxide levels; this may be a source of some rush hour traffic accidents in urban areas.

Cellular Respiration

All the processes thus far described, which are involved in getting oxygen into the cells and carbon dioxide out of them, are accessory to the process of cellular respiration.

The reader should now review the process of cellular respiration discussed in Chapter 5. It is a complex process, not of one step or reaction but of many, and a number of important points stand out. The whole series of reactions occurs with great speed. The energy stored in the glucose molecule is released in small packets in the step-by-step process, and this energy is trapped in the formation of ATP, where it is then available for the various energy-requiring processes in the cell. Each step in the process is catalyzed by one of a whole battery of enzymes. It is interesting to note again that many of these enzymes have as a part of their functional composition one of the vitamins. For instance, thiamin forms part of the prosthetic group of cocarboxylase, an important enzyme in the metabolism of pyruvic acid; riboflavin forms the prosthetic group of the flavoprotein enzyme; niacin forms part of many dehydrogenase enzymes; and pantothenic acid forms part of coenzyme A.

EXCRETION

In the metabolism that occurs in the cells of the body, various useless and injurious end products are formed. The removal of these from the body is *excretion.*

Several parts of the body are involved in this process. The role of the lungs in eliminating waste carbon dioxide has already been discussed. The sweat glands function in a very minor way in the removal of urea. Perspiration is much like dilute urine in composition, but with maximum perspiration a person would lose in a day no more than a tenth of the urea that is ordinarily eliminated in the urine. Some excretory substances are excreted through the digestive tract. The bile pigments that result from hemoglobin breakdown in the liver pass into the intestines. Excess calcium and iron are excreted as salts into the lumen of the large intestine. Such salts, together with the bile salts, are true excretions and they pass off from the body in the feces. Most of the fecal material, however, is made up of indigestible material and large masses of bacteria; so a distinction must be made between *excretion,* which is the elimination of the waste products of metabolism, and the elimination of fecal matter or *defecation.* Fecal matter is made up largely of material which has never entered the body cells. The *kidneys* are usually considered to be the primary excretory organs of the body. A most important function of the kidneys and associated structures is the elimination of nitrogenous wastes, especially *urea.*

Formation of Urea

In an earlier section the functions of amino acids in the body were discussed. Their principal use is in the synthesis of proteins for the growth and repair of cells. Those not used in this way undergo *deamination,* the removal of the amino group, and the remainder of the molecule containing carbon, hydrogen, and oxygen is converted to sugar and either burned directly in cellular oxidation or converted to glycogen or fat for storage. Deamination is effected by enzymes, and it results in the formation of ammonia, NH_3, illustrated as follows:

$$\underset{\text{alanine}}{\begin{array}{c} CH_3 \\ | \\ HCNH_2 \\ | \\ COOH \end{array}} + \tfrac{1}{2}O_2 \rightarrow \underset{\text{pyruvic acid}}{\begin{array}{c} CH_3 \\ | \\ C{=}O \\ | \\ COOH \end{array}} + NH_3$$

Ammonia, even in very low concentration in the body, is toxic and it is quickly converted to urea, a relatively nontoxic substance.

The liver is the organ primarily involved in both deamination and the formation of urea. The following experimental facts establish the role of the liver in urea formation.

1. Mann found that in dogs with the liver removed there is an increase in the amount of amino acids in the blood, and no urea is formed.
2. When certain amino acids are injected into the portal vein and circulated through the liver, there is an increase in the urea in the blood coming from the liver.
3. When ammonium carbonate is circulated through the liver, it is transformed into urea.
4. The addition of amino acids to liver pulp results in the liberation of ammonia.

The formation of urea, $CO(NH_2)_2$, may occur by more than one path. One possible path is

$$2NH_3 + CO_2 \rightarrow H_2N{-}\underset{\underset{O}{\|}}{C}{-}NH_2 + H_2O$$

Most of the urea formed in the mammalian liver involves a cycle, usually referred to as the ornithine or urea cycle, in which three of the amino acids participate. According to this concept, ornithine combines with ammonia and carbon dioxide to form citrulline; then citrulline combines with ammonia to form arginine; the arginine then decomposes to form urea and ornithine; and the ornithine is set free to repeat the cycle. The reactions for this cycle are given in Figure 10.6.

The urea that is formed in the liver and is the chief nitrogenous waste product of metabolism is carried in the blood to the kidneys where it is removed.

The Human Urinary System

The kidneys are dark red bean-shaped structures about 10 cm (4 inches) long; they are located in the back of the body cavity just below the stomach, one on either side of the middorsal line (Fig. 10.7). The kidneys lie outside of the body cavity and only one surface of each is covered with peritoneum. They are supplied with rather large renal arteries and renal veins. Leading from the indentation in each kidney is a *ureter,* which empties into the *urinary bladder.* A tube, the *urethra,* carries urine from the bladder to the outside. In the human female this duct is rather short, but in the human male it is much longer, leading to the outside through the *penis.* When a kidney is cut lengthwise, it shows that the ureter is expanded in the kidney to form a large cavity, the *pelvis* of the kidney. The gross appearance of the rest of such a sectioned kidney shows an outer layer, the *cortex,* and an inner region next to the pelvis, the *medulla.* The medulla is made up of a number of cone-shaped structures called *renal pyramids* which project into cuplike cavities of the renal pelvis called the *renal calyxes.*

Microscopic Structure of the Kidney

Microscopic examination of a sectioned mammalian kidney reveals that it is composed primarily of many tiny tubules. In the outer cortical

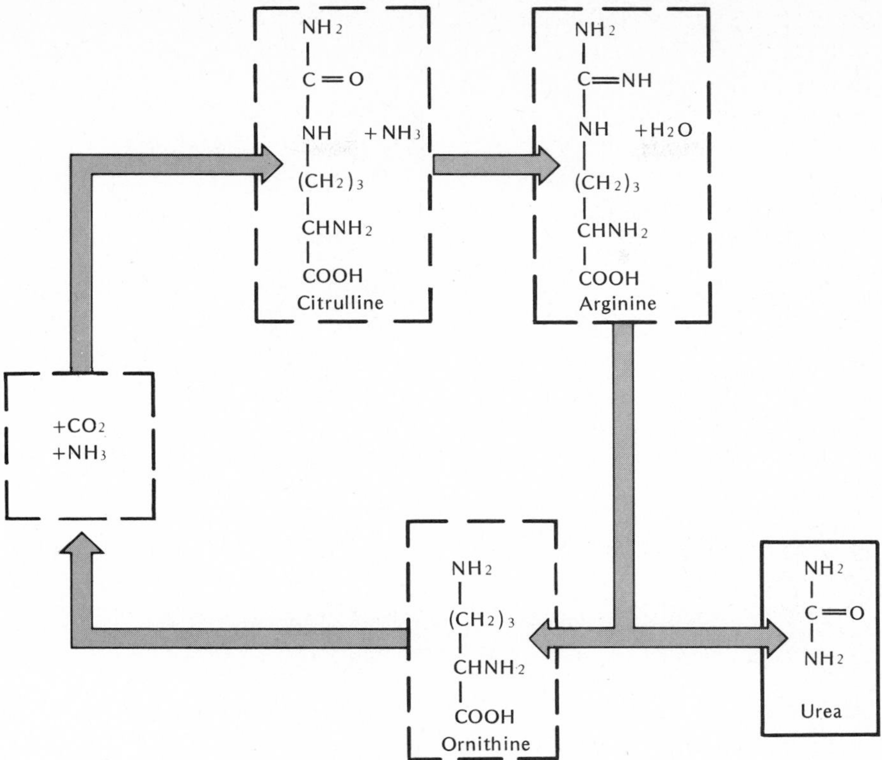

Figure 10.6 The ornithine or urea cycle.

portion of the kidney these tubules are coiled, whereas the tubules in the medullary portion are straight, giving that portion a striated appearance. The outer part of the kidney appears granular due to the presence of numerous bodies called *renal* or *malpighian corpuscles.* A single urinary unit is composed of one of these renal corpuscles and its associated *uriniferous tubule.* There are approximately 1 million of these urinary units in each human kidney.

An examination of Figure 10.7 shows the relationships of these parts. Each such unit begins with a renal corpuscle, which is composed of an outer, thin double-walled capsule, the *renal (Bowman's) capsule,* and this surrounds a tuft or knot of blood capillaries called a *glomerulus.* The capsule extends away from the renal corpuscle as the uriniferous tubule. The relationship of the capsule to the glomerulus can be better understood by trying to visualize what occurs during their development. A hollow blind tube approaches the knot of blood capillaries. When this tube touches the knot, it indents and surrounds the knot of capillaries as indicated in the diagram. A small *afferent arteriole* enters each glomerulus and an *efferent arteriole* leaves each one. The bore of the efferent vessel is smaller than that of the afferent vessel. Each coiled uriniferous tubule joins a larger *collecting tubule* that passes through the medulla of the kidney and empties into the pelvis. In humans each uriniferous tubule is differentiated into three regions: the *proximal convoluted tubule,* the *loop of Henle,* and the *distal convoluted tubule* (Fig. 10.7). The efferent vessel that leaves each glomerulus forms a capillary net around each uriniferous tubule. These capillaries form small veins that empty into the renal veins.

Functions of the Kidney

URINE FORMATION

Urine formation involves three different processes: *filtration, resorption,* and *tubular secretion.* As the blood passes through the glomerulus, it is under pressure because the afferent vessel leading into a glomerulus is a branch of an artery. This pressure is actually increased in the glomerulus because the efferent vessel is smaller than the afferent vessel. Due to this pressure, the liquid part of the blood is actually filtered out and passes into the cavity of the capsule as the *glomerular filtrate.* Experiments show that with an increase in the rate of the heart beat, which increases the blood pressure, the glomerular filtrate forms more rapidly.

In the early work on kidney functioning, one worker was able to obtain a sample of the glo-

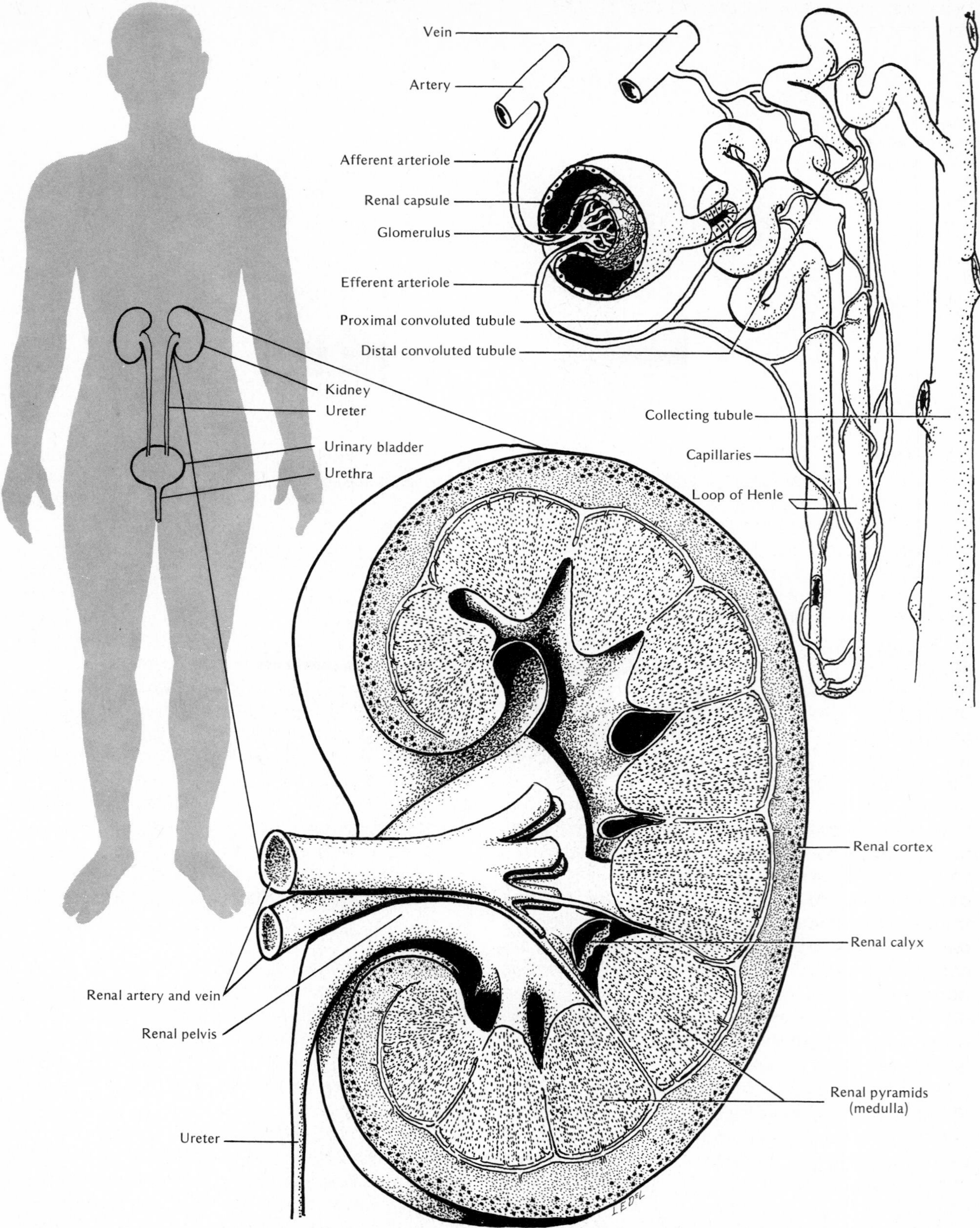

Figure 10.7 Human urinary system. **(Center)** Longitudinal section of kidney. **(Right)** Functional unit of kidney.

merular filtrate in a frog's kidney in a micropipette. When he subjected this sample to a microchemical analysis he found that the glomerular filtrate had the same composition as the blood plasma minus the blood proteins and other colloidal materials. When a sample of the fluid

Table 10.2 Composition of Blood Plasma and Urine*

	Blood Plasma (percent)	Urine (percent)	Times Increase in Concentration
Water	90-93	95	
Proteins and other colloids	7-9	None	
Glucose	0.1	None	
Na	0.30	0.35	Little or none
Cl	0.37	0.6	2
Urea	0.03	2.0	60
Uric acid	0.004	0.05	12
K	0.02	0.015	7
NH_4	0.001(?)	0.04	40
Ca	0.008	0.015	2
Mg	0.0025	0.006	2
PO_4	0.009	0.15	16
SO_4	0.002	0.18	90
Creatinine	0.001(?)	0.075	75

* Compiled from various sources.

taken from the beginning of a ureter was analyzed, it had the same composition as *urine.* This suggested that as the glomerular filtrate passes through the tubules it is changed in composition. Diffusion could be a part of the process, but it should be possible for the cells of the tubules to carry on active transport of some of the materials involved. It should be possible for the cells to resorb certain materials from the filtrate, and it should be possible for the cells to pass certain materials in the blood capillaries into the lumen of the tubules. It appears that both of these processes are involved, but in the mammalian kidney the process of resorption is the primary one. Table 10.2 shows the composition of blood plasma and urine.

In the course of 24 hours the human kidneys form from 75 to 150 liters of glomerular filtrate, yet the normal amount of urine produced each day is from 1 to 1½ liters. This means that a tremendous amount of water is resorbed each day in the tubules. The analysis shows that glucose is present in glomerular filtrate but not in normal urine. This also must be resorbed in the tubules. This resorption occurs primarily in the proximal convoluted tubules. Most of the water is resorbed and normally all of the glucose is too, but the amount of resorption of other substances varies. There seems to be a selective action by the tubule cells. Substances like urea, uric acid, and phosphates are resorbed to a minor extent while creatinine is not resorbed at all.

The earlier theories about the concentration of urine by the gradual reabsorption of water as the glomerular filtrate passes along the tubules had to be abandoned because of later experimental results. Several workers used micropuncture studies at different regions of both the uriniferous tubules and the collecting tubules. Such studies indicate that the fluid in the descending limb of Henle's loop becomes more and more concentrated, while the fluid in the ascending limb of Henle's loop becomes more and more dilute, with a concentration at the beginning of the distal tubule slightly less than the original glomerular filtrate. Then with the passage through the collecting tubule, the fluid becomes more and more concentrated.

Today this aspect of kidney functioning is explained on the basis of a *counter current exchange.* This is diagrammed in Figure 10.8 showing how the movement of water and sodium ions varies in different regions of the tubules. The permeability of the collecting tubules to water is controlled by the antidiuretic hormone (ADH) formed in the hypothalamus of the brain. An increase in the salts in the blood results in the release of ADH which, in turn, makes the collecting tubules more permeable to water. With a drop in the salt concentration in the blood, the release of ADH is shut off.

It would seem that the development of the loops of Henle in the evolution of the vertebrate kidney has made possible the formation of urine more concentrated than the blood. Only birds and mammals possess these loops, and in general, those with the longest loops of Henle produce the most concentrated urine. Desert animals have very long loops of Henle.

The fact that the creatinine concentration in urine is more concentrated than can be accounted for by water resorption indicates that a third mechanism, tubular secretion, is involved in urine formation. There is now quite a bit of evidence to support this concept. In the frog kidney the glomeruli are supplied with blood from the renal arteries while the tubules are

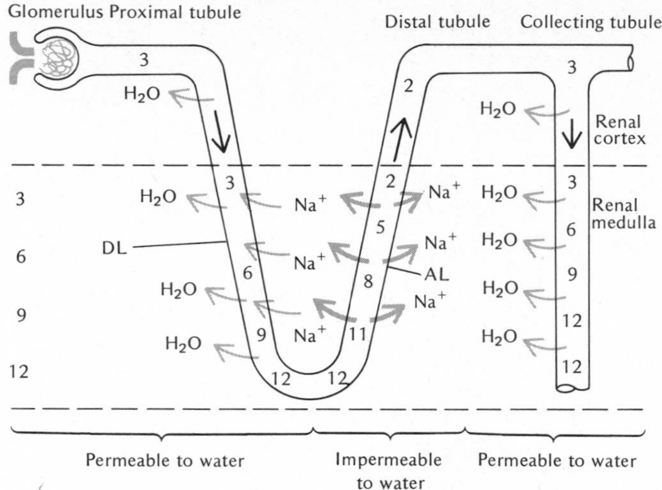

Figure 10.8 Diagram of the counter exchange mechanism in the human nephron which makes possible the conservation of water and the production of a concentrated urine. The numbers indicate the relative osmotic concentrations in the tubules and in the tissues of the renal medulla. The arrows indicate the movement of sodium ions and water: heavy arrows indicate active transport of sodium, light arrows the diffusion of sodium ions and of water. DL is the descending limb of Henle's loop and AL the ascending limb of Henle's loop. Sodium ions are actively transported from the ascending limb of Henle's loop, which is impermeable to water. Sodium ions leaving the ascending limb diffuse into the descending limb. In this way there is a counter current exchange of sodium ions between these limbs, causing sodium ions to become more and more concentrated in the descending tubule and in the extracellular fluids of the renal medulla. As the collecting tubules pass through this region, water is passively reabsorbed (diffusion) out of the tubule into the extracellular fluid.

supplied with blood from the renal portal vein. If the renal arteries are tied off in a frog, it still produces urine, and the only way this can happen is by tubular secretion. In certain experiments, where the dye, phenol red, was injected into the renal circulation of animals and the animals were killed at successive intervals after injection for examination, the following was found. When a kidney section from an animal that was killed shortly after the injection was examined, the dye was found in the ends of the tubule cells next to the blood capillaries; in an animal killed somewhat later, the dye was found in the center of these tubule cells; and in one killed still later, the dye was found passing from the cells into the lumen of the tubules.

It seems probable that not only is creatinine—a product of muscle metabolism—excreted in this way but that also other substances are so excreted. For instance, it has been shown that the drug penicillin is thus excreted. Although the actual extent of tubular secretion in humans is not known, it must be considered as a part of the process of urine formation.

When the excretion of carbon dioxide from the lungs was considered, it was described as a strictly passive process so far as the cells of the alveoli are concerned—the carbon dioxide simply diffuses through the cells. Here in the kidneys the process of filtration is a passive one so far as the cells of the capsule are concerned. The force for the filtration is supplied by the heart. The diffusion of ions and water that occurs in the tubules is a passive process, but the other activities that occur in the tubules, both in resorption and in tubular secretion, are different—the cells of the tubules actually perform most of these processes. As the glomerular filtrate passes down in the tubule, the concentration of glucose in the filtrate is the same as that in the blood of the capillaries surrounding the tubule. Thus there is no basis for a diffusion gradient. When the cells of the tubule take glucose out of the filtrate and pass it into the blood, they must perform work. In a similar way, when the cells of a tubule take dye molecules or those of penicillin out of the blood and pass them into the lumen of the tubule, they must perform work. This work performed by the cells of the kidney tubules is active transport. It has been shown that the kidney cells use oxygen at a faster rate than the cells of many of the tissues in the body. Electron micrographs show that the cells of the kidney tubules are richly supplied with mitochondria (Fig. 10.9).

OTHER FUNCTIONS OF THE KIDNEY

One ordinarily thinks of the kidneys as organs that are primarily involved in the elimination of urea and other wastes of protein metabolism from the body, but this is just a part of their more general function of helping to maintain the constancy of the internal environment. This constancy of the internal environment is called *homeostasis*. Many parts of the body are in-

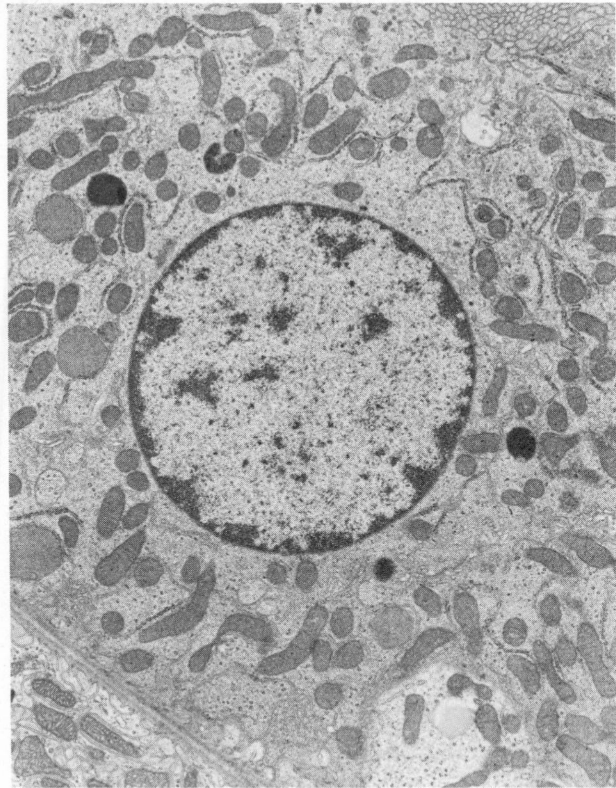

Figure 10.9 Portion of a kidney tubule cell, showing numerous mitochondria. (Courtesy of T. Kuwabara, Howe Laboratory of Ophthalmology.)

volved in homeostasis but the kidneys are especially important. It is obvious that the removal of harmful substances such as urea, uric acid, and creatinine is necessary if this constancy is to be maintained. When substances are present in the blood in excess of normal, they are removed by the kidneys. Thus if the concentration of glucose in the blood rises above 0.17 percent, the excess is promptly removed by the kidneys. The osmotic pressure of the body fluids is maintained by the kidneys by regulating the concentration of salts in the blood. An excess amount of water in the blood, which tends to lower the osmotic pressure, results in more water elimination by the kidneys. In a like manner, an excess of salts in the blood will be eliminated. The quantity of urine produced depends not only on the amount of liquid taken into the body, but also on the concentration of salts. When unusually salty food is eaten, the kidneys must excrete a proportionately larger amount of salt to maintain the osmotic pressure, and the urine volume is increased. In a similar manner, the kidneys aid in maintaining the proper pH of the blood. Any excess of acid or base produced during metabolism in the cells throughout the body will be given off by the kidneys.

Just how the cells of the kidney tubules carry out their selective activities involving different thresholds for different substances is not known, but active transport is involved.

ABNORMAL KIDNEY FUNCTION

The most common type of kidney disease, *nephritis,* is caused by a bacterial infection of the capillaries of the glomeruli, which may result in their destruction. Such destruction will reduce the filtering surface of the kidney. As a result of this destruction, proteins and even blood cells may pass into the urine. With this loss of blood proteins, *edema,* which is a swelling due to the accumulation of water, particularly in the legs, may result. In severe nephritis the individual may be unable to excrete water, urea, and other substances properly. The gradual accumulation of urea and other wastes in the blood, called *uremia*, produces a toxicity that may lead to death.

Occasionally the passage of urine is blocked by the formation of *kidney stones.* Some of the constituents of the urine, such as uric acid and calcium phosphate, when present in large amounts, may be precipitated to form these hard deposits. When one of these becomes large enough to block the passage of urine, it must be removed by surgery.

Damage to the glomeruli of the kidneys often occurs with chronic high blood pressure. Although it is known that the kidneys normally produce one or more vasodilator substances that counteract high blood pressure, recent work indicates that when the circulation to the kidneys is restricted, they produce a vasoconstrictor substance, called *renin,* that further aggravates the high blood pressure.

There are two endocrine disturbances that result in abnormal kidney function although the kidneys themselves are not deranged. One of the hormones released by the posterior lobe of the pituitary gland, the antidiuretic hormone, ADH or *vasopressin,* serves to regulate the resorption of water by the cells of the kidney tubules. When the posterior lobe of the pituitary is damaged and there is not enough of this hormone in the bloodstream, the tubules fail to resorb the normal amount of water and large quantities of very dilute urine are produced. In severe cases the amount of urine eliminated per day may be 20 to 30 liters instead of the normal 1 to 1½ liters. This disorder is known as *diabetes insipidus.* A victim of this disorder has an almost insatiable thirst. Although it cannot be cured, periodic injections of the hormone or of an extract of the posterior lobe will keep the condition under control.

As mentioned above, the kidney tubules normally resorb all of the glucose in the glomerular filtrate. However, the normal amount of sugar in the blood is regulated by the hormone *insulin* produced in the pancreas. This hormone regulates the storage of glucose as glycogen. When the hormone is not produced in normal amounts, the sugar is not taken from the blood and stored and a much larger amound of sugar is filtered out in the kidneys. The kidney tubules can resorb sugar only up to a certain amount and when this level is reached, the rest passes out in the urine. This is what happens when there is an undersecretion of insulin and large quantities of glucose are excreted in the urine. This disorder is known as *diabetes mellitus* and will be considered further in the chapter on endocrine glands.

Osmotic Regulation

In the discussion of the functions of the human kidney, it was pointed out that in addition to the removal of waste products, the uriniferous tubules play a very important role in returning water to the bloodstream. But not all vertebrates have the problem of water conservation; some need to get rid of water.

It is generally considered that marine invertebrates are in osmotic equilibrium with the sea and thus have no problem of water regulation. This must have been the situation with the chordate ancestors of vertebrates. However, there is strong evidence that the first vertebrates were freshwater forms. This means that they lived in water containing less salts than their blood and body fluids so that there would be a tendency for water continually to enter their bodies by osmosis. Accordingly, there was the problem of getting rid of excess water. It has been suggested that the glomerulus and capsule appeared in their evolution as an adaptation for filtering out and eliminating the excess water. It is further suggested that at the same time the tubules became capable of reabsorbing sugar, salts, and other useful materials that could be lost by this filtration process. Freshwater bony fish and frogs excrete a copious amount of dilute urine. Frogs excrete on the average about one third of their body weight in water each day, whereas humans excrete only one fiftieth of their weight per day. If the salt concentration is to be maintained higher than that of the surrounding water, work must be done in order to absorb salts against the concentration gradient, that is, active transport must be involved. Freshwater fish have special cells in their gills that carry out this function.

Some of the descendants of these first freshwater vertebrates became marine dwellers as well as land dwellers. The marine bony fishes have an osmotic concentration about one half as great as that of seawater and thus tend to lose water to their environment. Marine fishes, with the problem of conserving water, have few glomeruli in their kidneys. They swallow sea water but give off (active transport again) large amounts of salts through their gills. The marine cartilaginous fishes (sharks, rays) have "solved the problem" in a different way. These forms have about the same amount of salts in their blood as marine teleosts (the largest order of bony fishes) but in addition about 2 percent urea. A segment of their renal tubule absorbs urea from the glomerular filtrate. This added urea brings the total osmotic pressure of their blood to slightly above that of sea water. These organisms retain numerous large glomeruli like the freshwater forms, and they excrete large amounts of dilute urine.

Sea birds such as gulls and petrels are able to drink seawater without upsetting the osmotic balance, because they have salt-secreting glands in their bills. The secretion of these glands may contain more than 5 percent salt.

The nature of the nitrogenous waste excreted varies with the environment of the organism. Species with an abundant water supply excrete nitrogen primarily in the form of ammonia. This is the case with freshwater teleosts. Most of the ammonia is lost through the gills, so in these organisms the kidney functions primarily in maintaining proper water balance. Marine teleosts excrete considerable ammonia but also some urea. Terrestrial animals primarily excrete urea or uric acid. In humans the nitrogenous waste is primarily urea. Birds, snakes, and lizards excrete a semisolid urine containing uric acid crystals, thus minimizing water loss more than any of the other vertebrates. Uric acid is quite insoluble in water, and it can be excreted without the use of much water. The amount needed is that necessary to flush the uric acid into the cloaca. Within the cloaca most of the water is absorbed and the waste is given off in paste form. It is interesting to note that the frog as a tadpole eliminates about 40 percent of its nitrogenous waste as ammonia, but as an adult, when there is need for more water conservation, about 80 percent of its nitrogenous waste is urea.

SUGGESTIONS FOR FURTHER READING

Baldwin, E. B. *An Introduction to Comparative Biochemistry,* 6th ed. Cambridge: Cambridge Univ. Press. 1967.

Best, C. H., and N. G. Taylor. *The Living Body,* 5th ed. New York: Holt, Rinehart and Winston (1969).

Cantarow, A., and B. Schepartz. *Biochemistry.* Philadelphia: W. B. Saunders (1967).

Evans, L. *The Sciences,* **12** (Oct. 1972).

Grollman, S. *The Human Body.* New York: Macmillan (1968).

Guyton, A. C. *The Function of the Human Body.* Philadelphia: W. B. Saunders (1969).

Moffat, D. B. *The Control of Water Balance by the Kidney.* New York: Oxford Univ. Press (1971). (One of the Oxford Biology Readers distributed by Carolina Biological Supply House.)

Torrey, T. W. *Morphogenesis of the Vertebrates.* New York: Wiley (1967).

Wood, D. W. *Principles of Animal Physiology.* New York: American Elsevier (1972).

From Scientific American

Fenn, W. O. "The Mechanism of Breathing" (Jan. 1960).

Merrill, J. P. "The Artificial Kidney" (July 1961).

Smith, H. W. "The Kidney" (Jan. 1953).

11 Transport and Blood Functions

As animals increased in size, there was an accompanying development of internal conducting systems as the surface–volume relationships became relatively too different for exchange of gases, transport of wastes, delivery of nutrients to the interior, and so forth, to take place. The middle germ layer, the mesoderm, became modified for duct systems as well as to form organs such as gonads and kidneys. This generalization applies to all organisms, beginning with the flatworms. One of the most elaborately and elegantly evolved systems for internal conduction is the circulatory system. Circulatory systems exist in several phyla. Some are called *open* circulatory systems because they consist of a pump, the heart, and blood vessels from the heart that empty the blood to bathe the cells directly. The blood of these forms is often called hemolymph. Outstanding examples of open circulatory system are found in the many mollusks and arthropods. The other kind of circulatory system is said to be a *closed* one. Blood is carried from the heart by arteries which branch throughout the body. The outward flow terminates in *capillaries,* small vessels whose walls are one-cell thick. Blood is returned to the heart by *veins.* Closed circulatory systems are exemplified by earthworms, squid, and vertebrates. In the latter, only *lymph* bathes the cells; this is the fluid that leaves the capillary bed and enters the tissue spaces. It is returned to the closed circulatory system by lymph vessels after passing through lymph nodes and other lymphoid organs.

As in other chapters, the emphasis will be on the human (a mammal) as a model organism. However, among the vertebrates there have been a number of significant changes in the heart and the circulatory pattern. A better appreciation of the evolutionary changes will emerge from examining these differences.

THE CIRCULATORY SYSTEM

The Human Heart: Its Action and Its Control

The human heart is a double pump of four chambers (Fig. 11.1). It lies in the chest, between the lung cavities, in its own pericardial cavity.

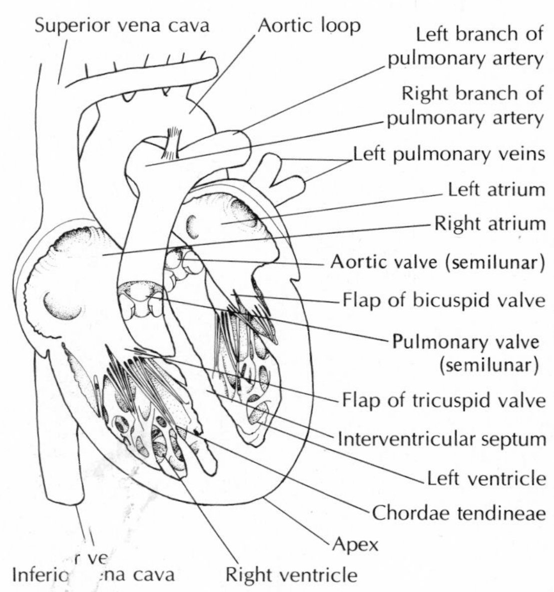

Figure 11.1 Longitudinal section of human heart.

This cavity contains lymph, a fluid that reduces friction as the heart beats. The blood is received in the heart by two thin-walled *atria* that contract together, and it is delivered from the heart by two thick-walled *ventricles* that contract together. Blood follows this course: oxygen-poor blood is received into the *right atrium* from two large veins, the *superior vena cava,* returning blood from the head and arms, and the *inferior vena cava,* returning blood from the trunk, viscera, and legs. Blood passes from the *right*

atrium to the *right ventricle* through the *tricuspid valve,* which is composed of three flaps whose edges are loosely anchored to the wall of the ventricle. The right ventricle delivers blood through the *pulmonary artery* to the lungs. The blood passes the three half-moon-shaped pockets of the *right semilunar valve* as it enters the pulmonary artery. After being oxygenated in the lung capillaries that surround the alveolar spaces of the lungs, oxygen-rich blood is returned to the *left atrium* by two pairs of *pulmonary veins.* The left atrium delivers the blood past the *bicuspid (mitral) valve,* composed of two flaps, to the left ventricle. The left ventricle delivers blood to the large thick-walled artery, the *aorta,* from which it is distributed throughout the body. In leaving the left ventricle, the blood passes the left semilunar valve; as on the right side, this valve prevents return of the blood to the ventricle.

The muscle tissues of the heart itself have an extensive system of vessels and capillaries called the *coronary* system. The coronary arteries arise from the aorta, and the coronary veins empty into the right atrium. Coronary occlusions, usually clots, which block the flow of blood in these heart vessels, lead to cardiac accidents that often result in death.

Chief Blood Vessels in Humans

Figure 11.2 shows the arrangement and distribution of the chief human blood vessels. In general for each artery that carries blood away from the heart, there is a corresponding vein that returns blood to the heart. Blood vessels that supply the head, arms, trunk, legs, kidneys, and gonads are bilaterally paired. The arterial supply to the digestive tract is by a few unpaired vessels. Veins from the various organs of the digestive tract—the stomach, the small intestine, and the large intestine—unite to form the *hepatic portal vein* which enters the liver. In the liver the vein distributes to many venous sinuses. These collect into *hepatic* veins which deliver into the main vein that returns blood from posterior to the heart, the *inferior vena cava.* The names of other major blood vessels and their distribution can be obtained from examination of Figure 11.2.

The Circulatory System in Other Vertebrate Classes

Of the other classes of vertebrates, only the class Aves has a completely separate double-pump system such as seen in humans and other mammals. (Because of its superficial similarity, no further details on avian hearts will be discussed here.)

One can trace a basic circulatory pattern to the invertebrate chordates, such as the cephalochordates as represented by amphioxus (Fig. 11.3). Stated simply, blood is delivered forward underneath the pharynx by a pulsating vessel or "heart" area that delivers to a ventral aorta. The ventral aorta sends numerous vessels upward around the pharynx where close proximity to the pharyngeal clefts allows for oxygenation of the blood. The oxygenated blood is collected into dorsal arteries, the aortae. Although paired in the region of the pharynx, the aortae unite into a common one that flows posteriorly, delivering tributaries to all parts of the body into a capillary network. Blood is collected and delivered back to the heartlike area by veins.

Fish have a pattern that is similar to that of amphioxus. However, there is a greater differentiation of the heart, and the pattern of circulation is more complicated. The fish heart has four chambers. It lies under the pharynx and its chambers are linearly positioned. The posterior chamber, the *sinus venosus,* receives blood from a pair of common cardinal veins that collect from the trunk and from hepatic veins that indirectly collect from the digestive tract by way of the liver. The sinus venosus delivers blood anteriorly successively into an atrium, a ventricle, and a truncus arteriosus. The truncus arteriosus passes blood into a ventral aorta (an artery) and through aortic arches around each side of the pharynx, during which time the blood is oxygenated. It is collected by paired dorsal aortae that unite posteriorly as a single artery. The aorta delivers oxygenated blood to tributaries in the trunk. Noteworthy in the fish is the presence of two *portal systems.* A portal system consists of blood being delivered to an organ, flowing through the organ as a complex of capillaries or sinuses, and then reassembling into a larger vessel of the same kind that entered the organ. There are both arterial portal systems (in the hypothalamus—pituitary area of mammals) and venous portal systems. The two portal systems of fish are venous. One kind, a hepatic portal system, exists in all vertebrates. Blood is collected from the walls of the alimentary tract into a single vessel, the *hepatic portal vein.* This vein takes venous blood, laden with the products of digestion, to the liver. The blood is distributed throughout the interior of the liver by liver sinuses; it is then collected into several large veins, the *hepatic veins,* which empty into the sinus venosus. The second portal system is associated with the kidneys. Here the blood from

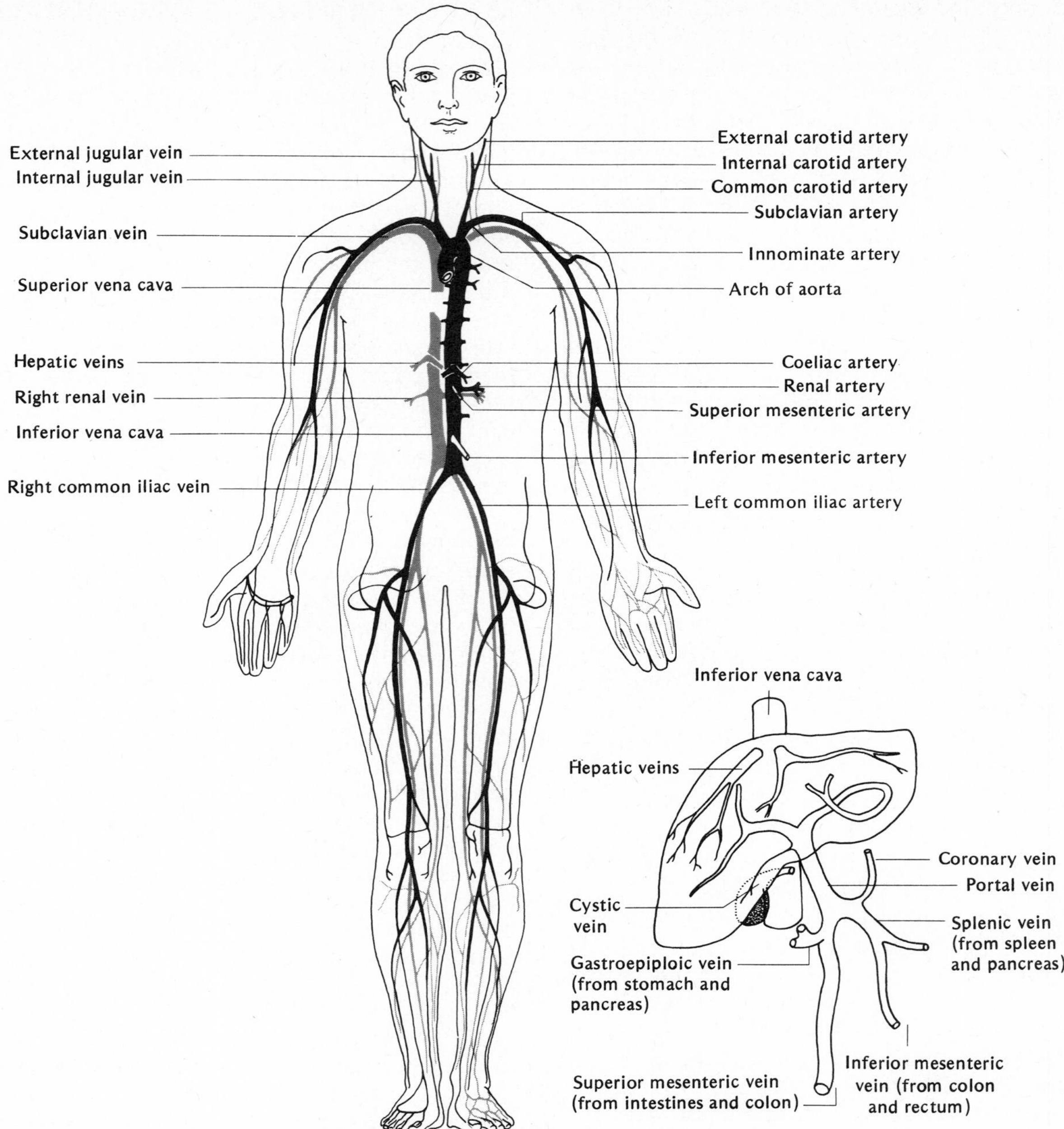

Figure 11.2 Chief blood vessels in humans. Systemic circulation. **(Lower right)** Details of hepatic portal system.

the tail region is delivered to each kidney by a renal portal vein. On entering the kidney, the venous blood is distributed through a series of capillaries and sinuses and then collected into the posterior cardinal veins; these flow into the common cardinal veins and thence into the sinus venosus of the heart.

Amphibia show a significant change in the organization of the heart. This is associated with the development of lungs. The amphibian heart actually has five chambers. The sinus venosus collects venous blood, as in fish. Now, however, the atrium has become partitioned into a right and a left side. The sinus venosus delivers into the right atrium. The left atrium receives blood from the lungs. When the two atria contract they deliver blood into the ventricle. By contrast with the relatively straight tube of fish, in amphibia the ventricle is strongly flexed and ventricular blood is forced past a spiral valve along the length of the truncus arteriosus. The blood passes into the aortic arches in essentially two streams. The blood from the right side of the single ventricular cavity tends to be shunted toward the rearmost aortic arches which, in amphibia and all other vertebrates, have tributaries that bring the blood to the lungs. The blood from the left side of the ventricular chamber is passed to the other aortic arches that deliver to the body. The point to be made here is that although an amphibian has a single ventricle, it functionally tends to separate the oxygenated and deoxygenated blood. Amphibia also have two venous portal systems, but a new blood vessel, the *inferior vena cava,* is present. It collects blood from the trunk and kidneys and delivers it to the sinus venosus. The hepatic veins empty into the inferior vena cava rather than directly into the sinus venosus. Another change between fish and amphibia is that the number of aortic arches around the pharynx becomes reduced.

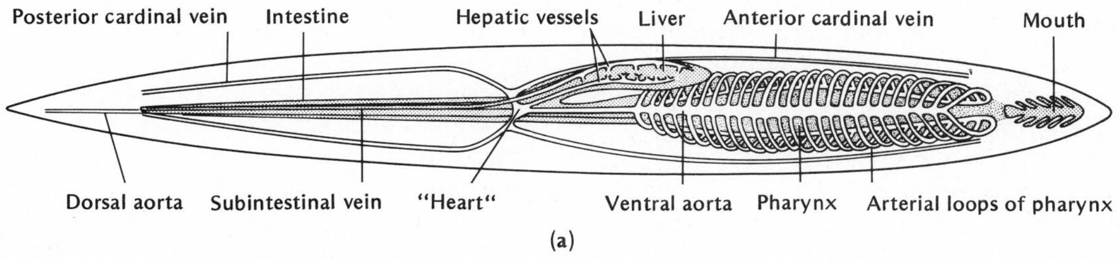

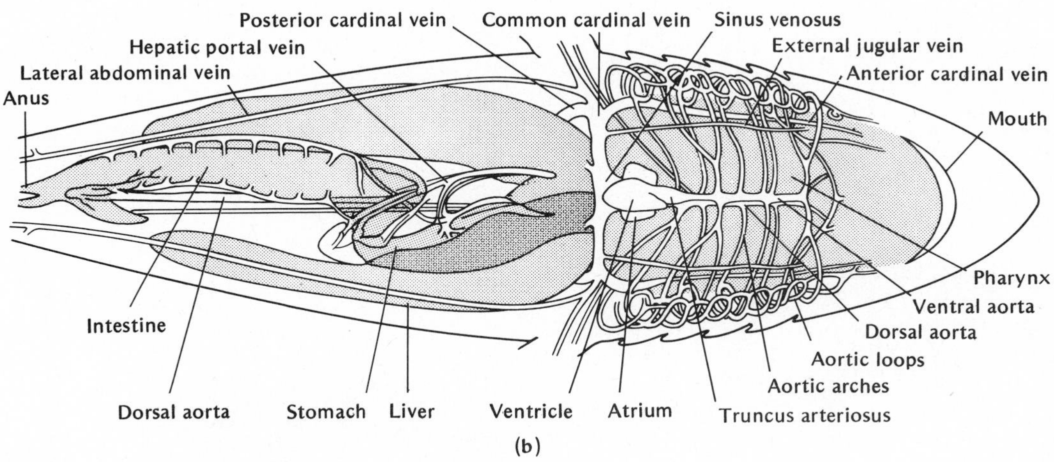

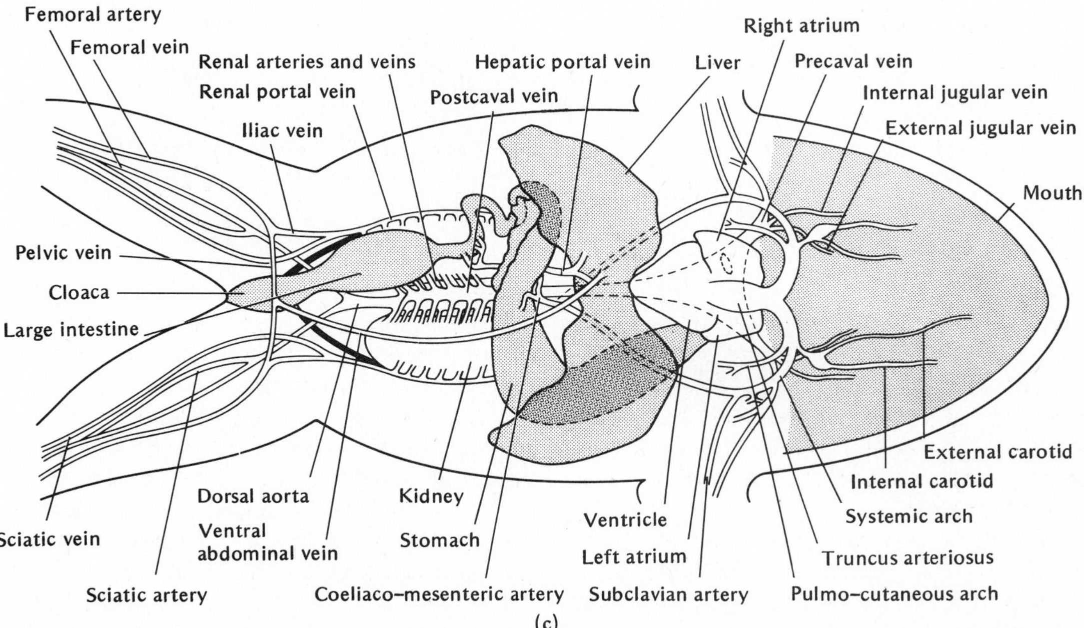

Figure 11.3 Schematized diagrams of the circulation in **(a)** amphioxus, a cephalochordate; **(b)** shark, a chondrichthyean fish; **(c)** frog, an anuran amphibian. Compare these patterns with human circulation, Figure 11.2.

The reptiles show several stages in evolutionary development of the circulatory structures. The sinus venosus becomes reduced in size, though still present and flowing into the right atrium. The ventricles range from having a small fencelike partition in some reptiles to complete partitioning in such as the alligator. The truncus arteriosus is much reduced and drawn into the ventral wall of the ventricle. Thus reptiles range between six- and five-chambered hearts. Even alligator hearts have a slight connection between the right and left systems. As with amphibia, there are two portal systems present.

In birds the heart consists of four chambers bearing the same names as those of mammals. While birds have the structural organization of a renal portal system, the flow of blood is through the kidney without significant sinus formation; therefore, functionally, the renal portal system is different from reptiles.

Blood Vessels

The arteries carry blood away *from* the heart, the veins *to* the heart. Capillaries are microscopic vessels that connect arterioles and venules, except in kidneys and gills, where they may connect arterioles, or in liver, where they may connect venules. Arteries are thick-walled and carry blood under greater pressure than that in veins or capillaries (Fig. 11.4). Their walls are three-layered. The inner layer, *tunica intima,* is formed of endothelial cells. The middle layer, *tunica media,* consists mainly of smooth muscle and elastic tissue. The outer layer, *tunica externa,* is connective tissue of a fibroelastic kind. The smallest arteries are called *arterioles.* The smooth muscle of the middle layer dominates, and these are the vessels whose innervation causes them to expand or contract. Capillaries connect arterioles to venules. Capillaries are composed only of endothelial cells. Sometimes their diameter is smaller than the red blood cells (7.5 μm), forcing these cells to bend in their passage. Veins, also with three layers, are thinner walled than arteries, and larger veins have valves along their length to prevent the backflow of blood.

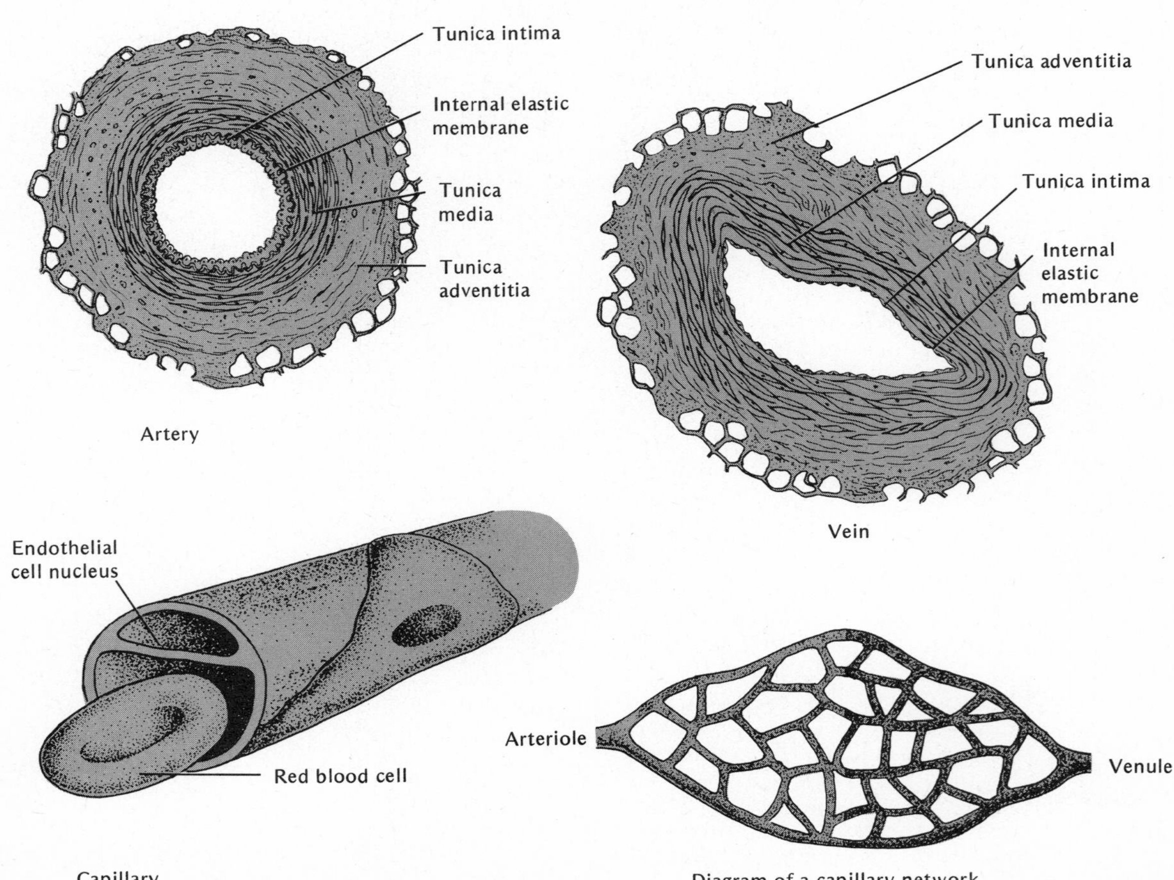

Figure 11.4 Structure of blood vessels.

Heart Beat, Heart Rate, and Blood Pressure

HEART SOUNDS

The cardiac valves determine the direction of flow in the heart. Their action creates the heart sounds. The heart beat begins as the atria contract, forcing residual blood that has not poured in from the veins into the ventricles. As the atria relax, the tricuspid and bicuspid valves snap shut. This creates the first heart sound, or "lubb." The ventricles next contract and the blood is forced into the pulmonary artery and the aorta. As the ventricles relax, the two semilunar valves close, creating the second heart sound, or "dup." This cycle of beats occurs, on the average, about 72 times a minute. Under resting conditions, nearly 60 ml (almost 2 ounces) of blood are delivered with each beat. During exercise, however, the heart rate accelerates, and the entire 6.6 liters (14 pints) of blood in the body may pass through the heart twice per minute.

THE PACEMAKER

Three places in the body are involved in the *control* of the heart rate and the blood flow: the *pacemaker* in the heart, the cardiovascular centers in the medulla of the brain, and the action of the peripheral blood vessels.

In the wall of the right atrium is a specialized patch of modified muscle tissue called the *pacemaker,* or sinoatrial (SA) node. It has an intrinsic capacity to initiate impulses that spread over the atrial walls. Impulses from the SA node stimulate a second patch of modified muscle tissue, the *atrioventricular (AV) node.* The impulses from the AV node travel to the ventricle by additional modified muscle. This distributes the impulse throughout the muscle of the ventricle. These events describe the *intrinsic* control of the beat of the heart. However, the pacemaker is influenced by impulses coming to it through nerves that exert *extrinsic* control on the rate of the heartbeat.

Nervous Control of the Pacemaker Two kinds of nerves of the autonomic nervous system, with opposite actions, innervate the pacemaker. Both of these pairs of nerves deliver impulses from a complex *cardiac center* in the medulla of the brain. The more active of the two pairs are branches of the tenth cranial nerve, the *vagus,* a parasympathetic nerve. Impulses from the vagus have an inhibitory (decelerating) effect on the heart. The other pair of nerves emerges from the spinal cord in the neck; they are sympathetic nerves and they accelerate the heart. Of the two nerves, changes in stimuli provided by the vagus nerves seem to be more important in regulating heartbeat than the accelerator nerves.

Nonnervous Control of the Pacemaker The heart muscle and the pacemaker can be stimulated directly by changes in temperature, pH, and hormones. Except in cases of fever, temperature is not a significant factor. As should be expected, high temperature increases the heart rate and low temperature decreases it. The heart is accelerated by a lowering of the pH, as can occur during exercise when the amount of carbon dioxide delivered to the blood increases and leads to the formation of carbonic acid. The hormone adrenalin also can act directly on the heart. This effect is applied sometimes in cardiac arrest, in which case adrenalin may be directly injected into the heart to stimulate it to resume beating.

NERVOUS CONTROL OF THE CARDIAC CENTER

The impulses reaching the heart are the result of sensory stimulation of the cardiac center. These impulses can be from many sources and can range from physical events, such as muscular activity, to emotional stimuli, such as the viewing of something very pleasant or very unpleasant. Two activities associated directly with blood flow illustrate how the cardiac center can be influenced through sensory impulses. If the blood pressure in the *vena cava* is increased because of muscular activity, for instance, this stretches the wall of this vessel. Receptors in the wall respond to the stretch and send impulses to the cardiac center that increase the rate of heartbeat by inhibiting the vagus nerve. This causes the blood to be moved through the heart faster and permits more oxygen-rich blood to flow to the body (Fig. 11.5).

Accelerated heartbeat increases the pressure in the aorta. Its walls also have receptors that are stimulated by the stretching to send impulses to the cardiac center. The action is different from the stretch reflex of the vena cava, however. In this case the cardiac center is stimulated to decelerate the heart. Thus by pressure on the veins entering the heart and on the aorta leaving the heart, stretch impulses send stimuli to the brain that adjust the heartbeat to the physiological situation confronting the person.

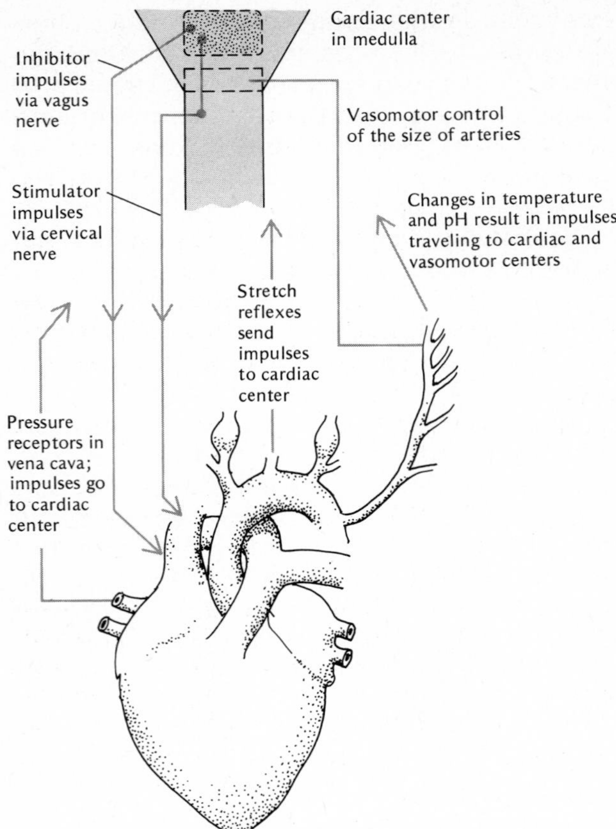

Figure 11.5 Nervous control of the heart and arteries. (The vagus and cervical nerves are shown, for simplicity, on one side only; they are really paired.)

INFLUENCE OF PERIPHERAL CHANGES ON THE HEART AND ON BLOOD PRESSURE

Blood pressure in humans is measured by a sphygmomanometer with an inflatable rubber cuff that is attached around the forearm and inflated. The cuff pressure is raised until the pulse cannot be heard in the arteries of the arm through a stethoscope. The first sound is heard when the pressure is reduced until the heartbeat, *systole,* is strong enough to force blood past the constriction; this is the *systolic pressure,* averaging 120 mm of mercury. The pulsing sound is heard until the pressure of the cuff drops to the point where it places no pressure on the vessel during the relaxation, *diastole,* of the heart; this pressure is the *diastolic pressure,* averaging about 80 mm of mercury. The *pulse pressure* is the difference between systolic and diastolic pressure, averaging 40 mm of mercury.

Blood pressure is the result of the force of contraction of the heart, the size of the blood vessels, and thus the amount of space the blood can occupy and the amount of blood in the body. Exercise can affect the force of heart contraction, and bleeding can influence the amount of blood, but under normal conditions the amount of space occupied by the blood is the most important in the control of blood pressure.

A center in the medulla of the brain that is important in controlling the space the blood can occupy is the *vasomotor center.* Nerves connect the vasomotor center to the walls of the small arteries. These can cause contraction or relaxation of the blood vessels. Generalized constriction raises the blood pressure; generalized dilation lowers it. Localized constriction can occur, as in instances of the hands and the face turning pale when cold. Localized dilation is exemplified by flushing or by the reddening of an overly warmed body part. As with the cardiac center, the vasomotor center is influenced by changes in the blood, such as by pH changes associated with CO_2 production.

The heart, the blood, the blood vessels, and the central nervous system operate together to meet the normal and unusual requirements of the body for an adequate blood supply. This can be visualized by tracing the flow of a portion of blood from the heart. The heart provides the principal force to propel the blood through the entire system until it returns to the heart. The pressure conveyed to the arteries is sustained through the contractions of their relatively thick walls. Of course, as the blood passes into successively smaller arterial tributaries, the walls of these smaller vessels provide increasing resistance, or "drag." On reaching the terminal arterioles and proceeding into the capillaries, the blood that left the heart in one pulse is now distributed through many small vessels whose collective cross sections exceed greatly the area of cross section of the aorta. This leads to a much slower flow and a much lower pressure (see below). The blood is gradually collected into fewer and fewer veins on its return to the heart. The veins have thinner walls, lack smooth muscle, and a number of veins also have valves. The pressure of the blood entering the heart is usually very low. However, muscular exercise provides a massaging action on the veins that assists the return of the blood.

It is at the tissue level where much of the significant functioning of the blood is achieved. Inasmuch as the blood does not normally flow out among the cells as it does in animals with "open" circulatory systems, the mechanisms providing exchange between the blood and the tissues are several. First, the capillary wall has selective permeability, filtering some dissolved matters and retaining others. Second, the plasma proteins, which do not leave the blood, contribute to the osmotic pressure of about 25 mm Hg, which helps to regulate the amount of fluid in the tissues. Third, difference in pressures between arterioles on the afferent side of the capillaries (about 40 mm Hg) and the small veins on the efferent side of the capillaries (about 15 mm Hg) creates a *filtration pressure,* resulting in the flow of water and dissolved substances from the blood into the tissues. Of course, under normal conditions the tissues do not become edematous ("waterlogged") because by the time the blood reaches the veins, the filtration pressure is lower than the osmotic pressure and an appropriate amount of water is returned to the blood. Normally, if the actions of the osmotic pressure and filtration pressure do not result in return of sufficient water to the blood, it eventually is returned through the action of the lymphatic vessels. We can summarize the simple pressure changes in the circulatory system as follows. At the aorta the pressure reaches about 120 mm Hg with the contraction of the heart. As the heart relaxes the pressure drops only to about 80 mm Hg due to the recoil of the elastic walls of the arteries from the systolic pressure. In smaller arteries the pressure gradually drops to an average of about 70 mm Hg. In the arterioles it tapers to about 40 mm Hg, and as the blood passes through the capillaries it drops to about 15 mm Hg. Venules and veins show an added drop so that as the blood in the vena cava enters the heart, the pressure is essentially zero, or even slightly negative. The blood is assisted in being returned to the heart by the valves in larger veins and by the syringing or pumplike activity of the abdomen and thorax. Exercising the legs also helps this process.

LYMPHATIC SYSTEM

The lymph that leaves the blood vessels for the tissues cannot accumulate in the tissues or edema, or dropsy, develops. This interferes with the delicate balance of exchanges between the cells and the environment. A system of collecting vessels, the lymphatics, normally prevents fluid accumulation (Fig. 11.6). In humans most of these vessels empty into one main lymphatic vessel, the *thoracic duct,* which runs along the

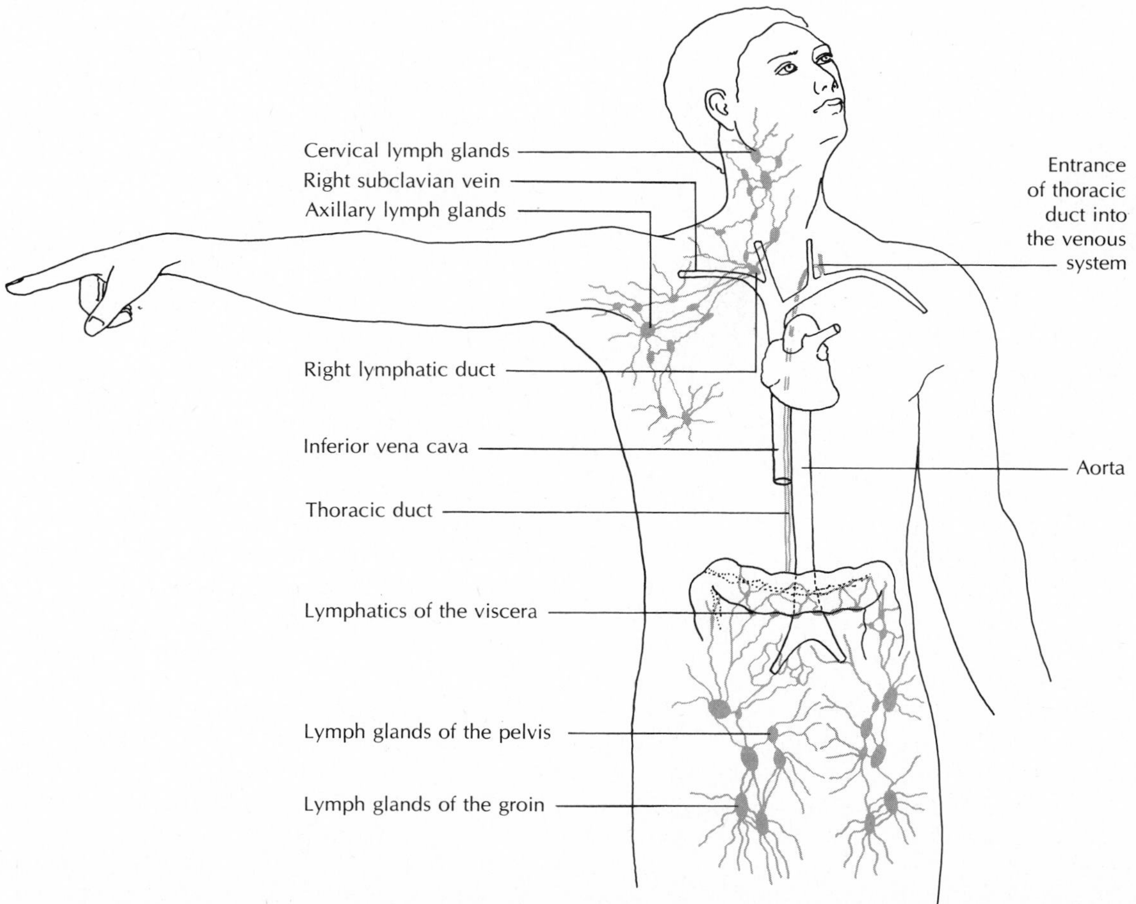

Figure 11.6 Some major lymph channels and beds in humans. Lymph vessels are distributed throughout the body in an extensive yet delicate meshwork.

back wall of the abdomen and thorax and empties into the left innominate vein near the neck. Another, smaller duct, the *right thoracic duct,* drains only the head and right arm and empties into the right innominate vein. The walls of the lymphatic vessels are quite thin. The larger vessels have valves in them like those of large veins; these control the direction of the flow of the lymph. Pressure on the vessels is produced by the various movements of the body, both voluntary and visceral, and this causes the lymph to move. At various points in the lymphatic system associated with the vessels are lymph nodes that act as filters of lymph. The significance of lymph nodes as protective structures is discussed later.

The spleen is a large filter associated with the lymphatics. It is a rather large [7.5–10 cm (3–4 inches) in diameter] irregular body, dark red in color, and located on the left of the stomach. In addition to serving as a reservoir of blood and red blood cells, it has important defense functions related to the entry of foreign substances. The *thymus* also plays an important immunological role (see below) and contributes cells to the lymphoid system.

Nature of Lymph

The blood in the arteries is under considerable pressure, and this pressure gradually decreases in passage through the arteries, capillaries, and veins, as previously described. At the beginning of a capillary net, however, the blood pressure is still relatively high. It is quite low at the venous end of the capillary net. This relationship is important. Because of the blood pressure at the beginning of a capillary net, some of the liquid part of the blood is actually filtered through the capillary walls. This is the lymph. It is often referred to as the internal environment between the blood and the cells. Toward the venous end of a capillary net where the blood pressure is much lower, the blood proteins exert an osmotic effect. As a result, some of the lymph diffuses back into the blood stream.

BLOOD

Composition of the Blood

Blood is a tissue of heterogeneous origin, with the plasma serving as its matrix. If a sample of blood is kept from clotting, it will be found to separate into about 45 percent formed elements and 55 percent plasma (Fig. 11.7).

All but one percent of the formed elements consist of red blood cells, or erythrocytes; the remainder consists of white blood cells, or leukocytes, and blood platelets (Fig. 11.8). This is the general case in all vertebrates, with some modifications in percentages. However, one difference is found in nonmammalian vertebrates: spindle-shaped cells called *thrombocytes* serve the same role as platelets.

RED BLOOD CELLS

These are biconcave disks in mammals, about 7–8 μm in diameter, lacking a nucleus at

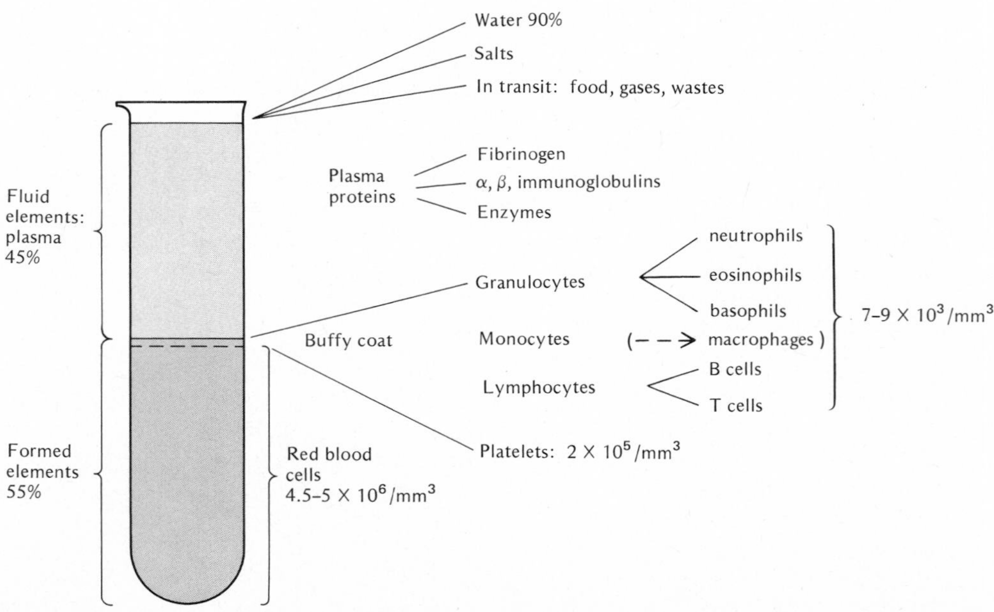

Figure 11.7 The composition of blood.

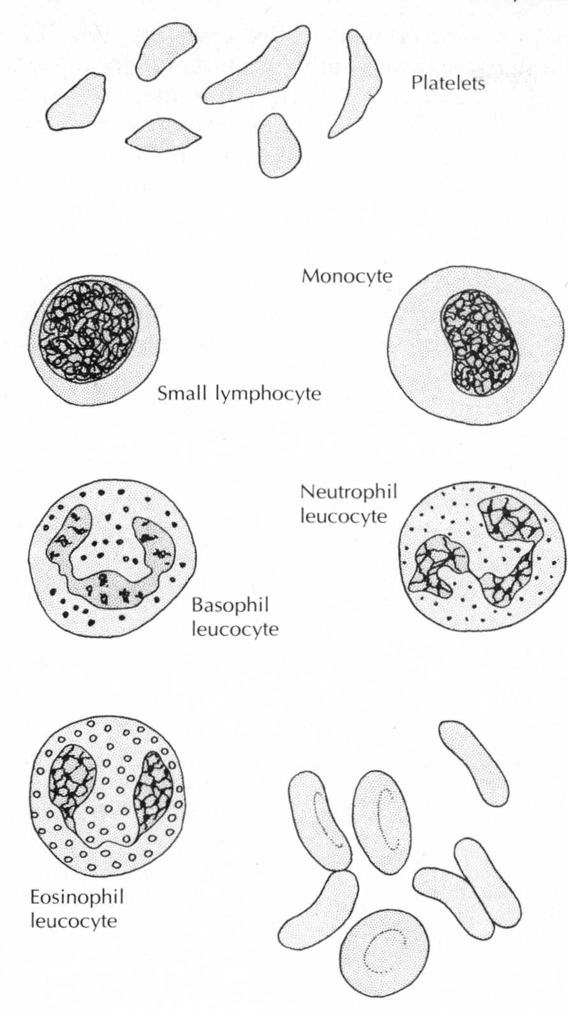

Figure 11.8 Different types of blood cells.

maturity, and having a large surface area that is associated with their high efficiency in performing one of their primary roles—oxygen transport and exchange. Except in mammals, the red cells (*erythrocytes*) have nuclei. These cells are manufactured at the rate of about two million per second in the marrow of flat bones such as ribs and in the red marrow of the ends of long bones; they are destroyed by the liver and spleen at about the same rate. The average lifetime of a red blood cell is 120 days. This leaves a relatively constant count of about 5,400,000 red blood cells per cubic millimeter for men and 4,800,000 for women under normal conditions. This concentration can rise to as high as 7½ to 8 million in people who reside at very high altitudes. It can be shown that this increase is due to the stimulus of lowered oxygen tension. In addition to the effect of altitude, the hormone erythropoietin, manufactured in the kidneys, plays a role in regulating erythrocyte production.

The role of hemoglobin in oxygen and carbon dioxide transport has already been discussed in Chapter 10. This pigment, in vertebrates, is found beneath the surface of the erythrocyte, whereas in some invertebrates, it or another oxygen carrier (for example, the blue blood protein of the lobster, hemocyanin) is found in suspension in the blood plasma. The *heme* part of hemoglobin contains iron and is responsible for the binding of oxygen. When red blood cells are aged and destroyed, much of the iron is either retained by the liver cells and reused or is lost to the circulation as heme of hemoglobin is converted to bile pigments. Under some conditions, normal numbers of erythrocytes are formed that lack an adequate amount of hemoglobin; this is a form of *anemia*. Another form is expressed by an inadequate number of red blood cells; a severe, and often fatal, expression of this kind of condition, called *pernicious anemia,* is associated with inadequate absorption of vitamin B_{12} from the digestive tract. Vitamin B_{12} is apparently essential in the synthesis of hemoglobin.

WHITE BLOOD CELLS

In humans, leukocytes occur at a concentration of 5000 to 9000 per cubic millimeter of blood, but they may increase sharply in severe infection. These cells fall into two classes: the *granular leukocytes,* which have lobed nuclei and cytoplasmic granules that take characteristic stains, either acid, neutral, or basic; and the *agranular leukocytes,* consisting of lymphocytes and monocytes, which have round or kidney-shaped nuclei, and a relatively small amount of cytoplasm. In some leukocytes the granules are packages of hydrolytic enzymes. This places these granules in the category of lysosomes (see Ch. 4). All leukocytes, except the lymphocytes, exhibit phagocytic activity, that is, they can surround and remove foreign materials. As will be discussed later, the lymphocytic leukocytes produce antibodies.

PLATELETS

Platelets, about 2 μm in diameter, lack nuclei and are formed from fragments pinched from the surface of large cells in the bone marrow. These structures play an important role in the clotting of blood. They number about 250,000 per cubic millimeter in the blood.

PLASMA

The plasma is the whole blood minus the formed elements. It is a very complex fluid and its composition probably fluctuates, within narrow tolerances, in performing its role for the rest of the body. Following are some of the materials that are transported by the plasma: (1) sugar, principally glucose; (2) lipids, which serve the body as an energy source or for synthesis into membrane systems of cells, for example; (3) amino acids from the intestine for synthesis; (4) gases, such as nitrogen, and carbon dioxide produced by cells and involved in the regulation of the pH of the blood (also, a small amount of oxygen is transported in the plasma); (5) hormones en route from site of synthesis to site of action; (6) waste products, such as urea.

In addition to the substances it transports, the plasma is about 90 percent water and inorganic salts and 7 percent proteins. The latter are of particular interest and can be subdivided into three basic categories: *serum albumins,* composing about 4 percent; *serum globulins,* over 2 percent; and *fibrinogen*, less than 1 percent. There are many other proteins, such as enzymes, in the blood, some of which are complexed with other substances, such as sugars, lipids, or metals. Although all proteins in the plasma are involved in maintaining the osmotic balance of the blood, a principal role for this function can be assigned to the albumins, which originate in the liver. The globulins are deeply involved in immune responses, and fibrinogen is a protein that contributes to clotting; both are considered in more detail below.

Before we proceed with functional aspects of some of the blood proteins, it is best to understand the classification given to various fluid parts of the blood. *Serum* is plasma minus fibrinogen and other factors associated with clotting. *Defibrinated blood* is whole blood that has had the clotting material, fibrinogen, removed. Students sometimes confuse another substance, lymph, with the blood itself. Lymph is filtered from the blood in the capillary beds and bathes cells directly. It resembles plasma but has a low concentration of protein.

Functions of Blood

There are three general functions of blood and lymph: transportation, regulation, and protection.

Transportation results in the movement of (1) *nutrients;* (2) *oxygen* for cell respiration; (3) *wastes,* such as carbon dioxide to the lungs and urea to the kidneys and skin; (4) *hormones* (see Ch. 12; the fact that blood carries secretin and cholecystokinin from the site of action in the pancreas and gall bladder, respectively, was mentioned in Chapter 9 and is representative of the blood's role in carrying hormones); and (5) *enzymes* in an active or inactive state, with the latter being represented by prothrombin.

There is regulation of *metabolism* and other body functions by the interaction of transported hormones with their targets and by the quantity of hormones (see Ch. 12); of *temperature* by changes in the bore of blood vessels under nervous, endocrine, and environmental control; and of *internal environment* in terms of salt balance, osmotic state, water concentration, and acidic or basic state of tissues.

Protection to the organism is provided from loss of blood through clotting, from invasion through the action of phagocytic cells, and from foreign substances or pathogenic organisms by developing immunity. While problems associated with transportation and metabolism have been discussed elsewhere in Part II, the problem of protection has not and is explored here.

Protection

CLOTTING AS A PROTECTIVE MECHANISM

The clotting mechanism in vertebrates prevents loss of precious blood through a very complex and intricate set of interactions that progressively convert a series of circulating, inactive enzymes in the plasma into active enzymes which ultimately leads to the conversion of the soluble blood protein, *fibrinogen*, into insoluble *fibrin* that entraps cells to form clots. The clot usually plugs the site of the wound until reparative processes occur. In an abbreviated form the following steps describe the process of clotting.

When wounding occurs, the injured tissues and the blood platelets release *thromboplastin* from thromboplastinogen. In the presence of *calcium* this converts the inactive enzyme *prothrombin* to an active enzyme, *thrombin.* Thrombin catalyzes soluble fibrinogen to insoluble strands of *fibrin* that create a meshlike plug at the wound. Thrombin is also an active agent in stimulating mitotic activity. In this way it probably also triggers reparative cell division at the wound site (Fig. 11.9).

Examples from the study of clotting illustrate two general biological principles. The first example illustrates that individual events in a series of steps are under genetic control. It is known that classical hemophilia is a sex-linked recessive gene (Ch. 19). The effect of the recessive gene is to prevent the formation of the antihemophilic factor (AHF), also known as factor VIII, that operates in conjunction with thrombin. Hemophiliacs can be protected by being ad-

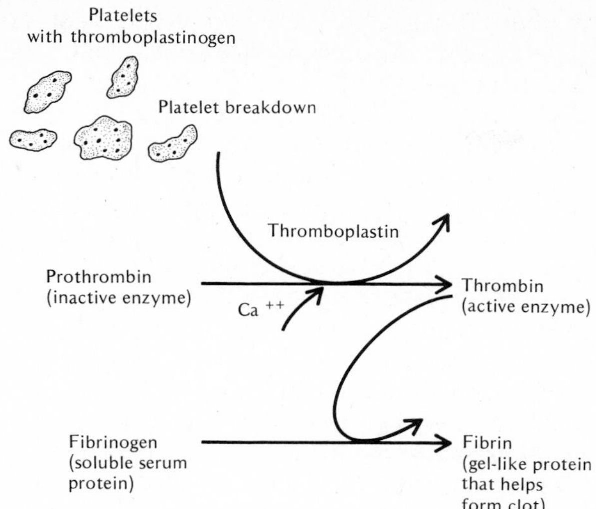

Figure 11.9 A simplified diagram of the steps in the clotting of blood.

ministered AHF obtained from normal people. The second example is more complicated but it illustrates another biological principle that biomolecules can be routed through more than one pathway. In this case another blood factor, XII, known to participate in blood clotting, is also involved in at least two other events. In one case factor XII can cause an inactive enzyme in the serum, plasminogen, to become active plasmin, to act on clots and assist in their dissolution. As repair takes place, the clot is also removed. The second event triggered by factor XII's activation is to release substances, presumably proteolytic enzymes, that through more than one step release fragments of alpha globulins in the blood. These fragments, usually 7–9 amino acids in length, are called *kinins.* They enhance vascular permeability by dilating small blood vessels and possibly also enhance the migrations of leukocytes. This is an adaptive advantage to marshal scavengers to a wound site.

BLOOD TRANSFUSIONS AND BLOOD GROUPS

Blood transfusion today is a common practice, thanks largely to the work of the Nobel Prize winner Karl Landsteiner at the beginning of the twentieth century. Prior to 1900 many transfusions had been tried, but the practice was not looked upon with much favor because death of the recipient often ensued. In some cases a blood transfusion saved a life, but in many cases death resulted. Landsteiner was able to show that such unpredictable results were caused by the mixing of two types of blood that were incompatible because they contained different natural *antibodies.* Antibodies are proteins that generally circulate in the blood after having been produced by special lymphoid cells under the stimulus of a foreign material or *antigen.* His results (see Fig. 11.10) form the basis for blood typing in hospitals today.

Human Blood Types In human blood Landsteiner found two antigens and two antibodies. The blood antigens are glycoproteins, polysaccharides associated with red cell surfaces, and are designated A and B. The antibodies are specific proteins in the plasma that are designated anti-A and anti-B. The antibodies are called *natural* antibodies. It used to be unclear how the human genetically controlled the production

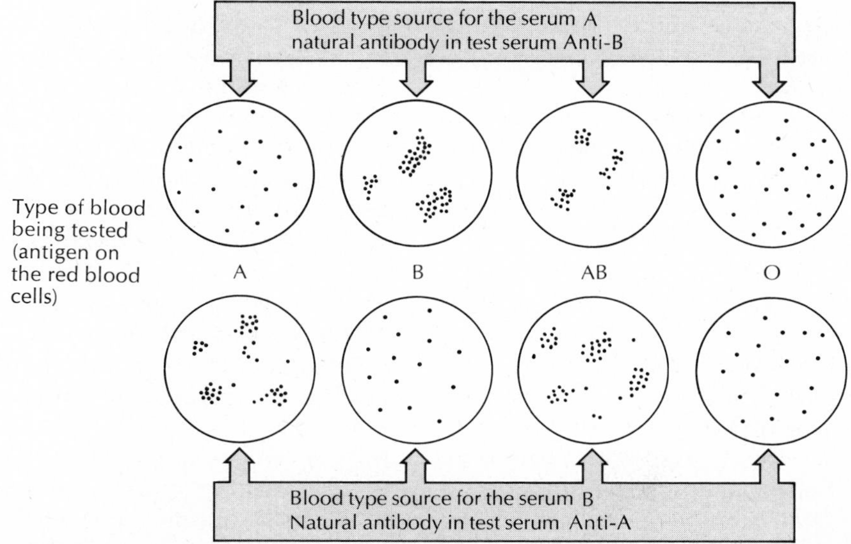

Figure 11.10 Typing of ABO blood groups by the agglutination method.

of these specific antibodies. While red cell antigens appear early in embryonic development, red cell antibodies appear and mature first during the first few months after birth. During this early postnatal period, the infant's intestinal tract becomes colonized with colon bacteria. These bacteria have polysaccharide antigens in their surface coats. The antigens are structurally similar to the glycoproteins on the surface coats of red blood cells. It is believed that these bacteria provide the antigens against which the antisera are developed. Thus a person with A blood type produces only anti-B antibody because their own A-type genetics make them tolerant to bacteria with A-like antigens. Two points tend to support this hypothesis for the development of the so-called natural ABO blood group antibodies. First, the genetics supports it. If there are no genes producing A or B antigens, as in a type O person, the individual can react to both classes of bacterial antigens, A-like and B-like, and produce both type A and type B antisera. If the person is type AB, tolerance to both bacterial antigen classes exists and no antisera are produced. Second, studies with chickens support the hypothesis. Chickens raised in germ-free environments lack bacteria and their sera are free of antibodies against the bacteria. If such chickens are experimentally contaminated with human colon bacteria, they will produce anti-A or anti-B antibodies in their sera. The situation is, nevertheless, an unusual one because humans usually do not otherwise build antibodies to molecules that so easily cross-react with antigens they carry on the surfaces of their cells.

The human population can be divided into four groups, depending on the presence, absence, or combination of A and B antigens and antibodies. Thus, we have the four blood groups: A, B, AB, and O. As Figure 11.10 shows, group A has anti-B antibody and not anti-A, group B has anti-A antibody and not anti-B, group AB has neither antibody, and group O has both antibodies. The nature of the inheritance of these blood groups is discussed in Chapter 19.

Normally, in transfusions the donor and recipient are of the same group. Group O people, however, are referred to as universal donors and their blood is sometimes used for people of other groups. In the blood of group O people there are no antigens, and the antibodies present during a transfusion are diluted so quickly that no ill effects result. From this it would appear that the ill effects that result when incompatible bloods are mixed in a transfusion, such as A (containing anti-B antibody) and AB (containing B antigen), stem from the clumping of red cells of the donor and the subsequent plugging of blood vessels.

Rh Factor About three decades ago another blood antigen was discovered in humans; this antigen must be recognized in blood transfusions performed today. It is known as *Rhesus antigen* (Rh) because it was first discovered in Rhesus monkeys. It is universally present on the red cells of these monkeys and is found in about 85 percent of the Caucasian population. People who have the antigen on their red cells are classified as Rh positive, and those lacking it are classified as Rh negative. Normally, Rh antibodies are not present in human blood. However, if blood from an Rh-positive individual is tranfused into an Rh-negative individual, Rh antibodies will then be formed in the Rh-negative individual. This would cause no immediate ill effects, but if that same person later received a blood transfusion from another Rh-positive individual, he or she might suffer a severe, or even fatal, reaction. The Rh condition is inherited in a more complex fashion than are other blood groups. It is now common procedure to type blood for the Rh condition as well as for the other groups.

The discovery of the Rh condition provided an explanation for the rather uncommon, but often fatal, ailment found in newborn babies, called *erythroblastosis fetalis.* Such infants are born in a severely anemic condition and are always of the Rh-positive type. The mothers are always of the Rh-negative type and the father must be of the Rh-positive type for this condition to arise. One out of every eight marriages in the United States is a potentially dangerous mismatch (Rh-wise). During *in utero* development, some of the red cells of the fetus may get into the maternal circulation through small breaks in the placental circulation, or more likely, the fetal red cells get into the maternal circulatory system at the time of parturition. The antigens then stimulate the formation of anti-Rh^+ antibodies in the mother. They may diffuse back into the blood of the next fetus and destroy the red cells containing the Rh antigen. The ailment does not always develop when an Rh-negative mother has several Rh-positive children. It may be in such instances that none of the Rh-positive blood ever gets into the mother's blood. Sometimes it is possible to save affected babies by prompt transfusion of the Rh-negative blood. The red cells in such blood would not be acted on by the antibodies in the baby's blood.

Recently Rh factor incompatibilities have been prevented by the following steps.

1. Rh-positive blood is injected into Rh-negative males (volunteers).
2. A fraction of the serum of these subjects is purified. This fraction contains anti-Rh antibodies and is called *Rho-Gam.*
3. Anti-Rh antibodies are injected into Rh-negative women that could have received Rh-positive cells from the infant at childbirth. These anti-Rh antibodies neutralize any fetal Rh-positive cells that may have crossed the placenta. Thus these cells are destroyed and do not induce anti-Rh antibodies in the mother. The inoculation must be repeated with each pregnancy, but it appears to be about 99 percent effective.

IMMUNE REACTIONS AS PROTECTIVE MECHANISMS

From an evolutionary standpoint it seems obvious that if there are going to be dependencies of one organism on another for energy sources, the success of the organisms will rely on the capability of the host species not to be eliminated in the process. This thesis is expanded in Part V. It is raised here to emphasize that animals have developed cellular and biochemical systems that protect them from attack by viruses, bacteria, fungi, and animal parasites. In both invertebrates and vertebrates these protective mechanisms are associated with factors in body fluids and with specialized cells. Coelomic fluids and hemolymph of invertebrates and the cells circulating in these fluids have been studied. While the research with invertebrates for their capabilities of rejecting invasion by microorganisms or of tissue transplants has not revealed the same kinds of molecules to be involved as in the vertebrates, the evidence is strong that they do have capability for rejection and also have what is called "memory" (see below).

Components of the Immune Response In vertebrates, and especially in birds and mammals, two components of the blood are involved in an animal's ability to recognize substances as foreign, to mount an immunity that specifically attacks and destroys the foreign substances. Also, a memory specific for the rejected substance is retained so that reexposure to the substance results in an accelerated and more efficient rejection than provided from the first exposure. These components are the circulating proteins, *immunoglobulins* (Ig), of the gamma-globulin blood proteins, the leukocytes and their descendants, and the macrophages found in both the blood and the tissues. Of the leukocytes, the lymphocytes are especially important as they contribute specifically to the production of immunity.

Cells of the Immune Response Lymphocytes arise, in birds and mammals, from the stem cells in bone marrow that also give rise to other kinds of blood cells. The fate of released lymphocyte percursor cells is best described in birds because there are two organs clearly involved in their becoming mature. One organ is the *thymus,* derived from the embryonic pharyngeal pouches (Ch. 16) and then located in the neck. The second organ is the *bursa of Fabricius,* derived as a dorsal epithelial outpouching of the cloaca. Cells from the bone marrow that pass through the vascular bed of the bursa of Fabricius are processed in a way not yet understood and then passed into the lymphatic system where they come to reside in specific parts of the lymph nodes and the spleen. Because of the route they travel, they are called B cells. As we shall see, when properly stimulated, their progeny produce antibodies, the Igs, that circulate in the lymph. While mammals lack a bursa of Fabricius, and its equivalent has not yet been clearly located, nevertheless mammals produce B-cell-equivalent cells that do the same things they do in birds. From this point the discussion is as relevant for mammals as for birds.

On the other hand, cells from the bone marrow that pass through or temporarily accumulate in the thymus in either birds or mammals before being circulated to the lymph nodes and spleen are called T cells. These cells do not necessarily release recognizable amounts of Igs in order to function.

A third cell type, the *macrophage*, may also be involved. Macrophages are large cells capable of endocytosis of both types (phagocytosis, pinocytosis, Ch. 4), located in tissues, lymph nodes, the spleen, and lining vascular channels. They originate by the enlargement of monocytes that roughly resemble lymphocytes.

The three cell types, B cells, T cells, and macrophages, participate in immune reactions.

Antibodies The serum protein component of some immune reactions, the immunoglobulins (Ig), are of five major classes that have molecular weights ranging from 150,000 to 900,000 daltons. We will not attempt to describe each. However, distinct functions are known for four of the five classes (see Suggestions for Further Reading). The most common class of immunoglobulin is IgG. IgG has a molecular weight of

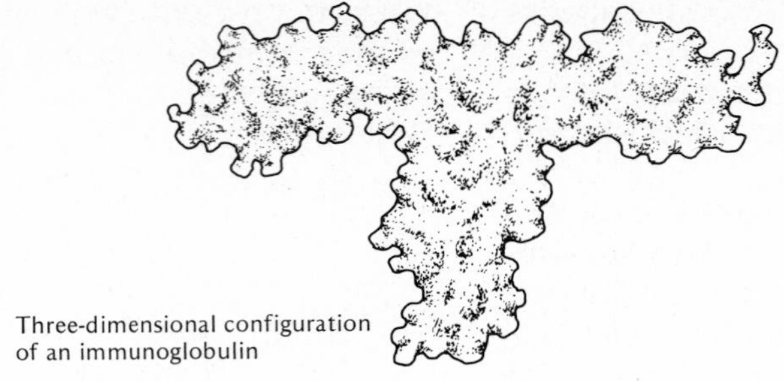

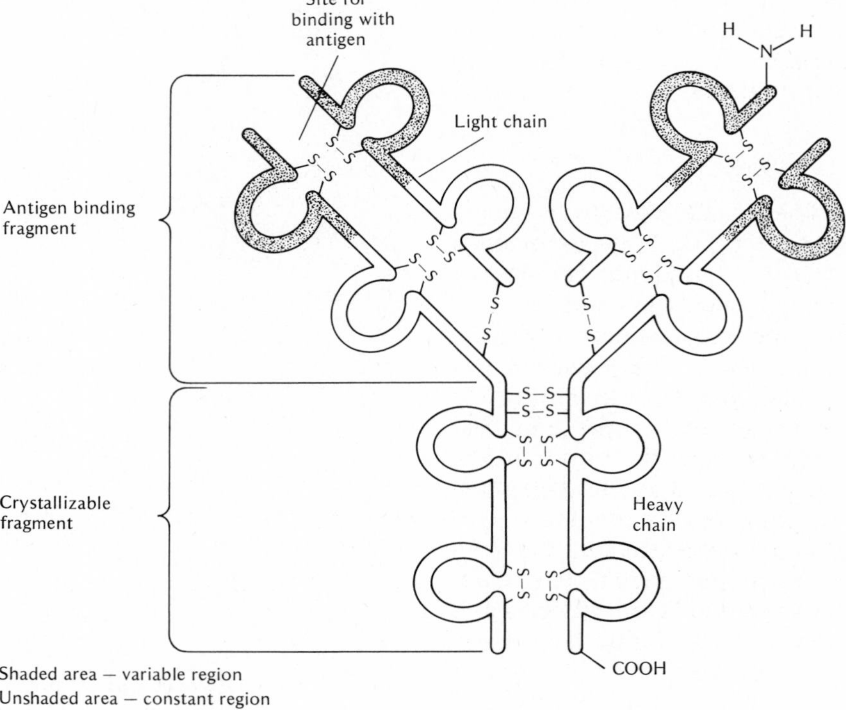

Figure 11.11 The structure of immunoglobulins (antibodies).

160,000 and is a protein with quaternary structure. It has two short *light* chains and two long *heavy* chains (Fig. 11.11). The amino acid sequence and the shape of aminoterminal ends of each light and heavy chain combination confer a specificity on each immunoglobulin that can be so precise that they will react or not with antigens with such small differences as the orientation of an OH group on a molecule. Put in practical terms, people produce IgGs that react with cat hair (dander) but not dog or horse hair, for instance. Or IgGs will be produced to one strain of influenza virus but do not react to another strain.

IgGs are produced and released by derivatives of B cells, called *plasma cells.* Plasma cells differ from B-cell lymphocytes principally in having more cytoplasm and a very well developed endoplasmic reticulum that is absent in unstimulated B cells.

Having described the principal components of immune reactions, let us now examine the development of immune reactions. Immune reactions fall into two general kinds: *humoral immunity,* in which a response to antigens leads to the proliferation of B cells, their transformation to plasma cells, and the release of circulating (humoral) immunoglobulins; and *cell-mediated immunity* in which the response to antigens is the proliferation of T cells, but no plasma cells and no detectable release of immunoglobulins. Both kinds of immunity are complex and have implications for understanding theories of development, disease, and aging that extend beyond the scope of this textbook. Each description to follow must be understood to be an oversimplified model.

Humoral Immunity An invading microorganism will have molecules of various shapes on its surface that are called antigens or immunogens. The part of the molecule to which an antibody is specific is called the *hapten.* Often, however, the isolated hapten will not induce production of the antibody, the whole molecule being required instead. When antigens enter the organism, they are recognized as foreign and usually attacked by granular leukocytes that destroy them. The products of this destruction are concentrated in lymph nodes where the imported immunogens are fixed in position, probably by macrophages. The presence of the immunogen in adequate concentration stimulates the nearby B cells to proliferation.

The response to the introduced immunogen, say the capsular coat of a bacterium, is for macrophages to process the immunogen so that "information" is relayed to the specific population of B cells capable of responding to the bacterial coat immunogens. This causes the B cells to undergo cell division to produce clones of cells. The cells produced undergo a maximum of about 6–9 divisions, thereby greatly increasing the number of cells capable of producing Igs. However, not all cells arrive at the Ig-producing state; instead, some stop dividing and are *memory cells* (Fig. 11.12). Also, there is growing evidence that some B-cell responses depend on T cells for initiation.

In the final stages of cell division as the cells approach the point of transforming into plasma cells, some begin releasing the Ig they have synthesized. The Igs unite with the immunogenic sites on the bacterium (our example) and lead to the destruction of the bacterium. Also, as all immunogenic sites are blocked, there is no longer a stimulus communicated to B cells by macrophages, and the stimulus to divide subsides. The example discussed here illustrates a minimum of five points of general biological interest: a recognition of foreignness by cells of an organism, stimulation of a very restricted population of cells to cell division, cell-to-cell communication, cellular transformation, and feedback regulation of cell division.

If the individual is later subjected to infection by the same kind of bacterium, rather than have the latent population of original B cells stimulated, the immunogenic stimulus is communicated to the population of memory cells derived from the first exposure. As the memory cell population is larger than the cells from which they were cloned, they can divide and rapidly transform into plasma cells. This is an excellent adaptive mechanism that helps the individual respond to a second exposure at an accelerated rate. (Memory cells continue to be retained; therefore, persons once immunized retain a long-lasting immunity in many instances.)

Not all humoral immunity is against microorganisms. Chemicals such as those found in hair dyes, some heavy metals, drugs such as penicillin, or insect stings or bites can cause the development of immunity. The manifestations of the immunity may be a rash or dermatitis in some milder cases but may also be vigorous and

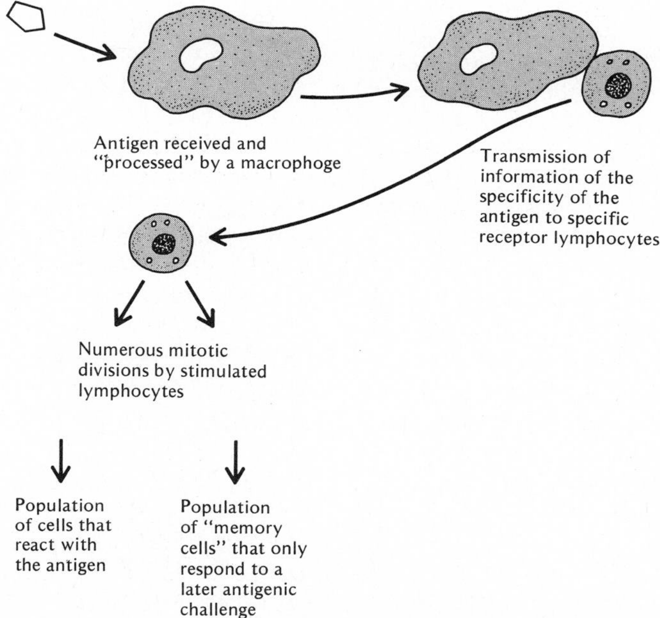

Figure 11.12 Simplified diagram of the general steps in the development of an immunity. If B cells are involved, the population that can react to the antigen converts to plasma cells that produce and release antibodies; this is humoral immunity. If T cells are involved, significant amounts of antibody are not released, and the cells must be present at the site of the reaction; this is cellular immunity.

severe. In some extreme cases, such as response of sensitive individuals to insect stings, the reaction can be sudden and massive leading to great vasodilation, loss of fluids from the blood, shock, and loss of consciousness leading to death; conditions such as these are called anaphylactic shock.

Cell-Mediated Immunity This form of immunity involves T-cells. Circulating Igs do not characterize the onset of this reaction. The protective effect of cell-mediated immunity is principally against viruses, fungi, and potentially cancerous cells that arise in the body. The reaction is usually relatively slow. With the development of surgical and prophylactic procedures in medicine that make a tissue and organ transplantation feasible, cell-mediated immunity has taken on added importance. Unless one is a twin, each person has unique biochemical characteristics not shown by any other individual. Consequently, two considerations must accompany transplantation surgery. The first is to match individuals genetically so that there are the fewest dissimilarities possible. In humans the white cells have distinctive, genetically determined immunogens on their surfaces called HL-A antigens. Specific antibodies can be produced to lymphocyte types that permit matching in the laboratory similarly to ABO blood typing. The second consideration is to provide some form of therapy to suppress adequately the person's immune response to the graft. The success of this suppression must not be so great that other necessary immunities cannot develop, nor can it be assumed that there will be an "adaptation" of the graft so that therapy can be eliminated. However, transplantation immunity is a very complex subject dealing not only with suppression of T-cells but also controlling the presence or absence of a number of factors such as "blocking" and "unblocking" antibodies that the interested student can pursue through the references quoted.

The principles of immunity outlined for humoral immunity are generally applicable here. Namely, immunogen causes specific T cells to proliferate. As they proliferate, some are set aside as memory cells. As the cells reach immunological maturity, however, they do not release Igs, and in order to act effectively T cells must be present at the site. Thus a foreign graft, such as skin, becomes highly infiltrated with lymphocytes.

Immunological Tolerance The successful transplantation of organs from one animal to another requires that the recipient be tolerant or not immune to the graft (organ) that it receives. Tissue transplantation can be done successfully into embryos before their immunological rejection systems become active and the embryo receiving the tissue will retain it throughout its life (Fig. 11.13). Animals that retain the grafts made during their early embryonic period are said to be tolerant of them. However, the immune machinery matures at different times in different organisms. Thus demonstration of immunological tolerance requires a knowledge of when the immune machinery matures in a species.

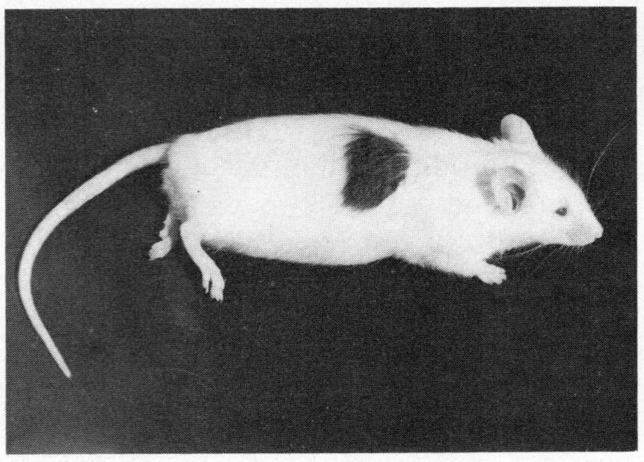

Figure 11.13 Immunological tolerance. A white mouse of strain A was injected before birth with spleen cells of black strain CBA. After birth a black patch was grafted onto the white mouse. The black was accepted rather than rejected. If the spleen cells had not been injected before birth, the patch would have been rejected. (From Fig. 3C, R. E. Billingham and W. K. Silvers, *Transplantation of Tissues and Cells*, p. 108, 1961; courtesy of the Wistar Institute.)

Allergy Allergies are responses in some people to substances that normally do not stimulate an immune reaction. A large segment of the population is afflicted by allergies of varying severities. Four different manifestions of allergy are common: hay fever type reactions usually confined to the nasal area; asthma, located in the lungs; food allergies most characteristically associated with the digestive tract; and rashes associated with the skin. The stimulating immunogen, called an *allergen*, combines with an antibody class of immunoglobulin, IgE, that triggers nearby cells in the tissues, *mast cells*, to release histamines. Histamines cause vasodilation, release of fluids, and inflammation, leading to the characteristic symptoms seen in hay fevers, for instance. If unable to avoid contact with allergens, afflicted individuals may control the effects of these reactions either by antihistamines or by corticosteroids.

The complexity and delicateness of some aspects of immunological protection provided by the blood is illustrated by the observation that the nature of these reactions can sometimes be influenced by the emotional status of the individual. Stress, therefore, seems to be a factor; this can be illustrated by the provoking of an asthmatic attack in a person under emotional stress and predisposed to asthma.

There are many other interesting aspects of immunity—immunity to cancer, autoimmunity, interferon, to name a few—and the student can find discussions about them in the Suggestions for Further Reading.

SUGGESTIONS FOR FURTHER READING

Beck, W. S. *Human Design.* New York: Harcourt, Brace, Jovanovich (1971).

Burnet, F. M. *Immunology, Aging, and Cancer. Medical Aspects of Mutation and Selection.* San Francisco: Freeman, 1976.

Florey, E. *General and Comparative Physiology* Philadelphia: W. B Saunders (1966).

Manning M. J., and R. J. Turner. *Comparative Immunobiology.* New York: Halsted (Wiley). 1976.

Schottelius, B. A., and D. D. Schottelius. *Textbook of Physiology,* 17th ed. St. Louis: C. V. Mosby (1973).

Vertebrate Adaptations. Readings from Scientific American (N. K. Wessels, Introduction). San Francisco: W. H. Freeman (1968).

From Scientific American

Adolph, E. F. "The Heart's Pacemaker" (Mar. 1967).

Aird, R. B. "Barriers in the Brain" (Feb. 1956).

Billingham, R. E., and W. K. Silvers. "Skin Transplants in the Hamster" (Jan. 1963).

Burnet, Sir M. "The Mechanism of Immunity" (Jan. 1961).

Burnet, Sir M. "The Thymus Gland" (Nov. 1962).

Clarke, C. A. "The Prevention of 'Rhesus' Babies" Nov. 1968).

Cooper, M. D., and A. R. Lawton, III. "The Development of the Immune System" (Nov. 1974).

Edelman, G. M. "The Structure and Function of Antibodies" (Aug. 1970).

Fox, H. M. "Blood Pigments" (Mar. 1950).

Hilleman, M. R. "The Induction of Interferon" (Aug. 1971).

Irving, L. "Adaptations to Cold" (Jan. 1966).

Laki, K. "The Clotting of Fibrinogen" (Mar. 1962).

Levey, R. H. "The Thymus Hormone" (July 1964).

Mayer, M. M. "The Complement System" (Nov. 1973).

Mayerson, H. S. "The Lymphatic System" (June 1963).

Perutz, M. F. "The Hemoglobin Molecule" (Nov. 1964).

Raff, M. C. "Cell Surface Immunology" (May 1976)

Warren, J. V. "The Physiology of the Giraffe" (Nov. 1974).

Wiggers, C. J. "The Heart" (May 1957).

Wood, J. E. "The Venous System" (Jan. 1968).

Zucker, M. B. "Blood Platelets" (Feb. 1961).

Zweifach, B. W. "The Microcirculation of the Blood" (Jan. 1959).

Coordination I: Hormones and Chemical Activity

The integration of the activities of the various parts of the body is brought about by the nervous system and by the products of the various *endocrine* (Gk. *endon,* within; *krino,* to separate) *glands.* The endocrine secretions are known as hormones (Gk. *hormon,* to rouse or excite) and the study of the structure and function of endocrine glands is called endocrinology. All the hormones that have been identified chemically are either proteins, peptides, amino acids or amino acid derivatives, fatty acid derivatives, or steroids. Hormones exert their effects in very small amounts, and in normal development and maintenance of an organism, a delicate balance of the hormones must exist. An excess secretion leads to *hyperfunction;* an inadequate secretion leads to *hypofunction* of each hormone (Table 12.1).

Table 12.1 Summary of Hormonal Activity in Vertebrates (Chiefly Mammals)

GLAND	HORMONE	CHIEF FUNCTIONS	DISORDERS RESULTING FROM MALFUNCTION
1. Adrenal cortex	Glucocorticoids Corticosterone Cortisone Hydrocortisone	Inhibit protein synthesis in muscle; stimulate glycogen formation; help maintain normal blood sugar level	
	Mineralocorticoids Aldosterone	Regulate sodium–potassium metabolism	
	Androgens Adrenosterone	Stimulate secondary sex characteristics	
2. Adrenal medulla	Epinephrine	Increases blood sugar by stimulating breakdown of liver glycogen; augments sympathetic division of autonomic nervous system; "fight or flight" reactions	
	Norepinephrine	Constricts blood vessels	
3. Cells in general	Prostaglandins	Antagonistic to AMP	
4. Damaged tissues	Histamine	Increases blood vessel permeability	
5. Hypothalamus (to anterior lobe of pituitary)	Specific releasing factors (RF):	RFs are transported from hypothalamus to anterior lobe via a vascular portal system where they control the release of specific hormones	
	Luteinizing hormone releasing factor (LH–RF) Follicle-stimulating hormone releasing factor (FSH–RF)		

GLAND	HORMONE	CHIEF FUNCTIONS	DISORDERS RESULTING FROM MALFUNCTION
	Corticotrophin releasing factor (CRF) Thyrotrophin releasing factor (TRF) Growth hormone releasing factor (GRF) MSH releasing factor (MRF) Melanocyte-stimulating hormone releasing factor (MIF) Prolactin release inhibiting factor (PIF)		
6. Hypothalamus (to posterior lobe of pituitary)	Oxytocin	Stimulates release of milk and contraction of uterine muscles	
	Vasopressin	Stimulates constriction of arterioles, increases blood pressure and water resorption	Hypofunction: diabetes insipidus
7. Intestine (duodenum)	Secretin	Stimulates secretion of pancreatic juice	
	Cholecystokinin	Stimulates contraction of gall bladder and release of bile; inhibits gastric secretion	
	Enterogasterone	Decreases mobility of stomach; secretion of HCl	
8. Kidney	Erythropoietin	Stimulates red bone marrow to produce new erythrocytes; controlled by oxygen-carrying capacity of blood	
9. Ovary	Estrogens	Initiate and maintain female secondary sexual characteristics; initiate periodic thickening of uterine mucosa; inhibit FSH	
	Progesterone	Cooperates with estrogens relative to female sexual characteristics; supports and glandularizes uterine mucosa; inhibits LH; stimulates hypothalamus	
	Relaxin	Stimulates relaxation of pelvic ligaments at end of pregnancy	
10. Pancreas	Insulin	Reduces amount of glucose in blood; important in formation and storage of glycogen	Hypofunction: diabetes mellitus
	Glucagon	Increases amount of glucose in blood by stimulation of glycogen breakdown	
11. Parathyroid	Parathormone	Controls calcium and phosphorus metabolism	Hypofunction: muscle twitch and parathyroid tetany
12. Pineal	Melatonin	Inhibits reproductive functions in mammals by decreasing secretion of hypothalamic gonadotrophin-releasing factors	
	Serotonin	Precursor of melatonin	
13. Pituitary (anterior lobe)	Human growth hormone (HGH), somatotrophic hormone (STH), or growth-stimulating hormone (GSH)	Stimulates bone and general body growth; affects lipid, protein, and carbohydrate metabolism	Hyperfunction in childhood: gigantism; hyperfunction in adults: acromegaly; hypofunction in childhood: dwarfism

GLAND	HORMONE	CHIEF FUNCTIONS	DISORDERS RESULTING FROM MALFUNCTION
	Follicle-stimulating hormone (FSH)	Stimulates follicle development in females and sperm and seminiferous tubule development in males	
	Luteinizing hormone (LH)	Stimulates ovulation, conversion of follicle to corpus luteum and progesterone secretion in female; stimulates testosterone production in male	
	Adrenocorticotrophic hormone	Stimulates adrenal cortex to produce corticoid hormones	Hyperfunctions: Cushing's disease, "bearded lady"
	Thyrotrophic hormone (thyroid-stimulating hormone, thyrotropin)	Stimulates thyroid to produce thyroxine and triiodothyronine	
	Prolactin (lactogenic hormone)	Stimulates milk production in mammary glands and corpus luteum to produce progesterone	
14. Pituitary (intermediate lobe)	Melanocyte-stimulating hormone (MSH)	Controls skin pigmentation in fish, amphibia, and reptiles; sebotrophic hormone in mammals—maintenance of hair lipids	
15. Pituitary (posterior lobe)		See hypothalamus, the posterior lobe is a storage-release center for hypothalamic hormones	
16. Placenta, chorionic	Gonadotrophins	Reinforce LH or substitute for it	
	Relaxin	Stimulates relaxation of pelvic ligaments	
17. Specialized cells and tissues	Pheromones	Released into the environment where they signal specific responses among members of a population; release is often dependent on hormonal stimulation	
18. Stomach (pyloric mucosa)	Gastrin	Stimulates secretion of gastric juice	
19. Testes	Testosterone	Initiates, maintains male secondary sexual characteristics	
20. Thymus	Thymosin	Retards aging by stimulation of lymphocyte proliferation	
	Thymin I and II	Stimulate bone marrow to produce T lymphocytes	
21. Thyroid	Thyroxin	Stimulates oxidative metabolism in all cells; inhibits thyroid-stimulating hormone (TSH)	Hypofunction in infants: cretinism; hypofunction in adults: myxedema; hyperfunction: exophthalmic goiter
	Triiodothyronine	Same as thyroxin	
	Thyrocalcitonin	Prevents excessive rise of calcium in blood; antagonistic to parathormone	

Because their hormones are secreted directly into the bloodstream, the endocrine glands also are called *ductless glands.* Although some of the endocrine glands have more than one function, the endocrine part of such an organ secretes directly into the bloodstream. The pan-

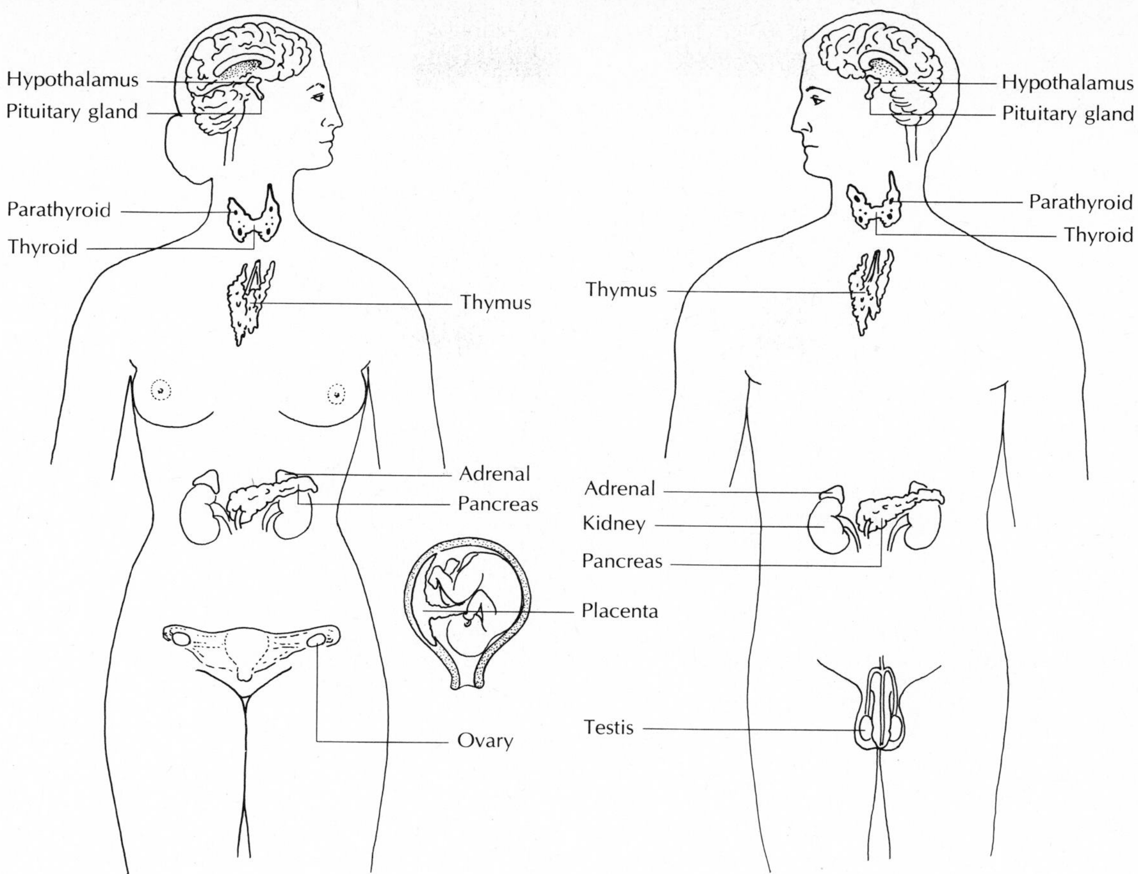

Figure 12.1 Distribution of endocrine structures in humans.

creas, the ovaries, and the testes are endocrine structures that, in addition, produce other materials (digestive enzymes in the case of the pancreas and sex cells in the case of the ovaries and testes) that leave the organ by way of a duct. Glands whose secretions are transported by ducts are *exocrine glands.*

In the discussion of digestion (Ch. 9) the functions of *secretin* and *cholecystokinin,* produced in the duodenum, as well as the function of *gastrin,* produced by glands in the stomach, were considered. Erythropoietin, produced by the kidney, was discussed in Chapter 10. Figure 12.1 shows the location in the human body of the main endocrine glands.

THE PANCREAS

As described in Chapter 9, the pancreas is an irregularly shaped organ in the mesentery between the stomach and the duodenum. When this organ is examined histologically, it is found to be composed of two kinds of glandular tissues. Scattered between the *acini,* which secrete digestive enzymes into ducts, are clusters of special cells called *islets of Langerhans.* This is the tissue that secretes the pancreatic hormones into the bloodstream (Fig. 12.2).

Insulin, produced by the beta cells, is considered first because of its historical significance. Physiologists have long been interested in the pancreas because of its digestive functions. In 1889 two German physicians, von Mering and Minkowski, were studying the digestive disturbances that resulted from the removal of the pancreas in dogs. In the course of their experiments the animal caretakers noticed that ants gathered in the cages of the experimental dogs from which the pancreas had been removed. An investigation showed that the urine from these dogs was loaded with sugar and this explained the presence of the ants. The two physicians found that after such an operation the dogs always died in from 10 to 30 days. The symptoms developed by these dogs were quite like those of the people afflicted with severe *diabetes mellitus.* This human disease had been known for centuries, but the cause was unknown and its control was very inadequate.

As a result of the observations in the laboratory of von Mering and Minkowski, other workers

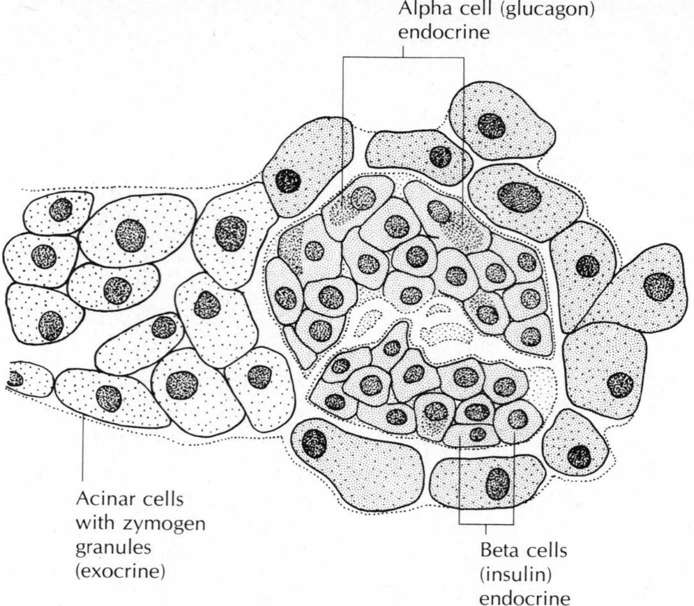

Figure 12.2 Section of pancreas showing exocrine acinar cells and endocrine islet of Langerhans cells. Alpha and beta cells have different endocrine functions.

became interested in the problem. They reasoned that if experimental diabetes could be produced in dogs, perhaps control would be possible. If the control was found for dogs, it might work also in humans.

The fact that the removal of the pancreas, not the operational procedure, was the cause of diabetes in dogs was demonstrated first. This discovery was accomplished by control operations without the removal of the pancreas. It was then possible to show that it was the removal of the islet tissue that produced the effects. When the ducts of the pancreas are tied off, the digestive part of the gland disintegrates, but the islet tissue remains intact. In dogs treated in this manner, no diabetes developed. The next step involved attempts to extract the hormone from the pancreas. For years many such attempts were made without success, and many of the extracts that were tried were injurious. The problem was finally solved in 1922 by Dr. Frederick Banting, working at the University of Toronto. Banting and associates reasoned as follows: the pancreas is a double organ and perhaps all the failures to extract the hormone resulted from the destruction of the hormone by the digestive enzymes in the pancreas, when the whole pancreas was ground before the hormone could be extracted.

Banting knew that in the embryo the islet tissue develops before the rest of the gland. So with the collaboration of his coworkers, Banting extracted embryonic pancreas and found that this extract would alleviate the symptoms in diabetic dogs.

Since 1922 methods of extracting the glands have improved and now quantities of purified insulin are available. Presently commercial insulin is extracted from bovine, sheep, or swine pancreas by an acidic alcohol method that rapidly inactivates the proteolytic enzymes. Thousands of people who otherwise would have died are able to lead rather normal lives through daily injections of insulin. Insulin cannot be taken orally because it is destroyed by enzymes of the digestive tract.

The primary function of insulin is the control of glucose metabolism. It is necessary for the proper oxidation of glucose to yield energy and also for the storage of glucose as glycogen in the liver. When there is a lack of insulin in the bloodstream, glucose accumulates in the blood and is passed off in the urine. The glucose content of the urine of a diabetic is often as high as 8 percent. If such a person abstains from eating carbohydrates, glucose continues to be formed; it can be formed from amino acids. Even tissue proteins are broken down in this process. A severe, untreated diabetic suffers from insatiable thirst, excessive urination, loss of weight, weakness, cataracts on the eyes, blindness, and finally a coma ending in death.

The lack of insulin or the proper amounts of it in the bloodstream is known as hypoinsulinism. The opposite condition, hyperinsulinism, also can occur. This may happen when a diabetic injects too much insulin and the level of the blood sugar drops so low that the cells of the brain are affected and the person suffers from *insulin shock.*

Insulin is a small protein composed of two polypeptide chains, A and B. The A chain contains 21 amino acids and the B chain contains 30; the total molecular weight of bovine insulin is 5733 daltons. The chains are held together by sulfur–sulfur bridges (—S—S—). Insulin was the first protein to have its primary structure determined. Within the past decade, insulin has been shown to be synthesized as a single-chain precursor form of 84 amino acids. This protein, proinsulin, is activated by the cleavage of a 33-amino acid "connecting peptide" (Fig. 12.3).

Treatment of diabetes includes stringent diet control. Some mild cases of diabetes can be controlled completely with regulated diet. Some cases of diabetes may be controlled with "oral insulins," which are synthetic compounds that stimulate sluggish beta cells to produce insulin. In the normal individual the level of blood sugar determines whether insulin will be released. This illustrates an interesting checks and balances system of a feedback mechanism (acting specifically on the beta cells) in which the blood sugar regulates its own concentration through insulin: a high concentration stimulates insulin secretion and the lowered blood sugar no longer stimulates insulin release.

Currently intensive research looks for an artificial pancreas. An artificial glucose sensor as small as the little fingernail has been developed; this device can measure variation of blood sugar in either direction. Ultimately the sensor will activate an artificial insulin-secreting "beta cell." The sensor will transmit fluctuations in blood sugar through a little battery to a computer which in turn will determine how much insulin is or is not needed. This will control a reservoir of insulin through another battery. The sensor, batteries, and insulin reservoir will all be inside the body, but near the surface for ease in handling and refilling. In 1974 the first artificial pancreas was sold to a medical center, but it is the size of a television set; miniaturization is yet to come. Diabetes is increasing as a health problem because the frequency of diabetics in the population is rising. Commercially available insulin has permitted many people, who normally would have died early in life, to survive and reproduce, thereby passing their "diabetic" genes to their children. Although the precise mechanism of inheritance of diabetes is not clear, diabetes does have a strong genetic component. It runs in families. There are 8 to 12 million diagnosed and undiagnosed diabetics in the United States and the annual medical bills for diabetes run into the millions of dollars. Diabetes mellitus may be divided into two general types: *infantile,* characterized by early onset, and *adult,*

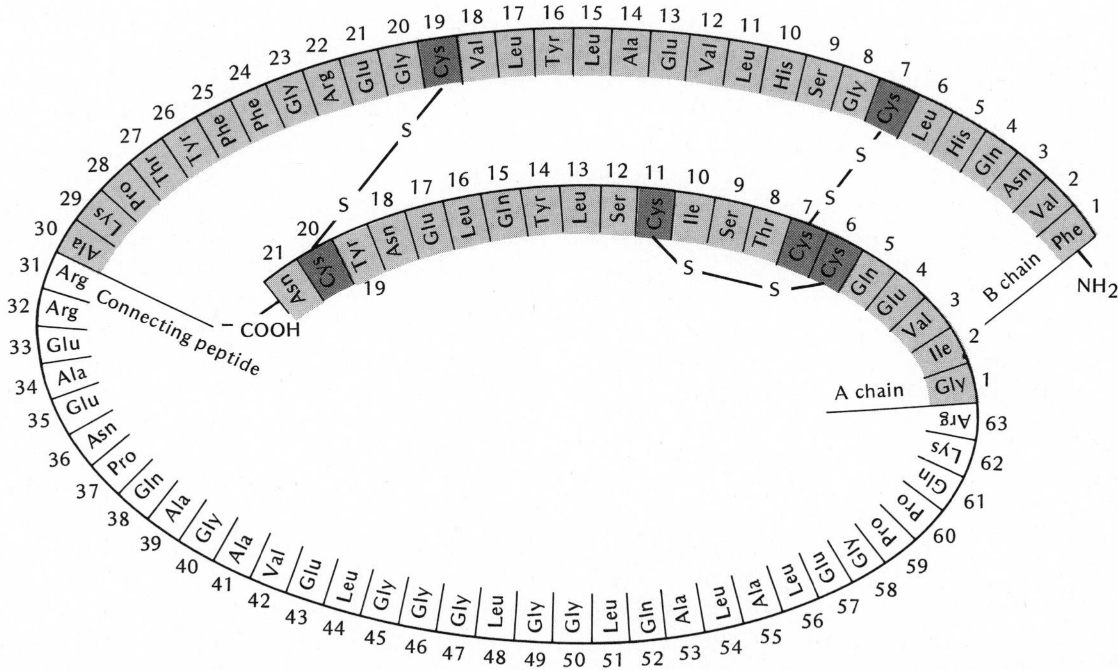

Figure 12.3 Insulin activation. The insulin molecule is synthesized as one protein chain, proinsulin. Removal of a segment, the connecting peptide, produces active insulin which has two chains. Insulin was the first protein to have its primary structure (amino acid sequence) determined completely.

which starts later. (Eighty percent of these cases are detected after the age of 40.)

The alpha cells of the islets of Langerhans produce glucagon, a single chain of 29 amino acids. It elevates the blood glucose by acting on the liver, apparently enhancing the conversion of glycogen to glucose. Glucagon serves as just one of several mechanisms whereby the body does this, since the removal of the pancreas does not interfere with this conversion.

The following general points can be gained from the example of the pancreas as an endocrine gland.

1. A specific cell type is associated with a specific hormone.
2. The endocrine cells are independent of the rest of the gland.
3. The functions of glucagon and insulin are antagonistic, but glucagon is dispensable because there are alternative routes for stimulation of glycogen breakdown, such as by epinephrine (see below).
4. The hormonal action at the cellular level can influence tissues or organs, for example, the development of cataracts.
5. Correction of the deficiency is possible.
6. The control of synthesis and the secretion of a specific hormone is regulated by a feedback mechanism.
7. The control is dependent on the concentration of a specific compound in the bloodstream.

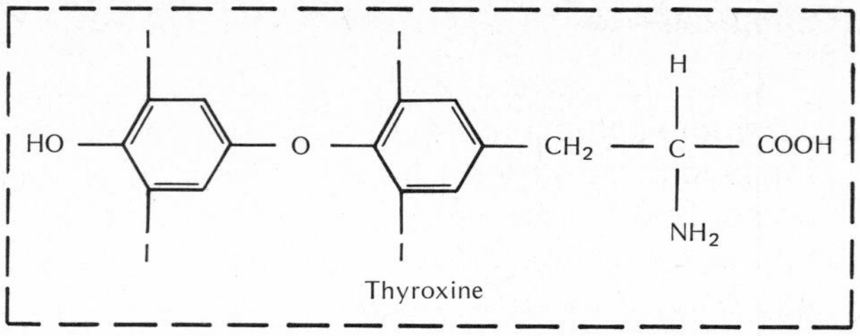

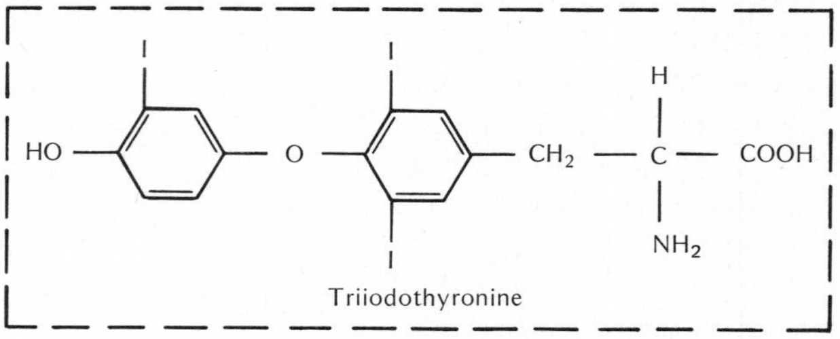

Figure 12.4 Thyroxine and Triiodothyronine

THE THYROID GLAND

In humans the thyroid is a bilobed structure located along the sides of the trachea just below the larynx. The two lobes are connected by a narrow band of tissue, the *isthmus.* In all vertebrates the thyroid develops as an outgrowth from the floor of the embryonic pharynx. A microscopic examination of a cross section of the gland shows that it is made up of numerous follicles that are lined by single layers of cuboidal epithelial cells.

Thyroxin, the classical thyroid hormone, is disseminated widely via the blood stream and acts on all cells where it determines the production of specific messenger RNAs associated with the production of enzymes that are involved with energy metabolism. Thus the general effect of thyroxin in the vertebrate animal is to regulate the rate of metabolism—the rate of burning foods. For normal metabolism this hormone must be present in definite amounts. An undersecretion lowers the rate of metabolism; an oversecretion increases the rate of metabolism.

The thyroid gland has an exceptionally rich blood supply and the follicle cells have a remarkable ability to accumulate iodine (in the form of iodide) from the blood. Iodide is used in the synthesis of the protein thyroglobulin, which is secreted into the follicular cavities and stored there. Proteolytic enzymes hydrolyze thyroglobulin into its constitutent amino acids, one of which is thyroxin. Thyroxin is derived from two molecules of tyrosine, a common amino acid, by iodination of the aromatic rings in four places. Thyroxin also is called tetraiodothyronin. Triiodothyronine (Fig. 12.4), another thyroid hormone, is three to five times more active than thyroxin in speeding up catabolism, but it is secreted in smaller amounts.

The production and discharge of the thyroid hormones are regulated by the thyroid-stimulating hormone (TSH) which is secreted by the anterior lobe of the pituitary gland. Control of the thyroid gland by the pituitary has been found in rats, humans, and other mammals. Secretion of TSH is regulated in turn by the amount of thyroxin in the blood. A decreased production of thyroxin stimulates the pituitary to release TSH (through a thyroid-stimulating hormone-releasing factor, TSH-RF) which then induces the thyroid gland to raise its output of thyroid hormones. When the thyroid hormone level is brought back to normal, secretion of TSH is decreased. Thus the feedback mechanism between thyroid hormones and TSH keeps the output of thyroxin relatively constant and the basal metabolic rate within the normal range.

Simple Goiter

The enlargement of the thyroid gland, known as *simple goiter,* has been recognized in medicine for several centuries. Long ago Greek physicians prescribed the drinking of seawater as a cure for it, and the Chinese found that eating seaweeds was helpful. Another interesting aspect of simple goiter, which has been known for some time, was its peculiar distribution around the world. Although occasional cases of simple goiter have been found in all parts of the world, the disorder was rather common until recent years in certain regions (the Alps, the Pyrenees, and the Carpathian mountain regions of Europe; the South American Andes; and the Great Lakes region and St. Lawrence River Valley of the United States and Canada).

Several discoveries ultimately led to the correlation of the unusual distribution of the disorder and the iodine in the medications (seawater and seaweeds, particularly) that had been used. Now it is known that these regions of the world where simple goiter has been prevalent for so many years have little iodine in the soil. In these regions the iodine was leached from the soil as a result of the action of glaciers during the last Ice Age. The use of seawater and seaweeds as a cure for simple goiter, which was done empirically, now makes good sense, because we know that both seawater and seaweeds have a high iodine content. All of these correlated facts have led to the use today of ordinary table salt fortified with potassium iodide.

Hypothyroidism

Simple goiter is actually one manifestation of undersecretion of the thyroid gland, or *hypothyroidism.* In this case the gland enlarges because of the lack of iodine in the food and water. This may result in normal or only slightly subnormal metabolic rates. A very low output of thyroxin in a new-born child will, if left untreated, lead to *cretinism.* This condition, once common in the "goiter regions" of the world, is now fairly rare. In a human cretin growth is stunted and mental development is retarded greatly. Other effects are bowed legs, defective teeth, enlargement and protrusion of the tongue, and a coarse and leathery texture of the skin. Fortunately these effects can be overcome or prevented if treatment is begun early.

If the thyroid atrophies in the adult with a resulting decrease in thyroxin secretion, a condition known as *myxedema* results. A person with myxedema has a basal metabolic rate much lower than normal. Associated with this lowering of metabolic rate are several other symptoms such as loss of vigor, reduced mental activity, the accumulation of fat, puffy skin, and a lessening of the sex drive. When proper doses of thyroxin are given to such an individual, the metabolic rate becomes normal again and the other symptoms disappear. Thyroxin, being a modified amino acid, is not affected by digestive enzymes and so may be taken orally.

Hyperthyroidism

In humans the disorder that results from an oversecretion of thyroxin is known as *exophthalmic goiter.* The gland may appear normal in size or it may be enlarged slightly, but in any case, the amount of thyroxin is greatly increased, sometimes up to twice the normal amount. This greatly increases the basal metabolic rate. Among the symptoms of this disorder are protrusion of the eyeballs (thus the name *ex-,* out; *ophthalmic,* eyes), high blood pressure, nervous tension and irritability, profuse sweating, loss of weight, and fatigue. This condition appears to result from an oversecretion of the thyroid-stimulating hormome of the pituitary. When this condition is severe, death usually results unless the individual is treated. As a rule, the metabolic rate of such an individual can be restored to normal by the surgical removal of just the right amount of thyroid tissue. Recently, radioactive iodine, ^{131}I, has been successfully used to correct this situation. Since iodine is concentrated in the thyroid cells for the synthesis of thyroxin, the rays that are given off by the radioactive iodine destroy some of the thyroid cells.

Thyrocalcitonin, discovered in 1961, is unrelated to the other thyroid hormones. The chief effect is the prevention of excessive calcium in the blood. Thus thyrocalcitonin is antigonistic to the action of parathormone, discussed next.

THE PARATHYROID GLANDS

In humans four very small parathyroid glands are imbedded in the tissue of the thyroid. Because of this intimate association of the parathyroids and the thyroid gland, considerable confusion existed in early experimental work on the thyroid. Now it is known that the secretion of the parathyroids, known as parathormone (PTH), is absolutely essential for life. In some of the early experimental work on thyroid removal, some workers reported death soon thereafter (in dogs), whereas others found no such drastic effects (in rabbits). Later it was understood that in those cases where death occurred both the parathyroids and the thyroids had been removed. In rabbits the parathyroids are located just outside of the thyroid; in dogs they are embedded in the thyroid, as in humans.

When the parathyroid glands of an animal are removed, the animal soon begins to suffer from severe muscular tremors (*parathyroid tetany*) and, finally, convulsions that lead to death. The injection of either parathormone or calcium will restore such an animal to the normal condition, but only for a short time. Repeated injections are necessary to keep the animal alive.

The hormone is essential for the maintenance of the proper level of calcium and phosphorus in the blood and for the control of calcium metabolism in the cells. Parathormone increases the concentration of calcium in the blood and decreases the concentration of phosphate by acting on the kidneys, the bones, and the intestine. It inhibits excretion of calcium by the kidneys and intestine, but stimulates release of calcium into the blood from the bones. Calcium in bones exists in the form of calcium phosphate, and the release of calcium also releases phosphate. Parathormone compensates for phosphate release into the blood by stimulating phosphate excretion by the kidneys. Actually, more phosphate is excreted than is added to the blood from the bones, and the net result is a decrease of phosphate in the blood as the parathormone concentration increases. Hyperparathyroidism sometimes results when the parathyroid glands develop tumors. The bones become weak and are easily bent or fractured because of withdrawal of calcium.

When the calcium content of the blood is too low because of too little parathormone, membrane functions are affected and the muscles and nerves become extremely irritable; this leads to the convulsive twitching. Complete absence of parathormone is usually fatal unless large quantities of calcium are included in the diet or injections of parathormone are taken.

THE ADRENAL GLANDS

There are two adrenal glands, and in humans one is located on the top of each kidney. The adrenal gland in mammals is composed of two distinct regions, an inner medulla and an outer cortex. The two regions have separate embryological origins. The cellular structure of the two regions as well as the hormones produced by them are quite different.

Two hormones have been identified from the medulla, but it appears that a number of different hormones are produced in the cortex. An animal can survive very well the complete removal of the medulla, but it will soon die if all the cortical tissue is removed. The two human glands have a richer blood supply per mass of tissue than any other organ in the body.

The Adrenal Medulla

The hormones produced in this part of the gland are epinephrine and norepinephrine (also called adrenaline and noradrenaline, respectively, Fig. 12.5). An injection of epinephrine produces a number of marked effects. There is an increase of heart rate, a rise in blood pressure, an increase in the glucose content of the blood, a decrease in the glycogen content of the liver and muscle, and an increase in muscular power and resistance to fatigue. In addition, the pupils of the eyes dilate, the hair "stands on end" due to the action of the arrector muscle, and the skin blanches due to constriction of the blood vessels.

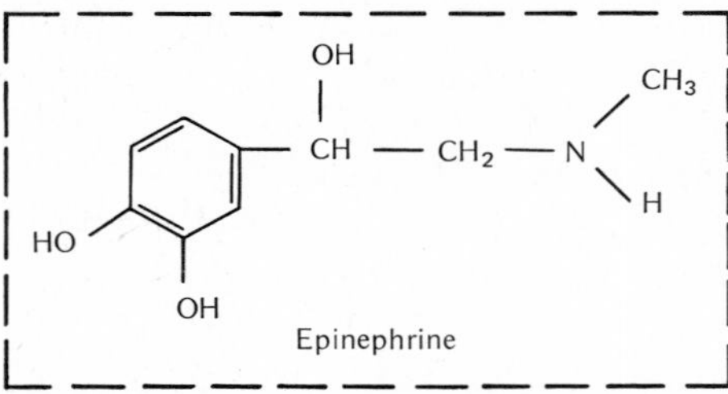

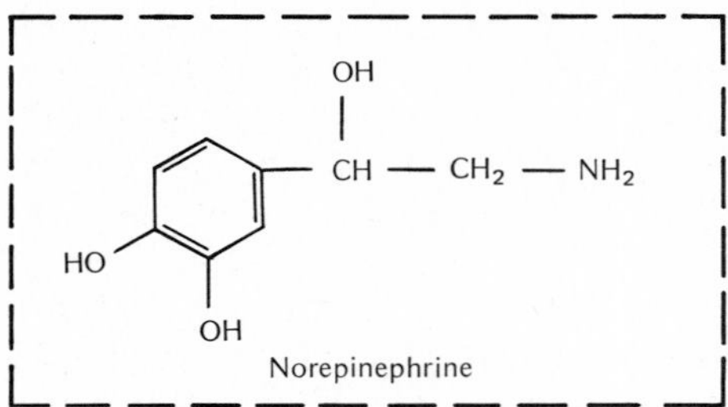

Figure 12.5 Epinephrine and norepinephrine.

In such a condition the body is ready to function to the maximum of its ability. Such conditions develop when there is great emotional stress, as in fear and rage, and apparently many of the reported unusual feats of strength are accomplished when the resources of the body are so rallied. The excited condition produced by epinephrine has been termed the "fight or flight" reaction.

Practically all the effects caused by epinephrine may be produced by the generalized stimulation of the sympathetic nervous system (Ch. 13). The adrenal medulla is innervated by sympathetic nerves. Inasmuch as an organism can live quite well with no medullary tissue, it may be that the function of epinephrine is merely that of augmenting the functions of sympathetic nerves. Norepinephrine has effects similar to epinephrine, but it causes more vasoconstriction and is less effective in stimulating the conversion of liver glycogen to glucose.

The Adrenal Cortex

This part of the gland is certainly one of the most complex endocrine structures in the body, and it has been, perhaps, the most difficult to analyze and understand. As mentioned above, the complete removal of all cortical tissue leads to death in a short time. *Addison's disease* is due to an atrophy of the adrenal cortex. The symptoms of this disease are loss of appetite, muscular weakness, loss of sodium chloride and water from the body, and a characteristic bronzing of the skin. In severe cases death results. The first extract prepared from the adrenal cortex that brought relief to sufferers of Addison's disease was referred to as *cortin.* Since that time, 1926, more than 50 different hormones have been isolated from the adrenal cortex. The one that is probably most familiar is *cortisone.* It has been widely used in the treatment of arthritis, leukemia, and certain skin conditions. The use of adrenocorticotrophic hormone (ACTH), produced by the anterior lobe of the pituitary, for the same purpose is explained on the basis that ACTH stimulates the adrenal cortex to release more hormone in the form of cortisol, which is converted to cortisone in the liver. ACTH increases the size of the adrenal cortex by stimulating RNA and protein synthesis. In turn, the secretion of ACTH is inhibited by cortisone; thus a feedback control exists between the pituitary and the adrenal cortex.

The hormones of the adrenal cortex are all steroids (Ch. 3) and may be grouped into three classes on the basis of their functions:

1. *Glucocorticoids* which act primarily in regulating carbohydrate and protein metabolism: conversion of proteins to carbohydrates; active as anti-inflammatory agents
2. *Mineralocorticoids* which act primarily in regulating salt and water balance: sodium and potassium metabolism

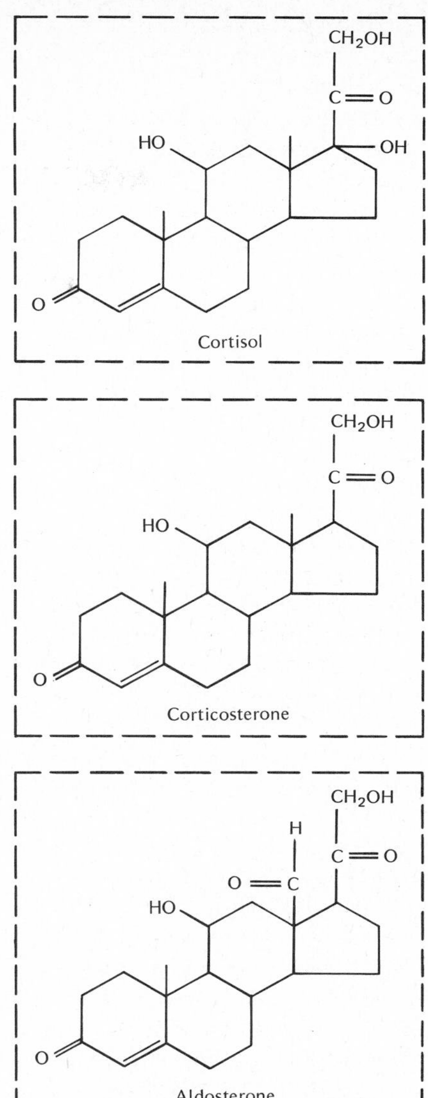

Figure 12.6 Hormones of the adrenal cortex. Cortisone is closely related to cortisol and corticosterone.

3. *Androgens* which act primarily as sex hormones: male sex activity

Most effects of cortical extracts have been attributed to the glucocorticoids (Fig. 12.6). In most tissues these hormones reduce amino acid uptake and stimulate the synthesis of glucose from amino acids and glycerol. Virtually every feature of lipid metabolism appears to be affected by the glucocorticoids.

Aldosterone (Fig. 12.6) is the most important mineralocorticoid in humans. It is absorbed preferentially by nuclei of kidney cells where it triggers the synthesis of selected messenger RNAs which in turn lead to increased production of oxidative enzymes in mitochondria and greater ATP production through oxidative phosphorylation. The greater availability of ATP at the site of the sodium–potassium pump in the cells of the convoluted tubules leads to sodium retention and potassium release. The retention of sodium leads to reabsorption of water and chloride ions, which increases both blood volume and blood pressure.

The androgens are very similar in structure and function to the sex hormones produced in the gonads and are involved in the development of secondary sexual characteristics. Although both male and female cortical androgens are produced, the male hormones (for example, adrenosterone) predominate.

THE PITUITARY GLAND

The pituitary gland, also called the hypophysis, is a little larger than a pea and is located in a depression of the sphenoid bone of the skull called the *sella turcica* (L. *sella,* saddle; *turcica,* Turkish). Thus the pituitary is one of the best protected and inaccessible endocrine glands. It is attached to the hypothalamus by a stalk, the *infundibulum.* The pituitary gland has a double origin, and in most vertebrates is composed of three distinct regions: the *anterior lobe* (*adenophypophysis*) and the *intermediate lobe* (*pars intermedia*) which arise as a pouch that grows dorsally toward the brain from the roof of the embryonic mouth; and the *posterior lobe* (*neurohypophysis*) which arises as a ventral outgrowth from the hypothalamus, a connection which is maintained in the adult. As will be seen, the hypothalamus has an important role in the functions of the endocrine system. In the young human the intermediate lobe is a distinct structure, but it gradually fuses with the posterior lobe and becomes obscured. However, in most vertebrates the intermediate lobe remains as a distinct structure. Sections of the whole gland show differences in the cellular composition of the different lobes.

When the pituitary gland is removed from a young rat, a number of striking effects are produced. The rat stops growing and never reaches sexual maturity. However, if instead of removing the pituitary, pituitary extracts are injected into the rat, it will grow rapidly, and ultimately reach a state of *gigantism.*

Hormones of the Anterior Lobe

From the above statement it is obvious that the pituitary secretes a hormone that regulates growth. This hormone is produced in the anterior lobe and is known as *human growth hormone* (HGH), *somatotrophin, somatotrophic hormone* (STH), or *growth-stimulating hormone* (GSH). The anterior lobe produces five other hormones. These are *thyrotropin* or *thyroid-stimulating hormone* (TSH) which stimulates the thyroid gland; *adrenocorticotrophic hormone* (ACTH) which stimulates the adrenal cortex; the two *gonadotropins* discussed in detail in Chapter 19—the *follicle-stimulating hormone* (FSH) and the *luteinizing hormone* (LH); and *prolactin,* the *lactogenic hormone* which stimulates the growth of mammary glands in pregnancy and initiates milk production by them.

If an oversecretion of the growth hormone starts after an individual reaches maturity, the condition known as *acromegaly* develops. By this time most of the parts of the body have lost their capacity for growth. Growth is possible only in the bones of the hands, feet, and face. The hands and the feet become much larger than normal, the lower jaw becomes abnormally long and broad, and there is a thickening of the nose and of the ridges above the eyes.

Chapter 15 describes FSH and LH, the gonadotrophic hormones of the pituitary. Both are necessary for the changes that take place at puberty in both males and females. FSH stimulates the development of the ovarian follicles in the female and the formation of sperm in the seminiferous tubules of the testes in the male. LH is necessary for ovulation to occur in the female, and it is necessary for the complete development of the interstitial tissue of the testes and the secretion of the male hormone, *testosterone.*

In addition to secreting hormones that regulate the development and function of the gonads, the anterior lobe secretes hormones that regulate the development of the adrenal glands and the thyroid (see discussion above).

Thyrotropin affects the development and functioning of the thyroid gland. A rat, deprived of its pituitary, shows an atrophy of the thyroid gland. When an extract containing this hormone is injected into a normal animal, there is a marked increase in size of the thyroid gland that leads to an increase in the basal metabolic rate.

Prolactin, the lactogenic hormone from the anterior lobe, is responsible for the secretion of milk by the mammary glands after the birth of young. When male animals are treated with female sex hormones (which causes their rudimentary mammary glands to enlarge), injections of prolactin will cause the secretion of milk. Prolactin also functions to induce the corpus luteum (Ch. 15) to produce progesterone and it also acts to prolong the functional life of the corpus luteum.

The preceding account indicates that the anterior lobe of the pituitary, through its different hormones, exercises considerable control over the functions of some of the other glands. This regulation, however, is not entirely a one-way affair. The stimulated glands may, in turn, cause some regulation of pituitary functioning. For instance, when more and more ACTH is secreted by the pituitary, the adrenal cortex secretes more and more of its hormones into the blood. However, a concentration of these hormones from the adrenal cortex is finally reached which feeds back to inhibit the further pituitary secretion of ACTH.

Hormones of the Posterior Lobe

Two posterior-lobe hormones have been isolated, structurally characterized and synthesized by Vincent duVigneaud. Both are peptides containing nine amino acids and they differ in only two amino acids (Fig. 12.7). Neither of these hormones is produced in the posterior lobe itself. Rather they are synthesized in the cells of the hypothalamus of the brain. From there they pass along nerve tracts in the infundibulum or stalk to the posterior lobe for storage prior to release into the bloodstream. The development of the posterior lobe from the floor of the brain explains the mechanism of this *neurosecretion* by the hypothalamus and notes the breakdown of a rigid distinction between the nervous and endocrine systems in coordination.

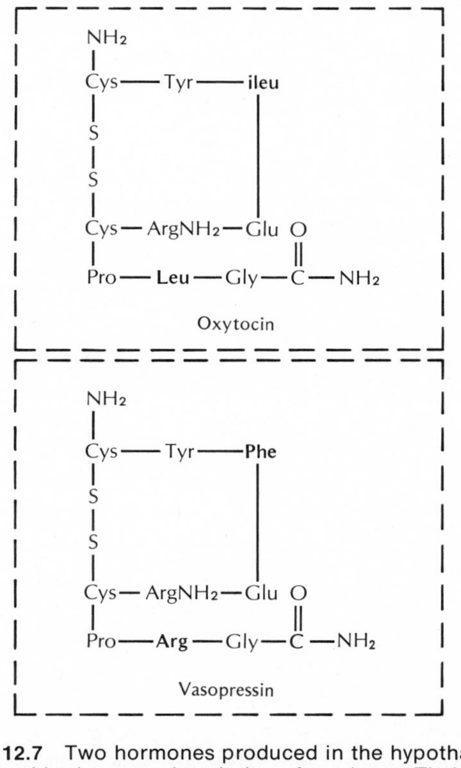

```
NH2
 |
Cys — Tyr — ileu
 |            |
 S            |
 |            |
 S            |
 |            |
Cys — ArgNH2 — Glu   O
 |                   ||
Pro — Leu — Gly — C — NH2

        Oxytocin

NH2
 |
Cys — Tyr — Phe
 |            |
 S            |
 |            |
 S            |
 |            |
Cys — ArgNH2 — Glu   O
 |                   ||
Pro — Arg — Gly — C — NH2

        Vasopressin
```

Figure 12.7 Two hormones produced in the hypothalamus and stored in the posterior pituitary for release. Their molecules differ only as shown by boldface type, but their functions are different. (See Table 12.1 and text.)

One hormone, *oxytocin,* causes strong contractions of the uterine muscles. It is sometimes injected into a pregnant female during the late stages of labor and also after childbirth to induce contraction of the uterus. Oxytocin also functions after childbirth to cause ejection of milk from the alveoli of the mammary glands. The other, *vasopressin,* causes a contriction of the small arteries throughout the body and thus a rise in blood pressure. This hormone also regulates the resorption of water by the kidney tubules. When the posterior lobe or the tract leading to it from the hypothalamus is damaged, the disorder known as *diabetes insipidus* develops. This condition is characterized by the passage of large quantities of dilute urine (as much as 40 liters daily). The condition is relieved for a period of several hours by the injection of vasopressin. Vasopressin is not a cure for this type of diabetes, but it is used to control it. Vasopressin is known also as the *antidiuretic hormone,* often abbreviated as ADH.

The Intermediate Lobe

In the frog the melanophore-stimulating hormone (MSH) causes the expansion of the pigment in melanophores or black pigment cells. These cells are responsible for the change in the color of frogs that occurs when they move from a light background to a dark one, and vice versa. This hormone regulates the chromatophores of fishes and reptiles as well as amphibians. MSH is formed in many birds and mammals, but for a long time no function was known for it. In 1973 some workers suggested that with the evolution of feathers and hair, the ability to change skin color was no longer important and, on the basis of experimental data, have suggested that MSH may be concerned with the maintance of hair lipid.

HORMONES OF THE REPRODUCTIVE SYSTEM

Discussions in this chapter have mentioned progesterone, follicle-stimulating hormone (FSH), luteinizing hormone (LH), prolactin, and testosterone. These hormones control various aspects of sexual development and reproduction in females and males. The details of reproduction and its controls, including the hypothalamic releasing factors, are described in Chapter 15 along with further emphasis on nervous and endocrine relationships which coordinate the complex set of feedback mechanisms. The hormones involved in reproduction, the menstrual cycle, and pregnancy are included in Table 12.1. Mating and contraception are discussed also in Chapter 15.

THE THYMUS

The thymus, a two-lobed gland located in the upper chest near the heart, has been an enigma for a long time, but the roles of the thymus as an endocrine gland are beginning to be unraveled. At birth the thymus is a prominent structure which begins to atrophy before maturity is reached.

Current research indicates that the thymus produces at least three hormones, thymosin, thymin I, and thymin II. Thymosin is a small acidic peptide that can prevent or modify many consequences of removal of the thymus gland. Thymosin stimulates the proliferation of lymphocytes (Ch. 11). Also, thymosin restores cell-mediated immunological functions such as the ability to reject first- or second-set skin grafts. Thymosin circulates in the blood of all mammals that have been tested. The amounts of thymosin circulating in the blood of humans decreases significantly with age. This observation suggests that many diseases of the aged may result from an inability of the thymus to produce normal amounts of thymosin which results in a decreased resistance to infection. Thymosin activity is considerably lower in the blood of patients with immunodeficiency diseases.

THE PINEAL

In mammals the pineal gland rests between the two cerebral hemispheres. It originates in the brain of the developing mammalian embryo, but loses direct connection with the brain soon after birth. The pineal produces two hormones, melatonin and serotonin. Production of melatonin is stimulated by darkness and is inhibited by light. Amphibian pineals have photoreceptive cells that can generate nerve impulses in direct response to environmental light, but the mammalian pineal does not contain photoreceptive cells. The mammalian pineal responds to light through the visual system and sympathetic (norepinephrine-releasing) nerves (Ch. 13) which innervate it. Light (and probably cold, stress, and other hormones) acts on the pineal to decrease melatonin synthesis by inhibition of specific enzymes, whereas darkness stimulates synthesis.

The pineal inhibits reproductive functions in mammals by controlling the regulation of the anterior pituitary, particularly the gonadotrophic hormones, by controlling the secretion of hypothalamic gonadotrophin-releasing factors, the luteinizing hormone releasing factor (LRF), and the follicle-stimulating hormone releasing factor (FSH-RF). Melatonin decreases the secretion of these releasing factors. Thus light, through melatonin depression, stimulates the secretion of the gonadotrophic hormones and, for example, the ovaries of the rat to enlarge.

During the past fifteen years, the pineal has been studied intensively because it provides a model for studying circadian rhythms (Chs. 13 and 43) and the regulation of endocrine organs by nerves. There are circadian rhythms in the pineal level of melatonin which persist in continuous darkness and which are abolished by exposure to light. These rhythms are generated by diurnal changes in the release of norepinephrine from the sympathetic terminals innervating the pineal. This ultimately results in melatonin production through a series of enzymatic reactions. An increased discharge of norepinephrine at night stimulates the pineal to produce melatonin.

PARAHORMONES

In addition to the substances that meet the classical definition of a hormone, other substances have hormonelike actions. Two major classes are the prostaglandins and pheromones.

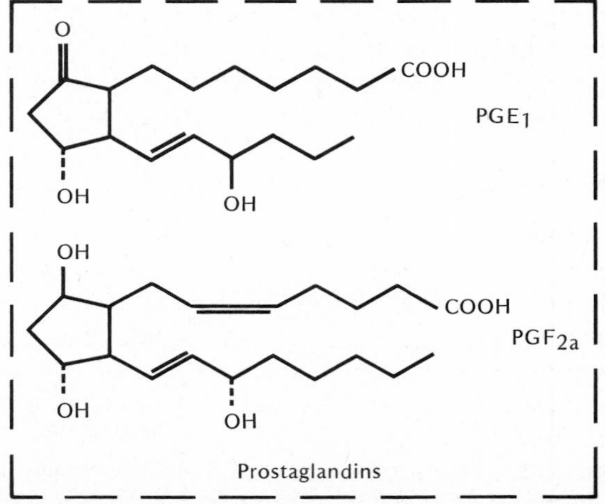

Figure 12.8 Two of several forms of prostaglandins.

Also, histamine, discussed in this chapter, and vitamin D, discussed in Chapter 9, have characteristics similar to hormones.

Prostaglandins

The prostaglandins were originally identified in human semen and found to be in the prostate glands in high concentration, but are now known to be present in most, if not all, mammalian cells. They have a wide range of actions, including potent oxytocic effects. For this reason they have been used to induce labor and have been proposed as contraceptive agents. Over twenty different prostaglandins have been isolated and characterized since the proper techniques were developed by the 1950s. They are cyclic oxygenated fatty acids that contain 20 carbon atoms (Fig. 12.8). The end results of their actions are varied.

A number of hormones, such as epinephrine, norepinephrine, glucagon, and ACTH, stimulate the breakdown of lipids by stimulating the synthesis of cAMP (see later section on the mechanism of hormonal action). Prostaglandins are known to antagonize the lipolytic effect of cAMP. It is thought that they do so by affecting adenyl cyclase, the enzyme responsible for cAMP synthesis from ATP. Since it is known that a variety of other hormonal events also mediated by cAMP are antagonized by prostaglandins, these substances may serve as general modifiers of cAMP in a number of tissues.

Damaged Tissues

Histamine (decarboxylated histidine, a common amino acid) is released from damaged tissues. This compound relaxes the muscles in the walls of blood vessels (Ch. 11) and increases their permeability. This allows the movement of white blood cells and antibodies into the damaged tissues to combat infection. Other biological actions of histamine were discussed with reference to the immune system in Chapter 11.

Pheromones

The definition of a pheromone (Gk. *pherein,* to carry; *hormon,* to excite) is a "chemical substance secreted by an individual that triggers either behavioral or developmental processes when perceived by other members of the same species." Pheromones also are called ecohormones. These terms were applied originally to sex attractants of insects (Ch. 30), but accumulated information has broadened the definition to include various agents released into the environment by all major groups of animals to signal and integrate the members of a population. The pheromones are not true hormones since they are products of exocrine glands; however, release from these glands often is dependent on hormonal stimulation. Endocrinologists consider pheromones for two other reasons:

1. Hormonal metabolites being eliminated from the organism may function as pheromones in some species
2. Pheromones initiate adjustments involving the pituitary gland and gonads

Pheromones are categorized into two types:

1. *Releaser* or *signaling* pheromones that produce rapid and reversible responses through the central nervous system or by neuroendocrine control mechanisms
2. *Primer* pheromones that activate a slow and long-acting series of neuroendocrine events involving prolonged stimulation

The question of whether pheromones are produced by humans is unanswered due to a lack of critical experimental data. However, in late 1974 a team of medical researchers reported the isolation of volatile chemicals from vaginal secretions of young women. These compounds include some simple carboxylic acids. The concentrations of these simple volatile acids were highest during the most fertile, middle part of the menstrual cycle and were lowest at the low-fertility beginning part of the cycle. Women who were taking oral contraceptive pills did not show the cyclic variation of these compounds. In Rhesus monkeys these compounds have been shown to be sex attractants. Whether they attract human males is not known. The researchers were cautious in interpreting their data.

MECHANISMS OF HORMONAL ACTION

In general, hormones comprise a regulatory device superimposed on the already existing internal controls of cellular metabolism. The mechanism of action of various hormones has challenged the imagination of biologists for many years and has spurred many speculations. Most evidence supports the idea that hormones act directly on their target cells. About half a century ago hormone action was proposed to affect membrane transport (cellular permeability). Many hormones change the rate of transport of inorganic ions, sugars, amino acids, and water. The permeability hypothesis is attactive since it can account for radically altered behavior of cells without a major change in normal control mechanisms. For example, if a hormone stimulates its target cells to grow and divide, the increased transport of nutrients into the cell as a result of membrane permeability changes leads to increased protein synthesis and energy metabolism. The mechanism of this permeability change can be related to the major theory for the mechanism of hormonal action: the second messenger hypothesis.

The second messenger hypothesis involves the cyclic nucleotide, cAMP. The hormone which is the first messenger stimulates adenyl cyclase, an enzyme in the membrane of the target cell (Ch. 4), to synthesize cAMP, the second messenger, from ATP:

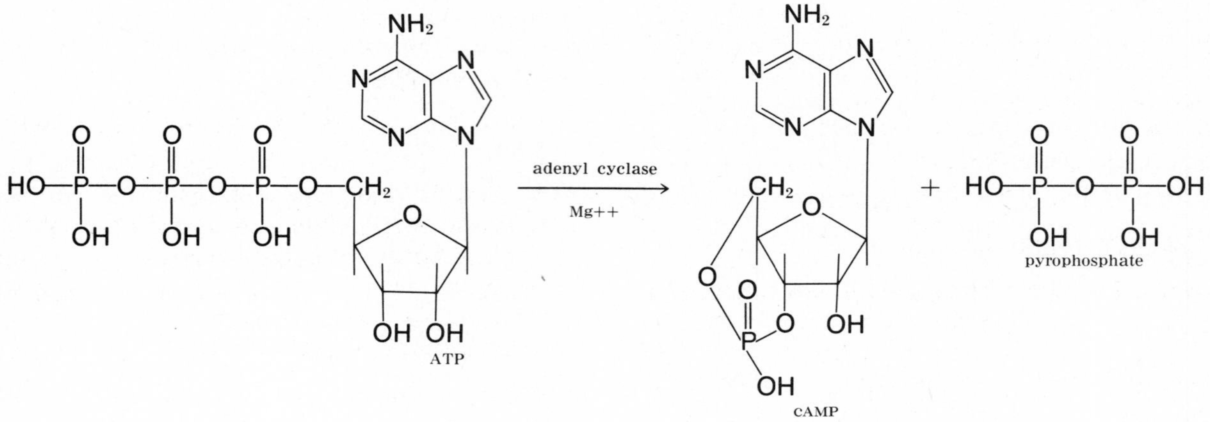

$$\text{ATP} \xrightarrow[\text{Mg}^{++}]{\text{adenyl cyclase}} \text{cAMP} + \text{pyrophosphate}$$

The second messenger, cAMP, then regulates the activity or synthesis of enzymes, permeability, and so forth of the target cell. The generalized role of cAMP as a second messenger in a variety of tissues suggests that the endocrine specificity must depend on the structure of the membrane receptor and the enzymatic content of the target cells. Cyclic AMP has been established as an intermediate in the action of many hormones. In fact, all hormones except the steroids are thought to be mediated through specific target cell membrane receptors and subsequent modulation of cAMP production. The majority of these hormones increase the activity of adenyl cyclase, whereas a few, such as insulin and some prostaglandins, decrease its activity.

Cyclic AMP must be destroyed after it is synthesized and released into the cell just as acetylcholine must be destroyed after it has traversed the synapse (Ch. 13). Specific enzymes, phosphodiesterases, are present in the cells, and these enzymes eventually destroy cAMP. Cyclic AMP stimulates the long-term production of phosphodiesterases and thereby initiates its own eventual destruction.

Cyclic GMP (cGMP) also has been shown to be involved with events concerning the second messenger concept. This cyclic nucleotide seems to have effects anatagonistic to cAMP. Overall, cGMP stimulates cell division and cAMP inhibits it. Cyclic GMP also stimulates the phosphodiesterases to breakdown cAMP. Probably cAMP likewise stimulates destruction or excretion of cGMP, thereby providing a parallel control system. The two cyclic nucleotides have been likened to the Yin and Y'ang of Chinese philosophy as opposing effectors of metabolism.

Steroid hormones pass easily through membranes and may act directly within the cell cytoplasm rather than through a second messenger. The initial step involves the complexing of the steroid molecule with a protein receptor in the cytoplasm. The steroid–receptor complex then passes through the nuclear envelope and binds with specific receptor sites on the chromatin, thereby inducing transcription of mRNAs from structural genes. Some recent evidence indicates that steroid–receptor complexes bind directly to both nuclear DNA and nonhistone chromosomal proteins. This leads to protein synthesis and gene expression (Ch. 6). The use of the term "messenger" in first messenger, second messenger, and messenger RNA should not be confused.

Elucidation of the details of the mechanism(s) of hormonal action provides many challenges to many biologists: endocrinologists, geneticists, biochemists, physiologists, and immunologists.

SUGGESTIONS FOR FURTHER READING

Banting, F. G., and C. H. Best. "The Internal Secretion of the Pancreas," in *Great Experiments in Biology,* M. L. Gabriel and S. Fogel, Eds. Englewood Cliffs, N.J.: Prentice-Hall (1955).

Cooper, J. R., F. E. Bloom, and R. H. Roth. *The Biochemical Basis of Neuropharmacology,* 2nd ed. New York: Oxford Univ. Press (1974).

Frieden, E., and H. Lipner. *Biochemical Endocrinology of the Vertebrates.* Englewood Cliffs, N.J.: Prentice-Hall (1971).

Gorbman, A., and H. A. Bern. *Textbook of Comparative Endocrinology.* New York: Wiley (1962).

Gordon, M. S. *Animal Function: Principles and Adaptations.* New York: Macmillan (1968).

Turner, C. D., and J. T. Bagnara. *General Endocrinology,* 5th ed. Philadelphia: W. B. Saunders (1971).

From Scientific American

Bonner, J. T. "Hormones in the Social Amoebae and Mammals" (June 1969).

Davidson, E. H. "Hormones and Genes" (June 1965).

Gardner, L. I. "Deprivation Dwarfism" (July 1972).

Gillie, R. B. "Endemic Goiter" (June 1971).

Guillemin, R., and R. Burgus. "The Hormones of the Hypothalamus" (Nov. 1972).

Li, C. H. "The ACTH Molecule" (July 1963).

Loomis, W. F. "Rickets" (Dec. 1970).

McEwen, B. S. "Interactions Between Hormones and Nerve Tissue" (July 1976).

Pastan, I. "Cyclic AMP" (Aug. 1972).

Pike, J. E. "Prostaglandins" (Nov. 1971).

Rasmussen, H., and M. M. Pechet. "Calcitonin" (Oct. 1970).

Wilson, E. O. "Pheromones" (May 1963).

Wilson, E. O. "Animal Communication" (Sept. 1972).

Wurtman, R. J., and J. Axelrod. "The Pineal Gland" (July 1965).

13 Coordination II: The Nervous System and the Reception of Stimuli

In the preceding chapter, coordination as it is produced by hormones was discussed. Coordination by hormones is a relatively slow process. When an animal requires great speed in coordination and response, it uses its nervous and muscular systems. In this chapter the nature and functioning of the nervous system are considered. A complete understanding of the functioning of the nervous system also involves an understanding of receptors (sense organs) and effectors (muscles and glands). Following the discussion of the nervous system *per se,* the receptors in humans are considered; a discussion of muscles was given in Chapter 8.

The nervous system is one of the most important systems in the body and certainly the most complex. Although quite a bit is known about the structure and the function of the individual cells that compose the nervous system, the *neurons,* the complexity of the system lies in the organization of these neurons.

All of our movements and what is called thinking, learning, experiencing sensations, and the like are involved with the functioning of the nervous system. The task of providing a satisfactory explanation for all of these processes is tremendous, and much work remains to be done. Even so, it is possible to obtain a partial understanding of the functioning of this system.

DIFFERENT KINDS OF NERVOUS SYSTEMS

The simplest type of nervous system is found in the coelenterates, such as *Hydra* (Ch. 23). Here the nervous system consists of a simple *nerve net* (Fig 13.1). There is no central nervous system. Impulses can move in either direction along the fibers. An impulse spreads from its point of origin in all directions; the stronger the impulse, the farther it spreads. Sensory or receptor cells are present, and the effectors are contractile elements in the epitheliomuscular cells. Such a simple system suffices for such a simple animal. *Hydra* has radial symmetry. Other radially symmetrical animals, such as jellyfishes and starfishes, have more complex nervous systems, but even these do not provide for the complex behaviors found in most bilaterally symmetrical animals.

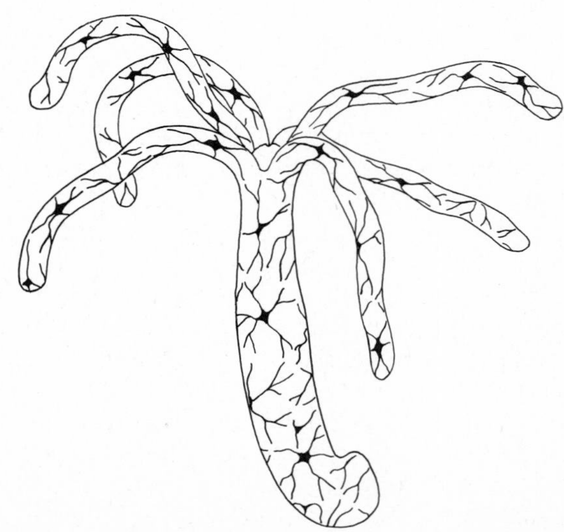

Figure 13.1 Nerve net system in *Hydra.*

Great advances in the evolution of central nervous systems are associated with the evolution of bilateral symmetry, which involves some kind of head formation. The most primitive of these still have a nervous system composed of a simple nerve net. However, in the planarians (Ch. 24) there is a central nervous system composed of two ventral nerve cords with joined enlargements at the anterior end. These two enlargements are often referred to as a "brain," but they are probably involved only in relaying the impulses that come in from the sense organs on the head to the cords. As pictured in Figure 24.*8*, this nervous system looks like a ladder because of the cross connections between the two cords.

In the annelids and arthropods the two ventral nerve cords have become fused or partly fused, and there is an aggregation of neuron cell bodies, or ganglion, in each segment in the primitive forms (Ch. 27–29). The most anterior pair of ganglia, which are located dorsally, is the brain. It shows some dominance over the other ganglia, but it is limited in this respect when compared to the vertebrate brain. In the evolution of the more specialized arthropods, some of the insects, there has been a fusion of some of the ganglia of the ventral nerve cord, with the result that some insects have only a few large ganglia.

The greatest development in nervous systems is found in the vertebrates. The vertebrate nervous system differs in a number of ways from those of annelids and arthropods. In development the central nervous system of vertebrates forms from foldings in the ectoderm (neural folds), which form a hollow tube in the dorsal region (Ch. 16). This tube remains hollow in contrast to the solid nerve cords of the invertebrates discussed. In early development there appear in the anterior region of the tube three swellings known as the *forebrain,* the *midbrain,* and the *hindbrain.* The rest of the tube becomes the spinal cord. In the development of all modern vertebrates the three brain regions undergo further differentiation: the forebrain differentiates into the *cerebrum* and diencephalon (the walls of the diencephalon become the *thalamus* and the floor, the *hypothalamus*); the midbrain remains undivided, but its roof forms the *optic lobes;* the ventral portion of the hindbrain becomes the *medulla,* and the anterior dorsal portion of the hindbrain becomes the *cerebellum.*

The central nervous system of vertebrates, with respect to overall size, is much larger in proportion to the corresponding system of invertebrates and it is dorsal in position. Furthermore more of the brain in vertebrates is devoted to the correlation and integration of information. This is particularly true of the more advanced vertebrates.

GROSS STRUCTURE OF THE VERTEBRATE NERVOUS SYSTEM

For convenience in study, the nervous system of vertebrates is often divided into three parts: the *central,* the *peripheral,* and the *autonomic* systems. These divisions are not completely distinct, either structurally or functionally, but they are useful nevertheless. The central system, composed of the brain and cord, is primarily the integrating part of the whole system. The peripheral system is composed of the *cranial* and *spinal nerves.* A nerve is a bundle of processes of neurons that are insulated from each other and that carry nerve impulses. None of the neuron cell bodies are located in the nerves. Nerves carry impulses to the central system from the sense organs (sensory nerves) and from the central system to the skeletal muscles (motor nerves) that effect the movements of the body. The autonomic system is associated with the glands of the body and the various organs of the viscera (internal organs). This system functions in connection with *visceral reflexes,* over which the organism has no conscious control. The autonomic system is divided into the *sympathetic* and *parasympathetic systems,* and they are considered in detail in a later section. The general pattern of the nervous system in all vertebrates is the same (see Fig. 13.2 and also Fig 13.7).

The Peripheral System

CRANIAL AND SPINAL NERVES

There are 12 pairs of cranial nerves in humans. Some of these nerves are purely sensory, some are purely motor, and some are mixed. The cranial nerves innervate the sense organs and muscles of the head. In addition, fibers of the tenth nerve (vagus) innervate many visceral structures such as the heart, stomach, intestine, esophagus, lungs, and aorta; fibers of the eleventh nerve innervate the muscles of the shoulder.

In humans there are 31 pairs of spinal nerves. In all vertebrates these nerves are mixed nerves with both motor and sensory fibers. The nature of the spinal cord and the attachments of the

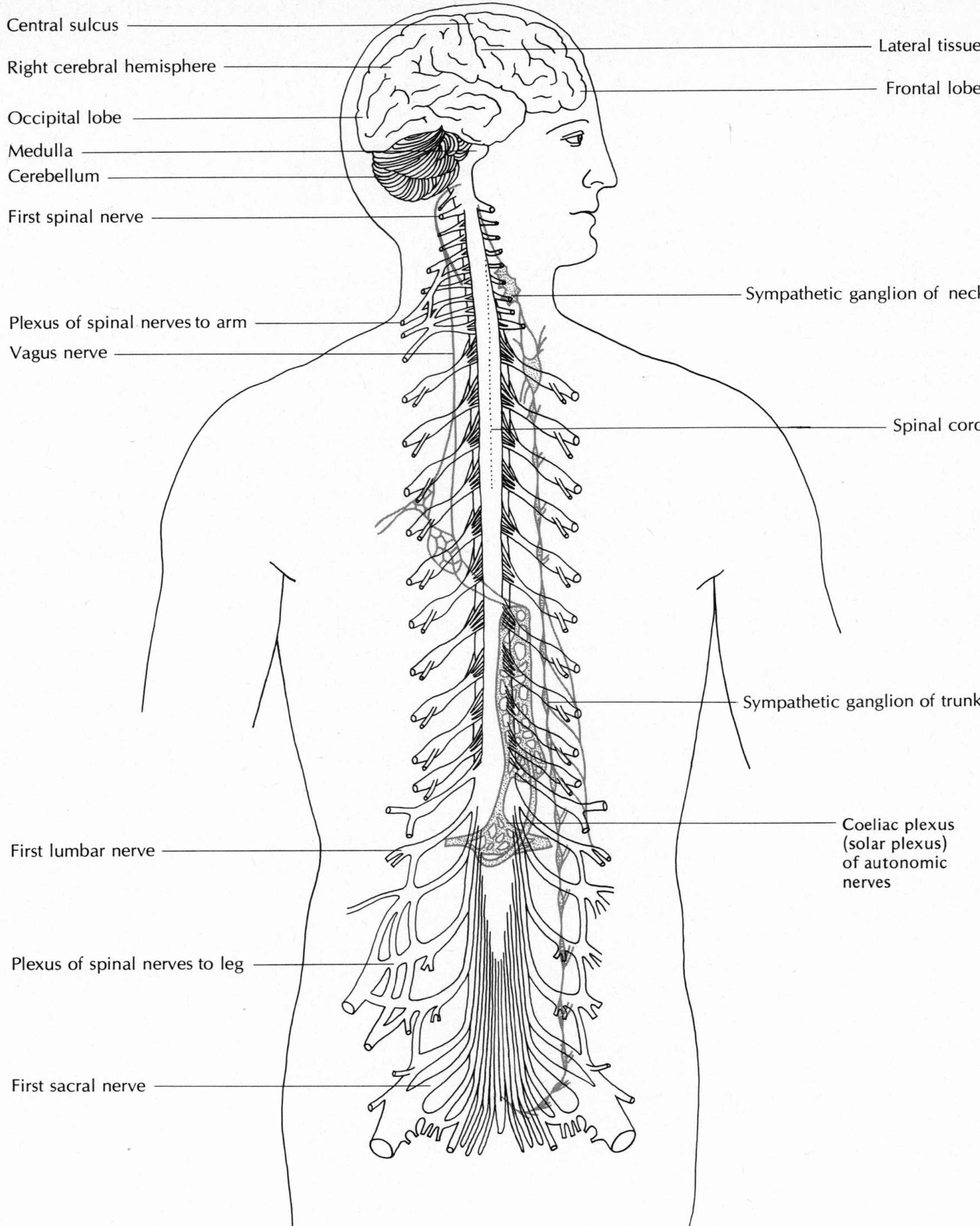

Figure 13.2 Human nervous system.

spinal nerves to it is the same in all vertebrates. Figure 13.3 illustrates the distribution of *white* and *gray matter* in the cord and the attachment of the spinal nerves. The white matter is composed of myelinated fibers, whereas the gray matter is composed of cell bodies of neurons, glial cells (small supporting cells which outnumber neurons about 10:1), and nonmyelinated fibers. Each nerve is attached to the cord by two roots: the *dorsal root* and the *ventral root.* The sensory fibers of the nerve enter the cord by way of the dorsal root, whereas the motor fibers of the nerve emerge from the cord by way of the ventral root. The swelling of each dorsal root, the *dorsal root ganglion,* is the location of the cell bodies of the sensory neurons.

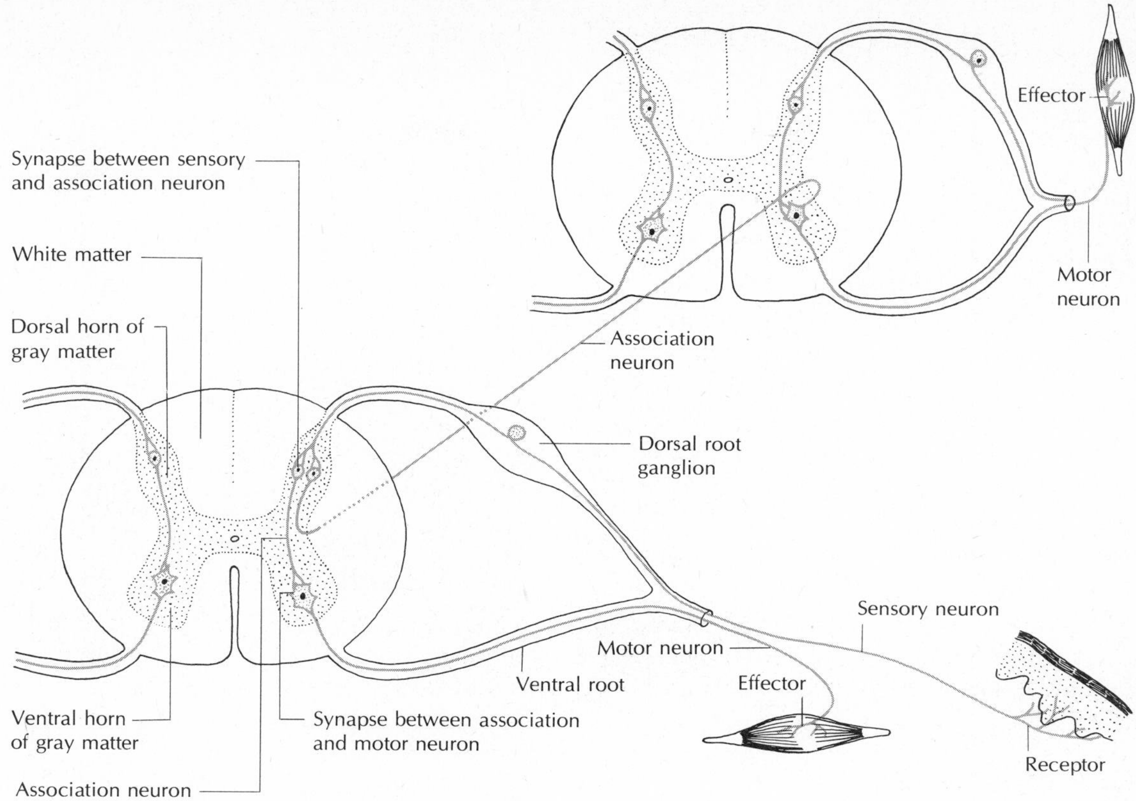

Figure 13.3 Parts of a reflex arc with two levels of the spinal cord shown.

NEURON—STRUCTURAL UNIT

As discussed in Chapter 4, the unit of structure in the nervous system is a specialized type of cell, the neuron.

Neurons vary greatly in architecture, size, and associated structures (Fig. 4.27). Here the motor neuron serves as a model of neuronal structure. The nucleus is localized in the main cytoplasmic mass, called the *cell body.* Each neuron has two kinds of thin processes. These processes are the *dendrites* and the *axons.* The dendrites carry impulses into the cell bodies and the axons carry them away. The axons of neurons are usually enveloped in a sheath composed of a series of cells called *Schwann cells.* Each Schwann cell forms a short tube around the axon. As a result, the fibers appear segmented; the points of segmentation are called *nodes of Ranvier.* When Schwann cells simply envelop an axon, the sheath is called the *neurilemma.* On some axons the Schwann cells have wrapped many times around the axial fiber so that there are many layers of the Schwann cell outer membrane between the axon and the enveloping cell's cytoplasm. This multilayered membrane is called the *myelin sheath* (Fig. 13.4).

Cell membranes have a high lipid content, and thus the myelin sheath gives a white or fatty appearance to fibers that possess it. It is generally believed that the myelin sheath serves to force the impulse to travel more rapidly down the nerve by allowing it to jump from node to node. The rate of conduction is much faster in myelinated fibers than in the nonmyelinated type.

Many consider that the neuron is the most specialized type of cell and we are still learning about its nature and functioning. Because of the very long axonal process the volume of the axon may be a hundred times the volume of the cell body. Students have been concerned about the origin of components of the axonal cytoplasm in growth, regeneration, and normal functioning. In 1948 Weiss and Hiscoe published a classic paper in which evidence was presented that the components of axonal cytoplasm are formed in the nucleated part of the cell body and then transported *(axonal flow)* throughout the axon. In recent years these observations have been confirmed and extended. There is evidence now that axonal transport of proteins occurs at different rates. Some of the faster moving components include the neurotransmitter-containing vesicles. The slower moving components include most

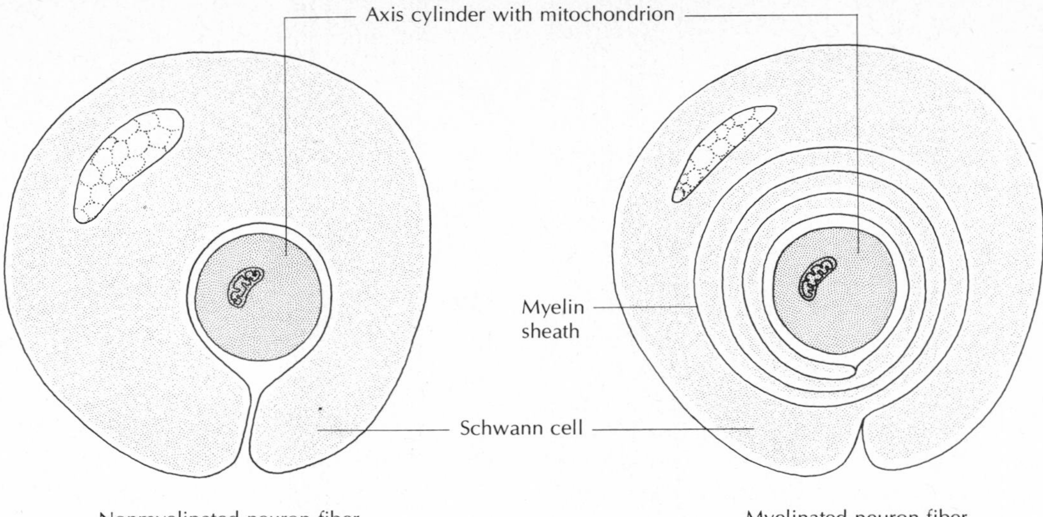

Figure 13.4 Cross sections of myelinated and nonmyelinated fibers.

of the soluble proteins in the axon. Many of these do not reach the nerve terminals which suggests local utilization in the axon.

REFLEX ARC—FUNCTIONAL UNIT

There are commonly five[1] parts to the basic reflex arc of vertebrates: receptor, sensory neuron, association neuron, motor neuron, and effector (Fig. 13.3). The transmission of an impulse begins with the stimulation of a receptor; for example, touch receptors in the skin. The receptor then relays the stimulus to the dendrite of a sensory neuron. These dendrites may be several feet long; for example, they may reach from the foot to the spinal cord. The cell bodies of sensory neurons are aggregated in clusters, ganglia, external to the spinal cord rather than within the tissues of the central nervous system, as are those of the association and motor neurons. The clusters to which sensory neurons go are specifically called dorsal root ganglia because of their position on the dorsal connection of the nerve to the central nervous system.

The impulse traveling the dendrite of a sensory neuron passes the cell body in the ganglion and proceeds along the axon. The axon enters the central nervous system and ends close to the dendrite of an association neuron. This close association between an axon and a dendrite is called a *synapse.* As will be discussed below, it is an important point of control of the passage of an impulse from one neuron to another. In the simplest situation, the association neuron forms a synapse with the dendrites of a motor neuron.

[1] The knee-jerk reflex in humans involves only four parts; there are no association neurons.

As Figure 13.3 shows, the spinal cord is composed of two parts, *white matter* and *gray matter.* White matter principally consists of fibers of neurons and forms the outer part of the spinal cord. Gray matter has a high concentration of nerve cell bodies and forms a fluted column throughout the center of the spinal cord; in cross section the gray matter is crudely H-shaped. Sensory axons enter the dorsal horns of this gray matter and motor-cell bodies lie in the ventral horns; association neurons form a bridge between these neurons. After the dendrites of the motor neuron are activated by the association neuron, the impulse travels away from the central nervous system along the axon of the motor neuron and finally stimulates the effector, a muscle or gland. Thus a response to a stimulus is completed.

A reflex arc is not usually so simple as just described. The middle cell, the association neuron, often branches to several motor neurons, and action may remain at one level of the central nervous system. Or the axon of the association neuron may travel up or down the length of the central nervous system to another level and there establish a synapse with motor neurons. An example of further complexity is that a reflex which flexes a muscle on one side of a joint usually inhibits the simultaneous action of opposing muscles. Some examples of reflex acts are the eye blink, the grasping reflex of infants, and the secretion of salivary fluid in response to visual, taste, or emotion-evoking stimuli.

NERVE IMPULSE, SYNAPSES, AND NEUROMUSCULAR JUNCTIONS

In spite of extensive research into the nature of the nerve impulse, there is much to be learned about it. At one time the nerve impulse was thought to be a simple electric current because the passage of the impulse can be detected with a galvanometer. But the speed of the impulse is too slow to be just an electric current. The change in electrical potential is known to be related to internal and external concentrations of sodium and potassium ions.

In the resting stage the membrane of the neuron is polarized; the inside is negative with respect to the outside. Stimulation of the membrane increases the permeability (along the membrane), and positive sodium ions rapidly enter the cell. As a result, the interior of the neuron actually becomes positive. Slightly later the membrane becomes more permeable to potassium ions, which move out of the cell. Thus in the initial resting state the charge on the inside is negative and the outside is positive; stimulation allows the inside to become positive, and then as the positive potassium ions move out, it becomes negative again but now has changed ion concentrations. The change in permeability continues down the length of the neuron; this change is the nerve impulse. Recovery of the neuron involves the return of potassium to the inside of the membrane and sodium to the outside. The return is accomplished by a mechanism involving active transport and has been called the *sodium–potassium pump* (Fig. 13.5).

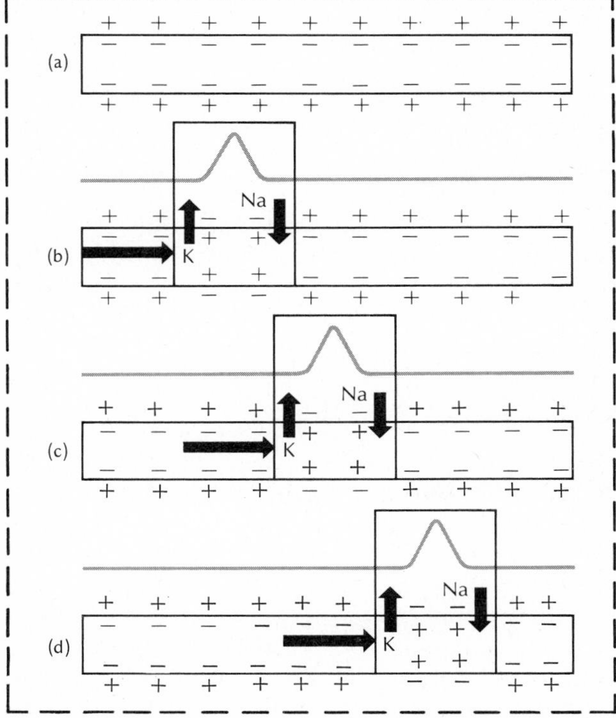

Figure 13.5 Diagram of membrane theory. **(a)** Resting fiber showing polarization of the membrane. **(b), (c), (d)** Fiber at different time intervals during the passage of an impulse showing, from left to right, the partially repolarized region behind the impulse, the depolarized region where the impulse is, and the polarized region ahead of the impulse.

The flow of sodium and potassium ions across the membrane temporarily depolarizes a section of it. This causes the neighboring region to be depolarized and allows sodium to flow in. In this way the propagation of the impulse along the fiber is achieved. When ATP production is blocked, the pump will not function. This shows that energy is used in the process.

The weakest stimulus which excites a neuron is the threshold stimulus of that neuron. Once started the impulse goes all the way, that is, the transmission of the impulse is an *all-or-none reaction.* Different neurons in a nerve have different thresholds; at a given level of stimulus, some axons in a nerve may be firing and some may not. More will fire if the stimulus is increased. Intensity of stimulus is also interpreted by the frequency of firing: the more intense the stimulus, the more times per second a neuron will fire.

An important place of control of the passage of the nerve impulse is the synapse. It can be demonstrated that the impulse normally travels only from the axon of one cell to the dendrite of another. When the nerve impulse reaches the terminations of the axon, it causes the release of a chemical, a transmitter substance, from numerous small synaptic vesicles located there (Fig. 13.6). A common transmitter substance is *acetylcholine.* If the acetylcholine crosses the synapse, special sensitive areas of the dendrite are stimulated. These areas trigger a change in the membrane of that neuron and an impulse is started.

There is a delicate control at the synapse. On the dendritic side the neuron produces the enzyme *cholinesterase.* After the acetylcholine has stimulated the membrane, it is inactivated by the enzyme. In a similar way other transmitter substances are inactivated.

Synapses, then, are points of resistance in pathways of neurons. The transmitter molecule apparently must reach a critical concentration to initiate depolarization of the next neuron. If one impulse does not produce a critical concentration of the transmitter substance, a series of

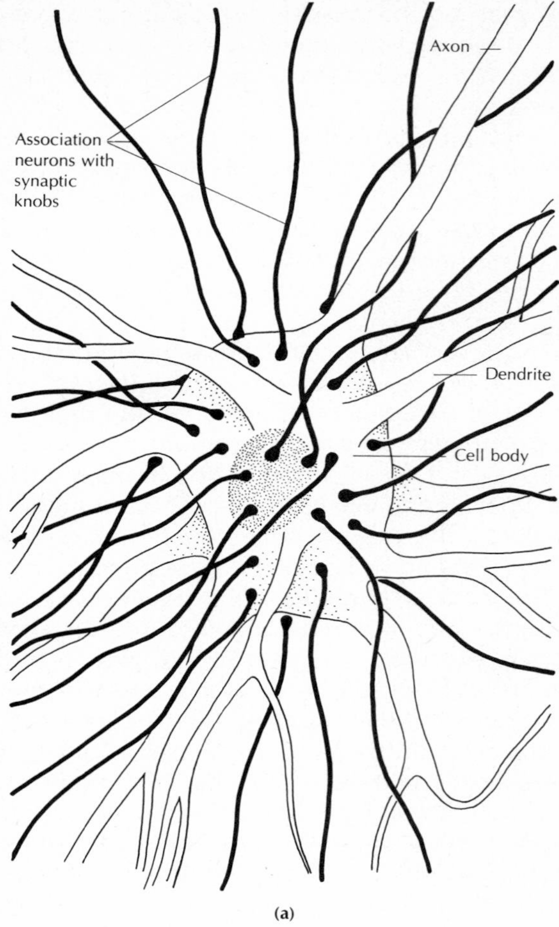

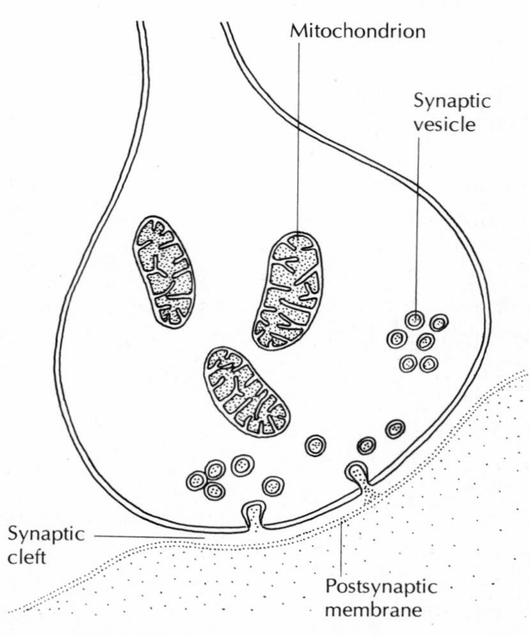

Figure 13.6 **(a)** Cell body of motor neuron showing synaptic junctions. **(b)** Drawing showing ultrastructure of synaptic region.

impulses may. This cumulative phenomenon is called *synaptic summation.*

While the most studied transmitter substances are acetylcholine and noradrenalin, recent work indicates that the following substances may serve as transmitters in different parts of the system: γ-aminobutyric acid, dopamine, 5-hydroxytryptamine, glutamic acid, and glycine.

The dendrites and cell bodies of association and motor neurons have many terminations of axons near them forming synaptic junctions. Electron micrographs show that these endings are knoblike. Within these knobs are very small *synaptic vesicles* (Fig. 13.6). It is presumed that these vesicles contain the transmitter substance that is given off into the very narrow synaptic clefts at the time the impulse reaches them. If the transmitter substance is excitatory, it diffuses to the membrane of the adjacent neuron, where it effects changes in the membrane, resulting in the influx of sodium ions and the start of an impulse in that neuron. Often the transmitter substance is inhibitory, producing changes so that the charge on the membrane becomes more positive, thus making the cell less easily stimulated. De Robertis has identified the receptor material as a special proteolipid.

The neuromuscular junction, or motor end plate, is the point of functional association with the muscle fiber (Fig. 8.10). Acetylcholine seems to be the transmitter substance at the neuromuscular junction, where the branch of the neuron delivers the transmitter to a single muscle fiber. This causes the depolarization of the membrane of the muscle fiber in the same way that a nerve fiber is depolarized. An impulse is produced that stimulates an interaction between the two significant components of muscle, *actin* and *myosin.* This interaction is *muscular contraction* (see Ch. 8).

THE CENTRAL NERVOUS SYSTEM

Cord

The cord is similar in structure and function in all vertebrates. The gray matter of the cord functions as centers for many reflexes. The white matter of the cord is made of pathways to and from higher levels and its fibers are segregated into bundles or tracts with specific functions—tracts that conduct impulses to higher levels of the cord and to the brain and tracts that carry impulses from the brain down the cord.

Brain

The brains of all vertebrates are similarly constructed, that is, the same basic parts are present in all of them. Figure 13.7 shows something of the evolutionary trend in the different classes of vertebrates. The greatest modifications are found in the cerebral hemispheres and in the cerebellum. The increase in size of the cerebral hemispheres is associated with an increase in the higher mental faculties of learning, memory, and thinking, whereas the difference in degree of development of the cerebellum is associated with differences in the degree of highly coordinated muscular activity. Another rather obvious difference is the larger proportionate size of the olfactory lobes in the lower forms where there is greater dependence on the olfactory sense.

The *medulla oblongata* most resembles the spinal cord in its organization. As in the spinal cord, it has ascending and descending association pathways present in the white matter. In addition to the pathways, the medulla contains the following vital reflex centers: the *respiratory center,* the *cardiac center,* and the *vasomotor center.* Also reflex centers for swallowing, coughing, sneezing, and vomiting are located there.

A network of neurons, the *reticular formation,* runs through the medulla and on into parts of the forebrain, including the thalamus and the cerebral cortex. This system functions in the arousal of parts of the brain, especially the cerebral cortex and the thalamus. Fortunately this system is very effective in that it does not make the cortex aware of every stimulus. A person with this network destroyed becomes permanently unconscious.

Next in tandem arrangement is the *cerebellum,* which interacts in the control of equilibrium and modulates the impulses from the cerebrum to the muscles. The cerebellum has its cell bodies concentrated as gray matter in the outer part, or cortex. The cerebellum does not initiate movements; it modulates impulses from the cerebrum resulting in very precise movements. If the cerebellum of a dog is destroyed, it can move but it lacks muscle tone and its movements are not coordinated. A pigeon with a damaged cerebellum can move its wings but cannot fly. A person with a damaged cerebellum may walk like a drunk person.

Just above the medulla is a region composed mainly of fiber tracts, called the *pons.* It carries impulses from one hemisphere of the cerebellum to the other, serves as a conduction pathway between lower and higher levels of the brain, and it is the location of a few centers for reflexes mediated by the fifth, sixth, seventh, and eighth cranial nerves.

The midbrain region in the human is significantly smaller in comparison with that of the lower vertebrates. This is a result of reduced function, which is limited to some eye and ear reflexes and to the conduction to and from the anterior parts of the brain. This contrasts with the lower vertebrate midbrain, which serves as a coordination center.

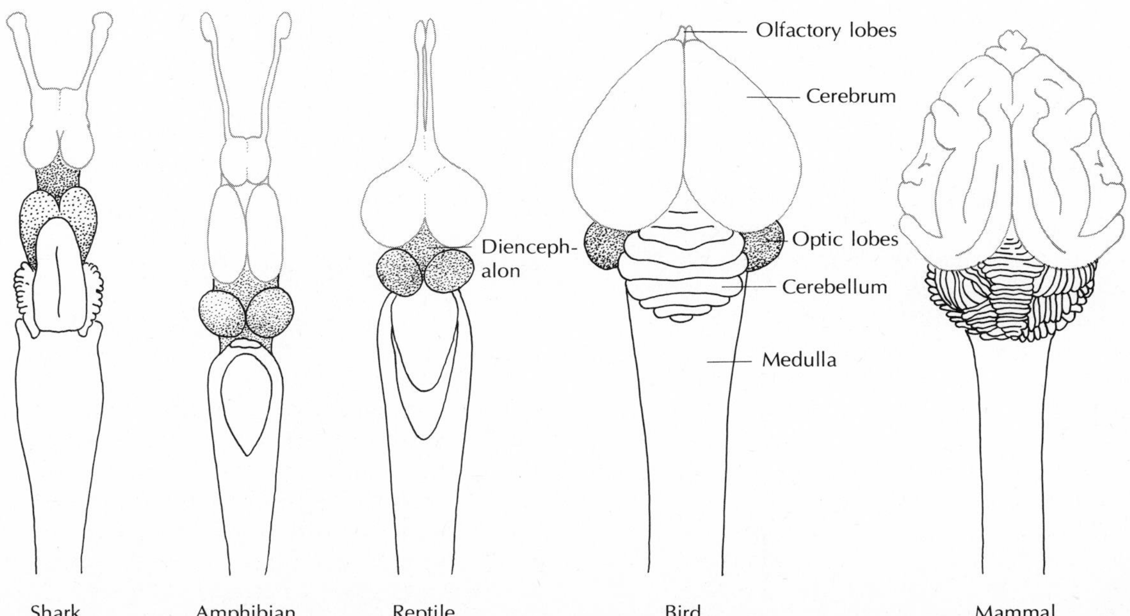

Figure 13.7 Representative vertebrate brains.

The forebrain is divided anatomically into the *diencephalon* and the *cerebrum.* The thick walls of the diencephalon form the *thalamus,* which serves as the relay center for sensory impulses. All sensory systems, with the exception of smell, funnel through it. Also, reflex centers for muscular and glandular activities are localized in the thalamus. Control of some levels of emotion, such as manifestations of rage, also have centers here. The region below the thalamus is the *hypothalamus* (Fig. 13.8). It contains centers for a number of functions such as water balance, fat metabolism, eating, blood pressure, body temperature, and sleep.

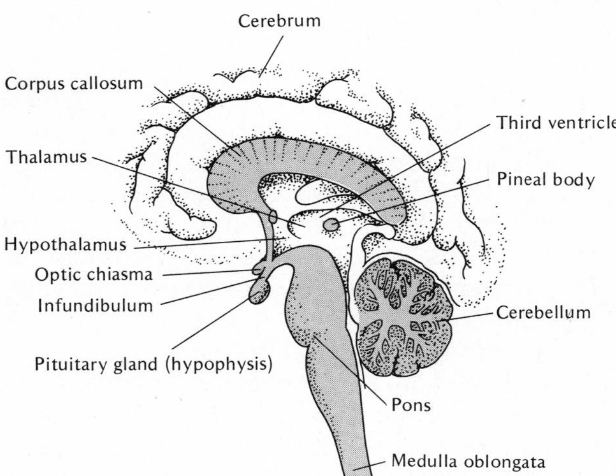

Figure 13.8 Sagittal section through the brain showing some of the principal regions.

We do not have a complete understanding of the neural mechanisms involved in sleep. Only the higher vertebrates with fairly well developed cerebral cortices sleep. Those with the larger and better developed cortices appear to require more sleep. All of these higher vertebrates have a basic rhythm of sleep alternating with wakefulness, and the pattern is regulated somehow by the hypothalamus. There is a sleep center in the anterior region and a wakefulness center in the posterior region of this part of the brain.

Also, the hypothalamus secretes hormones that are transported to the pituitary gland and stored there. It produces releasing factors that control secretions by the pituitary gland. The pituitary gland is attached to the ventral side of the hypothalamus (see Ch. 12).

The parts of the brain discussed so far are associated with involuntary functions, or unlearned behavior. The most anterior part, the *cerebrum,* is associated with conscious behavior, or learning and discrimination. This was not always the case. In the lower vertebrates the cerebrum served primarily as a pathway for olfactory impulses en route to the midbrain for interpretation. This basic pathway persists on the underside of the cerebrum in mammals, but it is inconspicuous because the evolution of the dorsal section has overshadowed it in size. The cerebrum consists of a thin outer layer, the *cortex,* which is gray matter and consists of about 5 billion nerve-cell bodies. Deep in the cerebrum are several aggregations of cell bodies, but most of the noncortical area consists of fiber pathways, some connecting different parts of a given hemisphere, some connecting the two hemispheres, and some connecting with lower parts of the brain and the cord.

By direct probing for function, by autopsy studies, and by studying accident victims with brain damage, the areas of the cortex have been mapped (Fig. 13.9). There is in the posterior part a center for vision and there are lateral centers for speech, hearing, and smell. On the anteriolateral surface lies a pair of elongated zones to the front and rear of a fissure, the *fissure of Rolando;* these are the somatic sensory and motor centers. Most of the remaining regions of the cerebrum are composed of association areas which connect different regions of the cortex.

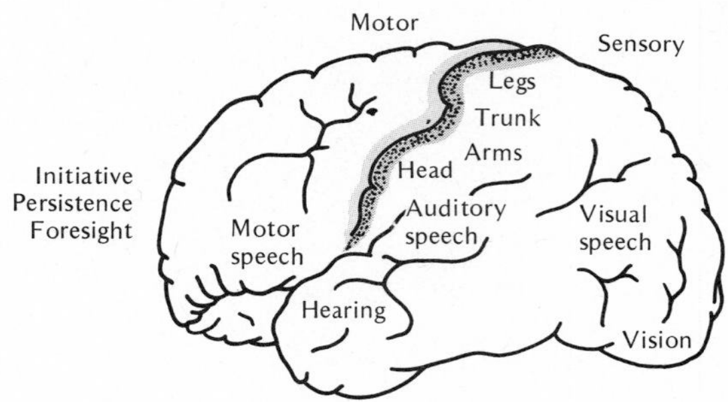

Figure 13.9 Localization of functions in cerebral cortex of humans.

The Autonomic System

Figure 13.10 illustrates the double nature of the autonomic system, with one part stimulating and the other part inhibiting a specific structure. The *sympathetic* division has its neurons connected with the central nervous system along the spinal cord (Fig. 13.2). The *parasympathetic* division connects principally with the brain but also with the sacral area of the spinal cord.

Both autonomic subdivisions deliver impulses to the structure they innervate by two neurons. The first neuron emerges from the central nervous system as a *presynaptic neuron* or fiber.

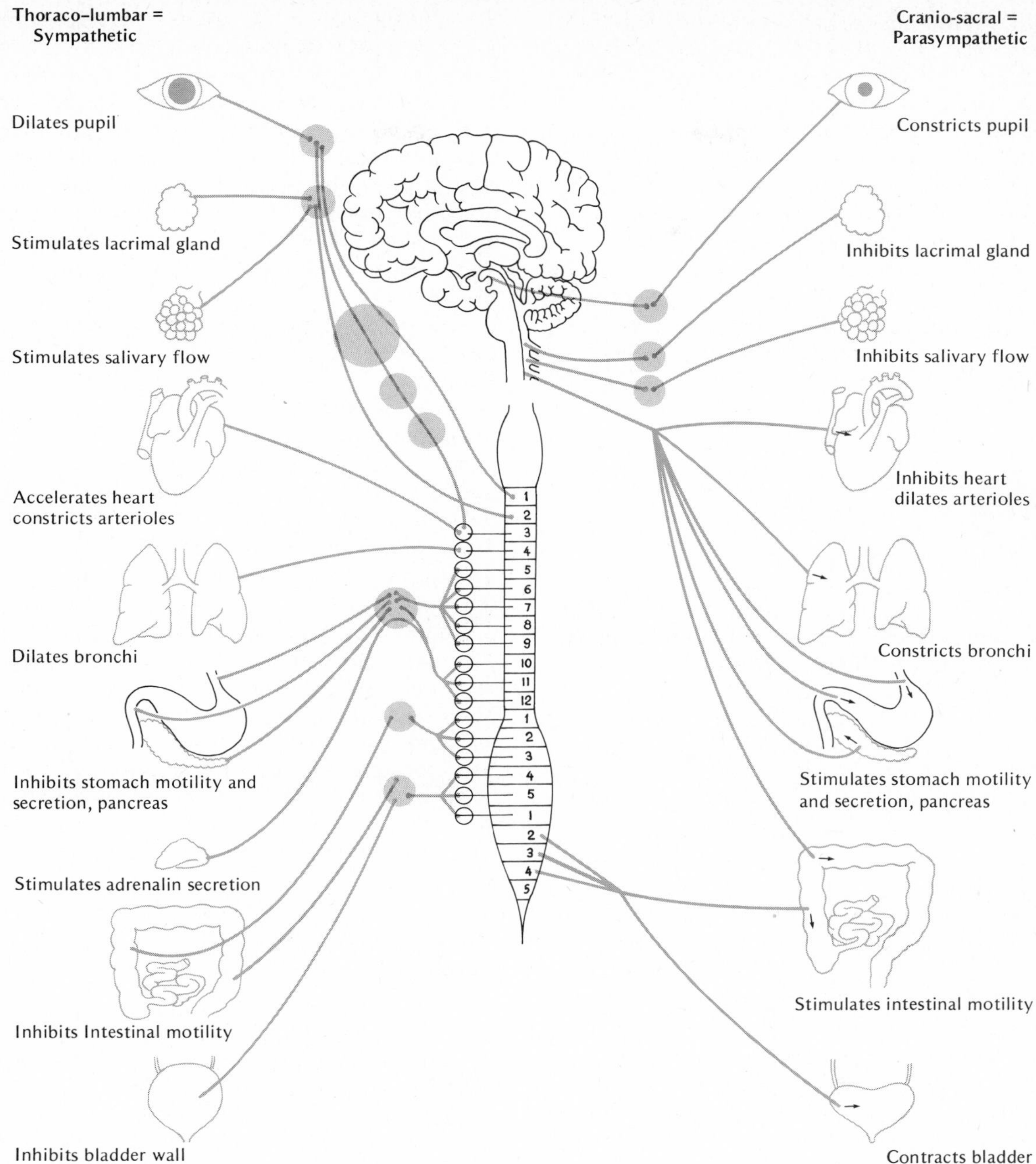

Figure 13.10 Diagram of autonomic system.

The second neuron, lying entirely outside the central nervous system, is called a *postsynaptic neuron* or fiber, and stimulates its effector. Presynaptic and postsynaptic fibers of sympathetic and parasympathetic divisions generally differ in the site where they form a synapse. Most parasympathetic pathways have relatively long presynaptic fibers, and the postsynaptic cell body is located in or adjacent to the structure innervated.

In the sympathetic division, presynaptic fibers establish synapses with postsynaptic fibers in specialized autonomic ganglia. These are of two principal types. *Sympathetic ganglia* lie close to the vertebrae and form two chains closely connected with spinal nerves. Other *visceral* ganglia are located in the mesenteries that support organs. An example of a cluster of these ganglia is the so-called solar plexus in the abdominal viscera.

The secretions of the terminal fibers differ in sympathetic and parasympathetic fibers. As in the somatic motor neurons, the parasympathetic division secretes acetylcholine. These fibers are *cholinergic.* The sympathetic fibers secrete *noradrenalin,* and they are *adrenergic.*

Most organs are innervated by both sympathetic and parasympathetic subdivisions of the autonomic division, and each has a function that is antagonistic to the other.

Even though the autonomic system operates at the involuntary or subconscious level, consequences of its function may reach the conscious level, and vice versa. Visceral autonomic functions may secondarily stimulate pain receptors, which then initiate impulses that proceed to the conscious level. Conversely, stimuli of pain or other sensation received at the conscious level can also be relayed to centers of autonomic or involuntary control. For example, the heart can involuntarily "quicken" in association with some conscious unpleasantness.

RECEPTION OF STIMULI IN HUMANS

General Nature of Receptors

With the evolution of the multicellular forms and the division of labor between the cells, certain cells, or groups of cells, have become specialized for the reception of particular kinds of stimuli. These specialized cells are sense organs or parts of sense organs. In a few instances, such as the pain receptors in the skin, the naked endings of sensory neurons serve as receptors. In general, however, the receptors are composed of specialized cells on which the fibers of sensory neurons end. A receptor must be acted upon by some particular environmental change to be stimulated. The specialization of each kind of receptor is associated with its high degree of irritability for a specific kind of energy change. The threshold for stimulating a given receptor is much below that of ordinary cells and even much below that of the sensory nerve fibers themselves.

All receptors share a common function; they generate nerve impulses. Each type of receptor functions as a transducer, converting one form of energy into another. At the present time all the details about how this is accomplished are not yet known for most receptors.

It cannot be assumed that the neurons leading from a particular receptor possess distinct patterns of neuronal activity. The specific nature of the reception is interpreted in the cerebral cortex as a particular conscious sensation.

For a long time a popular notion has existed that man has five senses: touch, taste, smell, sight, and hearing; but there are several others, as a little reflection will reveal. When the so-called sense of touch is analyzed, one realizes that stimulation of the skin may evoke sensations not only of touch, but of pain, pressure, heat, and cold. In addition to these, we have the important sense of *proprioception.* Here sense organs in the muscles and tendons furnish information about the tension of muscles and the position of body parts. Also the sense organs in the inner ear that are concerned with the balance of the body and with experiencing rotation add to knowledge about the position of parts of the body.

We experience sensations of hunger and thirst. Also, definite sensations are experienced with the stimulation of some of the organs involved in sexual behavior. In addition to the receptors that, upon stimulation, usually elicit conscious sensations, there are many others in the body that are continually functioning without any conscious sensations. The receptors in the blood vessels that function in the reflex dilation of the walls of the vessels are examples of this.

Receptors are often classified as *exteroceptors* and *interoceptors.* Exteroceptors receive information about the outside world, whereas interoceptors are affected by changes that take place within the body. Most of our discussion here is focused on exteroceptors. All that we know about the outside world is based upon information given us by these sense organs when it is interpreted by the sensory areas of the brain.

Cutaneous Receptors

The skin contains a number of kinds of sense organs involved in the sensations of touch, pressure, heat, and cold. The sensation of pain results from the stimulation of the nerve endings in the skin and other parts of the body. Figure 13.11 shows the structure of three of these cutaneous receptors.

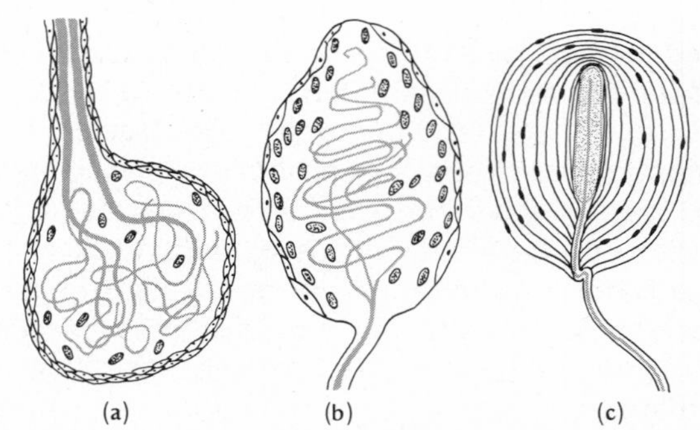

Figure 13.11 Some cutaneous receptors. **(a)** Cold (end bulb of Krause). **(b)** Touch (Meissner's corpuscle). **(c)** Deep pressure (Pacinian corpuscle).

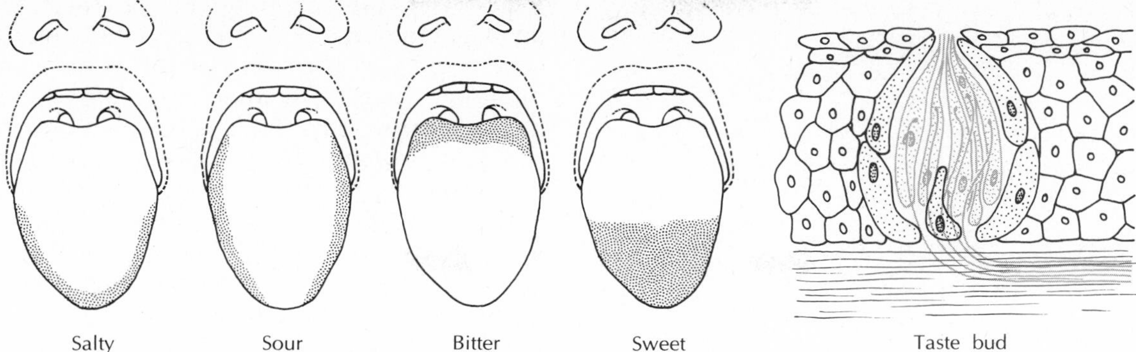

Figure 13.12 Details of taste bud and distribution on the tongue of receptors for true taste.

Taste and Smell

Taste is in reality a deceptive term. True taste receptors are confined to the tongue. They are clusters of specialized cells (Fig. 13.12): sweet on the tip; salty on the tip and in an arc along the sides; sour along the lateral edges; and bitter near the back. But taste is more than the reception of the four basic taste senses. Texture of the food is an important component. There are variations in different individuals; some are more receptive to sweet than sour, for instance.

Besides such individual variations, much that passes for taste is really olfactory reception (smell). Anyone with a head cold soon recognizes this as foods lose their characteristic flavor and are less enjoyable. Inflammation of the nasal lining, the mucosa, congests the sensory receptors of smell.

Smell is the result of direct stimulation of dendrites of neurons that are embedded in the nasal mucosa. Their axons pass through a perforated bone in the floor of the skull and proceed directly into the brain. These neurons form the *olfactory nerve.* There are variations in the sensitivity for olfactory reception in the vertebrates. This becomes obvious from observing, for example, dogs or cats. In humans, *olfactory fatigue* can develop. Continuous exposure to an odor, even a strong one, can lead to a lack of awareness or ability to smell that odor, but other odors still can be detected.

It is not clear how one odor can be discriminated from another. The theories that have been proposed have not met with universal acceptance. Amoore's theory holds that there are seven different primary odors (camphor, musk, floral, peppermint, ether, pungent and putrid), with other odors due to combinations or overlapping stimulation. In the reception of taste and smell it is quite certain that the stimulating substance must be in solution to be recognized. This can be demonstrated by drying the tongue thoroughly before applying the stimulus. Both taste and smell are examples of chemoreceptors.

The Eye and Vision

Light-detecting structures have evolved in a great variety of animal forms. The single-celled plant-animal *Euglena* has a red eye spot or stigma, by means of which it orients itself to light. The flatworm planaria has simple eyes for detecting the directions of light. They do not form images. Such simple eyes are found in many organisms. In the arthropod group, which includes such forms as insects and lobsters, in addition to simple eyes, compound eyes are found. Such eyes are composed of many visual units and produce a mosaic type vision, with each unit of the eye contributing a separate image to the mosaic. Compound eyes are especially effective in detecting the slightest motion. Among all the invertebrates, in only one group (that to which the squid and octopus belong) have eyes evolved that can focus for near and far vision like the vertebrate eye. Although the evolutionary pathways of these mollusks and the vertebrates have been quite far apart, the eyes in the two groups are similar in many respects. This is an example of convergent evolution, in which similar structures develop in organisms that are not closely related.

THE HUMAN EYE

The human eye is composed of three coats, as shown in Figure 13.13. The outer tough opaque layer of connective tissue is the *sclera,* which protects the inner structures and helps to maintain the shape of the eyeball. In front this layer becomes the thinner transparent *cornea* through which light enters. The middle layer, the *choroid coat,* is a black pigmented layer that absorbs imperfectly focused light rays. In the front part of the eye this middle layer becomes modified to form the *iris* and the muscle-containing

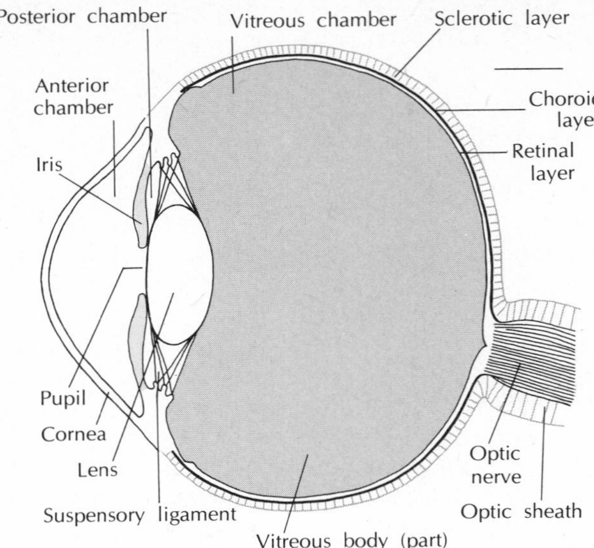

Figure 13.13 Human eye in sagittal section.

structure, the *ciliary body.* The colored iris extends around a central opening, the *pupil,* in the front part of the eye. The color of the eye is due to the pigments in the iris. The *lens* of the eye is attached to the ciliary body by the *suspensory ligament.* The inner layer, the *retina,* contains the light receptors, and it extends up to the points of attachment of the lens.

The cavity of the eye anterior to the lens is filled with a watery fluid, the *aqueous humor.* The *anterior chamber* is in front of the iris; and the one between the iris and the lens attachment is the *posterior chamber.* The large cavity behind the lens is filled with a jellylike material, the *vitreous humor,* which has the consistency of uncooked albumen. These substances fill the cavities of the eye and maintain a pressure within to keep the layers in place and prevent the eye from collapsing. Attached to each eyeball are six muscles that function in the movements of the eyeball in its bony socket.

The iris of the eye acts as a diaphragm to regulate the size of the pupil and thus the amount of light that reaches the retina. There are two sets of muscles in the iris, one set arranged circularly and one set arranged radially. They are reflexively controlled. Contraction of the circular muscles constricts the pupil, whereas contraction of the radial muscles dilates the pupil.

The eye perceives not only light but also form. Form perception is possible because the lens system reproduces true images on the retina. The images are oriented upside down, but the cerebrum interprets the impulses so that the impression of proper orientation is achieved.

In the eye the bending of the light rays to form an image on the retina is largely a function of the curved cornea. The lens is an important part of the total lens system, but it plays a minor role in the total refraction. The lens functions in changing the focusing power of the eye through changes in its curvature. Because of the changes in the lens, one is able at one second to focus on a distant object and in the next second to focus on a nearby object. This change is called *accommodation.*

When the eye is at rest, the fibers of the suspensory ligament are under tension, the lens is somewhat flattened, and the eye can focus on distant objects. When the ciliary muscles (located near the points of attachment of the suspending fibers) contract, the tension on the fibers is reduced and the lens bulges and becomes thicker. In this condition the eye is focused for near objects. The changes in the ciliary muscles, and thus the process of accommodation, are under reflex control. Accommodation depends upon the contractile power of the ciliary muscles and the elasticity of the lens. As the elasticity of the lens decreases with age, an older person must use convex glasses for near vision in reading to compensate for the reduced powers of accommodation.

The sensitive layer of the eye is the retina. This layer is composed of three layers of cells, as shown in Figure 13.14. The cells that form the layer next to the choroid coat are the *rods* and *cones,* the receptor cells of the eye. From their location it is obvious that light must pass

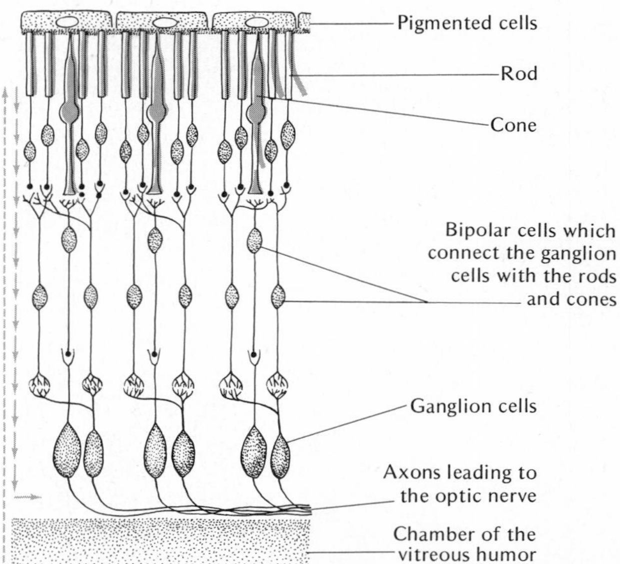

Figure 13.14 Structure of the retina (highly schematized and simplified). The dashed arrows show the pathway of the impulse after the rods and cones respond to light.

through the other parts of the retina before reaching the actual receptors.

The function of the eye as a photoreceptor is based on chemical changes of visual pigments located in the rods and cones. The proportion of these two cell types varies, depending on the vertebrate. For instance, in the chicken most of the retina is composed of cones. In humans, the cones are principally concentrated in a small spot, the *fovea centralis.* It is here that most of the light entering the eye is focused. The remainder of the retina is made up primarily of rods.

Rods and cones perform separate functions. Rods are receptive to light of low intensity and do not respond to differences in color. Cones respond to strong intensities of light and discriminate color.

A single eye, then, has two visual systems in its retina. Rods function in dim light and produce poorly defined images. Cones respond to color and produce detailed images. Rod vision involves the breakdown of the pigment, *rhodopsin,* which is present in rods, into the protein, *opsin,* and a carotenoid pigment, *retinene.* When rhodopsin is struck by light, it dissociates into the two components. This process leads to a change in membrane potential and initiates a nerve impulse. Retinene decomposes into vitamin A. Rhodopsin can be regenerated from vitamin A and opsin. The energy for joining the two is derived from ATP. Deficiency in vitamin A leads to night blindness, a failure to see in dim light.

Color vision has been shown to depend on three kinds of cone pigments. The cells bearing them are called blue, green, and red receptors. Different combinations of receptors are stimulated by wavelengths of different shades of color. The impulses are interpreted in the brain.

EYE DEFECTS

Earlier it was mentioned that many older people need to use convex glasses for near vision to compensate for the reduced powers of accommodation. The commonest eye defects are known as *farsightedness* and *nearsightedness.* In the normal eye the shape of the eyeball is such that the light rays converge on the retina. In a farsighted eye the eyeball is too short and the light rays strike the retina before converging, causing a blurred image. This condition may be corrected by the use of convex lenses that will cause the light rays to converge farther forward on the retina. In a nearsighted eye the eyeball is too long and the light rays converge in front of the retina, also causing a blurred image. Concave lenses, which cause the light rays to converge farther back in the eye, are used to correct this condition.

Astigmatism is another common eye defect. This condition is due to a cornea or a lens that is irregular in its curvature. To correct for astigmatism, lenses must be ground unequally to correct for the unequal curvature of the cornea or the lens.

Either as the result of disease or aging, the lens system of the eye may become opaque, resulting in blindness. When the lens becomes opaque, the condition is known as *cataract.* In the hands of a good surgeon such a lens can be removed. In such a case the individual must wear special glasses to substitute for the removed lens. An opaque cornea is even more serious. The only correction for this is the replacement of the opaque cornea with a cornea from a normal eye. This is a very delicate operation, but one that is becoming more common all the time. The cornea is unusual in being able to be transplanted to different individuals. As a result it is not uncommon now for people to "will" their corneas to others at the time of death, and in some centers cornea "banks" have been established.

The Human Ear

The ear (Fig. 13.15) is also an organ of double function. It serves for hearing and for equilibrium. The ear has a long evolutionary history. Appreciation of this leads to a better understanding of the function of the mammalian ear. Early chordates had specialized hair cells in

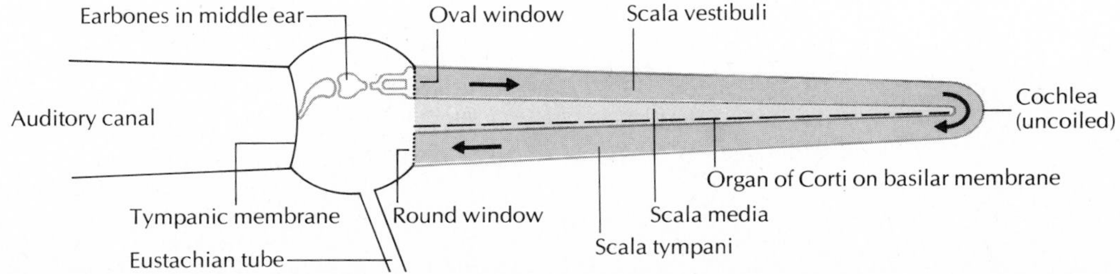

Figure 13.15 Diagram showing relationships of parts of the human ear. The cochlea is shown as though it were uncoiled.

grooves on the surface of the body. These appear to be responsive to pressure waves in the water. Some hair cells became locally restricted in an internal vesicle derived from embryonic skin, the ectoderm. This vesicle became subdivided, by differential growth rate, into a pair of sacs and three arching tubes or *semicircular canals.*

A fluid fills both the sacs and the canals. Specialized areas of these sacs and canals respond to changes in pressures. In the sacs, the *sacculus* and *utriculus,* are aggregations of crystals of calcium carbonate, called *otoliths,* which are embedded in a localized jellylike matrix. Changes of the head (or body) in space shift the otoliths on the hair cells, and this discharges impulses to the brain where orientation in space or *static equilibrium* is interpreted. At one end of each of the three semicircular canals is a bulblike expansion called an *ampulla,* which also contains hair cells. In movement, or *dynamic equilibrium,* the inertia of the fluid in the semicircular canals increases or decreases pressure on the hair cells in the ampullae, and this change of pressure causes impulses that are sent to the brain for interpretation of movement.

The use to which each canal is put in its orientation to the body is apparently a matter of training and experience. For instance, a rapid forward movement by a person is usually not sensed in any unusual way. However, a sudden vertical change, as in a rapidly accelerating elevator, can elicit a response that is communicated to the brain to create the "stomach dropping out" sensation. This can be avoided by bending the head to look up at the ceiling of the elevator, thus "fooling" the semicircular canals. Motion sickness is an excellent but unpleasant example of response to unusual forms of motion; some people can be trained or conditioned so that a particular form of motion ceases to be a problem.

The part of the inner ear that receives sound is a spirally coiled tube, the *cochlea* (Fig. 13.16). It lies deeply embedded in bone. The cochlea consists of three tapering, parallel canals. The central canal is filled with a fluid, *endolymph,* and possesses hair cells along a special basal membrane, a combination of structures called the *organ of Corti.* Flanking the central canal are two canals containing a fluid called *perilymph.* At the end of one of these canals, the *vestibular canal,* there is a membrane-covered *oval window.* This window holds the base of a small middle ear bone, the *stapes.* The other canal, or *tympanic canal,* also has a membrane, the *round window,* near the middle-ear chamber, but has no bone fitted to its surface.

The middle ear is a chamber between the inner ear and the tympanic membrane. It is connected to the pharynx by a tube, the *eustachian tube.* This helps to regulate the air pressure on

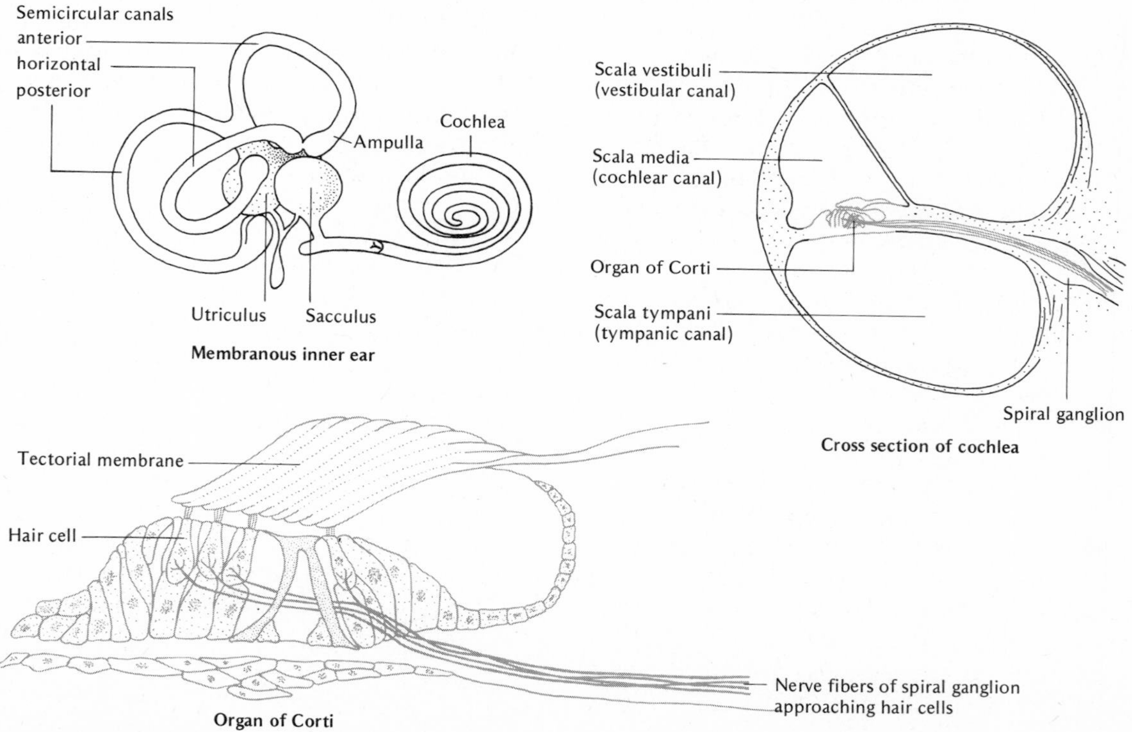

Figure 13.16 Details of inner ear.

the inner surface of the tympanic membrane. Bridging this membrane and the oval window are three middle-ear bones: the *malleus, incus,* and *stapes* (hammer, anvil, and stirrup) which form a set of levers.

The outer ear consists of the tube from the exterior to the tympanic membrane, the *auditory canal,* and the fleshy external part, or *pinna.*

Sound is transmitted in the ear as follows. Vibrations enter the auditory canal and cause the tympanic membrane to vibrate. The three middle-ear bones provide a leverage system to convey these vibrations to the oval window. The vibrations are passed through the perilymph of the vestibular canal. The pressure waves in this perilymph cause waves in the basilar membrane of the organ of Corti. When the hair cells along this membrane touch the overhanging tectorial membrane, the mechanical sound waves become translated into nerve impulses. The basilar membrane contains fibers, some 24,000 of them that increase progressively in length from the base to the apex of the cochlea, like the strings of a harp or piano. Short sound waves (high pitch) stimulate hair cells in the lower part of the spiral cochlea; long waves (low pitch) stimulate near the apex. The vibrations in the perilymph of the vestibular canal are carried to the apex and then travel in the reverse direction in the perilymph of the tympanic canal. They are damped out by the round window.

Stimulation of specific hair cells in the organ of Corti in its simplest sense determines pitch. Hearing, however, also involves interpretation of amplitude (loudness) and quality of tone. Amplitude is determined by the strength of the vibrations striking the organ of Corti. Quality of tone is interpreted by the overtones, which stimulate secondary areas of the organ of Corti. The impulses from different areas in the organ of Corti go to slightly different brain areas for interpretation. It becomes obvious that the interpretations made by the brain are influenced by the pattern of hair cells stimulated.

EVOLUTION OF THE VERTEBRATE EAR

A comparison of the ears of the different classes of vertebrates furnishes another interesting story of evolutionary change. In the fishes the ear is composed of the utriculus, the sacculus, and the semicircular canals. There is neither cochlea nor middle or outer ears; it is primarily an organ for equilibration. This nonacoustic part of the ear remains essentially the same in all the other vertebrate classes; the changes are found in the parts of the ear involved with hearing. In the amphibia, the first animals to live on the land, a number of changes occurred. They possess a small outgrowth of the sacculus, the *lagena,* that develops into the cochlea in mammals and is present as a very small diverticulum in some fishes. In the amphibia the middle ear cavity is formed from one of the embryonic pharyngeal pouches; the hyomandibular cartilage ceases to be a suspensory apparatus of the jaw and is incorporated in the middle-ear cavity as the *columella* (in higher forms called the stapes). With this, the thin membrane separating the middle ear from the outside becomes the ear drum. There is no external ear in the amphibians (Fig. 13.17).

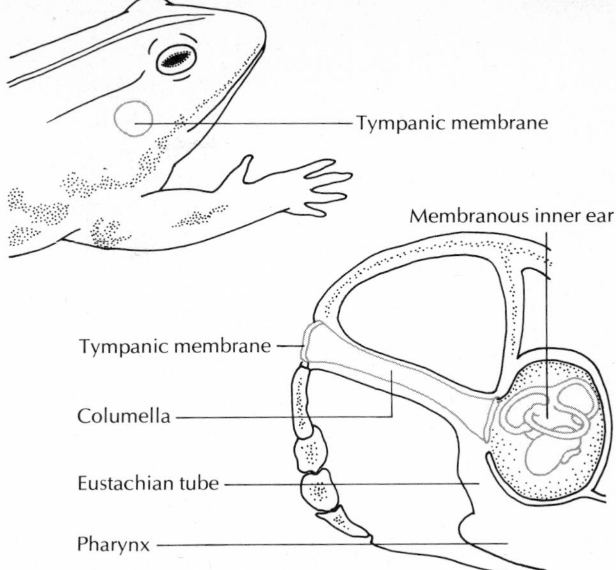

Figure 13.17 Ear of frog.

In reptiles the lagena becomes an elongated cochlea, and in mammals the cochlea reaches its greatest development, varying from a half turn in the primitive mammal, the echidna, to three and a half turns in the deer. In the different mammals the keenness of hearing varies with the degree of development of the cochlea. The pinna is a development found only in mammals. However, in reptiles and birds the ear drum may be in a depression, the auditory canal, below the surface of the head.

In this evolution many original structures have become modified for their new roles. The middle ear and the eustachian tube were originally parts of the respiratory system of the fish; as mentioned above, the columella or stirrup was a part of the suspensory apparatus of the jaw in fishes; and the malleus and incus were parts of the lower and upper jaws, respectively, of all vertebrates but mammals.

Proprioceptors

Many different muscles are involved in almost every act we perform. It has been estimated that over 50 different muscles are involved in a simple forward movement of one's leg in walking. What causes each muscle in such an involved group to contract just the right amount? Studies in the last century have shown that every muscle and every tendon that attaches a muscle to a bone are equipped with special nerve endings called *proprioceptors,* which are continually sending impulses to the brain as changes in the tension on the muscle or tendon occur. As a result of this stream of impulses going to the brain, the further activity of each muscle can be synchronized with the activity of other muscles in a reflex manner.

One is less aware of these sense organs than of any others in the body, yet they are probably the most numerous and most active of any of the sensory receptors. Without them, skilled movements would be impossible. Because of them one can perform such complicated operations as dressing in the dark and tying a knot behind the back. Without them there would probably be a continuous feeling of numbness, such as is experienced when an arm or leg "goes to sleep."

SUGGESTIONS FOR FURTHER READING

Bloom, W., and D. W. Fawcett. *A Textbook of Histology.* Philadelphia: W. B. Saunders (1968).

Calder, N. *The Mind of Man.* New York: Viking Press (1970).

DeRobertis, E. "Ultrastructure and Cytochemistry of the Synaptic Region." *Science,* **156,** 3777 (1967).

DeRobertis, E. "Molecular Biology of Synaptic Receptors." *Science,* **171,** 3975 (1971).

Giorgi, P. P., J. O. Karlsson, J. Sjorstrand, and E. J. Fields. "Axonal Flow and Myelin Protein in the Optic Pathway." *Nature New Biology,* **244** (1973).

Hoar, W. S. *General and Comparative Physiology.* Englewood Cliffs, N.J.: Prentice-Hall (1966).

Ochs, S., and R. A. Jersild, Jr. "Fast Axoplasmic Transport in Nonmyelinated Mammalian Nerve Fibers Shown by Electron Microscopic Radioautography." *J. Neurobiology,* **5** (1974).

Ranson, S. W., and Clark, S. L. *The Anatomy of the Nervous System,* 10th ed. Philadelphia: W. B. Saunders (1959).

Weiss, P., and Hiscoe, H. B. "Experiments on the Mechanism of Nerve Growth." *J. Exp. Zool.,* **187** (1948).

From Scientific American

Amoore, J. E., J. W. Johnson, and M. Rubin. "The Stereochemical Theory of Odor" (Feb. 1964).

Baker, P. F. "The Nerve Axon" (Mar. 1966).

Bekesy, G. Von. "The Ear" (Aug. 1957).

Eccles, J. "The Synapse" (Jan. 1965).

French, A. E. "The Reticular Formation" (May 1957).

Katz, B. "The Nerve Impulse" (Nov. 1952).

Lowenstein, W. R. "Biological Transducers" (Aug. 1960).

MacNichol, E. F. "Three-Pigment Color Vision" (Dec. 1964).

Michael, C. R. "Retinal Processing of Images" (May 1969).

Neisser, W. "The Processes of Vision" (Sept. 1968).

14 Behavior: Learned

Behavior is what an organism does. A behavior is a response of an organism to a stimulus or to stimuli. The stimuli may originate outside the organism or within the organism (secretions, changes in chemical state, and the like). Observations of animal behavior are probably as old as human existence. Early writers, such as Aristotle, included behavioral observations in their accounts of natural history.

All organisms exhibit behavior, even plants and lower animals that lack nervous systems. This chapter, however, is devoted to behavior in multicellular animals (primarily vertebrates) that have a well-developed nervous system. The basic equipment for behavior in such animals is a nervous system associated with receptors (sense organs) and effectors (muscles and glands). The nervous system, the receptors, and the effectors result from developmental processes and are determined very largely by the genetic makeup of the organism. The structures that form the basis of behavior are much alike in all members of a given species and can be considered as adaptations; they are the result of a long process of natural selection and evolution.

At the cellular level it is now known that all nerve cells are very much alike in basic structure and function from worm to mammal. Each has a cell body and two kinds of processes; they conduct nerve impulses, each of which is a wave of electrical negativity along the fiber; and they have transmitter substances secreted at the synapse or end plate. All these facts are essential to an understanding of the nervous system and behavior. It now seems clear that the differences in the behavior of a fish and a human, for example, are based in large part on the much greater complexity in the organization of human nerve cells.

Students of behavior often categorize two main kinds: *innate* or *stereotyped,* and *learned.* Innate behavior is behavior that is normally unaffected by experience. It occurs in essentially the same way in each member of the species when the stimulus is presented, regardless of past experience. It is due to an innate, or inherited, behavior pattern. Learned behavior, on the other hand, is behavior that is modified by experience. An animal records specific experiences and modifies its behavior in terms of these experiences.

This is a neat system of "either–or" classification. There has been considerable discussion in the past as to whether a particular behavior was innate or learned. Much evidence gathered in recent years indicates that many of the behavior types described earlier as completely stereotyped may be modified by experience. This is an area of active study today.

Many of the current students of behavior, instead of asking the question, "Is the behavior of the animal under consideration innate or learned?" are asking the question, "To what degree, if any, is the behavior of the animal in question modifiable by experience?" With this latter question in mind, it can be stated in a general way that the culmination of the annelid–arthropod line of evolution, the insect, exhibits *primarily* innate behavior, whereas the culmination of the echinoderm–chordate line of evolution, humans, exhibit *primarily* learned behavior. However, in a certain sense, all behavior is determined primarily ("primarily" because no organism develops in an environmental vacuum) by heredity. There is no question about this even in those cases where learning is demonstrable; the organism inherited the specific type of nervous system with a certain pattern of neurons that makes learning possible. So in this sense the potentialities for learned behavior are inherited.

The study of behavior is complex. Studies in cell physiology, comparative physiology, neurology, biochemistry, development, genetics, ecology, and evolution have all contributed to our understanding of behavior. However, two different schools of study of behavior—one in Europe and one in the United States—are generally recognized for their recent contributions, especially those of the last two decades. The members of the European school are known as ethologists and are concerned with instincts and behavioral adaptations as they occur in nature. The American school, composed of both psychologists and biologists, has focused on mechanisms of adaptations, motivational states, and the nature of learning.

This chapter focuses on learned behavior, with some examples of the methods used in its study. In Chapter 43 the basic types of innate behavior, together with some of the techniques that modern workers have used in their studies, are considered. In addition, in Chapter 43 the modification of innate behaviors through learning, the genetics of behavior, the evolution of behavior, and certain other aspects of behavior are discussed. However, before starting on learned behavior, it will be useful to consider a few examples of innate behavior.

EXAMPLES OF INNATE BEHAVIOR

The simplest types of innate behavior are the simple reflexes, and as was mentioned earlier, attempts to understand or partially understand the more complex types of behavior necessitate an understanding of the basis for simple reflex acts. The withdrawal of a stimulated leg by a brainless frog is a simple reflex act. The basis for this action, the reflex arc, has already been described in Chapter 13. The connections of the different neurons with sense organs and muscles forming the arc that makes such a response possible were made during the development of the frog. Until such connections are made in the process of development, such a response is not possible. That the responses of an animal are dependent upon the proper connections of the parts of the reflex circuit has been shown in experiments with developing salamanders. For a time the embryos are motionless and respond to no stimuli. After a while there is a little movement, and, finally, the animal begins to swim. A study of the developing nervous system shows that the movements begin at the time that the necessary connections have been completed.

But not all innate behavior is in the form of simple reflexes. Many complicated activities, in humans as well as in lower animals, are innate or unlearned. The human sucking reflex is a good example of a complicated and highly integrated activity that is unlearned. A newborn or even a prematurely born infant will carry out this series of movements without any help or training. In executing this series of reactions the baby bends its head to take the nipple, closes its lips, makes the characteristic movements of the lips, tongue, and mouth to suck, and finally swallows when the mouth is filled.

The behavior of many of the lower animals consists mainly of complicated actions that are often called instincts (see Ch. 43). Every reader is familiar with a number of striking examples. A young spider of a given species, without any contact with any other members of its species, will, in its first attempt, spin and weave a web characteristic for that species. In a similar way a caterpillar that has been kept isolated from all others of its kind will form a cocoon characteristic of its species. Also, a pair of young birds that have been kept apart from all other birds will build a nest characteristic for their species. One of the most complicated chains of activity found in some birds, which is instinctive in nature, or unlearned, is the migration phenomenon. Many species of birds that spend the summer in the northern part of this continent migrate to Central America or to other southern regions for the winter, and the place at which the members of a given species spend the winter is usually quite specific. In experiments that have been made using banded birds it was possible to determine that young birds, birds that had never made the trip south before, when caged and held back until all the other members of their species had migrated south, would, when released, make the trip south to the place occupied by that species for overwintering.

These are remarkable examples of how complex unlearned behaviors can be. Such complicated responses are dependent on circuits that are established in the organism during development.

Mammals vary considerably in their degree of development at birth. Newly born guinea pigs can walk; so can young guinea pigs that have been removed prematurely from the uterus. On the other hand, young rats do not walk until they are about three weeks old. Walking is not learned by young rats; they will walk as soon as the circuits are completed in their development. The growing together of nerves and muscles that makes such acts possible is called *maturation.* Maturation, so far as the basis for walking is concerned, occurs before birth in the guinea pig and after birth in the rat. Considering this and similar cases that have been studied, one should exhibit caution in attributing to learning rather than to maturation any complex activity in a young organism until one knows all of the facts.

A number of characteristics of unlearned behavior may now be enumerated. (1) The neural pattern for such behaviors is hereditary, and these patterns are laid down during development. (2) Unlearned responses are automatic in nature. (3) Unlike learned behavior, unlearned responses are not forgotten. (4) All unlearned responses must be looked upon as having survival value. To be sure, such responses may have survival value for the individual organism, but more important from the standpoint of evolution is that they have survival value for the species. In the evolution of each species the behavior patterns that have survival value have been selected and preserved in the present forms. (5) Although the stimuli that set off many unlearned responses originate in the outside world, it must be kept in mind that many of them are started by internal stimuli, hormones, changes in chemical states, and other internal changes.

LEARNED BEHAVIOR

Learning has been defined as a more or less permanent change in behavior that is recorded in the nervous system. Instinctive, or innate, behavior serves an organism well in a stable environment. Animals that can learn may be able to adjust to different and changing environments so that the behavioral flexibility provided by learning will give them an advantage over the fixity of some innate patterns. Learning is certainly one of the most important phenomena in living organisms. Attempts to understand learning have, to a considerable extent, dominated research in psychology and psychobiology for the past half-century. Much of the learning process has been recognized, but much remains to be identified. The ability to learn requires memory—but what is memory? There is no definite answer yet, but some of the current thinking is reviewed at the end of the chapter. Learning has been demonstrated in many kinds of animals, beginning with the flatworms in the phylogenetic series. It is most highly developed in mammals, with human beings at the pinnacle. Indeed it can be said that great learning capacity, together with the capability for the development and use of language has made possible human culture. This distinguishes a human from all other animals.

As mentioned above, learning causes more or less permanent changes in behavior. In studying learning, changes in behavior due to sensory adaptations, fatigue, and fluctuations in motivations must be recognized because such changes do not last long. Also, changes due to maturation (referred to earlier in connection with swimming of salamander larvae and walking in rats) must be recognized because these are not learned responses. Several kinds of learning situations are recognized and some of these are described.

KINDS OF LEARNING SITUATIONS

Habituation

Perhaps the simplest type of learning situation is habituation, by which the animal gradually learns not to respond to unimportant repeated stimuli. When the web of a spider is vibrated at a particular point by an experimenter, the spider runs out to investigate; finding nothing, it returns to the center of the web. After the same stimulus is applied several times, the spider no longer runs out. This is *habituation.* However, the spider will come out if a different spot on the web is vibrated or if a more intense stimulus is applied. In a like manner a cat may become habituated to a particular sound. It will orient toward the sound at first, but with each successive exposure it orients less and less until it finally does not respond. However, the cat will respond to an entirely different sound frequency.

Imprinting

Another type of simple learning situation is *imprinting,* originally described by Konrad Lorenz in his study of recently hatched goslings. This behavior is relatively permanent, and the phenomenon is limited to a short *critical period* in the young animal. Ordinarily a young goose or a young duck will start following the mother, which is the first moving object seen. However, if the eggs are hatched in the absence of an old bird, the young birds become imprinted by the first large moving and sound-producing object that they see in the first hours after hatching. For example, a young duckling exposed to a decoy

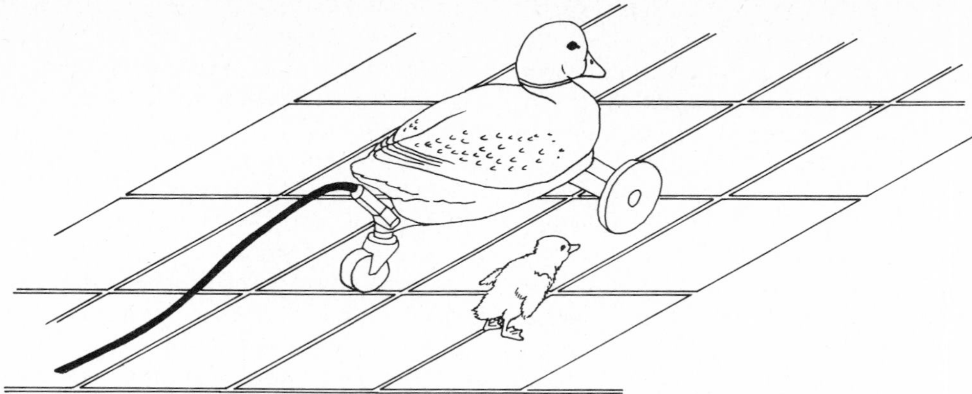

Figure 14.1 Young duck follows remote-controlled decoy.

containing a ticking clock followed the movements of the remotely controlled decoy (Fig. 14.1). After some exposure to the decoy, the duckling persistently followed the decoy instead of real older ducks; it had become imprinted by the decoy. Such a young bird may be imprinted by almost any object, animal, or person. Lorenz found that young birds would follow him around as faithfully as they would have followed the mother bird when they were exposed to him as the first moving object during this critical period. The critical period in young ducklings is the first 36 hours after hatching. If the duckling is not imprinted during this period, it may never imprint.

E. H. Hess has reported recently that the "unnatural" imprinting originally described by Lorenz is not as permanent as he described. In one experiment performed by Hess and his associates, they imprinted newly hatched ducklings to themselves. Then the ducklings were placed with a female mallard and after an exposure of one and a half hours to this female, the human-imprinted ducklings followed the female on the first exodus from the nest. Weeks later these ducklings followed the female mallard rather than the humans involved in their original imprinting. Hess has found that young ducklings begin to emit peeping sounds a few hours before hatching and that the female mallard clucks in response to these sounds. He believes that this auditory interaction is involved in the natural imprinting that occurs.

Imprinting has been described in a number of species of birds, mammals, and fishes. It seems to depend upon some special condition of the nervous system during this critical period. It would be very interesting to know what the effects of imprinting are on humans. Related to this phenomenon is an interesting discovery made by Harlow on young Rhesus monkeys. A group of young monkeys was reared in the complete absence of their mothers. They seemed healthy, but they were quite abnormal in their behavior—they acted frightened and withdrawn. Furthermore, they showed no interest in mating when they were sexually mature.

Classical Conditioning

One of the first great contributions to some understanding of the learning process was made by the Russian physiologist Ivan Pavlov in his study of conditioned reflexes. Pavlov's basic work was with dogs. Salivation, or the secretion of saliva by the salivary glands, is an innate reflex that occurs naturally when food is placed in the mouth. In his experiments Pavlov, with the dog lightly held in place by a harness, rang a bell just before he placed food in the experimental dog's mouth (Fig. 14.2). After doing this on several successive days, he placed no food in the dog's mouth at the time he rang the bell, yet salivation occurred. In this experiment Pavlov called the salivation response to the bell a *conditioned reflex,* the sound of the bell a *conditioned stimulus,* the salivation to the meat an *unconditioned reflex,* and the food itself an *unconditioned stimulus.*

This salivation to the sound of the bell persisted for some time. Pavlov had conditioned an innate reflex. He had substituted the sound of the bell for the natural stimulus for salivation. It must be assumed here that one afferent pathway, the one leading from the ear, had been substituted for the original afferent pathway leading from sensory endings in the mouth. Also, it must be assumed that association neurons connecting the new afferent pathway with the original motor pathway to the salivary glands were involved. These connections involve the cerebrum (Fig. 14.3). In a sense, then, Pavlov's

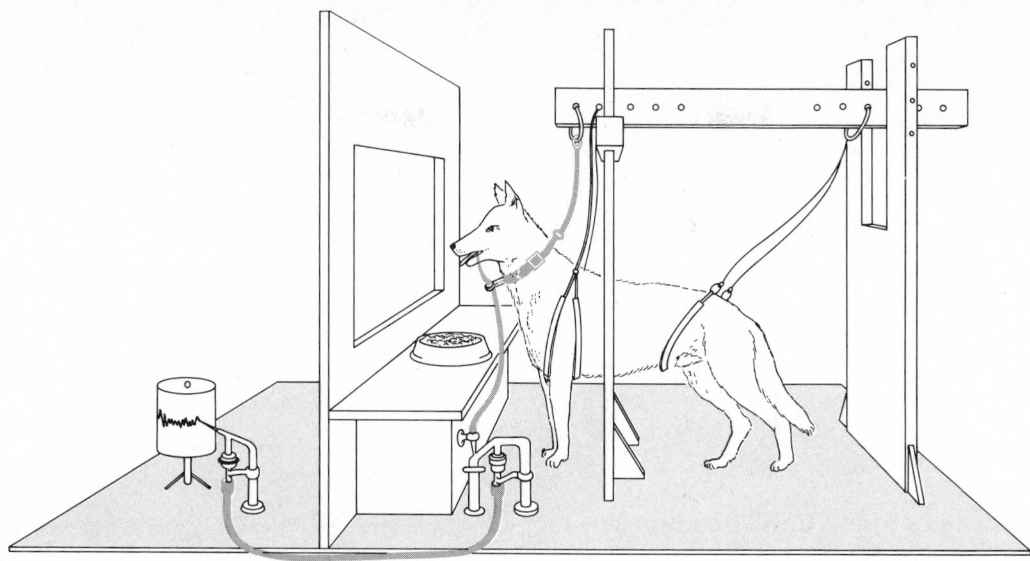

Figure 14.2 Pavlov's dog experiment.

dog *learned* to salivate at the ringing of the bell. Pavlov found that the conditioned response would persist for some time, but eventually, in the absence of the unconditioned stimulus (food in the mouth), the conditioned response disappeared. This he called *extinction.* He concluded that there was something about the unconditioned stimulus that was essential for the strengthening and the maintenance of the conditioned response, and this he called *reinforcement.*

Much work has been done on conditioning since Pavlov's time, and it has been possible to build whole chains of conditioned responses. Garrett described one such interesting chain (see Suggested Readings). In this experiment the dog was strapped in an apparatus, with one foot resting on a metal plate. When an electric circuit was closed, the dog was shocked through the metal plate, resulting in the reflex withdrawal of the foot. For a time the experimenter tried sounding a buzzer just before and during the time the current was on. After several trials the dog started to lift its foot with the sound of the buzzer, before the current was turned on. This was a conditioned response, a learned reaction. In still other trials an electric light was turned on before the buzzer was sounded. In

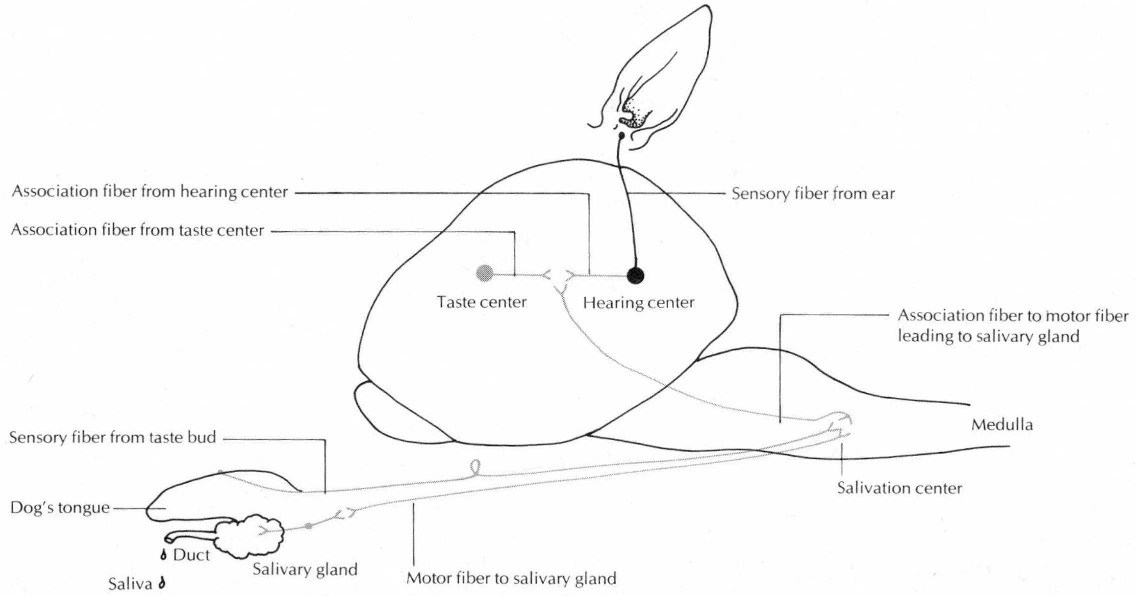

Figure 14.3 Diagram of neuronal pathways in conditioned reflex.

time, the dog was conditioned to this stimulus. The foot was lifted each time the light was turned on. In a similar way this dog was further conditioned to lift its foot when touched on the flank just before the light was turned on; and still later it was conditioned to respond to a blast of cold air on its nose just before being touched on the flank. In this experiment the dog was so thoroughly conditioned that for a period of 18 months it would respond to any one of the four experimental stimuli by lifting its foot.

For a time following Pavlov's work and the similar work of others some workers took the position that all learning might be explained on the basis of conditioned reflexes.

Instrumental, or Operant, Conditioning

In classical conditioning the animal has no control over the stimulation. In *operant* or *instrumental conditioning,* the animal, through its own activity, selects from the various stimuli in the environment a stimulus to which he reacts, and its behavior is instrumental in bringing about some significant change in its environment. Put in other terms, behavior performed by an animal is often called *operant,* that is, the animal *operates* on its surroundings. Such behavior may lead to a conditioned or learned response.

The simplest setup for such conditioning is a T maze or Y maze, where the animal must make a right or left turn. On one turn the animal may escape or find food or water; on the other it receives a shock. The facility in learning is indicated by the number of trials required before the animal always makes the right choice. In addition to simple mazes, more complicated mazes are used (Fig. 14.4). Another type of setup is

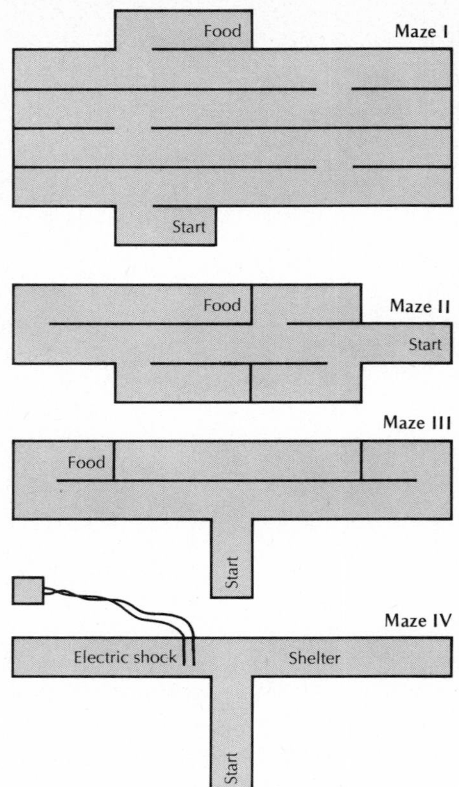

Figure 14.4 Different kinds of mazes. The upper three are of graded difficulty, as used by Lashley in studies on the effects of brain lesions on learning in rats. The lower one is a simple T maze, the type used by Yerkes for training earthworms to avoid electric shock.

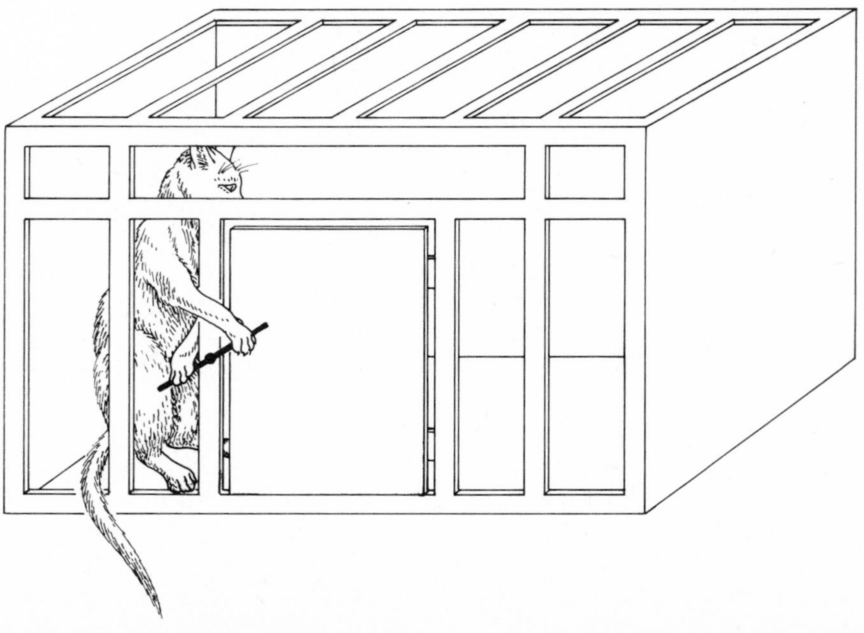

Figure 14.5 Cat trying to escape from puzzle box.

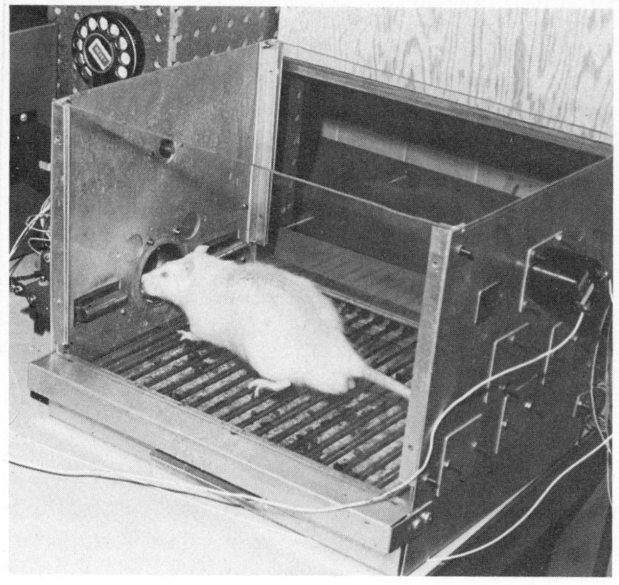

(a)

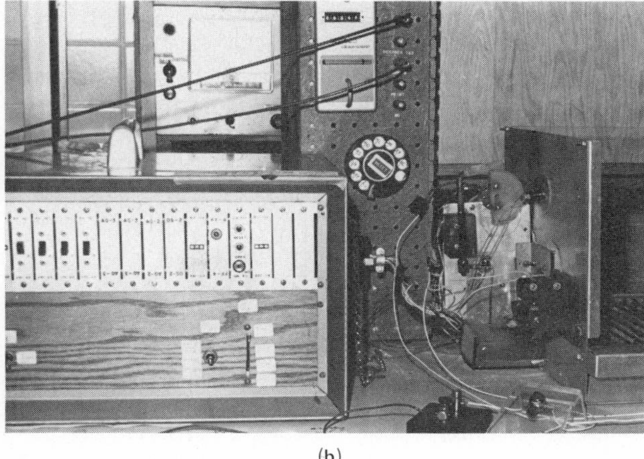

(b)

Figure 14.6 **(a)** Rat in Skinner box. **(b)** Wiring and recording instrument associated with Skinner box. (Courtesy of Psychology Department, Wabash College.)

found in the use of *Thorndike's puzzle box* (Fig. 14.5). This is a chamber shut by a complicated latch that can be operated from the inside. When a cat or other animal is placed inside, it will first perform many activities. It finally makes a move that opens the latch and allows the animal to escape and obtain food. In subsequent tests the behavior that led to the opening of the latch is used with increasing rapidity until a minimum time is reached for throwing the latch.

Still another type of experimental setup is a *Skinner box* (Fig. 14.6). This is a soundproof box with two transparent walls for observations. On one wall is a key and a closed food hopper. The key and the hopper are wired so that the food hopper will open temporarily when the key is touched. Also, the key is wired to a cumulative recorder, which plots a graph of the total number of times the key has been pushed as a function of time. When a hungry pigeon is placed in the box, it will make many different movements. Such behavior is often called *trial and error.* Ultimately the pigeon will peck the key. This results in the opening of the food hopper—long enough for the pigeon to pick up a grain of corn. Usually the pigeon will peck the key again immediately. When each peck results in food, the pigeon pecks at a faster rate. The food has reinforced this pecking behavior. However, if the box is set so that pecking the key no longer exposes food, the pigeon will keep on pecking for some time, but at a gradually reduced rate, until it finally quits; extinction occurs. The Skinner box can also be used effectively with such animals as rats.

Motivation

Many workers have studied what are called the motivational aspects of many innate behaviors. They start with the conception that the pattern can be analyzed into a *drive* directed to a *goal,* in the case of many types of behavior. With the attainment of the goal there is a reduction of the drive, or satiation. As an example, a hungry animal seeks food, finds it, eats, and then stops eating. The drive is the striving for food; the goal is the food. In Chapter 43 it is pointed out that numerous centers in the hypothalamus of the brain are associated with such drives and their satiation in a variety of animals involving several types of behavior. It seems reasonable to assume that these drives and the mechanisms that are responsible for them are the products of the evolutionary process.

Motivation, involving internal drives, is also important in many learning situations. If an animal without motivation is placed in a maze or one of the experimental boxes, it may just sit and do nothing; learning does not occur. Also, as has been pointed out, the learned behavior, if it is to persist, must be reinforced from time to time.

Insight

Simple examples of the insight phenomenon have been studied, using the *detour* problem. An example is illustrated in Figure 14.7. The animal is tied with a rope to a stake, with the rope around another stake as shown. Food is presented, but the animal cannot reach the food

Figure 14.7 Detour problem.

unless it detours around the second stake. When an animal does not go through a period of trial-and-error behavior, but surveys the problem and then acts correctly on the first try, it exhibits *insight or reasoning.* Many animals have been tested in detour problems, but only some monkeys and chimpanzees are successful in the first exposure. Many rats, dogs, and raccoons learn to do the correct thing rather rapidly. Fishes and birds may eventually learn the way around a barrier.

Wolfgang Köhler has described a more complex example of insight solution in a chimpanzee. The chimpanzee was placed in a cage with some bananas hanging out of reach. A number of empty boxes were also scattered about the cage. After surveying the situation, the chimpanzee piled the boxes on top of one another and then obtained the food.

The very complex processes found in humans, such as reasoning and the use of language, can be studied in part by analyzing other animals, such as chimpanzees. These processes and their influence on behavior properly belong in the province of psychology.

LEARNING IN DIFFERENT KINDS OF ANIMALS

To date, no convincing demonstrations have been made that learning occurs in any of the protozoans or in any of the radially symmetrical metazoans. The first real evidence of learning, for the present at least, is found in the bilaterally symmetrical worms. A planarian worm has been conditioned in the classical way. With a worm moving along in a container of water, a light was turned on, followed in 2 seconds by an electric shock. The shock caused the worm to contract longitudinally. It required 150 trials for the worm to contract to the light alone 90 percent of the time. Earthworms have been trained to go to one arm of a T-maze leading to a dark moist chamber instead of going to the other arm where they received an electric shock. In this case about 200 trials were required before the worm made the correct choice in 90 percent of the trials.

Among the invertebrates below the arthropods, the octopus, which has both well developed eyes and brain, exhibits the greatest learning potential. An octopus can be conditioned easily, and it quickly learns a maze. In one experiment an octopus learned to avoid food when a white card, signifying shock, was associated with it. At first the octopus was trained to come from its nest in the aquarium and seize a crab lowered into the aquarium by a thread. In the experiment on one half of the trials a white card was lowered with the crab, and in these trials the octopus was given an electric shock before it reached the crab. After 24 trials the octopus always remained in its nest when the card accompanied the crab, but it always came out to feed when the crab was presented alone. The octopus, however, was unable to solve a simple detour problem. When a transparent glass partition was lowered between the octopus and the crab, the octopus always swam straight into the glass.

Very complex behaviors are found in the insects, especially the social insects. Most of this behavior is innate and stereotyped; however, insects do exhibit learning. Most insects are capable of classical conditioning. Ants are adept at running mazes. Honeybees have been trained to associate food with a particular color or a particular shape. It has been demonstrated that young honeybees learn landmarks around the nest and recognize them by sight. Learning capacities are present in all classes of vertebrates. However, the greatest capacities are found in the birds and mammals, and some examples of learning in both groups have been described.

Memory

Learning involves a change in the nervous system that results in new or changed behavior. This change in the nervous system is the basis of what we call *memory.* Individual memory traces are often called *engrams.* For many years students have been asking these two questions: where in the brain are these memory traces located, and just what is involved in producing them? So far it has not been possible to obtain completely satisfactory answers for either of them. It has already been pointed out that it is possible to locate certain specific centers in the cerebral cortex for various conscious sensations and for various motor activities. It has not been possible to locate specific centers in the cerebral cortex for the changes involved in learning that result in memory. Since we have observed an increase in learning capacity in the vertebrates as we move from fish to human, and since this is paralleled by an increase in the development of the cerebrum and in the cerebral cortex in the higher forms, it seems logical to expect that memory would be located in the cortex.

Karl Lashley performed many experiments in which he tried to locate the sites of memory in the cortex of rats. In one series of experiments he first trained rats to run a maze. These animals were then operated on with incisions made all over the cortex. Such operated animals did not lose the ability to run the maze. Furthermore they were just as able to learn to run other mazes as normal rats. In another series of experiments he tested the effects of total removal of sections of the cortex on the ability of rats to learn to run mazes. These experiments indicated that the resultant defects in learning ability depend upon the amount of the cortex removed, not the region of the cortex. Experiments on a number of other mammals have produced similar results. However, in a study on monkeys different results were obtained. When lesions were made in different parts of the temporal cortex, there were defects in the learning of different sensory discriminations.

In an interesting clinical case the temporal lobes of the cerebral cortex and the underlying hippocampus were removed from a man for the treatment of epilepsy. There was no interference with old permanent memories, but new permanent memories could not be established. He could read an article, understand and discuss it, but as soon as he put it aside, he forgot it. This suggests that the hippocampal system is required for the establishment of new *long-term memories.*

The preceding case casts a new light on memory. It has been demonstrated in forms as different as an octopus, a rat, and a human being that memory involves at least two phases, which are referred to as *short-term memory* and *long-term memory.* In experiments with rats to demonstrate these two phases several groups of rats were trained to run from one compartment to another to avoid an electric shock. They were given one trial each day, and after each trial the rats were given an electric shock through the brain. (It is known that such shock does not affect permanent memories in a variety of forms, including humans.) One group was given the shock treatment 20 seconds after the trial; other groups received their shock after 1 minute, 4 minutes, 15 minutes, 1 hour, 4 hours, and 14 hours, respectively. In those that received the shock within an hour there was no learning, while those that received the shock after 4 hours learned as well as untreated rats. This experiment shows that there is a phase in the establishment of memory that is short and that can be wiped out by electric shock, and that if this phase is not interrupted, the second phase occurs.

It has been suggested that the short-term phase is physiological, that it involves the sustained and integrated firing of many nerve cells, and that it can be eliminated through shock, which simultaneously discharges large numbers of nerve cells. The second phase, which provides the permanent basis for memory, may involve an anatomical change. The clinical case mentioned above, in which the temporal lobes of the cortex and the underlying hippocampus were removed, resulted in the loss of ability to form new permanent memories but did not destroy either old permanent memories or the temporary acquisition of new ones. This suggests that there may be a third aspect to the total formation of memory, that is, the storage of long-term memories.

Although many suggestions have been made concerning the exact nature of the changes in the brain that result in learning and memory, there are not enough facts at present to prove any of them. Among the theories advanced are those that involve the formation of new neural interconnections, the permanent physiological or anatomical modification of existing synaptic pathways, and biochemical changes in the nerve cells themselves. Recently much interest has developed in experiments that implicate RNA in the process and in others that implicate protein synthesis.

In several experiments implicating RNA, the flatworm *Planaria* was used (see Corning and Ratner, Suggestions for Further Reading). Conditioned worms were cut and allowed to regenerate; the learned response was present in the worm that regenerated from both the head and tail sections. Then worms were conditioned and were fed to untrained worms. These worms that had fed on trained worms showed a shorter learning time than the controls. Tail sections of conditioned worms, after cutting, were placed in water containing ribonuclease (an enzyme that degrades RNA). Such tail sections, when regeneration was complete, did not retain the learned response that was present in the sections not treated with ribonuclease. Finally RNA obtained from conditioned worms was injected into unconditioned worms, and such worms showed a shorter learning period than untreated ones.

James McConnell, Allan Jacobson, and a number of others who have worked in this area feel strongly that RNA is somehow involved in learning and memory.

It has also been reported that the injection of ribonuclease into cats interferes with memory and sensory discrimination, that the injection of RNA into rats shortens conditioning time and lengthens the time to extinction of conditioned responses, and that the injection of 8-azaguanine, an inhibitor of RNA synthesis, slows the learning of rats.

Some workers feel that these experiments definitely implicate RNA in learning and memory; others criticize some of the work, saying that the effects are rather small and that the experiments were not adequately controlled. Some workers state that they have been unable to duplicate the results of some of the experiments. William Byrne and others (see Suggestions for Further Readings) have reported on attempts to show the transfer of learned behavior as a result of ingesting RNA-containing fractions from the brains of trained rats into untrained rats. Results from several laboratories gave negative results. Yet it seems clear that some of the supposed duplications did not duplicate all of the conditions of the original experiments. Thus the role of RNA is not settled.

More recently, experiments implicating protein synthesis have been reported. In one case goldfish treated with puromycin, an inhibitor of protein synthesis, exhibited short-term memory but did not develop long-term memory. In another, the injection of puromycin into the hippocampus of mice prevented learning. Louis Flexner and Josefa Flexner have studied the effects of acetoxycyclohexamide on memory in mice. This antibiotic, like puromycin, produced deep and prolonged inhibition of cerebral protein synthesis, yet it did not destroy memory of simple maze learning in mice. When they used acetoxycyclohexamide together with puromycin the memory was not affected. These workers propose that the acetoxycyclohexamide affords protection for polysomes and for their mRNA, and that this accounts for the failure to obtain loss of memory, either when acetoxycyclohexamide is used alone or in combination with puromycin.

In 1968 Ungar and coworkers reported on an experiment in which dark avoidance was induced in mice with extracts of brain taken from rats that were trained to fear the dark. Recently (1972) they have reported the isolation, identification, and synthesis of the active material. It is a peptide with a sequence of 15 amino acids and it has been named scotophobin. These workers obtained essentially the same results using the natural material and the synthetic material. However, this work, like that reported above, has been criticized by other workers. A few workers have obtained confirming results; others have failed. If this work turns out to be reproducible by several different workers, it will constitute a major discovery.

We know that RNA is involved in protein synthesis, so there may be a connection between these two types of experiments. More experiments are needed in this area. If RNA is involved in memory, how does the newly synthesized protein form the memory traces? We can expect more enlightenment from future experiments.

Split Brains

Everyone knows that he has two lungs, two kidneys, two eyes, and a number of other organs that exist in pairs. In addition, many people who have had a kidney or a lung removed can function in a normal fashion. It is also true that humans and other mammals have two cerebral hemispheres in their brains. Each hemisphere has a full set of centers for the various sensory and motor activities of the body, and each hemisphere is primarily associated with one side of the body. The right hemisphere is associated primarily with the left side of the body, and vice versa. However, it has been shown that when a region of one hemisphere has been damaged, the corresponding region in the other hemisphere can take over and control the functions for both sides of the body. Either hemisphere, then, can to a large extent serve as a whole brain.

The two hemispheres are connected both by the lower brain stem and by bundles of fibers, particularly by the very large fiber tract known as the *corpus callosum* (see Fig. 13.8). It has been known for some time by neurosurgeons that the severing of the corpus callosum produced little or no noticeable changes in the patients' ability to perform in an apparently normal way. Experiments on monkeys when the corpus callosum was cut also produced similar results. For many years surgeons cut the corpus callosum in some cases of severe epilepsy to prevent the spread of epileptic seizures from one brain hemisphere to the other.

Until the work of R. W. Sperry (see McGaugh and others, Suggestions for Further Reading) and his associates, that started in 1950, it had not been possible to determine the exact role of the corpus callosum in mammalian brains. Their work involving what is known as "split brain" experiments has pinpointed some of the roles of this huge fiber tract, and some of the important roles involve learning. They have studied and observed cats, monkeys, and chimpanzees in which the corpus callosum had been cut, and they have also studied and observed humans with split brains who had been operated on for severe epilepsy and had been freed of convulsive attacks.

In one type of operation not only was the corpus callosum of the experimental animal cut but also the optic chiasma. Half of the fibers in the optic nerve of each eye cross over to the opposite side in the optic chiasma. Thus when the optic chiasma is cut, each eye transmits impulses only to the hemisphere on the same side of the head.

Such an experimental animal was trained to solve a problem presented only to one eye, the other eye being covered. Food was the reward for solving the problem. After the animal had solved the problem it was then presented in the same way to the other eye, with the first eye covered. In this case the animal behaved as though it had never been presented with the problem before. The number of trials required for the learning indicated that there was no carryover of the learning that occurred when the first eye was used. However, the transfer of learning occurred quite readily in an animal with the corpus callosum intact. This means that the corpus callosum functions in the sharing of learning and memory in the two hemispheres. This sharing could occur in two ways: by transmitting the information at the time that learning takes place or by supplying the information later. Sperry and his associates have found that in the cat engrams are formed in both hemispheres at the time of learning, while in humans, where one hemisphere is usually dominant, a single engram is usually formed at the time of learning. Their experiments also show that learning is slower in some situations in split-brain individuals. The interested student should consult Sperry for some of the detailed effects found in people with split brains.

Gazzaniga's work on split brains in humans involves the separation of function between the right and left hemispheres in conceptual abilities. The right hemisphere controls spatial and patterning type relationships while the left hemisphere seems to control verbal and analytical relationships.

SUGGESTIONS FOR FURTHER READING

Byrne, W. L., et al. "Memory Transfer." *Science,* **153;** 658 (1966).

Carthy, J. D. *The Study of Behavior.* New York: St. Martin's Press (1966).

Corning, W. C., and S. C. Ratner, Eds. *Chemistry of Learning.* New York: Plenum (1967).

Dethier, V. G., and E. Stellar. *Animal Behavior,* 3rd ed. Englewood Cliffs, N. J.: Prentice-Hall (1970).

Garrett, H. E. *Great Experiments in Psychology.* New York: Appleton-Century-Crofts (1951).

Glassman, E., Ed. *Molecular Approaches to Psychobiology.* Belmont, Calif.: Dickenson (1967).

Lashley, K. S. "In Search of the Engram," *Symp. Soc. of Experimental Biology,* **4,** 454-482 (1950).

McGaugh, J. L., N. M. Weinberger, and R. E. Whalen, Eds. *Psychobiology: The Biological Bases of Behavior.* San Francisco: W. H. Freeman (1966).

McGill, T. E., Ed. *Readings in Animal Behavior,* 2nd ed. New York: Holt, Rinehart and Winston (1973).

Ungar, G., et al. "Isolation, Identification and Synthesis of a Specific-Behavior-Inducing Brain Pepotide," *Nature,* **238:** 5361 (1972).

Van der Kloot, W. G. *Behavior.* New York: Holt, Rinehart and Winston (1968).

From Scientific American

Gazzaniga, M. S. "The Split Brain in Man" (Aug. 1967).

Hess, E. H. "Imprinting in a Natural Laboratory" (Aug. 1972).

Rosenzweig, M. R. "Brain Changes in Response to Experience" (Feb. 1972).

Sperry, R. W. "The Great Cerebral Commissure" (Jan. 1964).

Continuity of Species

Part II, in discussing the problems that individual organisms face, concentrates on the vertebrate animal. In addition to the problems considered in Part II, the problem of how a species is maintained, generation after generation, is apparent. The solutions to this problem, involving reproduction, development, and how genes are transmitted from generation to generation, are discussed in Part III. The organisms which illustrate the fundamental solutions to these problems are not exclusively vertebrates, but include a variety of other organisms.

Reproduction involves gametogenesis, the formation of eggs and sperm, and fertilization, the joining of the gametes to produce a zygote. Protection, support, locomotion, nutrition, exchange of gases, elimination of wastes, blood transport of materials, endocrine control, nervous control, and behavior contribute to gamete production, mating, and fertilization.

The principles involved in the transmission of genes from generation to generation were first elucidated with the pea plant by G. Mendel, an Austrian monk, in the middle of the nineteenth century. Unrecognized at that time, these genetic principles languished until the beginning of the twentieth century when they were rediscovered independently by three European investigators. Both the pea plant and the fruit fly are used to illustrate Mendelian genetics and neo-Mendelian genetics, the elaboration and modification of the original statements of Mendel.

Genetics and its implications are pursued further at two levels: development and human welfare. How does the genetic information in the zygote produce the form of the fully developed organism? This is a central concern of developmental biology and developmental genetics, active fields of current research. Drawing upon information presented in Parts I and II, some of the knowledge obtained in recent years and some of the questions being investigated currently are presented.

By applying genetic principles to humans and agricultural stocks, both animals and plants, workers have been able to understand better the functioning of the human body and to produce agricultural stocks that have contributed to the well-being of humankind in many countries in addition to the developed countries of North America, Europe, and Japan. These topics are particularly relevant with respect to numerous articles and releases in the popular media on human genetics, the nature-nurture controversy, and the green revolution.

The chapters in Part III anticipate later discussions of human ecology, human evolution and population growth, food production, genetic engineering, and management of natural resources.

Karyotype of Down's syndrome (World Health Organization Photo)

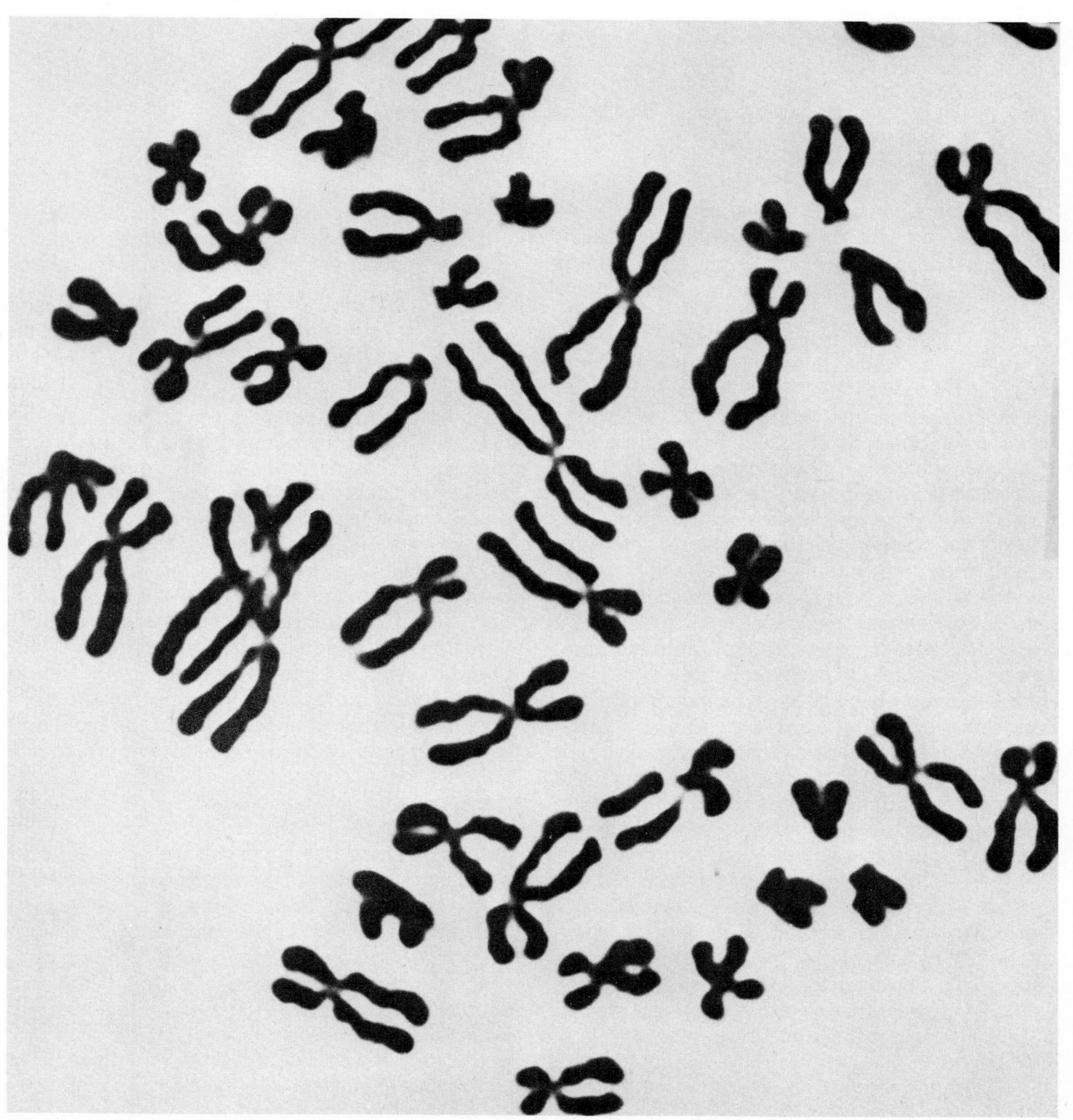

15 Reproduction

That "Life begets life" is a truism. The process for sustaining life from one generation to another is reproduction. Organisms have various adaptations that function in the reproductive process.

In the absence of true immmortality, here are a few of the ways in which continuity from one generation to the next is maintained. In some organisms, the entire animal undergoes division, much as do single cells in our bodies; this is especially true of protozoa. Other animals reproduce by developing localized buds that pinch off as new individuals; coelenterates characteristically often do this. Some animals increase their reproductive potential by having several stages in their life cycles, each of which can produce many copies of itself or the next stage; this method of reproduction is common among some of the flatworms, especially the parasitic ones where survival of the species depends on the production of large numbers. Other animals have the labor of feeding concentrated in one stage of a life cycle, a larval stage, and confine reproduction to another stage, adult, where feeding may be absent; some forms of insects practice this "division of labor."

A very important aspect of reproduction in most animals is the presence of two sexes—*bisexuality*—a very significant aspect of reproduction. Why is bisexuality considered to be so important? Certainly the capacity to reproduce asexually appears to be an economical way to produce new members of the species, and it is. But it ignores a condition in the natural world, namely, change. Asexually reproducing animals usually make almost exact copies of themselves; this offers little opportunity for variation and variation in organisms is essential in order to cope with change. For this reason nearly all living things have developed two sexes and reproduce sexually. Each sex produces gametes that are haploid. The union of gametes, eggs and sperm, results in the production of a zygote that has a genetic contribution from both parents. Thus the new, sexually produced individual will be different from either parent. There is variation. With variation there is an increased chance that some members of the new generation will be better able to cope with changes in the environment.

This explanation does not appear to come to grips with the possibility that eggs and sperm could be produced in the same animal, a condition called *hermaphroditism.* There are functional hermaphrodites among the animals, but they are relatively rare. Also, even though both sexes may be represented in some animals, there may be different timings at which the male and female organs become sexually mature. While self-fertilization can occur in some hermaphrodites, it is uncommon and usually cross-fertilization occurs. The more common condition of separate sexes has the adaptive advantage of creating a division of labor between the sexes. Just one manifestation of this is seen in the production of eggs versus the production of sperm. Eggs are usually produced in fewer numbers than sperm and tend to conserve their cytoplasm as food-storing reservoirs (yolk) while sperm are much more numerous, restrict their cytoplasm, and usually have large capabilities for movement.

In this chapter we will place strong emphasis on bisexual reproduction. However, you are reminded that a normal cell cycle (Ch. 7) is a reflection of one kind of reproduction. Also, meiosis (Ch. 7) is an essential part of preparing for the production of eggs and sperm.

Consistent with the approach developed in Part II, we shall examine reproduction in the human before proceeding to examine development in the next chapter.

REPRODUCTION OF AN ADVANCED ANIMAL: THE HUMAN

Mating in most animals is accompanied by physical and behavioral changes; many of these are under hormonal control and may be quite stereotyped. In humans, however, there is considerable variation in mating play, posture, and frequency of mating. The human female is subject to sexual arousal in no pattern and at relatively frequent intervals. It has been postulated that this is a naturally selected condition corresponding with the male's relatively continuous sexual drive.

In the copulatory embrace the erect penis is inserted into the vagina and approximately 3 ml of semen containing 300 million sperm are discharged from the male through a series of reflexive spasms of male ducts and glands. The erogenous areas of the penis and the clitoris are stimulated in the sex act to a peak of excitement that then ebbs.

Organs of the Human Male

The *scrotum,* a sac hanging from the lower wall of the abdomen (Fig. 15.1), houses the pair of *testes,* the sites of sperm production. The testes originate in the body cavity but descend into the scrotum through the inguinal canal. This route normally becomes blocked off by connective tissue but sometimes is a site of weakness at which an inguinal hernia can develop, in which case a part of the intestine may tend to push into the scrotum.

The testis (Fig. 15.2) is composed of coiled *seminiferous tubules* lined with epithelial cells that produce sperm cells; also interstitial cells around the tubules produce the male sex hor-

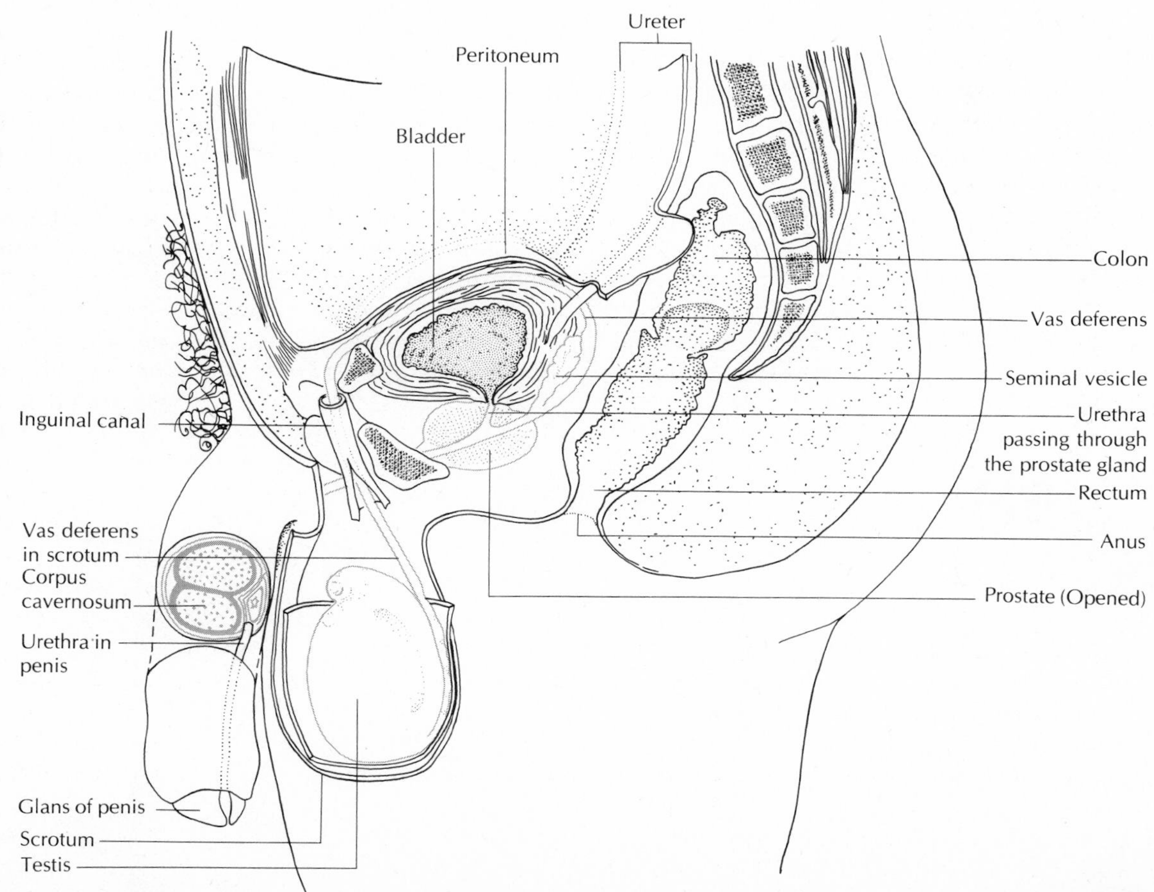

Figure 15.1 Reproductive organs of human male.

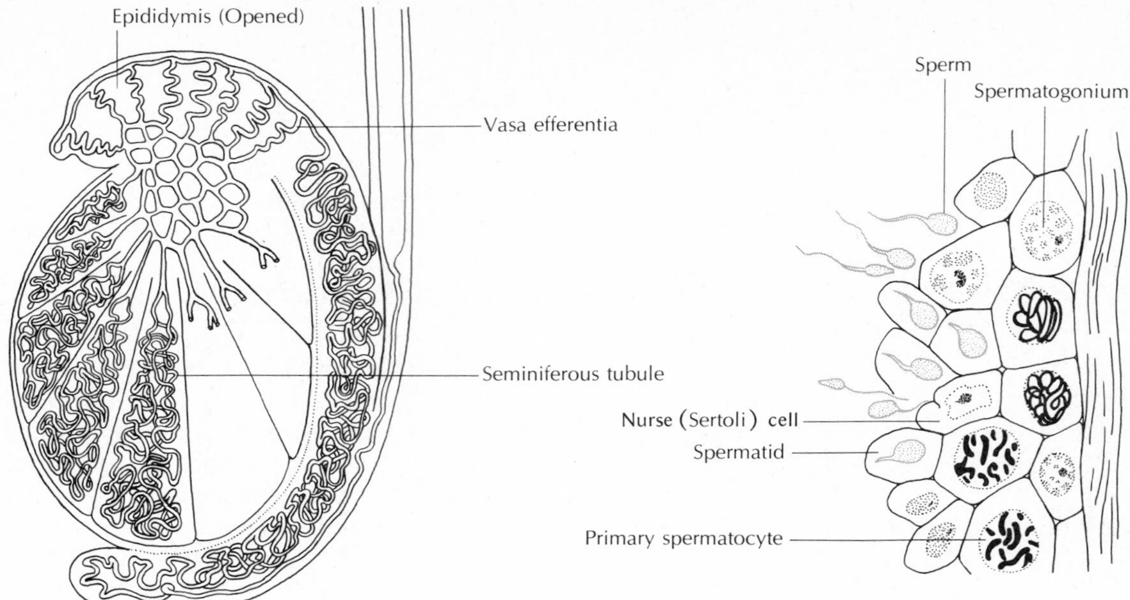

Figure 15.2 Section of human testis and the development of sperm in seminiferous tubules.

mone, testosterone, which controls male secondary sex characteristics.

Sperm leave the seminiferous tubules in the testes and pass through a set of ducts, from the *vasa efferentia* to the *epididymis,* derived from embryonic kidney, and into the *vas deferens.* The vas deferens leaves the scrotum by the inguinal canal and empties into the *urethra,* the duct that leads from the bladder. Fluids of the semen are added to the sperm en route from three sets of glands. A pair of *seminal vesicles* empties into each vas deferens before it connects to the urethra; a single *prostate* gland surrounds the urethra near the junction of the vasa deferentia, and *Cowper's glands* are connected to the urethra near the base of the spongy tissue of the penis.

In the penis the urethra is surrounded by three columns of spongy tissue permeated with blood sinuses that engorge to erect the penis.

Organs of the Human Female

At the midpoint of each menstrual cycle, the human female usually releases one egg from only one of a pair of almond-sized *ovaries* (Fig. 15.3) lying in the pelvic part of the body cavity (Fig. 15.4). Each ovary is surrounded partially by the expanded opening of an *oviduct* (Fallopian tube). The pair of oviducts lead to a central pear-shaped uterus of about 5 by 10 cm (2 by 4 inches). The uterine wall is thick and is the site where development will take place in the event of fertilization. The uterus is connected to the

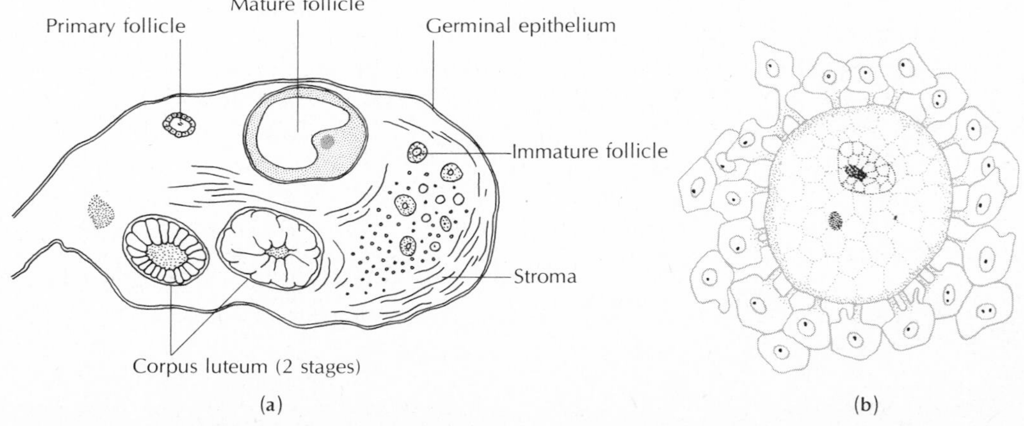

Figure 15.3 Section of human ovary. **(a)** Stages of follicular variation. **(b)** Released egg.

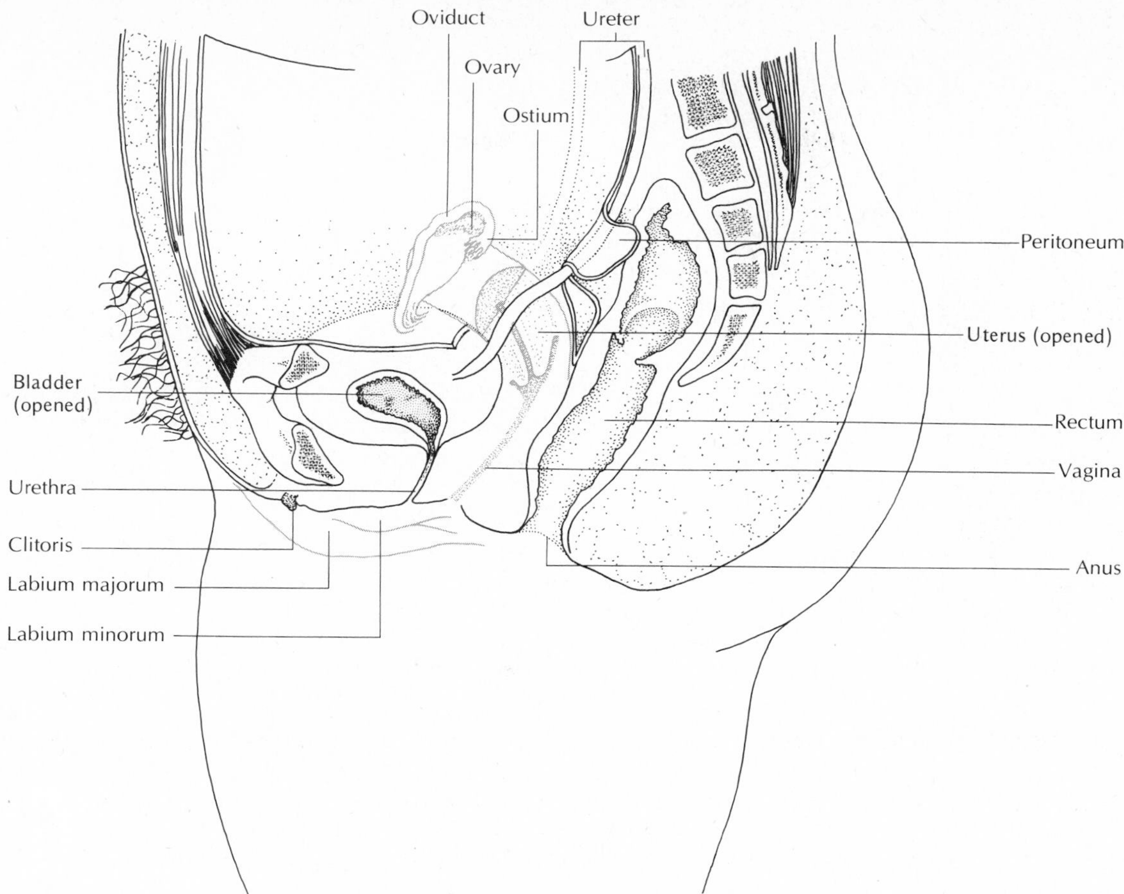

Figure 15.4 Reproductive organs of human female.

vagina, whose external opening is flanked by an inner pair of moist folds, the *labia minora,* enclosed with the fleshier hair-covered outer *labia majora.* These structures enclose the clitoris, mentioned earlier, and conceal the opening of the urethra that drains the bladder.

If semen is deposited in the vagina, the sperm rapidly travel through the uterus and into the oviducts. If an egg has been released recently from the ovary and has started down the oviduct, fertilization may occur to produce the zygote. In recent years the biological, psychological, and sociological problems associated with inducing or thwarting such fertilization have received much attention relative to family planning and population control.

SEX ORGAN HOMOLOGIES

While the structures of the male and female reproductive systems seem quite different, each mammal develops similarly up to a point, a period sometimes called the indifferent phase of development. During this period the following structures are in common. There are paired mesonephric kidneys (Ch. 10) whose mesonephric ducts (also called Wolffian ducts) drain into the cloaca. Adjacent to the mesonephric kidneys and supported from them by a mesentery, *genital ridges* give rise to the gonads. Running parallel to each of the mesonephric ducts is a paramesonephric (Müllerian) duct. Externally a small structure, the *genital tubercle,* develops just anterior and ventral to the part of the cloaca that persists in association with the urogenital structures. On each side of this opening are genital folds and genital swellings. The events between this indifferent phase and the development of the definitive male or female sex are complex, under hormonal control, and develop depending on the genetics of the embryo. Only a brief summary of these events is described and tabulated (Table 15.1).

If the embryo develops as a female, the mesonephric kidney and mesonephric ducts regress. The genital ridge areas become ovaries. The paramesonephric (Müllerian) ducts persist as the oviducts, the uterus, and part of the vagina; the urogenital sinus becomes the rest of the vagina. The genital tubercle undergoes relatively modest growth to form the *glans clitidoris* of the clitoris. The genital folds and swellings become, respectively, the labia minora and labia majora.

If the embryo develops as a male, part of the mesonephric kidney persists and ducts connect it and the developing testis. The part that was mesonephric kidney becomes the epididymis. The mesonephric ducts persist as the vas deferens that drains into a section of the original urogenital sinus. The genital tubercle becomes displaced to the tip of the developing penis as the *glans penis.* The shaft of the penis is produced normally by the extended fusing of the paired genital folds. The genital swellings enlarge in the male, incorporate a part of the body cavity, and are the site into which the testes grow to form the scrotum. Thus, as shown in Table 15.1, there are true homologies between male and female reproductive structures.

TABLE 15.1 Homologies between female and male reproductive structures.

FEMALE	INDIFFERENT STAGE	MALE
Ovary	Genital ridge	Testis
Vestigial	Mesonephric kidney	Epididymis
Vestigial	Mesonephric duct	Vas deferens
Lower vagina	Urogenital sinus	Proximal urethra
Oviduct, uterus, upper vagina	Paramesonephric duct	Vestigial
Labia minora	Genital folds	Dorsal shaft of penis and penile urethra
Labia majora	Genital swellings	Scrotum

VIVIPARITY

The Uterus

When an egg is fertilized high in the oviduct, it becomes a *zygote* with the diploid number of chromosomes reestablished. During a period of three or four days, as it passes down the oviduct, the zygote divides frequently and enters the lumen of the uterus as a *blastocyst* (Fig. 16.12). Its fate is partially dependent on the conditions existing along the uterine walls. Some blastocysts that may have developed abnormally never establish a relationship with the lining of the uterus and are discharged without the woman even recognizing that an egg may have become fertilized. The critical importance of the condition of the uterine lining is that it is the site where the blastocyst becomes embedded and which nourishes the development of the embryo. Humans are examples of animals that develop their young internally for a period of *gestation.* Thus we are viviparous animals. Viviparity is not a trait restricted to mammals, however. For instance, some sharks develop internally. In fact, even mammals have varying periods of development that they support in the uterus. For instance, mice, rats, and rabbits are born hairless, with their eyes closed, and quite helpless. Calves and colts, on the other hand, can stand and nurse very shortly after birth, and guinea pigs are so well developed that they resemble miniature adults shortly after birth. Human babies are better developed in a number of ways than rodents, but they are still quite dependent on adult care. A whole subclass of the mammals, the Metatheria (Ch. 38) or marsupials, remain in the uterus for only part of their development. When young marsupials are born, they may not have completed the development of all their appendages, for instance. However, they can reach the marsupial pouch where the mammary teats are located and complete development there.

The Structure of the Uterus and Its Hormonal Regulation

The uterus of mammals varies in its general shape. Two shapes are especially worthy of description. Most mammals have *bicornuate* uteri, that is, the two short oviducts lead into long uterine horns where the developing young are implanted. These horns join before they lead to the vagina. In humans, during the development of the female, only the oviducts remain paired and the uterus is fused into a single pear-shaped, *simplex* uterus.

The uterus is thick-walled and muscular. It is lined by an epithelium that, in humans, undergoes a rhythmic increase in thickness followed by a brief period of discharge of the uterine lining. This period of discharge is called *menstruation,* lasts four to five days, and reoccurs generally about every 28 days. Menstruation is a trait that occurs only in higher apes and humans. Its repeated appearances are called *menstrual cycles.* Nonprimate mammals do not show such cycles. Instead, they have *estrous cycles* that are spaced further apart. Some animals "come into heat" only once a year, others several times a year, and others may not be sexually receptive over more than the span of a year. Estrous cycles have in common with menstrual cycles a period when the lining of the uterus is in an appropriate condition for the implantation of the developing embryo. However, if implantation fails to occur, there is no sudden loss of uterine lining as there is in higher primates. Instead, the uterine lining returns to an inactive and nonreceptive state.

In animals with estrous or menstrual cycles, the regulation of the uterus is under hormonal control. While it is difficult to analyze critically the endocrine controls in humans, the following description seems to fit the combination of evidence drawn from experimental animals and from testing of humans.

The condition of the uterine lining, the *endometrium,* is controlled by the interaction with hormones produced by three other areas of the body, the ovaries, the pituitary gland (Ch. 12), and a section of the brain, the hypothalamus (Ch. 13). These organs not only control the condition of the uterus, but also influence the development of the gametes in both females and males and condition the responses of sexual partners. Because of the complexity of the interactions, the description of endocrine controls that ultimately influence the uterus will not begin with the uterus but will start with the pituitary and its controls. It should be emphasized that while the description emphasizes the female endocrine pattern, the male is controlled by the same glands and hormones.

The pituitary produces many hormones (Ch. 12), but the ones we consider here are called *gonadotrophins.* These are the *follicle-stimulating hormone* (FSH) and the *luteinizing hormone* (LH). (A third hormone, *prolactin,* may be involved in the complex steps but for our purposes it will be described as principally stimulating milk secretion from the mammary gland after birth.) Let us describe the events in a stepwise fashion.

When the hypothalamus releases a hormone to the pituitary through a small portal system (Ch. 11), this hormone, *follicle-stimulating hormone releasing factor* (or *hormone*) (FSH-RF), triggers the release of FSH into the blood. On arrival at its target organ, the ovary, development of a follicle is initiated. The follicle bears in it one egg. During the expansion of the follicle, its cells synthesize an *estrogen,* principally *estradiol.* This hormone enters the blood stream; its principal target organ is the uterine epithelium (mucosa). It stimulates this to proliferate. Thus during the ten days after the last menstruation, the estrogen increases with the expanding follicle and the endometrium thickens and develops narrow tubular glands.

When the estrogen in the blood reaches a certain concentration, it also acts on the pituitary to shut down FSH production and on the hypothalamus to result in the discharge of *luteinizing hormone releasing factor* (LH-RF) to the pituitary. The pituitary now release LH to the circulation. In combination with the residual FSH, it results in *ovulation,* the rupture of the follicle and the release of the egg, which normally is drawn into the open end of the oviduct. Here it may or may not be fertilzed by a sperm. The LH also causes the cells lining the follicle to alter, enlarge, and fill the old follicular cavity. The structure formed is the *corpus luteum.* The corpus luteum produces a hormone, *progesterone,* whose principal target is the thickened endometrium. Its action is to increase the vascularization and glandularization of the endometrium. Also, the level of progesterone in the blood feeds back information to the hypothalamus in the event no implantation occurs. Luteinizing hormone release is shut down, progesterone is shut down, and support of the uterine lining fades, leading to menstruation. However, if implanation occurs, the membranes surrounding the developing embryo embedded in the uterine lining contribute to the formation of the placenta and begin to secrete a progesterone-like hormone that finally replaces that produced by the corpus luteum. This prevents the uterine lining from sloughing and supports pregnancy. The interrelationship between organs and hormones in the menstrual cycle is illustrated in Figure 15.5.

The hormones estrogen and progesterone, produced by the developing placenta, also stimulate enlargement of the breasts. These hormones do not lead to actual milk secretion, which is regulated by the anterior pituitary hormone, *prolactin*, and the hormone stored in the posterior pituitary, *oxytocin.* These hormones are released after the inhibitions of the placental hormones are removed at birth.

Male Reproductive Capacity

The same hormones, FSH and LH, that regulate female reproductive capacity influence the development of the testes. The hormone produced by interstitial cells in the testes is *testosterone.* These hormones must function properly for males to be reproductively competent.

Puberty

In both males and females the arrival of sexual maturity, the onset of sex organ function, is achieved under the control of the pituitary–gonad axis. In females the changes in skeletal structure, fat deposition, breast development, and pubic and axillary hair are principally controlled by estrogens. In males, testosterone controls the development of the penis, the charac-

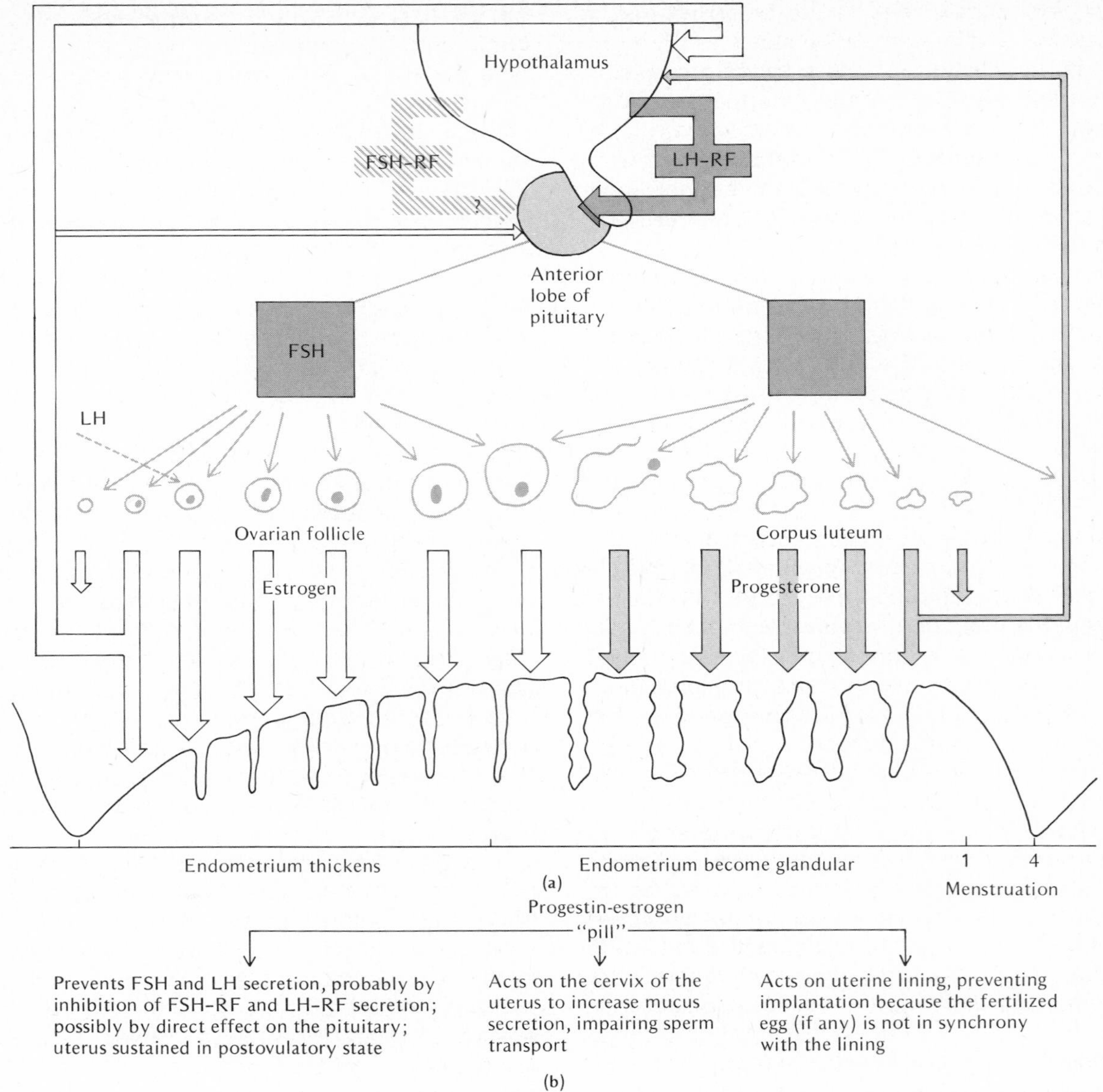

Figure 15.5 **(a)** Menstrual cycle showing interactions among organs, changing action within single organs, and feedback of information to organs. **(b)** The role of "the pill" in the reproductive cycle.

teristic hair pattern, and changes in the voice, in muscular, and in skeletal development. Hair pattern not only involves the axillary area and the pubic hair, where the pattern differs from the female in tending to extend in the midline toward the navel, but it also involves development on the face and chest to varying degrees, depending on the individual genetics of the man, as well as his race.

SEXUAL RESPONSE

With the development of the biological subscience of behavior (ethology) and the growing awareness of the human as another animal worthy of analysis, attention has in recent years been directed to the human sexual response. This has been examined as a strictly physiological problem and as a societal problem. As becomes obvious in Chapters 14 and 43, much of the societal aspects of animal populations revolve around reproductive drives as adaptations for survival.

Furthermore, the currently acceptable patterns of human behavior permit a franker and more objective examination of human sexuality. In past decades, sexuality was identified so closely with concepts of sin and with concepts of aggression that open discussion was inhibited. With the growing dignity given to women as an outgrowth of the women's liberation movement, sexual response has come to receive attention in a more complete and realistic light. Furthermore, with the development of more reliable contraceptive methods, such as "the pill" and the IUD (see below) in humans, sexual response has become somewhat divorced from concepts of procreation. Sex, therefore, is now viewed by many as are other human endeavors—a matter of choice on the part of both partners of the sexual act. In recent years the issue of choice has extended even to the choice of the sexes of the partners; this is not a new situation in human societies, but only one that has been openly practiced rather than covertly practiced or examined.

Male Sexual Response

In both males and females, sexual arousal usually follows preplay of varying forms and intensities. When fully aroused, the penis becomes engorged with blood and enlarges from an average of approximately 7.5 cm to 15 cm (3–6 inches) and achieves a diameter of about 3 cm (1 ¼ inches). This is the period of excitement. When in copulation with a female and with movements of the penis in the vagina, the male enters a plateau period when the breathing and heart rates accelerate. Also, the accessory reproductive glands, such as the seminal vesicles, prostate, and urethra, become especially well vascularized. When continued stimulation of the glans penis occurs, this triggers the spasm of muscles of the genital organs creating the orgasm and resulting in the discharge of semen from the urethra. Shortly thereafter the penis usually shrinks to near its unstimulated size. Under appropriate conditions after a refractory period, the man may be induced to reestablish an erect penis and repeat the sequence.

Female Sexual Response

While the female usually passes through the stages of excitation, plateau, orgasm, and often refractory period, her patterns are more variable and complex than are those of the male. The phase of excitement not only includes clitoral and genital fold swellings, but the vaginal wall secretes and creates a moist lubricant. Also, the breasts usually swell and the nipples become erect. On entering the plateau phase, her breathing and heart rates increase. It is now also recognized that clitoral and vaginal stimulation can both be elicited. The stimulation usually starts at the clitoris but spreads over the vagina and external genitals. Orgasm in the human female is usually associated with contractions of the vaginal wall. These may be spasmodic. Also, the result of an orgasm in the female is often to create a syringing action on the pool of semen deposited by the male near the cervix of the uterus; this acts to aspirate sperm into the uterus. In human females there may be two striking differences from the usual performance in males. First, under some circumstances the female may proceed rather rapidly from excitation to orgasm. Second, the female may have repeated orgasms without the usual refractory period seen in males.

These reviews of human sexual behavior can only be considered as brief surveys. Since the studies of Alfred Kinsey in the 1940s and the more recent studies of Masters and Johnson, much has been learned about the great variety of sexual expressions that exist in human populations and in different human societies. The noteworthy theme that is in common with most of these events is, as said before, the dissociation of sexual behavior from childbearing. Two things follow logically from this. One is that sexual partners need to understand each other and the implications of their activities. Among the larger problems that seem to exist are those revolving around whether orgasm is a minimal requirement for sexual satisfaction. A concensus of students of human sexuality is that the mental attitude of sexual partners is a very important aspect of a sense of sexual gratification and can produce gratification independently of orgasm. The second point that follows current attitudes is the implication that pregnancy can be a matter of choice. Controls on pregnancy, therefore, become important.

Contraception

Applied reproductive physiology has been under public scrutiny in recent years in relationship to the control of pregnancy, the right to determine family size, and the social responsibility for population control.

Some believe that true population control comes from the will of the people not to have children. It is agreed that continence rather than artificial means played a significant role in lowering the birth rate during the economic depression of the 1930s. Today, however, restraint is assumed to be an unrealistic approach. Therefore three major methods that allow intercourse and prevent childbearing are used: physical interference, restraint, or chemical inhibition of conception. Physical interference can be accomplished by the unreliable act of withdrawing the penis from the vagina before ejaculation, by ensheathing the penis in a condom, or by damming the route of passage of the sperm with a diaphragm. More radical is the tying and severing of sections of the ducts of males or females. In males this is accomplished by a small incision in the scrotum, and in females by small punctures either through the abdominal wall or through the vagina alongside the cervix of the uterus. In both males and females these procedures can be done quickly and without hospitalization. Intrauterine devices (IUDs), small inserts of loops or coils (usually of plastic) into the uterine cavity, can be used to prevent implantation of the developing embryo.

Restraint, or commonly "the rhythm" method, assumes that abstinence from intercourse will be practiced during the period of receptivity of the ovulated egg for sperm. If one assumes a three day variation in ovulation from the normal midpoint at 14 days, abstinence between approximately days 11 through 17 is required. This is an unreliable procedure because of error in memory, error in calculating the day of the cycle, or irregularity of the menstrual cycle.

Chemical interference has become a major means of employing or mimicking the female hormones. This has largely supplanted older chemical means that used douches, oils, and jellies. The most widespread means of chemical contraception in North American is oral contraception by use of synthetic estrogens and progesterones, commonly called "the pill." As Figure 15.5 shows, the estrogen–progestin "pill" prevents FSH and LH secretion among other things to sustain the postovulatory phase because the dosages of the estrogenlike and progesteronelike substances are controlled. In one popular form, the woman begins taking one pill a day for 21 days, then waits 7 days before taking them again. During the 21 days she takes the pill, she passes her normal period of ovulation without releasing an egg, yet this allows relatively normal uterine wall development. During the seven days after the 21st pill, the level of the synthetic hormones in the circulatory system drops and menstruation occurs. The clue to the success of the procedure lies in the balance between the two hormones in the pill and the physiology of the woman; in some cases pills containing only estrogens or only progesterone may be used.

Recently prostaglandins have been employed as postcoital contraceptives by causing contractions of the uterine wall to prevent implantation. At present this method does not have the same widespread public usage as does the progestin–estrogen pill.

Also, research in recent years has been directed to finding a male contraceptive.

SUGGESTIONS FOR FURTHER READING

Asdell, S. A. *Patterns of Mammalian Reproduction.* London: Constable & Co. (1965).

Austin, C. R., and R. V. Short (Eds.). *Reproduction in Mammals,* 5 vols. Cambridge: Cambridge Univ. Press (1972).

Berrill, N. J. *The Person in the Womb.* New York: Dodd, Mead (1968).

Bullough, W. S. *Vertebrate Reproduction Cycles.* London: Methuen (1961).

Carr, D. E. *The Sexes.* Garden City, N.Y.: Doubleday (1970).

Corner, G. W. *The Hormones in Human Reproduction.* New York: Atheneum (1963).

Crawley, L. Q. *Reproduction, Sex, and Preparation for Marriage.* Englewood Cliffs, N.J.: Prentice-Hall (1973).

Kinsey, A. C., W. B. Pomeroy, and C. E. Martin. *Sexual Behavior in the Human Male.* Philadelphia: W. B. Saunders (1948).

Kinsey, A. C., and the staff of the Institute for Sex Research. *Sexual Behavior in the Human Female.* Philadelphia: W. B. Saunders (1953).

Masters, W. H., and V. E. Johnson. *Human Sexual Inadequacy.* Boston: Little, Brown and Co. (1970).

McCary, J. L. *Human Sexuality,* 2nd ed. New York: Van Nostrand (1973).

Odell, W. D. *Physiology of Reproduction.* St. Louis: C. V. Mosby (1971).

Perry, J. S. *The Ovarian Cycle of Mammals.* New York: Hafner (1972).

Smith, A. *The Body.* Harmondsworth, Middlesex: Penguin Books (paperback), (1968).

Swanson, H. D. *Human Reproduction; Biology and Social Change.* New York: Oxford Univ. Press (1974).

Van Tienhoven, A. *Reproductive Physiology of Vertebrates.* Philadelphia: W. B. Saunders (1968).

Wickler, W. *The Sexual Cycle: The Social Behavior of Animals and Men.* Garden City, N.Y.: Doubleday (1972).

Wood, C. *Human Fertility: Threat and Promise,* World Science Library. New York: Funk and Wagnalls (1969).

From Scientific American

Berelson, B., and R. Freedman. "A Study in Fertility Control" (May 1964).

Dahlberg, G. "An Explanation of Twins" (Jan. 1951).

Gordon, M. J. "The Control of Sex" (Nov. 1958).

Jaffe, F. S. "Public Policy on Fertility Control" (July 1973).

Mittwoch, U. "Sex Differences in Cells" (July 1963).

Tietze, C., and S. Lewit. "Abortion" (Jan. 1969).

16 Development

The topic of development has no physical or temporal time span. In humans, it is true, it has been a convention for biologists and physicians to consider development up to birth. Psychologists, however, extend the study of development at least through the prepubertal years. Also, biologists sometimes vary in the scope of their study of development. For instance, students of amphibian or insect development may examine early development from the egg to the acquisition of body form and think of it as embryology. But development in these two examples extends beyond embryology to studies of problems associated with metamorphosis from tadpole to frog or salamander, or from larva to pupa to adult in insects. Thus development has a large scope and cannot be covered fully in one chapter. We will confine ourselves to a rather restricted survey that not only describes physical changes during development but also exposes some of the questions that remain to be examined in order to better understand this complex subject.

It should be emphasized at the outset that many, many events, usually under specific genetic control, must occur with exquisite precision in order for a normal individual to be produced. This becomes obvious if one simply thumbs through a treatise on medical genetics and observes the large number of kinds of abnormalities in development that can occur in humans. There is evidence that similar relationships exist in other organisms also.

From Chapters 7 and 15 we see that the cells that unite to form a new individual, a zygote, are the eggs and sperm. Two conditions are created when this occurs in any animal: the diploid chromosome constitution is reestablished and the egg is activated to develop into a new individual. In its development, cell divisions, called cleavages, occur. When the large number of cells are produced by cleavages, a stage called the *blastula* is reached. This is a round hollow sphere in many animals. However, some animals produce a blastula with a cell-filled center. The blastula stage is followed by one in which there is a striking rearrangement of cells through *formative movements* of these cells in relationship to each other. This process is called *gastrulation* and the product is a *gastrula*. In the gastrula stage, cells or groups of cells establish new relationships with each other. The subsequent development of the animal is often dependent on the influence of one group of cells on another, a process called *induction*. Our objective here will be to discuss development briefly from gametogenesis through fertilization and zygote formation, cleavage, blastula formation, gastrulation, and the establishment of body form (Fig. 16.1). How we discuss this for individual species of organisms depends on the kind of eggs one starts

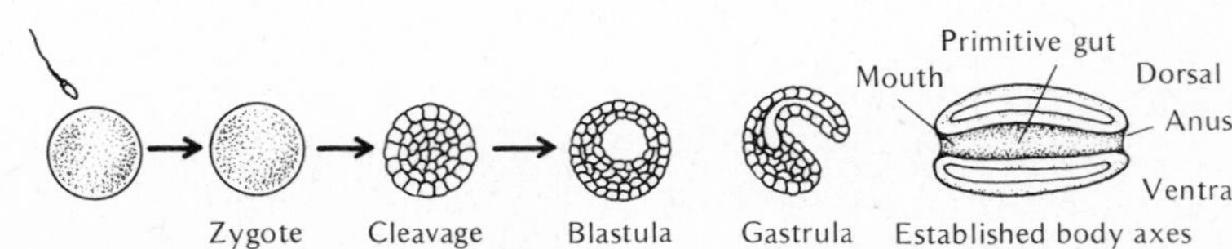

Figure 16.1 Representative stages of development from fertilized egg to establishment of basic body form.

with and the variations that occur during the stages outlined.

GAMETOGENESIS

The eggs and sperm mature in the ovaries and testes, respectively. Each gonial stage (oogonia or spermatogonia) ceases to proliferate by mitosis, and two meiotic cell divisions produce, respectively, the eggs and the sperm. Our attention is called here to the differences between these two kinds of gametes.

Eggs

Eggs are often large and laden with yolk (for example, the yolk of a hen's egg), although not all eggs are. Table 16.1 shows ways in which eggs are classified depending on their yolk concentration.

Table 16.1 Classification of Egg Types.

EGG TYPE	DESCRIPTION	EXAMPLES
Isolecithal, alecithal	Having a small amount of yolk distributed in the cytoplasm; or lacking visible yolk	Sea urchins, humans
Telolecithal	Having the yolk more highly concentrated at the vegetal pole than at the animal pole; cytoplasm concentrated toward the animal pole	Moderately telolecithal: frogs, salamanders Strongly telolecithal: birds, reptiles, some sharks
Centrolecithal	Having the yolk located in the center of the egg and surrounded by cytoplasm	Insects

We will see that the concentration and distribution of yolk influences the cleavage pattern following fertilization. One mature egg develops from each primary oocyte. Depending on the species, eggs may be liberated from the ovary at stages ranging from primary oocytes, where the first meiotic division has not started, to completion of meiosis. For instance, sea star eggs are released prior to onset of meiotic divisions while vertebrate eggs stop at metaphase of the second meiotic division.

The zygote must contain the nutrients to support development until nourishment can be obtained from the environment. Sea urchins and sea stars have only a small amount of stored nutrients in the zygote but they develop rapidly to freely feeding larvae. Frog larvae are late in developing, but the egg is supplied with a considerable store of yolk to support development until the tadpole can feed. Human eggs have little yolk, but the early embryo becomes embedded in the wall of the uterus, from which it draws nourishment until the placenta is established and food from the mother's blood passes across the placenta to the embryo's blood.

Where yolk is present, it is deposited in the developing egg while it is in the ovary. The source of the yolk is from surrounding follicle cells that assist in the transfer of yolk into the cytoplasm of the egg. The egg proteins and phospholipids that contribute to the formation of yolk are produced in the livers of vertebrates as can be shown by using radioactive phosphates.

Preparation of the egg for cleavage can be traced back to the prophase of the first meiotic division. During cleavages there are three important events that occur which depend on the appropriate biochemical machinery: increased amounts of mitotic spindle assembly, increased amount of plasma membrane, and increased number of nuclei. In the frog egg the efficiency for accomplishing these events is achieved by concentrating over 1000 times the normal somatic cell's amount of rRNA in the egg. This is achieved by repeated transcriptions, extra copies of rRNA being made during oogenesis and concentrated in multiple small nucleoli that deliver the rRNA to the cytoplasm (Fig. 16.2).

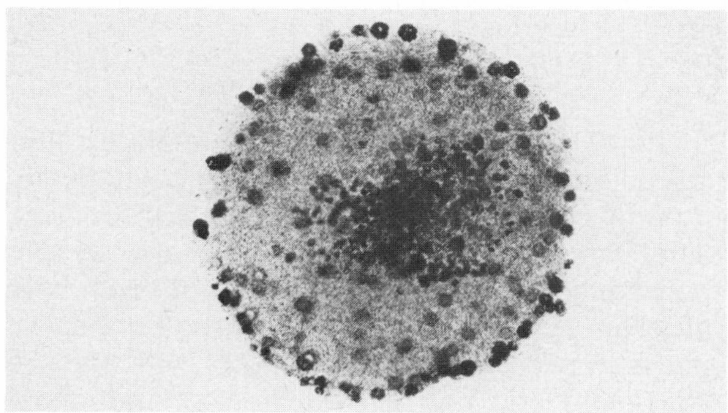

Figure 16.2 Isolated South African clawed toad germinal vesicle showing some of its hundreds of nucleoli (dark spots) stained with cresyl violet. The germinal vesicle is about 400 μm in diameter. (From D. D. Brown and I. Dawid, *Science*, **160,** 1968; © 1968 by the American Association for the Advancement of Science, with permission.)

This is a unique trait of oogenesis and does not occur in somatic cells. As a result of this, a mature frog's egg comes equipped with a reserve of rRNA that it uses in the translational processes which occur all through cleavage and blastulation. Thus while the frog's egg is proceeding rapidly to a multicellular stage, no energy is required for rRNA production, and this energy can be employed in other developmental events such as making DNA synthetase enzymes, manufacturing new plasma membrane, and so forth. Furthermore it now seems that mRNA may be present in an inactive or "informational" form which becomes active during development. In summary, the egg has a "dowry" of energy reserves in the form of yolk and an important part of its genetic equipment, rRNA, provided to it in the ovary prior to ovulation.

Sperm

Sperm differ from eggs in having completed meiosis by the time they are released. Furthermore, four sperm normally develop from each primary spermatocyte. After passing through the two meiotic divisions the spermatids produced undergo a metamorphosis in which they become highly elongated, have only a very small amount of cytoplasm, and in many animals can be subdivided into three regions: a head containing the nucleus and an acrosome; a middle piece containing centrioles, microtubules, and a high concentration of mitochondria; and a tail mostly made up of microtubules, seen in cross section to have a "9 + 2" arrangement (Ch. 4) (Fig. 16.3). Not all animals have sperm shaped as described. For instance, the sperm of arthropods have several configurations and are often immotile.

FERTILIZATION AND ZYGOTE FORMATION

The process of fertilization in animals is better appreciated if some adaptations associated with bringing eggs and sperm together are discussed. Many animals release their gametes directly into their aqueous environment. However, even here there are adaptations and synchronies that increase the chances that a sperm will reach a short-lived egg and activate its continued development. Some animals such as sea urchins and sea stars seem to release their eggs and sperm freely into the sea. However, usually the releasing animals are quite close to each other and the gametes are released in very large numbers. Some mollusks lying in beds will undergo a wave of gamete release presumably triggered by a pheromonelike stimulus emanating from nearby mollusks. Frogs and some fish also release their gametes directly into the water. Here, however, the males and females are very close to each other. In some fish, such as salmon and trout, as the female deposits eggs, the male swims over them and releases sperm (milt). In frogs the male clasps the female around the trunk in *amplexus,* and as she releases eggs, he releases sperm.

Other adaptations for fertilization involve internal deposition of the sperm. This is achieved in two ways. In many animals a penis is inserted into the genital tract of the female and sperm deposited. Other animals, such as some salamanders and squids, produce gelatinous packages of sperm, *spermatophores.*

In the case of salamanders, the male deposits a gelatinous cone topped by packaged sperm onto a leaf or stone in the water. The female passes over the spermatophore and picks sperm up with her cloacal lips. In squid, the spermatophore is transferred by specialized tentacles (arms) into the mantle of the female (Ch. 26).

The process of union of eggs and sperm has been thoroughly studied in only a relatively small number of species of animals. Most of the knowledge about the union of gametes comes from studies on invertebrates. There is still no major generalization about how sperm are attracted to eggs; random contact on the one hand or *chemotaxis,* chemical attraction, on the other have been proposed. However, there do appear to be cell surface conditions that hold the gametes together once they have met.

In a number of species, for instance, the acorn worm, *Saccoglossus* (Fig. 16.4), an *acrosome* on the tip of the sperm's head ruptures to liberate enzymes that lyse jelly coats around eggs. This is followed by the penetration of the jelly by an acrosomal tubule that contacts and penetrates the vitelline membrane around the egg. The tubule fuses the sperm plasma membrane with that of the egg. The egg's cytoplasm bulges into this point of union, and the sperm nucleus passes into the egg cytoplasm to become the

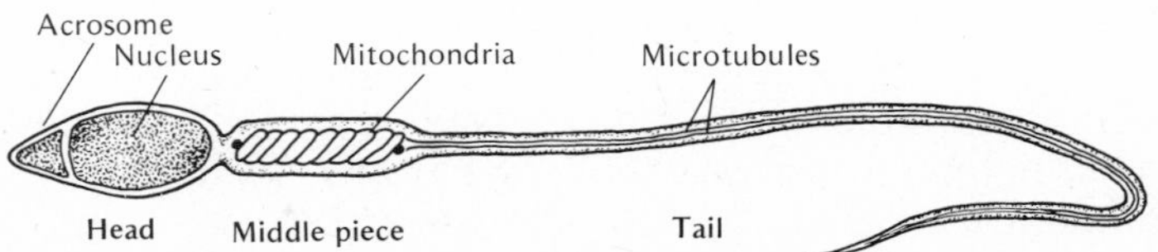

Figure 16.3 Diagram of the structure of a sperm.

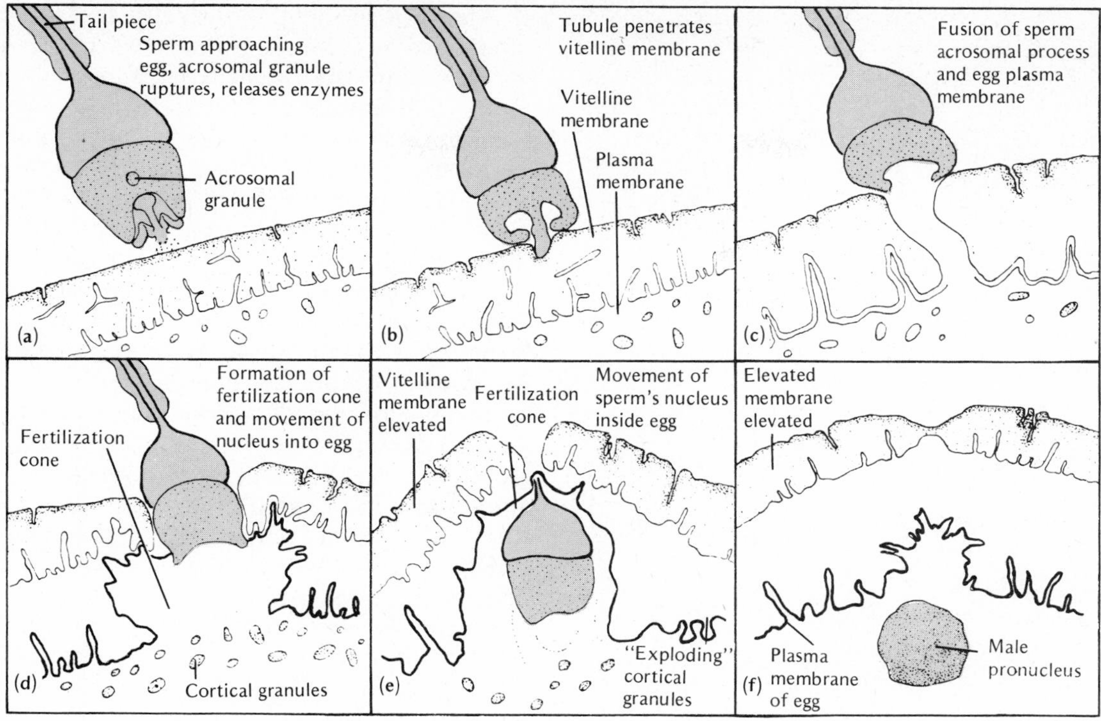

Figure 16.4 Stages in the union of a sperm and egg in a hemichordate. (Modified after Colwin and Colwin, 1963.)

male pronucleus. This moves to unite with the egg nucleus, the *female pronucleus,* to create the diploid zygote. It is conjectured that processes similar to this operate in mammals. For instance, mammalian eggs are surrounded by a mucoprotein and attached cells that are dissolved by an enzyme, hyaluronidase, which is found in the acrosome.

CLEAVAGE

The process of cell division, cleavage, leads to the development of numerous blastomeres (Fig. 16.5). The cleavage process itself occurs in several ways. Zygotes that have little to moderate amounts of yolk start to cleave at the animal pole and the cleavage furrow proceeds to the

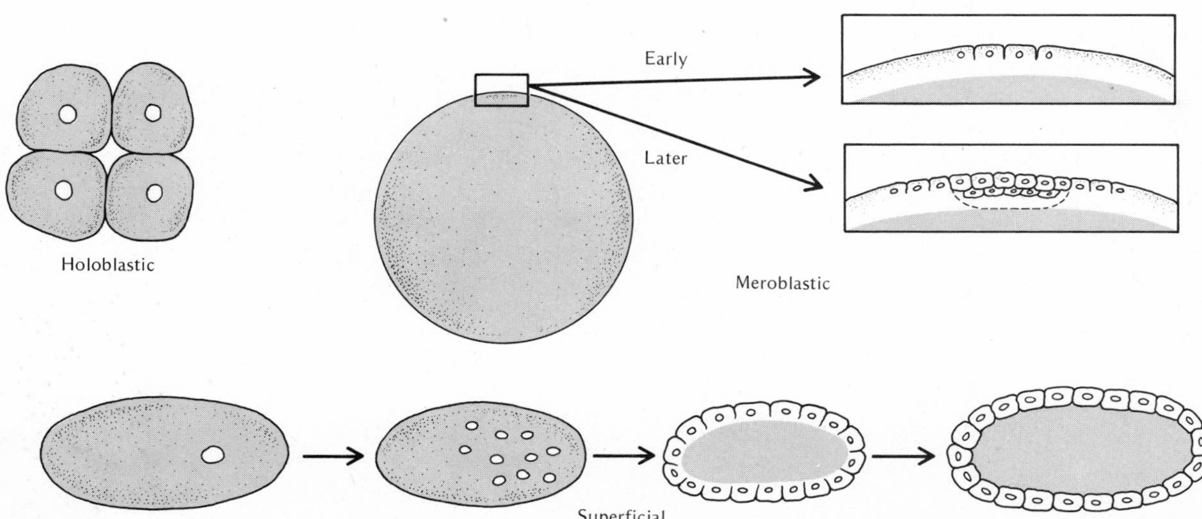

Figure 16.5 Types of cleavage.

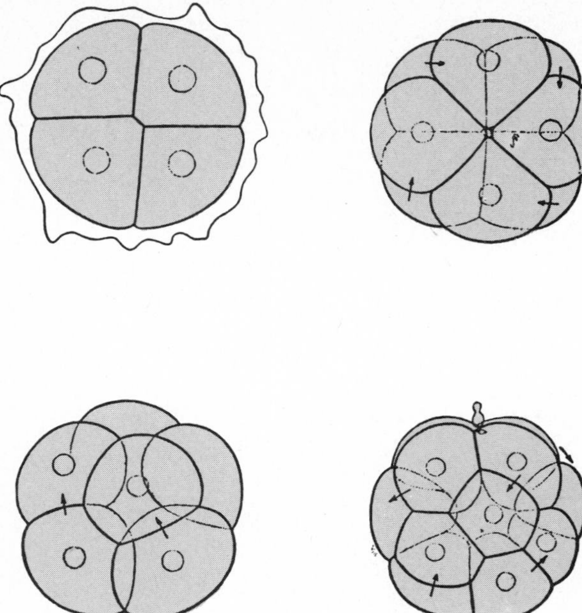

Figure 16.6 Spiral cleavage in *Polygordius.* (From Kellicott, *General Embryology,* Holt, Rinehart and Winston, 1913.)

vegetal pole, interposing plasma membranes between the two progeny cells. Such complete separation is called *holoblastic* cleavage. There are two kinds of cleavage that do not totally isolate the cytoplasm of individual cells from the surroundings. In bird eggs, for instance, the cleavage furrows extend through the cytoplasm to the yolk and stop, leaving the inner surface of the cell's cytoplasm exposed to the yolk. This is *meroblastic cleavage.* Later cleavages will isolate cells completely from the yolk to create a small disk of cells, a *blastodisk,* at the animal pole. The central cells of the disk become isolated from the yolk, the peripheral cells are exposed to it. Continued cleavages are centrifugal, thus expanding the size of the blastodisk. In centrolecithal eggs, found in insects, for example, the zygote nucleus divides many times and the nuclei migrate to the cytoplasm at the periphery of the egg. This stimulates a *superficial* cleavage in which cytokinesis creates a set of furrows between nuclei that resemble early stages of meroblastic cleavage. Soon the cells become completely separated from the yolk.

Many of the invertebrates of the protostome line have a form of holoblastic cleavage that differs from that seen in sea urchins or sea stars, for instance. This cleavage patttern is called *spiral* cleavage. The cleavage spindles tend to orient at angles to prior blastomere axes. The result is that blastomeres tend to lie in furrows between other blastomeres rather than on top of them (Fig. 16.6). In eggs with this form of cleavage there are often fairly well restricted differences in the cytoplasm; the blastomeres isolate different parts of the cytoplasm from each other. If a blastomere is lost, a part is usually lost from the developing embryo. This is also called *determinate* cleavage for this reason. It contrasts with *indeterminate* cleavage such as seen in sea urchins and sea stars. Here the loss of a blastomere does not necessarily mean the loss of a part later. Instead, the remaining blastomeres are said to *regulate* to produce a whole.

THE BLASTULA

Figure 16.7 shows some characteristic blastula stages. The important point is that at this time the embryo is about to undergo shifting movements of cells and enter gastrulation.

THE GASTRULA

To understand better the development of the body plan through gastrulation, four examples are used: sea stars, frogs, birds, and humans.

Sea Star

The zygote cleaves into two blastomeres from animal to vegetal pole (Fig. 16.8). The second cleavage is at right angles and produces four cells. The third cleavage is at right angles to the two prior cleavages creating an eight-cell stage of two tiers of four cells. Subsequent cleavages occur at rather well coordinated intervals to produce 16, 32, 64 cells, and so forth, until a blastula is formed. An early blastula has rounded

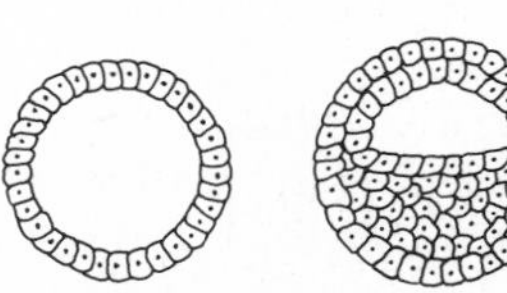

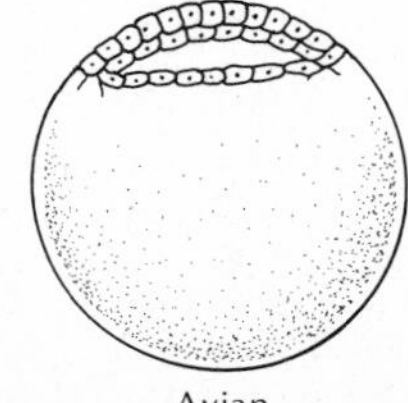

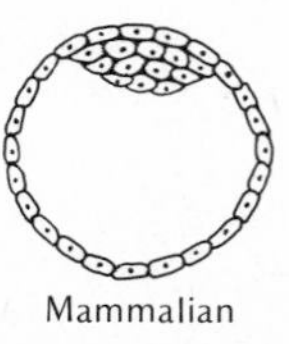

Figure 16.7 Forms in which blastula stages occur in several deuterostomes.

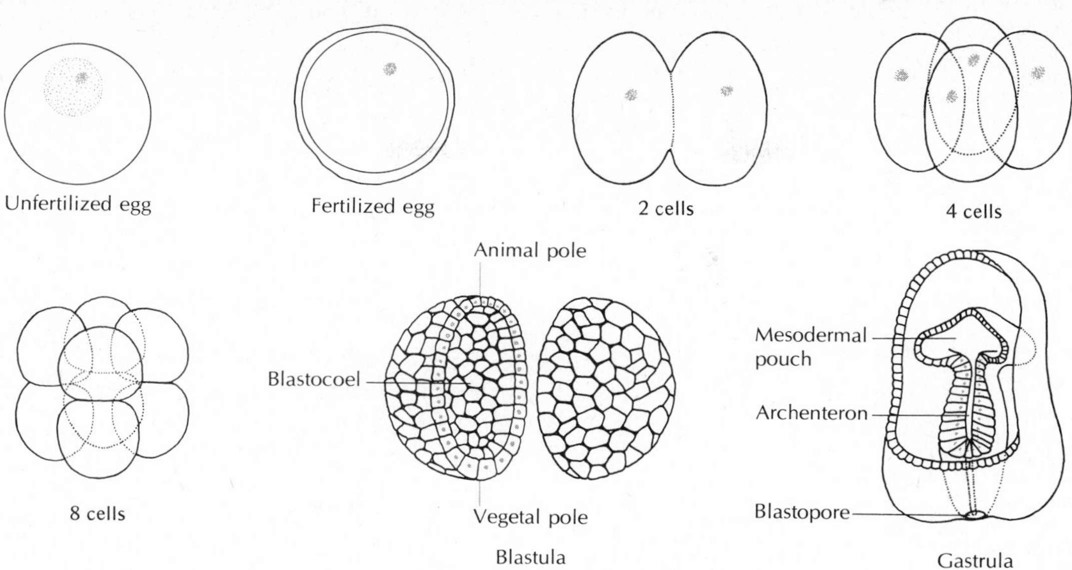

Figure 16.8 Some developmental stages in the sea star (starfish).

blastomeres composing its wall, but as cell division continues, the cells become compressed to each other in a single layer of columnar cells. This emphasizes the great increase in the amount of DNA that is produced for each new nucleus and the large amount of new plasma membrane that develops. The next event, characteristic of echinoderms, is for a group of small cells to separate from the surface at the vegetal pole. These are primary mesenchyme cells. They assist in the process of gastrulation. Gastrulation is initiated by a sinking in of the vegetal pole region. In association with this, mesenchyme cells develop pseudopodia which extend out as thin processes that explore the inner surface of the animal hemisphere. These processes establish contact and begin to contract. This assists in the movement of the vegetal area to the interior in the form of a blind tube. The point of movement to the interior is the blastopore. The invaginated tube is the *archenteron.* The innermost portion of the archenteron is destined to be mesoderm. It bulges outward into pouches that become separated into *mesodermal vesicles.* One or both of these, depending on the echinoderm, expands to line the old blastocoel and form the coelom. This kind of coelom formation is called enterocoelic because of its outpouching from the archenteron. When the mesoderm has separated from the archenteron, that which remains is the endoderm lining the future gut. The blind end of the archenteron now grows in a curved fashion to contact the outer wall of the embryo. At the point of contact, the layers break down and a mouth opening is formed. The blastopore is the site of the future anus.

Frog

Frog cleavage is similar to that in the sea star except that cleavage furrows running through the yolk of the vegetal hemisphere are retarded (Fig. 16.9). The result is that the blastula consists of small cells near the animal pole grading to larger cells at the vegetal pole and with the blastocoel displaced toward the animal pole. Gastrulation in frogs also is through a blastopore and with the formation of an archenteron. As with the sea star, the materials coming to line the archenteron are mesodermal and endodermal. While the pattern of movements from exterior to interior is more complicated than in sea stars and will not be expanded here, it is noteworthy that the blastopore does not lie at the vegetal pole, but is displaced toward the equator where the cells are smaller and more capable of formative movements. In the frog the mesoderm forms the roof of much of the archenteron. At the lateral edges of the archenteron, where mesoderm and endoderm have been in continuity, they now separate through a loss of tissue affinities. The freed endodermal edges grow upward and complete the lining of gut while the mesoderm grows downward between the outer ectoderm and the endoderm of the archenteron. The flanking mesoderm splits into an inner and outer sheet, and the coelom is formed between them. The forward end of the archenteron meets the outer

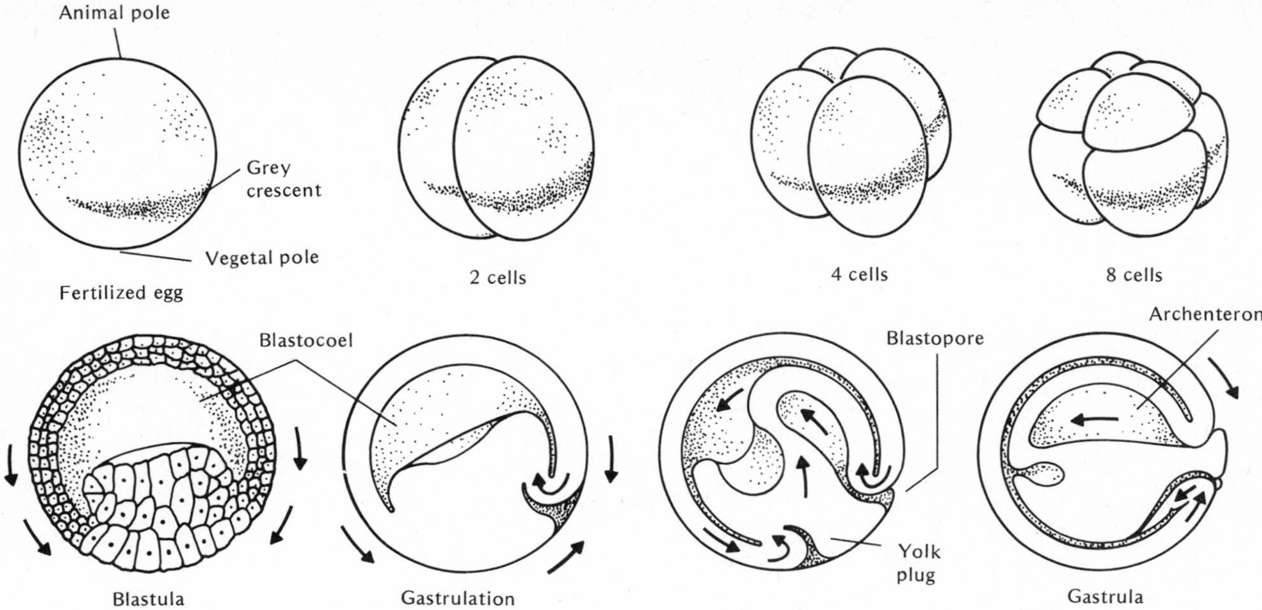

Figure 16.9 Stages in the formation of frog gastrula (gastrulation).

ectoderm and later perforates to form the mouth. In the frog and bird, another developmental principle is illustrated, tissue interaction. In these two vertebrates, the mesoderm underlying the ectoderm along the dorsal side of the embryo induces that ectoderm to thicken to become neural ectoderm (see below).

Chick

Recall that through meroblastic cleavage a blastodisk forms. The blastodisk expands and the central part, now free from the underlying yolk, thickens. The inner layer next to the yolk in this region separates away from the cells above to form the *hypoblast,* the endodermal germ layer. The space between the hypoblast and the surface tissues is the *blastocoel* (Fig. 16.7). On the surface, extending from the center toward the periphery, a thin pair of elevations appear with a depression between them; this is the *primitive streak.* It corresponds to the blastopore. Cells on the surface move in an orderly fashion from each side of the primitive streak, enter the streak, and emerge below the surface on each side as a middle mesodermal germ layer. This is gastrulation. In the area in front of the primitive streak where this has occurred, the underlying mesoderm also induces the overlying ectoderm to thicken as neural ectoderm in the form of a thickened horseshoe-shaped plate.

The neural plate folds to form a neural tube (Fig. 16.10). This tube is the future brain and spinal cord. Its presence shows that the long axis of the embryo is established. During later stages of development various structures along this axis appear. Eyes appear near the anterior part of the head, and a series of ridges, the pharyngeal arches, appear along the future neck. Also, in the chick the animal rotates onto its left side and flexures appear in the head and neck region. Thus gradually the definitive traits of the chick develop in an orderly progression.

Birds, along with reptiles and mammals, differ from other vertebrates in developing *extraembryonic membranes.* These are sheets of tissue which function during development but do not persist at hatching or birth. There is an evolutionary legacy associated with these membranes. Figure 16.11 illustrates these membranes. The *chorion* and *amnion* in reptiles and birds arise by common folds that extend from the embryo's body wall. They provide an outer moist surface at the boundary of the egg with its surrounding albumen and shell. Within the amnion a mildly saline fluid is secreted so that the embryo develops in a watery environment.

The yolk sac and allantois are associated with the primitive gut. The yolk sac of reptiles and birds is large and surrounds the fluid yolk. It is also important in that blood cells and blood vessels first appear in its walls. The allantois grows into the coelom at the rear of the primitive gut and expands into the extraembryonic coelom between the amnion and the chorion. It makes a close union with the chorion to produce the *chorioallantoic membrane.* This provides a large surface for the exchange of gases between the blood of the embryo and the exterior. It serves another function as a repository of metabolic waste products such as uric acid.

Egg yolk
Uncleaved blastodisc
Incomplete (meroblastic) cleavage
8 cells
Short primitive streak (stage 2)

(a)

Area opaca
Area pellucida
Notochord
Primitive streak
Stage 5 (19-22 hours)

Dorsal view
Longitudinal section
Head fold
Foregut
Notochord
Somites
Primitive streak
Stage 8 (26-29 hours)

(b)

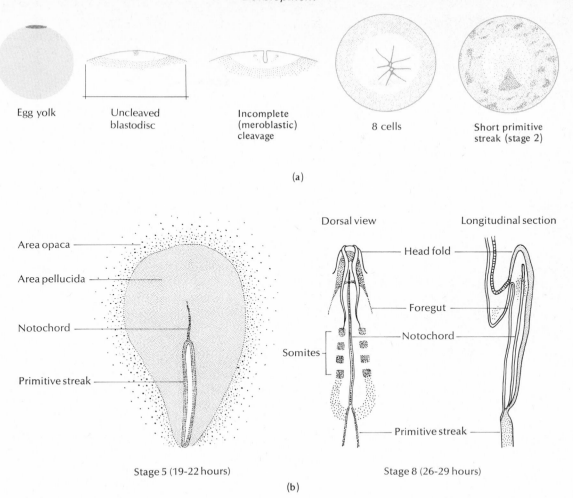

Figure 16.10 Early chick development. **(a)** Egg showing area of "active cytoplasm" (left), early cleavages (center), and early blastodisk (right). **(b)** Elongated blastodisk (left), surface view (center), and longitudinal section (right) of a chick after incubation for more than one day.

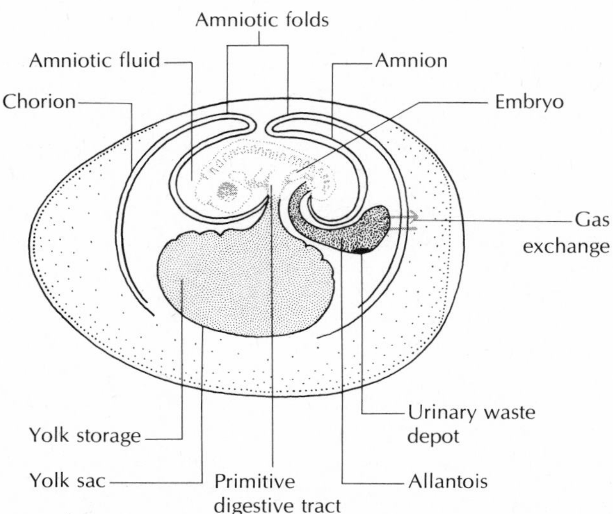

Figure 16.11 Diagram of the four extraembryonic membranes of a bird. All four membranes, chorion, amnion, allantois, and yolk sac, are present in the development of reptiles, birds, and mammals which are called Amniotes. Fish and amphibia are Anamniotes.

Human

The basic stages of development described above for the establishment of body form apply to humans but, in addition, we must understand the contributions of *extraembryonic membranes* to establishing relationships with the maternal tissues of the uterus.

An egg fertilized in the human oviduct undergoes cleavages as it moves along and enters the uterus in the equivalent of the blastula stage. It is called a *blastocyst* (Fig. 16.12) and differs from blastula stages of nonmammalians by having a solid cluster of cells, the *inner cell mass,* internally pendant from one wall.

The blastocyst invades the uterine wall through lytic activity of its outer cells which also proliferate vigorously in the invasion process. This outer proliferating area of the blastocyst, made up of ectoderm and a lining of mesoderm, is an extraembryonic membrane,

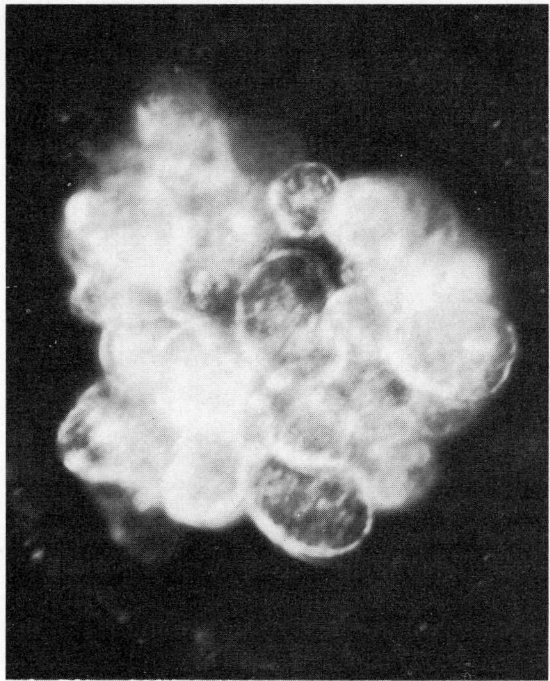

Figure 16.12 Mammalian blastocyst. (From L. B. Shettles, *Nature,* **229,** 1971.)

the *chorion.* The chorion will be the site of exchange of materials between the embryo or fetus and the mother.

The inner cell mass becomes divided into two small cavities (Fig. 16.13) whose linings make up, respectively, the *amnion* which lies closest to the chorion and the hollow *yolk sac* which lies closest to the cavity of the blastocyst. The amnion is destined to become fluid-filled, to expand, and to bathe the developing fetus that will be suspended in it, just as occurred in reptiles and birds. The yolk sac, while yolkfree in most mammals, is the site of early blood-cell formation.

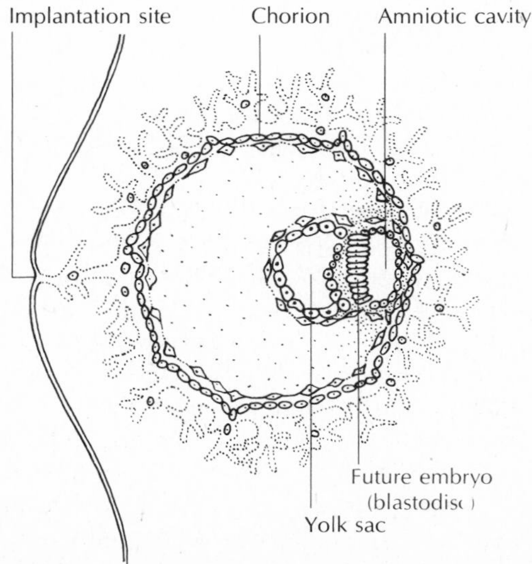

Figure 16.13 Implantation of embryo in wall of uterus.

BODY FORMATION

The formative movements of gastrulation take place in a disk of tissue, the *blastodisk,* at the confluence of the amnion and the yolk sac. Only the blastodisk contributes to the formation of the embryo; for this reason the other tissues are called extraembryonic. In addition to the formative movements that create the primitive gut, the outer ectoderm thickens, produces the *neural tube,* and becomes the brain and spinal cord. The formative movements have hardly begun, however, before a thin fingerlike endodermal evagination appears at the rear end of the blastodisk (Fig. 16.14). This projects into the *body stalk* which connects the developing embryo with the chorion. The outgrowth is the *allantois.* While the allantois itself remains small, it is vitally important because it is the route over which the embryo's blood vessels develop to grow into fingerlike outgrowths of the chorion, called *villi,* to provide for exchange of nutrients, wastes, and gases between the embryo and the mother. This is the region of the placenta.

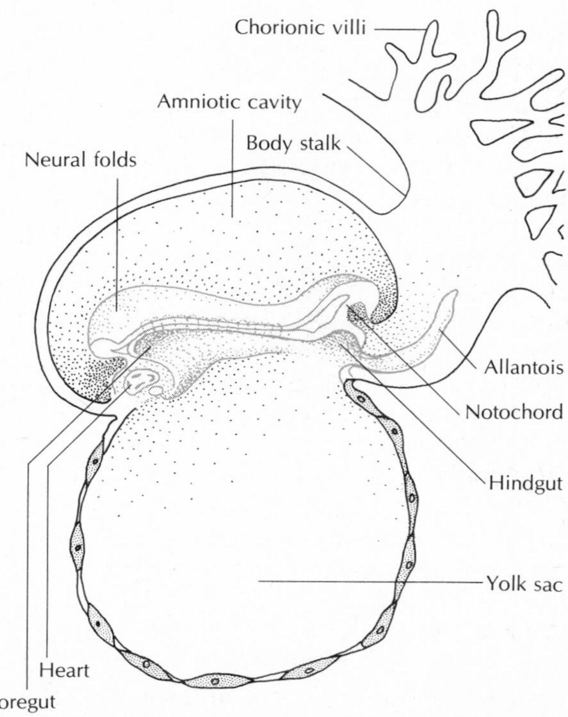

Figure 16.14 Human allantois in relationship to embryo. (Modified from Arey, *Developmental Anatomy,* Saunders, 1965.)

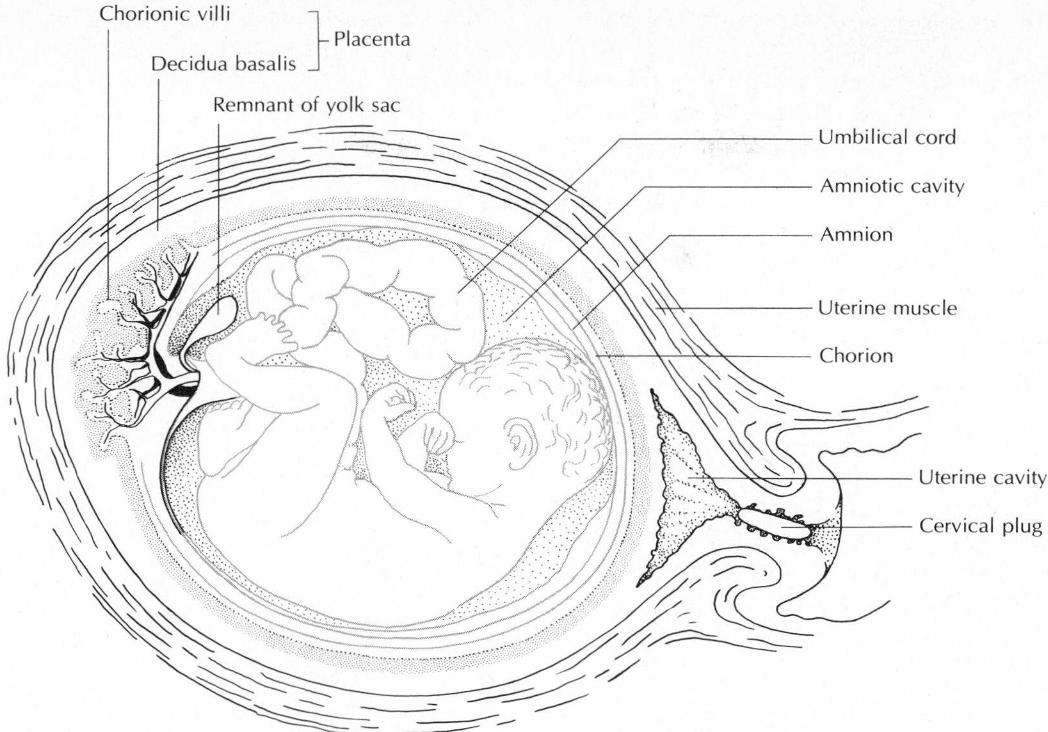

Figure 16.15 Diagram of advanced fetus shows its membranes and their relationship to the uterus. (From T. W. Torrey, *Morphogenesis of the Vertebrates,* Wiley, 1971.)

The embryo grows into a fetus within the amniotic cavity and this is accompanied by great enlargement of the uterus.

The Placenta

The placenta is formed by a combination of tissues from the embryo and from the mother. The region of the endometrium of the uterus into which the chorionic villi grow and to which the definitive unbilical cord is attached is called the *decidua basalis,* the site of the future placenta. The portion of the endometrium which surrounds the embryo just external to the chorion is the *decidua capsularis* and the remaining endometrium is the *decidua parietalis* (Fig. 16.15).

Placentas vary among different species of mammals. In humans it is a disk about 17.5 cm (7 inches) in diameter and about 2.5–3.5 cm (1–1½ inches) thick. Some animals, such as cows, have numerous isolated clusters of fetal-maternal associations, while others have diffuse fingerlike projections of the chorion. In carnivores, such as dogs and cats, the placenta forms a broad ring around the fetus.

Placentas also differ in the degree to which the chorion enzymatically erodes the surface of the endometrium. In some animals there is no erosion and the chorionic villi fit into crypts and crevasses of the endometrium. Other animals only partially erode the endometrium. In humans, however, the chorionic tissue erodes the endometrium until maternal blood pools are created between the chorion and the endometrium. The chorionic villi project into these pools of blood, making for an efficient exchange of gases, nutrients, and wastes between the blood vessels in the cone of the villi and the maternal blood pools.

Changes at Birth

A number of important changes occur in both the mother and the fetus at the time of birth, usually at nine calendar months from conception. Some of these changes are hormonal. A hormone *relaxin,* produced by the ovary, placenta, and uterine tissue, prepares the birth canal by relaxing the pelvic region. The pubic symphysis separates and the pelvic bones become less rigid. Relaxin also influences the condition of the uterus. The beginning of the birth process is marked by uterine contractions, influenced in part by a hormone of the posterior lobe of the pituitary, *oxytocin.* As the uterine contractions become stronger, the membranes surrounding the fetus break and the amniotic fluid is discharged. Eventually, accompanied by distention of the vagina, the uterine contractions, and the force of abdominal muscles, the fetus is expelled through the birth canal. Of course, the newborn baby still has the umbilical cord attached. This is tied and cut. Further uterine contractions force out the placenta, the remaining extraembryonic membranes, and much of the uterine lining as the *afterbirth.* (Some normally herbivorous animals, such as the cow, eat their afterbirth, presumably benefitting from nutrients and hormones.)

The newborn baby fills its lungs and starts breathing. Accompanying this is an increased flow of blood to the freshly expanded lungs. This change in the amount of blood flowing to the heart is reflected in an increased amount returning to the left atrium. This results in the closure of two flaps that had allowed blood to be exchanged between the right and left atria. This separates the right and left sides of the heart into a functionally double pump. Also, where the blood used to flow from the right ventricle to a shunt that took the blood to the aorta instead of to the lungs, it now flows to the lungs, and the shunt, known as the *ductus arteriosus,* reduces to a closed *ligamentum arteriosus.* Failures of the changes in circulation just described can lead to abnormalities creating so-called "blue babies."

Another event associated with birth is the release of milk when the baby suckles the breast. Of course, progesterone had caused enlargement of the breasts during pregnancy. *Prolactin,* from the anterior pituitary, however, causes the onset of the secretion of milk, and its ejection is influenced by oxytocin which is released from the posterior pituitary (though synthesized in the hypothalamus) through reflexive responses to the suckling of the infant.

From the four examples chosen, the following generalizations about early development can be made.

1. A zygote develops through predictable cleavage patterns to the blastula stage.
2. Formative movements shift cells from the surface to the interior in gastrulation.
3. The blastopore or primitive streak through which the cells migrate is at the future posterior end of the body.
4. The long axis of the embryo is thus created with the neural plate in vertebrates lying dorsally along the axis.
5. Changes in the cells on the surface are induced by underlying cells. This is the result of a tissue interaction.
6. The changed status of the cells that are induced is to move from a more general, less differentiated condition to populations of cells that become progressively specialized. This is a reflection of *differentiation* of cells, presently one of the poorly understood frontiers in biology.

EXPERIMENTS RELATED TO INDUCTION

In the case of induction of tissues, these briefly described experiments focus on some of the problems.

When a piece of the dorsal lip of the blastopore is transplanted to a site on a late blastula or early gastrula in frogs, the transplanted lip turns to the interior and causes the surrounding tissues also to become involved. The result of such an operation can be to produce an embryo with its normal archenteron and a secondary one arising from the transplant. Furthermore, this new archenteron's roof will induce the overlying ectoderm that normally would only form skin to form neural plate, neural tube, and finally central nervous system. The embryo then has two, one its own, the second induced. This clearly shows that external influences on the ectoderm have altered the direction of differentiation of the ectoderm (Fig. 16.16).

A second experiment shows that the nature of the response is under the genetic control of the responding cells. The experiment involves transplanting ectoderm from the oral region between frogs and salamanders. The frog ectoderm normally forms oral suckers; some salamander ectoderm normally forms rodlike projections, called balancers. Both suckers and balancers require the inductive influence of underlying tissues to develop at all. In this experiment the frog ectoderm lying over the tissue that normally induces balancers is induced by this tissue. But it responds according to its own genetic composition and produces frog suckers on the salamander. In the reciprocal experiment salamander balancers develop on the frog.

A third experiment shows that once the primary induction of the central nervous system is achieved, a succession of inductions follows. After the neural plate becomes a neural tube, at the anterior end a pair of lateral bulges appear. These are the optic vesicles. Their outer sufaces will fold in to form optic cups, the future lining of the eye. When these optic cups expand laterally and touch the lateral ectoderm, the ectoderm thickens locally, indents to form a vesicle that pinches from the outer ectoderm, and comes to lie within the rim of the optic cup; this is the future lens of the eye. By surgically

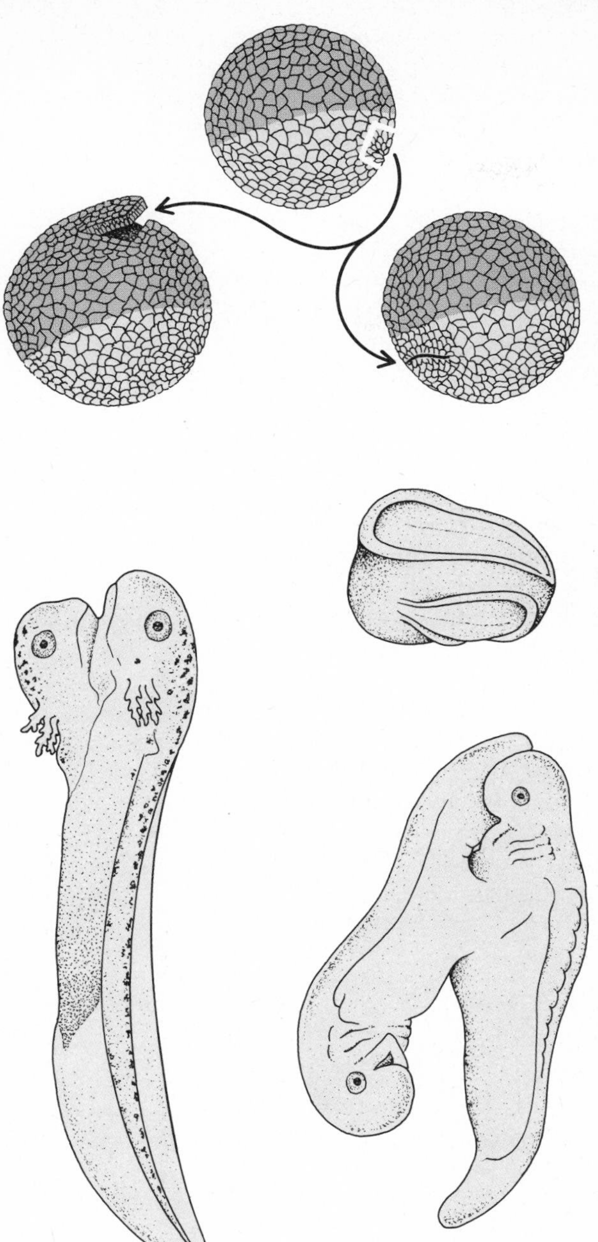

Figure 16.16 Role of the dorsal lip of the blastopore as an organizer of subsequent development. When the dorsal lip of the blastopore (outlined in white, top center) is removed and pushed through the opening into the blastocoel (upper left), the formative movements of the recipient push the piece into contact with tissue, which will, along with the transplant, differentiate into head structures (lower left). When the dorsal lip is grafted into the presumptive flank area of another gastrula (upper right), the dorsal lip graft undergoes formative movements of its own and also involves the cells of the recipient so that a second axis of organization (center right) forms and may differentiate into another nearly complete embryo (lower right).

removing only an optic cup without disturbing the outer ectoderm, no lens develops on the side of the operation. From this we can conclude that the optic cup induces the ectoderm to become lens. Since the optic cup came from the original neural plate that was induced from the underlying mesoderm, we see that there is a hierarchy of inductions. In this case neural plate formation is a primary induction, optic lens formation is a secondary induction. Other experiments can be done to show that the lens induces the differentiation of the cornea, a tertiary induction.

ESTABLISHING CHORDATE CHARACTERISTICS

Four things distinguish chordates from other animals: a dorsal hollow nervous sytem, a notochord, pharyngeal pouches (or clefts), and a postanal tail (Fig. 16.17). It has earlier been shown how a dorsal hollow nervous system is established. The notochord is a stiff rod of cells derived from mesoderm that develops in the midline in the roof of the archenteron before the archenteron becomes completely enclosed in endoderm. This tissue has a capablity for self-differentiating, that is, at the time the cells that comprise it move through the blastopore they are determined to differentiate as notochord. In the vertebrate chordates the notochord is either reinforced or replaced completely by vertebrae that develop from the adjacent mesoderm.

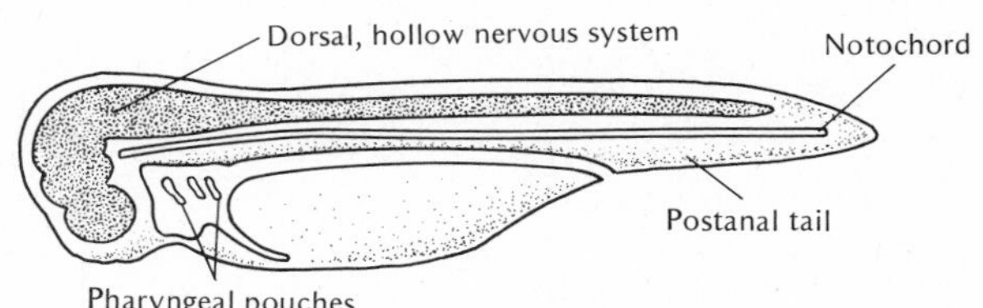

Figure 16.17 Chordate characteristics as seen in an early frog larva, tadpole (median sagittal section).

The third characteristic, the pharyngeal clefts, develops on the lateral walls of the anterior end of the archenteron, the area that expands to become the pharynx. Here a complex set of inductive relationships between outer ectoderm, intermediate mesenchyme (loose mesoderm), and inner endoderm results in endodermal pouches serially extending outward and ectodermal grooves indenting inward to meet the pouches. In the lower vertebrates, the fishes and amphibia, the tissue between the pouch and the groove breaks down and gill slits are formed. In higher mammals the perforation normally does not occur, but the endodermal pouches give rise to several pharyngeal derivatives such as the tubotympanic canal and the middle ear chamber, the thymus glands, and the parathyroid glands.

The fourth characteristic, a postanal tail, does not occur in other animals. In other words, in other animals the anal opening is terminal, although it may be concealed by such outgrowths as a telson in crayfish.

EMBRYONIC GERM LAYER DERIVATIVES

Later development of vertebrates goes beyond the scope of this textbook. However, the three embryonic germ layers contribute predictably to specific parts of the future animal. Table 16.2 itemizes these.

Table 16.2 Derivatives of Embryonic Germ Layers

DERIVATIVE	ECTODERM	MESODERM	ENDODERM
Skin and skin glands	×	—	—
Hair, feathers, parts of scales, covering of horns	×	—	—
Entire nervous system	×	—	—
Pituitary	×	—	—
Lining of mouth to rear edge of teeth	×	—	—
Part of lining of rectum	×	—	—
Adrenal medulla (from neural crest)	×	—	—
Connective tissue	—	×	—
Most cartilage (pharyngeal cartilages probably mainly from neural creast)	—	×	—
Bone	—	×	—
Muscles (a few exceptions)	—	×	—
Kidneys and ducts	—	×	—
Gonads and ducts, including uterus and part of vagina	—	×	—
Blood vessels, heart, lymphatics, blood (at least partly)	—	×	—
Lining of alimentary canal from pharynx to rectum	—	—	×
Thyroid and parathyroids	—	—	×
Thymus	—	—	×
Trachea, lungs, swim bladders of fish, liver, and pancreas (in the main)	—	—	×
Bladder	—	—	×
Cloaca (most) and perhaps some stem cells of white blood cells	—	—	×

METAMORPHOSIS

It was stated at the beginning of this chapter that development does not stop at birth, or at a free-feeding stage. An outstanding change that illustrates this is seen in the process of metamorphosis.

Metamorphosis occurs in postembryonic stages, and in a limited time morphological and physiological changes occur. In the discussions of development in various invertebrate groups a number of examples will be given. Here, two will suffice.

The tunicate, a urochordate, has a larva superficially resembling the tadpole of a frog. It has the typical chordate characteristics (see above). At a stage in its life cycle it rapidly resorbs its tail and notochord, most of the nervous system except for a ganglionic mass, and it settles to become sessile and attached. In this form it is difficult to see its chordate affinities.

The frog represents a better known metamorphosing form. From a tailed, totally aquatic form, possessing gills, the frog larva undergoes changes that lead to the resorption of the tail and gills, the development of limbs, and respiration by lungs. It has been clearly demonstrated, through thyroid or pituitary removal, or thyroid injections, that metamorphosis is under the control of the thyroid gland.

Insects, too, undergo metamorphosis, also under hormonal control, that causes them to pass through a few or several stages where reorganization of the anatomy and physiology occurs.

AGING

There is growing interest in the study of aging as a developmental process. Although there are several theories of aging, one that is noteworthy here is that proposed by Leonard Hayflick of Stanford University. He proposes that cells have built into them a maximum average number of cell divisions that they can undergo before they cease to divide and perhaps then begin to die. Hayflick developed his hypothesis from the results of tissue culture where he found that some cell lines could not be cultured past a certain number of generations. While there is much yet to be learned about aging, one observation that is not inconsistent with Hayflick's hypothesis is that central nervous system cells stop dividing after a period of time in an organism's development. Also the number of living central nervous system cells decreases as the individual ages. This very brief statement about aging perhaps makes it understandable why the United States National Institutes of Health now has an Institute for Aging.

SUGGESTIONS FOR FURTHER READING

Arey, L. B. *Development Anatomy.* Philadelphia: W. B. Saunders (1965).

Bodemer, C. W. *Modern Embryology.* New York: Holt, Rinehart and Winston (1968).

Brachet, J., and A. E. Mirsky. *The Cell,* vol. III. New York: Academic Press, 1961, (chs. by D. Mazia and M. M. Rhoades).

DeRobertis, E. O. P., W. W. Nowinski, and F. A. Saez. *Cell Biology.* Philadelphia: W. B. Saunders (1970).

Ebert, J. D., and I. Sussex. *Interacting Systems in Development.* New York: Holt, Rinehart and Winston (1970).

Saunders, J. W., Jr. *Patterns and Principles of Animal Development.* New York: Macmillan (1970).

Torrey, T. W. *Morphogenesis of the Vertebrates.* New York: Wiley (1970).

From Scientific American

Allen, R. D. "The Moment of Fertilization" (July 1959).

Ebert, J. D. "The First Heartbeats" (Mar. 1959).

Edwards, R. G. "Mammalian Eggs in the Laboratory" (Aug. 1966).

Edwards, R. G., and R. E. Fowler. "Human Embryos in the Laboratory" (Dec. 1970).

Etkin, W. "How a Tadpole Becomes a Frog" (May 1966).

Fischberg, M., and A. W. Blackler. "How Cells Specialize" (Sept. 1961).

Freiden, E. "The Chemistry of Amphibian Metamorphosis" (Nov. 1963).

Gray, G. W. "The Organizer" (Nov. 1957).

Gurdon, J. B. "Transplanted Nuclei and Cell Differentiation" (Dec. 1968).

Hayflick, L. "Human Cells and Aging" (Mar. 1968).

Jacobson, M., and R. K. Hunt. "The Origins of Nerve Cell Specificity" (Feb. 1973).

Lowenstein, W. R. "Intercellular Communication" (May 1970).

Mazia, D. "How Cells Divide" (Sept. 1961).

Moscona, A. A. "How Cells Associate" (Sept. 1961).

Reisfeld, R. A., and B. D. Kahan. "Markers of Biological Individuality" (June 1972).

Reynolds, S. R. M. "The Umbilical Cord" (July 1952).

Singer, M. "The Regeneration of Body Parts" (Oct. 1958).

Taylor, T. G. "How an Eggshell Is Made" (Mar. 1970).

Wessels, N. K. "How Living Cells Change Shape" (Oct. 1971).

Wessels, N. K., and W. J. Rutter. "Phases in Cell Differentiation" (Mar. 1969).

17 Genetics

Although man has long been aware that related individuals are generally more similar than unrelated ones, most of what is known about heredity today has been discovered in the twentieth century. However, the roots of genetics can be traced to antiquity. Tablets prepared by the Babylonians some 6000 years ago have been interpreted as horse pedigrees (breeding records); this suggests a conscious effort toward improvement of the horse stock. Other Babylonian stone carvings of the same period illustrate artificial pollination of the date palm. The early Chinese and the early American Indians improved their stocks of cereal grains by selection of desirable types. The Greeks were aware of inheritance also: Hippocrates (400 B. C.), Aristotle (350 B.C.), and other Greek philosophers considered that an individual's traits developed as a result of environmental influences. However, this idea was not stated precisely until the nineteenth century when Jean Baptiste de Lamarck proposed a theory of evolution (see Ch. 42).

It is obvious that the transmission of hereditary materials in sexually reproducing forms is by the gametes. Any real understanding of heredity would be dependent upon some knowledge of gametes and fertilization. Sperm were not described until the beginning of the eighteenth century; fertilization in plants was not worked out until 1761 by Joseph Kölreuter; and the nature of the mammalian egg was unknown until the middle of the nineteenth century. Genetics is actually a rather new discipline in biology. This is easy to understand since eggs, sperm, and fertilization came to be understood relatively recently.

Genetics includes two major areas of study: transmission genetics, which is concerned with how characteristics are passed from one generation to another; and cellular genetics, which is involved with how the bioinformation of the hereditary material is expressed and regulated in the cell. The basic tenets of cellular genetics and bioinformation were presented in Chapters 6 and 7. This chapter deals mainly with transmission genetics. Further aspects of genetics are taken up in Chapters 18, 19, and 42.

The first successful analysis of heredity was made by the Austrian monk, Gregor Mendel, in the middle of the nineteenth century. Mendel's primary work was the study of the inheritance of traits in the common garden pea. He studied the inheritance of seven different pairs of contrasting traits in pure strains (truebreeding) of the garden pea. A mating between individuals that differ in one trait is a *monohybrid cross.* Mendel made several monohybrid crosses, each involving a contrasting pair of traits. The results given in Table 17.1 are the totals for all crosses made. The first-generation individuals (F_1) were similar in all crosses. The term *dominant* describes the trait or characteristic that always appeared in the first generation, whereas the term *recessive* describes the characteristic that was not expressed in the first generation. The second generation (F_2), produced by mating two F_1s, showed that the recessive type is expressed, and in a definite ratio. The F_2 ratio in each experiment was essentially 3:1, that is, approximately $3/4$ of the F_2 generation in each experiment were the dominant type and $1/4$ were the recessive type. For example, plants with red flowers, when crossed to plants with white flowers, produced an F_1 generation of red-flowered plants. Mating of the F_1 plants gave an F_2 ratio of 3 red to 1 white. The 3:1 ratio is theoretical or ideal and will not always be obtained in the progeny from a single cross with relatively small numbers. This 3:1 ratio becomes apparent only

Table 17.1 F_2 Results of Mendel's Monohybrid Crosses

EXPERIMENT	DOMINANT CHARACTER	RECESSIVE CHARACTER	F_2 RESULTS	TOTAL NUMBER OF F_2 PLANTS	RATIO OF F_2 DOMINANTS TO F_2 RECESSIVES
1	Round seeds	Wrinkled	5474 round; 1850 wrinkled	7324	2.96:1
2	Yellow seeds	Green	6022 yellow; 2001 green	8023	3.01:1
3	Red flowers	White	705 red; 224 white	929	3.15:1
4	Axial flowers	Terminal	651 axial; 207 terminal	858	3.14:1
5	Inflated pods	Constricted	882 inflated; 299 constricted	1181	2.95:1
6	Green pods	Yellow	428 green; 152 yellow	580	2.82:1
7	Long stems	Short	787 long; 277 short	1064	2.84:1

when large numbers of individuals from similar crosses are analyzed.

Mendel also studied the inheritance of two pairs of contrasting traits at the same time (*dihybrid* crosses). For example, he crossed plants that were truebreeding for long stems and round seeds with truebreeding ones for short stems and wrinkled seeds. In such crosses all the F_1 individuals had long stems and round seeds, that is, both dominant traits were expressed in the F_1 generation. Mendel's analysis of several experiments carried to the F_2 generation showed that four kinds of plants were formed and in a definite ratio: 9/16 of the F_2 individuals were long stemmed with round seeds, 3/16 were long stemmed with wrinkled seeds, 3/16 were short stemmed with round seeds, and 1/16 were short stemmed with wrinkled seeds. Other dihybrid crosses using other combinations of characteristics gave similar results. The F_2 individuals were of four kinds in 9:3:3:1 ratios. In summary, the classical Mendelian ratios are 3:1 for a monohybrid cross and 9:3:3:1 for a dihybrid cross.

MENDEL'S POSTULATES

From the analysis of his results Mendel constructed four postulates.

1. Unit factors exist in pairs and are responsible for the inherited traits.
2. In the formation of gametes and spores the factors are separated or segregated so that each gamete gets only one factor of a given pair. This is often called Mendel's first law, or the *law of segregation.*
3. When two factors for the alternative expression of a trait, such as the factor for round seed and the factor for wrinkled seed, are brought together in fertilization, the resulting individual will show only the dominant trait. The recessive factor is not expressed. This is the principle of *dominance and recessiveness.*
4. When two or more pairs of factors are involved, they may assort or combine at random during the formation of gametes; or, stated another way, all possible combinations of factors will occur in equal numbers. This is often referred to as Mendel's second law, or the *law of independent assortment.*

These postulates may be illustrated in many genetic crosses. For example, in the cross of long-stemmed plants with short-stemmed plants the letter *s* represents short stems and *S* represents long stems. (Common genetic practice uses a small letter for the recessive factor and the corresponding capital letter for the dominant factor.)

	LONG STEMMED		SHORT STEMMED
Parent plants	*SS*		*ss*
Gametes	*S*	×	*s*
F_1 plants		*Ss* (long stemmed)	

When the F_1 plant is self-fertilized, or when two are crossed, the formation of gametes may be represented as follows:

F_1 plants	*Ss*	×	*Ss*
Gametes	Ⓢ ⓢ		Ⓢ ⓢ

Since gametes unite at random, all possible combinations occur when the gametes are numerous. The results of the random unions of gametes can be represented by the Punnett square, which was first used by the English geneticist R. C. Punnett. In this case the square is

		♂ Gametes		
		S	*s*	
♀ Gametes	*S*	*SS*	*Ss*	F_2 zygotes
	s	*Ss*	*ss*	

where the symbol ♂ represents male and the symbol ♀ represents female.

The results shown in the Punnett square represent the possible kinds of individuals that will be produced in the F_2 generation and their proportions—not the actual number of individuals. From this we see that three fourths of the F_2 plants will be long stemmed and one fourth will be short stemmed. Because of dominance of the *S* factor, the *Ss* type cannot be distinguished from the *SS* type. In this monohybrid cross the first three of Mendel's postulates have been illustrated: unit factors in pairs, segregation of the factors in gamete formation, and dominance *versus* recessiveness.

Before we consider Mendel's fourth postulate, certain terms should be defined. Individuals in which both factors of a given pair are alike, such *SS* or *ss*, are said to be *homozygous,* whereas individuals with a dominant and a recessive factor, such as *Ss*, are said to be *heterozygous.* The term *phenotype* indicates the appearance of an individual, whereas the term *genotype* designates the precise genetic makeup, for example, *SS* or *Ss*. Mendel's dihybrid experiment, in which he crossed plants homozygous for long stems and round seeds with plants homozygous for short stems and wrinkled seeds and carried to the F_2 generation, is represented in Table 17.2.

Table 17.2 One of Mendel's Dihybrid Crosses

Parental phenotypes	Long round		Short wrinkled
Parental genotypes	*SSWW*		*ssww*
Gametes	*SW*		*sw*
F_1 plants genotype		*SsWw*	
F_1 plants (phenotype)		(long round)	
Cross of F_1 plants	*SsWw*	×	*SsWw*
Gametes	*SW*		*SW*
	Sw		*Sw*
	sW		*sW*
	sw		*sw*

Punnett square

		♂ Gametes				
		SW	*Sw*	*sW*	*sw*	
♀ Gametes	*SW*	*SSWW*	*SSWw*	*SsWW*	*SsWw*	F_2 zygotes
	Sw	*SSWw*	*SSww*	*SsWw*	*Ssww*	
	sW	*SsWW*	*SsWw*	*ssWW*	*ssWw*	
	sw	*SsWw*	*Ssww*	*ssWw*	*ssww*	

In the formation of gametes to obtain the F_2 generation, independent assortment is illustrated. *S* assorts or combines with *W* and *w,* and *s* assorts with *W* and *w*, that is, all possible combinations are formed. Each gamete has one and only one of each pair of factors, that is, *segregation occurs in gamete formation.* It cannot be emphasized too strongly that these two basic phenomena, segregation and independent assortment, both occur during the formation of gametes.

Mendel obtained the 9:3:3:1 phenotypic ratio in all F_2 results for dihybrid crosses. Therefore his postulates, when applied in theoretical crosses, seem to explain his observed results. The significance of Mendel's conclusions has been made clear in the thousands of experiments that have been performed since 1900, using a great variety of organisms. Although it has been necessary to modify slightly and add to Mendel's concepts, the basic statements stand today as laws of heredity. The significance of his work is even more impressive when it is realized that Mendel did not know anything about the chromosomes and the physical basis of heredity.

Although Mendel published his results in 1866, his work received no attention until 1900, when three biologists, Carl Correns, Hugo de Vries, and Erich von Tschermak, independently rediscovered the principles of genetics. It has been said that the reason for the delay in recognition of Mendel's work was its publication in an obscure journal. However, it seems likely that his paper was read by a number of workers before 1900. Two other reasons have been suggested for the failure to appreciate Mendel's work at the time it first appeared. This was the period when the work of Charles Darwin was occupying the minds of most biological workers. Also, most of the biologists of this period took descriptive, not analytical, approaches to their work. Perhaps they were not ready for the analytical approach of Mendel.

Why did Mendel succeed in elucidating basic principles of heredity where others before him had failed? In the hundred years following the understanding of fertilization in plants (1761), many biologists had worked on the problem of heredity. A number of things contributed to Mendel's success.

First, he used favorable material. Normally, the pea plant is not cross-fertilized because of the structure of the flower. Thus Mendel was able to hand pollinate with confidence in the type of cross that he was making.

Second, his study focused on particular traits, as indicated above. The other students of heredity up to this time had been attempting to determine the nature of the process by considering the complete heredity of each organism.

Third, Mendel kept accurate data on all experiments, and from the analysis of his data he was able to formulate definite hypotheses that were testable.

Finally, Mendel was fortunate (lucky) in that the seven pairs of contrasting traits were located on different chromosomes. The probability of picking seven genes on seven different chromosomes is approximately 6 out of 1000.

Serendipity aside, the work of Mendel illustrates the scientific method in operation in a very striking way.

THE PHYSICAL BASIS OF HEREDITY

Although the interests of many biologists during the last half of the nineteenth century were focused on the problems of evolution that had been raised by Charles Darwin, this period did see many advances in the field of cytology. A great deal was learned about the nature and behavior of chromosomes. In this connection, the reader should review the discussion on mitosis and meiosis in Chapter 7. As is mentioned there, body or somatic cells of animals have a characteristic number of chromosomes: the diploid, or $2n$ number.

The constancy in the number of chromosomes in the body cells of sexually reproducing organisms poses a problem. In sexually reproducing organisms a reduction in the number of chromosomes would have to occur at some time in the life cycle in order to have a constant number maintained in each generation. If such a reduction did not occur, the number of chromosomes would double with each fertilization. That this reduction (meiosis) does occur also was determined by the students of cytology during the last part of the nineteenth century.

When the principles of genetics were rediscovered in 1900, most of the aforementioned facts about meiosis were known. Several workers recognized certain parallelisms between Mendel's postulates and the nature and behavior of the chromosomes.

1. The chromosomes exist in pairs; Mendel postulated factors in pairs.
2. The paired chromosomes are separated in meiosis, with each gamete receiving only one member of a given pair; Mendel postulated the segregation of the paired factors in gamete formation.
3. When more than one pair of chromosomes is involved, they assort at random in meiosis. This is the basis for Mendel's independent assortment of hereditary factors. (To verify this, examine Figure 17.1 and follow the explanation in the caption.)

Thus a firm basis for what Mendel observed and postulated was found in the nature and behavior of chromosomes during meiosis and fertilization. The later work on chromosomes has added further support to the concept. Many other parallelisms have been found between

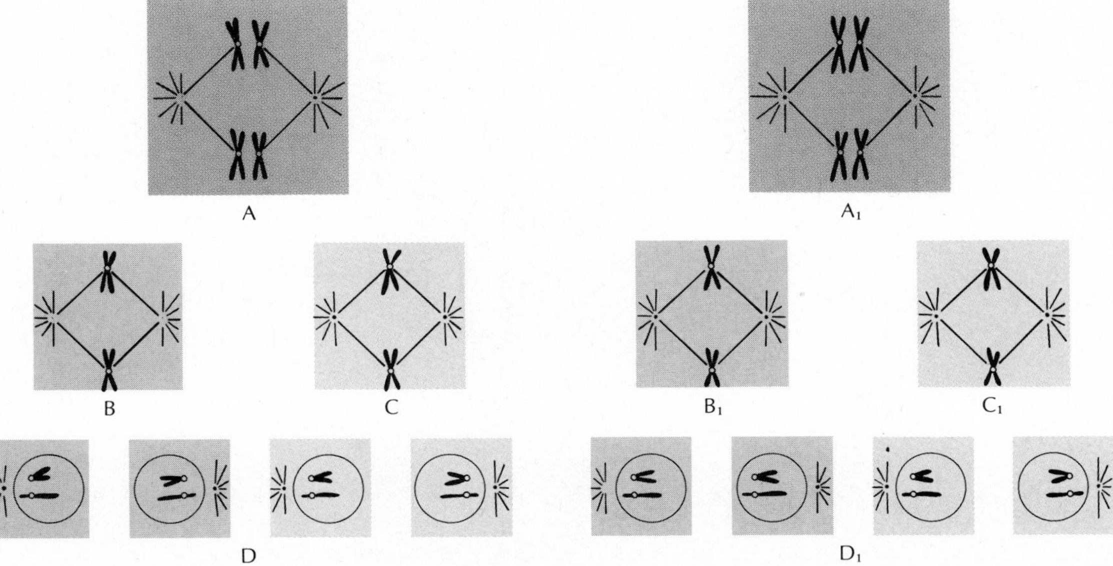

Figure 17.1 Diagram of meiosis including the two meiotic divisions. A, A_1, the two alternative arrangements of the chromosome pairs on the first meiotic spindle; B–D, B_1–D_1, the second meiotic divisions and the different types of reduced chromosome constitutions of the gametes.

genetic results and the nature and behavior of chromosomes (see the discussions of sex determination, linkage, and crossing over in a later section of this chapter). The chromosome theory of heredity has been established, and the particular entities (now termed *genes*) carried on the chromosomes are arranged in a linear order. Each point, or *locus,* on a chromosome occupied by a gene has a corresponding point or locus on the homologous chromosome that is occupied by the same gene or by an alternate form, an *allele,* which has a somewhat different effect on the same characteristic. For example, in the garden pea, *S* and *s*, for long and short stems, respectively, are alleles.

The Chromosome Concept Applied

With this knowledge of chromosomes it is helpful to diagram genetic crosses showing the genes on chromosomes. This is illustrated in a cross involving hair characteristics in guinea pigs. In these animals black hair color (*W*) is dominant over white (*w*). When homozygous black and white individuals are mated, the F_1 generation is all black and the F_2 generation shows the characteristic 3:1 ratio, that is, three fourths black and one fourth white.

The rough hair condition (*S*) is dominant over smooth hair (*s*). A cross between a homozygous black–smooth guinea pig and a homozygous white–rough guinea pig is shown in Figure 17.2. All F_1 individuals form four kinds of gametes. These unite at random to produce the F_2 generation, and this is shown in Table 17.3. An analysis of the F_2 generations shows four phenotypes in a 9:3:3:1 ratio: black–rough, black–smooth, white–rough, and white–smooth.

What happens when three pairs of alleles, each located on a different chromosome pair, are involved? A triple heterozygote might be written *AaBbCc.* Such an individual would produce eight kinds of gametes as follows: *ABC, ABc, AbC, Abc, aBC, aBc, abC, abc.* If two such heterozygotes were crossed, eight phenotypes would be present in the progeny in a ratio of 27:9:9:9:3:3:3:1. In working out this ratio, a Punnett square with 64 boxes is necessary.

The simple formula for determining the number of different kinds of possible gametes is 2^n, where *n* equals the number of heterozygous pairs of alleles located on different pairs of chromosomes. In a dihybrid cross each individual would produce 2^2, or 4, unique gametes; in a trihybrid cross each would produce 2^3, or 8; and in humans, with 23 pairs of chromosomes, there are 2^{23} (over eight million) when each of the 23 chromosome pairs has one heterozygous gene pair. If an organism has several heterozygous loci among the thousands of loci it pos-

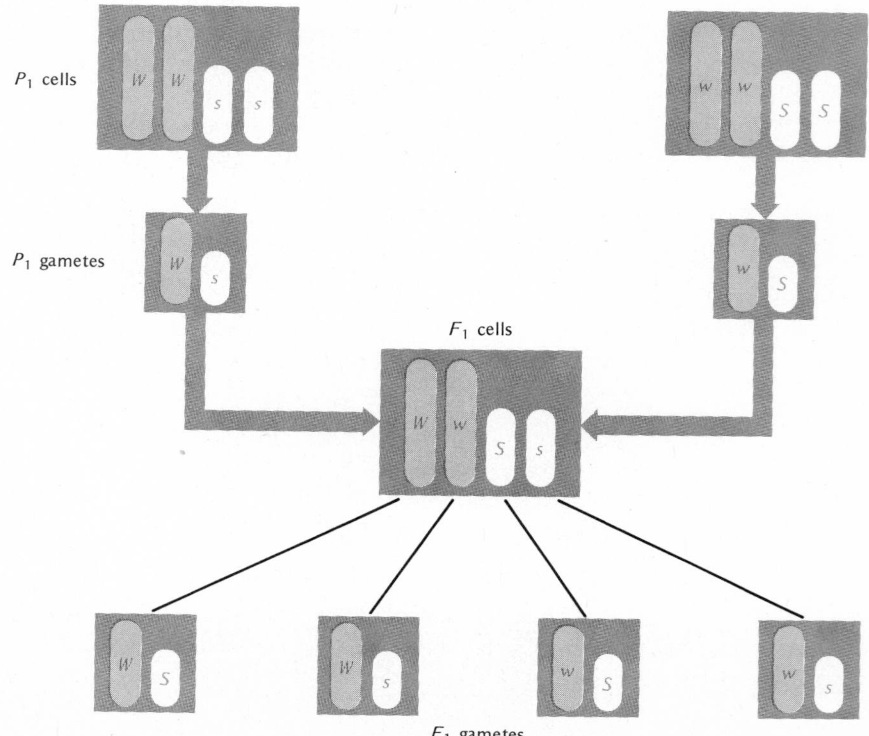

Figure 17.2 P_1 and F_1 gametes from dihybrid cross in guinea pigs.

sesses, then the number of different kinds of unique gametes is even larger.

The Test Cross

When the Punnett square showing the F_2 generation of the guinea pig cross is examined (Table 17.3), four different genotypes for black–rough individuals are found. These are (1) *WWSS*, (2) *WWSs*, (3) *WwSS*, and (4) *WwSs*. All these individuals look alike (have the same phenotype). In breeding work it is necessary to ascertain the genotype of each individual. This can be done by making a test cross, which is done by crossing the individual of the unknown genotype to one that is homozygous recessive. In this particular case the results that would be obtained in the four test crosses are shown below:

1. *wwss* × *WWSS*		
ws	*WS*	all black–rough (*WwSs*)
2. *wwss* × *WWSs*		
ws	*WS*	1/2 black–rough (*WwSs*)
	Ws	1/2 black–smooth (*Wwss*)
3. *wwss* × *WwSS*		
ws	*WS*	1/2 black–rough (*WwSs*)
	wS	1/2 white–rough (*wwSs*)
4. *wwss* × *WwSs*		
ws	*WS*	1/4 black–rough (*WwSs*)
	Ws	1/4 black–smooth (*Wwss*)
	wS	1/4 white–rough (*wwSs*)
	ws	1/4 white–smooth (*wwss*)

Thus by analyzing the results of a test cross, a breeder may determine the genotype of any given individual. This method is used widely and is, of course, applicable to cases involving one, two, or several pairs of genes.

Table 17.3 Phenotypes and Genotypes of the F_2 Generation of a Dihybrid Cross

		F_1 Eggs				
		Ws	*Ws*	*wS*	*ws*	
F_1 Sperm	*WS*	*WWSS* black–rough	*WWSs* black–rough	*WwSS* black–rough	*wwss* black–rough	F_2 zygotes
	Ws	*WWSs* black–rough	*WWss* black–smooth	*WwSs* black–rough	*Wwss* black–smooth	
	wS	*WwSS* black–rough	*WwSs* black–rough	*wwSS* white–rough	*wwSs* white–rough	
	ws	*WwSs* black–rough	*Wwss* black–smooth	*wwSs* white–rough	*wwss* white–smooth	

EARLY QUALIFICATIONS AND ELABORATION OF MENDELIAN GENETICS

Lack of Dominance

After the rediscovery of Mendel's work the concept of dominance soon had to be modified. In many different experiments with both animals and plants, the F_1 individuals are not like one of the parents but are intermediate in nature. This is true when homozygous red shorthorn cattle are crossed with homozygous white ones; the F_1 progeny are roan, a color intermediate between red and white.

Another example of this type of inheritance is found in the Andalusian breed of chickens. In this breed two stocks have been maintained for many years; one has white feathers and the other has black ones. That both stocks are homozygous for the feather color is shown by the fact that regular breeding maintains the stocks. When black is mated with black, the progeny are always black; likewise, the white stock also breeds true. However, when a black Andalusian chicken is mated with a white one, the F_1 individuals are blue. When the F_1 blue chickens are mated, the F_2 generation consists of black, blue, and white offspring in a 1:2:1 ratio. In this instance all heterozygous individuals (*Bb*) are blue, and the phenotypic ratio is the same as the genotypic ratio (Fig. 17.3). There are many cases of lack of dominance in plants and animals.

Factor Interaction

In all cases of inheritance discussed up to this point, only one pair of alleles has been involved in a given pair of contrasting traits. Although there are many instances of this kind of inheritance, there are many others where two or more pairs of genes are involved in the production of a given trait.

One of the first cases of this kind to be worked out involved the inheritance of flower color in the sweet pea. W. Bateson and R. C. Punnett had found that purple color was dominant over white and gave a typical 3:1 ratio in the F_2. The white strain that they used always bred true. However, when they crossed this white strain with another white strain from a different locality, they obtained unexpected results. The F_1 plants were all purple. When the F_1 plants were self-fertilized to obtain an F_2 generation, approximately 9/16 of the plants were purple flowered and 7/16 were white flowered.

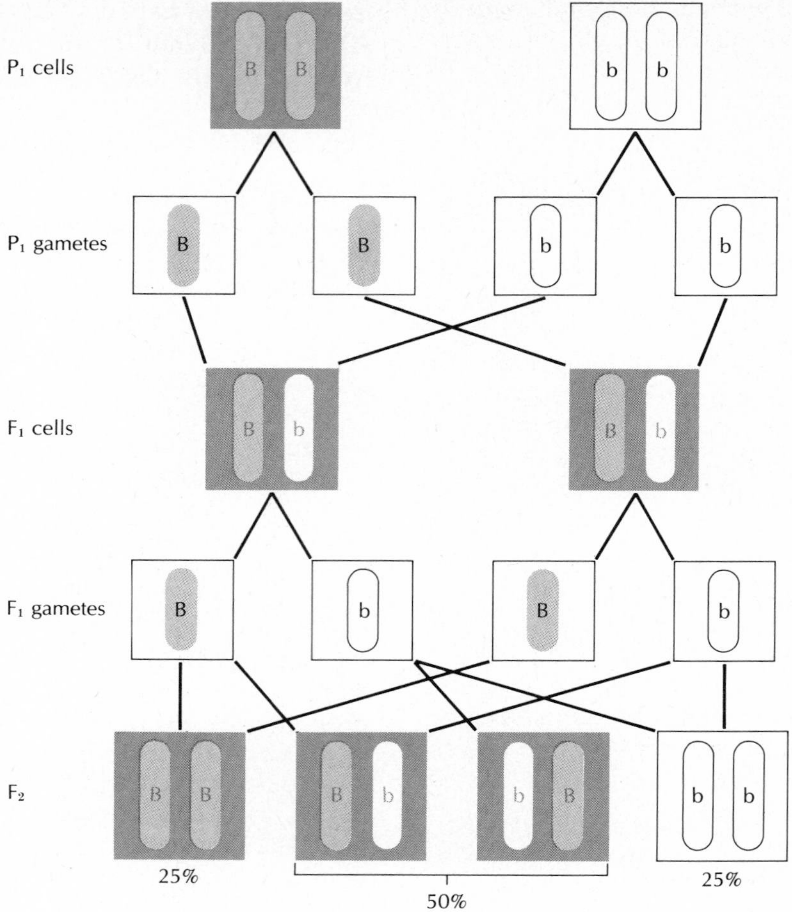

Figure 17.3 Cross of black and white Andalusians and inbreeding of their offspring (blue Andalusians).

They explained these results by suggesting that two pairs of genes are involved in producing flower color in the sweet pea. The purple flowered plants must have at least one dominant gene from each pair, and all individuals that do not have at least one dominant gene from each pair are white. Their explanation for these results is shown below:

Parents	white × white
	CCpp × *ccPP*
Gametes	*Cp* *cP*
F_1	purple
	CcPp

F_2		*CP*	*Cp*	*cP*	*cp*
	CP	*CCPP* purple	*CCPp* purple	*CcPP* purple	*CcPp* purple
	Cp	*CCPp* purple	*CCpp* white	*CcPp* purple	*Ccpp* white
	cP	*CcPP* purple	*CcPp* purple	*ccPP* white	*ccPp* white
	cp	*CcPp* purple	*Ccpp* white	*ccPp* white	*ccpp* white

Although at first glance, factor interaction seems quite different from that described by Mendel, only a slight modification of the Mendelian scheme is needed to fit the Mendelian principles to factor interaction. This, of course, is the importance and the significance of the Mendelian principles.

Multiple Factors

Another type of inheritance bothered geneticists for some time. In the cases described so far, the different kinds of individuals fall into a few alternative classes. In the study of inheritance of many traits (such as size, skin color, mental ability in humans, size of domestic animals, yield of fruits or eggs or milk), instead of a few alternative classes, there are often many small ones. Careful studies of this type of inheritance have revealed that it, too, can be explained on a Mendelian basis. The explanation involves the concept of multiple factors, or multiple genes, by which two or more pairs of genes contribute to the characteristic in a cumulative way.

An insight into this type of inheritance was provided in the studies on wheat by the Swedish geneticist H. Nilsson-Ehle. He found that when he crossed one variety that had white kernels with another that had very dark red kernels, the F_1 individuals produced kernels that were intermediate in color. When he obtained an F_2 generation, 15/16 of the individuals were red and 1/16 were white. However, a close analysis of the red kernels showed that they did not have the same shade of red. Only 1/16 were very dark red, 4/16 were medium dark red, 6/16 were intermediate red, and 4/16 were light red. This can be explained on the assumption that there were two gene pairs involved, *A* and *a, B* and *b*. The genes *A* and *B* caused redness; *a* and *b* caused whiteness. Only those individuals that received four of the "red" genes were very dark red, those that received three "red" genes were medium dark red, those that received two "red" genes were intermediate red, and those that received one "red" gene were light red. Finally those that received no "red" genes were the white individuals.

In another study Nilsson-Ehle crossed a red and a white variety that, in the F_2 generation, gave a ratio of 63/64 red to 1/64 white. This was due to three different genes for red color, and in this case an analysis of the F_2 red-kerneled plants showed a greater range in the intensity of red color.

The F_2 generation of a simple case involving two pairs of cumulative genes, such as described above for wheat, may be arranged to indicate the distribution of classes.

AaBb × *AaBb*

		AaBb		
		AaBb		
	AABb	*AaBb*	*Aabb*	
	AABb	*AaBb*	*Aabb*	
	AaBB	*AAbb*	*aaBb*	
AABB	*AaBB*	*aaBB*	*aaBb*	*aabb*
1/16	4/16	6/16	4/16	1/16

In this case there are five classes of individuals. In a situation involving three pairs of cumulative genes there would be seven classes with the following frequencies: 1/64, 6/64, 15/64, 20/64, 15/64, 6/64, and 1/64. With four pairs there would be still more classes, and so on. Actually, with a large number of gene pairs involved, the classes are so numerous and the differences between them so slight that it is difficult to show that the variation in F_2 progeny is not continuous. When the number of pairs of genes is large, the pyramid (as shown above) becomes a smooth bell-shaped curve. Such a bell-shaped curve is a normal distribution.

Multiple Alleles

In the early studies on heredity it was considered that a given locus on a chromosome could be occupied by one of two possible alleles. Here, again, the concept had to be modified because it was found that a gene can exist in more than two allelic forms. In the fruit fly, *Drosophila melanogaster,* many alleles have been found for the gene responsible for red eye color. The gene for white eye color is one allele, and there is a graded series from white to red. In this case the red gene is dominant over all others in the series, and the white gene is recessive to all. Obviously, any one fly can possess only two of the multiple alleles, but in the whole population of the species, many alleles, including red, are found for this locus. A simple case of three alleles is found in humans as the basis for the inheritance of the blood groups A, B, AB, and O. The importance of these blood groups in connection with blood transfusions was discussed in Chapter 11 on the circulatory system. The three alleles in this case may be designated as I^A, I^B, and *i*. Both I^A and I^B are dominant to *i*, but I^A and I^B show no dominance with respect to one another; this is called codominance. The genotypes of the four phenotypes are given below.

PHENOTYPES	GENOTYPES
A	I^AI^A
	I^Ai
B	I^BI^B
	I^Bi
AB	I^AI^B
O	ii

As a result of this knowledge, it is now possible to predict what kinds of offspring will be possible from various matings. This is used in many states in courts of law where cases of disputed parentage are involved. Such cases sometimes arise because of accidental exchanges of babies in hospitals or where there is disputed parentage involving illegitimacy or infidelity. A man of group A and a woman of group O could not be the parents of a child of group B. Neither could they be the parents of an AB child, but they could have children of the A or O type. A man and a woman, one heterozygous for group B and the other heterozygous for group A, could have all four blood groups represented in their offspring. Work out other possibilities, using the six different genotypes.

The genetic basis for several other blood groups in humans is known, and among these the Rh group is the most important. For a short time following the discovery of the Rh factor in 1940 a simple genetic mechanism was assumed. With further studies it became apparent that there are several alleles at the Rh locus. At the present time there is still dispute about the exact number of alleles and their effects.

Lethal Genes

One other type of seeming deviation from Mendelian inheritance should be mentioned. This involves genes that cause the death of the organism at some time during development. One of the earliest cases to be worked out involved the inheritance of yellow coat color in mice. Individuals with the "normal" or wild coat color (gray) breed true. However, yellow mice, when bred together, always produce both yellow and gray individuals in a ratio of 2 yellow to 1 gray. No pure-breeding yellow mice have ever been found. When yellow and gray mice are mated, approximately half of the offspring are yellow and half are gray. These two ratios suggest two things: that yellow is dominant over gray and that yellow mice are heterozygous. In studies on mating between yellow parents it was noted that the litters were only three quarters as large as other mice litters.

The final proof of the hypothesis that yellow mice are heterozygous came when investigators examined the uteri of pregnant yellow females that had been mated with yellow males. Approximately one fourth of the embryos found had stopped development and were being resorbed. The genotypes may be represented as follows:

PARENTS			OFFSPRING				
Yellow		Yellow	Dies		Yellow		Gray
Gg	×	*Gg*	1*GG*	:	2*Gg*	:	1*gg*

CHROMOSOMES AND SEX

Sex Determination

It is common observation that there are about as many men as women in the world and also that this is true for most populations of animals in nature. In general we find a 50:50 sex ratio. As early as 1902, while studying meiosis in the testes of grasshoppers, a cytologist noticed that there were 11 pairs of chromosomes and an odd one with no homologous mate. He suggested that this odd chromosome might be associated with sex determination. A study of the chromosomes in a great variety of species revealed that in one sex in many species there is either an odd chromosome or an odd pair of chromosomes in which the two members are unlike in size and shape. The situation found in *Drosophila* is typical.

The diploid number of chromosomes in this species is eight. There are two pairs of large V-shaped chromosomes and a pair of small dot-shaped ones in the cells of each sex; these are called *autosomes.* The fourth pair of chromosomes in the cells of females is rod-shaped. In the male the fourth pair consists of a rod-shaped chromosome, similar to those in female cells, and a smaller bent chromosome. These chromosomes are called the sex chromosomes, the rod-shaped one being designated as the X and the bent chromosome as the Y. Thus a typical female in this species has diploid cells with six autosomes and two X chromosomes, whereas the male cells have six autosomes, and one X and one Y chromosome. In the reduction that occurs in meiosis, all eggs are alike, but two kinds of sperm are formed, one half with an X chromosome and one half with a Y chromosome. With eggs and sperm uniting by chance, this gives a basis for the 50:50 sex ratio (Fig. 17.4).

The chromosomal situation in humans and many other animals is like that found in *Drosophila,* but the chromosomal mechanism for sex determination is not the same in all animals. In some forms, as described above for the grasshopper, there is no Y chromosome (XO). In butterflies, moths, birds, and some salamanders and fishes the situation is reversed from that found in *Drosophila,* the odd pair of chromosomes is found in the cells of females instead of males. Usually, the letters Z and W are used in such cases, that is, males are ZZ and females are ZW.

If all species had the XY situation, then it might be assumed that the Y chromosome contains a dominant gene or genes for maleness. However, since some species have no Y chromosome, this concept is not valid. What, then, is the reason for an XY or an XO individual being a male and an XX individual being a female? An answer to this has been found in studies on individuals produced from abnormal gametes. Occasionally in the formation of sperm cells in *Drosophila,* abnormal meiotic divisions occur

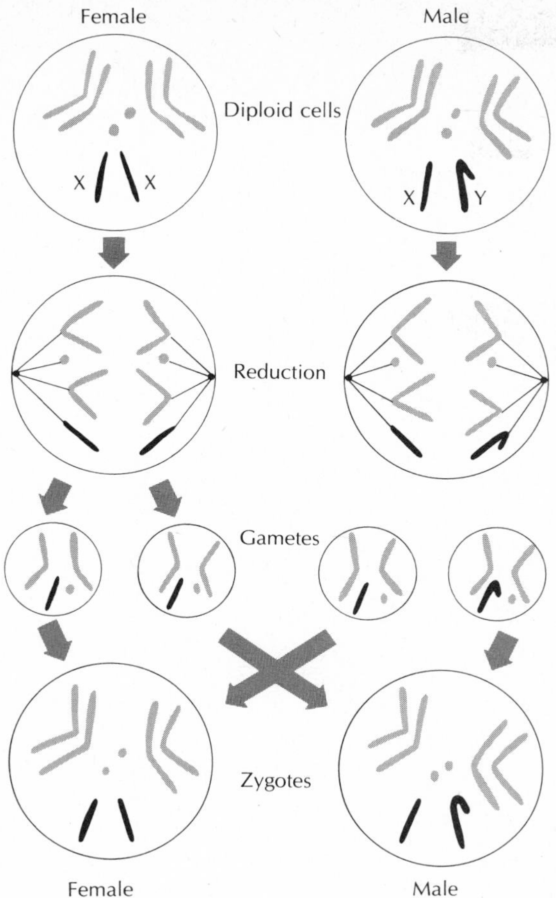

Figure 17.4 Distribution of sex chromosomes in *Drosophila melanogaster.* Sex chromosomes are in black; autosomes in gray.

which result in an abnormal distribution of the chromosomes. Sometimes a sperm is formed that contains both an X and a Y chromosome. When this happens, another sperm is formed that contains neither an X nor a Y. When a sperm of the XY type fertilizes an egg, the zygote formed has the normal number of autosomes plus two X and one Y. Such a zygote develops into a female. When a sperm without a sex chromosome fertilizes an egg, the zygote formed has the normal number of autosomes and only one X chromosome. Such a zygote develops into a male. So from these and similar observations involving irregular chromosome numbers, the hypothesis has been developed that sex is determined by the ratio between the X chromosomes and the autosomes. If there are two X chromosomes and the normal number of autosomes, a female is produced. If there is one X chromosome and the normal number of autosomes, a male is produced. (This genic-balance hypothesis does not hold for vertebrates, however, see Ch. 19.)

A male *Drosophila* without a Y chromosome is sterile. This suggests that the Y chromosome must be present for functional gametes to be formed in this species. Although some genes have been located on the Y chromosome in different *Drosophila* species, in general these genes are not alleles of genes on the X chromosomes. Therefore when considering the inheritance of traits controlled by genes on the X chromosome, described below, the Y chromosome generally may be considered to be a blank.

Sex Linkage

Characteristics due to genes located on the X chromosomes are said to be sex linked. Thomas Hunt Morgan, an American geneticist who pioneered the study of the genetics of *Dorsophila melanogaster,* first discovered sex linkage in 1910. A single white-eyed fly was found in his stocks of flies; the normal eye color is red. Morgan mated the white-eyed male with a virgin red-eyed female and carried the experiment to the F_2 generation. In the F_1 generation all flies had red eyes. In the F_2 generation three fourths of the individuals had red eyes and one fourth had white eyes. This seems like the characteristic monohybrid ratio, but Morgan noticed that all F_2 white-eyed individuals were males. By further breeding he was able to obtain white-eyed females. He then made the reciprocal cross, that is, he mated a virgin white-eyed female with a red-eyed male. In the F_2 generation one half of the females were red and one half were white, and one half of the males were red and one half were white.

Morgan reasoned that this type of inheritance could be explained by assuming that the alleles for eye color, red and white, were located on the X chromosomes. His explanations are shown in Figures 17.5 and 17.6.

Several hundred different sex-linked genes have been identified in *Drosophila,* all showing the same pattern of inheritance. Several sex-linked characteristics are known in humans, including red–green color blindness and hemophilia. In both these instances the genes that cause the defects are recessive to the genes for the normal condition. As in the case of white eyes in the fruit fly, a human male will show the defect if he has a single gene for it; the female, on the other hand, must have two of the recessive genes for the defect to appear. As a result, there are a great many more color-blind men than women. A distinctive characteristic of all sex-linked inheritance is the criss–cross type of inheritance shown in Figure 17.6. The males always receive their sex-linked recessive genes from their mothers. Heterozygous females are "carriers" of sex-linked recessive genes and do not express the trait.

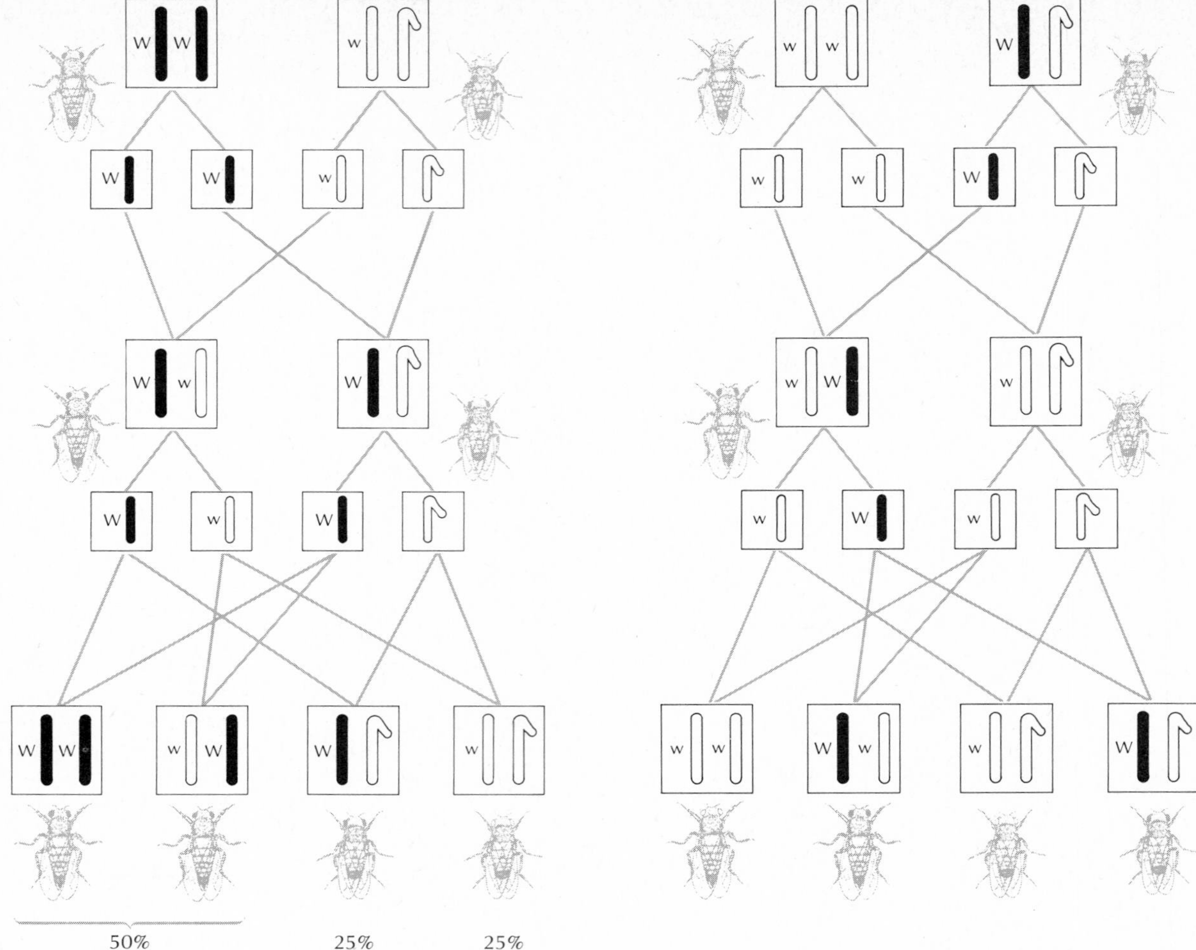

Figure 17.5 Cross of red-eyed female *Drosophila* and white-eyed male. F_1 inbred.

Figure 17.6 Cross of white-eyed female *Drosophila* and red-eyed male. F_1 inbred.

Linkage

The term "sex linkage" has been used above to denote those genes that are located on the X chromosomes. There is another concept that must be added to those developed after Mendel. He studied seven pairs of alleles in the pea and obtained independent assortments with all seven of them. Many more pairs of alleles have now been studied in the pea, and one does not always obtain an independent assortment when several pairs are studied together. Why? An independent assortment will occur only when the different gene pairs are located on different chromosome pairs. In an organism such as *Drosophila,* where several thousand genes have been studied, it is possible to have independent assortment occurring only when the gene pairs involved are on different chromosomes. The genes that are located on the same chromosome are said to be *linked;* they tend to stay together. Similarly, groups or blocks of characteristics tend to appear together.

The basic nature of linkage may be illustrated with a simple cross in the fruit fly. The normal or wild type fly has long wings (*V*) and a gray body (*B*). Vestigial wings are caused by a recessive gene (*v*) and black body by a recessive gene (*b*). If a homozygous gray–long fly is mated with a black–vestigial fly, the F_1 individuals are gray–long. If a heterozygous F_1 gray–long male is test crossed to a black–vestigial female, only two kinds of progeny are produced, black–vestigial and gray–long, in equal numbers. If these genes were located on different chromosomes, and thus assorted independently, one would obtain four kinds of offspring in equal numbers in such a test cross. If they were located on the same chromosome, one would obtain only the two kinds. This is shown in Figure 17.7.

Crossing Over

In the test cross just described, complete linkage is demonstrated. However, complete

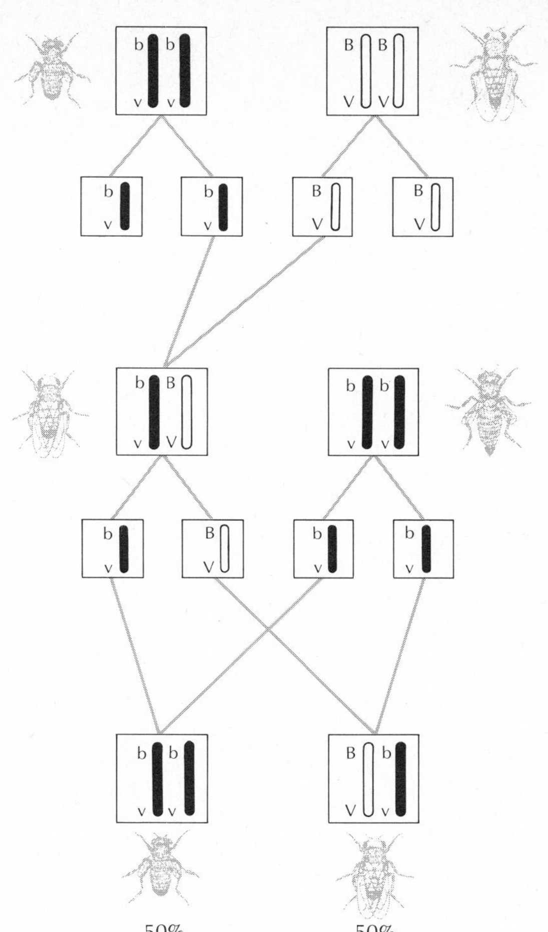

Figure 17.7 Complete linkage in *Drosophila;* cross of black–vestigial fly with gray–long. F_1 male test crossed to black–vestigial female.

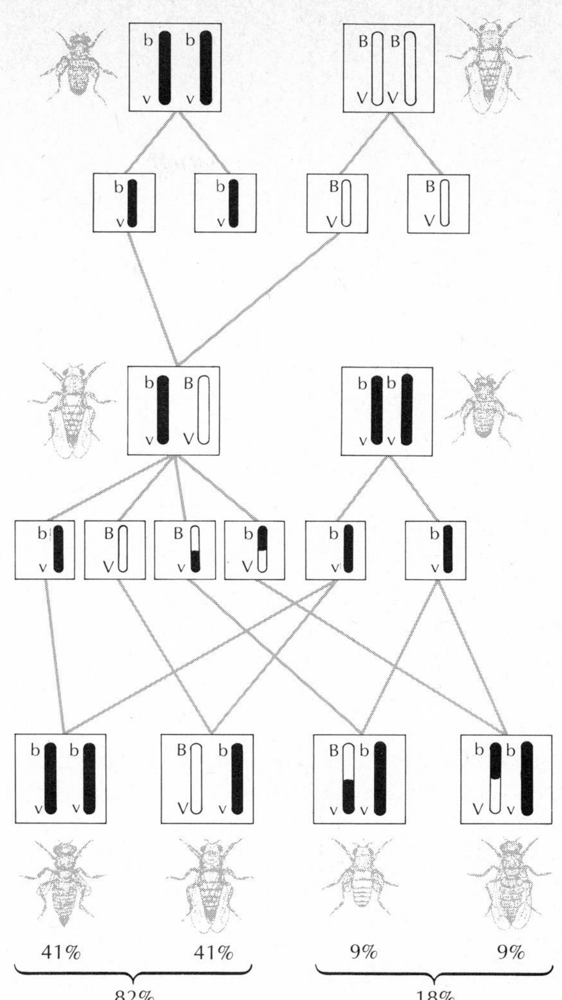

Figure 17.8 Incomplete linkage or crossing over in *Drosophila* cross of black–vestigial fly with gray–long. F_1 female test crossed to black–vestigial male.

linkage is found only rarely. It regularly occurs in the male fruit fly and a few other organisms. Just why this is true is not clear, but it appears that there is something about the relations of the chromosomes in the male *Drosophila* during meiosis that prevents *crossing over.* This phenomenon involves an equal exchange of genes between homologous chromosomes. When crossover occurs, the genetic results are different. Thus if an F_1 female from the experiment above is test-crossed to a black–vestigial male, four kinds of offspring are produced instead of two. This is shown in Figure 17.8.

To produce four kinds of offspring, the F_1 female has to form four kinds of gametes. The two crossover gametes result from the exchange of segments of the homologous chromosomes that contain the genes for body color and wing length. During meiosis two chromatids become twisted about each other, and when they separate breaks often occur, resulting in exchange of segments. Although crossing over may be explained as the exchange of segments of homologous chromosomes, the process is very complex and not well understood.

In this test cross using the F_1 female, the four kinds of offspring are not produced in equal numbers, as would be the case in independent assortment, but 41 percent are gray–long, 41 percent are black–vestigial, 9 percent are black–long, and 9 percent are gray–vestigial. Thus in 82 percent, linkage was complete (parental types) and in 18 percent crossover occurred (crossover types). Crossing over, then, provides for variations or new types, just as independent assortment does. A diagram of crossing over is shown in Figure 17.9. The parental types are easily recognized since their frequency is always greater than the crossover types.

CHROMOSOME MAPPING IN DIPLOID ORGANISMS

The crossover experiment involving the black locus and the vestigial locus has been repeated

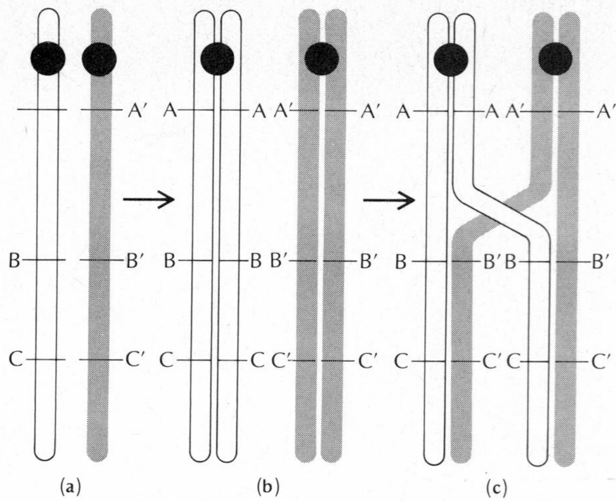

Figure 17.9 Diagram of crossing over. **(a)** Pair of *homologous* chromosomes. **(b)** Tetrad formation. **(c)** Crossing over between adjacent chromatids. Symbols represent hypothetical gene loci.

many times by many workers, and the result is always the same. When other linked genes are studied, crossover values varying from nearly 0 to nearly 50 percent are obtained. The crossover values obtained between a particular gene and a series of other linked genes are all different. In explaining this, the assumption is made that genes that are close together rarely cross over but that genes farther apart will cross over with greater frequency. It is assumed also that the genes are in a linear order on the chromosomes; no other arrangement would be compatible with the crossover results.

Using crossover data, geneticists have been able to construct chromosome maps. In chromosome mapping one percent of crossover is defined as one unit of map distance and is termed a *morgan,* after Thomas Hunt Morgan. Chromosome maps indicate the sequence of the genes as well as the relative distances apart. To illustrate, if three linked genes *A, B,* and *C* show 5 morgans between *A* and *B*, and 3 morgans between *B* and *C*, then the number of morgans between *A* and *C* will determine the order. If it is 8 morgans, then the order is *ABC*; if it is 2 morgans, the order is *ACB*. Fairly complete chromosome maps have been made for *Drosophila* and corn. Less complete maps have been made for a number of other organisms.

SUMMARY

This introductory chapter on genetics has emphasized how genetic characteristics are passed on from one generation to another: transmission genetics. Additionally the concepts involved in chromosome mapping were discussed. Genetic analysis, as seen by chromosome mapping, is a powerful tool.

Historically, genetics is a young discipline of biology. It should be apparent that the science of genetics lies at the very heart of biology. These basic concepts of genetics are pertinent to a good understanding of many areas of biology: development, evolution, population dynamics, ecology, and behavior.

Many students find genetics a difficult subject. It is necessary to understand the basic Mendelian postulates and the methods of working problems in Mendelian genetics. Then, as each variation of the theme of Mendelian genetics is presented, the student must be able to relate these topics to the fundamental principles. Further aspects of genetics are presented in many of the remaining chapters. Indeed, it has been said that genetics is the axial thread of biology.

SUGGESTIONS FOR FURTHER READING

Burns, G. W. *The Science of Genetics.* New York: Macmillan (1972).

Gardner, E. J. *Principles of Genetics,* 5th ed. New York: Wiley (1975).

Goodenough, U., and R. P. Levine. *Genetics.* New York: Holt, Rinehart and Winston (1974).

Levine, R. P. *Genetics.* New York: Holt, Rinehart and Winston (1972).

Merrell, D. J. *An Introduction to Genetics.* New York: W. W. Norton (1975).

Srb, A. M., R. D. Owen, and R. S. Edgar. *General Genetics.* San Francisco: W. H. Freeman (1965).

Strickberger, M. W. *Genetics,* 2nd ed. New York: Macmillan (1976).

From Scientific American

Benzer, S. "Genetic Dissection of Behavior" (Dec. 1973).

Benzer, S. "The Fine Structure of the Gene" (Jan. 1962).

Dobzhansky, T. "Genetics" (Sept. 1950).

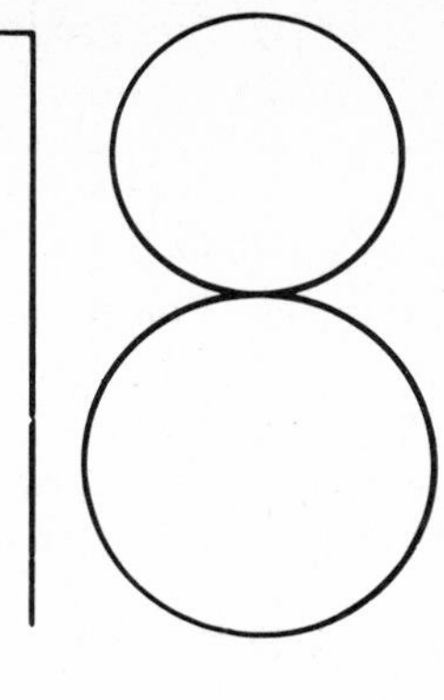

Genetics and Developmental Biology

Chapters 15 and 16 (Reproduction and Development) explored the maturation of sperm and eggs, fertilization, germ layer formation, organ formation, and, finally, the acquisition of the form of the organism. The mature sperm and egg, the genetic bridges between generations, are highly differentiated cells, specialized for their specific functions in fertilization. Zygote formation and restoration of the diploid number signal the beginning of a train of events which must be controlled by a fastidious set of mechanisms if a healthy mature organism is to be the result. The zygote, a single cell, contains all the genetic information required for the program of development, that is, all development ultimately is specified by genes, whether they be nuclear, cytoplasmic, or organellar genes. Since the mechanism of gene action is through specification of proteins, all cells derived from a zygote would be the same if differential gene activity did not occur among cells. The potential of a cell is related to the structure of that cell at a particular time and the protein population (including enzymes) which it contains. A human erythrocyte contains about 95 percent of one protein, hemoglobin, while the cells of the iris of the eye are characterized by specific pigments. Likewise, in muscle cells the genes for myoglobin, actin, and myosin are turned on while in fibroblasts the gene for collagen is expressed. Erythrocytes and iris cells are traceable back to their origins from the zygote and its genetic endowment. As development and differentiation proceeded, the iris cells and erythrocytes had some genes turned on and other (most) genes turned off. The levels of regulation of genetic and developmental control may be identified at three general points on a hierarchy of complexity:

1. *Intracellular,* within the cell
2. *Intercellular,* among cells
3. *Environmental,* through factors coming from the surroundings of the developing organism

INTRACELLULAR FACTORS IN DEVELOPMENT

The modern theory of animal development at the cellular level relies on two principal components: the nucleus and the cytoplasm. The general theme of transport of information (mRNA) and materials (tRNA, rRNA) as well as regulation of transcription according to the operon model were detailed in Chapter 6. Various cytoplasmic events and factors (accumulation of maternal ribosomes in the egg, the role of the grey crescent and nuclear transplantation) were discussed in Chapter 16.

Intracellular Processes and Their Controls

During interphase a eukaryotic cell does not show definite chromosomes since the chromosomal material or *chromatin* is in a diffuse or relaxed state (Ch. 7). This chromatin can be isolated and has been shown to contain DNA, RNA, and proteins. The proteins are of two major classes, basic histones and nonhistones. If these proteins are removed (with proteolytic enzymes, for example), the deproteïnized chromatin is more active in RNA synthesis than is the intact, "native" chromatin. Presumably deproteinization unmasks the DNA so that more of it is free to be transcribed by RNA polymerase. For example, using rat liver chromatin, a deproteinized sample provided with the enzyme RNA polymerase and the four ribonucleoside triphosphates (ATP, UTP, GTP, CTP) will produce five times more RNA than the native chromatin.

When "native" chromatin is isolated from different tissues and used as a template for transcription *in vitro,* each chromatin preparation produces RNA that is similar to the RNA produced by the corresponding tissue *in vivo.* This fact suggests that regulatory elements reside in the structures of the chromosomes (native chromatin) themselves. This conclusion came from experiments in which RNA produced *in vivo* in a given tissue competed with RNA transcribed *in vitro* for hybridization (Ch. 20) with DNA. Competition was greater when the two RNAs came from the same tissue type as compared to different tissues.

Since many more genes exist than types of histones, a one (or few) histone to one gene ratio is probably an untenable hypothesis. Some workers have suggested that chromosomal RNA (RNA found in chromosomes, Ch. 7) may be involved with histones in repression of chromosomal expression. Some evidence exists that chromosomal RNA is associated with the histones. If the chromosomal RNA recognized specific sites on the DNA by base pairing and the basic groups (positively charged) on the histones were attracted to the negative charges of the phosphates in the sugar–phosphate backbone, then specific interactions for repression and derepression of genes could be accounted for. On the other hand, the histones seem to be equally spaced over active and inactive regions of the chromosomes, a point consistent with a role as structural components of chromosomes rather than as genetic regulators. Recent evidence from experiments involving the cleavage of chromatin with certain nucleases (DNA hydrolyzing enzymes) suggests a repeating unit of 200 nucleotide pairs that is associated with six histone molecules. The extended length of 200 nucleotide pairs would be 68 nm, but the actual length is only 10 nm. Thus the complex of histones and DNA is wound into a ball. Chromatin fibers must be able to coil or fold extensively (as in metaphase chromosomes) and accordingly must be flexible. Electron micrographs of chromatin fibers show a structure that resemble beads on a string. This suggests a flexibly jointed chromatin chain. This model has been called the "kinky helix."

In the search for regulator proteins attention has been turned to nonhistone proteins, which are simply defined as chromosomal proteins other than the histones. Unlike the histones, there is no constant relationship between the amount of DNA and the amount of nonhistones in the chromatin. Like the histones, these proteins are synthesized outside of the nucleus. The histones are transported into the nucleus immediately after synthesis, but some of the nonhistones are transported immediately into the nucleus where they associate with DNA while others are transported more slowly. Thus the differential rate of transport of nonhistone proteins into the nucleus possibly could contribute to their specific regulatory function. Whereas histones and DNA occur in about a 1:1 constant ratio, at least part of the nonhistone proteins vary greatly.

While the histones fall into only five clearly defined types, the nonhistones show a great structural and functional diversity. The number of nonhistones is not known, but it is certainly large. They range in molecular weight from under 10,000 to over 150,000 daltons and include a variety of enzymes such as DNA and RNA polymerases, proteases, and DNA modification enzymes (addition and removal of acetate, methyl, and phosphate groups). Further, different tissues possess a different complement of nonhistones and these proteins recognize specific nucleotide sequences in DNA. Such specific interactions between regulatory proteins and DNA is what would be expected.

The mechanism of action of nonhistones in gene regulation may involve the *inhibition of the inhibition* of RNA synthesis (transcription) by the histones. R. Stewart Gilmour and John Paul isolated chromatin from rabbit thymus and bone marrow and dissociated the DNA, histone, and nonhistone components. Then these components were put back together; when all components were derived from the thymus, the transcribed RNA hybridized normally to thymus DNA. When the chromatin was reconstituted from bone marrow DNA and histone and thymus nonhistones, the transcribed RNA hybridized more efficiently to thymus DNA than to marrow DNA. Likewise, chromatin reconstituted from thymus DNA and histone and marrow nonhistone transcribed RNA that was bone marrow RNA: it hybridized more efficiently to marrow DNA than to thymus DNA. These experiments indicate that the presence of tissue specific nonhistones determines which genes will be transcribed in various tissues.

Heterochromatin and Euchromatin in Chromsomes

By staining technique chromatin may be divided into *heterochromatin* and *euchromatin.* Heterochromatin is highly compacted and does not disperse into highly diffuse interphase material as does euchromatin. The heterochromatic regions that maintain their compactness all the time are called *constituitive heterochromatin* and generally are considered to be inert genetically. Other chromosomal regions may be condensed (heterochromatic) only in certain situations. Such *facultative* or *functional heterochromatin* is thus sometimes inert and sometimes active (diffuse or relaxed); therefore, facultative heterochromatin often is thought to be associated with gene regulation. Constitutive heterochromatin is found as whole late-replicating chromosomes of Barr bodies (sex chromatin, Ch. 17) and in the centromeres which are known to serve as sites of attachment for mitotic spindle proteins.

The differences afforded by the heterochromatic (constitutive and facultative) and euchromatic regions suggest possibilities for differential gene activity and the basis of cellular differentiation. The ratio of histone to DNA is the same in heterochromatin and euchromatin, but in heterochromatin there are considerably fewer nonhistones than in active extended euchromatin. Here again, the presence of nonhistones is correlated with gene expression or active RNA synthesis.

Further Examples of Gene Regulation in Eukaryotes

PUFFS AND POLYTENE CHROMSOMES

The giant polytene chromosomes of *Drosophila* and other dipteran flies have been used in important work on differential gene activity. These polytene chromosomes have distinctive banding patterns, and the experienced cytologist can recognize even parts of chromosomes on the basis of the bands. At certain positions, puffs replace the sharp bands. Studies of the puffing pattern have revealed that the locations of puffs are different in different tissues at any one time, are different in the same tissues at different stages of development, and are the same in all cells of a given tissue type at any one time.

Wolfgang Beerman studied the puffs in salivary glands of two species of a dipteran fly, *Chironomus.* In one species, *C. pallidivittatus,* four cells of the salivary gland produce a granular secretion while the rest of the cells produce a clear secretion. One chromosome in the granule-secreting cells has a puff not present in the cells that produce the clear secretion. In another species, *C. tentans,* the four cells mentioned above are the same as all other cells and no differences in the puffing pattern exists. Hybrids between the two species possess the granule-secreting cells, but the amount of secreted granules is reduced. The hybrid chromosomes in the four cells have the puff inherited from *C. pallidivittatus* (Fig. 18.1). Thus this work correlates a gene product (the granular secretion) with the presence of a specific puff.

Studies with radioactive nucleobases have shown that the puffs are the sites of rapid RNA synthesis and thus of very active genes. Puffing can be prevented by actinomycin D, an antibiotic known to inhibit the synthesis of RNA. In 1973 evidence was obtained that RNA synthesized in a particular puff is transported to the cytoplasm (and therefore presumably functions as mRNA) and is translated into proteins. Other recent work in attempting to correlate puffing patterns with specific protein products has been carried out in the laboratory of Herschel Mitchell at the California Institute of Technology. Salivary

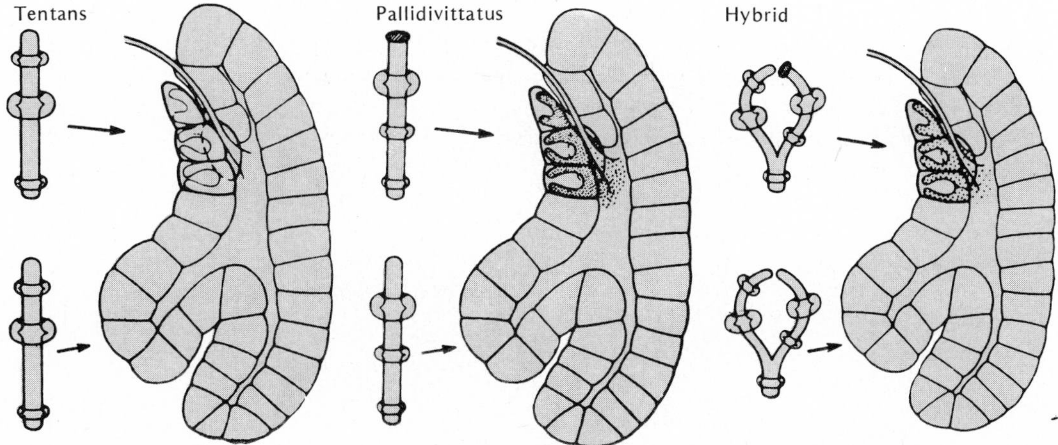

Figure 18.1 Pattern of chromosome puffs in the functionally different portions of the salivary gland in *Chironomus tentans, C. pallidivittatus*, and their hybrids. (From Beermann, *American Zoologist* **3**, 1963, with permission.)

glands of *Drosophila* larvae and early pupae were dissected out and incubated in a buffer containing ^{35}S-methionine. After labeling with this radioactive amino acid, the proteins of the glands were separated by electrophoresis. The changes in the kinds of proteins at different times between the larval and pupal stages could be correlated with the changes in patterns of chromosome puffs at the corresponding times.

Heat shock of the salivary glands changes the puffing pattern and likewise changes the kinds of proteins that are synthesized. Furthermore, modifications of puffs can be induced by the hormone ecdysone, and the protein patterns of ecdysone-treated salivary glands is changed also.

Gene Amplification

An equal increase of all genes in a genome, such as occur in polyploidy (Ch. 42), is not likely to alter the balance of gene expression greatly. Gene amplification refers to an increase in the number of one or more genes in a genome without a corresponding increase in other genes. Proof of specific gene amplification was provided in 1968 with the isolation of ribosomal genes by Brown and coworkers. In the South African clawed toad, *Xenopus laevis,* each haploid genome contains about 450 adjacent sets of ribosomal genes (Fig. 18.2). Early in oogenesis these gene sets are increased selectively so that each mature egg contains about 2 million ribosomal gene sets. The extra sets of ribosomal genes are accommodated in 1000–1500 extrachromosomal nucleoli (Ch. 16). Amplification seems to be peculiar to rRNA genes where the function is the production of maternal ribosomes that sustain early development of the animal (Ch. 16). In three well-studied cases, where a specialized cell produces a majority of one kind of protein, no gene amplification of the respective genes occurs:

1. Posterior silkgland of the silkmoth—silk fibroin
2. Duck erythrocytes—hemoglobin
3. Chick oviduct cells—ovalbumin

Nuclear Transplantation

The constancy of DNA in most cells of a given organism was first established in 1948 (haploid gametes and euploid liver cells have concomitant or corresponding decreases or increases in nuclear DNA content). This deduction is based on biochemical and cytological techniques and has been confirmed by nuclear transplantation experiments. J. B. Gurdon of Oxford University transplanted nuclei from differentiated intestinal cells of the clawed toad, *Xenopus laevis,* into eggs from which the original nucleus had been removed. (The technique of nuclear transplantation was first carried out successfully in *Amoeba* in 1939 and has been successful with the leopard frog, *Rana pipiens,* the Mexican salamander, the fruit fly, and the honey bee.) Using the O-nu (anucleolate) marker, Gurdon was able to show that eggs containing transplanted nuclei grew into adult frogs wholly derived from the donor nucleus from differenti-

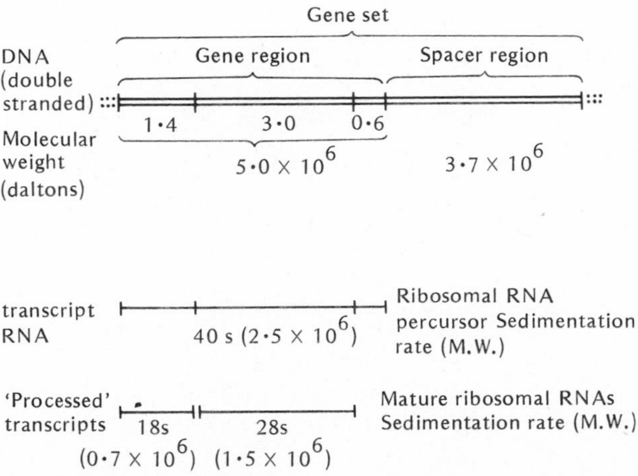

Figure 18.2 Diagram of a ribosomal gene set. The gene set includes DNA which codes for 18s and 28s rRNA as well as an untranscribed "spacer" region. (Redrawn from Gurdon, *The Control of Gene Expression in Animal Development,* Harvard Univ. Press, 1974.)

ated cells. An excellent review of gene expression in animal development, with emphasis on nuclear transplantation experiments, is given in Gurdon (see Suggestions for Further Reading).

INTERCELLULAR FACTORS IN DEVELOPMENT

As Chapters 15 and 16 (Reproduction and Development) clearly show, developmental processes are affected by specific signals coming from other cells; this section gives some examples.

Clonal Myogenesis

Embryonic muscle starts out as single undifferentiated cells, *myoblasts.* In skeletal muscle the myoblasts fuse to form myotubes and lose their cellular identities (a syncytium is formed). Myotubes ultimately show the typical cross striations characteristic of mature muscles (see Ch. 8). Because of the sequence of cellular interactions that produce the syncytium, some experiments were designed to determine if a single isolated myoblast can give rise to a colony of differentiated skeletal muscle tissue.

Muscle from the leg of an embryonic chick was dissociated into individual cells, and single cells were transferred into isolated culture chambers. Cell division over a four-day period produced clusters of cells, about four dozen in number. At about this time many of the cells would fuse into myotubes and there would be a reduction in the rate of cell division. Continued division was confined to isolated unfused cells. Successful development of differentiated striated fibers was, however, meager, without an added condition being taken into account.

Normal muscle not only has muscle fibers, but it also has fibroblasts, connective tissue cells. Fibroblasts are secretors of collagen, the important fibrous protein of connective tissue. I. Konigsberg found that the most success in getting truly differentiated striated muscles to grow in tissue culture involved putting isolated clones of myoblasts into old culture medium which had supported the growth of fibroblasts. This *conditioned* old medium was carefully filtered to remove the old cells before the clones of myoblasts were cultured in it. Therefore the fibroblasts had left something in the medium that signaled the myoblasts to differentiate into striated muscles. That collagen had been left in the medium by the fibroblasts seemed to be a reasonable hypothesis. Konigsberg finally obtained good development of myoblasts to the level of striated muscle fibers by coating the culture dishes with pure collagen. When this was done, fresh and unconditioned medium served to support differentiation of striated muscle from a single cell.

The inferences from these experiments for conditions in normal development are obvious. However, a definitive identification that collagen is the necessary factor for normal *in vivo* differentiation of muscle has not been established.

Transfilter Technique of Tissue Culture

This experiment involves the culture of pancreatic mesenchyme, a loosely organized mesoderm, on one side of a thin nitrocellulose membrane filter with micropores of a size known to exclude the passage of cells. Pancreatic endoderm is placed on the other side of the filter (Fig. 18.3). In this arrangement molecules from the mesenchyme migrate through the pores, and the endoderm is induced to form the characteristic secretory clusters of pancreatic cells, *acini,* that would form in normal pancreatic development.

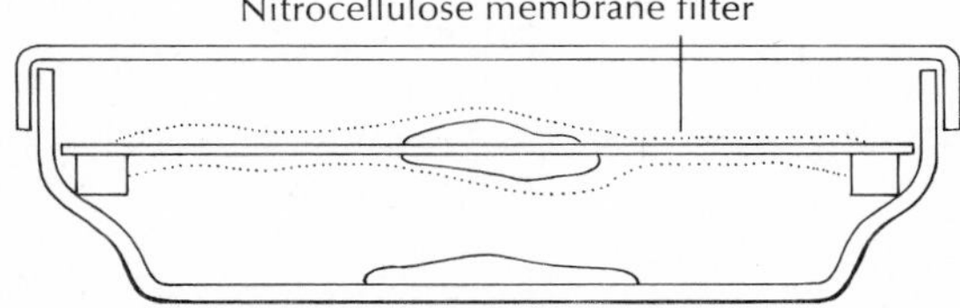

Figure 18.3 Transfilter technique of tissue culture. In covered dish, cellulose filter with pores of known size is supported on Plexiglass ring. Mesenchyme is placed on one side of the filter and epithelium on the other. Both tissues are covered by a plasma clot (dotted). Drying is prevented by providing moisture in the bottom of the dish. Differentiation of epithelium under such conditions occurs in predictable patterns.

In the absence of the mesenchyme on the other side of the filter, no such differentiation occurs. An amino acid (*proline*) is found in its hydroxylated form (*hydroxyproline*) in molecules of collagen. When the mesenchyme on one side of the filter is labeled with radioactive proline, the label appears later on the opposite side of the filter in a pattern characteristic of the collagen matrix, which lies around the differentiated acini. If the endoderm is labeled with the radioactive proline, the pattern of radioactivity does not correspond to the pattern of the collagen. This experiment produces strong evidence that the soluble tropocollagen, a precursor of collagen, is synthesized in the fibroblasts on the mesenchyme side, moves across the filter, and is aggregated under the influence of the epithelial cells of the acini into insoluble collagen fibers.

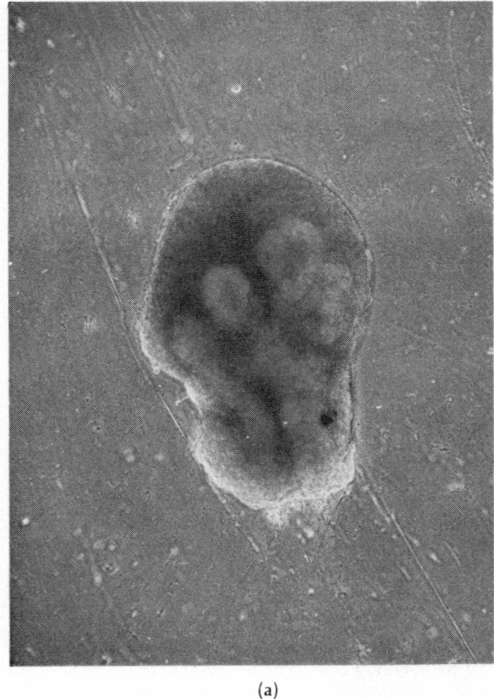

(a)

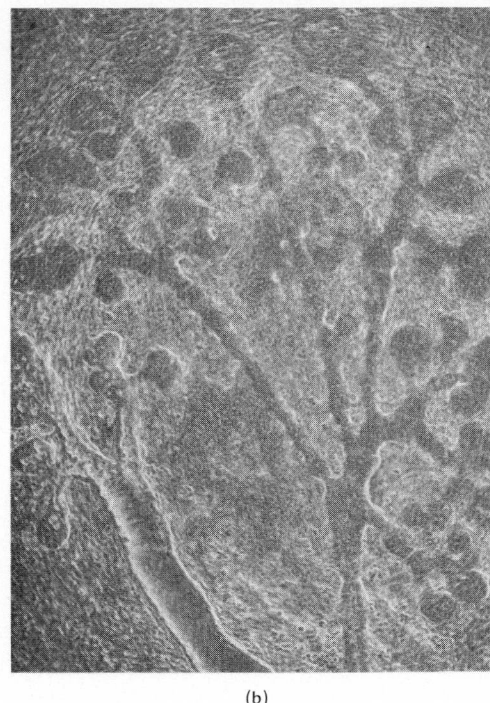

(b)

Figure 18.4 Salivary gland tissue culture. **(a)** Onset of culture; **(b)** 6 days later. Note differentiation of duct system. (Photos represent frames selected from a time-lapse cinematographic film, "Development of Mouse Submandibular Salivary Gland Rudiments in Organotypic Culture, W. D. Morgan and C. J. Dawe, Laboratory of Pathology, National Cancer Institute, Bethesda, Maryland.)

Here, then, is a form of organization of molecules that involves two kinds of tissues. This organization does not occur with either tissue alone. Figure 18.4 illustrates the differentiation of another organ composed of epithelium and mesenchyme, whose differentiation is dependent on interactions of the tissues. This is the salivary gland of the mouse.

Molecular Changes During Development

The proteins of an adult organism are not necessarily identical to the proteins that perform similar functions during earlier stages of development. For example, the enzyme lactase in infants and adults is controlled by different genes, and this is the basis for the galactosemic child outgrowing its intolerance of milk sugar (lactose). Another example is fetal and adult hemoglobins, these hemoglobins differ in their amino acid sequences of one pair of chains. The alpha chains of fetal and adult hemoglobins are identical but fetal hemoglobin contains gamma chains instead of beta chains. The gamma and beta chains are governed by separate genes.

There are also multiple forms of enzymes called *isozymes.* For example, lactate dehydrogenase (LDH) contains four polypeptide subunits that may be either the A or B type. Therefore five forms of the enzyme may exist: AAAA, AAAB, AABB, ABBB, and BBBB. When several kinds of tissues of one adult animal are examined, they are seen to vary in their pattern of isozymes. Also, when a single tissue is analyzed through several stages of development, the patterns or relative proportions of the isozymes shift.

The study of isozymes is of increasing importance since some developmental abnormalities can be identified with specific mutations, and shifts in enzymatic patterns associated with these mutations have been shown. Also the isozyme pattern of the blood changes with certain physical traumas; for example, the LDH isozyme pattern of the blood may be used to detect myocardial infarction (heart attack) in hospital laboratories.

Tissue Affinities

Johannes Holtfreter showed that the form of the neural tube could be influenced by orientations of tissue associated with the neural tissue. Figure 18.5 shows how isolated neural tube, lying on a somite bed, thickens on the side adjacent to the somites and not on the side away from the somites. With the development of the nervous system as an introduction, consider the great variety of interrelationships that exist in a developing embryo and how they may influence each other. The pancreas, for instance, be-

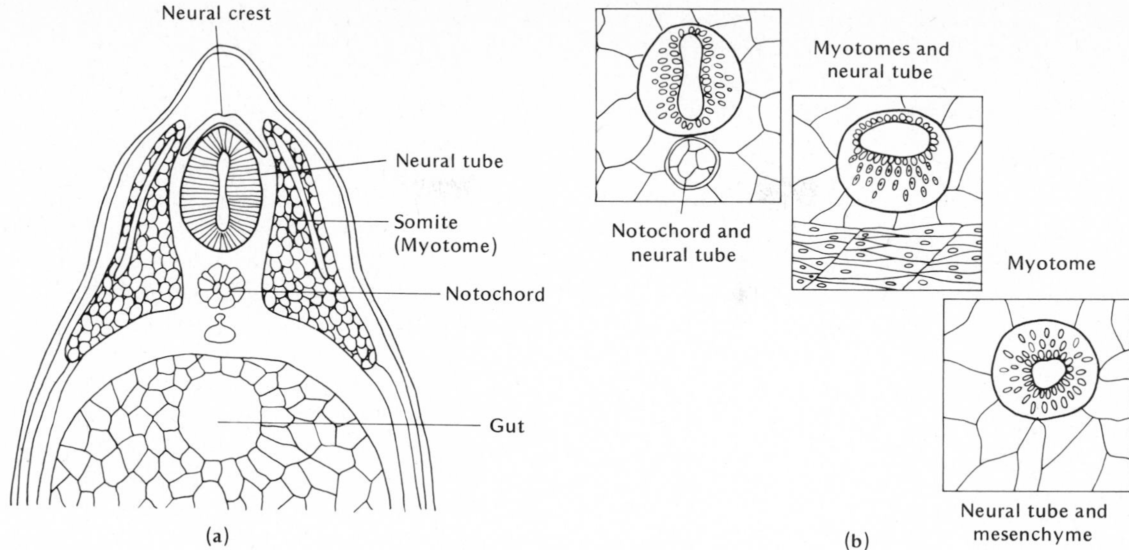

Figure 18.5 Influence of surrounding tissues on differentiation of neural tube. **(a)** Cross section through dorsal trunk of normal frog tadpole. **(b)** Effect of specific tissues on differentiation of neural tube under experimental conditions.

gins as a pair of precisely located tubular outgrowths from the developing embryonic gut, and the filter separation experiments discussed previously show that the pancreas will develop normally only if there is an association between the endodermal duct tissue and the mesoderm into which it penetrates. Pouches grow out from the pharynx into mesoderm and form various organs, such as the thymus and parathyroids. A normal adult mammalian kidney is produced by an outgrowth of a mesodermal duct from another and transient (mesonephric) kidney in the embryo and only develops normal nephrons if the outgrowing duct penetrates into a small region of mesoderm that responds to the inductive influences of the duct. These are but a few common examples where tissues establish spatial relationships with each other. In order to understand some of the spatial relationships, some developmental biologists have combined various tissues *in vitro,* have observed the affinities of these tissues for each other, and have noted whether one tissue tends to surround another or whether it is surrounded by another, and so forth. These studies have established hierarchies of differentiated cells that show selective associations among themselves.

One procedure is to dissociate cells of selected embryonic tissues by gentle enzymatic digestion or by using a calcium and/or magnesiumfree medium. Cells from two kinds of disaggregated tissue, when grown together in culture, will sort out with one kind of tissue covered by the other tissue. Thus when chick heart and chick cartilage cells are grown together, the cartilage cells sort to the interior, the heart cells sort to the exterior. Heart and liver cell combinations result in the heart cells to the interior and the liver cells to the exterior. From these two experiments, using only two tissues in each, it would be reasonable to predict that there is a hierarchy with cartilage cells having stronger association forces for themselves than do heart cells, and, in turn, heart cells have stronger association forces than do liver cells. A combination of the three kinds of disaggregates confirms this: The liver envelops the heart that envelops the cartilage (Fig. 18.6).

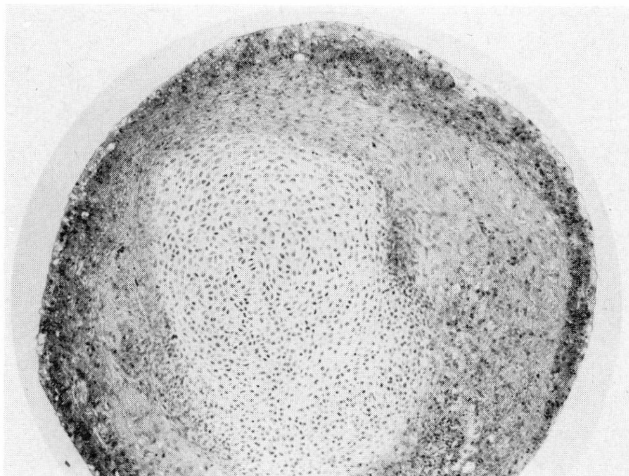

Figure 18.6 Reorganization of three tissues. Liver, heart ventricle, and precartilage tissues were apposed linearly *in vitro.* They have reorganized to produce concentric arrangement: liver covering heart ventricle covering precartilage. (M. S. Steinberg, from *Conference on Cellular Dynamics,* L. D. Peachey (Ed.), New York Academy of Sciences, 1968, with permission.)

The ramifications of such studies raise questions in the realm of biochemistry and biophysics. For instance, are the differences between cells only differences in numbers of binding sites in different tissues or are there qualitative differences as well? Are there differences in the developmental period in which the tissues change their ranks in the hierarchy of selective associations? Questions such as these are being tackled by developmental biologists, biochemists, and biophysicists.

ENVIRONMENTAL FACTORS IN DEVELOPMENT

The environmental factors which affect developmental processes are many and the course of animal evolution has produced a remarkable set of mechanisms for protecting the embryo, particularly in the transition of vertebrates from an aqueous to a terrestrial habitat. For the concerted developmental processes to occur properly, extreme variations of environmental factors must be minimized. The pinnacle of protective barriers is seen in the mammalian uterus and the amniotic cavity. If the protective barriers fail, gross abnormalities may result.

Proper development depends on proper nutrition. If essential amino acids or fatty acids, calories, vitamins, or gases are not supplied in the proper range of concentrations, damage to the developing embryo will be permanent. Likewise, a hormonal imbalance (Ch. 12) will seriously affect developmental processes.

If the membranes of an amphibian blastula are removed and the embryo is grown in a hypertonic solution, normal gastrulation does not occur. The prospective chordamesoderm and other mesodermal and endodermal cells do not invaginate, but grow outward instead. This *exogastrulation* is shown in Figure 18.7. Under these conditions, endodermal and mesodermal structures develop to a great extent [Fig. 18.7 (c)], but the ectoderm fails to form neural structures. This is a dramatic example of a physical factor which affects the relationships between tissues, thereby inducing the abnormality.

Another example of an environmental factor affecting development is the rubella (German measles) virus. Rubella is a relatively mild disease for children or adults. However, if a pregnant woman develops rubella during the first 2–3 months of pregnancy, the rubella virus can cross the placenta and damage seriously the developing embryo with resultant abnormalities such as eye and heart defects. On the other hand, more severe viruses for children and adults such as mumps or measles are not teratogenic (literally, monster producing).

A third category of an outside factor which may affect development is drugs. Some years ago in Germany many babies with abnormally short and malformed arms and legs were noted to have been born of mothers who had used a mild tranquilizing drug, *thalidomide.* The mother need not have used the drug for long. But if

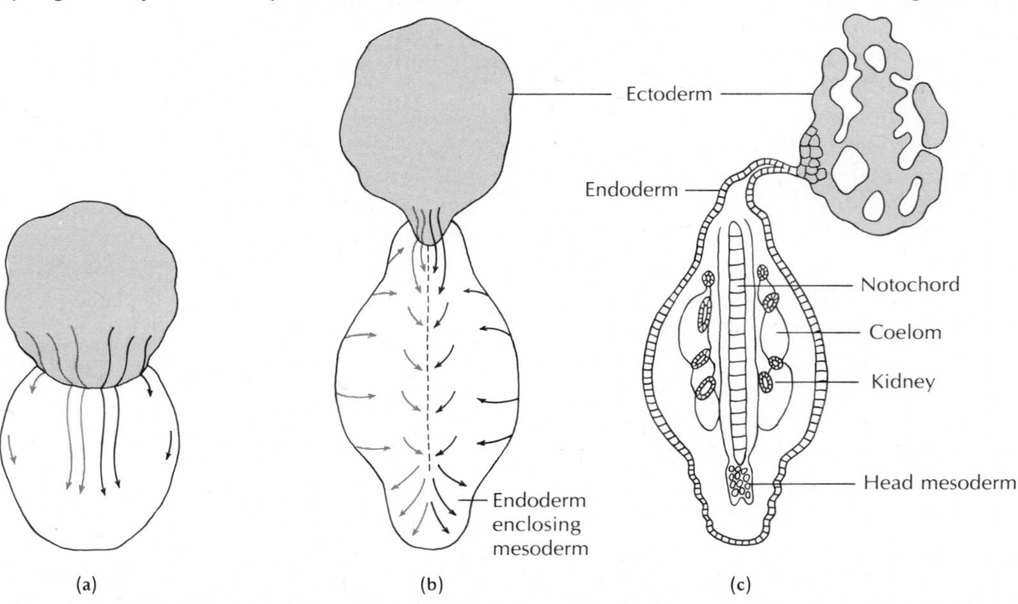

Figure 18.7 Exogastrulation in amphibian. **(a), (b)** Two stages in directions of regional migrations indicating movement of mesoderm and endoderm away from ectoderm rather than to interior as in normal gastrulation. **(c)** Diagrammatic longitudinal section showing differentiation of isolated mesoderm and endoderm and lack of differentiation of ectoderm. (After Holtfreter, *Biol. Zbl.,* **53,** 1933.)

she used it during a critical period of her pregnancy, it affected the growth of the infant's limbs. In this case the drug was removed from the market, but society is left with many deformed individuals who are dependent on others for their care.

SUMMARY

This chapter completes the fundamental statement of genetics and development (Ch. 15–17). In this chapter we have seen some special topics of genetics (genetic change, extrachromosomal inheritance, and biochemical genetics) and how the processes of development operate against the genetic framework of the organism. Parts I and III of this text are related closely in the first part of the hierarchy (Ch. 1) of biological organization: atom, molecule, organelle, cell, tissue, organ, organ system, and organism.

In animal development each differentiating cell has its own inner controls, but is subject to extrinsic factors from other cells and tissues as well as extrinsic environmental factors. One of the major tasks of developmental biologists is to identify the extracellular controls and determine the ways in which they influence the intracellular controls in development. The remaining chapter of Part III carries the themes established in previous chapters to systems which affect the human condition rather directly.

SUGGESTIONS FOR FURTHER READING

Bonner, J. *The Molecular Biology of Development.* New York: Oxford Univ. Press (1965).

Brachet, J. *Introduction to Molecular Embryology.* New York: Springer-Verlag (1973).

Cohen, S. C. "Mitochondria and Chloroplasts Revisited." *American Scientist,* **61**:4, 437 (1973).

Ebert, J. D., and I. M. Sussex. *Interacting Systems in Development.* New York: Holt, Rinehart and Winston (1970).

Ebert, J. D., A. G. Loewy, R. S. Miller, and H. A. Schneiderman. *Biology.* New York: Holt, Rinehart and Winston (1973).

Gurdon, J. B. *The Control of Gene Expression in Animal Development.* Cambridge, MA: Harvard Univ. Press (1974).

Hood, L. E., J. H. Wilson, and W. B. Wood. *Molecular Biology of Eucaryotic Cells.* Menlo Park, CA: W. A. Benjamin (1975).

Jacobson, M. *Developmental Neurobiology.* New York: Holt, Rinehart and Winston (1970).

Jinks, J. L. *Extrachromosomal Inheritance.* Englewood Cliffs, NJ: Prentice-Hall (1964).

Lash, J., and J. R. Whittaker (Eds). *Concepts of Development.* Stamford, CT: Sinauer Assoc. (1974).

Pasternak, C. A. *Biochemistry of Differentiation.* New York: Wiley (1970).

Saunders, J. W., Jr. *Patterns and Principles of Animal Development.* New York: Macmillan (1970).

Saxen, L., and J. Rapola. *Congenital Defects.* New York: Holt, Rinehart and Winston (1969).

Sussman, M. *Developmental Biology—Its Cellular and Molecular Foundations.* Englewood Cliffs, NJ: Prentice-Hall (1973).

Truman, D. E. S. *The Biochemistry of Cytodifferentiation.* New York: Halsted Press (1974).

From Scientific American

Beer, A. E., and R. E. Billingham. "The Embryo as a Transplant" (Apr. 1974).

Britten, R. J., and D. E. Kohne. "Repeated Segments of DNA" (Apr. 1970).

Brown, D. D. "The Isolation of Genes" (Aug. 1973).

Cohen, S. "The Manipulation of Genes" (July 1975).

Fischberg, M., and A. W. Blackler. "How Cells Specialize" (Sept. 1961).

Goodenough, U. W., and R. P. Levine. "Genetic Activity in Mitochondria and Chloroplasts" (Nov. 1970).

Guillery, R. W. "Visual Pathways in Albinos" (May 1974).

Gurdon, J. B. "Transplanted Nuclei and Cell Differentiation" (Dec. 1968).

Hadorn, E. "Transdetermination in Cells" (Nov. 1968).

Loewenstein, W. R. "Intercellular Communication" (May 1970).

Reisfeld, R. A., and D. Kahan. "Markers of Biological Individuality" (June 1972).

Sobel, H. M. "How Actinomycin Binds to DNA" (Aug. 1974).

Stein, G. S., J. S. Stein, and L. J. Kleinsmith. "Chromosomal Proteins and Gene Regulation" (Feb. 1975).

Taussig, H. B. "The Thalidomide Syndrome" (Aug. 1962).

Wessells, N. K., and W. J. Rutter. "Phases in Cell Differentiation" (Oct. 1971).

19 Human and Applied Genetics

How does genetics relate to the college student, the college graduate, or any involved citizen living in the latter part of the twentieth century? Human and agricultural genetics were set aside for this chapter to emphasize that genetics affects all people to a substantial degree, both directly and indirectly. A missing bit of a chromosome can mean the difference between mental deficiency, anatomical and physiological abnormalities, or a general inability to live a productive life—and a healthy, relatively happy existence. Much of the agricultural wealth of North America is the result of scientific breeding programs. Cattle, swine, wheat, corn, and many other animal stocks and plant crops are the results of genetic programs conducted by universities, governmental agencies, and private organizations or individuals. Human and agricultural genetics as "applied" or relevant genetics clearly demonstrate how this subject relates to all citizens of the twentieth century. Furthermore, these topics provide excellent review and teaching exercises for the genetics discussed up to this point. The fact that genetics pervades many areas of biology is emphasized by previous discussions of heredity in earlier chapters. More genetics is considered in conjunction with the topics of evolution, behavior, and ecology in succeeding chapters.

HUMAN INHERITANCE

Human Chromosomes

The normal diploid chromosome number of humans is 46. For many years this number was thought to be 48, but in 1956 workers who used cultured fibroblasts of human embryonic lung tissues found that the earlier reports had been in error. These unexpected findings were quickly confirmed by other investigators. This is a good example of how the "facts" sometimes change rapidly and how the workers in the field must be receptive to new data and new ideas.

As discussed in Chapter 7, the karyotype of the human female shows 23 homologous pairs of chromosomes; the male has 22 pairs plus an X and a Y chromsome. (A normal karyotype of human chromosomes is illustrated in Figure 7.6.)

The human is usually not thought of as a good organism for the study of genetics—the life cycle is long, controlled matings are not made, and the number of offspring is rather small for statistical treatments. Such organisms should be well studied biologically, or they should be of "simple" structure and function. Nevertheless in the past few years there has been an increase in appreciation of humans as organisms for genetic study. Medical histories and research have provided much knowledge about the anatomy and physiology of humans.

Chromosome Abnormalities

Two syndromes will illustrate X-chromosome abnormalities that are fairly well understood in humans. The *Turner's syndrome* individual has only one sex chromosome, an X. Such individuals are said to have an X0-sex chromosome constitution; they have only 45 chromosomes. The Turner's syndrome individual has external female genitalia, but the breasts are underdeveloped, the ovaries are small fibrous streaks, and the uterus is small. Thus the Turner's syndrome individual is a *phenotypic female.* The *Klinefelter's syndrome* results from an extra X chromosome in a regular male chromosomal consitution. Thus the Klinefelter's syndrome individual is XXY and has 47 chromosomes. In the

Klinefelter's individual the external genitalia are male, but the testes are small and the body hair is sparse. Although Klinefelter's individuals are *phenotypic males,* most have a femalelike breast development. Both Turner's and Klinefelter's individuals are sterile. These two syndromes indicate that the loss or gain of chromosomes has great effects on the development and well-being of the individual. Figure 19.1 shows how abnormal gametes may arise and lead to the two syndromes.

The Turner's and Klinefelter's syndromes are related to the presence of *sex chromatin* in the interphase nuclei. In 1949 Murray L. Barr discovered that the nuclei of most females have a deeply staining body attached to the inside of the nuclear membrane. The nuclei of most males lack this structure, which is known as the Barr body, or sex chromatin. This sex chromatin is thought to represent an inactive X chromosome. Normal females have one of the two X chromosomes inactivated and are sex chromatin positive. On the other hand normal males have only one X chromosome, which is not inactivated, and the male is sex chromatin negative. Turner's individuals have only one X chromosome and are sex chromatin negative; Klinefelter's individuals have two X chromosomes and are sex chromatin positive. Thus the sex chromatin in these syndromes is reversed with respect to the *apparent* sex. There is evidence that the number of Barr bodies always is one less than the number of X chromosomes.

XYY individuals are known also and they may be produced by fertilization of a normal egg by a YY-bearing sperm produced by nondisjunction. The *XYY syndrome* was associated with criminal tendencies in the 1960s and has come to be called the criminal syndrome. Studies on aggressive behavior of patients in mental institutions and prisons showed that a relatively large percentage of inmates (3 percent) were XYY. In the general population about 1 in 1000 (0.1 percent) males is XYY. The correlation studies of aggressive or antisocial behavior and the XYY condition were used as legal defense in some courts, and some reduced sentences resulted. Now the fact that a man is genetically XYY is no longer acceptable as a legal reason for acquittal since many nonviolent XYY in-

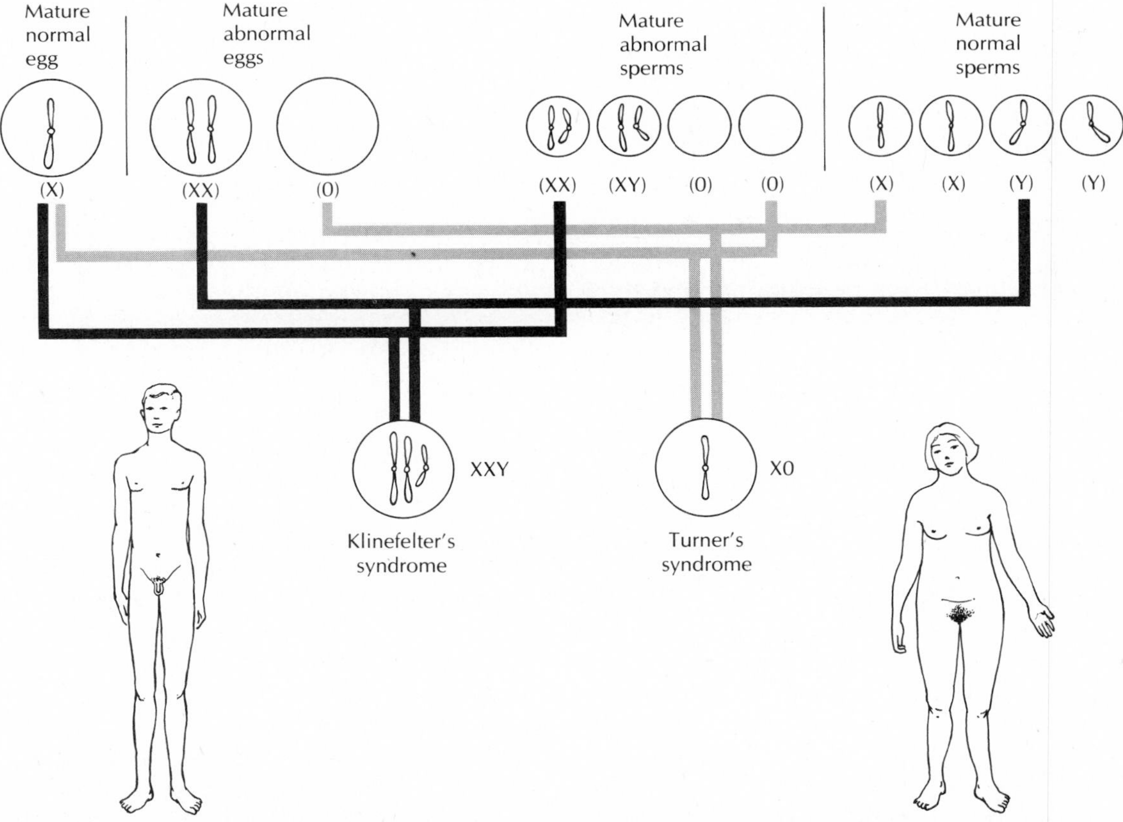

Figure 19.1 Formation of Klinefelter's and Turner's syndromes. Klinefelter's syndrome (XXY) results in a phenotypic male with reduced body hair, female-like breast development, and unusually long legs. Turner's syndrome (XO) results in a phenotypic female with short stature, webbed neck, and underdeveloped breasts. Only the X and Y chromosomes are shown; each individual has 22 pairs of autosomes. The total chromosome number for Klinefelter's and Turner's syndrome is 47 and 45, respectively.

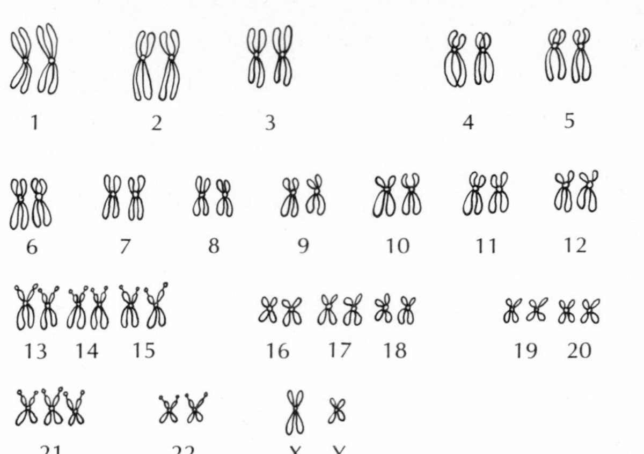

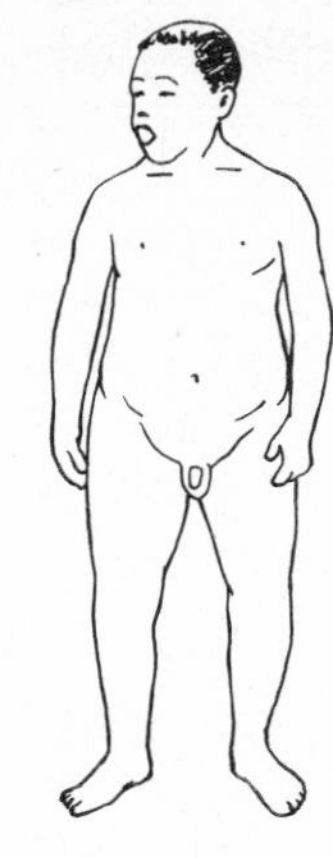

Figure 19.2 Down's syndrome. The individual possesses a third chromosome 21, is of low intelligence, short stature, and "mongoloid" due to the fold of the upper eyelid.

dividuals exist, and whether specific behavior patterns are due to genetic constitution or environmental factors is difficult or impossible to determine. In 1973 a report of studies in Denmark reported that 22 XYY syndrome males had greater truancy, vagabondage, impulsiveness, school difficulties, and antisocial behavior than a group of 42 male criminals with more or less similar behavior characteristics.

Other abnormal sex chromosome constitutions are known. XXX individuals are called triplo-X and may or may not show phenotypic abnormalities. The most common abnormality of the triplo-X person is low intelligence and sterility. The tetra-X (XXXX) syndrome is also known. The following additional sex chromosome abnormalities have been found, and they are all classified as Klinefelter's syndromes: XXXY, XXXXY, XXYY, and XXXYY.

Abnormal numbers of autosomes exist also. *Down's syndrome* results from an extra chromosome 21 (or a small chromosome in the same group as 21 according to the Denver classification system of 1960; possibly chromosome 22). It is also known as *trisomy 21 syndrome,* or *mongolian idiocy,* the former term being preferred. The individuals usually are mentally deficient and have short stature, a round face with epicanthal folds (a heavy skin fold above the eye), and a protruding tongue (Fig. 19.2). Chromosomal nondisjunction appears to be the basis for Down's syndrome. Older mothers are much more likely to have children with this condition than are younger mothers.

Also, Down's syndrome occurs by a mechanism other than nondisjunction of chromosome 21. This mechanism is translocation. The translocation involves chromosome 21 and chromosome 14 or 15. Such individuals have 46 chromosomes, which include two normal chromosomes 21, one normal 14 (or 15), and a large chromosome that is the translocated or fused chromosome 21 and 14 (or 15). Such cases are known as *translocation Down's syndrome.* As with the Down's syndrome resulting from nondisjunction, the genetic material of chromosome 21 is present in triplicate. Persons with translocation Down's syndrome are phenotypically indistinguishable from those with the more common trisomy 21 resulting from nondisjunction. Down's syndrome of all types occurs in 1 of 500 or 600 births. *This is not a rare or infrequent event.*

The incidence of Down's syndrome is much influenced by maternal age. For mothers of 25 the risk of having an affected child is about one in 2000–3000, but for mothers of 45 it rises to one in 40–50. Since ova reach the prophase stage of meiosis I in the fetus and may remain in that stage for 40–45 years before ovulation, aging of the ova might predispose them to nondisjunction. The incidence of Down's syndrome appears to be unrelated to the age of the father.

It is significant that trisomy of chromosome 21, a very small chromosome, has been found often, and trisomy of large chromsomes has not been found at all. Trisomies of chromosomes 13 (Fig. 19.3) and 18, slightly larger than number 21, have been found, but they lead to an early death. Many trisomies are probably not detected because their effects lead to an early death *in utero.* The large chromosomes apparently contain so much genetic information that overdosage is always lethal. Autosomal monosomy is rare, even for the smallest autosomes.

Besides additions and deletions of whole chromosomes and translocations, other human

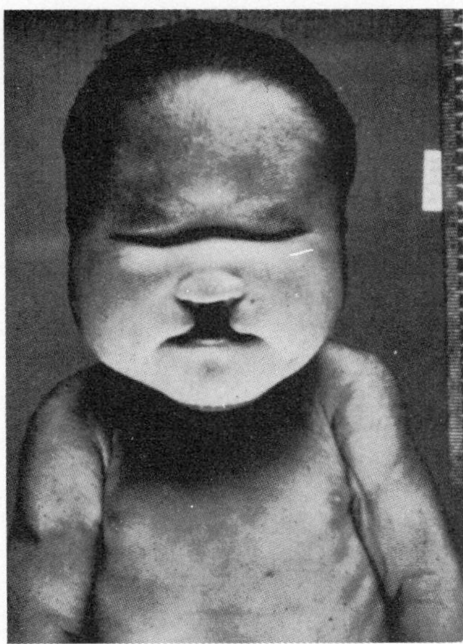

Figure 19.3 Trisomy-13. (Courtesy of Dr. Robert R. Eggen and Charles C Thomas, Publishers.)

chromosomal abnormalities are known. The best known example of a visible deletion of a chromosomal part (believed to belong to chromosome 5) was described in 1963. The individuals who lack the part have severe mental deficiency, a small head, wide spacing of the eyes, and a "moon face" appearance. Another feature of the condition is an abnormal larynx, which results in a plaintive cry, described as similar to the cry of a cat. The syndrome has been called the *cri du chat syndrome* for this reason. Figure 19.4 shows a photograph of an infant with the cri du chat syndrome.

Human Pedigrees

Since it has not been a practice to make controlled matings with humans, inbred strains and

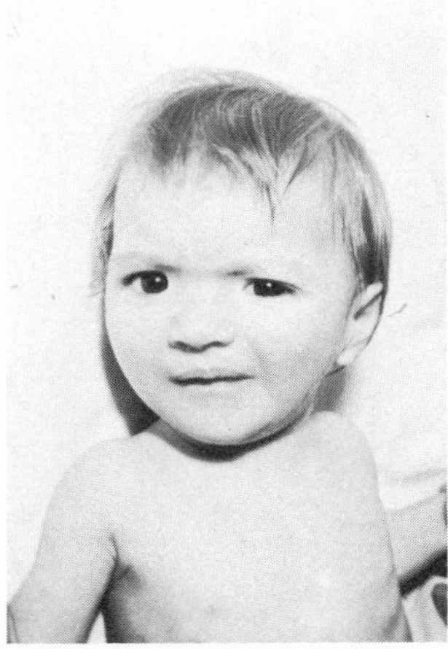

Figure 19.4 Infant with *cri du chat* syndrome. (From J. Le Jeune.)

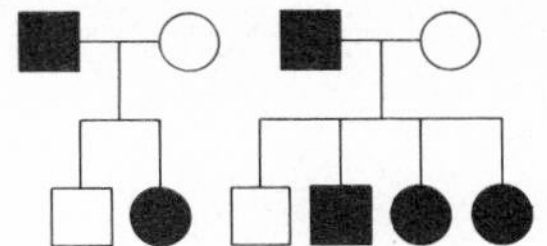

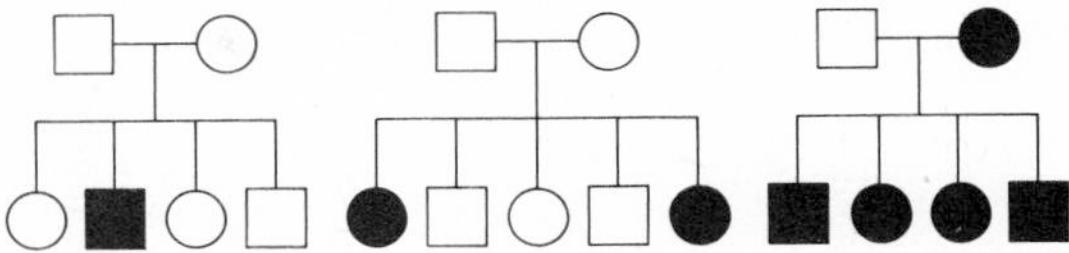

Figure 19.5 Five pedigrees of taste deficiency in man. Individuals represented by the shaded symbols are unable to taste crystals of phenylthiocarbamide. Those represented by the unshaded symbols find the crystals to be very bitter. Males are represented by squares, females by circles.

pure lines are not available. Students of human genetics have used the *pedigree method* for studying certain traits. A pedigree is a record of the distribution of one or more traits in a group of related individuals. Figure 19.5 shows five pedigrees with the distribution of tasters and nontasters for phenylthiocarbamide (PTC). About 70 percent of the general population find that PTC has a definite taste, usually bitter, whereas the remaining 30 percent find it tasteless. From these and many other pedigrees that have been studied, it has been determined that the inability to taste PTC depends on a recessive gene.

Although the nature of the inheritance of many human traits is now understood, the study of human inheritance, when compared to that of corn or *Drosophila*, is still in its infancy. Table 19.1 lists a few human traits and the mode of inheritance. In general the traits listed in Table

Table 19.1 Human Traits Inherited as Autosomal Characters

DOMINANT	RECESSIVE
Brown eyes	Blue or gray eyes
Farsightedness	Normal
Normal	Nearsightedness
Dark hair	Blond hair
Nonred hair	Red hair
Curly hair	Straight hair
Scaly skin	Normal
Thickened skin	Normal
Broad lips	Thin lips
"Roman" nose	Straight nose
Extra digits	Normal
Short digits	Normal
Allergy	Normal
Huntington's chorea	Normal
Normal	Congenital deafness
Normal	Tay-Sachs disease
Normal	Phenylketonuria

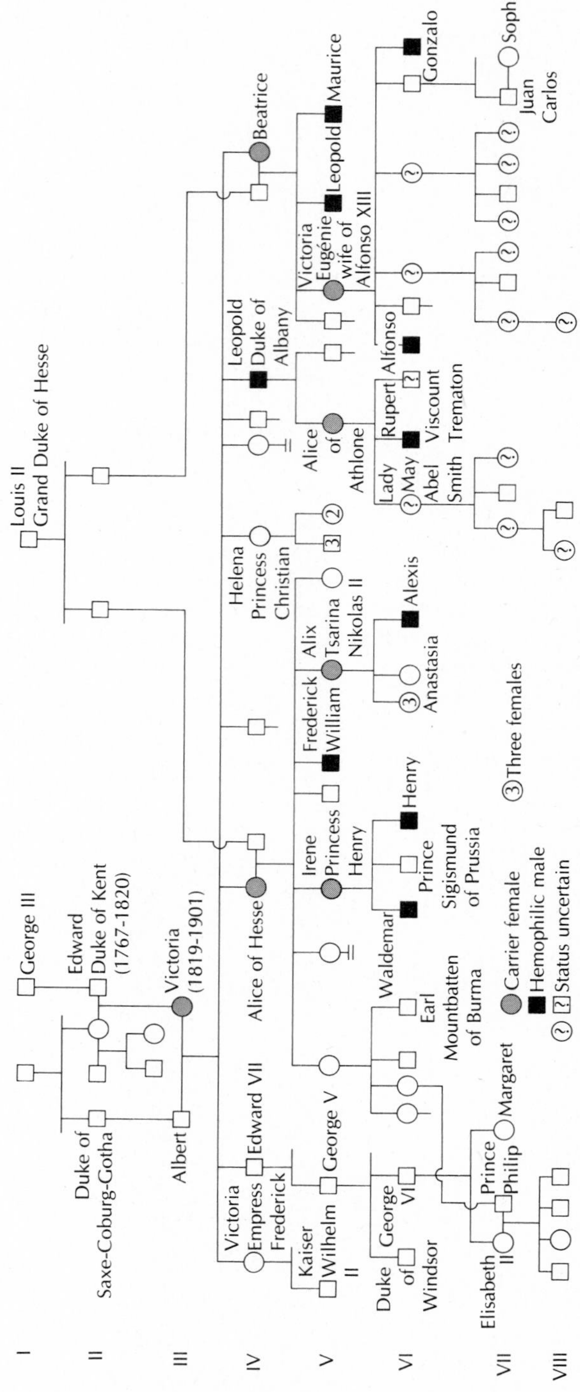

Figure 19.6 Pedigree of Queen Victoria and her descendants. The mutation for hemophilia, an X-linked recessive trait, probably occurred in Queen Victoria or in her father, Edward, Duke of Kent. (Redrawn from McKusick, *Human Genetics*, 2d ed., Prentice-Hall, 1969.)

19.1 are determined by genes that behave according to the simple dominant–recessive pattern.

Other modes of human inheritance are known. Like autosomal traits, those determined by X-linked genes may either be dominant or recessive. In X-linked dominant inheritance both males and females are affected and both transmit the disorder to their offspring. One of the best studied X-linked dominant traits is vitamin-D resistant rickets.

The genes for hemophilia and color blindness are sex-linked recessives. Hemophilia is a particularly interesting case, both biologically and historically. The royal houses of Europe have been afflicted with hemophilia since the days of Queen Victoria's children. Queen Victoria herself was a carrier, and since none of her ancestors and relatives were affected, the mutation probably occurred in one of her parents or in one of her own germ cells. Figure 19.6 shows the pedigree of Queen Victoria and her descendants for the hemophiliac trait. Leopold, Duke of Albany, died of hemophilia at the age of 31. Prince Albert, Victoria's consort, did not possess the trait. (If the argument is made that Albert possessed hemophilia in a mild undiscovered form, how does the pedigree disprove this possibility and exonerate Albert?) Two of Victoria's daughters were carriers, and several of their male descendants were "bleeders." The histories of Russia and Spain have been affected considerably by these genetic situations. Table 19.2 lists some of the traits thought to be controlled by genes on the X chromosome.

Other patterns of inheritance known in humans include multiple alleles (blood groups, Ch. 11) and multiple factors (skin color, height, and intelligence, Ch. 28).

Table 19.2 Some Human Traits Thought to Be Controlled by Genes on the X Chromosome

1. Partial color blindness, deutan series
2. Partial color blindness, protan series
3. Total color blindness; most cases are autosomal, however
4. Glucose-6-phosphate dehydrogenase structure
5. Xg blood group system
6. Hemophilia A (AHG deficiency)
7. Hemophilia B (Christmas disease)
8. Agammaglobulinemia
9. Diabetes insipidus, nephrogenic type
10. Diabetes insipidus, neurohypophyseal type
11. Absence of central incisors
12. Congenital deafness; there are autosomal forms also
13. Progressive deafness
14. Mental deficiency; a small portion of cases show X linkage
15. Hydrocephalus; most cases are not X linked
16. Ocular albinism
17. Congenital cataract with microcornea

Many of the known hereditary traits that are not listed in Tables 19.1 and 19.2 involve abnormalities that are of particular interest to the medical profession. As more and more is learned about human inheritance, it becomes necessary for people entering the practice of medicine to have a sound training in human genetics.

GENES AND DISEASE

Usually only rare diseases are inherited along simple genetic lines. But genetic factors are involved in many diseases. About 2000 human diseases are now considered to have a strong genetic component. Approximately 25 percent of all hospital beds and extended care facilities in the United States are occupied by persons who suffer from a "genetic disease." The relative contribution of intrinsic (genetic) factors and extrinsic (environmental) factors varies from disease to disease. Figure 19.7 represents this range for several human diseases. At one end of the scale there are diseases such as Down's syndrome, hemophilia, and so forth, which are exclusively genetic in origin and where the environment plays no direct role. The intermediate

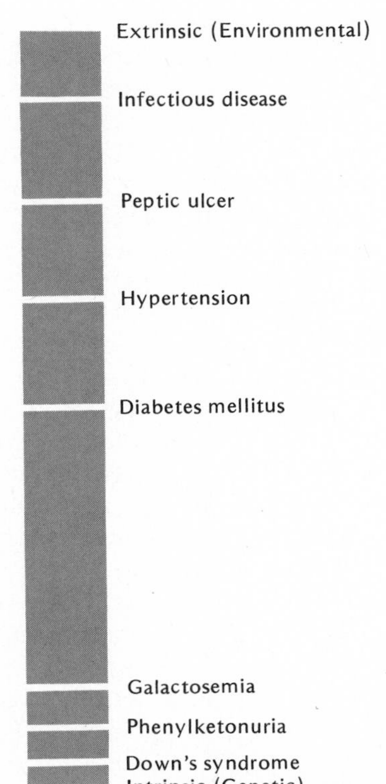

Figure 19.7 Relative contribution of genetic and environmental factors to various human diseases.

diseases, such as diabetes mellitus, hypertension, and peptic ulcer, have significant contributions from both sides. The far extreme of infectious diseases is almost entirely the result of the environment. Ease of treatment of the diseases is probably inversely proportional to their genetic factor. Infectious disease is relatively easy to control with modern medicine. Smallpox, polio, and scarlet fever are usually rare diseases in technologically advanced countries today. Genetic diseases, on the other hand, may be completely untreatable by modern medicine. Hemophilia and sickle-cell anemia[1] are good examples of "incurable" genetic diseases. Other genetic diseases may be controlled completely or to a large degree in some affected individuals. In galactosemia and phenylketonuria, where galactose and phenylalanine respectively, cannot be metabolized properly, elimination of galactose and reduction of phenylalanine in the diet at an early age can prevent irreversible damage. Certain individuals are sensitive or allergic to various drugs or chemicals. These "genetic diseases" can be controlled or avoided with proper medical attention. Thus as with environmental diseases, genetic diseases range over varying difficulties in treatment.

The question is sometimes asked, "Which is more important in the formation of traits in an organism, heredity or environment?" This question is actually unanswerable because an organism in its development always has an environment; a gene or a whole group of genes always has an environment. Many studies have shown that a particular genotype develops one way in one environment and another way in another environment (phenocopies). The common Chinese primula, when grown at temperatures of 13–18° C (55–65° F) will produce red flowers, generation after generation. The same plant, when grown at a temperature of 35° C (95° F) will produce white flowers as long as it is kept at this high temperature. However, if such plants are restored to low temperatures, they will produce red flowers. A mutant strain of *Drosophila* has been discovered with branched legs; this occurs only at low temperatures. This same strain, when grown at normal temperatures, has normal legs. Many similar examples have been reported. Such studies clearly indicate that what is inherited is the ability to react in a certain way to a certain environment. But the question may be asked, "Is it possible to set up an experiment in which it may be determined in a particular situation that heredity is more important than environment, or vice versa?"

In human society there are certain situations that, when combined, approach such an experimental setup. Human twins constitute the materials for this natural experiment. Ordinarily twins occur about once in 88 births in the United States, but the number has increased since the use of oral contraceptives became widespread. (Oral contraceptives, "the pill," have increased the frequency of multiple births by increasing the number of eggs released from the ovary when the pill is discontinued.) Human twins are of two kinds: one-egg (*monozygotic*), or *identical,* twins that originate from the division of a single fertilized egg, and thus have the same genotype; and two-egg (*dizygotic*), or *fraternal,* twins that, because they result from two different fertilized eggs, have different genotypes. Therefore fraternal twins are no more alike genetically than are ordinary brothers and sisters. Identical twins are always of the same sex, whereas fraternal twins may be of either the same or opposite sexes.

A number of investigators have realized the value of studying human twins in analyzing human heredity. One extensive study used 50 pairs of identical twins that had been reared together, 52 pairs of fraternal twins that had been reared together, and 19 pairs of identical twins that had been reared apart. Identical twins, whether reared together or apart, are very much alike in many physical traits, such as facial features, hair and skin color and texture, eye color, tooth characteristics, height, head length, and head width. With regard to these physical traits, differences in the environment have little effect.

An extensive consideration of twin studies is not feasible, but some reference should be made to an analysis of the relative effects of heredity and environment on intelligence in twins. Human intelligence should not be considered a single well-defined characteristic. However, intelligence tests have been devised; the results of these tests are expressed by an intelligence quotient, known as the IQ. This measure is a useful tool in discussing intelligence. A score of 90 to 110 is average, and theoretically denotes an individual who can make normal adjustments in everyday life. A score of 140 indicates a genius; one of 120 indicates a superior intellect; and one of 70 to 0 places the individual in the feeble-minded category. No one suggests that the IQ, as used generally, measures only innate intelligence. Psy-

[1] Sodium cyanate has been used clinically on a small scale, but it is suspected of being a toxicant of the central nervous system. In January of 1975 a new agent which inhibits sickling was reported, but clinical tests must await study of its action and possible toxicity.

chologists subdivide human mental abilities into distinct primary abilities, such as an ability to visualize objects in space, to memorize, or to reason inductively. It has been found that two individuals with the same IQ may vary considerably with respect to their primary abilities. Studies in the future undoubtedly will make more use of this new approach. As mentioned earlier, the distribution of the IQs in the whole population would seem to indicate that intelligence is determined by multiple factors. But the number of these factors, and precisely how each works, remain to be determined. Of course, IQ tests have been criticized because they do not take cultural backgrounds and experiences into account. The inadequacies of IQ tests are being brought to public attention more frequently than a few years ago.

One study found that the average pair difference for IQ scores of the 50 pairs of identical twins reared together was 5.9, the average pair difference for the 52 pairs of fraternal twins reared together was 9.9, and the average pair difference for the 19 pairs of identical twins reared apart was 8.2. The identical twins reared apart show a greater average difference than identical twins reared together, suggesting an environmental influence on IQ. The identical twins reared together show a smaller average pair difference than that found in the fraternal twins, suggesting that heredity is also important. It should be pointed out that the average difference between identical twins reared together is no greater than that obtained when the same individual takes the test twice. Furthermore the greater average pair difference for the identical twins reared apart was caused by four of these pairs. In the majority of the 19 separated pairs the individual differences were about the same as those found for the unseparated twins. In these cases the environments of the separated twins were not greatly different. However, in a few of the pairs the opportunities for schooling and the amount of formal education varied greatly.

A consideration of the case history of a pair of identical twins will illustrate the effects of great environmental differences. Gladys and Helen were separated at 18 months and did not meet again until they were 28 years old. Helen lived on a farm in Michigan, obtained a college education, and was then employed as a teacher. Gladys attended school for 2 years in Ontario, spent the next 2 years in the Canadian Rockies where there was no school, and when her family returned to Ontario, she never resumed her formal education. Prior to the time she was examined, she had worked at several occupations. Helen scored 116 on the IQ test and Gladys scored 92. This is the greatest difference found in any of the pairs studied, and this correlated with the greatest difference in educational experience of all the twins studied.

Any conclusions about the relative effects of heredity and environment from the twin studies must be tentative. Much more study is needed on the identification of these genes and factors and their specific roles in determining intelligence. However, it does seem fair to state that the hereditary components can be modified by the environment. Heredity seems to provide the individual with upper and lower limits between which some level of intelligence is realized. If this concept is correct, then it must follow accordingly that it is very important for each individual to have a favorable environment in which to develop. Without a favorable environment, a person with relatively favorable heredity for intelligence may fall short, in the achievement of intelligent behavior, of a person with relatively unfavorable heredity.

EUGENICS

Francis Galton coined the term *eugenics,* which means being well born. Galton used the term to cover the whole study of agencies under social control that may result in improving the hereditary qualities of future human generations. Since Galton's time many people have become interested in the eugenics movement, several eugenics societies have been formed, and many states have passed eugenics laws.

Several studies have pointed up the fact that there is a differential birth rate in the United States. In general these studies show that college graduates have fewer children than those who have finished only high school; these in turn have fewer children than those who finished only grade school; and the largest number of children is found in those families where the parents had less than seven years of formal education. Many people see a cause for alarm in our differential birth rates because they feel that the result will be a slow but continuous downward trend in the average intelligence. Professional geneticists, in the main, are a bit wary of drawing any definite conclusions at the present time because they still do not have adequate information about the many factors, both hereditary and environmental, that determine intelligence.

The approach of the eugenicists to the general problem of improving human heredity has been along two lines: negative eugenics and positive eugenics. The negative approach involves segregation or sterilization of individuals with undesirable traits. In either case the reproduction of such individuals would be prevented. The various state laws that have been passed provide for segregation or sterilization, or both. Most of the sterilization laws have pertained primarily to the feeble-minded and the insane. The sterilization operation is rather simple. In the male it involves cutting the sperm ducts; in the female the Fallopian tubes are tied. Such operations do not affect the endocrine functions; they only prevent the transport of gametes. Vasectomy, the cutting of the sperm ducts, is beginning to be a fairly popular means of birth control among couples who have the number of children that they desire.

However, there are great difficulties in such a negative program of eugenics. Not all feeble-mindedness is hereditary; some is environmentally produced. Also not all insanity is hereditary. Thus such a program must be in the hands of very competent people.

The other aspect of the eugenics program, positive eugenics, aims to raise the average intelligence of the population by encouraging those with desirable traits and talents to have more children.

With the present state of knowledge of human genetics, great caution must be followed in connection with specific eugenic measures. However desirable it may be to eliminate undesirable traits from the human population, this cannot be done until the exact nature and mode of operation of the defective genes are known. And in cases involving recessive genes and multiple genes, it must be realized that the problem of their elimination is enormous and that the time required may be many hundreds of years.

Genetics cannot be applied to humans as it is applied to livestock, for very obvious reasons. And in the case of humans, who is to say what the desirable type is? In the human population there are many desirable types, and society would not function for long if all people had the same talents and interests. Human diversity is needed for a society with many goals. This, however, does not mean that a sane program of eugenics based on practical and scientific methods should not be adopted.

Genetic Engineering

Another approach to the problem of human welfare may be made possible by altering the developmental processes of an individual destined to be afflicted with genetic defect. Joshua Lederberg, a Nobel Prize winner for his studies in microbial genetics, has proposed the term *euphenics* for such experimental modification of human development by physiological and embryological methods. Euphenics would have its effects only on the individual and not on future generations, since the germ cells would not be affected by such "developmental engineering."

Eugenics and euphenics are complementary; some traits may be best modified developmentally; others may be more readily changed by selection. Genetic engineering, developmental engineering, and cellular engineering are discussed in some detail in the summary chapter, Chapter 47.

Eugenics and Genetic Counseling

Presently eugenics does not have a united group presenting its perceived merits. Instead, a growing field of professional activity is that of *genetic counseling*.

In brief, a genetic counselor lays the evidence before individuals either contemplating childbearing or already pregnant so that they may make reasonable decisions about conceiving a child or aborting a fetus. The genetic counselor meets with the prospective parents, gathers as much information as possible for a pedigree analysis, and then discusses the probabilities (figures the odds) that certain traits will or will not appear in the offspring. For instance, if one parent has a single faulty dominant gene, there is a 50 percent probability that any child will also be faulty; if both parents have a single faulty recessive gene, the child has a 25 percent chance of showing the faulty trait; or if the genes appear to be sex linked, expectancies for the offspring can be worked out for the child just as they could be for *Drosophila* in Chapter 17. Sometimes in the case of multifactorial inheritance, the transmission pattern may be less defined and the genetic counsellor tries to make this obvious to the potential parents.

There are several positive things that can be done besides taking family histories. Sometimes carriers of traits can be tested quite simply. This is true, for instance, for carriers of sickle-cell genes. Sometimes a few cells from a potential carrier can be subjected to short-term tissue culture and these cells used to measure enzyme levels; over 60 metabolic diseases can now be identified by such procedures.

In cases where the woman is already pregnant, it is possible to obtain cells produced by the fetus through a technique called *amniocentesis.* Here, after the 14th week of pregnancy, amniotic fluid can be taken by hypodermic syringe through the abdominal wall of the woman. The fluid has cells from the fetus. These can be cultured and tested as above. The procedure also allows prenatal determination of sex by the presence or absence of Barr bodies.

Of course there are two principal alternatives after receiving genetic counseling. Either conception can be avoided or terminated, or the decision can be made to have the child. This often is a case of calculated risk that the child will not show the abnormal trait.

There is, however, another aspect to this problem that should not be ignored. The decision to have or not to have a child is a short-term decision. For instance, parents both of whom are heterozygous for a recessive trait have a 75 percent probability of producing children that appear normal. However, 66 percent or two thirds of those that do appear normal will be heterozygous. If the recessive phenotype is not as successful in reproduction and if given sufficient time, the human population will have a growing number of heterozygous persons. Conception from two heterozygotes will increase in frequency and the number of abnormal children appearing in the population will increase. The significance of this in terms of species survival will be considered apparent again in Chapter 42 where the Hardy–Weinberg concept and genetic drift are considered. Thus prospective parents who receive counseling have a threefold concern: concern for themselves, concern for the child, concern for society and the human species.

AGRICULTURAL GENETICS

The objective of agricultural genetics is the application of genetic knowledge to agricultural practice. The most economically important organisms to which this has been applied are the vertebrate animals and the flowering plants. Genetic manipulation with these organisms has shown that inherited traits that are agriculturally useful can be located, transferred, and utilized. In a world where the majority of people have too little food, only a few of the species that are used for food have been improved by genetic breeding programs. Proper breeding has produced disease-resistant varieties as well as varieties that give greatly increased yields.

In general, three procedures are commonly practiced by plant and animal breeders: selection, inbreeding, and outbreeding and hybridization. In modern breeding programs of agricultural genetics the principles of selection are used in various ways. Two main selection techniques used in plant-breeding practice are mass selection and pedigree selection. In mass selection the undesirable plants are destroyed, and the desirable ones are used for seed. In general, mass selection brings about a rather slow change. In the selection of offspring of naturally (somewhat at random) cross-pollinated plants there is no control over the pollen source. In pedigree selection individual plants and their progeny are maintained as separate lines. Such pedigree lines enable the plant breeder to follow closely the genetic behavior of his material. This technique reduces the amount of variability that might be attributed to environmental influences. Pedigree selection is often used to produce lines that do not have great agricultural or economic value in themselves, but when crossed to another pedigree line, produce superior offspring.

The animal breeder usually considers the individual characteristics and the characteristics of its genetic relatives—ancestral, collateral, and offspring. The selection of individuals to be parents of the next generation of a herd or flock, and so forth, usually requires some sort of measurement—the opinion of experienced breeders in the case of pedigreed dogs, for example. The ribbons and trophies won by the dog would be an index to its breeding potential, or data in the case of quantitative characteristics such as the number of eggs, pounds of meat, amount and quality of wool, and so on. Figure 19.8 shows a hypothetical record of a trait being selected. Sometimes the individuals of a generation will be closer to the original population mean than their parent; such a phenomenon is called *regression.* This emphasizes the complexities involved in studying traits that may be controlled by many genes. The nature of gene action in

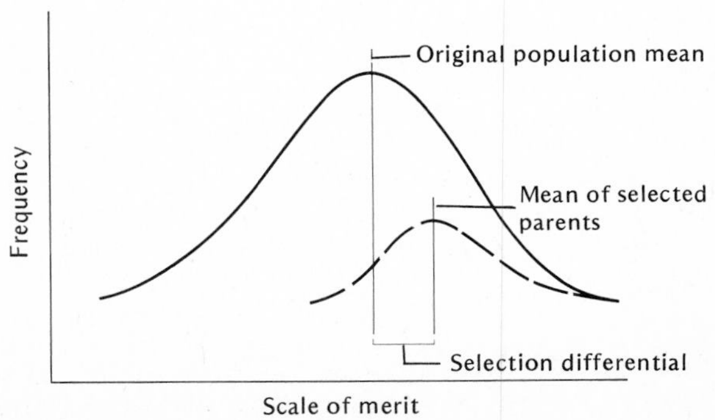

Figure 19.8 Hypothetical record of trait being selected.

multiple-gene (quantitative inheritance) systems remains rather obscure presently. It is clear that multiple genes usually do not operate through simple additive effects.

The systems of mating in natural populations are usually considered to be random (Hardy–Weinberg equilibrium and population changes, Ch. 42). Other systems of mating are found in nature and in human-controlled populations. Inbreeding is the mating of closely related individuals. Extreme inbreeding occurs naturally in self-fertilizing plants, such as beans and peas. (Mendel's experiments with pure lines of the garden pea depended on this system of mating.) The effect of prolonged inbreeding is to greatly reduce the number of heterozygous individuals. It is doubtful whether complete homozygosity is ever achieved, but for practical breeding purposes, lines subjected to inbreeding for long periods of time are pure lines.

Self-fertilization occurs only among a few plant types and does not occur at all among higher animals. In animals homozygosity can be approached quickly by brother–sister matings. Sometimes progeny are backcrossed to their parents. Inbreeding combined with selection over the years has resulted in many valuable breeds of domestic animals. Thomas Jefferson was an early proponent of genetic improvement of agricultural stocks in the United States. Jefferson was influential in bringing the merino sheep to this country. The merino sheep themselves make an interesting story. The strain was developed in Spain during the seventeenth century. The original stock had two coats of wool fibers–long coarse ones and short fine ones. Intensive selection for fine wool and inbreeding produced the merino sheep which have a more uniform production of short-fibered wool. For a time, Spain had a monoply on these sheep, but when France invaded Spain, some of the breed was taken back to France, from where these sheep were distributed to many parts of the world.

Intensive inbreeding has led to some unfortunate genetic situations. For example, inbreeding of Hereford beef cattle has brought together recessive genes for dwarfism. Dwarfs are of little economic value. The dwarf gene has been brought out also in the breeding of the Angus and Shorthorn cattle lines. Some dog breeders have so selectively inbred their lines that undesirable traits have been established; for example, hip dysplasia is a rather common disease in pedigreed boxers.

Hybridization is performed in two ways: outbreeding and crossbreeding. Outbreeding is the crossing of unrelated individuals, whereas crossbreeding involves mating of individuals of different races or even different species. Corn is a good example of a plant that has been improved by the techniques of outbreeding to produce hybridization. In the case of corn it has not been possible up to the present time to produce inbred lines that have all the desired characteristics. In 1968, 99 percent of the corn planted in the United States was hybrid corn. It is produced as shown in Figure 19.9. The four lines used in the production of the commercial hybrid corn seed are maintained by inbreeding. The final product obtained greatly increases the quality and yield. This phenomenon, which is frequently associated with hybridization, is called *heterosis,* or hybrid vigor. Heterosis, in general, may be explained on this basis. Each of two inbred lines is homozygous for certain desirable traits and also for some undesirable recessive traits, but the two lines are homozygous for different genes, and each has dominant genes to mask the undesirable recessive genes of the other. In the case of corn double crosses are used to produce commercial seed. Here when two inbred lines are crossed, the maternal parent, as is characteristic of most inbred corn lines, is small and produces small low-yielding ears. Double crosses result in the production of seed on large uniform ears of single-cross plants. It is not possible, however, to obtain good results by using the seed produced by plants grown from hybrid corn grains. Such grains result in a large variety of different kinds of plants and ears as a result of independent assortment, that is, such hybrids will not breed true to type, and continual crossing is practiced to produce the commercial seed. As is readily seen, both inbreeding and outbreeding are involved in the production of hybrid corn.

Several plant species other than corn have shown considerable hybrid vigor when two or more lines were crossed. Sorghum, onions, and sugar beets—all important agricultural crops—are good examples. Outbreeding has been used in animal husbandry, also. Poultry, rabbits, cattle, and sheep are such cases.

Crossbreeding is much more common with animals than with plants for agricultural purposes. The mule is the result of a successful cross between the horse and the donkey. The mule is superior in many ways to either parent. It is faster, stronger, and larger than the donkey and, at the same time, is more durable and more resistant to disease than the horse. The mule is usually sterile since the chromosomes do not

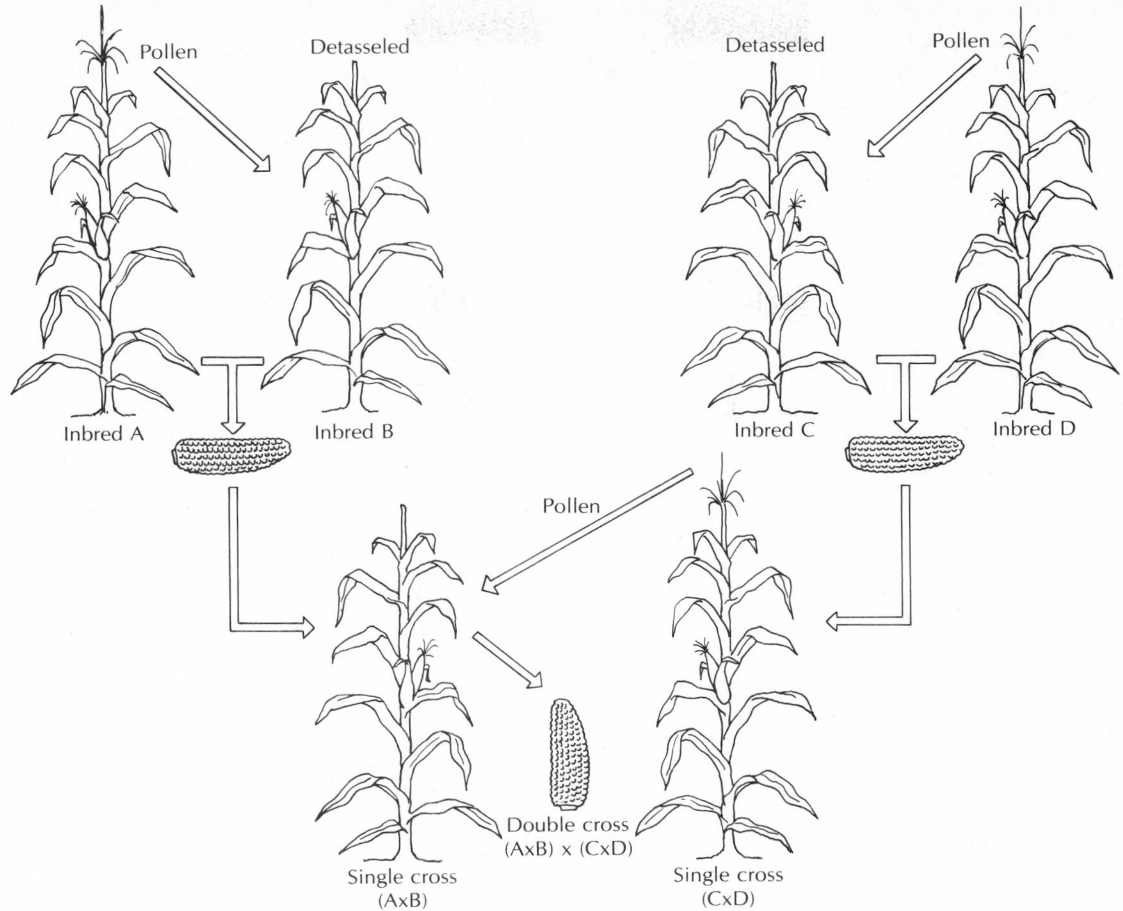

Figure 19.9 Production of hybrid corn. Plants from four inbred strains of corn are crossed pairwise: A × B, C × D. The Progeny of these two crosses are then used to produce the hybrid corn that is used in commercial planting.[Double cross: (A × B) × (C × D).]

have homologs for pairing in meiosis. Strictly speaking, a *mule* is the offspring of a male donkey (a jackass) and a mare. The offspring of a stallion and a female donkey is a *hinny.*

Successful crosses between zebus, the native cattle of India, and European cattle have produced hybrids that are adaptable to warm humid climates, with good production of milk and meat. Probably a great potential exists for breeding other desirable hybrids for food and work in many areas of the world.

In summary, selection, inbreeding, and hybridization are powerful techniques for improving stocks of plants and animals for agricultural and other purposes.

Although corn is a staple in the diet of large numbers of people and has been improved by genetic techniques, it is deficient in the amino acid lysine. A strain of corn that is high in lysine has been studied at Purdue University. When corn is the sole or major source of protein in the diet, such as in Latin America, this *opaque-2* strain is superior in support of growth. The discovery of the high lysine content of opaque-2 corn has stimulated a search for other high-lysine mutants in corn and other cereal grains.

An example of application of basic biology to food production by animal is the egg of the domestic fowl. The egg that you may have had for breakfast this morning probably was laid by a Single Comb White Leghorn hen, which had 13 vitamins and 13 minerals added to its soybean meal–corn–wheat diet, has spent all of its life in a windowless building, produces about 20 eggs per month from age 5 to 25 months, and will be in the soup can after its peak of egg production is past. The Single Comb White Leghorn has more traits desired by the egg rancher than any other breed: good egg production, large eggs with sound shells and high interior quality, resistance to disease, and it starts laying at a young age. The genetic makeup of the modern Leghorn strain is the result of many poultry breeding experiments. In the 1920s vitamin D was discovered; it became possible to raise chickens in confinement without sunshine. The minerals and other vitamins are added to the diet in the proper amounts for optimum growth, health, and productivity of the chicken. In addition protein and caloric (energy) requirements were determined. Soybeans are good sources of protein, but raw soybeans contain a protein which inhibits the action of trypsin, a digestive enzyme; therefore raw soybeans interfere with digestion. The tryspin inhibitor is destroyed by heating, so heated soybean meal is now the bulk of the protein source in chicken feed. Chickens have biological clocks which are set by the intensity and duration of light. Normally chickens lay more eggs in the spring than during any other time of the year. The modern chicken ranch uses artificial light cycles to influence the biological clock and thereby maximize hen development and egg production. Maturing pullets (young hens) are exposed to 8-hour (short) days to prevent them from laying until their bodies and oviducts have reached the proper size. If the pullet starts laying too soon, permanent damage to the oviduct may occur. At the proper age and size the light regimen is switched to 16 hours on and 8 hours off, a schedule which induces maximum egg production.

In this one example of food production by animals, biological principles of several disciplines are used: genetics, development, nutrition, biochemistry, and behavior.

RECOMBINANT DNA

In closing this chapter on human and applied genetics, we turn to a technique that has been discussed increasingly since 1972 when a new class of enzymes, *restriction endonucleases,* was discovered. The technique, joining of unrelated DNAs by enzymatic methods, involves the cleavage of one kind of DNA with restriction endocucleases and then joining the fragments to the ends of other DNAs (cleaved with the same endonuclease) with *DNA ligase,* the "joining enzyme" (Ch. 6). The restriction endonucleases cut both strands of the DNA double helix at points of specific base sequences in such a way that a short, single-stranded region is left at each end of the DNA fragments. Often called "sticky" because they can form hydrogen-bonded associations with the ends of other pieces, these ends are covalently joined when the DNA ligase is added.

This technique is basically powerful and simple because, in many cases, the two types of DNA may be mixed, digested and added to specially treated bacteria that take up the fragments. Once inside a bacterium which contains the DNA ligase, the fragments are joined and the cells carrying recombinant DNA only need to be isolated with standard culture techniques of microbiology to obtain strains which are "factories for manufacturing" the recombinant DNA.

Some of the events of the continuing discussion have been

1. In 1974 several scientists called for a moratorium on recombinant gene research.
2. A conference in February, 1975 at Asilomar, California—attended by persons from several countries besides the United States—produced temporary guidelines to govern this type of research.
3. Several guidelines meetings were held and the last one, a public input session at Bethesda, Maryland, where the National Institutes of Health (sponsors of much recombinant DNA research) are located was climaxed with long and boisterous public response in February, 1976.
4. The NIH guidelines were released in June, 1976. Designed to ensure that recombinant DNA research can proceed without harm to laboratory workers, the general public or the environment, the guidelines ban certain types of experiments, require justification that the same knowledge cannot be obtained with traditional techniques, establish levels of safety precautions commensurate with potential hazards and set responsibilities for investigators, institutions and the NIH.

This topic has come to national attention in several specific instances: The city council of Cambridge, Massachusetts, voted to ban these experiments at Harvard University and the Massachusetts Institute of Technology; the attorney general of the state of New York held a hearing in the fall of 1976 for the purpose of deciding what action to recommend to the state legislature; and the University of Michigan and Indiana University have felt the impact of public concern over establishment of recombinant gene laboratories on their campuses.

Succinctly, the possible dangers of recombinant DNA research involve the fact that with "synthetic biology" we leave the web of natural evolution, and the production of new strains of microorganisms might break down the existing barriers against genetic exchange between prokaryotes and eukaryotes. The possible beneficial applications include transfer of genes for nitrogen fixation into common food plants, facilita-

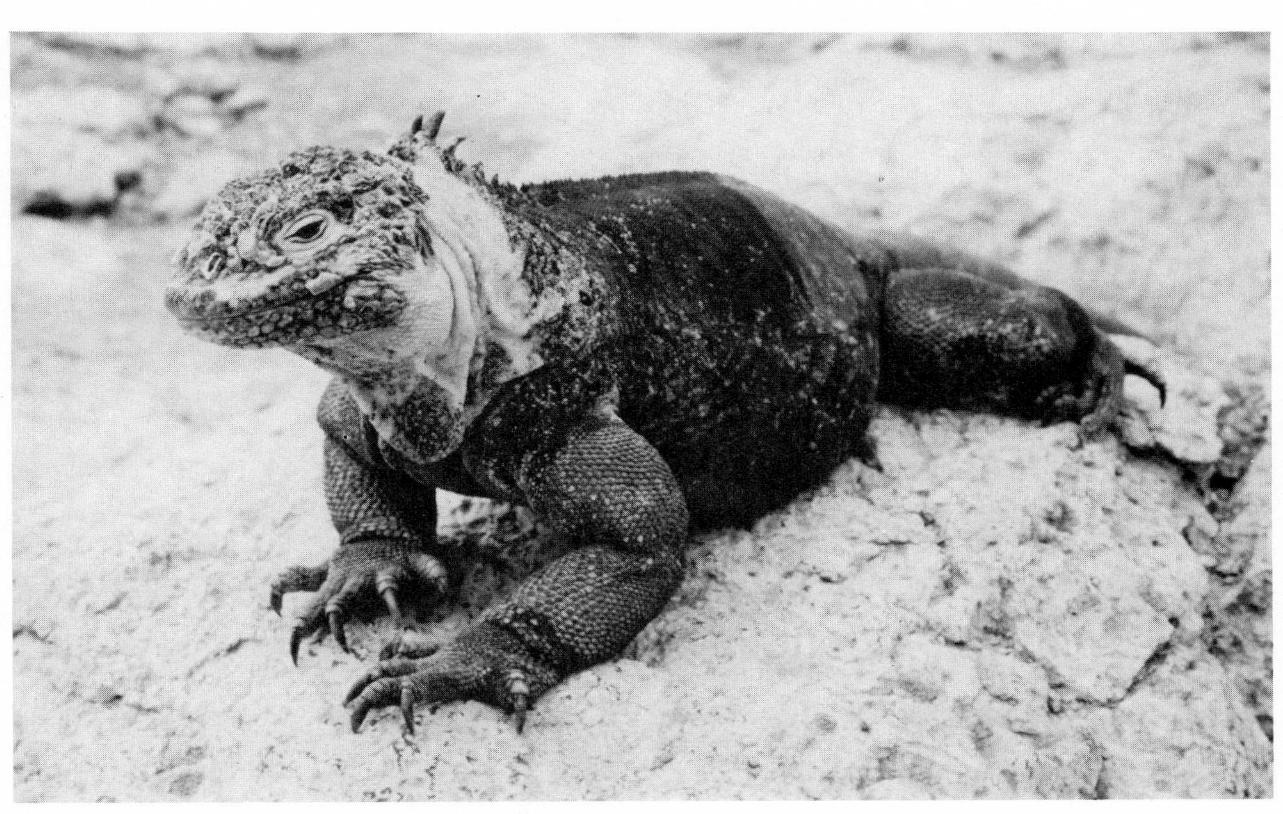

Conolophus subcristatus, the Galápagos Island land iguana.
(Photo — E. C. Williams)

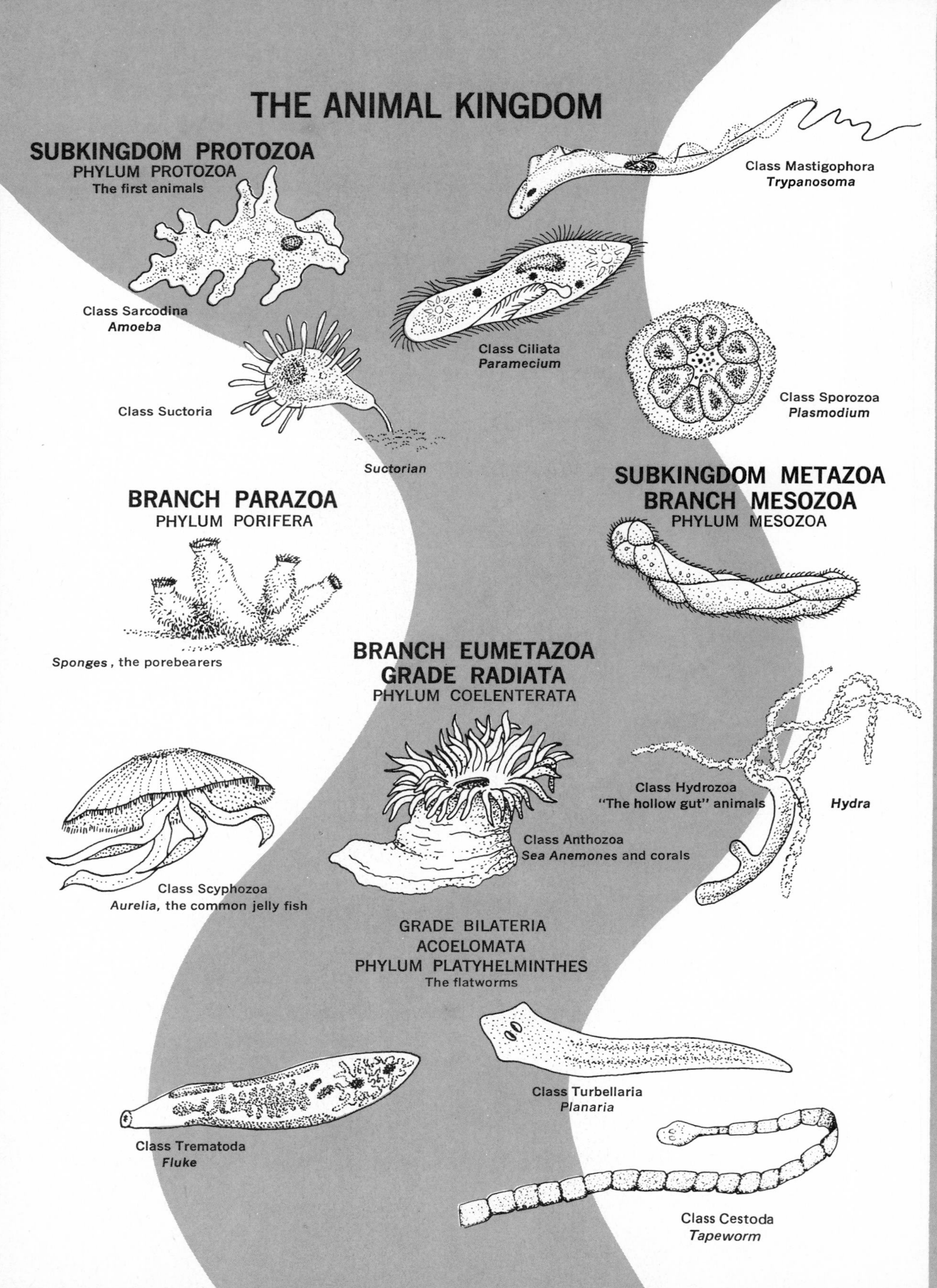
THE ANIMAL KINGDOM
SUBKINGDOM PROTOZOA
PHYLUM PROTOZOA
The first animals
Class Mastigophora
Trypanosoma
Class Sarcodina
Amoeba
Class Ciliata
Paramecium
Class Sporozoa
Plasmodium
Class Suctoria
Suctorian
BRANCH PARAZOA
PHYLUM PORIFERA
SUBKINGDOM METAZOA
BRANCH MESOZOA
PHYLUM MESOZOA
Sponges, the porebearers
BRANCH EUMETAZOA
GRADE RADIATA
PHYLUM COELENTERATA
Class Hydrozoa
"The hollow gut" animals
Hydra
Class Anthozoa
Sea Anemones and corals
Class Scyphozoa
Aurelia, the common jelly fish
GRADE BILATERIA
ACOELOMATA
PHYLUM PLATYHELMINTHES
The flatworms
Class Turbellaria
Planaria
Class Trematoda
Fluke
Class Cestoda
Tapeworm

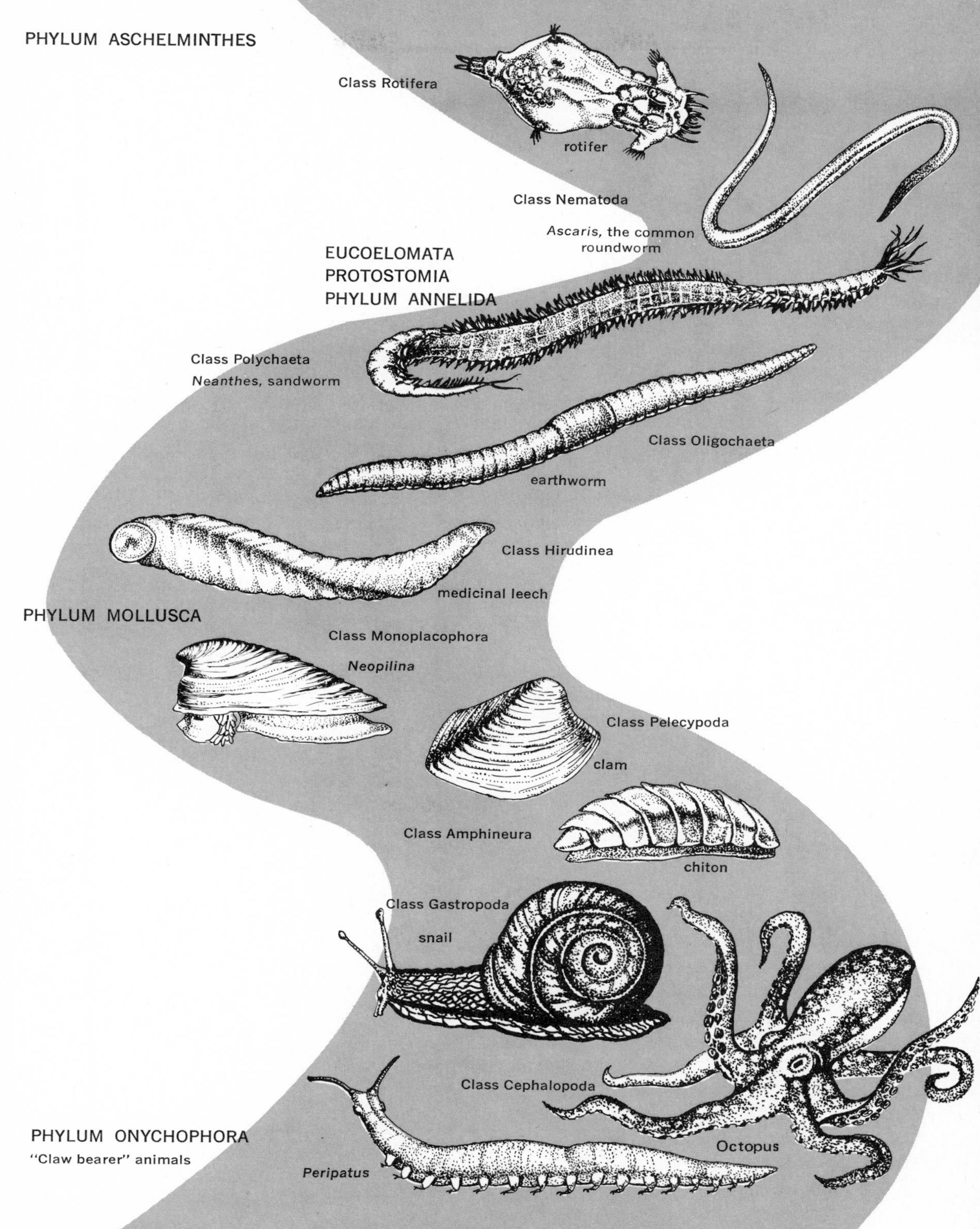
PHYLUM ASCHELMINTHES
Class Rotifera
rotifer
Class Nematoda
Ascaris, the common roundworm
EUCOELOMATA
PROTOSTOMIA
PHYLUM ANNELIDA
Class Polychaeta
Neanthes, sandworm
Class Oligochaeta
earthworm
Class Hirudinea
medicinal leech
PHYLUM MOLLUSCA
Class Monoplacophora
Neopilina
Class Pelecypoda
clam
Class Amphineura
chiton
Class Gastropoda
snail
Class Cephalopoda
Octopus
PHYLUM ONYCHOPHORA
"Claw bearer" animals
Peripatus

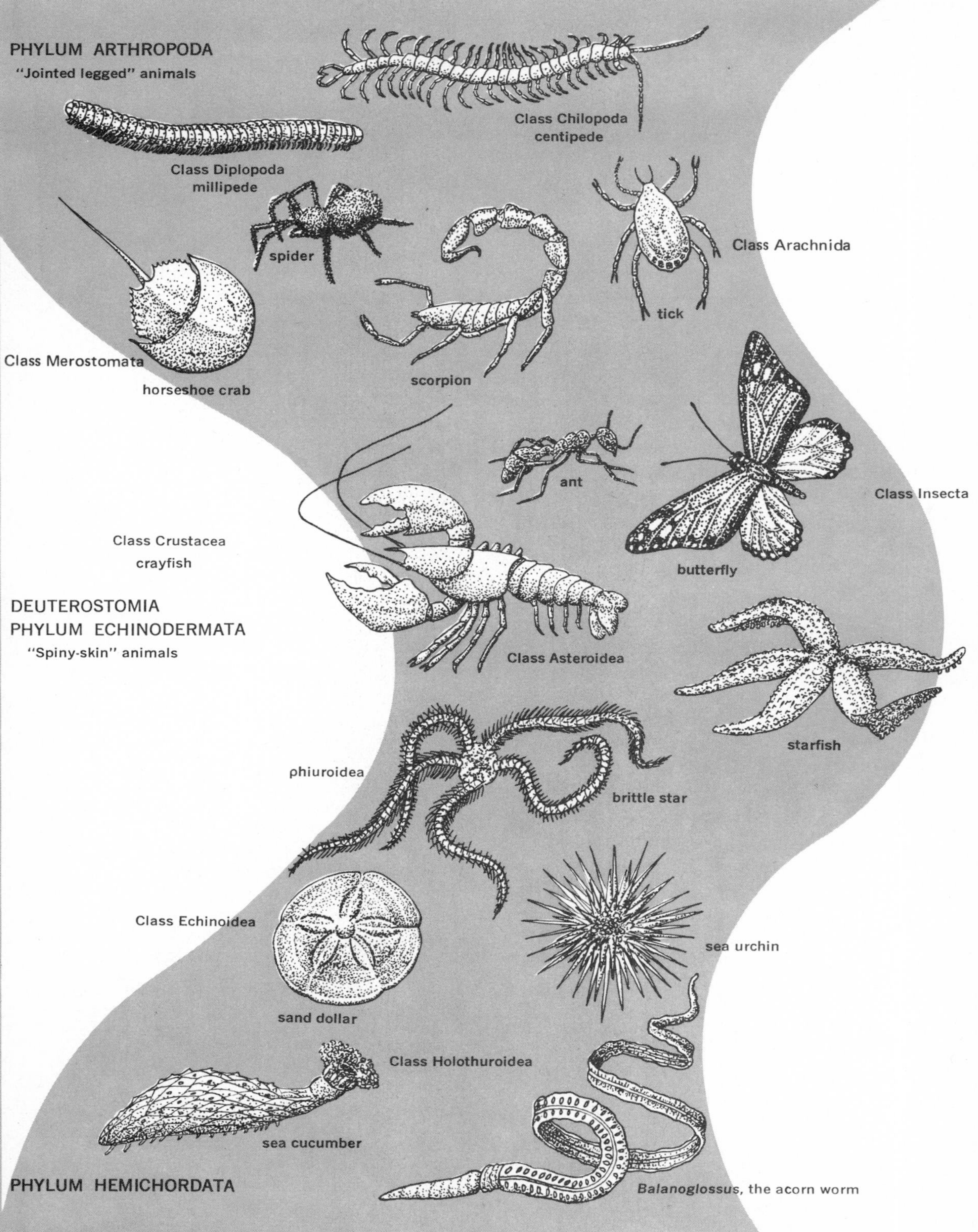
PHYLUM ARTHROPODA
"Jointed legged" animals
Class Chilopoda
centipede
Class Diplopoda
millipede
spider
Class Arachnida
tick
Class Merostomata
horseshoe crab
scorpion
ant
Class Insecta
butterfly
Class Crustacea
crayfish
DEUTEROSTOMIA
PHYLUM ECHINODERMATA
"Spiny-skin" animals
Class Asteroidea
starfish
phiuroidea
brittle star
Class Echinoidea
sea urchin
sand dollar
Class Holothuroidea
sea cucumber
PHYLUM HEMICHORDATA
Balanoglossus, the acorn worm

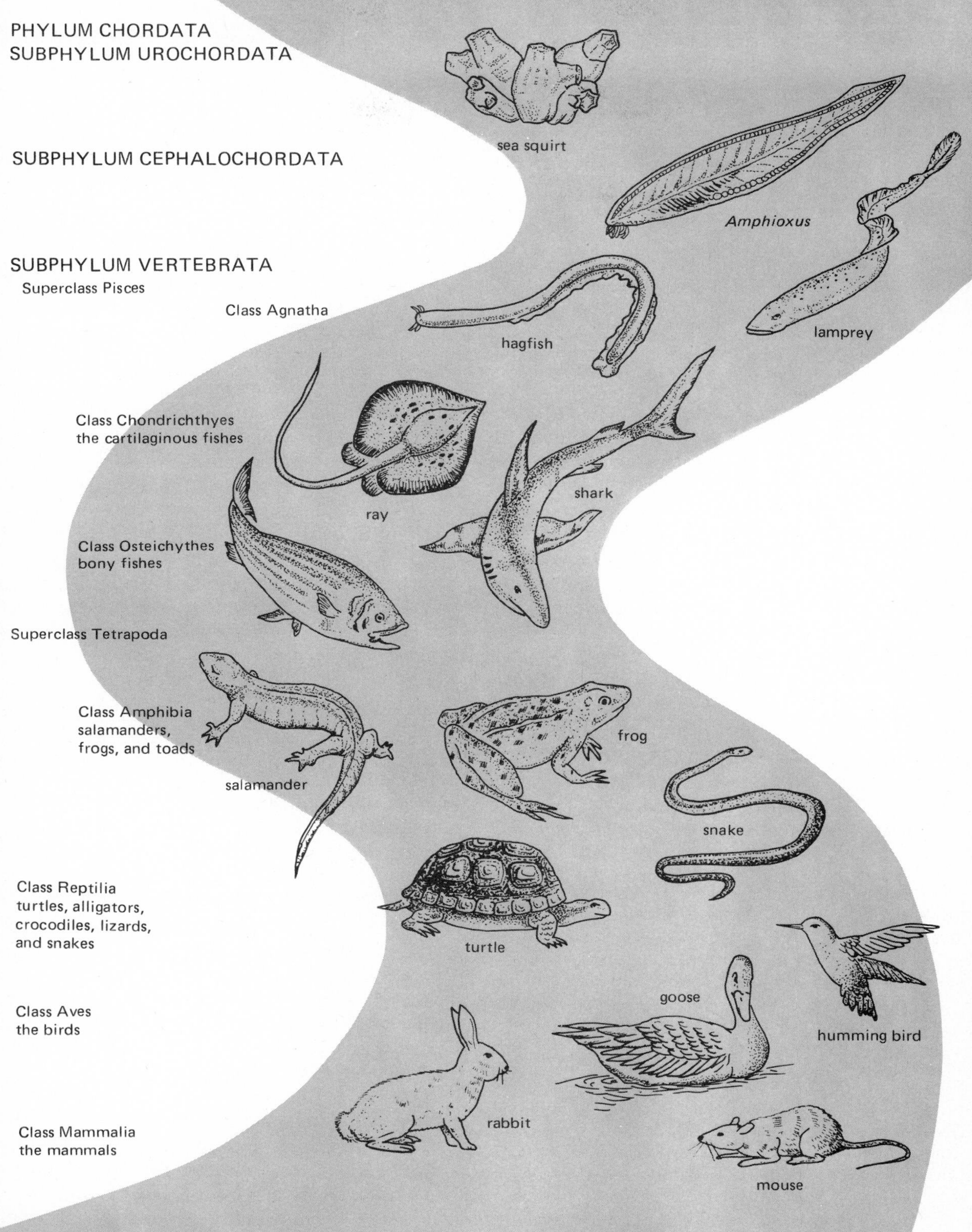
PHYLUM CHORDATA
SUBPHYLUM UROCHORDATA
SUBPHYLUM CEPHALOCHORDATA
SUBPHYLUM VERTEBRATA
Superclass Pisces
Class Agnatha
Class Chondrichthyes
the cartilaginous fishes
Class Osteichythes
bony fishes
Superclass Tetrapoda
Class Amphibia
salamanders,
frogs, and toads
Class Reptilia
turtles, alligators,
crocodiles, lizards,
and snakes
Class Aves
the birds
Class Mammalia
the mammals
sea squirt
Amphioxus
hagfish
lamprey
ray
shark
frog
salamander
snake
turtle
goose
humming bird
rabbit
mouse

20 The Naming and Classification of Animals

Every species of organism is a unique genetic system, differing from all other species in structural, physiological, behavioral, and other characteristics. Every biologist must determine the identity of the organism with which he is working to make sure that his observations and experiments are meaningful. A classification of the diverse species is an indispensable information retrieval system. The name of an organism is like the index number of a file; it gives immediate access to all the information existing about it. The scientific name is the key to the entire literature that deals with a particular species or higher taxon. Evolutionary relationships among organisms, revealed by classification, represent an important source of information for every kind of biological study.[1]

HISTORICAL BACKGROUND

The naming of animals has been going on since language first appeared in human civilization. We cannot talk about the different things in our environment without having names for them. In the course of human civilization over one million different animals have been named, and the job is by no means completed. This is not saying, however, that only a million names have been used for the million different kinds of animals. The same kind of organism has been given a different name (so-called common name) in different countries where different languages are used and even different names in regions where there is a common language. In different parts of the United States the same snake is known as the common water snake, the water moccasin, the water copper, and the banded water snake. The same thing is true for a great many other animals.

The mere naming of animals without organization becomes an unwieldy hodgepodge. For scientific use of the names there must be some kind of classification. Actually, science is sometimes defined as accumulated knowledge systematized and formulated with reference to the discovery of general truths, or the operation of general laws.

Since ancient times, humans have devised systems of classifying animals as well as giving them names. Perhaps the first significant attempt to classify was made by the Greek scholar, Aristotle. He divided animals into air dwellers, water dwellers, and land dwellers. This classification was quite artificial and, for our purposes today, quite unsatisfactory; yet it was a beginning. Nothing better was developed during the Roman period. St. Augustine divided all animals into three groups–useful, harmful, and superfluous. A more effective system of classification was finally begun with the work of John Ray in England in the seventeenth century and the work of Linnaeus (Karl von Linnè) in Sweden in the eighteenth century. Ray's great contribution was his clear-cut formulation of the concept of *species,* a term that had been used previously to indicate various assemblages of animals or plants. Ray defined a species as an assemblage of individuals derived from similar parents and having the ability to pass parental characteristics to the next generation (Ch. 42). Linnaeus gave us the binomial system of naming organisms which meant that the same organism would go

[1] From P. Handler, (Ed.). "The Diversity of Life," in *Biology and the Future of Man.* New York: Oxford Univ. Press (1970), Ch. 13.

by the same name in all parts of the world in scientific writings since the name in each case was in Latin. His *Systema Naturae,* the tenth edition of which was published in 1758, is regarded as the beginning of present-day taxonomy. He arranged many plants and animals in such a way that they could be identified easily. One of the rules he established was that each organism should have a name of two words—the first, the *genus* name, and the second, the *specific* name. (The genus name is always placed first and is capitalized, while the specific name is placed after the generic name and is not capitalized. Both are printed in italics. If the scientific name is handwritten or typed with a conventional typewriter, underlining is used to indicate italics. The species name designates a group of individuals that resemble each other very closely (more on this later in the chapter). Several species that are much alike constitute a genus. The generic name *Felis* is used for house cats and other smaller members of the cat family. The name of the house cat is *Felis domestica;* the ocelot, *Felis pardalis;* the bobcat, *Felis rufa.* The large cats are in the genus *Panthera,* the lion being *Panthera leo,* the tiger *Panthera tigris.*

The name of the person who first described the animal, or his initial, completes the scientific name, as in *Canis familiaris* L. The initial here is for Linnaeus who first used this name to denote the common dog.

The institution of the binomial system was a great advance, but as more and more species were described and named, problems did arise. It was inevitable that different workers would use different names for the same species and also the same name for different species. In time, both botanists and zoologists in international congresses formulated codes of nomenclature to settle such problems, and today international commissions interpret these codes and rule on cases called to their attention. In general the rule of priority operates: the first published name is the valid name for a group.

Linnaeus classed together animals that were similar in structure, but to him this did not mean genetic relationship. At the time that Linnaeus made his contributions, almost all biologists believed in the special creation of each species. It was not until more than a century later, after the writings of Darwin had appeared, that belief in evolution became common among biologists. From the evolutionary point of view, the presence of fundamental structural similarities in two species is explained on the basis that both species were derived from a common ancestor. This is in contrast with the position of Linnaeus, which was that the Creator had just made them similar. However, the fact that Linnaeus used structural similarities as the basis for his system of classification made it possible for later workers to take over much of it. The system used today is one that reflects the actual relationships of the organisms classified. To Linnaeus the presence of the same bones in the forelimbs of such diverse forms as bats, frogs, birds, dogs, and man was due to the plan of the Creator at the time each was separately created. To the taxonomist or zoologist of today these bones are *homologous structures* (ones that have the same parts, the same arrangement of parts, and similar embryonic origins, but not necessarily the same functions) and their presence in all of these different forms indicates relationship.

TAXONOMIC CATEGORIES

Linnaeus divided all living things, as is done by some today, into two great *kingdoms*—the animal kingdom and the plant kingdom. In referring to the two great divisions, Linnaeus had this to say by way of differentiation. Plants "are bodies organized, and have life and not sensation," while animals "are bodies organized, and have life, sensation, and the power of locomotion." Today there are no absolutely satisfactory criteria for deciding whether an organism is a plant or an animal. In most cases it is easy, but in some cases, as will be pointed out in later discussions, it is not.

As mentioned above, Linnaeus grouped into a genus different species that showed great similarity. In addition to the genus and species, he used two other categories in his system—*order* and *class.* One or more genera were placed together to form an order, and one or more orders were placed together to form a class. Each grouping or category was formed on the basis of certain similarities. In his classification of animals, he used only six classes—mammals, birds, amphibians, fishes, insects, and worms. He included only seven orders in class Mammalia, and only one of these as such, order Primates, is retained in our present system.

In the modern system of classification, other categories have been added. Between the genus category and the order category, the *family* category is now used. Also, the highest category in the modern system is not the class but the *phylum.* In addition, suborder, subclass, subphylum, and so on, are used. By adding either prefix sub- or super- to a category, that is, subphylum, superorder, the taxonomist can add refinements to the evolutionary history of a group.

A few examples will help to clarify the system in use today. The phylum Chordata includes all animals that possess, at some time in their existence, the following four characteristics: (1) a rodlike skeletal structure in the mid-dorsal line called the *notochord*; (2) a dorsal, hollow nervous system; (3) lateral pouches from the pharynx (pharyngeal pouches) that may open to the outside by *pharyngeal clefts*; (4) post anal tail.

There are three subphyla included in this phylum, one of which is the subphylum Vertebrata. An obvious characteristic of all vertebrates, as the name implies, is the possession of a vertebral column. There are seven living classes of vertebrates, as follows:

Class Agnatha	Lampreys and hagfishes (no jaws or limbs)
Class Chondrichthyes	Cartilaginous fishes (all the fishes have gills and have fins as appendages)
Class Osteichthyes	Bony fishes
Class Amphibia	Salamanders, frogs, and toads (smooth skin, gills in larval stage, lungs in adults)
Class Reptilia	Turtles, lizards, snakes, crocodiles (scaly skin, breathe by lungs)
Class Aves	Birds (skin covered with feathers, forelimbs are wings)
Class Mammalia	Mammals (mammary glands, skin covered with hair)

In the Amphibia three orders are recognized as given below:

Order Gymnophiona (Apoda)	Legless amphibians
Order Urodela (Caudata)	Salamanders (amphibians with legs and a tail)
Order Salientia (Anura)	Frogs and toads (tailless amphibians)

There are several families in the order Salientia, the Bufonidae and Ranidae being the largest. The family Bufonidae includes the toads. *Bufo americanus,* the American toad, is a common species. The family Ranidae includes over 20 species of frogs. Among the familiar species are *Rana pipiens* (common grass frog); *Rana catesbeiana* (bullfrog), and *Rana sylvatica* (wood frog). The members of the toad family differ from the frog family in a few important respects. The skin of toads is somewhat rough and warty, while the skin of frogs is smooth; toads are terrestrial and nocturnal in habit and enter water only to spawn; frogs spend their life in and near water. The different species of frogs are not hard to recognize. The bullfrog is greenish in color, quite large [15–20 cm (6–8 inches) in length], and has very large ear drums. The grass frog is brownish green in color on the upper surface with two rows of brownish spots, each narrowly margined with white, and 7.5–12.5 cm (3–5 inches) in length. The wood frog is only 5–7.5 cm (2–3 inches) in length with a pale to dark brown back and with the ear regions dark.

The mammals that are living on the earth today are placed in 18 different orders. The order Primates, which includes humans (Table 20.1), and the order Carnivora, which includes the different flesh-eating mammals, are two orders familiar to everyone. Among the families in the order Carnivora are the Felidae (cats), Canidae (dogs), Ursidae (bears), and Hyaenidae (hyenas).

Table 20.1 The Classification of Man

Kingdom	Animal
Phylum	Chordata
Subphylum	Vertebrata
Class	Mammalia
Order	Primates (superior nervous system and nails on digits)
Family	Hominidae (no tail or cheek pouches and nonopposable big toe)
Genus	*Homo* (humanlike)
Species	*Homo sapiens* (present-day human)

Different species of the cat family have already been mentioned. In the dog family are such species as *Canis familiaris* (common dog), *Canis nubilis* (timber wolf), *Canis latrans* (coyote), and *Vulpes vulpes* (red fox). Each species is separate and distinct from the others in the same genus because of certain structural differences and also because it is reproductively isolated from the other species, that is, its members cannot or do not normally mate with members of other species and produce fertile offspring. The problem of what constitutes a species is a very important one in the study of evolution as well as in classification and we shall return to it in the discussion on evolution.

CRITERIA FOR THE CLASSIFICATION OF ANIMALS

Knowing that about one million different kinds of animals have been named and classified might, on first thought, make the classification of several animals appear as a very difficult task. Actually, however, the identification to the proper phylum (and also class) of most animals is not difficult. In most cases this can be done by recognizing a few basic characteristics. Classification beyond the class to order, family, genus, and species is more difficult. Below are listed some of the more important structural criteria used in classifying an animal into its proper phylum.

Number of Cells

The phylum Protozoa is composed of animals that exist as a single cell or as a colony of undifferentiated cells. All other animals are multicellular with some differentiation and division of labor between and among the cells. All such organisms are grouped in the subkingdom Metazoa. About 28 metazoan phyla are recognized.

Number of Germ Layers

An important feature in differentiating the phyla of Metazoa is whether the body is formed from two or three embryonic germ layers. In one class of the phylum Coelenterata (class Hydrozoa) there are only two embryonic germ layers, the *ectoderm* and the *endoderm,* which give rise to all the adult structures. This condition is called *diploblastic.* The Porifera (sponges) have two layers of cells, but their embryological origin is such that these layers cannot be homologized with ectoderm and endoderm of higher forms. Hence the use of the term *diploblastic* for the sponges is not advisable. The small group of parasites which comprise the phylum Mesozoa also does not qualify, since the body is made up of an external layer of cells with a single reproductive cell or multinucleate mass within. These layers are definitely not ectoderm and endoderm. In all the higher phyla, as well as in the other two classes of the phylum Coelenterata, three embryonic germ layers are formed, *ectoderm, endoderm,* and *mesoderm.* This condition—possessing three germ layers—is known as *triploblastic.*

Nature of the Digestive Structures and the Openings to the Digestive Tract

In all the Metazoa above the sponges there is a digestive cavity (except in certain parasitic organisms). In sponges digestion is entirely an intracellular process. In some of the animals which possess a digestive cavity, there is only one opening to the cavity, the *mouth.* This situation is found in the Coelenterates and in the Platyhelminthes (flatworms). The members of the other phyla have, in addition to a mouth, an *anal opening.*

The Presence of a Coelom

In all the triploblastic organisms above the flatworms there is a cavity between the body wall and the gut. In the Acanthocephala, Aschelminthes, and Entoprocta the cavity is not a cavity in the mesoderm and it is called a *pseudocoelom.* In all the other metazoans above these phyla, the body cavity is a cavity in the mesoderm and is thus a true *coelom.* In the Mollusca (snails, clams, and so on), the arthropods, and in three small phyla closely related to the arthropods the coelom is greatly reduced.

Segmentation

In the three most highly organized animal phyla, annelids, arthropods, and chordates, there is a linear repetition of parts that is called *metamerism* or *segmentation.* This is quite obvious in an external examination of an earthworm (phylum Annelida) or a grasshopper (phylum Arthropoda). The segments of an earthworm are essentially alike, while many of those in the grasshopper and other arthropods are quite modified. In Chordates the segmentation is largely internal, involving primarily muscles and nerves and, in the vertebrates, the vertebrae. The segmentation in chordates shows up clearly in the embryo stage in the segmentally arranged *mesodermal somites.* The mesodermal somites are blocks of tissue that differentiate on either side of the middorsal line during development and subsequently give rise to muscles and skeletal elements.

Skeleton

In most of the large multicellular animals and in the fast-moving smaller ones, some kind of skeleton is found that serves for protection and support and in locomotion. In chordates the skeleton is internal and is called an *endoskeleton.* In some of the other phyla, such as the mollusks and arthropods, the skeleton is external and is called an *exoskeleton.* In one major group of the chordates the basic chordate supporting structure, the notochord, has been replaced by a segmented vertebral column, leading to the use of the term *Vertebrata* for this group of animals, and to the use of the term *Invertebrata* for all other members of the animal kingdom. The term Invertebrata, however, does not have category significance in classification as does the term Vertebrata. Vertebrata is a subphylum of the phylum Chordata; Invertebrata is simply a term used in referring to all other animals besides the vertebrates.

Appendages

The structures that protrude from the body of an animal and that are used in locomotion and feeding are called appendages. Among the different kinds found are the *tentacles,* which encircle the mouths of coelenterates, the fleshy *parapodia* of annelids, the *muscular foot* of mollusks, the *legs* of arthropods, and the *fins, legs,* and *wings* of vertebrates. The legs of all arthropods and the legs and wings of vertebrates are characteristically *jointed.*

Symmetry

Most animals are so constructed that they exhibit either *radial symmetry* or *bilateral symmetry,* although some are *asymmetrical.* The Coelenterata have radial symmetry; they are constructed like a bottle. In the case of a rubber bottle it is possible to cut it into two equal halves by making cuts through any number of planes that pass through the central axis from the mouth to the base. A cake or a pie has radial symmetry, and each may be cut into equal halves through any number of planes. The same thing may be done to any animal with radial symmetry. Animals that have bilateral symmetry are quite different. Their body parts are so arranged on two sides with reference to a central plane that that they can be cut into two equal halves (which are mirror images) by only one particular cut running along the central plane.

With the development of bilateral symmetry come *anterior* and *posterior* ends, *right* and *left* sides, *dorsal* and *ventral* surfaces. Most bilaterally symmetrical animals move with one particular part of the body always in front. Thus a definite head is found in such organisms. The development of bilateral symmetry in the evolutionary process led to more efficient locomotion, among other things.

Radial symmetry is found in some sponges, but it is most characteristic of coelenterates and ctenophores. Such organisms have *oral* and *aboral* ends. They do not have anterior and posterior ends; they do not have sides, nor heads, nor dorsal and ventral surfaces. Radial animals often lead a sessile life, and their locomotion is usually quite slow.

All the animals above the ctenophores have bilateral symmetry except the members of phylum Echinodermata (starfishes, sea urchins, and so on). It is known, however, from embryological studies of echinoderms that the radial symmetry of the adult forms has been superimposed upon a basic bilateral symmetry that is present in echinoderm embryos.

THE ANIMAL KINGDOM

Although it is usually not possible to place a given animal in its proper phylum by using just one of the above criteria, it is possible to do the job with relative ease by using these criteria in certain combinations.

Table 20.2 gives an outline of the breakdown of the animal kingdom into phyla, and Table 20.3 gives the estimated numbers of species in the various major groups. In Table 20.4 the distinguishing characteristics of the various subdivisions of the animal kingdom are given.

Table 20.2 An Outline of Animal Classification

Kingdom Animalia	
Subkingdom Protozoa	
	Phylum Protozoa
Subkingdom Metazoa	
Branch Mesozoa	
	Phylum Mesozoa
Branch Parazoa	
	Phylum Porifera
Branch Eumetazoa	
Grade Radiata	
	Phylum Coelenterata
	Phylum Ctenophora
Grade Bilateria	
Acoelomata	Phylum Platyhelminthes
	Phylum Rhynchocoela
	Phylum Gnathostomulida
Pseudocoelomata	
	Phylum Acanthocephala
	Phylum Aschelminthes[a]
	Phylum Entoprocta
Eucoelomata	
Protostomia	
	Phylum Mollusca
	Phylum Priapulida
	Phylum Sipunculida
	Phylum Echiurida
	Phylum Annelida
	Phylum Onychophora
	Phylum Tardigrada
	Phylum Pentastomida
	Phylum Arthropoda
	Phylum Ectoprocta
	Phylum Phoronida
	Phylum Brachiopoda
Deuterostomia	
	Phylum Chaetognatha
	Phylum Echinodermata
	Phylum Pogonophora
	Phylum Hemichordata
	Phylum Chordata

[a] This phylum includes the following classes: Rotifera, Gastrotricha, Kinorhyncha, Nematoda, and Nematomorpha. Many workers consider each of these as a separate phylum.

Table 20.3 Estimated Numbers of Known Species of Recent Animals[a]

Protozoa	30,000
Mesozoa	50
Porifera	4,500
Coelenterata	9,000
Ctenophora	90
Platyhelminthes	15,000
Rhynchocoela	750
Gnathostomulida	100
Acanthocephala	300
Aschelminthes	
Class Rotifera	1,500
Class Gastrothricha	175
Class Kinorhyncha	100
Class Nematoda	10,000
Class Nematomorpha	100
Entoprocta	60
Mollusca	100,000
Priapulida	9
Sipunculida	250
Echiurida	60
Annelida	7,000
Onychophora	65
Tardigrada	280
Pentastomida	70
Arthropoda	
Chelicerata	35,000
Crustacea	25,000
Insecta	850,000
Other arthropods	13,000
Phoronida	15
Ectoprocta	3,300
Brachiopoda	250
Chaetognatha	50
Echinodermata	5,700
Pogonophora	125
Hemichordata	80
Chordata	
Urochordata	1,600
Cephalochordata	20
Vertebrata	
Fishes	18,000
Amphibia	2,500
Reptilia	6,000
Aves	8,590
Mammalia	4,500
Total	1,153,189

[a] Modified after Lord Rothschild. *A Classification of Living Animals.* Longman's Green & Co., Ltd., and Wiley (1961).

PROBLEMS OF CLASSIFICATION

The system of classification used today has enabled us to organize living things in an orderly and systematic way. This is important to all workers in the field of biology. The validity of all biological research is dependent on repeatability. Unless a worker can accurately identify the organism used in his research, no one else will be able to repeat, and thus verify, his work.

Table 20.4 Classification of the Animal Kingdom [1]

SUBKINGDOM PROTOZOA. Unicellular (acellular) animals. Amoeba, paramecium, and so on.

Phylum Protozoa

SUBKINGDOM METAZOA. Cellular animals, composed of cells, which may lose their boundaries in the adult state.

Branch A. Mesozoa. Cellular animals having the structure of a stereogastrula composed of a surface layer of somatic cells and interior reproductive cells.

Phylum Mesozoa

Branch B. Parazoa. Animals of the cellular grade of construction with incipient tissue formation, interior cells of several different kinds, without organ systems, digestive tract, or mouth; porous with one to many internal cavities lined by choanocytes. Sponges.

Phylum Porifera

Branch C. Eumetazoa. Animals of the tissue or organ system grade of construction, with mouth and digestive tract (except when lost by parasitic degeneration), interior cells reproductive only in part, not porous, without body spaces lined by choanocytes.

Grade I. Radiata. Eumetazoa with primary radial symmetry, of the tissue grade of construction with incipient organ systems, incipient mesoderm mesenchymal in nature and mostly of ectodermal origin, digestive cavity the sole body space, no anus.

1. Symmetry radial, biradial, or radiobilateral, mouth usually encircled by tentacles armed with nematocysts, no rows of ciliated plates. Hydra, jellyfishes, corals, and anemones.

Phylum Coelenterata

2. Symmetry biradial, tentacles when present not encircling the mouth, no nematocysts, eight radial rows of ciliated swimming plates. Comb jellies.

Phylum Ctenophora

Table 20.4 Classification of the Animal Kingdom *(Continued)*

Grade II. Bilateria. Eumetazoa with bilateral symmetry or secondary radial symmetry, of the organ-system grade of construction, mostly with a well-developed mesoderm of endodermal origin, mostly with body spaces other than the digestive cavity, anus generally present.

Division I. Protostomia. Bilateria in which the mouth arises from the blastopore or from the region of the blastopore; cleavage is spiral and determinate for the most part; mesoderm typically arises from a special cell set aside in early cleavage (the 4d cell); coelom, when present, arises typically as a split in the mesoderm—a schizocoel; the characteristic larval form is the trochophore, although all protostomes do not possess it.

Subgrade Acoelomata. Region between digestive tract and body wall filled with Mesenchyme, excretory organs of protonephridial type with flame bulbs, unsegmented, or if segmented, youngest segments nearest the head.

1. Anus absent

a. Jaws absent. Flatworms.

Phylum Platyhelminthes

b. Jaws Present. Gnathostomulids.

Phylum Gnathostomulida

2. Anus present, eversible proboscis in hollow sheath above the digestive tract. Ribbon worms.

Phylum Rhynchocoela

Subgrade Pseudocoelomata. Space between the digestive tract and body wall, but this space is a pseudocoel—the unlined remnant of the blastocoel—and is not a true coelom, anus present, with or without protonephridia, flame bulbs may be present.

1. Endoparasitic, vermiform, without a digestive tract, anterior end an invaginable proboscis armed with hooks. Spiny-headed worms.

Phylum Acanthocephala

2. Intestine more or less straight, anus posterior, no anterior ciliated projections except in a few rotifers.

Phylum Aschelminthes

(The Phylum Aschelminthes includes the following classes: Rotifera, Gastrotricha, Kinorhyncha, Nematoda, and Nematomorpha. Many authors treat each of these classes as a separate phylum.)

3. Intestine looped, bringing anus near mouth, mouth and anus encircled by a circlet of ciliated tentacles.

Phylum Entoprocta

Subgrade Coelomata. Have a true coelom and usually a well-developed endomesoderm, excretory organs when present are protonephridia, with solenocytes or metanephridia, with or without nephrostomes, anus normally present.

1. Without a lophophore, coelom a schizocoel.

a. Mostly unsegmented.

a′. Visceral mass covered by a body fold, the mantle, which in most species secretes a calcareous shell of one or more pieces; coelom usually reduced; one small class shows evidence of segmentation in arrangement of some organs. Clams, snails, squids, and so on.

Phylum Mollusca

b′. Without a mantle, naked, wormlike, coelom spacious.

a″. Bulbous spiny anterior end is eversible; body shows superficial segmentation; anus posterior.

Phylum Priapulida

b″. Eversible proboscis, never segmented, anus anterior and dorsal. Peanut worms.

Phylum Sipunculida

c″. Proboscis not eversible; adults unsegmented but there is some segmentation in larval states; anus posterior and terminal. Spoon worms.

Phylum Echiurida

b. Segmented.

a′. Without jointed appendages.

a″. No appendages, or if unjointed appendages are present they have no terminal claws. Earthworms, leeches, and so on.

Phylum Annelida

b″. Unjointed appendages bearing claws in a terminal position present at some stage of the life cycle.

a‴. Open circulatory system, respiration by trachea, excretion by nephridia. Peripatus.

Phylum Onychophora

b‴. No circulatory system, no trachea, no nephridia present.

a⁗. Minute, free-living forms, poorly segmented. Water bears.

Phylum Tardigrada

b⁗. Wormlike endoparasites of vertebrates; no appendages in the adult; larvae with two pairs of unjointed clawed appendages. Nose worms, tongue worms.

Phylum Pentastomida

b′. With jointed appendages. Insects, spiders, crabs, and so on.

Phylum Arthropoda

Table 20.4 Classification of the Animal Kingdom *(Continued)*

2. With a circular or crescentic or doubly spirally coiled ridge, the lophophore, bearing ciliated tentacles; intestine looped, bringing anus near the mouth; coelom originates either as a schizocoel or as an enterocoel; cleavage radial or spiral.

a. Solitary with one pair of metanephridia, and a closed circulatory system.

a′. Wormlike, secretes leathery tube.

Phylum Phoronida

b′. With a bivalve shell, valves dorsal and ventral. Lamp shells.

Phylum Brachipoda

b. Colonial with gelatinous, chitinous, or calcareous encasements; no nephridia or circulatory system, anus outside lophophore ring. Moss animals.

Phylum Ectoprocta

Division II. Deuterostomia. Coelomate bilateria in which the anus arises from the blastopore or from the region of the blastopore; cleavage is radial and indeterminate; mesoderm arises from outpocketing of the archenteron as does the coelom which is thus an enterocoel; dipleurula larva characteristic of some.

1. With a secondary, usually pentamerous, radial symmetry; water vascular system present. Starfish, sea urchins, crinoids, sea cucumbers.

Phylum Echinodermata

2. Bilateral symmetry retained throughout life.

a. Without pharyngeal clefts or endoskeleton.

a′. Tube-dwelling deep sea forms with no digestive tract at any time in their life cycle. Beard worms.

Phylum Pogonophora

b′. Small pelagic, predaceous marine forms with horizontal fins and spiny jaws. Arrow worms.

Phylum Chaetognatha

b. With pharyngeal clefts, endoskeleton, or both.

a′. Without a typical notochord in adult or embryonic stage. Acorn worms.

Phylum Hemichordata

b′. Embryo with a notochord, adults with pharyngeal clefts or vertebral column, or both. Tunicates, amphioxus, and vertebrates.

Phylum Chordata

Adapted from Hyman, L. H. *The Invertebrates,* McGraw-Hill, 1940, with permission.

But the modern system of classification is very important for another reason: it is based on relationships of organisms, which is the core of evolutionary studies. In a very real sense, then, taxonomic studies are studies in evolution.

There are problems in taxonomic studies and a few of them should be mentioned. There are even problems with respect to the number of kingdoms of organisms. Organisms have been separated into two kingdoms, plant and animal, which has been common practice for a long time. But some workers today use a third kingdom, *Protista* ("the first ones"). This proposal was first made by Haeckel in the last century. Some workers who use this third kingdom place all the protozoa (unicellular animals) and all the thallophytes (algae, fungi, bacteria, and slime molds) in this kingdom. Others who use this kingdom are more restrictive and include only the protozoa, the plantlike flagellates, the bacteria, and the slime molds.

The reasons for setting up a third kingdom include the following. Most of the organisms included are unicellular (sometimes called *acellular*), they all have very ancient origins, and their relationships are vague and not well understood. Furthermore, some of them have characteristics of *both* plants and animals.

A few workers use a fourth kingdom, *Monera.* Those who use this system also use Protista, but they separate the bacteria and the blue-green algae from the Protista and place them in the fourth kingdom. There seems to be little doubt that these two groups represent the most primitive types of living organisms. There are problems here. The relationships between these groups are vague. It is true that in the flagellates some are found that are truly plantlike, some are truly animallike, and some have characteristics of both plants and animals. Many workers consider that the ancestry of both plant and animal lines involves ancient flagellated forms. In an evolving world, perhaps what is found in the flagellates is to be expected. The typical mode of plant nutrition is photosynthesis as found in the higher green plants, but many of the organisms classed in Protista and Monera are capable of photosynthetic activity. The typical method of animal nutrition is holozoic (ingestion of solid organic matter), and many of the organisms placed in the Protista are of this type. So it would appear that no basic problems are solved by the creation of one or two more kingdoms; their creation, however, does call attention to the problems. To some this may seem to have some advantages over the two-kingdom system. (Some students of classification use a fifth kingdom for the fungi.) Each student will have to decide for himself.

The species is the fundamental unit of classification and there are even problems at this level. One problem involves the transition from the system of Linnaeus to the modern system. Under the system of Linnaeus each species was a fixed entity and was thought of in terms of a *type* or *archetype.* Variations between members of a species were either ignored or considered nuisances. After the evolutionary viewpoint was accepted, the idea of a type, however, was still retained. The type in this case was not an abstraction but rather was a specific unique specimen housed in a museum or herbarium. In practice, other organisms were compared with these type specimens to determine which type each most closely resembled. This practice, however, does not take into account the great variability that may occur within a group of organisms that function in nature as a species. What is a species? Not all workers agree, and no one simple definition seems to be applicable to all kinds of organisms. The following definition of a species, advanced by Mayr, is considered by many biologists to be a good working definition for sexually reproducing organisms. *Species are groups of actually or potentially interbreeding populations, which are reproductively isolated from other such groups.* This definition emphasizes that a species is a *population* of organisms. In characterizing a species, therefore, it is necessary to know the range of variation in the different characteristics of the group. The definition emphasizes also that members of a species do or can interbreed and produce fertile offspring and that each species is reproductively isolated from other species. [The student interested in uniparental species, which reproduce asexually, should consult G. G. Simpson (see Suggestions for Further Reading).]

But this definition is not truly universal. There are some cases where members of two recognized species do interbreed and produce fertile offspring. More cases of this are known in plants than in animals. Then there is the practical matter of knowing whether a given specimen does or can mate with members of more than one group. Often a specialist is asked to place a dead specimen, far removed from its natural habitat, in the proper species category. In such instances he must depend largely on the structural features of the specimen and his knowledge of similar organisms. Before reaching a decision he must compare the structural features of the specimen with the ranges in the different characteristics found in the population that constitutes the species, and find if the specimen fits.

There are still other problems and perhaps one more example should be mentioned. Each reader is probably acquainted with the common grass frog, *Rana pipiens.* This species ranges over most of the eastern half of the United States, from Minnesota to Texas and from New England to Florida. Specimens of this species look very much alike wherever they are collected. John Moore has found that individual frogs from Minnesota when mated with individuals from Texas do not produce viable offspring; yet individuals from closely adjacent regions throughout this whole geographic range can mate and produce viable and fertile offspring. There are other cases of a similar nature known. In other words, in species that occupy a very wide geographic range it sometimes happens that the subpopulations at the extremes of the range, when brought together, are not able to function as members of the same species, according to the definition given previously. This indicates, for one thing, that a population which we call a species is in a dynamic state and not a static one. It further suggests how new species might arise. Later on, after the discussions on genetics, this subject will be taken up again in Chapter 42.

While similar structural features, indicating relationship, are very important in the work of taxonomists, more and more taxonomists today are no longer primarily type oriented. They take into account many aspects of the organisms with which they work: the habitat and ecological adaptations, behavior, development and life cycles, physiology, chromosome morphology, and biochemistry. A few examples will illustrate this approach.

For a long time there were differences of opinion as to the proper placement in the scheme of classification of such animals as barnacles, sea squirts, and the parasite, *Sacculina,* which lives in crabs. Those who have been to the seashore are familiar with barnacles. These sessile animals in their adult form show no obvious resemblances to common crustaceans (lobsters, crabs). A careful study of their early development and life cycle, however, revealed that they pass through a larval stage that is very similar to many other crustaceans. The same thing can be said about *Sacculina.* In the adult form it has no hard parts, no appendages, and no gut; in fact, it resembles no other animal, yet it was found to have a characteristic crustacean larval development (Fig. 30.14). Sea squirts are also common on the seashore. These sessile and anomalous creatures look like cellophane bags with two openings in them for the entrance and exit of water. As adults their affinities are certainly not clear. Yet in their development they have a swimming larval stage that is tadpolelike, and during this stage in their life cycle they have a distinct dorsal hollow nervous system with an underlying notochord. These features clearly make them members of the phylum Chordata (Fig. 33.6).

Chromosome morphology may also play an important role in differentiating between groups. There are many examples of this and one will suffice to illustrate. Most biology students become well acquainted with the common fruit fly, *Drosophila melanogaster.* There are a number of closely related species in the genus *Drosophila.* Individuals of *Drosophila melanogaster* have four pairs of chromosomes (Fig. 17.4), each pair with a characteristic shape, in each of the body cells. In another species three pairs of chromosomes are found; in still another, five pairs. Chromosome morphology alone will not differentiate between species, but it can be an important factor.

MORE RECENT APPROACHES

Biochemical similarities and differences are also used to indicate relationship, and, as such, they can be of help in problems of classification. Remember that every structural feature of an organism ultimately can be related to its biochemistry. Biochemical criteria, as previously mentioned, have long been used to identify bacteria, but only in recent years have the techniques of the biochemist been adopted by the taxonomist.

Perhaps most of the chemical taxonomy used with animals involves the nature of proteins and nucleic acids. Protein characterization may involve the primary sequencing of a particular protein, the rate of protein migration in an electric field (electrophoresis), or the antigenic similarities of proteins, to name a few.

One widely used technique is called double diffusion. First antiserum (Chapter 11) is prepared against a particular protein. The antiserum is placed in a depression made in a firm gel. Test proteins, from organisms of uncertain taxonomic position perhaps, are placed in depressions around the one with the antiserum. Each preparation diffuses into the gel, and a zone of precipitation results from the antigen–antibody interaction (Fig. 20.1). The more intense the reaction, the more closely related the organisms are said to be.

Electrophoresis of proteins on paper and on gels of various kinds is also very useful for classification. For example a particular enzyme from a variety of organisms may be electrophoresed, then stained, and the relative positions may be noted and recorded. The pattern of bands or spots represents an electrophoretogram; the patterns may be compared for differences and similarities. So it can be shown by this technique that the multiple forms of the enzyme esterase, from a cow, are distinct from those of a rat. And those of the rat are similar to the esterases of the mouse, and so on.

One of the most exciting developments in the general area of taxonomy is the use now being made of molecular hybridization techniques to establish the degree of relationship between organisms of a particular genus, family, or other taxonomic category. These hybridization techniques were originally developed by the biophysicists collaborating with Ellis T. Bolton and Brian J. McCarthy of the Carnegie Institution of Washington. Since its development, the technique has been modified in several ways.

With the original technique, high-molecular-weight DNA prepared from one organism is heated to make it single stranded; then it is put into molten agar. The hot agar is solidified by cooling. This lump of DNA-containing agar is

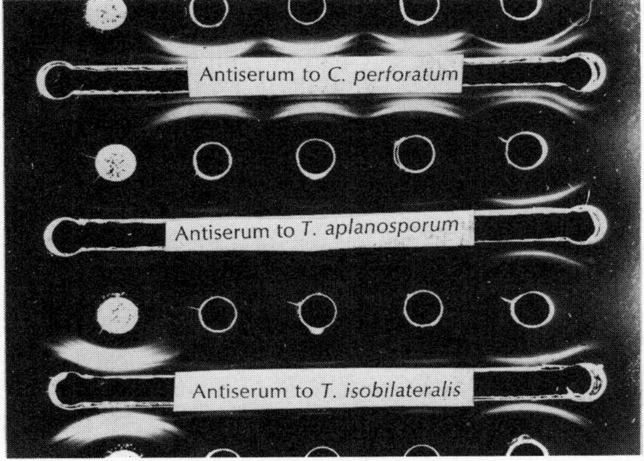

Figure 20.1 Double diffusion. Trough contains antisera to *Chlorococcum perforatum, Tetracystis isobilateralis,* and *T. aplanosporum.* Wells contain, from left to right, antigens of *T. aplanosporum, C. diplobionticum, C. perforatum, C. echinozygotum,* and *C. ellipsoidem.* The more intense the precipitin line, the closer the relationship between the test organisms. For example, of the *Chlorococcum* species, *C. ellipsoideum* is most closely related to the *Tetracystis* species. (From R. M. Brown, University of North Carolina.)

then forced through a sieve so small that fine chunks of agar result. The agar is put into a glass tube having a screen over one end, and then it is washed with a salt solution to eliminate any high-molecular-weight DNA not immobilized in the agar.

In the meantime radioactive DNA is prepared from a species that has been grown in a medium containing an isotope such as ^{32}P or ^{14}C-thymidine (a nucleoside of DNA). The resulting radioactive DNA is first fragmented into short segments and then converted into single strands by heating to a high temperature and then cooling quickly. This radioactive DNA is put into the solution that covers the agar. After about 18 hours at 60° C the agar is washed repeatedly with a solution having a high salt concentration. This treatment removes radioactive DNA that has not hybridized with the agar-bound DNA. Then by using a salt solution of low concentration the hybrid molecules can be removed from the agar. The method is shown in Figure 20.2. Since these molecules contain a radioactive strand they are easy to recognize. A hybrid then represents a molecule in which there was a complementarity in the nucleotide sequence of the immobilized and radioactive DNAs. The greater the similarity between the DNAs, the greater the relatedness and the greater the radioactivity.

In the past few years the basic technique has been modified for use with DNA–RNA hybrids. Also the agar method is technically rather difficult and often requires considerable amounts of DNA. For this reason the DNA–RNA filter method has been applied by many workers to DNA–DNA hybridizations. DNA is immobilized on a membrane filter and then incubated with radioactive DNA as described for the agar method. After being washed, the filters are counted for radioactivity. The major advantage of the filter technique is that it requires little manipulation and small amounts of DNA. Unfortunately the hybrid molecules are not easily recovered from the filters.

Using the basic technique of DNA hybridization, workers have made many interesting findings. For instance, the studies show that there

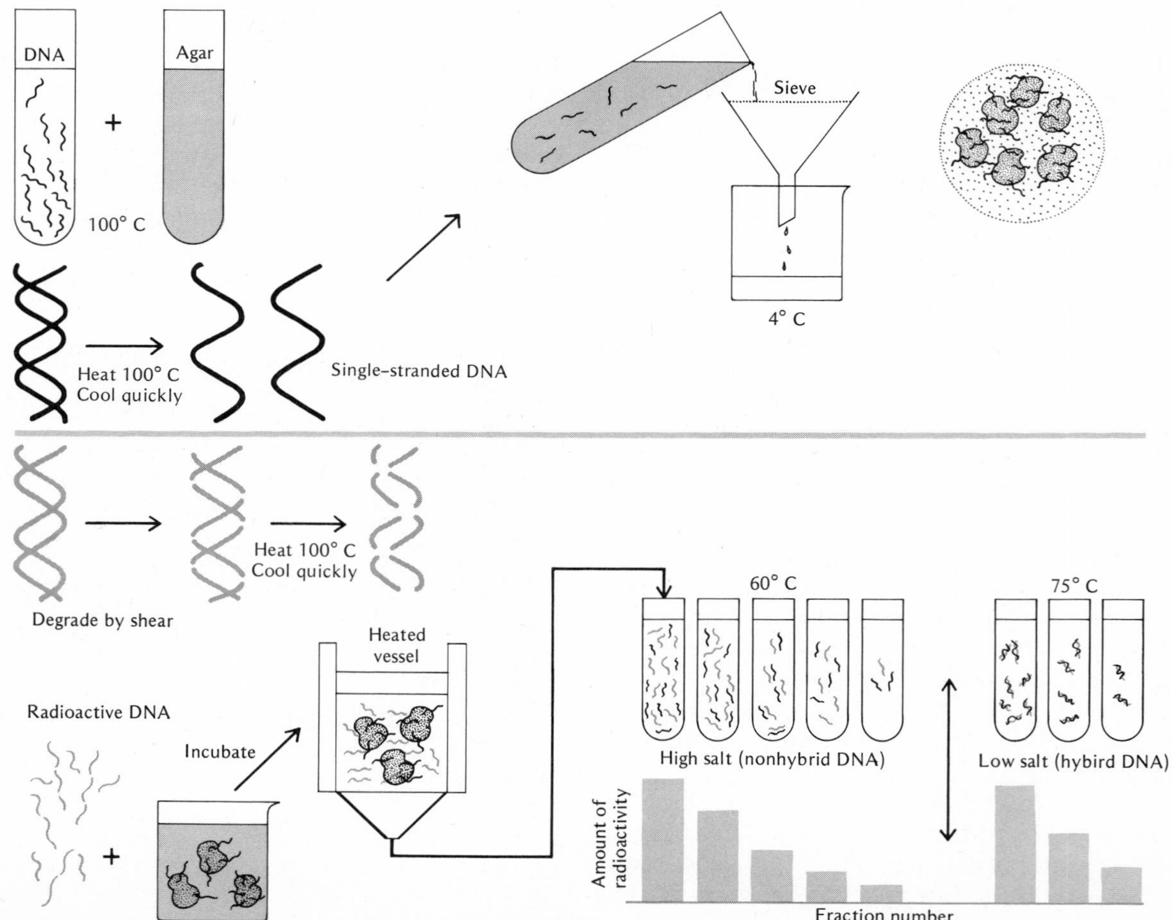

Figure 20.2 DNA–DNA hybridization technique using agar columns. (Upper right) DNA immobilized in agar.

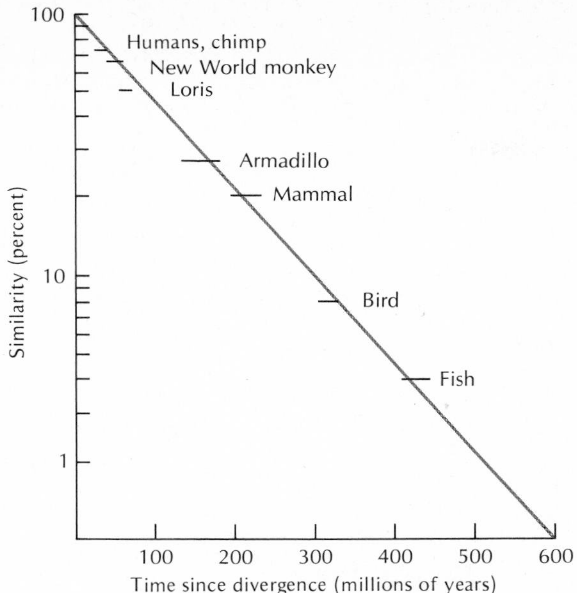

Figure 20.3 Relationship between polynucleotide similarity and time of evolutionary divergence. (From E. T. Bolton, Carnegie Institute of Washington, Year Book 63, 1963–1964, with permission.)

is a thread of hereditary continuity among all vertebrates (Fig. 20.3). In the primate group the experiments found that the Rhesus monkey and the baboon have polynucleotide sequences that are indistinguishable. They found that human and chimpanzee DNAs are very similar and that the very primitive primates, the lorises, share only about half of their polynucleotide sequences with the Rhesus monkey. About 20 percent of all mammalian DNA is shared in common.

Since these early hybridization experiments it has been shown that a large part of the DNA of eukaryotes (not prokaryotes) consists of repeated nucleotide sequences. Many individual DNA sequences in DNAs of higher organisms are repeated a large number of times, in some instances as much as one million times, while other DNA sequences in the same genome are unique (not repeated). In experiments the repetitious DNA reacts quickly while the nonrepeated DNA reacts very slowly. It is now clear that the early experiments (Fig. 20.3) were dealing with repetitious DNA. However, it is possible to show degrees of relationship in experiments with nonrepeated DNA. Kohne (see Suggestions for Further Reading), for example, has shown in *in situ* hybridization that a very high percentage of labeled nonrepeated DNA of the mouse reacts with rat DNA, while very little mouse DNA reacts with cow DNA. This is what one would expect on other grounds. Some 40–60 percent of mammalian DNA is the nonrepeated type. At present the significance of repetitious DNA in the evolution of higher organisms is not clear.

Although these sorts of studies have not significantly altered the major taxonomical categories, they have helped greatly to confirm certain suspected relationships.

Finally the involvement of mathematics in classification should be mentioned. In recent years taxonomists have made great use of the computer not only for data storage and retrieval but also for solving problems involving organisms of uncertain taxonomic position. The subdiscipline, numerical taxonomy, by assigning numerical values to morphological and biochemical characteristics or organisms, has as its goal a more quantitative approach to taxonomy.

SUGGESTIONS FOR FURTHER READING

Alston, R. E., and B. L. Turner. *Biochemical Systematics.* Englewood Cliffs, NJ: Prentice-Hall (1963).

Kohne, D. E. "Taxonomic Applications of DNA Hybridization Techniques," in *Chemotaxonomy and Serotaxonomy,* G. G. Hawkes (Ed.). London: Academic Press (1968), pp. 117–130.

Kohne, D. E., J. A. Chiscon, and B. H. Hoyer. "Evolution of Primate DNA Sequences," *J. Human Evolution,* **1,** 627–644 (1972).

Mayr, E. *Animal Species and Evolution.* Cambridge, MA: Harvard Univ. Press (1963).

Mayr, E., G. E. Linsley, and R. L. Usinger. *Methods and Principles of Systematic Zoology.* New York: McGraw-Hill (1953).

Simpson, G. G. *Principles of Animal Taxonomy.* New York: Columbia Univ. Press (1961).

From Scientific American

Britten, R. J., and D. E. Kohne. "Repeated Segments of DNA" (Apr. 1970).

Sokal, R. R. "Numerical Taxonomy" (Dec. 1966).

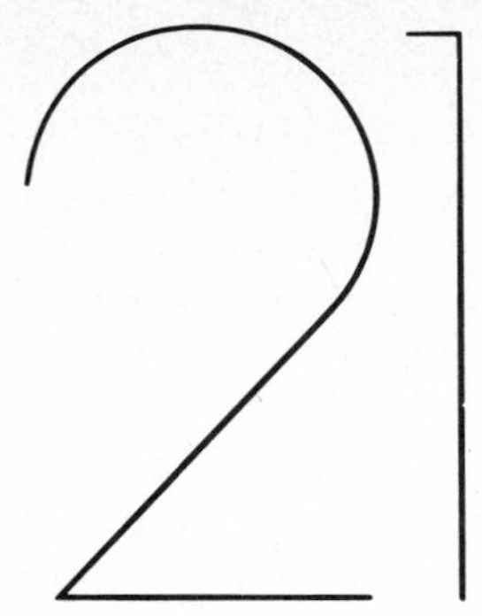

The Protozoa

PHYLUM PROTOZOA

Protozoans are animals which consist of a single cell or colony of cells. They possess no tissues or organs and the colonies differ from metazoans in that their components are all alike except when engaged in reproductive activities. More than 30,000 species have been described.

The phylum is made up of two subphyla and includes four classes.

Classification

SUBPHYLUM **Plasmodroma** Members of the subphylum Plasmodroma have but one kind of nucleus; if they reproduce sexually, it is by *syngamy,* the union of gametes; they move either by pseudopodia or flagella, if at all.

CLASS **Mastigophora** This class includes those protozoa which possess one or more flagella as locomotor organelles either throughout life or at some time during their life. *Trypanosoma gambiense* (Fig. 21.9).

CLASS **Sarcodina** The sarcodina are characterized by the presence of pseudopodia. *Amoeba proteus* (Fig. 21.11).

CLASS **Sporozoa** All the sporozoa are internal parasites. Asexual reproduction in this phylum involves *schizogony,* multiple fission producing many daughter cells. In many there is a second type of multiple fission following sexual reproduction which includes meiosis. This type of multiple fission is called *sporogony. Plasmodium vivax* (Fig. 21.16).

SUBPHYLUM **Ciliophora** The subphylum contains a single class, the class Ciliata. Therefore the characteristics of the class are the characteristics of the subphylum.

CLASS **Ciliata** Ciliates move by means of cilia. There are two types of nuclei, a compact macronucleus and one to many micronuclei in each individual; sexual reproduction is by the union of two individuals in a process called *conjugation* during which there is a mutual exchange of nuclear material. *Paramecium caudatum* (Fig. 21.18).

General Characteristics

The name Protozoa means "first animals." However, we were not aware of the existence of these animals until the work of the early microscopists in the seventeenth century, particularly the work of the Dutchman, Anthony van Leeuwenhoek (see Dobell, Suggestions for Further Reading). Leeuwenhoek made a number of simple microscopes, some of which would magnify objects over 200 times. He had great curiosity and examined materials and objects of many kinds. Many of the things he had examined he described in letters to the Royal Society in London. His descriptions were accurate and Dobell's translations of them are quite interesting to read. He called all of the motile things that he saw "little animals." One knows now that some of them were bacteria, but many of them were protozoans.

These tiny organisms are of interest to all students who desire to know the different kinds of organisms in the world and their possible relationships; they have been and still are the objects of considerable research. Most of them may be grown in large numbers, and important discoveries have been made in researches on protozoans in connection with the nature of the living material, nutrition and other aspects of metabolism, differentiation, movement, heredity, and the development of sexuality. As humans and all other animals are parasitized by one or more kinds of these organisms, a knowledge of the Protozoa is essential for the prevention and control of human and animal diseases. The Protozoa are important in the food cycles in nature (see Ch. 45). Many protozoans feed on bacteria and other small plants; in turn, the protozoans are eaten by larger animals. Although the Protozoa as a group are the smallest animals, they are certainly the most numerous. It has been estimated that the waters of the Mediterranean average 4 million protozoans per quart and that some soils may contain that many per pound.

SIZE AND DISTRIBUTION

In size, protozoans range from 3 μm to over 1000 μm in length, with a majority of the species being less than 100 μm in length. The larger ones are visible to the naked eye. In the majority of the species, each individual is a single cell, although in some species several daughter cells adhere together, forming a colony.

Protozoans are cosmopolitan—the same species may be found on every continent. They are found in a great variety of habitats: in fresh water, in the ocean, in the soil, in hot springs, in the snow on mountains, and living as parasites in the bodies of most animals. In the active phase the different species of Protozoa are found in a moist or fluid environment. In many species, during adverse conditions the organism is able to secrete and accumulate a tough protective membrane around itself. Such a stage is known as a *cyst,* and such cysts may often withstand extremes of dryness and other adverse conditions. When placed in favorable conditions, the organism excysts and becomes active again.

NATURE OF THE PROTOZOAN CELL

The body of a protozoan is morphologically a single cell. All the characteristics of living organisms—metabolism, growth, irritability, and reproduction—are carried on by the single cell. In multicellular metazoans there is a division of labor between the cells, but in the protozoans all the activities are accomplished in this small mass of material. A protozoan is a complete organism. In some forms, like *Amoeba*, there is little differentiation within the cell, whereas in others, like *Paramecium,* numerous differentiated structures are found. The specially differentiated parts or structures within the body of a protozoan are known as cell organs or *organelles.*

A number of biologists now refer to the Protozoa as noncellular or *acellular* organisms. They argue that because many of the Protozoa are very complex in organization and have more than one nucleus, and because they are all complete organisms, we should reserve the term cell for the units of structure of multicellular organisms. However, it seems to the authors that use of the term unicellular is valid, especially for the beginning student. All biologists support the concept of organic evolution. According to this concept the multicellular forms were derived in the process from unicellular forms. We have in some of the Protozoa living today organisms that, by ordinary standards, are simple cells. Many of them, on the other hand, are quite complex in structure and are not "simple" cells in any sense of the word. No one would suggest that any of the present-day forms are just like the ancestral forms. They have all undergone a long evolution; yet some of them still retain a body form that is a single cell with little specialization. The mistake that is often made is to refer to the Protozoa generally as "simple unicellular animals."

Because most of the Protozoa have soft bodies without any skeletal structures, their fossil forms are unknown. However, two orders of the class Sarcodina, the Foraminifera and the Radiolaria, form conspicuous shells that are found as fossils (Figs. 21.1 and 21.2). The calcareous shells of foraminiferans are found in the Cambrian rocks, the oldest strata of rocks that show good fossils, and the siliceous shells of radiolarians are found in rocks of the Paleozoic and up to the present time. The shells of foraminiferans may be brought up undamaged in the borings in well drilling, and the presence of these fossils is used by oil companies in locating oil. The compressed shells of these shell-forming protozoans make up portions of the earth's crust as seen in the chalk cliffs of Dover, the Indiana limestones, and certain flints. The time involved and the number of organisms included in the formation of such deposits is almost beyond our comprehension.

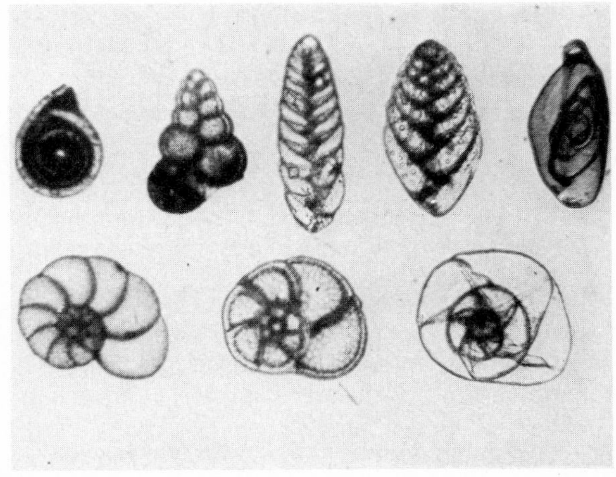

Figure 21.1 Different types of Foraminifera.

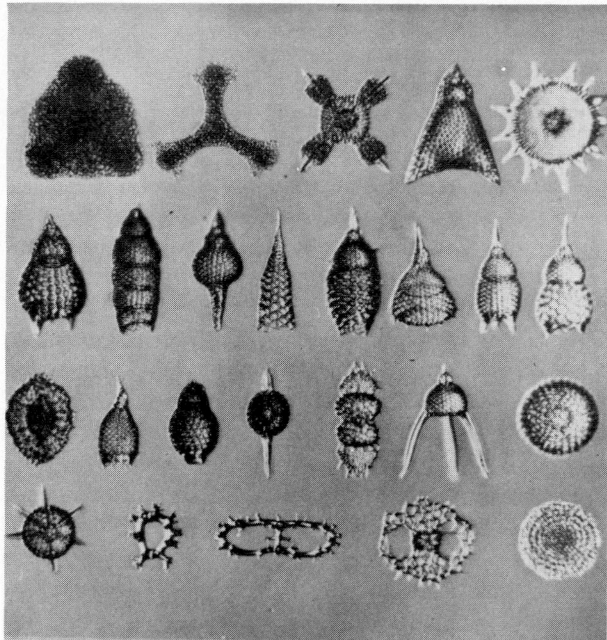

Figure 21.2 Different types of Radiolaria. (From Turtox/Cambosco, Macmillan Science Co., Chicago.)

TYPES OF NUTRITION

Several different types of nutrition are found in the Protozoa. Many of them ingest complex organic foodstuffs in the form of other organisms. This is the characteristic animal method of nutrition and is known as *holozoic* nutrition. Others absorb their food material from decaying organic matter and are said to have *saprozoic* nutrition. Those containing chlorophyll are able to synthesize their food in essentially the same manner as green plants and thus have *holophytic* nutrition. Large numbers of protozoans that live as parasites absorb their food materials from the fluids in the bodies of their hosts. Some species may exhibit more than one method of nutrition.

CLASS MASTIGOPHORA

CLASSIFICATION

SUBCLASS **Phytomastigina** These are the plantlike flagellates and most of them possess chlorophyll.

ORDER **Chrysomonadida** These are small flagellates with yellow or brown plastids and usually two flagella. They form cysts encased in silicon. Both marine and freshwater species. *Chromulina, Synura.*

ORDER **Cryptomonadida** Small with two flagella and with an anterior invaginated reservoir. Both marine and freshwater species. *Chilomonas.*

ORDER **Dinoflagellida** Flagellates with transverse and longitudinal flagella in grooves. Body naked or covered with cellulose plates or by a cellulose membrane. Brown and yellow plastids present, mostly marine. *Noctiluca, Gonyaulax, Ceratium.*

ORDER **Euglenoidida** Green or colorless flagellates which are elongated and have one or two flagella. The flagella arise from a reservoir. Stigma or eye spot present in pigmented species. Primarily freshwater. *Euglena, Peranema.*

ORDER **Chloromonadida** This is a little-known group of flagellates with pale green plastids. Two flagella are present. Food is stored in the form of oil, not carbohydrate. *Gonyostomum.*

ORDER **Phytomonadida** Green plastids and stigma present. Two flagella of equal length. Colonial species are common. Mostly freshwater. *Chlamydomonas, Gonium, Eudorina, Volvox.*

SUBCLASS **Zoomastigina** These are the animallike flagellates which lack plastids. One to many flagella present.

ORDER **Protomonadida** Small flagellates with one or two flagella. Mostly parasitic. The free-living species include the choanoflagellates such as *Proterospongia.* Among the parasites are *Trypanosoma* and *Leishmania.*

ORDER **Polymastigida** Three to eight flagella present. Nuclei vary in number from one to many. Mostly found in the gut of vertebrates and arthropods. *Chilomastix, Pyrsonympha.*

ORDER **Trichomonadida** One to many nuclei. Three to six flagella. A central stiffening rod called the axostyle present. Parasites and commensals in vertebrates and in insects. *Trichomonas.*

ORDER **Hypermastigida** Many flagella, some of them often very long. One nucleus present. All live in the gut of termites and roaches. *Trichonympha.*

ORDER **Opalinida** Leaflike flattened forms found only in the gut of frogs and toads. Many nuclei present. Body uniformly covered with very short flagella. *Opalina.*

ORDER **Rhizomastigida** A single flagellum present. Body also capable of ameboid movement. *Mastigamoeba.*

The classification of the flagellated protozoans has always presented some difficulty. Some of them contain chlorophyll and are definitely plantlike, whereas others are colorless and animallike. Attempts to classify all the green flagellates as algae produce difficulties because in some cases closely related forms differ only in the presence and absence of chlorophyll. The fact that we find forms that show characteristics of both plants and animals should not be disturbing. We should expect to find such forms if there has been an evolution.

Euglena is definitely an "in-between" type of organism. It has green chloroplasts but no cellulose cell wall; astaxanthin, a pigment found in the eyespot, is definitely an animal carotenoid.

That the large group of unicellular flagellates grades into the algae on the one hand and the protozoa on the other is evidence of their primitive position. Most workers agree that primitive green flagellates gave rise in evolution to the green algae and through them to the other higher plants, and also that they gave rise to the animallike flagellates and through them to the rest of the animal kingdom.

For a long time members of the class Sarcodina were considered to be the most primitive of the Protozoa. This concept is no longer accepted because it is generally believed that plantlike organisms came into existence before animals. The Mastigophora are now considered to be the most primitive. It is possible to derive the Sarcodina from the flagellated Protozoa because of the existence of a number of species that possess flagella at some stage in their life cycle. Also, some of the Sporozoa have flagellated microgametes, which suggests their derivation from flagellated organisms. Electron micrographs show that both cilia and flagella have the same basic "9 + 2" arrangement of their fibrils or tubules (Fig. 4.17). This suggests that the ciliates were derived from flagellated ancestors.

Anatomy and Physiology of *Euglena*

HABITAT

Euglena is a common green flagellate found in freshwater ponds. The individuals are quite small (35–60 μm), but they often occur in such numbers on the surface of the water that they form a green scum.

STRUCTURE

As shown in Figure 21.3(**b**), this organism is somewhat spindle shaped. The shape of *Euglena* is maintained by an outer covering or *pellicle.* Digestion experiments have shown that this pellicle is not composed of cellulose but that it is protein in nature. There are parallel thickenings in the pellicle that run diagonally over the body. The *reservoir* is a flask-shaped cavity at the anterior end from which emerges a single long *flagellum.* The flagellum originates at the bottom of the reservoir in two *blepharoplasts.* A light-sensitive red pigmented *eyespot* or *stigma* is located near the reservoir, and a single *contractile vacuole* empties into the reservoir. The large *nucleus* is located just posterior to the center of the cell. Numerous *chloroplasts* are found in the cytoplasm, and a single *pyrenoid* or starch-forming center is usually associated with each chloroplast. Stored carbohydrate in the form of *paramylum bodies* is also found in the cytoplasm.

LOCOMOTION

Euglena swims toward ordinary daylight but away from the direct rays of the sun. Normally locomotion is effected by lashing movements of the flagellum. These movements of the flagellum draw the organism after it in a characteristic spiral path. It is often difficult to see the flagellum in the living organism, but a drop of iodine solution added to a small sample of a culture makes the flagellum stand out when viewed under the high power of a microscope. In addition to the normal flagellar movements, these organisms may, when trapped, move by means of wormlike contractions and expansions, called *euglenoid movement.* The elasticity of the pellicle permits this activity.

NUTRITION

Normally *Euglena* carries on holophytic nutrition like a green plant, synthesizing its food from inorganic substances in the presence of light. However, some species of *Euglena* have been maintained in cultures in total darkness for long periods of time, when the medium contained certain necessary food materials in solution. Under these conditions the chloroplasts and pyrenoids disappear and the organisms live by saprozoic nutrition, absorbing through their

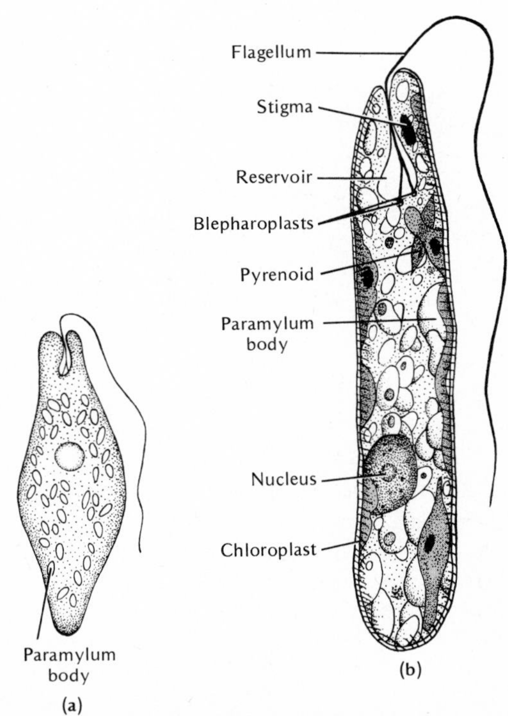

Figure 21.3 **(a)** *Astasia* sp., a colorless flagellate. **(b)** *Euglena*, a holophytic flagellate.

cell membranes the necessary nutrients from the surrounding medium. Such organisms, after a period of growth and multiplication in darkness, will develop chloroplasts again when returned to the light. Older accounts of the nutrition of *Euglena* described the ingestion of solid food and the formation of food vacuoles, but numerous observations in recent years have failed to verify this. However, some of the closely related colorless flagellates, such as *Peranema,* do ingest bacteria, yeast, and small protozoans, thus exhibiting holozoic nutrition. Some species of *Euglena* require vitamin B_{12} as an outside nutrient. They may be used in bioassays of different substances for the detection of this vitamin.

RESPIRATION AND EXCRETION

The exchanges of oxygen and carbon dioxide in respiration occur by diffusion through the cell membrane. Soluble waste products of metabolism diffuse through this membrane to the outside. It has often been suggested that the contractile vacuole serves as an excretory organ, but in *Euglena,* as in other freshwater Protozoa, this organelle serves primarily to eliminate excess water.

REPRODUCTION AND DEVELOPMENT

The life cycle of euglenas involves both an *active* or *trophic phase,* during which the organism moves about, and an *encysted phase,* in which the organism is rounded up with a protective cyst membrane surrounding it. Asexual reproduction by cell division occurs in both the active and encysted stages. The division is typically a longitudinal binary fission (Fig. 21.4). The nucleus divides by mitosis, and then the cytoplasm divides lengthwise. Sexual reproduction is unknown in *Euglena.* It is generally considered that some unfavorable condition or conditions in the environment induce encystment. With the return of a favorable environment, the encysted organism excysts and becomes active again. The most noticeable characteristic of the behavior of euglenas is their reaction to light. When a large container of euglenas is placed so that part of it is shaded, part subjected to intense sunlight, and part subjected to moderate light, the organisms will collect in the latter region. The red eyespot, already mentioned, is known to function in this reaction. For a long time it was assumed that the stigma was the photoreceptor. The consensus now is that the stigma functions as a light-absorbing shield that can shade the flagellar swelling, the true photoreceptor. As *Euglena* rotates, the stigma intermittently shades the photoreceptor (Fig. 21.5). In general, the reactions of euglenas are adaptive, that is, they move away from harmful stimuli and toward beneficial stimuli.

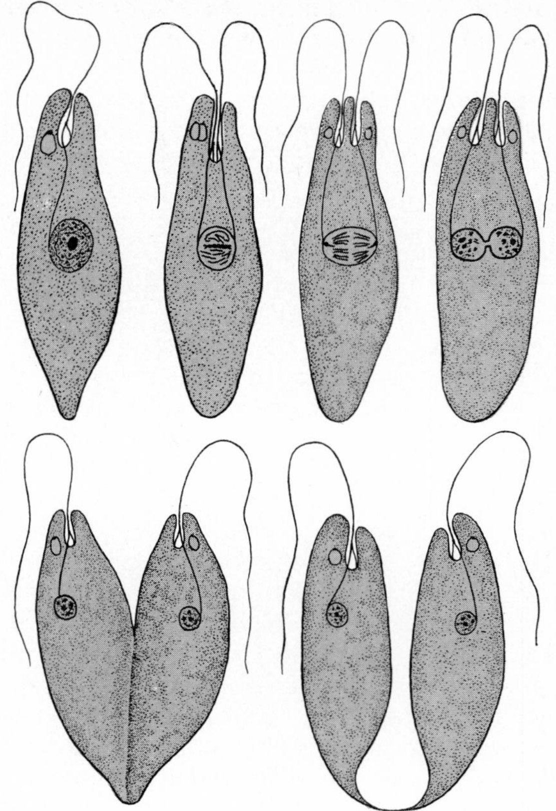

Figure 21.4 Longitudinal binary fission in *Euglena.* Note the intranuclear mitotic figure.

When *Euglena* is treated with streptomycin, or certain other antibiotics, the chloroplasts disappear. Cultures of such colorless forms have been maintained on appropriate media for a long time. *Astasia* is a naturally occurring colorless

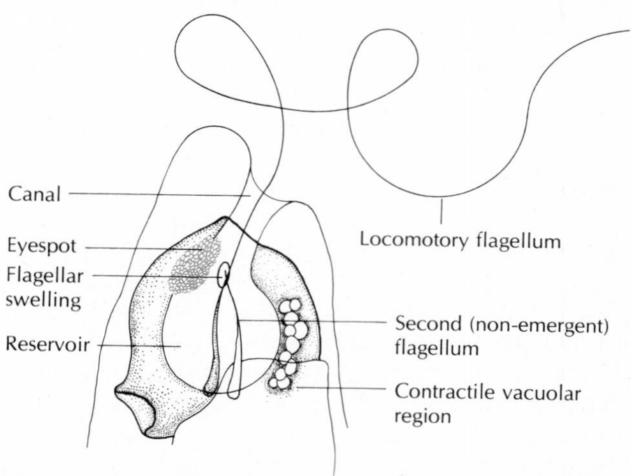

Figure 21.5 Details of reservoir region in *Euglena.* (After G. F. Leedale, *Euglenoid Flagellates,* Prentice-Hall, 1967, with permission.)

organism quite similar to *Euglena* except that it lacks the chloroplasts [Fig. 21.3 (**a**)].

Other Mastigophora

There are many species in this class and they are placed in two subclasses. One subclass, the Phytomastigina, is made up of organisms, like *Euglena,* that have chlorophyll and the other, the Zoomastigina, is composed of colorless forms that are distinctly animallike in their nutrition.

Among the Phytomastigina, the dinoflagellates are an interesting group. Members of this order have two flagella, one of which originates near the center of the cell and extends posteriorly in a longitudinal groove or *sulcus.* The other flagellum lies in a groove, the *annulus,* which encircles the midregion of the cell. *Ceratium,* a typical genus with both freshwater and marine species, is shown in Figure 21.6.

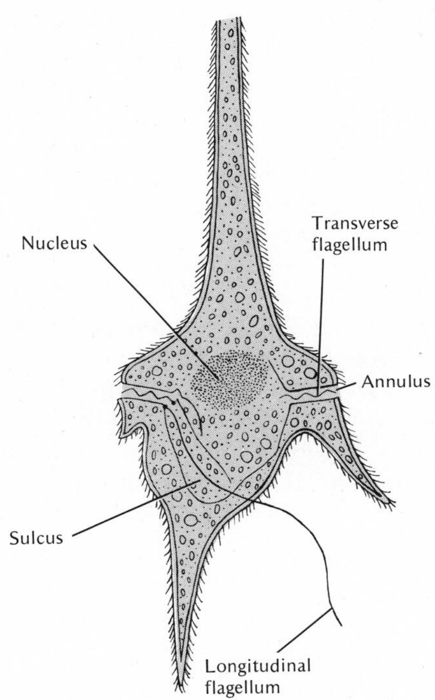

Figure 21.6 *Ceratium,* a dinoflagellate.

Members of several genera of dinoflagellates, including *Gymnodinium* and *Gonyaulax,* contain a red pigment and also produce a highly toxic metabolite. On several occasions off the coast of Florida and Central California, a sudden increase in the numbers of these forms to densities of 50,000,000 per liter and more caused the water to appear red in color—a condition known as the "red tide." The toxic metabolite caused the death of millions of fish and other marine animals and indirectly caused the death of humans who ate fish or mussels which had ingested the poisonous protozoans.

Another well-known plantlike flagellate is the colonial form, *Volvox* (Fig. 21.7). The colony is a hollow sphere with hundreds of small flagellated cells, connected by thin protoplasmic bridges, embedded in a gelatinous matrix. Asexual reproduction occurs by the budding of numerous small colonies to the interior of the sphere with the eventual breakdown of the parent colony, releasing the new ones. In sexual reproduction, eggs and sperm are formed in cell masses which also bud into the interior portion of the sphere.

Volvox was once thought to be important phylogenetically because it is structurally much like the blastula stage of development of many higher animal forms. Haeckel hypothesized that a form similar to *Volvox* was the link between the Protozoa and the Metazoa. In Haeckel's original theory this stage was called the blastaea and the hollow blastula of many metazoans was considered to be a recapitulation of this stage. According to Haeckel the blastaea invaginated to produce a two-layered organism, the gastraea. This he considered to be the hypothetical ancestral organism equivalent to the gastrula stage in metazoan development. Haeckel pointed out the similarity of his postulated gastraea and some of the simple metazoans such as the coelenterates.

Gastrulation in the coelenterates, however, is not by invagination but by the proliferation of cells from the blastular wall to the interior. Because of this Metschnikoff proposed that the gastraea was a solid rather than a hollow organism, and modern proponents of the theory share this view. The gastraea is then considered to have been an ovoid radially symmetrical organism with flagellated cells on the outside and a solid interior. This is very much like the planula larva of coelenterates and it is often referred to as the *planuloid* ancestor of metazoans. This would account for the basic radial symmetry found in coelenterates; the bilateral symmetry of the flatworms would have to be considered as a later development (see Ch. 23).

One difficulty with this theory is that *Volvox* has been thought to be surrounded by a cellulose membrane, and cellulose production is normally considered to be a plant characteristic. If such forms actually were the ancestors of metazoans, sometime in their evolution it would have been necessary for them to lose their chloroplasts and their cellulose membrane. From what has already been said it is not too difficult to imagine the loss of chloroplasts, but the loss

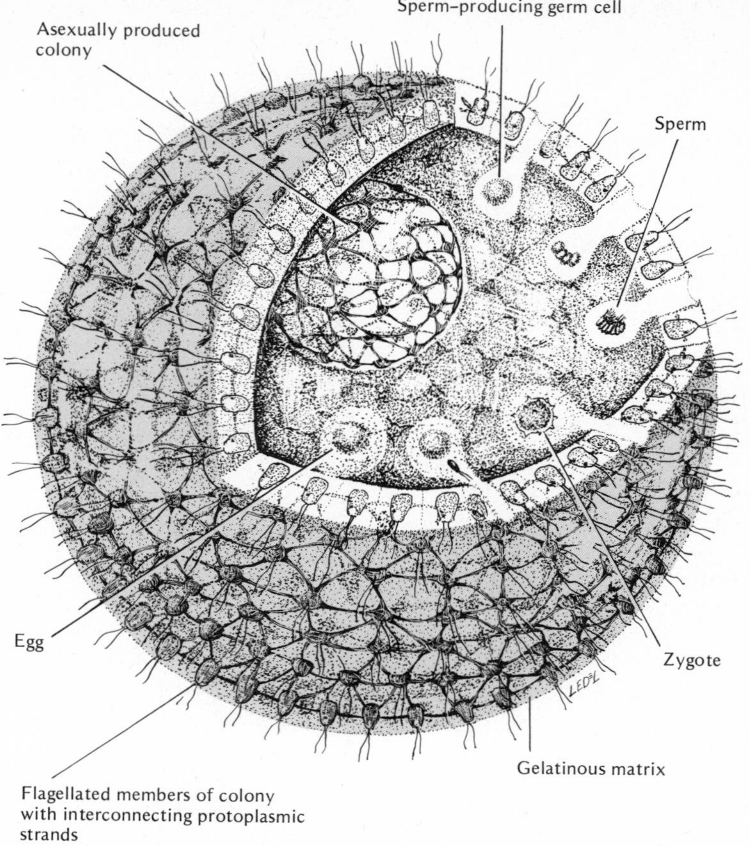

Figure 21.7 *Volvox,* a colonial plantlike flagellate.

of cellulose presents more of a problem. However, Norma Lang has indicated that her electron micrographs of *Volvox* do not show cellulose walls. If this is verified, it would add additional support to the colonial theory. Some workers have proposed that the metazoans arose from an extinct group of animal flagellates that possessed a colonial organization similar to that found in the colonial plantlike flagellates.

Among the Zoomastigina a group of colonial flagellates with thin, transparent, funnel-shaped collars, extending from the surface of the cell and surrounding the flagellum, is also important in phylogeny. A representative example, *Proterospongia,* is shown in Figure 21.8. The phylogenetic significance of this group lies in the fact that the *collar cell,* or *choanocyte*, is found only in the so-called choanoflagellates and in the sponges. The generally accepted interpretation is that the sponges evolved from ancestral choanoflagellates. The sponges are a "dead end" as far as evolution of any higher groups is concerned. The origin of other multicellular animals was from some other protozoan stock. Some theories of metazoan origins will be discussed later.

Many of the Zoomastigina are parasites. Perhaps the best known are the trypanosomes, for example, *Trypanosoma gambiense,* which causes African sleeping sickness in humans (Fig. 21.9). In the trypanosomes the flagellum is attached by means of a thin membranous structure to the edge of the cell. The flagellum extends free at

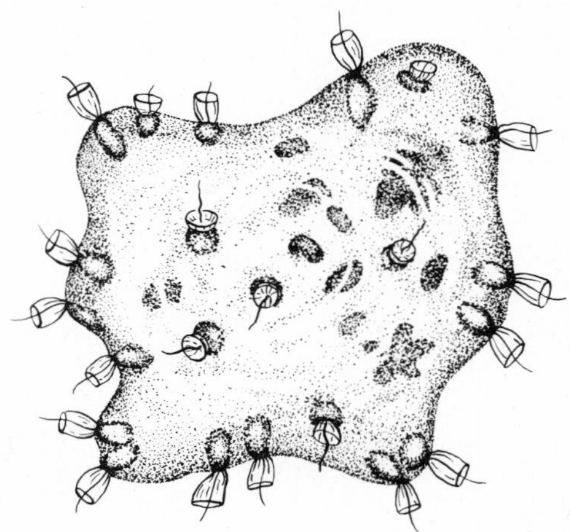

Figure 21.8 The colonial protozoan, *Proterospongia.* The colony has collar cells, like those in sponges. These cells are imbedded in a gelatinous material through which amoebocytes wander.

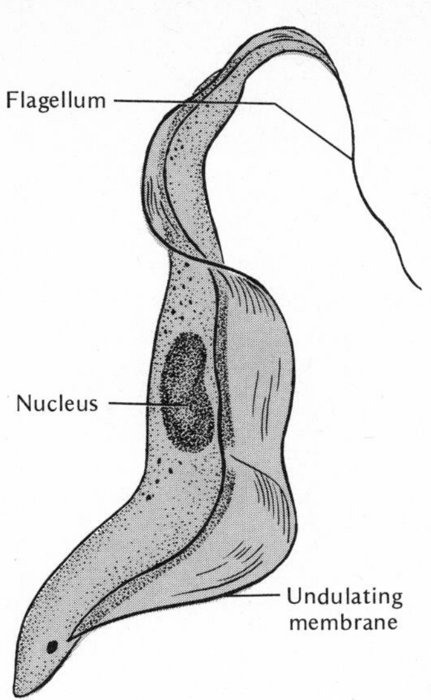

Figure 21.9 *Trypanosoma gambiense,* the cause of African sleeping sickness. The edge of the undulating membrane bears the flagellum which extends free at one end of the animal.

one end of the organism. The structure formed by the flagellum and the membrane is called an *undulating membrane* and it is used in locomotion. *T. gambiense* and several related species which cause serious illness in humans are transmitted by the *tsetse fly, Glossina* sp. When a tsetse fly bites an infected animal or a human, some of the parasites are sucked with the blood into the fly's intestine. Here they undergo changes and finally invade the salivary glands of the fly. In these glands they multiply and undergo further changes and are ready to be injected into another vertebrate host. In the blood of humans and other vertebrate hosts, they multiply and liberate their poisonous by-products of metabolism. Fever starts when they become numerous in the bloodstream, and if they are not destroyed by treatment with certain drugs (arsenic compounds have been used), they invade the cerebrospinal fluid and ultimately cause death. Control of this disease is complicated, for many wild animals can serve as hosts. The best control would seem to be ultimate destruction of the tsetse fly, a very difficult task. Closely related species of the genus *Trypanosoma* cause the death of domestic cattle, and there are vast areas of central Africa which as yet cannot be inhabited by humans because of this danger to both people and their domestic animals.

Other trypanosomes cause serious tropical diseases in various areas of the world. Some of the diseases due to these parasites are Chagas disease in Central and South America, transmitted by several kinds of blood-sucking insects of the order Hemiptera; and kala azar and Oriental sore caused by members of the genus *Leishmania* and transmitted by sand flies, of the genus *Phlebotomus.*

Parasites are not limited to the trypanosomes but are found in most of the other orders of Zoomastigina as well. One group, the order Opalinida, is found only in the lower gut of frogs and toads. *Opalina* is a flattened leaflike organism with many nuclei, all of the same type. The cell surface is covered with many short flagella, actually so short as to be comparable in length with cilia. For many years these organisms were considered as ciliates, but the presence of a single type of nucleus was unlike the rest of the ciliates and they are now considered by most workers to be flagellates.

The most complex of all the flagellates belong to the order Hypermastigina, distinguished by the presence of many flagella, some of which are often very long. Members of this order inhabit the intestines of termites and cockroaches. One of the forms found in termite intestines is *Trichonympha,* shown in Figure 21.10. Cleveland demonstrated that the termites were dependent upon the flagellates to digest the wood which is the main food of the termite. He found that the protozoans were sensitive to increased oxygen tension and temperature and could be killed without killing the termite. Termites which had lost their flagellates would eat wood as usual, but it passed through the digestive tract unchanged and the termites soon died of starvation. The protozoans are provided with a suitable environment and a steady source of food and in return the termites also receive nourishment. Such a relationship, in which individuals of two species live together in an association of mutual benefit, is called *mutualism.*

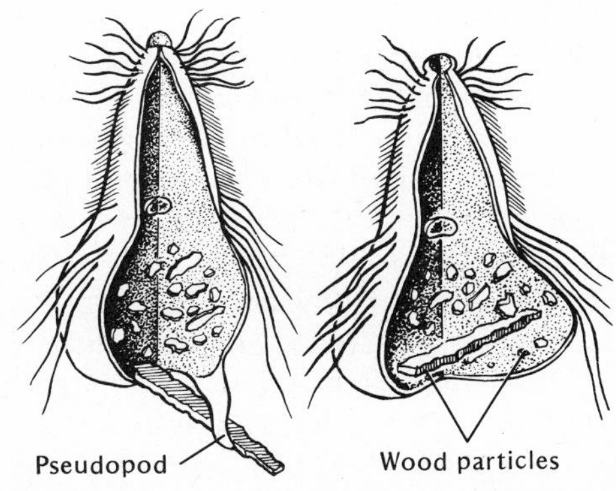

Figure 21.10 *Trichonympha* engulfs wood particles by pseudopodia extended from the lower part of the animal.

CLASS SARCODINA

Classification

SUBCLASS **Actinopoda** Floating or sessile Sarcodina. Pseudopodia are always axopodia.

ORDER **Helioflagellata** A small group of marine and freshwater species with one or more flagella in addition to pseudopodia. *Tetradimorpha.*

ORDER **Heliozoida** Largely freshwater species. Body spherical with radiating axopods. *Actinophrys, Actinosphaerium.*

ORDER **Radiolarida** Entirely marine. Body consists of a central perforated capsule of tectin with an outer cortex of cytoplasm. Outer skeleton of silicon or in some cases of strontium sulfate. *Collozoum, Sphaerozoum.*

SUBCLASS **Rhizopoda** Pseudopodia are lobopodia, filopodia, or reticulopodia. Axopodia never present. Largely creeping forms.

ORDER **Amoebida** Naked species which move by lobopodia or by flowing of the entire body. Mostly free-living species in marine, freshwater, and soil habitats. Some parasites. *Amoeba, Entamoeba.*

ORDER **Testacida** Freshwater species which secrete a single-chambered shell or test in which foreign particles may be imbedded. There is usually only a single opening in the test through which the pseudopodia are protruded. *Arcella, Difflugia.*

ORDER **Foraminiferida** Largely marine species with multichambered shells of calcium carbonate. Shells are perforated by many pores. Pseudopodia are reticulopodia. *Globigerina, Orbulina, Spirillina.*

Anatomy and Physiology of Amoeba

HABITAT

The common amoeba of freshwater ponds and streams, *Amoeba proteus,* is one of the most fascinating objects of study for a beginning zoology student (Pl. 25). This unicellular organism is about 250 μm in diameter (although some specimens may reach a length of nearly 1 mm during locomotion) and its movements are so sluggish (2 or 3 μm per second) that its study under the microscope is relatively easy. The common amoeba is found living on decaying leaves and other vegetation at the bottom of ponds where it feeds on small algae, bacteria, and other protozoans. Each amoeba is a small mass of jellylike material, slightly grayish in appearance, without any definite shape. Unlike some of the other protozoans described in this chapter, the amoeba has few specialized parts or *organelles.* It is simple in structure, yet it carries out all the vital functions of higher animals—metabolism, growth, reproduction, locomotion, and response to stimuli.

CELL STRUCTURE

The cell of an amoeba, like most cells, is differentiated into cytoplasm, nucleus, and a cell or plasma membrane. In *Amoeba* this cell membrane is often referred to as the *plasmalemma.* As in other cells, the plasma membrane determines what substances enter and leave the cell. The *nucleus* occupies no fixed position and is a biconcave disk, often with a folded or wrinkled appearance. The *cytoplasm* is differentiated into a clear thin outer layer, the *ectoplasm,* and the inner granular *endoplasm.* The outer region of the endoplasm is stiff or jellylike and is called the *plasmagel,* whereas the inner part of the endoplasm is fluid in nature and is called the *plasmasol.* The endoplasm contains a large clear *contractile vacuole, food vacuoles,* and numerous *granules* and *crystals.* The extensions from the body of the amoeba are *pseudopodia* ("false feet") and are used in locomotion (Fig. 21.11).

LOCOMOTION

The locomotion of *Amoeba,* commonly called "amoeboid movement," is one of the most interesting aspects of the study of this organism. Amoeboid movement is considered to be the simplest type of animal locomotion. Amoeboid movement is also found in human white blood corpuscles, sponge amoebocytes, and a number of other kinds of cells.

A moving amoeba may be observed to send out a projection from its body, and then this *pseudopodium* may advance through the flowing into it of some of the endoplasm. Two or three pseudopodia may be formed simultaneously, but ultimately one of these will become dominant for a time. As new pseudopodia are formed, the old ones withdraw into the general body region. In its locomotion, the amoeba often alters its course by forming a new dominant pseudopodium on the opposite side, thus moving in a very irregular fashion. There are no permanent anterior or posterior ends.

At the present time there is no complete explanation of the chemical and physical changes involved in amoeboid movement. One explanation of this type of movement is based on changes in the sol and gel state of the cytoplasm (Fig. 21.12). At the point where a pseudopodium starts to form, the outer gelated region of the amoeba liquefies, probably due to some local chemical change, and the fluid plasmasol flows through this weakened region. As the plasmasol flows forward, it turns back and becomes a part

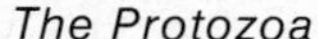

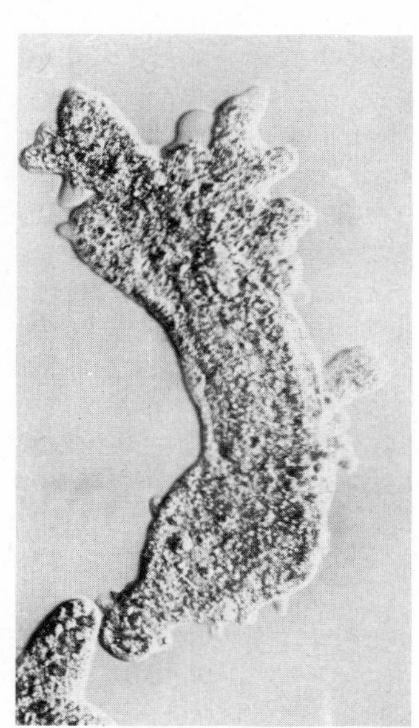

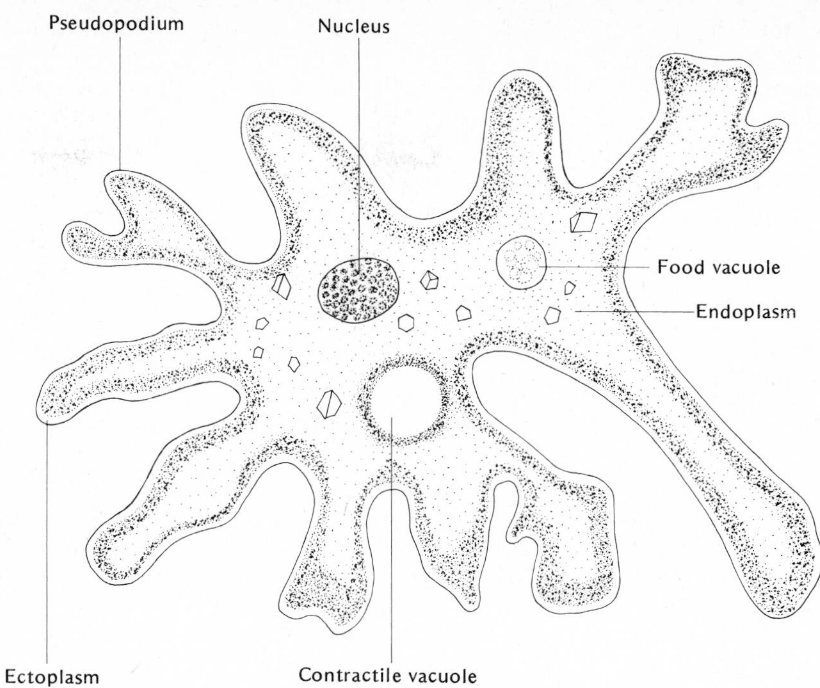

Figure 21.11 *Amoeba proteus.* Drawing and photograph of a living specimen. (Photo from Carolina Biological Supply Company.)

of the outer plasmagel, while more plasmasol flows into the advancing tip. At the temporary rear end of the amoeba, plasmagel changes into plasmasol. It is assumed, in this explanation, that the outer layer of plasmagel contracts and forces the fluid plasmasol forward into the pseudopodium where the plasmagel is thinnest. If the tip of an advancing pseudopodium is touched, the plasmagel thickens at this point and the direction of movement is changed. When an amoeba is stimulated all over, as it is when it is squirted out of a pipette onto a slide, it rounds up into a motionless sphere for a short time. Apparently, when so stimulated the amoeba forms a heavy layer of plasmagel on all sides. When a microscope is so arranged that a moving amoeba is viewed from the side, it is found that only the tips of the pseudopodia are in contact with the substrate.

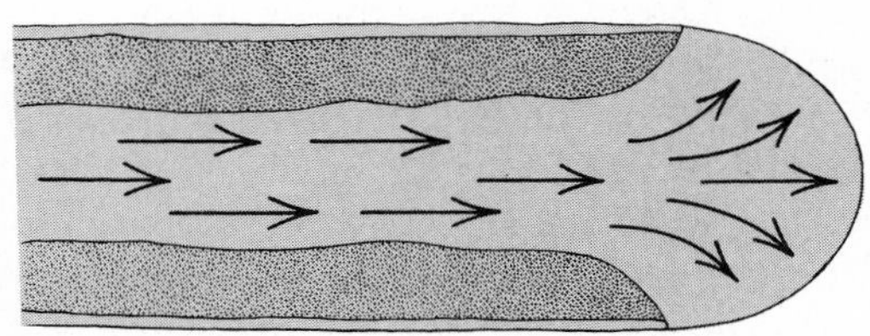

Figure 21.12 Tip of a pseudopod showing one theory of the mechanism of amoeboid movement. The central, more fluid plasmasol flows to the tip of the pseudopod where it turns posteriorly and becomes a more rigid material, the plasmagel. According to this concept the plasmagel changes to plasmasol at the temporary posterior end of the amoeba.

The above explanation, or perhaps it would be better to say description, of amoeboid movement was first proposed by S. O. Mast over fifty years ago. It is sometimes referred to as the Mast theory of amoeboid movement. In this explanation the force for pushing the endoplasm along comes from the posterior contraction of the ectoplasmic tube which surrounds the endoplasm. In 1962 R. D. Allen (see Suggestions for Further Reading), after a number of observations and experiments, announced that the force which causes the movement is a contraction in the fountain zone of the advancing pseudopodium, and this forward contraction pulls the endoplasm toward the forward end of the ectoplasmic tube. Many workers have accepted Allen's explanation, but a detailed molecular explanation of amoeboid movement is yet to come.

NUTRITION

As mentioned above, the amoeba feeds on small plants and animals. In the process of capturing and ingesting food objects, pseudopodia are used. In food getting, however, pseudopods are thrown out on all sides of the food object, forming what has been called a *food cup.* The lips of the food cup gradually come together, trapping the algal cell or protozoan inside the body of the amoeba. The food object and the surrounding water that was trapped now constitute a *food vacuole.* When an amoeba ingests an algal cell, which is motionless, only a small amount of water is taken into the food vacuole. However, when it ingests a fast-moving protozoan, a larger food cup is formed and more water is taken into the food vacuole. *Digestion* occurs within the food vacuole. When an actively moving protozoan is ingested, it can be observed to move within the vacuole for a short time. However, acid secreted into the food vacuole by the cytoplasm of the amoeba soon immobilizes the ingested organism. It has been possible, by using certain color indicators, to follow the pH changes in food vacuoles. In the early stages the contents of the vacuole are acid, but they later become alkaline. In *Amoeba,* as in humans, digestion is effected by enzymes, and different enzymes act at definite hydrogen-ion concentrations. The enzymes that function in the vacuoles of an amoeba enter by fusion with lysosomes and they cause the change of proteins into amino acids, fats into fatty acids and glycerol, and carbohydrates into simple sugars. These end products of digestion are *absorbed* through the vacuolar membrane and *assimilated* in the cytoplasm of the amoeba. These end products provide the materials for the release of energy in respiration and for the synthesis of new cellular material in growth.

Once a food vacuole is formed, it circulates about the body of the amoeba in the fluid endoplasm. Materials that are not digested in the food vacuoles are finally *egested* from the body. In *Amoeba* there is no definite spot where egestion occurs; food vacuoles containing undigested material may break through the surface at any point and the residue is left behind.

RESPIRATION AND EXCRETION

There is no special mechanism for gaseous exchange in *Amoeba.* The oxygen dissolved in the surrounding water passes through the plasma membrane into the cytoplasm by diffusion. Likewise, the carbon dioxide formed in cellular respiration diffuses out of the amoeba into the surrounding water.

Not only must an organism rid itself of carbon dioxide, but it must also rid itself of the waste products of protein metabolism (ammonia and so forth). The *contractile vacuole* of *Amoeba* was long associated with this function, but that position has been largely abandoned. Since it was possible to remove samples from the contractile vacuole of an amoeba with a micropipette and make a chemical analysis of the contents, it was found that the concentrations of urea and ammonia were not high enough to indicate that all the nitrogenous wastes of an amoeba are given off by this vacuole. It is now the position of most workers in the field that the vacuole functions in a hydrostatic way—to eliminate excess water; the nitrogenous wastes apparently diffuse from the amoeba as does the carbon dioxide. This position is supported by a number of facts. An excess of water does enter the body of an amoeba. In addition to the water formed in cellular oxidations and the water that enters in food vacuoles, a large amount of water is continually entering the body of the amoeba by diffusion because the surrounding pond water is hypotonic to the cytoplasm of the amoeba. In contrast with this, marine amoebae do not have contractile vacuoles. However, marine amoebae form contractile vacuoles when transferred by stages to fresh water and freshwater amoebae lose their contractile vacuoles when additional salt is added to their normal pond water.

REPRODUCTION

Reproduction in *Amoeba* is by the asexual process of *binary fission.* The amoeba rounds up and the nucleus divides by mitosis which, as in most protozoans, differs from the mitosis of higher forms by having the spindle form within the confines of the nuclear membrane. After the nucleus divides, the cytoplasm constricts and two daughter cells are produced. As in all organisms that reproduce by simple fission, there is no parent left after the process; the parent has become two individuals. For this reason it is often said that *Amoeba,* and other organisms that reproduce in this way, are "immortal"; that is, every amoeba that now exists is in a direct continuum through the ages with the first amoeba. However, this concept does not imply that a single atom present in the first amoeba is present in any amoeba now living, because there is a constant turnover of all the constituents of living cells in the metabolic processes.

Some species of *Amoeba* are known to form protective cysts under unfavorable conditions. In the encysted state such an organism, with its metabolism reduced to a very low level, may exist for a considerable period of time until it is again surrounded by favorable conditions.

BEHAVIOR

Although there are no special structures differentiated in an amoeba for the reception of stimuli, for conduction, and for coordination,

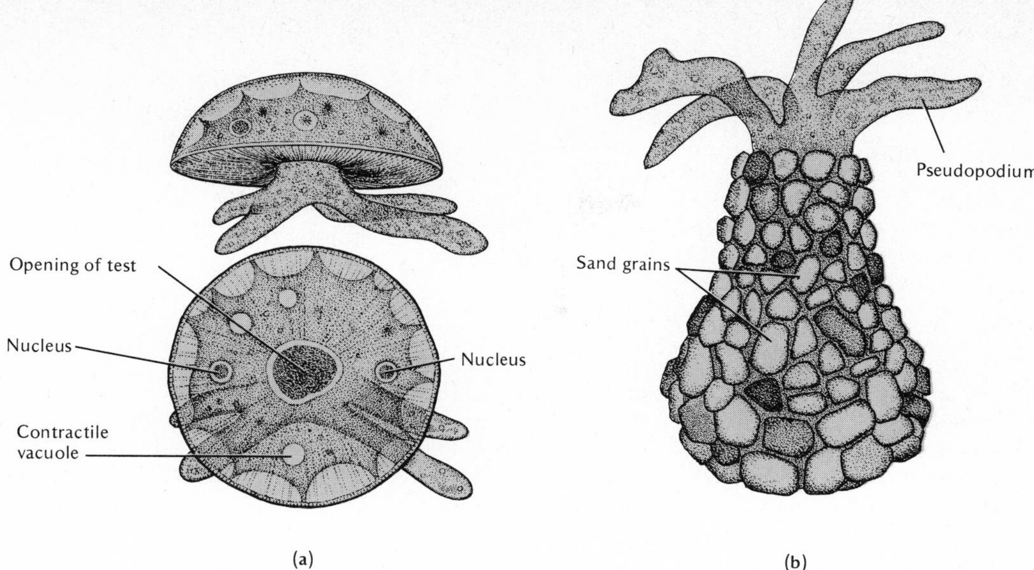

Figure 21.13 **(a)** *Arcella,* of the order Testacida, with a test of tectin. **(b)** *Difflugia,* another testacid, with a test of tectin in which sand grains are imbedded (an arenaceous test).

the organism reacts very definitely to different stimuli. In general its behavior is adaptive. It is able to distinguish food objects from those with no food value. It moves away from strong light, strong chemicals, and mechanical prodding; thus it avoids injury.

Other Sarcodina

There are six orders in the class Sarcodina with many species which are quite different from *Amoeba.* Several kinds of pseudopodia are found in addition to the simple *lobopodia* described for *Amoeba* (Fig. 21.11). The most complex type of pseudopod is the *axopodium,* found in Heliozoa and Radiolaria. The axopodium is a more or less permanent structure in which there is a central *axial filament* and sometimes a basal granule. A *filopodium* is a fine filamentous extension of the ectoplasm with no axial filament. The filopod may branch, but the branches do not *anastomose* (join) with other branches. Finally there are *reticulopodia* which are similar to filopodia except that the branches of the fine filaments do anastomose to form a netlike structure.

Many sarcodinans secrete a shell or *test* (Fig. 21.13). *Arcella,* a freshwater form, secretes a chitinlike test composed of *tectin. Difflugia* secretes a similar one, but before the material hardens, sand grains become imbedded in it to form what is called an *arenaceous test.* Foraminifera secrete a shell of calcium carbonate, while the Radiolaria secrete an outer shell of silicon dioxide, or in some cases, strontium sulfate, and an inner perforated *central capsule* of tectin. As previously mentioned, the shells of radiolarians and foraminiferans have formed large deposits in the earth's crust and the shells of the foraminiferan *Globigerina* form a thick ooze which covers vast areas of the ocean floor.

There are many species of parasitic amoebae, and one of these is of special interest to us. The only amoeba living in the body of humans that is quite harmful is *Entamoeba histolytica.* This is the organism that causes *amoebic dysentery.* It lives in the intestine of humans where it may do great damage by the formation of abscesses and bleeding ulcers. In this country, where modern sanitation now prevails in most places, the incidence of this disease is rather low, but in other regions, such as some of the places in the Pacific where troops were sent in World War II, the incidence may be very high. The disease is transmitted by food and drink contaminated by the excreta of infected persons. This amoeba does not live outside of the human body in the trophic or active stage; it is transmitted in the *cyst* form. Although the disease may be treated with broad-spectrum antibiotics (against associated bacteria) and with other drugs which act directly on the amoeba, its control lies in proper sanitation (Fig. 21.14).

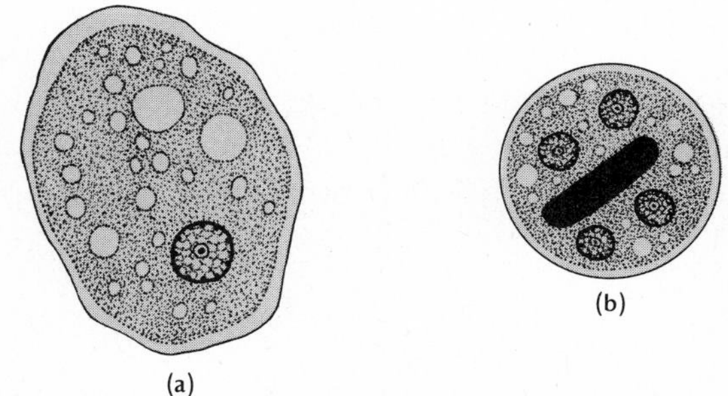

Figure 21.14 *Entamoeba histolytica.* **(a)** Feeding stage (trophozoite) from human intestine. Note ingested red blood cells. **(b)** Cyst. Note the characteristic four nuclei and the darkstaining chromatoid body which disappears in time as food reserves are exhausted.

CLASS SPOROZOA

Classification

SUBCLASS **Telosporidia** Spores naked or enclosed in a resistant spore membrane. Polar capsules absent.

ORDER **Gregarinida** Unicellular spores develop directly from the zygote. Extracellular parasites of invertebrates, mostly arthropods and annelids. Reproduction primarily by sporogony, although some species also have schizogony. *Monocyctis, Gregarina.*

ORDER **Coccidia** Intracellular parasites of epithelial cells in vertebrates and higher invertebrates. Complex life cycle involves alternate schizogony and sporogony, often with a change of host. *Eimeria, Isospora.*

ORDER **Haemosporidia** Intracellular parasites of blood corpuscles or other parts of the circulatory system of vertebrates where schizogony occurs. Sporogony following union of gametes occurs in blood-sucking insects or ticks. *Plasmodium.*

SUBCLASS **Cnidosporidia** Spores contain *polar capsules,* oval bodies resembling nematocysts and containing a coiled thread which is released due to the action of digestive juices and attaches the spore to the gut wall of the host.

ORDER **Myxosporidia** Mostly parasites of fish, in which they are found in various hollow organs or in liver, spleen, kidney, and many other tissues. *Myxobolus, Leptotheca.*

ORDER **Actinomyxidia** Parasites of coelom or digestive tract of freshwater oligochaetes and marine sipunculids. Spores have three polar capsules. *Triactinomyxon.*

ORDER **Microsporidia** Intracellular parasites of fish and arthropods with single spore case. *Nosema bombycis* causes a serious disease, called pebrine in silkworms. It was discovered by Pasteur in 1865. This was the first scientific study of a protozoan which resulted in a practical method of controlling infection.

SUBCLASS **Sarcosporidia** Parasites of muscle in mammals, including humans, as well as birds and reptiles. Life history and method of transmission are not known. Form a crescent-shaped cyst containing many spores. *Sarcocystis.*

Anatomy and Physiology of Plasmodium

Plasmodium is the genus of the sporozoan which causes the disease *malaria* in people. Four species affect us, but the most common one is *Plasmodium vivax.* The most dangerous species is *P. falciparum,* largely restricted to tropical and subtropical regions. Mortality may be as high as 25 percent in some places. Recently strains of this species resistant to chloroquine, an important antimalarial drug, have appeared in Viet Nam. Historically, malaria has played a very important role. It is believed to have been a major factor in the fall of ancient Greek and Roman civilizations. In the Civil War, as well as in the Spanish-American War, the disease was more destructive than sabers and bullets. In World War II and in the Viet Nam war it has presented many problems. Although widespread efforts at control have greatly eased the problem in many places, malaria is still a very important cause of death, either directly or indirectly, in many parts of the world. In India alone, over 1 million people die of malaria every year. At one time malaria occurred commonly in the southern part of the United States, but through treatment and preventive measures the disease is now kept under control in most areas.

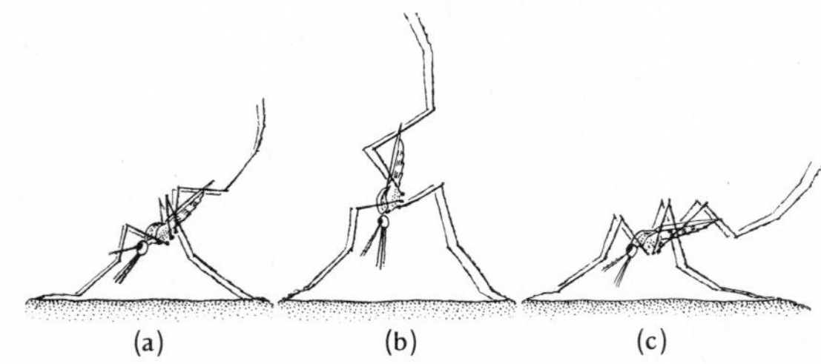

Figure 21.15 Biting positions of mosquitoes. **(a), (b)** *Anopheles.* **(c)** *Culex.*

This sporozoan is injected into the bloodstream of humans with the bite of an infected mosquito (Fig. 21.15). Only female mosquitoes of the genus *Anopheles* transmit the sporozoans that cause malaria in humans. The stage in the life history of *Plasmodium* that is injected into humans with the bite of the mosquito is a very minute spindle-shaped form, called a *sporozoite* (Fig. 21.16). The sporozoites pass from the bloodstream into various tissue cells of organs associated with the circulatory system, characteristically in the liver. Within these cells the sporozoite feeds, enlarges, and undergoes *schizogony*—multiple division of the nucleus with the eventual production of many small cells, called *merozoites,* which may enter other liver cells to repeat the schizogony with more merozoites produced. This phase of the life cycle is called the *exoerythrocytic* phase of the cycle since it occurs outside of the erythrocytes. From 10 to 18 days after the original introduction of sporozoites by the bite of the mosquito, merozoites from the exoerythrocytic stage appear in the bloodstream and enter red blood cells where they increase in size—a stage called the *trophozoite*—and then undergo schizogony with the production of more merozoites. The merozoites in turn infect more red blood cells, and this process is repeated until billions of cells

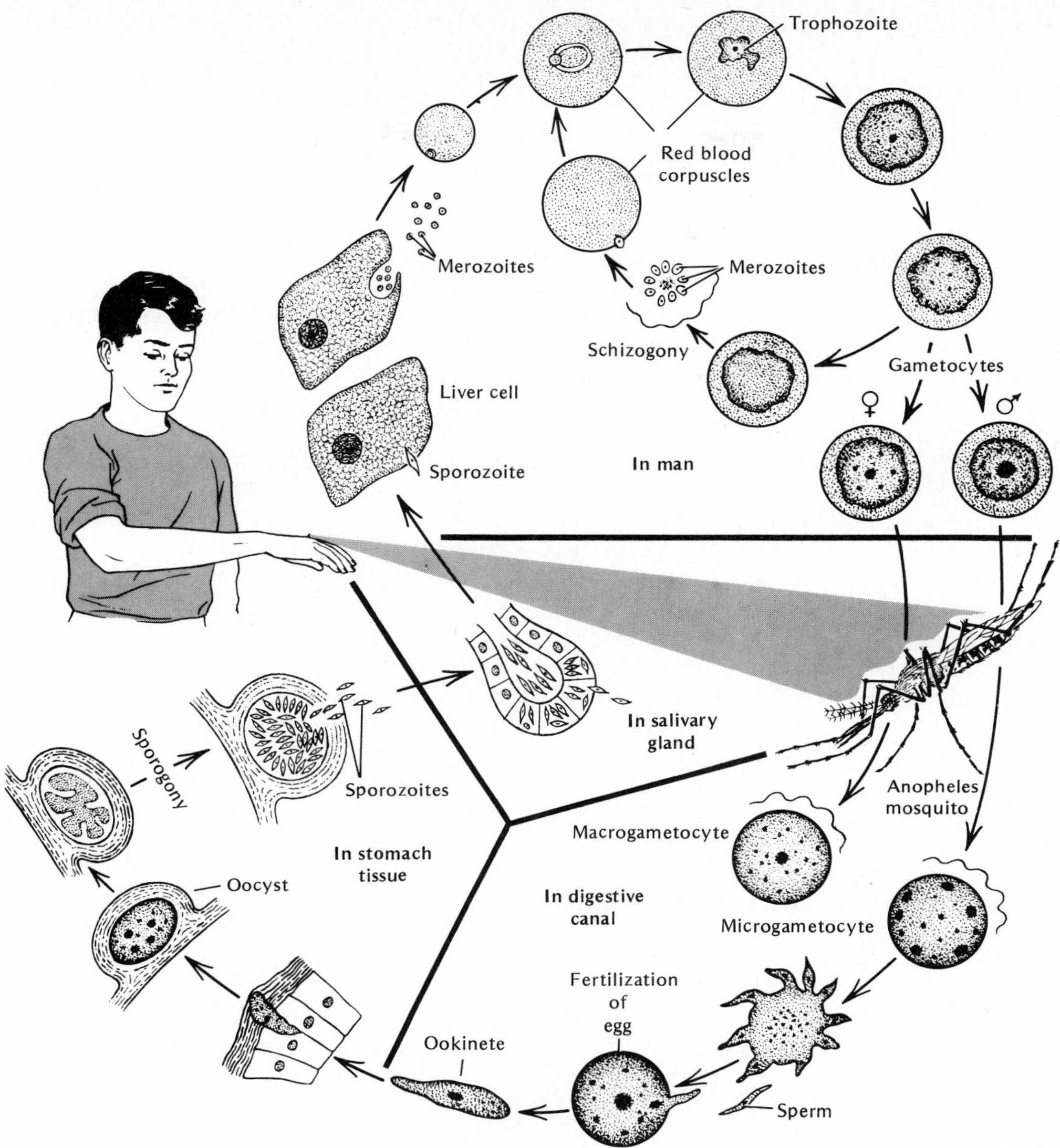

Figure 21.16 Life cycle of the malarial parasite, *Plasmodium.*

are infected. For *P. vivax* it takes 48 hours for the merozoite which enters a red blood cell to grow and undergo schizogony. When large numbers of red blood cells are bursting with the release of merozoites at about the same time, there is also a release of significent amounts of toxic metabolites from the developing parasites. It is the release of these metabolites that causes the characteristic chills and fever which recur every 48 hours in cases of infection with *Plasmodium vivax.* (The cycle is also about 48 hours in *P. falciparum* and *P. ovale* and it is 72 hours in *P. malariae.*)

After the asexual process of schizogony has progressed for some time, certain trophozoites do not undergo multiple division but develop into sexual forms, some becoming *female gametocytes* and others becoming *male gametocytes.* Unless these cells are taken into the stomach of a mosquito with her blood meal, there is no further development. However, if they are ingested by the mosquito, the sexual phase of the *Plasmodium* life cycle begins. The female gametocyte enlarges to become the female gamete or egg. The male gametocyte divides to produce four to eight slender flagellated sperm. Meiosis occurs in the processes of egg and sperm formation. After fertilization, which occurs in the stomach of the mosquito, the zygote becomes an elongate wormlike *ookinete* which works its way through the stomach wall and forms an *oocyst* on the outer side of the stomach. Within the oocyst there is a second type of multiple division, called *sporogony*, which results in the production of numerous spindle-shaped sporozoites which burst out of the oocyst and pass via the hemolymph to the salivary glands of the mosquito where they are ready to infect a new host when next the mosquito takes a blood meal. The developmental period in the mosquito varies somewhat according to the temperature and the species, about 16 days at 20° C, 10 days at 25° C, and 8 days at 30° C in *P. vivax.*

The life cycle of *Plasmodium* illustrates a feature found in many parasites—more than one host is necessary for the completion of the life cycle. In this case it is a human and the female *Anopheles* mosquito. If all the mosquitoes of this species were destroyed, that would be the end of the human malarial parasite. Unfortunately it is probably impossible to eradicate this species completely, but its control has been the chief way of controlling the disease. In regions of the world where the breeding places of mosquitoes have been destroyed and where measures are used for killing the larvae of the mosquitoes, this disease is kept under control. For many years the drug *quinine* has been used in the treatment and prevention of malaria, and since World War II *atabrine* and *chloroquine* have been used. Although malaria has been under control in the United States for some time, many soldiers in the Pacific area in World War II and in Viet Nam suffered from the disease.

Other Sporozoa

Members of the class Sporozoa are all parasites, and there are host species in most of the animal phyla. Except for some of the male gametes, there are no specific locomotor organelles in members of the class. The name sporozoa is derived from the fact that infective spores are formed at some stage in the life cycle. Hosts are infected by passage through an intermediate host as in the case of *Plasmodium* or by ingestion of infective spores.

Monocystis is an example of a sporozoan which has only one host, the earthworm. The sporozoite enters the seminal vesicles where it grows at the expense of the developing sperm cells. Two trophozoites unite, but do not fuse or lose identity, in a process called *syzygy*, and secrete a cyst wall around themselves. Each of the cells then undergoes multiple fission to produce many gametes. The gametes pair and each pair forms a *sporoblast* within which eight infective sporozoites are formed. The sporoblast may reach the outside by way of the sperm ducts, or it may be ingested with its host by a bird and pass unharmed through the digestive tract of the bird. When it is eventually ingested by an earthworm, the spore case is digested away and the sporozoites move to the seminal vesicles to start the cycle over again. The life cycle of *Monocystis* is shown in Figure 21.17.

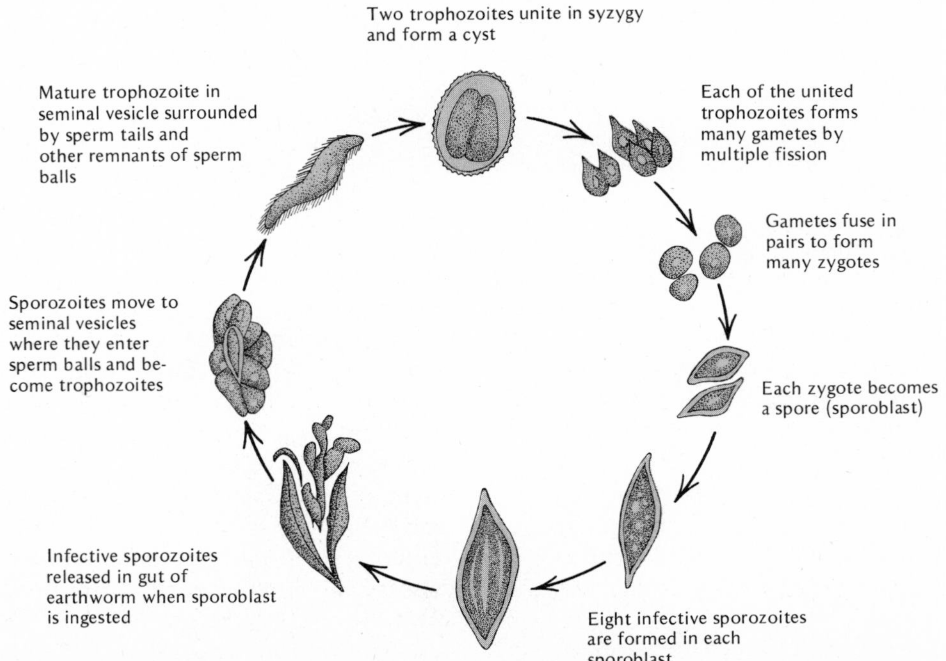
Two trophozoites unite in syzygy and form a cyst
Each of the united trophozoites forms many gametes by multiple fission
Gametes fuse in pairs to form many zygotes
Each zygote becomes a spore (sporoblast)
Eight infective sporozoites are formed in each sporoblast
Infective sporozoites released in gut of earthworm when sporoblast is ingested
Sporozoites move to seminal vesicles where they enter sperm balls and become trophozoites
Mature trophozoite in seminal vesicle surrounded by sperm tails and other remnants of sperm balls

Figure 21.17 Life cycle of *Monocystis lumbrici*, a common sporozoan parasite of the seminal vesicles of earthworms.

CLASS CILIATA

Classification

SUBCLASS **Holotricha** Simple cilia over all or part of body, no aboral cilia.

ORDER **Gymnostomatida** Mostly large species with no cilia around the mouth. Mouth opens directly to the outside with no gullet. *Coleps, Lacrymaria, Prorodon, Didinium.*

ORDER **Trichostomatida** Gullet with a vestibule provided with cilia. *Colpoda, Balantidium.*

ORDER **Chonotrichida** Vase-shaped species which lack cilia on the body in the adult. A funnel-shaped region at the anterior end is ciliated. Mostly marine. *Spirochona.*

ORDER **Suctorida** Sessile, stalked forms with sucking tentacles. Adult stage lacks cilia. *Podophrya, Tokophrya.*

ORDER **Apostomatida** Cilia on the body arranged in a spiral fashion. Marine parasites or commensals with complex life cycles. *Foettingeria.*

ORDER **Astomatida** Commensals or parasites in the gut and body cavity of oligochaete worms. Uniform ciliation on the body. *Anoplophrya.*

ORDER **Hymenostomatida** Small forms with uniform ciliation. Buccal cavity present with undulating membrane and a zone of membranelles around the anterior end. *Colpidium, Tetrahymena, Paramecium.*

ORDER **Thigmotrichida** A small group of marine and freshwater forms. Associated with bivalve mollusks. Tuft of sensory cilia at anterior end. *Thigmophrya.*

ORDER **Peritrichida** Cilia normally lacking on adult body. Conspicuous cilia associated with the mouth. Mostly attached forms with a distinct stalk. *Vorticella, Zoothamnium.*

SUBCLASS **Spirotricha** Reduced body cilia and well-developed cilia associated with the mouth.

ORDER **Heterotrichida** Body uniformly covered with cilia or encased in a secreted test and lacking cilia. *Stentor, Spirostomum, Folliculina.*

ORDER **Oligotrichida** Small in size with body cilia reduced or absent. Conspicuous membranelles associated with the mouth. *Halteria.*

ORDER **Tintinnida** Free-swimming forms with a secreted case. Conspicuous membranelles associated with the mouth. Chiefly marine. *Tintinnus, Favella.*

ORDER **Entodiniomorphida** Commensal forms found in the digestive tract of herbivorous mammals. Body cilia reduced or absent. *Entodinium.*

ORDER **Odontostomatida** A small group of laterally compressed forms with reduced body and oral cilia. *Saprodinium.*

ORDER **Hypotrichida** Dorsoventrally flattened forms with the body cilia restricted to cirri on the ventral surface. *Euplotes, Stylonychia.*

Anatomy and Physiology of Paramecium

HABITAT

Paramecium is one of the most common protozoans, frequently found in water containing bacteria and decaying organic matter. There are several species of *Paramecium*, ranging in length from 120 to over 300 μm. The appearance and behavior of a paramecium stand in striking contrast to the appearance and behavior of an amoeba. A paramecium has a definite shape and fixed anterior and posterior ends. Its movements are quite rapid in contrast to the sluggish amoeba. By contrast with as simple a type of protozoan as an amoeba, in terms of organelle variety, a paramecium is quite complex.

CELL STRUCTURE

Figure 21.18 shows the structure of a paramecium. It has often been called the "slipper animalcule" because of its shape. The anterior end is narrow and rounded; the posterior end is broad but pointed. The outer surface is a stiff but flexible *pellicle,* which accounts for its permanent shape. This pellicle is composed of numerous hexagonal. plates that are slightly concave. From the bottom of each concavity, a short hairlike cytoplasmic process, a *cilium,* emerges. These are the locomotor organelles in this class of protozoans. In *Paramecium* the cilia are arranged in rows all over the body. A conspicuous groove, the *oral groove,* runs diagonally across the anterior part of the animal. At the posterior end of this groove a *mouth pore* or *cytostome* is located. From this mouth opening a funnellike tube, the *gullet,* or *cytopharynx,* extends down into the cytoplasm. *Food vacuoles* are formed at the base of the gullet. The cytoplasm is differentiated into an outer *ectoplasm* and an inner more granular *endoplasm.* Lying in the ectoplasm, just beneath the pellicle, are numerous rodlike structures, the *trichocysts.* The contents of the trichocysts may be discharged as long threads. This discharge may be induced by adding a small amount of acetic acid to the water near a paramecium. Sometimes paramecia are observed to discharge their trichocysts when they are attacked by carnivorous protozoans, but sometimes they do not. It has been suggested that the trichocysts serve as defense organelles and this may be one of their functions. However, it is also known that the trichocysts are used for anchoring the animal during feeding.

In *Paramecium* there are two kinds of nuclei, a *macronucleus* and one or more *micronuclei.* This unusual nuclear condition characterizes the class Ciliata and is not found elsewhere. In *Paramecium caudatum,* one of the most common species studied, there is one macronucleus and one micronucleus. Metabolic activities of the cell are controlled by the macronucleus, and

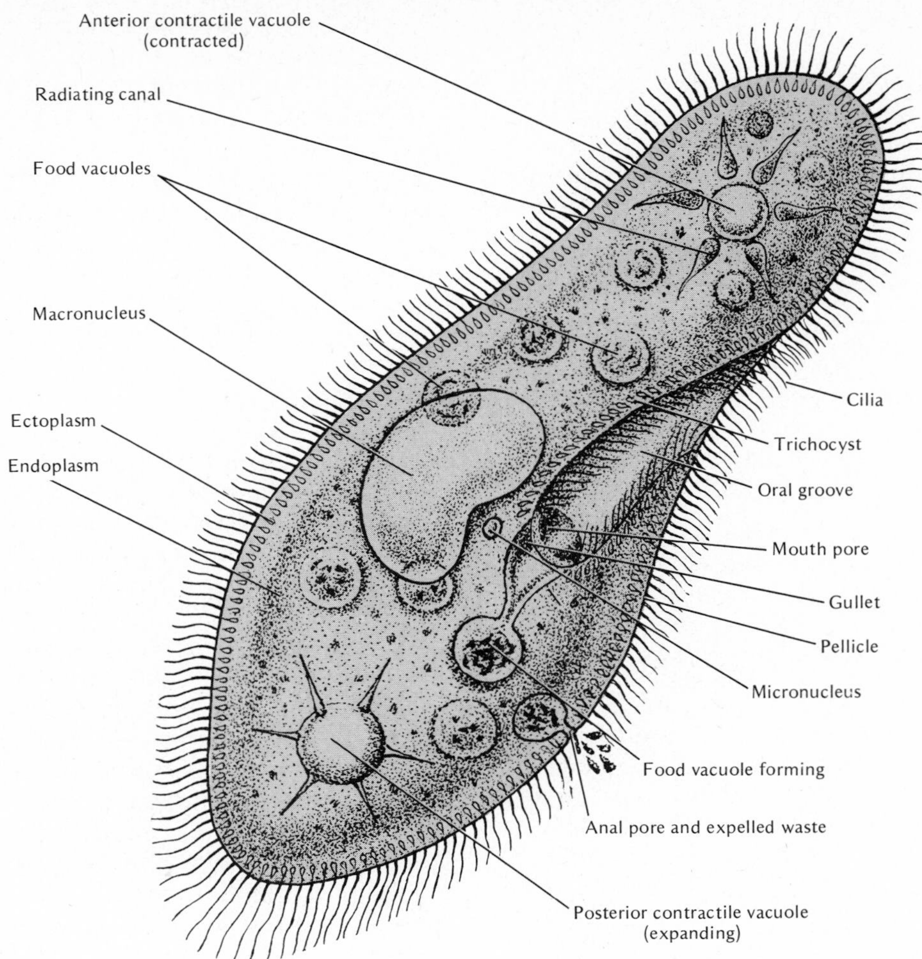

Figure 21.18 Structure of *Paramecium*.

the reproductive and hereditary functions by the micronucleus.

In *Paramecium,* ordinarily, there are two *contractile vacuoles.* They differ from the contractile vacuoles in *Amoeba* in that they are fixed in position (each has a permanent pore to the outside), and associated with each vacuole are several *radiating canals.* It can be observed that the canals gradually fill with fluid and then contract, filling the vacuole, which in turn empties to the outside through a pore. Under ordinary conditions, the two vacuoles alternate in contraction. The function of the contractile vacuoles is the same as that described for amoeba—the elimination of excess water in the cytoplasm.

Numerous food vacuoles are present in the endoplasm in different stages of digestion. Near the posterior end is a permanent *anal pore* where undigested particles are given off from the body.

LOCOMOTION AND COORDINATION

The ciliates are very rapid swimmers in many instances. Each movement of a cilium consists of an effective stroke in which the cilium is rigidly extended, and a recovery stroke in which the cilium is bent and thus presents less resistance (Fig. 21.19). The beat of the cilia is obliquely backward. This causes the animal to revolve on its long axis as it moves. However, because the cilia in the oral groove are longer and their beat is stronger than those on the rest of the body, the anterior end describes a circle, with the result that the animal swims in a spiral path. The cilia beat in successive waves which start at the anterior end and progress backward. A surface view of the cilia in action appears like a field of wheat in a slight breeze. This phenomenon of successive coordinated waves of action is called *metachronism.* As mentioned before, the basic structure of cilia and flagella is the same (9 + 2 fibrils).

Satir (see Suggestions for Further Reading) has recently reviewed the sliding microtubule hypothesis to explain ciliary action. The so-called fibrils of a cilium are microtubules composed of two kinds of proteins called tubulins. The collection (9 + 2) of microtubules in a cilium,

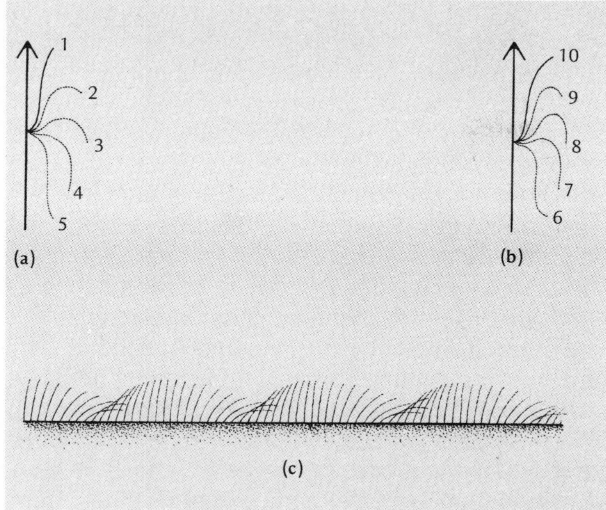

Figure 21.19 Ciliary movement. **(a)** 1–5. the effective stroke. **(b)** 6–10, the recovery stroke. **(c)** Metachronism, waves of ciliary movement from front to back of an organism.

apart from the surrounding membrane, constitutes an *axoneme.* This membrane has been removed, both by glycerin and detergents, and the naked axoneme will move when ATP and certain ions are added. It has been determined that the enzyme which breaks down ATP in cilia is a protein called *dynein.* The microtubules in each of the nine doublets are of unequal lengths. This enzyme, dynein, forms arms between the paired microtubules, and the dynein arms cause the microtubules to slide past one another. The interested student should read Satir's article and examine the figures. Figure 21.20 explains in part this sliding mechanism. No complete explanation for the reversal of ciliary beat is available at present. However, it is known that ciliary reversal is caused by an influx of calcium ions into the cell.

When a paramecium strikes an obstacle, it stops, reverses its ciliary stroke, and swims backward for a short distance, then pivots and swims forward again. This is called the *avoiding* reaction and is the basis for the *trial-and-error* behavior seen in this organism.

How does this unicellular creature, with over 2000 cilia, coordinate its movements? Each cilium ends in a basal granule or *kinetosome* (Fig. 21.21). The basal granules of a given row of cilia are all connected by coordinating fibers called *kinetodesmata* just beneath the pellicle. In addition, the different coordinating fibers are connected with adjacent ones at intervals, forming the *kinety system.* In *Paramecium* these fibers converge in the region of the gullet to form a coordinating center. When this center is experimentally destroyed, the cilia no longer beat in a coordinated manner.

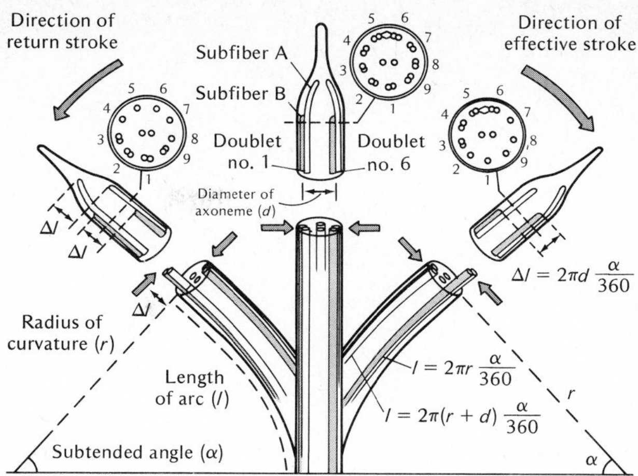

Figure 21.20 Sliding of microtubules is manifested by differences in the pattern of subfibers near the tip of the cilium. Here the relations among sliding, bending, and tip pattern are shown for two doublets, 1 and 6, in three stroke positions: straight up, bent in the effective-stroke direction (right), and bent in the return stroke direction (left). The short arrows mark equal distances along the microtubules from the beginning of the bend. Because the microtubules are flexible but of constant length, the doublet on the inner concave side of a bend must slide tipward. The resulting displacement Δ can be measured at the tip, at the end of subfiber B, or at the end of the bent region. The length l of the arc of the bend is given by the formulas. Because the length of the inner arc is proportional to the radius r and the length of the outer arc is proportional to that radius plus the diameter of the axoneme d, the difference in length Δ is proportional to the axoneme diameter d. As the three base-to-tip cross-section diagrams of the indicated part of each cilium show, in the straight position all subfibers B should be present, but in the bent positions subfiber B of the doublets on the outer side of the bend will be missing. When the dynein arms cause microtubules to slide past one another, shear resistance in the cilium changes sliding to bending. (From P. Sater, "How Cilia Move;" © 1974 by *Scientific American,* with permission.)

Because of the large number of cilia possessed by *Paramecium,* it is not a very good organism to experiment on in this connection. C. V. Taylor used to advantage the ciliate *Euplotes* to demonstrate the role of these coordinating fibers. *Euplotes* is not ciliated all over and, in addition, its locomotor organelles are

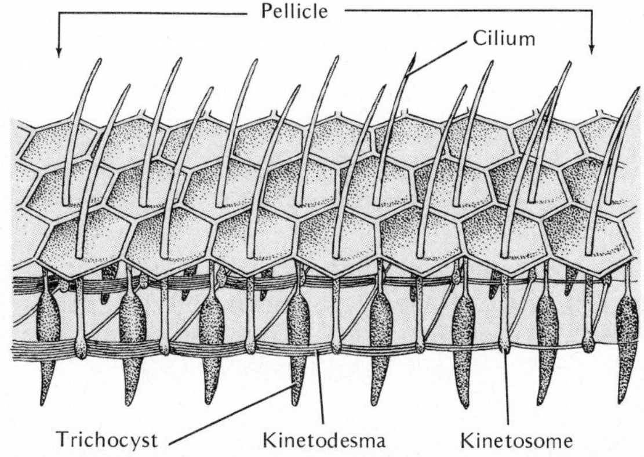

Figure 21.21 Portion of the pellicle of *Paramecium* with associated organelles and the kinety system.

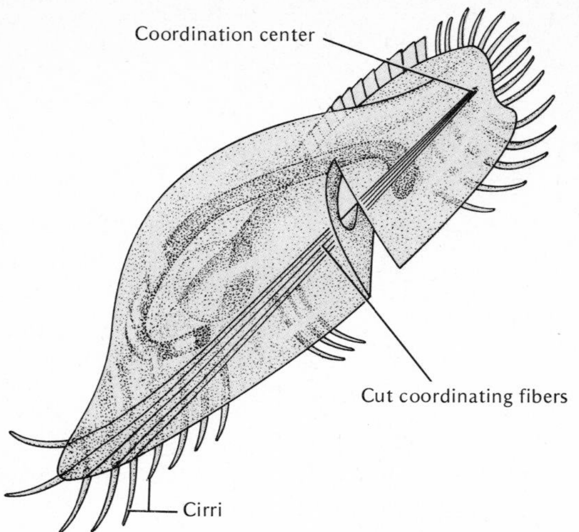

Figure 21.22 Taylor's experiment on *Euplotes.*

composed in each case of several fused cilia, forming what is called a *cirrus.* Figure 21.22 shows the nature of Taylor's experiment. When he cut through the coordinating fibers that lead from the coordinating center to the five posterior cirri, he found that the cirri would no longer move in coordination.

The role of these fibrils in coordination has been challenged by Naitok and Eckert (see Suggestions for Further Reading). Using cinematography, they found that sectioning of the fibrils in *Euplotes* did not destroy coordination between widely separated organelles. Their evidence indicates that ciliary coordination results primarily from electrotonic spread of spontaneous and evoked membrane potentials. Additional studies are needed before a definitive answer can be given to the question raised above.

NUTRITION

The natural food of *Paramecium* is bacteria and yeast. A paramecium may be observed to anchor itself alongside a clump of bacteria and feed. Through the beat of the cilia in the oral groove, the bacteria are swept into the mouth opening. The food particles are passed down the gullet into a food vacuole forming at its base. When a food vacuole attains a certain size, it breaks away from the gullet and, owing to cyclosis or streaming, slowly passes around the endoplasm.

During this movement of the food vacuoles through the endoplasm, digestion occurs and the end products of digestion diffuse from the vacuoles into the cytoplasm where they are utilized in metabolism. As in *Amoeba,* the enzymes that cause digestion are secreted into the food vacuoles by the cytoplasm.

Although ciliates like *Paramecium* are normally holozoic organisms, it has been possible in recent years to culture one species, *Tetrahymena pyriformis,* in a completely sterile (free of other organisms), chemically known, fluid medium. In this case a solution of salts, sugar, vitamins, amino acids, and nucleic acid derivatives gave good growth. Attempts have been made to culture *Paramecium* in the same way. It has been possible to obtain growth of *Paramecium* in the absence of other organisms using a medium containing inorganic salts, a carbon source, five B vitamins, 14 amino acids, a purine and a pyrimidine, the sterol stigmasterol, either stearic or oleic acid, and an unknown fraction or fractions from yeast. There is now evidence that at least one protein is required. Thus *Paramecium* will grow in a fluid medium devoid of particulate food material. In this situation the food substances enter in part through the plasma membrane and in part through the fluid food vacuoles that it continues to form.

RESPIRATION AND EXCRETION

These processes are accomplished in *Paramecium* in much the same way as they are in *Amoeba.* The movement of oxygen into the animal and of carbon dioxide out of the animal is by diffusion through the pellicle and plasma membrane. The pellicle is quite permeable to these gases and is not a barrier to this exchange. In addition to carbon dioxide, the *Paramecium* must excrete water and nitrogenous compounds. As already discussed, water is given off by the contractile vacuoles. Most of the nitrogenous waste in *Paramecium* is ammonia. This waste diffuses from the animal as it is formed in the same way that carbon dioxide leaves the animal.

REPRODUCTION

In contrast with *Amoeba, Paramecium* reproduces both *asexually* and *sexually.* Asexual reproduction occurs by transverse binary fission (Fig. 21.23). In this process the macronucleus divides by simple constriction (amitosis), but the micronucleus divides mitotically. The numerous very small chromosomes (30–40 in *P. aurelia*) reduplicate and are equally distributed to the two micronuclei which are formed. The nuclear membrane does not break down in this mitotic division of the micronucleus in *Paramecium.*

Under favorable conditions of food supply and temperature, paramecia will divide two or more times every 24 hours. If all these animals lived, prodigious numbers would be present after a short period of time.

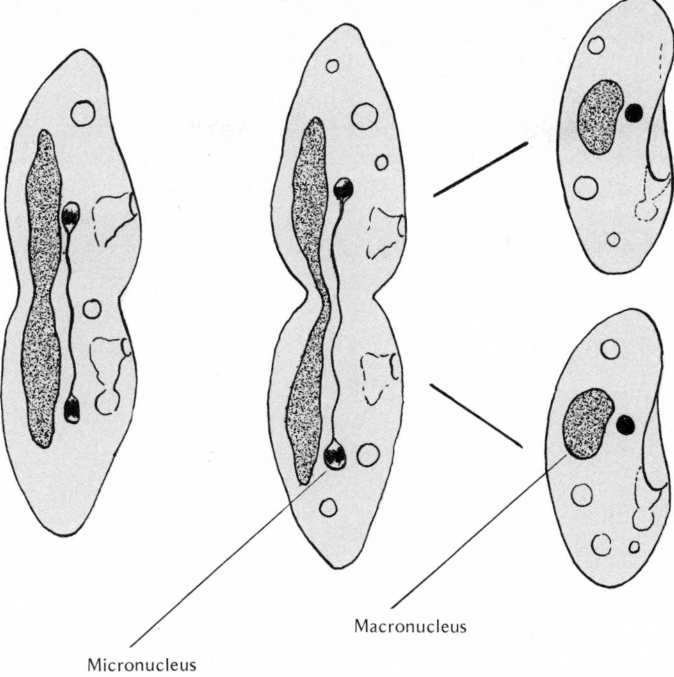

Figure 21.23 Transverse binary fission in *Paramecium caudatum*. Note that the micronucleus divides mitotically with an intranuclear spindle and that the macronucleus divides amitotically.

Paramecium is cosmopolitan in its distribution. The widespread distribution of many unicellular organisms is usually associated with cyst formation. However, attempts to demonstrate a cyst stage in *Paramecium* have been unsuccessful.

The process of binary fission may continue without interruption for long periods of time. However, at times the process of *conjugation,* a sexual process, occurs. In conjugation two individuals come in contact with each other in the region of the oral grooves and remain attached for some time. Figure 21.24 shows the series of nuclear changes that occur in the process. In the early stages of the process the micronuclei in each organism undergo meiosis. Three of the nuclei formed disintegrate in each conjugant, and the remaining one in each divides unequally. The smaller micronucleus in each one migrates across the cytoplasmic bridge that joins them and fuses with the larger micronucleus in the other conjugant. This exchange of nuclei is a sexual phenomenon and the diploid fusion nuclei are equivalent to zygote nuclei in other forms. While these changes have been taking place in the micronuclei, the macronucleus in each individual conjugant has been breaking down and has disappeared in the cytoplasm.

When fertilization is completed, the two conjugants separate. Then in each exconjugant, the single micronucleus (the product of fertilization) undergoes three divisions to produce eight micronuclei. Of these, four enlarge and become macronuclei and three of the others disintegrate. Then each exconjugant undergoes two binary fissions to produce four individuals. With each division the micronucleus divides so that each new individual receives a single micronucleus and a single macronucleus. The four macronuclei are simply distributed to the four individuals in the two divisions. Conjugation is then completed with the formation of eight paramecia from the two original conjugants.

In this process not only are individuals produced that have hereditary materials from two different individuals, but also, the macronucleus in each individual is recreated out of micronuclear material. That the periodical formation of new macronuclei (the nuclei that regulate vegetative functions) is important is indicated by another process called *autogamy* or self-fertilization.

Although paramecia may reproduce asexually by fission for many generations without conjugation occurring, several workers have shown that periodically the process of autogamy occurs. In this process a series of micronuclear divisions occurs, the macronucleus degenerates, and two micronuclei (each with half the normal number of chromosomes) fuse. After this the process is like that described in conjugation. In autogamy there is no exchange of

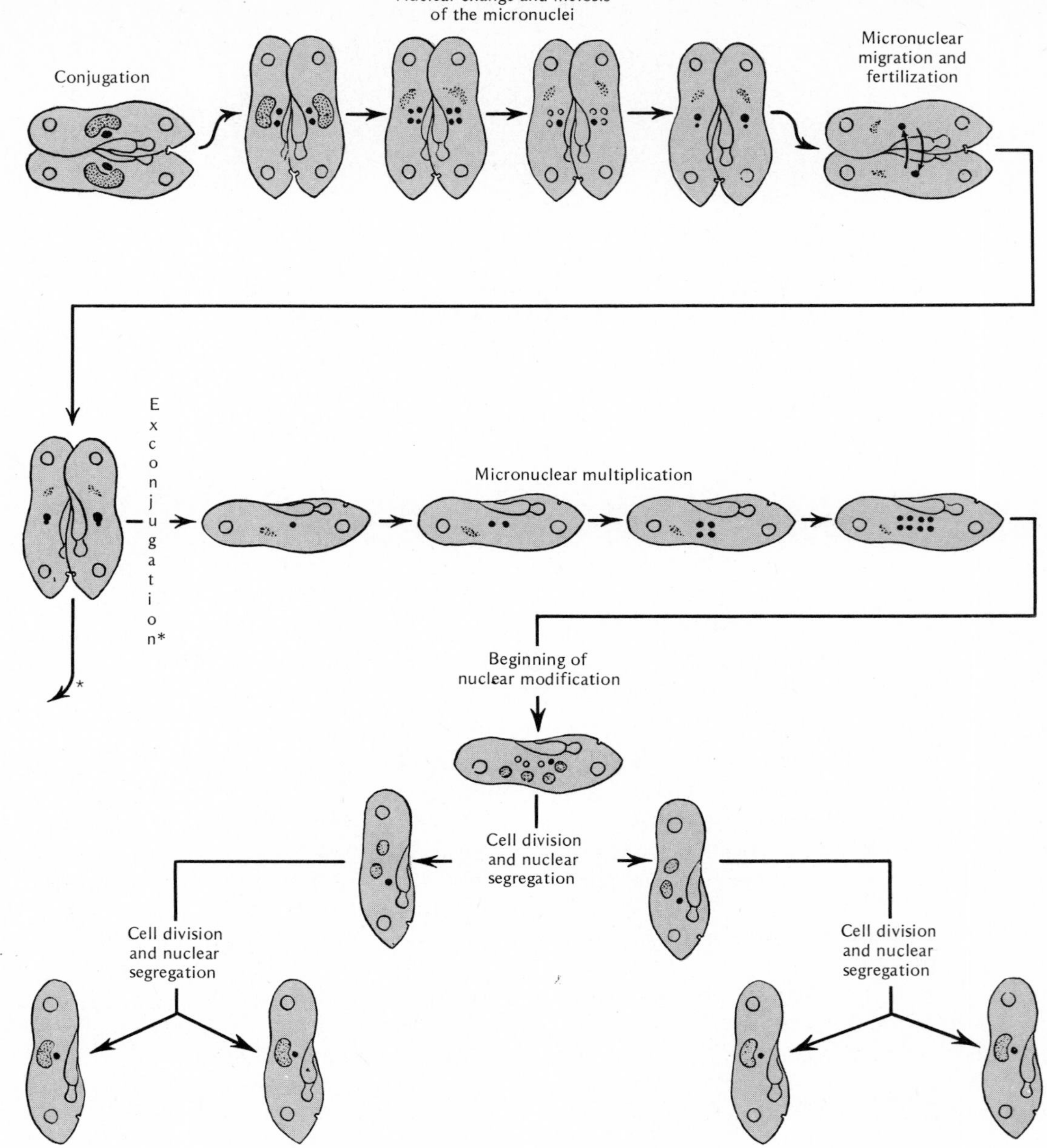

Figure 21.24 Conjugation in *Paramecium caudatum.* Follow the arrows. After the conjugants separate,* only one of the exconjugants is followed. Note, however, that eight new individuals result from one conjugation. See text for details.

nuclear material between two individuals, but the process results in the formation of new macronuclei.

What causes paramecia to conjugate? Are there sexes in this organism? Studies by Jennings, Sonneborn, and others have shown that sex as we ordinarily think of it—with a male and a female in each species—is not involved. The situation is more complicated. Table 21.1 shows the situation in one species, *Paramecium aurelia,* as worked out by Sonneborn. He describes seven varieties in this species. In each of the first six varieties there are two mating types; in variety 7 he has found only one. In variety 1, mating type I will conjugate with mating type II, and with no others. The table shows the rest. In this species the different mating types are not distinguishable on a morphological basis. However, it is assumed that there are differences of a physiological and chemical nature that cause the behavior described. This phenomenon of mating types has been found in a number of other ciliates.

BEHAVIOR

The basis for the responses of *Paramecium*—the coordination of ciliary movements in the avoiding reaction—was pointed out earlier. A

Table 21.1 Varieties and Mating Types in Paramecium aurelia[a]

		VARIETY												
		1		2		3		4		5		6		7
VARIETY	MATING TYPE	I	II	III	IV	V	VI	VII	VIII	IX	X	XI	XII	XIII
1	I	–	+	–	–	–	–	–	–	–	–	–	–	–
	II	+	–	–	–	–	–	–	–	–	–	–	–	–
2	III	–	–	–	+	–	–	–	–	–	–	–	–	–
	IV	–	–	+	–	–	–	–	–	–	–	–	–	–
3	V	–	–	–	–	–	+	–	–	–	–	–	–	–
	VI	–	–	–	–	+	–	–	–	–	–	–	–	–
4	VII	–	–	–	–	–	–	–	+	–	–	–	–	–
	VIII	–	–	–	–	–	–	+	–	–	–	–	–	–
5	IX	–	–	–	–	–	–	–	–	–	+	–	–	–
	X	–	–	–	–	–	–	–	–	+	–	–	–	–
6	XI	–	–	–	–	–	–	–	–	–	–	–	+	–
	XII	–	–	–	–	–	–	–	–	–	–	+	–	–
7	XIII	–	–	–	–	–	–	–	–	–	–	–	–	–

[a] + indicates that conjugation occurs; – indicates that it does not. (From Kudo, *Protozoology,* Charles C Thomas, 1954.)

number of conditions in the environment elicit the avoiding reaction; mechanical objects, excessive heat or cold, irritating chemicals, or predacious enemies are examples. Their avoidance of extremes of heat or cold, of strong chemicals, and of solid objects may be easily demonstrated in the laboratory. Paramecia are negatively geotropic—they may be seen to collect near the top of a glass container. They respond positively to dilute acids. This response is probably associated with feeding, because bacteria are apt to be near decaying organic matter that makes the surrounding water more acid. Paramecia are not able to discriminate easily between food and nonfood particles. They may be observed to ingest carbon and carmine particles and form numerous food vacuoles. However, after a time they will reject the inert particles and take in only food material. In some experiments it has been shown that paramecia may discriminate between different kinds of bacteria.

T. M. Sonneborn at Indiana University has used strains of *P. aurelia* in studies on cytoplasmic inheritance. "Killer" strains, when mixed with sensitive strains, cause the death of the sensitive organisms. The killer effect is due to particles, called "kappa," present in the cytoplasm of the killers, which release into the medium a toxin that kills the sensitive organisms. By appropriate matings kappa can be transferred to sensitives, making them killers. The viruslike kappa particles are self-reproducing. The maintenance of kappa is determined by a specific nuclear gene.

Other Ciliates

There are 15 orders of ciliates, placed in two subclasses. The major basis for distinguishing the various orders is the nature and distribution of the cilia. The cilia may be individual entities as described above for *Paramecium,* but there are also various fusions of cilia to produce more complex locomotor organelles. The leglike cirri formed by the fusion of many cilia into a single structure have been mentioned above. In *membranelles* the cilia are fused side by side to form paddlelike structures, and a long wavy membrane formed in the same manner is an *undulating membrane.* Figure 21.25 shows these ciliary modifications.

Paramecium feeds on small bacteria and yeasts and has a mouth located at the base of

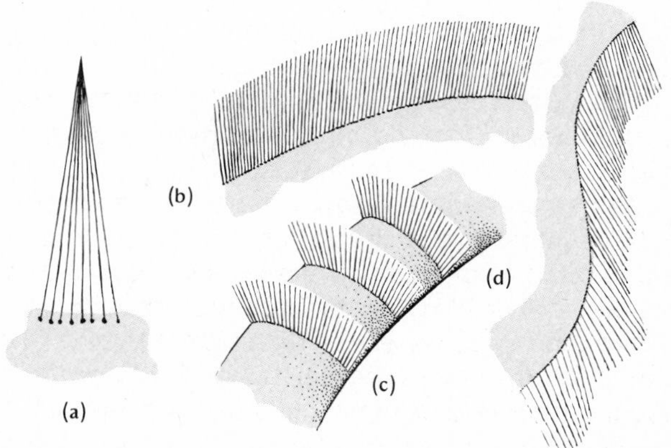

Figure 21.25 Ciliary modifications. **(a)** Cirrus. **(b)** Membranelle. **(c)** Paddlelike membranelles as in *Stentor.* **(d)** Undulating membrane.

its gullet. Other ciliates feed on larger food particles or capture other protozoa in some cases. In these forms the mouth becomes more specialized; for example, in *Didinium*, which captures and engulfs paramecia, there are small stiffening rods, called *trichites,* around the mouth.

The order Suctorida is made up of organisms which are stalked and sessile and lack cilia as adults. They do have a ciliated developmental stage and are thus included in the class Ciliata, although formerly they were set aside as a separate class. The name Suctorida is derived from the fact that the adults possess sticky knobbed tentacles which entangle and paralyze prey. Subsequently the prey is sucked into the body of the suctorian through the central portion of the tentacle. (Fig. 21.26).

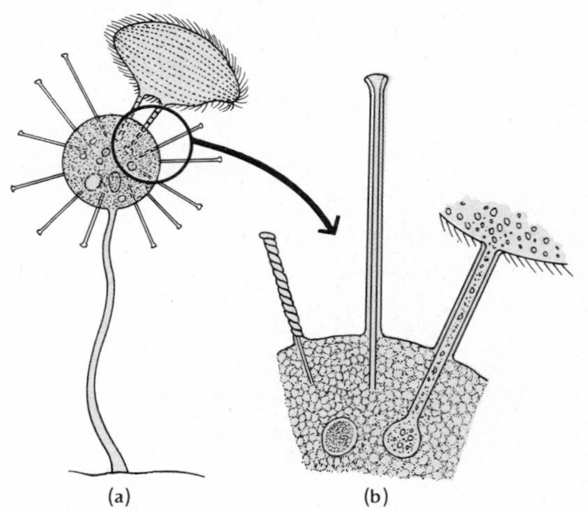

Figure 21.26 Structure of a typical suctorian, *Podophyra.* **(a)** Adult suctorian which has caught a small ciliate. **(b)** Detail of structure and function of the sucking tentacles.

The macronucleus is varied in shape in different types of ciliates. In *Paramecium* it is kidney shaped, in *Vorticella* it is sausage shaped, and in *Stentor* and *Spirostomum* it is beaded. The micronucleus is normally rounded.

Although most ciliates are free-swimming forms and generally rapid swimmers at that, there are some sessile species. The suctoria have been mentioned above. *Vorticella* is another sessile ciliate, but it has contractile elements in its stalk, called *myonemes,* which enable it to pull down close to the substratum and then extend upward for some distance (Pl. 24). This capacity greatly increases its feeding area, and the rapid contraction of the stalk may also serve a protective function. Several of the ciliates mentioned are shown in Figure 21.27.

There are relatively few parasitic ciliates and only one, *Balantidium coli,* sometimes is reported as a parasite in the human intestine. Figure 21.27(**c**) shows a trophozoite of this species. Note the cytopharynx and the permanent anal region called a *cytopyge.* The trophozoite contains red blood cells and other cell fragments from damaged host tissue.

PROTOZOANS AND THE ORIGIN OF MULTICELLULAR ANIMALS

All zoologists agree that multicellular animals or metazoans were derived in evolution from unicellular organisms, but there is no general agreement about the ancestral group or groups. It should be stated, as Hyman has pointed out, that on the basis of present information the problem is insoluble, and therefore we can only speculate. Because there are no fossils, the theories presented must be based on inferences made from studies of living forms. Some propose that the ancestral metazoan arose from a multinucleate ciliate; others that the ancestral metazoan arose from a colonial flagellate; and still others propose that the metazoans had multiple origins from different unicellular groups. The first two of these theories will be considered briefly.

The first theory has been called the *syncytial theory.* According to this theory, the ancestral metazoan was derived from a multinucleate ciliate. It was at first syncytial in structure, that is, a multinucleate mass of cytoplasm. Later multicellular forms appeared as the result of the development of cell membranes around the nuclei. According to this theory the primitive metazoans were something like the living acoel flatworms. These acoels have cilia but tend to be syncytial and are in the general size range of some ciliates. There are a number of objections to this theory. The acoel flatworms are not syncytial during early development; the syncytial condition arises later. A ciliate ancestor does not explain the flagellated sperm found in metazoans, unless these flagella were derived secondarily from cilia. The acoel flatworms are bilaterally symmetrical. This theory, then, would make bilateral symmetry a primitive condition in metazoans. On this basis the radial symmetry of coelenterates would have to be derived secondarily from the flatworms. All the evidence, however, indicates that the radial symmetry of the coelenterates is primary.

The second or *colonial theory* is the older and more classical one. It is based on the assumption that the ancestral metazoan probably arose

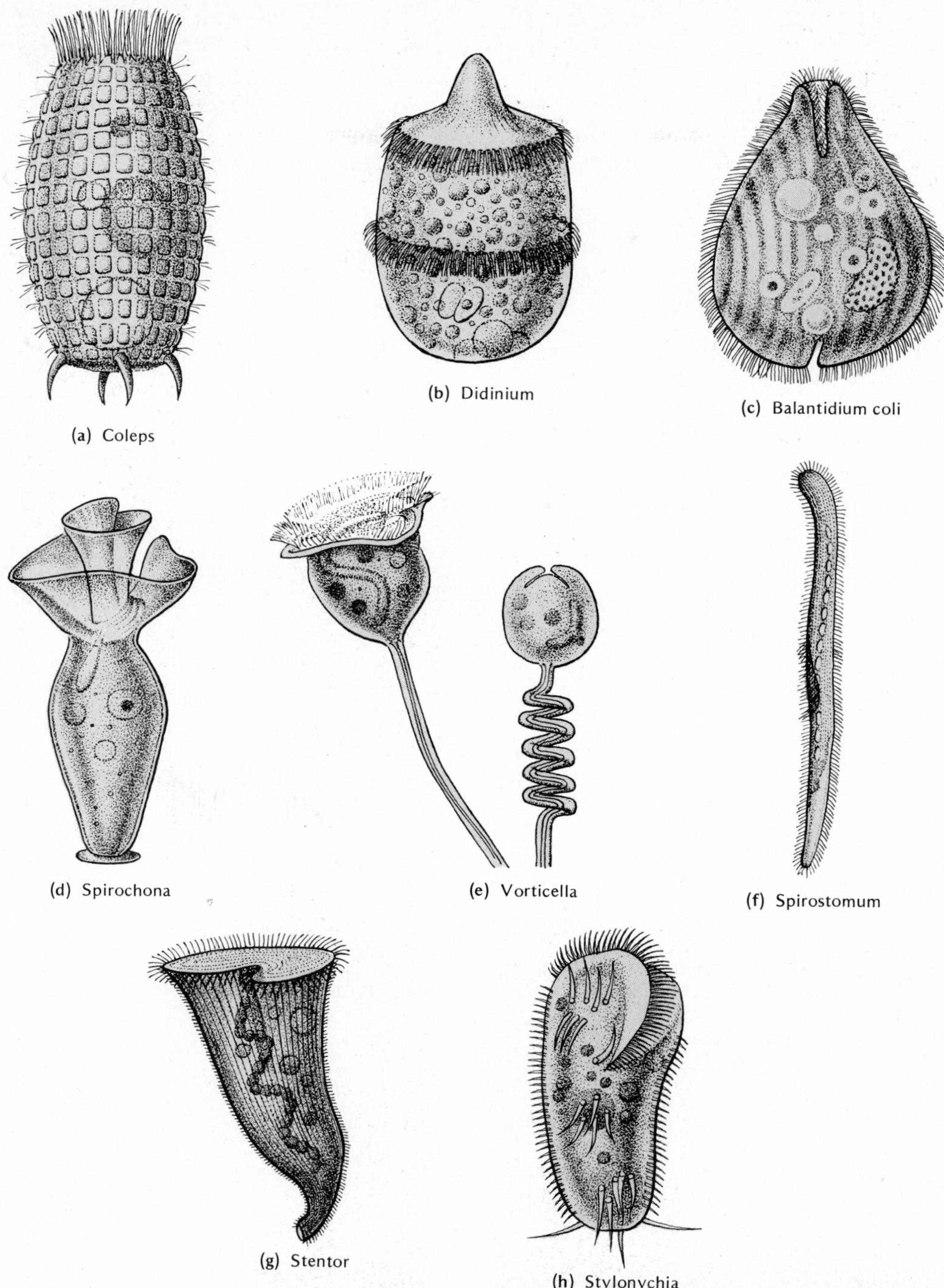

Figure 21.27 Representative ciliates. **(a)** *Coleps,* a species with thick plates in the pellicle. **(b)** *Didinium,* feeds on *Paramecium.* **(c)** *Balantidium coli,* intestinal parasite of man. **(d)** Spirochona. **(e)** *Vorticella;* note the contractile myonemes in the stalk. **(f)** *Spirostomum.* **(g)** *Stentor.* **(h)** *Stylonychia.*

from a hollow colonial flagellate. This theory and its modern interpretation have been discussed in some detail above in connection with the colonial plantlike flagellate, *Volvox.*

The possible origin of the sponges from the choanoflagellates has also been discussed. In Chapter 22 the sponges will be dealt with in some detail.

SUGGESTIONS FOR FURTHER READING

Corliss, J. O. *The Ciliated Protozoa.* New York: Pergamon (1961).

Dobell, C. *Anthony van Leeuwenhoek and His Little Animals.* London: Staples Press (1932).

Edmondson, W. T. (Ed.). *Ward and Whipple's Fresh-Water Biology,* 2nd ed. New York: Wiley (1959).

Hutner, S. H., and A. Lwoff (Eds.). *Biochemistry and Physiology of the Protozoa,* vols. 1 and 2. New York: Academic Press (1951–1955).

Hyman, L. *The Invertebrates,* vol. 1: *Protozoa through Ctenophora.* New York: McGraw-Hill (1940).

Jahn, T. L., and F. F. Jahn. *How to Know the Protozoa.* Dubuque, IA: William C. Brown (1949).

Johnson, W. H. "Nutrition of Protozoa," *Annual Review of Microbiology,* **10** (1956).

Kudo, R. R. *Protozoology.* Springfield, Il: Charles C. Thomas (1966).

Leedale, G. F. *Euglenoid Flagellates.* Englewood Cliffs, NJ: Prentice-Hall (1967).

Naitok, Y., and R. Echert. "Ciliary Orientation; Controlled by Cell Membrane or Intracellular Fibrils?" *Science,* **166**:3913 (1969).

Pennak, R. W. *Fresh-Water Invertebrates of the United States.* New York: Ronald Press (1953).

Pitelka, D. R. *Electron-Microscopic Structure of Protozoa.* New York: Pergamon (1963).

Wichterman, R. *The Biology of Paramecium.* New York: McGraw-Hill (1953).

From Scientific American

Allen, R. D. "Amoeboid Movement" (Feb. 1962).

Alvarado, C. A., and L. J. Bruce-Schwatt. "Malaria" (Feb. 1962).

Hawking, F. "The Clock of the Malaria Parasite" (June 1970).

Satir, P. "How Cilia Move" (Oct. 1974).

22 Mesozoa and Porifera

PHYLUM MESOZOA

Characteristics

1. Solid two-layered construction with inner cell or cells reproductive in function, not digestive. These layers are not homologous with germ layers of higher forms.
2. Microscopic in size.
3. All are internal parasites of marine invertebrates and have complex life cycles.
   Figure 22.1

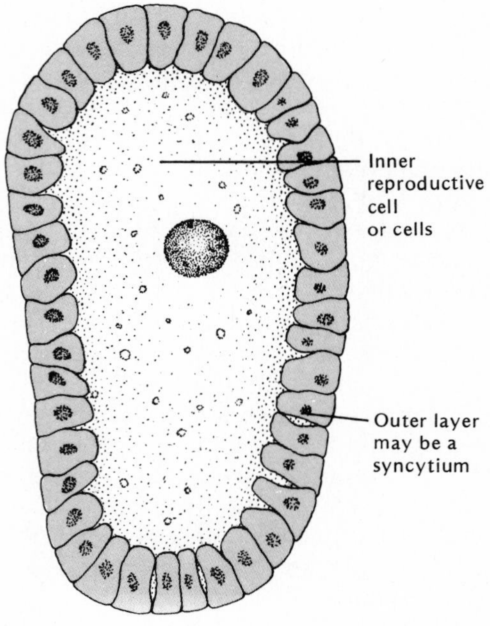

Figure 22.1 Basic mesozoan organization.

Classification

CLASS **Moruloidea**

ORDER **Dicyemida** These are common parasites of the kidney of octopuses and squids. Each species has a constant small number of cells surrounding a single reproductive cell in the adult stage of the life cycle. *Dicyema, Dicyemennea, Conocyema.*

ORDER **Orthonectida** Rare parasites of the internal spaces and tissues of various marine invertebrates, including flatworms, nemerteans, brittle stars, annelids, and a clam. The asexual stage is a multinucleate amoeboid plasmodium. In the sexual phase the central reproductive region is multicellular. No intermediate host in the life cycle. *Rhopalura, Stoecharthrum.*

Anatomy and Life Cycle of Order Dicyemida

There has been a good deal of mystery about the life cycle of the dicyemids. Several workers have published conflicting accounts and it was formerly assumed that there was a sexual stage in the life cycle in some organism other than a cephalopod. Recently Lapan and Morowitz have found dicyemids in the kidneys of young octopuses (cephalopods) hatched in aquaria which contained no possible source of infection other than adult octopuses.

The vermiform stage, shown in Figure 22.2, is the mature parasite found in the cephalopod

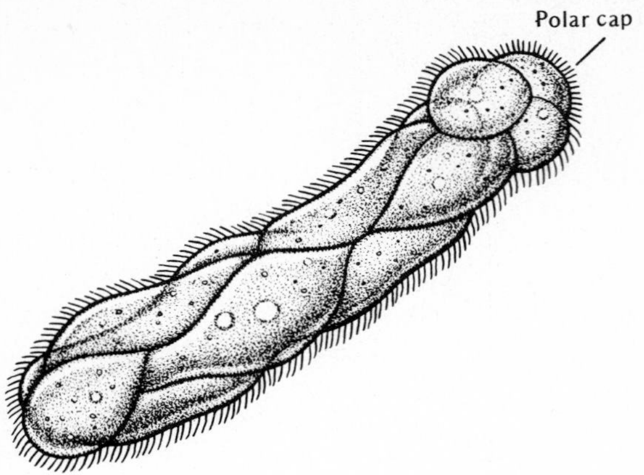

Figure 22.2 Order Dicyemida, *Pseudodicyema* sp., vermiform stage.

kidney. Ranging in size from 0.5 to over 7 mm, the adult is a long slender organism. It is made up of a constant number of ciliated jacket cells, usually about 25, surrounding a long cylindrical *axial cell.* Within the axial cell are a number of small reproductive cells, called *axoblasts* or *agametes.* Axoblasts may either give rise to adult vermiforms or remain within the axial cell to become a hollow structure which produces both eggs and sperm. This structure, called an *infusorigen*, is actually a hermaphroditic gonad. The eggs, fertilized by sperm from the same or a nearby infusorigen, develop into minute ciliated *infusiform* larvae. The infusiform larva escapes into the kidney tubule sytem and is shed into the sea with the urine.

Settling to the bottom, the infusiform larva is swallowed by an octopus or a squid and moves by an uncertain pathway, perhaps the circulatory system, to the host kidney. Cells released from the infusiform larva develop into the vermiform adults and the cycle starts over again in the new host.

The factor which determines whether axoblasts become vermiforms which remain in the host or undergo the unusual changes resulting in the production of infusiform larvae seems to be the size of the population of vermiform adults within the kidney. Infusiform larvae are produced only when a large population of vermiform adults builds up.

The immature vermiforms swim freely in the kidney fluids of the cephalopod host. When mature, they attach to the spongy tissue of the kidney by means of the cilia which cover the rounded polar cap. There is little or no evidence of damage to the kidney tissue and they are quite probably harmless to the host. Food substances such as albumin and amino acids are absorbed from the kidney fluids. In line with their habitat, all stages of the life cycle except the free-swimming infusiform larva are anaerobic, metabolizing glycogen with the liberation of lactic acid.

Anatomy and Life Cycle of Order Orthonectida

In view of the new details of the life cycle of the dicyemids, it may turn out that the orthonectids with their more complex life cycle are not closely related to the dicyemids. Perhaps they will be removed from the phylum Mesozoa and placed somewhat higher in the phylogenetic sequence.

Orthonectids occur in a variety of invertebrate hosts. The plasmodial stage, shown in Figure 22.3, develops within the host. The plasmodial stage is clearly parasitic, causing considerable damage to host tissues. It reproduces by simple fragmentation for some time and then produces agametes which become males or females. In most cases a given plasmodium produces only males or females. These ciliated forms escape into the sea. Fertilization occurs within the female following insemination by the male via the female genital opening. Development of a ciliated larval form occurs within the female. When liberated, the larva seeks out its proper host species, enters, and gives rise to a new plasmodium which starts the cycle over once again.

Phylogenetic Relationships

Although these animals were first discovered in 1839, they still constitute a puzzle to students of phylogeny. The name Mesozoa implies a primitive condition intermediate between protozoa and metazoa, and structurally there is

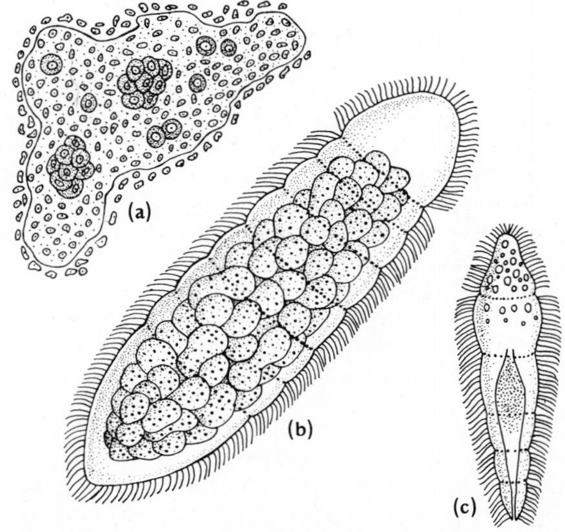

Figure 22.3 Order Orthonectida, *Rhopalura,* sp. **(a)** Plasmodial stage. **(b)** Female. **(c)** Male.

much to support this hypothesis. The structural features of this group are so distinctive that they cannot rightly be placed with any other phylum. However, the mesozoans are all internal parasites of marine invertebrates, and the distinctive features may be the result of parasitic degeneration. In subsequent chapters, examples of such degeneration will be seen in several groups of parasites.

Van Beneden, the famous Belgian cytologist, took the view that the mesozoans were in fact primitive forms, intermediate between the protozoans and the metazoans, and it was he who introduced the term Mesozoa in 1877. Hatschek, a German authority on phylogeny, placed the mesozoans in the phylum Coelenterata because of their resemblance to a coelenterate larval form, the *planula* larva. More recently there have been those who regard the mesozoans as degenerate flatworms and append them to the phylum Platyhelminthes.

If the structural features of the mesozoans could be shown to be original and not the result of parasitic degeneration, the group would be of great phylogenetic significance as a logical stage bridging the gap between the protozoans and the metazoans. They would furnish solid evidence that the first step in metazoan organization had a solid, not hollow, construction and that the gastraea theory of Haeckel is definitely erroneous. However, in the absence of fossil evidence the question of the origins of the mesozoans may never be answered.

The most promising evidence for a close relationship of the dicyemids with the protozoans has recently come from biochemical studies. Lapan and Morowitz examined the DNA of *Dicyemennea,* a common genus of dicyemid, and found that it contained 23 percent guanine and cytosine. Comparison of this value with those from many other animal groups reveals that it is a very low percentage. Values this low have been found in only a few other organisms, all protozoans. Species of the ciliate *Tetrahymena* have values ranging from 23 to 30 percent and species of *Paramecium* also have a guanine plus cytosine content as low as 23 percent. The few guanine–cytosine values available for platyhelminthes range from 35 to 50 percent.

Although the evidence above is the most important yet found to indicate a primitive position for the dicyemid mesozoans, it is perhaps too early to place them firmly in this place on the phylogenetic tree.

PHYLUM PORIFERA

Characteristics

1. No true tissues or organs.
2. No digestive tract or mouth.
3. Body wall pierced by many pores through which water enters.
4. One to many internal cavities, at least some of which are lined with choanocytes.
5. Internal skeleton of spicules or spongin or both.
6. Two cell layers, but these layers are not homologous with the germ layers of higher forms.
7. Radially symmetrical or asymmetrical.
8. Free-swimming larval form, but sessile as adults.
9. All are marine except for the small freshwater family Spongillidae.
   Figure 22.4

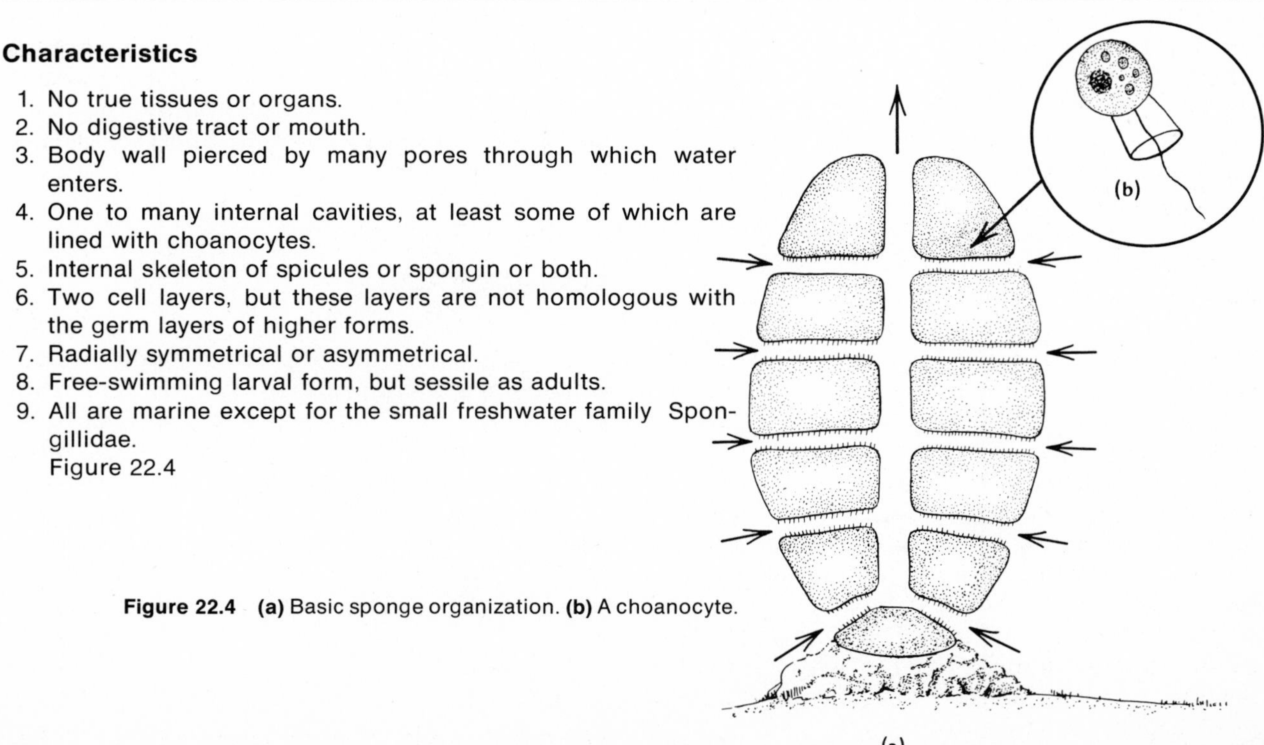

Figure 22.4 **(a)** Basic sponge organization. **(b)** A choanocyte.

Classification

BRANCH **Parazoa** Because of the distinctive nature of their organization, the sponges are separated from all other metazoans as the branch Parazoa. It seems probable that the sponges constitute the simplest living multicellular group of animals. The sponges are a dead end as far as evolution is concerned. Evolution occurred within the phylum with the development of a variety of types, but no higher phylum can be directly related in any way to the sponges.

CLASS **Calcarea** These are sponges in which the internal skeleton is of calcium carbonate spicules. They are generally of small size and many are solitary or, if colonial, they retain their individuality (Fig. 22.5). *Leucosolenia Sycon, Scypha.*

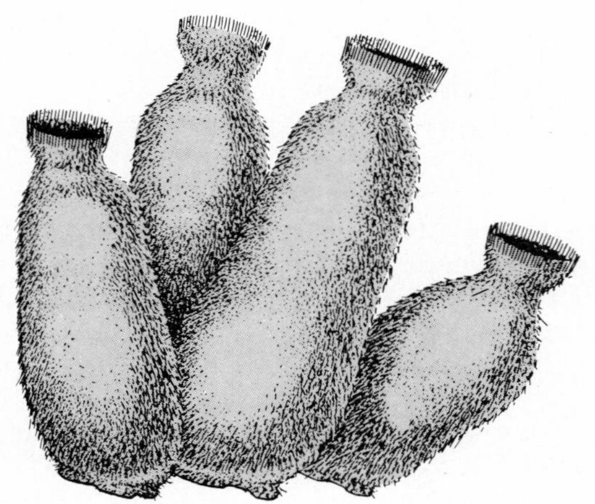

Figure 22.5 Class Calcarea, *Scypha* sp.

CLASS **Hexactinellida** Glass sponges are distinguished by their skeleton of silicon dioxide spicules which are triaxon six-rayed structures. They are of moderate size, inhabit deeper portions of the sea, and are mostly solitary (Fig. 22.6).

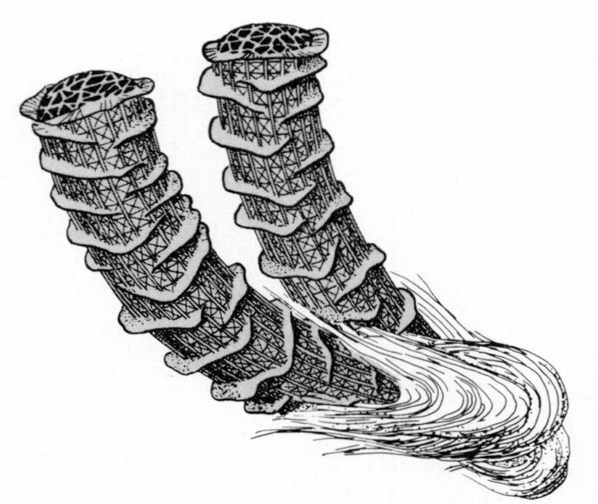

Figure 22.6 Class Hexactinellida, *Euplectella* sp., Venus's flower basket.

CLASS **Demospongiae** A few members of this group have no skeleton but most have one either of silicon spicules which are single-rayed or four-rayed; or of an organic substance called spongin; or of both. This is the largest class in terms of numbers of species. Many form complex colonies in which boundaries between individuals are indistinct. The bath sponge, whose skeleton of spongin was in great demand before the advent of cellulose sponges, is in this class (Fig. 22.7). *Euspongia, Haliclona, Spongilla.*

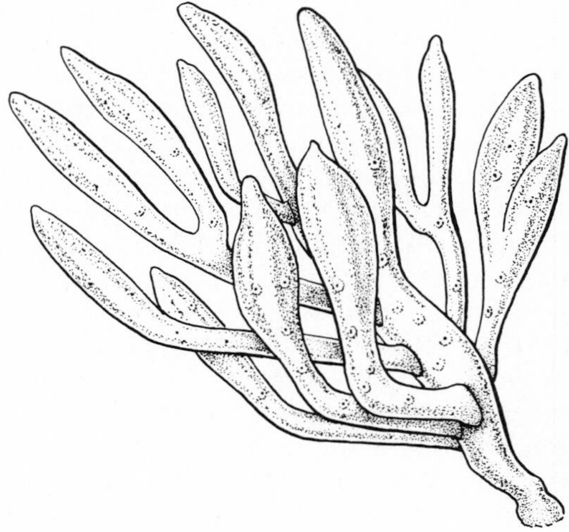

Figure 22.7 Class Demospongiae, *Haliclona* sp.

Since the sponges form a rather homogeneous group and have quite similar anatomical features, the discussion will be organized for the phylum as a whole without a separate discussion for each of the classes. A detailed breakdown to the order level would not be profitable since the differences depend primarily on fine distinctions between spicule types.

Anatomy and Physiology

BASIC BODY PLAN

The simplest type of anatomy is seen in the *olynthus* stage, a developmental stage through which some sponges pass (Fig. 22.8). The olynthus has a vase-shaped structure with an outer layer of flattened *pinacocytes* and a single large inner cavity, the *spongocoel,* which is lined with choanocytes. Between these two cellular layers is a jellylike *mesoglea* in which a number of types of cells are found. At intervals there are special cells, called *porocytes,* which contain an intracellular pore connecting the spongocoel with the outside. Water enters by way of the porocytes and is passed through the spongocoel and out again by way of the single large opening, the *osculum* (Pl. 26). The water is moved by

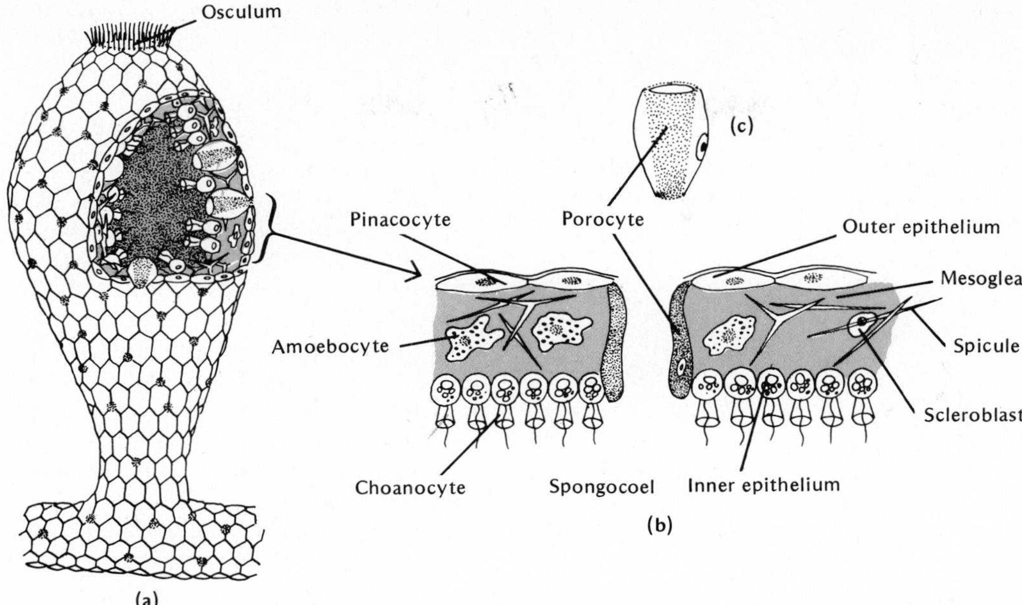

Figure 22.8 Olynthus stage of development. **(a)** Gross structure. **(b)** Diagrammatic section of portion of body wall. **(c)** Diagram of a porocyte showing the intracellular pore.

the action of the flagella of the choanocytes. The porocytes are contractile and may close off the pore in the event of unfavorable conditions in the water.

The cells in the mesoglea are *amoebocytes, scleroblasts,* and *archeocytes.* In addition, the spicules are located in the mesoglea. The amoebocytes are wandering cells which carry nutritive materials from one part of the sponge to another and also gather excretory wastes. In some cases they become so filled with food materials that they can be considered storage cells.

The scleroblasts form the skeletal elements. Figure 22.9 shows the method of formation of the different skeletal types. A monaxon spicule is started as an organic core within a single scleroblast which soon divides to form two cells, the *founder* and a smaller *thickener.* The spicule is shaped by the founder cell and the thickener cell moves back and forth over the surface laying down more material ($CaCO_3$ or SiO_2) and enlarging the spicule. A triradiate spicule forms from three scleroblasts which divide to produce founders and thickeners as outlined above. Similarly, a tetraxon spicule (four radii) starts with four scleroblasts.

The spongin fibers are formed within *spongioblasts* which line up side by side and finally disintegrate after the adjacent portions of the

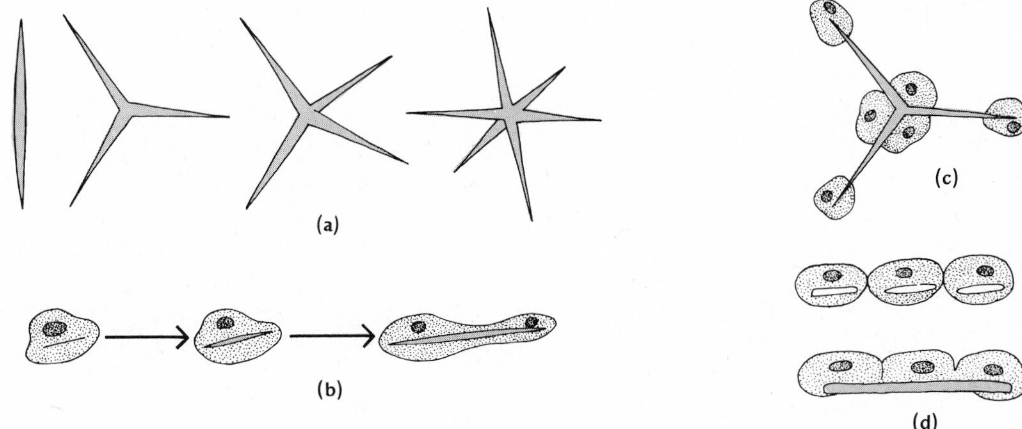

Figure 22.9 Sponge spicules, types and methods of formation. **(a)** Monaxon, tri-radiate, textron, and hexaxon (left to right). **(b)** Formation of a monaxon spicule. Scleroblast secretes an organic core; silicon dioxide or calcium carbonate laid down, scleroblast divides to form founder and the smaller thickener cell (left to right). **(c)** Triaxon spicule formed by union of three scleroblasts which have each divided to form a founder and thickener. **(d)** Formation of spongin fiber. Three spongioblasts form portions of fiber which fuse. Spongioblasts later disintegrate.

spongin fiber have fused. Spongin is a protein related to keratin (the material found in mammalian hair), silk, and collagen fibers. In addition to the fibrous form, the spongin may be secreted as an amorphous mass and used to cement the sponge to the substratum. The archeocytes are the source of gametes and play an important role in regeneration. These functions will be discussed later.

MODIFICATIONS OF THE BASIC OLYNTHUS BODY PLAN

There are three types of body plan found in mature sponges. The simplest is the *ascon type* which is found in *Leucosolenia,* a common calcareous sponge. This is essentially the same as the olynthus stage with a thicker body wall. Choanocytes line the spongocoel and incurrent openings are the porocytes with intracellular canals described above. Figure 22.10 shows a typical ascon sponge.

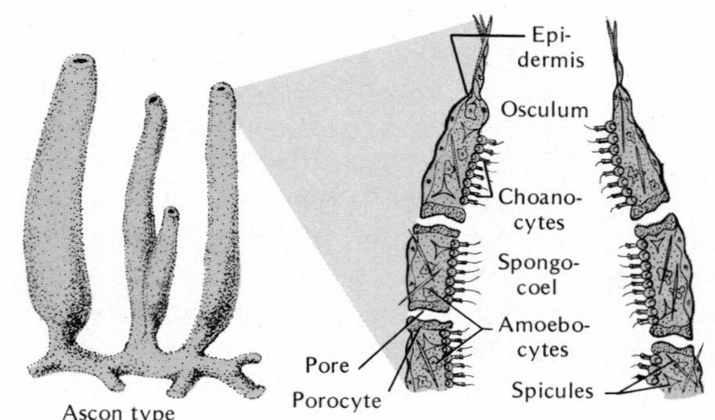

Figure 22.10 Ascon type of sponge body, *Leucosolenia.*

Folding of the ascon type results in the *sycon type* (Fig. 22.11), which is characteristic of *Scypha.* The folding results in a central spongocoel, as in the ascon, with lateral extensions, called *radial* or *excurrent canals.* The choanocytes are restricted to the lining of the radial canals. The folding which produces the lateral radial canals also results in infoldings between them, called *incurrent canals.* Water enters the opening of the incurrent canal, called an *ostium,* and passes through an opening in the sponge wall, called a *prosopyle,* into the radial canal. It leaves the radial canal by way of an opening, called the *apopyle,* and enters the spongocoel from whence it passes out through the osculum. The intracellular pores characteristic of the ascon type are absent in the sycon sponges. All of the openings are between cells, although they may originate as openings in porocytes which ultimately degenerate, leaving a space in the epidermis.

Continued complex folding results in the *leucon* type of body wall (Fig. 22.12). The choanocytes are now confined to numerous small cavities, called *flagellated chambers.* Water passes through an ostium into an incurrent canal and enters the flagellated chamber by one of a number of prosopyles and leaves via a single apopyle to enter an excurrent canal. The excurrent canal joins others to form a main excurrent canal which eventually leads to the osculum. In most of the leucon sponges there is no well-defined spongocoel. The common bath sponge, *Spongia,* as well as freshwater sponges (*Spongilla*) and most other members of the class Demospongiae are of the leucon type.

FEEDING, DIGESTION, RESPIRATION, AND EXCRETION

The currents set up by the flagellar action of the choanocytes bring nutrients into the sponge body. Most of the food is captured and ingested by the choanocytes in thin regions at the base

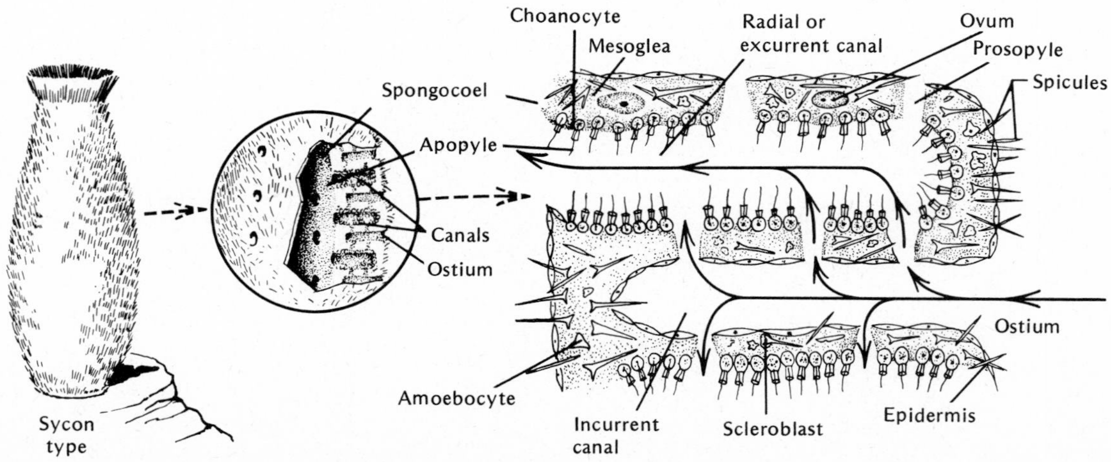

Figure 22.11 Sycon type of sponge body, *Scypha.*

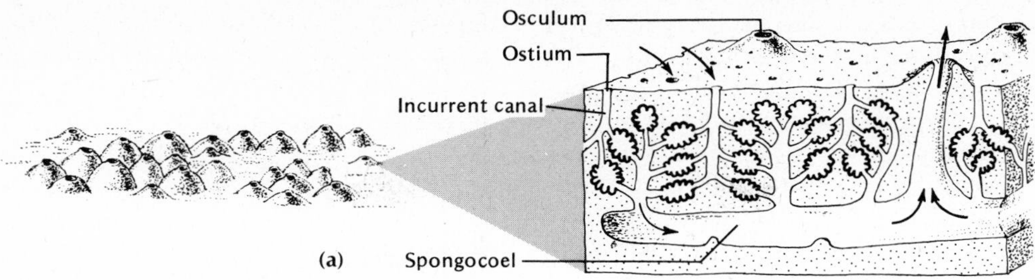

Figure 22.12 Leucon type of sponge body. **(a)** Encrusting form. **(b)** Upright vaselike form.

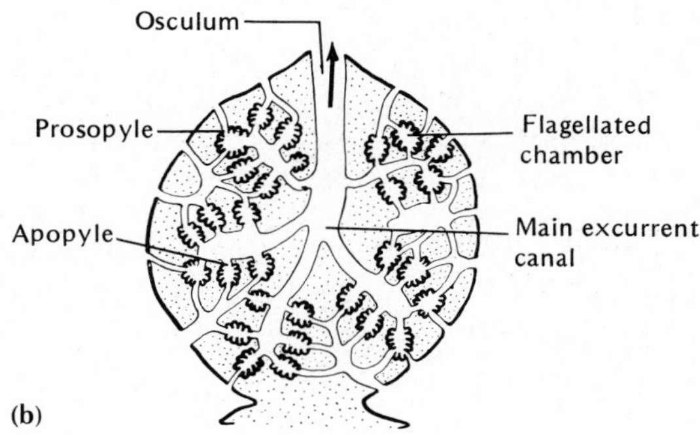

of the collar. The food may be digested by the choanocyte, but often it is passed to an amoebocyte which completes the digestion and delivers nutrients to other cells in the body. Some food may be captured and ingested by other types of cells lining the various passages through which the water passes. Other necessary materials dissolved in the water are readily absorbed by the sponge cells. Respiratory exchange occurs between the cells and the water. Excretion is primarily by diffusion, although the amoebocytes may pick up excretory wastes and take them to the sponge surface for elimination. One result of the increasing complexity of body form from ascon to leucon, described above, is the great increase in number of incurrent openings and in surface area. All of the functions described in this paragraph are greatly enhanced by this increase in surface as well as the increase in volume of water passing through the sponge body.

PROTECTION AND RESPONSE

The spicules prevent most animals from molesting sponges. In addition, many sponges have a most disagreeable odor and many produce toxic substances which also serve as a protection.

Sponges have no nervous system and all responses are by individual cells which function as independent effectors. As mentioned earlier, the porocytes may contract and close as a result of unfavorable stimulation. Similarly there are certain cells, called *myocytes,* in the outer epidermis which contain contractile fibers. Such cells associated with ostia and with the osculum can cause the closure of these openings as a result of local stimulation. However, there is little or no transfer of such stimuli to other regions of the sponge—it is entirely a local response.

REPRODUCTION AND DEVELOPMENT

Asexual Sponges reproduce asexually by budding, and large colonies are formed from small beginnings. Some of the large loggerhead sponges *Spheciospongia* of warm tropical seas reach a diameter of 2m. Fragmentation results in the formation of new individuals, and sponges show a great capacity for *regeneration*—the replacement of lost or injured parts. In 1907 H. V. Wilson demonstrated that when a sponge is squeezed through fine-meshed silk cloth, the individual cells will collect in small groups and each group will reorganize into a new sponge like the original one. This phenomenon is still being studied (see Moscona, Suggestions for Further Reading). If two sponges of contrasting colors are used, it is easy to see that cells from each type of sponge will aggregate only with others of the same kind. This capacity for regeneration is used in a practical way in the sponge industry. Commercial sponges are cut into many small pieces which are "planted" in a favorable locality. In five or six years these pieces will grow large enough to harvest.

A special type of asexual formation is found in freshwater sponges and in some marine forms. In the autumn when conditions become unfavorable for the freshwater sponges, many archeocytes gather in a mass in the mesoglea and become filled with food reserves. Other amoeboid cells move in and secrete a thick membrane around the archeocytes. A layer of special spicules, called *amphidisks,* is next laid down by scleroblasts, and finally an outer thin membrane is added. The layers which make up the wall of the globular structure, called the *gemmule,* are largely spongin, but there is a small fraction which is chitin. At one end of the gemmule there is an opening, called the *micro-*

pyle. The parent sponge dies and the gemmules are able to withstand the rigors of winter. The following spring the archeocytes flow out through the micropyle and organize a new sponge in much the same way as did the masses of cells squeezed through the bolting cloth by Wilson. Figure 22.13 shows the structure of a typical gemmule formed by the freshwater sponge, *Ephydatia.*

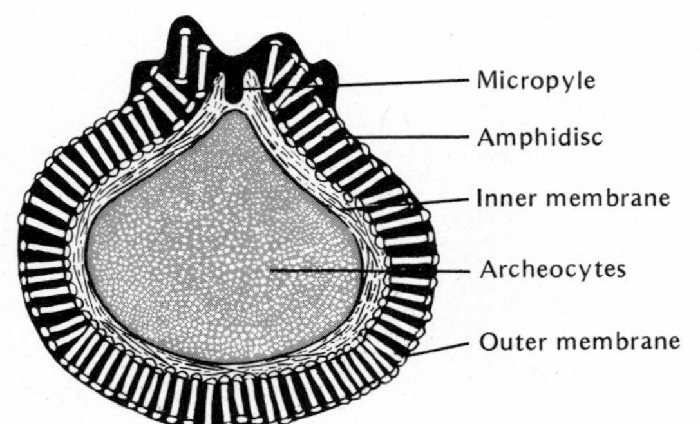

Figure 22.13 A gemmule of the freshwater sponge, *Ephydatia.*

Sexual The eggs and sperm are formed either by archeocytes or choanocytes. Most sponges are hermaphroditic, but self-fertilization seldom occurs since eggs and sperm are produced at different times in any given sponge. The egg cells are amoeboid when first formed, and they move to the base of the choanocyte layer. Sperm enter the sponge with the water currents. A sperm does not enter the egg directly. It enters the opening in the collar of a choanocyte and is absorbed (Fig. 22.14). Subsequently the choanocyte loses its collar and flagellum, moves inward, and comes close to an unfertilized egg. The sperm, which has lost its tail and has become surrounded by a capsule, now penetrates the egg (Fig. 22.15).

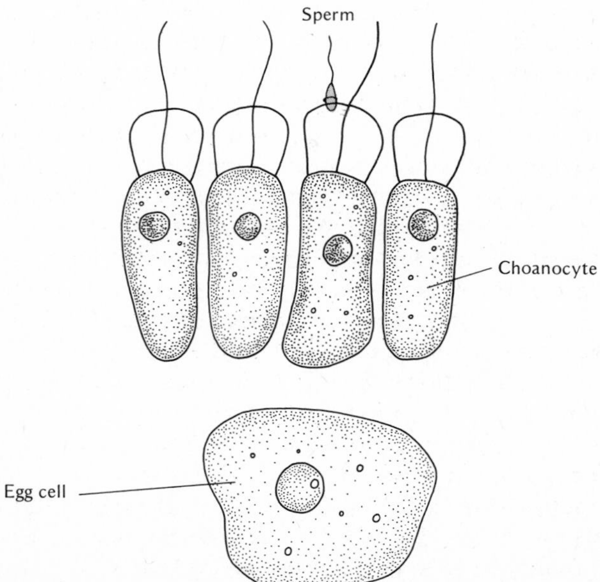

Figure 22.14 Sperm entering the collar of a choanocyte.

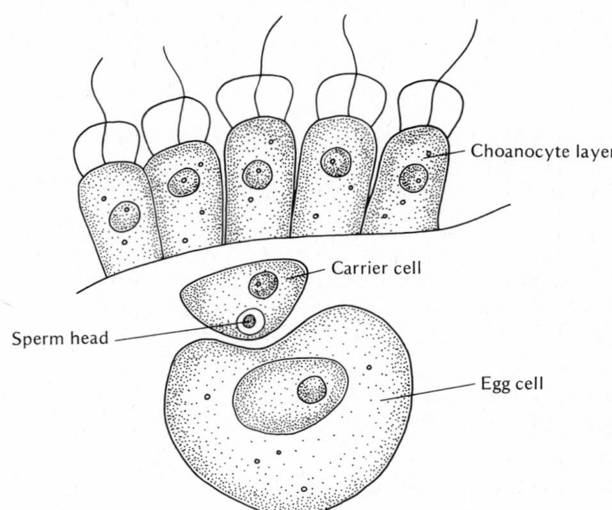

Figure 22.15 Carrier cell delivering sperm to an egg.

Cleavage begins and at the 16-cell stage the flattened disk-shaped embryo lies just beneath the choanocyte layer. The eight macromeres adjacent to the choanocytes will become the epidermal cells and the rest will become choanocytes. The latter cells continue to divide, but the eight presumptive epidermal cells do not divide for a time. An opening appears at the center of the macromeres and the embryo proceeds to engulf nearby choanocytes. Eventually a flagellated blastula stage is reached. The larva now breaks through the choanocyte layer into the canal system and leaves by way of the osculum. After a brief free-swimming period the larva settles down to form a new individual.

The larvae of sponges are of two kinds. In most calcareous sponges the larva is an *amphiblastula* (Fig. 22.16) with a reduced blastocoel

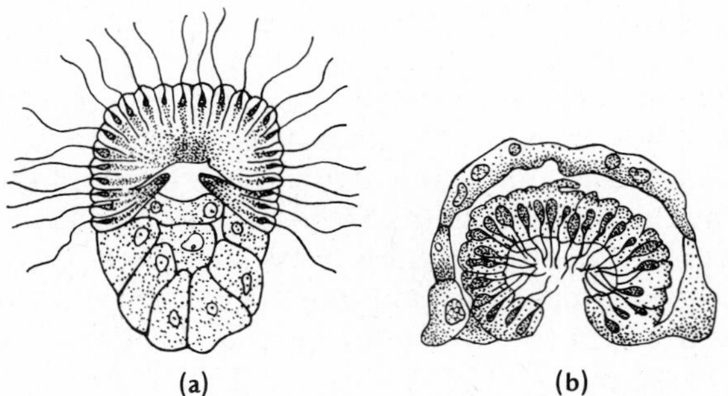

Figure 22.16 Amphiblastula larva. **(a)** Free-swimming larva as it leaves parent sponge. **(b)** Larva after it settles to the bottom and the posterior cells have overgrown the anterior flagellated cells.

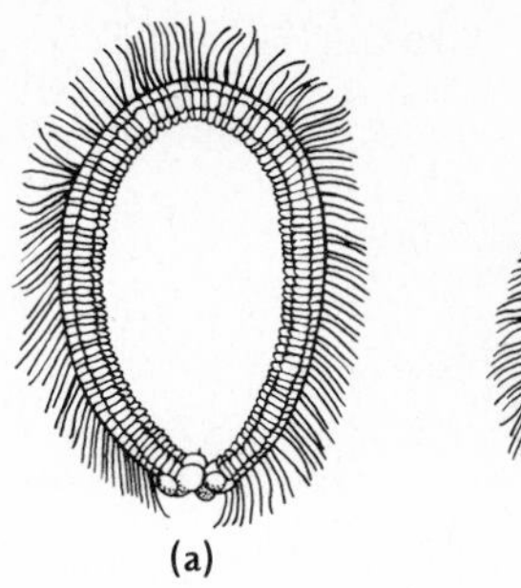

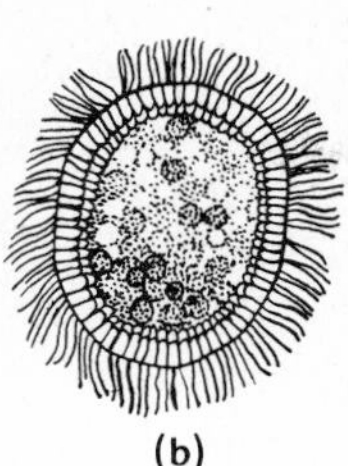

Figure 22.17 Parenchymula larva. **(a)** Early stage with blastocoel and a few nonflagellated cells at the posterior. **(b)** Free-swimming larval stage in which the blastocoel has been obliterated by division of the nonflagellated cells which migrated into it. When this flagellated larva settles to the bottom, the internal cells will migrate out and cover the flagellated cells.

and flagellated cells covering the anterior portion of the larva. After a brief free-swimming existence, the larva settles to the bottom and the flagellated cells are overgrown by the posterior nonflagellated cells and come to be on the inside of the new sponge where they eventually will be the choanocytes.

The second type of larva, characteristic of Demospongiae, is a solid stereogastrula, called the *parenchymula larva* (Fig. 22.17). By the time the larva escapes, the blastocoel has become obliterated by cells which have moved into it and divided repeatedly. The entire surface is covered with flagellated cells and after a short time the larva settles to the bottom. The outer flagellated cells migrate to the inside and become the choanocytes.

Ecology

Sponges are entirely marine with the exception of the small family Spongillidae. There are about 50 species of freshwater sponges and they inhabit lakes and streams throughout the world. They are quite inconspicuous, a flat encrusting or slightly branching growth on twigs, stems, and other submerged objects. Many are green in color due to the presence of intracellular algae, called *zoochlorellae,* in their archeocytes. This is an example of the relationship which is called *mutualism* since both sponge and alga benefit by the association. Others are brightly colored, red, orange, purple, and so forth.

Marine sponges are found worldwide and in both shallow and deep waters. Most sponges are attached to rocks or other hard surfaces such as coral (Pl. 27). Hexactinellids like Venus's flower basket, *Euplectella,* live in deep water where the bottom is a soft ooze. They have a long rootlike tuft of silicon fibers which anchors them.

Sponges provide an excellent home for many organisms. A large loggerhead sponge, *Spheciospongia,* in the Tortugas Islands was found to contain over 16,000 small shrimp. There were also many other kinds of animals living in the canal systems of this sponge. Other shrimps live in pairs in the central cavity of Venus's flower basket. They enter when young and soon they are too large to pass through the sievelike plate which covers the osculum.

Hermit crabs inhabit empty snail shells, and certain encrusting sponges grow over the surface of the shell, eventually covering it. Later the sponge dissolves away the snail shell and the hermit crab now lives within a cavity in the sponge.

The crab *Dromia* breaks off pieces of sponges and holds them over its back with its last pair of legs. The spider crab has many hairlike bristles on its carapace to which are attached bits of sponge as well as algae, hydroid coelenterates, and other small sessile animals. These attached organisms usually continue to grow, and soon the crab looks like a portion of the surrounding substratum, an obvious protective advantage.

Phylogenetic Relationships

As mentioned in discussions of the flagellated protozoa, it is generally believed that the sponges were derived from choanoflagellate ancestors. These highly specialized cells are found nowhere else in the animal kingdom except in these two groups. The entire plan of organization of the sponge is built around these cells, and the sponge is dependent upon them for its very existence. The sponges are considered as a separate subkingdom because of their distinctive features. In addition to the choanocytes, the complete absence of a digestive system and the embryological origin of the cellular layers are also distinctive. The outer layer of the adult is derived from inner cells of the larva, and the inner layer of choanocytes in the adult is derived from the outer flagellated cells of the larva. Thus these layers cannot be homologized with ectoderm and endoderm of other metazoan forms. For all these reasons, therefore, it is concluded by most biologists that the sponges gave rise to no higher groups.

SUGGESTIONS FOR FURTHER READING

PHYLUM MESOZOA

Florkin, M., and B. T. Scheer (Eds.). *Chemical Zoology,* vol. II, *Porifera, Coelenterata,* and *Platyhelminthes.* New York; Academic Press (1968).

Hyman, L. H. *The Invertebrates,* vol. 1; *Protozoa through Ctenophora.* New York: McGraw-Hill (1940).

Hyman, L. H. *The Invertebrates,* vol. 5: *Smaller Coelomate Groups.* New York: McGraw-Hill (1959).

McConnaughey, B. H. "The Life Cycle of the Dicyemid Mesozoa." *Univ. of California Publ. in Zoology,* **55** (1951).

McConnaughey, B. H. "The Mesozoa," in E. C. Daugherty (Ed.). *The Lower Metazoa.* Berkeley, CA: Univ. of California Press (1963).

Stunkard, H. W. "The Life History and Systematic Relations of the Mesozoa." *Quart. Rev. of Biology,* **29** (1954).

From Scientific American

Lapan, E. A., and H. Marowitz. "The Mesozoa" (Dec. 1972).

PHYLUM PORIFERA

Barnes, R. D. *Invertebrate Zoology,* 3rd ed. Philadelphia: W. B. Saunders (1974).

Fry, W. G. (Ed.). "The Biology of the Porifera," *Symposia of the Zoological Society of London,* no. 25. New York: Academic Press (1970).

Hegner, R. W., and J. G. Engemann. *Invertebrate Zoology.* New York: Macmillan (1968).

Hyman, L. H. *The Invertebrates.* vol. 1: *Protozoa through Ctenophora.* New York: McGraw-Hill (1940).

Hyman, L. H. *The Invertebrates,* vol. 5: *Smaller Coelomate Groups.* New York: McGraw-Hill (1959).

Pennak, R. W. *Fresh-Water Invertebrates of the United States.* New York: Ronald Press (1953).

Wilson, H. V. "On Some Phenomena of Coalescence and Regeneration in Sponges." *J. of Experimental Zoology,* **5** (1907).

Wilson, H. V., and J. T. Penney. "Regeneration of Sponges from Dissociated Cells." *J. of Experimental Zoology,* **56** (1930).

From Scientific American

Moscona, A. A. "Tissues from Dissociated Cells" (May 1959).

Moscona, A. A. "How Cells Dissociate" (Sept. 1961).

The Radiates–

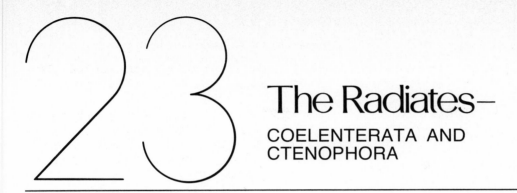

COELENTERATA AND CTENOPHORA

SUBKINGDOM METAZOA– BRANCH EUMETAZOA– GRADE RADIATA

The first phylum of the branch Eumetozoa, subkingdom Metazoa, is the phylum Coelenterata, also called Cnidaria. The branch Eumetazoa includes all the rest of the animal kingdom. The coelenterates, along with members of the phylum Ctenophora, are set apart as the grade Radiata since these two phyla include the only basically radially symmetrical Eumetazoa. It is true that the echinoderms are radially symmetrical as adults, but they start life as bilaterally symmetrical larvae and their radial symmetry is thus secondary.

PHYLUM COELENTERATA (Cnidaria)

Characteristics

1. Basically radially symmetrical.
2. The digestive cavity, called a gastrovascular cavity, has but a single opening, the mouth, and is the sole internal cavity.
3. Body wall with two layers, epidermis and gastrodermis, with mesoglea in between (diploblastic); mesoglea contains cells and connective tissue (ectomesoderm) in some (triploblastic).
4. Although there is incipient organ development, especially in the class Anthozoa, the coelenterates are essentially at the tissue level of development.
5. One to several whorls of tentacles associated with the mouth.
6. Special cells, called cnidoblasts, produce distinctive stinging structures, called nematocysts.
7. Nervous system is a simple network of primitive nonpolarized neurons with no centralized coordinating center.
8. No special excretory or respiratory system.
9. Polymorphism is characteristic of the classes Hydrozoa and Scyphozoa with polyp and medusa being the two body forms. Many species are colonial.
10. Entirely aquatic and mostly marine—only a few Hydrozoa are found in freshwater.
Figure 23.1

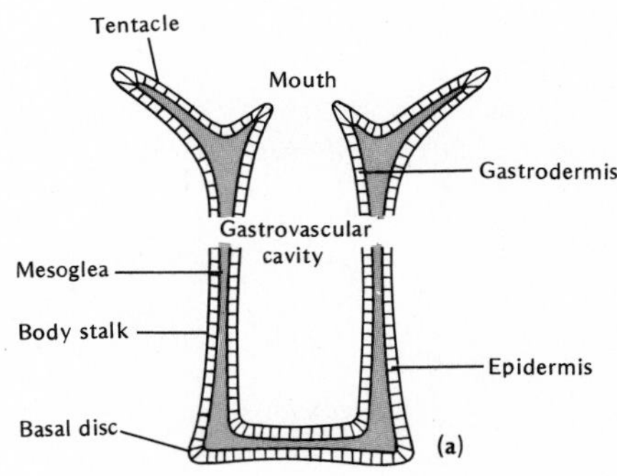

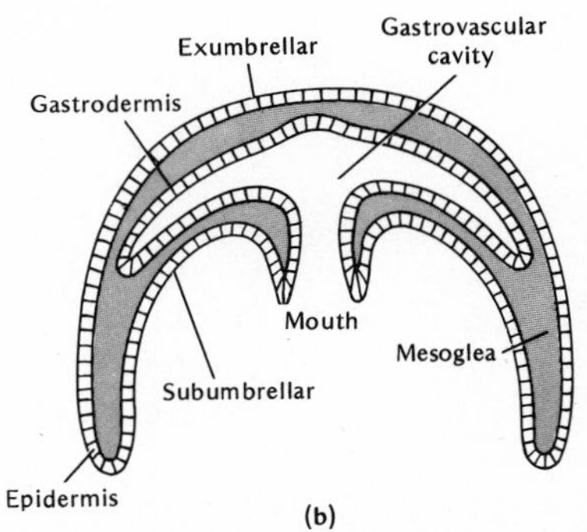

Figure 23.1 Coelenterate body plan. **(a)** Polyp or hydroid body form. **(b)** Medusa body form.

Classification

CLASS **Hydrozoa** This is generally considered to be the most primitive class of the phylum. The body wall is made up of two layers of cells, and the polyp stage, which is often colonial, has a single open gastrovascular cavity with no partitions (mesenteries). No stomodaeum present. The sex cells ripen in the epidermis and alternation of generations with an asexual polyp stage and a sexual medusa stage is the rule. The medusa stage possesses a velum. In most species the polyp stage has a noncellular secreted exoskeleton. Mostly marine, but a few freshwater species also belong to this class. *Hydra, Obelia* (Fig. 23.2).

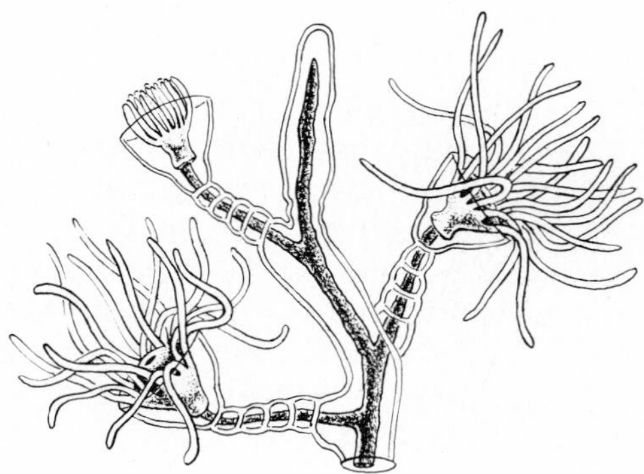

Figure 23.2 Class Hydrozoa, *Obelia* sp.

CLASS **Scyphozoa** Exclusively marine coelenterates in which the medusa stage is the prominent one with the polyp stage being rudimentary or absent. Mesenchymatous layer between the epidermis and gastrodermis. Gonads endodermal in origin. No skeletal structures present. Medusa lacks a velum. True jellyfishes. *Aurelia* (Fig. 23.3).

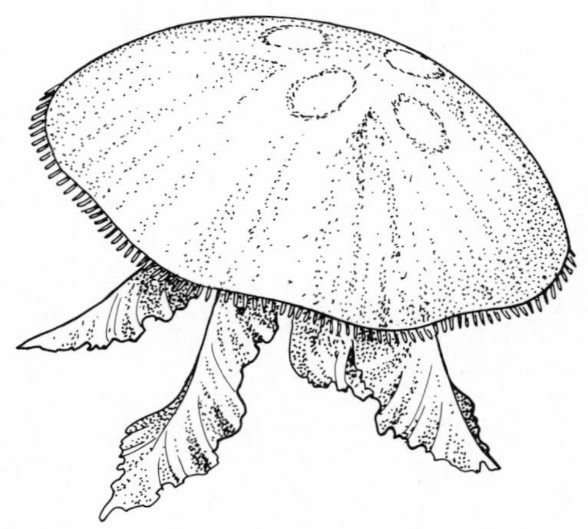

Figure 23.3 Class Scyphozoa, *Aurelia* sp., a jellyfish.

CLASS **Anthozoa** Exclusively marine coelenterates with a polyp stage only. Mesenchymatous layer, including muscle fibers, between the epidermis and gastrodermis. Gastrovascular cavity has vertical partitions, called mesenteries. Stomodaeum present. Gonads endodermal in origin. Exoskeleton may or may not be present. Anemones and corals. *Metridium* (Fig. 23.4).

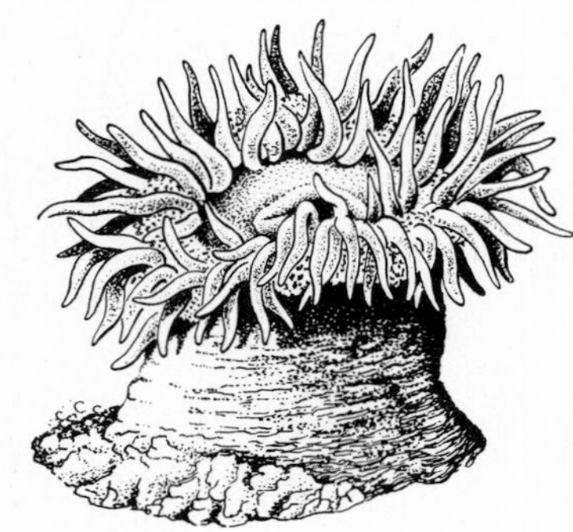

Figure 23.4 Class Anthozoa, *Metridium* sp., a sea anemone.

CLASS HYDROZOA

Classification

ORDER **Hydroidea** Polyps usually well developed, either solitary or colonial. Medusae budded from polyps and possessing eyespots and statocysts. Some produce sessile degenerate medusae or none at all. *Hydra, Obelia, Gonionemus.*

ORDER **Milleporina** Colonial hydroids which form a massive porous calcareous mass. Two kinds of polyps, feeding gastrozoids and protective dactylozoids, occurring in cavities in the calcareous mass. Small degenerate medusae which lack digestive system and tentacles. A single genus, *Millepora.*

ORDER **Stylasterina** Similar to the millepores but with small dactylozoids usually arranged in systems with the gastrozoids. Medusae degenerate and sessile. *Stylaster, Allopora.*

ORDER **Trachylina** Polyp stage reduced or absent. Tentacles usually attached above the margin of the umbrella. *Liriope, Cunina, Tetraplatia.*

ORDER **Chondrophora** Free-floating colonies of polyps with a high degree of polymorphism. A gas-filled float, modified from the perisarc, keeps them afloat. Some have a flattened saillike structure on the upper surface. *Velella, Porpita.*

ORDER **Siphonophora** Highly polymorphic free-swimming or floating colonies with reduced medusae which are rarely free-swimming. *Physalia, Halistemma.*

Anatomy and Physiology of *Hydra*

GENERAL STRUCTURE

Only a few coelenterates live in freshwater and the most common form is *Hydra.* Various species of this genus are found in freshwater streams all over the world. They are quite small, usually not more than 15 mm in length. They spend much of their time attached to the substrate or some object in the water, held in place by secretions from the cells of their basal ends. The body of a hydra is a simple tubular polyp with the basal end closed and with the free end drawn out into a slight elevation, the *hypostome*, in the center of which is the *mouth*. Surrounding the hypostome are extensions of the body, called *tentacles*, that function primarily in obtaining food. The central cavity is called the *gastrovascular cavity* since it serves both the functions of digestion and circulation.

CELLULAR STRUCTURE

Figure 23.5 illustrates the two-layered nature of *Hydra*. The outer layer is the *epidermis* and the inner layer is the *gastrodermis*. Between the epidermis and the gastrodermis there is a rather

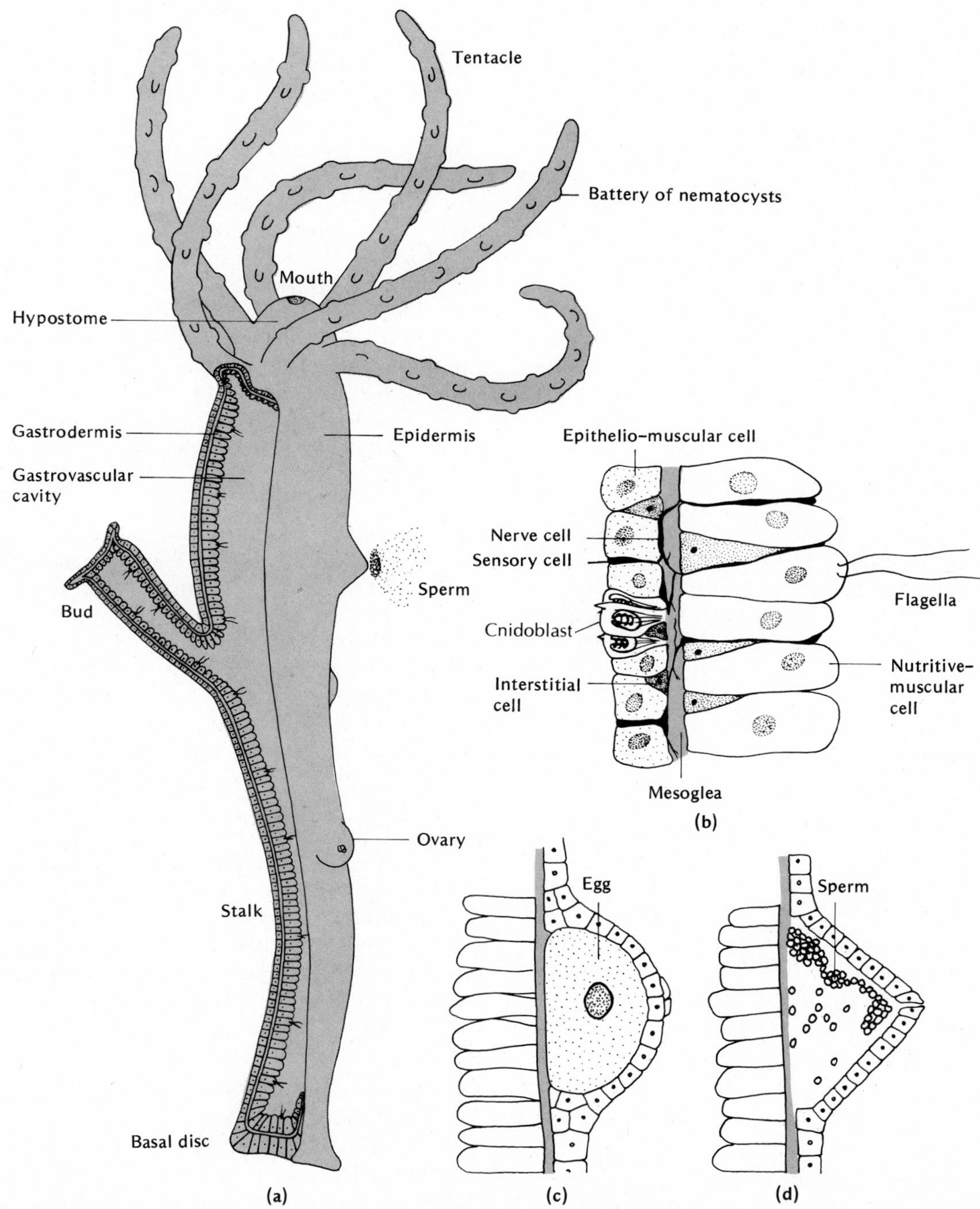

Figure 23.5 *Hydra.* **(a)** Budding hydra showing its gross anatomy. **(b)** Detail of cellular structure of the body wall. **(c), (d)** Egg and sperm in the gonads.

thin layer of jellylike material, called the *mesoglea.*

In the epidermis most of the cells have elongated bases with contractile fibers. These cells are known as *epitheliomuscular* cells. They function both for protection and in contraction, that is, they are partly epithelial and partly muscular in nature. The fibrils in these cells run lengthwise, whereas their counterparts in the gastrodermis have the fibrils running at right angles to them, or circularly. The cells in the gastrodermis that have these fibrils are known as *nutritive-muscular cells,* and they constitute most of the gastrodermis. Contractions of the fibrils in the outer layer cause a shortening or bending of the body, whereas contractions of the fibrils in the inner layer constrict and lengthen the body. Movements of the tentacles are effected in the same way.

Scattered here and there in the epidermis are the *cnidoblasts* (Fig. 23.6). These cells contain the characteristic stinging structures of all coelenterates, called *nematocysts.* There are numerous kinds of nematocysts but basically they are alike. A nematocyst is a fluid-filled capsule containing a long spirally coiled hollow thread. On the surface of each cnidoblast is a fine projection, the *cnidocil*, which may serve as a trigger. When the cnidoblast is stimulated, the coiled hollow thread of the nematocyst is turned inside out, much as the fingers of a rubber glove are everted when one blows into the glove. The cnidoblast is an independent effector, that is, it responds to external stimuli independent of the nerve net system. Many chemicals will cause the nematocysts to be discharged. Mechanical stimulation will not cause discharge unless food is present. The concensus is that the nematocyst discharge is caused by mechanical stimulation only after its threshold is lowered by chemical stimulation from a food object. Thus a nematocyst will not discharge when the food object is too far away to be reached by the nematocyst thread. Most of the nematocysts contain a poison, and they are discharged with such force that the thread will penetrate the body of a food animal, and, thus, inject poison into it. The threads of other types of nematocysts do not penetrate the prey but are used to entangle small organisms.

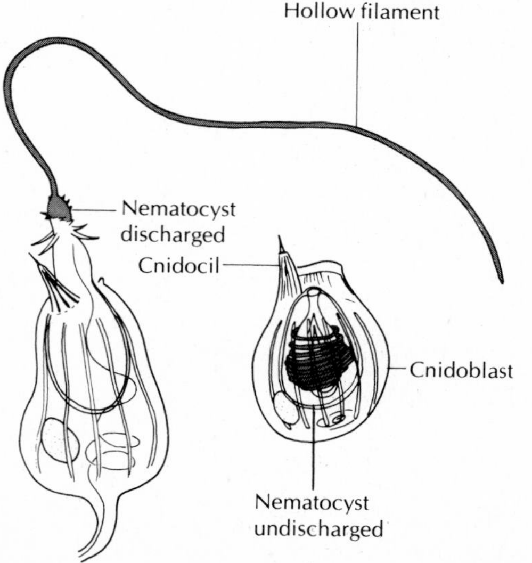

Figure 23.6 Cnidoblasts, cells which contain and discharge nematocysts.

The discharge of a nematocyst seems to be due to an increase in pressure within the capsule of the nematocyst. According to one theory, a rapid intake of water causes an increase in hydrostatic pressure in the capsule; according to another theory, the increase in pressure is cause by contractile fibers that surround the capsule. Once a nematocyst is discharged, it can never function again. Cnidoblasts are continually being replaced by new ones that develop from interstitial cells. Although cnidoblasts are found scattered all over the epidermis, they are most concentrated on the tentacles, where they are found in groups or clumps known as *batteries.*

Scattered throughout both layers are elongate sensory or *receptor* cells. The base of a receptor cell has one or more neurofibers that connect with *nerve cells* lying in the inner part of the epidermis next to the mesoglea. These nerve cells form a network throughout the body known as the *nerve net.*

Numerous small, somewhat rounded cells are found between the bases of the cells of both layers. These are the *interstitial cells.* These cells retain the properties of embryonic cells and are capable of transforming into any of the other cell types. Thus an interstitial cell may become a cnidoblast or an epitheliomuscular cell; these cells participate in the formation of *gonads* and of buds. The coelenterates as a group show great powers of regeneration, and the interstitial cells are important in this process.

The nutritive-muscular cells are not all alike. Some of them secrete into the gastrovascular cavity the enzymes that start the disintegration of food organisms. Others send out pseudopodia and ingest particles of such broken-down organisms. Still others have flagella on their free surfaces, and the movements of these flagella keep the materials in the gastrovascular cavity in circulation.

In addition to the cell types mentioned above, there are gland cells in the epidermis. The epidermis of the basal disk of *Hydra* contains many of these gland cells. In a number of coelenterates a horny covering is formed by secretions of gland cells in the epidermis. In corals a calcium carbonate cup is formed within which the coral polyp lives.

FEEDING AND DIGESTION

Hydra feeds on small aquatic animals. The hydra first paralyzes its prey with nematocysts and then pushes the small crustacean or worm into its mouth with its tentacles. The body wall is capable of great distension, and a hydra may often swallow a worm much larger than itself.

As mentioned above, some of the cells lining the gastrovascular cavity secrete enzymes that start the breakdown of the food mass. The enzymes that are active in this process are proteases. After the food mass has been reduced to small particles, other nutritive-muscular cells ingest the particles in food vacuoles and continue the digestion intracellularly. In a certain sense these cells function like an amoeba in digestion. When the digestion occurs in food vacuoles, it is *intracellular digestion,* whereas digestion that occurs in the gastrovascular cavity is termed *extracellular digestion.*

Enzymes which effect the hydrolysis of different kinds of food materials—proteins, fats, and carbohydrates—are secreted into the food vacuoles. The end products of digestion are then either used in the cells in which they are formed in synthesis or in respiration, or they may diffuse into the adjacent cells of the inner layer or across the mesoglea into the cells of the outer layer for utilization. In any case, the organism is of such simple construction that no circulatory system, as found in higher forms, is needed. Materials that are not digested in the cavity are ultimately expelled by way of the mouth.

GAS EXCHANGES AND EXCRETION

Because of the small size and simple structure of *Hydra,* the exchange of gases in respiration and the removal of nitrogenous wastes take place by diffusion just as they do in the protozoans. There are no special structures for gas exchange or excretion.

LOCOMOTION

A hydra may glide along the region of attachment due to amoeboid movements of the basal cells. More rapid movements are effected by a sort of somersaulting. In this type of locomotion the animal bends over and attaches its tentacles to the substrate; then it loosens its basal attachment, swings the basal region completely over, and attaches it again. By a series of such somersaults a hydra is able to move fairly rapidly. In some instances a hydra may be observed to move after the fashion of a measuring worm. In this method of locomotion the hydra reaches out and attaches its tentacles to some object; the base is loosened and is then moved near the place where the tentacles are holding on, for a new attachment.

Although a hydra is able to move from place to place, it spends most of its existence attached to some object, its body and tentacles extended and slowly swaying about.

COORDINATION AND BEHAVIOR

It is easy to demonstrate that a hydra will respond to stimuli. If a tentacle is touched with a needle, all of the tentacles and the body will contract into a compact mass. After a time the body and the tentacles will expand and the tentacles will sway back and forth. If a small food organism is brought into contact with one of the tentacles, all the tentacles will work together to push the food organism into the mouth. Glutathione, a sulfur-containing tripeptide present in most tissues, seems to stimulate *Hydra,* bringing on the vigorous feeding activity. As pointed out in the discussion of cellular structure, *Hydra* has sensory cells, nerve cells, and contractile elements in cells of both the epidermis and the gastrodermis. These form the basis for the types of response and coordination mentioned above. The nervous mechanism in *Hydra* is much simpler than that found in higher animals. The nervous system is simply a nerve net with no central system present. Although the nerve net is more concentrated around the mouth than elsewhere, this region is in no sense a brain. When a sensory cell is stimulated, an impulse is relayed to a nerve cell, which in turn relays the impulse to other nerve cells. The transmission is slow and it is nonpolarized, that is, it goes in all directions. There are no synapses in this type of system, and the cells are called *protoneurons* to distinguish them from the neurons of more advanced forms which do have synapses that give polarity to the nervous system. The coelenterate type of system is often called a *receptor–effector* type of nervous system, and although it is very primitive in comparison to the vertebrate type of nervous system, it is adequate for these organisms.

REPRODUCTION AND DEVELOPMENT

Asexual Hydras reproduce asexually by budding. A well-fed hydra may have two or three buds in different stages of development. As shown in Figure 23.5, a bud involves both of the body layers. In the beginning there is only a slight elevation at the point where the bud is starting. This elevation grows and elongates and in a few days looks like a little hydra with tentacles and mouth. The cavity of the bud is continuous with that of the parent hydra, but when the bud is mature, a constriction forms near its base, separating it from the parent.

Sexual Under certain conditions hydras develop testes and ovaries. Loomis has been able to show that increased carbon dioxide tension in the water stimulates the formation of sex organs (see Suggestions for Further Reading). In some species both testes and ovaries are found in the same individual, a condition known as *hermaphroditism.* In other species the sexes are separate. Figure 23.5 shows both an ovary and a testis. These organs are formed in the outer layer from interstitial cells causing bulges in the body wall. The testes form on the upper half of the body, while the ovaries develop nearer the basal disk. The interstitial cells in a testis divide a number of times to form small sperm cells. In the ovary a number of interstitial cells fuse, with all the nuclei but one degenerating, forming a single large egg cell. When mature, the egg breaks through the surrounding epidermis, but it remains attached for a time. Sperm are released from a testis and they swim in the water until they reach an egg. Only one sperm will penetrate the egg and effect fertilization.

Development The early stages of development take place while the egg is still attached. The fertilized egg or zygote undergoes cleavage; then a hollow ball of cells, the blastula, is formed. In the formation of the gastrula stage, the cells of the blastula divide and some of them fill the central cavity. In this stage the outer layer secretes a protective shell and the gastrula separates from the parent and drops to the substrate. In this condition it may remain over winter, or if the conditions are right, it may start development again in a week or so and emerge from the shell as a young hydra. In this development within the shell, the central cell mass, or endoderm, becomes hollowed out to form the gastrovascular cavity, a mouth forms, and tentacles develop.

Regeneration The process of replacing lost or injured parts of an organism, or even of forming a whole organism from a small part, is known as *regeneration.* The capacity for regeneration is greatest among organisms lowest in the phylogenetic scale, that is, among organisms that are least specialized. *Hydra* has great powers of regeneration. When the body of a hydra is cut into a number of pieces, most of them will regenerate into complete but miniature hydras. The name "hydra" was given to this organism because one of the early workers saw a resemblance in this regenerative capacity to the mythical nine-headed Hydra which grew two heads in the place of one that had been cut off. The regenerative powers of *Hydra* and other coelenterates apparently are associated with the capacity of the interstitial cells to develop into all the kinds of cells found in a complete hydra.

Ecology

Hydra is typically found in shallow waters of lakes and streams. It has been reported from waters as deep as 450 m (1400 feet). It can also be found on rocks in swift streams or wave-pounded lake shores.

As in some of the ciliates and freshwater sponges, the cosmopolitan green hydra, *Chlorohydra viridissima,* has a mutualistic association with intracellular green algae. The alga has been isolated from the hydra and it will grow independently. The cells seem to be identical with those of members of the genus *Chlorella.* Some marine Hydrozoa have similar relationships with algae, not only green algae but also yellow and brown algae.

Two ciliates, *Kerona* and *Trichodina,* are commonly found on the external surface of *Hydra,* gliding about and feeding on small bits of food. Somehow these ciliates do not trigger the discharge of the nematocysts even though other ciliates are readily captured by discharging nematocysts. This type of interaction between two species in which one benefits, the protozoans, and the other is neither harmed or benefitted, the hydra, is called *commensalism.*

Although *Hydra* does not have a medusa stage, there are freshwater coelenterates with a medusa stage. The most common one is *Craspedacusta sowerbyi* which has been found in the majority of areas of the United States as well as on most of the other continents. It is about 15 to 20 mm in diameter when mature and delicately transparent in appearance. Interestingly, it occurs in widely scattered bodies of water, and all the medusae in a given place are generally of the same sex. This peculiar situation reflects the fact that *Craspedacusta* is distributed in the hydroid stage, a minute nontentacled polyp. The mode of distribution is probably by migrating waterfowl with the polyps being carried on the feet or feathers of the birds. Thus one polyp may be introduced and it multiplies asexually to establish the species. Since polyps will produce only males or females, all of the descendants of the original polyp will produce medusae of one or the other sex. The result is that in most instances sexual reproduction does not occur.

Anatomy and Physiology of *Obelia*

GENERAL STRUCTURE

Polyp There are numerous species of marine coelenterates that are closely related to *Hydra* but that exist as colonies. *Obelia* is one of the best known of these (Fig. 23.7). This organism is found attached to different kinds of objects in the intertidal zone and down to a depth of 30 m (90 feet) or more. In many respects an *Obelia* colony looks like a number of hydras connected at their bases to a common stalk. However, in a mature *Obelia* colony all the polyps or zooids are not alike. Many of them have tentacles and are much like individual hydras; these are the *nutritive polyps.* Others lack tentacles and are *reproductive polyps.* An examination of an *Obelia* colony and the substrate to which it is attached shows that the colony is anchored to the substrate by a rootlike structure, the *hydrorhiza.* This structure gives off at intervals upright branches, known as *hydrocauli.* The nutritive and reproductive polyps are attached to a hydrocaulus.

The soft parts of the *Obelia* colony are surrounded by a chitinous covering called the *perisarc,* which is secreted by the epidermis. The cup-shaped part of the perisarc surrounding each nutritive polyp is called the *hydrotheca,* while that part of the perisarc surrounding each reproductive polyp is called the *gonotheca.* The perisarc is ringed at different places, allowing for movement of the colony. All the soft parts—the actual living parts of the colony—constitute the *coenosarc.* In structure the coenosarc is like the body of a hydra, consisting of an outer epidermis, an inner gastrodermis, with mesoglea inbetween. Thus the gastrovascular cavity is continuous, connecting all parts of the colony.

Medusa The medusa of *Obelia* is umbrella shaped with a fringe of tentacles on the edge. Hanging down from the center, like a short handle to the umbrella, is the *manubrium* with the mouth at the end. The gastrovascular cavity extends out from the cavity of the manubrium into four radial canals on which are located the

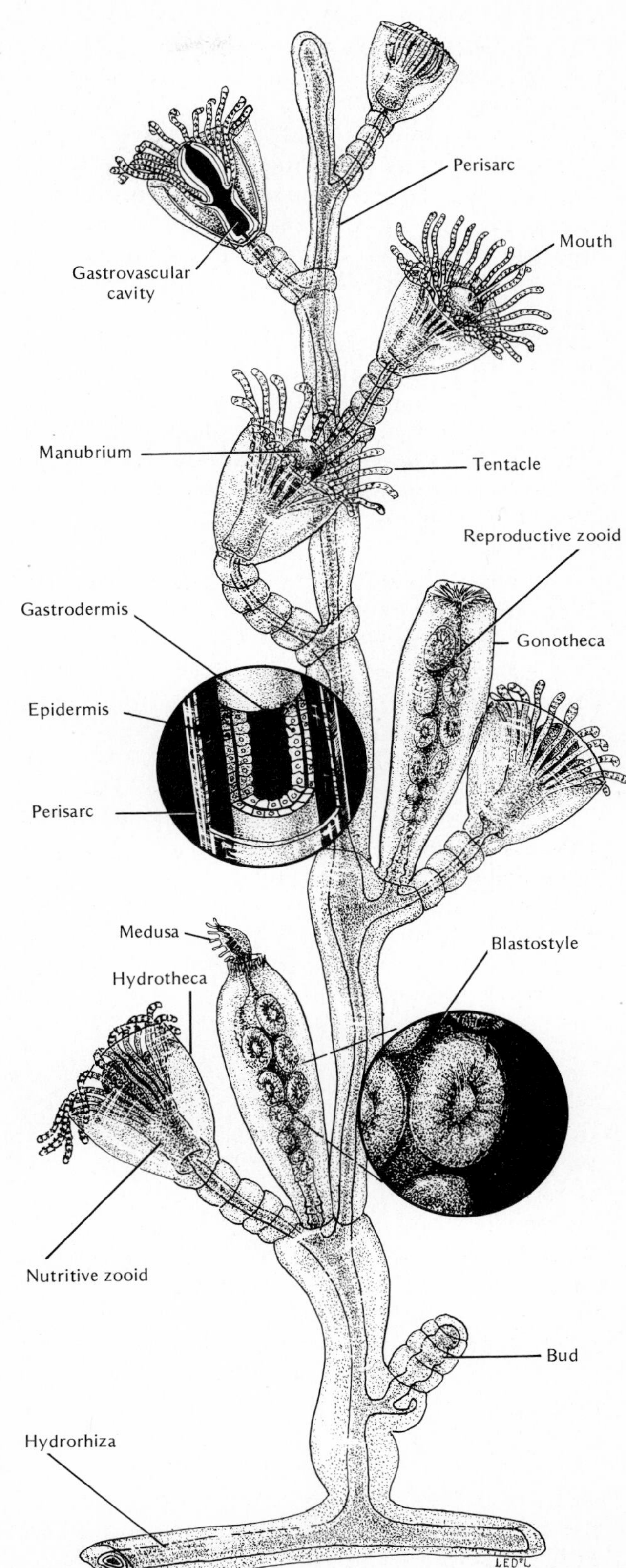

Figure 23.7 Structure of *Obelia.* **(a)** Colony. **(b)** Section of stalk. **(c)** Medusa bud.

reproductive organs. The radial canals connect with the circular canal at the margin of the umbrella. Balancing organs, called *statocysts,* are located at intervals along the margin. The statocysts are spherical cavities lined with sensory hairs upon which rests a small calcareous body. Any change in position causes the calcareous body to stimulate the sensory hairs and appropriate righting reactions follow. Many medusae also possess light-sensitive structures, called *ocelli,* on the margin of the umbrella. The medusae of *Obelia* are quite small, but those of the related genus *Gonionemus* average about 20 mm in diameter (Pl. 31). In general form it is similar to the medusae of *Obelia* and it is commonly used as a laboratory example of a hydrozoan medusa. As Figure 23.8 shows, the subumbrella is partly closed by a membrane called the *velum*. This structure, along with other features, distinguishes hydrozoan medusae from the true jellyfishes of the class Scyphozoa. When they swim, water is forced from the subumbrella cavity through the opening in the velum by contractions of the body. This propels the animal in the opposite direction and in this way it moves about. The suctorial pads shown on the tentacles of *Gonionemus* are used to anchor the organism to seaweeds.

PHYSIOLOGY

The physiology of an *Obelia* colony is similar to that of *Hydra.* The nutritive polyps of *Obelia* are much like individual hydra. At the free end of the polyp is an elevation, the *manubrium,* through which the mouth opens into the gastrovascular cavity. Numerous solid tentacles surround the manubrium. When fully extended, the manubrium and tentacles are outside of the hydrotheca, but when disturbed, these parts are withdrawn into the cup of the hydrotheca. Food is captured by the tentacles and digestion occurs in the gastrovascular cavity and in the cells of the gastrodermis. Food ingested by one polyp may be passed along the common gastrovascular cavity by the action of flagella and be ultimately utilized in other parts of the colony. Respiration and excretion are accomplished by simple diffusion as in *Hydra.* Also, the nervous system is similar to that found in *Hydra.*

REPRODUCTION AND DEVELOPMENT

Asexual The *Obelia* colony starts as a single nutritive polyp like those in Figure 23.7. By asexual budding this polyp gives rise to a number of others like itself, and they all remain attached together. After a number of nutritive polyps have been formed, reproductive polyps begin to appear. These are also formed by asexual budding and they form in the angle between a nutritive polyp and the hydrocaulus. A single hydrocaulus may support many polyps. Also, as was mentioned above, the *hydrorhiza,* which continues to grow horizontally, gives off upright buds at intervals, and these will, by budding, form other units.

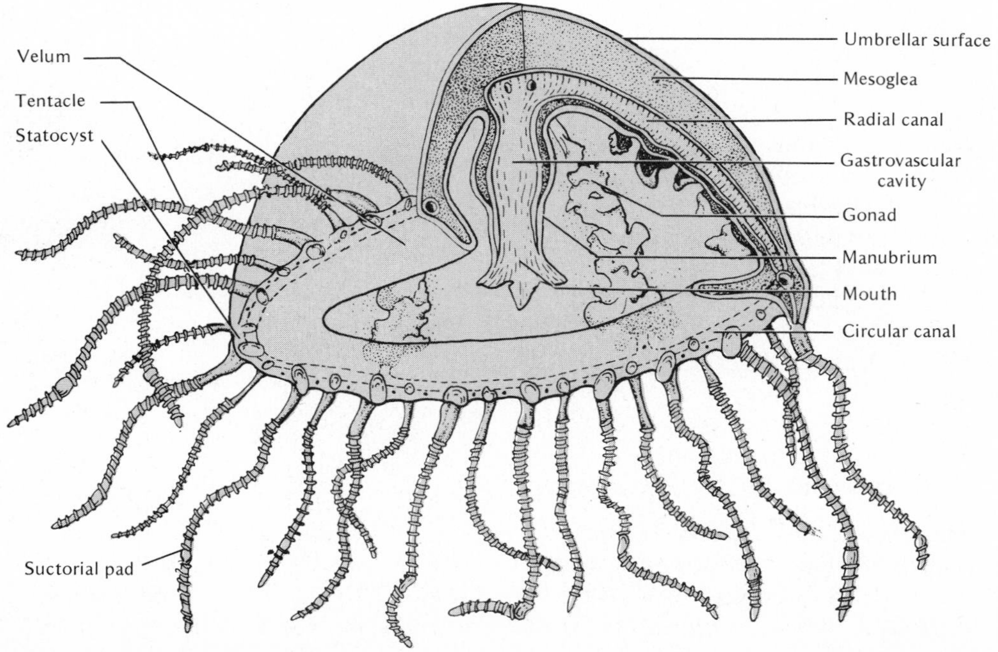

Figure 23.8 Medusa of *Obelia.*

The central core of a reproductive polyp is the *blastostyle.* Each blastostyle produces a series of button-shaped buds, which break off when mature and swim out of the opening in the free end of the reproductive polyp as free-swimming medusae.

Sexual The medusae provide for dispersal or spread of the species. The medusae of *Obelia,* and in most other species as well, have separate sexes; ovaries are formed in some and testes in others. Sexually mature medusae release their eggs or sperm into the water, where fertilization occurs. The zygote undergoes cleavage, blastula, and gastrula formation like that described for *Hydra.* The gastrula elongates and develops into a ciliated swimming larva, called a *planula.* After swimming about, this larva develops a central cavity, becomes attached to the substrate, and proceeds to form a new colony by asexual budding. Figure 23.9 shows the life cycle of *Obelia.* The alternation of a polyp and a medusa stage is often called *alternation of generations* or *metagenesis.*

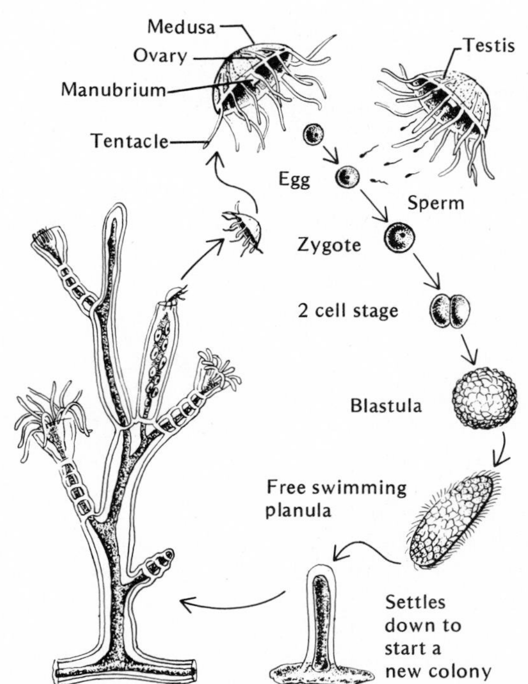

Figure 23.9 Life cycle of *Obelia.*

Ecology

Marine hydrozoa such as *Obelia* have many definite relationships with other organisms. Several hydroid species are found only on the surface of the floating alga *Sargassum.* Other species are always associated with particular sponges, crabs, mollusks, or fish. *Hydractinia* is usually found on snail shells—in temperate regions on shells inhabited by hermit crabs and in the arctic on shells of living snails.

Hydrichthys is an ectoparasite on fish. Its hydrorhizae spread over the surface and send rootlike stolons into the tissues of the host. The polyps lack tentacles, but they bend down and suck blood and tissues with their mouths.

A number of hydroids, including *Obelia,* are luminescent. The light is produced by granules found in the polyps and throughout the stalk. The adaptive value of this feature is not understood.

One of the most bizarre interactions is seen in the relationship of *Obelia* to certain nudibranch mollusks—sluglike organisms which lack a shell. One of these nudibranchs, *Hermissenda crassicornis,* feeds on *Obelia* and somehow, in the digestion of the *Obelia* tissue, it prevents the discharge of the nematocysts. The nematocysts move from the stomach through special ducts to the tips of small fingerlike projections on the dorsal surface of the sea slug. Here they are stored in small sacs and later squeezed out to discharge in a normal manner, protecting the nudibranch.

Polymorphism

The occurrence of more than one body form in the life cycle of an animal is a phenomenon called *polymorphism.* This phenomenon is quite obvious in many species of coelenterates, but it is found in other groups also, such as the social insects. Actually there are three different body forms or kinds of individuals in the life cycle of *Obelia:* the nutritive polyps, the asexual reproductive polyps, and the sexual medusae. In most species of organisms each individual in the species must perform all the work that is done. In those species with polymorphism, different kinds of individuals are produced, each specialized for a different function. In some of the coelenterates still other kinds of individuals are formed. For instance, in the Portuguese man-of-war, *Physalia,* there are polyps specialized for protection, others for sensory perception, and the gas-filled float called a *pneumatophore,* is actually a modified medusa (Fig. 23.10, Pl. 12). Although the sting of most coelenterates may cause temporary annoyance to humans, the poison associated with the nematocysts is not particularly dangerous. *Physalia* and some related species are an exception in that the toxin produced is quite dangerous to humans. Serious injury and even fatalities have resulted from contact with large specimens. It is interesting

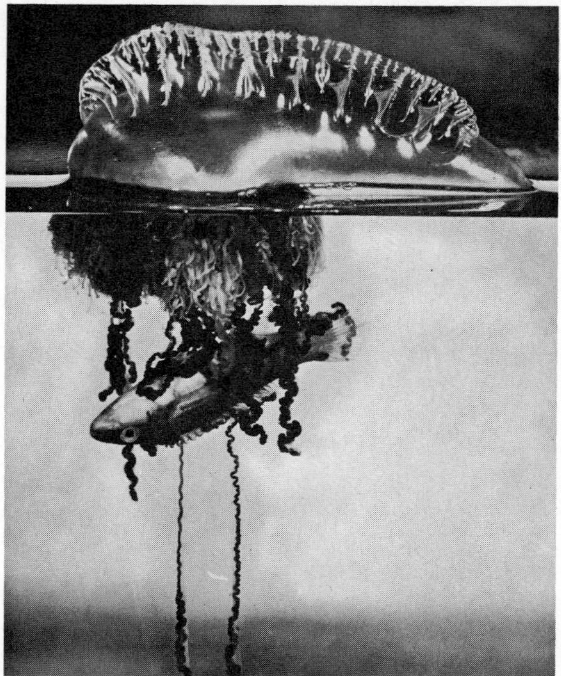

Figure 23.10 *Physalia physalis*, the Poruguese man-of-war, with a recently captured fish in its tentacles. (From P. Wilson, Marine Biological Laboratory.)

that a small fish, *Nomeus,* is able to live among the tentacles of *Physalia* with perfect freedom from injury. It is reported that this fish lives nowhere else but in close association with *Physalia.* Another siphonophore, *Porpita,* is shown in Plate 38. The flattened oval stem forms a float with many dactylozoids on its margin. A large, central gastrozoid encircled by gonozoids is on the underside of the float.

Although at first glance the body of a polyp and the body of a medusa look quite different, a careful examination shows that they are constructed on the same plan. Figure 23.11 shows such a comparison. Not all coelenterates have polymorphism. The hydra does not and neither do any of the members of class Anthozoa, to which the corals and anemones belong. The question naturally arises as to which body form developed first and how did the two types of body form arise. There have been many discussions of this question and opinions have varied. Hyman has examined the evidence and considers that it favors the view that the primitive coelenterates were medusalike forms. According to this view, the polyps were derived from the polyplike planula larvae produced by the medusa. Students interested in further details on the evolution of the different coelenterate types should consult Hyman (see Suggestions for Further Reading).

CLASS SCYPHOZOA (JELLYFISHES)

Classification

ORDER **Stauromedusae** Adults sessile, attached by an aboral stalk and developing directly from the scyphistoma. Marginal sense organs are modified tentacles if present at all. *Haliclystus.*

ORDER **Cubomedusae** Cubical medusae with four trailing tentacles, alternating with four groups of sense organs located in rhopalia. Stomach divided by septa. *Carybdea, Chiropsalmus.*

ORDER **Coronatae** Medusae with a scalloped margin. Tentacles born on flattened leaflike structures, called pedalia. Stomach divided by septa. *Nausithoe, Periphylla.*

ORDER **Semaeostomeae** Corners of the mouth prolonged into four long frilled oral arms. Stomach not divided by septa. *Aurelia, Cyanea.*

ORDER **Rhizostomeae** Oral lobes fused, obliterating the mouth. Numerous small mouths on the oral lobes which also contain many canals. Stomach not divided by septa. *Cassiopeia, Rhizostoma.*

Anatomy and Physiology of *Aurelia*

GENERAL STRUCTURE

The class Scyphozoa includes the true jellyfishes, such as *Aurelia,* a common form found along both the Atlantic and Pacific coasts. These

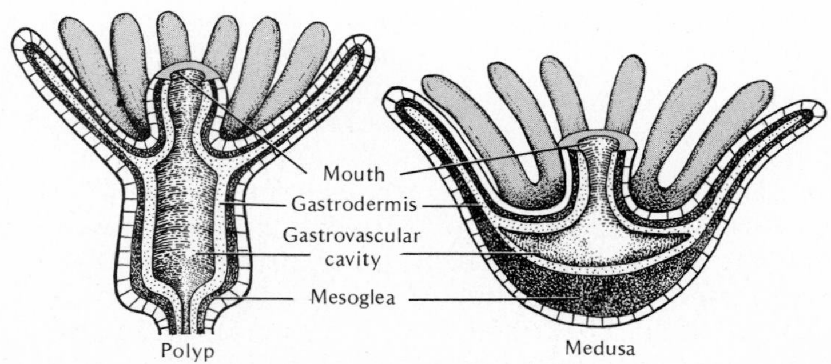

Figure 23.11 Comparison of polyp and medusa.

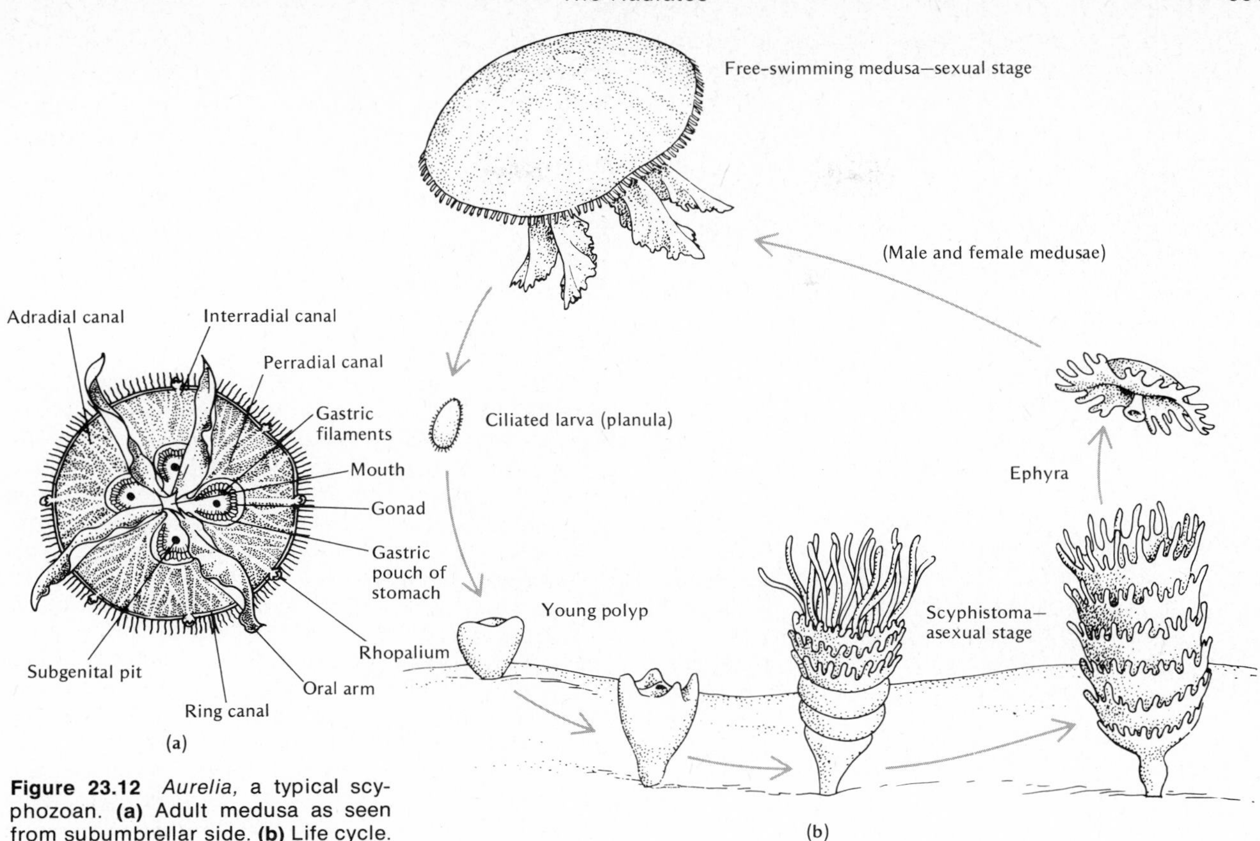

Figure 23.12 *Aurelia,* a typical scyphozoan. **(a)** Adult medusa as seen from subumbrellar side. **(b)** Life cycle.

jellyfishes may be distinguished from the medusa of hydrozoans by their size and by the absence of a velum. *Aurelia* ranges in diameter from around 7.5 cm to over 30 cm (3–12 inches). The medusa stage is the conspicuous stage in the life cycle of *Aurelia,* the polyp stage being quite reduced. The structure and life cycle of this organism are shown in Figure 23.12.

The manubrium is very short and the corners of the mouth are drawn out into four trailing *oral arms* that possess the nematocysts and function in capturing small animals. The mouth leads into the large centrally located stomach which has four diverticulae, called *gastric pouches.* The gonads are located on the walls of these pouches and originate from the lining which is of endodermal origin. Small tentaclelike structures, called *gastric filaments,* protrude from the lining of the gastric pouches. The gastric filaments bear nematocysts and are also the source of some digestive enzymes. Numerous canals extend from the stomach to the edge of the umbrella. Some are branched and others are not. These canals are indicated in Figure 23.12. A small invagination from the subumbrellar surface into the floor of each gastric pouch is called a *subgenital pit.* These pits do not open into the pouch and their function seems to be largely respiratory. The margin of the umbrella bears numerous short tentacles. The ring of tentacles is interrupted by eight equally spaced notches in which are located club-shaped structures, called *rhopalia.* The rhopalia, located between two protective *marginal lappets,* contain calcareous statoliths, an ocellus, and two sensory pits which may serve as chemoreceptors. The rhophalium or tentaculocyst functions in maintaining equilibrium. The calcerous statoliths at the tip of the club weight it so that any change of position will shift the club and thus stimulate sensory hairs on the epidermis (Fig. 23.13).

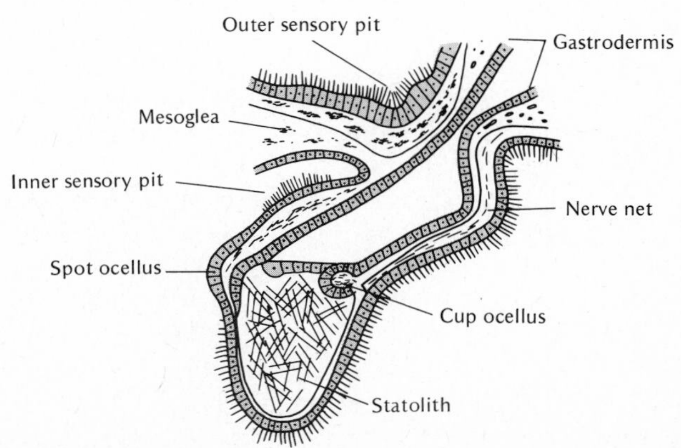

Figure 23.13 Sectional view of rhopalium of *Aurelia.*

Analyses show that most jellyfishes are composed of 94–96 percent water. In *Cassiopeia* the water content of the jelly is 94.6 percent and of the cellular parts, 93.8 percent. In spite of this amazingly high water content, they carry on all of the activities of other members of the phylum. The cellular layers are comparable with those described for hydra, but the mesoglea now contains cells, derived from the ectoderm, and constitutes a third cellular layer.

REPRODUCTION AND DEVELOPMENT

The sexes are separate in *Aurelia,* as in most jellyfishes. Sperm are discharged by way of the mouth. The eggs are fertilized within the female in the stomach, and after passing out of the mouth, they are held on the oral arms during early development. The ciliated planula larvae leave this position and soon settle down on the substrate where they develop into small polyps, called *scyphistoma.* After a period of growth, the polyp undergoes a series of transverse divisions, a process called *strobilization* that is peculiar to the scyphozoa, forming a pile of saucerlike structures. In time these saucerlike structures develop tentacles, break away as *ephyra larvae,* and develop into adult aurelias. Figure 23.12 shows the stages in the life cycle of *Aurelia.*

Ecology

Scyphozoans normally live in the open ocean. If they are carried into the coastal surf zone, they are rapidly torn to bits. They float at the mercy of the currents, swimming only slowly and erratically by sudden contractions of the bell-shaped body (Pl. 33). The swimming motions aid in keeping them at the surface. Some species periodically stop pulsating and sink, later swimming back up to the surface. During the sinking period the outstretched marginal tentacles capture small food organisms.

Some jellyfishes, for example, *Cyanea arctica,* reach a diameter of over 2 m (7 feet) with tentacles 37 m (120 feet) long. Entanglement in the tentacles of a specimen that large can be quite serious even to humans.

Like the hydrozoans, some scyphozoans also have commensal zoochlorellae in their cells. Many immature crustaceans enter the bell of a jellyfish and gain protection, a free ride, and probably some food particles. Although most animals give jellyfishes a wide berth, the giant ocean sunfish, *Mola mola,* lives almost exclusively on jellyfishes and siphonophores. Considering the fact that jellyfishes are 95 percent water, the sunfish must spend most of its time feeding. One specimen of sunfish caught off the coast of California weighed over 1140 kg (2500 pounds).

CLASS ANTHOZOA (Sea Anemones and Corals)

Classification

SUBCLASS **Alcyonaria** Polyps with eight pinnate tentacles and eight complete septa in the gastrovascular cavity. One siphonoglyph present. All colonial with a secreted endoskeleton.

ORDER **Stolonifera** Polyps not fused, connected by a basal sheet of tissue or by stolons. Skeleton of separate spicules or tubes made of fused spicules. *Tubipora, Clavularia.*

ORDER **Telestacea** Long axial polyps with lateral polyps as side branches. *Telesto.*

ORDER **Alcyonaceae** Lower portions of polyps fused with only oral ends protruding. Skeleton of separate calcareous spicules. *Alcyonium;* soft corals.

ORDER **Coenothecalia** Skeleton massive with erect cylindrical cavities for the polyps. *Heliopora,* the blue corals.

ORDER **Gorgonaceae** Axial skeleton of calcareous spicules or of a horny substance, called gorgonin, or of both. Polyps short and borne on the sides of the skeletal axis. *Gorgonia,* the horny corals, sea fans, sea whips, sea feathers.

ORDER **Pennatulacea** Colony made of a single long axial polyp with many lateral polyps on its side. Lower portion of the axial polyps lacks lateral polyps. *Pennatula, Renilla,* the sea pens and sea pansies.

SUBCLASS **Zoantharia** Tentacles simple and rarely branched and other than eight in number. Septa in pattern other than that found in the alcynonarians. Skeleton, when present, not of loose spicules. Anemones and stony corals.

ORDER **Actiniaria** Paired septa often in multiples of six, some of which extend from body wall to pharynx. One or more siphonoglyphs usually present. Always solitary and lacking a skeleton. *Metridium,* sea anemones.

ORDER **Madreporaria** Compact calcareous exoskeleton, polyps solitary or colonial and lacking a siphonoglyph. *Astrangia, Orbicella,* the true corals.

ORDER **Zoanthidea** Solitary or colonial, lacking a skeleton and a pedal disk. Usually growing on sponges or other animals. *Epizoanthus.*

ORDER **Antipatharia** Deep-water colonial forms with an axial skeleton of a black thorny hornlike material. *Antipathes,* the black or thorny corals.

ORDER **Ceriantheria** Long solitary polyps which lack a pedal disk and are adapted for burrowing in the sand. Numerous simple tentacles in two whorls, one siphonoglyph present. *Cerianthus.*

Anatomy and Physiology of *Metridium*

GENERAL STRUCTURE

The class Anthozoa includes the anemones and the corals. *Metridium* is a common sea anemone that may be seen at low tides in many places along the Atlantic and Pacific coasts. The members of this class exist only in the polyp form, but it is not difficult to distinguish the large fleshy sea anemones from the small polyps of hydrozoans. Many of the West Coast anemones are quite large and brightly colored, often looking more like flowers than animals.

The basic structure of an anemone is shown in Figure 23.14. The body or column of a sea anemone is quite stout, with the upper end somewhat expanded into a disk with a central mouth surrounded by several circles of hollow tentacles. The lower end is differentiated into a pedal disk which attaches the animal to the substratum. Internally, anthozoan polyps may be distinguished from hydrozoan polyps by the presence of vertical partitions, called *septa* or *mesenteries,* which divide the gastrovascular cavity. Some of the septa are complete and extend from body wall to pharynx, but others are incomplete with their edges free in the gastrovascular cavity. Longitudinal retractor muscles are located in the septa, and filamentous structures, called *acontia,* are located on the lower free edges. The acontia bear nematocysts and probably aid in dispatching organisms which reach the gastrovascular cavity alive. They also serve in defense, since they may be protruded through the mouth or through small pores in the column and pedal disk, called *cinclides.*

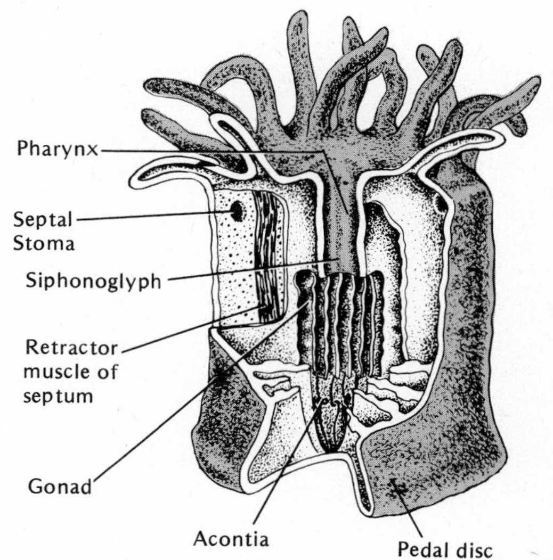

Figure 23.14 Sectional view of an adult sea anemone, *Metridium.*

Another feature found in anthozoan polyps and not in hydrozoan polyps is that the epidermis turns in at the mouth and lines the pharynx. Such a structure, lined with ectoderm, is called a *stomodaeum.* In *Metridium* there are two longitudinal grooves in the pharynx, called *siphonoglyphs,* in which there are long cilia whose effective beat is downward. Other groups of anthozoa may have but a single siphonoglyph, while others lack them entirely. The mesenchyme located between the epidermis and the gastrodermis is a cellular layer with well-developed muscles in some places.

FEEDING AND DIGESTION

Like other coelenterates, *Metridium* is a carnivore. Its tentacles are well provided with powerful nematocysts and it feeds on a variety of animals, including fish, crustacea, and mollusks. It can ingest organisms almost as large as itself, greatly extending its body to accommodate the prey.

Metridium has a remarkable ability to distinguish between food substances and inert material. If a piece of filter paper is dropped on the oral disk, ciliary currents carry it out to the tentacles which bend out and drop it. On the other hand, a similar piece of paper soaked in clam juice delivered to the tentacles will cause them to bend toward the mouth and the paper is swallowed.

Food materials pass down the pharynx and are enveloped in the folds of the septa which divide the gastrovascular cavity into compartments. Proteases and lipases are secreted by the gastrodermal cells, and by both intracellular and extracellular digestion the food materials are quickly digested. Anemones, like most if not all other coelenterates, are unable to digest starches. Undigested materials are egested through the mouth.

RESPIRATION AND EXCRETION

As in other coelenterates both respiration and excretion are carried out by simple diffusion. The ciliary currents of the siphonoglyph carry oxygenated water into the gastrovascular cavity while cilia on the rest of the pharynx beat in the opposite direction and thus force water out of the cavity, carrying with it carbon dioxide and nitrogenous wastes.

REPRODUCTION AND DEVELOPMENT

Asexual Anemones reproduce asexually both by longitudinal fission and by budding. In *Metridium* a peculiar type of asexual reproduction, called *pedal laceration,* occurs. Small pieces, a few millimeters in diameter, separate from the edges of the pedal disk. Each piece soon reorganizes into a complete little anemone.

Sexual The gonads originate in the gastrodermis and are located on the septa in the gastrovascular cavity. When mature, the eggs or sperm are released, passing to the outside via the pharynx and mouth. Anemones are either male or female or hermaphroditic. In the latter case, however, eggs and sperm are seldom produced at the same time in one individual. Thus cross-fertilization is the rule. Following fertilization, cleavage results in a typical planula larva which soon settles down to form a new anemone.

Ecology

Sea anemones are found from the tidal zone all the way down to the deepest portion of the ocean from which living organisms have been collected – 10,500 m (34,448 feet). It was long believed that living organisms could not withstand the great pressures at such depths [over 1,000 kilograms per square centimeter (15,000 pounds per square inch)]. However, they obviously can and the reason lies in the fact that the pressure within the organism equals that which is on the outside.

Along the central California coast, a species of sea anemone, *Anthopleura elegantissima,* occurs in dense aggregations on rocky surfaces in the high intertidal zone where it is exposed to the air for relatively long periods of time. Lines of sucking tubercles cover its stalk, and sand, gravel, and bits of shell adhere to the tubercles. This covering effectively protects the organism from desiccation during its twice-daily exposure to the air.

Like the sponges, many sea anemones are also carried about on the shells of hermit crabs, a given species of hermit crab always carrying a certain species of anemone. When the hermit crab outgrows its shell and moves to a new one, it carefully pulls the anemone loose from the old shell and transfers it to the new one. If it somehow loses its anemone, it finds a replacement and holds it against its shell until the anemone becomes firmly attached. Other crabs carry small anemones around in their pincers, holding them out defensively when danger threatens.

Certain small fish called clown fish, are able to live as commensals among the tentacles of anemones, even to swim in and out of the gastrovascular cavity. One report tells of such small fish driving other fish into the tentacles of the host anemone to be summarily captured and swallowed (Fig. 23.15).

Figure 23.15 *Amphiprion* sp., a clown fish, among the tentacles of an anemone, Lizard Island, Queensland, Australia. The clown fish is immune to the dangers of the anemone tentacles and may actually lure other less adapted species of fish to their death. This would be an example of a mutualistic relationship between the anemone and the clown fish. (Photo – E. C. Williams)

Anatomy and Physiology of Corals

The stony corals are much like anemones, except that they secrete delicate limestone cups, called *corallites.* The small coral polyps can retract into the cups when danger threatens (Pls. 13, 30). Most corals are colonial, but there are some solitary species. The mushroom coral, *Fungia,* shown in Figure 23.16, is an example of a solitary coral. Each individual corallite has extending inward from its wall a series of radially arranged vertical plates that fit into corrugations in the polyp located between the septa. The material of these cups is secreted by the epidermis and is what is ordinarily called coral. The differences in the form of coral masses are due to the differences in the patterns of budding in the various species. In the brain corals (Pl. 15) the colonies build up large masses whose surface is covered with long, curved depressions which are occupied in life by a row of small polyps. Many kinds of solitary corals grow in the temperate and even cold ocean waters, but the extensive masses of colonial corals grow only in tropical and subtropical waters. These colonial organisms have built up great deposits of limestone during the past history of the earth as seen in some of the islands and reefs of the South Pacific, the Great Barrier Reef of Australia, and the Keys of southern Florida (Fig. 23.17).

Figure 23.16 *Fungia* sp., a mushroom coral from the Great Barrier Reef of Australia. Although most corals are colonial, this is an example of a solitary coral. (Photo—E. C. Williams)

Formation of Islands and Reefs

Borings taken on some of the coral islands indicate that in some places the coral formation

Figure 23.17 The Great Barrier Reef at low tide, Heron Island, Queensland, Australia. (Photo—E. C. Williams)

extends to depths of 300 m (1000 feet). It is known that these colonial corals cannot live at depths below 46 m (150 feet) because of the low temperatures in deeper water; this has posed a problem about the presence of coral formations at greater depths. Charles Darwin was one of the first workers to become interested in this problem, and the theory he offered to explain it is still quite widely accepted. There are three main types of coral reefs: the fringing reefs, the barrier reefs, and the atolls. The fringing reefs are found in shallow waters, bordering the coast closely; the barrier reefs parallel coasts but are separted from them by a deep channel of water; the atolls are ring-shaped coral formations enclosing central lagoons. Darwin reasoned that all these formations started the same way, as fringing reefs. In cases where there are barrier reefs today, there has been a gradual sinking of the coast with a continued growth of the coral, leaving a wide and deep channel between the two. Finally, in the formation of atolls, Darwin postulated that in the beginning there was an island with a fringing reef, which passed through a stage when it was surrounded by a barrier reef: with eventual complete submergence of the island, the coral formation alone was left above the water (Fig. 23.18).

Another plausible theory to account for the great thickness of coral formations is one that assumes that coral formations started on submerged platforms in the ocean—platforms that may have resulted from volcanic activity—and then with a gradual increase in the depth of the ocean, the corals kept pace with their growth, resulting in these very thick formations. That the oceans were shallower during periods of great glacier formation, and also that the oceans increased in depth with the melting of the glaciers, are known facts.

Ecology of Coral Reefs

Although one generally speaks of coral reefs as though they were indeed built up entirely by the coral animals, many other reef-dwelling animals secrete calcium carbonate shells or tubes and thus contribute to the buildup of the reef. Many reefs are as much alga reefs as coral reefs. Certain marine algae, called *coralline algae*, have large amounts of calcium carbonate in their cell walls, and their contributions to the reef are often at least equal to those of the corals.

Corals form gigantic communities stretching for hundreds of kilometers in many instances. They provide a habitat for a specialized flora and fauna found nowhere else.

The reef is a firm base of attachment for sessile organisms, and the countless nooks and crannies provide homes and refuges for animals large and small. Predatory animals find the coral reef an ideal hunting ground.

Few if any beauties in nature surpass those of a tropical coral reef with its kaleidoscope of colors from corals, algae, sponges, anemones,

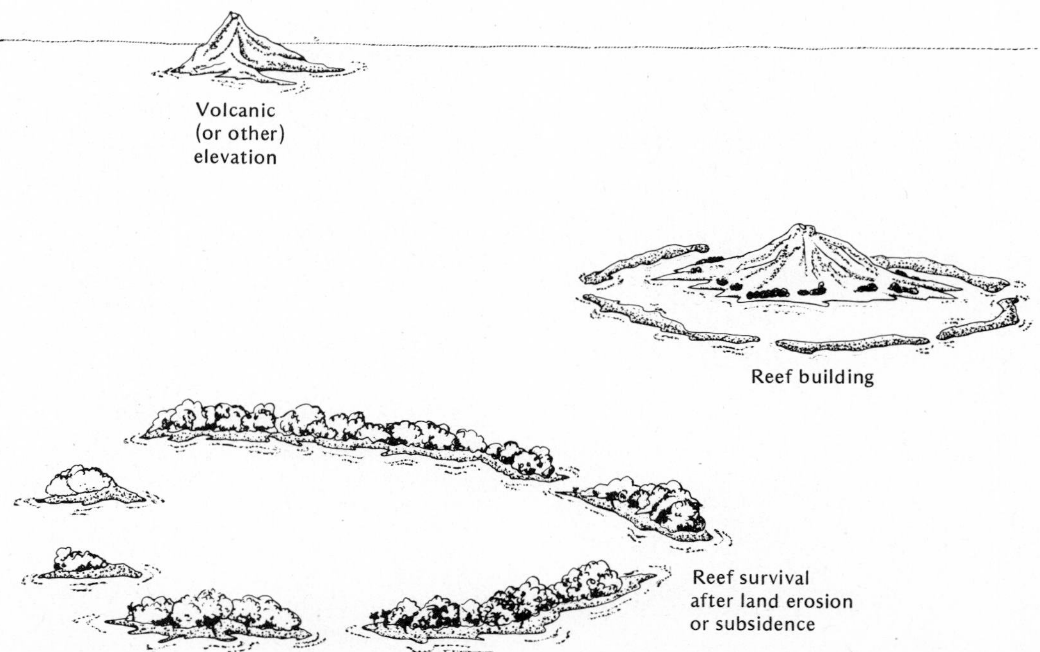

Figure 23.18 Darwin's postulate of the method of coral reef formation.

and fishes, delicate and grotesque. To spend an hour or two floating above such a reef with face mask and snorkel tube is to enter a veritable fairyland.

Many animals have special adaptations which enable them to bore into the coral and make their own refuges. Bivalve mollusks such as *Lithophaga* are aided in the process by acidic secretions of their salivary glands which dissolve the calcium carbonate. The female coral gall crab, *Harpalocarcinus,* while still immature, settles in the fork between two branches of coral. The respiratory currents set up by the crab cause the branches to broaden out and curve around the crab until a chamber is formed from which the crab cannot escape. The crab in turn is modified so that it can subsist on minute organisms drawn into its chamber. Male gall crabs do not form chambers. They are minute in size, able to enter the chamber occupied by a female and fertilize the eggs.

Corals also have their intracellular algae, usually yellow-green algae called *zooxanthellae.* The algae gain protection and remove carbon dioxide and nitrogenous wastes from their hosts, a mutually beneficial activity.

In recent years the crown of thorns starfish, *Acanthaster planci,* has caused serious damage to some coral reefs in the Pacific. This species feeds on corals, and for some unaccountable reason its numbers increased dramatically in certain areas with resultant increased pressure on the coral reefs. Many feared that this increase was related to some unknown environmental pollutant and advocated extermination of the starfish. Others hold that this is just another example of a well-known phenomenon, the periodic population explosion of certain species. They believe that the pendulum will swing in the other direction, with no permanent irreparable damage to the coral reefs (Fig. 23.19).

Figure 23.19 *Acanthaster planci,* the crown of thorns starfish, One Tree Island, Capricorn Group, Queensland, Australia. This species is a serious predator of reef corals in some areas where populations have grown very large in recent years. (Photo—E. C. Williams)

PHYLUM CTENOPHORA

Some authors have placed the ctenophores within the phylum Coelenterata as a subphylum. The distinctive features of this group are sufficient to warrant their position as a separate phylum. Although there are less than 100 species, these organisms are an important component of the marine planktonic community where they feed, in many cases quite voraciously. The organisms which make up the plankton are plants and animals which may or may not have the capacity to move, but which are at the mercy of the water currents and cannot move against them.

Although most ctenophores are spherical or oval in shape, they do have structures such as paired tentacles which make it impossible to speak of them as strictly radially symmetrical; hence they are said to have biradial symmetry.

Characteristics

1. Biradial symmetry.
2. Locomotion by means of eight rows of ciliated comb plates, or ctenes.
3. Mesenchymatous ectomesoderm present.
4. Gastrovascular cavity with a single opening, the mouth, leading into a large pharynx which is lined with ectoderm and is thus a stomodaeum.
5. Usually with paired retractable tentacles which bear specialized adhesive cells, called colloblasts. With one exception, nematocysts are lacking.
6. Gonads originate from the endoderm. All species are hermaphroditic and have mosaic development quite unlike that of the coelenterates.
   Figure 23.20

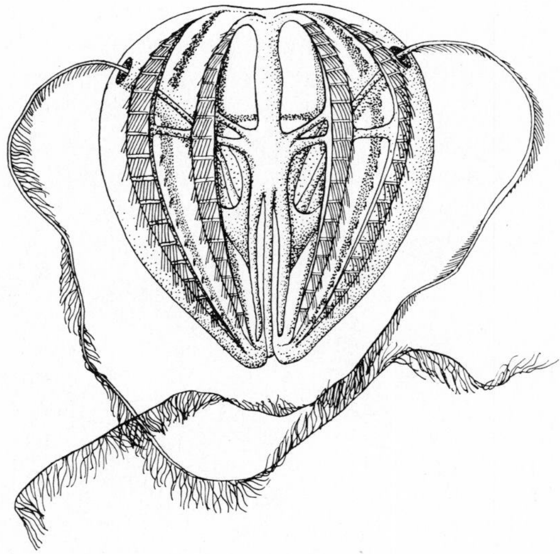

Figure 23.20 *Pleurobrachia,* a typical ctenophore.

Classification

CLASS **Tentaculata** Two or more tentacles present.

ORDER **Cydippidea** Spherical or oval body form with two retractile tentacles provided with a special sheath. Branches of the gastrovascular cavity end blindly. *Pleurobrachia.*

ORDER **Lobata** Body with two large oral lobes and four smaller folds associated with the mouth, called auricles. Tentacles lack sheaths and gastrovascular canals unite in mouth region to form a ring canal. *Mnemiopsis.*

ORDER **Cestida** Body compressed and ribbonlike, comb rows rudimentary and four in number. Two main tentacles reduced but provided with sheaths. Two rows of small tentacles around mouth. *Velamen.*

ORDER **Platyctenea** Strongly flattened creeping forms with two tentacles in sheaths. Comb rows present only in the larval forms in some species. *Ctenoplana.*

CLASS **Nuda** Tentacles absent.

ORDER **Beroida** Conical in shape with wide mouth and pharynx. *Beroë.*

Anatomy and Physiology of *Pleurobrachia*

GENERAL STRUCTURE

A common ctenophore of the North Atlantic coast, often used for laboratory study as a representative of this phylum, is the sea gooseberry, *Pleurobrachia.* Figure 23.21 shows the anatomical details of this organism. It has a transparent spherical body about 18 mm in diameter. The body wall is made up of an outer ciliated epidermis, an inner gastrodermis lining the gastrovascular cavity, and in between is a gelatinous, mesenchymatous layer containing scattered cells, connective tissue fibers, and numerous muscle fibers. It is a true cellular layer, derived as in the coelenterates from the ectoderm—it is an ectomesoderm. Eight rows of meridional comb plates pass from a point near the aboral sense organ to the mouth. These comb plates, or *ctenes,* are a distinctive feature of the ctenophores. Each ctene is made up of a transverse row of long fused cilia (Fig. 23.22).

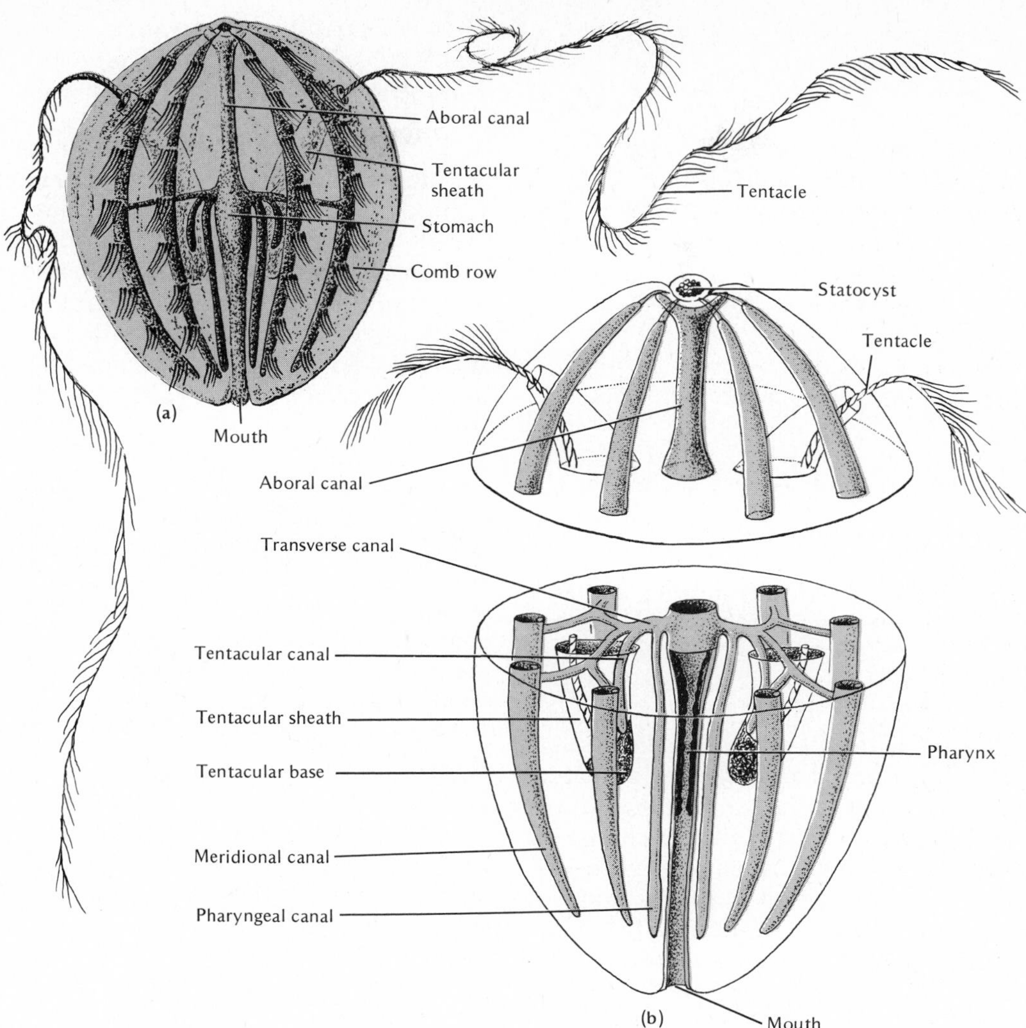

Figure 23.21 Anatomy of *Pleurobrachia*. **(a)** General anatomy. **(b)** Bisected specimen showing some details of internal anatomy.

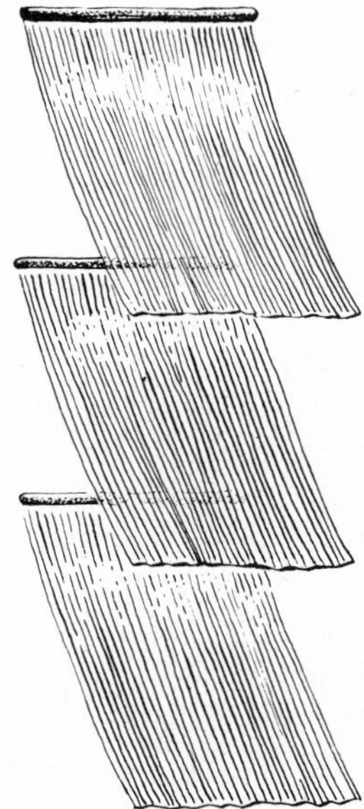

Figure 23.22 Detail of ctene structure. Each comb row is made up of a transverse row of fused cilia.

The beating of these paddlelike structures in successional waves which start at the aboral pole moves the organism through the water. The effective beat is toward the aboral pole so that the organism moves with its mouth in front. The synchrony of beat as well as the direction are controlled by nervous impulses originating in a statocyst in the aboral sense organ (Fig. 23.23).

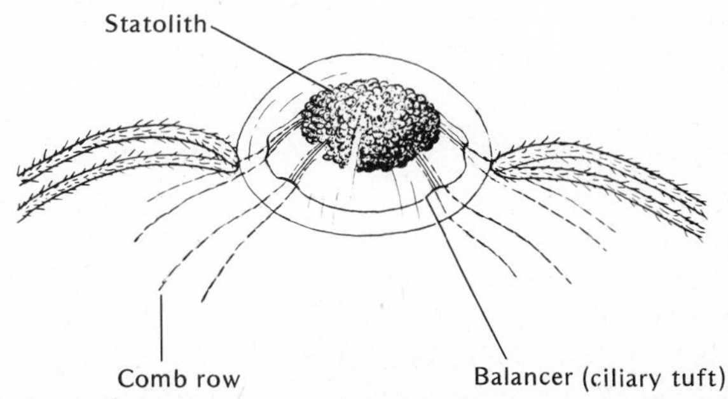

Figure 23.23 Detail of aboral sense organ.

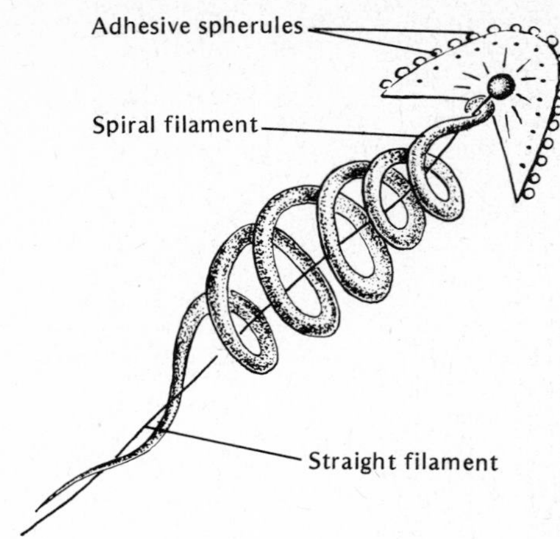

Figure 23.24 A single colloblast with its filaments.

FEEDING AND DIGESTION

Two retractable tentacles originate in blind sacs which open near the aboral end. The tentacles when fully extended trail out behind the organism for some distance. There are feathery lateral branches on the tentacles and the surface is provided with specialized cells, called *colloblasts* (Fig. 23.24). The colloblasts have a sticky secretion which functions in capturing the small organisms upon which the ctenophore feeds. When an organism is entangled in the adhesive, the straight filament ruptures and the contractile spiral filament functions to retain the prey. Meanwhile the tentacle moves to the mouth region and the prey is removed.

The mouth leads into the pharynx, which is a stomodaeum, and the stomach is located near the aboral pole. Five canals extend from the stomach: two toward the mouth along the pharynx; two transverse canals; and one aboral canal. Each transverse canal branches to form a tentacular canal and four meridional canals which lie beneath the rows of comb plates.

NERVOUS SYSTEM

The nervous system of ctenophores is essentially a nerve net as in the coelenterates. There is some condensation of fibers associated with the coordination of the ctenes, but the elements which make up the system are still protoneurons.

RESPIRATION AND EXCRETION

Respiration is by simple exchange between the cells and the environment. Excretion probably is also by simple diffusion, but specialized structures, called *cell rosettes* (Fig. 23.25), may also be involved. These structures are ciliated cells which surround a small opening between the gastrovascular cavity and the mesoglea. If they are indeed excretory in function, they would constitute the earliest phylogenetic appearance of specialized excretory structures.

REPRODUCTION AND DEVELOPMENT

Ctenophores are all hermaphroditic. The gonads originate from endodermal tissue. On either side of each meridional canal there are bands of tissue, one of which is a testis and the other an ovary. It was formerly believed that eggs and sperm were shed into the sea, leaving via the meridional canals and the mouth. Recent studies indicate that eggs are fertilized internally and that they reach the sea by passing through the epidermis.

Development in ctenophores is quite different from that seen in coelenterates. In fact, there is a unique type of cleavage pattern, called biradial, found only in this phylum. At the eight-cell stage a biradial symmetry is established which persists throughout development. The blastula stage is a solid structure with numerous micromeres at the future aboral surface and with a smaller number of macromeres at the future oral end. Gastrulation is by invagination and epiboly with the micromeres growing down over the larger macromeres to form the ectoderm, while the macromeres invaginate to form the endoderm. The mesoderm is derived by delamination from the inner surface of the ectoderm. The fate of the blastomeres is determined very early in development and thus the cleavage is said to be *determinate*. Experimental destruction of cells in early cleavage stages will result in deficiencies in the larva which develops. If the blastomeres are separated at the two-cell stage, a half larva will result from each blastomere. Similarly, one-quarter larvae result from separation of the blastomeres at the four-cell stage. The larval

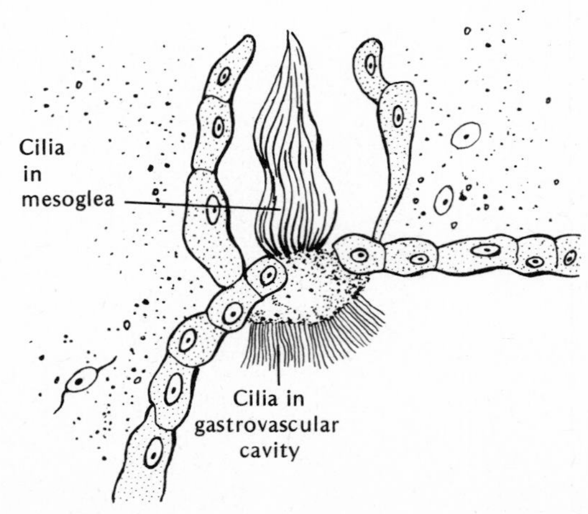

Figure 23.25 A cell rosette.

form characteristic of the ctenophores is the *cydippid larva,* resembling an oval or spherical adult.

Most ctenophores are capable of light production, called *bioluminescence.* They are capable of luminescing very early in development and continue to possess this faculty throughout life. Regenerative capacity is high in this group, as in the coelenterates.

PHYLOGENETIC RELATIONSHIPS OF RADIATA

Haeckel's concept of the gastraea and the modification of that concept by Metschnikoff have been discussed in connection with the colonial flagellate *Volvox.* According to this concept, the ancestral coelenterate was a solid mass of ciliated cells much like the characteristic planula larva of present-day coelenterates. Derivation of the coelenterate medusa from this so-called planuloid ancestor involved development of a gastrovascular cavity and tentacles. The significant differences between ctenophores and coelenterates—lack of nematocysts and the distinctive cleavage type of ctenophores—leads students of phylogeny to believe that the ctenophores and the coelenterates arose from a common ancestral type, perhaps related to the primitive medusa of the order Trachylina.

Two aberrant flattened ctenophores belonging to the genera *Coeloplana* and *Ctenoplana* have certain superficial resemblances to polyclad turbellaria, and some authors have taken this resemblance to indicate a phylogenetic relationship between the phylum Ctenophora and the phylum Platyhelminthes. Current thinking casts serious doubt upon this concept. It postulates the origin of the phylum Platyhelminthes from other sources, and considers the radiate phyla as an evolutionary dead end.

SUGGESTIONS FOR FURTHER READING

Buchsbaum, R. *Animals Without Backbones.* Chicago: Univ. of Chicago Press (1949).

Crowell, S. (Ed.). "Behavioral Physiology of Coelenterates." *American Zoologist,* **5** (1965).

Hickman, C. P. *Biology of the Invertebrates.* St. Louis: C. V. Mosby (1973).

Hyman, L. H. *The Invertebrates: Protozoa through Ctenophora.* New York: McGraw-Hill (1940).

Lenhoff, H. M., and W. F. Loomis (Eds.). *The Biology of Hydra and Some Other Coelenterates.* Coral Gables, FL: Univ. of Miami Press (1961).

Lentz, T. L. *The Cell Biology of Hydra.* New York: Wiley (1966).

McConnaughey, B. H. *Introduction to Marine Biology.* St. Louis: C. V. Mosby (1974).

Miller, R. L. (Ed.). "The Developmental Biology of the Cnidaria." *American Zoologist,* **14,** (1974).

Pennak, R. W. *Fresh-Water Invertebrates of the United States.* New York: Ronald Press (1953).

Stoddart, D. R. "Ecology and Morphology of Recent Coral Reefs." *Biological Rev.,* **44,** (1969).

Tardent, P. "Regeneration in Hydrozoa." *Biological Rev.,* **38** (1963).

From Scientific American

Berrill, N.J. "The Indestructible Hydra" (Dec. 1957).

Lane, C. E. "The Portuguese Man-of-War" (Mar. 1960).

Loomis, W. F. "The Sex Gas of Hydra" (Apr. 1959).

The Acoelomates– PLATYHELMINTHES, RHYNCHOCOELA, AND GNATHOSTOMULIDA

INTRODUCTION TO THE BILATERIA

Most of the animal kingdom is included in the grade Bilateria of the branch Eumetazoa. These organisms are all basically bilaterally symmetrical in their body plan. Within the Bilateria there are several large subdivisions which set the pattern of organization for this section of the book. The presence or absence of a body cavity other than the digestive cavity and the nature of that cavity if it is present form the basis of subdividing the rest of the animal kingdom into three *subgrades.* Those bilaterally symmetrical animals which have only a digestive cavity are placed in the subgrade Acoelomata. Three phyla, the Platyhelminthes, Rhynchocoela, and Gnathostomulida, are in this category. Three other phyla, the Acanthocephala, Aschelminthes, and Entoprocta, do have a definite space between the gut and the body wall. This space, which is a remnant of the embryonic blastocoel and is not completely lined with mesoderm, is called a pseudocoel, and the three phyla are in the subgrade Pseudocoelomata. Finally, the other phyla all possess a true coelom—a cavity completely lined with mesoderm. They are placed in the subgrade Coelomata.

One further breakdown is made of the bilaterally symmetrical animals. It is based on the fate of the embryological structure, the blastopore. In all the Acoelomata and the Pseudocoelomata the blastopore region becomes the mouth. These phyla, along with a large group of coelomate forms in which the mouth also originates from the blastopore region, are placed in the *division Protostomia.* (The phyla, in addition to the Acoelomata and the Pseudocoelomata, which make up the division Protostomia, are Mollusca, Sipunculida, Echiurida, Priapulida, Annelida, Onychophora, Tardigrada, Pentastomida, Chaetognatha, Ectoprocta, Brachiopoda, and Phoronida.)

In a smaller group of coelomate phyla, Echinodermata, Chaetognatha, Pogonophora, Hemichordata, and Chordata, the region of the embryonic blastopore becomes the anus in the adult. These phyla are placed in the *division Deuterostomia.*

PHYLUM PLATYHELMINTHES

Two important groups of parasites which affect humans and domestic animals, the flukes and the tapeworms, as well as a group of free-living species, the planarians and their allies, make up this phylum. The phylum Platyhelminthes ranks among the "top ten" phyla in terms of numbers of living species. Since the Platyhelminthes are generally considered to be the most primitive group of bilateria, they are significant as being the phylum which introduces the pattern upon which evolutionary change operated and produced all the higher phyla. Their bilaterial symmetry with the beginnings of concentration of nervous tissue at the anterior end, called *cephalization,* is a very significant development. The third germ layer, the mesoderm, is a clearly defined layer with important organs and organ systems differentiating in it. The bulk of the mesoderm in the platyhelminthes, as well as in the succeeding phyla, is derived from the endoderm as opposed to the ectodermally derived mesoglea of the preceding groups. The digestive system still retains the single opening and is thus a gastrovascular cavity.

The coelenterates and ctenophora are considered to be organized on the *tissue level* of organization, but the flatworms, and all the higher animal phyla, are organized on the *organ level* or *organ-system level* of organization.

Characteristics

1. Bilateral symmetry with the beginnings of cephalization.
2. Mouth but no anus.
3. Triploblastic.
4. No coelom.
5. Body soft and usually flattened dorsoventrally.
6. Protonephridial excretory system with flame bulbs.
7. Early development characterized by spiral, determinate cleavage with all or most of the mesoderm arising from a single special cell.
   Figure 24.1

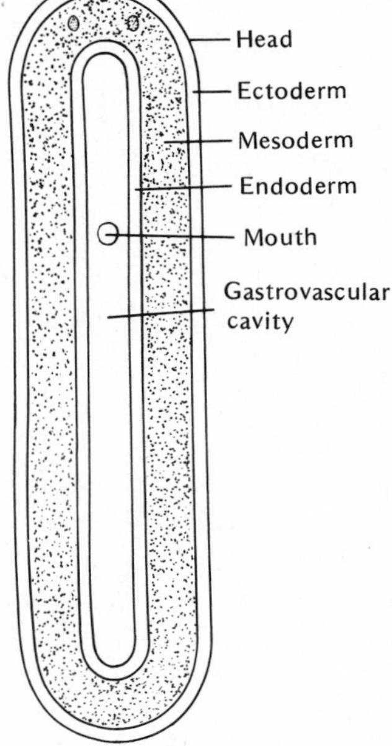

Figure 24.1 Platyhelminth body plan.

Classification

CLASS **Turbellaria** These are mostly nonparasitic forms, lacking hooks and suckers. An intestine is present in all but the most primitive order (*Acoela*) which has a mouth and pharynx but no well-defined intestine. The outermost surface is a cellular or syncytial epidermis which is usually ciliated, at least in some regions (Fig. 24.2).

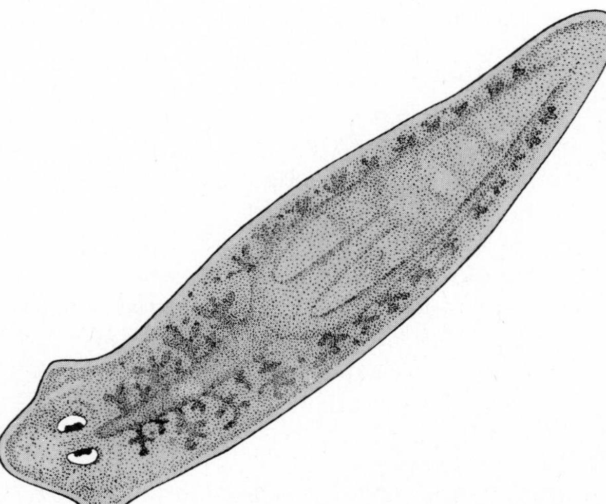

Figure 24.2 Class Turbellaria, *Dugesia* sp.

CLASS **Trematoda** The flukes are all parasitic, mostly internal parasites. One or more suckers present. A mouth, pharynx, and intestine are present, the latter usually Y-shaped. Body covered with a nonciliated living tegument which contains mitochondria and endoplasmic reticulum but lacks nuclei (Fig. 24.3).

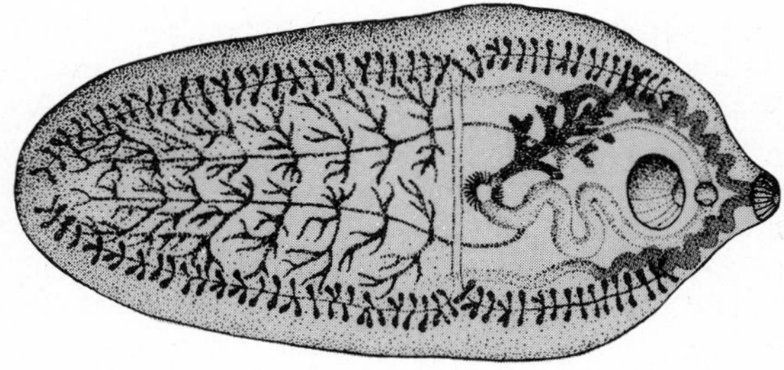

Figure 24.3 Class Trematoda, *Fasciola hepatica,* the sheep liver fluke.

CLASS **Cestoda** The tapeworms are all internal parasites, provided with hooks or suckers, or both. No digestive system present. Body usually divided into a head or scolex which bears the organs of attachment, a short neck, and a series of similar segments, called proglottids, which contain both male and female reproductive structures. Body covered with a nonciliated living tegument which

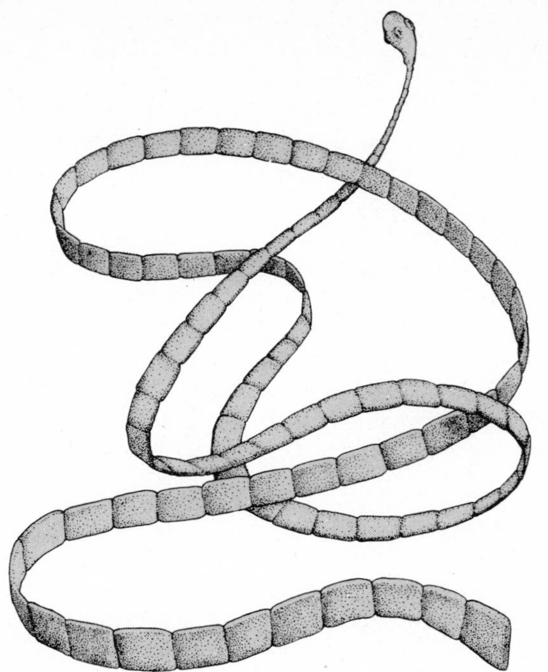

Figure 24.4 Class Cestoda, *Taenia saginata,* the beef tapeworm.

contains mitochondria and endoplasmic reticulum but lacks nuclei. Microvilli project from the surface of the tegument (Fig. 24.4).

CLASS TURBELLARIA

Classification

ORDER **Acoela** Small worms with mouth, pharynx, but no intestine. Exclusively marine. No protonephridia or well-defined gonads. *Convoluta, Polychoerus.*

ORDER **Rhabdocoela** Small worms with mouth, pharynx, and digestive tract which is a straight tube without diverticulae. *Stenostomum, Mesostoma.*

ORDER **Alloeocoela** Moderate in size. Intestine with short diverticulae. *Plagiostomum, Pseudostomum.*

ORDER **Tricladida** Intestine with one anterior and two posterior branches. *Dugesia, Bdelloura.*

ORDER **Polycladida** Intestine with many branches. Entirely marine. *Notoplana, Stylochus.*

Anatomy and Physiology of *Planaria*

GENERAL STRUCTURE

The typical representatives of the phylum Platyhelminthes are small flatworms, about 12 mm long and gray or black in color, called planarians. Planarians are abundant all over the world in uncontaminated spring-fed streams. If a piece of liver is placed in such a stream, in a few hours it may be covered with hundreds of these flatworms.

Many of the common freshwater flatworms belong to the genus *Dugesia* (or *Euplanaria*), and are commonly called planaria. The head of planaria is bluntly triangular in shape with a pair of conspicuous eyespots. The lateral projections of the head are known as *auricles* and they serve as tactile receptors and chemoreceptors. If the animals have fed recently, the outline of the gastrovascular cavity will show through the thin body wall. Although some planarians are almost colorless, most of them show brown or black pigmentation. The *mouth* opening is on the ventral surface, and through this opening the muscular pharynx is protruded during feeding. A smaller opening, the *genital pore,* is located posterior to the mouth.

The outer layer of a planarian, as shown in Figure 24.5, the *epidermis,* is a cuboidal epithelium that is ciliated on the ventral and lateral surfaces. Numerous gland cells open on the surface of the body, secreting a mucous material upon which the worm moves. The slow gliding of these worms results from the beating of the cilia in this bed of mucus.

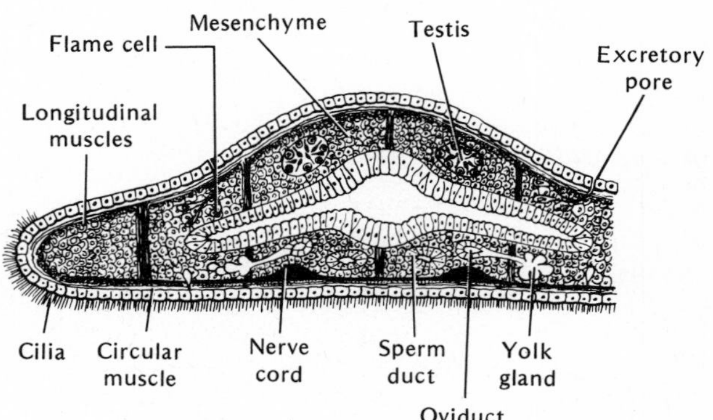

Figure 24.5 Cross section through a planarian worm.

MUSCULAR SYSTEM

In the planarian worm there are three kinds of muscles differentiated in the mesoderm: an outer *circular layer,* just beneath the epidermis; an inner *longitudinal layer;* and *dorsoventral* muscles. Contractions of the circular muscles cause an elongation of the body, while contractions of the longitudinal muscles cause constriction of the body. Other alterations of body shape are produced by contractions of the dorsoventral muscles.

DIGESTIVE SYSTEM

The general nature of the digestive system is indicated in Figure 24.6. As pointed out above, it lacks an anal opening.

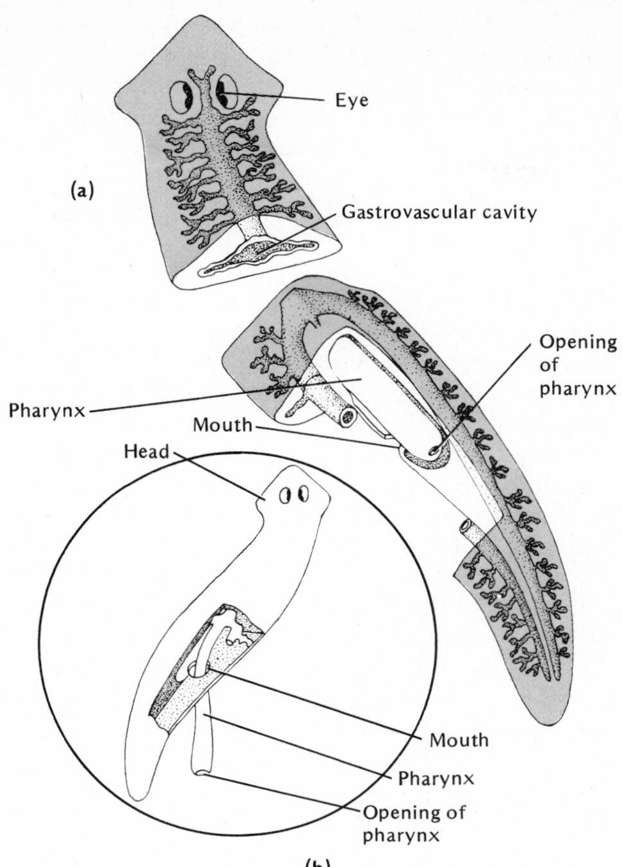

Figure 24.6 Digestive system of a planarian. **(a)** Complete system with pharynx withdrawn. **(b)** Diagram of worm with pharynx extended. Note that the mouth opening is in the body wall, not at the tip of the pharynx.

Although the origin of the flatworms is not revealed by paleontological studies, it seems possible that their beginning was an ancestral form that was planulalike with a gastrovascular cavity and a mouth directed downward, which started to live on the substrate and moved with one part of the body always in advance. The ventral position of the mouth would seem to suggest such a possibility.

The mouth of a planarian opens into a cavity that contains the muscular pharynx. The food of this organism consists of small worms, crustacea, and insect larvae as well as dead animal matter. In feeding, the planarian moves over its food object and traps it with its body. The pharynx is then extended, attached to the food material, and, by sucking movements produced by the muscles of the pharynx, the food is torn into bits and swallowed. The pharynx leads into the three-branched *gastrovascular cavity.* (Thus the order name "triclad" for this group of turbellarians in contrast to the order name "polyclad" of certain marine turbellarians that have a many-branched gastrovascular cavity.) There are one anterior branch and two posterior branches; each of these three main branches has numerous side branches. This branching of the gastrovascular cavity provides for the distribution of the end products of digestion to all parts of the body.

Most of the digestion in planarians is intracellular. The end products of digestion diffuse throughout the tissues of the body. Indigestible particles are eliminated through the mouth.

Some of the cells of the endoderm are used for storage. As a result of this ability to store food reserves, a planarian may live a long time without feeding. Planarians have been observed to shrink to one tenth of normal size during several months of starvation. In such extreme instances not only are the stored reserves used up, but the tissues of the body, such as the reproductive organs, are also resorbed and utilized.

RESPIRATION

There is no respiratory system in this organism. Its size permits the diffusion of oxygen from the surrounding water into the cells throughout the body and the diffusion of carbon dioxide to the outside.

EXCRETORY SYSTEM

A network of tubules running the length of the body on each side, as shown in Figure 24.7, is usually called an excretory system. This system is located in the mesenchyme or parenchyma, and it communicates with the outside by minute excretory pores. There are side branches of the tubules that consist of *flame cells,* also shown in Figure 24.7. The hollow cavity of each flame cell connects to an excretory tubule, and the beat of the long cilia in each flame cell causes the fluid, which enters from the surrounding mesenchyme, to move along the tubes to the excretory pores. Such a system, in which the ducts originate in closed hollow cells or bulbs and open only at the body surface, is called a *protonephridial* system. In later phylogenetic developments the excretory system is open, both internally and externally, and is called a *metanephridial* system.

Much water is eliminated via the flame cells and their ducts. Although some nitrogenous wastes may be passed out with this excess water, a primary function of this system seems to be the elimination of excess water. Much of the nitrogenous waste is eliminated by way of the gastrodermis and mouth, and the carbon dioxide diffuses from the surface of the animal. Nevertheless, the appearance of flame cells and

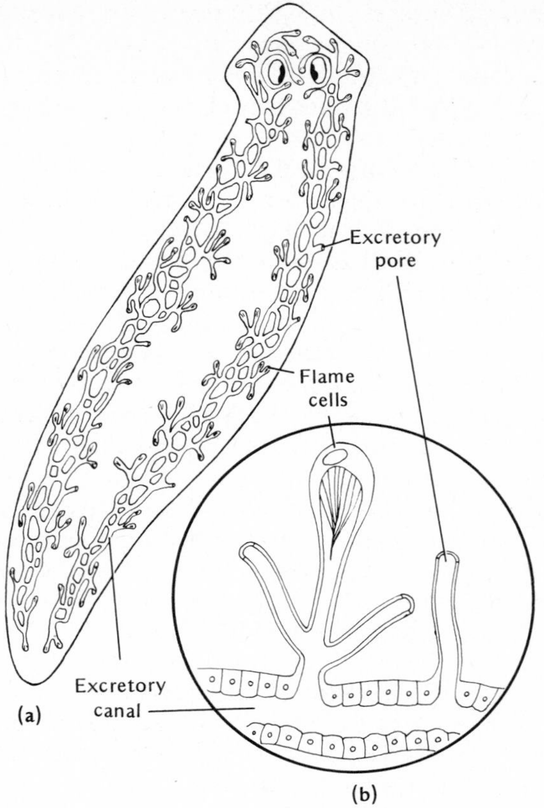

Figure 24.7 Excretory system of a planarian. **(a)** Entire system. **(b)** Detail of flame cell and duct.

their ducts is significant in that they are the forerunners of the well-developed excretory systems of higher phyla.

NERVOUS SYSTEM AND COORDINATION

The planarian nervous system is illustrated in Figure 24.8. This system is composed of paired ganglia in the head region, the *brain,* two longitudinal *ventral nerve cords* with cross connectives, and numerous peripheral nerves. Because the cross connectives give the appearance of rungs on a ladder, this type of nervous system is called the "ladder type." The auricles are highly innervated, and these sensory projections are especially sensitive to touch, chemicals, and to currents of water. The eyes are specialized for light reception. Each eye consists of a cuplike structure lined with black pigment and filled with receptor cells, the ends of which continue as nerves and enter the brain. The epithelium over each eye cup is not pigmented so that light can enter. The sensory cells in the eye cups are so shaded by the pigmented lining that the animal is able to respond to the direction of light. Numerous other sensory cells are scattered all over the body in the epidermis.

If the head of a planarian worm is removed, coordinated movements are still possible. This would tend to indicate that the brain serves largely as a sensory relay station that receives stimuli and transmits them to other parts of the body.

The planarian central nervous system is quite an advance over the nerve net found in *Hydra.* As a result, the responses of planarians are more rapid and more varied than those of *Hydra.* Planarians react negatively to light; they are usually found in dark places. They respond positively to contact, tending to keep the ventral surface in touch with the substrate. They respond to chemicals in the water and their response to food is quite rapid. Many planarians respond positively to water currents.

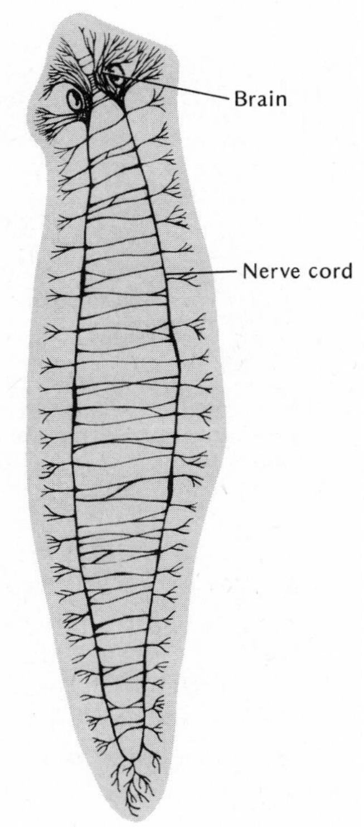

Figure 24.8 Nervous system of a planarian.

LEARNING

Turbellarians have some capacity for retention of reactions which are brought about by previous experience. They can be trained to return to the dark at a light–dark boundary if they receive an electric shock when they fail to do so. Retention time is relatively short.

Some experiments seem to indicate that feeding tissue of trained planarians to an untrained one transfers some of the training to the latter. If true, it would seem that information storage, at least in planarians, is at the molecular level, independent of an intact nervous system.

REPRODUCTION AND DEVELOPMENT

Asexual Some species of planarians may go for long periods without reproductive organs, during which time reproduction is asexual by fission. In this process the worm constricts at the region behind the pharynx. Then follows a "tug of war" between the anterior and posterior halves, and eventually the two halves break apart. Following this the head piece regenerates a new tail and the tail piece regenerates a new head.

Regeneration Planarians, like *Hydra,* have great powers of regeneration and because of their larger size, they have been studied more in this connection. If a planarian worm is cut in two, each part will regenerate into a whole worm, that is, a new tail will be formed on the head piece and a new head will be formed on the tail piece. Also, if a worm is cut into four pieces, as shown in Figure 24.9, three whole worms will be regenerated. The capacity for this regeneration is related to certain cells of the mesenchyme, called *neoblasts,* that retain their embryonic nature and can proliferate and differentiate into the different tissues found in planarians.

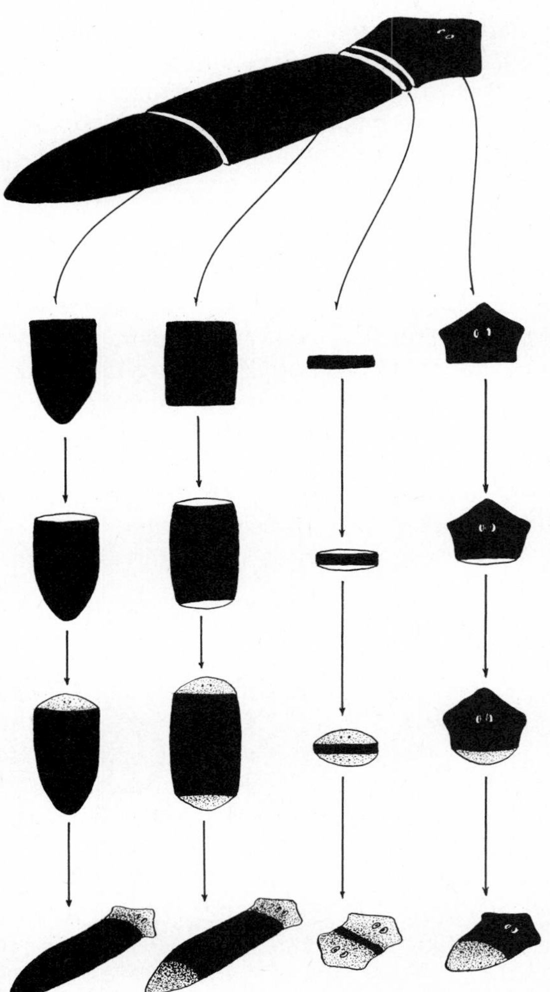

Figure 24.9 Regeneration in planarians.

But, you may ask, why does the large middle piece in Figure 24.9 develop a new head on the cut surface next to the original head end, and a tail on the cut surface next to the original tail end? In such experiments this virtually always happens.

There is an antero–posterior metabolic gradient in planarians with the rate of metabolism highest in the head region and decreasing toward the posterior end. If several worms are cut into three pieces, the head, middle, and tail pieces are put together in three groups, and the rate of respiration of the three groups is determined, the head pieces consume the most oxygen, the tail pieces the least, and the consumption of the middle pieces is intermediate. Further evidence of the presence of a gradient is gained by placing planarians in solutions of certain poisons. In this situation cytolysis and death do not occur all over the body at once, but rather start at the head region and progress backward.

It is possible to cut pieces of planarians so narrow that there is little or no difference in metabolic rate between the two cut surfaces. When such a narrow piece is cut from the head region, two heads may regenerate on the cut surfaces (Fig. 24.9). Similarly, if a narrow piece is isolated from the tail region, tails may regenerate from the two cut surfaces.

It was long thought that these results indicated some sort of control over the structures which regenerate by the difference in metabolic rate at the two ends of the cut piece. Recent studies indicate that the dominant anterior regions produce both inductive and inhibitory substances. The brain is particularly important in this regard, inducing head and eye formation and inhibiting duplication of these parts.

Many other interesting experiments may be made with these animals. It is possible to cut the head region down the middle, keep the cut surface apart, and have two heads regenerate.

The striking degree of regeneration that may be seen in hydra or planarians gradually disappears higher in the phylogenetic scale. A starfish may regenerate its arms and an earthworm can regenerate a new head, but when we come to human beings and other warm-blooded vertebrates, regeneration is restricted to the healing of wounds and broken bones. This loss of regenerative capacity is associated with greater specialization, the details of which have not been completely worked out by students in the field.

Sexual By far the most complex organ system found in planaria is the reproductive system. As mentioned above, during periods of starvation, the reproductive organs may be completely resorbed. However, following such a loss, when in a well-fed condition, the worm can regenerate new reproductive structures.

Most planarians are hermaphroditic, each individual producing both eggs and sperms, but self-fertilization rarely if ever occurs. The parts of this complicated system are shown in Figure 24.10. There are two *ovaries* located near the head region, and these are connected with the genital chamber by two *oviducts.* Numerous *yolk glands* connect with each oviduct. Several *testes* are found on both sides of the body, and each one connects by a short duct to one of the *sperm ducts.* Each sperm duct leads into a *seminal vesicle,* and the two seminal vesicles terminate in a pear-shaped *penis.* The penis projects into the *genital chamber,* which communicates with the outside by the genital pore. Also connecting with the genital chamber is a *seminal receptacle.*

During copulation the penis of one worm is inserted into the genital pore of the other, and sperm are deposited in the seminal receptacle of the other worm. After this the worms separate. The sperm deposited in the seminal receptacle become activated and move up the oviducts where they fertilize the eggs released from the ovaries. As the fertilized eggs pass down the oviducts, they are mingled with yolk cells from the yolk glands. This is a rather unusual arrangement, for in most animals the yolk is stored within the egg.

When the yolk cells and fertilized eggs reach the genital chamber, several eggs and many yolk cells are enclosed in a capsule which forms from droplets present in the yolk cells. As the capsule passes out through the genital pore, an adhesive secretion is added. This secretion is drawn out into a short stalk by which the capsule is attached to a suitable object, usually the underside of a stone. Development occurs within the capsule, and in two or three weeks minute planarian worms emerge.

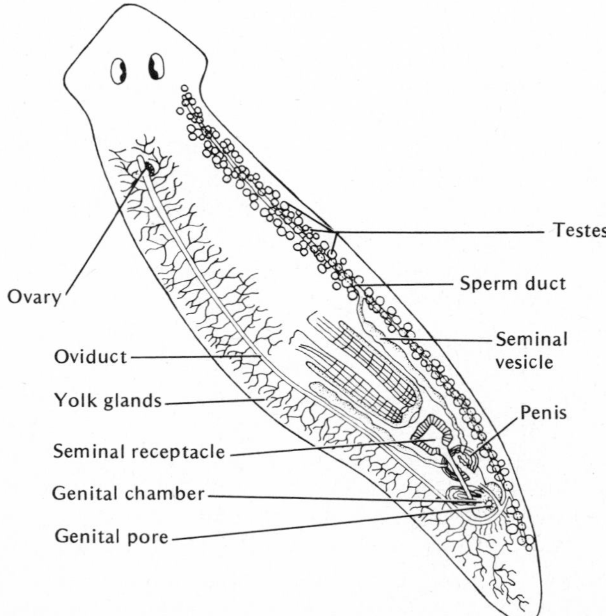

Figure 24.10 Reproductive system of a planarian. For clarity the testes and duct are shown on the right, ovary and duct on the left.

Sexual Reproduction in Marine Polyclads As indicated above, planarian development occurs in a cocoon. The development is complicated by the fact that the yolk is in yolk cells which surround the egg. There are no larval stages and the worms hatch from the cocoon as miniature planarians.

As an example of turbellarian development, it is better to examine that of a marine polyclad since this is the type of development considered to be the original method for the class. Figure 24.11 shows the development of a typical marine polyclad turbellarian. The first two cleavages, at right angles to each other, result in four equal-sized cells which are designated as A, B, C, and D. The third cleavage is unequal and at right angles to the first two. As a result, four small cells, called *micromeres,* are produced at the animal pole. The micromeres are designated la, lb, lc, and ld. The larger cells, now called *macromeres,* are 1A, 1B, 1C, and 1D.

The cleavage plane which produces the first micromeres is at a 45-degree angle to the right so that the micromeres lie within the furrows

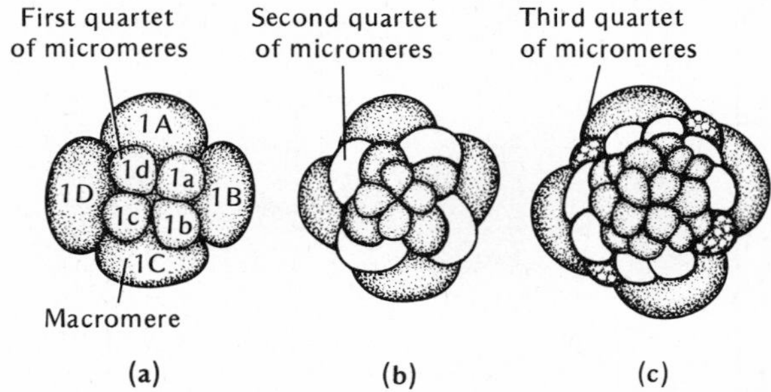

Figure 24.11 Spiral cleavage in polyclad turbellarians. **(a)** Third cleavage resulting in four macromeres and four micromeres. **(b)** Second quartet of micromeres has formed and the first quartet has divided also. **(c)** Third quartet of micromeres has formed and all of the previously formed micromeres have also divided.

between the macromeres rather than directly above them. The next division is also unequal and the cleavage plane is 45 degrees to the left so that the second quartet of micromeres is not above the first. Third and fourth quartets of micromeres are produced in the same way with cleavage planes alternating from right to left, resulting in cells which alternate in position with one another. The pattern produced is the basis of the term *spiral cleavage.* As the succeeding sets of micromeres are being formed from the original macromeres, the earlier formed micromeres continue to divide. An exact method of identifying these cells has been worked out in cell lineage studies, and the student is referred to an embryological text for more detailed information.

By the time the fourth quartet of micromeres has been formed, the blastula stage has been reached. At this time it is possible to demonstrate by experimentation that one cell of the fourth quartet—the 4d cell, often called the mesentoblast cell—is the source of all the endoderm and all the endomesoderm, which constitutes the bulk of the mesoderm of the adult worm.

The fate of the other blastomeres is also determined at this stage. The first quartet of micromeres forms the future nervous system, eyes, and dorsal ectoderm. The second and third quartets form the rest of the ectoderm and the ectomesoderm which forms the muscles of the pharynx. All the macromeres as well as the other three cells of the fourth quartet are used as food by the developing embryo. This very early establishment of the fate of specific cells is the reason for the term *determinate cleavage* used to describe this type of development. Spiral determinate cleavage is typical of many of the phyla in the division Protostomia, including the large phyla Mollusca and Annelida. The cleavage pattern and developmental history of the early blastomeres is essentially that described above, except for the fate of the macromeres and the cells of the fourth quartet other than the 4d cell. In most cases these cells are the source of most of the endoderm.

In many polyclads there is a free-swimming larval form, called a *Müller's larva,* which is considered significant as a possible foreshadowing of the trochophore larva of several higher phyla (Fig. 24.12).

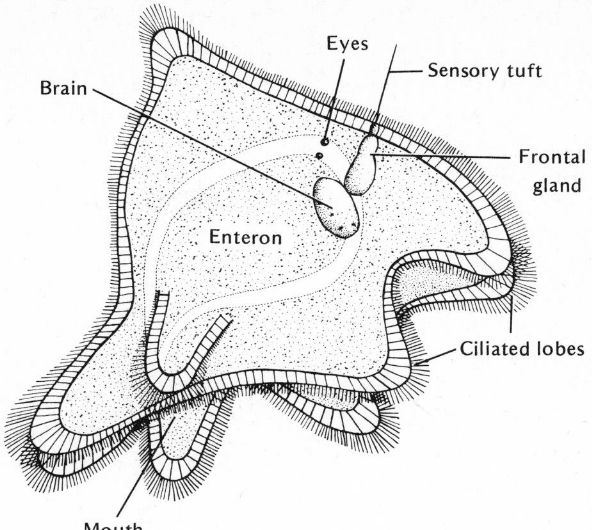

Figure 24.12 Müller's larva of a polyclad turbellarian.

Ecology

Most turbellarians live in the sea. It seems clear that they originated in a marine environment. Later some moved to freshwater and even moist terrestrial habitats.

Freshwater species are most often found in quiet waters of lakes and ponds, but there are some species which live only in the flowing waters of streams. A few occur in cold alpine streams. Although most species are *stenothermal,* living in a narrow temperature range, a few are *eurythermal.* The latter term is used for organisms which are able to live over a wide range of temperatures. *Microstomum lineare,* for example, ranges from cold alpine lakes to hot springs with temperatures as high as 47° C.

One family, *Kenkiidae,* occurs only in north American caves. In this eternally dark habitat they, like many other cave-dwelling animals, lack eyes and pigment.

Land planarians are for the most part limited to tropical and subtropical rain forests. One species, *Bipalium kewense* is sometimes found in greenhouses where conditions are similar (Fig. 24.13).

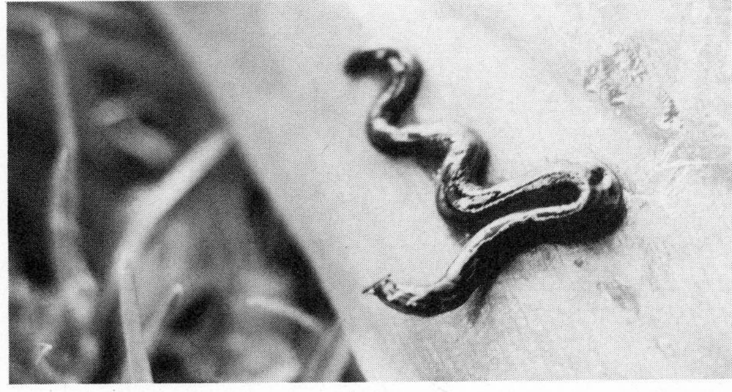

Figure 24.13 *Bipalium kewense,* a land planarian first described from Kew Botanical Gardens in England. Its actual native habitat is not known. This specimen was found in Hawaii National Park, Hawaii. (Photo—E. C. Williams)

Marine turbellarians are mainly bottom dwelling, living in shallow coastal waters. No species have been collected at great depths in the sea. Larger species live on hard rocky substrata, while small and microscopic ones live on sand and mud bottoms. The latter are modified for existence in this difficult environment with long slender bodies, many sensory hairs, loss of eyes, and specialized food-getting mechanisms.

There are some pelagic species, that is, they live in the open sea. They are broadly oval in shape and often quite transparent. Species found in the floating sargassum have brown and white patterns which match the background in which they live.

Many turbellarians harbor commensal algae in the muscle layers just below the epidermis. One marine acoel, *Convoluta roscoffensis,* is said, when mature, to live entirely by the digestion of its zoochlorellae.

Most turbellarians are carnivorous, feeding on entire organisms of appropriate size or on recently dead larger forms. Some species of turbellarians live in or on other animals, but the relationship seems to be commensal and not parasitic. However, this may be indicative of the manner in which the parasitic classes of the phylum began their evolutionary path to parasitism. Members of the genus *Bdelloura* are found on the horseshoe crab, *Limulus.* A posterior sucking disk has evolved in *Bdelloura,* enabling it to maintain its position on the smooth chitinous carapace of its host.

Turbellarians have few predators other than members of their own kind. The mucous secretions found in all members of the class are objectionable, if not actually toxic, to potential predators.

CLASS TREMATODA

The flukes belong to the class Trematoda and they are all parasites. Some of them are ectoparasites, living on the surface of their hosts. Such forms usually have well-developed holdfast structures such as suckers and hooks, and their life history is rather simple. However, most of the flukes are endoparasites, living within the host, and they require more than one host to complete their life cycles. For parasites that require more than one host for the completion of the life cycle, we refer to the host which harbors the sexually mature or adult stage as the *final* or *definitive host,* whereas the host or hosts that harbor the immature stages are called the *intermediate hosts.* The final host of most flukes is a vertebrate, whereas the intermediate host, or at least one of them, is a mollusk. There are many species of flukes that infest humans and domestic and wild animals. Some of them are blood flukes, some are lung flukes, some are liver flukes, and some are intestinal flukes.

Classification

ORDER **Monogenea** Oral sucker weak or absent, anterior end usually with two adhesive structures. Posterior end with sucker and usually with hooks. Mostly ectoparasites. A few are found in structures opening to the exterior, such as nasal and pharyngeal cavities or the urinary bladder. Hosts are usually lower vertebrates, but a few occur on crustaceans or cephalopod mollusks. Life cycle with only one host. *Gyrodactlylus, Polystoma.*

ORDER **Aspidobothria** Oral sucker and anterior adhesive structures absent. Large ventral sucker or row of suckers on ventral side, hooks absent. Simple life cycle with only a single host. Endoparasites of freshwater and marine mollusks, fish, and reptiles. *Aspidogaster, Cotylapsis.*

ORDER **Digenea** Usually with an anterior sucker which surrounds the mouth, a ventral sucker, hooks absent. Endoparasites with complex life cycles involving two or more hosts. *Fasciola, Opisthorchis, Schistosoma.*

Anatomy and Physiology of *Opisthorchis sinensis,* the Chinese Liver Fluke

Perhaps the best form to use for illustrating flukes in general is the Chinese liver fluke, *Opisthorchis* (= *Clonorchis*) *sinensis.* It has one of the most complex life histories, and for laboratory study the organs in the adult of this species are easier to identify than are those in many other fluke species.

GENERAL STRUCTURE

The adult of this species is about 12 mm long and the body is quite flat. It lives in the bile passages of the liver where it attaches itself by suckers and feeds on bile. Figure 24.14 shows the anatomy of *Opisthorchis.* Until recently it was believed that the trematodes and also the cestodes were covered by a nonliving cuticle secreted by the underlying cells. Electron microscope studies have revealed that the protective covering is actually living material (Fig. 24.15). Mitochondria and endoplasmic reticulum are present along the lower surface and cytoplasmic strands connect the layer with nucleated cell bodies which lie beneath the muscle layers. Since the term cuticle is normally used in in-

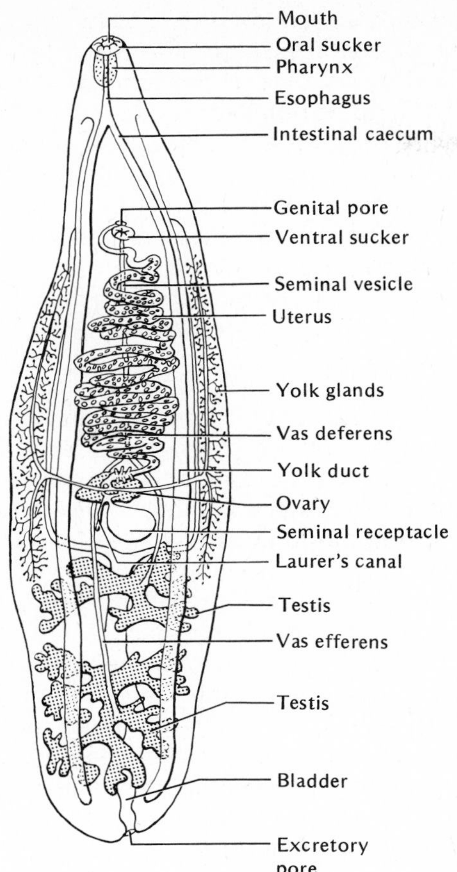

Figure 24.14 Anatomy of *Opisthorchis sinensis,* ventral view.

vertebrates for a nonliving secreted covering, the term *tegument* seems more appropriate for this living protective covering.

The tegument is an adaptation for parasitism for it protects the worm from enzymatic action. With the development of this tegument, the worm has lost the cilia and sense organs found on the surface of free-living forms. The *mouth* is located at the anterior end and is surrounded by the *oral sucker.* Another sucker, the *ventral sucker*, is located some distance posterior to the mouth. The mouth opens into a muscular pharynx which is followed by a short *esophagus.* To the back of this the alimentary tract branches into two long *intestinal caeca* that extend posteriorly almost the whole length of the worm. The excretory system consists of a pouchlike *bladder* that empties at the posterior end and two protonephridial ducts that empty into the bladder. These protonephridial ducts connect with flame cells in the anterior part of the worm lateral to the intestinal caeca. The nervous system of *Opisthorchis* is essentially like that found in planarians.

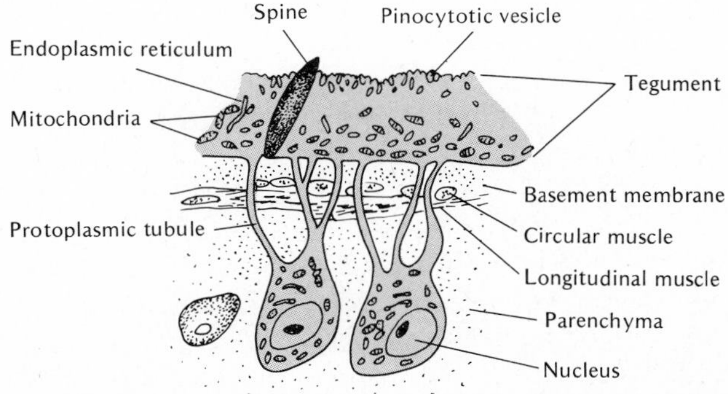

Figure 24.15 Tegument and underlying tissues of the body wall of a trematode.

By far the most complicated system in these worms is the reproductive system. This fluke is also hermaphroditic, like planarians. The two branched *testes* are conspicuous in the posterior third of the body, with one lying in front of the other. The two *vasa efferentia,* one leading from each testis, join anterior to the ovary to form the *vas deferens* that enlarges into the *seminal vesicle* before it terminates at the *genital pore* just anterior to the ventral sucker.

The trilobed *ovary* lies in the midplane, in front of the anterior testis and between the rounded *seminal receptacle* and the much-coiled *uterus.* A short *oviduct* connects the seminal receptacle and ovary and continues forward as the uterus. There are numerous *yolk glands* in the lateral regions of the body, and the two *yolk ducts* connect with the oviduct at the level of the ovary. A small delicate duct, *Laurer's canal,* connects with the oviduct just anterior to the seminal receptacle.

REPRODUCTION, DEVELOPMENT, AND LIFE CYCLE

As there is no copulatory structure in this species, self-fertilization is common. Sperm that reach the genital pore may then pass back down the uterus. *Laurer's canal,* found in some flukes, is considered by some workers to be a vestigal vagina. In *Opisthorchis* it opens by a pore on the dorsal surface. It seems possible that sperm from another worm might be deposited in the seminal receptacle by way of this canal.

When the worm is mature, egg formation starts. The uterus of an adult worm is always filled with developing eggs, and these ovoidal structures with their light brown shells are quite conspicuous within the uterus. The eggs are given off at the genital pore and then pass with the bile into the host's intestine and leave with the feces. Once begun, the process of egg-laying seems continuous. The average daily output per parasite is over 1000. The probable life expectancy of this worm living in man is between 15 and 30 years. Ova of these Asiatic parasites have been recovered from the feces of Chinese immigrants 25 years after leaving China. Eggs develop as they pass through the uterus and each egg that leaves the liver with the bile contains a fully formed ciliated embryo known as a *miracidium* (Fig. 24.16).

Further development in this species occurs only when the eggs are eaten by certain species of freshwater snails. Hatching occurs in the digestive tract of the snail, and the miracidium penetrates the tissues of the snail where it transforms into a *sporocyst.* By asexual means the sporocyst produces a number of *rediae* within its body. When the rediae are fully developed, the wall of the sporocyst ruptures and they escape into the sinuses of the snail where they grow and produce asexually a second generation of redia. Within each of these second-generation redia six to eight *cercariae* are formed, again asexually, and these ultimately break out of the mother redia into the lymph sinuses of the snail. When mature, the cercariae escape from the body of the snail into the water.

The mature cercaria has an ellipsoidal body with a long tail that it uses in swimming. Essentially an immature adult worm, the body has both an oral and a ventral sucker and a primitive digestive tract. The cercaria leads a free-swimming existence for a period of 24–48 hours, and unless it encounters a fish belonging to one of several species within this time it sinks to the bottom and dies. If it encounters a fish of the right kind, it penetrates under the scales of the fish and into the muscles, loses its tail, and forms a cyst called the *metacercaria.* If some mammal (human being, dog, or cat) eats the fish containing the cyst, the cyst wall will be digested in the stomach and duodenum and the freed larva will migrate through the bile duct into the liver where it matures into an adult fluke.

CHARACTERISTICS OF PARASITES

The life cycle of *Opisthorchis,* just described, illustrates some of the common characteristics of parasites in general. Associated with the parasitic habit are changes in morphology in relation to the morphology of related free-living forms. In *Opisthorchis* a protective tegument is present, cilia and sense organs are absent, suckers are present, and the digestive tract is reduced. In a great many endoparasites more than one host is required for the completion of the life cycle. In *Opisthorchis* this involves three hosts. In many instances there is a high degree of specificity in the host relationship. Although the life cycle of *Opisthorchis* does not involve the extreme specificity of hosts found in some parasites, there are only certain species of snails

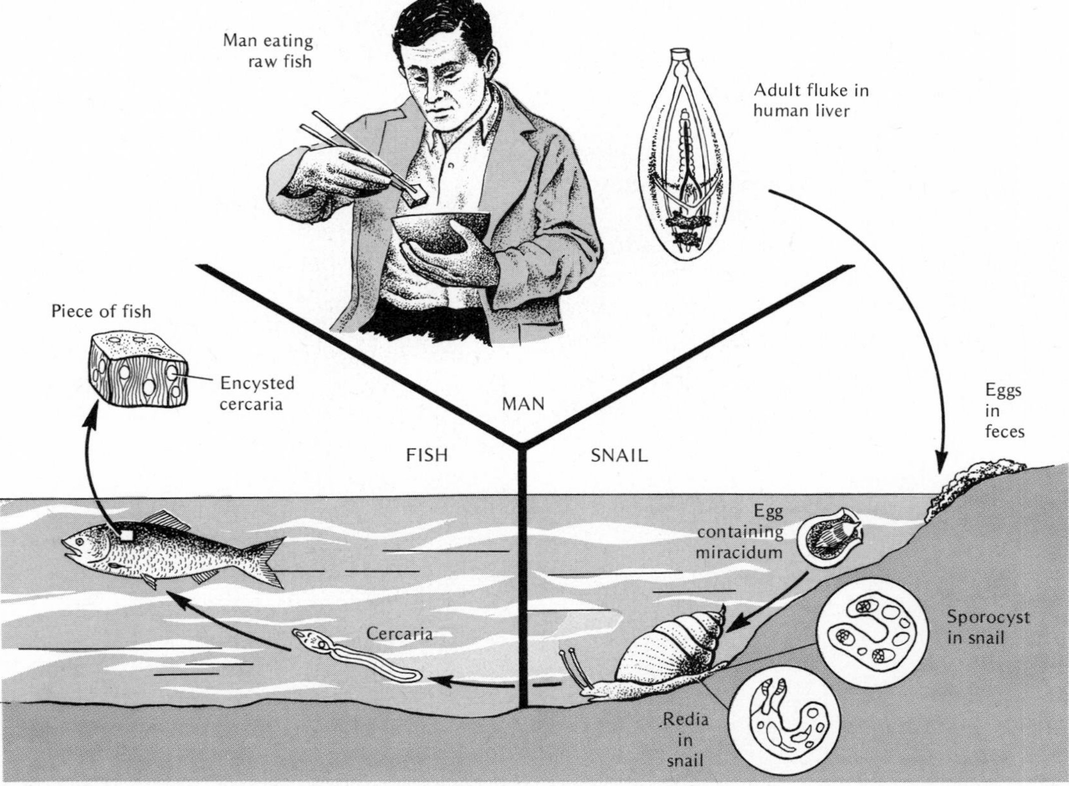

Figure 24.16 Life cycle of *Opisthorchis sinensis,* the oriental liver fluke.

and fish that will serve as hosts, and for the definitive host a mammal that eats raw fish is required.

However, the greatest modification associated with parasites generally is the great increase in reproductive potential over free-living forms. This is very well illustrated in *Opisthorchis.* Not only is the egg-laying capacity of the adult tremendous, but also the ability of the larval stages in the snail to reproduce asexually greatly increases the reproductive potential. That this adaptation is advantageous seems obvious when one considers the improbability of any one stage in the life cycle ever going on to completion.

Human infection by *Opisthorchis* is high in southern China, Japan, and Korea. This parasite, in heavy infections, will cause impairment of liver functions and, in extreme cases, may lead to death. Treatment has not been particularly effective, although some success has resulted from the administration of chloroquine. Control of this parasite is more likely to come from preventive measures. Its high incidence is associated with two practices that should be eliminated: the eating of raw freshwater fish and the use of human feces as fertilizer. The thorough cooking of all fish before eating and the mixing of ammonium sulfate with human feces to destroy the eggs would go a long way in cutting down the incidence of this infection.

Other Trematodes

The sheep liver fluke, *Fasciola hepatica,* infects many other vertebrates, including humans, and its distribution is worldwide. The intermediate host is a snail and infection of the definitive host occurs by ingestion of plants on which the cercariae encyst, growing in or near the water.

Fasciolopsis buski is a large intestinal fluke infecting humans and the pig in the Far East. Man is infected when he eats various water plants such as the water chestnut on which the encysted metacercaria stage occurs.

The lung fluke, *Paragonimus westermani,* lives as an adult in a cystlike capsule in lung tissue. Eggs are coughed up with the sputum and may either leave by expectoration or, after passing unharmed through the digestive system, in the feces. After as long two weeks in the water, the eggs hatch and the miricidia seek out the proper snail host. Cercariae leaving the snail enter a crab or crayfish and encyst. Humans are infected by eating raw crab or crayfish meat. The metacercariae excyst in the intestine, move through the gut wall, and migrate through the body until they reach the lungs where they become sexually mature within a fibrous capsule produced by the host tissues.

The most serious flukes affecting human beings are the blood flukes belonging to the genus *Schistosoma.* These are an exception to the usual condition in the phylum since there are separate sexes. It is also unusual in that no cyst is formed. The cercaria invades human skin directly. The adults inhabit blood vessels. The female lays eggs in capillaries of either the intestinal wall or the wall of the urinary bladder, depending upon the species. The eggs are provided with sharp spines, and they gradually pass through the capillary wall and the intervening tissues and fall into the gut or bladder from whence they pass to the outside with the wastes. The secondary host is a snail and humans are infected by direct entry of the cercariae through the skin while bathing in or drinking water which contains cercariae. The great damage caused from infection is caused by eggs that fail to make their way into the bladder or intestinal lumen. The resulting reaction of the host tissues produces the conditions of the disease recognized as *schistosomiasis.*

Schistosoma haematobium is endemic in the greater part of Africa, where it is commonly called bilharzia. It also occurs in some southern Mediterranean countries. *S. japonicum* is confined to the far east, while *S. mansoni* is found in parts of Africa, Brazil, Surinam, and a number of Caribbean islands. Introduction to the Western Hemisphere seems to be related to the slave trade in the eighteenth and nineteenth centuries.

CLASS CESTODA

Tapeworms belong in the class Cestoda. They are among the most highly modified of all parasites. They are covered by a tegument like that of the flukes, but the gastrovascular cavity is completely absent. Thus they depend on the host not only for food but also for its digestion. The adult stage of all tapeworms lives in the intestine of the host and it is provided with holdfast organs, hooks or suckers or both, to keep it in place.

Classification

There are nine orders in the class, six of which are parasitic in fishes or other lower vertebrates and are of no medical importance. One order is found only in swans. The two orders which contain species that infect man are as follows.

ORDER **Pseudophyllidea** Body either segmented or not; usually with a scolex provided with two slit-like structures, called bothria, which are used for attachment. Many yolk glands present. Two intermediate hosts. *Dibothriocephalus, Eubothrium.*

ORDER **Cyclophyllidea** Segmented tapeworms with four suckers on the scolex and usually with hooks at its apex. Yolk gland single. One intermediate host. *Taenia, Echinococcus, Hymenolepis.*

Anatomy and Physiology of *Taenia saginata,* the Beef Tapeworm

GENERAL STRUCTURE

A typical cestode is the beef tapeworm of humans, *Taenia saginata.* An individual consists of a head or *scolex,* a short *neck,* and numerous *proglottids.* The scolex is provided with suckers by which the worm attaches to the wall of the intestine. In many species, such as the pork tapeworm of humans, *Taenia solium,* there are hooks on the scolex in addition to the suckers. The proglottids are budded off from the neck region; thus the youngest proglottids are nearer the neck. The adult worm may vary from 4 to over 9 m (12–30 feet) in length, and the number of proglottids may be as many as 2000.

From Figure 24.17 it can readily be seen that the mature proglottid is composed primarily of reproductive structures. There are two *nerve cords* that extend the length of the body. Also,

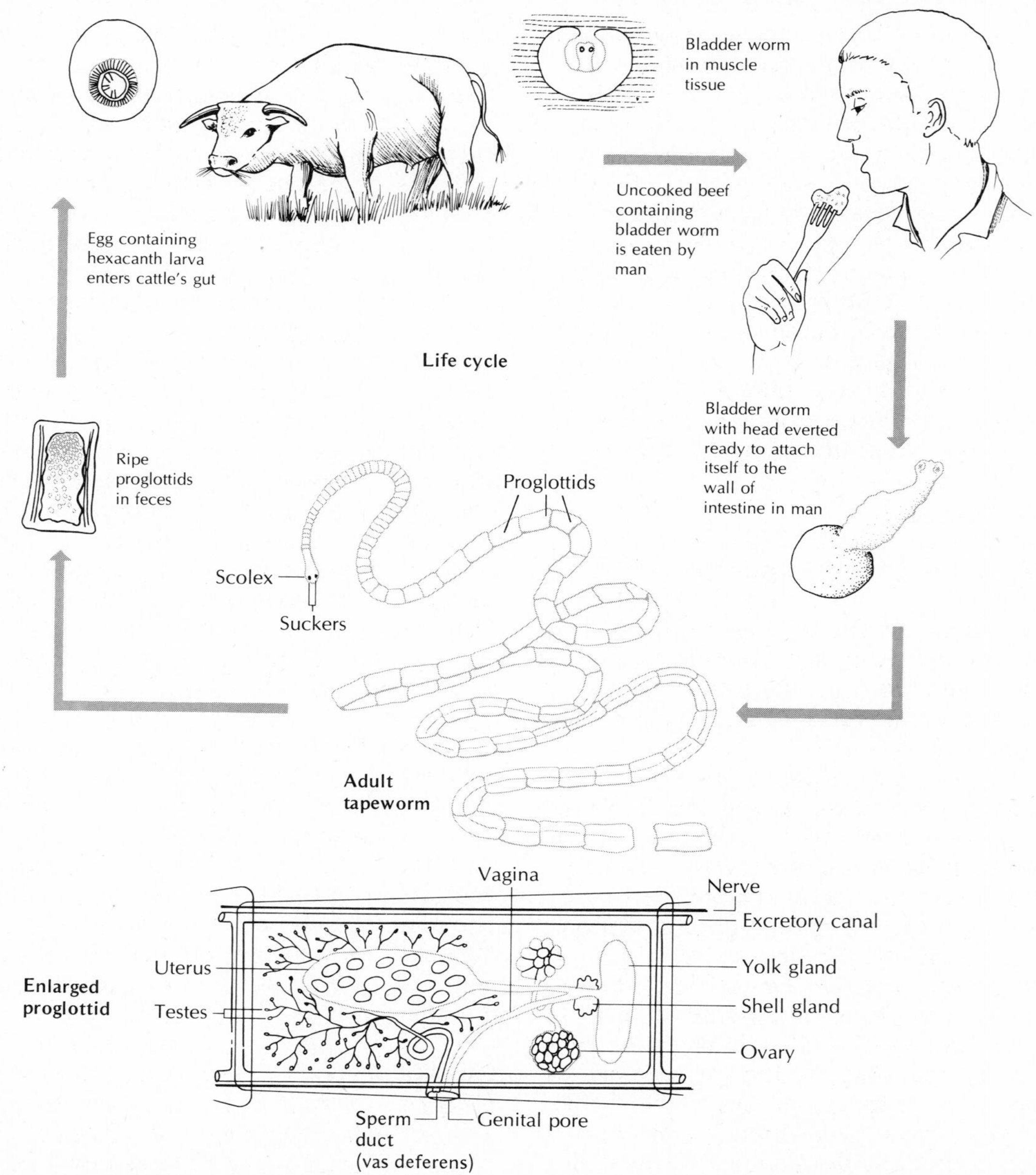

Figure 24.17 Life cycle of the beef tapeworm, *Taenia saginata.* Detailed anatomy of a mature proglottid is shown in the lower drawing.

there are two *excretory canals* that parallel the nerve cords and are connected by a transverse duct in the posterior part of each segment.

The tapeworms are hermaphroditic and each proglottid contains a set of male and female reproductive organs. As shown in Figure 24.17, the male system consists of numerous *testes* that are connected by small ducts to a large *sperm duct* that, in turn, leads to the *genital pore.* The female system consists of a pair of *ovaries,* a short *oviduct,* a *yolk gland,* a *shell gland,* a *vagina,* and a *uterus.*

REPRODUCTION AND LIFE CYCLE

Self-fertilization within a proglottid may occur. However, cross-fertilization, either between different proglottids of the same worm or between proglottids of different worms, usually occurs. When eggs are released from the ovary, they are fertilized by sperm that have been stored in the enlarged part of the vagina. The fertilized eggs receive yolk from the yolk gland and are then covered with a shell secreted by the shell gland. The eggs then pass into the uterus for storage. As the embryos develop, the uterus becomes distended, while the testes and the other parts of the female system degenerate. Thus a "ripe" proglottid consists essentially of a greatly distended uterus filled with developing eggs. These ripe proglottids break off from the worm and pass out of the body of the host with feces. It has been estimated that a ripe proglottid contains about 124,000 eggs and that the annual output of one worm is 594 million eggs; the worm may continue this production for well over 20 years.

For the life cycle to be completed, a cow must eat food contaminated with human fecal material containing the eggs. Under favorable conditions the eggs may remain viable on exposed pasture land for eight weeks or more. In the gut of the cow the shell is digested away and the *hexacanth* (*six-hooked*) embryo emerges. This embryo bores its way through the gut wall into the bloodstream where it is carried to muscle tissue. It becomes embedded in the muscle and develops into the *cysticercus* or *bladderworm* stage shown in Figure 24.17. If uncooked beef containing bladderworms is eaten by humans, the bladderworms hatch. In this process the tiny scolices evert and become attached to the intestinal wall. Once attached, the process of budding off proglottids begins.

Improved sanitary conditions and the inspection of meat have greatly reduced the incidence of this parasite in this country. Thorough cooking of all beef eaten will prevent infection. In cases where an infection occurs, the worms may be dislodged by the use of certain drugs. In *T. saginata* it is often extremely difficult or virtually impossible to eliminate the worm.

Other Tapeworms

The pork tapeworm, *Taenia solium*, of humans has a similar cycle except that the pig is the intermediate host.

Human beings also serve as a rare intermediate host for *T. solium* when eggs do not pass out of the gut, but hatch, penetrate the gut wall, and circulate to tissues where they produce cysticerci. Some results of such infections include blindness and death in cases of brain involvement, serious muscle damage, and pain when localized there.

Echinococcus granulosus is a minute tapeworm, 3 to 6 mm long, which has the dog as its normal definitive host. This small tapeworm consists of a scolex, neck, and two or three proglottids and may number in the thousands in the intestine of a dog. The normal secondary host is an herbivorous animal such as a cow, sheep, or pig. Humans may become the unwitting secondary host with rather disastrous results. Instead of forming single bladderworms as in the case of *Taenia,* this species forms large cysts, called *hydatid cysts,* containing multiple scolices and often reaching considerable size. If the site of the cyst is in the brain, it is comparable to a serious brain tumor and usually results in death of the patient. Wherever these large cysts occur, liver, lung, or other organs, they cause very serious symptoms. Infection may occur from contaminated food or water or from the friendly lick of a household pet. It is normally not a serious health problem except in those areas where sheep and cattle are raised in large numbers.

Many tapeworms have more than one intermediate host. The broad fish tapeworm, *Diphyllobothrium* (=*Dibothriocephalus*) *latum,* with three hosts in its life cycle is an example. First the eggs must be eaten by a copepod, a small crustacean in which a *procercoid* stage develops; then a fish must eat the copepod and the infective *plerocercoid* stage develops; a human being finally gets the parasite by eating the uncooked fish. This fish tapeworm has been known for centuries in the Baltic regions of Europe, but it has now spread to this country, apparently by immigrants to the Great Lakes region. Fish from many of the lakes of Northern Minnesota, Northern Michigan, and Canada have been found to be infected. So to be safe, all fish from these regions should be thoroughly cooked before they are eaten. This fish tapeworm is the largest of all the human tapeworms; records are known of its reaching a length of 18 m (60 feet) and a width of 2 cm (3/4 inch).

PHYLUM RHYNCHOCOELA

The second phylum in the subgrade Acoelomata, sometimes called the phylum Nemertina, is made up of about 750 species of worms, mostly marine but with a few freshwater and terrestrial species. The major advance over the flatworms is the appearance of a second opening in the digestive tract, the anus. The resultant pattern, sometimes called the "tube-within-a-tube body plan," is characteristic of all the higher phyla. The first appearance of a circulatory system is also a noteworthy phylogenetic advance.

Characteristics

1. Bilaterally symmetrical, unsegmented vermiform body.
2. Triploblastic.
3. No coelom.
4. Digestive tract complete, with mouth, gut, and anus.
5. Eversible proboscis in a sheath dorsal to the digestive tract and usually with a separate opening.
6. Closed circulatory system present.
7. Sexes normally separate.
   Figure 24.18.

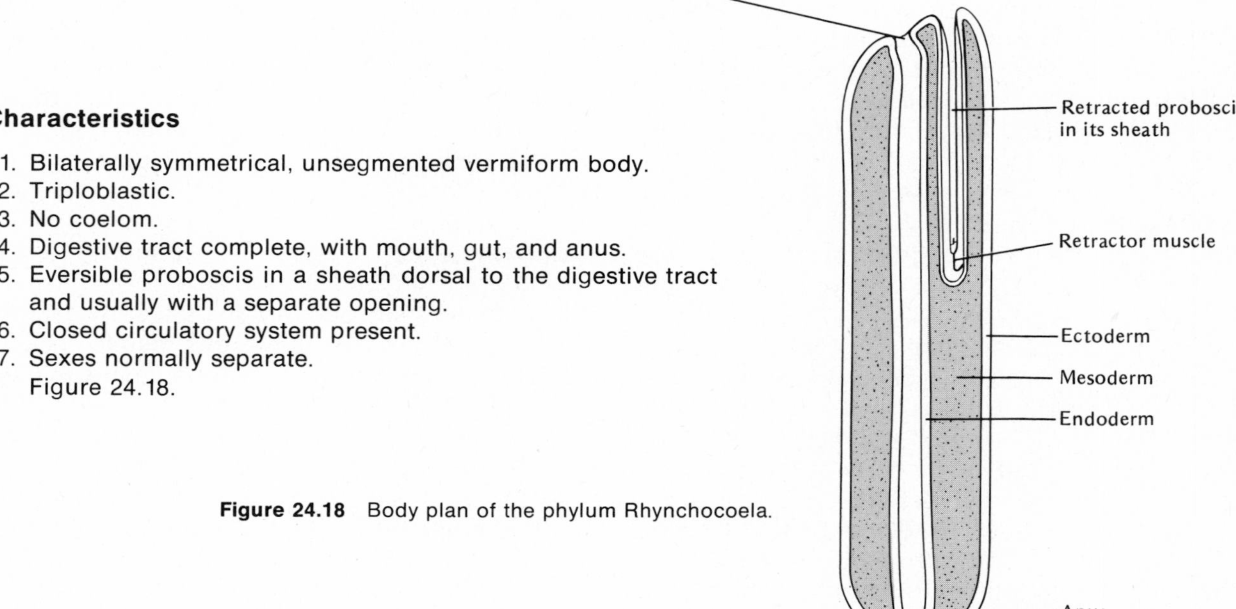

Figure 24.18 Body plan of the phylum Rhynchocoela.

Classification

CLASS **Anopla** Mouth posterior to the brain. Proboscis without stylets. Central nervous system external to or located within the muscle layers.

ORDER **Palaeonemertini** Body musculature in two or three layers. Dermal layer poorly developed, gelatinous. *Tubulanus, Carinoma.*

ORDER **Heteronemertini** Body musculature in three layers, innermost layer is longitudinal. Dermal layer fibrous. *Cerebratulus, Lineus* (Fig. 24.19).

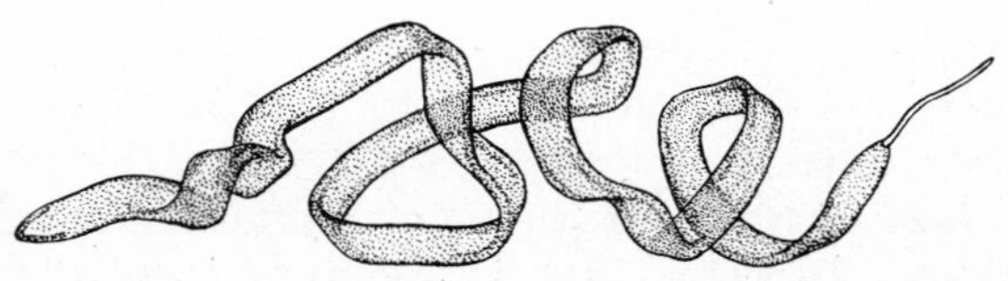

Figure 24.19 *Cerebratulus,* a representative nemertean.

CLASS **Enopla** Mouth anterior to the brain. Proboscis may be armed with stylets. Central nervous system in the parenchyma, internal to the muscle layers.

ORDER **Hoplonemertini** Proboscis armed with one or more stylets. Intestine straight with lateral diverticulae. *Tetrastemma, Prostoma.*

ORDER **Bdellonemertini** Proboscis without stylets. Intestine sinuous and without diverticulae. Commensals living in the mantle cavity or pericardial sac of mollusks. *Malacobdella,* the only genus.

Anatomy and Physiology

GENERAL STRUCTURE

The size range of nemertean worms is enormous, ranging from a few millimeters to as much as 25 m. Many species are brightly colored, often

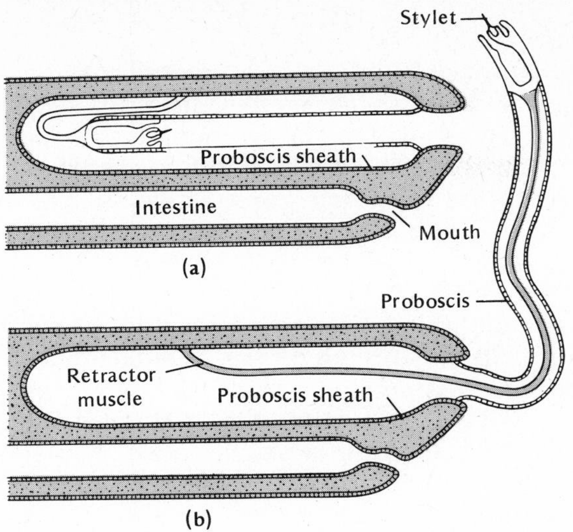

Figure 24.20 Nemertean proboscis. **(a)** Retracted. **(b)** Extended.

with striking striped patterns. The proboscis is probably the most distinctive feature of the phylum. It functions as a tactile organ, as an organ for offense and defense, and since it is prehensile, it is used in grasping food. Figure 24.20 shows the retracted proboscis in its cavity, the *rhynchocoel.* Contraction of the circular muscles of the proboscis sheath causes the proboscis to evert due to hydrostatic pressure on the fluid within the rhynchocoel. Retraction of the proboscis is accomplished by contraction of the retractor muscle. Stylets are present on the tip of the proboscis in many species. Poison glands at the base of the stylet make it a very effective weapon. The surface of the proboscis is covered with a sticky mucuslike secretion which also may contain a poison.

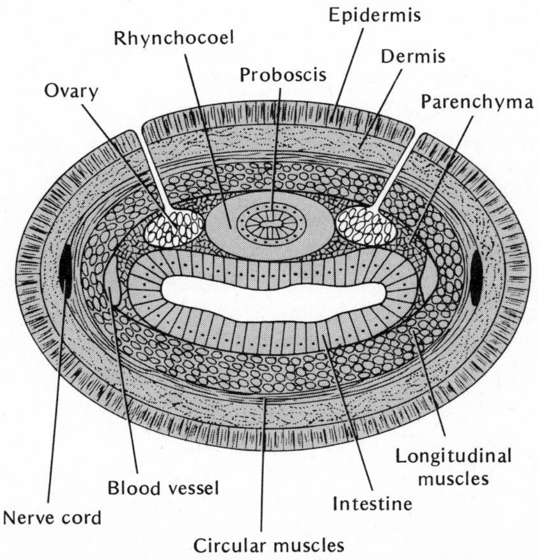

Figure 24.21 Cross section of a nemertean worm.

The body wall is quite similar to that in turbellaria with a well-defined ciliated glandular epidermis, a dermis, and circular and longitudinal muscles. The number and arrangement of the muscle layers varies in the different orders (Fig. 24.21). The space between the muscles and the gut is filled with parenchyma.

As already pointed out, the digestive system is complete (Fig. 24.22). The mouth is subterminal and leads to a simple esophagus which passes to an intestine, which may or may not have lateral diverticulae, and extends posteriorly to the terminal anus.

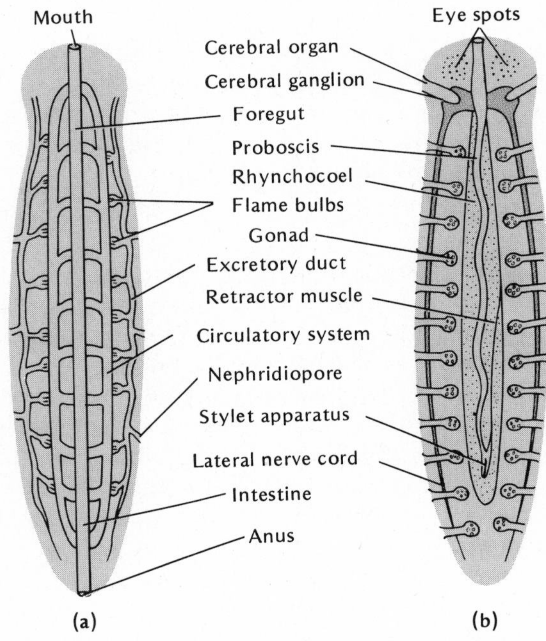

Figure 24.22 Structure of a nemertean worm. **(a)** Digestive, excretory, and circulatory systems. **(b)** Nervous and reproductive systems.

The nervous system consists of two pairs of ganglia at the anterior end, dorsal and ventral to the rhynchocoel and connected by a commissure. Lateral nerves pass posteriorly from the ventral ganglion and a pair of nerves also passes posteriorly along the proboscis sheath. Sense organs consist of eyespots, statocysts, and a pair of ciliated invaginations at the anterior end, called the *cerebral organs.* The latter probably serve as chemoreceptors.

The circulatory system, appearing for the first time in the phylogenetic sequence, is a well-developed closed system. There are three principal contractile vessels with numerous connections between them. The blood is a colorless fluid containing corpuscles which may or may not be pigmented. In some species the pigment is hemoglobin.

Excretion is by means of a protonephridial system in which the flame bulbs are often multicellular or at least multinucleate. Respiration is by simple diffusion through the body wall.

Most nemerteans are dioecious. The gonads are numerous, located laterally along the trunk, and each is provided with its own duct. In forms with lateral diverticulae on the intestine, the gonads lie between the diverticulae.

REPRODUCTION AND DEVELOPMENT

Asexual Nemerteans in general have great powers of regeneration, as do the flatworms. When handled or placed in unfavorable conditions they are apt to break off posterior parts of the body. In such cases the anterior portion will regenerate the lost posterior region.

In the genus *Lineus* asexual reproduction occurs regularly by a process of fragmentation. The posterior region breaks into many small pieces, each of which regenerates into a perfect small worm.

Sexual Fertilization occurs either externally in the sea or internally. In the latter case the male and female worms lie close to one another, sometimes in a secreted mucus tube. Eggs are deposited in a gelatinous mass and there is no parental care. In a few species, development occurs within the female in the ovaries and the juvenile worms leave via the oviduct.

Development is by spiral determinate cleavage as in the polyclad turbellaria. In many forms development is direct, but some species have a characteristic helmet-shaped larval form, called a *pilidium larva* (Fig. 24.23), which is in some ways similar to a trochophore larva.

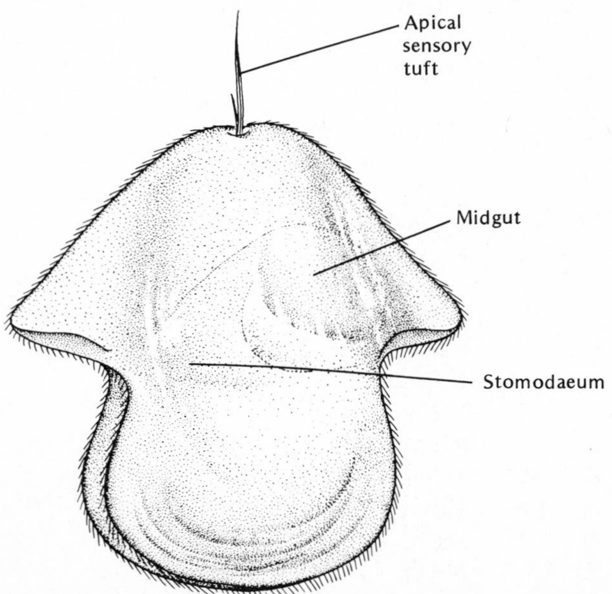

Figure 24.23 Pilidium larva of a nemertean.

Ecology

Nemerteans are mostly marine, living in coastal areas under stones, in mud, sand or gravel, or in algal beds. Although they occur in tropical and arctic regions, they are most abundant in the temperate zone. Some species are pelagic, living at depths from 200 to 3000 m. These forms have broad flat bodies.

A number of species of the genus *Prostoma* are found in freshwater habitats throughout the world. In the United States there is but one species, *Prostoma rubrum,* a small slender worm reaching a length of 20 mm. It is not common and where found it is usually present in small numbers, creeping on rocks and submerged vegetation. The genus *Geonemertes* contains a number of terrestrial species, all of which are tropical or subtropical in their distribution.

The nemerteans are carnivorous, feeding on live or dead animals, usually annelid worms which they capture by shooting out the proboscis which then follows the trail of an unwary worm. The sticky proboscis wraps around the prey, which soon ceases to struggle due to the paralyzing toxins in the mucus, and the toxin from the stylet in the case of those species which possess stylets.

Although there seem to be no truly parasitic nemerteans, a number of species live within other organisms in a commensal relationship. All members of the order Bdellonemertini are in this category. A number of forms in other orders are also commensals, for example, *Carcinonemertes,* order Hoplonemertini, lives on the gills and egg masses of crabs. The worm does not harm the host, but it does receive food in the form of small animals, and oxygen—all brought in by ciliary currents—and excellent shelter as well.

Phylogenetic Relationships of Platyhelminthes and Rhynchocoela

The origins of the bilaterally symmetrical phyla are the subject of much speculation, and a number of lines of descent have been postulated. The acoel flatworms are generally conceded to be the most primitive bilaterally symmetrical animals. It is possible to derive the other free-living turbellarians from an acoel ancestor. The origins of the parasitic trematodes and cestodes are obscured because of the great changes associated with parasitism. Hyman believes that both the trematodes and the cestodes evolved from an ancestral rhabdocoel turbellarian. Certain rhabdocoel species are commensals, living in close association with mollusks or echinoderms. A commensal organism benefits by the association but does not harm its host. Since it is generally believed that parasitic species evolved from free-living ancestors, a commensal relationship is a logical step in the direction of a parasitic mode of life.

Several theories concerning the origin of the flatworms have been developed. One which appeals to many is that of Hyman. She contends that the planula larva of present-day coelenterates is much like a hypothetical ancestral form—the so-called planuloid ancestor—which gave rise to the radiate phyla, coelenterata and ctenophora, and also to the acoel flatworms. An acoel flatworm is similar in size and appearance to a planula larva. Figure 24.24 shows how Hyman believes the derivation could have occurred.

In spite of the advances which the rhynchocoela show over the platyhelminthes—the appearance of an anus and a circulatory system—the nemertean worms are quite clearly related to the flatworms. There is little doubt that both groups either had a common ancestral type or that the nemerteans evolved from an ancestral free-living flatworm.

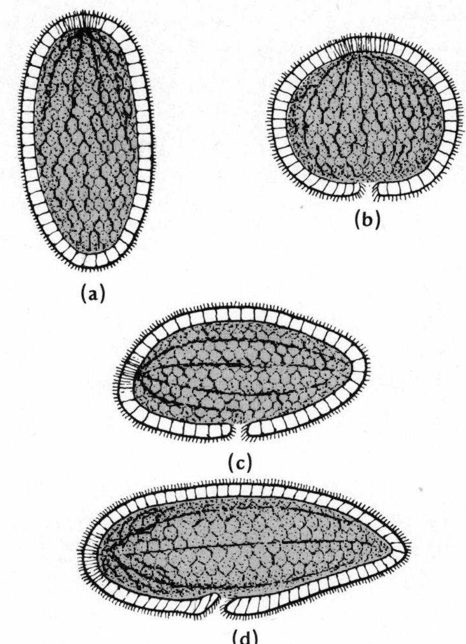

Figure 24.24 Diagrams illustrating the possible derivation of the Bilatera from a planuloid ancestor. **(a)** Planula larva without a mouth. **(b)** Mouth appears and organism shortens and thickens. **(c)** Body elongates and nervous system shifts to what will become the anterior region. **(d)** Bilaterally symmetrical form similar to a present-day acoel flatworm.

PHYLUM GNATHOSTOMULIDA

In 1928 Remane was the first to observe these interesting organisms. It was not until 1956, however, that the first two species were described. Since then, species have been found in many parts of the world and it seems likely that they are worldwide in their distribution. In establishing the phylum in 1972, Wolfgang Sterrer indicates that there are 18 genera and over 80 species described or in press. He estimates that the number of species will exceed 100.

Characteristics

1. Minute, wormlike, acoelomate, marine animals usually less than 1 mm long.
2. Muscular pharynx usually with paired jaws and a basal plate.
3. Anus absent, parenchyma poorly developed.
4. Protonephridial excretory system.
5. Single layer of monociliated epithelial cells.
6. All monoecious.
   Figure 24.25

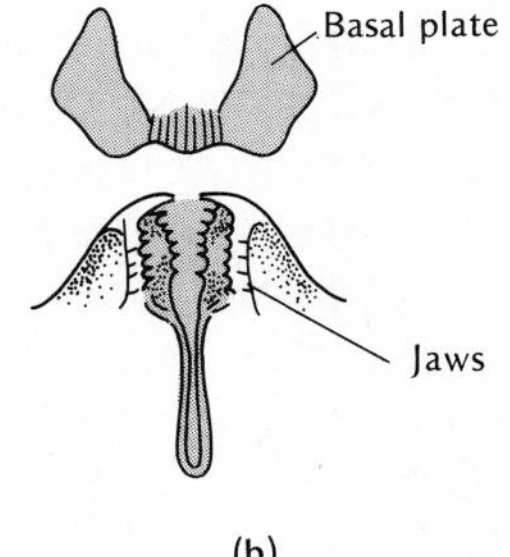

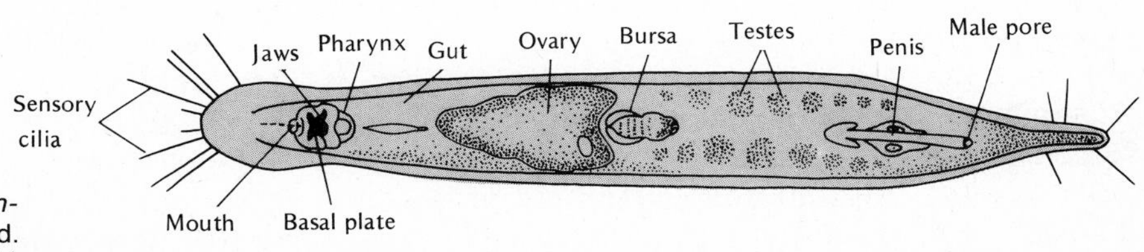

Figure 24.25 *Gnathostomula jenneri*, a representative gnathostomulid. **(a)** Sexually mature individual. **(b)** Detail of jaws and basal plate.

Classification

ORDER **Filospermoidea** Bursa, vagina, and penis absent. Sperm filiform (threadlike) in shape. *Haplognathia, Pterognathia.*

ORDER **Bursovaginoidea** Bursa always present, vagina usually so. Male opening with a penis. Sperm not filiform in shape, lacking flagella but may have a group of short filaments. *Gnathiella, Gnathostomaria, Gnathostomula.*

Anatomy and Physiology

GENERAL STRUCTURE

The gnathostomulids have elongate, often threadlike, cylindrical bodies, usually 0.5–1.0 mm long. The body is made up of three distinct regions: a short head or *rostrum* anterior to the mouth, an elongate body, and a tapered tail. Often a slightly constricted neck or collar region separates the rostrum from the body. The tapered tail varies a good deal in length in different species.

There is no cuticle and the single layer of epithelial cells is unique in that each cell bears a single cilium. By means of the cilia, the animals move by gliding over the substrate or slowly swimming in loops and spirals. Beneath the epithelium lies a thin layer of circular muscle followed by strong longitudinal muscle fibers arranged in three or four paired groups allowing for body contractions and lateral nodding of the rostrum. Parenchyma tissue is poorly developed except in the rostrum and around the gonads. The nervous system is located at the base of the epidermis with a brain at the anterior end. Ciliated sensory pits and single sensory cilia are concentrated in the rostrum.

Excretion seems to be by a protonephridial system with a modified type of flame cell, found in groups of two or three in the parenchyma near the reproductive organs. The digestive system has but one opening, the mouth, located ventrally in the collar region, often far from the anterior end. The mouth is provided with a pair of lateral jaws [Fig. 24.25 (**b**)], a hardened base plate, and sometimes a similar plate on the upper lip. A complex musculature operates the jaws and plates, enabling the animal to scrape up the fungi and bacteria upon which it feeds. The simple midgut is made up of large cells which surround the saclike gut cavity. There is no anus and unused solid materials pass out through the mouth opening. Circulatory and respiratory systems are lacking.

REPRODUCTION AND DEVELOPMENT

The gnathostomulids are all hermaphroditic, although some individuals may have only male or female systems. The male organs consist of paired posterior testes, a copulatory organ which may be poorly developed, and a ventral male pore near the tip of the tail. The female system is made up of the single ovary located dorsally in the midbody region, a sperm storage sac or bursa in some forms, and in some cases a vagina. In copulation, one worm transfers a packet of sperm surrounded by mucus to the body of another worm. Subsequently the sperm migrate to the bursa or region of the ovary where fertilization occurs. Single fertilized eggs are released by rupture of the body wall, which soon regenerates.

Cleavage is spiral and development is direct without a larval stage. Whether the cleavage is determinate is not yet known. One would suspect that it is determinate.

Ecology

The late discovery of this phylum can be attributed to the small size of its members and their specialized habitat. They live in the interstitial spaces of fine sand and silt in shallow marine waters. There must be at least 50 percent fine sand, usually 70–85 percent. Special extraction methods are required for their collection, but they are often extremely abundant, over 6000 specimens per liter of sediment. Usually these sediments smell strongly of hydrogen sulfide and are rich in iron bacteria which indicates anaerobic conditions. Thus these organisms must be able to carry on respiration anaerobically.

Phylogenetic Relationships

The gnathostomulids show a mixture of characteristics which seem to place them in an intermediate position between the acoelomates and the pseudocoelomates. Like the turbellarians they are acoelomate, lack an anus, possess a ciliated epithelium lacking a cuticle, and are hermaphroditic with much similarity in reproductive systems. The excretory system of the gnathostomulids is basically similar to that of the turbellarians. The ventrally located mouth, far from the anterior end in simpler groups, is also a turbellarianlike feature.

The monociliated epithelium is a unique feature, and the poorly developed parenchyma sets them apart from the turbellaria as well. In the platyhelminthes flagellated sperm with two flagella are the rule, whereas flagellated sperm found in gnathostomulids bear only one flagellum.

Relationships with the pseudocoelomates are seen in the jaws which are very rotiferlike. The ciliated pits of the head region are like those of both rotifers and gastrotrichs, and the sensory cilia on the gnathostomulid rostrum are much like those of the gastrotrich head region.

The characteristics of the gnathostomulids set them apart as a separate phylum, but their affinities with turbellaria and some of the aschelminthes seem clear. As more information becomes available, it is likely that the relationships of the gnathostomulids will be clarified.

SUGGESTIONS FOR FURTHER READING

Bils, R. F., and W. E. Martin. "Fine Structure and Development of the Trematode Integument." *Trans. American Mircroscopical Society,* **85** (1966).

Chandler, A. C., and C. P. Read. *Introduction to Parasitology.* New York: Wiley (1961).

Cheng, T. C. *The Biology of Animal Parasites.* Philadelphia: W. B. Saunders (1964).

Grasse, P. O. (Ed.). *Traite de Zoologie,* vol. 4. Paris: Masson (1961).

Hyman, L. H. *The Invertebrates,* vol. 2. New York: McGraw-Hill (1951).

Noble, E. R., and G. A. Noble. *Parasitology.* Philadelphia: Lea and Febiger (1964).

Pennak, R. W. *Fresh-Water Invertebrates of the United States.* New York: Ronald Press (1953).

Riedl, R. "Gnathostomulida from America. First Record of the New Phylum from North America." *Science,* **163** (1969).

Riser, N. W., and M. P. Morse (Eds.). *Biology of the Turbellaria.* New York: McGraw-Hill (1974).

Smyth, J. D. *The Physiology of Trematodes.* Edinburgh: Oliver and Boyd (1966).

Smyth, J. D. *The Physiology of Cestodes.* Edinburgh: Oliver and Boyd (1969).

Sterrer, W. "Systematics and Evolution within the Gnathostomulida." *Systematic Zoology,* **21** (1972).

Wardle, R. A., and J. A. McCleod. *The Zoology of Tapeworms.* Minneapolis: Univ. of Minnesota Press (1952).

Yamaguti, S. *Systema Helminthum.* vols I–IV. New York: Wiley (1958–1963).

The Pseudocoelomates—

ACANTHOCEPHALA, ASCHELMINTHES, AND ENTOPROCTA

The general overall organization of the animal kingdom into the major categories above the phylum level has been discussed. Recall that animals in the subgrade Pseudocoelomata are those in which a remnant of the embryonic blastocoel remains in the adult as a space between the gut and the body wall. In these forms no mesoderm covers the gut. Other characteristics of the group include spiral and determinate cleavage; an adult mouth formed at or near the embryonic blastopore; no skeletal, respiratory, and circulatory systems; excretion usually by means of protonephridia. It is generally conceded that the pseudocoelomate phyla arose from ancestral acoelomate stocks, but that they are not in the main line of evolution leading to the true coelomate forms. They represent a terminal offshoot from the main line of evolution, and the pseudocoelom is not a stage in the evolution of the true coelom.

PHYLUM ACANTHOCEPHALA

Characteristics

1. Parasitic, unsegmented worms.
2. Retractable proboscis with numerous recurved spines at the anterior end.
3. No digestive system at any stage in the life history. Figure 25.1

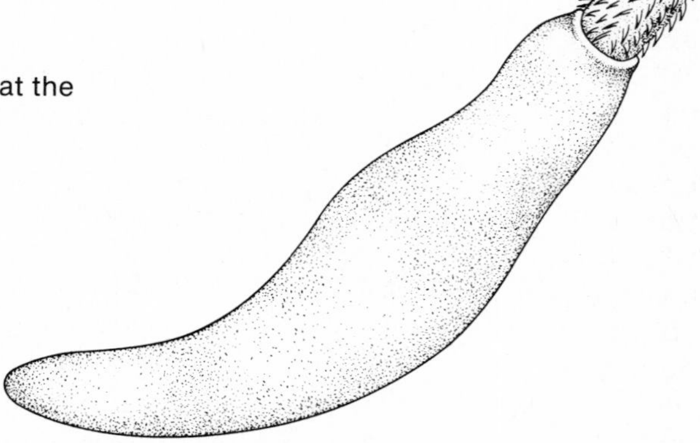

Figure 25.1 Phylum Acanthocephala, a spiny-headed worm.

Classification

The spiny-headed worms constitute a small phylum of parasites in which the definitive host is a vertebrate and the intermediate host is an arthropod. Some authors include these forms with the nematode worms, but they are sufficiently distinctive to warrant their position as a separate phylum.

ORDER **Archiacanthocephala** Concentrically arranged spines on the proboscis. Protonephridial excretory system present. Hosts terrestrial. *Macranthorhynchus, Moniliformis.*

ORDER **Palaeacanthocephala** Spines on proboscis in alternating rows. No excretory system. Mostly aquatic hosts. *Centrorhynchus, Gorgorhynchus.*

ORDER **Eoacanthocephala** Spines on proboscis

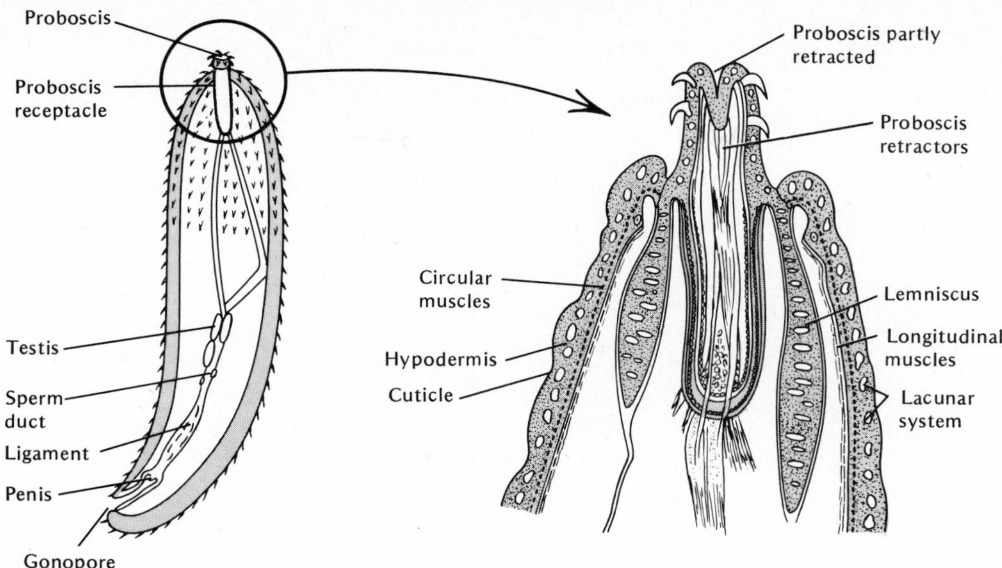

Figure 25.2 Acanthocephalan anatomy.

radially arranged. No excretory system. Hosts aquatic. *Neoechinorhynchus, Quadrigyrus.*

Anatomy and Physiology

GENERAL STRUCTURE

The anatomy of a typical acanthocephalan is diagrammed in Figure 25.2. The body consists of two parts, an anterior *presoma* which includes the proboscis and neck, and the main body or *trunk.* The most characteristic feature of the phylum is the retractable proboscis with its recurved spines by means of which the adult worm attaches to the intestinal wall of its host. There is no digestive tract at any time in the life cycle and the worm obtains its nourishment by absorption of nutrients through its body wall. The surface of the worm is covered by a delicate cuticle which is secreted by the underlying syncytial hypodermis. The hypodermis has a unique fibrous structure with three layers of fibers, an outer set being radial, a middle set in which the fibers run in different directions, and an inner set again radial. The thick inner layer contains the nuclei which are very large and the number is constant for a given species. In many species the total number of cells or nuclei, or both, in the organism is constant, a condition called *eutely.* This feature is also found in members of the phylum Aschelminthes. A peculiar system of fluid-filled interconnecting channels, called the *lacunar system,* is located in the hypodermis. These channels have no definite walls. It is thought that this system aids in distribution of absorbed nutrients. Anteriorly there is a pair of glandular infoldings of the hypodermis into the pseudocoel, called the *lemnisci.* They are well supplied with lacunar channels and may serve as a reservoir for the lacunar fluid of the presoma when the proboscis is retracted. Beneath the hypodermis lie two layers of muscles, an outer circular layer and an inner longitudinal one. There is no lining internal to the longitudinal muscle layer. Specialized circulatory or respiratory structures are absent.

The nervous system is made up of an anterior ventral brain with two or more longitudinal chords. Sense organs are poorly developed.

Within the pseudocoel a tubular membranous sheath, called the *ligament,* extends from the proboscis to the posterior end where it opens to the outside as a gonopore. The gonads, either testes or ovaries since all species are dioecious, are attached to the ligament.

REPRODUCTION AND DEVELOPMENT

Males are provided with a penis and fertilization is internal. Early development occurs within the pseudocoel of the female and the young larvae, called *acanthors* (Fig. 25.3), are enclosed in a shell and released via the gonopore into the gut of the host whence they pass out with the feces. The intermediate host, an insect or crustacean, is infected when it ingests the larval stage enclosed in its shell. Development to a stage much like the adult, except that it lacks gonads, occurs within the intermediate host. The definitive host—fish, bird, or mammal—acquires this juvenile stage when it ingests the intermediate host. In some cases one or more so-called transport hosts may be involved before

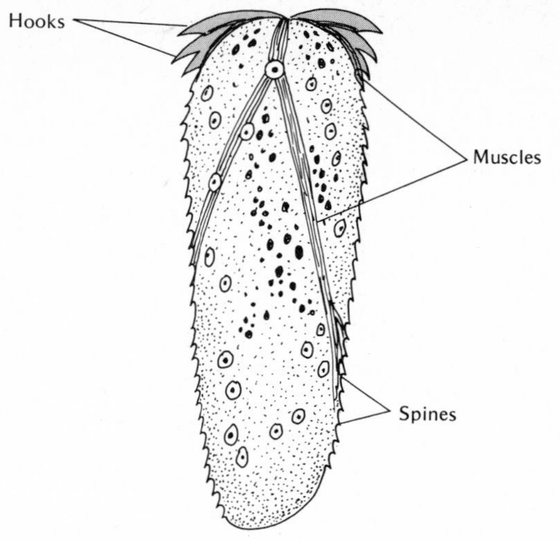

Figure 25.3 Acanthor larva of an acanthocephalan as it appears when it hatches from the egg in the gut of the arthropod intermediate host.

the definite host is reached. The juvenile worm is passed from host to host in the food chain but remains encysted until it reaches a proper definitive host.

Phylogenetic Relationships

The structural peculiarities of the acanthocephala coupled with the extreme changes associated with their parasitic existence make the determination of their phylogenetic position difficult. The presence of the pseudocoel, even though its embryological origin (by withdrawal of the central cells of the solid early larval stage from the body wall) is peculiar, is the main reason for placing the group in the grade Pseudocoelomata. Many of the structural features of this phylum are more nearly like those of the platyhelminthes.

PHYLUM ASCHELMINTHES

The bulk of the pseudocoelomate animals are grouped together within the phylum Aschelminthes. The five classes are quite distinct from one another but there are also significant characteristics held in common. Some schemes consider each of these classes as a separate phylum, and there are other schemes which group some together and separate others as phyla.

Characteristics

1. Pseudocoelomate.
2. Unsegmented or superficially segmented body.
3. Body covered by a noncellular secreted cuticle.
4. Digestive tract consists of a mouth, well-developed pharynx, digestive tube which lacks muscles, and an anus which is posterior and usually terminal.
5. Usually a constant number of cells or nuclei, or both, in each species.
   Figure 25.4.

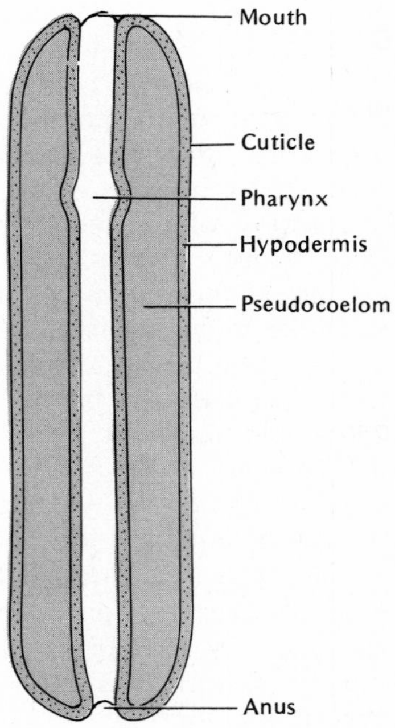

Mouth
Cuticle
Pharynx
Hypodermis
Pseudocoelom
Anus

Figure 25.4 Aschelminth body plan.

Because of the diverse nature of the classes of the phylum Aschelminthes the usual format with a complete classification of the phylum will not be used. Instead, a classification will be given for each class in connection with the treatment of the class.

CLASS ROTIFERA

Characteristics

1. Microscopic, aquatic, mostly in fresh water.
2. Anterior end with a specialized ciliary disk, the corona.
3. Possess a highly developed muscular pharynx, called a *mastax,* which contains specialized jaws. Figure 25.5.

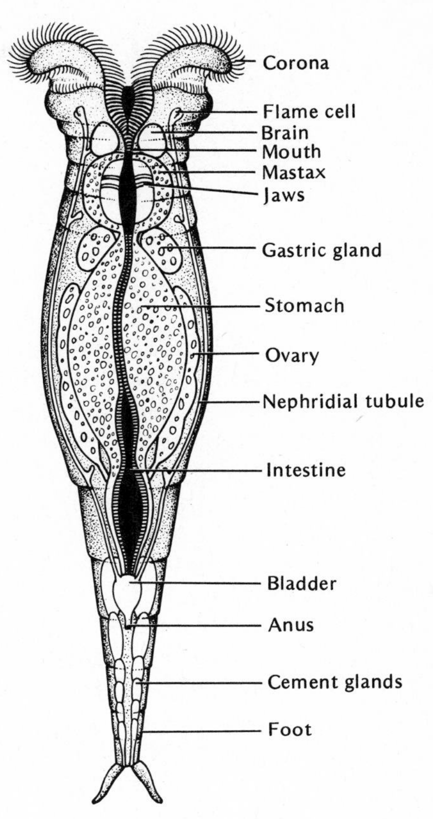

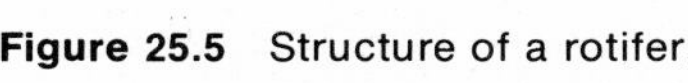
Figure 25.5 Structure of a rotifer.

Classification

ORDER **Seisonacea** Greatly elongated forms which are epizoic, that is, they live on the surface of other animals. Poorly developed corona. Males and females both well developed, gonads paired, no specialized yolk glands in the female. *Seison,* the only genus.

ORDER **Bdelloidea** Swimming or creeping forms with a retractable corona. Paired ovaries with specialized yolk glands, males unknown. *Philodina, Habrotrocha.*

ORDER **Monogononta** Swimming or sessile forms. One ovary with associated yolk gland, males small and degenerate. *Asplanchna, Brachionus.*

Anatomy and Physiology

GENERAL STRUCTURE

A rotifer body is made up of three regions: head, trunk, and tail or foot. A cuticle, secreted by the underlying syncytial hypodermis, covers the body.

The head bears a crown of cilia, or *corona.* These cilia are often arranged in disk-shaped lobes and, beating in succession, they give the appearance of revolving wheels. This unique feature is the source of the name Rotifera, which means wheel bearer. The corona serves both as a means of locomotion and to bring food to the mouth which is located ventrally in the corona.

The middle trunk region contains the visceral organs and the tapering foot completes the body. There are usually spines or toes at the end of the foot. Cement glands located within the foot open through the toes, and the secreted material anchors the rotifer during feeding.

The shapes of rotifers vary largely with their mode of life (Fig. 25.6). Some free-swimming forms and those which creep on the substrate are wormlike. Other free-swimming forms and those which merely float are oval, spherical, or saclike, often secreting a thickened cuticle, called a *lorica.* The lorica may be in ringed sections, allowing for greater freedom of movement. Sessile forms are vase shaped, often living within tubular cases secreted by glands in the foot region. The cases are often covered with sand grains or other foreign materials. Some sessile species live within a secreted gelatinous matrix.

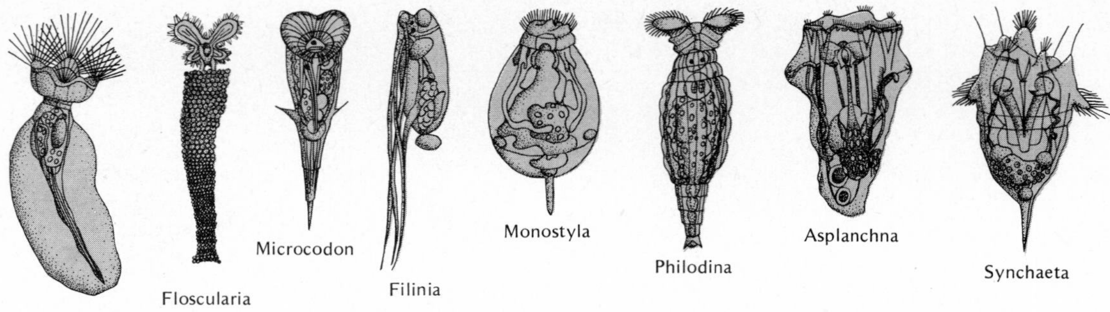

Figure 25.6 Various types of body form found in the class Rotifera.

Beneath the hypodermis there are groups of circular and longitudinal muscles arranged in a rather irregular fashion. The pseudocoelom contains a noncorpuscular fluid which circulates only as a result of body movement.

The mouth leads into the specialized pharynx or *mastax* with its jaws which crush the food. In some species the pharynx can be everted and the jaws are used for seizing prey. Following the mastax, food passes into a stomach which is often provided with gastric glands, thence into the short intestine and cloaca, leaving by way of the anus which opens dorsally at the base of the foot. Movement of food through the digestive tract is facilitated by the ciliated lining.

Excretion is protonephridial with a single pair of branching tubules. There are usually two to eight flame bulbs associated with each tubule. The tubules unite in a posterior reservoir, the urinary bladder, which empties into the cloaca. The rate of excretion is very great in most rotifers. Some expel water equal to the total body weight in 10 minutes. Such a rapid output indicates that here, as in the protozoan contractile vacuole, a very important function is the elimination of excess water which enters by osmosis. Specific evidence of an excretory role is lacking, but it is assumed that at least some excretory wastes do leave by this pathway.

The nervous system consists of a dorsal ganglion, or brain, with fibers leading to sense organs and other structures in the head region and two ventral cords which lead posteriorly to the various internal organs. Sense organs are usually an eyespot, a small dorsal antenna, believed to be a tactile receptor, and a number of different kinds of chemoreceptors associated with the head region.

Rotifers show the interesting characteristic of *eutely* better than any other group. Each species has its own characteristic number of cells, or more often nuclei, since most adults are syncytial. The number of nuclei in adults ranges between one thousand and two thousand in most species. This constancy is carried to the level of each organ and organ system so that the brain or the hypodermis of any given species will have a characteristic number of nuclei, for example.

REPRODUCTION AND DEVELOPMENT

The sexes are separate in rotifers, usually with a distinct sexual dimorphism. Males are small and degenerate in some groups and in others they are lacking entirely. In only a few species are the males well developed and comparable in size with the female. Males possess a single testis with a duct, penis, and gonopore. Since the male usually lacks a digestive system, there is no cloaca in most species. Fertilization may involve copulation in which sperm are placed in the cloaca of the female. However, more often the penis punctures the body wall of the female and sperm are released into the pseudocoel where they swim directly to the ovary. This process is called hypodermic impregnation. The female possesses one or two ovaries which are usually closely associated with a yolk-producing gland.

In forms which reproduce only parthenogenetically the female releases eggs singly and they develop into new females. In those species which have males, early in the season the females produce large eggs which undergo only the first meiotic division and thus remain diploid ($2n$). These diploid eggs develop parthenogenetically into females. Late in the summer or with the onset of unfavorable conditions, a second, smaller type of egg in which both meiotic divisions occur is produced. These haploid (n) eggs develop parthenogenetically into males which then impregnate females containing haploid eggs. When the haploid eggs are fertilized, they develop a heavy shell. They must remain dormant for a period, often requiring a period of freezing and drying before they will begin development. The following spring these thick-shelled eggs develop into female rotifers which produce more females parthenogenetically, thus starting the cycle over once again.

In review, there are two types of eggs produced by rotifers: large thin-walled diploid eggs which produce only females parthenogenetically; and small thin-shelled haploid eggs which produce males parthenogenetically or, if fertilized, produce thick-walled winter eggs which require a dormant period before developing into females.

Development in rotifers is direct, there is no larval stage. Cleavage is spiral and determinate. In later cleavage stages a rather unique type of bilateral cleavage pattern is seen. There is a stereoblastula and a single cell moves internally in gastrulation. This cell becomes the source of the gonads. The outer layer of cells is the source of all the rest of the cells or nuclei, including the mesoderm.

Ecology

Although rotifers are found in many habitats, including benthic and pelagic marine and freshwater, terrestrial, on and within other organisms, most of them are found in relatively shallow freshwater situations. Like many other microscopic organisms, they are cosmopolitan in their distribution. The same species are found in similar ecological situations worldwide. This reflects their ability to exist in a desiccated state, either as eggs or adults, and thus to be carried great distances by winds or animals.

Many pelagic species undergo a seasonal change in body proportions, particularly in the length of various projecting spines—a phenomenon called *cyclomorphosis.* This change is believed to be associated with the lowered density of warmer water. In the summer months the forms with longer spines are more buoyant.

Sessile species are often found only on particular species of plants. Some select specific regions of the plant, such as the distal or proximal portions of the axis. Terrestrial species live in mosses or lichens and other similarly damp environments. When water is absent, the animals dry up to a small cystlike mass. Most of them have no real protective cyst wall, just the normal body wall. They may remain viable in this dried-up state for as long as four years, swelling to normal size and becoming active again in a short time when water is again available. In addition to withstanding drying, many rotifers can survive great changes in temperature and pH.

Phylogenetic Relationships

Early workers made much of the superficial resemblance of some rotifers to the trochophore larva, characteristic of a number of phyla higher on the phylogenetic tree. This resemblance is now believed to be a simple case of convergence related to the pelagic existence of the trochophorelike rotifers and the trochophore larvae of annelids and mollusks.

Both anatomy and embryology point to a relationship between the rotifers and primitive turbellarians; the protonephridial excretory system of both is essentially identical. The organization of the ovary is the same in both groups.

CLASS GASTROTRICHA

These microscopic aquatic animals have often been included with the rotifers, but they are sufficiently distinct in their characteristics to be placed in a class by themselves.

Characteristics

1. Microscopic, aquatic forms with a cuticle which is usually differentiated into scales, bristles, or spines.
2. Cilia on the head and on the ventral surface in specific patterns of longitudinal or transverse bands or in patches.
3. Single pair of protonephridial tubes, each with a single flame bulb in freshwater forms.
4. Some marine species lack an excretory system. Figure 25.7

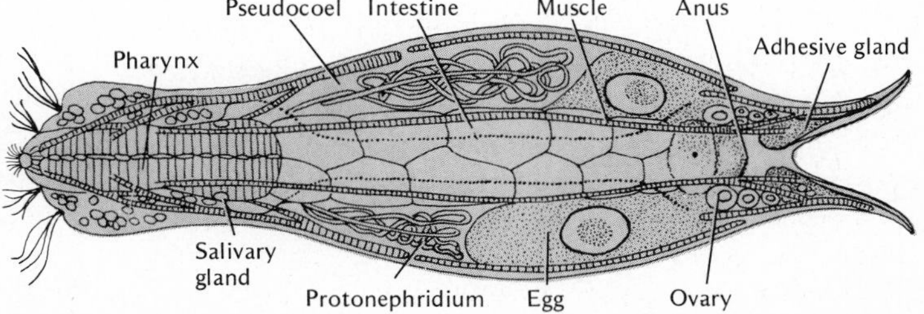

Figure 25.7 Class Gastrotricha, *Chaetonotus* sp.

Classification

ORDER **Macrodasyoidea** Exclusively marine. Lack protonephridial system. Hermaphroditic with both male and female systems. *Cephalodasys, Macrodasys, Platydasys.*

ORDER **Chaetonotoidea** Primarily freshwater forms, only a few marine representatives. Protonephridial system present. Only parthenogenetic females except in the marine genera *Neodasys* and *Xenotrichula* which have both male and female systems. *Chaetonotus, Lepidermella, Dasydytes.*

Anatomy and Physiology of *Chaetonotus*

These are minute aquatic forms, both freshwater and marine, with the largest being only a few hundred micrometers in length. *Chaetonotus* is a typical example (Fig. 25.7) with its ciliated tufts on the head, two ventral bands of cilia, and bifurcated posterior end. Locomotion is accomplished by the beating of the ventral cilia, much as in the turbellaria. Paired adhesive glands opening at the tip of the posterior end are a characteristic feature of the phylum. In other species there may be a series of laterally placed adhesive glands, each leading to the outside by its own tube. The cuticle is differentiated into scales which may or may not overlap, or spines which in many cases are very highly developed, or both (Fig. 25.8). Beneath the cuticle is a syncytial hypodermis and the rest of the body wall is essentially like that of the rotifers. There is only a very small pseudocoel between the body wall and the gut. The digestive system is quite simple, consisting of an anterior terminal mouth, a pharynx which is remarkably similar to that of the nematodes, and an intestine with a posterior anus. The tract is not ciliated, another feature in common with the nematodes. The excretory system, when present, consists of a single pair of nephridia, each with a simple multinucleate flame bulb. The nervous system consists of a two-lobed brain located on either side of the pharynx, and a pair of lateral trunks. Sensory structures are similar to those of rotifers, but eyespots are rare.

The gastrotrichs are basically hermaphroditic, but in *Chaetonotus* and most of the rest of the order Chaetonotoidea, which includes all the freshwater forms, the male system does not develop and reproduction is parthenogenetic. Development is by a modified spiral cleavage and is determinate, with no special larval form.

Ecology

Both fresh- and salt-water species are known, and most of them are bottom dwellers, gliding about over the substrate by means of their ciliary bands. Many of them are part of the interstitial fauna, living in the spaces between sand grains. They feed upon bacteria, protozoans, and unicellular algae.

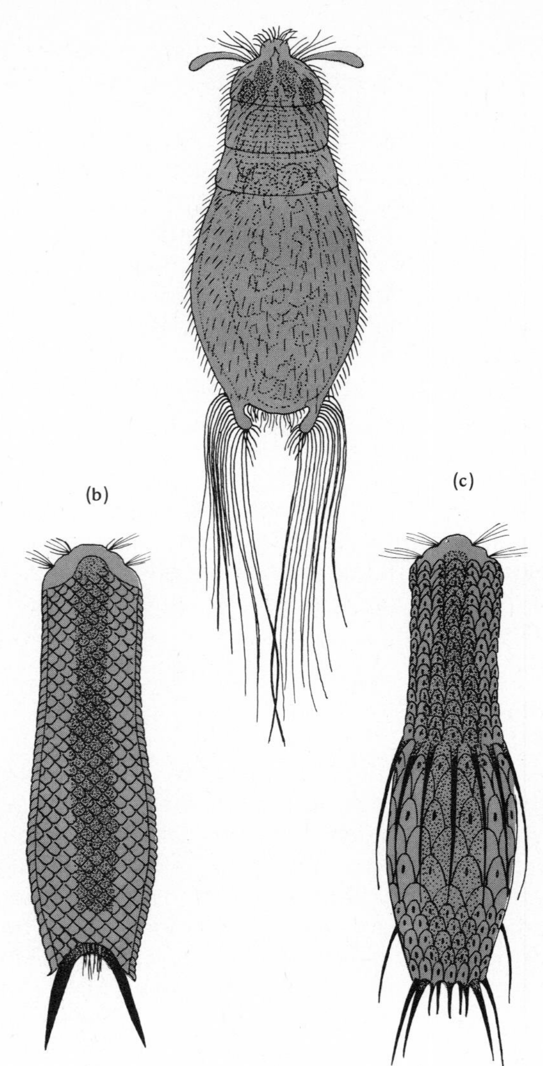

Figure 25.8 Examples of cuticular modifications in the class Gastrotricha. **(a)** *Neogossea* sp. **(b)** *Aspidiophorus* sp. **(c)** *Chaetonotus* sp.

Phylogenetic Relationships

Some place the gastrotrichs close to the rotifers because of similarities in muscles, nervous system, protonephridial system, and adhesive glands. The digestive system, however, is quite unlike that of the rotifers and resembles closely that of the nematodes. It is probable that the gastrotrichs are most closely related to the nematodes and that both nematodes and gastrotrichs as well as the rotifers descended from the turbellaria.

CLASS KINORHYNCHA (Echinodera)

Characteristics

1. Microscopic and entirely marine.
2. Cuticle clearly divided into 13 or 14 segments.
3. Head provided with spines and completely retractable into the neck region.
   Figure 25.9

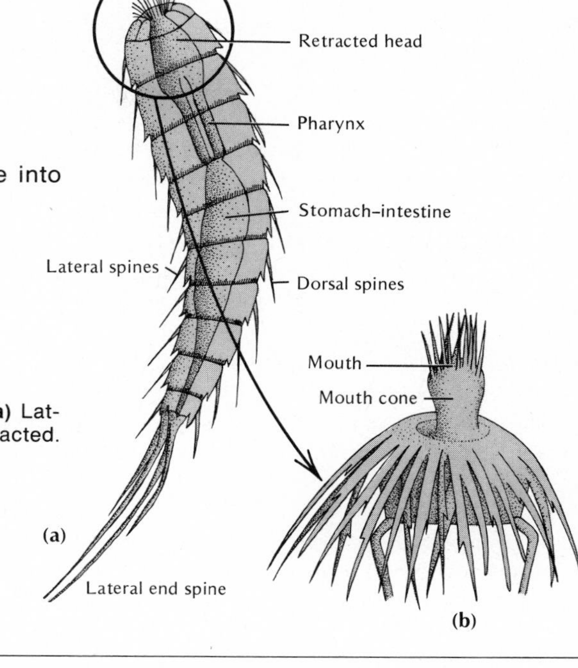

Figure 25.9 Class Kinorhyncha, *Echinoderella* sp. **(a)** Lateral view. **(b)** Detail of anterior end with head protracted.

Classification

ORDER **Cyclorhagida** First zonite retracted with the head and cuticular plates or placids of the second zonite form a protective cover. *Echinoderes, Echinoderella, Centroderes.*

ORDER **Homalorhagida** First two zonites retracted with the head. Protective cover formed by one dorsal and three ventral placids of the third zonite. *Pycnophyes, Trachydesmus.*

ORDER **Conchorhagida** First two zonites retracted with the head. Protective cover formed by a pair of lateral shells on the third zonite. *Semnoderes,* the only genus.

Anatomy and Physiology of *Echinoderella*

Echinoderella, a typical kinorhynch, is diagrammed in Figure 25.9. The cuticle is well developed and clearly segmented, with one segment or *zonite* covering the head, a second the neck, and the other 11 covering the trunk region. Most of the zonites bear well-developed movable spines. The mouth is anterior and terminal and is surrounded by a circlet of spines. The entire head region may be retracted into the neck or trunk, and when it is so retracted, the cuticular plates or *placids* of the second zonite close down to cover the entire anterior end. The body wall is essentially like that of rotifers and gastrotrichs, and there is a well-developed pseudocoel between it and the gut. A special feature of the body wall is the thickened longitudinal cords in the hypodermis, one dorsal and two lateral, which are quite similar to thickenings found in the nematodes. The digestive system is a simple nonciliated tube with a pharynx much like that of the nematodes. Excretion is by a pair of nephridia, each with a single flame bulb containing a long flagellum. Sexes are separate but little is known of the early development. The kinorhynchs inhabit shallow marine waters in mud or sand where they feed on algae, diatoms, or detritus.

Phylogenetic Relationships

Because of the external segmentation and superficial resemblance to an insect larva, the group was thought by some to be related to annelids and arthropods, perhaps intermediate between the two. However, the external nature of the segmentation, the protonephridial excretory system, and the absence of a coelom indicate far too primitive an organization to place them that high on the phylogenetic tree. Similarities in the pharynx and in the cuticular modifications point clearly to a relationship with both nematodes and gastrotrichs with which they probably share a common ancestry.

CLASS NEMATODA

The nematodes, or roundworms, constitute the largest class of the phylum Aschelminthes and are among the most numerous of all animals. Most of the nematodes are very small. There are a number of very important parasites, of both plants and animals, among the nematodes, and probably because of this, many texts leave the impression that it is primarily a parasitic group. But there are many free-living forms. They are found in almost every kind of habitat: in the soil, in fresh and salt water, in aquatic vegetation, and in moist places generally. In numbers of organisms, the roundworms are exceeded only by the protozoa and the arthropods. So numerous are the roundworms that a great student of the group, Cobb, has said, "If all the matter in the universe except the nematodes were swept away, our world would still be dimly recognizable, and if, as disembodied spirits, we could then investigate it, we should find its mountains, hills, vales, rivers, lakes and oceans represented by a film of nematodes. The location of towns would be decipherable, since for every massing of human beings there would be a corresponding massing of certain nematodes. Trees would still stand in ghostly rows representing our streets and highways. The location of the various plants and animals would still be decipherable, and, had we sufficient knowledge, in many cases even their species could be determined by an examination of their erstwhile nematode parasites."

Characteristics

1. Elongated, cylindrical, unsegmented worms.
2. Well-developed, noncellular cuticle present.
3. Three-angled pharynx.
4. Cloaca present in males but not in females where the genital pore is anterior.
   Figure 25.10

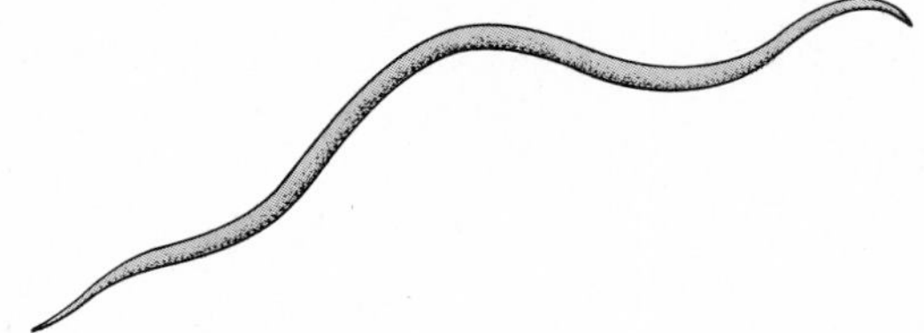

Figure 25.10 Class Nematoda, *Ascaris lumbricoides.*

Classification

SUBCLASS **Aphasmida** Phasmids (paired glandular pouches at the posterior end) are absent.

ORDER **Enoplida** Free-living, mostly marine. Three circles of anterior sensory papillae. Single-celled renette gland. *Enoplus, Thoracostoma, Halalaimus.*

ORDER **Dorylaimida** Common soil and freshwater species. Hollow stylet present, used in sucking plant and animal juices. *Dorylaimus, Actinolaimus, Xiphinema.*

ORDER **Mermithida** Thin threadlike worms. Immature forms parasitic in terrestrial or freshwater mollusks and arthropods. Adults free-living in soil or fresh water. *Mermis, Tetradomema, Aproctonema.*

ORDER **Trichiurida** Pharynx nonmuscular, enclosed in a stichosome—two longitudinal rows of large cells which represent a pharyngeal gland. Mouth lacks lips. All parasitic in vertebrate animals. Includes the whipworm, *Trichiuris trichiura,* and the causitive agent of trichinosis, *Trichinella spiralis,* both of which affect humans. *Trichiuris, Trichinella, Capillaria.*

ORDER **Dioctophymida** Male with a bell-shaped tail. Mouth lacks lips but is encircled by one or two rows of sensory papillae. Only three genera, all endoparasites of birds and mammals. None are known to affect human beings. *Dioctophyme renale,* with an aquatic annelid worm and a fish as first and second intermediate hosts, affects dogs. *Dioctophyme, Eustrongyloides, Hystrichis.*

SUBCLASS **Phasmida** Phasmids, a pair of glandular cuticular pouches with ducts opening near the anus, always present.

ORDER **Rhabditida** Protrusible stylets absent. Mostly terrestrial, both free-living and parasitic species, none of which affect humans. The vinegar eel, *Turbatrix aceti,* which feeds on bacteria in fermenting vinegar, is in this order. *Rhabditis, Turbatrix, Cephalobium.*

ORDER **Tylenchida** Buccal cavity with a protrusible stylet. Free-living soil dwellers and parasitic in plants. Many economically important plant parasites are found in this order, including the sugarbeet eelworm, *Heterodera schachtili;* the golden potato nematode, *H. rostochiensis;* and the wheatgall eelworm, *Anguina tritici.* The latter can remain viable for as long as 28 years as a dormant second-stage larva in dried stored grain. *Heterodera, Anguina, Ditylenchus.*

ORDER **Rhabdiasida** Complicated life cycles, with alternating free-living terrestrial generations and parasitic generations in a vertebrate host. *Strongyloides stercoralis,* an intestinal parasite of humans, is in this order. Infection is by direct penetration of the skin by the larva, or by its ingestion.

ORDER **Oxyurida** All are obligatory parasites, mostly in vertebrates, with a simple life cycle involving only one host. The pinworm, *Enterobius vermicularis,* is a common intestinal parasite of humans, particularly of children. Gravid females pass out with the feces, or they lay their eggs during the night around the anus causing intense itching. Autoinfection often occurs because of scratching. Infective eggs may infect entire families, spread by shaking of bedding used by an infected child. *Enterobius, Oxyuris, Thelandros.*

ORDER **Ascarida** Obligatory parasites of the intestine of vertebrates. Many species are of large size. Mouth surrounded by three lips followed by a well-developed muscular pharynx. Simple direct life cycle with only one host. *Ascaris lumbricoides,* which affects humans, will be used as an example of the class. *Ascaris, Toxicara, Ophidascaris.*

ORDER **Strongylida** Lips absent, mouth with teeth around the edges. Males with an expanded pouchlike bursa at the posterior end. All are obligatory parasites of vertebrates, mostly mammals. Life cycle is direct with free-lving juvenile stages. Infection is by ingestion or, in the case of the hookworms, direct penetration of the skin by the infective larval stage. The hookworms, *Ancylostoma duodenale* and *Necator americanus,* are among the most injurious of human parasites. *Necator, Ancylostoma, Strongylus.*

ORDER **Spirurida** Slender worms with a cuticularized buccal capsule. All are obligatory parasites of vertebrates and all have complex life cycles involving at least one arthropod intermediate host. Some species of the genus *Thelazia* infect the eyes of humans in the adult stage. Mode of infection in these cases is not known. *Thelazia, Habronema, Spirura.*

ORDER **Dracunculida** Obligatory parasites of the body cavities and of connective tissues of vertebrates. These slender worms lack lips and a cuticularized buccal capsule. The life cycle involves an intermediate host, often a copepod. The guinea worm of humans, *Dracunculus medinenis,* is a spectacular species in which the adult female, often more than a meter long, lives in the subcutaneous tissue where it can be readily seen. *Dracunculus, Philometra, Micropleura.*

ORDER **Filarioida** Slender worms of fair size, with the females much larger than the males. All are obligatory parasites of the circulatory system, coelom, muscles, or connective tissues of vertebrates. They are similar in structure to members of the order Spirurida, differing in the method of transmission which is through the skin by a bloodsucking insect, the intermediate host. The causitive agent of the dread disorder elephantiasis, *Wuchereria bancrofti,* is in this order. The heart worm of the dog, *Dirofilaria immitis,* is a filarial worm transmitted by mosquitoes and perhaps by ticks. *Wuchereria, Dirofilaria, Onchocera.*

Anatomy and Physiology of *Ascaris*

GENERAL STRUCTURE

A large specimen of a free-living nematode would not be over 1 cm in length and, therefore, not very satisfactory for detailed study. Some of the parasites, such as *Ascaris,* which lives in the intestine of humans, pig, horse, and a number of other mammals, may attain a length of 25–30 cm (10–12 inches). In *Ascaris* the female is larger than the male. The male has a characteristically incurved tail with a pair of projecting *reproductive spicules.* The body is elongated, pointed at both ends, and light in color. The *mouth* is a triangular opening at the anterior end and is surrounded by three lips. The *anus* is located a short distance from the posterior end. When the body wall is cut open, as shown in Figure 25.11, the digestive tract and reproductive organs are exposed in the unlined body cavity, or pseudocoel.

The digestive tract is quite simple. The mouth opens to a short muscular three-angled pharynx, and the tract continues on through the length of the worm as the intestine. Just anterior to the anus the intestine narrows to form a short rectum. In the male this region is actually a cloaca since the duct from the testis opens into it. Food material is pumped into the ascaris intestine from the intestine of the host by the action of the muscular pharynx. The wall of the intestine is formed of a single layer of columnar cells, and the absorbed nutrients are passed by this layer into the body fluids in the pseudocoel where they are distributed to all parts of the body.

The nervous system consists of a nerve ring around the pharynx and two large nerve cords, one dorsal and one ventral. The body wall is composed of an outer noncellular cuticle, a syncytial hypodermis that secretes the cuticle, and a layer of large muscle cells. In many respects these cells are a very primitive type of muscle cell. The cell bodies, in cross section, appear as rounded masses projecting into the body cavity (Fig. 25.11), and the contractile portion lies immediately under the hypodermis. The contractile fibers all run longitudinally and, as a result, the movements of the animal are limited to bendings of the body from side to side. The elasticity of the cuticle causes the body to snap back after contraction of the muscles, and the worms move in a characteristic whiplike manner. The muscle layer is interrupted by four longitudinal thickenings of the hypodermis that extend the length of the body. Two of these are known as lateral lines, one the dorsal line, and the other the ventral line.

The excretory system is unique, consisting of two longitudinal tubes, one in each lateral line, which open on the ventral surface near the anterior end by a single excretory pore. In marine

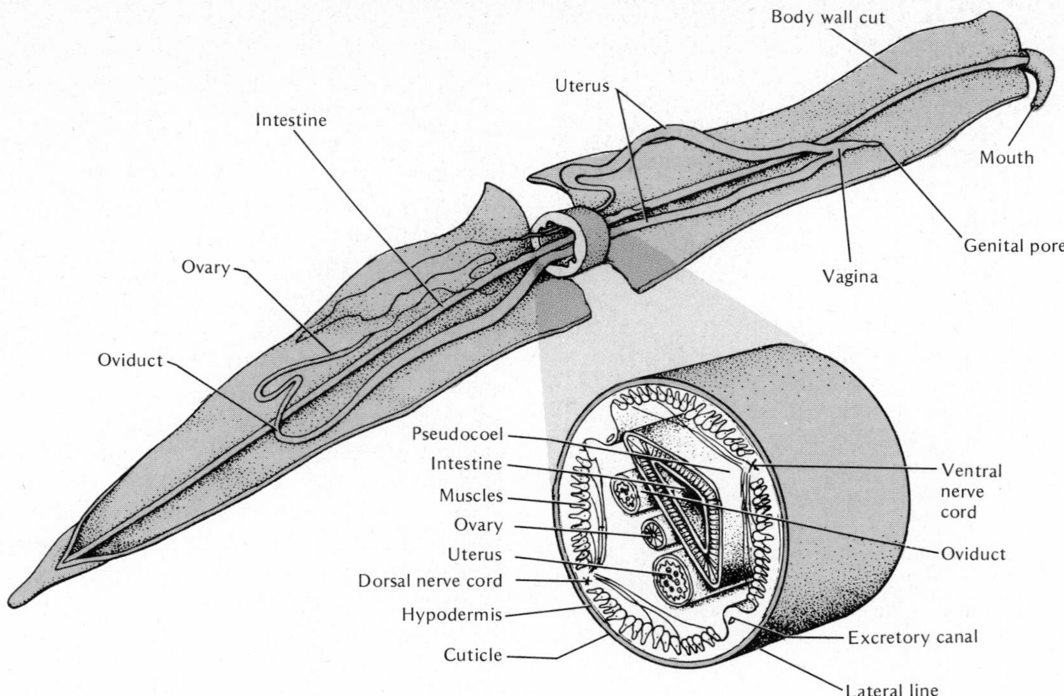

Figure 25.11 Structure of *Ascaris* showing internal organs of female and a cross section.

nematodes excretion is accomplished by means of a simple saclike *renette cell.* Its narrow neck opens to the outside at the excretory pore. A pair of similar cells is found in the cephalic region of some terrestrial and parasitic forms in addition to the tubes in the lateral lines. There are no cilia in the system. It was formerly believed that cilia were completely lacking in the class Nematoda. Roggen and his coworkers showed by electron microscopy that cilia with a typical 9 + 2 pattern of tubules were present in cephalic sensory structures of a nematode.

REPRODUCTION AND LIFE CYCLE

In the female the reproductive system consists of a genital pore, located about one third the length of the body from the anterior end, a single vagina, and two long convoluted tubes, each differentiated into a uterus, an oviduct, and a coiled threadlike ovary. In the male the reproductive organs do not have a separate opening, but rather open into the cloaca. The male system is composed of a single coiled threadlike testis, a vas deferens, a seminal vesicle, and a short muscular ejaculatory duct that opens into the cloaca. The male worm is believed to utilize a pair of horny bristles in transferring the unusual amoeboid sperm to the female. Fertilization takes place in the uterus and the zygote is surrounded by a thick shell. It has been estimated that a single mature female *Ascaris* lays about 200,000 eggs per day.

These eggs leave the body of the host in the feces and remain viable for a long time. Infection of a new host will occur by the swallowing of the eggs. Cleavage follows a modified spiral pattern and it is determinate. The first molt occurs within the egg shell. Hatching occurs in the intestine of the new host, and the young worms then start on a strange trip through the body. They burrow through the intestine into the blood vessels and are carried to the pulmonary artery. When they reach the lungs, they bore into the bronchial tubes and after a second molt they eventually reach the throat where they are swallowed. They finally reach the intestine again, molting twice more en route, where they remain and grow to adult size.

Other Parasitic Nematodes

There are many species of roundworms which parasitize plants and animals as well as a number which infect humans. Some important parasites are mentioned in the synopsis of the class. The most important ones affecting humans are the hookworm, the trichina worm, and the nematode that causes elephantiasis.

Hookworm disease is a tropical and semitropical disease that causes much human misery and even loss of life. There are two species of hookworm and one of these, *Necator americanus,* is widely distributed in our southern states. The adult worms, which are about 12 mm

1. *Holocanthus ciliaris,* the queen's angelfish. The blue ring on the head which resembles a halo is responsible for the common name of this colorful tropical fish. (Ch. 34)
2. *Equus burchelli,* Burchelli's or common zebra, at a water hole in Ngoro Ngoro Crater, Tanzania. (Chs. 38, 45)
3. *Crocuta crocuta,* the spotted or laughing hyena, Ngoro Ngoro Crater, Tanzania. (Chs. 38, 44)
4. *Alopex lagopus,* the arctic fox, in summer coat, Northern Sweden.

   The spotted hyena and the arctic fox are examples of ecological equivalents since the spotted hyena feeds on the leavings of lion kills and ostrich eggs and the arctic fox feeds on the leavings of polar bear kills and the eggs of a bird, the guillemot. (Chs. 38, 44)

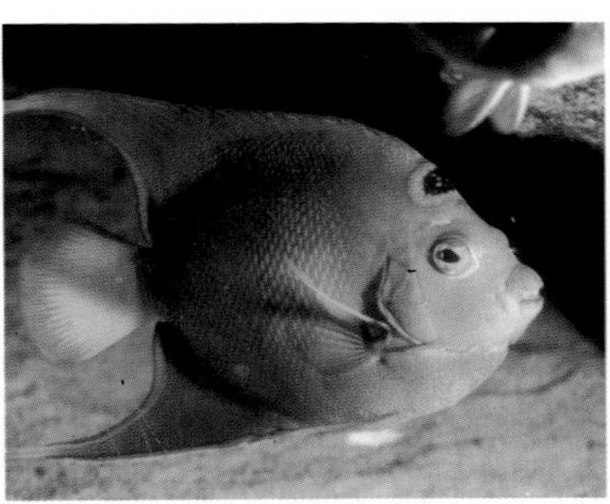

1

2

3

4

5

6

5. *Loxodonta africana,* the African elephant, Queen Elizabeth Park, Uganda. (Chs. 38, 45)
6. *Procavia capensis,* a rock hyrax, Serengeti Plains, Tanzania. This small mammal, which is found on rocky outcrops and mountain slopes throughout much of Africa, belongs to a small order whose members are the closest living relatives of the elephant. (Ch. 38)
7. *Balearica pavonina,* the crowned or crested crane, the national bird of Uganda. (Ch. 37)

7

8. *Panthera leo,* African lion, Serengeti Plains, Tanzania. (Chs. 38, 45)
9. *Rana pipiens,* the common leopard frog. (Ch. 35)
10. *Chrysemys picta,* the painted turtle, a familiar sight in lakes and ponds of Eastern United States. (Ch. 36)
11. *Hippopotamus amphibius.* Queen Elizabeth Park, Uganda. (Ch. 38)

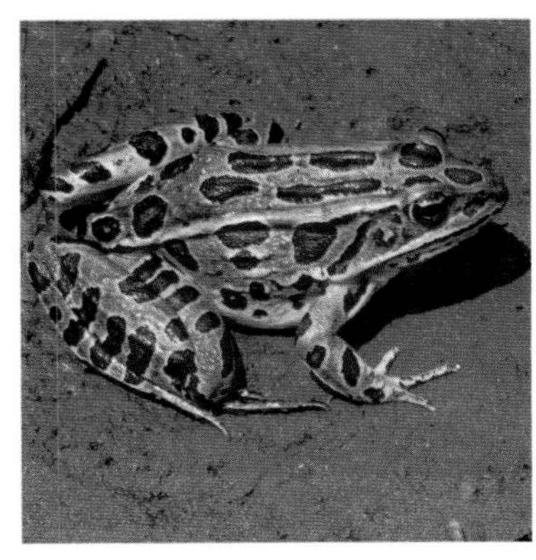

9

8

10

11

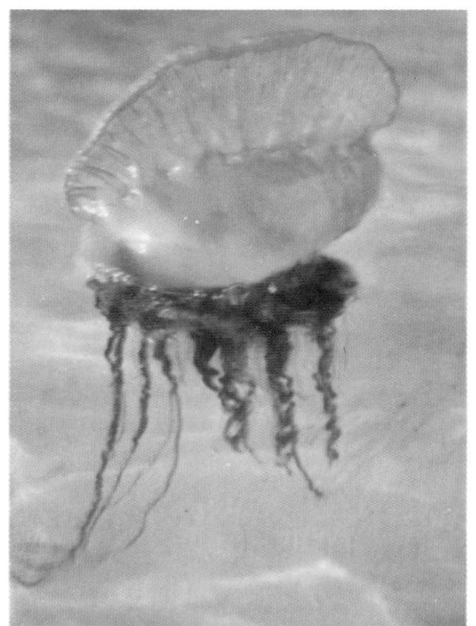

12 13

14

15

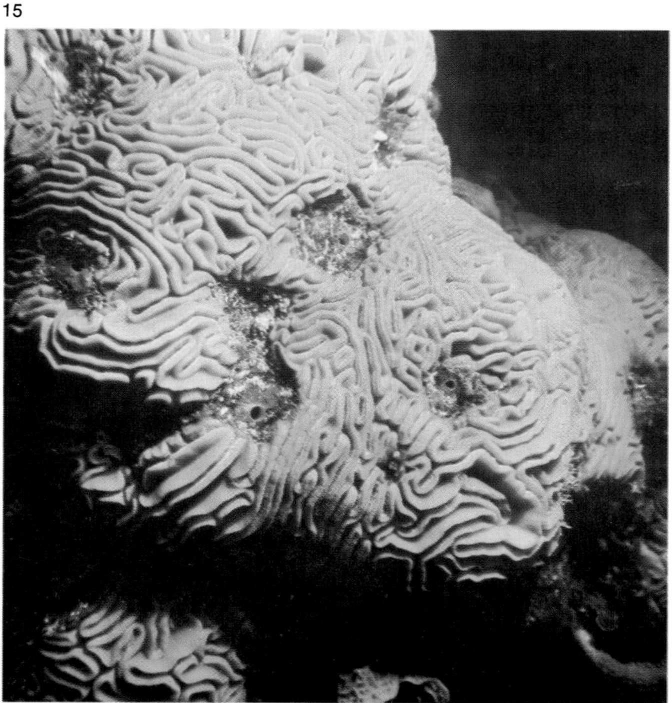

12. *Physalia* sp., the Portuguese Man-of-War. (Ch. 23)
13. Gorgonian corals, 170 feet deep in the La Jolla Submarine Canyon, California. Note the many small polyps that extend from the surface of the colony. (Ch. 23)
14. Fan Worm, a polychaete annelid. (Ch. 27)
15. *Diploria* sp., brain coral, Grand Bahama Island. Note the small red sponges growing on the surface of the coral colony. (Ch. 23)
16. *Hermissenda crassicornis,* a nudibranch mollusk, feeding on *Obelia* polyps. This shell-less gastropod digests the *Obelia* without causing discharge of the nematocysts and subsequently uses them for its

16

own defense in small glandular structures at the tips of the tentacle-like cerata which cover its back. (Ch. 26)

17. *Eudystylia polymorpha,* a feather duster worm. This polychaete lives in a secreted leathery tube and feeds on microscopic organisms that are trapped in the mucus which covers the feathery gills and are carried to the mouth by ciliary currents. (Ch. 27)

18. *Strongylocentrotus franciscanus,* a West Coast sea urchin found in deep tidal channels and pools. (Ch. 32)

17

18

19

20

21

19. *Achatina achatina,* the giant West African land snail that reaches a length of over ten inches. A related species has been introduced to many tropical areas of the Pacific, including the Philippines and Hawaii, and it has become a serious pest animal in most places. (Ch. 26)
20. *Mopalia* sp., a chiton, firmly attached to a rock by means of its suction cup-like foot. (Ch. 26)
21. *Modiolus modiolus,* the horse mussel, a common East Coast species. Note the stout byssus threads that the animal secretes as a holdfast mechanism. (Ch. 26)
22. *Pecten irradians,* the common scallop. Note the large adductor muscle, which is the edible portion, and the series of bright blue eyes on the mantle edges. (Ch. 26)
23. *Octopus* sp., swimming. Note the nature of the suckers, which are typical of all cephalopod mollusks. (Ch. 26)

22

23

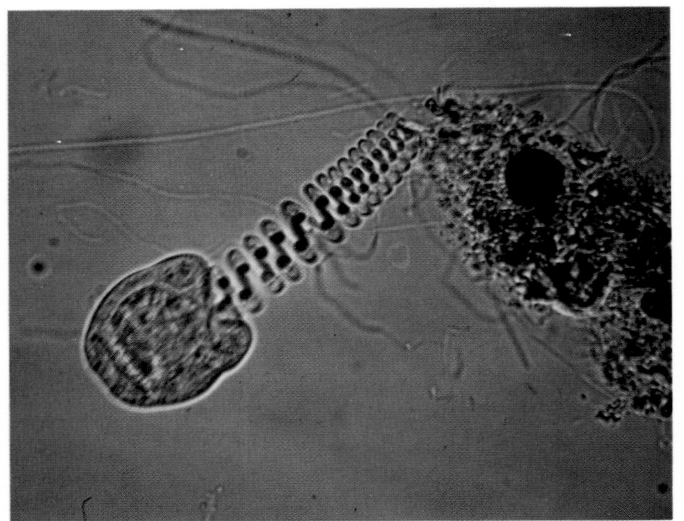

24

25

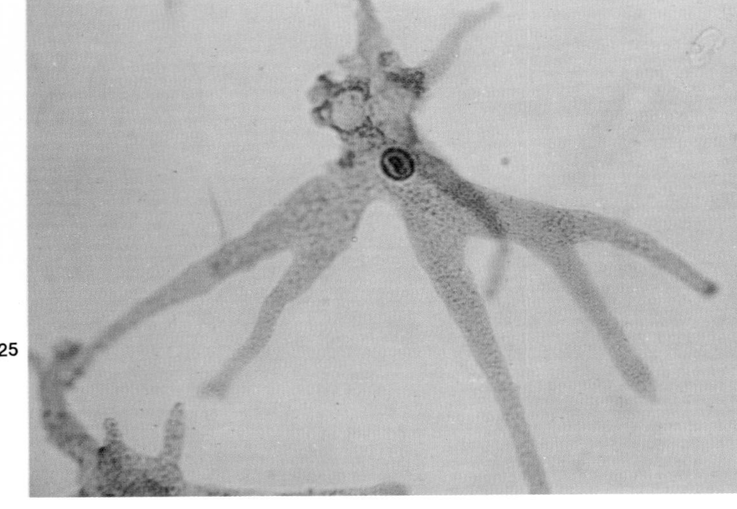

26

27

28

24. *Vorticella*, sp., living specimen, contracted. Note the myonemes in the spirally coiled stalk. (Ch. 21)
25. *Amoeba proteus*, stained preparation. Note the chromatin material in the stained nucleus. (Ch. 21)
26. *Craniella gravida*, a sponge of the class Demospongiae. Note the prominent osculum, the excurrent opening of a sponge. (Ch. 22)
27. *Microciona prolifera*, a common East Coast sponge encrusting stones and shells. (Ch. 22)
28. *Chaetopterus variopedatus*, a highly specialized, tube-dwelling polychaete annelid worm. This photo was taken of an animal in a transparent tube. (Ch. 27)
29. *Nereis succinea*. Close-up view of the head of this polychaete showing its eyes, tentacles, and palps as well as a number of parapodia. Note the red color of the dorsal vessel due to the presence of the pigment hemoglobin dissolved in the blood. (Ch. 27)

29

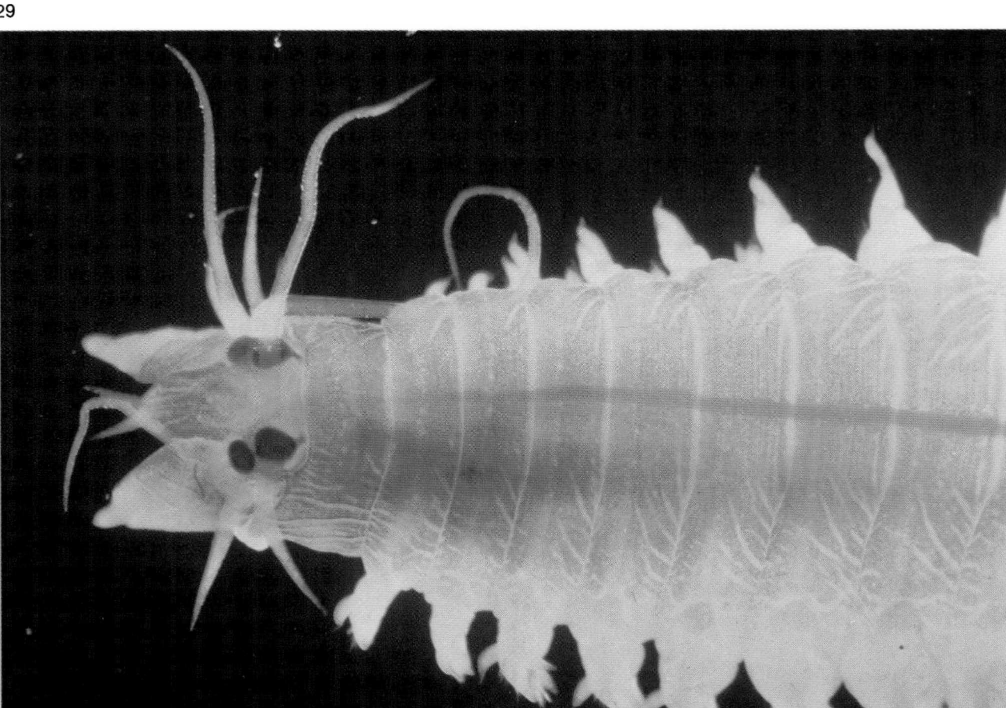

30

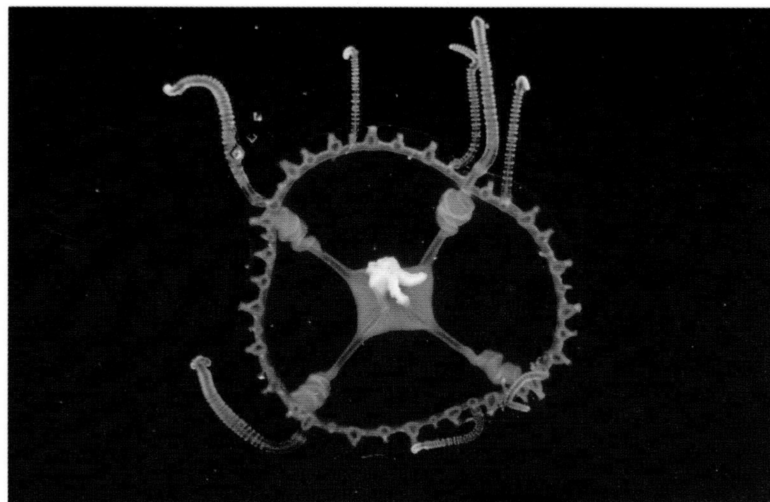

31

32

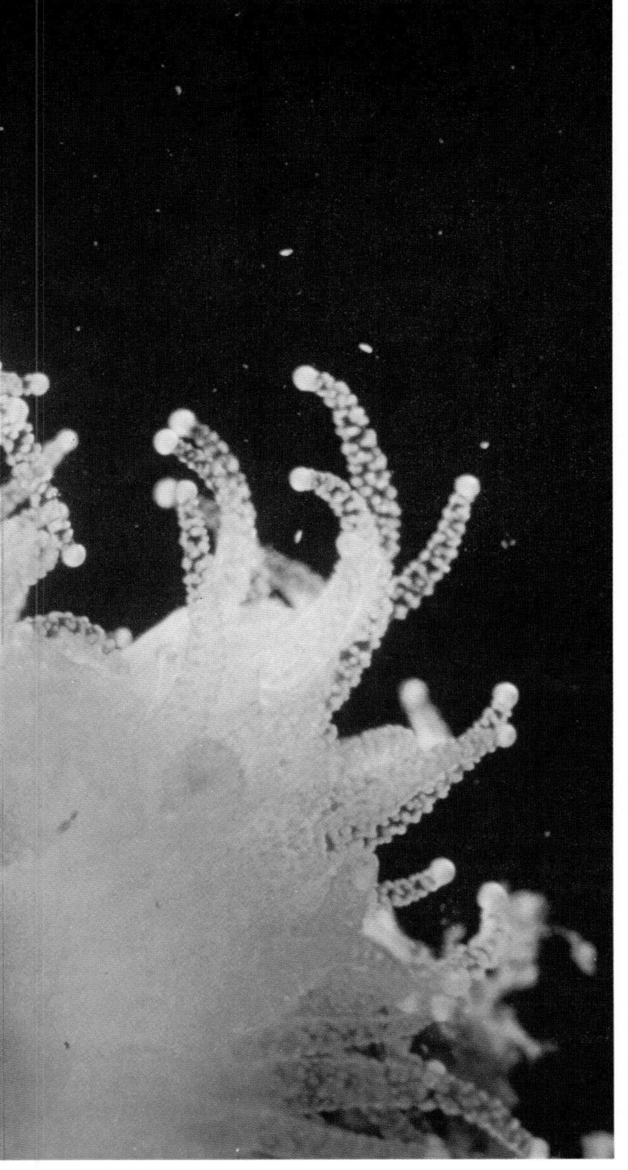

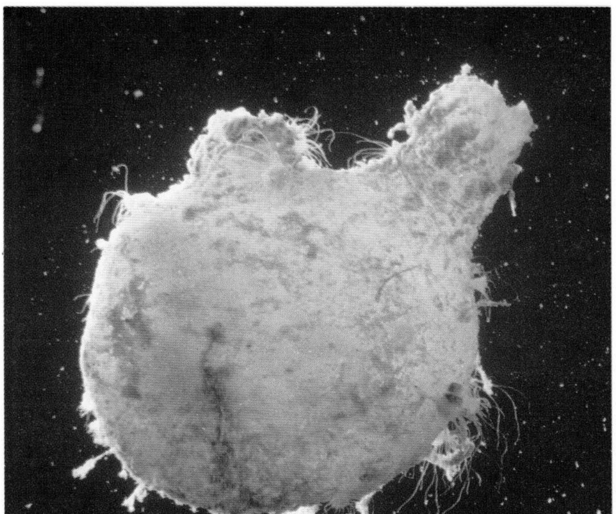

30. *Astrangia danae,* found on the East Coast from Florida to Cape Cod. Note the coral polyps extending from their stony, corallite cups. (Ch. 23)
31. *Gonionemus vertens,* a hydrozoan medusa, shown from the subumbrellar side. Note the white manubrium and the gonads that are located on the four radial canals. Most of the tentacles are extended straight upward and only their bases show in this view. (Ch. 23)
32. *Molgula manhattensis,* a sea squirt, member of the phylum Chordata, subphylum Urochordata, class Ascidiacea. Note the small scyphozoan polyps (scyphistomae) attached to its surface. (Ch. 33)

33

34

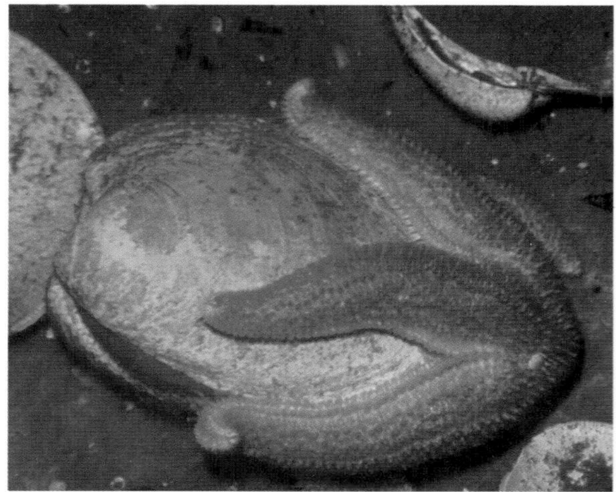

33. *Chrysaora quinquecirrha,* a scyphozoan medusa which is exceptional in that it is hermaphroditic. (Ch. 23)

34. Starfish, *Asterias forbesi,* attacking a clam. (Chs. 26, 32)

Some planktonic animals: (35) *Glaucus* sp., a gastropod mollusk of the order Nudibranchia; (36) *Parapeneus longpipes,* a crustacean of the order Decapoda; (37) *Pontella securifer,* a crustacean of the subclass Copepoda, order Calanoidea, and (38) *Porpita* sp., a hydrozoan coelenterate of the order Siphonophora. (Chs. 23, 26, 30)

39. The tidepool habitat. Crabs, anemones, and mussels in a tidepool on the California coast. (Ch. 45)

35

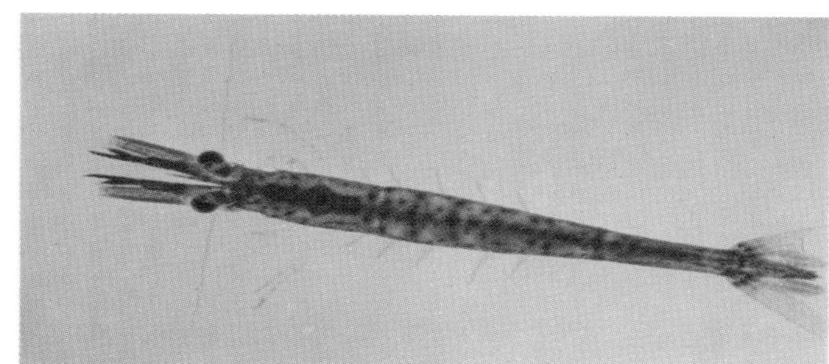

36

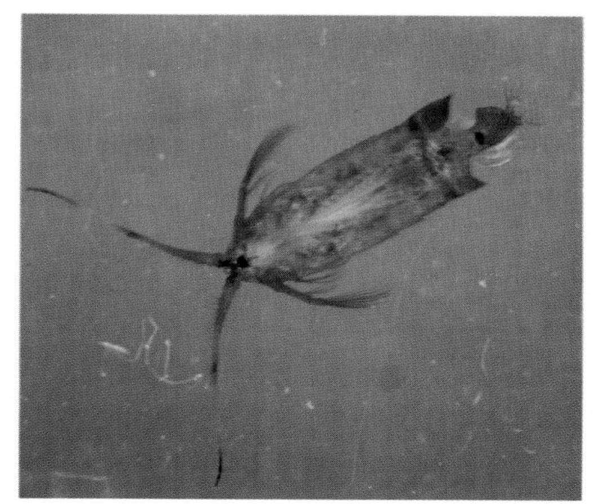

37

38

39

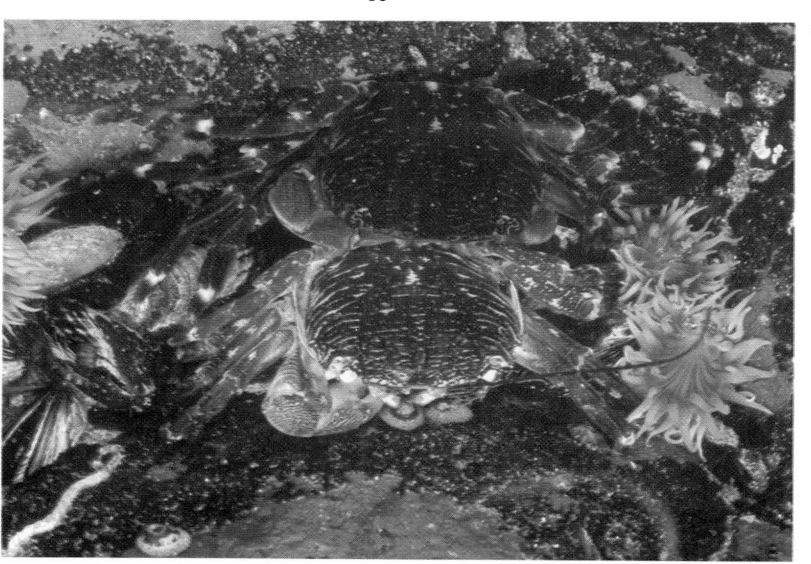

40

40. The temperate grassland biome. Reconstituted prairie at the University of Wisconsin Arboretum. Few if any stands of natural prairie have been left untouched by man. However, this stand developed in Madison, Wisconsin, is probably quite typical of the natural prairie. (Ch. 45)
41. The northern coniferous forest (or taiga) biome. Winter in the taiga of interior Alaska. (Ch. 45)
42. The chaparral biome, a typical example in Southern California. This biome is characterized by abundant winter rainfall and dry summers. The climax vegetation consists of shrubs and small trees with hard, thick, evergreen leaves. Similar communities are also found on the shores of the Mediterranean and this biome type is often called the Mediterranean biome. (Ch. 45)
43. The savannah biome, Serengeti plains, Tanzania. The flat-topped acacia trees and the broad expanses of grassland are typical of the East African savannah. In total numbers and in variety of species, the ungulate population of the African savannas is unequalled anywhere in the world. (Ch. 45)
44. The desert biome, the Sonoran Desert of northwest Mexico. Among the plants seen here are the organ pipe cactus, mesquite, and brittle bush. (Ch. 45)

41

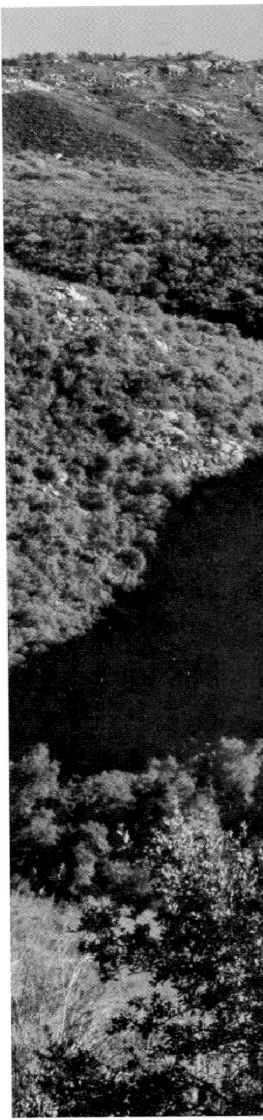

42

43

44

45

46

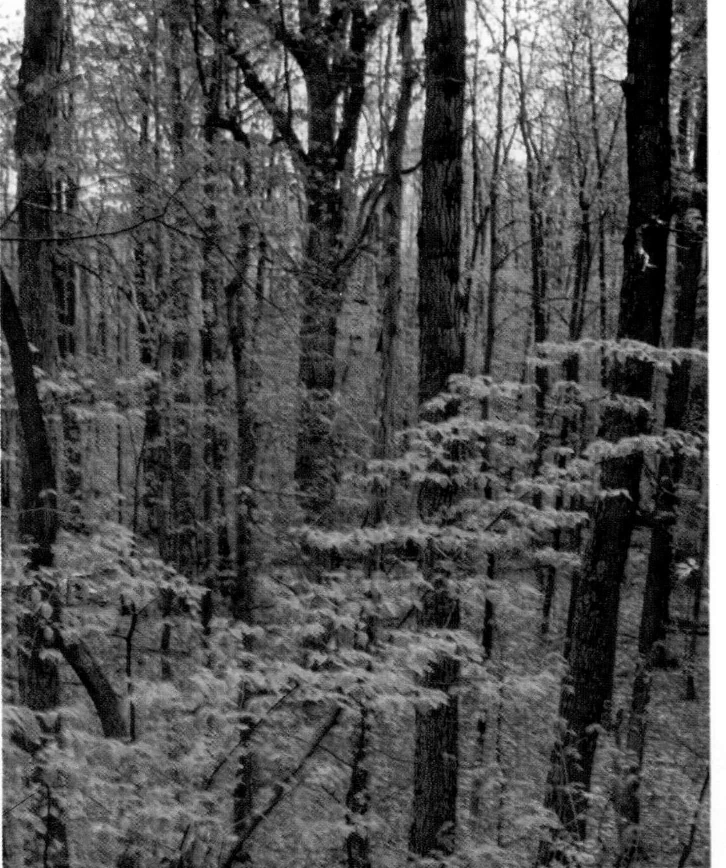

45. The tropical rain forest biome. A giant mora tree with its buttressed trunk in the rain forest on the island of Trinidad. (Ch. 45)

46. The temperate deciduous forest biome. Midsummer in a typical deciduous forest of eastern United States. (Ch. 45)

47. The tundra biome, Mt. McKinley Park, Alaska. (Ch. 45)

47

in length, live in the small intestine where they suck blood and lymph from the intestinal wall. Eggs are given off into the intestine and pass out with the feces. It has been estimated that a single worm may release as many as 10,000 eggs per day. The eggs soon hatch in the soil. The larvae feed on bacteria, grow and molt twice to form an infective nonfeeding third-stage larva in 1–2 weeks. When these worms come in contact with the bare skin of a human being, they bore through the skin and enter the bloodstream. Once in the body, they make the same tour as described for *Ascaris.* The debilitating effects of hookworm infection are largely due to the loss of blood. This loss is high because the worms continuously pump out blood and move to new sites, leaving the old ones still bleeding. It is estimated that the loss of blood is 0.2–0.5 ml per worm per day.

In the past, many people in the southern states were infected with hookworm. Educational programs, treatment, preventive measures such as wearing shoes, and improved sanitation have greatly improved this situation. However, there are still significant numbers of people in the South and large numbers in other areas of the world who are infected with hookworm.

The trichina worm, *Trichinella spiralis*, causes the disease trichinosis in humans, hogs, mice, and rats. Unlike infections of *Ascaris* and hookworm, in which eggs leave with the feces, this disease is transmitted when one animal eats the infected flesh of another animal. Humans obtain their infections from uncooked pork. Pigs usually become infected by eating garbage containing scraps of uncooked pork, or by eating diseased rats or mice. Thus because human flesh is almost never eaten by any kind of animal, when a trichina worm enters the human body it terminates the life cycle.

When people eat infected pork, the encysted worms emerge from the cysts in the intestine and soon develop into mature worms about 1 mm long. After mating, the female worms, which are viviparous, burrow into the intestinal wall and deposit their larvae, several thousand each, in blood or lymph vessels. These larvae are carried throughout the body where they leave the systemic capillaries and encyst in striated muscles, especially those of the diaphragm, chest, and tongue (Fig. 25.12). Infections produce rheumatic muscular pains, swelling, and other symptoms. In very heavy infections, death occurs. Ingestion of ten larvae per gram of body weight of the host is usually fatal. It is estimated that about 16 percent of the people in this country are lightly infected with *Trichinella.* It is impossible to inspect pork for the encysted worms because they are microscopic in size. Therefore the only sure way to prevent human infections is to cook pork thoroughly before eating it. Prohibition of the feeding of raw garbage to hogs, or requiring that such garbage be cooked prior to feeding, would greatly reduce the incidence of trichinosis. At this time there is no known treatment for an infection after the worms have encysted in muscle tissue. It is reassuring, however, that mild infections provide a strong immunity against later heavy and possibly fatal infections.

The filaria worm, *Wuchereria bancrofti,* is of great medical importance in many tropical and subtropical countries. Charleston, South Carolina, is the only locality in this country where human infections have been found. The adult worms live in the lymph glands and ducts. The females, which are from 7 to 10 cm (3–4 inches) in length, are threadlike; the males are only about half that size. The females give birth to tiny larvae, known as *microfilariae*, which enter the bloodstream. For the life cycle to be completed, these microfilariae must be sucked up by a mosquito. They undergo further development in the body of the mosquito and are then transferred back to humans when the mosquito bites. In some cases, after multiple infections over a considerable period of time, the worms block the lymph ducts in parts of the body. The result of this is the disease *elephantiasis,* in which the legs or other parts of the body swell to enormous proportions.

Phylogenetic Relationships

Nematodes have many characteristics in common with other members of the phylum Aschelminthes. The pseudocoelom; well-developed

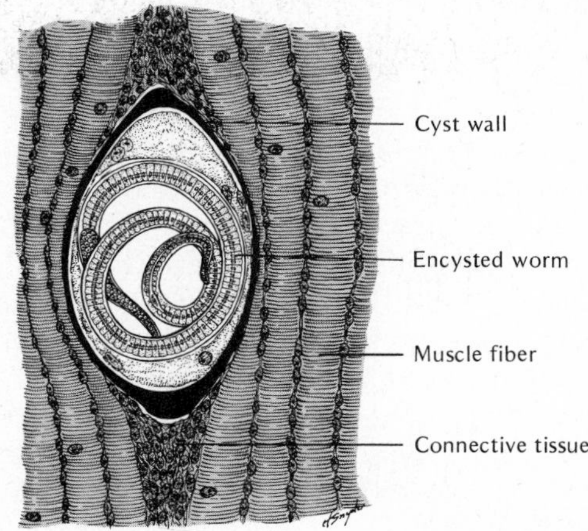

Figure 25.12 Trichina worm encysted in muscle. (From Turtox/Cambosco, Macmillan Science Co., Chicago.)

cuticle with various spines, bristles and scales; superficial segmentation; and molting of the cuticle are quite general among the rest of the classes. The pharynx of nematodes and gastrotrichs is strikingly similar. Caudal adhesive glands are found in free-living nematodes, rotifers, and gastrotrichs. The absence of protonephridia in nematodes is a problem, but marine gastrotrichs have a structure which resembles the renette cells of nematodes. Probably the closest relationship of the nematodes is with the kinorhyncha.

CLASS NEMATOMORPHA

This small class of "horsehair snakes" was formerly placed in the class Nematoda, to which they are closely related. However, the nematomorpha are sufficiently different from the nematodes to warrant placing them in a class by themselves.

Characteristics

1. Long, slender, round, unsegmented worms.
2. Cloaca present in both sexes.
3. No specialized excretory system.
4. Alimentary tract atrophied in adults.
5. Young stages parasitic in arthropods, adult-free living. Figure 25.13

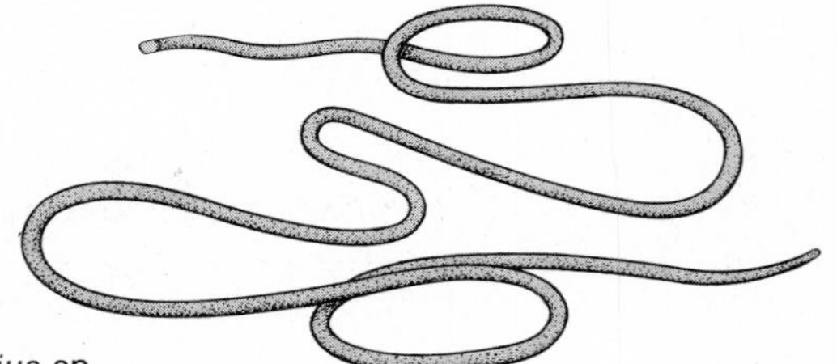

Figure 25.13 Class Nematomorpha, *Gordius* sp.

Classification

ORDER **Gordioidea** Freshwater forms. Parasitic stage in terrestrial or aquatic arthropods especially of the insect order Orthoptera. Pseudocoel much reduced by mesenchymal tissue. *Paragordius, Chordodes, Gordius.*

ORDER **Nectonematoidea** Marine forms with parasitic stage in crustacea; pseudocoel not reduced. *Nectonema,* the only genus.

Anatomy and Physiology

GENERAL STRUCTURE

Members of the class are very long and slender, opaque milky white to brown in color. The body wall is similar to the nematodes with a cuticle, cellular hypodermis, and a longitudinal muscle layer with cells resembling those of nematodes. They lack the longitudinal thickening of the hypodermis found in nematodes and do not have an excretory system. The simple alimentary tract becomes atrophied in the free-living adult stage. Even in juvenile forms it is more or less degenerate and it seems unlikely that food is ever ingested by these animals. The condition of the pseudocoel varies. Some have a well-defined cavity between the body wall and the gut while in others the space is more or less filled with mesenchyme cells. The nervous system is just below the hypodermis with a ganglion around the gut at the anterior end and a ventral nerve cord. There are few sensory structures.

REPRODUCTION AND DEVELOPMENT

Pairs of gonads run the length of the body in members of the order Gordioidea, while the marine genus *Nectonema* has but a single gonad. The genital ducts of both males and females open into the posterior part of the digestive tract and eggs and sperm reach the outside by way of the anus. Copulation involves the coiling of the posterior portion of the male around the female. Sperm are released near the female anus and they migrate into the female cloaca. Eggs are laid in long strings, held together by secretions from special glands. Early development is similar to that in nematodes, and the young larvae must enter the proper host or die. Within the host, for example, a grasshopper or cricket in the case of many freshwater species, development continues over a period of several weeks to several months. The larval stages have a retractible proboscis which is lost when they leave the host in or near water and undergo a final molt to the adult condition. A cricket falling into a watering trough would be relieved of its parasite, and the long thin worms, looking much like an animated horse hair, became the source of the legend about the "horse hair snake." Supposedly a horse hair changed into a moving, living organism.

PHYLUM ENTOPROCTA

Characteristics

1. Adults stalked and sessile, either solitary or colonial.
2. Small in size, most species are microscopic.
3. U-shaped digestive tract with both mouth and anus opening within a circlet of ciliated tentacles. Figure 25.14

   The phylum is not broken down into orders. *Loxosoma, Barentsia, Urnatella* are representative genera.

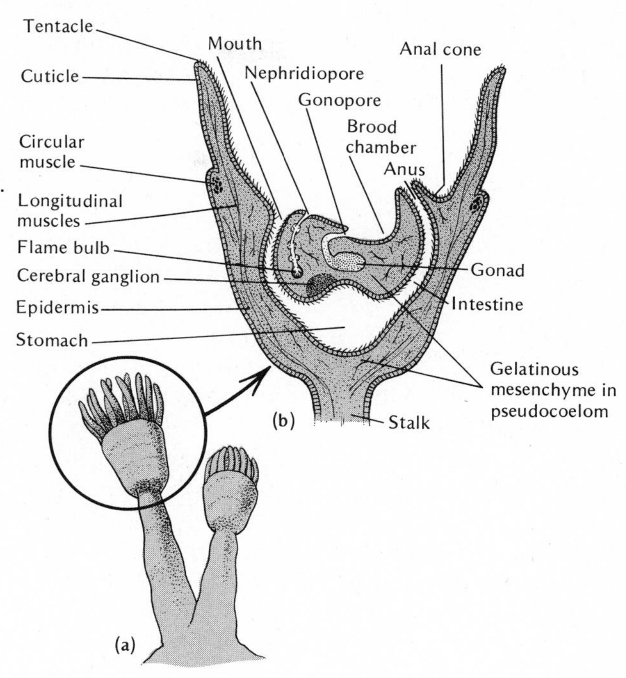

Figure 25.14 Phylum Entoprocta, *Urnatella* sp. **(a)** External appearance. **(b)** Diagram of cross section through an individual.

Anatomy and Physiology

GENERAL STRUCTURE

These small animals are all sessile as adults, have a crown of tentacles, a main body called the *calyx*, a stalk, and a basal attachment disk (Fig. 25.14). They appear superficially much like hydroid polyps, but the ciliation of the tentacles as well as the internal structure clearly distinguish them.

A secreted cuticle covers the outside of the calyx and the stalk. The tentacles have cuticle only on the outside surface, while the inside of the tentacles and all the surface within the *vestibule*, which the tentacles surround, lack cuticle and are largely ciliated. The ectoderm is cellular, and a layer of longitudinal muscles lies beneath it. The extensive pseudocoel between the body wall and the gut is filled with a gelatinous material containing numerous mesenchyme cells. A single ganglionic mass with nerves leading to the various structures constitutes the nervous system. Sense organs are simple tactile receptors, mostly on the tentacles.

The U-shaped digestive tract with mouth and anus both opening within the tentacular ring is ciliated and consists of a single layer of cuboidal or columnar cells. The entoprocts are filter feeders and currents set up by the tentacular cilia pass water through the tentacular crown where microscopic organisms are filtered from the water by the cilia. They are then moved to the mouth by special ciliated pathways.

Excretion is accomplished by a single pair of flame bulbs whose ducts meet and lead to the outside by a single nephridial pore located just posterior to the mouth.

REPRODUCTION AND DEVELOPMENT

Asexual Budding is very common among the entoprocts. Buds arise from the horizontal stolon or the upright branches in colonial species and from the calyx in solitary ones. In the latter case, the bud detaches from the parent and settles to the substrate. Regenerative capacity is high among the entoprocts. In some species when conditions become unfavorable, the calyces die and drop off but the stalks remain alive. With the return of favorable conditions, new calyces are regenerated.

Sexual Some entoprocts are monoecious and others are dioecious. In the monoecious species there are two testes and two ovaries, all of whose ducts unite and utilize a single gonopore. Dioecious forms have a pair of testes or ovaries whose ducts unite near the gonopore.

Fertilization is internal, either in the ovary or its duct. Early development occurs in the *brood chamber,* a depression in the vestibule between the gonopore and the anus. There is a ciliated free-swimming larva which varies with the species. Some are quite rotiferlike while others bear at least a superficial resemblance to a trochophore.

Ecology

There are less than 100 species in the phylum Entoprocta, and all are marine except for members of the small genus *Urnatella.* Freshwater forms have thus far been found only in eastern United States and in India, usually in running waters, attached to the undersides of rocks.

Marine entoprocts live in coastal waters, including the brackish waters of bays and inlets. One family, the Loxosomatidae, is made up of solitary individuals, all of which live on the surface of other animals. They are commonly found on sponges, annelid worms, and hydrozoan colonies. Often they live within the secreted tubes of annelid worms. The other family, the Pedicellinidae, is made up of colonial forms which grow on rocks and pilings or on the surface of algae, worm tubes, mollusk shells, and hydroid colonies.

Phylogenetic Relationships

The entoprocts were formerly united with the Ectoprocta in a single phylum, called the phylum Bryozoa. For a number of reasons this union is unacceptable. The presence of a pseudocoelom in the entoprocts and a true coelom in the ectoprocts is the most significant reason. The apparent structural similarity between the two groups is quite probably associated with their sessile existence. The most reasonable affinity of the entoprocta is with the rotifers. Features such as pedal glands which are found in some entoprocts, the general nature of the digestive system, and the preoral organ of some entoprocts which bears resemblance to the antennae of the rotifers are prominent reasons.

SUGGESTIONS FOR FURTHER READING

Bird, A. F. *The Structure of Nematodes.* New York: Academic Press (1971).

D'Hondt, J. L. "Gastrotricha," *Ann. Rev. of Oceanography and Marine Biology,* **9** (1971).

Edmondson, W. T. (Ed.). *Fresh Water Biology.* New York: Wiley (1959).

Goodey, T. *Soil and Freshwater Nematodes.* New York: Wiley (1951).

Grassé, P. (Ed.). "Nemathelminthes, Rotifers, Gastrotriches, et Kinorhynques," in *Traité de Zoologie,* vol. 4, pts. 2 and 3. Paris: Masson (1965).

Hyman, L. H. *The Invertebrates,* vol. 3. New York: McGraw-Hill (1951).

Hyman, L. H. *The Invertebrates,* vol. 5. New York: McGraw-Hill (1959).

Kaestner, A. *Invertebrate Zoology,* vol. 1. New York: Wiley-Interscience (1967).

Marshall, A. J., and W. D. Williams (Eds.). *Textbook of Zoology Invertebrates.* New York: American Elsevier (1972).

Noble, E. R., and G. A. Noble. *Parasitology.* Philadelphia: Lea and Febinger (1972).

Pennak, R. W. *Fresh-Water Invertebrates of the United States.* New York: Ronald Press (1953).

Zuckerman, B. M., W. F. Mai, and R. A. Rhode. *Plant Parasitic Nematodes,* vol. 2. New York: Academic Press (1971).

26 The Mollusks

INTRODUCTION TO THE SUBGRADE COELOMATA

Most of the animals in the groups covered thus far have been of relatively small size. One of the problems presented by increase in size is adequate distribution of materials to all parts of the body. With the appearance of the pseudocoelom, some distribution of materials was possible by way of the pseudocoelomic fluids. However, the full development of complex animals seems to stem from the appearance of a true coelom and the consequent development of two distinct layers of mesoderm with greater potential for differentiation of organ systems, including distribution mechanisms. The two main lines of evolutionary development of more complex animals are distinguished in several ways, one of which is the method of coelom formation. In protostome coelomates of the annelid–arthropod line, a solid mass of mesoderm originates from a single cell which can be identified early in development. This mesoderm splits to form the coelom, called a *schizocoel* (Gk. *schizo,* split; *koilos,* cavity). In this group the blastopore or region thereof becomes the mouth. Many of the schizocoels have a similar larval form, the trochophore, whose mouth region is derived from the blastopore.

In deuterostomes of the echinoderm–chordate line, the mesoderm and subsequently the coelom develop from outpocketings of the primitive gut. This type of coelom is called an *enterocoel* (Gk. *enteron,* intestine; *koilos,* cavity). The blastopore or region thereof becomes the anus in the deuterostomes (Fig. 26.1).

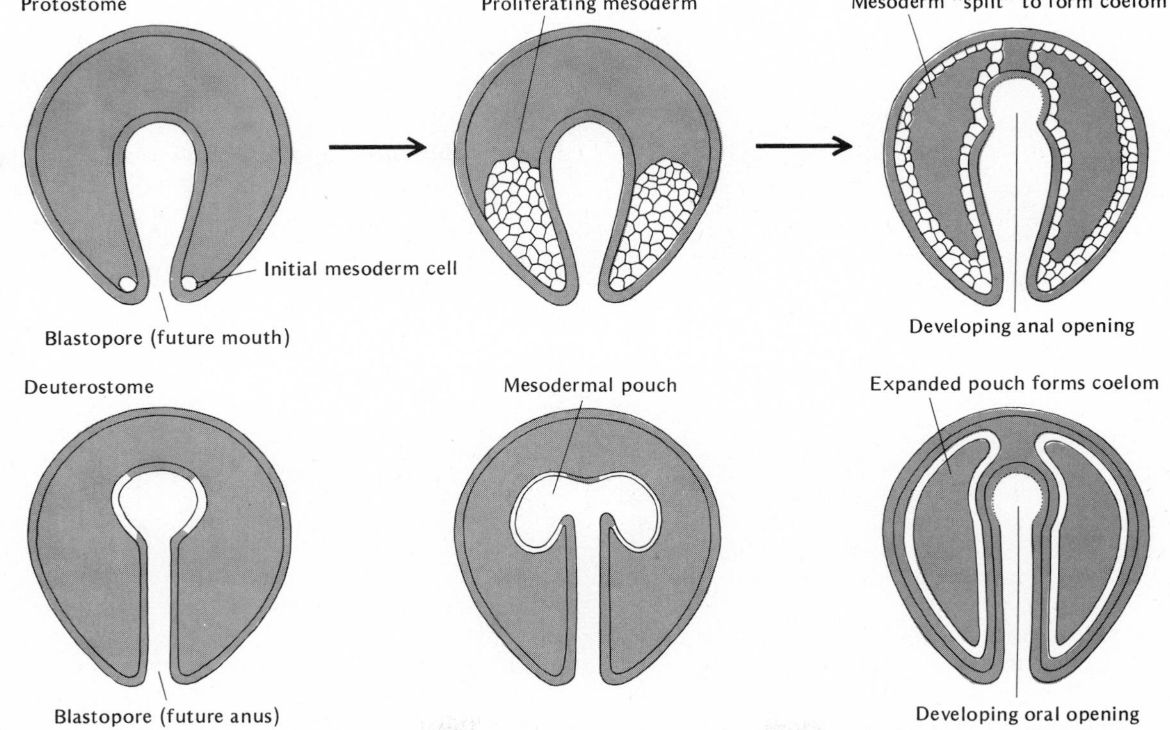

Figure 26.1 Schematic comparison of the early development and origin of the coelom in protostomes and deuterostomes.

PHYLUM MOLLUSCA

The large and ubiquitous phylum Mollusca is second to the phylum Arthropoda in number of described species. It is made up of six diverse classes having the same basic plan of organization. Representatives are found in marine, freshwater, and terrestrial environments.

Characteristics

1. Soft-bodied, mostly unsegmented.
2. Coelom reduced to the pericardial cavity, lumen of gonads and nephridia.
3. A ventrally located, highly muscular structure, called the foot, is present.
4. A *mantle* is present: an extensive evagination of the body wall which covers all or part of the body and usually secretes a protective shell.
5. Anterior mouth, typically provided with a characteristic rasping organ, called a *radula.*
6. Open circulatory system, except in some cephalopods. Figure 26.2

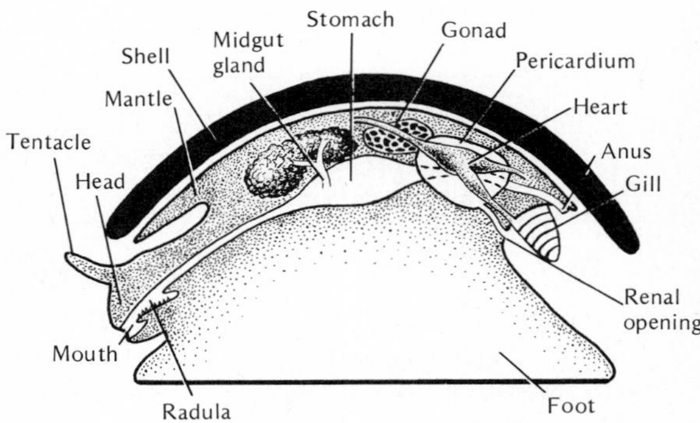

Figure 26.2 Molluscan body plan.

Classification

CLASS **Monoplacophora** Bilaterally symmetrical with a ventrally located foot. Single dome-shaped shell. Segmentation evidenced by the arrangement of internal organs: five pairs of gills, five pairs of nephridia, eight pairs of foot retractor muscles, ten pairs of pedal nerves. Radula present. Entirely marine, only four living species known, many extinct species (Fig. 26.3).

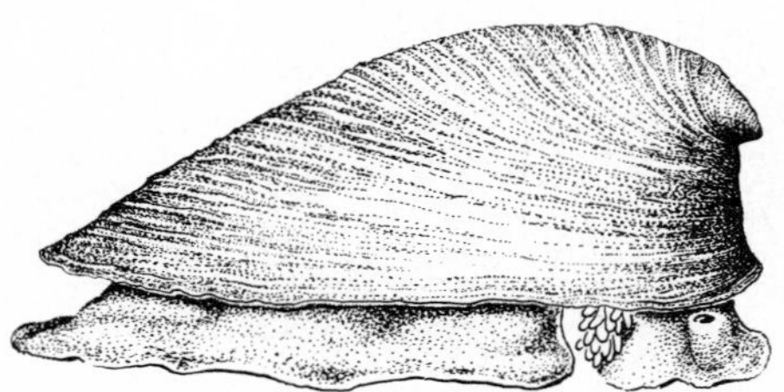

Figure 26.3 Class Monoplacophora, *Neopilina galathea.*

ORDER **Tryblidioidea** Only order of living forms.

CLASS **Amphineura** Dorsoventrally flattened and elongated. Extensive mantle and usually provided with a dorsal shell made up of eight serially arranged plates. Numerous external gills. Radula present, but head is poorly developed. Entirely marine (Fig. 26.4 Pl. 20).

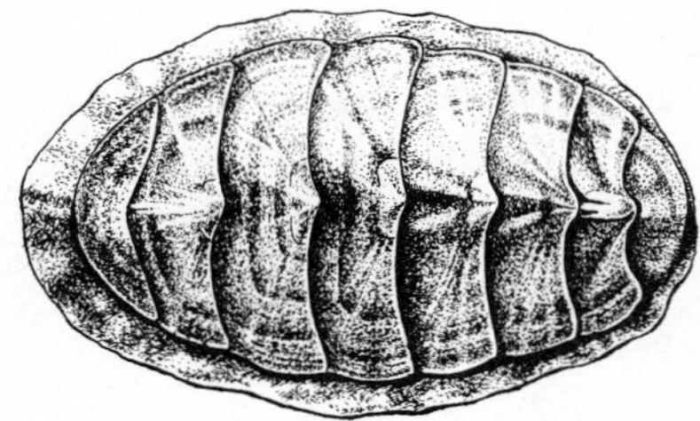

Figure 26.4 Class Amphineura, a typical chiton.

ORDER **Aplacophora** Wormlike forms with shell reduced to calcareous spicules; foot also reduced. *Chaetoderma.*

ORDER **Polyplacophora** Flattened elliptical forms with shell consisting of eight overlapping valves. *Chiton.*

CLASS **Gastropoda** Asymmetrical body, usually with a one-piece spirally coiled shell. Well-developed head with eyes and tentacles. Radula present. Ventral portion of foot flattened for creeping. Marine, freshwater, and terrestrial (Fig. 26.5).

Figure 26.5 Class Gastropoda, a typical snail.

CLASS **Scaphopoda** Bilaterally symmetrical with a single tubular tusklike shell which is open at both ends. Gills absent. Foot elongated. Mouth surrounded by lobed outgrowths. Radula present. Specialized food-gathering structures called *captacula* on head. Entirely marine (Fig. 26.6). The class is not broken down into orders.

Figure 26.6 Class Scaphopoda, a typical tooth shell.

CLASS **Pelecypoda** Bilaterally symmetrical with a bivalve shell, lateral in position. No head, jaws, or radula. Paired well-developed gills. Marine and freshwater (Fig. 26.7).

Figure 26.7 Class Pelecypoda, a scallop.

CLASS **Cephalopoda** Bilaterally symmetrical. Shell spiral chambered in a few, more often internal or absent. Head very well developed and provided with large complex camera eyes, jaws, and a radula. Foot modified to form armlike tentacles usually provided with suckers. One or two pairs of well-developed gills. Circulatory system mostly closed. Entirely marine (Fig. 26.8).

Figure 26.8 Class Cephalopoda, an octopus.

CLASS MONOPLACOPHORA

Discovery and Fossil History

Prior to 1957 this class was believed to have been extinct since early Devonian times, about 350 million years ago. However, the Danish research ship Galathea dredged some material from the ocean bottom at a depth of 3570 m off the coast of Costa Rica in 1952 which was found, when the samples were sorted in 1957, to contain specimens of a mollusk which clearly belonged in this group. The animal had certain features quite similar to a long-extinct genus *Pilina,* and it was given the name *Neopilina galathea* by Lempche who discovered the specimen. Since that time three other species have been collected in other areas.

Anatomy and Physiology of *Neopilina*

The structure of *Neopilina* is shown in Figure 26.9. It is a small animal, less than 40 mm in length. In ventral view one sees a small head, large circular foot with a ciliated surface and surrounded by the pallial groove. The segmentally arranged gills are found in the pallial groove. The foot has eight pairs of segmentally arranged retractor muscles. The scars of these muscles on the fossil monoplacophoran shells were the sole indication of the segmentation, which was quite apparent when the soft parts became available for study.

Neopilina is apparently a detritus feeder since the intestines contained skeletons of diatoms, radiolarians, and foraminiferans. It has an anterior mouth with a flattened *velum* on both sides and a series of tentacles to its rear. A well-developed radula, the characteristic molluscan rasping organ, is found in the mouth. The radula is a strip of chitin provided with recurved teeth which is pulled back and forth by muscular action over a protrusible cartilaginous rod located beneath it. Next comes the pharynx with two diverticulae and a single salivary gland, followed by the esophagus and stomach with a pair of digestive glands. The coiled intestine leads to the posterior rectum and anus. The excretory system consists of six pairs of segmentally arranged metanephridia, of which two pairs also serve as ducts for the gonads. Metanephridia have two openings, one into the coelom and the other to the outside. This contrasts with the more primitive protonephridium which has but one opening, to the outside. There is a circumoral nerve ring with two pairs of longitudinal cords, the pedal nerve in the foot, and the lateral nerve in the roof of the *pallial groove.* Segmentally arranged connections join the two sets of cords. There is an open circulatory system. The heart is divided by the rectum, resulting in a ventricle and two auricles on each side (Fig. 26.9). The ventricles empty into the dorsal

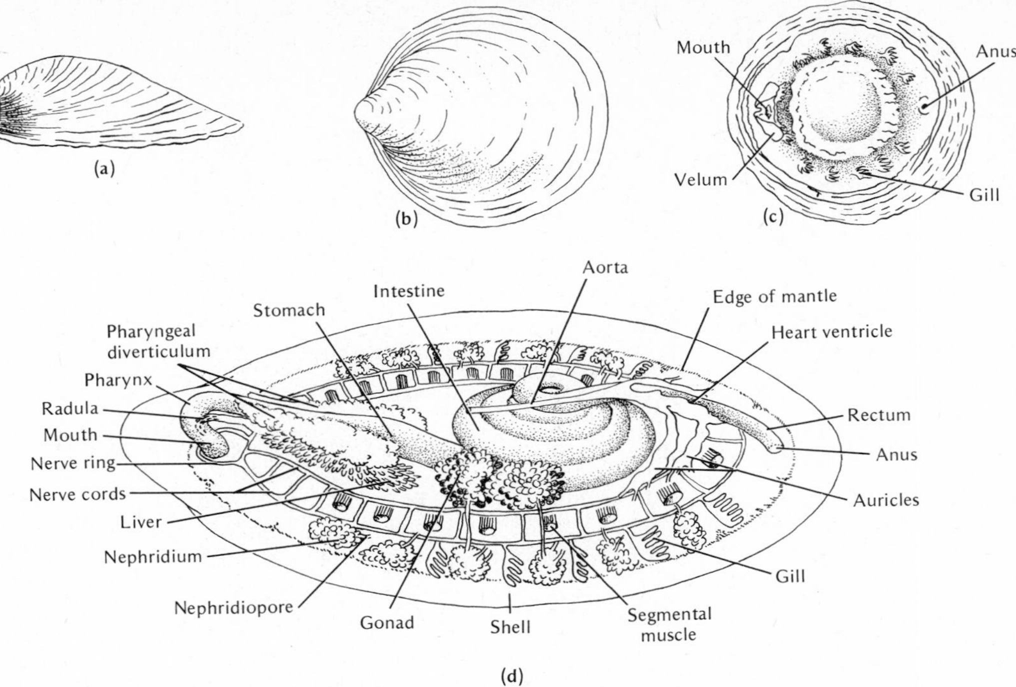

Figure 26.9 *Neopilina galathea.* **(a)** Left view of shell. **(b)** Dorsal view of shell. **(c)** Ventral view. **(d)** Internal structure.

aorta which extends to the anterior blood sinuses. Blood moves through the sinuses to the gills where it is aerated and sent back to the auricles. There are two pairs of gonads, located below the intestine, and the sexes are separate. Sex cells pass from the gonads to a nephridium by a short gonoduct and to the outside via the nephridial ducts.

Phylogenetic Relationships

The phylogenetic significance of *Neopilina* rests in the clearly segmental arrangement of several of its systems. The segmental arrangement serves as additional evidence of common ancestry for the mollusks and annelids. This concept of common ancestry has long been accepted because of the practically identical spiral, determinate cleavage patterns, and the trochophore larvae of these two phyla.

CLASS AMPHINEURA

Anatomy and Physiology of *Chiton*

GENERAL STRUCTURE

The chitons (Fig. 26.10) are common intertidal forms which cling tenaciously to rocks. Muscles in the foot pull the central portion of the foot away from the rock surface, creating a very effective suction cup. If the suction is broken by gently prying up the edge of the foot, chitons are easily removed. Members of many species roll up into a ball when they are taken from a flat substratum.

A typical chiton is elongate oval in shape and bears eight calcareous overlapping *valves* on its convex dorsal surface. Lateral to the valves is a thickened portion of the mantle, often covered with spines or bristles, called the *girdle.* In some species the entire dorsal surface is covered by the mantle with the shell often reduced or entirely absent. In ventral view, most of the surface is that of the large flat foot. The small poorly developed head is anterior, and there is a pallial groove between the foot and the mantle. Numerous fingerlike gills hang down in the pallial groove. The digestive system is quite similar to that described for *Neopilina.*

The heart, located in the dorsal pericardial cavity at the posterior end, is made up of two auricles and a ventricle which leads into the dorsal aorta. A pair of metanephridia connect the pericardial cavity with the outside. The metanephridia are more than just ducts. They are well-developed highly vascularized excretory structures in which filtration and resorption occur. A circumoral nerve ring with a pair of pedal nerves and a pair of pallial nerves make up the nervous system. There are serially arranged connections between these cords.

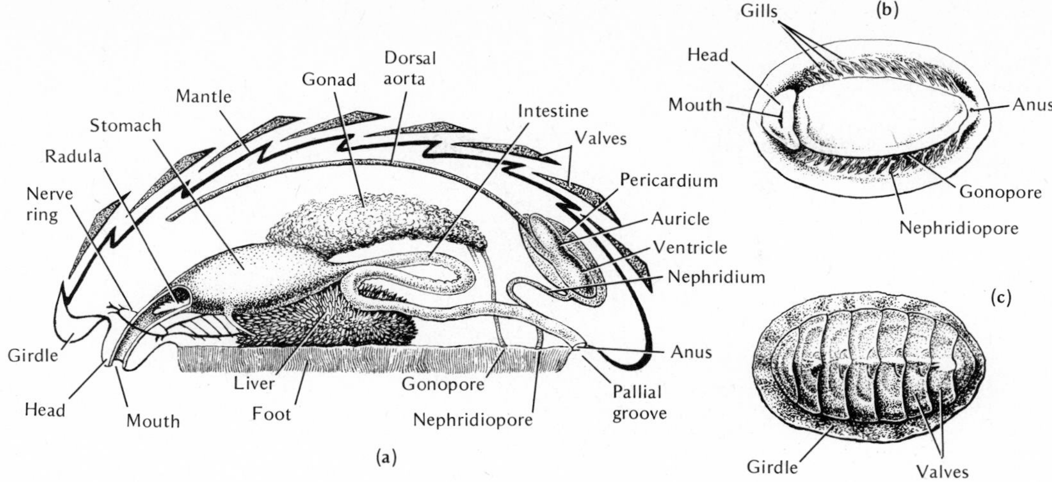

Figure 26.10 A typical chiton. **(a)** Sagittal section showing internal structure. **(b)** Ventral view. **(c)** Dorsal view.

REPRODUCTION AND DEVELOPMENT

Chitons are dioecious with a single dorsally located gonad and paired gonoducts opening into the posterior portion of the pallial groove. There is no copulatory apparatus and there is no special mating behavior. Eggs and sperm are shed into the sea, where fertilization occurs. Secretions released with the sperm stimulate the females to release their eggs, often within a minute. Females may be stimulated to release eggs by males that are as much as 7.5 m (25 feet) away.

In some species the females brood their eggs in the pallial groove, carrying them until they are well-developed small chitons. One viviparous species was reported from Chile. A female had fifteen small chitons within her ovary. How fertilization is accomplished is unknown.

Development is by spiral determinate cleavage with a typical trochophore larva (Fig. 26.11).

Ecology

Chitons are primarily animals of the intertidal zone in regions where the shore is rocky. As mentioned above, they cling firmly to the rocks, remaining in one secluded spot during the day (Pl. 20). At night they come out to feed, primarily on algae which they scrape from the surface of the rocks with the radula. They normally range not much more than a meter from home base, usually returning to the same spot after feeding. If taken not more than 1.5 m from their normal feeding areas, they easily return to their normal resting place. When taken a greater distance, they establish a new territory.

Being sluggish animals, the chitons are often covered by a variety of plants and sessile animals. As many as 26 different species of plants and animals have been found on a single chiton. In addition, a number of commensal forms utilize the pallial groove as a sheltered residence.

CLASS GASTROPODA

This largest of the molluscan classes is made up of the familiar snails and slugs as well as limpets, nudibranchs, and other marine forms which may be less well known. As an example

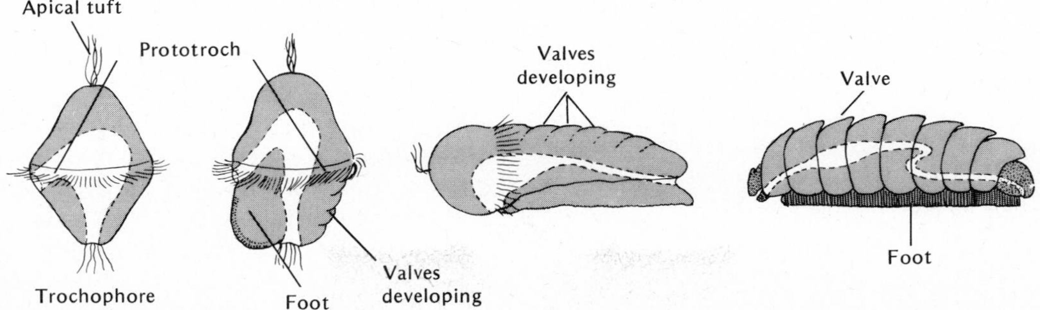

Figure 26.11 Metamorphosis of trochophore to adult chiton.

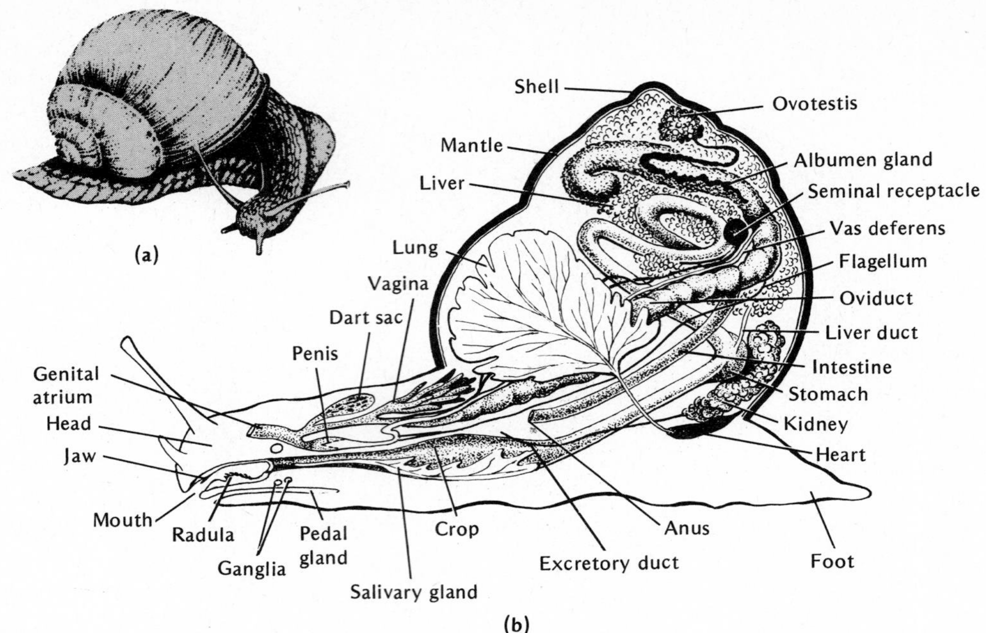

Figure 26.12 *Helix,* the garden snail, a typical pulmonate gastropod. **(a)** External appearance. **(b)** Internal anatomy.

of a typical gastropod, the common garden snail *Helix* will be discussed (Fig. 26.12).

Classification

SUBCLASS **Prosobranchia** Visceral hump with 180-degree torsion. Gills anterior to heart. Mantle cavity opens anteriorly. Sexes separate with a single genital opening. One pair of tentacles. Mostly marine.

ORDER **Archeogastropoda** Siphon, proboscis, and penis absent. Gills with two rows of plumelike filaments. Nervous system not concentrated. *Patella* (limpet), *Haliotis* (abalone).

ORDER **Mesogastropoda** Only one gill, auricle, and nephridium. Gill with a single row of filaments. Penis and operculum, a portion of shell on the foot which covers the shell aperture when the animal withdraws, are usually present. *Littorina* (periwinkle), *Strombus* (conch), *Cypraea* (cowrie).

ORDER **Stenoglossa** Shell with a siphonal canal. Nervous system highly concentrated. *Busycon* (whelk), *Urosalpinx* (oyster drill).

SUBCLASS **Opisthobranchia** Shell reduced or absent. Hump partially or completely uncoiled. Two pairs of tentacles. Hermaphroditic. Exclusively marine.

ORDER **Tectibranchia** Shell usually present, but reduced. Mantle cavity and gills present. *Aplysia* (sea hare), *Cavolinia* (sea butterfly).

ORDER **Nudibranchia** No shell, gills, or mantle cavity in adult. *Hermissenda, Doris* (sea slugs).

SUBCLASS **Pulmonata** Mostly small. Shell a simple spiral or absent. Mantle cavity anterior. Gills absent and respiration occurs in pulmonary sac, the modified mantle cavity, with a contractile pore on the right side. Mostly freshwater and terrestrial.

ORDER **Basommatophora** One pair of tentacles with eyes at the base. Freshwater snails. *Planorbis, Physa, Lymnaea.*

ORDER **Stylommatophora** Two pairs of tentacles with eyes at tip of posterior pair. *Helix* (garden snail), *Limax* (slug).

Anatomy and Physiology of *Helix,* the Garden Snail

GENERAL STRUCTURE

The flattened ciliated foot, with which the snail moves on a bed of mucus secreted by the pedal gland, is a most characteristic feature. There is a well-developed head with two pairs of retractable tentacles. The eyes are located on the tips of the posterior pair, and both pairs also have tactile functions. The dorsal surface, which is elongated and coiled, is called the visceral hump, and it contains most of the internal organs. A coiled shell covers the visceral hump, and the entire animal can retract into the shell primarily by the action of the columellar muscle which connects the animal with the apex of the coiled shell. The mouth opens into the pharynx on the dorsal surface of which is located a chitinous jaw. The radula, on the floor of the pharynx, may be protruded through the mouth opening for rasping off bits of plant material upon which the snail feeds. A pair of salivary glands open into the pharyngeal cavity. Next is an esophagus followed by a thin-walled crop and muscular stomach. The intestine is long and coiled, returning to the anus in the anterior part of the mantle cavity. A large digestive gland, often called the *liver,* is located in the dorsal part of the visceral hump and is connected to the stomach by means of a duct. Digestion occurs both extracellularly in the stomach and intracellularly within the liver.

Although most gastropods carry on respiration by means of gills, the land snails as well as most freshwater species have a modified mantle cavity called a *lung*. A small opening allows air to enter and leave, and the walls of the cavity are highly vascular. The heart, consisting of a single auricle and ventricle, receives blood from the mantle and pumps it through arteries to sinuses in the head, foot, and visceral hump from whence it passes to the lung. One kidney leads from the pericardial cavity to the mantle cavity. The nervous system consists of a cerebral ganglion, dorsal to the pharynx, and pedal, pleural, and combined visceral and parietal ganglia close together and ventral to the pharynx.

REPRODUCTION AND DEVELOPMENT

Like other pulmonate snails, *Helix* is hermaphroditic. The gonad, called an *ovotestis,* is located in the apex of the coiled visceral hump. Sperm pass from the ovotestis through the hermaphroditic duct and into the vas deferens which leads to the penis in the anteriorly located *genital atrium.* Sperm are enclosed in small packets, called *spermatophores,* in the long tubular *flagellum* which extends posteriorly from the base of the penis. Eggs leave the ovotestis and pass through the hermaphroditic duct to the albumen gland, through the oviduct to the vagina and genital atrium, and out through the genital pore. A reciprocal exchange of sperm between two snails follows a rather elaborate mating procedure. Each snail shoots into the other one or more sharp calcareous *darts* which are produced and stored in the *dart sac,* an evagination from the genital atrium. Presumably this stimulates sexual activity. Next the penis of each snail, in turn, is placed in the genital atrium of the other and spermatophores are deposited therein. Following copulation, eggs are fertilized, covered with albumen, and deposited in large masses in holes in the soil. Some species of pulmonate snails retain the eggs until hatching, a condition called *ovoviviparity.* Development is direct, with minute snails hatching after several weeks.

Development of Marine Gastropods

Development of marine gastropods includes the characteristic trochophore larva with later appearance of a typically molluscan larva, called a *veliger.* The veliger is characterized by a swimming organ called the *velum,* made up of two large ciliated semicircular folds (Fig. 26.13).

Torsion of the Gastropod Shell

The degree of coiling of the visceral hump varies among the various groups of gastropods. In the opisthobranchs some forms become highly coiled, and during development there is a loss of one gill and one auricle. It is believed that the ancestral gastropods were uncoiled. In addition to the coiling seen in many gastropods, there is another type of twisting in many, which is called *torsion.* The members of the subclass Prosobranchia show 180-degree torsion bringing the gills and anus to an anterior position. The former left visceral ganglion ends up on the right side and the right visceral ganglion on the left. In the subclass Opisthobranchia there are some forms which undergo the usual 180-degree torsion during early development, but later the torsion is reversed either partially or completely. The nudibranchs show complete detorsion. Figure 26.14 diagrams the condition believed to exist in the hypothetical gastropod ancestor and shows the effects of torsion in the three subclasses.

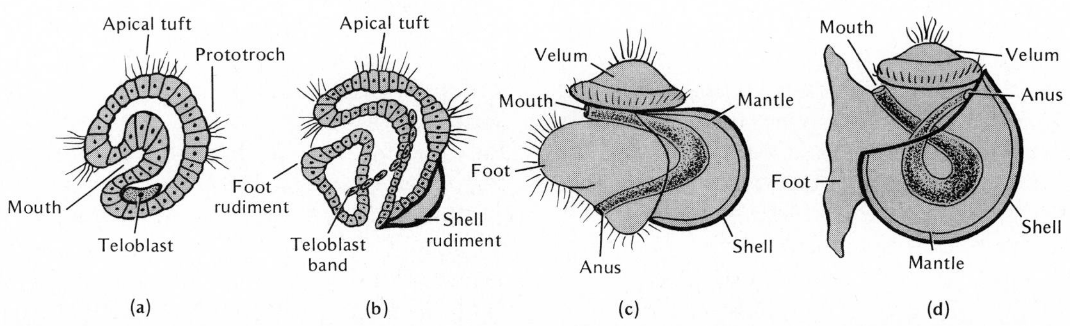

Figure 26.13 Development in a marine gastropod. **(a)** Section through a trochophore larva. **(b)** Early veliger larva. **(c)** Mature veliger prior to torsion. **(d)** After torsion.

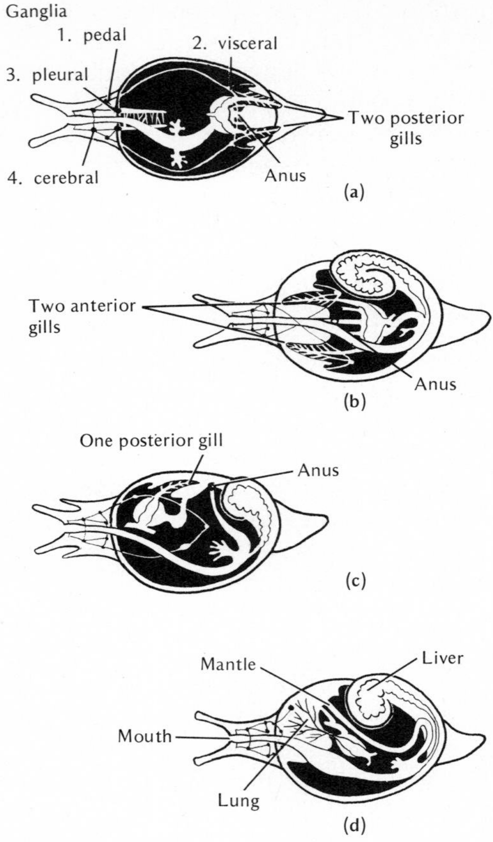

Figure 26.14 Diagram of hypothetical gastropod ancestor and the general organization in the three subclasses. Note the changes in the nervous system as a result of torsion as well as its effects on the gills and on the heart. **(a)** Hypothetical ancestor. **(b)** Prosobranchia. **(c)** Opisthobranchia. **(d)** Pulmonata.

Ecology

While many gastropods are algal grazers like the chitons or feeders on other plant material like the garden snail, others have evolved a variety of feeding habits. Some, like the slipper shell, *Crepidula,* are plankton feeders. Thick mucous secretions of the mantle trap the plankton, and ciliary pathways carry it to the mouth. Others live in tubelike shells and capture plankton with the tips of their gills which protrude from the open end of the tube. The giant West African land snail, *Achatina achatina,* which is shown in Plate 19, is herbivorous. Such snails have become serious pests in many parts of the Indo-Pacific region into which they were introduced because of their food value.

Many gastropods are carnivorous predators. In the genus *Conus,* the radula apparatus is modified to produce sharp poison-covered darts which are shot into the prey. The poison of some species is potent enough to kill humans. In the oyster drill, *Urosalpinx,* an accessory gland secretes materials which soften the shell, enabling the radula to bore a neat round hole and gain access to the soft flesh within. This serious pest to the commercial oyster beds, a native of our eastern seaboard, has been introduced all over the world in the transfer of young oysters—the oyster spat.

Respiration in marine gastropods is by means of one or two specialized outgrowths of the mantle surface called *gills* or *ctenidia.* Some gastropods, such as the pelagic heteropods, have a much reduced shell and the familiar terrestrial slugs have a reduced internal shell or none at all. The sea slugs or nudibranchs also lack shells. Certain nudibranchs of the family Eolidae, such as *Hermissenda crassicornis,* have fingerlike evaginations on their dorsal surface, called *cerata* (Pl. 16). The cerata contain small ducts which are actually diverticulae from the gut. *Hermissenda* feeds on hydrozoan coelenterates such as *Obelia* and somehow inhibits the discharge of the nematocysts which are freed by digestive action and pass into the ducts in the cerata to be stored in small muscular sacs at the tip. Upon stimulation the sacs contract and squeeze out the nematocysts which now discharge and function in protection of their "new master"—a truly remarkable adaptive feature. A pelagic nudibranch, *Glaucus* sp. which is shown in Plate 35, has a reduced head and tentacles, a taillike foot and three pairs of lateral projections which are tipped with tentaclelike cerata. As in *Hermissenda,* the cerata have small muscular sacs which contain nematocysts derived from siphonophores such as *Porpita* (Pl. 38) upon which *Glaucus* feeds.

CLASS SCAPHOPODA

The tooth shells comprise a small class of less than 300 species. They are entirely marine, living partly buried in the substrate from shallow coastal waters down to depths as great as 5000 m. *Dentalium* (Fig. 26.15) is a typical scaphopod. The body is greatly elongated and the tubular mantle secretes a toothlike shell which is open at both ends. The pointed foot protrudes from the larger opening and is used for digging and pulling the animal down into the sand or mud. The head is very poorly developed with many delicate filamentous tentacles, called *captacula,* surrounding the mouth. These ciliated structures are prehensile and bear a knoblike expansion at the tip. They function in food capture and in sensory perception. Gills are absent and the extensive tubular mantle serves for respiratory exchange.

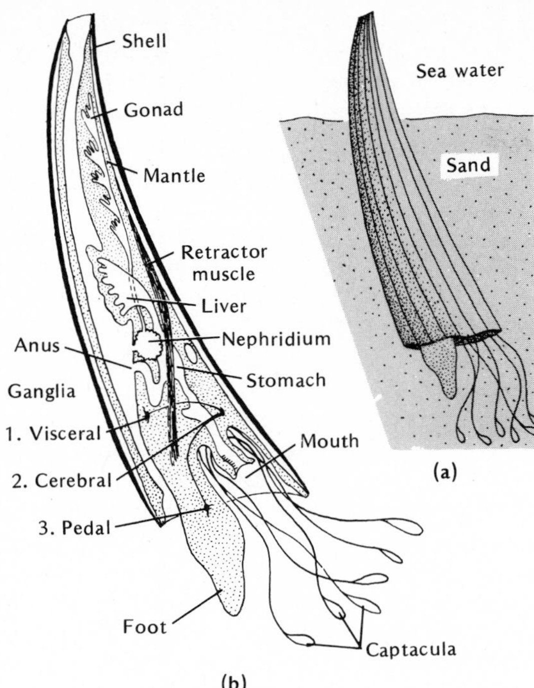

Figure 26.15 *Dentalium,* a typical scaphopod. **(a)** Adult organism in its normal habitat. **(b)** Anatomical detail.

The digestive system is typical with a radula in the mouth cavity which can be protruded. The circulatory system is reduced to a few sinuses and there is no heart. Two nephridia, with nephridiopores in the mantle cavity near the anus, are present, but they do not open internally since there is no pericardial cavity. The nervous system has the typical molluscan ganglia, but specialized sense organs are lacking. Sexes are separate and the unpaired gonad makes up most of the apex of the body. It makes use of the right nephridiopore for release of eggs or sperm. Development is typical with a trochophore and veliger larva.

CLASS PELECYPODA

Classification

ORDER **Protobranchia** Gill with two divergent rows of flat filaments arising from a central axis. Foot flattened on ventral surface. *Nucula* (nut clam), *Solemya* (awning clam).

ORDER **Filobranchia** Gill filaments elongate and reflexed, usually with adjacent lamellae attached at intervals by ciliary tufts. Often attached to substrate by byssus threads. *Mytilus* (mussel), *Pecten* (scallop), *Ostrea* (oyster).

ORDER **Eulamellibranchia** Gill filaments elongate and reflexed with adjacent lamellae firmly attached at regular intervals by vascularized tissue. *Unio* (freshwater clam), *Mercenaria* (edible clam), *Venus* (hard-shelled clam), *Ensis* (razor clam), *Teredo* (the shipworm).

Anatomy and Physiology of a Freshwater Clam

GENERAL STRUCTURE

Clams are often studied as representative mollusks because they are fairly large in size and easy to obtain. The clam, oyster, and other mollusks with two lateral shells are called *bivalves,* and they are placed in the class Pelecypoda. This name means "hatchet foot," referring to the shape of the foot in many species. The clam is an atypical mollusk since it lacks a head and a radula.

The two equal shells or valves, one on the right and one on the left side, are hinged along the dorsal edge by an elastic ligament. Near the anterior edge of this ligament each valve bears an elevated hump, the *umbo.* This is the oldest region of the shell. As the clam grows, the mantle secretes successive layers of shell, and the concentric lines on the shell represent successive growth increments. The thin outer layer of the shell, called the *periostracum*, is made up of a horny protein similar to keratin, called *chonchiolin* (Fig. 26.16). The middle, *prismatic layer* is of calcium carbonate crystals with the crystals oriented perpendicular to the

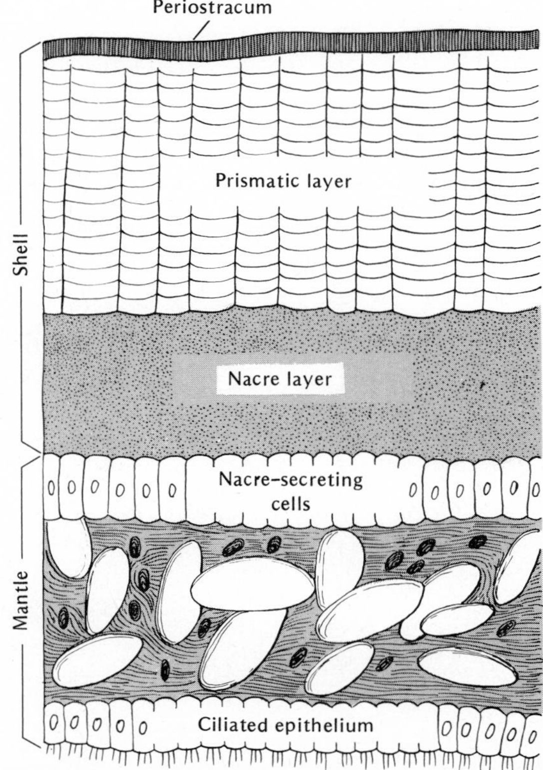

Figure 26.16 Diagrammatic cross section of a clam shell and the underlying mantle.

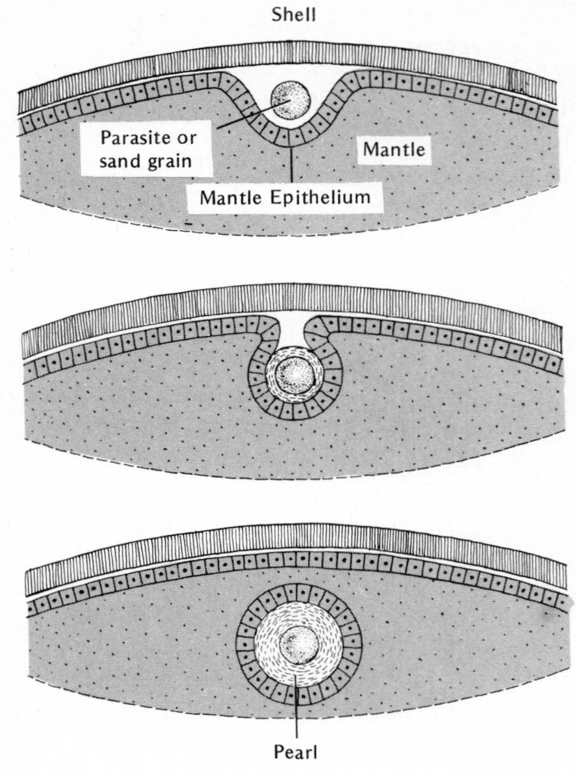

Figure 26.17 Pearl formation.

mantle. The innermost layer, just beneath the mantle, is the *nacreous layer* or mother-of-pearl. Here the calcium carbonate crystals are parallel with the surface of the mantle. The periostracum and prismatic layers are secreted by the edge of the mantle while the nacreous layer is secreted by the entire outer surface of the mantle. Foreign materials such as sand grains or even parasites which get between the mantle and the shell are covered by successive layers of nacre, and thus pearls are formed (Fig. 26.17).

Within the two valves, as shown in Figure 26.18, are the *mantle lobes* surrounding the *mantle cavity.* At the posterior end the mantle lobes are modified to form the *incurrent* and *excurrent* siphons. Projecting into the mantle cavity, on each side, are two large platelike gills. The foot occupies the central part of this cavity and the *visceral mass* is continuous with the foot dorsally. Two large muscles, the *anterior* and *posterior adductors,* are attached to the inner surfaces of the valves. Contraction of these muscles brings the edges of the valves together. When these muscles are relaxed, the valves separate due to the elasticity of the ligament, and the muscular foot may be protruded. Locomotion in the clam is a very slow process. When the foot is extended into the mud or sand, the tip of the foot is filled with blood and swells, forming an anchor. The foot is then contracted and this draws the body of the clam forward.

The inner surface of the mantle and the surfaces of the gills are ciliated. The beating of these cilia causes water to enter the incurrent siphon. The water passes over the sievelike gills

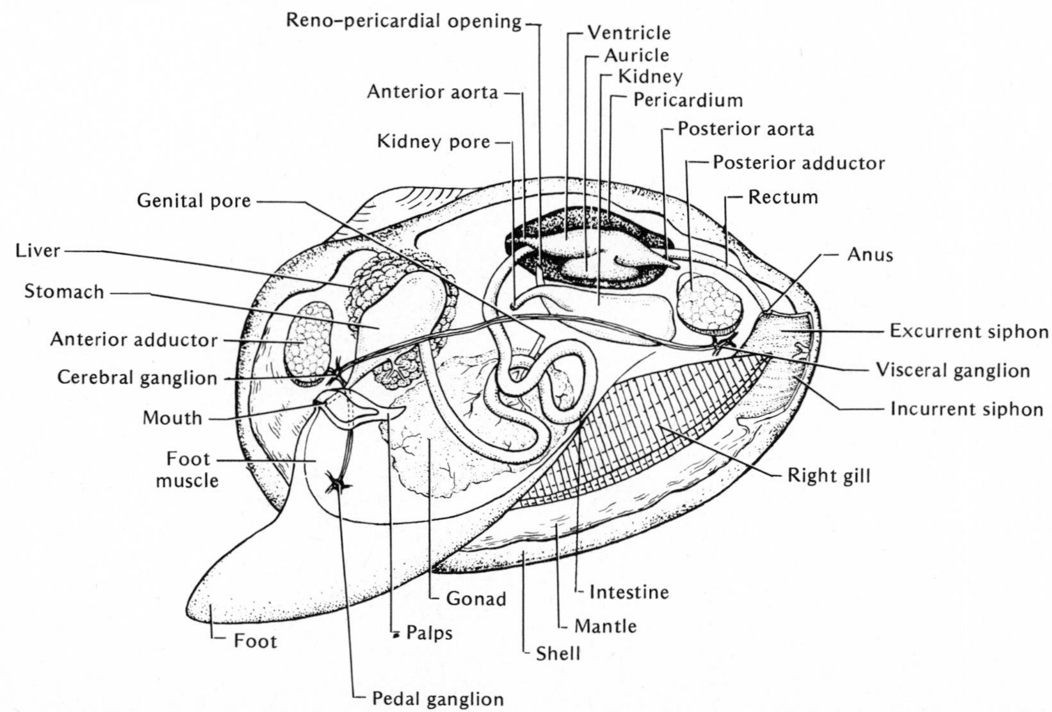

Figure 26.18 Internal anatomy of a clam.

and enters water tubes within them. The water flows from the tubes into a chamber at the dorsal edge of the gills and then out the excurrent siphon. As the water passes over the gills, dissolved oxygen is taken up by the blood in the gills and carbon dioxide is given off. The current of water, which is continually passing over the gills, also provides the clam with its food. Minute organic particles become entangled in mucus secreted by the gills and are passed by ciliary action to the mouth. On either side of the mouth are two ciliated flaps, the *labial palps.* From the mouth a short esophagus leads into the stomach, which in turn leads into a long coiled intestine. The anus opens near the excurrent siphon. A large digestive gland is connected with the stomach, and there is evidence that much of the digestion occurs intracellularly in this gland.

Associated with the digestive gland is a structure called the *crystalline style,* found also in other groups of mollusks. This transparent gelatinous structure is formed in a ciliated diverticulum of the stomach. The cilia beat and force the tip of the crystalline style out into the stomach where it provides an amylase used in starch digestion.

The coelom in the clam is reduced to the pericardial cavity surrounding the heart. The intestine passes through this cavity and through the heart. The circulatory system is an open one. The heart, which consists of a ventricle and two auricles, pumps blood over the body through the arteries, which empty into sinuses where the exchanges occur. The blood is returned to the heart by veins. Most of this returning blood goes to the kidneys where wastes are removed. The two kidneys are located just below the pericardial cavity on each side. Each kidney is a bent tubular structure, differentiated into a ventral glandular region where wastes are removed from the blood by filtration, and a dorsal bladder where wastes are stored. The glandular region connects with the fluid-filled pericardial chamber. The bladder, which narrows into a ureter empties into a chamber in the gills that leads to the excurrent siphon.

The nervous system in this essentially headless animal is reduced. There are three pairs of ganglia, one pair for each of the main regions of the body: the *cerebral ganglia* are located on either side of the mouth and are joined by a connective that runs over the esophagus, the *pedal ganglia* in the foot are connected with the cerebral ganglia by two connectives, and the *visceral ganglia,* located just below the posterior adductor muscle, are also connected with the cerebral ganglia by a pair of connectives. Sense organs are poorly developed. An organ of equilibrium, the statocyst, is located in the foot near the pedal ganglia. Sensory cells located near the visceral ganglia apparently function as chemoreceptors, as there is evidence that they are sensitive to chemicals in the water that enters the incurrent siphon. The edges of the mantle and siphons contain sensory cells that are affected by touch and light (Pl. 22).

REPRODUCTION AND DEVELOPMENT

Reproduction is sexual, and in most species the sexes are separate. The gonads surround the coils of the intestine that lie in the foot, and the genital ducts empty into the dorsal gill chamber near the points where the ureters empty. In the freshwater clam the female releases eggs through a short oviduct into the mantle cavity, and the eggs lodge on the interior of the gills. Sperm, released by a male, are carried by the water entering the incurrent siphon into the mantle cavity of the female where fertilization occurs. The zygotes develop, while attached to the gills of the female, into bivalved larvae known as *glochidia* (Fig. 26.19).

Figure 26.19 Glochidium larva of a freshwater clam.

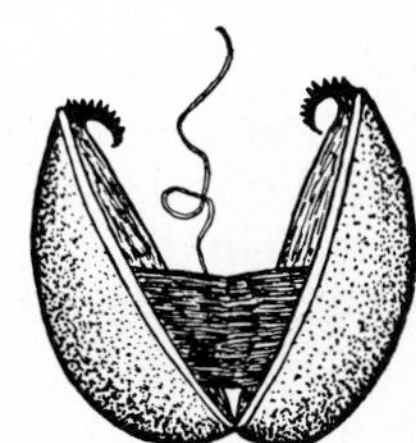

These glochidia are released into the water and further development will occur only if they become attached to the gills or fins of a fish within a few days. After attachment the skin of the fish grows over the glochidia, forming "blackheads," which live as parasites in this position for a period of weeks. When they have developed into young clams, they drop from the fish and assume an independent existence. This stage in the life history of the freshwater clam is an adaptation for the spread of the species.

Marine bivalves, on the other hand, shed both their eggs and sperm directly into the water, and development occurs there without any parasitic stage. The zygote develops into a trochophore larva, which in turn becomes a veliger larva. This stage sinks to the bottom and becomes a miniature bivalve.

Ecology

Many clams are sessile, attached to rocks, pilings, and other suitable surfaces, often in large numbers. A bathing float in the Bay of Kiel in the Baltic Sea had 30,000 mussels on each square meter of surface after a few months in the water. The competition for "a space to sit down" is the main problem facing such sessile animals. In terrestrial situations it is only the plants that have this problem. The mussels attach firmly to the substratum by means of tough fiberlike *byssus threads* secreted by special glands in the foot (Pl. 21).

Clams of the intertidal zone are subject to extreme variations in environmental conditions. They are often exposed for several hours twice a day at low tide. Such species in northern areas must withstand great extremes of temperature. Along the Massachusetts coast they can survive temperatures −10 to −20° C when 60–65 percent of the tissue water freezes.

The wood-boring clam, *Teredo,* which is called the shipworm because of its elongate wormlike shape, has small filelike shells which cover only a small part of the anterior portion of the body. Movement of the shells by action of the foot cuts a tubelike cavity in which the clam lives. The cavity is lined with shell material secreted by the naked mantle which covers most of the body. Small wood fragments are carried by ciliary currents to the mouth, giving the animal its main food. Whether the clam secretes a cellulose-splitting enzyme or depends on mutualistic microorganisms for this function is not definitely known.

The giant clam, *Tridacna,* which is found in the coral reefs of the Indo-Pacific region, has an interesting mutualistic relationship with algae which live within amoeboid cells circulating in the blood. These clams, which may reach a length of over 1 m and a weight of 200 kg, are largely dependent upon the algae for nourishment. The mantle is exposed by the gaping of the great valves and has special pigments which reduce the light intensity to a suitable level for photosynthesis. Sinuses in the mantle receive the blood with its alga-containing cells. Later the algae are taken to the digestive gland.

Oysters and scallops as well as many types of clams are important as a source of human food. The economic importance of pearls from oysters and clams as well as pearl buttons from clam shells is well known.

CLASS CEPHALOPODA

Classification

SUBCLASS **Tetrabranchiata** Two pairs of gills. Shell external, coiled in one plane and chambered. Open circulatory system. Ink sac absent.

ORDER **Nautiloidea** Single genus exists today, *Nautilus.*

ORDER **Ammonoidea** Ammonites, all extinct.

SUBCLASS **Dibranchiata** One pair of gills. Shell internal or absent. Ink sac present. Closed circulatory system.

ORDER **Decapoda** Ten tentacles present. Well-developed coelom. Shell internal. *Loligo* (squid), *Archeteuthis* (giant squid), *Sepia* (cuttlefish).

ORDER **Octopoda** Eight tentacles present. Coelom reduced. Shell reduced or absent. *Octopus* (octopus), *Argonauta* (paper nautilus).

Anatomy and Physiology of the Squid

GENERAL STRUCTURE

Some of the most highly specialized mollusks are found in this class, including the largest of all invertebrate animals, the giant squid, *Architeuthis,* which may be over 15 m in length and lives at depths of up to 5000 m. *Loligo,* a typical cephalopod, is shown in Figure 26.20. The squid's body is greatly elongated in the dorsal–ventral plane. The large head with its mouth surrounded by ten sucker-bearing *arms,* which represent the molluscan foot, is located ventrally. The conical body is covered by the muscular mantle with a triangular fin at the dorsal end. The edge of the mantle at the ventral open end is called the *collar,* and it articulates with cartilaginous ridges on the neck, making possible the effective closure of the mantle cavity except for the opening in the muscular *siphon.* The expansion and contraction of the mantle force water in and out of the mantle cavity between the collar and the neck, enabling respiratory exchange to occur between the water and the pair of gills located on the inner wall of the mantle. However, a sudden contraction of the mantle can close the collar opening around the neck and force the water out of the funnellike siphon, moving the animal rapidly by jet propulsion. Muscles control the position of the siphon so that if the jet is in the direction of the tentacles, the animal moves with the tip of the body first; or the siphon can be bent to send the jet in the opposite direction so that the animal moves with its tentacles first. By making use of its fins and tentacles the organism is also able to swim more slowly.

The outer surface of the mantle contains *chromatophores,* elastic pigment cells surrounded by attached muscle cells which can rapidly contract, enlarging the pigment cell and thus exposing more pigments (Fig. 26.21). The muscles can relax just as quickly and reduce

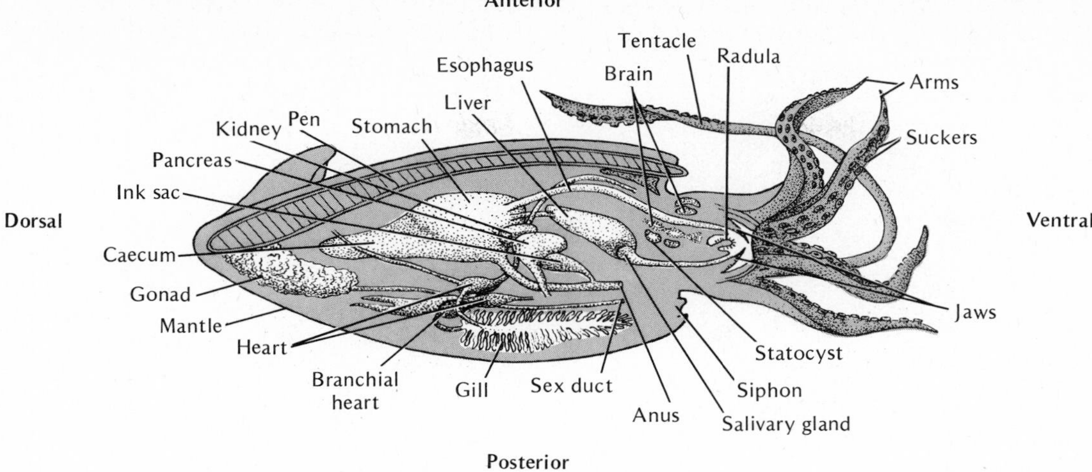

Figure 26.20 *Loligo,* a typical cephalopod, in sagittal section.

the color. This mechanism enables the squid to change color very rapidly, which it does when excited or in response to a change in background. Below the mantle on the anterior surface is the horny *pen,* the reduced shell which acts as a stiffening rod for the body.

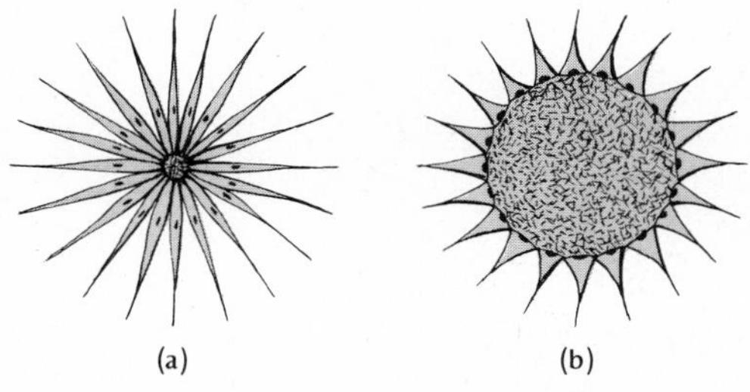

Figure 26.21 Squid chromatophore. **(a)** Muscles are relaxed and the chromatophore is in a contracted state with little pigment exposed to view. **(b)** The surrounding muscles have contracted, pulling the edges of the chromatophore out, thus exposing much pigment to view.

Squids are predaceous animals, capturing fish, crustaceans, or other mollusks with their arms, and drawing them rapidly to the mouth where they are quickly dispatched by the sharp beaklike jaws. There is a radula in the pharynx, but it is probably little used. Food passes next to the narrow esophagus and on to the muscular stomach with its long thin-walled *caecum.* A complex valve allows only liquid materials to pass into the caecum. The tract now turns back on itself with the intestine and rectum leading to the anus located in the mantle cavity near the inner opening of the siphon. Two pairs of salivary glands which open into the pharynx and a large liver and smaller pancreas which open into the stomach aid in digestion. The characteristic *ink sac* is located above the rectum and opens into it near the anus. When a squid is attacked by an enemy, contractions of the ink sac sends the dark pigmented fluid, containing melanin granules, through the anal opening into the mantle cavity and out through the siphon. The resultant black cloud enables the squid to seek shelter from its enemy.

The circulatory system is well developed and is almost a closed one. The heart has two auricles which receive oxygenated blood from the gills and send it to the ventricles which pump the blood through arteries to all parts of the body. The blood returns to the gills by veins, and a *branchial heart* at the base of each gill pumps the blood out into the gills whence it returns to the auricles. Excretion is by a pair of kidneys which connect the pericardial cavity with an opening into the mantle cavity.

The nervous system is highly concentrated with most of the typical ganglia fused in the brain region around the pharynx. The eye of the squid is a remarkable example of convergent evolution since it is strikingly similar in structure and function to the vertebrate eye, yet these two eyes evolved independently. Figure 26.22 shows the structure of a squid eye. The parts are named in accordance with the terminology used for the vertebrate eyes, even though they are not in any sense homologous. It is interesting that focusing is accomplished not by change in shape of the lens as in the human eye, but by moving the lens forward or backward as in an ordinary camera.

REPRODUCTION AND DEVELOPMENT

Squids are dioecious, as are all cephalopods. The single gonad is located at the apex of the

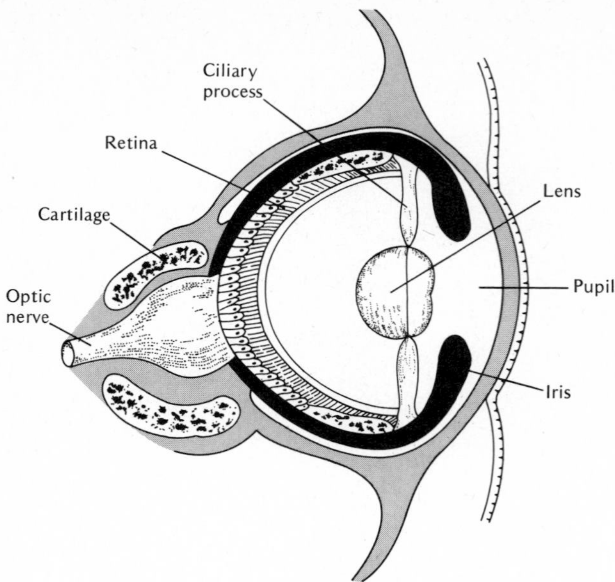

Figure 26.22 Section through the "camera eye" of a cephalopod.

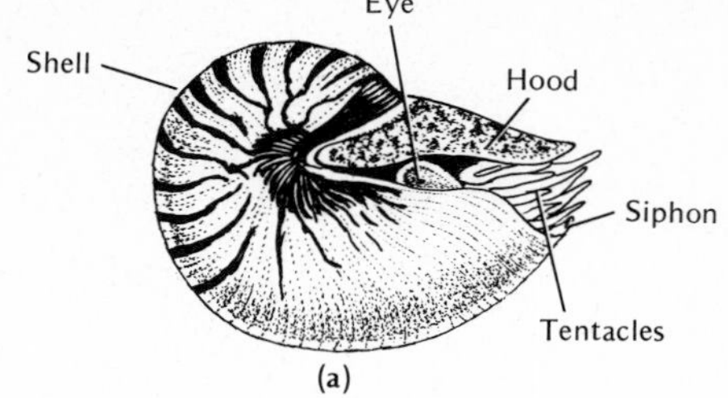

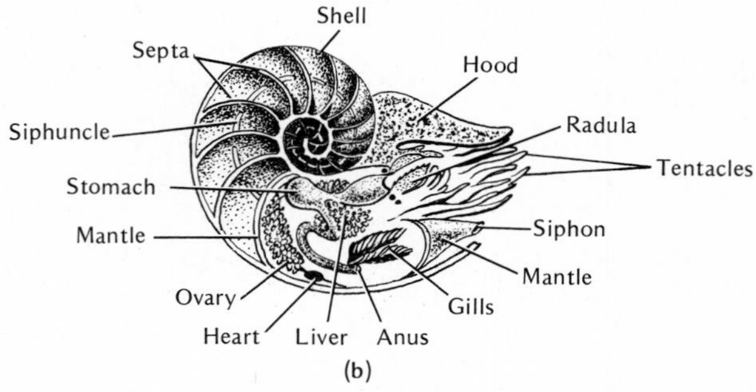

Figure 26.23 Nautilus, a shell-bearing cephalopod. **(a)** External appearance. **(b)** Sagittal section showing details of internal anatomy.

mantle cavity with a duct opening into the mantle cavity near the anus. Sperm are packed into a specialized spermatophore in the *spermatophoric gland.* At mating they are taken by one of the arms, called a *hectocotylus,* which has its tip modified for this purpose, and placed in the mantle cavity of the female. The large heavily yolked eggs are fertilized in the mantle cavity and then surrounded by a capsule secreted by the *nidamental glands.* Development is modified by the large amount of yolk, and cleavage is superficial as in birds. Coelom formation is schizocoelic, as in other mollusks, but there is no larval form. Young squids hatch as miniature adults capable of swimming and feeding immediately.

Other Cephalopods

In addition to squids and the very similar cuttlefish with a calcareous internal shell, the class includes the octopi which lack a shell, and the chambered nautilus which inhabits the last chamber of a coiled chambered shell. The nautiloids were dominant organisms in Paleozoic seas, and the sometimes enormous ammonites [over 2.5 m (8 feet in diameter)]were common during Mesozoic times. Only the genus *Nautilus* with about six species remains today (Fig. 26.23). Octopi range in size from a few centimeters to over 8 m (25 feet) in spread of their arms. They live mostly on rocks and in tide pools along the coast. Science fiction films and stories to the contrary, the octopus is relatively harmless to people (Pl. 23).

Nearly 90 percent of the squid species and almost 10 percent of the octopus species are luminescent. In some there are open pockets in the epidermis which contain luminescent bacteria, while others have specialized light-producing organs.

PHYLOGENETIC RELATIONSHIPS OF MOLLUSCA

Several features point to a relationship between the mollusks and annelids. The trochophore larvae of marine forms in both phyla are almost indistinguishable at early stages. The spiral, determinate cleavage pattern is also very similar in the two groups. The discovery of *Neopilina* with well-defined segmentation is strong evidence for a relationship as well.

The basic ladder-type nervous system of turbellarian flatworms is seen in primitive mollusks. Such a system does not occur in adult annelids or arthropods, although some forms have a similar condition in developmental stages.

The similarities between Müller's larva of the marine turbellarians and the trochophore lead one to speculate on the possible derivation of both mollusks and annelids from a turbellarian ancestor. The question remains as to when the ancestral mollusks and annelids diverged.

SUGGESTIONS FOR FURTHER READING

Abbott, R. T. *American Sea Shells.* Princeton, NJ: Van Nostrand (1954).

Hyman, L. *The Invertebrates,* vol. 6: *The Mollusca.* New York: McGraw-Hill (1967).

MacGinitie, G. E., and N. MacGinitie. *Natural History of Marine Animals.* New York: McGraw-Hill (1967).

McConnaughey, B. H. *Introduction to Marine Biology.* St. Louis: C. V. Mosby (1974).

Meglitsch, P. A. *Invertebrate Zoology,* 2nd ed. New York: Oxford Univ. Press (1972).

Wilbur, K. M., and C. M. Yonge (Eds.). *The Physiology of Mollusca,* vol. 1. New York: Academic Press (1964).

Young, J. Z. "Learning and Discrimination in the Octopus," *Biological Rev.,* **36** (1961).

From Scientific American

Boycott, B. B. "Learning in the Octopus" (Mar. 1965).

Cadart, J. "The Edible Snail" (Aug. 1957).

Keynes, R. "The Nerve Impulse and the Squid" (Dec. 1958).

Willows, A. O. D. "Giant Brain Cells in Mollusks" (Feb. 1971).

Annelids and Annelid Allies—

ANNELIDA, PRIAPULIDA, SIPUNCULIDA, AND ECHIURIDA

PHYLUM ANNELIDA

With the annelid worms, segmentation, which has been seen heretofore only in its incipient stages, becomes firmly established as a basic feature of all the rest of the protostome animals. It will also be seen later that segmentation as a basic feature of organization evolved independently in the phylum Chordata. It is felt that the high degree of development seen in the chordates and the arthropods is correlated with the appearance of segmentation. With segmentation there is a greatly increased opportunity for division of labor among the different regions of the body. Although division of labor between different segments of annelid worms is not great, in the course of evolution the higher phyla capitalized upon this condition.

Characteristics

1. Segmented vermiform animals.
2. Extensive coelom, divided by septa.
3. Closed circulatory system.
4. Nervous, circulatory, excretory, muscular, as well as other systems are segmentally arranged.
5. Body covered by a thin cuticle, chitinous setae usually present.
6. Larval form, when present, is a trochophore. Figure 27.1.

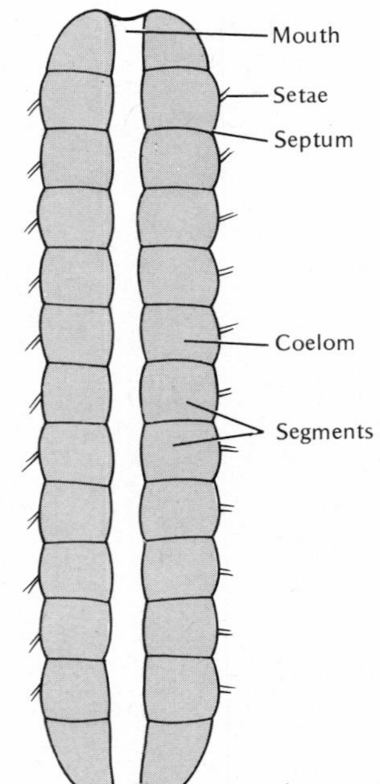

Figure 27.1 Annelid body plan.

Classification

CLASS **Polychaeta** Numerous well-developed setae, usually on fleshy outgrowths, called parapodia. Well-developed head with tentacles usually present. Sexes separate. Trochophore larva. Entirely marine. (Fig. 27.2).

CLASS **Oligochaeta** Well-defined external and internal segmentation. Few setae and no parapodia. Head poorly developed. Clitellum present at sexual

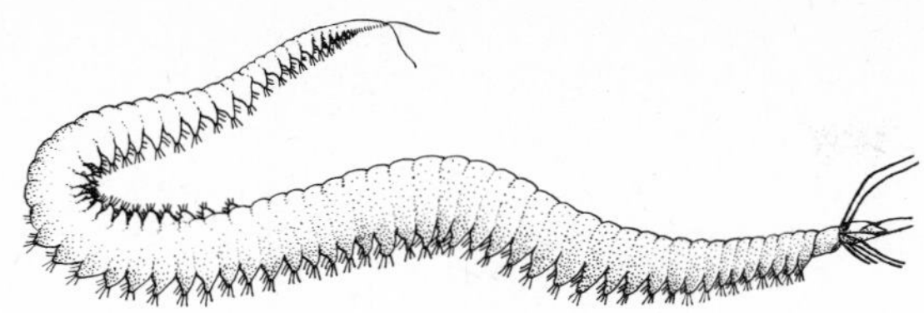

Figure 27.2 Class Polychaeta, *Nereis* sp.

maturity. Monoecious, eggs deposited in a cocoon. Direct development, no trochophore larva (Fig. 27.3).

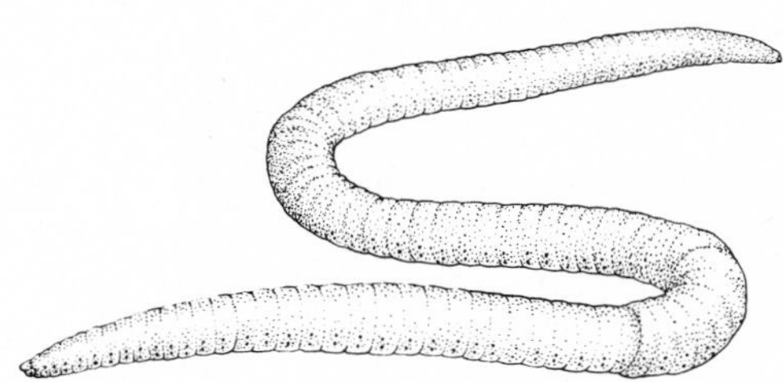

Figure 27.3 Class Oligochaeta, *Lumbricus terrestris.*

CLASS **Hirudinea** Dorsoventrally flattened ectoparasites with a sucker at each end. No parapodia. Setae absent except in one species. Constant number of segments (externally the apparent segmentation is actually a superficial annulation and does not reflect the true number of segments). Coelom reduced, filled with mesenchyme. Monoecious with direct development. Terrestrial, freshwater, and marine (Fig. 27.4).

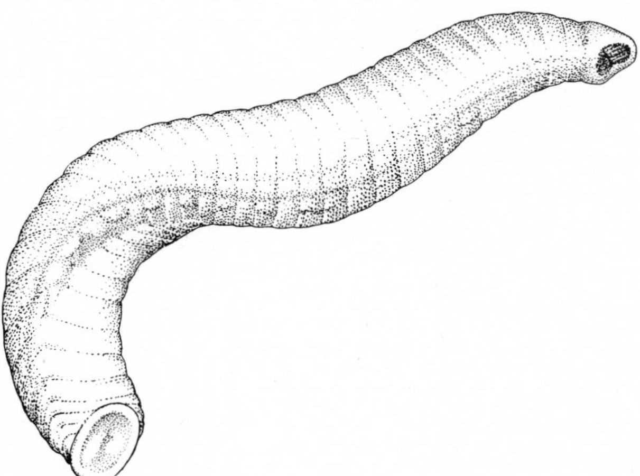

Figure 27.4 Class Hirudinea, a typical leech.

CLASS OLIGOCHAETA

Since the annelid worm most familiar to the majority of students is the earthworm, the discussion of this phylum will begin with the class Oligochaeta, even though the most primitive class is undoubtedly the class Polychaeta.

Classification

ORDER **Lumbriculida** Male ducts open in testis-bearing segments. Four setae per segment. Entirely freshwater. *Lumbriculus.*

ORDER **Moniligastrida** Testes and their funnels in paired dorsal sacs. One family. All terrestrial species in tropical Asia. *Moniligaster.*

ORDER **Hoplotaxida** Sperm ducts pass through at least one septum.

SUBORDER **Hoplotaxina** Two pairs of testes. *Hoplotaxis.*

SUBORDER **Tubificina** One pair of testes. *Tubifex.*

SUBORDER **Lumbricina** Four pairs of setae per segment. One or two pairs of ovaries. Sperm ducts pass through two or more segments. *Lumbricus.*

Anatomy and Physiology of the Earthworm

GENERAL STRUCTURE

Associated with the burrowing habits of the nocturnal earthworm are a degeneration of the head and sense organs, the loss of the segmentally arranged lateral appendages, and the development of a complex reproductive system. Earthworms can be found in burrows in the soil ranging in depth from several centimeters to almost 2 m when there is abundant moisture and humus. The number of segments in the average full-grown specimen may vary from 110 to 180. The body of the worm is protected by a thin cuticle. In warm weather the worms are found near the surface if the soil is moist, but in dry periods they move to deeper regions. During the winter they are found below the frost line. In loose soil they are able to make their burrows simply by pushing the soil away on all sides, but in compact soil they actually eat their way through. From this activity the earthworm deposits the soil, which passes through its alimentary tract, as *castings* at the mouth of its burrow. The soil that it eats provides food as any organic matter in the soil, such as seeds, eggs, small animals, and the decaying bodies of plants and animals, will be digested and utilized. The body of the worm must be moist at all times. When night comes the worm emerges to explore the surrounding area. With the exception of the first few and last few segments, the others bear four pairs of slender bristles, or setae. These not only assist in locomotion but also help to anchor the tail of the worm in the burrow when it is extended, making quick retreat possible.

NUTRITION: FOOD ACQUISITION AND ASSIMILATION

The foraging worm obtains its food from the soil or from debris, such as dead leaves lying on or in the soil. The mouth is located at the anterior blunt end just beneath a small lobelike projection, the *prostomium.* The first complete segment is the *peristomium.* There are no jaws, teeth, or other special structures associated with the mouth. The anus is a vertical slit in the last segment. The food materials taken into the mouth through the alimentary canal, which is suspended in the coelom by the septa that form partitions between segments. Anteriorly the alimentary canal is differentiated into the following regions: a mouth cavity, a muscular pharynx, an esophagus, a thin-walled crop for storage, a thick-walled muscular gizzard, and an intestine. From segment 19 the intestine continues without modification to the anus. In the esophagus the *calciferous glands,* three swellings on each side of the esophagus, secrete calcium carbonate into the lumen. This is mixed with the food material and tends to neutralize any organic acids present. From the crop the food passes into the thick-walled muscular gizzard, where it is thoroughly ground. Particles of sand taken in with the food aid in this grinding. Most of the digestion and absorption takes place in the intestine. The intestine is not a simple cylindrical tube; its dorsal wall is infolded, forming the *typhlosole* (Fig. 27.5), which greatly increases the surface for absorption and secretion. The mucosa of the intestine is composed of cells that are secretory and cells that function in absorption. The secretory cells secrete a digestive fluid that contains enzymes capable of breaking down proteins, carbohydrates, and fats. The end products of digestion pass into blood vessels in the intestinal wall. The undigested material is voided through the anus in the forms of castings.

TRANSPORT OF MATERIALS WITHIN THE ORGANISM—THE CLOSED CIRCULATORY SYSTEM

The blood of the worm is red because of the hemoglobin dissolved in the plasma. A median *dorsal vessel,* visible through the skin, is contractile and has rhythmic peristaltic waves. This vessel, which lies on top of the alimentary tract, is the main collecting vessel, and in it the blood flows forward. A median ventral vessel, suspended from the alimentary tract by a mesentery, is the main distributing vessel. In the region of the esophagus five pairs of enlarged vessels, the hearts, connect the dorsal and ventral vessels, and force the blood through the ventral vessel. The flow in the ventral vessel is forward, anterior to the hearts, and backward, posterior to the hearts. A subneural vessel also runs the length of the worm. The blood flows posteriorly in this vessel. In each segment a pair of parietal vessels connects the subneural vessel with the dorsal vessel. The ventral vessel gives off branches in each segment to the body wall, the alimentary tract, and the nephridia. The body wall and segmentally arranged excretory structures, the nephridia, are drained by small vessels that empty into the parietals. Blood from the alimentary canal is drained into the dorsal vessel. Valves in the hearts and dorsal vessel control the direction of the blood flow (Fig. 27.5).

GAS EXCHANGE AND ELIMINATION OF WASTES

There is no specialized respiratory system in the earthworm; the highly vascular skin serves as the respiratory surface. Oxygen and carbon dioxide are exchanged through the entire surface of the worm that is kept moist by the secretions of cells of the skin. The thin cuticle is quite permeable to both oxygen and carbon dioxide. The hemoglobin dissolved in the blood plasma aids in the transportation of oxygen.

The primary excretory structures in the earthworm are coiled tubes, called *nephridia,* which are found in pairs in every segment except the first three and the last. The position of these structures in the body is shown in Figure 27.5. Each nephridium occupies parts of two successive segments. A ciliated funnel, or *nephrostome,* in one segment is connected, through a septum, by a thin tube, with the major part of the structure in the segment posterior to it. The nephridium is coiled into three loops, and it expands into a bladderlike region just before it empties to the outside through the *nephridiopore.* The cilia of the nephrostome cause materials in the coelom to enter the tubule; also in the glandular region of the nephridium, wastes in the blood are removed.

In addition to the nephridia there are specialized cells of the outer layer of the intestine (visceral peritoneum), *chloragen* cells, that have an excretory function. Wastes extracted from the blood are deposited in these large and glandular cells. Chloragen cells become detached and float in the coelomic fluid, where they disintegrate. The detritus is either expelled through the nephridia or is engulfed by amoeboid cells in the coelomic fluid which, in turn, migrate into the outer tissues, disintegrate, and leave their contents as deposits of protective pigment.

COORDINATION

The earthworm has basically a ventral ladder-type nervous system with ganglia in each seg-

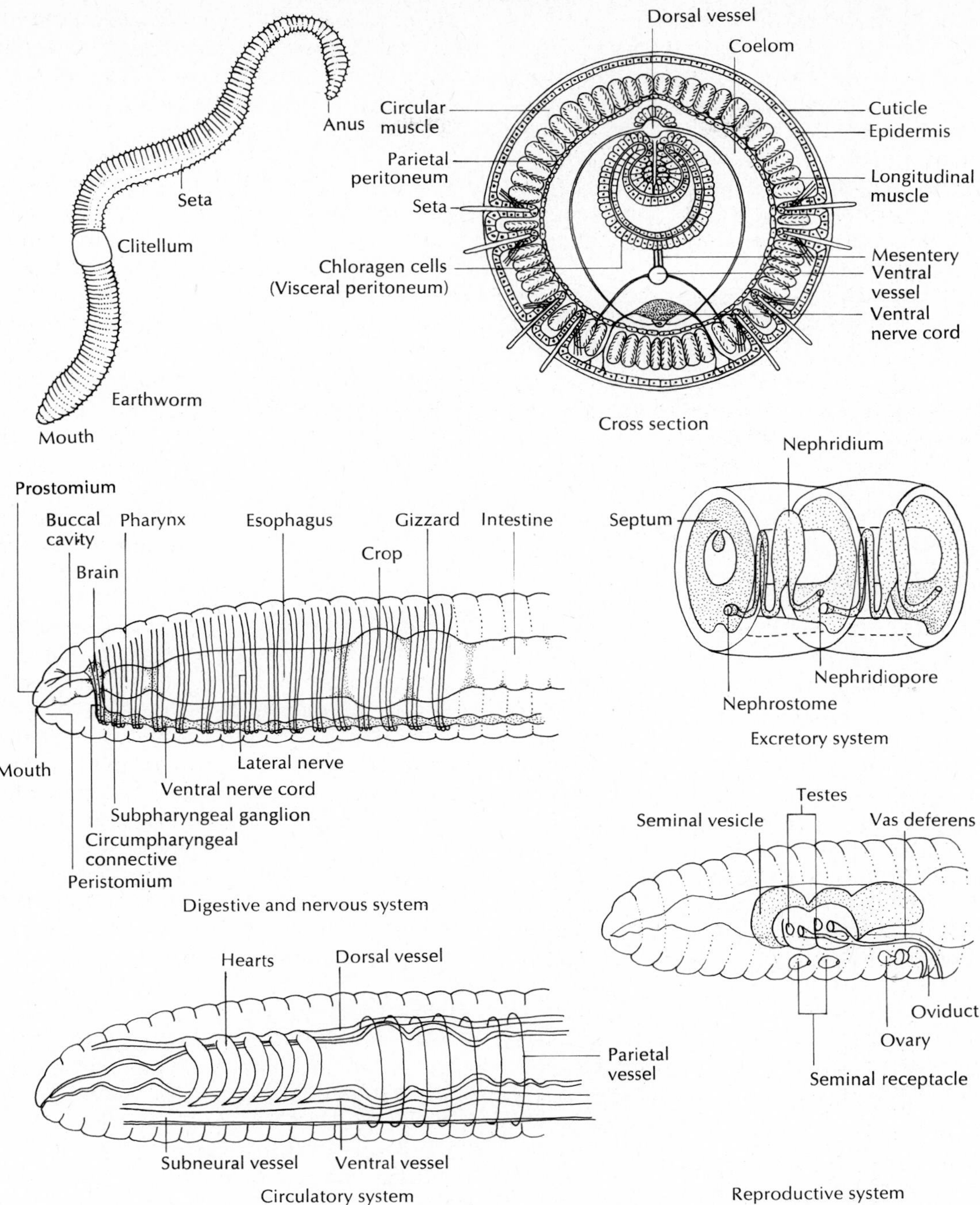

Figure 27.5 Structure of the earthworm showing external features, body plan, organ systems and cellular detail.

ment. The double cord and the ganglia are fused in the adult worm, but the double nature is still evident in the brain, or *suprapharyngeal ganglion,* located dorsally in segment three, as a bilobed mass representing the partial fusion. The brain is connected with the first ventral ganglion, the *subpharyngeal ganglion,* located in segment four, by two *circumpharyngeal connectives* that extend around the pharynx. From this first ventral ganglion the *ventral nerve cord* extends along the floor of the coelom to the last segment. The nerve cord enlarges in each segment to form a ganglion, and it branches into three pairs of lateral nerves. These lateral nerves contain both sensory and motor fibers. In the ventral nerve cord are giant nerve fibers that extend over long distances, even throughout the length of the cord. The speed of transmission in these fibers is 7 to 12 m per second and in the median giant fiber, 17 to 25 m per second. This compares with a speed of 0.025 m per second in the fine nerve fibers. It is assumed that rapid movements, such as the sudden retreat into the burrow, involve these giant fibers.

A simple experiment demonstrates an interesting feature of coordination in the earthworm. If one cuts an earthworm in half and ties the two halves back together loosely with thread—one loop on each side will do—the two halves will still move in a coordinated fashion. When a wave of contraction reaches the end of the front half, the tug on the severed posterior half causes the contraction to continue on down just as it does in an intact worm.

Associated with the burrowing habit of the earthworms is the degeneration of the head and sense organs. However, numerous sense organs are scattered in the epidermis. Light receptors are concentrated at the two ends of the worm. Other receptors scattered in the epidermis respond to touch, chemicals, and changes in temperature. These worms do not respond to sound, but they are quite sensitive to vibrations in the substrate.

LOCOMOTION AND SUPPORT

Two layers of muscle in the body wall and the four pairs of short bristles or setae and their associated muscles in each segment are the structures involved in locomotion (Fig. 27.5). The outer layer of muscles is the *circular* muscle layer, and just beneath it is the *longitudinal* muscle layer. Contractions of these two sets of muscles account for the changes in shape of the worm. The body of the worm is filled with coelomic fluid that is relatively incompressible, so the contraction of either layer of muscles results in a change not of size, but of shape. Contraction of the circular muscles makes the worm slender, but correspondingly longer, while contraction of the longitudinal muscles makes it shorter and thicker. It is the contraction of these latter muscles that enables the worm to withdraw into its burrow when it is disturbed. The setae function both to anchor the worm in its burrow and to determine the direction of locomotion.

BEHAVIOR

Simple observations on living worms indicate the nature of their basic responses. They react negatively to light, as everyone who has hunted "night crawlers" with a flashlight knows. The earthworm is sensitive to changes in temperature; it reacts negatively to extremes of temperature and responds positively to moderate temperatures. The earthworm cannot survive desiccation; it therefore responds positively to moisture. Earthworms react negatively to volatile chemicals, such as acetic acid. Charles Darwin made some simple observations on the chemical sense of the earthworm in connection with its selection of food. He found that when the worms were offered a variety of foods they showed a definite preference for certain types.

REPRODUCTION AND DEVELOPMENT

Both in its reproductive organs and in the process of reproduction the earthworm is quite different from the marine annelids. In the marine annelids the sexes are separate and the gametes are budded off from the lining of the coelom in many segments of the body. The earthworm is hermaphroditic. When any two individuals meet, cross-fertilization is possible.

The female system consists of a pair of ovaries, a pair of oviducts with their associated egg sacs, and two pairs of small seminal receptacles. The ovaries (Fig. 27.5) are in segment 13, and the funnel-shaped openings of the oviducts penetrate the septum between segments 13 and 14 and open to the outside by small pores on the ventral surface of segment 14. Ripe eggs are stored in the egg sacs prior to laying. The sperm receptacles receive the sperm from the opposite worm.

The male possesses two pairs of testes, two large seminal vesicles with lateral pouches, and two vasa deferentia that open to the outside in segment 15. The tiny testes are housed within the seminal vesicles. Sperm are discharged from the testes into the vesicles, where they are stored. The sperm during copulation enter the funnel-shaped openings and pass to the outside through vasa deferentia.

During copulation two worms are united, ventral surface to ventral surface, with the anterior ends in opposite directions and with the anterior quarter of the length of the bodies overlapping (see Fig. 27.6). They are held together in this position in part by secretions from a saddlelike swollen glandular region, the *clitellum,* located on the dorsal sides of each worm from segments 32 to 37. The mucus from the clitellum forms a sleeve about the animals. During copulation each worm discharges sperm that pass from the openings of the vasa deferentia in segment 15 to the openings of the seminal receptacles in the opposite worm through temporary longitudinal furrows that form in the skin of the worms. Following this exchange of sperm, the worms separate.

After this the clitellum of each worm secretes a mucous ring that slides forward over the body of the worm as it contracts. As the mucous ring passes segment 14, eggs are deposited in it. When it reaches the openings of the seminal receptacles, sperm are discharged over the eggs

Figure 27.6 Earthworms copulating. (From Turtox/Cambosco, Macmillan Science Co., Chicago.)

to fertilize them. Finally as it passes over the anterior end of the worm, it becomes closed at both ends to form a sealed capsule or cocoon. These small cocoons are deposited in damp soil, where development takes place. Development is direct, that is, there is no larval stage. In the laboratory young worms hatch from cocoons in two or three weeks, but in nature the process probably requires a longer time.

The adult earthworm has considerable ability to regenerate lost segments, but its regenerative capacities are much more limited than those in planaria. Loss of tail segments is usually followed by complete regeneration. However, regeneration of losses at the anterior end is limited. No more than four or five segments will be regenerated, and if more than 15 segments are removed no head is reformed.

ECONOMIC IMPORTANCE

In addition to the use of the earthworm as bait for fishing, these organisms exert definite long-term effects on the nature of the soil. The significance of their effects on the soil was first recognized by Charles Darwin in his treatise on "The Formation of Vegetable Mould," published in 1881. As a result of observations by Darwin and others it is known that the earthworm castings deposited on the surface of the earth may amount to as much as 18 tons per acre per year in certain regions.

Other Oligochaetes

There is an earthworm found in Australia, *Megascolides,* which may be as much as 3 m long and 2.5 cm in diameter. Many aquatic oligochaetes, such as *Aeolosoma,* are microscopic in size, less than a millimeter in length.

Ecology

In addition to the familiar earthworm, whose ecology has been discussed above, there are many species of aquatic oligochaetes. Most of them are found in shallow waters up to a meter in depth. Members of the family Tubificidae, however, are found at the bottom of deep lakes in great abundance. Here they are able to live even when the oxygen levels fall very low. Several families of oligochaetes have species which live in brackish water or in the intertidal region of the sea. These families are the Tubificidae, Enchytraeidae, and Megascolidae.

Parasitism is not found in the class except in the case of the family Branchiobdellidae whose members live in the gill chambers of freshwater crayfish. Even this relationship may be one of commensalism.

CLASS POLYCHAETA

Classification

Most workers believe that the polychaetes are too diverse to group the various families into orders, and they divide the class into two subclasses, but do not use orders. A few representative families are given for each subclass.

SUBCLASS **Errantia** Segments numerous and essentially similar except in the head region. Parapodia alike and provided with internal stiffening rods, called aciculae. Swimming, crawling, burrowing, or tube-dwelling.

Family **Aphroditidae** Sea mice. Many long setae form a feltlike pad on the dorsal surface. *Aphrodite.*

Family **Glyceridae** Burrowing worms with a conical prostomium and with an eversible proboscis armed with four jaws. *Glycera.*

Family **Nereidae** Prostomium with four eyes, peristomium with four pairs of cirri. *Nereis.*

Family **Eunicidae** Free living or tube dwelling with an elongate body. *Eunice.*

SUBCLASS **Sedentaria** Body with two or more regions with unlike parapodia and segments. Head appendages reduced or absent. Parapodia reduced and lacking aciculae. Usually tube-dwelling forms.

Family **Sabellariidae** Head modified to form an operculum which closes the tube. Tube dwellers in sandy tubes. *Sabellaria.*

Family **Arenicolidae** Lugworms. No head appendages. Construct a U-shaped burrow. *Arenicola.*

Family **Terebellidae** Tube dwellers with many long filiform prostomial tentacles. *Terbebella, Amphitrite.*

Family **Serpulidae** Fan worms, feather duster worms. Secrete a calcareous tube within which they live. *Serpula.*

GROUPS OF UNCERTAIN POSITION Two groups of annelidlike forms have been treated in a variety of ways in the past, either as separate phyla or as classes of the phylum Annelida. On the basis of present evidence it seems more appropriate to include them in the class Polychaeta.

Myzostomaria, family Myzostomidae, of the subclass Errantia. These are flattened rounded commensals or parasites of echinoderms, especially crinoids. *Myzostoma.*

Archiannelida. A heterogeneous group of minute polychaetes, perhaps three or four families. Some lack setae and/or parapodia; some have a ciliated epidermis. Most of them live in coastal marine waters in the interstitial water of sandy bottoms. Some occur in brackish and fresh water. *Polygordius.*

Anatomy and Physiology of *Nereis virens*

GENERAL STRUCTURE

The common sandworm or clamworm, *Nereis virens,* is a typical representative of the largest class of annelids, class Polychaeta. This class is made up of marine annelids with many bristles. *Nereis* lives in burrows in sand or mud at tide level and is active mostly at night.

Although most of the annelid features are shown in the earthworm, there are three others that are found in *Nereis* and other polychaetes that should be considered: a well-developed head, parapodia, and a trochophore larval stage during development.

The prostomium and peristomium form a distinct head with conspicuous sense organs (see Fig. 27.7, Pl. 29). The peristomium surrounds the ventrally placed mouth and bears four pairs of *peristomial tentacles* dorsally. On the prostomium are two short *prostomial tentacles* medially, a pair of stubby conical *palps* laterally, and two pairs of eyes dorsally. The eyes of *Nereis* are more advanced than those found in planaria. There is a layer of light-sensitive cells and a gelatinous lens that concentrates the light in each eye. Food organisms, crustaceans, and other small animals are captured by a pair of strong chitinous jaws that are everted with a part of the pharynx when *Nereis* is feeding.

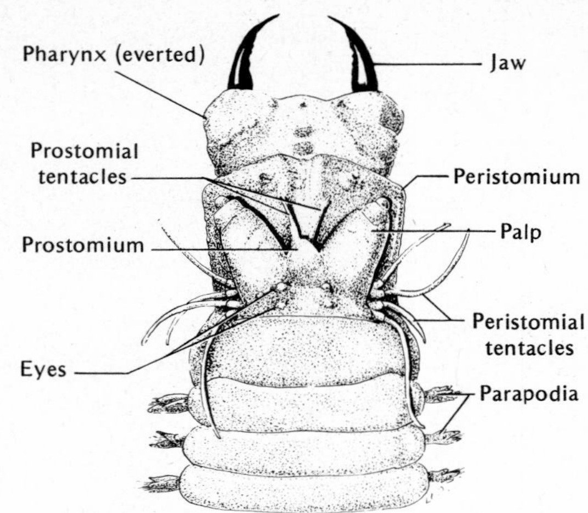

Figure 27.7 Head of *Nereis* with pharynx everted.

Each segment, with the exception of the first and last, bears a pair of lateral extensions, the *parapodia.* The parapodia serve for locomotion in both creeping and swimming, and also as respiratory structures. A large part of the gaseous exchange in external respiration in these worms is effected through the capillaries in the parapodia. Figure 27.8 shows a cross section through *Nereis,* including a parapodium. Note the stiff bristles, called *aciculi,* which act as internal skeletal elements for the parapodium.

REPRODUCTION AND DEVELOPMENT

The rest of the internal anatomy of *Nereis* is quite similar to the earthworm except for the reproductive system. The sexes are separate and the gonads develop in the coelomic peritoneum only during the breeding season. The gonads develop in the posterior portion of the body and various structural changes also occur in this part of the body. The color may change and the parapodia may become greatly enlarged. At certain specific times, apparently associated with external factors such as the phases of the moon, worms rise to the surface and release their gametes either through the nephridial ducts or by rupture of the body wall. The vast reproductive powers, synchrony, and abundance of ma-

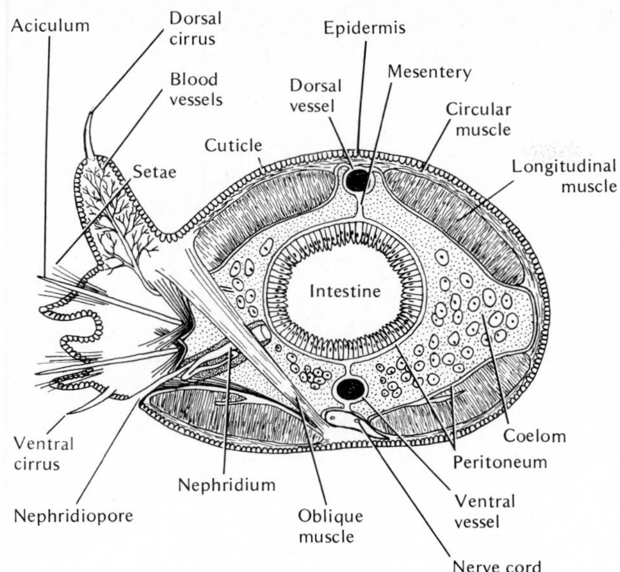

Figure 27.8 Cross section of *Nereis.* On the left a parapodium is included.

rine polychaetes are shown by the palolo worm, *Eunice viridis,* of the South Pacific. Every year during the last quarter of the moon at the lowest tides during October and November the natives are on hand to fill buckets for a feast when the posterior portions of these worms break off and rise to the surface by the millions. Following fertilization in the sea, development begins with spiral determinate cleavage and a trochophore larva which is quite indistinguishable from a molluscan trochophore in its early stages. Metamorphosis to the adult results in the formation of the prostomium from the apical portions of the larva and body segments from the rest of it with new segments being added at the rear.

To those familiar only with the earthworm, it may be surprising to learn that many marine polychaetes are beautifully colored animals—bright reds, greens, or blues in various combinations and often coupled with a lovely irridescence. The tube dwellers often have delicate feathery gills or tentacles which protrude from their dwelling places and are quickly withdrawn at the first sign of danger.

Ecology

Polychaetes are important members of marine sand and mud bottom communities at all levels from the intertidal to the greatest depths that have been sampled. Many are carnivorous, but some feed on plant material and others are ciliary filter feeders.

Some of the tube dwellers have evolved highly sophisticated feeding mechanisms. *Chaetopterus* lives in a permanent U-shaped tube lined with a parchmentlike material (Pl. 28). Water is drawn in through one opening, passes through the tube, and leaves by a second opening. Modified parapodia form fanlike structures which move water through the tube. Just behind the fans, a pair of long winglike parapodia secretes a baglike net of mucous in which particles of food are entangled. The narrow tip of the mucous net is periodically rolled up by a cup-shaped ciliated structure and passed forward to the mouth in a ciliated groove. This unique feeding method even allows for some selectivity on the part of the worm. If undesirable material passes over sensory structures anterior to the mucous net, the net is lowered to allow these unwanted substances to pass on through the tube. Another colorful tube-dweller, *Eudystylia polymorpha,* is shown in Plate 17.

Commensal relationships are found in some groups where worms inhabit the burrows or tubes of other animals, including other polychaetes. Some live within the mantle cavity of clams and oysters and others in the ambulacral grooves of starfish. Several species are commonly found sharing the snail shell with a hermit crab. Many of the tube-dwelling fan worms live within coral masses with only the fanlike tentacles protruding. They quickly retract the tentacles to the safety of their tubes when danger threatens. Plate 14 shows a group of fan worms with tentacles extended and in Plate 15 the small holes surrounded by the red encrusting sponges are the external opening of fan worm tubes as they appear when the worm is retracted.

Parasitism is rare among polychaetes. Members of the family Histriobdellidae are parasitic in the gill chambers of crustaceans and the family Ichthyotomidae are blood-sucking parasites on the fins of marine eels.

The Myzostomaria (Fig. 27.9), a group of uncertain position, are all parasites or commensals on echinoderms. Most of them creep over the

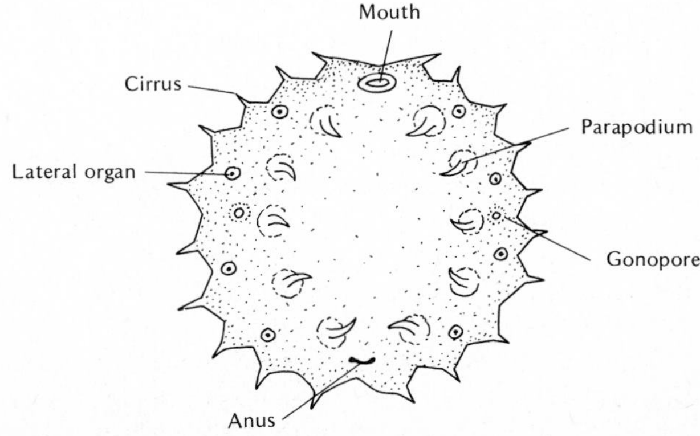

Figure 27.9 *Myzostoma,* ventral view. Length 4.5 mm.

surface of the host or live within gall-like outgrowths which they induce. Some few are internal parasites in the digestive tract or coelom.

CLASS HIRUDINEA

Classification

ORDER **Acanthobdellida** Setae present in five anterior segments. Coelom not completely obliterated. One Russian species parasitic on salmon. *Acanthobdella.*

ORDER **Rhynchobdellida** Jaws lacking, proboscis present. All aquatic, either marine or freshwater. *Glossiphonia, Placobdella.*

ORDER **Gnathobdellida** Noneversible pharynx with three jaws, five annuli per segment. Aquatic or terrestrial. *Hirudo.*

ORDER **Pharyngobdellida** Nonprotrusible pharynx which lacks jaws and teeth. Freshwater or semiterrestrial, carnivorous. *Erpobdella.*

Anatomy and Physiology

GENERAL STRUCTURE

The body of the leech is usually flattened in the dorsoventral plane, and in most species there are two suckers, an anterior one which surrounds the mouth and a posterior one. There are always 34 segments in the leech body. Externally there appear to be many more due to the superficial annulations of the body wall. The determination of the true number is based on the internal structure, particularly of the nervous system. With the exception of a single genus, *Acanthobdella,* setae are absent in the class.

The sharp jaws of a leech make a Y-shaped cut in the skin of the host. The secretion of the salivary glands contains a substance, called *hirudin,* which prevents the coagulation of the blood of the host and also acts as an anesthetic so that the host is unaware of the leech's presence. The medicinal leech, *Hirudo medicinalis,* will be used as an example of the class (Figs. 27.10 and 27.11). This organism is called the medicinal leech because it used to be employed in blood-letting when it was believed that many disorders could be cured by reducing the volume of blood in human patients. A jar of living leeches was part of the stock and trade of the corner pharmacy. Blood is sucked into the alimentary canal by the muscular pharynx and a short esophagus leads into a capacious *crop* or *stomach* which is provided with up to 11 pairs of lateral *caeca* in which blood is stored. These capacious caeca allow the worm to go for several months between meals. Following the caeca there is an intestine, which may be a straight tube or may also have lateral storage sacs. A short rectum leads to the anus located just in front of the posterior sucker. Excretion is by means of paired metanephridia located in the midsegments of the body. Much of the coelom in leeches is filled with a spongy connective tissue, and the nephridial tubules are imbedded in it. The circulatory system consists of sinuses in the connective tissue which fill the coelom. There are two lateral sinuses and a dorsal and ventral one. Pulsations in the sinuses circulate the blood. The nervous system is made up of a dorsal bilobed brain, circumpharyngeal connectives, and a ventral nerve cord in which there has been some fusion of ganglia at the anterior and posterior ends. Sense organs include eyespots on the head, chemoreceptors in the mouth region, and tactile organs on the body surface.

REPRODUCTION AND DEVELOPMENT

Like the oligochaetes, all leeches are hermaphroditic. On each side there is a series of

Figure 27.10 *Hirudo medicinalis,* the medicinal leech. This European species in the past was used to remove blood from patients in the belief that such removal had therapeutic effects. (From Turtox/Cambosco, Macmillan Science Co., Chicago.)

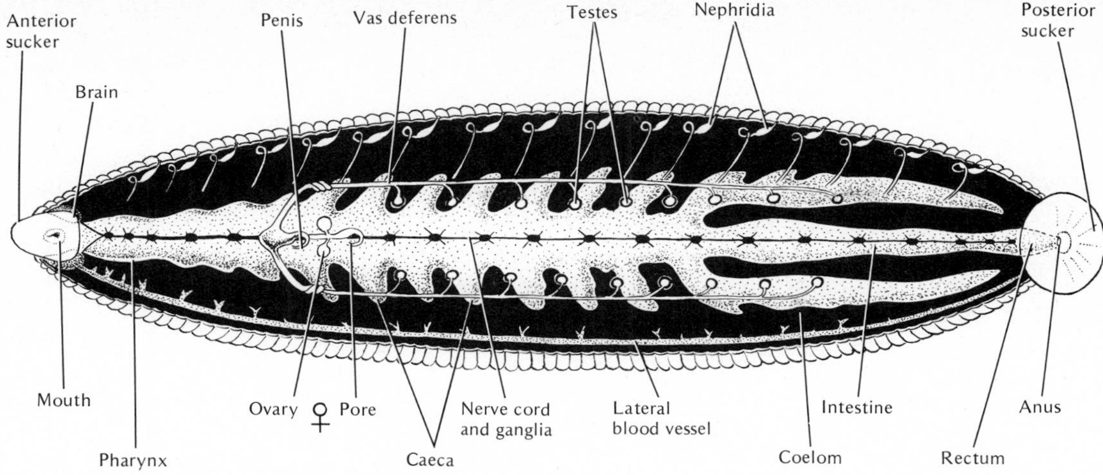

Figure 27.11 Internal structure of *Hirudo medicinalis,* ventral view.

paired testes attached to a vas deferens which joins its mate anteriorly in a median muscular penis. Two ovaries connect in a median vagina which opens posterior to the male pore. Copulation with mutual exchange of sperm occurs. Later a cocoon is formed by a clitellar region, and fertilized eggs are placed in it along with nutritive secretions of clitellar glands. Development is direct with no larval stages.

Ecology

While a few species of leeches are marine and terrestrial, most of the species live in fresh water. They generally occur in shallow waters of lakes, ponds, and slow-moving streams. In some cases they may be very abundant—many thousands per square meter.

In most cases the parasitic leeches are free living except when feeding. In the family Piscicolidae, the fish leeches, they remain attached to the host except when breeding. Usually leeches will feed on a variety of hosts, only a few being restricted to a specific host. Many do confine their feeding to members of a single vertebrate class.

Terrestrial leeches are largely confined to humid rain forests of the tropics where they prefer bird and mammal hosts, dropping on the host as it moves through the lush vegetation.

About one fourth of the leeches are not parasitic. They are all predators on a variety of invertebrate animals including worms, snails, and insect larvae.

PHYLOGENETIC RELATIONSHIPS OF THE ANNELIDS

The annelids have many typical protostome characteristics which relate them to the more primitive flatworms and to the mollusks as well. Their cleavage and early development are quite similar, and the trochophore larvae of annelids and mollusks are remarkably alike at early stages. A hypothetical trochophore ancestor has often been suggested, but since this is only a larval form in present-day animals, it seems unlikely that it ever existed as an adult animal.

Within the phylum Annelida, the polychaetes seem to be the most primitive and they probably gave rise to the oligochaetes. The latter in turn seem to be the source of the leeches.

THE ANNELID ALLIES

The three small phyla to be discussed in this section offer a transition between the unsegmented protostomes and the segmented ones. The segmentation of the Monoplacophora such as *Neopilina* foreshadowed this significant phylogenetic advance.

While the priapulids are coelomate, their affinities are not clearly with the annelids but more with certain acoelomate groups. Although the sipunculids show no segmentation even in their larval stages, many aspects of their development and the general nature of their nervous system clearly indicate close annelid relationships. The echiurids are more closely related to the annelids as indicated by the appearance of segmentation in their larval forms, which is largely lost in the adult. It is not believed that the relationship between the mostly unsegmented mollusks and the segmented annelids can be traced in a direct line via the priapulids, the sipunculids, and the echiurids.

PHYLUM PRIAPULIDA

Characteristics

1. Cylindrical, marine worms less than 10 cm long.
2. Bulbous anterior end lacks tentacles and is eversible.
3. Surface of the body is superficially segmented and covered with warts or spines.
   Figure 27.12
   The phylum Priapulida is not broken down into classes and orders. There are six genera including *Priapulus* and *Halicryptus* and less than ten described species.

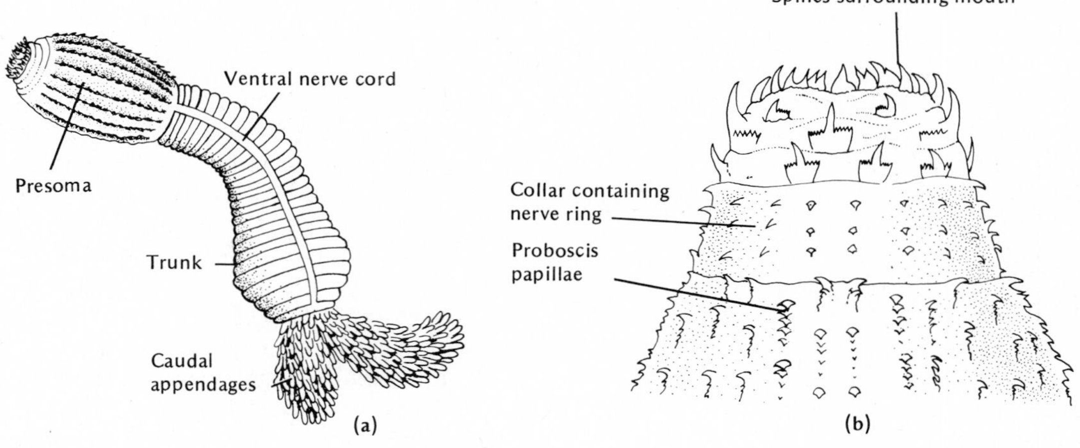

Figure 27.12 Phylum Priapulida, *Priapulus bicaudatus.* **(a)** Ventral view. **(b)** Detail of anterior end with mouth everted.

Anatomy and Physiology of *Priapulus*

GENERAL STRUCTURE

The priapulid body is made up of a bulbous anterior portion, called the *presoma,* which is eversible, and a cylindrical trunk. Some species have warty appendages on the posterior end. The mouth is located at the tip of the presoma (Fig. 27.12). A series of concentrically arranged recurved spines is associated with the mouth. When the mouth is everted, these spines function in grasping prey and pulling it in. Posterior to the spine-covered region is the proboscis which is covered with longitudinal rows of papillae. A constriction marks the beginning of the trunk region which appears segmented due to superficial annuli numbering from 30 to 100, depending on the species. The trunk region is covered with warts and spines. The anus and two urogenital pores are located at the posterior end.

The cuticle is secreted by the underlying cellular hypodermis. The nervous system, consisting of a circumesophageal ring without a brain enlargement and of a midventral nerve cord, lies in the hypodermis. Beneath the hypodermis are two well-defined muscle layers, an outer circular and an inner longitudinal layer. Internal to the longitudinal muscles is a thin lining membrane which was formerly said to be noncellular. Recently Shapeero has shown it to be cellular.

The alimentary canal is a straight tube leading from the mouth to the muscular pharynx and through a tubular gut to the posterior anus (Fig. 27.13). Excretion is by means of protonephridia in the paired urogenital organs in which the collecting cells are cellular solenocytes rather than flame cells. A *solenocyte* has long flagella instead of cilia which are characteristic of flame cells. Specialized circulatory and respiratory systems are lacking in the phylum.

REPRODUCTION AND DEVELOPMENT

The sexes are separate with the gonads enclosed within the paired urogenital organs. Each urogenital organ contains a duct with the gonad lying along one side and the solenocytes on the other. Each duct opens to the outside by a pore near the anus.

Fertilization is external in the sea. Development is direct with the larval stage similar to the adult, except that it is encased in a chitinous

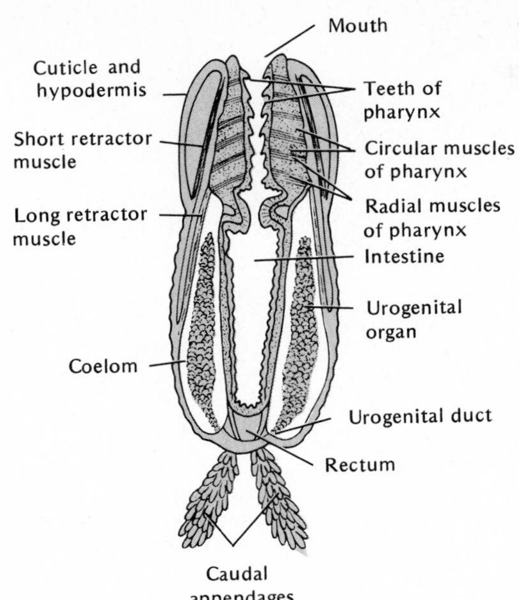

Figure 27.13 Internal anatomy of *Priapulus.*

lorica which it retains for two years before molting to become an adult.

Ecology

Priapulids live buried in soft muddy bottoms of cold northern and southern seas from the littoral zone down to depths of 500 m. Occurring as far north as Greenland and Siberia and in the south to Antarctica, they are not found in the intervening warmer latitudes. This peculiar distribution presents a problem to the student of animal geography. The priapulids are predaceous on soft-bodied polychaete worms and other priapulids, seizing their victim with their spiny teeth and drawing them into the pharynx to be swallowed whole.

Phylogenetic Relationships

The priapulids were long thought to lack a true cellular lining to the body cavity and thus were placed in the pseudocoelomate group as a class in the phylum Aschelminthes. Electron microscope studies by Shapeero have shown that the body cavity is lined with a thin cellular layer and it thus must be considered a coelom. Unfortunately, nothing is known about the method of coelom formation.

Priapulids are not closely related to the annelids or their allies, the sipunculids and echiurids. The retractable introvert is similar to that of the acanthocephalans and kinorhynchs. The pharynx and nervous system are also much like those of the kinorhynchs. More details of the development of priapulids may clarify their relationships with other groups.

PHYLUM SIPUNCULIDA

Less than 250 species, ranging in size from 2 mm to over 50 cm, constitute this phylum of burrowing marine worms which are found at depths from the shallow intertidal zone to as much as 5000 m. Their most distinctive characteristic is the *introvert,* the long slender anterior end which is rapidly, and often continuously, rolled in and out.

Characteristics

1. Unsegmented marine worms.
2. Anterior end an extremely extensible and retractable introvert.
3. Digestive tract highly coiled and recurved so that the anus is anterior and dorsal.
4. No circulatory or respiratory systems.
5. Spacious coelom which is not divided by septa.
6. No setae present.
7. Sexes separate, trochophore larva.
   Figure 27.14
   The phylum Sipunculida is not broken down into classes and orders. Representative genera are *Sipunculus* and *Phascolosoma.*

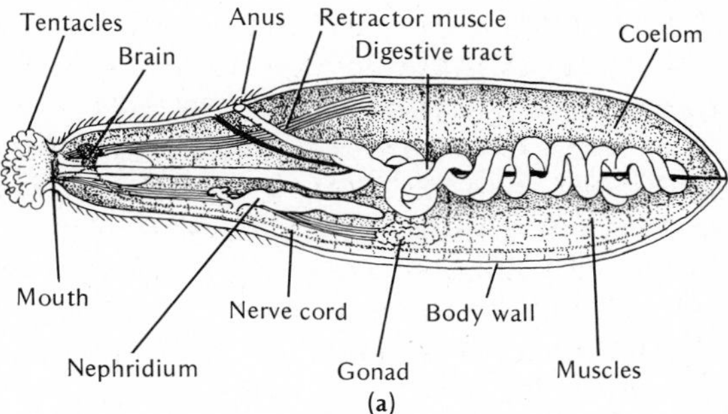

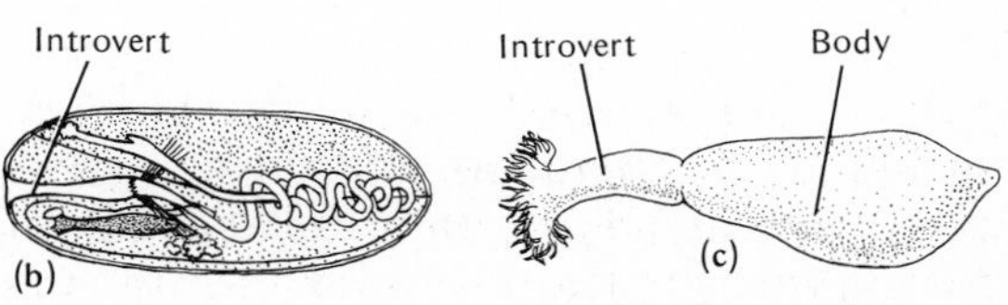

Figure 27.14 Phylum Sipunculida. **(a)** Diagrammatic section of body with the introvert extended. **(b)** Introvert retracted. **(c)** External appearance.

Anatomy and Physiology of *Sipunculus*

GENERAL STRUCTURE

Sipunculus, diagrammed in Figure 27.14, will be described as a representative of the phylum. The eversible introvert is perhaps the most interesting feature of the group. Approximately the anterior third of the animal can be drawn into the trunk region of the body. This retraction is possible because of the well-developed retractor muscles which extend from the body wall of the trunk to the esophagus. The rolling outward of the introvert is brought about by the contractions of the circular muscles of the body wall which force the coelomic fluid into the introvert, thus rolling it outward by hydrostatic pressure. The body wall is covered by a thin cuticle which is often modified on the introvert in the form of scales or spines. The cuticle is secreted by the underlying hypodermal cells beneath which are connective tissue cells of the dermis. In *Sipunculus* and some other genera there is a system of *coelomic canals* in the dermis which communicate by pores with the coelom. They are lined with peritoneum and contain coelomic fluid, including corpuscles with a respiratory pigment, *hemerythrin.* They undoubtedly serve some of the functions of a circulatory system, which is lacking. Beneath the dermis are the muscle layers: an outer circular and inner longitudinal layer with a thin diagonal layer in between in some groups including *Sipunculus.*

The mouth, surrounded by tentacles, is located at the tip of the introvert. The digestive tract is a thin-walled tube without much specialization. Correlated with the habitat in burrows and crevices, the anus is located anteriorly and dorsal at the base of the introvert. The tract is highly coiled, resulting in an extensive surface area for absorption. The lining cells of the tract are ciliated, and a distinctive *ciliated groove* extends the length of the tract.

The spacious coelom contains a variety of cells within its fluid, including red corpuscles, amoebocytes of several kinds, and a type of structure peculiar to sipunculids, called an *urn.* These vaselike urns (Fig. 27.15) with their ciliated opening originate as fixed urns attached to the peritoneum. Their function either in the fixed state or floating in the coelomic fluid seems to be related to excretion since they collect bacteria and various kinds of cellular debris and may deposit it at the nephridia. There is a pair of elongate recurved tubular metanephridia located ventrally in the anterior portion of the trunk. The nephrostome is located at the point of attachment close to the metanephridial pore.

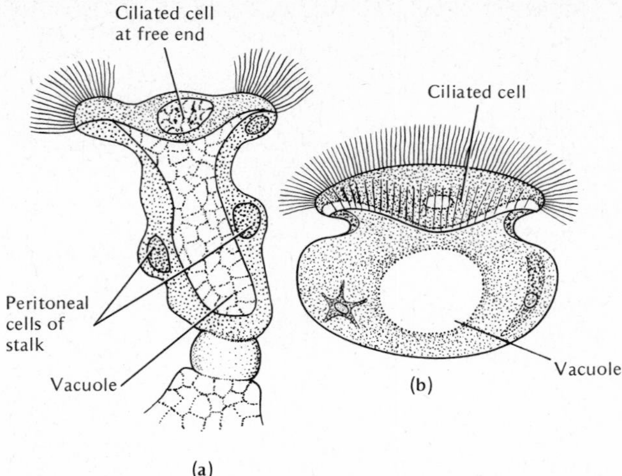

Figure 27.15 Urns of *Sipunculus.* **(a)** Fixed urn attached to peritoneal lining of the coelom. **(b)** Free urn from the coelomic fluid.

The nervous system consists of a dorsal bilobed brain at the anterior end with circumesophageal connectives which meet below to form the ventral nerve cord. The ventral nerve cord extends the length of the worm in the midventral line, giving off many lateral nerves. This type of nervous system, as will be seen later, is much like that of an annelid worm and strengthens the argument in favor of a phylogenetic relationship between these phyla. Sense organs include a chemoreceptor, called the *nuchal organ,* located near the mouth and made up of ciliated cells; and pigmented spots imbedded in the brain which seem to function as light receptors.

REPRODUCTION AND DEVELOPMENT

Sipunculids are dioecious with gonads developing on the wall of the coelom. The gametes reach the outside by way of the nephridia. Fertilization is external in the sea where development ensues with typical spiral determinate cleavage, schizocoelous coelom formation, and a trochophore larva. After about a month as a free-swimming organism, the larva settles to the bottom and metamorphoses into an adult.

Ecology

Sipunculids occur at all latitudes and from the intertidal zone down to depths of 5000 m. Some burrow through the mud, others live in holes in rocky substrates or in abandoned snail shells, while yet others construct mucous-lined burrows in sand. Some small species inhabit the burrows of other animals such as burrowing anemones and tube-dwelling annelids.

In the East Indies an interesting mutualistic relationship exists between sipunculids of the genus *Aspidosiphon* and solitary corals of a number of genera. The sipunculid inhabits a snail shell and the coral becomes attached to the shell, eventually growing down over it and actually lengthening the tubular cavity as the worm grows.

Phylogenetic Relationships

The sipunculids are clearly related to the annelids by reason of their typical spiral cleavage, method of mesoderm formation from teloblast cells, coelom formation by a split in the mesoderm, and the nature of the trochophore larva. The adult worm has a nervous system much like that of annelids. Since segmentation is lacking both in the larval stages and in the adult, the sipunculids cannot be placed in the phylum Annelida. They occupy an intermediate position in the main line of protostome evolution, leading to the segmented annelids and arthropods.

PHYLUM ECHIURIDA

About 150 species of sausage-shaped worms, living mostly in mud or sandy bottoms of shallow marine waters, make up this phylum. They have a contractile, but not retractable *proboscis* which is usually short and spoon shaped, but in some species it is long, with a broad tip.

Characteristics

1. Sausage-shaped marine worms with an anterior prostomial proboscis.
2. Unsegmented as adults, but with segmentally arranged somites in late larval life.
3. One pair of stout setae on the anterior ventral surface and often with a circle of setae around the anus at the posterior end.
4. Closed circulatory system.
5. No respiratory system.
6. Sexes separate, trophophore larva.
   Figure 27.16
   The phylum Echiurida is not broken down into classes and orders. Representative genera include *Urechis, Echiurus,* and *Bonellia.*

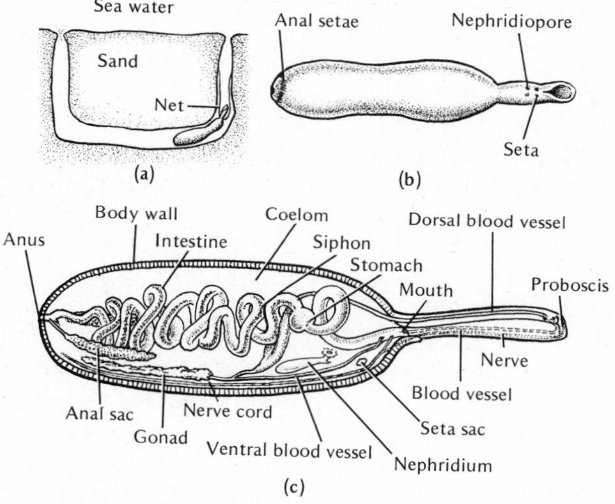

Figure 27.16 Phylum Echiurida, *Urechis caupo.* **(a)** Adult worm in its burrow. **(b)** External appearance of adult. **(c)** Sagittal section of an adult.

Anatomy and Physiology

GENERAL STRUCTURE

Figure 27.16 diagrams a typical echiurid, *Urechis caupo.* The spoonlike anterior proboscis is quite mobile and contractile but it cannot be drawn into the body as in the case of the sipunculid introvert. The thick fleshy body wall is covered with a thin cuticle secreted by the epidermal cells. The muscle layers are similar to those in the sipunculids. The single pair of recurved movable setae at the anterior end arise in a muscular seta sac. *Urechis* also has a circle of setae at the posterior end. The mouth, located at the base of the proboscis, opens into a short stomach and a long greatly coiled intestine. There is a ciliated groove in the intestine as in the sipunculids and also a unique structure, called the *siphon.* This is a small tube which extends the entire length of the intestine, enclosed within the intestinal peritoneum and opening into the intestine only at the anterior and posterior ends in the region of the ciliated groove. Its function is not known.

A pair of diverticula, called the *anal sacs,* which secrete mucus and may also aid in excretion since they have small ciliated funnels which open into the coelom, are attached to the posterior end of the gut near the anus. Although there is no respiratory system, some species draw water in and out of the anal opening by contractions of the gut muscles. Respiratory exchange occurs through the thin-walled gut. A pair of nephridia similar to those in sipunculids open to the outside near the anterior setae. The closed circulatory system is made up of dorsal and ventral vessels which connect at the anterior end. The blood, containing amoeboid corpuscles, is colorless. The extensive coelom is not divided by septa and its fluid must also play a role in circulation since hemoglobin-containing cells are found in it.

The nervous system is on the same pattern as in the sipunculids and annelids, but lacks ganglionic enlargements. A loop around the esophagus and a midventral cord with lateral branches make up the system. There are no specialized sensory receptors.

REPRODUCTION AND DEVELOPMENT

The sexes are separate with a single gonad developing on the peritoneum in the posterior portion of the body cavity. Gametes are released into the coelomic cavity in an immature stage and after ripening they leave via the nephridia. Fertilization occurs in the sea followed by spiral determinate cleavage and a trochophore larva. As the trochophore matures, there is a transitory appearance of segmentation with 15 mesodermal somites and an equal number of ganglionic swellings with lateral branches on the ventral nerve cord. All evidence of segmentation is lost by the time metamorphosis to an adult is complete.

Ecology

Urechis lives in a permanent U-shaped burrow through which it keeps a current of water flowing by peristalsislike contractions of its body wall [Fig. 27.16 (**A**)]. The animal secretes a thin net of mucus in one portion of the tube which traps bacteria and other particulate material. Periodically the mucus net with its trapped material is eaten and a new net is formed. A number of organisms, including several species of annelids, crabs, and clams, commonly take advantage of the protected burrow with its constant source of food and oxygen.

Other Echiurida

Although echiurids are dioecious, there usually is no morphological difference between the sexes. The genus *Bonellia,* however, shows an extreme dimorphic condition with females as much as 1 m in total length and males only 1 or 2 mm long. These small males are internal parasites living in an *egg storage sac* within the female. In this form, sex is determined in an interesting manner. An isolated larva settling to the bottom will develop into a female worm. However, if a larva settles on the proboscis of a female it eventually enters her mouth and comes to live within the egg storage sac, a fold in the nephridium, as a small degenerate ciliated parasite whose sole function is to fertilize the eggs of its host. It appears that some secretion from the female influences the direction of the development of the indifferently sexed larva.

Phylogenetic Relationships

As in the case of the sipunculids, many features point to a close relationship between the echiurida and the annelids and therefore a close relationship between the echiurids and the sipunculids as well. These features include the circulatory, excretory, and nervous systems as well as the nature of the body wall of the adult. The segmentation which appears during development of some species suggests an origin of the echiurids from early ancestral polychaetes or from the line leading to the polychaetes.

SUGGESTIONS FOR FURTHER READING

Barnes, R. D. *Invertebrate Zoology.* Philadelphia: W. B. Saunders (1974).

Brinkhurst, R. O., and B. G. Jamieson. *Aquatic Oligochaeta of the World.* Toronto: Toronto Univ. Press (1972).

Borradaile, L. A., F. A. Potts, L. E. S. Eastman, and J. T. Saunders (rev. by G. A. Kerkut). *The Invertebrata.* London: Cambridge Univ. Press (1961).

Darwin, C. *The Formation of Vegetable Mould through the Action of Worms.* London: John Murray (1881). Reprinted, New York: Appleton (1898).

Hyman, L. H. *The Invertebrates: Smaller Coelomate Groups,* vol. 5. New York: McGraw-Hill (1959).

Laverack, M. S. *The Physiology of Earthworms.* New York: Pergamon (1963).

Mann, K. H. *Leeches (Hirudinea), Their Structure, Physiology, Ecology, and Embryology.* New York: Pergamon (1962).

Pennak, R. W. *Fresh-Water Invertebrates of the United States.* New York: Ronald Press (1953).

Sawyer, R. T. "North American Freshwater Leeches,

Exclusive of the Piscocolidae, with a Key to all Species." *Illinois Biological Monographs,* no. 46. (1972).

Stephenson, J. *The Oligochaeta.* Oxford: Clarendon Press (1930).

From Scientific American

Nicholls, J. G., and D. Van Essen. "The Nervous System of the Leech." (Jan. 1974).

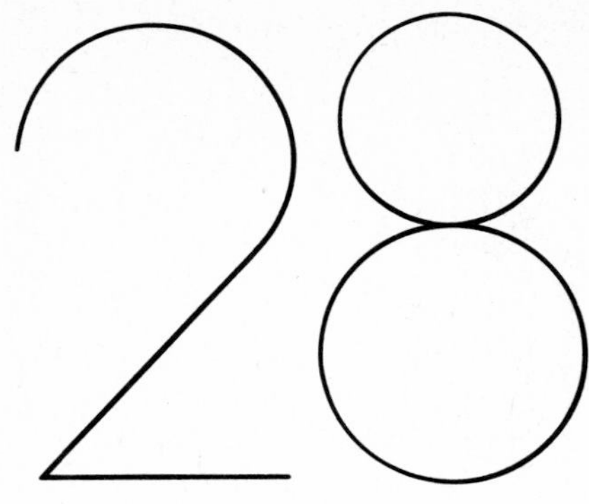

The Arthropod Allies—

ONYCHOPHORA, TARDIGRADA, AND PENTASTOMIDA

The three small groups to be discussed in this chapter have been variously treated in terms of their taxonomic position. Their characteristics indicate relationships with both the annelids and the arthropods. Some have included them as subphyla within the phylum Arthropoda. At least one of the phyla, the Onychophora, is sometimes included in the phylum Annelida. Other workers unite the three phyla in a single phylum, called phylum Oncopoda. However, they are quite distinctive groups and it seems most logical to consider each as a separate phylum.

PHYLUM ONYCHOPHORA

This small phylum of less than 75 species is in many ways one of the most interesting groups of animals from the point of view of phylogeny. The combination of annelid and arthropod characteristics found in these organsims clearly indicates that there is a close phylogenetic relationship between the annelids and the arthropods. Onychophorans are found in damp habitats under logs and leaf mold in tropical, subtropical, and south temperate forests of both the eastern and western hemispheres. In order to emphasize the significant annelid and arthropod characteristics of these organisms, the phylum characterization below is divided into three parts: one giving the annelidlike characters; a second arthropodlike ones; and in a third section the distinctive characters found only in the phylum Onychophora.

Annelidlike Characteristics

1. Paired, segmentally arranged nephridia.
2. Gonoducts lined with cilia.
3. Nervous system quite annelidlike, although it lacks distinct ganglia.
4. General body form is annelidlike.

Arthropodlike Characteristics

1. Respiration by means of tracheal tubes with spiracles opening to the outside.
2. Open circulatory system with a dorsal tubular heart provided with ostia.
3. Body cavity not a true coelom. The coelom is reduced to the cavities of the gonads and to small sacs associated with the nephridia.
4. Jaws are modified walking legs.

Characteristics Distinctive to the Onychophora

1. Only one pair of mouth parts while arthropods typically have more than one pair.
2. Only three segments in the head, whereas the annelid head is a single segment and six segments are involved in most arthropod heads.
3. Numerous fleshy unjointed pairs of appendages—much like annelid parapodia which bear a pair of claws on the tip—as most arthropod appendages.
4. A pair of oral papillae which shoot out an adhesive material.
   Figures 28.1 and 28.2

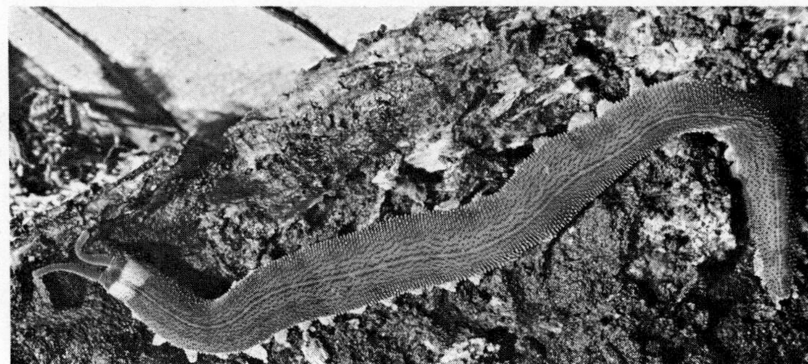

Figure 28.1 Phylum Onychophora, *Peripatus* sp. (From Nicholas Smythe, National Audubon Society.)

Figure 28.2 Onychophoran anatomy. **(a)** Ventral view of anterior region. **(b)** Ventral view of posterior region. **(c)** Internal anatomy, lateral view. **(d)** Detail of spiracle and tracheal tubes.

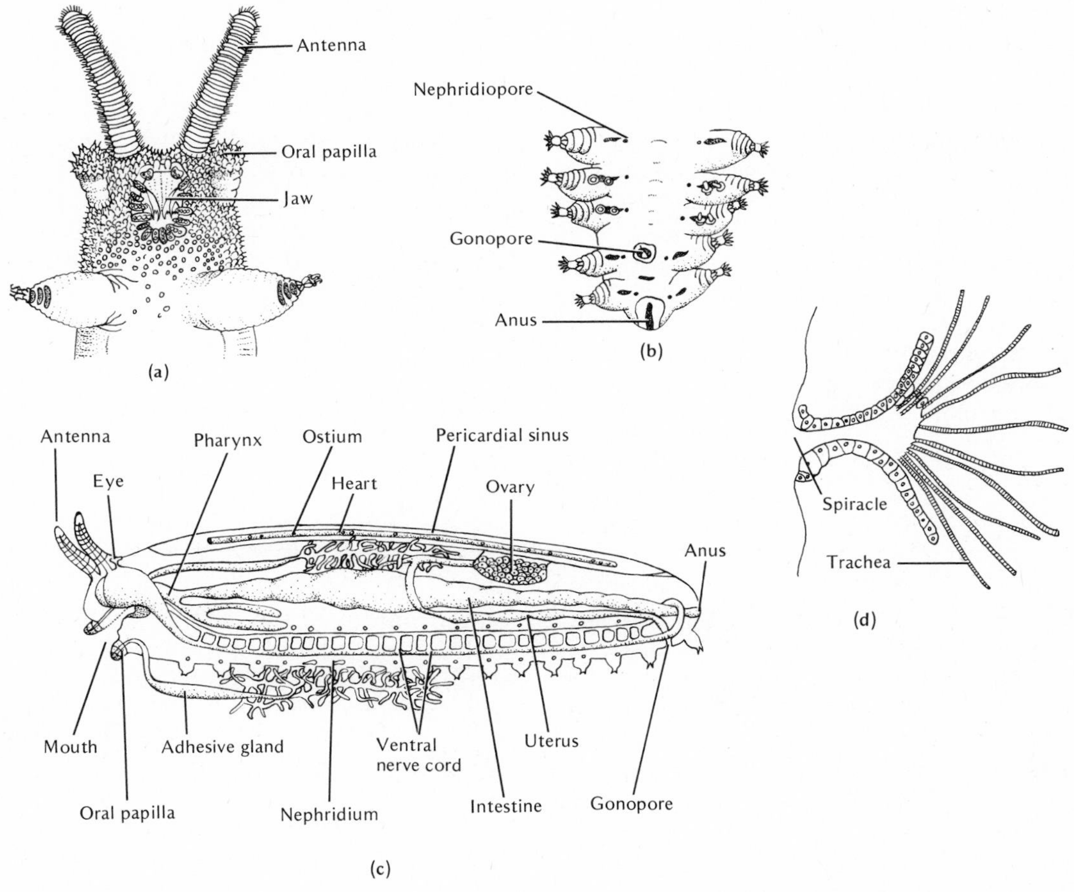

Anatomy and Physiology

The cylindrical bodies of onychophorans are covered with a velvety chitinous cuticle. The entire body is covered with tubercles and papillae, the large ones being provided with a sensory bristle. The three segments of the head are evidenced by the antennae and eyes on the first segment, the *oral papillae* on the second, and the mouth with its jaws on the third. The antennae are not retractable and serve a sensory function. The oral papillae contain openings of the adhesive glands. They can be aimed at prey or an enemy, followed by the explosive release of sticky mucus which effectively entangles the target at distances as great as half a meter. The only external evidence of segmentation is the paired appendages which are unjointed but bear a pair of claws at the tips. The openings of the nephridia are located ventrally at the base of the legs.

The body wall is quite annelidlike with a single layer of epidermal cells and circular, diagonal and longitudinal muscles below. The muscles are not striated, another annelid feature since arthropod muscles are striated. The digestive system consists of an anterior mouth within which are two lateral movable jaws which are modified appendages and a single dorsal median tooth. The straight tubular system ends posteriorly in the terminal anus. Excretion is by segmentally arranged paired metanephridia, each of which opens internally into a small coelomic sac and externally at the base of the legs. The circulatory system is a simple dorsally located tubular heart which is open at both ends and has a pair of *ostia* in each segment. The colorless blood which contains amoeboid cells moves anteriorly in the heart and out into the open blood sinuses. The *trachea* are thin unbranched tubules, many of which open into each *spiracle* [Fig. 28.2 (**d**)]. The many small spiracles are not arranged in segmental fashion as in the arthropods, but open all over the body surface. The annelidlike nervous system consists of a bilobed brain which connects around the pharynx with a pair of ventral nerve cords. There are connections between the cords but no conspicuous segmental ganglia.

Onychophorans are dioecious. The paired gonads are attached to the dorsal body wall in the posterior region. The gametes reach the gonopore which is located just anterior to the anus, by ciliated ducts which are modified nephridia. The ducts in the female are enlarged to form a uterus in which the fertilized eggs are retained in most species until development is complete. The eggs are like those in arthropods with a mass of yolk in the center, and cleavage occurs only in the superficial cytoplasm; subsequent development is similar to that in arthropods.

Phylogenetic Relationships

The obvious blend of arthropod and annelid characteristics seen in the onychophorans would tempt one to say that they are the "missing link" between annelids and arthropods.

Most students of phylogeny reject this view, however, because of basic structural features unique to the Onychophora. It is more likely that the onychophorans descended from a line of annelids earlier than that which later evolved into the arthropods. Thus there is common ancestry but not direct descent.

PHYLUM TARDIGRADA

The tardigrades, commonly called water bears, are microscopic animals less than a millimeter long. The presence of four pairs of unjointed legs, each with claws at the tip, led early workers to place them in the class Arachnida of the phylum Arthropoda. However, the unjointed nature of the legs and other characters such as the lack of chitin leads most students of the group to place them in a separate phylum. The characteristics are such that they, like the onychophorans, are evidence of a relationship between annelids and arthropods.

Characteristics

1. Microscopic animals with a body of six segments, four of which bear unjointed clawbearing legs.
2. Body covered by a nonchitinous secreted cuticle which is periodically molted.
3. Cell numbers constant in each species.
4. Circulatory, respiratory, and excretory systems absent. Figure 28.3

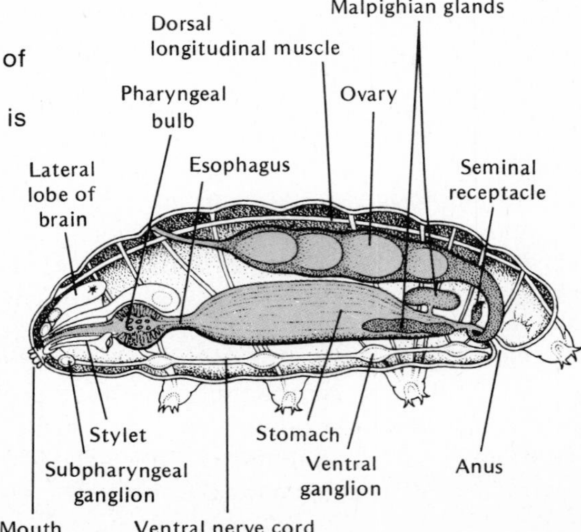

Figure 28.3 Phylum Tardigrada, anatomical detail.

Anatomy and Physiology

The body is composed of a head and five trunk segments, four of which are provided with short fleshy appendages at the tip of which there are claws (Fig. 28.3). The thin nonchitinous cuticle, secreted by the epidermal cells, not only covers the external portions of the body but also lines the anterior and posterior portions of the gut. Various sorts of spines or other modifications of the cuticle are often present. Beneath the epidermis the muscles of tardigrades are not arranged in wormlike fashion, but are thin bands which extend from one part of the exoskeleton to another, a relationship which, it will be seen, is found also in the arthropods. The extensive space between the body wall and the gut is not a true coelom but is a *hemocoel.* Although *coelomic sacs* appear during development, all but the cavity of the gonad and its ducts disappear in the adult.

Tardigrades are mostly herbivores, feeding upon plant juices which are obtained by piercing the plant cell wall with a sharp *stylet.* The muscular pharynx then sucks out the contents of the cell. A short esophagus connects the pharynx with the stomach and on to a short hindgut which leads to the anus on the midventral surface anterior to the fourth pair of legs. A pair of glandlike diverticulae of unknown function, which are sometimes called the *Malpighian tubules,* is located at the beginning of the hindgut. There are a dorsal brain above the pharynx, circumpharyngeal connectives, a subpharyngeal ganglion, and a ventral nerve cord with four ganglia.

Tardigrades are dioecious with a single dorsally located ovary or testis. The gonoducts open either into the hindgut or to the outside by a separate gonopore. Development is external, often in the newly shed cuticle of the female. Five pairs of coelomic sacs evaginate from the gut during early development. This is typical enterocoelic coelom formation, not the usual pattern for the protostome line. As mentioned above, the coelomic sacs disappear in the adult except for the cavities of the gonads and their ducts.

Of the less than 300 species of tardigrades, some are aquatic, both freshwater and marine; but most of them live in moist terrestrial situations such as in mosses or in lichens. Like rotifers, many of which live in similar situations, the tardigrades show a remarkable capacity to withstand desiccation. There are records of animales surviving over four years in a shriveled-up condition in which minimal metabolic activity goes on. Experimentally they can be revived and subjected to drying again and again without apparent harm.

PHYLUM PENTASTOMIDA

This small phylum, sometimes called the phylum Linguatulida, of less than 70 species is made up entirely of forms in which the adults are parasitic in the respiratory passages of carnivorous vertebrates, mostly reptiles. They are largely tropical in their distribution but a few species do occur in the United States. The presence of the two stubby, unjointed, claw-bearing legs in the larva is believed to indicate arthropod relationships since the larvae of some mites have similar appendages. The degeneration associated with parasitic existence has effectively obscured any other significant phylogenetic evidence.

Characteristics

1. Adults are wormlike, unsegmented parasites of carnivorous vertebrates, usually in the respiratory passages.
2. Two pairs of claws adjacent to mouth.
3. Circulatory, respiratory, and excretory systems absent.
4. Larvae with two pairs of fleshy unjointed legs armed with claws at the tip.
   Figure 28.4

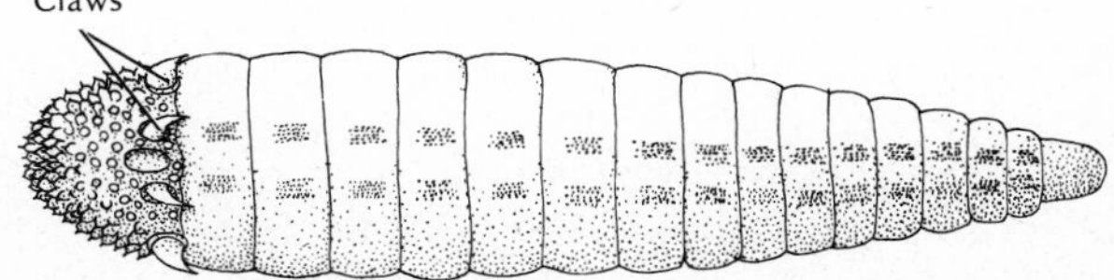

Figure 28.4 Phylum Pentastomida, *Porocephalus* sp. Ventral view of an adult.

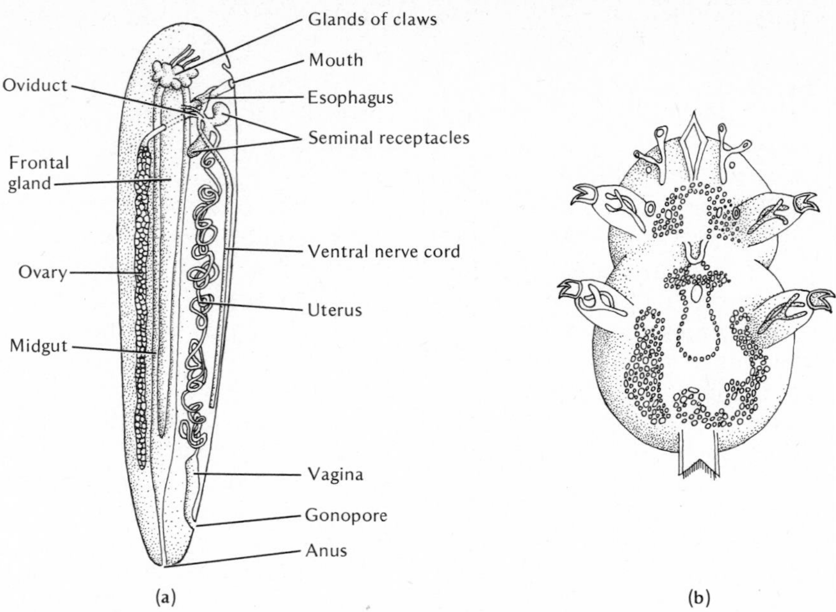

Figure 28.5 Phylum Pentastomida, *Porocephalus* sp. **(a)** Internal anatomy, lateral view. **(b)** Larva.

Anatomy and Physiology

The adult worms show a superficial annulation and the body is covered by a thin chitinous cuticle. The mouth is located at the anterior end, and two pairs of retractible claws, sometimes on stubby outgrowths, are located on either side (Fig. 28.4). The early observers mistook the four claws for separate mouths—hence the phylum name, Pentastomida. The worms attach to the lining of the respiratory passages and the claws are used to cause bleeding of the host. Glands associated with the claws secrete an anticoagulant and the blood is sucked into the mouth by activity of the muscular pharynx. Following the pharynx is a large midgut extending most of the length of the body, and a short rectum leading to a posterior anus. The nervous system is a simple ladder type with three segmentally arranged ganglia. There is a circumpharyngeal nerve ring but no brain enlargement. The circulatory, respiratory, and the excretory systems are lacking, as are cilia (Fig. 28.5).

The sexes are separate and dimorphic with the female being larger than the male. Fertilization is internal and zygotes leave the host and must be ingested by a suitable intermediate host which is a fish in the case of a species infecting crocodiles, for example, or an herbivorous mammal for species with terrestrial definitive hosts. The egg hatches in the gut and the larva bores through the gut wall, enters the blood or lymph, and is transported to various parts of the body. The larva which hatches from the egg has two pairs of unjointed claw-bearing legs and has, as mentioned earlier, some resemblance to the larvae of certain mites. The larvae encyst in the tissues of the intermediate host and undergo several molts. When the intermediate host is eaten by a carnivorous vertebrate, the larvae excyst and move by way of the esophagus to the respiratory passages where they attach and mature.

SUGGESTIONS FOR FURTHER READING

Barnes, R. D. *Invertebrate Zoology.* Philadelphia: W. B. Saunders (1974).

Meglitsch, P. A. *Invertebrate Zoology.* New York: Oxford Univ. Press (1972).

Pennak, R. W. *Fresh-Water Invertebrates of the United States.* New York: Ronald Press (1953).

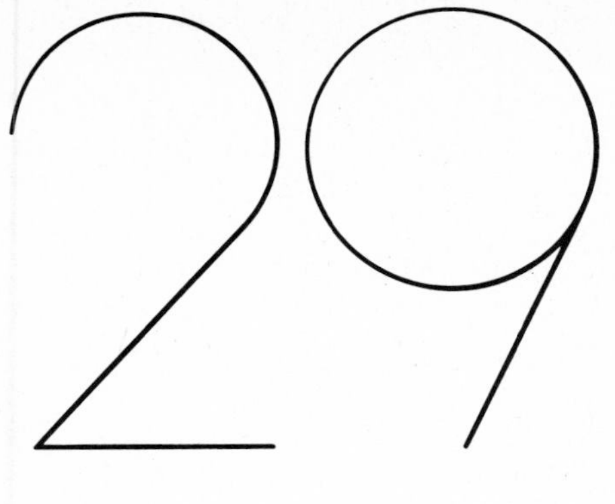

The Trilobites and the Chelicerate Arthropods

PHYLUM ARTHROPODA

The jointed-legged animals comprise the phylum Arthropoda, the largest phylum in the animal kingdom. With almost a million species there are more arthropods than all other kinds of plants and animals combined. Not only are there great numbers of species in the phylum, but many species are represented by countless numbers of individuals. If sheer numbers are a measure of success of any group of organisms, the arthropods are by far the most successful animals. All habitats and all climates, from the vast frozen wastelands of the poles to the depths of the sea, have their arthropod residents. Some species are parasitic, but most are free living. Among the latter are found some very highly organized societies with a differentiation into castes and considerable division of labor. Some members of the phylum are of immense economic importance. The oldest known arthropods in the fossil record are the extinct trilobites that flourished from the Cambrian to the Silurian, from 550 to 360 million years ago.

Following the listing of distinctive arthropod characteristics and a general discussion of arthropod features, this chapter will deal with the subphyla Trilobitomorpha and Chelicerata. The following chapter will discuss the third and largest subphylum, the Mandibulata.

Characteristics

1. Segmented body.
2. Jointed legs with at least one pair serving as jaws.
3. Chitinous exoskeleton.
4. Well-developed head, usually with compound eyes.
5. Coelom greatly reduced, main body cavity a hemocoel derived from the embryonic blastocoel. Figure 29.1

Head
Eye
Segmented appendage
Segmented body
Chitinous exoskeleton

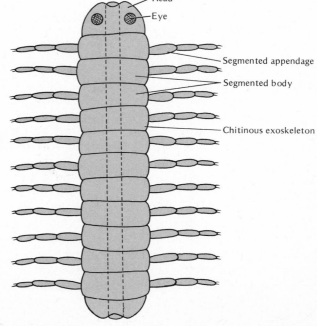

Figure 29.1 Arthropod body plan.

Classification

Because of the large size and diverse nature of the phylum Arthropoda, the following outline is only a listing of the subphyla, classes, and subclasses. Detailed characteristics and a breakdown into orders will be given in the treatment of each class.

Phylum Arthropoda
SUBPHYLUM **Trilobitomorpha**
CLASS **Trilobita,** the trilobites
SUBPHYLUM **Chelicerata**
CLASS **Merostomata**
SUBCLASS **Xiphosura,** the horseshoe crabs
SUBCLASS **Eurypterida,** the eurypterids, all extinct
CLASS **Pycnogonida,** the sea spiders
CLASS **Arachnida,** spiders, scorpions, mites, and ticks
SUBPHYLUM **Mandibulata**
CLASS **Crustacea,** crabs, lobsters, shrimps, barnacles, and pill bugs
CLASS **Diplopoda,** the millipedes
CLASS **Pauropoda,** the pauropods
CLASS **Chilopoda,** the centipedes
CLASS **Symphyla,** the symphylids
CLASS **Insecta,** the insects

General Anatomy and Physiology

JOINTED APPENDAGES

The jointed appendages are one of the most distinctive features of the phylum, and it is this characteristic which gives the phylum its name. The success of the arthropods is in large measure due to the wide variety of adaptive uses which resulted from the evolutionary changes in these important structures. Not only did they serve as walking legs, they became modified for swimming, breathing, and reproductive activity. They also produced specialized structures which are used in defense, food procurement, chewing, and sensory perception (Fig. 40.2).

CHITINOUS EXOSKELETON

Besides the jointed appendages, the next most conspicuous characteristic of arthropods is their *chitinous exoskeleton.* In the annelids only a thin cuticle covers the epidermis and it is in no sense a skeleton. The cuticle of the arthropods is quite different; it is much thicker and harder than the cuticle of annelids. When examined carefully, the exoskeleton is found to be composed of an outer waxy layer, an intermediate layer of chitin, a horny flexible substance often infiltrated with lime salts to make it rigid, and an inner more flexible chitinous layer. The hard middle layer is not present at the joints and the flexible layer is thin. This allows for the hinge action at the joints in the legs and between the segments of the body. Thus this exoskeleton serves as a protective coat of armor without sacrificing mobility.

The muscles of arthropods are attached to this exoskeleton across the joints. The arrangement here is quite different from that found in the vertebrates, as shown in Figure 29.2. The skeletal muscles of arthropods, like those of vertebrates, are striated and, thus, they are adapted for strong and rapid contractions.

In addition to serving as a protective coat and as a surface for the attachment of muscles, the exoskeleton has made it possible for the arthropods to live on land. Land life for any kind of organism requires a relatively impermeable outer covering to prevent desiccation, and the arthropod exoskeleton certainly does this.

Although there are several advantages to this tough lightweight exoskeleton with its great area for muscle attachment, there are also some disadvantages. The arthropod exoskeleton is just like a suit of armor and it allows very little growth. As a consequence, all arthropods, during their growth, must undergo a number of molts. The process of molting, or *ecdysis*, is a complicated one stimulated and controlled by a complex system of hormones, but in this process

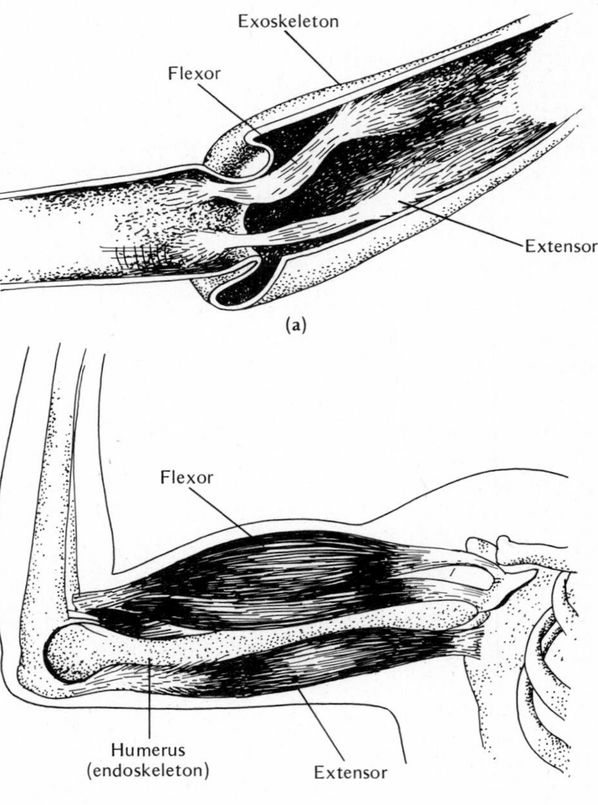

Figure 29.2 Diagram showing attachment of muscles in an arthropod and in a vertebrate. In the arthropod **(a)** the muscles are attached to the exoskeleton; in the vertebrate **(b)** they are attached to the endoskeleton.

the exoskeleton splits open and the organism moves out, leaving it behind. Following this there is a period of rapid expansion before the new outer layer becomes hard and armorlike. During the period of the formation of the new exoskeleton, the arthropod is quite vulnerable, and in this stage many arthropods are destroyed by their enemies. An exoskeleton also puts some limitation on size and somewhat restricts development of external sensory receptors to specialized appendages.

NERVOUS SYSTEM

Still another advance that the arthropods show over the annelids is an increase in cephalization. There is a primitive sort of head in planaria and a much better developed head in *Nereis,* but even in these annelids the head is really not a distinct body region. In the arthropods the head is a distinct body region. Associated with this distinct head development, there is found in some of the arthropods a further division of the body into *thorax* and *abdomen.* In some arthropods the head becomes fused secondarily with the thorax to form a cephalothorax. Also correlated with this head development is a higher development of the nervous system in the anterior segments and a greater specialization of sense organs associated with the head. The *compound eye,* described in a later section, is an important and rather complex sense organ found in many arthropods.

DIGESTIVE SYSTEM

Arthropod digestive systems consist of a fore-, mid-, and hindgut. Both the fore- and hindgut are lined with chitin and are embryologically a stomodaeum and a proctodaeum, respectively. An esophagus and stomach are usually present in the foregut while the midgut is the intestine. The posterior hindgut opens in the anus which is usually terminal. Salivary glands as well as other digestive glands are commonly present.

EXCRETION, CIRCULATION, AND RESPIRATION

Excretion is accomplished in two different ways in the phylum Arthropoda. In aquatic forms a glandular structure, which is probably a modified nephridium, collects excretory wastes and opens to the outside at the base of an appendage. Horseshoe crabs have *coxal glands* which open at the base of walking legs and crustacea have the so-called *green glands* which open at the base of certain head appendages, usually the second antennae. Terrestrial arthropods, such as insects and spiders, possess excretory organs, called *Malpighian tubules,* which open into the gut at the juncture of the mid- and hindgut. The open circulatory system consists of a dorsal heart with openings, called *ostia,* through which blood enters the heart. Arteries take the blood, which may be colorless or contain various types of respiratory pigments in the plasma, and deliver it to various parts of the body where it enters the large open sinuses through which it returns to the cavity around the heart. Respiration in the aquatic forms is usually by means of gills or *gill books,* while most terrestrial forms have *lung books* or tracheal systems, or both.

REPRODUCTION AND DEVELOPMENT

Most arthropods are dioecious. Fertilization is internal in terrestrial forms and may be either internal or external in aquatic species. Eggs are well provided with yolk which is centrally located with a thin layer of cytoplasm surrounding it. Cleavage involves division of the centrally located nucleus, but there are no cytoplasmic divisions. Later the nuclei migrate to the periphery where cell membranes form around them (Fig. 30.56). At this point the blastula stage has been reached. Gastrulation occurs by invagination or by delamination. Subsequent development varies widely within the phylum with larval stages of different types occurring in many of the groups.

SUBPHYLUM TRILOBITOMORPHA, CLASS TRILOBITA

Characteristics

1. Ancestral marine arthropods, extinct since Permian times.
2. Body divided by two anterior–posterior furrows into three lobes.
3. Body made up of a distinct head and a thorax fused to the posterior pygidium.
4. Head with antennae, usually two well developed eyes, and four pairs of appendages.
5. Each remaining body segment except the last with similar biramous (two-branched) appendages.
Figure 29.3.

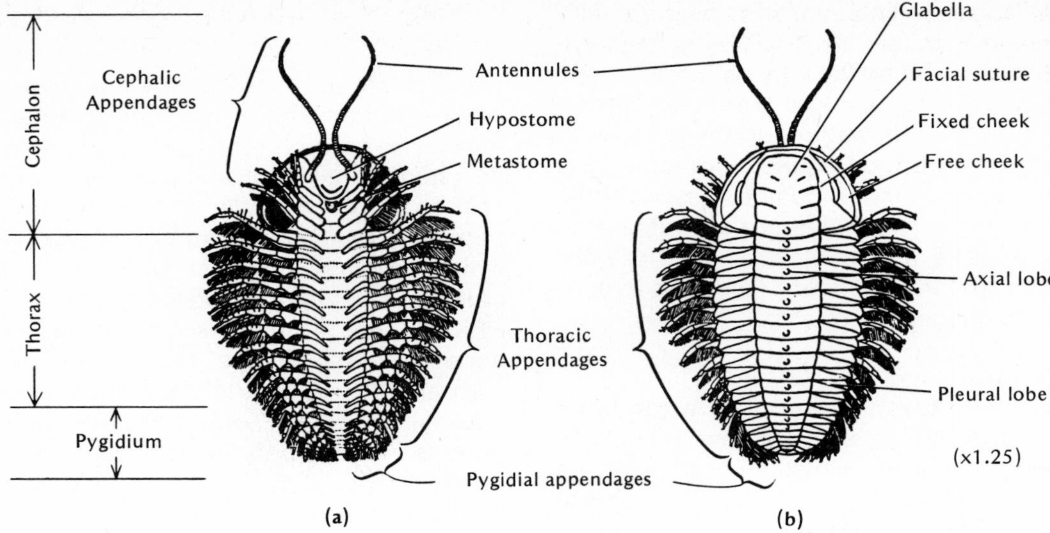

Figure 29.3 Class Trilobita. **(a)** Ventral and **(b)** dorsal views of a completely restored specimen. (From Shrock and Twenhofel, *Principles of Invertebrate Paleontology,* McGraw-Hill, 1953, with permission.)

Anatomy and Physiology

GENERAL STRUCTURE

The trilobite body was divided into three *lobes,* a median axial lobe and two lateral pleural lobes, by two longitudinal furrows, hence the name of the group (Fig. 29.3). The head was distinctly separated from the rest of the body and bore a pair of jointed antennae near the ventral mouth and four pairs of similar *biramous appendages.* Dorsally, many species had a pair of well-developed compound eyes. The four pairs of head appendages were probably used in pushing food toward the mouth.

The thorax and fused tail region, or *pygidium,* were made up of similar segments, each one except the last with a pair of biramous appendages bearing fringelike gills. The evolutionary relationship of these appendages to those of present-day arthropods will be discussed in some detail in Chapter 40.

Fossil Record

The first arthropods of which there is any record were the trilobites. Fossil remains of these organisms are abundant in the rocks of the Cambrian period, dating back approximately 550 million years. They were the dominant animal forms during the early part of the Paleozoic era, but they declined gradually and became extinct by the end of that era, some 200 million years ago.

Ecology

Some 4000 species of trilobites have been described. Most of them were of modest size, from 3 to 6 cm long, although a few species reached sizes of over 60 cm. Most trilobites seem to have been bottom dwellers, but the young larvae and some adult species were *pelagic,* that is, they were swimmers and probably spent much of their time in the open ocean.

SUBPHYLUM CHELICERATA

The body of chelicerates is made up of two regions, an anterior prosoma and a posterior opisthoma. The prosoma bears one pair of chelicerae which have either claws or pincers, a pair of pedipalps, and four pairs of walking legs. The opisthoma is quite variable having up to 13 somites and a telson and various types of modified and reduced appendages.

The subphylum Chelicerata is composed of three classes. Each of these classes will be discussed in turn below.

CLASS MEROSTOMATA

The prosoma is covered dorsally by a single continuous carapace. One pair of compound eyes and one pair of simple eyes are present on the dorsal surface. The opisthoma bears flattened appendages used in respiration. There is a pointed telson at the posterior end.

There are two subclasses in the class Merostomata, subclass Xiphosura and the subclass Eurypterida.

SUBCLASS XIPHOSURA

Characteristics

1. Semicircular, horseshoe-shaped carapace covers prosoma.
2. Opisthoma wide and unsegmented.
3. Chelicerae have three segments.
4. Pedipalpi and legs have six segments, the former being essentially walking legs.
   Figure 29.4

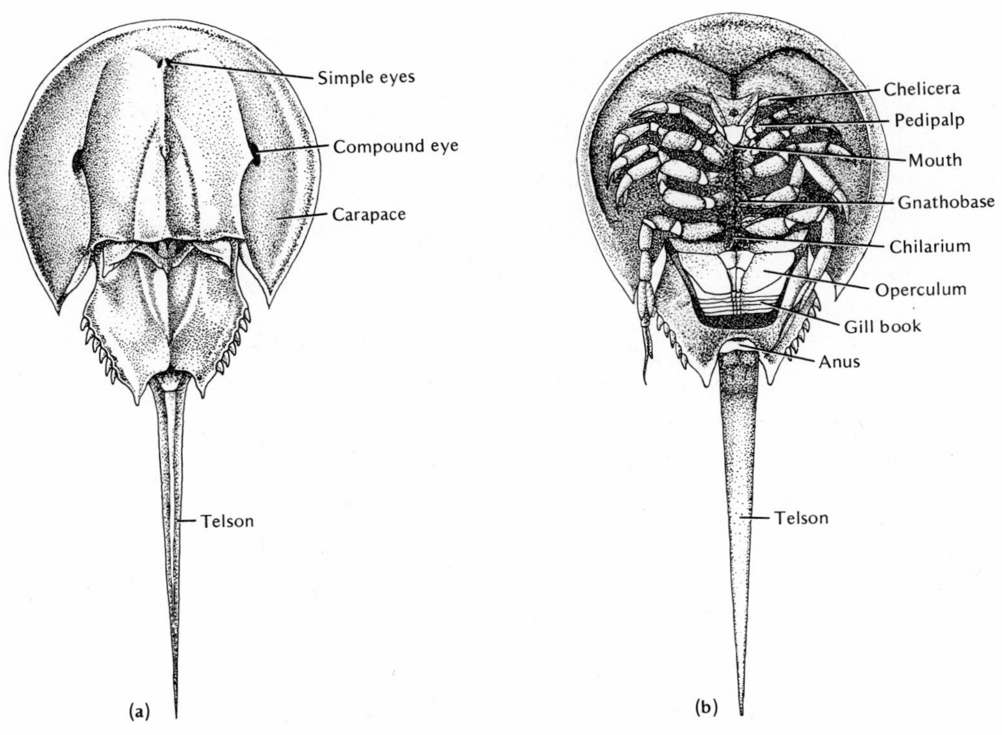

Figure 29.4 Subclass Xiphosura, *Limulus polyphemus,* the horseshoe crab. **(a)** Dorsal view. **(b)** Ventral view.

Anatomy and Physiology of *Limulus*

GENERAL STRUCTURE

The horseshoe crab is a peculiar relic of ancient times, the present-day genus *Limulus* going back over 200 million years to Triassic times. *Limulus* is often rightly called a living fossil. Anyone who has ever seen this armored tanklike creature with all its appendages safely tucked beneath its spiny *carapace,* can easily see that it is well protected against enemies and can understand its continued presence long after its ancient contemporaries became extinct.

Limulus polyphemus, the common horseshoe crab of the Atlantic seashore, reaches a length of 50 cm. A pair of lateral *compound eyes* and a median pair of *simple eyes* can be seen on the dorsal surface of the carapace which covers the *cephalothorax.* Ventrally, the mouth is surrounded by the anteriormost appendages. There are two small, three-jointed, pincerlike chelicerae anterior to the mouth. The pedipalps and the first three pairs of walking legs are six segmented and quite similar in appearance with small pincers at the tip. At their bases are located spiny processes, called *gnathobases,* which aid in crushing food. The larger posterior pair of walking legs has flattened tips used to clean mud off the abdominal appendages. The next posterior pair of appendages are the *chilaria,* small and greatly modified. Whether these are a posterior pair of prosomal appendages or the first pair of opisthomal appendages is debatable.

The six pairs of opisthomal appendages are flattened, leaflike, and highly modified. The first pair form a protective cover for the rest. The reproductive openings are on this appendage, which is called the *operculum.* The other five pairs have many flat highly vascularized gill plates on their posterior surface, forming the so-called gill books. Posterior to the anus, the long pointed *telson* projects to the rear. The telson is used to push the body forward when the animal burrows in the mud and also aids in righting the animal if it is overturned. Excretion is by means of four pairs of coxal glands which are located at the bases of the walking legs and open to the outside at the excretory pore, located at the base of the posterior pair of legs. Other internal structures are much like those described earlier in the discussion of general arthropod anatomy.

REPRODUCTION AND DEVELOPMENT

Sexes are separate and the males are smaller than the females. They become sexually mature after about three years. In the spring, large numbers of horseshoe crabs come to shallow waters to mate. The smaller males ride on the back of a female, maintaining their position with their modified second appendages. The female digs a hole in the sand in the intertidal zone and deposits as many as 1000 eggs. The male releases sperm into the hole. The returning tide fills in the hole with sand. In some species the female actively covers the holes. After several months the young larvae hatch and for a time these trilobite larvae (Fig. 29.5) live a free-swimming existence.

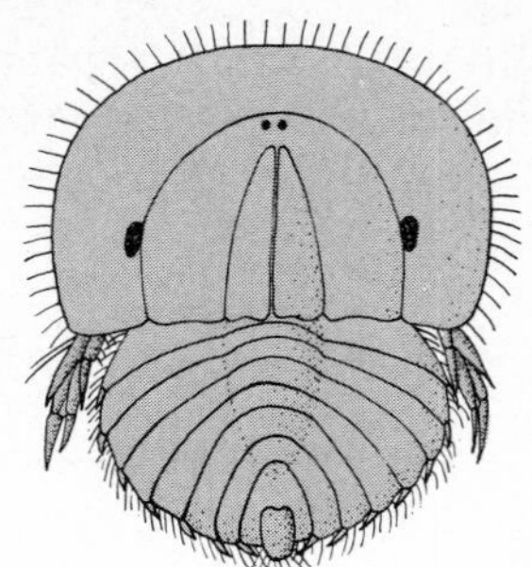

Figure 29.5 Trilobite larva of *Limulus polyphemus.*

Ecology

While generally marine, some forms are occasionally found in brackish or even fresh water. Horseshoe crabs are usually nocturnal in habit, living in shallow waters of the continental shelf on muddy or sandy bottoms where they feed largely on annelid worms, nudibranchs, and other soft-bodied animals. The turbellarian *Bdelloura* is a common commensal resident in the gills of *Limulus.* The life span of *Limulus* may be 14–19 years, rather long for an invertebrate.

Phylogenetic Relationships

This ancient animal is the closest relative of the original arthropod, the trilobite. Its body organization has not changed since Silurian times, and the genus *Limulus* goes back to the Triassic. The trilobite larva is good evidence of a relationship between the horseshoe crabs and the trilobites. Unfortunately there is no fossil record of the transitional forms between these two groups.

SUBCLASS EURYPTERIDA

These animals, often called the giant water scorpions, have been extinct since the Permian.

Characteristics

1. Compound eyes present as well as ocelli.
2. Appendages of the prosoma were a pair of chelicerae and four pairs of walking legs, the fifth pair often being modified as paddles for swimming.
3. Opisthoma of 12 segments with respiratory appendages on segments 2–6.
4. A terminal telson which may have been a poisonous sting in some species.
   Figure 29.6

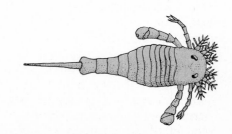

Figure 29.6 Subclass Eurypterida, *Eurypterus* sp.

The eurypterids differed from the Xiphosura in having a segmented opisthoma and a generally more elongate body. Some of them reached a length of almost 3 m, making them the largest arthropods ever known. While the Xiphosura were marine, the eurypterids invaded brackish and fresh water and perhaps even land. They probably were predators on other invertebrates, and the larger species may have fed upon vertebrates.

Phylogenetic Relationships

It was long thought that the eurypterids gave rise to the scorpions. A tapering tail, with a pointed telson at the tip, gives some species the appearance of a scorpion and strongly suggests a relationship. In addition, the concealed book lungs of the eurypterids are remarkably like those of modern scorpions. However, the primitive eurypterids were much less scorpionlike. By the time scorpionlike species of eurypterids did evolve, the scorpions themselves were already on the scene. Students of phylogeny are not in agreement that the scorpions are the most primitive arachnids.

While the picture is not clear, there seems to be little doubt that the arachnids were in some way derived from the eurypterids.

CLASS PYCNOGONIDA

These peculiar marine creatures seem to be all legs. They are quite spiderlike in general appearance but are, in fact, only distantly related to the spiders. They occur from the intertidal zone to depths of over 4000 m and are quite abundant in cold polar waters.

Characteristics

1. Prosoma is short and cylindrical with a minute, peglike opisthoma.
2. Large anterior proboscis has the mouth at its tip.
3. Usually there are four pairs of legs, but species with as many as 12 pairs are known.
   Figure 29.7

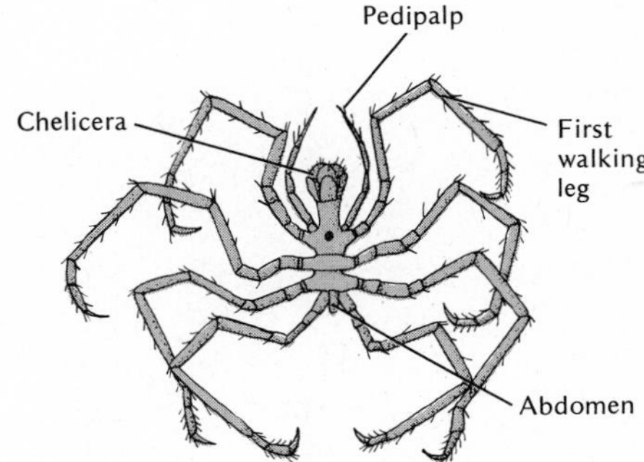

Figure 29.7 Class Pycnogonida, *Nymphon* sp.

Anatomy and Physiology

Most pycnogonids are of modest size, under 5 cm in length. The head has a long cylindrical proboscis at the end of which the mouth is located. The number of head appendages varies, but many include pairs of pedipalps, specialized legs in the male with which the eggs are carried during their development, and a pair of walking legs. Posterior to the head there are typically three segments, each with a pair of walking legs. Additional segments, each with a pair of legs, may also be present. Posteriorly there is a minute opisthoma.

Pycnogonids are carnivorous, feeding largely on sponges, coelenterates, and ectoprocts. Internal anatomy is quite simple and there are neither respiratory nor excretory systems present. Sexes are separate and in most species the eggs are carried by the male during early development.

Phylogenetic Relationships

Structurally the pycnogonids are clearly chelicerate arthropods, and they appear to be much more like the arachnids than the Merostomata. They are, however, so much modified that any clear relationships are not available. Their larvae are quite unique and their primitive nature has led some to suggest a very early origin for the group.

CLASS ARACHNIDA

By far the most successful of the chelicerate arthropods, numbering over 30,000 species, the arachnids were the first arthropods to take up a terrestrial mode of life. Most present-day forms are terrestrial and familiar members of the class include spiders, scorpions, mites, and ticks.

Characteristics

1. Mostly terrestrial with respiration by lung books, trachea, or both.
2. Prosoma has paired chelicerae, pedipalpi, and four pairs of walking legs.
3. Compound eyes are absent.
4. Opisthoma lacks external gills and locomotor appendages.

Classification

ORDER **Scorpionida** The scorpions (Fig. 29.8). Body elongate with the prosoma broadly joined to the opisthoma. Chelicerae small. Pedipalpi large and provided with well-developed pincers. Opisthoma long and tapering to a narrow posterior portion at the end of which is a poison sting. Featherlike pectines on ventral surface of second opisthomal segment. Four pairs of book lungs present. Size range from 2.5 to over 17 cm. *Centuroides.*

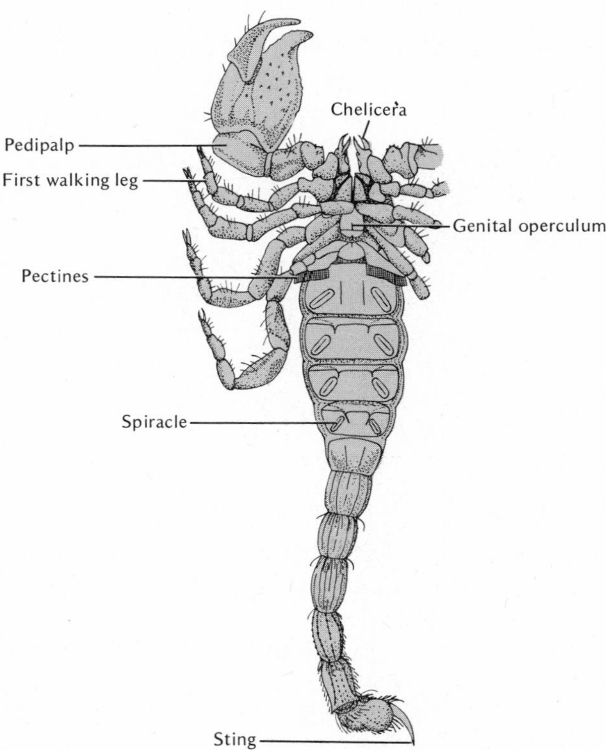

Figure 29.8 Class Arachnida. Ventral view of a scorpion.

ORDER **Uropygi** The whip scorpions (Fig. 29.9). Prosoma narrowly joined to the opisthoma. Chelicerae small and chelate. Pedipalpi well developed, stout, and provided with pincers. Tarsus of first pair of legs modified into a long feeler. Opisthoma with a long, slender, many-jointed flagellum at its tip. Size range from 2 to over 60 mm. Tropical and subtropical in distribution. *Mastigoproctus.*

ORDER **Amblypygi** (Fig. 29.10) Similar to the uropygids, but lack the long flagellum at the tip of the opisthoma, and have much longer whiplike modifications of the first pair of legs. Size range from 4 to 45 mm. Tropical and subtropical in distribution. *Acanthophrynus.*

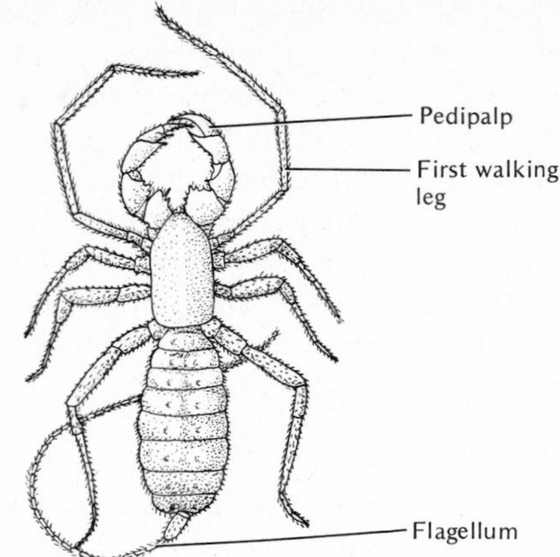

Figure 29.9 Order Uropygi, a whip scorpion.

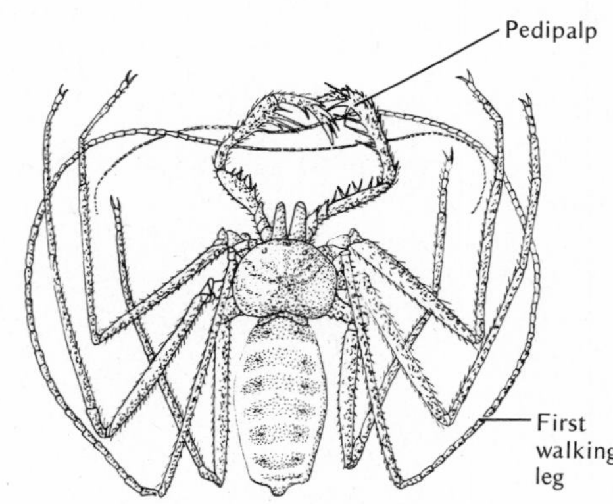

Figure 29.10 Order Amblypygi, a typical amblypygid.

ORDER **Palpigradi** (Fig. 29.11) Minute in size, less than 3 mm. Stout chelicerae. Pedipalpilike walking legs. First pair of walking legs used for sensory perception rather than locomotion. Opisthoma oval with a slender segmented tail at its tip. Tropical and subtropical in distribution. *Koenenia.*

ORDER **Araneae** The spiders (Fig. 29.16). Body made up of prosoma and opisthoma, both unsegmented and joined together by a narrow peduncle. Chelicerae with poison fang. Pedipalpi leglike, basal joint enlarged for crushing prey. Opisthoma with spinnerets which extrude silk used in spinning webs and cocoons. Size range from 0.5 mm to over 10 cm. *Latrodectus, Miranda, Eurypelma.*

ORDER **Solpugida** Sun spiders (Fig. 29.12). Prosoma broadly joined to the segmented opisthoma. Chelicerae long and stout and joined to the carapace which covers the anterior portion of the prosoma. The leglike palpi and the first pair of legs are both tactile structures. Size from 10 to 50 mm long. Found in warm, dry regions of the world. *Eremobates.*

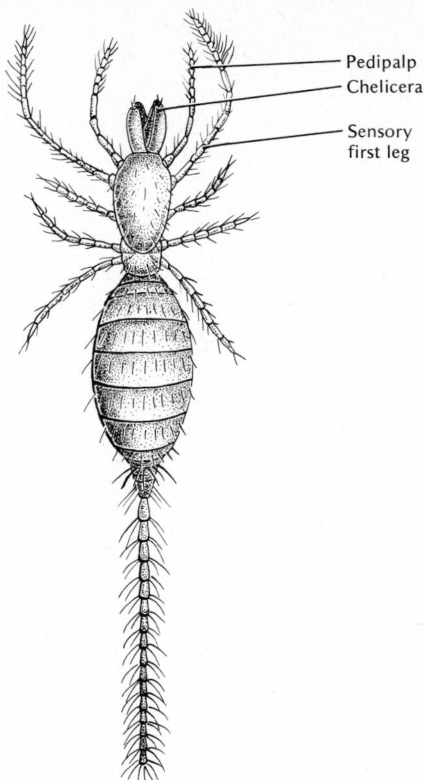

Figure 29.11 Order Palpigradi, *Koenenia,* a representative palpigrade.

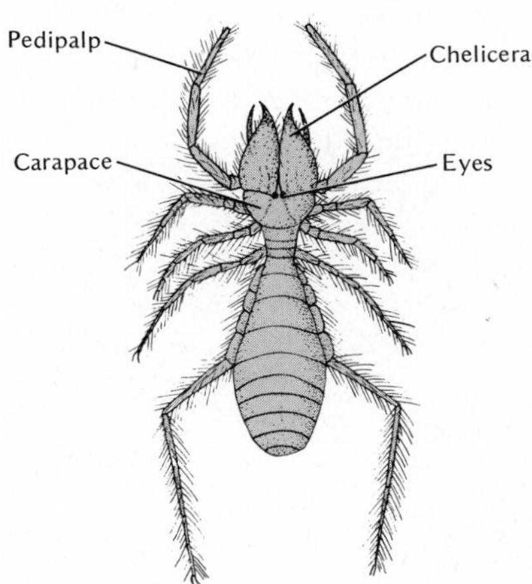

Figure 29.12 Order Solpugida, *Eremobates,* a sun spider.

ORDER **Pseudoscorpionida** The false scorpions (Fig. 29.13). Small forms, between 4 and 10 mm long, with prosoma broadly joined to the segmented opisthoma which lacks a tail and sting. Chelate chelicerae have comblike structures, called serulae, which are connected with ducts from silk glands and are used to construct a nest for the young. Pedipalpi large and provided with well-developed pincers which contain poison glands that open at the tips of the pincers. *Chthonius.*

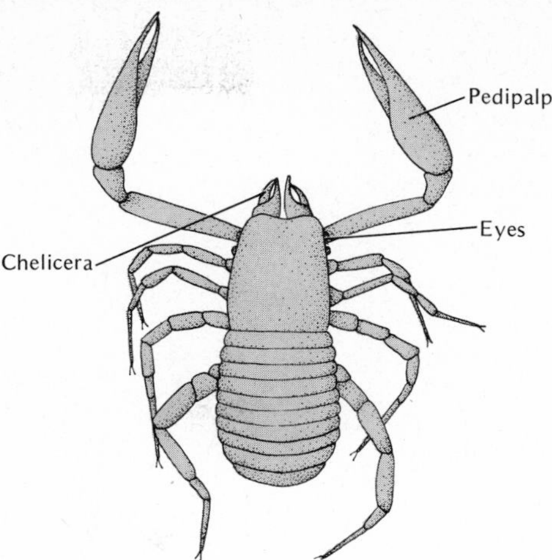

Figure 29.13 Order Pseudoscorpionida, a representative false scorpion.

ORDER **Ricinulei** (Fig. 29.14). Small forms, 4 to 10 mm in length, with a movable hood on the front of the prosoma which covers the chelicerae. Short pedicle connects the prosoma with opisthoma. The opisthoma has at its tip a small tubercule bearing the anal opening. Pedipalps shorter than the legs and provided with small pincers. Tropical and subtropical in distribution. *Ricinoides.*

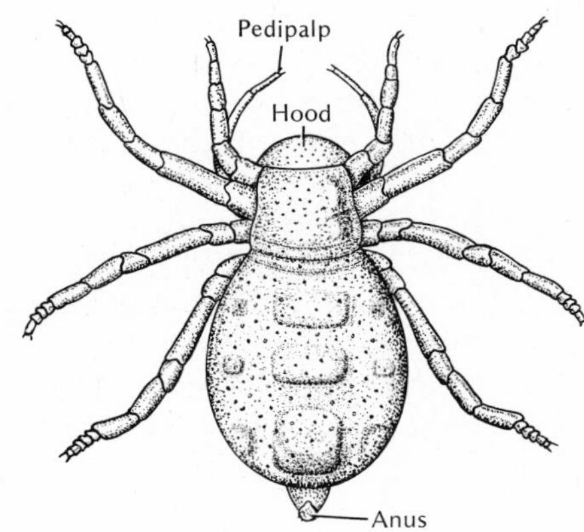

Figure 29.14 Order Ricinulei, *Ricinoides,* a representative ricinuleid.

ORDER **Phalangida** The harvestmen or "daddy-long-legs" (Fig. 29.15). Body quite short with prosoma broadly joined to the opisthoma which is segmented. Chelicerae small. Pedipalps without pincers. Legs very long and slender. Stink glands open along edge of prosoma. No silk glands or book lungs present. Body length 1 to 15 mm. *Liobunum.*

ORDER **Acarina** The mites and ticks (Fig. 29.17). Body very compact and rounded. Small to microscopic in size, 1 to 30 mm long. Prosoma and opisthoma fused and unsegmented. Chelicerae and pedipalpi quite varied in form and function. Legs usually widely separated. Free-living and parasitic species. *Dermacentor, Trombicula.*

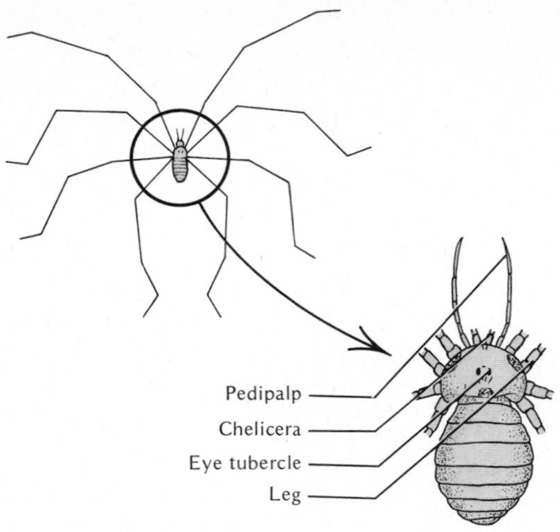

Figure 29.15 Order Phalangida, a daddy-long-legs or harvestman.

Anatomy and Physiology of Representative Arachnids

ORDER SCORPIONIDA

Scorpions live in arid tropical and subtropical regions, for the most part. The primitive segmentation pattern is most pronounced in the scorpions and the appendages also seem to be quite primitive. The elongate body is divided into a prosoma and a segmented opisthoma (Fig. 29.8). Anterior to the mouth one finds a pair of appendages, the short, three-jointed chelicerae. Behind the mouth are the pedipalps, greatly enlarged and modified with strong pincers at the tip. The four pairs of walking legs complete the appendages of the prosoma. The tapering opisthoma has an operculum covering the opening of the reproductive system on the ventral side of its first segment. Ventrally on the second segment is a pair of comblike structures called *pectines* which seem to be tactile receptors. Each of the next four segments has a pair of slitlike spiracles which open into the book lungs. The opisthoma now tapers sharply to form a cylindrical tail, the last segment of which, the telson, bears the sharp-pointed *sting.* Poison glands in the telson open near the tip of the sting. Most scorpions are not deadly to humans, but some genera contain species known to have caused fatalities, especially in young children.

Scorpions feed on small arthropods which they grasp with their pedipalps and sting by arching the tail over the prosoma. The mouth sucks the fluids of the victim and the chitinous portions are discarded. Internally, scorpions are quite typical of arthropods in general. Excretion is by means of Malpighian tubules and respiration by lung books. Fertilization is internal, following a courtship dance. The young develop within the mother and after birth they cling to her for some days.

ORDER ARANEAE

The spiders are the most familiar arachnids, occurring in almost all types of terrestrial habitats. The body is made up of an unsegmented prosoma which is joined by a narrow peduncle to the unsegmented opisthoma (Fig. 29.16). Dorsally, on the anterior portion of the prosoma, there are up to four pairs of simple eyes. The six pairs of appendages include the chelicerae, which are sharp-pointed *fangs* equipped with poison ducts; the pedipalpi with enlarged bases used in crushing food; and four pairs of walking legs. The poison ducts, mentioned above, are present in all spiders but as in the scorpions

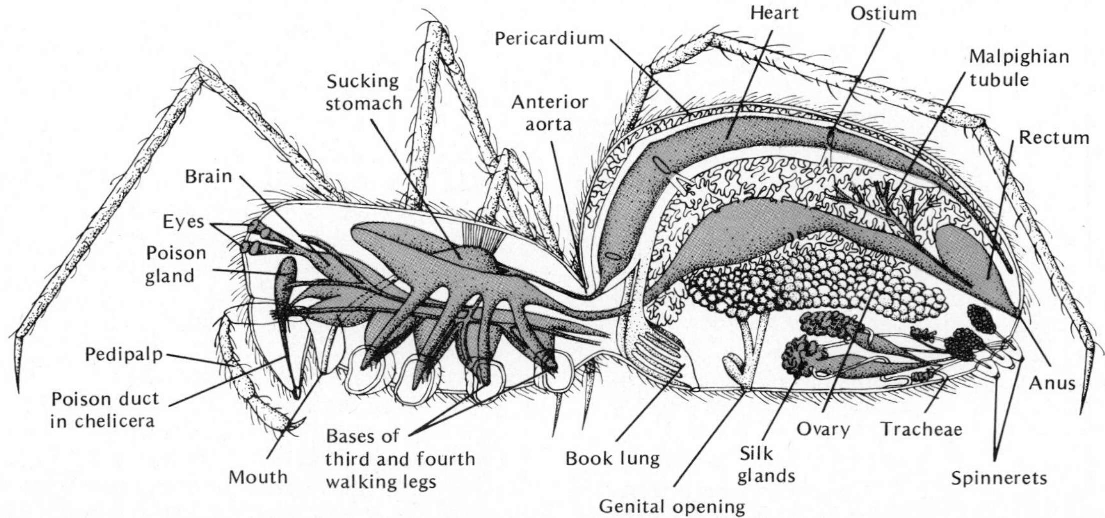

Figure 29.16 Order Araneae. Structure of a typical spider; the left side of the body has been removed.

there are relatively few spiders which are dangerous to humans. The black widow spider, *Latrodectus mactans,* and other members of the same genus have a very potent toxin which causes severe pain and even death to humans. The bite of the brown recluse spider, *Loxosceles reclusa,* causes severe pain and ulcerating sores.

The opisthoma has openings to the book lungs just behind the *peduncle.* The book lungs are unique respiratory structures which are found only in the class Arachnida. Each consists of a series of flat leaflike plates which are highly vascularized. Gaseous exchange occurs as air circulates over the flat plates, of which there may be as many as 20 in each book lung. Just behind the opening to the book lungs is the genital opening, which is covered by a heavy plate, called the *epigynum* in females. In many forms a second pair of spiracles at the posterior part of the opisthoma opens into a tracheal system similar to that found in the insects. This gives the spider two types of respiratory organs. Two or three pairs of *spinnerets* are also seen at the tip of the opisthoma just in front of the anus. They contain the openings of fine tubules which connect to the *silk glands.* Spider silk is a protein which hardens into a solid thread as soon as it comes in contact with air. Some of the threads are sticky and others are not, the sticky ones being used in portions of the web where prey becomes entangled. Many diverse uses of silk are seen among the spiders, including the large orb weaver webs which capture flying insects, lining of burrows in trapdoor spiders, cocoons in which young develop, and long gossamer threads spun out by young spiders which are caught by the wind and "balloon" the young spiders great distances, giving dispersal to these wingless arthropods.

As can be seen in Figure 29.16, the internal anatomy of a spider is quite typical of arthropods in general.

Sexes are separate and there is often considerable difference in size and color between them, with the female being larger than the male. The pedipalps of the male are modified for transfer of sperm, and with them the male places the sperm in the female genital pore. Often a complex courtship precedes this precarious operation—precarious because the female often kills and devours the male. Eggs are laid in a cocoon which may or may not be carried by the female until the young hatch.

ORDER ACARINA

The mites and ticks are important to humans because some species are vectors for serious diseases which affect humans and domestic animals, and others attack many crops. The mites are by far the most numerous of all arachnids—soil and surface litter-dwelling forms may number many millions per hectare. The majority of acarinids are small to microscopic in size. The body has all its parts fused into a single unit, and external evidence of segmentation is absent in most forms (Fig. 29.17). The body may be membranous or covered dorsally with thickened plates. All ticks and most mites lack eyes. The chelicerae vary from chelate types in most free-living species to pointed sucking types in parasitic ones. The pedipalps also vary from simple leglike forms to stout structures with well-developed pincers. Each of the four pairs of walking legs bears a pair of claws on the terminal segment. Internal anatomy is typical of that of most arthropods. In most forms the circulatory system is reduced, and although most species have trachea for respiration, there are some which lack any specialized system and carry out respiratory exchange through the general body surface. Excretion is by means of coxal glands, varying from one to four pairs, or Malpighian tubules, or both. In one group, the Trombidiformes, excretion occurs in a specialized excretory organ which develops from the hindgut.

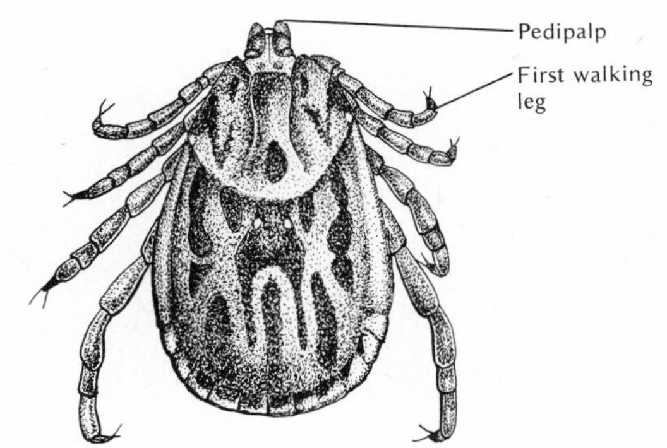

Figure 29.17 Order Acarina. *Dermacentor andersoni,* the vector for Rocky Mountain spotted fever; a representative tick.

Sexes are separate and in most cases there is a six-legged larval form which feeds and molts into a sexually immature eight-legged form, called a *nymph.* The nymph feeds and molts a number of times before becoming a sexually mature adult. In parasitic species which feed upon blood, the molts usually occur following a blood meal.

Several serious diseases of humans are transmitted by the bite of ticks, one of the most serious being Rocky Mountain spotted fever which is transmitted by the wood tick, *Dermacentor andersoni.* Tularemia, relapsing fever, and Q fever are also carried by ticks. The chigger is the larval stage of a mite which attacks vertebrates, including humans. It causes severe itching, and in the Far East it is the vector for scrub typhus.

SUGGESTIONS FOR FURTHER READING

Baker, E. W., and G. W. Wharton. *An Introduction to Acarology.* New York: Macmillan (1952).

Barnes, R. D. *Invertebrate Zoology.* Philadelphia: W. B. Saunders (1974).

Borradaile, L. A., and F. A. Potts. *The Invertebrata,* 4th ed., rev. by G. A. Kerkut. Cambridge: Cambridge Univ. Press (1961).

Cloudsley-Thompson, J. L. *Spiders, Scorpions, Centipedes, and Mites.* New York: Pergamon (1958).

Gertsch, W. J. *American Spiders.* Princeton, NJ: Van Nostrand (1949).

King, P. E., *Pycnogonids.* New York: St. Martin's Press (1973).

From Scientific American

Petrunkevitch, A. "The Spider and the Wasp" (Aug. 1952).

Savory, T. H. "Spider Webs" (Apr. 1960).

Witt, P. "Spider Webs and Drugs" (Dec. 1954).

30 The Mandibulate Arthropods

Subphylum Characteristics

The body may consist of a head and trunk; or it may have a head, thorax, and abdomen; or the head and thorax may be fused into a cephalothorax, so that the main divisions are cephalothorax and abdomen. There are usually one or two pairs of antennae and one pair of mandibles (jaws). One or more additional pairs of appendages, called maxillae, are associated with the mouth. The trunk appendages are mainly used in walking or swimming, but they are highly specialized in function in some groups. Abdominal appendages may or may not be present.

CLASS CRUSTACEA

Characteristics

1. The body is composed of a head of five segments which may or may not be fused with the thorax, and an abdomen with a terminal telson.
2. Head appendages include two pairs of antennae, one pair of jaws, and two pairs of maxillae.
3. Compound eyes are usually present.
4. Appendages are mostly biramous.
5. Respiration is by gills or by the general body surface.
6. The crustacea are mostly aquatic, both freshwater and marine.
   Figure 30.1

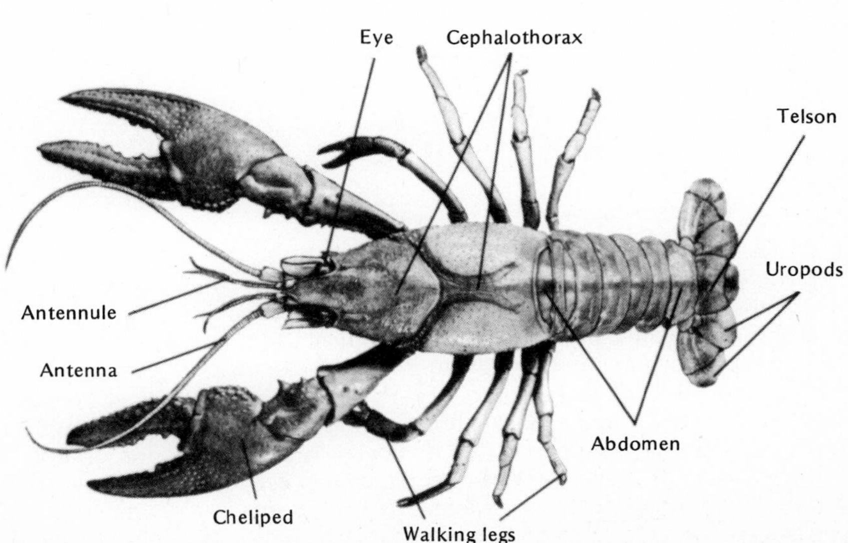

Figure 30.1 Class Crustacea, *Cambarus* sp. External features of crayfish anatomy. (From American Museum of Natural History, New York, with permission.)

Classification

SUBCLASS **Cephalocarida** (Fig. 30.2) The most primitive of living crustacea. Body shrimplike with horseshoe-shaped head bearing short antennae and antennules. Elongated, tapering trunk of 19 or 20 segments of which the first nine bear triramous appendages. Less than 4 mm long. Only a few species known, all of which are hermaphroditic. *Hutchinsoniella.*

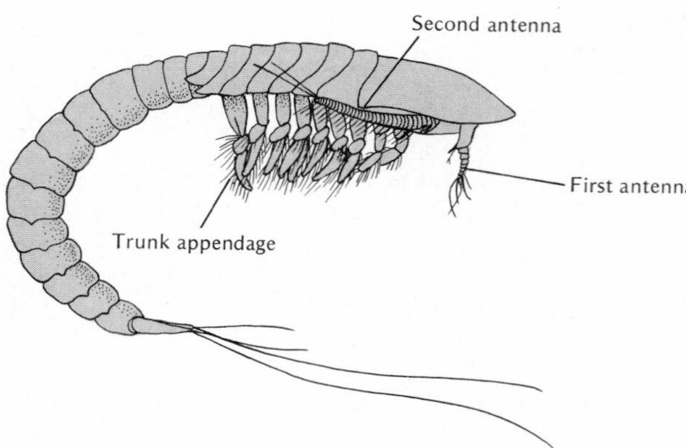

Figure 30.2 Subclass Cephalocarida. *Hutchinsoniella,* one of the most primitive living crustaceans.

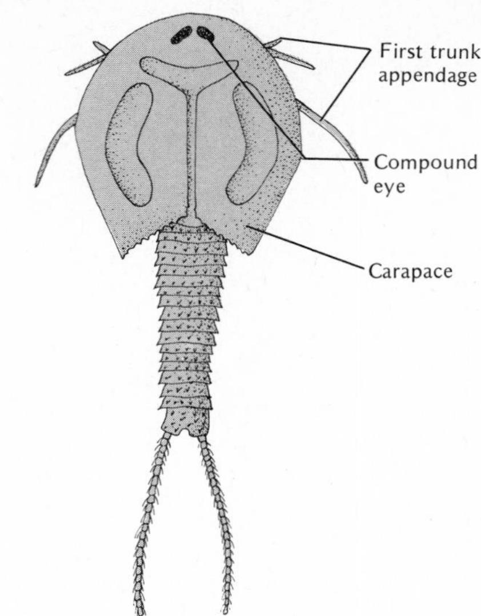

Figure 30.4 Order Notostraca. A tadpole shrimp.

SUBCLASS **Branchiopoda** Antennae adapted for swimming or reduced in size. Four or more pairs of thoracic appendages which are flattened, leaflike, and provided with gills on their margins. Usually a carapace is present. Mostly freshwater in distribution.

ORDER **Anostraca** Fairy shrimps (Fig. 30.3). Elongate, slender, delicate forms without a carapace. 11 to 19 pairs of leaflike trunk appendages. Eyes stalked. 5 to 180 mm long. *Eubranchipus,* the fairy shrimp, appears in temporary pools formed by melting snow in early spring. *Artemia salina,* the brine shrimp, is familiar to those who maintain aquaria.

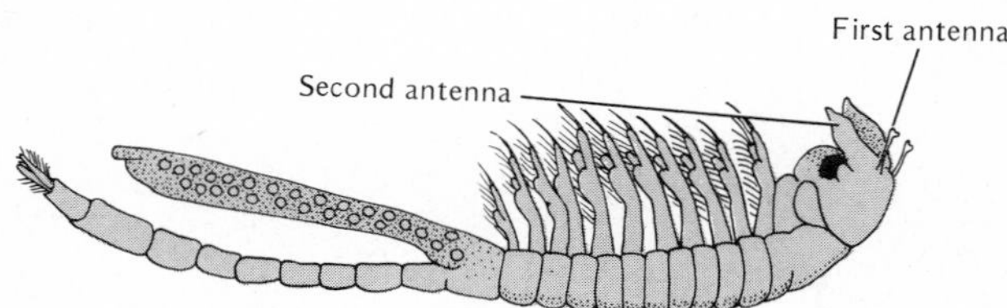

Figure 30.3 Order Anostraca. A fairy shrimp.

ORDER **Notostraca** Tadpole shrimps (Fig. 30.4). Carapace flattened and oval, covering the head and anterior half of the trunk. Eyes not stalked. Appendages number from 35 to 70 pairs; 20 to 30 mm long. *Lepidurus, Apus.*

ORDER **Chonchostraca** Clam shrimps (Fig. 30.5). Body laterally compressed and enclosed in a bivalve carapace. 10 to 32 pairs of appendages. Eyes sessile. Antennae used in swimming. 3 to 15 mm long. *Limnadia.*

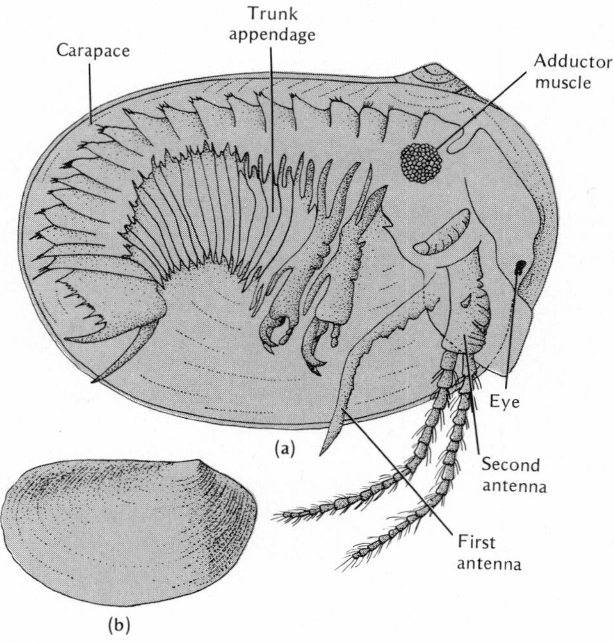

Figure 30.5 Order Chonchostraca. A clam shrimp. **(a)** Left valve of carapace removed. **(b)** Left valve of carapace.

ORDER **Cladocera** Water fleas (Fig. 30.6). Bivalve carapace covers body but not the head. Four to six pairs of trunk appendages, second antennae enlarged and used in locomotion. 0.15 to 3 mm long. *Daphnia.*

SUBCLASS **Ostracoda** Seed shrimps (Fig. 30.7). Laterally compressed, poorly segmented body enclosed in a bivalved carapace. Two pairs of trunk appendages present often reduced in size. Disproportionately large head with well-developed antennae used in locomotion. Some species are luminescent. Mostly minute, up to 8 mm long but one marine species reaches 3 cm.

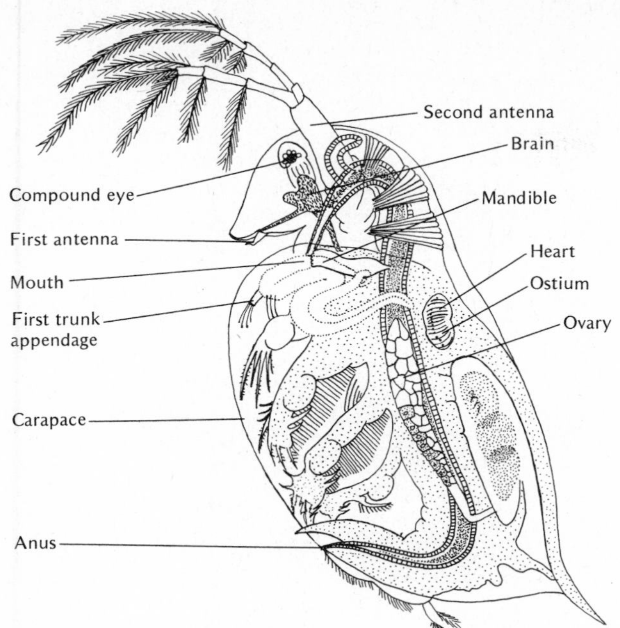

Figure 30.6 Order Cladocera. *Daphnia pulex*, the water flea.

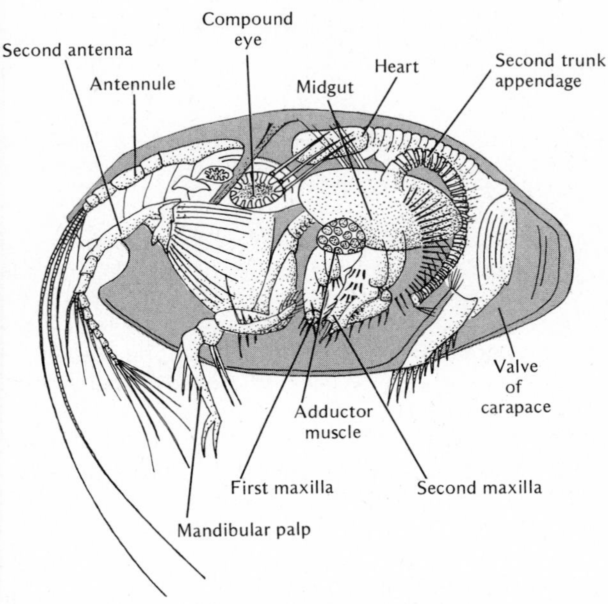

Figure 30.7 Subclass Ostracoda, order Myodocopa. *Cypridina*, a representative seed shrimp, with left valve of the carapace removed.

ORDER **Myodocopa** Carapace with a notch allowing extension of the second antennae which are used for swimming. Marine. *Cypridina.*

ORDER **Cladocopa** Carapace without notch. Both pairs of antennae used in swimming. Marine. *Polycope.*

ORDER **Podocopa** Carapace without notch. Second antennae leglike. *Darwinula* in fresh water. *Cythereis* marine.

ORDER **Platycopa** Carapace without notch. The two pairs of well-developed antennae are not used in locomotion. Marine. A single genus, *Cytherella.*

SUBCLASS **Mystacocarida** (Fig. 30.8) Microscopic less than 5 mm long, wormlike forms. Cylindrical body. Both pairs of antennae well developed. Compound eyes absent, four pairs of simple eyes. Each of the four thoracic segments bears a pair of legs. Abdomen of six segments lacks appendages. There are only a few known species in the group. Their habitat is damp sand in the intertidal zone.

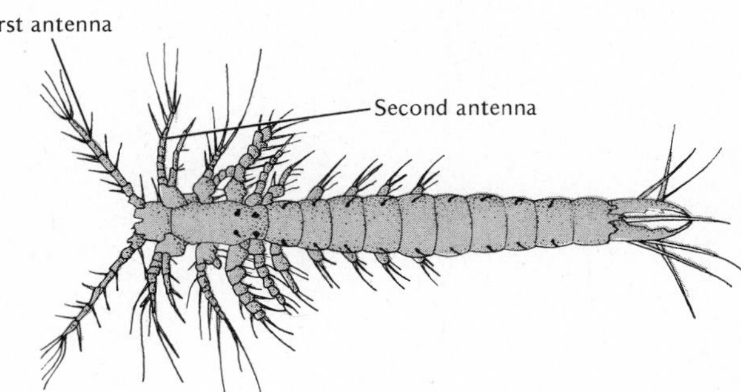

Figure 30.8 Subclass Mystacocarida. *Derocheilocaris*, a representative mystacocardian.

SUBCLASS **Copepoda** Antennae often well developed and used in swimming. Compound eyes absent, ocelli often fused to form a single, median eye. Thorax usually with six segments, each with a pair of biramous appendages. Abdomen of four segments, without appendages. Many highly modified parasitic species. Small to microscopic in size, 1 to 4 mm long.

ORDER **Calanoidea** (Fig. 30.9) Body constricted behind the fifth leg. Second antennae biramous, freshwater and marine. *Calocalanus, Diaptomus.*

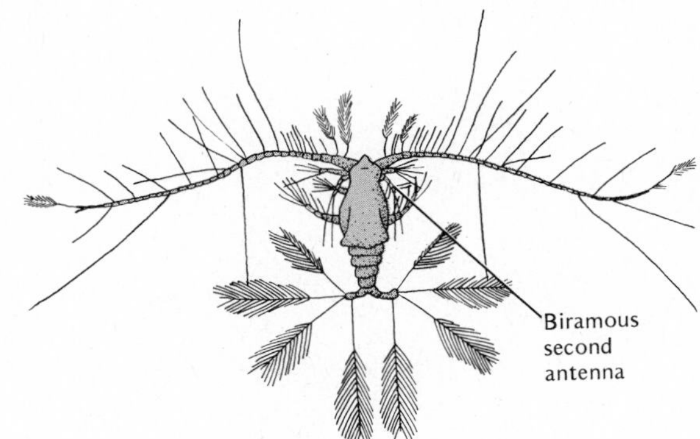

Figure 30.9 Order Calanoidea. *Calocalanus*, a representative calanoid copepod.

ORDER **Harpacticoida** Body elongate and cylindrical. Body slightly constricted behind the fourth leg. First antennae short, second antennae biramous. Marine and freshwater, usually bottom dwellers. *Harpacticus.*

ORDER **Cyclopoida** (Fig. 30.10) Body constricted behind the fourth leg. Second antennae uniramous. Freshwater and marine. *Cyclops, Macrocyclops.*

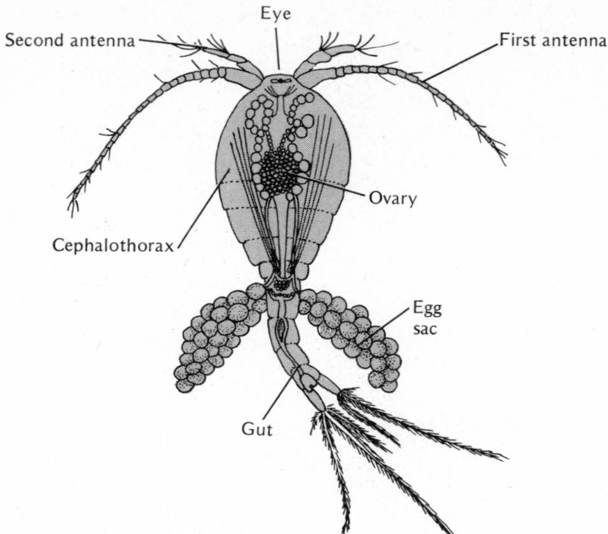

Figure 30.10 Order Cyclopoidea. *Macrocyclops,* a representative cyclopoid copepod.

ORDER **Notodelphyoida** Body hinged between the fourth and fifth thoracic segments in the male and between the first and second abdomen somites in the female. Live as commensals in tunicates. *Doropygus.*

ORDER **Monstrilloida** Adults lack second antennae and mouth parts. Larvae parasitic in polychaetes. *Monstrilla.*

ORDER **Caligoida** Body hinged between the third and fourth thoracic somites. Ectoparasites on gills of freshwater and marine fishes to which they attach by means of the second antennae. *Caligus.*

ORDER **Lernaeopodoida** (Fig. 30.11) Highly modified ectoparasites on marine and freshwater fishes with segmentation reduced or absent. Second maxillae modified as attachment organs. Thoracic appendage reduced or completely lacking. *Lesteira, Lernanthropus.*

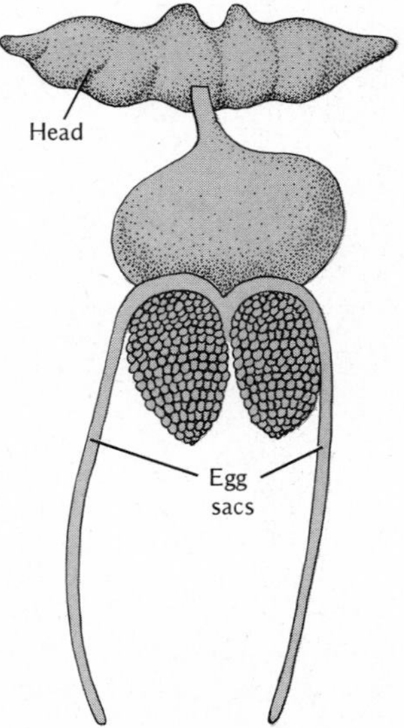

Figure 30.11 Order Lernaeopodoida. *Leisteira,* a highly modified ectoparasite of fish. The animal imbeds its head in the skin of the fish and the rest of the body hangs free.

SUBCLASS **Branchiura** (Fig. 30.12) Fish lice. Body flattened dorsoventrally. Head and thorax covered with a large disclike carapace. Sessile compound eyes present. Appendages modified as suckers and claws used in attachment to the marine and freshwater fish which they parasitize. Less than 6 mm long. *Argulus.*

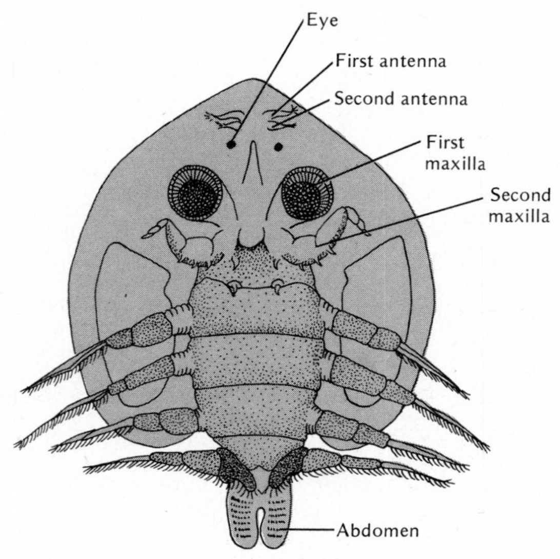

Figure 30.12 Subclass Branchiura. *Argulus,* the fish louse, attaches to its host by the suckerlike first maxillae and the claws of the second maxillae.

SUBCLASS **Cirripedia** Barnacles. Adults sessile and exclusively marine. Body enclosed in a mantle formed by the carapace which in many forms secretes calcareous plates. Abdomen vestigial. Second antennae lacking. Thoracic appendages biramous and used in feeding. Usually monoecious. Many highly modified parasitic species.

ORDER **Thoracica** (Fig. 30.13) Six pairs of thoracic appendages, mantle usually secretes calcareous plates. Some species with a fleshy stalk, others lack a stalk. Up to 25 cm long. *Lepas, Mitella, Balanus.*

ORDER **Acrothoracica** Boring barnacles which live in burrows cut in molluscan shells and coral. Mantle present but calcareous plates absent. Four pairs of thoracic appendages. Usually less than 10 mm long. *Trypetesa.*

ORDER **Ascothoracica** Digestive tract has branches extending into the mantle which does not secrete calcareous plates. Six pairs of thoracic appendages. Parasitic in soft corals and echinoderms. *Laura.*

ORDER **Rhizcephala** (Fig. 30.14) Mantle present. Shell, appendages and digestive system lacking. Animal saclike parasite on crabs which sends rootlike absorptive processes into the tissue of the host. *Sacculina.*

SUBCLASS **Malacostraca** Body usually of 19 segments. Five in the head, eight in the thorax and

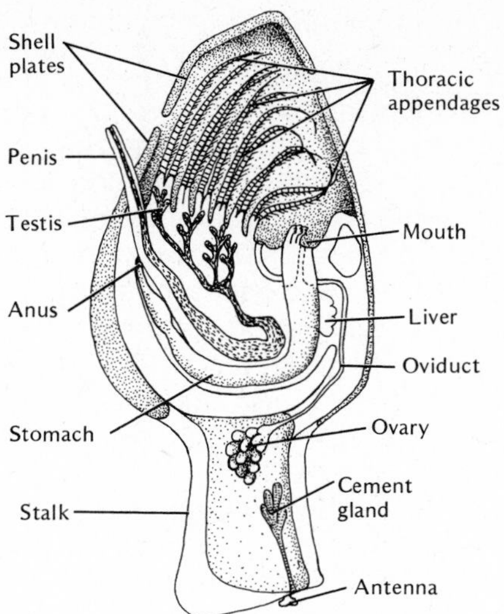

Figure 30.13 Subclass Cirripedia, order Thoracica. *Lepas,* a goose barnacle.

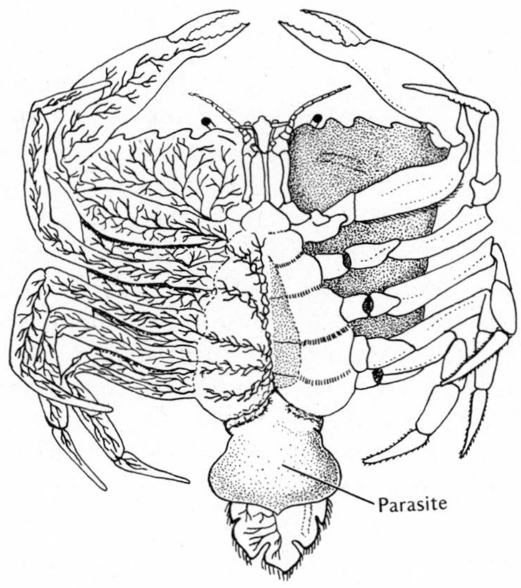

Figure 30.14 Subclass Cirripedia, order Rhizocephala. *Sacculina,* a barnacle parasitic on crabs. The adult is reduced to a saclike mass with branching internal parts which extend throughout much of the body and absorb food. Left side of a parasitized crab illustrated with exoskeleton removed to show the rootlike branching body of the parasite.

six in the abdomen. Carapace often present. Head usually fused with one or more thoracic segments. Appendages present on the abdomen.

ORDER **Nebaliacea** Abdomen with seven segments. Carapace bivalved and covering the head, thorax and first abdominal segment. Less than 10 species. Entirely marine. 4 to 12 mm in length. *Nebalia.*

ORDER **Anaspidacea** Carapace lacking. First thoracic segment fused with the head. Up to 5 cm long *Anaspides.*

SUPERORDER **Peracarida** When present the carapace is fused to anterior thoracic segments only, four or more always distinct.

ORDER **Mysidacea** The opossum shrimps (Fig. 30.15).

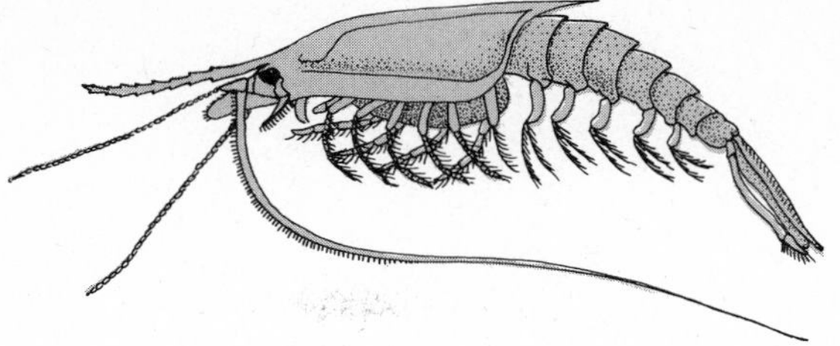

Figure 30.15 Order Mysidacea. *Gnathophausia,* a representative mysid shrimp.

Carapace covers entire thorax, but unfused to last four segments. Pair of biramous appendages on each segment. Uropods and telson form a tail fan. Mostly marine but *Mysis* is an important genus found in freshwater lakes such as the Great Lakes. Mostly 1.5 to 3 cm long, one species reaches 15 cm. *Gnathophausia.*

ORDER **Cumacea** (Fig. 30.16) Mostly burrowers in sand and mud. Carapace often extending anteriorly and covering the head. Abdomen slender and lacking tail fan. Small, marine forms. 1 to 35 mm long. *Diastylis.*

Figure 30.16 Order Cumacea. *Diastylis,* a filter-feeding cumacean.

ORDER **Tanaidacea** Carapace small. Second thoracic appendages enlarged and chelate. Burrow in mud or live in specially constructed tubes. Entirely marine. Usually microscopic, 2 mm or less. *Apseudes.*

ORDER **Isopoda** Wood lice, pill bugs (Fig. 30.17). Carapace absent, body flattened dorsoventrally. Abdominal segments may be fused. Marine, freshwater and terrestrial. Mostly 5 to 15 mm long, one marine species 40 cm. *Asellus, Idotea.*

ORDER **Amphipoda** Sand hoppers. Carapace absent. Body usually laterally compressed. Abdomen flexed ventrally between the third and fourth segments. Marine and freshwater. Mostly 3 to 12 mm long, one marine species 28 cm. *Gammarus.*

SUPERORDER **Hoplocarida**

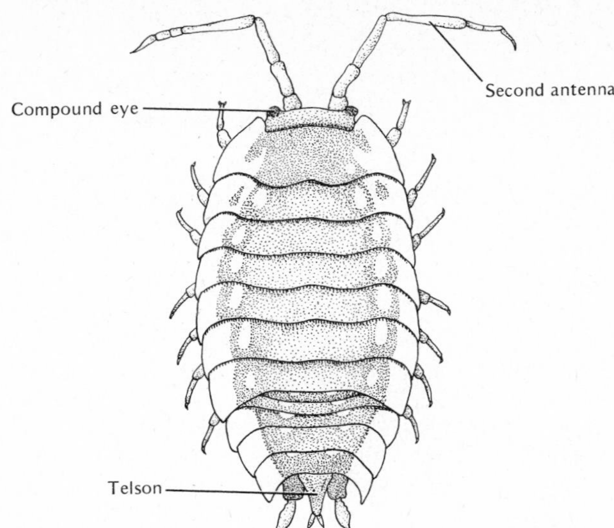

Figure 30.17 Order Isopoda. *Oniscus,* a representative terrestrial isopod.

ORDER **Stomatopoda** Mantis shrimps. Body elongate and dorsoventrally flattened. Carapace covers first two thoracic appendages and does not extend to the anterior end but stops short of the eyes and antennules which are on movable segments. Second thoracic leg greatly enlarged and modified for seizing prey. Marine. 20 mm to 30 cm in length. *Squilla.*

SUPERORDER **Eucarida** Carapace large, fused with all thoracic segments. Eyes stalked. Gills in thoracic region.

ORDER **Euphausida** Biramous thoracic appendages. Entirely marine. 3 to 6 cm long. *Euphausia.*

ORDER **Decapoda** Crabs, lobsters, crayfish. First three pairs of thoracic appendages modified as maxillipeds and used in feeding. Remaining five pairs of thoracic appendages are walking legs (hence the name "decapoda") although the first pair is often modified as pincers. Largest group of crustacea. Most of them are marine but there are some freshwater forms and a few terrestrial ones. Range in size from 1 mm to the Japanese spider crab 45 cm long with 4 meters between the tips of the chelipeds. *Homarus, Cambarus, Uca.*

Anatomy and Physiology of the Crayfish

All the crayfish found west of the Continental Divide are placed in the genus *Astacus.* Those in waters east of the Continental Divide are placed in one of six different genera, the genus *Cambarus* being one with a number of common and widely distributed species. The crayfish is very similar in structure to its much larger marine relative, the American lobster, *Homarus americanus,* found along the East Coast. Crayfish live in freshwater streams, lakes, and swamps and a few species live in burrows along the shore.

EXTERNAL FEATURES

The body of the crayfish (Fig. 30.1) is covered with a hard exoskeleton and consists of an anterior cephalothorax and a posterior abdomen. The abdomen is clearly segmented, consisting of six segments with appendages and a terminal telson, on the ventral side of which is the anus. The part of the exoskeleton that covers the dorsal and lateral surfaces of the cephalothorax forms a single large shield and is known as the *carapace.* The carapace masks the segmental nature of the head and the thorax, but this segmentation is evident on the ventral surface in the paired appendages. A groove on the carapace, the *cervical groove,* marks the division between head and thorax. The carapace extends anteriorly as a median, pointed *rostrum.*

The appendages of the crayfish are interesting because they illustrate in a striking way how a primitive type of structure has become modified, in the process of evolution, for several different functions. The extinct ancestral arthropod, the trilobite (Fig. 29.3), had paired appendages on most segments of the body very similar to the paired appendages on the abdomen of the crayfish, the *swimmerets.* These appendages are called *biramous* because they are composed of two branches or rami. Each biramous appendage consists of a basal portion, the *protopodite,* an outside branch, the *exopodite,* and an inside branch, the *endopodite* (Fig. 30.18).

In the extinct trilobite it is assumed that these appendages served in swimming. This type of appendage is found throughout the Crustacea, and it is considered to be the fundamental plan of all crustacean appendages. These appendages arise in development from similar origins and are, thus, homologous structures. When homologous structures are arranged in series on successive segments of the same animal, they are said to be *serially homologous.* In the description of the crayfish appendages that fol-

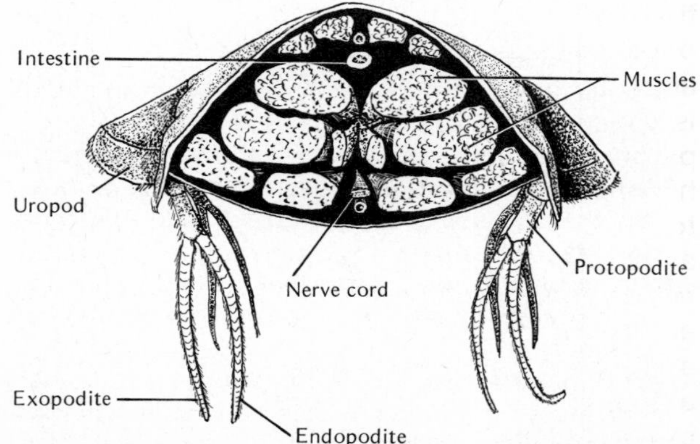

Figure 30.18 Cross section of the abdomen of a crayfish (seen toward the tail).

lows, it will be noted that many of the appendages are no longer biramous in the adult condition. However, their homology with the ancestral biramous appendage is clearly shown during development. In an early stage of a developing lobster, a marine relative of the crayfish, all the appendages are biramous.

The most anterior appendages of the head region are the stalked compound eyes and the branched *antennules* (first antennae) that are sensory in function. Both of these pairs of appendages are probably not homologous with the other appendages of the crayfish. These structures develop from the nonsegmented prostomial region of the embryo. Furthermore, the branched condition of the antennules does not represent the endopodite and exopodite of the other appendages; the branching occurs late in development and is a secondary matter. The other appendages of the head region, the antennae (second antennae), the *mandibles*, the *first maxillae*, and the *second maxillae*, are homologous structures. Figure 30.19 shows the modifications from the basic type of these appendages. The antenna is composed primarily of a long endopodite that is sensory in function; the exopodite is a small scalelike projection. Each mandible has a small segmented palp that is a modified endopodite; the exopodite is absent. The heavy jawlike part of the mandible is derived from the protopodite. The mouth is located just above the mandibles. The first and second maxillae pass the food toward the mouth. The exopodite of the second maxilla is modified to form a thin indented plate, called the *bailer*, which circulates the water through the gill chamber.

There are eight pairs of appendages on the thorax, one pair for each thoracic segment. The first three pairs—the *first, second*, and *third* maxillipeds—tear and pass food on to the mouth. The next pair of appendages is the large pinching legs, or *chelipeds*, (sometimes called first walking legs) that are used in fighting and obtaining food. The last four pairs of thoracic appendages are the walking legs. No exopodite is present in the adult in any of the last five pairs of thoracic appendages. When a crayfish has a leg injured or caught, it has the capacity to shed the appendage by a process of self-amputation, called *autotomy*. There is a specific region near the base of the leg, called the *autotomy line*, where special muscles contract and pinch off the distal portion of the leg. After autotomy, a valve and diaphragm close the end of the stump, thus minimizing the loss of blood. The lost appendage is later replaced by regeneration. Attached to the bases of most of the

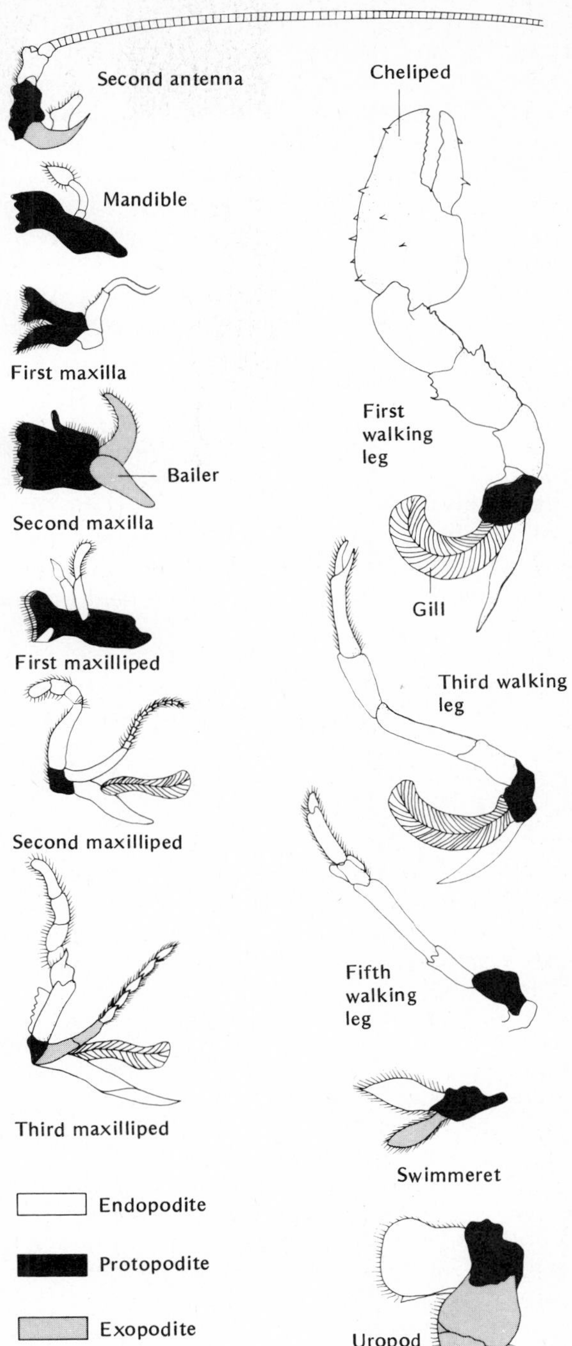

Figure 30.19 Crayfish appendages showing homology of parts with basic biramous appendages.

thoracic appendages are gills. In the female the openings of the reproductive system are at the bases of the third pair of walking legs, and in the male the openings are at the bases of the fifth pair.

As mentioned above, the abdominal appendages are the swimmerets. In the male the first two pairs are modified for transferring sperm to the female; in the female these first two pairs are reduced in size. In both sexes the next three pairs are typical biramous appendages that function for the movement of water and, in the female, for the attachment of eggs. The sixth pair of swimmerets is greatly enlarged, forming the *uropods*. These uropods with the telson form a powerful tail fan, used in backward swimming.

The appendages of the crayfish have been described in some detail because they illustrate the great specialization that has occurred in serially homologous structures during the course of evolution.

THE DIGESTIVE SYSTEM

The crayfish uses its chelipeds to catch and crush its food, which is then passed on to the mouth with the aid of the maxillipeds and maxillae. The food material is somewhat broken up by the action of the maxillipeds and still more by the action of the mandibles before it enters the mouth.

Figure 30.20 shows the parts of the digestive system of the crayfish. The mouth leads into a short tubular esophagus. From the esophagus food passes into the stomach that consists of a large thin-walled anterior part, the *cardiac* region, and a smaller posterior *pyloric* region. The pyloric region of the stomach leads into a short midgut that continues as the intestine to the anus. Two large multilobed digestive glands connect by ducts to the midgut. The cardiac part of the stomach serves for storage. In addition, this region has, projecting into its cavity, three toothlike structures, one median and two lateral, which constitute the *gastric mill.* The parts of the gastric mill are moved by muscles outside the stomach, reducing the food material to very fine particles. The entrance to the pyloric chamber is guarded by folds containing many hairlike setae that serve to filter out all but very fine particles. These minute particles of food material are digested and absorbed in the midgut region and in the cavities and tubules of the digestive glands. These digestive glands then not only secrete the proteases, lipases, and carbohydrases necessary for digestion, but they are responsible for a large part of the absorption of the end products of digestion. The fact that so much of the digestion and absorption takes place in the digestive glands explains how such a large animal as the crayfish is able to get along with such a short uncoiled intestine.

In the crayfish only the midgut region and the digestive glands have an endodermal lining. The lining of the other parts of the alimentary canal is ectodermal in origin, and these parts are in turn lined with a thin cuticle that is continuous with the exoskeleton and is shed with each molt. Quite often two calcareous deposits, the gastroliths, are present in the wall of the cardiac region of the stomach. The calcium salts in these deposits are absorbed during molting, and it is thought that they are used for the hardening of certain parts of the newly formed exoskeleton.

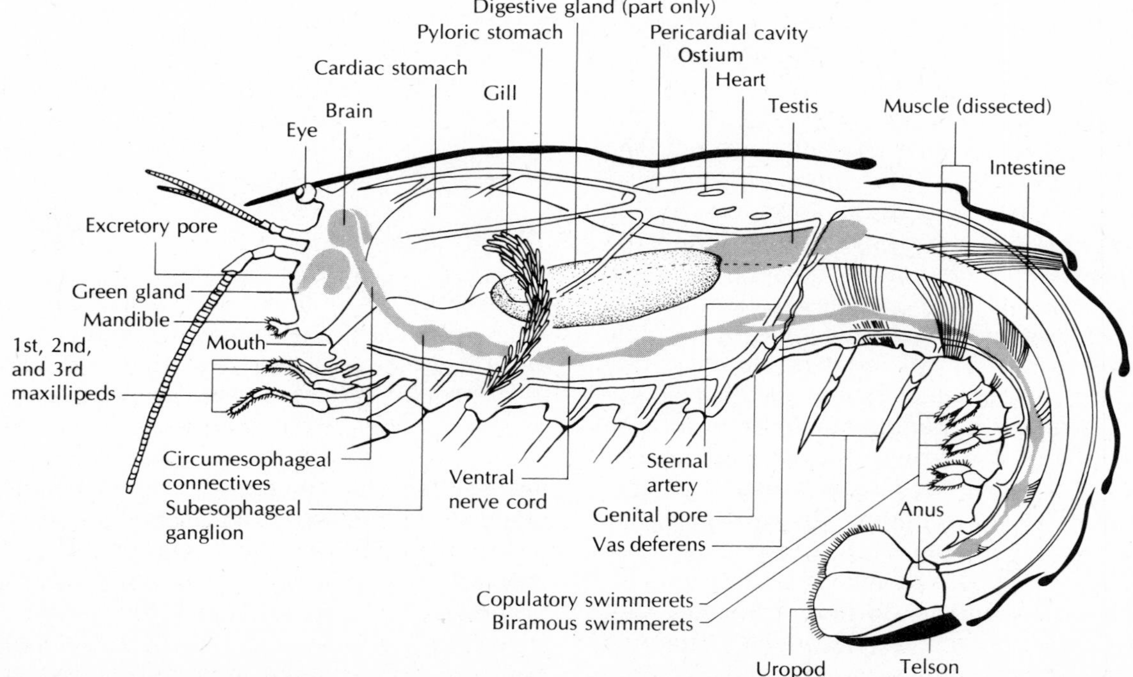

Figure 30.20 Crayfish. Internal anatomy.

THE RESPIRATORY SYSTEM

When a lateral wall of the carapace is cut away, numerous featherlike gills are exposed. The gills are attached to the bases of the thoracic appendages, to the membranes between these appendages, and to the wall of the thorax. The cavity in which they are located is the gill chamber. Water enters the gill chamber under the free edge of the carapace and passes upward and forward over the gills. The modified second maxilla, on each side, forms a bailer that moves up and down and causes the water to flow in this direction. The gills are highly vascularized, and through their thin walls the exchanges in external respiration occur.

THE CIRCULATORY SYSTEM

The circulatory systems of the frog, the human, and the earthworm, which have been described in some detail, are closed systems consisting of a heart (or hearts), arteries, capillaries, and veins. The circulatory system in the crayfish, and other arthropods, is an open one. The heart is located dorsally in a blood-filled chamber, the *pericardial sinus.* The heart has three pairs of tiny openings, the *ostia,* which allow the blood to enter from the pericardial cavity. Upon contraction of the heart, valves prevent the blood from going out the openings, and it is forced into the seven arteries that leave the heart and carry blood throughout the body. The smallest branches of these arteries empty into cavities in the tissues, the sinuses, and the blood bathes the cells directly. From these tissue sinuses the blood collects in a large ventral sinus, the *sternal sinus,* and from here it passes through afferent vessels into the gills where the exchanges in external respiration occur. From the gills the blood is returned to the pericardial sinus around the heart by way of efferent vessels.

The blood of the crayfish is nearly colorless, but it contains a dissolved respiratory pigment, *hemocyanin,* which has a slight bluish color (when completely oxygenated, it is a deep blue, but the blood of larger crustaceans is never fully oxygenated *in vivo*) and serves in the same capacity as hemoglobin in the vertebrates, that is, as an oxygen carrier. There are free amoeboid corpuscles in the blood. Associated with the frequent loss of appendages by the crayfish is a strong clotting capacity of the blood.

THE EXCRETORY SYSTEM

Two large *antennary* or *green glands,* located at the bases of the antennae, were formerly believed to be the major excretory organs of the crayfish. Recent work reveals that the green glands are mainly osmotic regulators, secreting a urine that is hypoosmotic to the blood and which contains little nitrogenous waste. Whether the green gland functions by secretion or filtration is still in question.

No agreement has been reached as to the major site of nitrogenous waste removal. Most of the nitrogenous waste in a crayfish is in the form of ammonia. Suggested sites include special gland cells in the gills and parts of the digestive tract.

THE NERVOUS SYSTEM

In the crayfish, and in some other arthropods, this system bears a general resemblance to that of the earthworm. The brain in the head receives nerves from the eyes, antennules, and antennae. From the brain a pair of circumesophageal connectives pass ventrally, one on either side of the esophagus, and fuse to form the subesophageal ganglion. This large ventral ganglion actually results from the fusion of six pairs of ganglia that are separate in the embryo. The ventral nerve cord, which runs posteriorly from this ganglion, is double in nature, but this is clearly evident only where the sternal artery passes through the cord. Posterior to the subesophageal ganglion, the nerve cord enlarges into ganglia in each segment. Nerves connect these ganglia with the appendages, muscles, and other organs of the body.

SENSE ORGANS

The crayfish is well equipped with sense organs. Many parts of the body are sensitive to touch. The receptors for touch are *tactile hairs* that are quite numerous on the chelipeds, the mouth parts, the underside of the abdomen, and the edge of the telson. The chemical sense (like our taste and smell) is associated with special hairs on the antennules, antennae, and mouth parts.

The sense organs for equilibrium are the statocysts, located at the base of the antennules. Each statocyst is a chitin-lined sac with a small opening to the outside. Within each statocyst are numerous sensory hairs to which sand grains are attached by mucus to serve as statoliths. Sensory nerve fibers lead from these hairs to the brain. When the body of the crayfish is tilted or inverted, the changed gravitational pull on the statoliths initiates impulses in these sensory fibers that pass to the brain and result in a righting reflex. Proof of the function of the statocysts was demonstrated many years ago by Kreidl in experiments on a species of shrimp. He used a shrimp that had just molted. During molting, the linings of the statocysts as well as the sand grains are shed. He placed the animal in filtered water and provided it with iron filings instead of sand. During the formation of the new exoskeleton, the shrimp introduced the iron filings into its statocysts by scraping its antennules on the bottom of the aquarium. Thus the iron filings were substituted for the sand grains in the statocysts. When the experimenter held a magnet over this animal, it responded by turning over on its back.

The compound eye of the crayfish is an interesting structure, and such an eye is characteristic of many of the arthropods (Fig. 30.21). It is quite unlike the vertebrate eye. These eyes are on movable stalks on either side of the rostrum. The outer rounded surface of each eye, the *cornea,* is divided into over 2000 tiny sections, the *facets.* The facet indicates the outer part of one unit of the compound eye, an *ommatidium.* Each ommatidium is composed of a number of parts. The outer facet is the *lens,* followed by the *crystalline cone* with its elongate tapered stalk, which functions as a second lens and focuses the light on the *retinula.* The center of the retinula, the translucent *rhabdome,* is the actual photoreceptor. It is formed by portions of the surrounding retinular cells. These parts of the retinular cells, which go to make up the rhabdome, are called *rhabdomeres.* The number of retinular cells contributing to each rhabdome varies in different species, but it is often seven or eight. The combined rhabdomeres function as a single unit, and their transmitted impulse represents a single point of light. Evidence is accumulating which points to the fact that at least in some insects each rhabdomere is sensitive to a different portion of the spectrum. This gives each individual ommatidium a very sensitive color vision with excellent hue discrimination. The nerve impulse generated in the rhabdome leaves the ommatidium via the axons of the retinular cells which join to form a nerve leading to the optic ganglion.

The retinular cells also contain the *proximal pigments. Distal pigment cells* surround the crystalline cone. In bright light the pigment is dispersed so that each ommatidium is isolated from the neighboring ones. Thus each unit of the eye is like a very small telescope directed at a point in space. Light passing axially through the lens from a small area on an object reaches the rhabdome and stimulates the nerve endings, while rays entering at an angle are refracted into the pigment and absorbed. The image that results from the functioning of many ommatidia is an *apposition (mosaic) image.* Thus the compound eye in bright light is well adapted for the detection of moving objects [Fig. 30.21 (**a**)].

In dim light the pigment becomes concentrated at the two ends of the ommatidia with the result that light rays can affect adjacent ommatidia. In this condition the resulting image is a continuous or *superposition image* that is probably less distinct than the apposition image [Fig. 30.21 (**b**)].

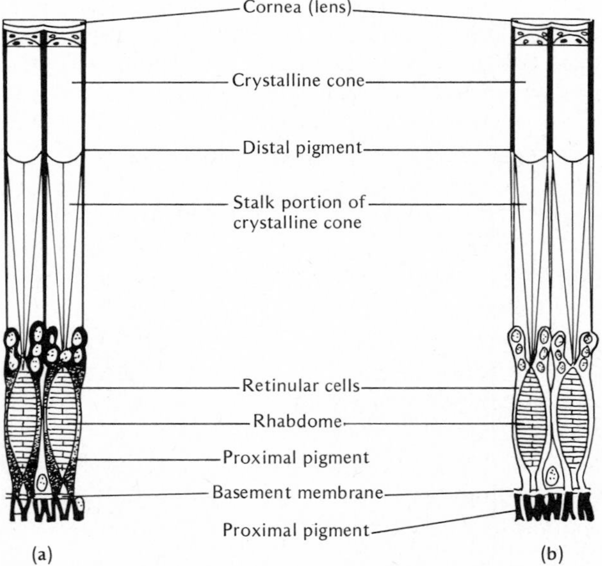

Figure 30.21 Structure of ommatidia in the eye of a crayfish. **(a)** Light adapted state. **(b)** Dark adapted state. Note that the pigment in the light adapted state **(a)** is spread out, shielding each ommatidium while in the dark adapted state **(b)** the distal pigment has migrated outward and the proximal pigment has moved below the basement membrane.

THE MUSCULAR SYSTEM

The abdomen of the crayfish is largely composed of segmentally arranged muscles. The largest of these are flexor muscles, the contraction of which flexes the abdomen ventrally with great force, causing the animal to shoot backward through the water. Others function in the movements of the swimmerets and in the extension of the abdomen. All these muscles are of the striated type that contract very rapidly. Similar muscles are found in the cephalothorax for the movement of the different appendages. The abductor muscles of the chelipeds are quite large. In the crayfish, as in the vertebrates, smooth muscles are found in such organs as the blood vessels, and the digestive tract.

THE REPRODUCTIVE SYSTEM AND REPRODUCTION

The sexes are separate in the crayfish. In both sexes the gonads are three lobed and are located in the dorsal part of the body between the alimentary canal and the pericardial sinus (Fig.

30.20). The gonads are hollow. In fact, the cavities in the gonads and those in the excretory glands are the only remnants of the coelom that persist in the adult crayfish. In the female the thin transparent oviducts pass ventrally to the openings at the bases of the third walking legs. In some species the female has a cuticular fold between the bases of the fourth and fifth pair of walking legs that serves as a seminal receptacle. In the male the coiled sperm ducts pass ventrally to openings in the bases of the fifth walking legs. The walls of the sperm ducts are thick and glandular and secrete a sticky gelatinous material over the amoeboid sperm, aggregating them into sperm packets or spermatophores.

During copulation the spermatophores leave the sperm ducts and pass along the grooves in the modified abdominal appendages of the male to the seminal receptacle of the female where they are stored. After this the male and female separate. Egg laying does not occur until some time after copulation, and in the interim the sperm remain alive, protected by the gelatinous covering. At the time of egg laying, the female thoroughly cleans the ventral side of the abdomen and turns over on her back. As the eggs emerge from the oviducts, they are fertilized and then pass backward to the swimmerets. Glands in these structures secrete a sticky material that causes the eggs to adhere to them. The eggs remain in this position until they hatch, which is between five and eight weeks, depending upon the temperature.

The young crayfish at hatching resembles the adult except for minor details. Many of the marine crustaceans, however, have several larval stages, the earlier of which are quite unlike the adult. In the lobster there is a larval stage known as the *mysis larva* in which there are exopodites on all the walking legs.

ENDOCRINE GLANDS

For a long time most of the work on animal hormones dealt with vertebrates. Studies have now shown that hormones are produced in crayfish, insects, and many other invertebrates. Work on the crayfish has revealed a tiny neurosecretory gland, called the *X-organ,* located at the base of the eyestalk. The X-organ produces a number of hormones which pass via axons to a storage-release center, the *sinus gland,* located in the eyestalk. These hormones influence a number of processes, including the distribution of retinal pigment in the process of light and dark adaptation described above, regulation of deposition of calcium in the gastroliths at time of molting, the process of molting itself, and color change, by affecting the distribution of pigment granules in the chromatophores located in the epidermis.

Ecology

Most crustaceans are aquatic. Those which live in terrestrial situations are still confined to damp environments or those which are not far from water. Aquatic crustaceans are found in fresh and salt water at all depths from pole to pole. Two pelagic crustaceans are shown in Plates 36 and 37. Specialized forms occur in hot springs and others in waters with salinities up to eight times that of the sea. All the larger groups are cosmopolitan in distribution. In many situations the crustaceans are the most important members of the lower portions of the food chain.

With the exception of the barnacles, most crustaceans are motile. Free-swimming planktonic animals and benthic walking or crawling types are all represented. Most crustaceans are probably predators or scavengers, but some have adopted symbiotic relationships with other organisms ranging from the complex internally parasitic barnacles through external parasites to commensal species living in worm burrows, the mantle cavity of oysters, and other similar situations.

Phylogenetic Relationships

Crustacea have an ancient lineage with a fossil record dating back to Cambrian times. In the Cambrian rocks there are fossil forms which have some features of both trilobites and crustaceans. These so-called Pseudocrustacea may indicate a close phylogenetic relationship between these two groups. All the major crustacean groups appeared so early in the fossil record that little can be said about evolution within the class.

CLASS DIPLOPODA

Characteristics

1. Body long and usually cylindrical with a distinct head and trunk, size range from 2 mm to over 20 cm.

2. Head with many small simple eyes; one pair of antennae (seven segments); jaws and maxillae; the latter are usually fused to form a platelike *gnathochilarium.*
3. Thorax of four segments the first of which is legless while each of the other three has a single pair of legs.
4. The abdomen makes up most of the length of the body with up to 100 or more segments, each of which bears two pairs of legs.
5. The reproductive ducts open in the middle of the ventral side of the third segment.
   Figure 30.22

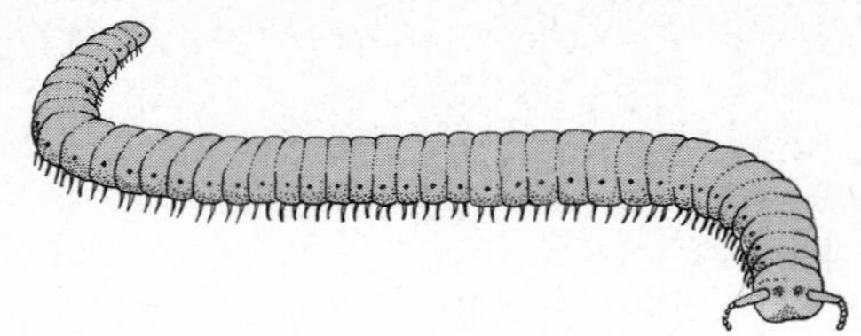

Figure 30.22 Class Diplopoda, *Julus* sp.

Classification

ORDER **Pselaphognatha** Minute forms with 10 to 12 body segments. Back and sides with tufts of bristles. Second maxillae leglike, not fused to form a gnathochilarium. *Polyxenus.*

ORDER **Limacomorpha** Eyeless tropical forms with 22 body segments. Male with gonopods, legs modified for clasping the female, on the last segment. *Glomeridesmus.*

ORDER **Oniscomorpha** 11 to 13 body segments, flattened on the ventral side. Repugnatorial glands absent. Last two pairs of legs in male modified as gonopods. Body capable of rolling up into ball. Found in the Old World tropics. *Glomeris.*

ORDER **Polydesmoidea** 19 to 22 flattened body segments. Repugnatorial glands on alternate segments. Male gonopods on seventh segment. Eyes absent. *Polydesmus.*

ORDER **Nematomorpha** 26 to 32 body segments. Repugnatorial glands absent. Male gonopods on seventh segment. Several pairs of silk-spinning glands open at posterior end. *Chordeuma.*

ORDER **Juliformia** Forty or more body segments, each with a pair of repugnatorial glands. Male gonopods on seventh segment. No spinning glands. *Julus, Spirobolus.*

ORDER **Colobognatha** 30 to 70 body segments. Male gonopods are second pair of legs of seventh segment and first pair of legs of eighth segment. Repugnatorial glands present. Mouth parts often modified for sucking. *Siphonophora.*

Anatomy and Physiology

The most obvious feature of the Diplopoda is that which gives the class its name—the two pairs of legs per segment—and it is also the basis for the common name for the group, millipedes (1000 feet). Each of the abdominal segments is quite obviously the result of the fusion of two, since in addition to the two pairs of legs each bears two pairs of spiracles opening into the tracheal respiratory system, and internally has two ganglia in the ventral nerve cord and two ostia in the long tubular dorsal heart which extends much of the length of the animal.

Millipedes are mostly herbivorous creatures, feeding on living and decaying vegetation. Some will also eat dead animal matter and one small group preys upon earthworms. The digestive tract is for the most part a straight tube with little or no modifications, the anus being located in the terminal segment. Salivary glands aid in digestion, and excretion is by means of Malpighian tubules.

REPRODUCTION AND DEVELOPMENT

There is a single gonad in the midventral line which opens on the ventral side of the third segment. Fertilization is internal, the male using modified appendages, *gonopods,* usually located on the seventh segment, to transfer the sperm to the female. In some species a special courtship behavior precedes mating. In many species the eggs are laid in a nestlike structure in the soil, often lined with fecal matter. In some cases the female remains near the eggs until they are hatched. Development is direct with a small animal with only three pairs of legs and seven segments hatching after two or three weeks. As the animal grows and molts, additional segments are added at the posterior part of the body just in front of the terminal segment. This method of growth is called *anamorphosis.*

Ecology

Millipedes live in dark moist environments beneath surface litter, logs, or stones. Only a few climb above the surface on trees and shrubs. Of the more than 8000 described species, many are tropical in their distribution. Some species have taken up a subterranean existence in the dark, damp constant environment of cave systems.

On occasion, very large numbers of millipedes will migrate from their birthplace, perhaps because of insufficient food after a very favorable period for reproduction. Such mass migrations of millipedes have been reported from all over the world. In at least one instance the millipedes were so numerous that a train was unable to move due to the crushed animals on the tracks.

Although the millipedes do not bite, they may possess a formidable protective mechanism—a type of chemical warfare. Many species possess *repugnatorial glands* ("stink glands"), one in each segment with lateral pores. Others have these glands in alternate segments. Although there are species which can spray the secretion by action of muscles on the saclike glands, in most cases it merely oozes out. In some of the polydesmids, the glands produce the highly toxic gas, hydrogen cyanide (oil of bitter almonds). These glands consist of two compartments, separated by a duct which can be closed by a muscular valve. The innermost larger compartment produces a precursor, *mandelonitrile.* The outer compartment produces an enzyme. When the valve opens, the precursor mixes with the enzyme and benzaldehyde and hydrogen cyanide are produced. One of these polydesmids in a jar will serve as a source of cyanide for killing insects which may be added. The animal itself is not affected by the hydrogen cyanide which is very volatile. The benzaldehyde evaporates more slowly, and it is also noxious to potential enemies. Other types of secretions produced by millipede repugnatorial glands include *p*-benzoquinone which causes pain and also stains human skin, aldehydes such as *trans*-2-dodecenal, *p*-cresol, and an unknown compound which smells like camphor.

CLASS PAUROPODA

Characteristics

1. Pauropods are minute millipedelike animals with a cylindrical body of 11 or 12 segments rarely over 2 mm long.
2. The head bears one pair of branched antennae and lacks eyes.
3. There are eight or nine pairs of legs.
4. The reproductive ducts open on the third trunk segment. Figure 30.23

The class Pauropoda is not broken down into orders.

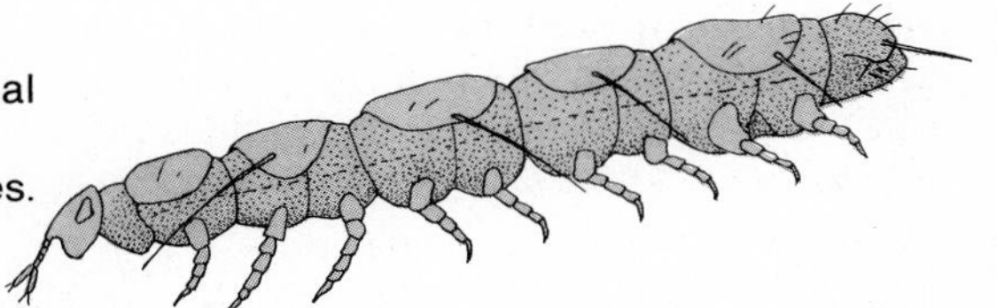

Figure 30.23 Class Pauropoda, *Pauropus* sp.

Anatomy and Physiology

Pauropods are unfamiliar to most people largely because of their small size and their restricted habitat. They live in moist soil and surface litter in temperate and tropical regions and may actually be quite numerous in some situations. They seem to be quite primitive forms and are probably most closely related to the millipedes. The distinctive *branching antennae* are found in no other arthropods except the crustaceans. The head also bears a pair of mandibles and a liplike structure which is probably homologous with the gnathochilarium of the millipedes. Although there are no eyes, a fluid-filled sensory structure of unknown function occupies the position where the eyes would normally be located. The reproductive ducts open in the third trunk segment as in the millipedes. Excretion is by Malpighian tubules, but specialized respiratory and circulatory systems are absent.

CLASS CHILOPODA

Characteristics

1. Centipedes possess an elongate body ranging from 3 to 25 cm in length, which is flattened in the dorsoventral plane.
2. Head with one pair of antennae; one pair of jaws; two pairs of maxillae, the first pair forming a lower lip; eyes simple, compound, or absent.

3. The first pair of body appendages is modified with a sharp poison claw at the tips.
4. The rest of the body segments, numbering from 14 to over 180, each with a single pair of legs.
5. The reproductive ducts open on the ventral side of the next to last segment.
Figure 30.24

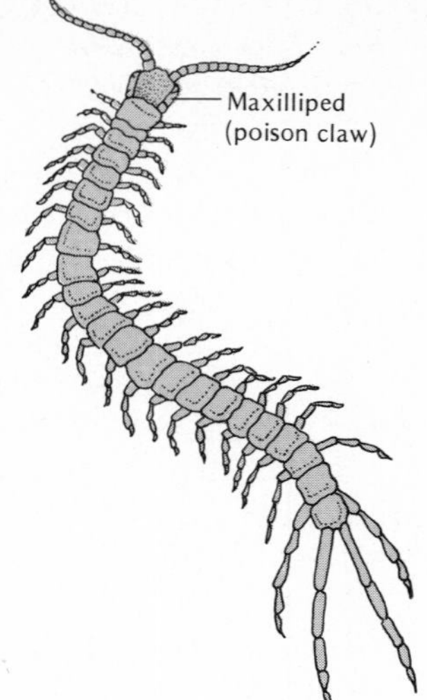

Figure 30.24 Class Chilopoda. A representative centipede.

Classification

ORDER **Scutigeromorpha** Antennae and legs long with 15 pairs of the latter. Spiracles unpaired and middorsal in position. *Scutigera forceps,* house centipede.

ORDER **Lithobiomorpha** Fifteen pairs of short legs. Antennae variable in length. Spiracles located laterally. *Lithobius.*

ORDER **Scolopendromorpha** 21 to 23 pairs of legs. Includes species which reach lengths of over 20 cm. *Scolopendra.*

ORDER **Geophilomorpha** Very slender forms with 31 to over 180 pairs of legs. *Geophilus.*

Anatomy and Physiology

The centipedes are readily distinguished from the millipedes by the presence of a single pair of legs on each of the clearly distinguishable body segments. The head is provided with a pair of antennae, often very long. In some groups there are no eyes. Others have a number of small simple eyes, and some have very well developed eyes much like the compound eyes of insects and crustaceans. The mouth parts consist of a pair of jaws and two pairs of maxillae, the first of which forms a lower lip, while the second pair varies in form in the various groups. The first pair of legs is also involved in feeding since they are modified as *poison claws.* Internally the centipedes are provided with a typical arthropod digestive system. Respiration is by means of a tracheal system with a pair of spiracles in each body segment. A pair of Malpighian tubules functions in excretion, and the tubular heart extends the length of the body and has ostia and lateral arteries in each segment.

REPRODUCTION AND DEVELOPMENT

The gonads consist of a single dorsally placed ovary in the female and from 1 to 24 testes dorsal to the midgut in the male. Development is direct and the care given to the eggs varies from species which guard the eggs until the young hatch to those which bury the eggs in the soil and leave them. In some orders the young have all their segments when they hatch, while in others the young have a small number of legs and new segments are added at the posterior end of the body.

Ecology

Centipedes are found in both temperate and tropical areas and are represented by over 3000 species. Like the spiders and scorpions, these predatory animals all possess poison fangs, but most species produce a painful but not serious bite. The bite of a few species is more serious and may prove fatal to young children.

During the day, centipedes remain under logs and stones, in leaf litter, or in the soil. At night they emerge to hunt for prey. Some species climb up the lower portions of tree trunks. Several groups have representatives that have adapted to life in caves, an environment that is well suited to their moisture requirements.

Several species are found on ocean beaches, a few being able to live in the intertidal zones where they are periodically covered at high tide. Remaining quiet, a small air bubble in each spiracular opening seems to enable gaseous exchange to take place.

CLASS SYMPHYLA

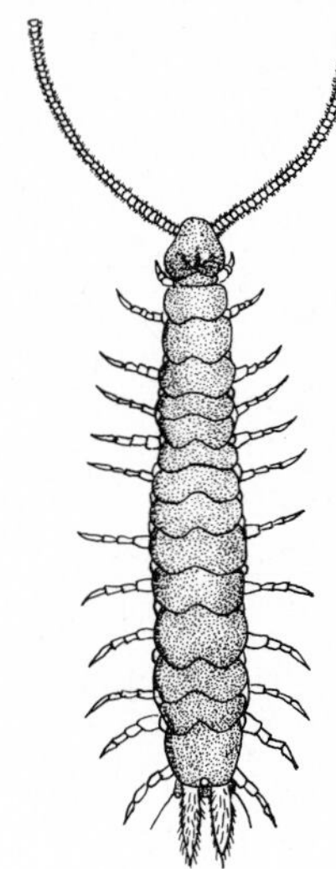

Characteristics

1. Symphylids are white, eyeless animals less than 10 mm long.
2. They have 12 pairs of legs and the reproductive ducts open ventrally on the third segment.
   Figure 30.25
   The class Symphyla is not broken down into orders.

Figure 30.25 Class Symphyla. *Scutigerella immaculata,* the garden centipede.

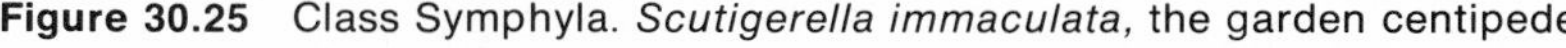

Anatomy and Physiology of *Scutigerella immaculata*

These secretive organisms which live in soil and surface litter are believed by many to be the arthropod group closest to the insects. The head, which lacks eyes, bears a pair of long antennae, jaws, and two pairs of maxillae with the second pair fused to form a *labium.* The labium is remarkably similar to that of insects. Also on the head are the two spiracles which open into the tracheal system limited to the first three trunk segments. Near the base of each leg on the body wall is a small unjointed appendage of unknown function, called the *stylus.* On the penultimate (next to last) segment a pair of pointed appendages, called *spinnerets,* are located. Internally the symphyla are much like centipedes and millipedes.

Eggs are laid in small clusters attached to the tip of short stalks. As in millipedes and centipedes, the young have less than the full complement of legs at birth, and new segments are added at the posterior end with successive molts. *Scutigerella immaculata* is cosmopolitan in its distribution and often attacks vegetables and flowers, both in the field and in greenhouses.

CLASS INSECTA

The culmination of the evolutionary process in the annelid–arthropod line is reached with the class Insecta. Without doubt the insects represent the most successful single group of organisms ever to evolve. Each of the 26 insect orders will be introduced by a figure and a brief characterization. Then a detailed consideration is given of the grasshopper. The chapter concludes with a general treatment of molting, hormones, pheromones, the importance of insects to humanity, and the phylogenetic relationships of arthropods. Space does not permit greater coverage. Entomology courses are usually available for those wishing more detailed information.

Characteristics

1. The insect body is composed of head, thorax, and abdomen.
2. The head bears one pair of antennae (except in the order Protura), one pair of mandibles, one pair of maxillae, and a labium.

3. The thorax consists of three segments, each usually with a pair of legs and each of the last two typically with a pair of wings.
4. The abdomen usually has 11 segments and lacks locomotor appendages.
Figure 30.26

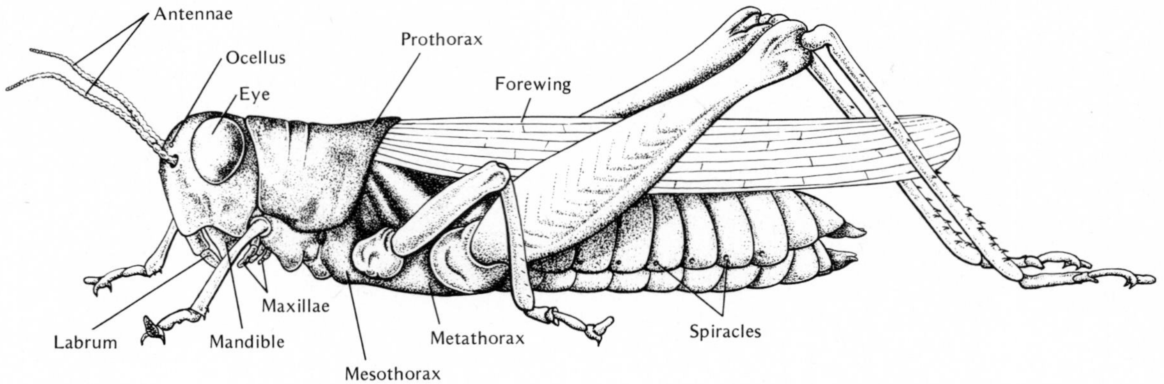

Figure 30.26 Class Insecta. External features of the grasshopper.

Classification

SUBCLASS **Apterygota** Primitively wingless insects without true metamorphosis, abdomen with ventral *styli* and usually with cerci at the tip.

ORDER **Protura** (Fig. 30.27) Minute in size, less than 1.5 mm. Eyes, antennae, and trachea absent. Piercing mouth parts. Abdomen with 12 segments, the first three bearing small ventral appendages, and lacking cerci. Young hatch with only nine abdominal segments and add one posteriorly in successive molts until the full number of 12 is reached. *Eosentomum.*

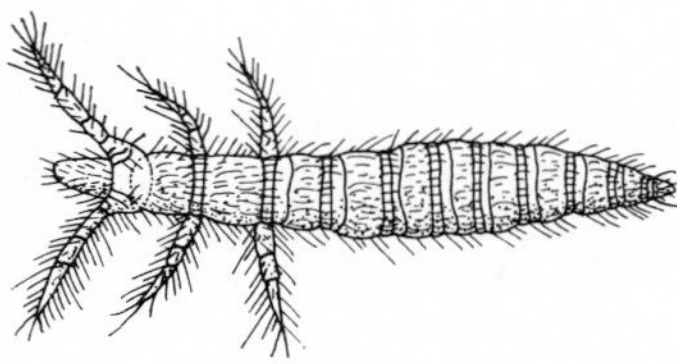

Figure 30.27 Order Protura. *Acerentulus,* a representative proturan.

ORDER **Diplura (Entotrophi)** (Fig. 30.28) Small in size, usually 8 to 10 mm. Head bears long antennae but no eyes. Biting mouth parts deep within head capsule. Abdomen with 10 visible segments and possessing a pair of cerci or a pair of pincers at the tip. *Japyx.*

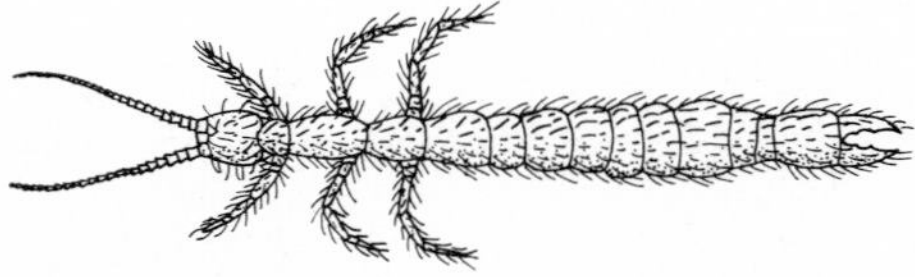

Figure 30.28 Order Diplura. *Japyx,* a representative dipluran.

ORDER **Thysanura**—Bristletails or silverfish (Fig. 30.29) Small in size, up to 30 mm. Antennae long. Biting mouth parts readily visible. May have compound eyes or ocelli, or both, or may lack eyes entirely. Abdomen with 11 segments and three long segmented filaments at tip. Body usually covered with scales. Live under logs and stones and some species invade homes where they may damage book bindings. *Lepisma, Thermobia.*

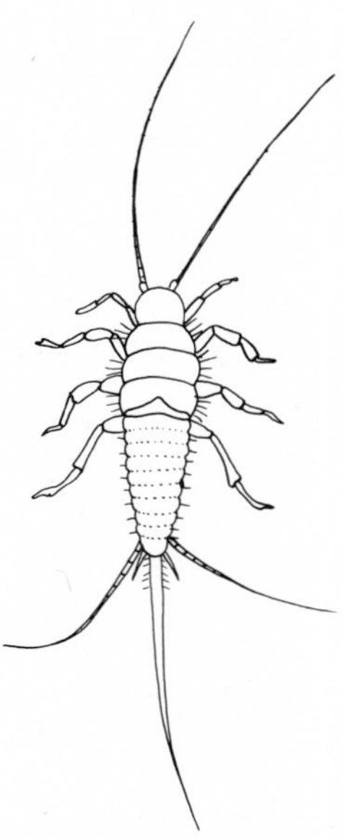

Figure 30.29 Order Thysanura. A representative silverfish.

ORDER **Collembola** (Fig. 30.30) Mostly very small in size, up to 5 mm. Head bears a pair of four- to six-segmented antennae. Eyes absent or represented by a few scattered ommatidia. Biting mouth parts. Abdomen of six segments, usually with a jointed

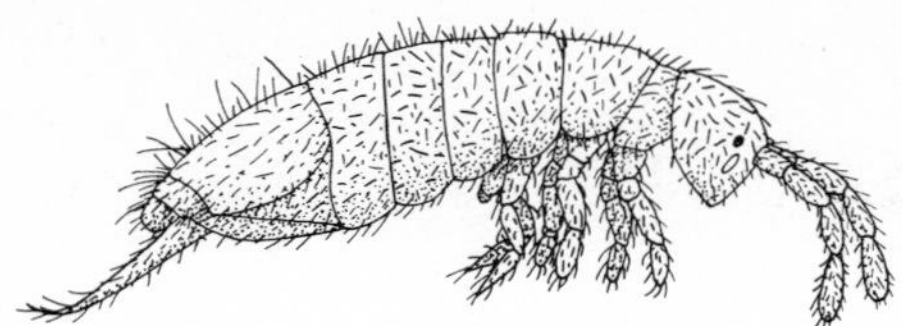

Figure 30.30 Order Collembola. A springtail.

springing structure, called a *furcula,* on the ventral side of the fourth segment and a catching device for the furcula on the ventral side of the third segment. Cerci and Malpighian tubules absent. These small primitive insects inhabit moist damp places and may be quite numerous in some cases. Some species, called snow fleas, may appear in large numbers on the surface of the snow in early spring. *Achorutes, Sminthurus.*

SUBCLASS **Pterygota** Insects with wings or secondarily wingless. Some type of metamorphosis present. No abdominal appendages except cerci and genitalia.

SUPERORDER **Exopterygota (Hemimetabola)** Young stages, all of which have compound eyes, are nymphs (terrestrial), or naiads (aquatic). Wings develop externally from pads in late instars. The immature forms resemble the adults.

ORDER **Ephemeroptera**—Mayflies (Fig. 30.31) Moderate in size, up to 25 mm. Chewing mouth parts which are vestigial in the adult. Eyes large. Antennae inconspicuous and hairlike. Two pairs of membranous net-veined wings, the posterior pair small. Wings held vertically over the back when at rest. Abdomen with a pair of long cerci and often a long median filament at the tip. Most of life is spent as an aquatic naiad, in some species as much as 3 years. The adult lives for a very short time, a few days. *Ephemera.*

Figure 30.31 Order Ephemeroptera. A mayfly.

ORDER **Odonata**—Dragonflies and damselflies (Fig. 30.32) Large in size, up to 18 cm wing span. Chewing mouth parts. Head with very large compound eyes. Antennae reduced to minute hairlike structures. Two pairs of similar net-veined wings. Abdomen long and slender. Both adults and aquatic naiads are predaceous, the former catching insects in flight. *Anax, Agrion.*

ORDER **Plecoptera**—Stoneflies (Fig. 30.33) Medium to large sized, 15 to 50 mm, soft-bodied forms. Chewing mouth parts, but often none in the adult. Immature stages aquatic. Two pairs of membranous wings held flat against the back when at rest, with the larger hind pair pleated and folded beneath the front pair. Abdomen usually with two long cerci. Immature stages may last as long as 3 years, but adult life usually short. *Perla.*

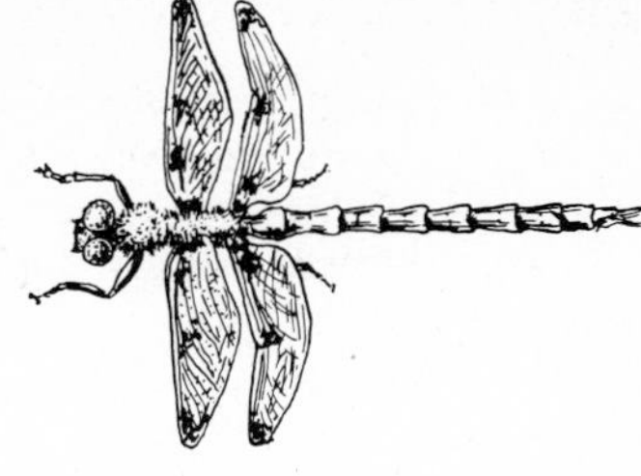

Figure 30.32 Order Odonata. A dragonfly.

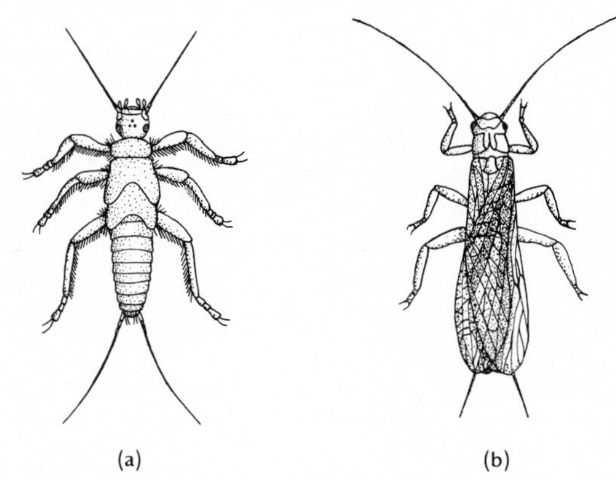

Figure 30.33 Order Plecoptera. Stonefly. **(a)** Aquatic naiad. **(b)** Adult.

ORDER **Orthoptera**—Grasshoppers, cockroaches, crickets, walking sticks, mantids (Fig. 30.34). Medium to large in size, up to 8 cm, walking sticks

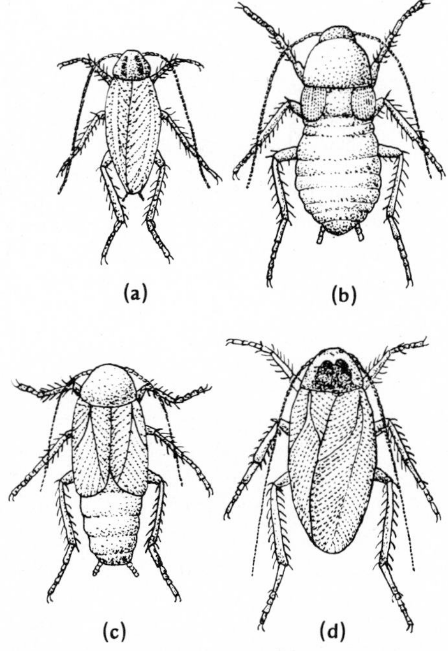

Figure 30.34 Order Orthoptera. Three common household pests. **(a)** *Blatella germanica,* the German cockroach. **(b)** *Blatta orientalis,* the Oriental cockroach, female. **(c)** Oriental cockroach, male. **(d)** *Periplaneta americana,* the American cockroach.

to 17 cm. Chewing mouth parts. Forewings narrow and leathery covering the folded membranous hindwings when not in flight. Some species lack wings. Cerci usually present. Herbivorous and predaceous. *Blatta, Gryllus.*

ORDER **Dermaptera**—Earwigs (Fig. 30.35) Small to medium-sized elongate forms, 2 to 20 mm long. Some species lack eyes. Chewing mouth parts. Hindwings large and membranous fold under short leathery forewings when at rest. Some species wingless. Cerci at tip of abdomen form a pair of sharp-pointed forceps. Some species become garden and household pests. *Forficula.*

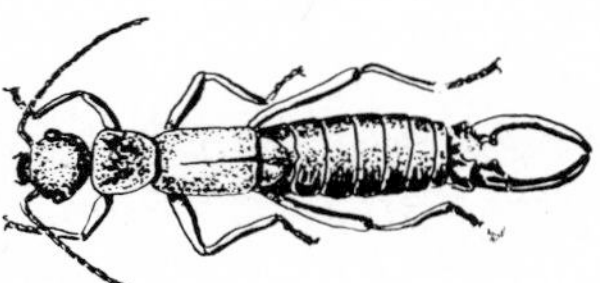

Figure 30.35 Order Dermaptera. An earwig.

ORDER **Embioptera**—Embiids, web spinners (Fig. 30.36) Small elongate soft-bodied forms, 4 to 7 mm long. Chewing mouth parts. Females wingless, males usually with two pairs of similar membranous wings. Short cerci present, front tarsi enlarged and contain silk glands used in spinning silken tunnels in which large numbers live together in a colony. Largely tropical in distribution. *Embia.*

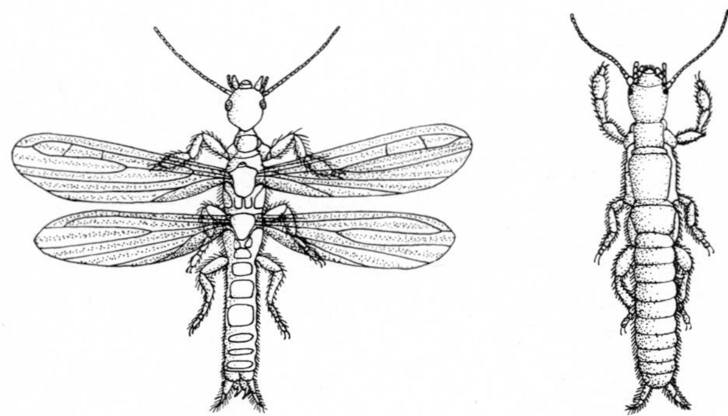

Figure 30.36 Order Embioptera. Webspinner, winged male and wingless female.

ORDER **Isoptera**—Termites (Fig. 30.37) Small social insects with chewing mouth parts. Workers up to 1 cm, queens may reach 10 cm or more. Reproductive forms have two pairs of similar membranous wings which are shed after a nuptial flight. Abdomen broadly joined to the thorax. Eyes present only in the reproductives; other castes—workers and soldiers—lack them. Only the reproductives are pigmented. Live in large colonies either underground or in large aboveground nests. Largely tropical but several species are found in temperate regions where they do extensive damage to wooden structures. *Reticulitermes, Calotermes.*

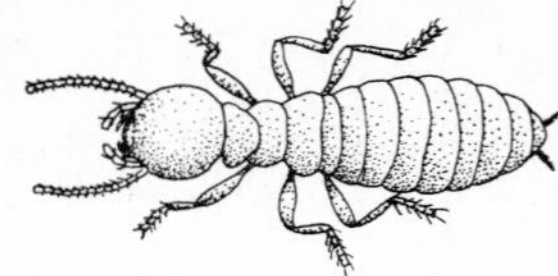

Figure 30.37 Order Isoptera. A worker termite.

ORDER **Zoraptera** (Fig. 30.38) Rare minute insects with chewing mouth parts. Up to 2.5 mm long. Blind and wingless or with compound eyes, ocelli, and two pairs of delicate membranous wings with only a few veins which are shed after mating. Antennae long and nine jointed. Abdomen with ten segments and a pair of one-segmented cerci. Usually live in small colonies of less than 100 individuals with no evidence of division of labor. *Zorotypus.*

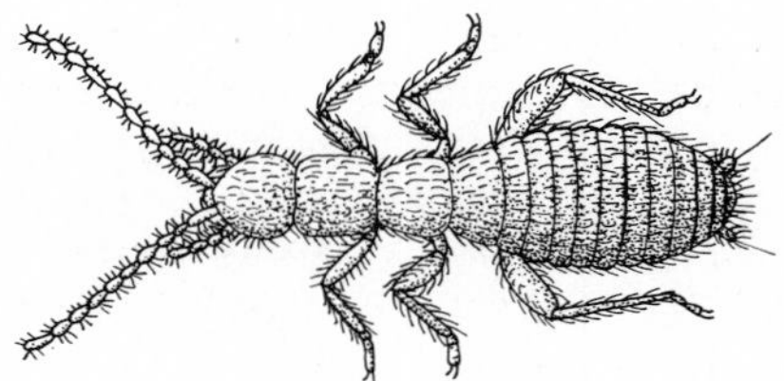

Figure 30.38 Order Zoraptera. An eyeless and wingless zorapteran.

ORDER **Psocoptera (Corrodentia)** Book lice (Fig. 30.39) Minute to small size, up to 10 mm. Chewing mouth parts. Two pairs of wings, if present, are similar, membranous, and roofed over the abdomen when at rest. Forewings much larger than hindwings. Ten-segmented abdomen lacks cerci. Labial palps spin silken webs in which eggs are laid. These insects may be quite numerous in soil and surface litter. Some species invade homes and museums where they attack book bindings, starches, and other organic materials. *Psocus, Liposcelus.*

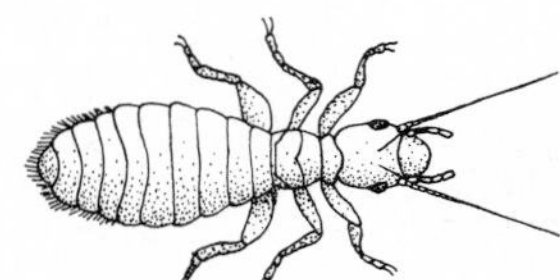

Figure 30.39 Order Psocoptera. A wingless book louse.

ORDER **Mallophaga**—Chewing lice (Fig. 30.40) Small, up to 5 mm. Dorsoventrally flattened ectoparasites of birds and mammals, the majority of species being found on birds. Chewing mouth parts. Eyes reduced. Antennae short. Legs each with a pair of claws used for clinging to feathers or hair. Feed on sloughed-off skin, hair, and feather fragments and other organic debris. Heavy infestations materially decrease production in poultry, cattle, sheep, and goats. *Menopon, Trichodectes.*

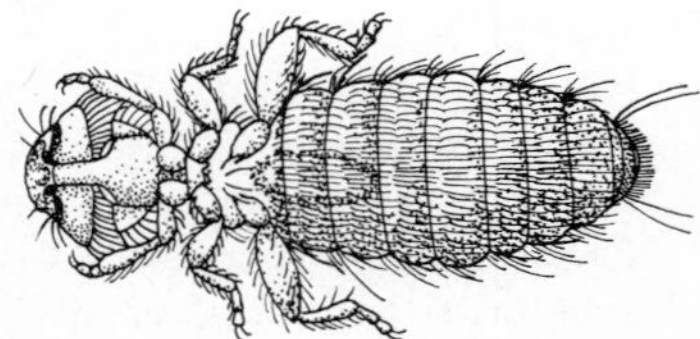

Figure 30.40 Order Mallophaga. A chicken louse.

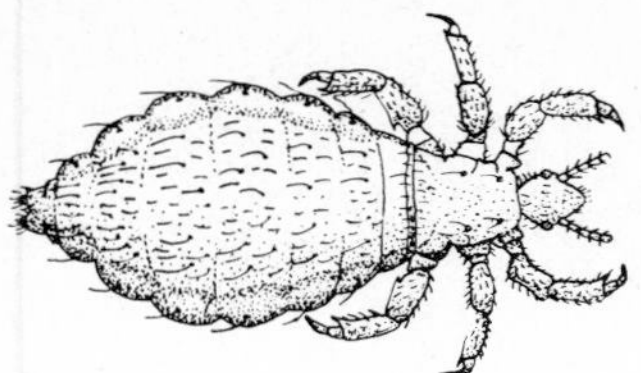

Figure 30.41 Order Anoplura. A human body louse.

ORDER **Anoplura**—Sucking lice (Fig. 30.41) Minute, up to 6 mm. Dorsoventrally flattened wingless ectoparasites of mammals. Mouth parts modified for piercing skin of host and sucking blood. Eyes reduced or absent. Antennae short. Legs each with a single claw used in holding on to a hair. Eggs, called *nits,* are glued to hairs. The Anoplura are the vectors of a number of serious human diseases, including typhus fever, trench fever, and relapsing fever. *Pediculus humanus* is the head and body louse.

ORDER **Thysanoptera**—Thrips (Fig. 30.42) Mostly minute in size, 0.5 to 14 mm. Mouth parts form a lacerating-sucker cone. Wings, if present, are very long and narrow with at most two veins and fringed with long hairs. The front pair of wings is often larger and the wings are laid flat along the body when at rest. Tarsi are one or two jointed and terminate in a blunt tip with an eversible bladderlike pad. Feeding habits include plant feeders and those which are predaceous on mites and small insects. In some species males are unknown and all reproduction is by parthenogenesis. Although metamorphosis is gradual, the last stage nymph forms a cocoon and enters a pupal stage prior to the final molt to the adult condition. Many species are quite destructive to crops and some are vectors for plant diseases. *Thrips, Leptothrips.*

Figure 30.42 Order Thysanoptera. A thrip.

ORDER **Hemiptera**—True bugs (Fig. 30-43) Medium to large in size, 2 to over 100 mm. Piercing–sucking mouth parts forming a jointed beak which arises from the front portion of the head. Forewings thick and leathery at the base and membranous toward the apex, crossed flat on the back when at rest with the membranous hindwings folded and hidden beneath them. Prominent triangular plate, called the scutellum, located between the bases of the forewings. This very large order of insects includes terrestrial and aquatic forms, plant-feeding species, predaceous species, and blood-sucking ectoparasitic species. Many are of great economic importance, including the chinch bug, the bedbug, and the cone-nosed bug which is the vector of a protozoan parasite causing Chagas' disease in Central and South America. *Blissus, Cimex, Triatoma.*

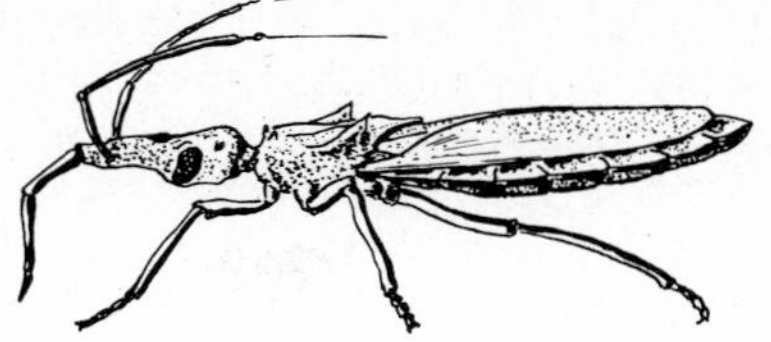

Figure 30.43 Order Hemiptera. A cone-nosed bug.

ORDER **Homoptera**—Aphids, cicadas, scale insects, leafhoppers (Fig. 30.44) Mostly small in size, but cicadas may reach a length of 5 cm. Piercing–sucking mouth parts forming a beak, which arises ventrally on the posterior portion of the head. Wings uniformly membranous when present, held roofed over the abdomen when at rest. All the members of this large order are plant feeders, many being very destructive. Some species are vectors for serious plant diseases. *Cicada, Aphis.*

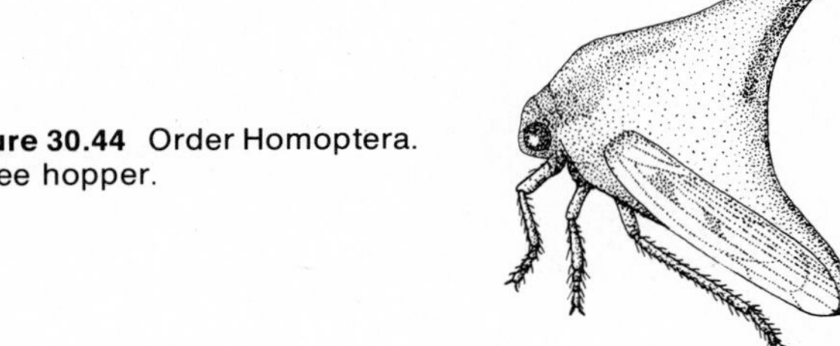

Figure 30.44 Order Homoptera. A tree hopper.

SUPERORDER **Endopterygota (Holometabola)** Young stages, the larvae, do not resemble the adult and lack compound eyes. Wings develop within the larval body. Metamorphosis to the adult condition occurs within the pupal case and results in an abrupt and drastic change in form from the preceding larval stage.

ORDER **Neuroptera**—Dobsonflies, lacewings (Fig. 30.45) Minute to large in size, from 2 mm to the dobsonfly with a wing span of over 12 cm. Chewing mouth parts. Long many-segmented antennae and large eyes. Two pairs of similar membranous wings with many veins and cross veins which are

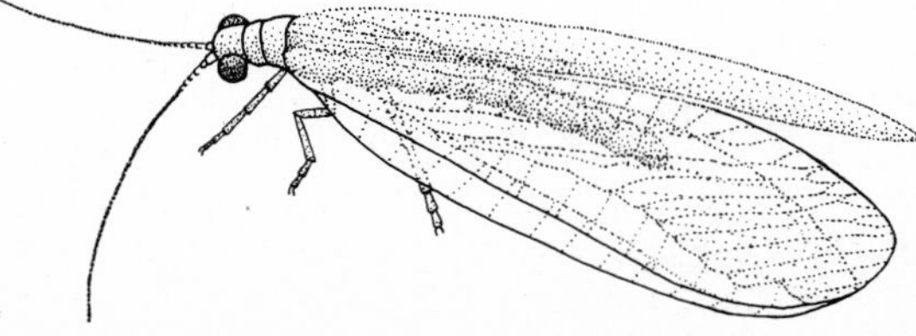

Figure 30.45 Order Neuroptera. A lacewing.

roofed over the body at rest. Adults and larvae are predaceous on mites and insects. The familiar ant lion or "doodle-bug" is a neuropteran larva. *Myrmeleon, Chrysopa, Corydalis.*

ORDER **Mecoptera**—Scorpion flies (Fig. 30.46) Small to medium-sized insects, 1 to 2.5 cm. Chewing mouth parts, often located at the tip of a beaklike downward-projecting elongation of the head. Eyes large. Antennae long. Two pairs of slender net-veined membranous wings which are roofed over the body at rest. Wings absent in some. Males of some with bulbous recurved genitalia which give an appearance a little like the abdomen of a scorpion, hence the common name of the order. *Boreus, Panorpa.*

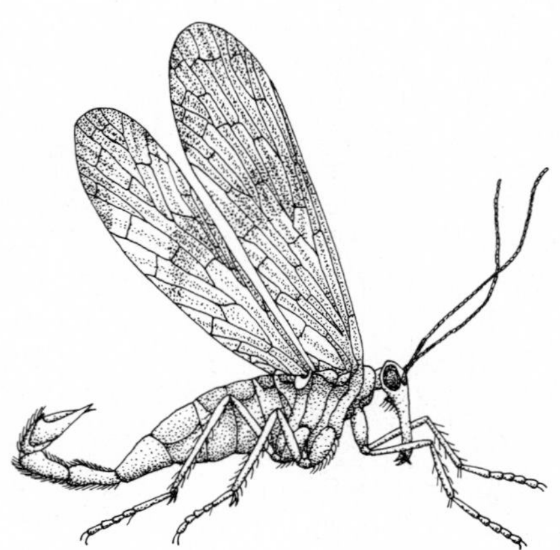

Figure 30.46 Order Mecoptera. A scorpion fly.

ORDER **Trichoptera**—Caddisflies (Fig. 30.47) Minute to medium sized, 3 to 25 mm, soft-bodied mothlike insects. Mouth parts rudimentary in adults but of chewing type in aquatic larvae. Antennae long and many-segmented. Large compound eyes. Two pairs of large membranous wings with many longitudinal but few cross veins. Wings and body hairy, in some species they are also provided with patches of scales. Wings roofed over body when at rest. The aquatic larvae are carnivorous and many species live in distinctive cases of stones, twigs, leaves, or other materials. *Limnephilus, Helicopsyche.*

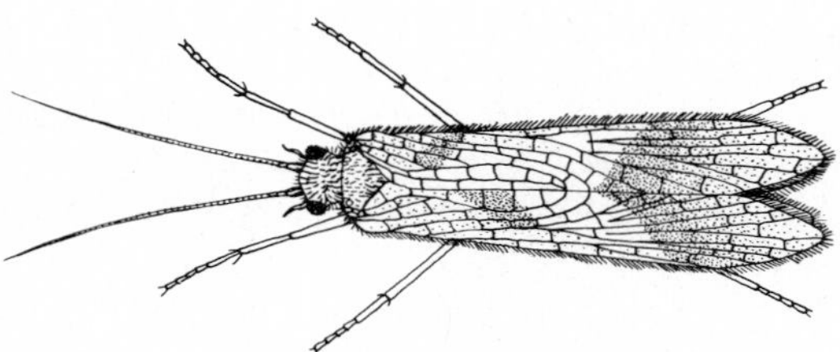

Figure 30.47 Order Trichoptera. A caddisfly.

ORDER **Lepidoptera**—Butterflies and moths (Fig. 30.48) Small to large in size, wing spread from 2 to over 300 mm. Chewing mouth parts in the larvae, sucking mouth parts in the form of a long coiled proboscis in the adults. Large compound eyes and long antennae. Membranous wings and body covered with scales. In moths wings held horizontal when at rest, in butterflies, vertical. Larvae are the familiar caterpillars which possess three pairs of thoracic legs and usually five pairs of abdominal prolegs. This large order includes many species in which the larvae are destructive pests to crops, stored foodstuffs, and cloth. In part this is offset by the sole source of natural silk, the silkworm, *Bombyx mori.* Other genera include *Sphinx, Papilio, Danaus.*

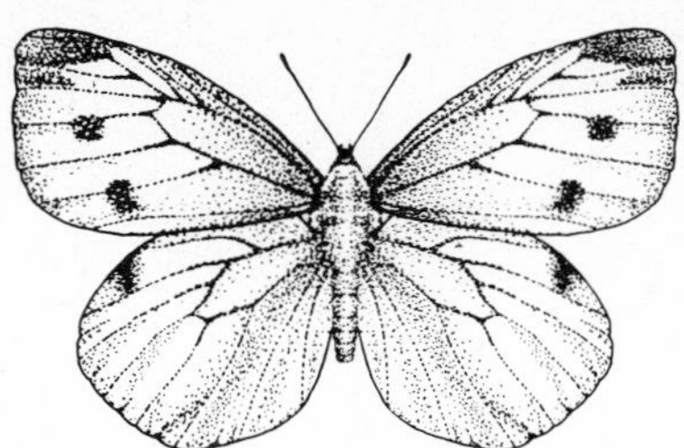

Figure 30.48 Order Lepidoptera. *Pieris rapae,* the cabbage butterfly.

ORDER **Diptera**—True flies, mosquitoes, and gnats (Fig. 30.49) Small to moderate-sized insects, 1mm to over 25 mm. Mouth parts either of the piercing–sucking or of the rasping–lapping type. One pair of wings normally present, the hind pair reduced to slender knobbed balancing organs, called *halteres.* This very large order includes a diverse group of insects with many different feeding habits. The food of the adult is usually quite different from that of the wormlike larva. In addition to species which are destructive to crops and injurious to domestic animals, there are many which are vectors for serious diseases affecting man. Included in the latter are malaria, yellow fever, equine encephalitis, African sleeping sickness, and many others. Flies carry disease-bearing filth and can thus transmit such diseases as cholera, typhoid fever, and several kinds of dysentery. Representative genera include *Musca, Anopheles, Drosophila, Glossina, Tabanus.*

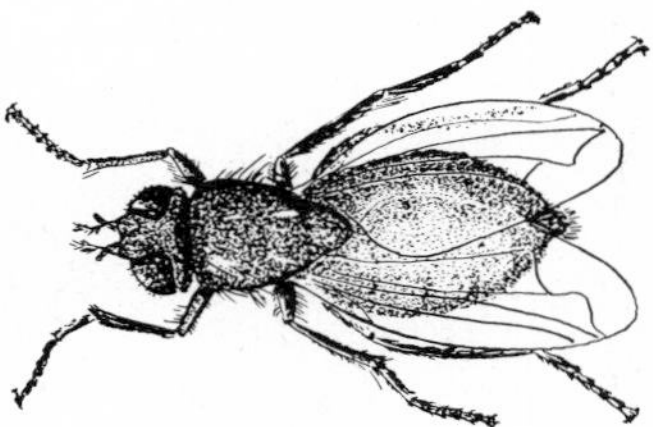

Figure 30.49 Order Diptera. A housefly.

ORDER **Siphonaptera**—Fleas (Fig. 30.50) Minute wingless insects, 3 to 4 mm. Laterally compressed. With piercing–sucking mouth parts used to obtain blood from their bird or mammal hosts. Human fleas (*Pulex irritans*) as well as dog and cat fleas

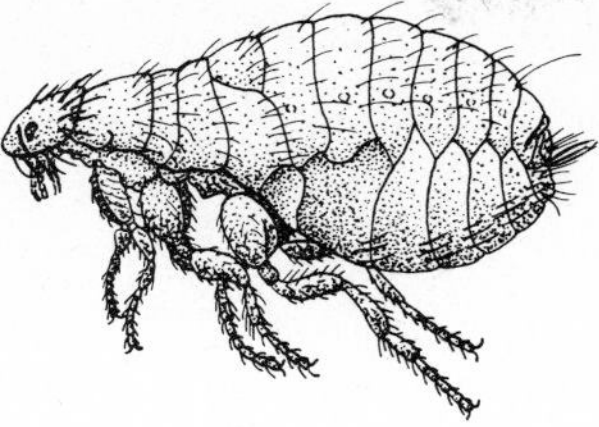

Figure 30.50 Order Siphonaptera. A flea.

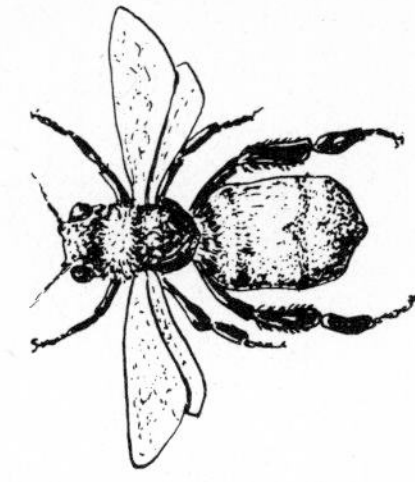

Figure 30.52 Order Hymenoptera. A bee.

cause much discomfort to humans since their bite causes itching and may result in secondary infection due to scratching. The rat flea, *Xenopsylla cheopis,* is the vector for bubonic plague and thus presents an everpresent serious threat to human populations.

ORDER **Coleoptera**—Beetles and weevils (Fig. 30.51) Size varies from minute to very large, less than 1 mm to 13 or 14 cm. Mouth parts of the chewing type. Head bears well-developed antennae and compound eyes. Usually with two pairs of wings, the anterior pair being veinless hard wing covers, called *elytra,* under which the membranous second pair is folded when not in use. Entire body is usually compact and hard due to the thick chitin in the cuticle. This largest of all insect orders (containing more than 40 percent of all described species of insects) must also be said to be the most successful in terms of diversity of habitat, feeding habits, and interactions with other animals. Many economic pests of agriculture, household goods, and stored foods are found among the beetles. Others are predaceous and may be important agents of biological control of noxious insects. Still others play an important role in removal of carrion and other animal wastes and decaying vegetation. Representative genera include *Lucanus, Scarabaeus, Cicindela, Tenebrio.*

Figure 30.51 Order Coleoptera. A ground beetle.

ORDER **Hymenoptera**—Ants, bees, and wasps (Fig. 30.52) Size varies from minute to large, 0.5 to 50 mm. Mouth parts may be modified for chewing, sucking, or lapping. Two pairs of membranous wings usually present, the smaller hind wings coupled to the front wings by hooks when in flight. In all but the sawflies, the first abdominal segment is fused with the thorax followed by a narrow waist-like pedicel between it and the rest of the abdomen. Ovipositor at tip of abdomen in the female may be modified as a saw or a sting. This very large order of insects includes minute parasitic species, species which are destructive to crops and trees, and the familiar honeybees which are beneficial not only because of the production of honey but because of their role in pollination. Although most species are solitary, the order includes many species with highly developed social organization. Representative genera include *Formica, Bombus, Vespa.*

Anatomy and Physiology of the Grasshopper

A lubber grasshopper, so called because it is large, stout, and clumsy, is often studied as a representative insect. It is large enough to be dissected with little difficulty, and is a relatively generalized insect, representing the insect body plan better than some of the more specialized ones. In the grasshopper, as in all insects, the body is encased in a chitinous exoskeleton. As in the crayfish, this exoskeleton is thin and flexible at the joints, thus allowing movements of the different parts of the body (Fig. 30.26).

EXTERNAL FEATURES

The body of any insect is divided into three regions: head, thorax, and abdomen. The segments of the abdomen and thorax are easily made out, but those of the head are not. The head is actually composed of six fused segments, the identity of which may be determined only during embryonic development. The paired head appendages of the adult, in part, indicate the segments of the embryo. There are two large compound eyes on the head of the grasshopper similar to those described for the crayfish. In addition, the grasshopper has three simple eyes or *ocelli.* Each simple eye is much like a single ommatidium of a compound eye. Attached to the head, medially between the eyes, are two long segmented sensory antennae. The mouth parts (Fig. 30.53) consist of an upper lip or *labrum,* a pair of toothed mandibles, a pair of maxillae, and the lower lip or *labium.* This labium represents the fusion of a pair of structures that arise in embryonic development and are homologous to the second maxillae of the crayfish. Beneath the labrum is a tonguelike organ, the *hypopharynx.* The grasshopper has *chewing* mouth parts of a primitive or generalized type. Grasshoppers feed on leafy material,

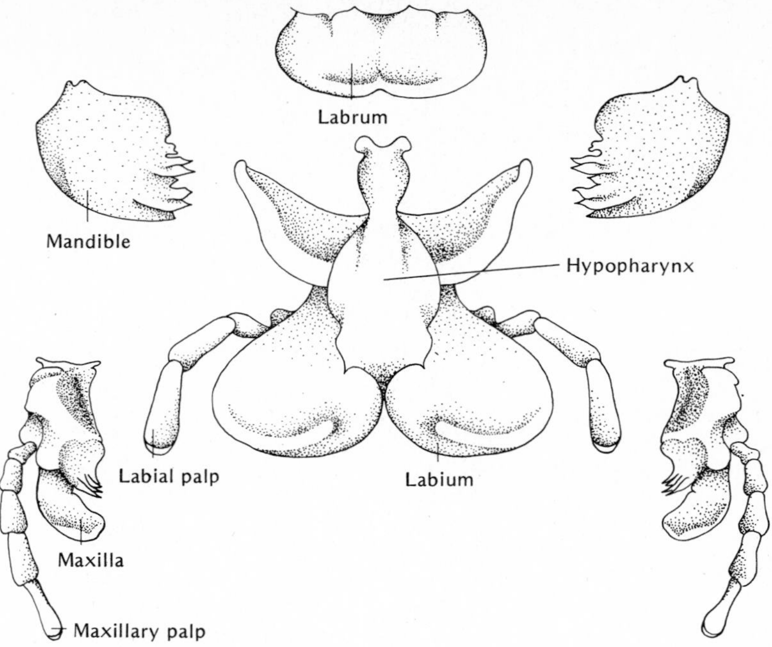

Figure 30.53 Mouth parts of the grasshopper.

and, as everyone who has driven through a grasshopper-infested region knows, when present in large numbers, they can quickly denude a field of corn.

The thorax is composed of three segments. Although it is partly covered dorsally by a posterior projection of the covering of the first segment, the three segments, with a pair of legs on each, are clearly visible at the sides. These segments, named from anterior to posterior, are the *prothorax,* the *mesothorax,* and the *metathorax.* The three pairs of thoracic legs, as found in the grasshopper, characterize all insects. In fact, the old name for this group was "Hexapoda." The grasshopper legs are all quite similar in structure. Each leg consists of five regions. The first two segments nearest the body are short and are known as the *coxa* and the *trochanter.* Next to these are two long segments, the *femur* and the *tibia.* The last region, the *tarsus,* is composed of three segments and ends in two *claws* with a pad between them called the *pulvillus.* The first two pairs of legs are relatively simple walking legs, but the third pair, with the femur much enlarged, is modified for jumping.

The forewings are attached to the dorsal side of the mesothorax, and the hindwings are attached to the corresponding part of the metathorax. The two pairs of wings are different from each other, and in this respect the grasshopper is specialized over some of the more primitive insects with both pairs of wings alike. In the grasshopper the first pair of wings is hardened and serves only as a cover for the hindwings that do the actual flying. The hindwings are quite broad and are folded like a fan under the forewings when not in use.

The development of flight in insects was a very important evolutionary advance. Insect wings are attached to both the dorsal and ventral chitinous plates (tergites and sternites). In all but the dragonflies, which have direct flight muscles, the up and down movement of the wings is accomplished indirectly by dorsoventral muscles which cause the body wall to bulge and flatten alternately. The wing moves to a horizontal position and then suddenly "clicks" the rest of the way up or down. Direct muscles, usually four pairs, move the wing anteriorly and posteriorly, causing it to rotate around its long axis as it moves up and down. The result is either a figure-eight pattern or an ellipse.

In many insects the rate of wing beat is moderately slow, ranging from eight beats per second to 70 or more. In such cases the rate of beat varies with the rate of the nerve impulses. In the Diptera and Hymenoptera the wing may beat 300 to 400 or more times per second. In these cases, the rhythm of contraction originates within the muscle, it is *myogenic* rather than *neurogenic.* Nerve impulses arrive at a rate of 100 or more per second, but the muscle fibers oscillate several times for each impulse received.

The abdomen of the adult grasshopper has no paired appendages except for the two *cerci,* small rudimentary appendages found near the

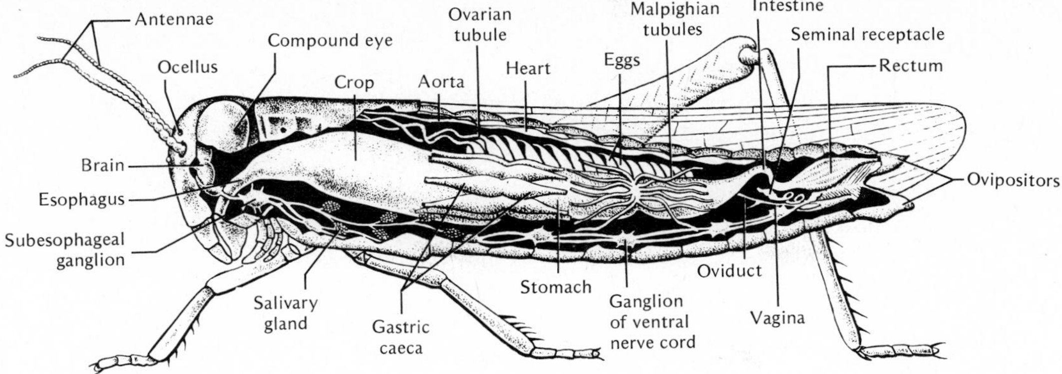

Figure 30.54 Internal features of the grasshopper.

tip of the abdomen in both sexes. The cerci are considered to be serially homologous with the legs and mouthparts.

THE DIGESTIVE SYSTEM

As in the crayfish, the digestive tract of the grasshopper is divided into three general regions: foregut, midgut, and hindgut (Fig. 30.54). The foregut and the hindgut are ectodermal in origin and are lined with chitin. The foregut consists of the pharynx, the tubular esophagus, the large thin-walled *crop*, and the muscular *gizzard.* The mouth, which receives secretions from the salivary glands, opens into the pharynx. The crop is primarily a storage region and the gizzard, which is lined with chitinous teeth, serves to grind the food thoroughly before it passes into the stomach. Digestion probably begins in the crop since the food is mixed with the salivary secretion in the pharynx. The stomach with six paired cone-shaped *gastric pouches,* or *caeca,* makes up the midgut. These pouches secrete a digestive fluid into the stomach, and absorption occurs through the stomach wall. The hindgut or intestine consists of the small intestine, the large intestine, and the rectum, which terminates in the anus. The junction of the stomach with the intestine is marked by the attachment of several long excretory tubules.

THE EXCRETORY SYSTEM

The tubules just referred to are called *Malpighian tubules* after their discoverer. Their closed distal ends lie in blood sinuses and extract nitrogenous wastes from the blood which are passed into the hindgut in the form of uric acid crystals. Water and nutrients are removed in the proximal portion of the tubule and returned to the body fluids. Thus the nitrogenous wastes leave the body with the fecal material as dry excretions. The elimination of wastes in a dry form is an adaptation found in a number of land animals that live in habitats where the water supply is limited.

THE RESPIRATORY SYSTEM

The characteristic respiratory system of insects consists of a network of tubes, the *trachea,* the fine branches of which terminate in the tissues throughout the body (Fig. 30.55). The tracheal tubes connect with the outside openings, the *spiracles,* which are conspicuous on the sides of the thorax and abdomen (Fig. 30.26). Each spiracle is guarded by a valve that can be opened or closed to regulate the flow of air. The walls of the larger tracheal tubes are reinforced with spiral chitinous threads that prevent their collapse. In the grasshopper several large *air*

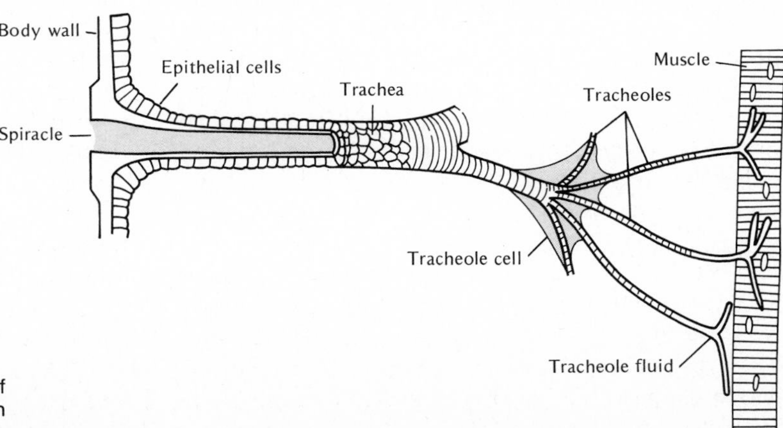

Figure 30.55 Diagram of the tracheal system of an insect showing the relationships between spiracle, trachea, tracheoles, and muscle tissue.

sacs, which serve as reservoirs of air, are associated with the tracheal tubes. The finer tracheal tubes are filled with fluid. Oxygen, in the air of the larger tubes, dissolves in this fluid and diffuses to the surrounding cells. Carbon dioxide in the cells leaves by the reverse route. Movement of air in this system is caused, in part, by contractions and expansions of the abdomen. During inspiration the first four pairs of spiracles are open and the other six are closed. In expiration this is reversed.

THE CIRCULATORY SYSTEM

The nature of the respiratory system indicates that the blood of the grasshopper, in contrast to the situation in the other complex animals studied up to this point, plays little role in the process of respiration. This system in the grasshopper, like that in the crayfish, is an open one. It consists of a single tube located in the middorsal line just under the body wall. The posterior part of this tube is a contractile heart, and it continues anteriorly as the aorta. The heart is partially constricted into a series of chambers and a pair of ostia open into each chamber. The heart lies in the pericardial sinus. When the heart contracts, the ostia are closed and the blood is forced forward through the aorta into communicating sinuses throughout the body. The blood serves chiefly to carry food and wastes; it is composed of a colorless plasma and phagocytic white blood cells.

THE NERVOUS SYSTEM

The nervous system is similar to that found in the crayfish. In the grasshopper there is a ganglion for each segment, but in the adult some of the ganglia are fused. The first three embryonic head ganglia fuse to form the brain, while the last three fuse to form the subesophageal ganglion. These two structures are joined by two connectives around the esophagus. The ventral nerve cord contains a large ganglion for each of the thoracic segments and five ganglia in the abdomen. This reduced number of abdominal ganglia results from fusions that occur during development. In some of the more complex insects there is considerably more fusion of these ganglia of the ventral cord, and in some cases all the thoracic and abdominal ganglia fuse into one mass.

SENSE ORGANS

The grasshopper has well-developed receptors for sight, touch, taste, smell, and sound. The compound eyes are basically similar to those described for the crayfish. The ocelli, or simple eyes, are considered to be simple organs for light perception. Fine hairs on various parts of the body, but especially on the antennae, are touch receptors. The receptors for smell and taste are small pits located on the antennae (smell) and around the mouth (taste). The auditory organs consists of a pair of structures located on the sides of the first abdominal segment. Each consists of a circular *tympanum* beneath which are sensory cells that, when stimulated, send impulses to the central nervous system.

Everyone is probably familiar with the sounds made by grasshoppers. When at rest, they make their noises by drawing the femurs of the hindlegs across the hindwings. While flying, the sharper noises are made by rubbing the forewings and the hindwings together. Most of the "singing" is done by males. It functions primarily in attracting a female, but some sounds are alarm or distress signals and others serve to intimidate other males.

MUSCLES

As in the crayfish, the muscles of the grasshopper that are associated with the exoskeleton are of the striated type and they are very numerous. The muscles of insects are powerful and efficient structures. It has been estimated that the wings of a honeybee make 400 strokes per second. An ant can carry an object heavier than itself up a vertical wall at a rapid rate. One worker has estimated this achievement as equivalent to a man climbing a vertical cliff at the rate of 20 miles an hour with a 100-kg (220-pound) load on his back.

THE REPRODUCTIVE SYSTEM AND REPRODUCTION

Female grasshoppers are usually larger than the males, and they can be distinguished from the males by the *ovipositor,* the four-pronged terminus of the abdomen. The internal parts of this system are shown in Figure 30.54. The female has a pair of ovaries and a pair of oviducts that unite to form the vagina. The opening of the seminal receptacle, which receives the sperm from the male at the time of copulation, connects with the vagina just within the genital opening.

In the male the paired testes occupy a position similar to that of the ovaries in the female. The two vasa deferentia that lead from the testes unite to form an ejaculatory duct, at the end of which is the penis. After copulation the female drills a hole in the ground with her ovipositor and then deposits a mass of eggs in the hole.

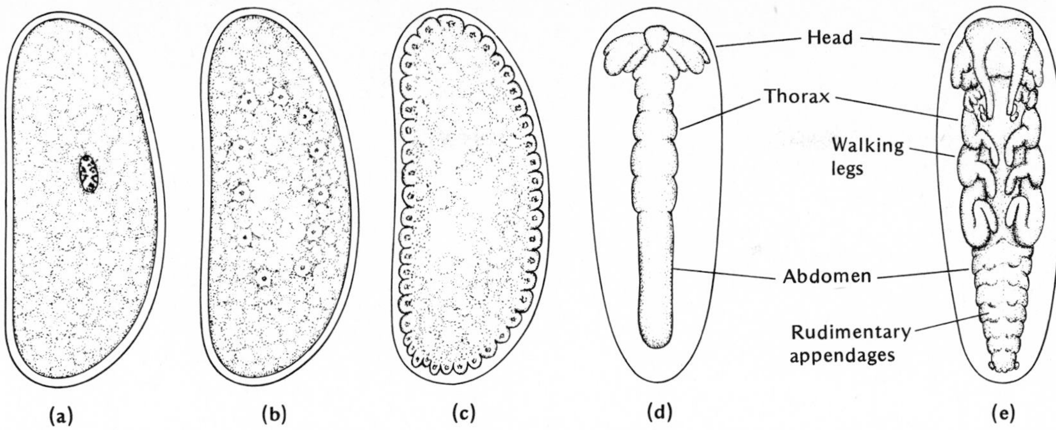

Figure 30.56 Embryology of an insect. **(a)** Fertilization. **(b)** Cleavage nuclei scattered throughout the cytoplasm migrate to the periphery of the egg. **(c)** Cell boundaries appear between nuclei, producing a single layer of cells. **(d)** Segmentation develops. Three segments are involved in forming the primitive head, and appendage primordia form on each segment. **(e)** Body regions are differentiated.

The early development of an insect is quite unlike that described for the starfish and the frog, probably because of the large amount of yolk in the insect egg. At first the zygote nucleus undergoes a number of divisions without cell membranes being formed (Fig. 30.56). These nuclei then migrate to the periphery, and cell membranes form between them, making an outer layer of cells. Subsequently the three primary germ layers are formed and morphogenesis takes place. During this development coelomic sacs form in the mesoderm, but these break down so that ultimately a coelom is not present in the adult, except in the cavities of the gonads.

On hatching the young grasshopper looks like the adult, except for its disproportionately large head and the absence of wings. This form is known as a *nymph*. It is a voracious feeder and grows rapidly. A nymph undergoes five molts in reaching the adult condition. During this process the wings are gradually developed from external wing buds. This type of development is called *gradual metamorphosis.*

Insect Metamorphosis

Three categories are recognized in the insect group as a whole with respect to metamorphosis.

No Metamorphosis Primitive insects, such as the silverfish and springtails, undergo no metamorphosis. They hatch from the egg much like adults, except for size.

Gradual Metamorphosis A number of species, among them the various kinds of grasshoppers, crickets, and cockroaches, undergo the gradual changes described above for the grasshopper.

In mayflies, stoneflies, and dragonflies the body form of the immature individuals in general resembles that of the adult, but the young forms are adapted for life in an entirely different environment from that of the adult. The immature forms, known as *naiads,* live in the water and feed on aquatic plants or animals. For the life in the water, the naiad has tracheal gills for external respiration. The major change in the metamorphosis of these forms involves the loss of the gills for terrestrial existence. Insects which undergo gradual metamorphosis are grouped together in the superorder Exopterygota (Hemimetabola).

Complete Metamorphosis The changes that take place in the insects having complete metamorphosis are striking. Among the insects exhibiting this kind of metamorphosis in their life cycle are the moths, butterflies, flies, beetles, and bees. Four stages are recognized in the life cycle of such forms: *egg, larva, pupa,* and *adult* (Fig. 30.57). The egg hatches as a segmented wormlike animal variously referred to as caterpillar (butterfly and moth), maggot (fly), and grub (beetle). The larval stage is a period of feeding and growth. Each larva undergoes several molts and finally becomes the pupa. In many moths the caterpillar spins a cocoon around the pupal case. In the fly the fullgrown larva ceases to feed, becomes motionless, and the outer covering of the larva becomes the pupal case. The pupal stage is sometimes referred to as a quiescent stage. Externally this is true, but internally it is certainly not. Within the pupal case the tissues of the larva change

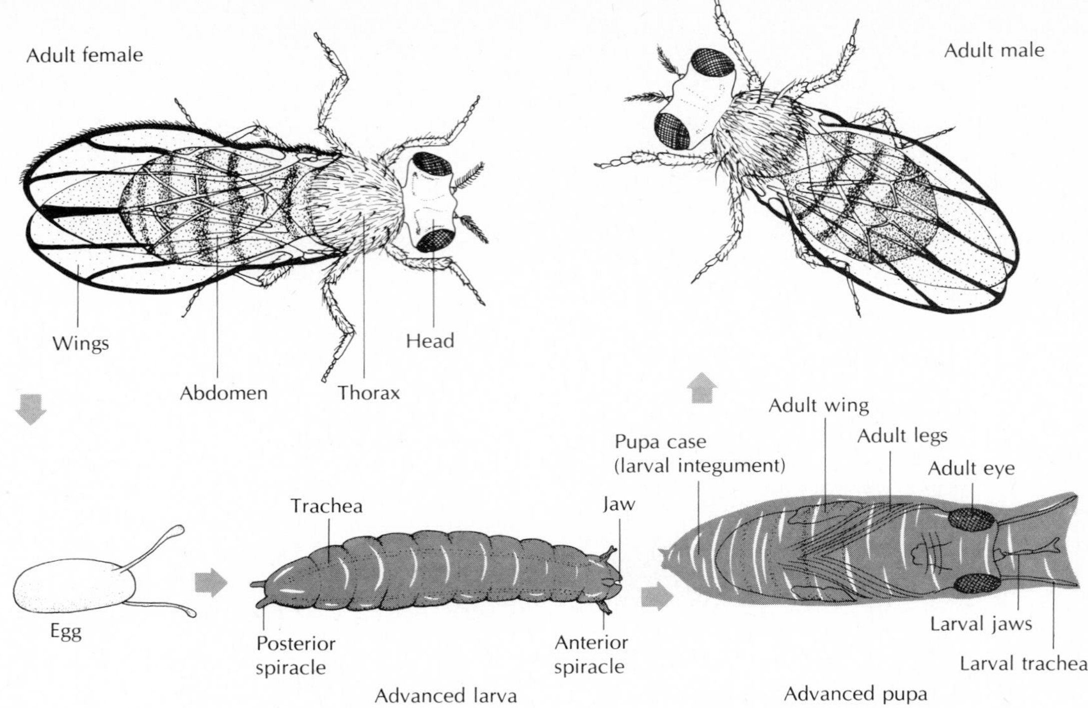

Figure 30.57 Life cycle of the fruit fly.

into those of the adult. The nervous system reorganizes, compound eyes form, musculature modifies, reproductive organs become functional structures. When these changes are complete, the pupal case splits and the adult emerges. After the wings dry a little, it flies away.

Insects which undergo complete metamorphosis are grouped together in the superorder Endopterygota (Holometabola).

Mechanism of Molting

Within the animal kingdom the term molt describes a dramatic shedding of skin or feathers by birds, reptiles, and arthropods, a process followed by a new growth. As mentioned above, molting in insects occurs in the larval or nymphal stage: the insect produces a new skin or cuticle underneath the old one, which splits along the dorsal side and the insect crawls out. Insects differ in the number of molts.

The mechanism of molting has been studied in a number of insects including the fruit fly, *Drosophila melanogaster,* and the silkmoth, *Antheraea polyphemus.* The fruit fly undergoes two molts, thereby dividing the larval stage into three periods or instars: egg $\longrightarrow$ first instar larva $\xrightarrow{\text{first molt}}$ second instar larva $\xrightarrow{\text{second molt}}$ third instar larva $\xrightarrow{\text{metamorphosis}}$ pupa. Figure 30.58 shows electron micrographs of cross sections of the fruit fly during the second molt. The events seem to be as follows.

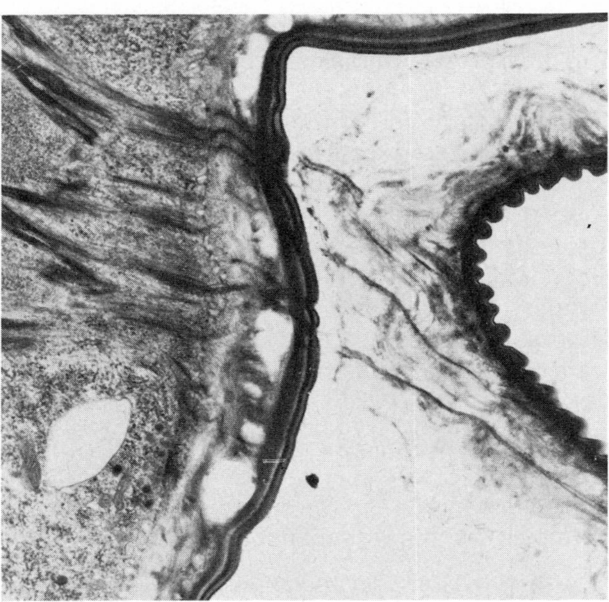

Figure 30.58 Larval cuticle of second instar of *Drosophila melanogaster* 70 hours from egg laying at 25° C. New epicuticle of the third instar larva is almost complete (center of photo). The space between the new epicuticle and old epicuticle is filled with molting fluid which has digested almost all of the chitin and muscle fibers. At 72 hours the old epicuticle will split and the larva, now a third instar, will crawl out of the old "skin." (Photo courtesy of Professor H. K. Mitchell, The California Institute of Technology.)

1. Epidermal cells secrete digestive enzymes which begin the destruction of the old cuticle, composed mainly of proteins and chitin.
2. As digestion of the old cuticle occurs, the epidermal cells then begin building up the new cuticle.
3. By the time that the old cuticle is thin and ready to split open, the new cuticle is laid down sufficiently to protect the third instar larva.

The events involved in shedding and synthesis of the old and new cuticles provide an interesting system of cellular regulation. What are the events in the epidermal cells that permit them to destroy the old cuticle and, soon afterwards, reverse their roles and begin the synthesis of new cuticle?

Insect Hormones

The process of molting and metamorphosis in insects is dependent upon several hormones. Work on the *Cecropia* moth has shown that three different hormones are involved in its growth and metamorphosis (Fig. 30.59). This insect undergoes a long period of dormancy (diapause) in the pupal stage. Carroll Williams found that, after chilling the pupae for 6 weeks and then returning them to normal temperature, they would metamorphose in about a month. Two hormones are demonstrated to be involved. The first is produced by certain cells in the brain and stored in the *corpus cardiacum,* located ventral to the brain in association with the hypocerebral ganglion. When released from the corpus cardiacum, the brain hormone stimulates the prothoracic glands to secrete the growth and differentiation hormone called *ecdysone.* This hormone causes the epidermis to secrete a molting fluid so that molting and metamorphosis occur. These hormones are not secreted in diapause.

Both of these hormones are secreted during the larval period; a third hormone also functions

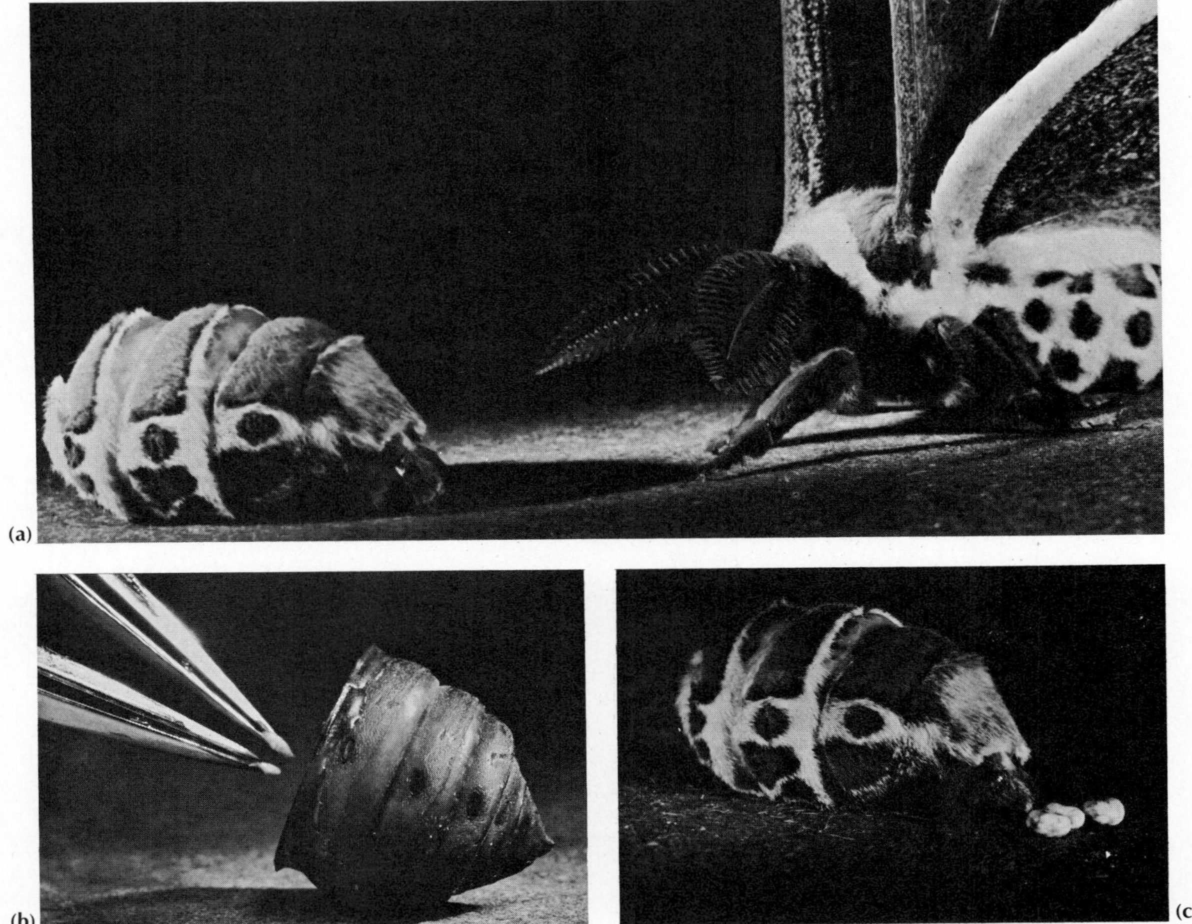

Figure 30.59 Hormone experiments on *Cecropia* moths. C. Williams of Harvard University has determined that two hormones are necessary for metamorphosis. When he transplanted the hormone-producing tissues (brain and prothoracic glands) separately into tail sections of female pupae, no changes occurred. But when both were implanted **(b)** it not only matured but also attracted a male **(a),** was fertilized, and laid eggs **(c).** (Photographs by R. Vishniac and C. Williams.)

in this period. It is known as the *juvenile hormone* and is produced in two tiny glands just posterior to the brain, the *corpora allata*. This hormone stimulates larval development, but inhibits metamorphosis. Figure 30.60 shows the scheme of action of the hormones involved in molting and metamorphosis. As long as the juvenile hormone is present in adequate amounts, ecdysone promotes larval growth. When the amount of the juvenile hormone is reduced, ecdysone causes the development of the pupa, and when the juvenile hormone is absent, the adult form differentiates. There is cessation of the secretion of this hormone in the last larval stage. When the corpora allata are removed from a young larva, it undergoes pupation at the next molt. When extra corpora allata are implanted in a mature larva, it does not metamorphose but, instead, molts and forms an extra large larva. These hormones have been identified in a number of insects, and the mechanism seems to be much the same in all insects.

Ecdysone has been chemically identified as a sterol and it has been obtained in pure form. In one experiment in which small amounts of ecdysone were injected into a midge, chromosome puffs were observed to form within 15 minutes at certain sites on salivary gland chromosomes. Puff formation is taken as an indication of the activation of a certain gene or genes. This is one bit of evidence suggesting that certain genes may be activated by hormones.

Pheromones

As mentioned in Chapter 12, insects produce hormonelike substances which have their action outside the body of the animal that produces them. These substances, called *pheromones,* affect other members of the same species in a number of very specific ways.

Some pheromones are used as trail markers leading members of a colony to food sources or to the nest. Many ants produce such pheromones. The mandibular glands of the queen honey bee produce a so-called "queen substance," identified as 9-oxo-*trans*-2-deconic acid. This material serves to integrate the ac-

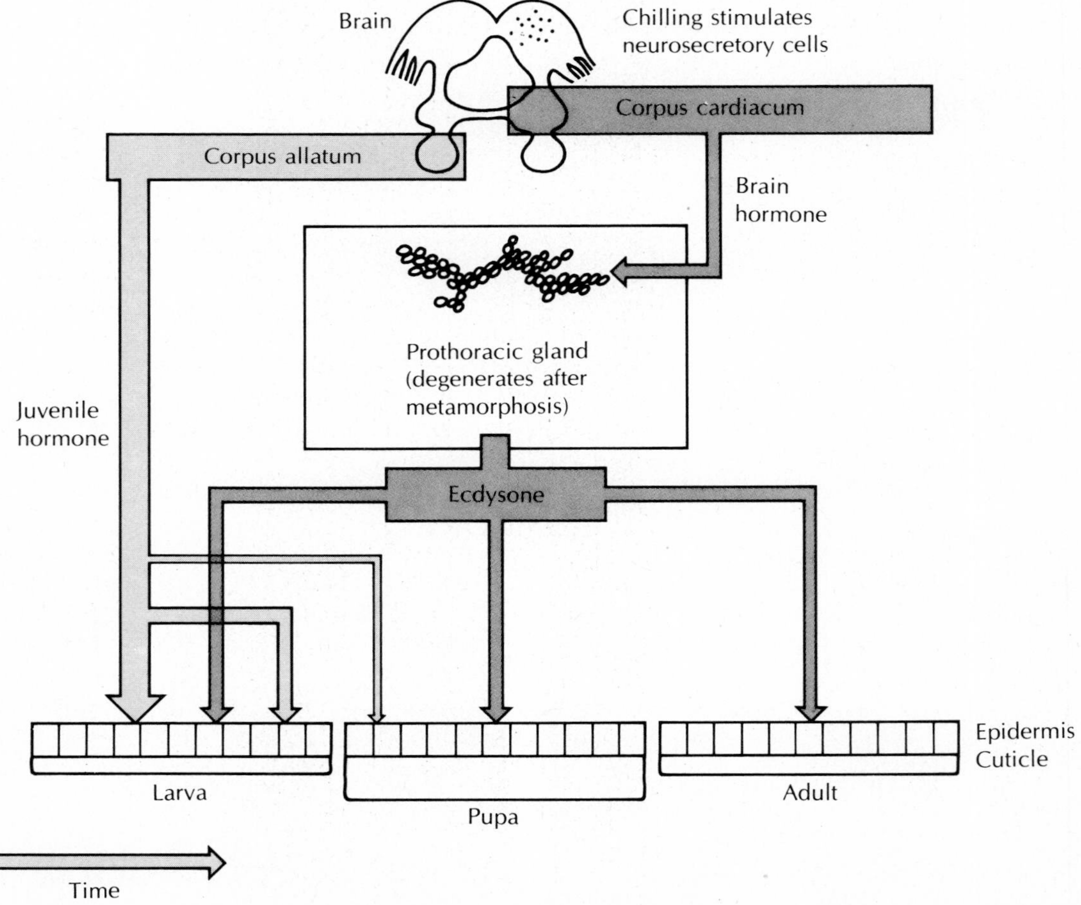

Figure 30.60 Scheme of hormone action in metamorphosis of the silkmoth. Varying proportions of juvenile hormone and ecdysone influence the kind of structures that will be synthesized by the cells.

tivities of the worker bees who obtain it by licking the body of the queen. The queen substance is also the sex attractant which brings the drones to the queen in her nuptial flight.

The other social insects, ants and termites, also produce pheromones which are important in integrating the activities of the colony. Each colony has its own identifying pheromone, and other ants of the same species but a different colony are usually viciously attacked. Many types of highly specialized guests have evolved numerous mechanisms which enable them to live unharmed in these complex societies. Some beetles living in ant colonies produce a pheromone quite similar to that produced by the host species and thus ward off attacks.

When honeybees sting an object, they release a pheromone which causes other bees to attack the same object. Bees have another pheromone which releases a strong alarm behavior in other bees.

Many insects in a large number of orders have been shown to produce a pheromone which attracts members of the opposite sex; usually the female attracts the male, often from great distances. Males also have sex pheromones, usually exciting the female to induce copulation. The sex-attracting pheromones have been the object of intensive study because of their great potential in the control of destructive insects. One of the most successful has been the control of the gypsy moth, *Porthetria dispar.* A compound, *cis*-7,8-epoxy-2-methyloctadecane, named "disparlure," which effectively attracts males of this species to traps, has been widely used.

Importance of Insects to Humanity

Reference has already been made to some of the ways in which insects affect people. Some insects are beneficial, but most are injurious. Among the beneficial ones are those that produce honey and silk, those that pollinate flowers, and those that destroy injurious insects. Bees are probably the most important insect pollinizers. Many plants are completely dependent upon the activity of insects for cross-fertilization. Some of the common forms important in the destruction of harmful insects are ground beetles, tiger beetles, ladybird beetles, and ant lions.

The harmful insects may be divided roughly into those that damage agricultural crops and livestock, those that cause damage or annoyance in homes, and those that transmit diseases. The United States Department of Agriculture estimates that the annual damage of insect pests to crops, stored foods, forests, and domesticated animals amounts to over 4 billion dollars.

The list of insects that contribute to the damage of plants and plant products includes the following: grasshopper, army worm, potato beetle, European corn borer, Japanese beetle, Mediterranean fruit fly, Hessian fly, cotton boll weevil, wood-boring beetles, alfalfa weevil, corn-ear worm, flour moth, gypsy moth, browntail moth, plant lice, leaf hopper, stink bug, scale insect, and chinch bug. Damage to domestic animals is caused by biting lice, sucking lice, horn flies, the larvae of botflies, and the larvae of ox warble flies. The following are common household pests: termites, cockroaches, clothes moths, ants, flies, bedbugs, and mosquitoes.

Although the damage to possessions due to injurious insects is tremendous, the harmful effects to society from the insects that transmit disease is as great. Someone has said that, with respect to the whole course of human civilization, malaria is the most important human disease. As mentioned earlier, this disease is caused by a protozoan transmitted by mosquitoes. Although it is now known that the disease can be controlled by draining swamps and other places where the mosquito lays its eggs, in certain places in the Orient this disease still claims the lives of many people each year. It is estimated that in India over a million people die because of malaria annually. The works of Ross and Manson, in establishing the role of the mosquito in this disease, form interesting chapters in the contribution of biology to human welfare.

Important, too, was the work of Walter Reed and others in the conquest of yellow fever. Through their researches it was established that the virus that causes yellow fever is transmitted by the stegomyia mosquito. There seems to be little doubt that the building of the Panama Canal was only made possible through this work that led to the control of yellow fever.

There are several other human diseases that are known to be transmitted by insects, the most important of which are typhoid fever, elephantiasis, bubonic plague, typhus fever, and African sleeping sickness. The bacterium that causes typhoid fever may be carried by the common housefly. Elephantiasis, a disorder caused by a parasitic roundworm that clogs the lymphatic circulation in the body, is transmitted by the bite of a mosquito. Bubonic plague, or Black Death, is caused by a bacterium carried by fleas. The causative agents of typhus fever are the tiny intracellular parasites, rickettsias, transmitted by lice. Flagellated protozoans, trypanosomes, are the cause of African sleeping sickness, and they are introduced into the human body in the bite of the tsetse fly.

Partly because of the great economic importance of insects to humans, *entomology,* which focuses on the study of insects, is a well-developed branch of biology. The United States Department of Agriculture, the health services, both state and national, and the pest control industry employ many entomologists who devote their talents to the control of insects. In addition to entomologists, many research chemists are participating in the fight to control insects by developing newer and safer insecticides (insect poisons) to use in this fight. Every reader is probably acquainted with the use of DDT, chlordane, and lindane, all of which have become so effective in the control of insects. As strains of insects appeared which were resistant to some or all of the above materials, newer types of insecticides, including a whole family of organic phosphate compounds, were developed. Because of evidence of the buildup of certain insecticides in the soil and concentration of these insecticides in animal tissues, there has been great concern over the indiscriminate use of insecticides in recent years and some, such as DDT and dieldrin, have been banned for most uses. The pressure for the prohibition of certain insecticides because of potential adverse environmental effects is in certain respects paradoxical. For example, some years ago, as an economic measure, the government of Ceylon decided to stop using DDT for mosquito control because malaria seemed to be well under control. Shortly thereafter the incidence of malaria rose sharply with a concomitant increase in the human death rate. In these days of worldwide food shortages it is improbable that the removal of effective insecticides from the market can continue without serious results, unless alternate methods of control are developed. Great emphasis is now being placed on nonchemical methods of insect control, including biological control through the use of predators and pathogens which affect pest species of insects and sex pheromones to attract insects to traps. However, these methods have not been brought to a state of perfection capable of handling the insect problem to the exclusion of proven chemical methods.

PHYLOGENETIC RELATIONSHIPS OF ARTHROPODS

The arthropods are clearly a well-defined assemblage of animals, but the actual origin and evolution of the major groups is not well understood. There are a number of points of view and no general agreement on the question. A major problem hinges on the basic arthropod origins. Is the phylum monophyletic, that is, does it stem from a single ancestral type; or is it polyphyletic with more than one ancestral type?

The evidence is pretty clear that the arthropods arose from wormlike ancestors very similar in organization to polychaete annelids. The ancient hypothetical progenitors had a body consisting of a series of similar segments. The head was simple and probably bore sensory bristles. The ventral mouth was located between the head and the first segment of the body. The appearance on each segment, except the last, of a pair of walking legs, unsegmented in the primitive forms, and the development of eyes and antennae on the head produced an organism having characteristics much like those of *Peripatus.* From such an ancestral form it is conceivable that the myriapods and insects evolved. Embryological evidence as well as structural features such as trachea and an open circulatory system point to a relationship among these groups.

Their aquatic habitat as well as many strucstructural and developmental features make the position of the crustaceans in close relationship to the myriapods and insects difficult to defend. The origin of the crustaceans and the homogeneous chelicerate group from a trilobitelike ancestor is appealing to many zoologists. Others look to a common ancestor for all three groups: the trilobites, the crustaceans, and the chelicerates. Some prefer to remove the peculiar pycnogonids from the chelicerates and place them in a separate subphylum.

The above concepts involving polyphyletic origins of the arthropods are gaining more favor. They necessitate the acceptance of the independent appearance of *"arthropodization"* at least twice in the course of the evolution of these groups. Arthropodization is the development of a more or less rigid exoskeleton and jointed appendages. Figure 30.61 shows two of the current concepts of arthropod origins and evolution.

Based on the concept of polyphyletic origins of the major arthropod groups, several workers have suggested that the phylum Arthropoda should be broken into three separate phyla: the phylum Uniramia (insects and myriapods), the phylum Crustacea, and the phylum Chelicerata. Such a major change will necessitate much further study and discussion before any general agreement is reached.

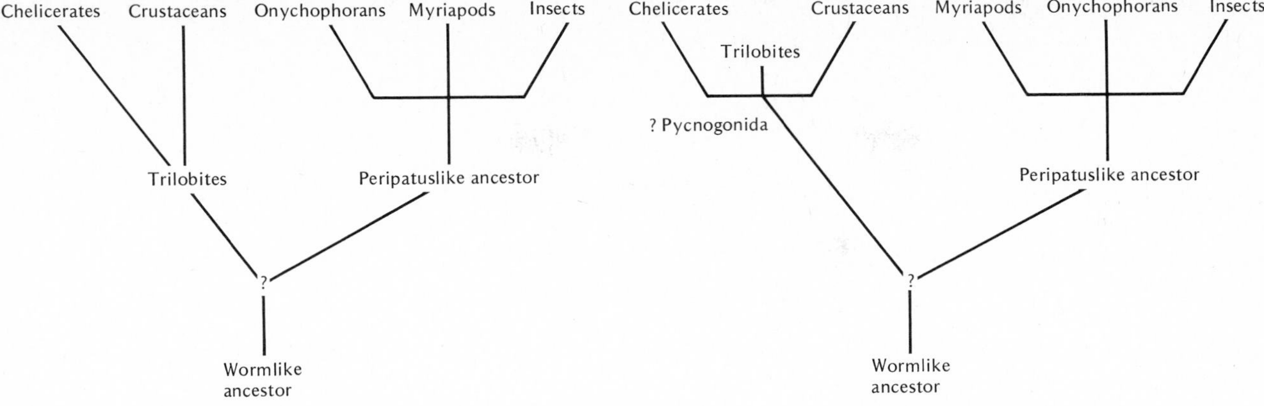

Figure 30.61 Two concepts of the origin of the arthropods.

SUGGESTIONS FOR FURTHER READING

Anderson, D. T. *Embryology and Phylogeny in Annelids and Arthropods.* New York: Pergamon (1973).

Barnes, R. D. *Invertebrate Zoology,* 3rd ed. Philadelphia: W. B. Saunders (1974).

Borror, D. J., D. M. DeLong, and C. A. Triplehorn. *An Introduction to the Study of Insects,* 4th ed. New York: Holt, Rinehart and Winston (1976).

Buchsbaum, R. *Animals Without Backbones.* Chicago: Univ. of Chicago Press (1949).

Herms, W. B., and M. T. James. *Medical Entomology,* 4th ed. New York: Macmillan (1961).

Jacobson, M. *Insect Sex Pheromones.* New York: Academic Press (1972).

Metcalf, C. L., and W. P. Flint. *Destructive and Useful Insects,* 4th ed. New York: McGraw-Hill (1962).

Romoser, W. S. *The Science of Entomology.* New York: Macmillan, (1973).

Ross, H. H. *A Textbook of Entomology,* 3rd ed. New York: Wiley (1965).

Wheeler, W. M. *The Social Insects.* New York: Harcourt, Brace and World (1928).

Wigglesworth, V. B. *Insect Physiology,* 6th ed. New York: Wiley (1966).

Williams, C. *Morphogenesis and the Metamorphosis of Insects,* Harvey Lectures, vol. 47. New York: Academic Press (1953).

Wilson, E. O. *The Insect Societies.* Cambridge: Harvard Univ. Press (1971).

From Scientific American

Beck, S. D. "Insects and the Length of Day" (Feb. 1960).

Edwards, J. S. "Insect Assasins" (June 1960).

Hocking, B. "Insect Flight" (Dec. 1958).

Hovanitz, W. "Insects and Plant Galls" (Nov. 1959).

Jacobson, M., and M. Berozoa. "Insect Attractants" (Aug. 1964).

Johnson, C. G. "The Aerial Migration of Insects" (Dec. 1963).

Lekh, R., and S. W. T. Batra. "The Fungus Gardens of Insects" (Nov. 1967).

Milne, L. J., and M. J. Milne. "Insect Vision" (July 1948).

Smith, D. S. "The Flight Muscles of Insects" (June 1965).

Smith, R. F., and W. W. Allen. "Insect Control and the Balance of Nature" (June 1954).

Williams, C. M. "The Metamorphosis of Insects" (Apr. 1950).

Williams, C. M. "Insect Breathing" (Feb. 1953).

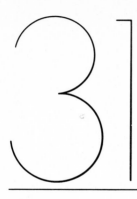

Lophophorates–

PHORONIDA, ECTOPROCTA, AND BRACHIOPODA

The three coelomate phyla to be discussed in this chapter, the Phoronida, Ectoprocta, and Brachiopoda, are clearly related because they all possess a unique structure, the *lophophore.* A lophophore is a tentaculated extension of the anterior portion of the body which embraces the mouth and not the anus, and which also contains an extension of the coelom. This excludes other types of tentacles associated with the mouth, in particular those of the pseudocoelomate Entoprocta, and restricts the term lophophore to the three phyla under discussion in this chapter.

PHYLUM PHORONIDA

Less than 20 species make up this phylum of marine tube-dwelling worms.

Characteristics

1. Small unsegmented marine worms which live in secreted chitinous tubes.
2. True coelom present.
3. Horseshoe-shaped lophophore.
4. Closed circulatory system.
5. U-shaped digestive tract with mouth located in the lophophore ring, but anus located outside of it. Figure 31.1

The phylum is not broken down into classes and orders.

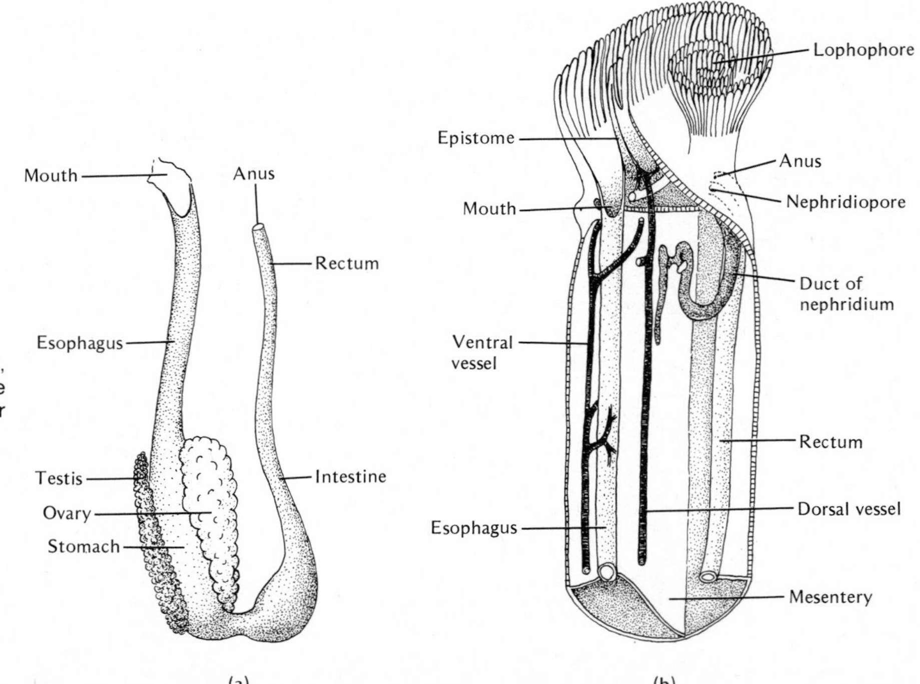

Figure 31.1 Phylum Phoronida, anatomy. **(a)** Diagram of digestive system. **(b)** Cut through anterior portion of an adult worm.

Anatomy and Physiology

The phoronids live in shallow water in tubes made of chitin which are secreted by the ectodermal cells. The tubes are either buried in sand or mud or attached to submerged objects. The characteristic circle of ciliated tentacles, called the *lophophore,* protrudes from the open end of the tube.

The body wall is made up of a cellular epidermis beneath which are a basement membrane, a layer of circular muscles, and a layer of longitudinal muscles. Internal to the longitudinal muscles is the peritoneum which is a syncytium. The coelom is divided into an anterior *mesocoel* and a posterior *metacoel* by a peritoneal septum. The nervous system consists of a nerve ring in the epidermis at the base of the lophophore with an enlarged portion which is the main neural center. Nerves lead from the nerve ring to the tentacles and the body wall. The digestive system is U-shaped, starting with the mouth in the lophophore ring which receives food filtered from the water by the tentacles and passed to it by strong ciliary currents. Following the mouth, food passes by means of the ciliated lining to the esophagus, stomach, and intestine. The anus is anterior and outside of the lophophore ring. Digestion is largely intracellular. The wall of the gut is made up of the mucosal lining, muscles and connective tissue, and an outer peritoneum.

There is a well-developed closed circulatory system with dorsal and ventral contractile vessels and red corpuscles containing hemoglobin. Excretion is by a pair of metanephridia which, it will be recalled, differ from the protonephridia of lower forms in having two openings—one into the coelom and one to the outside. The opening into the coelom is a ciliated funnel, called the *nephrostome,* and the two nephridial pores are located near the anus. The nephridial ducts also serve as pathways for the eggs and sperm. Most phoronids are hermaphroditic, with eggs and sperm being formed from cells which line the coelom. Fertilization may be internal or external. Cleavage varies, with some species showing spiral cleavage, others radial, while others have an irregular cleavage pattern. Mesoderm arises by outgrowth of cells from the endoderm, and the coelom arises from mesenchyme cells which migrate and form a hollow saclike structure. There is a trochophore-like larva called an *actinotroch* (Gk. *aktis,* ray; *trochos,* wheel) (Fig. 31.2). The ciliated lobes at the apical end resemble the spokes of a wheel.

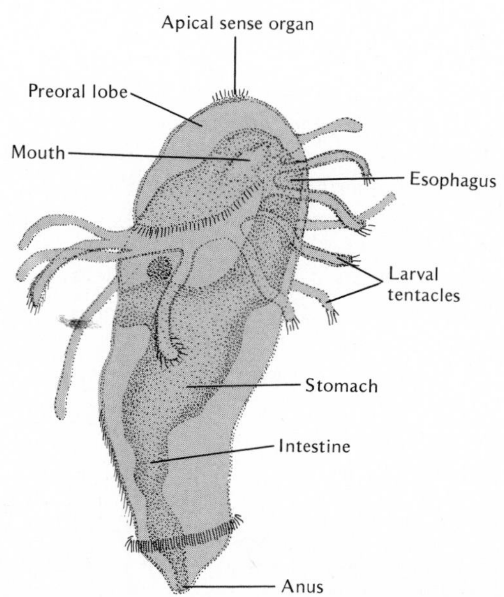

Figure 31.2 Actinotroch larva of a phoronid worm.

PHYLUM ECTOPROCTA (BRYOZOA)

This rather large phylum of microscopic colonial animals is mostly marine in distribution; less than 50 species are found in fresh water.

Characteristics

1. Microscopic, sessile, and colonial.
2. Lophophore present.
3. Circulatory and excretory systems absent.

Classification

CLASS **Gymnolaemata** Marine with few exceptions. Lophophore circular. No direct connections between the coelomic cavities of zooids. Exoskeleton calcareous or membranous (Fig. 31.3).

ORDER **Ctenostomata** Exoskeleton membranous. Orifice with pleated collarlike closing device. Without brood chambers. *Clavopora, Victorella.*

ORDER **Cheilostomata** Boxlike zooids, usually with calcareous exoskeleton. Orifice with a hinged operculum. Usually with avicularia or vibracularia, or both. *Bugula, Membranipora.*

ORDER **Cyclostomata** Tubular zooids. Calcareous exoskeleton. No operculum or avicularia. *Tubulipora.*

CLASS **Phylactolaemata** Entirely freshwater. Lophophore horseshoe shaped. Connections present between coelomic cavities of the zooids. Exoskeleton gelatinous. This small class is not broken down into orders. Representative genera, *Pectinatella, Plumatella* (Fig. 31.4).

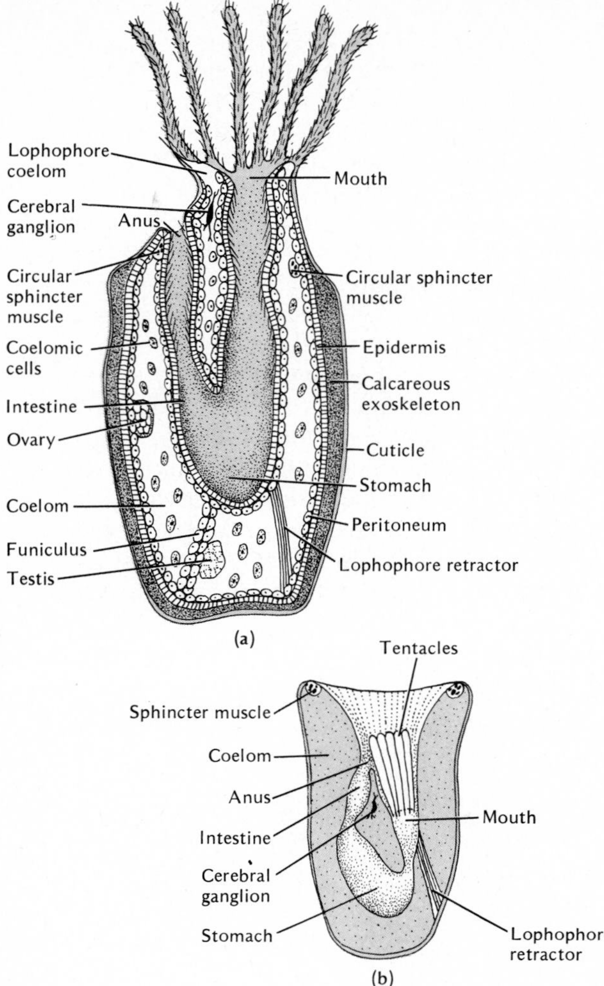

Figure 31.3 Class Gymnolaemata. A marine ectoproct. **(a)** Detailed anatomy of a single zooid. **(b)** Diagram showing retracted zooid.

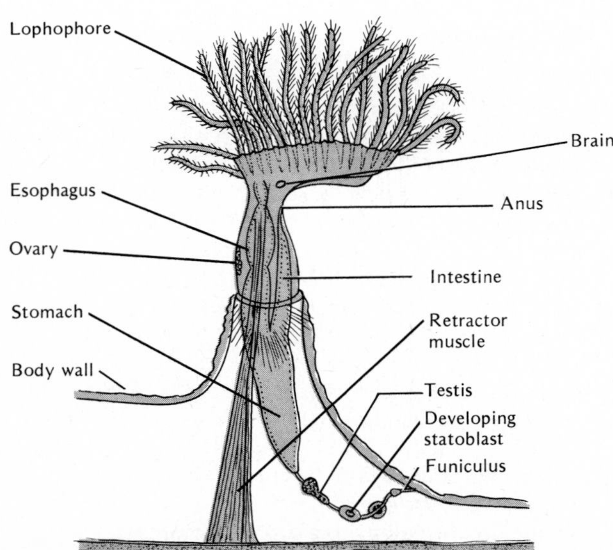

Figure 31.4 Class Phylactolaemata. *Plumatella* sp., a freshwater ectoproct.

Anatomy and Physiology of Class Gymnolaemata

The bulk of the phylum is found in this class, that is, about 4000 living species and more than 15,000 described fossil forms. The colonies may reach considerable size and vary in form from branching upright types to flattened mats encrusting surfaces of rocks, algae, or other animals. Many species are polymorphic with several kinds of *zooids* making up the colony. The basic feeding zooid is diagrammed in Figure 31.3. The exoskeleton secreted by the zooids is a boxlike structure with an opening on one side through which the lophophore protrudes. The body wall consists of an epidermis that secretes the skeleton, which may be entirely of chitin or may have a calcium carbonate layer below the thin chitinous cuticle, and a lining peritoneum. Body wall musculature is lacking in members of this class.

The tentacles of the lophophore are arranged in a single circular row around the mouth opening. The lophophore can be retracted completely by action of the powerful retractor muscles. Various mechanisms for protraction of the lophophore are found in members of the phylum. They all utilize hydrostatic pressure from the coelomic fluid. Cilia on the inner surface of the tentacles set up currents which take food particles to the mouth. The pharynx and intestine are ciliated, but the esophagus and stomach are not. Movement of food is accomplished by action of the muscular fibers in the gut wall and ciliary action. Digestion is both intracellular and intercellular, with enzymes secreted into the lumen by the gland cells in the lining epithelium. The gut is covered with a thin peritoneum, and the extensive fluid-filled coelom lies between it and the body wall. The stomach is anchored to the body wall by a cordlike structure, called the *funiculus.* The nervous system is quite simple with a cerebral ganglion dorsal to the pharynx, a nerve ring around the base of the lophophore, and a plexus beneath the epidermis.

Although there is no excretory system, an interesting feature of most marine species is the periodic formation of a *brown body.* Most of the soft parts of the zooid degenerate into a brownish shapeless mass within the coelom of the former individual. The remaining body wall regenerates the lost organs with the brown body coming to lie within the stomach of the new individual. Upon completion of the new zooid, the brown body is passed out via the anus. It is believed that excretory wastes build up in the tissues of the zooid, particularly in the stomach epithelium, and thus this peculiar mechanism plays a role in excretion. There is no circulatory system and respiration takes place by simple diffusion.

As mentioned above, many species are polymorphic. Several types of specialized zooids are found, including the *avicularium,* a structure much like a bird's beak which seems to function largely in protection of the colony from other organisms which might settle on it; and the *vibracularium* with a long vibratile extension which moves over the surface of the colony and probably serves a function similar to that of the avicularium (Fig. 31.5).

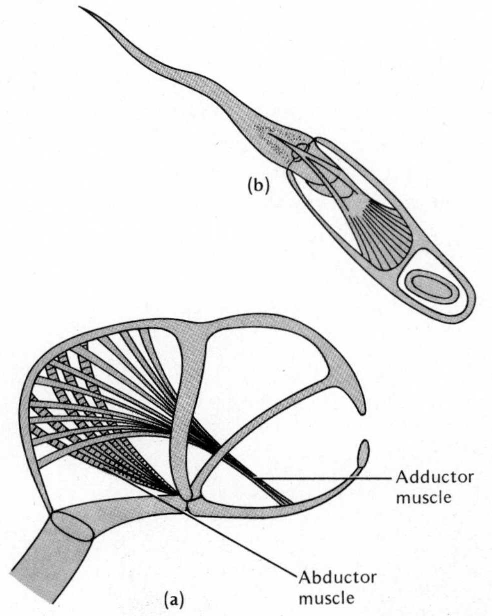

Figure 31.5 Polymorphism in marine ectoprocts. **(a)** Avicularium of *Bugula.* **(b)** Vibracularium of *Scrupocellaria.*

Most species are hermaphroditic. Gonads develop on the peritoneal lining, and there are no special ducts or openings to the outside. Release is by rupture of the body wall. Fertilization occurs internally and some species release the fertilized eggs into the sea, but most of them brood their eggs in the coelom or in invaginated areas near the tentacles. Cleavage is radial, and a highly modified trochophore larva results (Fig. 31.6). The larva is called a *cyphonautes larva,* and the form varies considerably in different species. After a free-swimming existence of as much as 2 months, the larva settles down on a suitable substrate and undergoes a very extensive metamorphosis to become an adult zooid. The larva practically disintegrates, leaving only a small heap of ectoderm-covered cells which form the new individual. Thus only the larval ectoderm carries through to the adult and produces all the adult structures. The original zooid, called an *ancestrula,* is the source of the new colony which enlarges by asexual budding.

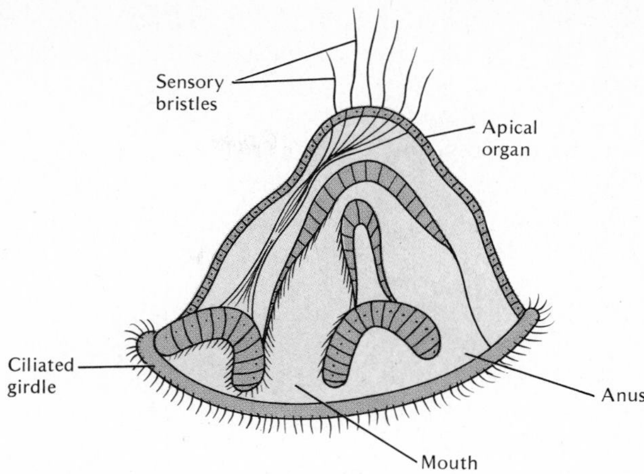

Figure 31.6 Cyphonautes larva of a marine ectoproct.

Anatomy and Physiology of Class Phylactolaemata

Anatomically the freshwater Phylactolaemata do not differ greatly from the gymnolaemata (Fig. 31.4). The exoskeleton is always gelatinous, the coelomic cavities of adjacent zooids are interconnected, and the lophophore is horseshoe shaped rather than circular. The body wall also differs in that there are circular and longitudinal muscles. Development takes place in an *embryo sac* which invaginates into the coelom. A small ciliated colony is produced which is released and soon settles to the bottom and grows by asexual budding.

The members of this class have an additional method of asexual reproduction which occurs at the onset of unfavorable conditions such as drought or cold weather. A resistant structure, called a *statoblast,* is formed on the funiculus (Fig. 31.7). Epidermal cells migrate to this region and surround a mass of peritoneal cells. The entire mass then secretes a chitinous covering, often furnished with hooks or containing air

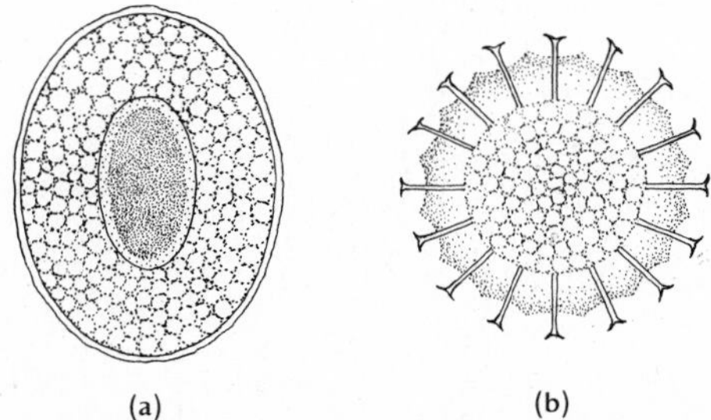

Figure 31.7 Statoblasts of freshwater ectoprocts. **(a)** Floating type with air-filled sacs (*Hyalinella*). **(b)** Statoblast with hooks (*Cristatella*).

spaces which cause them to float. Each species has a characteristic type of statoblast, and they are used in the classification of the group. Following the unfavorable period during which the statoblast remains dormant, the statoblast opens and a new zooid forms from the cells within. By asexual budding, a new colony results.

PHYLUM BRACHIOPODA

This phylum has a very ancient lineage and has been very prominent in the fossil record. Over 30,000 fossil species have been described, but there are less than 300 species alive today.

Characteristics

1. Lophophore present.
2. Bivalve shells, dorsal and ventral in position, with an extensive mantle on the inner surface of each valve.
3. Poorly developed, open circulatory system.
4. Sessile as adults, attached to the substratum by a stalklike *peduncle.*
5. Entirely marine.

Classification

CLASS **Inarticulata** Valves held together by muscles only; lophophore lacks an internal skeleton. Anus present (Fig. 31.8).

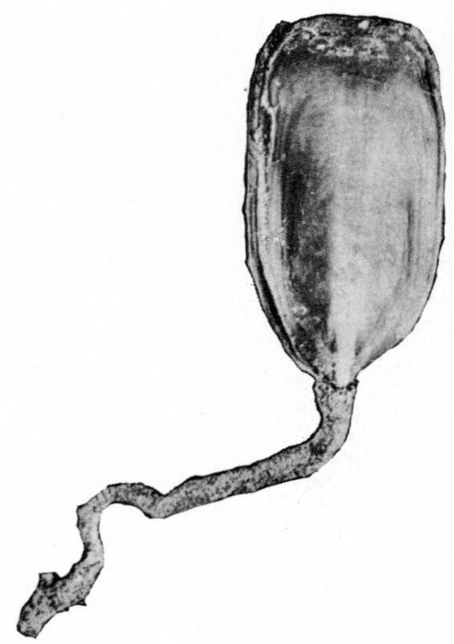

Figure 31.8 Class Inarticulata, *Lingula* sp. (From Smithsonian Institution, with permission.)

ORDER **Atremata** Peduncle attached to the ventral valve, but both valves are involved in its passage to the outside. *Lingula.*

ORDER **Neotremata** Peduncle emerges from an opening confined to the ventral valve. *Crania.*

Figure 31.9 Class Articulata, *Terabratulina* sp. (From Smithsonian Institution, with permission.)

CLASS **Articulata** Valves held together by a tooth and socket arrangement. Lophophore with an internal skeleton. Anus absent. *Terebratulina* (Fig. 31.9). This class is not broken down into orders.

Anatomy and Physiology

The anatomy of a typical articulate brachiopod is diagrammed in Figure 31.10. The bivalve shell differs from that of the pelecypod mollusks in having dorsal and ventral valves rather than lateral as in the clams and oysters. As in the pelecypods, the shell is secreted by thin extensions of the body wall, called the *mantle.* The shell has a thin external layer of chitin under which is a thicker layer containing calcium carbonate. The main body of the animal fills the posterior third of the space within the valve, and the anterior *mantle cavity* contains the two coiled arms of the lophophore. A pair of calcium carbonate supports for the lophophore arm extend from the dorsal valve. Between the bases of the lophophore arms is lo-

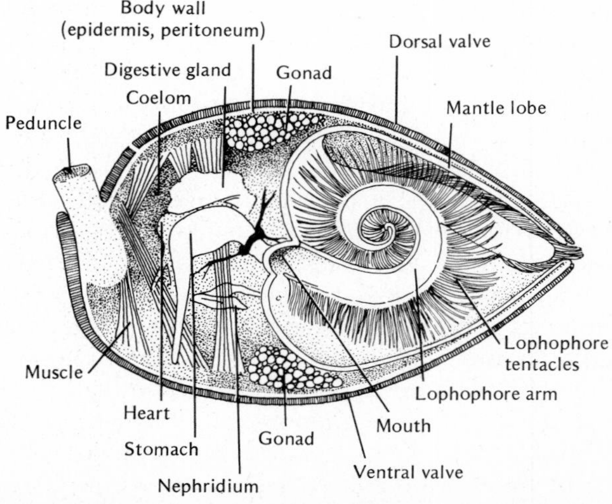

Figure 31.10 Anatomy of an articulate brachiopod.

cated the slitlike *mouth* which leads into the ciliated digestive tract that consists of a saclike stomach with a large digestive gland and an intestine which ends blindly. The open circulatory system is little more than a contractile *heart* and an anterior and posterior channel which empty into open sinuses. The blood lacks pigment and contains few corpuscles.

The nervous system consists of an esophageal nerve ring with dorsal and ventral ganglionic thickenings. Nerves extend anteriorly to the mantle and lophophore as well as posteriorly to the valve muscles. Specialized sensory structures are lacking in most species. Excretion is by means of one or two pairs of metanephridia with nephrostome funnels opening into the coelom and short tubules leading to the nephridial pores located lateral to the mouth opening. The sexes are usually separate, and gonads, usually four in number, develop in the coelomic cavity attached to the peritoneum. The nephridial tubules form the exit pathway for the gametes. Fertilization may occur within the mantle cavity or externally in the sea. In the former case, early development may occur within the mantle cavity. Cleavage is radial and mesoderm and coelom are formed by *enterocoelic pouches*—evaginations of the primitive gut. The larva is quite distinctive. It swims around briefly, settles down on its *peduncular lobe,* and the mantle folds reverse position to cover the mouth (Fig. 31.11).

The inarticulata differ from the articulata in the nature of the valves. The valves are chitin or chitin and calcium phosphate. They also lack a specialized hinge. In addition an anus is present, opening into the mantle cavity, and there are no skeletal elements in the lophophore. *Lingula,* an inarticulate brachiopod (Fig. 31.8), has the distinction of being a member of one of the oldest living genera. Fossil species of this genus are known at least as far back as Ordovician times, over 400,000,000 years ago.

PHYLOGENETIC RELATIONSHIPS OF LOPHOPHORATES

The lophophorates seem to be in a transitional position since they have some protostome characteristics and some deuterostome characteristics. In echinoderms and hemichordates the larval coelom is divided into three compartments, the *protocoel, mesocoel,* and *metacoel.* These three compartments are also found in the coelom of lophophorates, although the protocoel is reduced in correlation with the absence of a well-developed head region. Coelom formation is also intermediate since in the phoronids and ectoprocts it is neither schizocoelous nor enterocoelous, while in the branchiopods it is enterocoelous in some and schizocoelous in others. Cleavage is radial in ectoprocts and brachiopods and spiral in some phoronids. The 4d cell, which is the source of all or most of the mesoderm in protostomes, is not found in any of the lophophorates. The blastopore becomes the mouth in phoronids and brachiopods, and these two phyla also have trochophorelike larvae. The cyphonautes larva of the ectoprocts is much less trochophorelike.

For all the reasons given above it is believed that the lophophore phyla were derived from a line which gave rise to both the protostome coelomates and the deuterostomes. It will be seen in the hypothetical phylogenetic tree developed in Chapter 39 that the branching into the two main lines of coelomate phyla takes place at the level of the lophophore phyla.

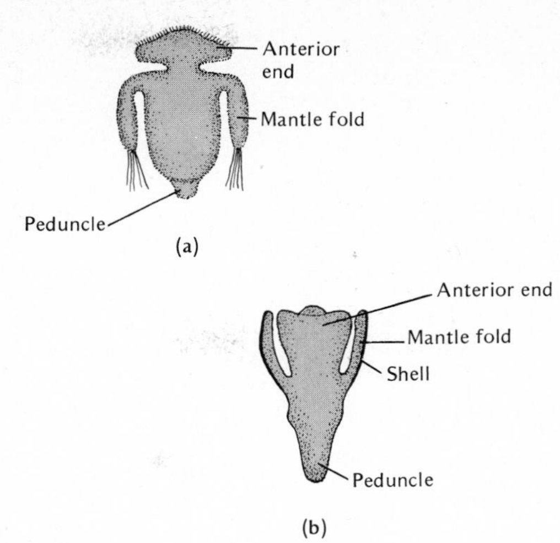

Figure 31.11 Metamorphosis of a brachiopod larva. **(a)** Free swimming stage. **(b)** Folding of mantle over the anterior end when the larva settles to the bottom.

SUGGESTIONS FOR FURTHER READING

Barnes, R. D. *Invertebrate Zoology,* 3rd ed. Philadelphia: W. B. Saunders (1974).

Hyman, L. H. *The Invertebrates: Smaller Coelomate Groups,* vol. 5. New York: McGraw-Hill (1959). (All three lophophorate phyla are covered in the volume.)

MacGinitie, G. E., and N. MacGinitie. *Natural History of Marine Animals.* New York: McGraw-Hill (1967).

McConnaughey, B. H. *Introduction to Marine Biology.* St. Louis: C. V. Mosby (1974).

Meglitsch, P. A. *Invertebrate Zoology,* 2nd ed. New York, Oxford Univ. Press (1972).

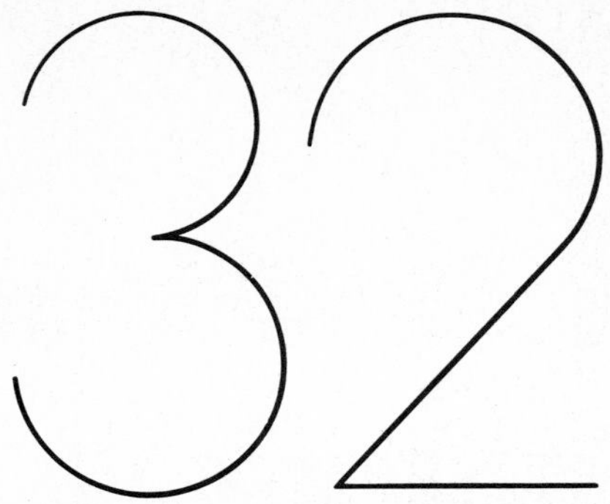

32 Echinoderms

THE DEUTEROSTOMIA

The common characteristics uniting the Deuterostomia, which at first glance may appear to be quite unrelated, are as follows. The cleavage pattern is basically radial and indeterminate, that is, the fate of the blastomeres is not determined at a very early stage. The mesoderm is derived from cells of the archenteron which evaginate to form the enterocoelic coelom. The anus arises from or near the blastopore. If larval forms are present, they are not trochophores. All these features are in contrast to those of the protostomous coelomates.

This chapter will deal with the phylum Echinodermata; the next will discuss three small phyla, the Chaetognatha, Pogonophora, and Hemichordata, and the protochordates—invertebrate members of the phylum Chordata. In Chapter 39 an overall summary of phylogeny will be given.

PHYLUM ECHINODERMATA

Characteristics

1. Bilaterally symmetrical in larval stages, radially symmetrical as adults; usually with five radii or some multiple thereof.
2. Calcareous endoskeleton, usually with spines which protrude through the skin.
3. Highly developed coelomic cavities, including a unique complex of water-filled canals, called the water-vascular system.
4. The water-vascular system usually has characteristic finger-like extensions, called the tube feet.
5. Mouth and anus present in all but one class in which the anus is lacking.
   Figure 32.1

Figure 32.1 Echinoderm body plan.

The phylum is made up of two subphyla, one of which has four present day classes while the other has but one.

Classification

SUBPHYLUM **Eleutherozoa** Members of this subphylum are never attached by means of a stalk, and the oral surface normally faces downward or laterally, never upward. Five classes make up this subphylum, one of which is entirely extinct. The other four classes, which include most of our present-day echinoderms, are characterized below.

CLASS **Asteroidea** Star-shaped body with arms not sharply marked off from the central disk. Tube feet, with suckers protruding from open ambulacral grooves on the oral surface of each arm, are used in locomotion. The oral surface faces downward. Anus and the opening of the water-vascular system, the madreporite, are on the upper or aboral surface (Fig. 32.2).

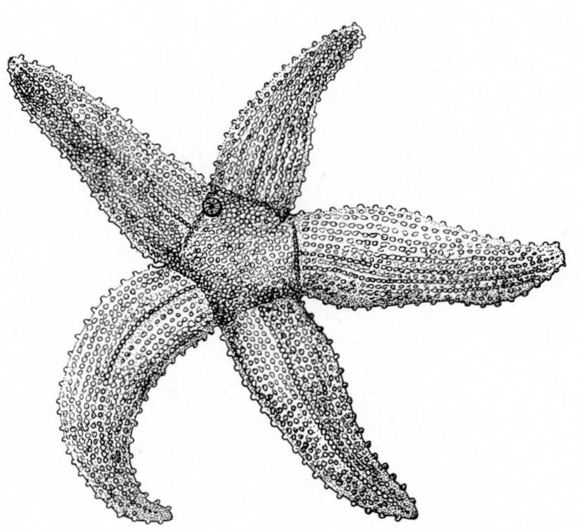

Figure 32.2 Class Asteroidea. A common starfish.

CLASS **Ophiuroidea** Body with long, flexible arms which are sharply marked off from the disk. No open abulacral grooves and the tube feet lack suckers. The madreporite is on the oral surface and an anus is lacking. (Fig. 32.3).

CLASS **Echinoidea** Arms are absent, and the body is spherical or flattened and disklike. Ambulacral grooves are covered, and the tube feet possess suckers. Anus and madreporite are on the aboral surface. Endoskeletal plates are fused to give an inflexible body. Well-developed movable spines are present (Fig. 32.4).

CLASS **Holothuroidea** Wormlike, almost bilaterally symmetrical forms with a leathery skin. Endoskeleton and spines are reduced or absent. The mouth is surrounded by branching tentacles. Anus is aboral and madreporite internal. Ambulacral grooves are closed and the tube feet possess suckers (Fig. 32.5).

SUBPHYLUM **Pelmatozoa** Members of this subphylum have the oral surface facing upward, and they are usually attached to the substratum by means of a stalk growing out from the middle of the aboral surface. Both mouth and anus are located on the oral surface. Of the six classes in this subphylum, five are extinct. The sole remaining class is the class Crinoidea.

CLASS **Crinoidea** There is a cuplike body or calyx with five radii. Arms are branched and usually bear small lateral projections, called *pinnules.* Mouth and anus are on the membranous oral surface. They are usually attached to the substratum by an aboral stalk, but some forms lack a stalk (Fig. 32.6). Of the four orders in the Class Crinoidea, only one has living representatives.

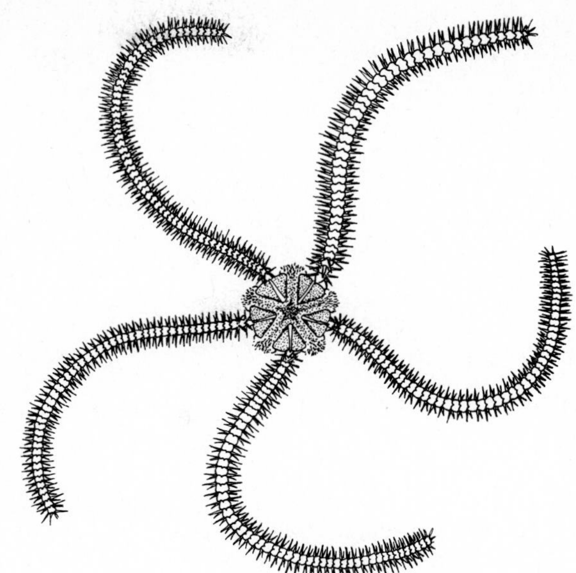

Figure 32.3 Class Ophiuroidea. A brittle star.

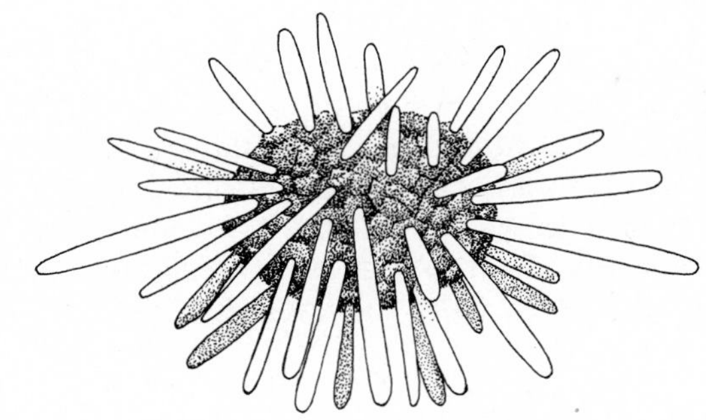

Figure 32.4 Class Echinoidea. A sea urchin.

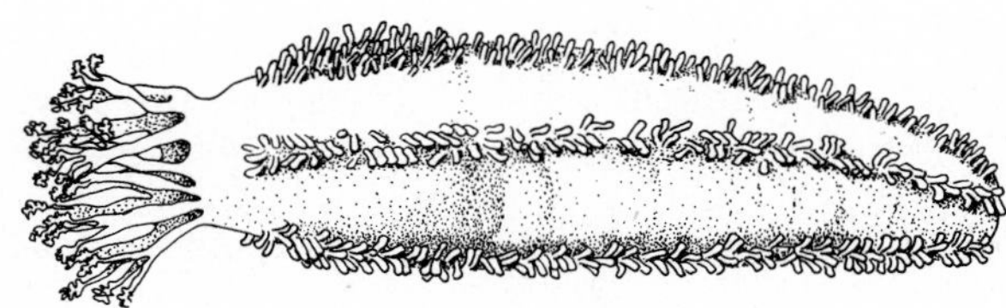

Figure 32.5 Class Holothuroidea. A sea cucumber.

Figure 32.6 Class Crinoidea. A sea lily.

ORDER **Articulata** Lower arm ossicles are incorporated into the calyx. Calyx flexible. Oral surface leathery with small calcareous plates imbedded in it. Mouth and ambulacral grooves not covered. *Antedon, Bathycrinus.*

Echinoderm Body Plan

Figure 32.7 shows the orientation of the echinoderm body relative to the substratum. The body plan indicates the location of mouth, anus, and madreporite, the distribution of tube feet, and whether or not the tube feet possess suckers. All of these features are significant in distinguishing among the classes.

CLASS ASTEROIDEA

Classification

ORDER **Phanerozonia** Arms with two rows of conspicuous marginal plates. Tube feet in two rows. When present, pedicellaria lack stalks. *Luidia, Dermasteria.*

ORDER **Spinulosa** Arms without conspicuous marginal plates. Pedicellaria rarely present. Tube feet with suckers. *Henricia.*

ORDER **Forcipulata** No conspicuous marginal plates. Pedicellaria stalked. Tube feet with suckers and usually arranged in two double rows. *Asterias, Pisaster.*

Anatomy and Physiology

BASIC BODY PLAN

The starfishes (sea stars) are probably the most familiar members of the phylum and perhaps more than any other group they conjure up visions of the seashore. The common starfish of the eastern shore, *Asterias forbesi,* will be used as a representative of the class (Fig. 32.8). It is composed of a central region, or *disk,* with *arms,* or *rays,* which are not clearly marked off from the central disk. The mouth is in the center of the underside, or *oral surface.* In contrast, the upper side is termed the *aboral surface.* The calcareous plates of the endoskeleton are so arranged that the arms are capable of movement, both laterally and up and down. The numerous spines which project from the surface are covered with a thin epidermis. In addition to the spines, there are numerous small pincers, *pedicellariae,* and small fingerlike processes, *dermal branchiae,* projecting from the surface. The former are used in food handling and policing the surface of the body, whereas the latter serve a respiratory function.

WATER-VASCULAR SYSTEM

On the oral surface of each ray there is a deep groove, the *ambulacral groove,* which extends from the mouth to the tip of the ray. Each one of these grooves contains two double or single rows of tube feet. These feet are part of the *water-vascular system* (Fig. 32.9), and they function in locomotion. In this system, which is actually a part of the extensive coelom, water enters through the sievelike *madreporite* on the aboral surface and passes through the *stone canal* to

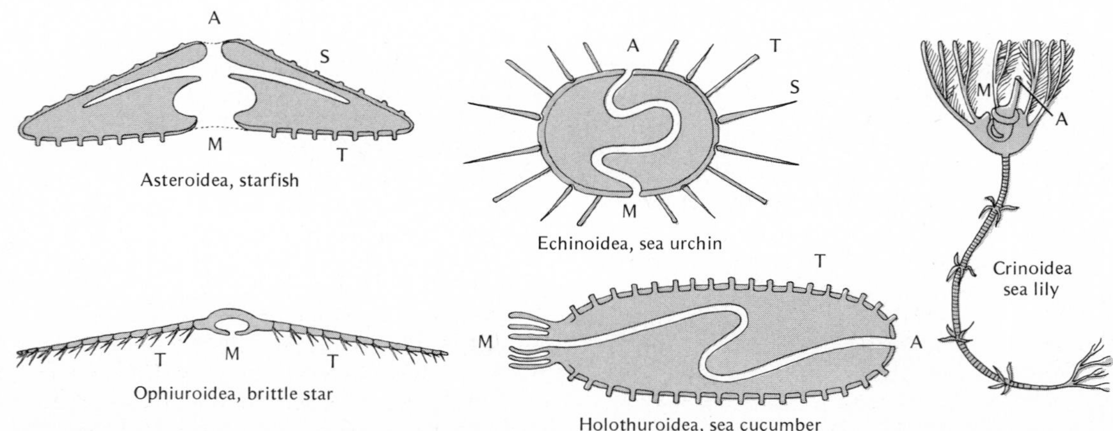

Figure 32.7 Diagram showing the main echinoderm features in each of the five classes. M—mouth; A—anus; T—tube feet; S—spines.

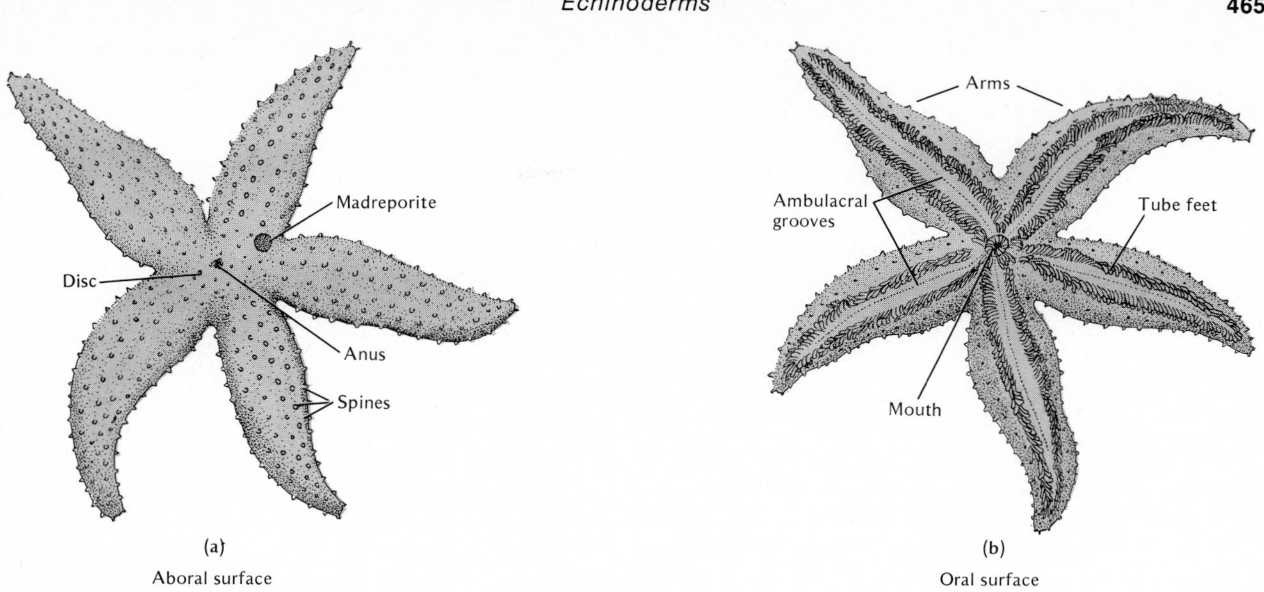

Figure 32.8 *Asterias forbesi,* the starfish. Oral and aboral views.

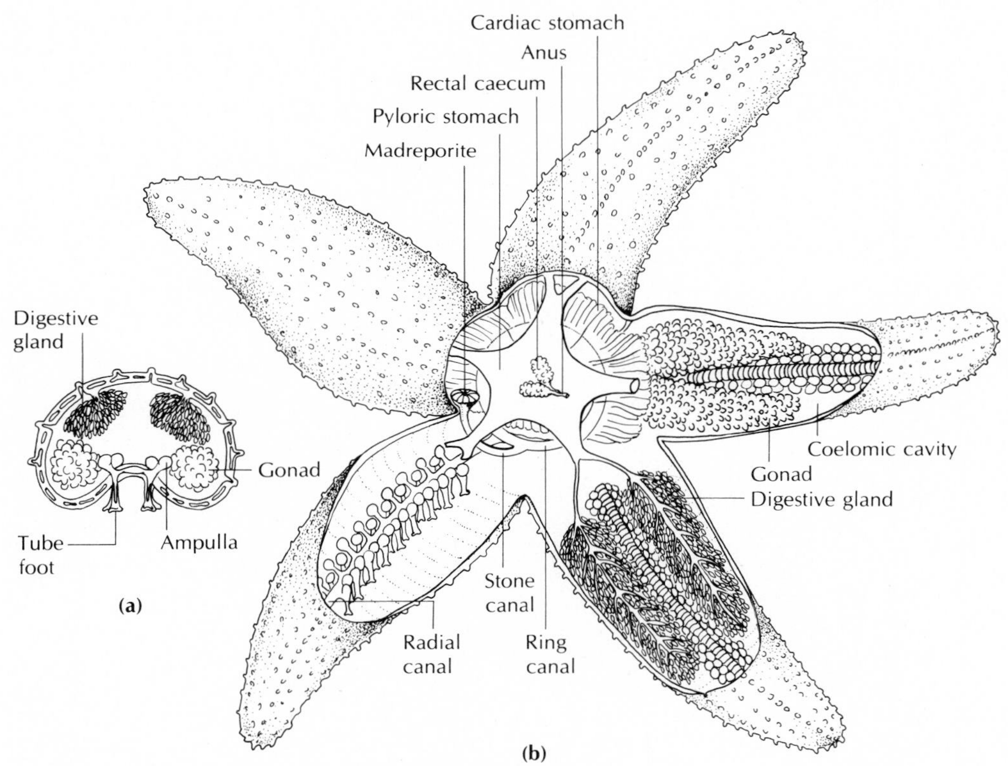

Figure 32.9 Internal anatomy of the starfish. **(a)** Cross section of an arm. **(b)** Aboral view, dissected.

a *ring canal* which surrounds the mouth. Five pairs of sacs, called *Tiedemann's bodies,* are located on the inner margin of the ring canal. They are the source of amoeboid cells which are found in the fluid of the water-vascular system. A *radial canal* extends from the ring canal into each arm just above the ambulacral groove. At intervals, lateral canals lead to muscular bulbs at the base of the tube feet. Each tube foot is about 1 cm long, and at its tip is a sucker. When the muscular bulb, called an *ampulla,* contracts, water is forced into the tube foot, extending it and causing it to attach to the substratum by means of its sucker. Contraction of the muscles of the foot forces some of the water back into the ampulla and thus shortens the foot. This action, combined with that of hundreds of other similar tube feet, pulls the animal forward slowly. The pull of the tube feet is also used by the starfish to open oysters and clams that often serve as its food, making the starfish a serious problem to the shellfish industry.

DIGESTIVE SYSTEM

The mouth leads to the stomach of the starfish which consists of two parts, the *cardiac region* and the *pyloric region.* The cardiac region can be everted to surround the soft parts of the bivalves which are its main source of food. The arms are attached by tube feet to opposite valves, and their slow, steady pull eventually causes the valves to gape enough for the starfish stomach to gain entry (Pl. 34). A space less than 1 mm wide is sufficient for this purpose. Digestive enzymes are secreted by the *hepatic caeca,* a pair of which is located in each arm. The partly digested material passes to the pyloric stomach and into the hepatic caeca where absorption occurs. The short intestine, with two *rectal caeca,* extends to the anus which is located in the middle of the aboral surface.

CIRCULATORY SYSTEM

There is a very much reduced blood-vascular or *hemal system.* It probably does not play a very important role. There is an oral *hemal ring* with a *radial hemal sinus* extending into each arm under the radial canal of the water-vascular system. Circulatory functions are probably served by the extensive coelomic cavity with its ciliated epithelium. The coelom is filled with fluid containing amoeboid cells which gather wastes and move to the outside through dermal branchiae which contain evaginations of the coelom. The dermal branchiae also provide an extensive surface through which respiratory exchange occurs.

NERVOUS SYSTEM

The nervous system is quite simple. There is a nerve ring around the mouth from which extend five radial nerves, one into each arm. These nerves lie in the epidermis in the ambulacral groove. At the end of each arm is a light-sensitive eyespot. Two other sets of nerve fibers in each arm complete the system: a pair of nerves below the oral epidermis and a single cord in the peritoneum on the aboral side.

REPRODUCTION AND DEVELOPMENT

Sexes are separate in the starfish, and in each sex a pair of gonads is located at the base of each arm. Each gonad opens by way of a small pore on the aboral surface of the central disk. Eggs and sperm are released into the sea where fertilization occurs. The early cleavage stages were described and shown in Chapter 16. Refer to that discussion for details. Following the gastrula stage, characteristic bands of cilia develop which are used in locomotion and food capture [Fig. 32.10 (**a**)]. The free-swimming bilaterally symmetrical larva is like the hypothetical *dipleurula* believed to be the ancestor of the echinoderms and the hemichordates. Three lobes develop on each side of the larva, which is now called a *bipinnaria larva* [Fig. 32.10 (**b**)].

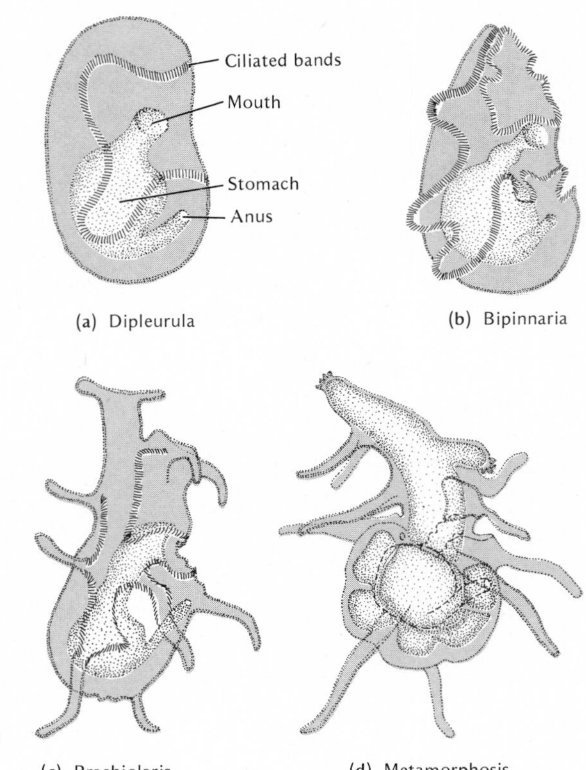

Figure 32.10 Later developmental stages in the starfish.

Additional lobes develop, and all of them become elongated and ciliated. This small larval form, about 3 mm long, is now called a *brachiolaria larva.* After several weeks of free-swimming existence, the brachiolaria larva settles to the bottom where it attaches by a specialized attachment disk on the anterior end. The posterior region enlarges and bends to the left while the larval arms are being resorbed. The posterior region rounds up, developing five lobes on the right side which will be the future aboral surface. The left side becomes the oral surface. A complex internal reorganization goes on at the same time with the water-vascular system developing as an outgrowth of the larval coelom.

Starfishes are noted for their powers of regeneration. A single arm with part of the disk will regenerate a whole organism. Before this was known, oyster fishermen, who must continually combat the starfish which feeds on the oysters, would break starfishes into pieces and dump them back into the ocean. Instead of destroying their enemy, they were unwittingly increasing its numbers.

Ecology

About 1700 species of starfish have been described, ranging in size from the tropical *Linkia* which may reach a diameter of 1 m to minute forms about 1 cm in diameter. In nature, starfishes are brightly colored, ranging from yellow to orange, green, blue, brown, and bright red. Starfish are usually predaceous, feeding on most any animal they can swallow or engulf in their protruded stomach. They seem to prefer shellfish, but are by no means restricted to that diet. As mentioned above, the starfish are a serious pest to the shellfish industry. Deep sea forms ingest the soft bottom ooze and utilize the organic detritus which it contains.

Starfish are mainly littoral forms, and are found from pole to pole. They are most numerous in tropical waters, associated with coral reefs. The threat to Pacific coral reefs by the starfish *Acanthaster* was discussed in Chapter 23. Some starfish do occur at great depths, some species having been taken at 6000 m.

CLASS OPHIUROIDEA

Classification

ORDER **Ophiurae** Arms simple and unbranched, moving in a horizontal plane, unable to twine around objects. Central disk and arms usually covered with well-developed plates. *Ophiura.*

ORDER **Euryalea** Arms simple or branched, capable of twining about objects and of moving vertically. Plates often poorly developed on disk and arms. *Gorgonocephalus,* the basket star.

Anatomy and Physiology

The brittle star has a circular disk and its five arms are quite distinct (Fig. 32.11). The slender arms are very flexible and they are readily broken off. Hence the two common names, serpent stars and brittle stars. The arms are not largely hollow as in the asteroids but contain a series of calcareous *vertebral ossicles,* which articulate with one another and are provided with muscles which make the characteristic movement

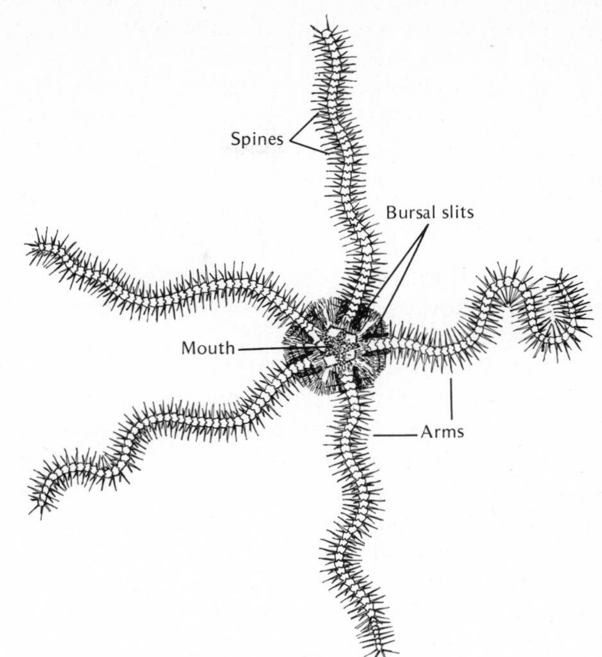

Figure 32.11 Oral view of a brittle star.

possible. The oral surface of the arms lacks the ambulacral groove seen in the asteroids. The small radial canal of the water-vascular system is internal, and it leads to small tube feet which lack both suckers and ampullae. They function in respiration and as sensory receptors but do not play a role in locomotion. This function is served by the flexible arms themselves. The arms also contain a much-reduced extension of the coelom, a radial nerve, and a hemal sinus. In addition to the radial canal, the water vascular system has a small madreporite on the oral surface and a stone canal which leads to a ring canal around the mouth from which the radial canals branch out into the arms.

The mouth opens into a simple saclike stomach which fills most of the cavity of the disk. There is no intestine or anus, and undigested material passes out through the mouth opening. Food consists of various small invertebrates and organic debris.

REPRODUCTION AND DEVELOPMENT

The gonads, located in the central disk, open into the small saclike *genital bursae* of which there are five pairs. Each bursa, in turn, opens to the outside by way of a narrow slit near the mouth. In addition to receiving the openings of the reproductive ducts, the bursae serve in respiration. Although the sexes are usually separate, there are species which are hermaphroditic. Fertilization occurs in the sea in most cases, but in some species fertilization and development take place within the bursae. In such cases development is direct, omitting all free-swimming larval stages. In other ophiuroids the early developmental stages are similar to those described for the starfish. Following the gastrula, the larva becomes conical in shape and its left side flattens out. Outgrowths of this flattened side, which will become the oral surface of the adult, form the long arms of the *pluteus larva,* which is characteristic of the ophiuroid. As many as four pairs of these arms, which contain a calcareous skeletal element, are present in older larvae. The pluteus larvae undergo metamorphosis while they are still free swimming. There is no attached stage as in the asteroids. Regenerative capacity is high in the brittle stars. The arms are readily cast off if the animal is handled roughly. These lost arms are soon replaced.

Ecology

Ophiuroids occur in all seas, mostly in shallow coastal waters, but some forms occur at depths as great as 6000 m. Often the numbers of ophiuroids are very large, up to millions per hectare. They may be brightly colored, but most are patterned and tend to match their environmental backgrounds. A few species are luminescent, with light-producing cells at the bases of spines along the arms. The luminescence is not continuous but results from stimulation, usually mechanical.

Some species are predaceous, feeding on small organisms, including some which browse upon coral polyps. Many species are largely scavengers, sorting out edible materials from the detritus of the ocean floor.

CLASS ECHINOIDEA

Classification

SUBCLASS **Regularia** Globular body, usually circular but may be oval. Radially symmetrical with a basic pentamerous body plan. Two rows of interambulacral plates in all living species. Aristotle's lantern well developed. Anus in middle of aboral surface.

ORDER **Lepidocentroida** Test flexible. Ambulacral plates extend to the edge of mouth. *Phormosoma.*

ORDER **Cidaroidea** Rigid globular test. Ambulacral and interambulacral plates, each in two rows and each extending to edge of mouth. Each interambulacral plate bears a large spine surrounded by smaller spines at the base. Sphaeridia absent. *Cidaris.*

ORDER **Diadematoida** Rigid globular test. Ambulacral and interambulacral plates, each in two rows, but they do not extend to the edge of the mouth, they stop at the edge of the leathery peristome. *Diadema, Arbacia, Strongylocentrotus.*

SUBCLASS **Irregularia** Test flattened, oval, or circular in outline. Tends to assume bilateral orientation with anus located away from the center of the aboral surface and the mouth often eccentrically placed on the oral surface.

ORDER **Holectypoida** Only one living family in which Aristotle's lantern is absent in the adult. Test is oval in shape with simple ambulacral patterns and anus is on oral surface not far from mouth. *Echinoneus.*

ORDER **Cassiduloida** Round or oval test. Ambulacral system in petal-shaped pattern on aboral surface. Lantern absent in adults. *Cassidulus.*

ORDER **Clypeastroida** Very flat test which is oval or circular in outline (sand dollars). Ambulacral system in petal-shaped pattern on aboral surface. Lantern present. *Clypeaster.*

ORDER **Spatangoida** Oval or heart-shaped test. Four of the aboral ambulacral areas petal shaped, but the fifth one not. Lantern present. *Echinocardium.*

Anatomy and Physiology

The spherical sea urchins and the disklike sand dollars still retain the basic echinoderm pattern (Fig. 32.12, Pl. 18). The common sea urchin, *Arbacia,* will be described as representative of the class. The sea urchin is enclosed in a rigid shell, or *test,* formed from rows of fused plates. There are ten distinct rows of these plates, five of which correspond with the arms of the starfish and bear rows of long slender tube feet. The rows bearing tube feet alternate with five wider rows which lack them. The former are called *ambulacral areas* and the latter, *interambulacral areas.* The plates of both regions

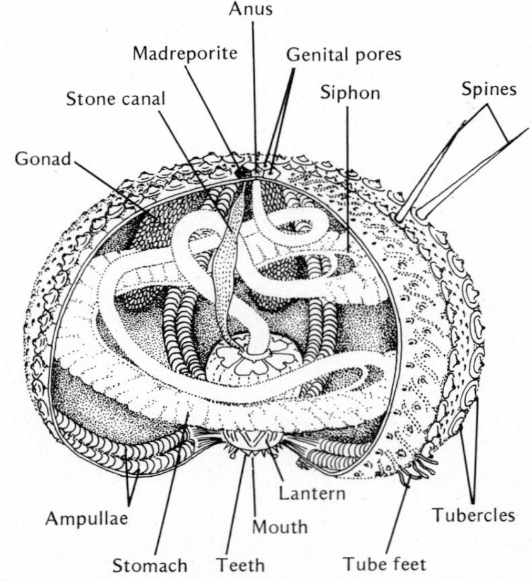

Figure 32.12 Anatomy of a sea urchin.

bear rounded *tubercles* with which the characteristic long spines articulate. The spines are moved by pairs of muscles at the base. In some species the sharp-pointed spines are provided with a toxic secretion which can cause painful injuries to humans. Locomotion is accomplished by action of the spines and of the tube feet which bear suckers as in the asteroids. Pedicellaria on very long stalks are present and function in keeping the body surface free of debris and other organisms. They may also capture small food organisms. Spherical sensory structures, called *sphaeridia,* which are actually modified spines, are found on the oral surface, and they function in maintaining equilibrium.

The mouth is provided with five large pointed teeth which have a complex internal structure. The entire unit is called *Aristotle's lantern.* The long coiled esophagus leads from the mouth to the saclike stomach, which in turn connects with a coiled intestine. A short rectum leads to the anus located near the center of the aboral surface. A peculiar tubelike structure, called the *siphon,* extends along the edge of the stomach connecting the esophagus and the intestine. Its lining is ciliated and currents set up may function to remove excess water from the food and bypass the stomach.

The madreporite on the aboral surface opens into a stone canal which descends to the ring canal which lies above the Aristotle's lantern complex. Five radial canals pass under the ambulacral plates and connect to the tube feet, each of which possesses a large ampulla and a well-developed sucker. The tube feet are quite long in some urchins, waving freely among the spines and extending beyond their tips.

The nervous system and hemal system each has a typical central ring with branches in each of the five ambulacral regions.

REPRODUCTION AND DEVELOPMENT

Echinoids are normally dioecious. The gonads, five in number, are located in the aboral portion of the spacious coelom. Each one opens by means of a small duct and a pore in one of the centrally located plates on the aboral surface. Eggs and sperm are usually released into the sea where fertilization occurs. Following typical cleavage stages, a pluteus larva similar to that of ophiuroids develops. After several weeks it metamorphoses into a small urchin. There are species of echinoids in which the zygotes develop on the body surface, held in place by spines and pedicellariae. In these forms the free-swimming pluteus larva is omitted and development is more or less direct.

The flattened sand dollars and partially flattened heart urchins have essentially the same structure as that described above, except that the complex Aristotle's lantern is modified, reduced, or absent entirely.

Ecology

The bottom-dwelling echinoids are found in all seas and on all sorts of substrata. Although they are primarily coastal dwellers, some do occur to depths as great as 5000 m. The regular or rounded sea urchins are more common on rocky bottoms while the irregular or flattened forms are more often on sandy bottoms. Many species of regular urchins are found in tropical coral reefs. Regular urchins are usually found in crevices and depressions in rocks, and several species actively excavate the cavities in which they reside, some of them remaining permanently in the cavities.

Most regular echinoids are omnivorous, but some are primarily herbivorous, feeding on algae, while others are carnivorous, feeding for the most part on sessile encrusting organisms such as hydroids, sponges, bryozoans, and barnacles. The efficient Aristotle's lantern is able to crush the calcareous shells where necessary. The flattened sand dollars feed on plant and animal detritus. protozoans, copepods, and other small organisms. The absence of jaws in the irregular urchins restricts their diet to small soft-bodied food sources.

Echinoids are subject to predation by many other animals including starfish, crabs, fish, birds, and mammals. Birds often seize an urchin and drop it from aloft to crack the shell. The ripe gonads are considered a delicacy by humans in many parts of the world. They are either eaten raw or roasted.

In recent years, large populations of sea urchins have denuded the kelp beds in many areas on the Southern California coast with disastrous effects on other members of the community.

CLASS HOLOTHUROIDEA

Classification

ORDER **Aspidochirota** Branching tentacles number from 15 to 30, usually 20. Tube feet numerous. Respiratory trees present. *Stichopus.*

ORDER **Elasipoda** Branching tentacles number from 10 to 20. Mouth normally ventral. Tube feet few in number. Respiratory trees absent. *Pelagothuria, Benthodytes.*

ORDER **Dendrochirota** Greatly branching tentacles,

often 10 in number. Tube feet numerous. Respiratory trees present. *Cucumaria, Thyone.*

ORDER **Molpadonia** Tentacles small and fingerlike, usually 15 in number. Tube feet absent except as anal papillae. Respiratory trees present. *Molpadia.*

ORDER **Apoda** Wormlike forms. Tube feet and respiratory trees absent. *Leptosynapta.*

Anatomy and Physiology

The sea cucumbers (Fig. 32.13) are animals with a soft body elongated in the oral–aboral axis. The characteristic calcareous endoskeleton and spines of other echinoderms are largely absent. They are represented only by microscopic *ossicles* imbedded in the leathery body wall. The animals usually lie on or burrow in the substratum with one particular side downward. For convenience, this side is called the ventral and the upper side dorsal. Using the same terminology, the oral end becomes anterior and the anus posterior. The mouth is surrounded by retractile branching *tentacles* which may number from 10 to 30. These tentacles are actually modified tube feet. Five rows of tube feet with well-developed suckers extend from the mouth to the anus. In some species the distribution of these tube feet is quite irregular.

The leathery body wall consists of an epithelial layer covered by a thin cuticle, under which is a thick dermal layer within which the ossicles are found. Next is a layer of circular muscle followed by five thick bands of longitudinal muscles. The epithelial lining of the body wall is ciliated. The large coelom is undivided by septa and is filled with a fluid which contains a number of types of amoebocytes, including some which contain hemoglobin.

The mouth, located in the middle of the membranous buccal membrane, opens into a muscular pharynx. The pharynx is surrounded by a ring of calcareous ossicles, ten in number, which may be homologous with Aristotle's lantern of the echinoids. The ossicles support the pharynx and serve as a site for muscle attachment. A short esophagus, which may be absent in some forms, leads into the relatively small stomach. The long coiled intestine is suspended from the body wall by mesenteries. It leads to a short muscular cloaca at the posterior end, which opens to the outside at the terminal anus. Two highly branched tubular structures, the respiratory trees, extend anteriorly into the coelom from the cloaca. Muscular action of the cloaca draws water in and forces it up into the respiratory trees. Respiratory and excretory exchanges take place between the coelomic fluids, thus forcing the water into the trees. Reversal of the process sends the water back out to the anus.

Some species of sea cucumbers have collections of long tubules, the organs of Cuvier, attached to the base of one or both respiratory trees. Upon adverse stimulation, these tubules are everted and cast out through the anus to produce a sticky viscous mass which effectively diverts most enemies. A more common reaction, seen in many species, is the passing out through the anus of part or almost all of the internal organs—cloaca, respiratory tree, intestines, gonads. Organs thus cast out are later replaced by regeneration.

The water-vascular system is basically similar to that in other classes with a ring canal surrounding the digestive tract near the beginning

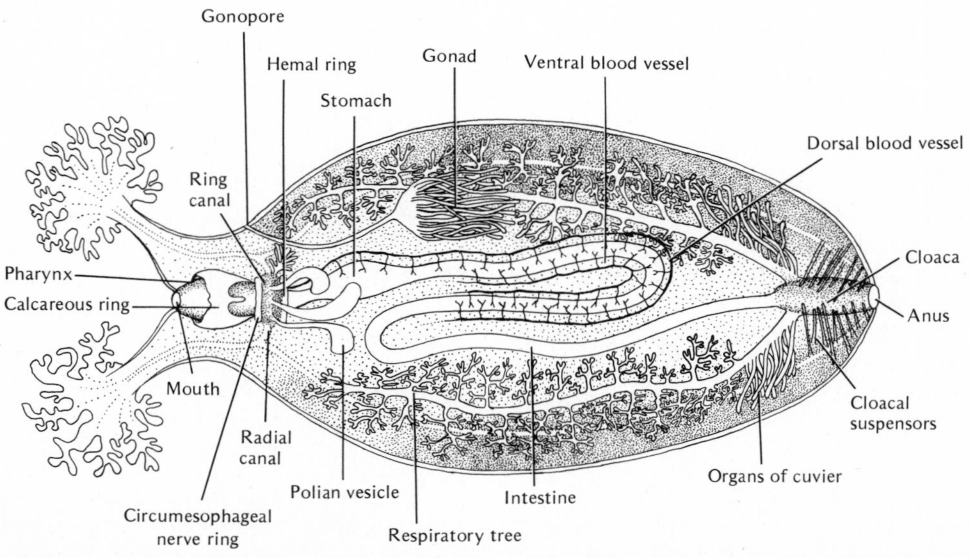

Figure 32.13 Anatomy of a sea cucumber.

of the esophagus. A stone canal leads from the ring canal to the madreporite. Unlike other classes of echinoderms in which the madreporite opens to the outside, it is internal in the coelomic cavity in the sea cucumber. Instead of taking in sea water through the pores in the sieve plate, it uses coelomic fluid. Large saclike structures, called *polian vesicles,* extend from the ring canal. The number varies from one to many, and they seem to serve as reservoirs for the fluid of the water vascular tissue. Five radial canals branch from the ring canal and extend the length of the body. Lateral canals connect the radial canals with the tube feet which may have well-developed suckers and ampullae. Some species have well-developed tube feet only on the functional ventral surface, and there are others which lack them entirely. Just after they leave the ring canal, the five radial canals send branches to the tentacles which surround the mouth. As mentioned earlier, these tentacles are modified tube feet.

The hemal system, which is rather poorly developed in other echinoderm classes, is well developed in the holothuroids. It has the same general pattern as in other classes, but the vessels are larger and more generally distributed. The most prominent vessels are the dorsal and ventral vessels associated with the intestine. The nervous system is typical with a circumesophageal nerve ring and five radial nerves.

REPRODUCTION AND DEVELOPMENT

Sexes are usually separate with a single gonad made up of a large number of fine tubules attached to the body wall. A single duct leads to a genital pore near the base of the tentacles. Fertilization is most often external in the sea, but some species brood their eggs in special pockets of the body wall, and in a few, development is internal. In brooding species, development is more or less direct. In those where development occurs in the sea, it proceeds as in the asteroids, and a larval form, called the *auricularia,* results. The auricularia is quite similar to the asteroid bipinnaria larva. As development continues, the larva assumes a barrellike shape with from three to five bands of cilia encircling the body. This larva is called a *doliolaria larva* and it soon metamorphoses into a young sea cucumber.

Ecology

Holothuroids are in general bottom dwellers, and they occur in all seas and at depths from the intertidal zone to the greatest depths yet sampled. They are more common on muddy or sandy bottoms, but some species occur in rocky crevices. There are some bizarre deep-sea pelagic species, placed in the order Elasipoda. They are highly modified with flattened bodies provided with various flotation and swimming structures such as large papilli, marginal extensions, and taillike appendages. Most sea cucumbers are quite dull in their coloration, being gray, dark green, or black.

Sea cucumbers feed in two ways, depending on whether they have highly branched tentacles or not. Those with the branching tentacles are planktonic feeders, entrapping small organisms and detritus on their extended mucus-covered tentacles. Periodically the tentacles are moved into the mouth where the collected material is removed. Holothurians with poorly developed tentacles simply ingest large quantities of the soft substrate in which they burrow. Useful materials are digested and the residue passes out through the anus.

In addition to the eversion of Cuverian tubules and parts of the digestive tract to entangle and confuse potential predators, some species of sea cucumbers are highly toxic to other animals. In the Pacific islands the natives squeeze the juices from certain species of sea cucumber into pools and soon the fish float to the surface where they are easily collected for food. In the Indo-Pacific region dried sea cucumbers of a number of species are considered a delicacy, particularly by the Chinese. The food material is called *trepang* or *beche de mer.*

Many other organisms have come to live in or on the sluggish sea cucumbers as commensals or as parasites. The classic case of the pearl fish is described in Chapter 44.

CLASS CRINOIDEA

Anatomy and Physiology

The sea lilies and other stalked echinoderms were very important components of the fauna in Paleozoic times. Of the five distinct classes in the subphylum Pelmatozoa, all but the class Crinoidea became extinct by the end of the Paleozoic era. Present-day forms have a cup-shaped body or *calyx* with five flexible arms which branch and may rebranch several times (Fig. 32.14). Each of the branches is fringed with numerous small appendages, called *pinnules.* The sessile forms have a long jointed stalk extending toward the substratum from the middle of the aboral surface. Rootlike *cirri* at the distal

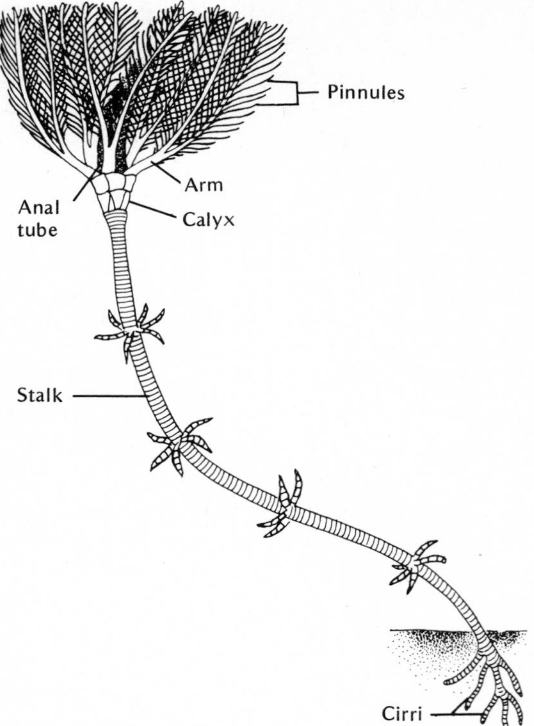

Figure 32.14 Anatomy of a crinoid.

end of the stalk serve to anchor the animal in the sand or mud. Some crinoids, the feather stars such as members of the genus *Antedon,* lack a stalk as adults. They do possess a ring of jointed cirri on the aboral surface with which they grasp the substratum or slowly walk upon it. They may also swim by waving their long feathery arms. The stalkless forms, all in the suborder Comatulida, make up the largest group of existing crinoids.

The mouth is in the center of the upper oral surface, and the anus also on the same surface, often near the base of one of the arms on a cone-shaped protuberance. An open ambulacral groove is located on the oral surface of each arm. The ciliated lining of the groove sets up currents which carry microscopic organisms and detritus to the centrally located mouth. Small tube feet are located in the ambulacral groove. They lack suckers and function in pushing food particles into the ambulacral groove and in respiratory exchange. There is no madreporite in the crinoids and water enters the water-vascular system by way of canals connected to the coelom. The coelom receives water through minute ciliated funnels which open to the outside.

REPRODUCTION AND DEVELOPMENT

The sexes are separate and the gametes are formed in the pinnules. Rupture of the pinnules releases the mature eggs and sperm. After fertilization there is considerable variation in the fate of the fertilized eggs. In some species they are shed into the sea, and in others they are attached to the pinnules, while in others the eggs are actually brooded in specialized brood chambers. Early development occurs within the egg membrane, and a barrel-shaped doliolaria larva hatches after five or six days. Following a brief free-swimming existence, the doliolaria larva settles to the substratum and attaches by means of an adhesive pit located on the anterior portion of the larva. This region will become the aboral portion of the adult into which the larva grows with several distinct stages intervening.

Ecology

The stalkless comatulids such as *Antedon* are mostly shallow water inhabitants with the greatest numbers occurring in tropical areas. Some are found in arctic and antarctic seas, so they are not restricted to warm waters. There are also some which live at depths as great as 1500 m, but it is the stalked crinoids which are more abundant in deeper waters, mostly between 200 and 5000 m. The stalked crinoids are most abundant in Indo-Pacific waters, but there are species in both the arctic and antarctic.

These sluggish or sessile animals have many kinds of commensal or parasitic guests living in or on their bodies. The stalk is often used by sessile polyps as a handy substrate. One group of aberrant annelids, the myzostomarians, are all ectoparasites on echinoderms, mostly on crinoids. They have elaborate suckers and hooks which enable them to hold fast to their host. Some species induce the growth of gall-like swellings on the pinnules, arms or disk and then take up residence within them. Many kinds of crustaceans find a haven among the arms and pinnules or even within the digestive tract.

PHYLOGENETIC RELATIONSHIPS OF ECHINODERMATA

There has been much speculation on the origins and relationships of the echinoderms and space does not permit an elaboration of all the proposals. Suffice it to say that no one theory enjoys general acceptance.

It does seem clear that the secondarily radially symmetrical echinoderms evolved from a motile bilaterally symmetrical ancestor which had a coelom which was divided into three sections—

protocoel, mesocoel, and metacoel. One theory suggests an ancestral form, called a *pentacula,* a bilaterally symmetrical organism with five hollow tentacles surrounding its mouth. Like a lophophore, these tentacles contained extensions of the coelom. The tentacles became the radial canals of the echinoderms, and the coelomic cavities within them formed the basis of the water-vascular system. Assumption of a sessile existence led to the radial condition seen in adult echinoderms. Subsequently the nonsessile echinoderms evolved from the sessile ancestor, retaining the radial symmetry which is actually more appropriate to a sessile organism.

An older theory postulated a hypothetical bilateral ancestor, called a *dipleurula,* which had features much like those of present-day echinoderm larvae.

There is much firmer ground when the relationship between the echinoderms and the hemichordates is examined. Here we find larval resemblances and developmental patterns which are unmistakably similar. More will be said on this score in Chapter 39.

SUGGESTIONS FOR FURTHER READING

Barnes, R. D. *Invertebrate Zoology,* 3rd ed. Philadelphia: W. B. Saunders (1974).

Borradaile, L. A., and F. A. Potts. *The Invertebrata,* 4th ed., rev. by G. A. Kerkut. Cambridge: Cambridge Univ. Press (1961).

Hyman, L. H. *The Invertebrates,* vol. 4: *Echinodermata.* New York: McGraw-Hill (1955).

MacGinitie, G. E., and N. MacGinitie. *Natural History of Marine Animals.* New York: McGraw-Hill (1967).

Marshall, A. J., and W. D. Williams. *Textbook of Zoology, Invertebrates.* New York: American Elsevier (1972)

33 Minor Deuterostomes and Protochordates

Three small deuterostome phyla, the Chaetognatha, Pogonophora, and Hemichordata, as well as the invertebrate chordates, subphyla Urochordata and Cephalochordata, will be discussed in this chapter. The subphylum Vertebrata will be discussed in Chapters 34-38.

PHYLUM CHAETOGNATHA

Characteristics

1. Transparent unsegmented planktonic marine organisms.
2. Head provided with groups of large sickle-shaped setae.
3. Cylindrical body divided into three regions: head, trunk, and tail.
4. Trunk with one or two pairs of *lateral fins;* tail with a fin around the tip; all fins are in the horizontal plane.
5. Circulatory, respiratory, excretory systems absent.
6. Hermaphroditic with direct development. Figure 33.1

   This small phylum of about 50 species is not broken down into classes and orders.

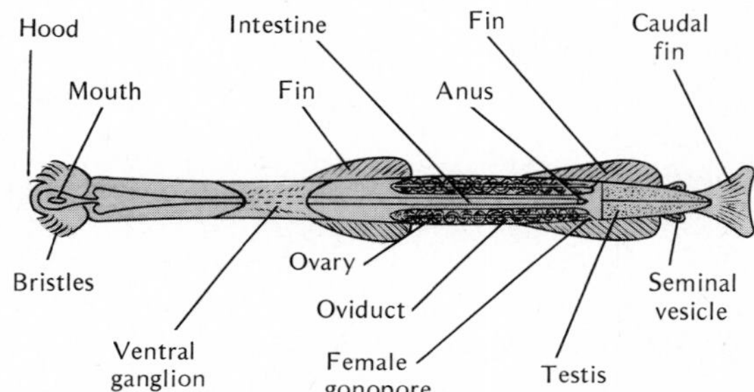

Figure 33.1 Phylum Chaetognatha. Anatomical features of *Sagitta,* the arrow worm.

Anatomy and Physiology

The torpedolike arrow worms are very common planktonic marine organisms. They are transparent and range in size from 2 to 3 cm long to as much as 10 or 12 cm. The body is made up of three regions: head, trunk, and tail. The phylum gets its name from the rows of large sickle-shaped chitinous bristles located on either side of the head. These large spiny structures are provided with muscles so that they serve as jaws for grasping the small crustaceans and other planktonic forms on which the animal feeds. Smaller chitinous teeth around the subterminal ventral mouth complete the food-getting adaptations of these voracious organisms. When present in great numbers, as they often are, these efficient predators must have a significant effect on plankton populations. There is a pair of eyes on the dorsal surface of the head. A peculiar fold of tissue at the base of the head, called the *hood,* is capable of extension to encompass the entire head when the animals swim rapidly about. Its function seems to be to streamline the body and thus reduce friction between the body and the water.

On each side of the trunk are one or two pairs of horizontal fins. The fins have raylike internal supporting elements, but they lack muscles and do not aid in movement. Their role as well as that of the tail fin seems to be largely in floatation. The midventral anus marks the end of the trunk region, and the short pointed tail, with a horizontal fin encircling the tip, completes the animal.

The body cavity of the adult is not lined with peritoneum, and thus the use of the term coelom for this cavity is debatable. In early development a true enterocoelous coelom develops by evagination of the wall of the primitive gut. It is not certain, however, that this original cavity is the source of the definitive body cavity of the adult. The cavity in the adult is divided into head, trunk, and tail compartments by transverse partitions. The trunk and tail compartments are further divided into two lateral portions by median mesenteries.

The body wall is surprisingly similar to that seen in most aschelminthes. There is a thin cuticle underlaid by a cellular epidermis which is one cell thick in most areas. The epidermis is separated from the underlying muscle layer by a thin basement membrane. As in the nematodes, the body wall muscles are all longitudinal. The muscles are cross-striated as in vertebrate muscle. They are arranged in four thick bands, a pair on the dorsal side and a pair on the ventral side. As mentioned above, the adult worms lack a peritoneal lining to the body wall. Movement of chaetognaths is very rapid and analysis of the exact mechanism is difficult. It seems probable that it is accomplished by rapidly alternating contractions of the dorsal and ventral muscle bands. In addition to the muscle bands just described, there are complex muscles in the head region associated with the spines, teeth, and hood.

The digestive system is quite simple, with the mouth leading into a muscular pharynx which in turn empties into a long straight intestine with no specialization other than a pair of lateral diverticulae near the anterior end of the trunk. The intestine terminates in the midventrally located anus at the juncture of the trunk and tail regions.

The nervous system is centered in a single cerebral ganglion in the dorsal portion of the head. Nerves pass from this center to the eyes and head musculature and a pair of nerves encircles the pharynx, meeting on the ventral side and passing in the midventral line to a large ganglion located in the anterior portion of the trunk. Nerves from this ganglion pass on to the rest of the body. In addition to the eyes, sensory structures include tactile receptors on the head and a U-shaped tract of cilia, called the *ciliary loop* and believed to function in chemoreception, extends from the head to the anterior end of the trunk.

REPRODUCTION AND DEVELOPMENT

All chaetognaths are hermaphroditic. A pair of elongated ovaries is located in the posterior part of the trunk cavity while paired testes fill much of the tail cavity. Immature sperm cells are released into the tail cavity where spermatogenesis is completed. The mature sperm enter a ciliated funnel and pass through a sperm duct into one of the laterally located seminal vesicles. Sperm, packed together in a single spermatophore, are released by rupture of the seminal vesicle. A pair of female gonopores is located laterally near the end of the trunk region. They function as a means of entry for the sperm which swim up the oviduct and fertilize the eggs internally. In some species self-fertilization occurs, while in others it does not. The zygotes reach the outside by temporary ducts which develop at the time of reproduction. In some forms the eggs develop while floating either singly or in small masses. Others cement their eggs to seaweed, other objects, or to their own body. Development is direct by radial indeterminate cleavage with a hollow blastula which gastrulates by invagination exactly as in the starfish. The blastopore closes, but the anus subsequently develops in the same area. The coelom and mesoderm are formed by typical enterocoelic pouches which evaginate from the wall of the archenteron. For a brief period the cavities of the coelom are obliterated and later the body cavities of the adult appear. Thus the mesoderm clearly arises as in other deuterostomes, but the exact nature of the coelom is obscured because the cavities which appear in the adult lack the usual peritoneal lining. The young worms which hatch from the egg resemble the adults and they mature without any specialized larval stages.

Phylogenetic Relationships

Because of their early developmental stages, the chaetognaths are quite clearly deuterostomes. They also show some affinities with more primitive groups. The nature of the body wall is quite like the pseudocoelomate nematodes as is the absence of the peritoneum in the adults. As will be seen in Chapter 39, the phylogenetic position of this group is at the very beginning of the deuterostome line. The chaetognaths seem to represent a line which developed from a hypothetical ancestor, often called the dipleurula, soon after it evolved from an ancestral protostome stock.

PHYLUM POGONOPHORA

Characteristics

1. Solitary tube-dwelling marine worms.
2. Body divided into three regions: *forepart, trunk,* and *opisthosoma.*
3. One to many fringed tentacles on the forepart.
4. Digestive tract and gill slits are absent. Figure 33.2

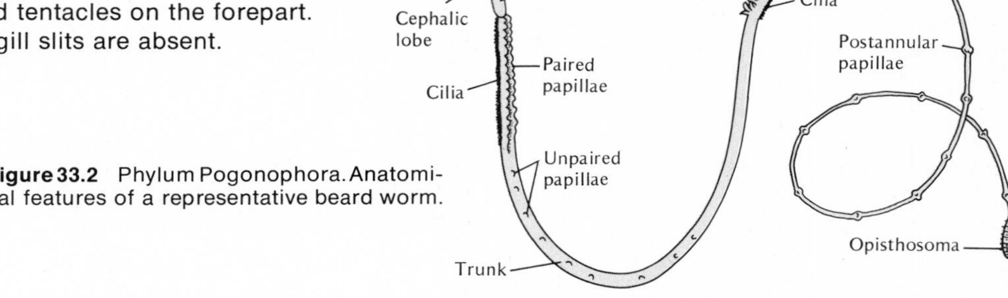

Figure 33.2 Phylum Pogonophora. Anatomical features of a representative beard worm.

Classification

ORDER **Athecanephria** Forepart clearly marked off from the trunk. One tentacle or a number of separate tentacles. Pericardial sac present. *Oligobrachia.*

ORDER **Thecanephria** Forepart not clearly marked off from the trunk. Tentacles numerous and either separate or fused at their bases. Pericardial sac absent. *Heptabrachia.*

Anatomy and Physiology

The Pogonophora or beard worms have only recently been generally accepted as a phylum. The foremost student of the group, Ivanov, has clearly demonstrated their distinctive features and was single-handedly responsible for the acceptance of this group as a phylum—a rather unique accomplishment in the mid-twentieth century. About 100 species in 14 genera have been described thus far. The animals are extremely delicate worms ranging from 10 to 30 cm or more in length, but having a diameter of a millimeter or less. They live in laminated tubes composed of protein and chitin which are secreted by multicellular glands. The tubes are imbedded in an upright position in mud and ooze. They are found in all major oceans, generally at depths of from 150 to 3000 m. Some species do occur in shallow cold waters of 10 to 12° C and lower.

The body, which is covered by a thin secreted cuticle, is made up of three regions: a short forepart, a long trunk region, and a short, segmented opisthosoma. The forepart bears from one to many ciliated tentacles which bear *pinnules,* small extensions of individual epidermal cells.

The trunk region bears papillae, and two girdles in its mid-region bear stout bristles. Some regions of the trunk are ciliated, and it is believed that these regions may have some sensory functions. The opisthoma is segmented, and its short setae probably aid in anchoring the worm in its tube. The opisthoma was not discovered until recently. This fragile tail is easily broken off in dredging specimens from the bottom.

The coelom is divided by transverse septa in each of the three regions of the body. Extensions of the coelom enter the tentacles and the opisthoma has internal septa in each segment which further compartmentalize the coelom in that region.

There is no trace of a digestive tract at any time of the life history. This is a unique feature for a free-living multicellular animal. As yet, feeding in these animals has not been observed. Ivanov suggests that the tentacles are arranged to form a food-catching net. Suspended food particles may be trapped on the pinnules. Digestion could occur by enzymes secreted by gland cells which flank the pinnules. End products could then be absorbed and picked up by the tentacular blood vessels. Pinocytosis has also been suggested as a means of obtaining food. It has been demonstrated that dissolved organic materials such as amino acids can pass through the wall of the tube and be absorbed by the body wall. In addition to playing a role in feeding, the vascular tentacles are also thought to be the site of gaseous exchange.

The nervous system consists of a network of nerves in the epidermis with a main middorsal nerve cord and a brain in the cephalic lobe near the base of the tentacles. In the opisthoma there are segmental ganglia.

The closed circulatory system is well-developed with middorsal and midventral vessels, an afferent and efferent vessel in each tentacle, and a contractile region in the ventral vessel which propels the blood. Hemoglobin is present in the blood.

The excretory system consists of a pair of long, coiled ciliated ducts in the coelom of the forepart. The ducts do not have a ciliated nephrostome, but merely open into the coelom. The dwelling tube may be a depository for non-soluble excretory wastes.

REPRODUCTION AND DEVELOPMENT

Sexes are separate with the paired gonads located in the trunk coelom. In males, the testes are elongate bodies occupying the posterior half of the trunk with sperm ducts opening anteriorly. A gland produces a complex spermatophore which contains the very elongate sperm. Ovaries are located in the anterior portion of the trunk. The duct is muscular near the external opening. The method of sperm transfer is unknown, but fertilization is known to occur within the female tube. From 5 to 100 eggs are laid in the tube where early development occurs. The cleavage pattern is unique, starting off as a modified radial type but the synchrony is lost by the 4- or 8-cell stage. Gastrulation is by delamination and epiboly. The method of coelom formation is not clear. The larvae leave the tube and live a free-swimming life for a few days before settling to the bottom.

Phylogenetic Relationships

There are three schools of thought on the phylogenetic position of the Pogonophora. Before the discovery of the segmented, seta-bearing opisthosoma, the Pogonophora were generally believed to be deuterostomes related to the hemichordates. The segmented opisthosoma is quite annelidlike and some prefer to consider the pogonophorans as protostomes related to the annelids. A third suggestion, prompted by the nature of the anterior end with its tentacles, is that they are related to the lophophorates. Further study of the group should clarify its phylogenetic position.

PHYLUM HEMICHORDATA

Until recently, the hemichordates were included as a subphylum within the phylum Chordata. Some hemichordates do possess *gill slits,* a characteristic chordate character. However, the main basis for including them in the phylum Chordata was a structure believed to be a *notochord.* It is now generally agreed that this structure is not homologous with the notochord of chordates, and thus they have been removed from the phylum Chordata and placed in a phylum by themselves.

Characteristics

1. Solitary or colonial wormlike marine animals.
2. Body and coelom divided into three successive regions: protosome, mesosome, and metasome and protocoel, mesocoel, and metacoel.
3. Gill slits may or may not be present. Figure 33.3

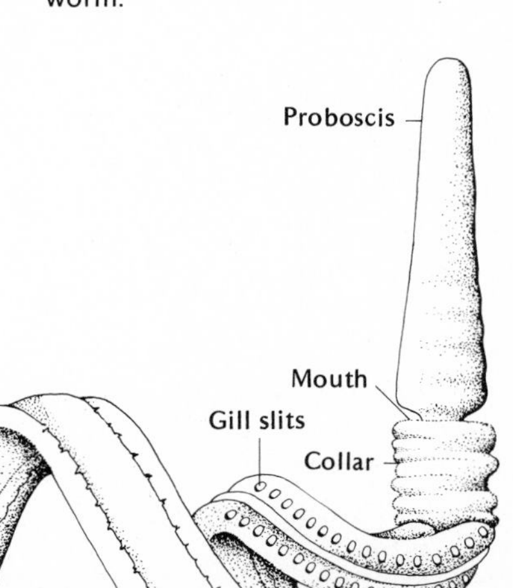

Figure 33.3 Phylum Hemichordata, class Enteropneusta. *Balanoglossus,* the acorn worm.

Classification

CLASS **Enteropneusta** Wormlike solitary organisms. Numerous gill slits present. Intestine a straight tube. *Balanoglossus,* acorn worms. The class is not broken down into orders (Fig. 33.3).

CLASS **Pterobranchia** Small, usually colonial forms living in secreted cases. Gill slits, two in number or absent. Mesosome with tentaculated arms. U-shaped digestive system.

ORDER **Rhabdopleurida** (Fig. 33.5) Animals in true colonies with organic connections between the members. Each member of the colony in a secreted tube. Gill slits absent. Two tentaculated arms present. A single genus, *Rhabdopleura.*

ORDER **Cephalodisca** Individual organisms aggregate in a common secreted case, but they are not organically connected with one another. One pair of gill slits present. Four to nine pairs of tentaculated arms. Two genera make up this order, *Cephalodiscus, Atubaria.*

CLASS **Planctosphaeroidea** This group is known only from its transparent pelagic larva which is spherical in shape and has branched ciliated bands on its surface. Its digestive tract is U-shaped and there are coelomic sacs in its gelatinous interior. The class is not broken down into orders.

CLASS ENTEROPNEUSTA

Anatomy and Physiology

The acorn worms are the most familiar hemichordates. Most of them range in size from 10 to over 40 cm (Fig. 33.3). Most species live in shallow waters in U-shaped burrows lined with mucus. Some deep-sea species have also been reported. The tripartite body consists of a short conical *proboscis,* a narrow *collar* region, and a long *trunk.* The proboscis is very agile. It is used in digging the burrow and is well provided with tactile receptors and chemoreceptors. The ventral mouth is located on the anterior edge of the collar. Food, which consists largely of small organisms and detritus, is caught in mucus secreted by the epidermis along with many sand grains and moved to the mouth by cilia, which cover the proboscis as well as the rest of the animal. The mouth opens into a buccal cavity followed by a pharynx. Along each side of the pharynx there is a series of U-shaped gill slits which open into the surrounding branchial chamber. The branchial chamber communicates with the outside by means of a series of gill pores (Fig. 33.4).

An anterior diverticulum extends from the dorsal part of the buccal cavity into the proboscis. For a long time this structure was believed to be homologous with the chordate notochord. Close examination reveals, however, that it is histologically similar to the buccal tube. It is thus just an anterior extension of the gut and not a notochord. On this basis the hemichordates are clearly separated from the chordates as a distinct phylum. Posterior to the pharynx, the long straight intestine leads posteriorly to the terminal anus.

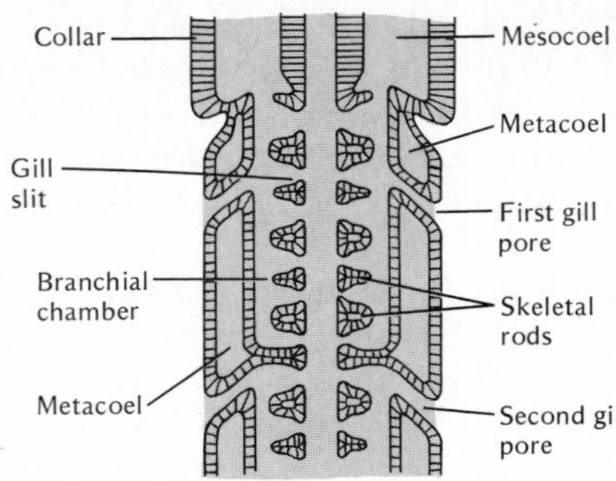

Figure 33.4 Diagram of the pharyngeal region of an acorn worm.

The circulatory system is an open one with two main vessels—one in the mid-dorsal line in which blood flows anteriorly and one in the midventral line in which it flows posteriorly. The two connect anteriorly in a large central sinus, sometimes called a heart. A small contractile pericardial sac presses against the central sinus and by its contraction effects movement of the colorless blood through the central sinus which is itself noncontractile. Branches from the main vessels lead to various parts of the body with extensive blood supply going to capillary beds in the tissues surrounding the gill slits.

Excretion seems to be accomplished by a pair of structures, called *glomeruli,* located on either side of the central sinus. The glomeruli are made up of blind tubular evaginations of the coelom. The internal cavities of the glomeruli are in direct connection with the central sinus, and all the blood flows through the glomeruli. Wastes are presumed to pass into the coelomic evaginations and to the outside via openings in the coelom.

The nervous system is rather primitive, located in the lower part of the epidermis with a mid-dorsal and a midventral cord. The dorsal cord sinks below the epidermis in the collar region. In some species this *collar cord* is hollow and thus it is comparable with the dorsal, hollow nervous system of the chordates. There is little or no centralization to form anything like a brain.

REPRODUCTION AND DEVELOPMENT

Sexes are separate and the gonads are located in the body wall of the pharyngeal region lateral to the branchial chamber. Each has its separate opening into the branchial chamber. Fertilization is external in the sea with typical deuterostome early developmental patterns. Some species have eggs which are laden with yolk and they develop more or less directly. In others there is a ciliated larval form, the *tornaria* (Fig. 39.4), which is very similar to some echinoderm larvae. This similarity in larval forms is an important piece of evidence of relationship between these phyla. The tornaria are small transparent forms, mostly less than 1 mm long. They lead a planktonic existence for some days and then gradually elongate and assume the wormlike adult form.

CLASS PTEROBRANCHIA

Anatomy and Physiology

There are but three genera in the class Pterobranchia. These organisms are all small to minute in size. They are basically like the enteropneusta described above, but differ in being colonial and possessing two or more arms with many tentacles. The digestive tract is U-shaped with the anus located dorsally. Gill slits, two in number, are present in two of the genera but are absent in the third. Two of the genera secrete a protective housing while the third does not. The colonies develop from one individual, known appropriately as the *ancestrula* (Fig. 33.5).

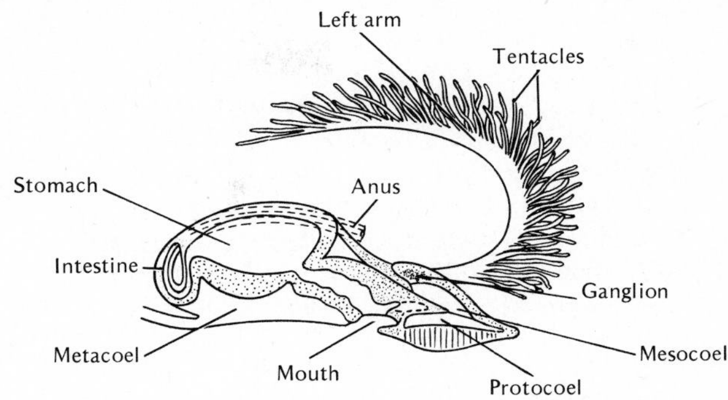

Figure 33.5 Phylum Hemichordata, class Pterobranchia. *Rhabdopleura*, a representative pterobranch.

THE PROTOCHORDATES

Two of the subphyla of the great phylum Chordata, Urochordata and Cephalochordata, are invertebrates since their members lack a vertebral column. They are often called the *protochordates.* The remaining subphylum, Vertebrata, will be discussed in some detail in Chapters 34–38. The phylum name Chordata reflects one of the basic characteristics of its members, the presence of a *notochord* (Gk. *noton,* back; and L. *chorda,* cord).

PHYLUM CHORDATA

Characteristics

The following are present at some time during the life cycle:

1. notochord
2. dorsal, hollow nervous system
3. paired gill slits
4. a postanal tail.

Figure 33.6

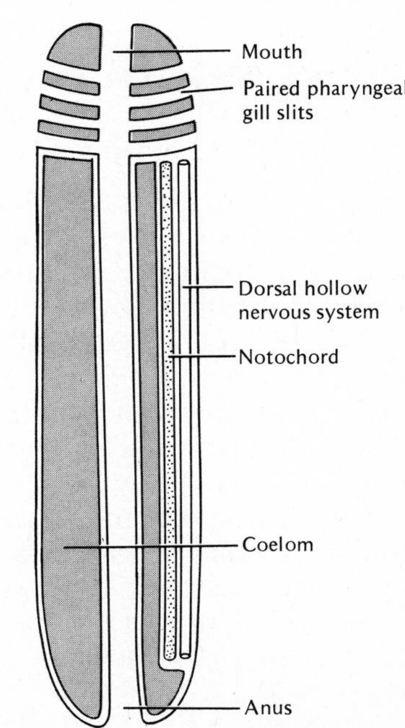

Figure 33.6 Phylum Chordata. Body plan.

SUBPHYLUM UROCHORDATA

Characteristics

1. Entirely marine, either sessile or pelagic.
2. Larva free living with a notochord and a dorsal hollow nerve cord in the tail region.
3. Gill slits present in both larvae and adults.
4. Adult rounded, tubular, or irregular in shape and covered with a secreted tunic.
5. Adult usually lacks notochord, coelom, segmentation, and nephridia.
6. Adult with nerve cord very much reduced.

The subphylum Urochordata includes a rather diverse group of marine animals. Some are solitary, others are colonial; most of them are sessile but a rather significant number are pelagic. In size, they range from microscopic to colonies over 30 cm in diameter.

Classification

CLASS **Ascidiacea** (Fig. 33.7) Solitary or colonial, usually sessile as adults. Larva tadpolelike with notochord and dorsal hollow nerve cord in the tail region. Both larva and adult with gill slits. Adults lack notochord, nerve cord, and tail. Adults have a U-shaped digestive tract. *Molgula.*

CLASS **Larvacea** (Fig. 33.8) Very small, less than 5 mm. Transparent pelagic organisms. Tail contains a notochord and nerve cord persists throughout life. One pair of gill slits are present. *Oikopleura.*

CLASS **Thaliacea** (Fig. 33.9) Planktonic forms, solitary or colonial, of varying sizes. Adults lack tail, notochord, and nerve cord. Gill slits few or numerous. Incurrent and excurrent siphons at opposite ends. *Pyrosoma.*

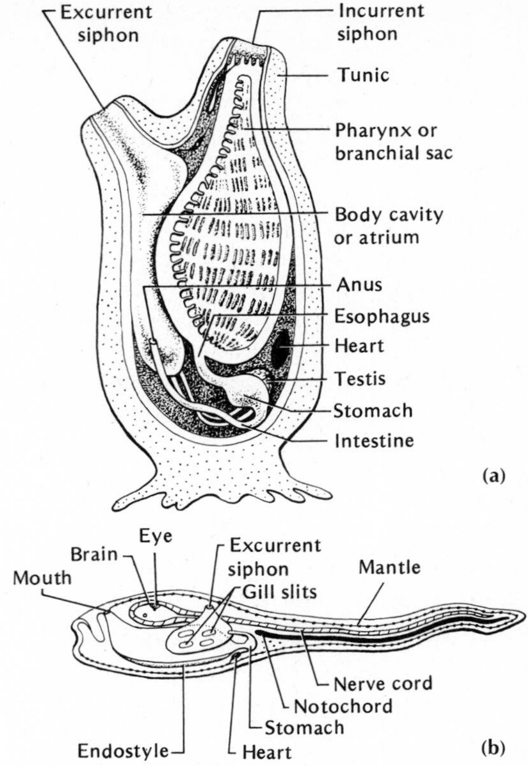

Figure 33.7 Class Ascidiacea. Anatomy of a tunicate. **(a)** Adult. **(b)** Larva.

CLASS ASCIDIACEA

Anatomy and Physiology

The bulk of the 1600 or so species in the subphylum are in the class Ascidiacea. The adult sea squirt is often vase shaped and is attached to the substratum by a flattened base or by a stalk (Fig. 33.7, Pl. 32). Externally there is a secreted *tunic* or *test* made of a type of cellulose, called *tunicine.* Cellulose, the main skeletal element in plants, is a rather unusual component for an animal. The tunic also contains proteins and inorganic compounds. Beneath the tunic is a membranous *mantle* with an outer epidermis which secretes the tunic, a connective tissue layer containing blood vessels and both circular and longitudinal muscles, and an inner lining epithelium.

Cilia, which line the internal chambers, draw water into the *incurrent siphon.* The water enters the enlarged pharynx and leaves by way of the numerous gill slits in the wall of the pharynx, passing into the *atrium* and out through the *excurrent siphon.* Food, in the form of minute planktonic organisms, is trapped in the mucus which lines the pharyngeal cavity. Ciliary currents carry the food into the esophagus and on into the stomach where digestion occurs. A narrow curved intestine leads to the anus at the bottom of the atrial chamber. Respiratory exchange takes place as the water passes through the pharynx and gill slits.

Most tunicates have no specialized excretory system and get rid of their nitrogenous waste by simple diffusion. Wandering cells, called *nephrocytes,* may deposit wastes in specialized areas where they seem to accumulate through the life of the organism. A rather poorly developed open circulatory system consists of a heart with blood vessels leading to the main parts of the body where the blood leaves the vessels and enters tissue spaces. A peculiar feature of the system is the periodic reversal of direction of flow, accomplished by a change in the direction of the beat in the heart itself. The blood contains numerous corpuscles which may be colorless or pigmented red, blue, and even green in one species.

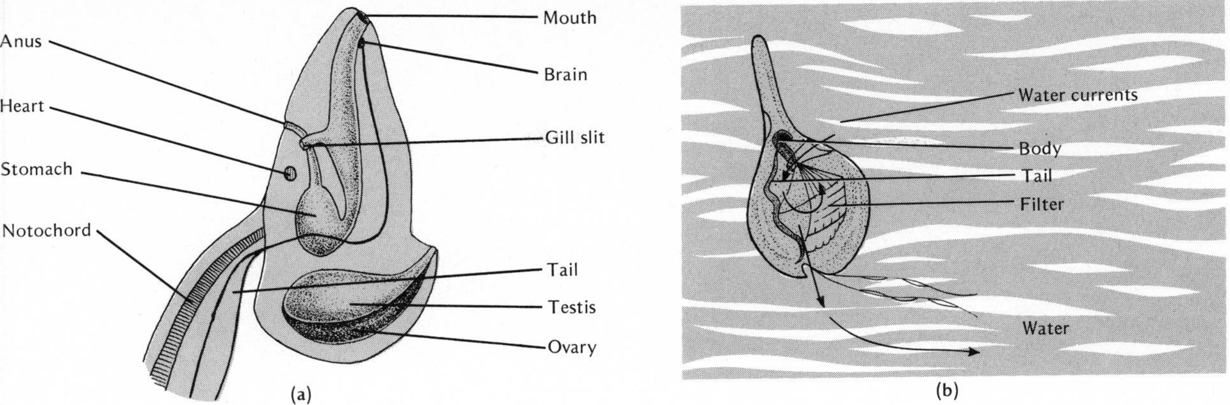

Figure 33.8 Class Larvacea. *Oikopleura,* a representative larvacean. **(a)** Diagram of an adult organism showing its anatomy. **(b)** Diagram showing an adult in its large gelatinous housing which contains a filtration system used in obtaining food. Movement of the tail sets up currents that draw water through as indicated by arrows. Periodically the animal leaves its "house" and secretes a new one.

The nervous system is reduced to a single ganglion located in the mantle between the two siphons. Nerves lead from this ganglion to the various parts of the body.

REPRODUCTION AND DEVELOPMENT

Most tunicates are hermaphroditic with a single ovary and testis whose ducts open into the atrial chamber. Fertilization and subsequent development usually occur externally in the sea. Cleavage is holoblastic, radial, and determinate. The latter feature is unusual since most deuterostomes have indeterminate cleavage. A hollow blastula is followed by gastrulation which occurs by invagination and epiboly. Elongation produces a tadpolelike larva (Fig. 33.7) with a dorsal hollow nerve cord and a notochord in its tail region. There are also a few pairs of segmentally arranged muscles. Following a very brief free-swimming existence—a few hours at the most—the larva becomes attached to the substratum by means of *adhesive papillae* at the anterior end. Metamorphosis begins with resorption of the tail with its neural tube and notochord. Rapid growth of the body between the attachment papillae and the mouth results in a 180-degree rotation, bringing the mouth to its upright position opposite the point of attachment. In addition to sexual reproduction, many colonial tunicates reproduce asexually by budding.

CLASS LARVACEAE

As the name of this class implies, many of its members retain some of the larval features throughout life. The adults are small and transparent, living in a secreted gelatinous housing as shown in Figure 33.8.

CLASS THALIACEAE

These pelagic forms may be either solitary or colonial (Fig. 33.9). The solitary *salps* are a very important component of the pelagic planktonic community and are often present in large numbers. The adults lose their tails, as in the ascidians, and after metamorphosis the incurrent and excurrent siphons are at opposite ends of the body. The water currents which pass through are thus used for locomotion in a kind of jet propulsion, in addition to the usual feeding and respiratory functions.

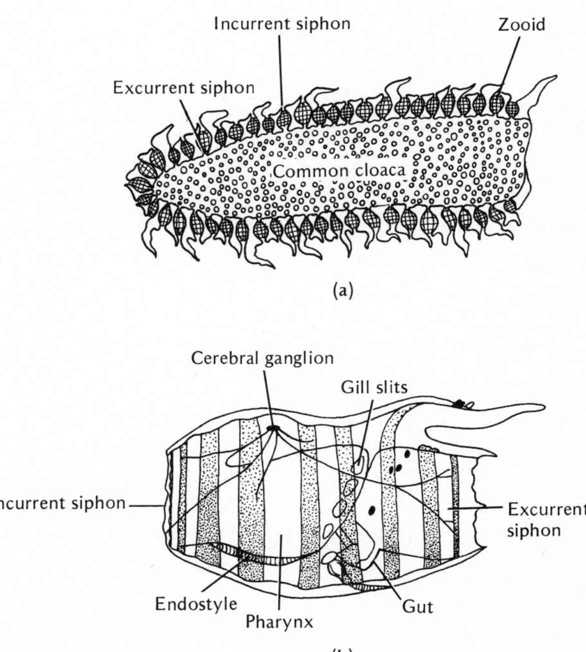

Figure 33.9 Class Thaliacea. **(a)** *Pyrosoma,* a colonial form. **(b)** *Doliolum,* a solitary form.

SUBPHYLUM CEPHALOCHORDATA

Characteristics

1. Small, slender, laterally compressed fishlike marine animals.
2. The segmented body lacks scales.
3. No head or brain is present.
4. Notochord and dorsal hollow nervous system present throughout life.
5. Pharyngeal gill slits present. This subphylum is not broken down into classes and orders.
   Figure 33.10

Figure 33.10 Subphylum Cephalochordata. Anatomy of amphioxus, the lancelet (*Branchiostoma* sp.).

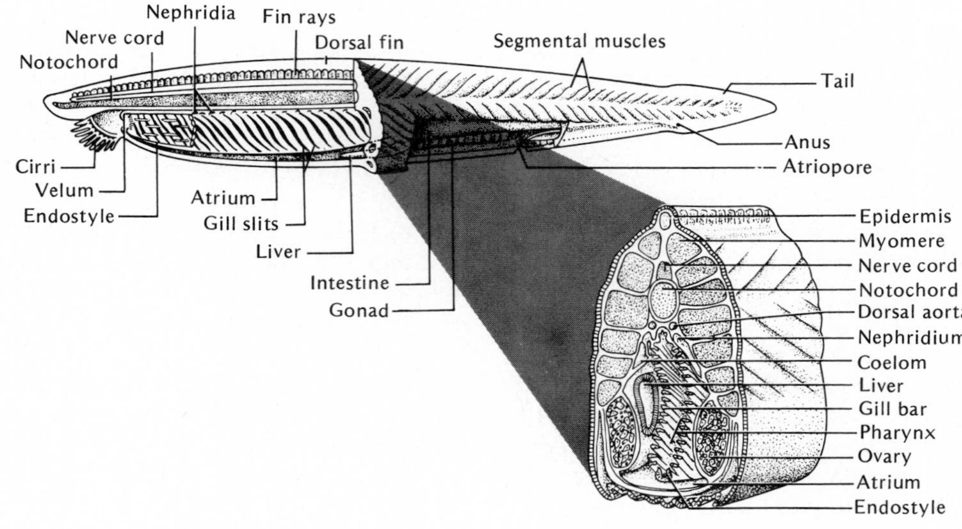

Anatomy and Physiology

There are only about 20 species in this subphylum. Its members live in shallow marine waters with sandy bottoms in tropical and temperate regions of the world. Amphioxus, the lancelet (genus *Branchiostoma*) will be used as an example of the group. This small (about 5 cm in length) fishlike chordate has been the object of much study because of the almost diagrammatic way in which it illustrates the basic chordate characteristics. Amphioxus has a long notochord which persists throughout life, a dorsal hollow nerve cord, and many gill slits in the pharynx.

The long slender body lacks a head. It is laterally compressed and pointed at both ends. A continuous dorsal fin expands into a tail fin and then continues ventrally to the *atriopore* as the ventral fin.

The mouth is an opening in a thin membrane, called the *velum,* which is located at the rear of a preoral chamber, called the *vestibule.* The opening into the vestibule, surrounded by a ring of fleshy fingerlike *cirri,* is located ventrally near the anterior end. Behind the mouth, which is surrounded by a fringe of *velar tentacles,* the digestive tract enlarges as the spacious pharynx with many gill slits on each side. A narrow straight intestine leads from the pharynx to the subterminal anus. A small diverticulum on the ventral side of the intestine, called the *liver,* is believed to secrete digestive enzymes.

There is a space between the pharyngeal wall and the body wall, called the *atrium.* It opens to the outside at the midventral atriopore anterior to the anus. Water drawn into the mouth and pharynx by action of the cilia which line the passages, passes through the gill slits into the atrium, and then out by the atriopore. Respiratory exchange occurs as the water passes through the very vascular gill slits. Some workers believe that most of the gaseous exchange occurs in the lacunae which lie just beneath the folded epidermis (*metapleural folds*) on the ventral side of the pharynx. Food is trapped by mucus in a groove, called the *endostyle,* on the floor of the pharynx and is carried back into the intestine by ciliary action.

The open circulatory system lacks a definite heart. Numerous contractile vessels, arranged much as in primitive vertebrates, force blood anteriorly in a series of ventral veins and posteriorly in a dorsal aorta. Capillaries are absent and the colorless blood, which lacks corpuscles, moves from arteries to veins through tissue spaces.

The body is supported mainly by the notochord, a rod of gelatinous material lying just beneath the dorsally located nerve cord. The nerve cord, with its small central canal, extends the length of the animal with a very slight anterior enlargement, called the *cerebral vesicle.* Each of the *myomeres* which lie along the sides of the animal receives a pair of nerves from the nerve cord. There is a dorsal nerve which is sensory and a ventral one which is motor. This is comparable with the situation seen in the spinal nerves of vertebrates.

Excretion is by means of numerous pairs of ciliated protonephridia which connect the coelom with the atrium. The presence of protonephridia in a group so high on the phylogenetic scale is difficult to explain.

REPRODUCTION AND DEVELOPMENT

Sexes are separate with a series of paired ovaries or testes bulging from the lower portion of the muscle segments into the atrium. Mature eggs and sperm, released by rupture of the gonads, pass out of the atrium via the atriopore. Fertilization is external in the sea.

Early cleavage is radial and indeterminate with a typical hollow blastula and gastrulation by invagination. Figure 16.8, showing early starfish development, would serve equally well to illustrate that of amphioxus. Development continues to be similar to that of the starfish with mesoderm and coelom arising by formation of enterocoelic pouches. The larva which hatches is quite asymmetrical, with its mouth on the left-hand side and gill slits only on the right. As the larva matures, it gradually loses this asymmetry.

SUGGESTIONS FOR FURTHER READING

Barnes, R. D. *Invertebrate Zoology,* 3rd ed. Philadelphia: W. B. Saunders (1974).

Hickman, C. P. *Biology of the Invertebrates.* St. Louis: C. V. Mosby (1967).

Hyman, L. H. *The Invertebrates,* vol. 5: *Smaller Coelomate Groups.* New York: McGraw-Hill (1959).

Ivanov, A. V. *Pogonophora,* transl. by D. B. Carlisle. New York: Consultant's Bureau (1963).

Meglitsch, P. A. *Invertebrate Zoology,* 2nd ed. New York: Oxford Univ. Press (1972).

34 The Fishes

INTRODUCTION TO THE VERTEBRATES

As identified in Chapter 33, the diagnostic characteristics of the phylum Chordata are the notochord, the dorsal hollow nervous system, and the pharyngeal clefts. Possession of a post-anal tail at some time in the life cycle also is sometimes used as a chordate characteristic. The subphylum vertebrata possesses these traits and an additional one, a vertebral column, which is present in members described in this and the four chapters that follow.

The vertebrates include the fishes, the amphibians, the reptiles, the birds, and the mammals. The classification of fishes has some problems, to be discussed below, but the remaining vertebrates fall reasonably well into classes.

The basic vertebrate trait is the presence of vertebrae or parts of vertebrae that either cap, surround, or replace the notochord which develops in each vertebrate embryo. The individual units of the vertebral column are either cartilage or bone, or combinations of both. In the trunk region of the animal a simplified vertebra is composed of two parts, a heavy centrum under the spinal cord, either surrounding or replacing the notochord, and a neural arch consisting of two cartilaginous or bony plates in contact or fused with the centrum and meeting above the spinal cord (Fig. 34.1). In the tail region of some vertebrates a second arch, the *hemal arch,* encloses the extension of the aorta from the trunk.

While the vertebral column is used as the major diagnostic characteristic for the subphylum, several other characteristics always accompany it (Fig. 34.2). An endoskeleton of cartilage or bone is always present in two principal and significant ways. First, the brain is always supported in a pan of cartilage or replacement bone.

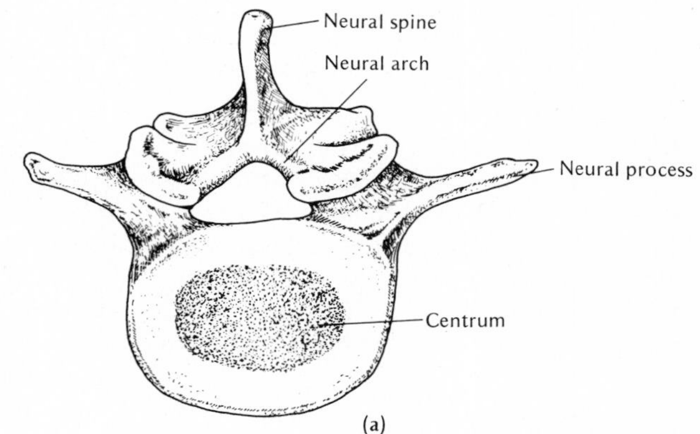

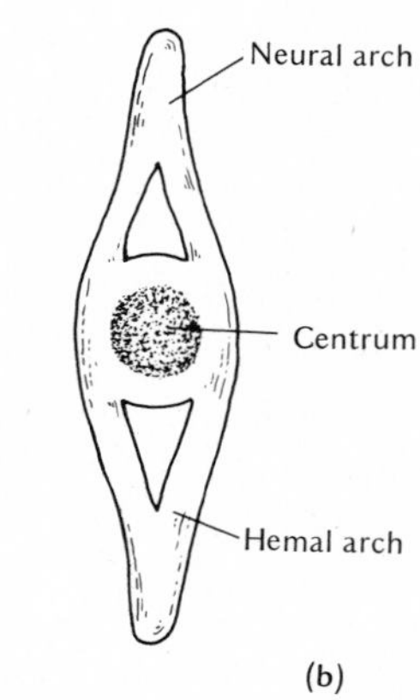

Figure 34.1 Schematic structure of vertebrae. **(a)** Trunk. **(b)** Tail.

Second, there are, with a few exceptions such as some snakes (Ch. 36), anterior and posterior pairs of appendages (limbs or fins). These are supported by an endoskeleton of cartilage or bone. Many bones developed among the muscles in some bony fish. However, in land animals, with few exceptions, such skeletal reinforcement of muscular structures is seen principally in the thoracic areas in the form of ribs and as

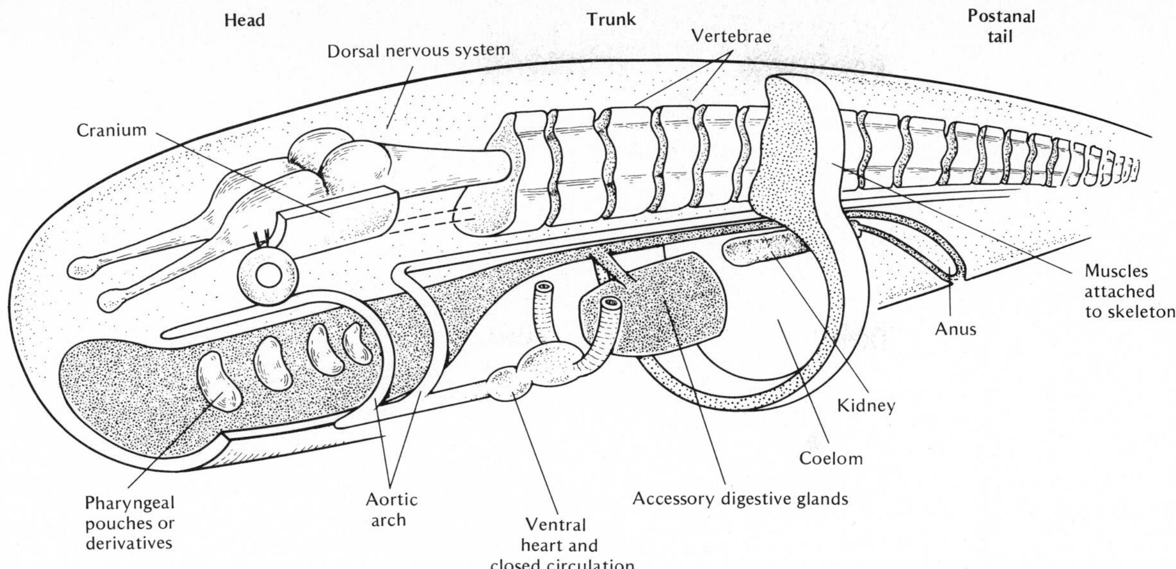

Figure 34.2 Phantom sketch of a hypothetical vertebrate to show chief diagnostic features.

special pharyngeal bones of the throat. One of the dramatic evolutionary events is the development of internal joints in the skeleton of the appendages. Accompanying this was the emergence of some vertebrates onto land and their radiation into various types related to their modes of locomotion.

The skeleton of vertebrates is not entirely derived from internal differentiation of cartilage as in the case of the vertebrae. Early in the evolution of some of the fishes, bone developed in the skin as plates or scales. While this armor has generally been reduced or lost in most vertebrates, some bones that are presumably derived from this ancient surface origin, called *dermal bones,* remain just below the surface. They form the vaults of bone over the brain, some parts of the face and lower jaw, and a bone in the girdle supporting the anterior appendages of many vertebrates.

Another characteristic of this subphylum is the relatively large coelom with internal organs loosely suspended by mesenteries.

The subphylum is also unusual in the expansion of epithelial tissues both internally and externally into cords, folds, and sacs. The sacs are exemplified by lungs, swim bladders, and urinary bladders, all structures capable of considerable change in shape and volume. Folds are noteworthy in the ways the lining of the digestive tract is convoluted, greatly increasing its surface area, in the folds or tubulations that produce special glands for digestion such as the pancreas or that occur on the surface to produce scales, feathers, hair, sweat glands, mammary glands, and poison glands. Cords of cells develop in all vertebrate embryos to produce a relatively large liver.

The circulatory system of vertebrates is a closed one (Ch. 11). There is a central pump, the heart, that develops under the pharynx and delivers blood through the walls of the pharynx to other parts of the body through arteries. The arteries terminate in capillaries with single cells forming their walls where gaseous exchanges occur. The blood is then collected into veins that return the blood to the heart. In the fish and larval amphibian modifications of the pharynx, called *gills,* are sites for oxygenation of the arterial blood. Soon in evolution, however, the lungs assume this function. Associated with the development of the lungs is the evolution of the circulation into two circuits, one to the lungs and back to the heart, the other to the body and back to the heart. As will be seen in the discussion of the tetrapod (four-legged) classes of vertebrates, the heart itself evolves into two separate pumping channels, although the beat of the heart in the two pumps is closely synchronized.

Another vertebrate characteristic is the evolution of segmental muscles into integrated sheets around the coelom and pharynx and into long columns flanking the vertebrae. Muscles associated with the appendages evolved in such a way that strictly two-dimensional hingelike action is superceded by greater flexibility and with rotational capabilities for the appendages. Accompanying the extensive development of muscle groups has been the expansion of two significant parts of the brain, first the cerebellum and later the cerebrum.

The combination of characteristics discussed above has permitted the subphylum Vertebrata to undergo radiation and diversity unequaled in multicellular organisms except for the arthropods. The vertebrates not only occupy the seas, where they form billions of tons of the biomass and often are at the end of major food chains (Ch. 45), but they also occupy the land in many habitats. Some of these are burrowing subterranean vertebrates such as moles, some are of limited motility and mobility such as turtles or sloths, some are speedy such as horses, some depend on plants for food and some are carnivores. As with most insects, one entire class of vertebrates has developed wings and usually can fly; however, with the insects, the birds are not completely emancipated from the land.

Emerging from the evolution of the vertebrates, of course, is the human species, a newcomer whose special modifications of the vertebrate characteristics have given its members an unusual and somewhat frightening position on this planet.

THE FISHES

The term "fish" is an imprecise one. It includes aquatic true fishes and fishlike vertebrates. It was once taught that the fish belonged to a class or superclass, called Pisces. This term is generally ignored today, however. This chapter will discuss four classes of vertebrates that are fishes or fishlike: Agnatha, jawless vertebrates (ostracoderms and cephalapsids), significant in the early evolution of vertebrates but now with only a few species surviving; Placodermi, an extinct class of generally armored, jawed fish; Chondrichthyes, the cartilaginous fish such as sharks and rays; and Osteichthyes, the bony fish. The fishes, in their aquatic environment, have adaptively radiated into an immense number of specializations. In fact, their evolution has been as diverse and amazing as that of the tetrapod classes taken collectively.

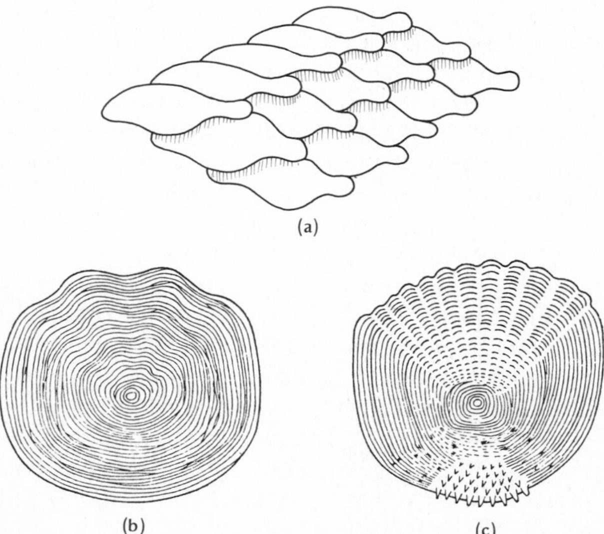

Figure 34.3 Three scale types in fish. **(a)** Ganoid, as in gars. **(b)** Cycloid, as in suckers. **(c)** Ctenoid, as in perch.

Structural and Functional Adaptations of the Fishes

While some insight can be obtained from the examination of the classification of the fishes at the end of this chapter, some more generalized statements about the adaptations of the fishes and fishlike vertebrates will be useful to an understanding of their biology.

Anatomy and Physiology

The form of a vertebrate is principally achieved through the interactions between the skeletal and muscular systems and the covering integument and its derivatives (Fig. 34.3) as adapted to its particular habitat. The aquatic environment has usually resulted in selection for a streamlined form.

THE SKELETAL SYSTEM

The *axial skeleton* of fishes principally consists of a cranium, a spinal column and associated ribs, and secondary ribs or ossifications in the *myocommata* or connective tissue segments between the *myotomes* or muscle segments. The *appendicular skeleton* consists of a pectoral girdle, usually arranged in the form of a U, and a pelvic girdle that is commonly a bar or plate. To these girdles are articulated the pectoral and pelvic fins, respectively. In addition there may be several unpaired pectoral and pelvic fins. The pectoral girdle may remain cartilaginous or it may ossify. In its simplest form the segment of this girdle ventral to the articulation of the pectoral fin is the *coracoid* and dorsal to it is the *scapula.* In bony fishes dermal bones often join to the pectoral girdle and articulate it with the bones of the skull; this is a strengthening arrangement. The pelvic girdles are never attached to the axial skeleton in fishes as they are in the tetrapod classes of vertebrates.

The bones of the skull can be very numerous, but they remain divisible into a cartilage-derived *chondrocranium* making up the floor and the membrane bone-derived *dermatocranium* forming the roof of the skull, the face and the bones covering the gill chamber.

THE MUSCULAR SYSTEM

The musculature of the fish is relatively simple. In the cartilaginous and bony fishes the segmentally arranged musculature is divided into dorsal *epaxial muscles* and ventral *hypaxial muscles.* In the trunk region the hypaxial muscles flank the body cavity, but in the tail they are as solid as is the epaxial group. Epaxial and hypaxial muscles are innervated respectively, by dorsal and ventral branches of spinal nerves. In the pharyngeal region there are opposing sets of branchial muscles between the cartilaginous or bony supports of the gill apparatus. These consist of dorsal and ventral constrictors and levators; they are derived from the wall of the pharynx and are innervated by cranial nerves. In the head region of the cartilaginous and bony fishes are the six muscles that move the eye and are present in all vertebrates evolved from the fish. They are believed to be segmental in origin and are innervated by the third, fourth, and sixth cranial nerves.

THE INTEGUMENT

The integument of fishes is usually not very thick. It principally serves as a site for the production of scales. Scales are of several kinds (Fig. 34.3). Cartilaginous fish have *placoid scales* consisting of a basal plate of bone with a point, consisting of dentine, projecting through the skin. The tip is capped with a hard material claimed to be homologous with enamel. Another analogy with teeth is that the placoid scales have a central pulp containing blood vessels. As has been said before, the placoid scales and the teeth of an elasmobranch grade into each other in size. The *ganoid scales* are characteristic of the primitive ray-finned fishes. Often these scales are close together and overlapping to provide a very hard armor consisting of a basal plate for each scale and a covering of layers of enamellike ganoin. The scales of the higher bony fishes lie in pockets in the integument at their anterior edges and overlap the next scales at the posterior edge. These scales are entirely dermal and have no outer coating. In shape they are either round *cycloid* scales, or *ctenoid* scales with toothlike projections on their posterior edges. Both cycloid and ctenoid scales may show growth rings.

In addition to the scales, fish skins are usually noteworthy for their pigmentation and the presence of mucous glands. The principal pigments are the brown and black melanins and the yellowish, orange, or reddish carotenoids; melanocytes contain the melanin pigments, lipophores contain the carotenoid pigments. In addition, iridiocytes may contain crystals of guanine and help to create iridescence. Fish may change color quite rapidly, both in response to the background with which the fish is associated and in response to excitatory stimuli such as threat or sex.

LOCOMOTION

Although some fish may have well-developed pectoral and/or pelvic fins, these are rarely used for locomotion but serve principally as stabilizers or for steering. Locomotion is produced in most fishes by alternating lateral movements of the body. There are exceptions to these generalizations, however. The large pectoral fins are used for swimming in such flattened fish as rays and in fishes that are heavily armored with spines such as puffer fish, even though their pectoral fins may not be very large.

One of the noteworthy locomotor adaptations among fish is the capacity to jump or sail. Common examples are seen in game fish such as marlin, tuna, and sailfish. The flying fish do not truly fly, but gain sufficient speed to leave the water and then spread their pectoral fins to plane for distances sometimes exceeding 300 m (1000 feet). Outstanding jumpers are the salmon in their drive from the sea to inland spawning waters. They may compulsively make repeated leaps to overcome waterfalls impeding their upstream migration.

The tails of some fish add significantly to their speed in the water. In fact, the tail is often a diagnostic trait for the classification of the fishes. Tail forms are of four principal types; three are illustrated in Figure 34.4.

In addition to the classical streamline shape of most fish, they are modified in a number of ways. Some are compressed laterally so that they are almost as deep as they are long. The trunk is often shortened, and accompanying this is a

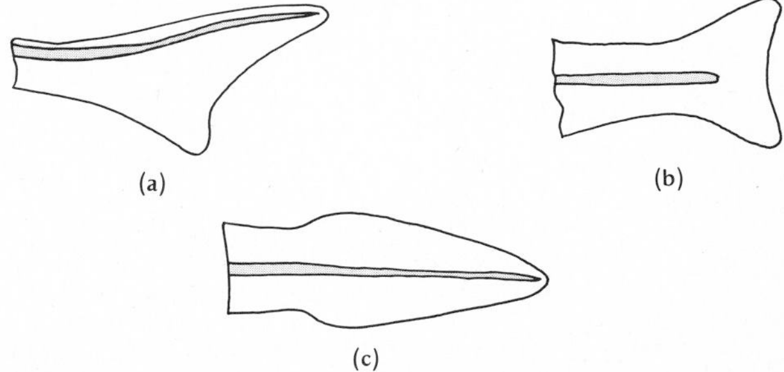

Figure 34.4 Three tail types in fish. **(a)** Heterocercal, as seen in sharks, sturgeons, paddlefish. **(b)** Diphycercal, as seen in lungfish. **(c)** Homocercal, as seen in teleost fish.

displacement of the pelvic girdle and anal opening as far forward as the throat area. Other fish become snakelike as in the case of the eels, or greatly elongated as with pipe fish that resemble the grasses they inhabit. Protective spines are not uncommon in both cartilaginous and bony fish. The rays often have a poisonous spine associated with a dorsal fin on the tail; the poison is sufficiently venomous to cause the death of unwary persons who step on them. Some fish seem to have some organs that are poisonous if eaten, but other parts of the body may be edible. It is wise to know what fish one is eating especially in areas of the Pacific Ocean.

While it is not common, some fish can use their pectoral fins for walking. For instance, the lungfish moves from one area to another in search for water during periods of drought. Also, recently in Florida accidentally released imported walking catfish have become a threat to native fish because of their capacity to spread by walking overland on their pectoral fins. They possess a respiratory area above the gills that makes it possible for them to use air directly, thus this contributes to their success.

An important part of the locomotory equipment is the structure of the pectoral girdle and pectoral fin of the fleshy-finned crossopterygian fish. The architecture of this assemblage seems to be ancestral to the primitive amphibian limb. The girdle itself is a cartilaginous scapula which articulates with the fin, supported by a dermal cleithrum and clavicle. Articulating with the scapula in the fin is a single bone, the humerus, distal to which are two articulating bones, a radius and an ulna; to these are attached several bony rays. There is a striking similarity between the appendage of the crossopterygian and the primitive fossil amphibian (Fig. 34.5).

ELECTRIC ORGANS

An interesting modification of muscle tissue into electric organs has developed in both the elasmobranchs in electric rays and in bony fish in electric eels and electric catfish (Fig. 34.6). The electric organs are columns of plates, called *electroplaxes,* that all face in the same direction. A network of nervous tissue innervates one side of each plate. When neural discharge triggers the electric organ, the current flows in one direction, from the head to the tail in the electric rays and from the tail to the head in electric eels. The eels are freshwater fishes and discharge up to 600 volts (V) in the high-resistance medium of the Amazon and Orinoco River systems of South America in which they live. Electric rays live in low-resistance salt water and only produce up to 50 V, but they are powerful and can produce over 5 kilowatts (kW) of power.

In addition to the protection and predation attributed to the use of the electric organ, those with low discharge may serve for orientation by having their flow patterns distorted by surrounding objects. Specialized neuromast cells (see below) in the lateral line system may assist in interpreting the form of these currents.

FEEDING AND DIGESTION

The most primitive food-getting device was filter-feeding or a slow vacuumlike sucking of organic material from broad-bladed algae, rocks, and so forth. This mode was apparently used by the primitive agnaths. Modern agnaths can suck and rasp at the flanks of other fish or even use

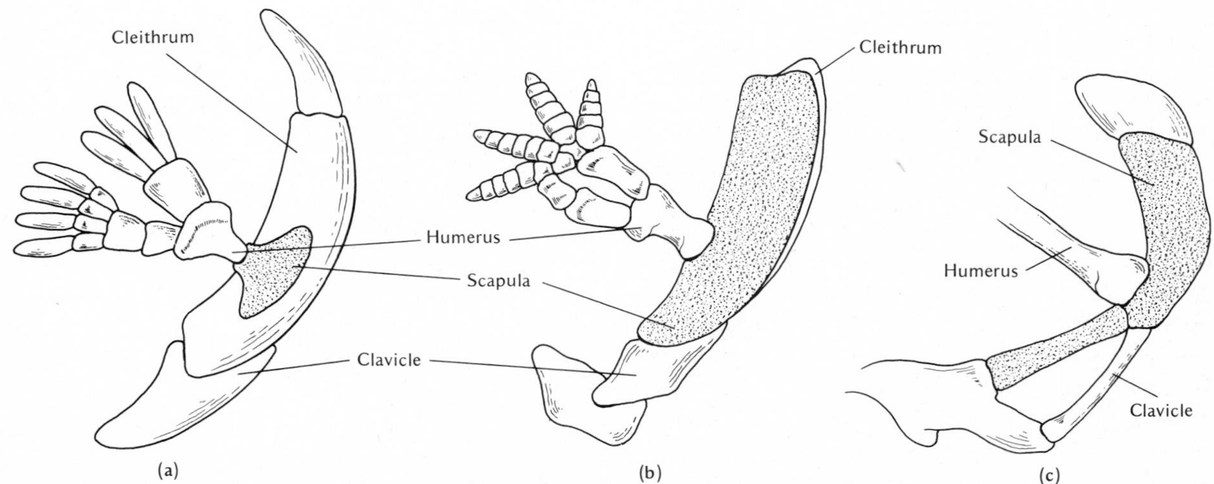

Figure 34.5 Evolution of the pectoral girdle as seen from the medial (internal) view of the girdle. **(a)** A crossopterygian. **(b)** A primitive (fossil) amphibian. **(c)** A frog.

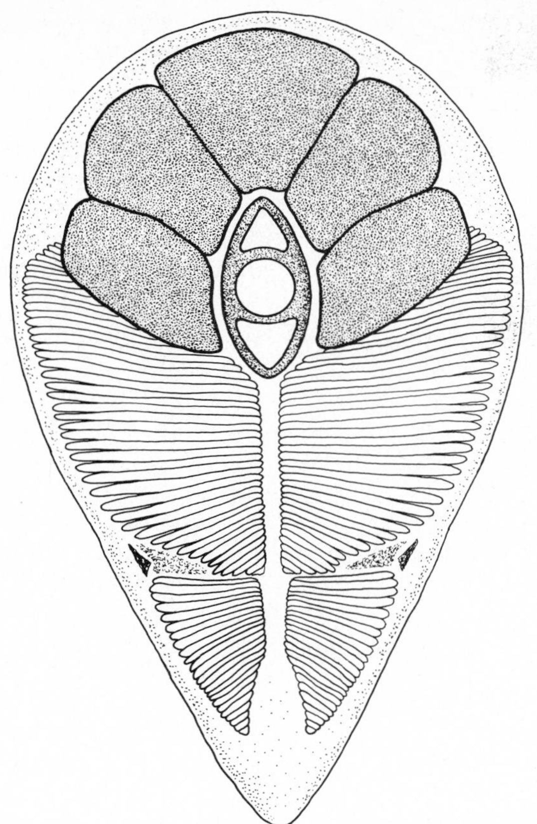

Figure 34.6 Electric organs in the tail of an electric eel. Note the plates, electroplaxes.

the horny ridges at the edge of the mouth to pull away pieces of flesh. With the evolution of a jaw system and with several forms of suspension, considerable flexibility in the aquisition of food evolved. The development of teeth made it possible to secure food in a predatory way. Some fish, such as the deep-sea gulper eels, have enormous jaws that can open to take in large amounts of food. A number of higher fish have teeth on the palate and pharynx as well as on the jaws. Some fish are adapted for eating heavy-shelled mollusks. Noteworthy are some of the rays and the chimaeras whose teeth are flattened grinding plates for this function. Other fish have lost teeth around the mouth, and the mouth itself is protrusible. Goldfish and other carp members have this organization, and these fish are browsers or detritus eaters.

There are a few mucous glands in the mouths of some fish, but mostly food is passed through the pharynx to a short esophagus leading into the stomach. The intestine of primitive animals like the cyclostomes has its internal surface increased by a longitudinal flap that is spirally suspended in an essentially straight intestine. Cartilaginous fish and a few primitive bony fish have a spiral valve through most of the length of the intestine, increasing the absorptive area. In the higher bony fish the spiral valve is missing, but the surface area is increased by lengthening and coiling of the gut. The fish have a standard array of proteolytic and lipolytic enzymes derived from the intestinal wall and the pancreas.

GASEOUS EXCHANGE AND INTERNAL TRANSPORT

Gaseous Exchange Fish employ two structures in gaseous exchange with their environment. The first and most common is exchange in the area of the pharyngeal walls associated with pharyngeal clefts. The respiratory surfaces associated with these clefts may be small, as in the case of the agnaths. When they are attached to prey, they draw water into the pharynx and expel it through the gill clefts. At the gill surfaces oxygen is removed from the water to the circulation and carbon dioxide is released to the water.

In cartilaginous fish, and even more extensively in bony fish, the surfaces between pharyngeal clefts become greatly increased through the formation of gill lamellae for the exchange of gases (Fig. 34.7) Water passes over them from the pharynx either through intake from the mouth or from the spiracle, a modified first gill cleft. The water is expelled through individual gill clefts. Several structural and behavioral adaptations have evolved that enhance the aeration of

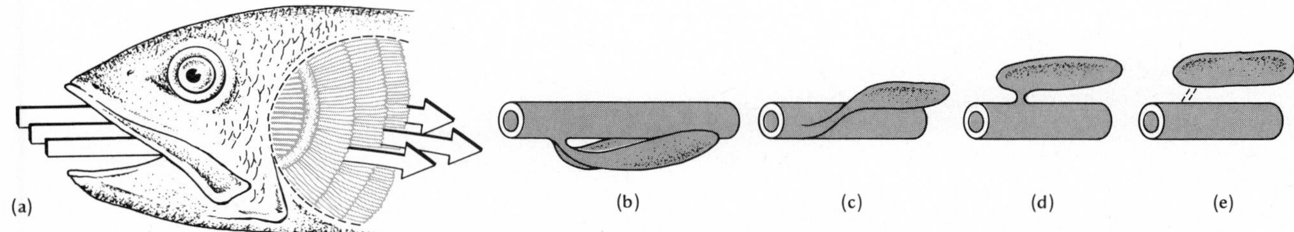

Figure 34.7 Fish gills and gas bladders. **(a)** Passage of water in modern teleost fish (opercular covering not drawn). **(b)** Generalized primitive fish lung. **(c)** Lung of Australian lung fish. **(d)** Dorsal lung with pneumatic duct as seen in sturgeon, some teleosts. **(e)** Gas bladder lacking pneumatic duct in adult form as seen in most teleosts.

gills. For instance, some fish swim with the mouth open, while others use the mouth to pump water past the gills.

In the higher fishes the separation between two sides of a gill is reduced and the water is expelled into a common opercular chamber and discharged through the opercular opening. The *operculum* is a large bony area which opens and closes to exit water to the rear.

In addition to gills, fish may have a second adaptation involving gaseous exchange. This is an air bladder (swim bladder) derived as an outgrowth from the developing gut. One chondrostean, *Polypterus,* in Africa uses a pair of such air bladders as lungs in times of drought.

The lungfish also have air bladders serving as lungs and that are subdivided into subsidiary chambers, increasing the respiratory surfaces.

Sometimes the air bladder is connected to the gut ventrally, sometimes dorsally, and in some advanced fish it loses its connection with the esophagus in the adult. Functionally, the air bladder assists the fish in achieving neutral buoyancy so that it can remain at specific depths rather than swim actively or rest on the bottom as the sharks must do, since they lack air bladders. The appropriate buoyancy is achieved either by gulping air into bladders connected to the esophagus or by a specialized secretory process. The former method is used by relatively unspecialized fish, while the latter method is used by those fish that have lost a connective pneumatic duct between the bladder and the gut. These fish have a remarkable and imperfectly understood gas secretion and resorption system related to the bladder. Gas glands can secrete gases into the bladder from a network of blood vessels, the *rete mirabile,* with arteries and veins running in a counter-current system. It is presumed that the gas moves from the blood to the air bladder through an active transport mechanism. Gas is absorbed through a special oval area on the bladder. The combination is remarkable in being able to concentrate gas in the bladder at pressures in excess of 225 atmospheres.

Air bladders are used for the reception of sound or for the production of sound. For instance, grunts produce a sound by grinding their pharyngeal teeth, and this is resonated by the air bladder.

Oxygen is distributed through the body in oval nucleated hemoglobin-containing erythrocytes. However, one antarctic ice fish has no erythrocytes. Its habitat is deep, cold, and with abundant food; the conditions apparently permit adequate oxygen uptake through absorption.

Blood formation is in the liver, spleen, and kidneys. In addition to erythrocytes, the usual population of leukocytes function in protection against disease, promotion of growth, and so forth. The vertebrates are the only organisms that have lymphocytes that function in the production of immunoglobulins (Ch. 11). While this activity is not fully functional in cyclostomes, it is well developed in cartilaginous and bony fish.

Internal Transport The blood is pumped by a linear flowing, but folded, heart consisting of a sinus venosus that receives all venous blood, delivers it through the atrium into the principal pumping chamber, the ventricle, that passes it through the conus arteriosus into a ventral aorta. This artery runs forward under the pharynx, branches bilaterally into aortic arches passing dorsally between the gill pouches. In the gills there is an afferent artery connected by arterial capillaries in the gills to an efferent artery that delivers blood to the dorsal aorta. Oxygenated blood is then delivered to the body and the viscera. Venous blood is collected from the trunk by bilateral common cardinal veins that empty into the sinus venosus. The common cardinals receive blood from paired anterior and posterior cardinals draining the head and trunk. Blood from the viscera is collected by a hepatic portal vein and carried to the liver from which it is delivered by the hepatic veins to the sinus venosus. The venous sinuses in the liver form a portal system (Ch. 11). Another portal system exists in the kidney where blood is distributed into sinuses as it passes toward the heart; it leaves the kidneys to drain into postcardinal veins. Another venous return is either by paired lateral abdominal veins as seen in cartilaginous fishes or unpaired ventral abdominal veins as present in some bony fish. These veins are of interest in tracing their evolution. They are associated with such extraembryonic structures as the allantois or placenta in reptiles, birds, or mammals, as will be seen in Chapters 36–38.

OSMOREGULATION, WATER BALANCE, AND EXCRETORY CONTROL

Osmoregulation and water balance are especially important to fishes because their aqueous environment usually is not isosmotic with the tissues and the blood. It may be hypotonic if freshwater or hypertonic if seawater. The tendency of freshwater fish to take in water and for marine fish to become dehydrated is principally compensated in fish by the structure and physiology of the kidney and by the presence of special cells in the gills.

The kidneys of fish are best understood by briefly reviewing their development. During the early embryogeny of a fish pronephric kidneys develop (Ch. 10). Soon moderately coiled ducts join the pronephric duct along its length. These coiled ducts are expanded and indented at their ends with a glomerular tuft of blood capillaries and are not connected to the coelom. This embryonic replacement of the pronephric kidney is a mesonephric type; as it utilizes all the available kidney area in development, it is called an *opisthonephric* kidney. This functional adult kidney is involved in osmoregulation and water balance. In freshwater fishes the kidney removes excess water, and the large renal capsules and glomeruli are helpful in this respect. Also, salt-absorbing cells in the gills add salts to the blood.

The problem of the marine fish is the reverse of that described above. As the blood's salt concentration is lower than that of sea water, the fish tend to lose water to the surroundings and risk dehydration. Compensation for this is achieved in several ways. The bony fish drinks sea water and then gets rid of excess salt through salt-secreting cells in the gills. Also, the renal capsules may be very small or absent, thus decreasing the tendency for loss of water through filtration (Ch. 10). This is not true of sharks, who have normal sized renal corpuscles. However, they have another compensation for retaining appropriate osmotic balance. They retain urea in the blood to increase its osmotic pressure.

Some fish are capable of adjusting quite easily to shifts in the salinity of their medium. The exact mechanisms for this are not fully understood. Such fish are called *euryhaline* fish while those with narrow salinity tolerance are *stenohaline* fish. Some euryhaline fish, such as salmon, migrate from the sea to fresh water to spawn; these are *anadromous.* Those of reverse behavior, migration from fresh water to the sea, as in the case of eels, are *catadromous.*

REPRODUCTION

Mating in fishes may be seasonal or not depending on the species. It may result in males simply spreading sperm (milt) over the eggs deposited by the female. Or there may be complex sets of stimuli that release the mating event; these may be courting patterns, nest buildings, feinted aggressions, or color displays between the sexes. Some fish such as the male sharks have claspers on their pelvic fins; these assist in the coupling of pairs and the deposition of sperm in the cloaca of the female.

There are a few fish that are naturally hermaphroditic. Functionally, there is variation in the expression of this hermaphroditism. In hagfishes, for instance, the anterior end of the gonad is ovary and the posterior end is testis. However, self-fertilization does not occur in the hagfish because the two parts mature at different times. There are some fish that make a change from one sex to another. If the change is from male to female, the fish are *protandrous* species, if from female to male, they are *protogynous.* Then there are one or two families where the sperm from the ovotestis will fertilize the eggs in the same organ before the eggs are released.

Testes are usually lobated, elongate structures, each lobule having sperm all at the same stage of development. In agnaths the sperm are released into the coelom and escape through genital pores, there being no sperm ducts. In cartilaginous and bony fish there is an association established between the testis and the opisthonephric (mesonephric) kidney. Such a relationship is retained with modification among all the vertebrates except the agnaths.

The ovaries in fishes often extend the length of the body cavity. Some fishes such as sharks ovulate into the coelom. In the sharks the oviducts are fused at the anterior end and have a common flared opening to receive the eggs. There are a number of variations in the genitourinary systems of fishes, ranging from those described above to cases where the oviducts or sperm ducts may be absent. In general, the Chondrichthyes have internal fertilization while in the Osteichthyes it is external.

While the *oviparous* condition exists in many fishes and they lay their eggs into the water, others are *ovoviviparous.* The latter fishes retain the eggs within the body, usually at posterior expansions of the oviduct, until they have undergone most or all of their major developments. Also, there are some fishes that are *viviparous;* these have intimate associations between the walls of the female genital tract and structures of the developing young. (Placenta formation in mammals is the common example of viviparity; Chs. 16, 38.) In fishes the association that is usually established is between folds in the walls of the oviduct and the yolk sac of the developing young; this facilitates the exchange of gases, at least.

In fishes, as among other vertebrates, there are numerous degrees of difference between oviparity, ovoviviparity, and viviparity.

SENSORY RECEPTION AND NEURAL COORDINATION

As was seen in Chapter 13, sensory reception is commonly achieved through interpreting touch, heat, cold, pain, taste, light, and sound. The fishes have these capabilities through the presence of skin receptors, taste structures associated within and sometimes around the mouth; their eyes have the general vertebrate origin and structure; and they are unusual, in our frame of reference, in interpreting sound principally through pressure exerted on the surface of the body rather than through specialized auditory structures such as found in humans. In view of the aquatic environment of fishes, their utilization of their sense organs has evolved in ways characteristic for that environment. For instance, inasmuch as taste interpretation is a chemical event requiring substances to be in solution, terrestrial vertebrates have this sense confined to the moist area of the mouth and nasopharynx. Fish have no such restrictions and taste receptors may exist on the body and on the fins as well as in the oral region.

Smell is interpreted by specialization of several forms of organization. In the jawless fish there is a single median specialization that either is a pit (lampreys) or a tube (hagfish) connected to the pharynx. Elasmobranchs have paired nasal pits lined with folds or ridges of sensory epithelium and called *choanae.* The dipnoan fishes have both internal and external choanae. This trait is believed to have existed in the crossopterygians and is one of the evidences for the origin of the tetrapods from these fishes.

The eyes of fishes have a number of interesting adaptations. These range from complete blindness in some cave fishes to very specialized lens systems that permit the fish to focus light traveling either through the water or through the air when the eyes are projected above the surface as seen in the South American "four-eyed fish" *Anableps* (Fig. 34.8). The lens of the *Anableps'* eyes is divided, and the lens for use in air and the lens for use in water are different distances from the retina. When the fish floats at the surface, it can see in both the air and water. If the habitat of a fish species is deep, two variations in the eye can occur. Some are blind, as in cave fish, but others have relatively large eyes. This latter variation occurs in all the classes of vertebrates when the organisms inhabit areas of low light intensity.

Fishes lack external ears, middle ears, and there is no cochlea present in the inner ear, it being confined to a utriculus and sacculus associated with static equilibrium and the semicircular canals. Although some may be able to interpret sound, the principal function of the inner ear is equilibrium. Closely associated with the auditory equipment of fishes is a system restricted to them and to larval amphibia, the lateral line system. This consists of a minimum of single rows along each side of the body where sensory receptors, *neuromasts,* may lie on the surface or be buried in canals that have sequences of pores that expose them to the surface. Presumably the lateral line system functions through the interpretation of pressure changes or currents in the water.

The central nervous system of fishes is relatively unspecialized. The brain is linear, consisting of five parts, the anterior *telencephalon* that gives rise to the cerebrum and the olfactory lobes; the former is primitive in fish while the latter are usually large in view of the strong olfactory functions of the sensory receptors.

Figure 34.8 The four-eyed fish, *Anableps.* The upper part of the eye is used for seeing above water, the lower part accommodates to underwater vision. (New York Zoological Society Photo)

The cerebrum is relatively trivial. It may be divided into two areas that will be important in later evolving vertebrates, the thin dorsal *pallium* and the basal ganglionic mass known as the *corpus striatum.* The *diencephalon* is important in fishes in that it gives rise to the *thalamus,* an important center for relaying visual and olfactory impulses. As dorsal outgrowths of the diencephalon, the agnaths have two structures, a parietal body and a pineal body; only the pineal body is present in most fish, however. The pineal body seems to be involved in the rhythms of "biological clocks" (Ch. 12). The midbrain, *mesencephalon,* is the center for nervous coordination, and its two optic lobes function in visual interpretation. The *metencephalon* in fishes varies in size, as it does in other vertebrates, depending on the degree of bodily activity of the fish. Those fish with slow and sluggish locomotion have small metencephalons, those that are active have larger ones. The hindbrain, *myelencephalon,* forms the medulla. In the fish this area has an importance that is shared with all other vertebrates. The medulla is very important as the center for many vital functions such as control of the heart, vascular control, metabolism, and so forth. In fishes and larval amphibia it serves as a center for the lateral line system and for the inner ear.

The peripheral nervous system has many similarities to other vertebrates (Ch. 13). The brain has only ten cranial nerves, a trait it shares with amphibia, instead of the twelve seen in the other classes. The agnaths are unusual in that their sensory and motor roots are not joined in the spinal nerves, but they are in elasmobranchs and bony fishes just as they are in the other vertebrates.

The autonomic nervous system of agnaths is little understood, is irregularly organized in the elasmobranchs, but in bony fishes it is organized into chains of ganglia in the sympathetic division as it is in other vertebrates. However, here the parasympathetic subdivision of the autonomic nervous system is not fully developed.

In summary, we see that the fishes have the basic neural mechanisms and sensory receptors to function as well integrated organisms, even though the pattern of organization is relatively simple by contrast with the tetrapod classes.

Chemical coordination is also important in fishes as it is in the other classes. The pituitary occupies a position under the diencephalon just as it does in other vertebrates. Also, it functions in the several ways that are characteristic of other vertebrates (Ch. 12). The thyroid gland is well developed. However, the parathyroid glands do not seem to exist, even in the bony fishes. The thymus appears first in cyclostomes as poorly developed tissue; it is present as several outgrowths above the gill pouches in the other fishes. This is an important set of organs since it is the source of the T cells (Ch. 11) that play such an important role in the immune capacities of vertebrates. The pancreas is unremarkable by contrast with other vertebrates and performs the same double functions. The adrenal glands differ in that the two functional areas seen combined in most vertebrates are completely separate in fishes.

BEHAVIOR

Noteworthy behaviors in fish relate to their reproduction, their sense of territoriality, and their migrations.

Reproductive behavior patterns have been mentioned above (see also Ch. 43). Often associated with reproduction and mating are strongly developed senses of territoriality. The stickleback (Ch. 43) aggressively defends a localized area in the water where conditions for nest building are favorable and continues to be aggressive toward males after the nest is built. Other fish, such as the bitterling, even have specific territories, a single freshwater mussel, which the male selects and protects because the gills of this mussel are where the female will deposit her eggs. Further, fish fanciers know that males of *Betta,* the Siamese fighting fish, cannot be kept together due to their strong sense of territoriality and their aggressiveness.

Migrations in fishes are still not fully understood, but they are real and can affect the activities of fishing fleets. These migrations may be related to feeding or to breeding. The migrations of salmon are noteworthy. Salmon eggs are laid in tributaries of a few major river systems in the northeastern and northwestern United States. The young of these anadromous fish migrate to the sea. Several years later, when they reach adulthood, they return to the same place where they developed from eggs. Strikingly different is the story of the Atlantic and the Pacific salmons. The Atlantic salmon, *Salmo salar,* makes repeated yearly runs to the spawning tributaries. The Pacific salmon return to their stream of origin just once, spawn, and die. This applies to North Pacific species of *Oncorhynchus:* the sockeye (*O. nerka*), the coho (silver) (*O. kitsutch*), the humpback (pink) (*O. gorbuscha*), and the chum salmon (*O. keta*). However, a close Pacific relative, the steelhead, a sea-going rainbow trout, *Salmo gairdinerii,* like the Atlantic salmon, does not necessarily die after its upstream spawning run.

A quite remarkable migratory pattern is seen in North American and European freshwater eels. The American eel lives in the waters of coastal streams of northeastern North America while the European eels inhabit streams in the British Isles and Europe. When mature, they migrate to sea and do not return. They proceed to the warm Sargasso Sea, near Bermuda, where they spawn and die. The eggs hatch into small, flat, transparent larvae that drift northward. When they arrive at the coastal waters they transform into elongate elvers, about 5–6 cm (2–2½ inches) long and move upstream where they grow to adulthood. As eels have commercial importance, their life histories are being explored thoroughly. Among the data recently accumulated is the suggestion that there is no true species difference in the eels on the two sides of the Atlantic but only that the environmental and temporal influences on the young larvae as they drift northward from the Sargasso Sea determine their characteristics. While not yet confirmed, this would simplify the problem of how European eels seemed always to migrate to Europe and American eels to America.

There is some evidence accumulating for fish, as for birds (Ch. 37), that celestial cues are used in their migrations.

Importance of Fish in the Biosphere

One of the major areas of importance of fishes is their presence in food chains. Fishes may not only be primary consumers (Ch. 45) of plant material, but they also exist as series of secondary consumers, with larger predatory fishes occupying the ends of the food chain. Thus in the seas alone, the cycling of the energy may be strongly routed through several kinds of fish. In addition, the fish themselves are the natural energy source for a number of birds and mammals. Seals and sea lions, for instance, are dependent on fish for food. Some fish establish strong relationships with other animals. The arrangement may be a loose one as in the case of pilot fish that accompany sharks or other fish. More intimate relationships may exist as with the remoras, medium-small fish, related to the perch, that have the anterior dorsal fin modified to form a suction disk. These fish attach themselves to the undersides of sharks. This provides them with a ready place from which to detach to pick up morsels from the shark's meal. Some fish establish similar relationships with invertebrates. There are small fish that spend much of their time in the oral and gut cavities of sea anemones without triggering the stinging assault from the anemone's tentacles (Ch. 23). They also have some territoriality as specific fish occupy specific individual anemones. Another fish occupies the cloaca of a sea cucumber (Ch. 32). The ultimate in interrelationship in fish is intraspecific. The oceanic angler fish (*Photocorynus*, Fig. 34.9) has a small male that is free-swimming for awhile but eventually each male grasps a female with its mouth and the tissues of the male and female become fused. The male is now a parasite, depending on the female for nutrients. Several deep-sea fish have such sexual parasitism by the male, seemingly an adaptation to assure close proximity of the male gametes at breeding times. It has been suggested that the female determines the reproductive activity of the male through endocrine control.

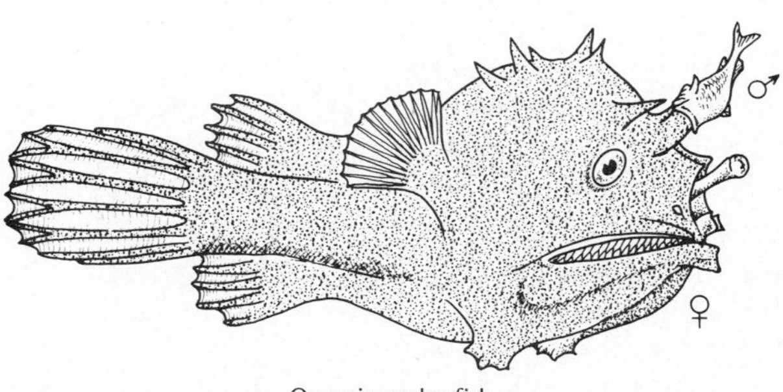

Figure 34.9 Female Angler fish with male as a fused parasite. After the male grabs the female the tissues grow together.

In addition to the role of the fish in the biosphere, the fish have had a great importance for humans. Whole societies have depended on fish as a principal source of food. Also, the fishing industry has been important to nations and has often been the key factor in the establishment of sea rights, determining naval action, or influencing the cultural patterns of peoples. Fish provide a rich protein source that is used for both human and animal consumption and for fertilizer.

Classification

CLASS **Agnatha** The jawless fish are the oldest and most primitive vertebrates. There are two subclasses, the extinct Ostracodermi and the Cyclostomata consisting of the lampreys and hagfish. (Some would further separate the lampreys and hagfish because they have a number of differences in structure and modes of life; it is probably true that their evolutionary roots in common are quite distant.)

The ostracoderms (Fig. 34.10) had an exoskeleton of dermal plates or denticles. This armor was present sometimes as a shield over the head and thorax and was protective in a way analogous to

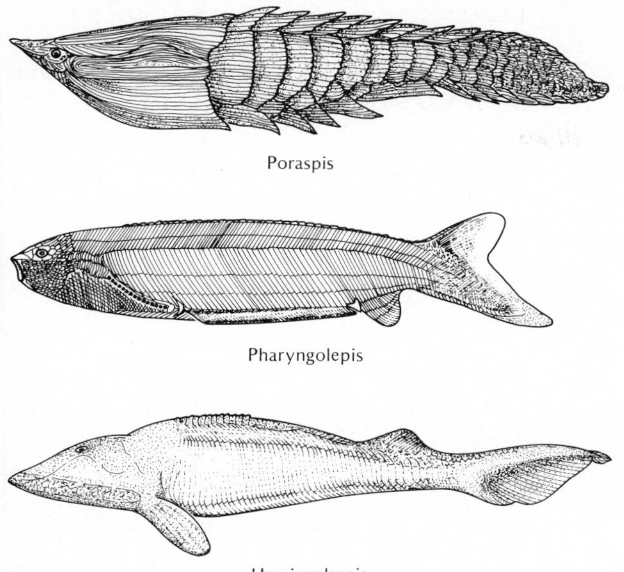

Figure 34.10 Three representative fossil ostracoderms. Bony coverings of the body characterize these fish. (From Romer, *The Vertebrate Body,* 4th ed. © 1970 by W. B. Saunders Company, Philadelphia.)

a similar, but chitinous, shield forming the carapace of the horseshoe crab (Ch. 29). While living agnaths are parasitic, the ostracoderms could not have been parasitic on other fish as they were the first. Their funnellike jawless mouth served them in filter feeding on detritus and small life in the mud. The ostracoderms also did not show paired appendages as a universal trait.

The cyclostomes lack external skeletal support, and the endoskeleton consists of a notochord with a few neural plates alongside the neural tube plus a fibrocartilaginous reinforcement of the walls of the pharynx. The cyclostomes are the only parasitic vertebrates. Their round mouth serves as a sucker. Lampreys attach to other fish and with a rasping tongue erode the walls to suck out blood and body fluids (Fig. 34.11). Then the animals detach and seek another host. Both freshwater and marine lampreys exist. Both, however, migrate by well-established routes to freshwater streams to spawn. At one time there were no lampreys in the Great Lakes. With the construction of the Welland Canal near Buffalo, New York, a marine lamprey, *Petromyzon marinus,* gained entrance to the Great Lakes, proliferated, and caused extensive damage to the fishing industry. For instance, between 1935 and 1949 the lake trout catch dropped from over 450,000 metric tons (500,000 short tons) to less than 1 ton. Rigorous chemical controls on lamprey spawning in tributary streams has, in recent years, reduced the lamprey population and the fishing industry has improved.

Hagfishes are all marine. Sometimes they actually penetrate the bodies of unhealthy fishes they attack. They function more as scavengers than as parasites.

CLASS **Placodermi** These fishlike vertebrates possessed jaws created by a modification of the first and second gill arches. The first arch became folded into a contribution to the upper jaw and a lower part that created the support for the lower jaw. The second arch became modified so that it gave support to the upper and lower jaws. Another trait of this class is that the cranium became partly ossified to serve as a support for the brain. Subclasses of the placoderms are partly determined by the degree to which a bony vault of bone lies over the brain; this is called the *dermatocranium* inasmuch as the bone is directly derived from membrane and is not preceded by cartilage. Ostra-

Figure 34.11 Lamprey parasitic on catfish. (Courtesy of the American Museum of Natural History, New York.)

Figure 34.12 West Indian shark. (Courtesy of the American Museum of Natural History, New York.)

coderms became prominent between the Silurian to the Permian periods and were the predatory vertebrates of those times. As happens in other vertebrates, these forms became adapted to several modes of existence. Some were sharklike while others evolved with broader bodies, somewhat resembling the rays, and they apparently were bottom dwelling.

With the appearance of the bony and cartilaginous fishes with their greater flexibility and variability through the loss of heavy dermal armor, the ostracoderms became extinct. This is the only total extinction of a class among members of the successful subphylum Vertebrata.

CLASS **Chondrichthyes** While the evolution of the bony fishes occurred slightly earlier in the Devonian period than that of the cartilaginous fishes, cartilaginous fish are discussed first inasmuch as they do not serve as ancestors for other vertebrates, the amphibians, as do the bony fishes.

The sharks, rays, and chimaeras make up this class. Their common characteristics are that they are jawed fishes with a cartilaginous skeleton. There is evidence that the class evolved from primitive bony fishes. While elasmobranchs, as the cartilaginous fishes are often called, presently are mostly marine, they evolved from freshwater ancestors. As has been discussed (Ch. 10), their successful adaptation to marine existence was accompanied by their capacity to retain large amounts of urea dissolved in the blood to achieve an osmotic equilibrium with the environment. There are two subclasses.

SUBCLASS **Elasmobranchii** This subclass is composed of sharks and rays. Usually sharks are streamlined, fast-swimming, voracious feeders (Fig. 34.12). One kind, the whale shark, is the largest known fish and may attain a length of over 15 m (50 feet). The sharks are characterized by pectoral and pelvic fins, the latter being modified in males to form claspers for mating; by two prominent dorsal median fins; by the external skeleton being reduced to placoid scales that, in the oral area, grade into the form of teeth; by a heterocercal tail in which the trunk turns upward into the large upper lobe of the tail; and by a subterminal mouth under a rostrum.

The rays (Fig. 34.13) are flattened dorsoventrally, and their pectoral fins are expanded to extend along the length of the trunk. They usually are bottom dwelling. Some rays have heavy flattened teeth that are used for crushing mollusks.

Figure 34.13 Cow-nosed rays. Note the flattened bodies and extended pectoral fins. Movement strongly resembles aerial flight (*top*). (New York Zoological Society Photo) Dorsal view of skate (*bottom*). (North Carolina Biological Supply).

SUBCLASS **Holocephali** This subclass is composed of the chimaeras or ratfishes (Fig. 34.14). The living members occupy deep oceanic waters. They lack placoid scales, have a different jaw suspension than the elasmobranchs, flat plates instead of teeth, and the number of gill slits is reduced and covered by a fold.

CLASS **Osteichthyes** The characteristics of the Osteichthyes are a bony skeleton, at least to some

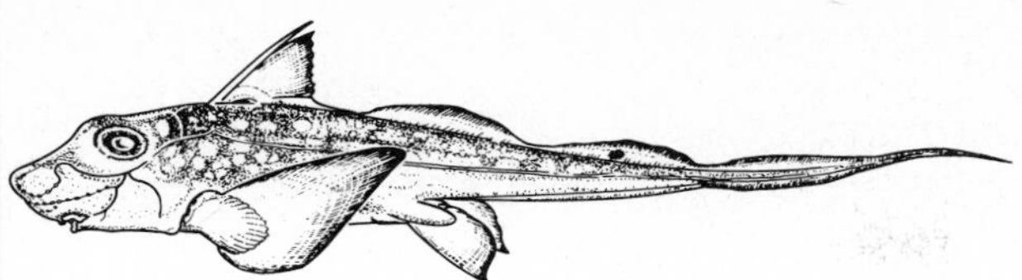

Figure 34.14 Chimaera or ratfish. This is a relatively rare cartilaginous fish. Though not flattened as are some rays that are mollusk-eaters, mollusks are the principal food of chimaeras. (From Romer, *The Vertebrate Body,* 4th ed. © 1970 by W. B. Saunders Company, Philadelphia.)

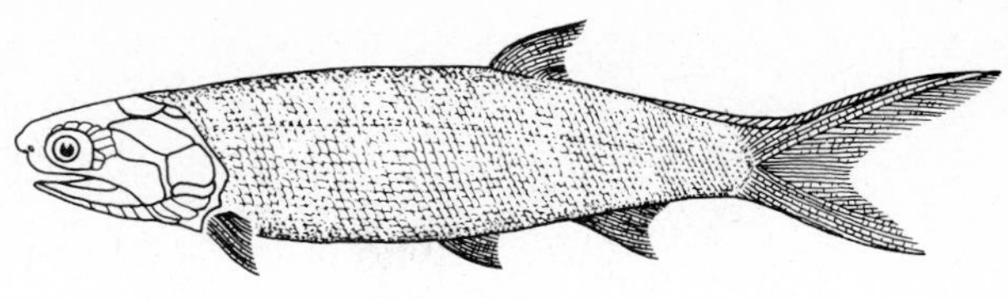

Figure 34.15 A paleoniscoid. An extinct bony fish.

degree; usually homocercal tail; terminal mouth; gills covered by an operculum; scales of three types, ganoid, cycloid, or ctenoid, when present; mucous glands in skin.

There are differences of opinion about the classification of the bony fish. In one scheme there are two subclasses, the ray-finned fish, subclass Actinopterygii; and the fleshy-finned fish, subclass Sarcopterygii. The major diagnostic criteria for these two subclasses are as follows. In Actinopterygii the fins are supported by numerous skeletal rays but there is no fleshy base to the fins; they also lack olfactory structures, called *choanae.* In Sarcopterygii the fins have a fleshy base, and usually the proximal portion of the fin has individually identifiable bones; they possess choanae.

SUBCLASS **Actinopterygii** In this classification the actinopterygians are classified into five superorders with, collectively, as many as 60 orders, living or extinct, recognized by some zoologists.

SUPERORDER **Paleonisci** The paleoniscoids are ancient extinct fish whose body, including the tail, resembled that of sharks. They lived in both fresh and marine waters (Fig. 34.15). In view of their existence from the Devonian to the Cretaceous, they appear to have been unsuccessful competitors with the sharks.

SUPERORDER **Polypteri** This group and the following two are called ganoid fish due to the form of the scales. Ganoid scales have a hard enamel-like covering, *ganoin,* over bone. This superorder is represented by one order, Polypteriformes. Only two living genera survive. *Polypterus,* commonly called a bichir is an example. The presence of numerous finlets derived from the dorsal fin is characteristic. The swim bladder functions as a lung. The primitive nature is illustrated by the presence of a spiral valve in the intestine.

SUPERORDER **Chondrostei** These animals have heavy ganoid scales as is seen in sturgeons (*Acipenser*) (Fig. 34.16), but much of the skeleton is cartilaginous. The notochord persists, and the vertebrae lack centra. A spiral valve in the intestine is present. Another representative of this superorder is the paddlefish (*Polyodon*). Though quite large, some over 8 m (25 feet) being recorded, the two living genera occupy freshwater of the Great Lakes and two major rivers, the Mississippi and the Yangtze.

SUPERORDER **Holostei** These ganoids have skeletons often resembling the true bony fish and are called the bony ganoids. They once had both marine and freshwater adaptations, but the two prominent living orders of the six, represented, respectively, by the bowfin (*Amia*) and the gar (*Lepidosteus*) (Fig. 34.17), are lake dwellers.

SUPERORDER **Teleostei** These true bony fish have been remarkably successful and diversified in their evolution. Characteristics in common among them are presence of well-developed vertebral centra, strong cranial ossification, usually small scales lacking ganoin, tail usually symmetrical (homocercal), air bladder dorsal to the gut and sometimes losing its connection with the gut, gut lacking spiral valve but elongate and coiled. Space does not permit the description of the traits of each of the over 40 orders that some ichthyologists

Figure 34.16 Chondrostean fish. Sturgeon (*above*). (Carolina Biological Supply Company) Paddlefish (*below*). (Steinhart Aquarium, San Francisco).

Figure 34.17 Alligator gar, *Lepidosteus spatula,* a holostean ganoid-scaled fish. Some skeletal features resemble those of true bony fish. (Steinhart Aquarium, San Francisco)

recognize. Instead, a few are selected to illustrate the remarkable diversification in this superorder (Fig. 34.18).

ORDER **Clupeiformes** This includes the herrings, tarpons, anchovies, salmon, and their relatives. Relatively, these are primitive teleosts in view of their soft-rayed fins, presence of a central aperture through the centrum, and lack of any association between the pectoral and pelvic girdles. They range in size from 2 cm (1 inch) to over 2 m (7 feet). Their air (swim) bladder is connected to the gut. They are both marine and freshwater and are widespread. This order is important in the fishing industry and economy of many people.

ORDER **Cypriniformes** This order shows fair diversity. While having soft-rayed fins, they possess an unusual chain of bony ossicles lying between the swim bladder and the inner ear (Webberian apparatus), an unusual sound reception device. Representatives of this order are the minnows, gold fish, and catfish on the one hand and the gymnotid eels on the other hand. Minnows have protrusible mouths, teeth in the pharynx, and the anal opening is typically posterior. The gymnotid eels have eel-like bodies, and they develop so that the anal opening is anterior on the throat.

ORDER **Anguilliformes** These are the true eels, and their elongate body has the dorsal and anal fins continuous with the caudal fin. They lack pelvic fins and are mostly marine.

ORDER **Beloniformes** The flying fish, needlefish, and halfbeaks make up this order. In the flying fish, of course, the pectoral fins are enlarged and capable of planing when the fish jumps from the water.

ORDER **Gadiformes** The codfish and hakes, while being soft-rayed, show a trait that develops in several orders, the pelvic fins are anterior to the pectoral fins.

ORDER **Percopsiformes** Consisting of pirate perch and trout perch, but not true perch, this order is included because it illustrates fish intermediate between primitive soft-rayed and advanced spiny-rayed fish. Although the pelvic fins are posterior to the pectoral ones, the pectoral fin has spines and the unpaired dorsal fins, the pelvic, and the anal fins have spines preceding soft-fin rays.

ORDER **Perciformes** The perch are spiny-rayed fish with a ligamentous connection between the pectoral and pelvic girdles. The swim bladder is not connected to the gut in the adult.

ORDER **Muligiformes** The barracudas, mullets, and silversides are members of this order. The pelvic and pectoral girdles are connected by a ligament. There is an anterior spiny-rayed dorsal fin and a soft-rayed fin behind it.

ORDER **Pleuronectiformes** These are the flatfishes that develop asymmetrically, such as the flounder, sole, and the halibut. Their bodies are laterally compressed and both eyes are on one side. The dorsal and anal fins extend along much of the edge of the body. They lack swim bladders and tend to be bottom dwelling with a pale-pigmented side down and pigmented side up.

ORDER **Zeiformes** These fish have their bodies lat-

Figure 34.18 (*opposite*) Variation among teleost fishes. **(a)** Tambaqui, a flattened solelike bottom dweller; note rotation of the eyes to one side of the body. **(b)** Morrish idol. **(c)** Flashlight fish. The white spot below the eye produces intense light from bioluminescent bacteria in the organ. An opaque curtain can be raised to obscure the light. **(d)** False stonefish. The eye and mouth can be seen but note how the rest of the fish blends with the rocks. **(e)** Tropical angler fish. **(f)** Sharksucker or remora. This fish is anchored to the shark just below the shark's four gill slits; it does not parasitize the shark but survives on residues from the shark's feeding. **(g)** Turkey fish. **(h)** Guinea fowl puffer fish. (All photos, Steinhart Aquarium, San Francisco)

(a) (b) (c) (d) (e) (f) (g) (h)

Figure 34.19 Cast of the coelacanth, *Latimeria.* (Courtesy of the American Museum of Natural History, New York.)

erally compressed but retain true bilaterality and thus are broad dorsoventrally.

ORDER **Lophiformes** The angler fishes are quite modified. Their pectoral fins are strengthened and can be used for walking, the pelvic fin is far forward on the throat, and the dorsal fin has its first ray elongated as a lure to attract other fishes.

ORDER **Tetraodontiformes** These fish, ranging from small to very large, include the trigger fishes, puffers, and ocean sunfishes. They all have relatively small mouths and gill openings. The puffer fish is noted for its ability to swell to nearly spherical dimensions and for its poisonous spines.

The preceding listing does not do full justice to the remarkable superorder Teleostei, but it is hoped that it conveys some of the great diversity that has occurred among its members.

SUBCLASS **Sarcopterygii** These fleshy-finned fishes consist of two orders.

ORDER **Crossopterygii** Believed to be extinct for millions of years until 1939 when *Latimeria* was caught off the coast of Madagascar, there are now two known living genera (Fig. 34.19). The order has two groups, the coelocanths and the extinct rhipidisteans. All have a fleshy lobe at the base of the fins, including the unpaired dorsal fins. Of considerable significance relating these fish to the amphibians is the presence of internal choanae, channels from the exterior to the mouth for the passage of air. Also, another interesting feature to be seen in the primitive amphibians is that the teeth have many folds in the enamel.

ORDER **Dipnoi** These are the lungfishes represented in living forms by three genera, one each in Australia (*Neoceratodus*), South America (*Lepidosiren*), and Africa (*Protopterus*) (Fig. 34.20). The swim bladder serves as a respiratory structure during times when the fish are encased in a mud cocoon during droughts. The dipnoi also have internal choanae and once were thought to be the ancestors of the land vertebrates.

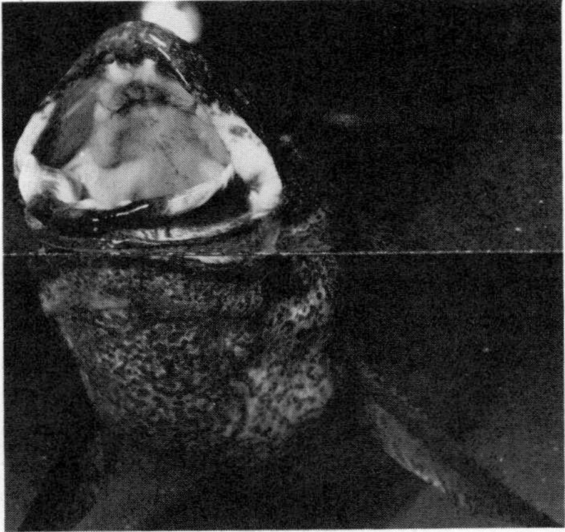

Figure 34.20 African lungfish, *Protopterus aethiopicus.* Lungfish coming to surface to breathe air (*right*) (New York Zoological Society Photos)

SUGGESTIONS FOR FURTHER READING

Berg, L. S. *Classification of Fishes, Both Recent and Fossil.* Travaux de l'Institut Zoologique de l'Académie des Sciences de l'URSS, vol. 5, pt. 2 (1940).

Brown, M. E. (Ed.). *The Physiology of Fishes,* 2 vols. New York: Academic Press (1957).

Colbert, E. H. *Evolution of the Vertebrates.* New York: Wiley (1955).

Gilbert, P. W. (Ed.). *Sharks and Survival.* Boston: D. C. Heath (1963).

Herald, E. S. *Living Fishes of the World.* Garden City, NY: Doubleday (1961).

Hoar, W. S., and D. J. Randall (Eds.). *Fish Physiology,* 6 vols. New York: Academic Press (1969–1972).

Lagler, K. F., J. E. Bardach, and R. R. Miller, *Ichthyology.* New York: Wiley (1962).

Lineaweaver, T. H., III, and R. H. Backus. *The Natural History of Sharks.* Philadelphia: J. B. Lippincott (1970).

Marshall, N. B. *Exploitations in the Life of Fishes.* Cambridge: Harvard Univ. Press (1971).

Romer, A. S. *The Vertebrate Body,* 3rd ed. Philadelphia: W. B. Saunders (1962).

Young, J. Z. *The Life of Vertebrates,* 2nd ed. New York: Oxford Univ. Press (1963).

From Scientific American

Applegate, V. C., and J. W. Moffett. "The Sea Lamprey" (Apr. 1955).

Brett, J. R. "The Swimming Energetics of Salmon" (Aug. 1965).

Carey, F. G. "Fishes with Warm Bodies" (Feb. 1973).

Gilbert, P. W. "The Behavior of Sharks" (July 1962).

Gray, J. "How Fishes Swim" (Aug. 1957).

Grundfest, H. "Electric Fishes" (Oct. 1960).

Hasler, A. D., and J. A. Larsen. "The Homing Salmon" (Aug. 1955).

Jensen, D. "The Hagfish" (Feb. 1966).

Johansen, K. "Air-Breathing Fishes" (Oct. 1968).

Leggett, W. C. "The Migration of the Shad" (Mar. 1973).

Lissmann, H. W. "Electric Location by Fishes" (Mar. 1963).

Lühling, K. H. "The Archer Fish" (July 1963).

Millot, J. "The Coelacanth" (Dec. 1955).

Ruud, J. T. "The Ice Fish" (Nov. 1965).

Scholander, P. F. "The Wonderful Net" (Apr. 1957).

Shaw, E. "The Schooling of Fishes" (June 1962).

Todd, J. H. "The Chemical Languages of Fishes" (May 1971).

35 The Amphibians

The Amphibians evolved from lung-bearing crossopterygian ancestors whose fins had the basic architecture from which the tetrapod limb developed. A unique distinction between crossopterygians and amphibians is the presence of two distinct modes of life in the amphibian life history. The early mode is aquatic and represented by a larval (tadpole) stage. The second mode is terrestrial and adult. Thus this class of vertebrates is transitional to the totally terrestrial capabilities of reptiles, birds, and mammals, and the name of the class, amphibia, is descriptive of this (Gk. *amphi,* double; *bios,* life).

Water is important in both phases of the life history of most amphibians. Their eggs are usually laid in ponds or streams, the larvae develop there, and after metamorphosis they must usually remain close to water for mating, for reproduction, and also to assure that they do not become desiccated.

Living amphibians fall into four orders that share these characteristics. They have mostly bony skeletons, sometimes with ribs present. They do not retain the notochord in the adult stage. The choanae are connected with the oral cavity through the presence of external and internal nares (nostrils). Like the fish, they are ectotherms and obtain their source of heat from the environment. Their skin is smooth, rarely has any scales, and it is often characterized by striking pigmentation. Except in the specialized caecelians, they are mostly pentadactyl tetrapods, meaning that their feet are built on a plan of five digits and there are four limbs. They all have one added chamber in the heart, the atria being paired, to create a heart of five chambers. Metamorphosis usually occurs.

Anatomy and Physiology

PROTECTION AND LOCOMOTION

Protection in amphibians involves three problems. One relates to prevention of desiccation; the other two are prevention of predation and the acquisition of food. For those amphibians that are not permanently aquatic, desiccation is prevented by two interrelated but different conditions. First, most amphibians possess numerous glands in the skin that produce a sugar-protein mucous secretion that deters loss of water. Also, there is an absence of water-releasing glands such as sweat glands. The skin also acts to deter predation since many amphibians have toxic secretions from poison glands in the skin. Some Central American frogs, for instance, can kill animals at least as large as dogs who come in contact with enough of the secretions from the frogs' skin. Newt secretions also seem to be unpalatable to most mammalian predators.

The behavioral pattern of most amphibians also protects them because most occupy shaded or secluded (cryptic) habitats. Only for a few desert toads is there usually a minimal opportunity to find shaded cooler locations, and these toads are unusual in the thickness and wartiness of their skins.

Protection from enemies is also related to the structural and behavioral traits of the animals. Salamanders and newts, being elongate and close to the ground, usually can scurry to a protective spot inaccessible to predators. The large order Anura accomplishes protection from predation by the unusual development of the hind limbs, to be described below, that permits them to leap large distances and also by a behavioral trait of "freezing" and remaining immobile in the presence of a potential enemy; the latter trait is only useful, of course, in the event the predator does not see the frog or toad.

THE SKELETAL SYSTEM

While the skull of fishes has a basic pattern of bones seen in all vertebrates, it is complicated by numerous bones not seen in the tetrapods. Therefore, a more detailed discussion of skull bones has been deferred until this point.

Most amphibians have a broad flat skull in contrast with the strong arch of bones overlying the head of fishes and the extinct amphibians. The dermal bones over the roof of the skull are reduced in number. From front to back these bones are the paired *premaxillaries, maxillaries, nasals, frontals, parietals,* and *squamosals* (Fig. 35.1). Some of the area over the brain actually lacks a bony covering in some amphibians, especially the urodeles. The back of the skull usually consists of a pair of *exoccipitals* that flank the entry of the spinal cord into the skull through the *foramen magnum.* Each has a condyle that articulates with the first vertebra. The floor of the braincase is cartilaginous and consists of the median *presphenoid* and *basisphenoid,* and the paired and flanking *alisphenoids.* These structures have a dermal bone underlying them, the *parasphenoid.* Furthermore, the parasphenoid also serves as the primary palate inasmuch as the amphibia lack a secondary palate. In front of the parasphenoids a pair of *vomers* lie medial to the internal nares.

The jaw pattern established in fish is retained by the amphibians; namely, a pair of cartilages, the *palatoquadrates,* create the upper jaw scaffolding, and a pair of cartilaginous rods, *Meckel's cartilages,* form the lower jaw and are ensheathed in a series of dermal bones. These two cartilages deserve some attention as they show interesting evolutionary changes in the tetrapod classes of vertebrates. The anterior part of the palatoquadrate produces the *epipterygoid* which does not ossify in the amphibians; it will be of some interest in later vertebrates. The posterior part of the palatoquadrate forms the *quadrate;* this ossifies in anurans but is only incompletely ossified in urodeles. The quadrate is the upper segment of the joint between the upper and lower jaws. The lower segment of this joint is formed by the posterior tip of Meckel's cartilage and is called the *articular;* it ossifies in urodeles but is cartilaginous in anurans. The anterior part of Meckel's cartilage remains cartilaginous or does not persist.

Teeth in modern amphibians are small or lacking. When present they may occupy these bones: premaxillaries, maxillaries, palatines, vomers, parasphenoids in the upper jaw area, and the *dentary,* a dermal bone in the lower jaw.

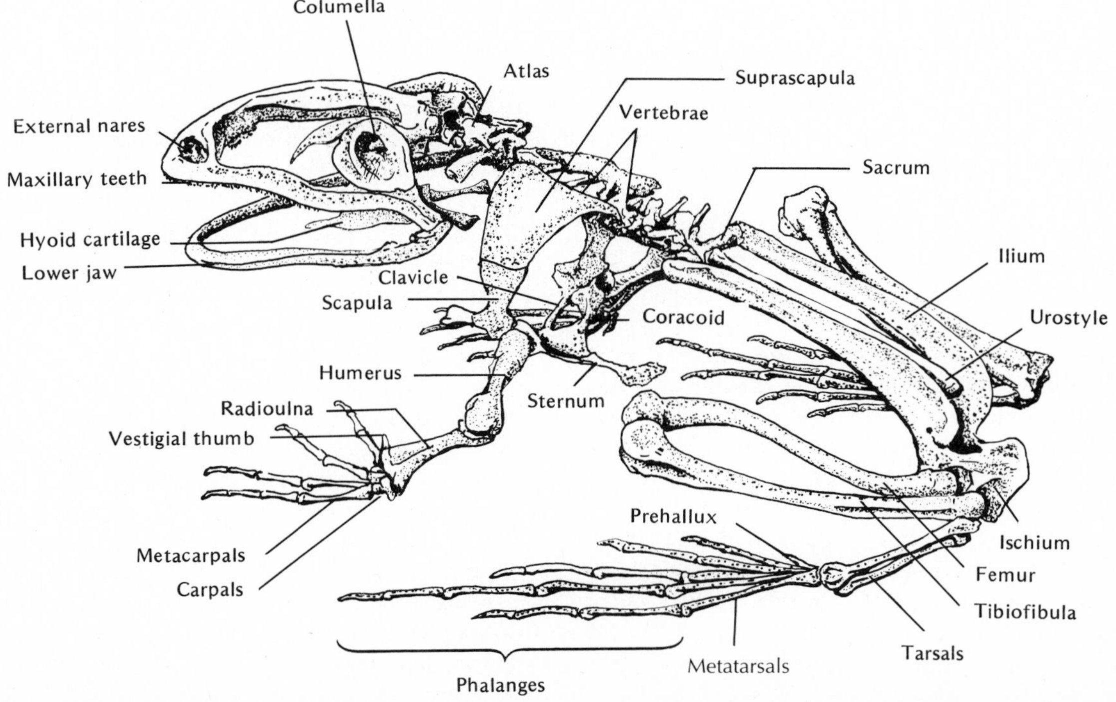

Figure 35.1 Frog skeleton.

The total effect of the changes in the amphibian skull relative to that of fishes is a lighter skull. This seems to be an adaptive advantage for movement on land.

The vertebrae in amphibians are variable in number, ranging from ten in the Salientia to about 200 in the Gymnophiona.

Ribs may be present but are poorly developed or fused to vertebrae. Certainly, they do not show the ventral connections we will see in the reptiles, birds, and mammals.

The two girdles of amphibians, pectoral and pelvic, show modifications not seen in fish, especially in the jumping anurans. There is no connection between the skull and the pectoral girdle, for instance. The pectoral girdle has become more closely associated with the functions of the forelimbs rather than the gill apparatus as in fishes. The pectoral girdle, from dorsal to ventral, consists of a *suprascapula,* a *scapula,* and a *coracoid.* Articulation of the forelimb is between the scapula and the coracoid. A sternum may be present in some amphibians, but it has no ribs articulating with it as it does in later vertebrates.

The pelvic girdle has a single pair of articulations with the sacral vertebra. Each dorsal *ilium* makes this connection. The ventral–posterior member on each side is called an *ischium,* and the anterior–ventral member, the *pubis.* The hind limb articulates at a cavity, the *acetabulum,* created by the union of the three bones. As we shall see below, the pelvic girdle has undergone very specialized evolution in the anurans in association with their jumping capabilities. However, both the pectoral and pelvic girdles reflect the fish ancestry of the amphibians in that these girdles articulate to the side of the vertebral column, making movement undulatory.

THE MUSCULAR SYSTEM

Amphibian musculature is transitional between that of fishes and reptiles. In this class one sees the shift from a primary development of the trunk musculature for lateral movements of swimming seen in the fish, or for gill and mouth movements, to the expansion of the muscles associated with the appendages that accompany terrestrial existence. The trunk musculature retains its segmented character in the dorsal trunk where most muscle fibers run in an anterior–posterior direction. The flank musculature is modified into sheets, layered in opposing directions. This gives support to the visceral contents in the absence of the buoyancy of water. The flank layers, the external and internal obliques, and the transverse muscle, established in this class, remain clearly evident in all other tetrapod classes. A strip of anterior–posterior muscle, the rectus abdominis, strengthens the midventral trunk.

After metamorphosis, gill musculature is reduced to functions associated with the throat where it helps to pump air into the lungs. It is in the appendicular musculature that the greatest changes occur. In fish this musculature is primarily confined to dorsal muscles opposed by ventral muscles to provide the small amount of movement associated with the stabilizer functions of most fins. The details of the muscle arrangement in amphibian limbs and those of other tetrapods are more than is warranted for discussion here. However, it is noteworthy that sets of muscles have developed that give great flexibility of movement associated with terrestrial existence. Furthermore, there are muscles that are intrinsic—they lie within the limb rather than bridging between the limb and the trunk.

A brief discussion of the skeletal–muscular functions of the frog will help to apply some of the anatomical characteristics in general and show the unusually specialized variation that has developed with the evolution of the anurans. The frog forelimbs function principally to stabilize the frog at the end of a jump, or to clasp a female during sperm deposition. Thus these limbs are short and not highly specialized. As jumpers, the trunk is short with only ten vertebrae. Terminally there is a long *urostyle,* probably a modification of caudal vertebrae, that lies in the middorsal plane following the vertebrae. It is flanked by two long processes of each ilium that lie parallel to the urostyle and attach to the transverse processes of the last, sacral, vertebrae. These structures provide a strong and stable anchorage for the well-developed thigh muscles. The hind limb is highly modified for lightness and for strength. Whereas the "typical" vertebrate hind limb consists, from proximal to distal, of femur, parallel tibia and fibula, tarsals, metatarsals, and phalanges, the frog has had reduction and fusion of these bones. In the same sequence, the frog hind limb consists of a femur, a tibiofibula, two tarsal bones, metatarsals, and phalanges. The frog's leg has intrinsic muscles, a strong *gastrocnemius* (calf) muscle, opposed by a smaller *tibialis* muscle. The calf muscle not only assists in a jump, but the two lower leg muscles aid the swimming movements of the frog, accomplished almost entirely by the hind limbs.

FEEDING AND DIGESTION

Adult amphibians are all carnivores, but their diet is varied. The burrowing Gymnophiona subsist on grubs, earthworms, and other subterranean organisms. Salamanders, newts, and anurans will eat nearly anything that moves and can be swallowed. Frogs will even eat nearby members of their own species, sometimes downing animals half their own size. Larval urodeles are also carnivores, beginning their eating with small zooplankton and tackling larger aquatic organisms, such as small fish, as they increase in size. By contrast, larval anurans are usually herbivores. Their diet is algae adhering to the surfaces of aquatic plants and rocks. These animals have paired suckers ventral to the mouth that also secrete an adhesive. The young larva attaches to the surface and rasps its food from the surface with a set of horny (keratinized) epidermal teeth that it loses as it metamorphoses.

In most amphibians the tongue is free at the posterior end and is quite rapidly protrusible. Its tip is glandular and adhesive. There are few glands in the mouth except that many terrestrial amphibians have an intermaxillary gland at the roof of the mouth that produces an adhesive.

Amphibians use their teeth only for anchorage of captured prey; they do not chew or cut. After passing the oral cavity, food enters a short esophagus, often hard to delimit, into the stomach. The stomach may be quite large and curved as with frogs, or straight and not distinct from other areas as with caecelians. In most amphibians the gut is relatively short, a trait of carnivores. It may or may not be clearly distinguishable into small and large intestine. It empties into the cloaca. The herbivorous anuran larvae have a relatively long and coiled gut; this adaptation permits microorganisms to work on the food (fermentation) to assure its utilization. At metamorphosis the gut becomes shortened proportionately to the animal's carnivorous existence (Fig. 35.2).

The enzymes produced by the alimentary canal and the pancreas are those characteristic

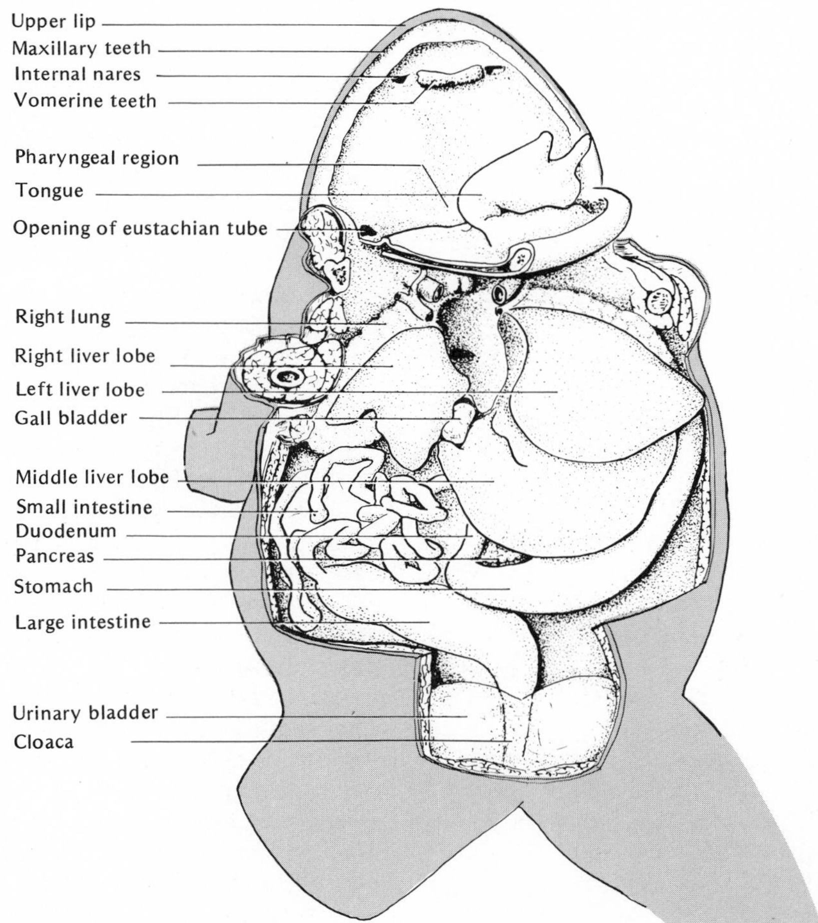

Figure 35.2 Digestive system of frog.

for the digestion of proteins, fats, and carbohydrates as discussed in Chapter 9.

GASEOUS EXCHANGE AND INTERNAL TRANSPORT

Gaseous Exchange Many amphibians face the problem of gaseous exchange in two ways. The first occurs as aquatic embryos and larvae; the second occurs when they are terrestrial adults. It is generally assumed that larval salamanders and anurans exchange gases between the skin and the water or between specialized external gills and the water. In fact, it has been asserted that larvae with especially large gills are symptomatic of conditions where the amount of oxygen in solution is low. However, the absolute relationship can be questioned, as salamander embryos from the same spawning have been raised in large numbers under identical conditions, and considerable variation in the size of the gills of individual animals has been noted. Furthermore, it has been possible to prevent gills from developing through some embryonic transplantation techniques, yet some of the salamanders have developed and metamorphosed to adults without ever possessing gills. Compensatory exchange between the blood vessels in the skin and the environment probably developed. The gills, however, do increase the surface area and provide close association between the blood vessels and the environment. The two external sites for gaseous exchange are augmented by the membranous surfaces of the mouth and by the presence of lungs. While the lungless terrestrial salamanders (Plethodontidae) exchange gases principally through the vascular skin, the other amphibians have lungs of varying degrees of complexity. In some salamanders each lung is a reticularly vascularized, elongated sac with a relatively smooth interior, except for having a ciliated surface. In anurans the lungs are shorter and probably more efficient in the amount of gas exchanged because they are divided into segments by septa and have small alveoli (Ch. 10) that increase the surface area. This increased complexity and efficiency corresponds well to the more active life style of anurans in contrast to urodeles (Fig. 35.3).

The lungs are connected to the larynx, sometimes by a short trachea. As the air enters the larynx from the pharynx it passes over vocal cords in some anurans. Vocalization is principally exercised by males during the breeding season. It is usually possible to distinguish the song of one species of frog from another.

Breathing in amphibians is achieved by a different means than is used by later vertebrates. Amphibians use a positive-pressure system whereby air is forced into the lungs rather than sucked in as by humans, for instance. The procedure can be observed in frogs quite easily.

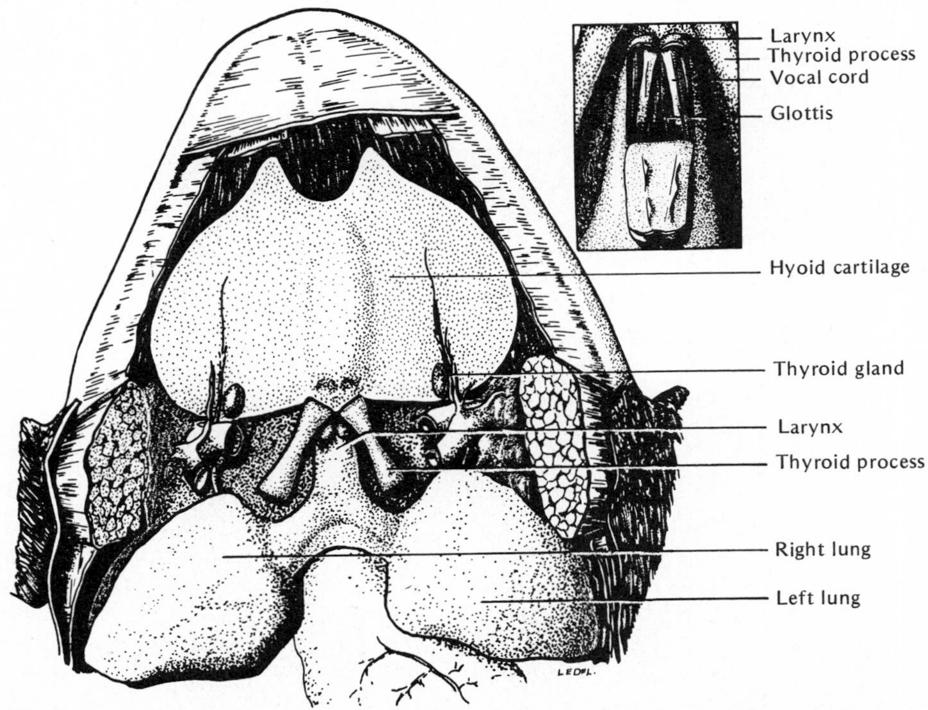

Figure 35.3 Respiratory system of frog.

The floor of the mouth is depressed while the mouth is closed. This draws air into the oral chamber through the external and internal nares (nostrils). The nostrils can be closed in a valve-like manner. The entry to the larynx is opened and the floor of the mouth is elevated, forcing air through the larynx into the lungs. Periodically the muscles of the body wall contract to force air from the lungs. In the interim oxygen is delivered from the lungs to the blood and carbon dioxide moves from the blood. (Probably more carbon dioxide is exchanged at the surface of the skin than at the surface of the lungs. Under some conditions as much as 50 percent of the oxygen is obtained through the skin.) The amphibia's form of breathing is not usually found in the later vertebrates. However, it is very similar to the buccal pumping seen in teleost fish.

The lungs are supplied by arteries from the sixth aortic arch of the embryo (Ch. 10), a trait retained by all other classes of vertebrates.

Internal Transport The pulmonary arteries delivering blood to the lungs arise from the anterior chamber of the heart, the conus arteriosus. The pulmonary artery is just one branch of the sixth aortic arch, another major branch being the cutaneous artery that delivers blood to the skin for cutaneous respiration. Before branching into the pulmonary and the cutaneous arteries, the sixth aortic arch is called the *pulmonocutaneous artery.* In most amphibians the distal part of the sixth aortic arch is lost, but in urodeles it may persist in its original embryonic connection to the dorsal aorta. This section is called the *ductus arteriosus* and its persistence allows some mixing of oxygenated and nonoxygenated blood.

Of the six embryonic aortic arches that emerge anterior to the heart, usually only the third, fourth, and sixth persist, although some urodeles retain the fifth aortic arch. The third aortic arches form the bases of the internal carotid arteries that supply the head and brain. The fourth aortic arches persist on both sides as the systemic arches that have branches to the arms and connect to the dorsal aorta to supply blood to the rest of the body.

The hearts of amphibians are five-chambered (Fig. 35.4). Blood is received from the veins (see below) into the sinus venosus that delivers blood to the right atrium. It is in the atria that the separation of heart functions into two pumps is seen to have begun in the evolution of vertebrate circulatory systems. The single atrium of fishes is divided in amphibians into right and left parts by the development of an interatrial septum that isolates the inflow of blood from the body to the right atrium. The left atrium receives blood from the lungs through pulmonary veins. The ventricle is a single chamber, but its interior has deep pockets that tend to isolate blood from each atrium and prevent mixing. When the ventricle contracts, blood is delivered to the fifth chamber, the conus arteriosus. In amphibians, this chamber has become modified to restrict the currents emerging from the ventricle through the development of spiral ridges and valves. Injection studies show a tendency for blood from the right atrium, after entry into the ventricle, to be channeled through the conus arteriosus to the pulmonocutaneous artery, while left atrial blood mostly is delivered to the systemic and carotid arteries. The lungless salamanders are exceptional in that the interatrial septum is incomplete and no pulmonary veins enter the left atrium.

The venous return from the body is similar to that possessed by the ancestral fish. Common cardinal veins enter each side of the sinus venosus and return blood from the head (anterior cardinal veins) and trunk (posterior cardinal veins). Lateral abdominal veins, seen in fish, are united into a single *ventral abdominal vein* that empties into the hepatic portal system. The latter receives blood from the viscera.

A comparison between the fishes and the amphibia clearly shows that the advances in evolution shown by the amphibians over fishes are accompanied by retained similarities. Thus the transitional nature of the class Amphibia is nicely illustrated by the respiratory and circulatory functions.

WATER BALANCE AND EXCRETORY CONTROL

Problems of retaining correct water balance in amphibians are more severe than in the fishes since the amphibians still retain many fishlike traits yet must cope with the physiologically hostile terrestrial existence. This is achieved with some success, but it appears that the amphibians are strongly limited in their habitats because of the problems of water balance.

What success there is is attributable to the organization and use of the skin and the urinary system, plus some natural behavioral traits.

The integument of amphibians consists of a dermis and an epidermis (Fig. 35.5). The dermis is principally composed of collagen secreted by the widely scattered fibroblasts. However, the inner zone of the dermis is composed of a set of dense collagen layers. The epidermis not only consists of epithelial cells in a covering sheet

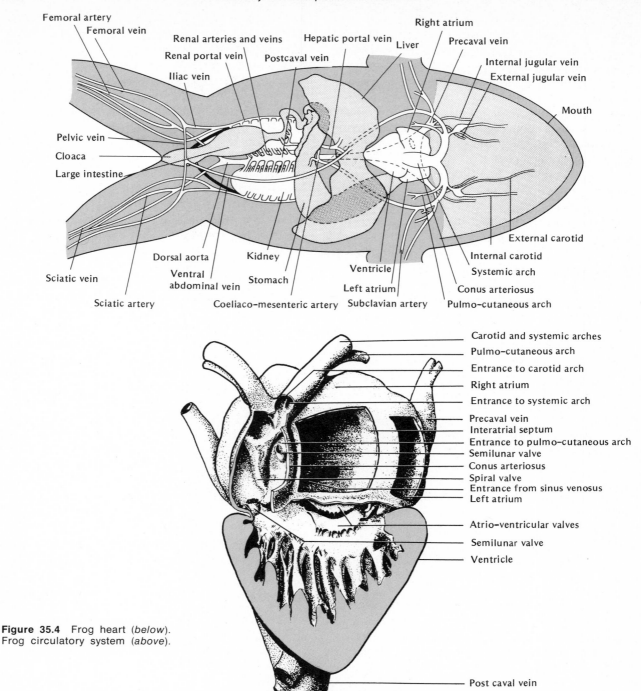

Figure 35.4 Frog heart (*below*). Frog circulatory system (*above*).

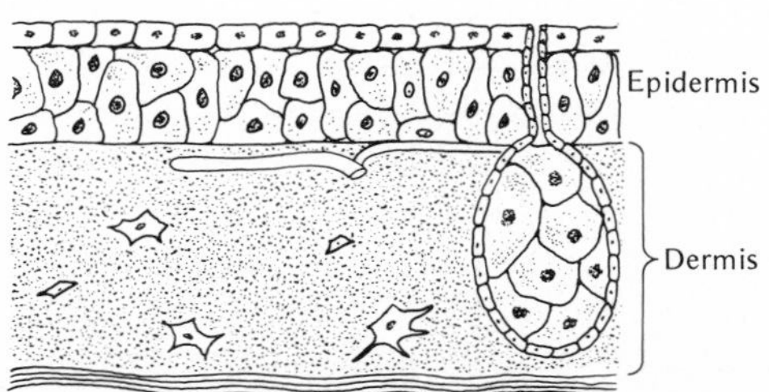

Figure 35.5 Skin of a salamander.

but also possesses specialized glands that lie embedded in the dermis. One kind of gland, mucous glands, produces a slimy secretion that helps to keep the skin moist and prevent water loss. Another kind of gland is the large poison glands also found widely distributed embedded in the dermis. These glands also secrete a thick whitish fluid that is irritating or even lethal to other animals. The epidermis has a basement layer of columnar cells that proliferates the overlying layers of cells; these flatten, keratinize, and may be shed in sheets. Those amphibians that are mostly aquatic have a thin epidermis; those that live on land, especially in desert regions, have thick, rough, and warty skins.

Pigment cells found in the skin are associated with protective coloration and control the penetration of light. Some amphibians have definite pigment patterns of spots as in frogs and some salamanders, or there may be banding patterns of striking beauty in some salamanders. In addition, it is possible for amphibians to alter their degree of pigmentation. Three kinds of pigment cells in the skins of amphibians are usually more concentrated just below the epidermis than elsewhere. Collectively called *chromatophores,* the three kinds of pigment cells are the *melanophores* that contain black or brown pigments, *guanophores* that contain whitish crystals, and *lipophores* (also called *xanthophores*) that possess yellow or red pigments. Melanophores create the variations in the skin pigmentation depending on the degree to which pigment granules are dispersed in processes extending from the cells or are concentrated in the body of the cell near the nucleus. This is controlled by a pituitary hormone, MSH (see below).

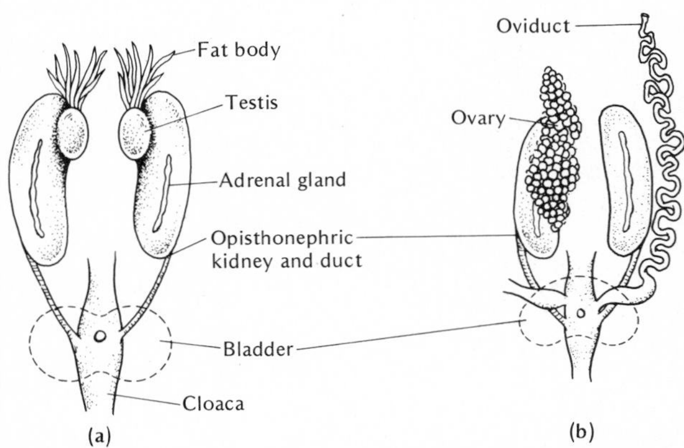

Figure 35.6 Urogenital system of frogs (ventral views). **(a)** Male. **(b)** Female. The ovary is removed on the left side of the frog to show the oviduct and is drawn as a small immature ovary on the frog's right to permit the kidney to show.

In addition to the skin, the kidneys influence water regulation. The amphibian kidney, like that of fishes, is opisthonephric (Fig. 35.6). It is an elongated straplike organ in all but the anurans where it forms a pair of ovoid kidneys. Although the kidneys of males may be associated with the testes and sperm movement (see below), the urinary functions are performed in a way similar to that described for fishes. However, there is no variation in the size of the renal corpuscles; they are all large and assist in the elimination of water. Fresh water, of course, is hypotonic to amphibian body fluids and there is a tendency for water to accumulate. The renal corpuscles compensate for this through the large filtration capacity they provide. The duct leading from the kidney to the cloaca is variously called the archinephric, opisthonephric, or mesonephric duct, or simply the ureter, although the latter term leads to confusion if equated with the ureters of mammals, for instance. In some amphibian males, where the anterior part of the kidney functions for sperm transport, there may be secondary urinary ducts that connect the kidney and the cloaca; these ducts more closely resemble the ureters of later vertebrates.

The bladder in amphibians is not a posterior enlargement of the archinephric duct as it is in some fishes, but is an evagination from the floor of the cloaca. Urine, therefore, must cross from the archinephric ducts to the bladder for storage. Storage capacity, rather than immediate release, serves as an adaptation in terrestrial amphibians so that resorption of water can occur from the bladder when water loss from the skin becomes excessive. In a neotenic amphibian (one that is permanently aquatic) it was noted that a vital blue dye injected into the coelom appeared as two thin streams of blue emerging from the cloacal opening within two minutes after injection. This illustrated, first, that the clearance function of the kidney is very rapid, and second, that delivery of fluids from the archinephric ducts can be to the exterior even though the salamander possesses a functional bladder. It is not fully understood what controls the delivery to the bladder rather than to the exterior.

In addition to water balance functions, the kidneys also eliminate nitrogenous and other wastes. In larval amphibians and in permanently aquatic adult amphibians the wastes are eliminated as ammonia and the form of excretion is called *ammonotelic.* In terrestrial adults the nitrogenous wastes are urea and the excretion is called *uretelic.*

REPRODUCTION

In most amphibians, but not all, fertilization is external. Because of the close association of the male reproductive system with the urinary system, just discussed, the structure of the male system is described first.

The testes are paired. They may be ovoid yellowish-white structures as seen in common leopard frogs, or they may consist of many lobes, each lobe having sperm development at one stage, but differing among the lobes. The testes deliver sperm either through the tubules of the anterior kidney region or directly into the archinephric ducts. Sperm are delivered to the cloaca for direct release over eggs being deposited by anuran females; or by way of a *spermatophore,* in urodeles, that is a package of sperm in a capsule atop a conical mass of jelly that is deposited on leaves, sticks, or stones for the female to retrieve; or by internal fertilization in Gymnophiona through the male's protrusible cloaca into the cloaca of the female during copulation.

A structure found anterior to the testis in some male toads, *Bidder's organ,* may develop into an ovary under some circumstances. Also, "sex reversal," in a sense, is seen in some old female toads that may produce sperm.

The ovaries are paired saclike structures with numerous eggs in the thin walls of the sac. On ovulation the eggs are moved in the coelom via ciliary and other movements to the paired oviducts that receive them. They either pass to the exterior via the cloaca or they may be temporarily stored in expansions, often called uteri or ovisacs, at the posterior ends of the oviducts. From here they may be released to the water or, in a few ovoviviparous species, they may be retained there for development of the young. The amphibian oviduct also contains glands that deposit a jelly around the eggs as they pass. The pattern of jelly coats can be used to determine the species laying the eggs.

Sexual recognition and courtship play an important role in amphibians as they do in other vertebrates. In anurans the vocalizations of the males during the breeding season play an important role. The chorus of frog or toad calls at ponds or streams in the early spring can be very loud. When a male has attracted a receptive female, a coupling in the water, called *amplexus,* is made (Fig. 35.7). The male clasps the female around the trunk, and as she releases eggs from the cloaca, the male releases sperm over them. The fertilized eggs, with their jelly coverings, form characteristic clusters, such as clumps, strings, or sheets, that are attached to plants or may float on the surface.

Some salamanders have courtship patterns wherein the males stimulate the females by a series of body movements, nudges, rubbings, and so forth. The male usually then deposits one or more spermatophores. Receptive females move over the spermatophores and with the cloacal lips pick them up to retain the sperm internally. The sperm fertilize the eggs as they are deposited. The number of eggs laid is variable with the species. Some frogs may lay a single egg and as few as three eggs are reported for a plethodontid salamander. On the other hand, the Woodhouse toad has been reported to lay 25,000 eggs.

Development in amphibians follows a similar pattern up to the establishment of the larval stage (Ch. 16). Zygotes cleave and form blastulae that undergo formative movements of gastrulation and the establishment of the body axis. Neurulation and elongation occur and major organogenesis is established. The young embryo survives on yolk stored in the cells lying in the wall of the gut. In time the young larvae are released from the jelly, probably through the secretions of a hatching enzyme. The released anuran young temporarily attach to plants, sticks, and other surfaces by their suckers to begin browsing for algae. Urodele young, on the other hand, being carnivorous, begin hunting for food.

Figure 35.7 Amplexus in toads. (From Rugh's *Experimental Embryology,* Burgess Publishing Company, with permission.)

The development of amphibians, because of the ease with which it can be followed, has contributed some of the basic knowledge about cellular movements and cellular interactions that have led to our appreciation of development in general, to understanding of abnormalities in organisms, and as a foundation from which the modern subdiscipline of developmental genetics has emerged.

Metamorphosis occurs in most amphibians (Fig. 35.8). In all cases it is under the influence of the hormones of the pituitary and the thyroid glands; these in turn are controlled through the hypothalamus (Ch. 12). Anuran metamorphosis leads to dramatic changes in body form and in internal anatomy. The larval tail is resorbed through the selective destruction of cells by lysosomal enzymes (Ch. 4), the horny teeth are lost, the gills disappear, new blood channels develop, and the gut becomes proportionately shorter. As an anuran tadpole begins to metamorphose, its hind limbs begin to develop, and by the time the tail is completely resorbed, the typical hindlimbs of a frog are functioning. Urodeles do not show such great changes. The gills are lost, as is the dorsal fin, there are changes in the eye, and lids develop. The skin begins to take on the adult glandular pattern and pigmentation.

Care of the young after spawning is nonexistent in many amphibians. On the other hand, some remarkable adaptations for their care have also developed. For instance, some tree frogs of South America create nests for the eggs; this is done in different ways by different species. Some simply beat the egg mass with their hind feet to create a frothy crust which dries and the eggs are protected within. Other tree frogs glue leaves together, deposit their eggs between them, then lay eggless jelly capsules over them to protect and keep them moist. One tree frog species has the male building a mud fence behind which the female lays her eggs. Other amphibians retain their eggs to carry young on the body. In an entire family of frogs, the male's vocal pouch is enlarged to hold the young; these are the mouth-breeding frogs. An unusual example of parental care is seen in the Surinamn toad where the female lays eggs on her own back through the presence of a long ovipositor, after which the male fertilizes the eggs and presses them into her skin so that each offspring develops in a pit of its own. Several kinds of frogs possess deep pouches on the backs of the females in which the eggs develop.

Guarding the eggs also occurs in amphibians. One South American frog has the male guard the eggs during development and he then later carries the larvae around attached to his back. Some plethodontid salamander females guard their eggs until they hatch.

Figure 35.8 Some stages in the metamorphosis of a frog.

SENSORY RECEPTION AND NEURAL COORDINATION

Larval amphibians, as in fish, receive stimuli associated with pressure through the lateral line system. However, at metamorphosis the lateral line system degenerates except in neotenic salamanders. With the shift to terrestrial existence, reception through the senses of smell, sight, and hearing play an important role, and the organs associated with these functions are described below.

The nervous system of the amphibians is very similar in its level of development to that seen in fishes (Fig. 35.9). The five parts of the brain, telencephalon (cerebrum), diencephalon (thalamic region), mesencephalon (optic lobe region), metencephalon (cerebellum), and myelencephalon (medulla), are arranged in linear order and do not show significant specialization. Although the telencephalon is principally olfactory, close study shows that some fibers enter the area from the rear and terminate in the upper part, called the *pallium.* This is a minor achievement, but it is the onset of a development that progresses through the reptiles and mammals and results in this region becoming a well-developed cerebral cortex with primary importance in nervous integration. Amphibians generally are animals with little spontaneous motility, and their metencephalic region is not well developed. The myelencephalon is important in retaining the centers of vital function and in serving as a center for hearing, a development more important in anurans than in urodeles.

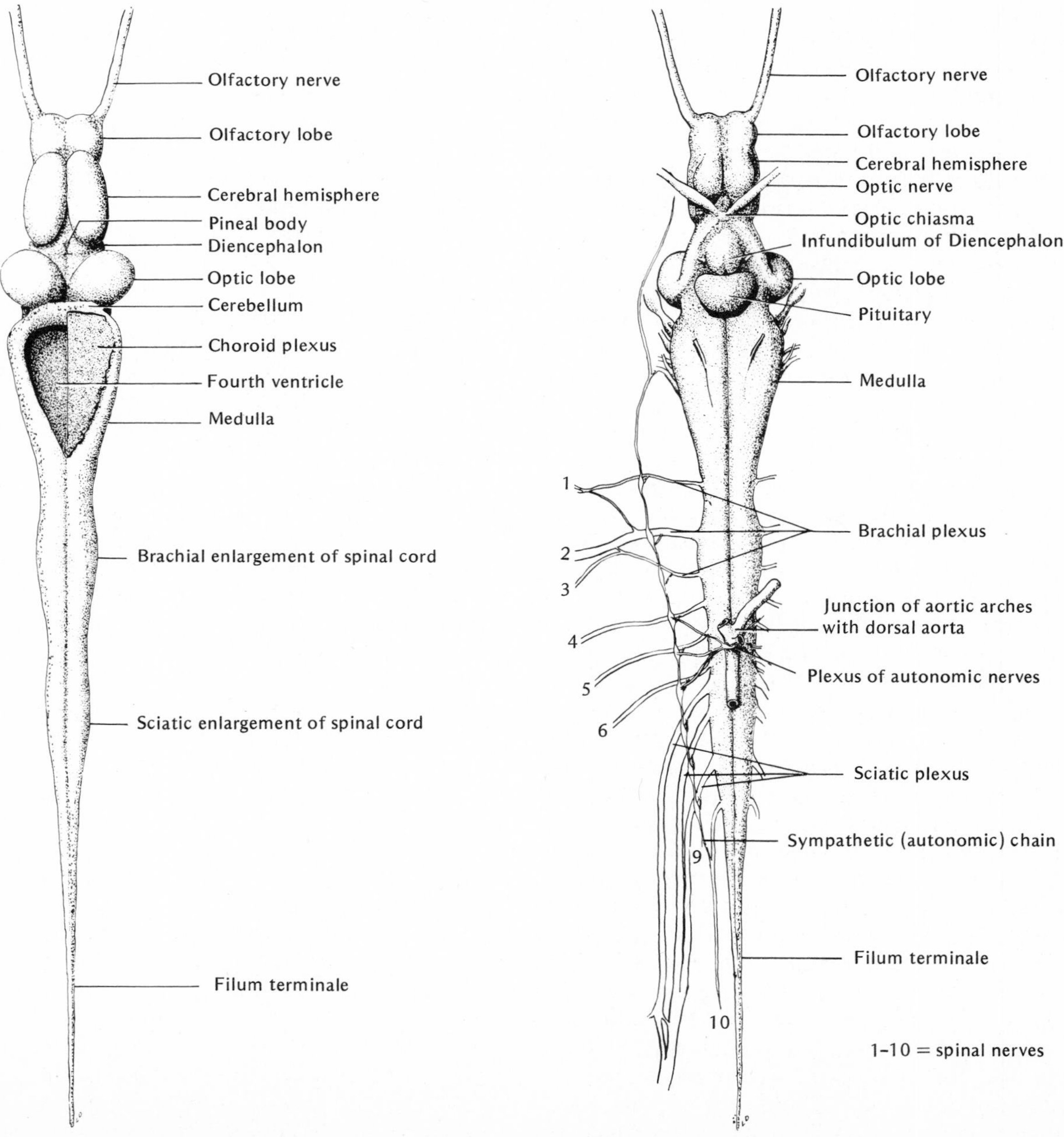

Figure 35.9 Frog brain and cord. Dorsal view (*left*); ventral view (*right*).

There are ten cranial nerves and a variable number of spinal nerves, depending on the kind of amphibian, together making up the somatic portion of the peripheral nervous system. The autonomic peripheral nervous system consists of two chains of special ganglia flanking the sides of the spinal column, and the parasympathetic division of this system is poorly developed.

Olfactory reception is in the dorsal part of the choanae, and nerve fibers from there penetrate the telencephalon. There is a new structure in the roof of the mouth, the *vomeronasal* (*Jacobson's*) *organ.* It arises from an evagination from the nasal area and is believed to function in taste as it is lined with chemoreceptors.

Vision in amphibians is received through an eye quite similar to that of fishes, except that its lens is farther from the cornea and is fixed for relatively distant vision. It is capable of some accommodation for near vision by small muscles. The pupils of amphibians vary in shape; they may be round, horizontal, vertical, three-cornered, or four-cornered. In terrestrial animals the upper eyelid is well formed and the lower lid is folded into a semitransparent *nictitating* membrane. Retention of moisture on the surface of the eye is important, and amphibians may have tear glands, but poorly developed. However, a new gland, the *Harderian gland,* is present and helps water retention by producing an oily secretion.

Hearing mechanisms vary in the amphibians. Salamanders do not have well-developed ear drums or middle ears, although a rod, the *columella,* may extend between the skin and the skull. Anurans have a middle ear, an ear drum, and a tubotympanic canal (Eustachian tube) connecting the middle ear to the pharynx. Bridging across the middle ear between the ear drum and the skull is the columella. The anuran ear represents a condition that we will see elaborated in the later vertebrates. The columella is derived from the supporting elements for the gills from the second pharyngeal arches of the fishes. In later vertebrates the columella becomes the stapes. The inner ear is still mainly for equilibrium and consists of two saclike structures, the *utricle,* from which arises the three semicircular canals, and the *saccule,* from which extends a small diverticulum, the *lagena.* We will see the lagena develop into the cochlea in later vertebrates. It can be demonstrated that the lagena has sensory receptors for the conversion of mechanical sound waves into nerve impulses for the interpretation of sound. The wave lengths for which the receptors are sensitive correspond to the wave lengths of the voices of the males of the species.

CHEMICAL COORDINATION AND HORMONAL CONTROL

Most conditions of endocrine development to be seen in later vertebrates are present in the amphibians, including parathyroid glands which are absent in fishes. An interesting point related to the interaction of endocrine glands has been noted in *neotenic* amphibians. Neoteny is a condition where adult amphibians, especially urodeles, retain gills, dorsal fins, and a number of traits usually associated with larval existence. Neotenic animals are not necessarily truly immature, however, as they reproduce and have well-developed immune systems as is shown by their capability to reject foreign skin grafts. Some urodeles, such as the mudpuppy, *Necturus,* cannot be caused to metamorphose by hormonal treatment. Others, such as the neotenic form of the tiger salamander, may metamorphose if the environment is altered as, for instance, by increasing the iodine content of the water. The Mexican axolotl, *Ambystoma mexicanum* normally does not metamorphose, apparently due to inadequate production of thyroid-stimulating hormone, TSH, by the pituitary, and not due to inadequacy of iodine provided to the thyroid gland. Furthermore, that this failure can be attributed to the pituitary recently has been reinforced by studies of the hypothalamus of axolotls. They contain as much thyrotropic-releasing hormone, TRH, needed to trigger pituitary activity as is found in some mammalian hypothalami. Also, some breeding experiments between tiger salamanders and Mexican axolotls point to a simple Mendelian genetic control of the pituitary's capability to produce TSH because F_2s of such a cross segregated with approximately $3/4$ metamorphosing, as in tiger salamanders, and $1/4$ remaining neotenic as in Mexican axolotls.

The thyroid gland not only directly regulates metamorphosis, as can be shown by removing it, but it probably also influences the molting of the skin.

The regulation of skin pigmentation, described above, is under the control of the pituitary hormone, melanocyte-stimulating hormone, MSH. The degree to which this is released depends on the amount of light that strikes the upper or lower parts of the retina. Most light, coming from above, strikes the lower half of the retina with greater intensity than it does the upper half of the retina. When the amphibian is on a light background, more light is reflected to the upper half. This increased illumination results in inhibition of MSH release, and the melanin granules in the melanocytes remain centrally aggregated, producing a pale-colored animal. If the animal is on a dark background, most illumination is from above and the MSH is released to cause dispersion of melanin granules to the processes of the melanocytes and creating a darker animal.

The adrenal glands in amphibians are compound glands as they are in all later vertebrates, being composed of intermixed cortical-type cells and medullary-type cells.

BEHAVIOR

The behavioral traits of some amphibians associated with their mating and nesting habits have been described above. In addition, a trait we have seen in fishes, migration habit, exists in some amphibians, and they possess homing traits. Several studies on homing in salamanders have clearly shown that amphibians may return to the stream or pond where, as eggs, they were first deposited. It is still not clear whether one or more senses play roles in this movement. Experiments with West Coast salamanders suggest that olfaction is of primary importance because blinded animals could still home. However, experiments with Fowler's toad suggest that celestial navigation is employed in orienting animals toward their home sites.

A behavioral trait exhibited by some amphibians is hibernation. Most frogs bury themselves in the mud of ponds and streams where they remain from fall to spring. Of course, their metabolism is reduced to minimal levels. In the Midwest as many as twenty salamanders have been found together in icy burrows approximately 15–30 cm above the winter water and ice of streams. Spring flow covers these sites and stimulates the release from hibernation.

Importance of Amphibians to Other Animals

Amphibians play a role in food chains as they eat various arthropods and small vertebrates and thus are, at least, secondary consumers (Ch. 45). They also serve as natural food for such animals as snakes, some turtles, aquatic birds, hawks, and racoons. Some adaptations associated with predator–prey relationships are remaining still, long jumps, feigning death, or in some instances even aggressive counterattack. If caught, it is not uncommon for frogs to void urine, probably disconcerting to the predator.

Importance of Amphibians to Humans

Froglegs have been delicacies in a number of countries for many years. It has been reported that Mexican Indians eat the large salamanders of their region. However, although not realized by most laymen, frogs have served significantly in the expansion of scientific knowledge. The nerve–muscle preparation often used by beginning biology students serves as more than just a teaching device. The frog gastrocnemius muscle and the accompanying sciatic nerve have contributed greatly to our understanding of muscle contraction and to nerve impulse conduction, to the roles of drugs, and to the impact of various chemicals on normal function. The simplicity of the nervous system has also made for useful studies on central nervous system function. As indicated earlier, studies on the amphibians have made great contributions to our knowledge of normal development and to factors leading to abnormal development.

Whether amphibians will be able to play a continuing role in our quest for knowledge is uncertain due to two things. First, shifting trends in the teaching of biology have led to a great increase in the use of frogs in secondary education as well as in colleges. This has placed a great demand on the collection of frogs. Also, natural habitats of frogs, marshy and swampy environments, have suffered greatly from drainage for urban and suburban development and for agriculture through man's continued consumption of land.

Classification

There are four modern orders of Amphibia: Urodela or Caudata (salamanders), Trachystoma (sirens), Gymnophiona or Apoda (caecelians), and the Anura (frogs and toads). The fossil record is unclear and paleontologists still debate the evolution of these modern orders. It appears that amphibians are derived from lobe-finned Crossopterygii. The early amphibians split into several groups, and the evolution of the modern orders from these is what principally is debated. Some feel that all modern orders evolved from a single Paleozoic stock and group them in a single subclass, Lissamphibia. Here we adopt the position that there are two subclasses based on the origins of the vertebrae.

SUBCLASS **Aspidospondyli** The centra are formed first from blocks of cartilage that replace the notochord. These blocks may or may not ossify. There are two superorders.

SUPERORDER **Labyrinthodontia** (Fig. 35.10) These extinct amphibians represent the stem stock of Amphibia. They were dominant during the latter half of the Paleozoic, or "age of amphibians." Some achieved lengths of 3 m or more and they had broad flat bony heads, elongate tailed bodies, dermal body coverings, and their limbs extended laterally in a primitive pattern not especially efficient in bearing the weight of the body.

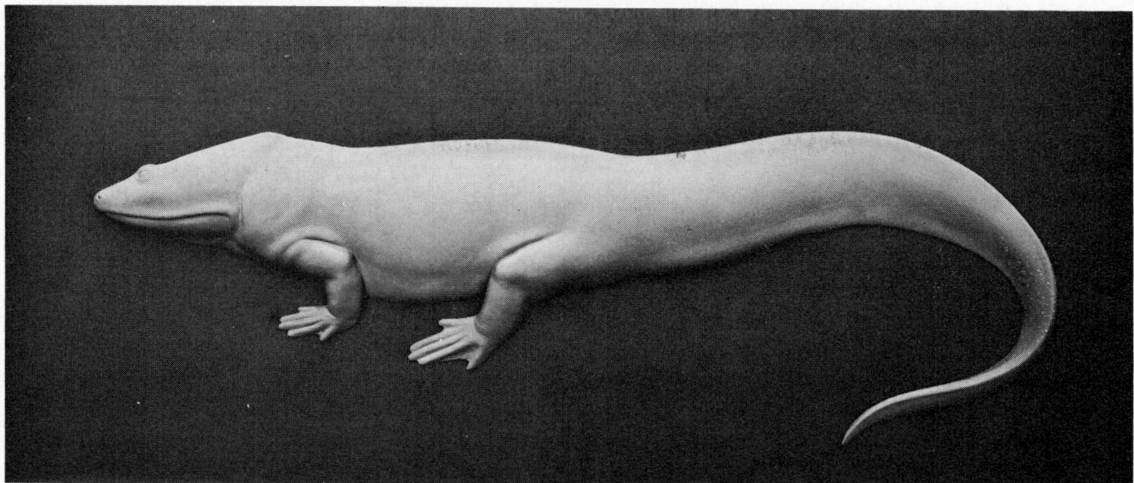

Figure 35.10 The labyrinthodont amphibian, *Diplovertebron*. (Courtesy of the American Museum of Natural History, New York.)

SUPERORDER **Salientia** These are specialized amphibians, and the vertebrae usually lack true centra, the bodies of the vertebrae being produced from fusion of the lower part of the neural arch. Two orders; one extinct.

ORDER **Proanura** Extinct. Possessed tails and ribs.

ORDER **Anura** Represented by about 1900 species and consisting of the frogs and toads (Fig. 35.11). Worldwide except arctic and antarctic regions.

SUBCLASS **Lepospondyli** The vertebral centrum usually results from direct deposition of bone around the notochord. Six orders; three extinct.

ORDER **Aistopoda** Extinct. Limbless, elongate.

ORDER **Nectridia** Extinct. Reduced limbs, elongate.

ORDER **Microsauria** Extinct. Limbs present.

ORDER **Trachystoma** Living. Medium sized. Gills and lack lungs, teeth, hindlimbs. These are the sirens (Fig. 35.12). Eastern N. America, Southern Europe.

Figure 35.11 Anurans Bullfrog (*above*). Large grass frog (*below*). (Carolina Biological Supply Company)

Figure 35.12 Siren or mud eel. Note the presence of fingered forelimbs, absence of hind limbs. (New York Zoological Society Photo)

ORDER **Gymnophiona** or **Apoda** Living. Elongate wormlike bodies. Limbs lacking. These are the caecilians (Fig. 35.13). Mexico, South America, Asia, Africa.

ORDER **Caudata** or **Urodela** Living. Elongate with tail and unspecialized limbs. The seven families consist of the salamanders, newts, mudpuppies (Fig. 35.14). Americas, Europe, Africa, Asia.

Figure 35.13 (*left*) Caecilian. Note absence of any limbs; an apodan. (New York Zoological Society Photo)

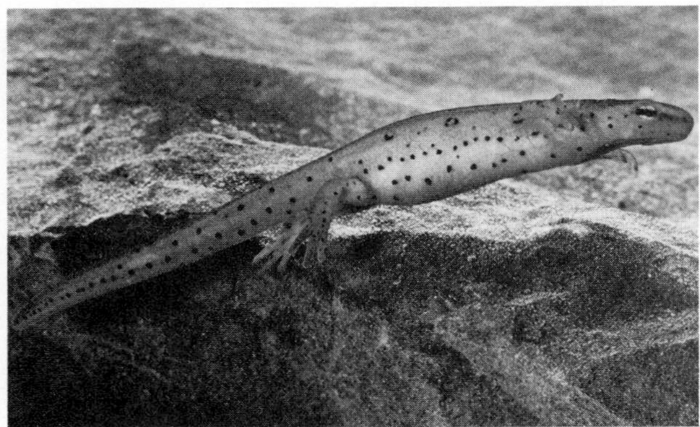

Figure 35.14 Urodeles. Fire salamander (*left*), common or red-spotted newt (*right*). (New York Zoological Society Photo)

SUGGESTIONS FOR FURTHER READING

Barbour, T. *Reptiles and Amphibians, Their Habits and Adaptations,* 2d ed. Boston: Houghton Mifflin (1934).

Bishop, S. C. *Handbook of Salamanders . . . of the United States, of Canada, and of Lower California.* Ithaca, NY: Comstock Publ. (1943). Reprinted: New York: Hafner Publ. (1962).

Conant, R. *A Field Guide to Reptiles and Amphibians of Eastern North America.* Boston: Houghton Mifflin (1958).

Dunn, E. R. *The Salamanders of the Family Plethodontidae.* Northampton, MA: Smith College (1926).

Dunn, E. R. *The American Caecilians. Museum of Comparative Zoology Bull. 91* (1942).

Goin, C. C., and O. B. Goin. *Introduction to Herpetology.* San Francisco: W. H. Freeman (1962).

Moore, J. A. (Ed.). *Physiology of the Amphibia.* New York: Academic Press (1964).

Noble, G. K. *Biology of the Amphibia.* New York: McGraw-Hill (1931). Reprinted: New York: Dover (1955).

Oliver, J. A. *The Natural History of North American Amphibians and Reptiles.* New York: Van Nostrand (1955).

Schmidt, K. P. *A Checklist of North American Amphibians and Reptiles* (American Society of Ichthyologists and Herpetologists). Chicago: Univ. of Chicago Press (1953).

Stebbins, R. C. *Amphibians and Reptiles of Western North America.* New York: McGraw-Hill (1954).

Wright, A. A., and A. H. Wright. *Handbook of Frogs and Toads of the United States and Canada,* 3d ed. Ithaca, NY: Comstock Publ. (1949).

From Scientific American

Etkin, W. "How a Tadpole Becomes a Frog" (May 1966).

Frieden, E. "The Chemistry of Amphibian Metamorphosis" (Nov. 1963).

Muntz, W. R. A. "Vision in Frogs" (Mar. 1964).

36 The Reptiles

If, from the human perspective, the development of a partially terrestrial habitat in the amphibian life history was a step of significance in the evolution of the vertebrates, the contributions to the evolution of subsequent vertebrates by the class Reptilia are of very great magnitude. To confine one's study to modern living reptiles does great disservice to the class and to the understanding of vertebrate biology. We will, therefore, place nearly as much emphasis on those reptiles that are extinct as we do on the relatively obscure living remnants of this great class.

The reptiles are the first fully terrestrial vertebrates. Their achievement of this status required emancipation from dependence on water, especially during early developmental stages. It is often said that this was achieved by the development of the extraembryonic membrane, the amnion (Ch. 16). What the amnion achieves is to provide a "little sea" in which the embryo can develop. This is a dramatic development in itself, but it is not the only one. The amniotic egg does not simply add one membrane to the amphibian egg. Instead, there are four membranes, each of which plays an important role as an adaptation for survival. This is a remarkable number of changes to be selected for. Let us examine these four membranes in a slightly different perspective than they were in Chapter 16. The *amnion* and the *chorion* develop as parts of single sets of folds that arise from the surface of the egg, encompass the embryonic zone of the egg, and then become separated. The function of the amnion has been given, to produce and retain a saline fluid surrounding the embryo. But what about the chorion? The reptilian egg could not stand the rigors of a terrestrial environment either, and it is covered by a semifluid proteinaceous jelly produced in a way similar to the amphibian's by secretions of the oviduct. Reptilian oviducts developed a new capacity, unknown in amphibians, to enclose the egg and its jelly in a leathery or calcified shell. This, however, still does not define the functional value of the chorion. A third adaptation, not seen in amphibians, helps give significance to the chorion. This is the development of an expanded membrane from the floor of the cloaca, the *allantois.* Whether embryonic conditions related to bladder formation in ancient amphibians served as a preadaptation for allantoic outgrowth is uncertain. Nevertheless, the allantois in reptiles and birds (but not in mammals, Ch. 38) expands greatly beyond the embryo and joins and fuses with the chorion to form the chorioallantoic membrane, or CAM. The allantoic cavity serves as a reservoir for wastes from the embryo's metabolism and the fused CAM as the site for exchange of gases between embryonic circulation and the surroundings. This is achieved because the expanded allantois is highly vascularized. In an analogous way, the CAM provides a respiratory function for the amniote embryo similar to expanded gills or vascularized skin in amphibian and fish embryos. Another adaptation established in reptiles, and associated with their survival on land, was the production of large yolky eggs to provide sustenance between the time of laying and hatching. This adaptation was valuable because the amniotes develop relatively longer in the egg membranes than do the anamniotes (fishes and amphibians) and lack a free-living, food-acquiring larval stage. Instead, the amniote develops to a stage where it can either fend for itself in seeking food, or receive parental care and sustenance until it can cope. The yolk of amniote eggs, when present, is usually quite fluid. It differs from the stored reserves in amphibians in being extracellular. It is sus-

Figure 36.1 Snapping turtle emerging from the shell. (Photo—Hal H. Harrison, from National Audubon Society)

pended below the developing embryo in a *yolk sac,* the fourth extraembryonic membrane of the amniotes.

Another adaptation that developed in many amniotes, not otherwise equipped to cope with the problem, is the temporary presence of a so-called "egg-tooth," a special modification on the face or bill of reptiles and birds, that permits them to escape through the surrounding egg shell when they are ready to hatch (Fig. 36.1).

Accompanying the development of these several adaptations, as the reptiles emerged from the amphibians, is the means for internally fertilizing the eggs of all reptiles. While various ways have developed in a few fishes and amphibians for internal fertilization, the development of an intromittant organ, the *penis* (or hemipenes, see below) occurs in all reptiles.

In summary, the advent of the reptiles as a class involved a group of adaptations each of which functioned to achieve survival for this group. It is not clear in which order the adaptations occurred, but it is remarkable that they occurred with sufficient synchrony to sustain the transition from the partly aquatic existence of amphibians to the totally terrestrial existence of reptiles, for even the eggs can be laid on land and develop there.

Origin of the Reptiles

The reptiles arose during the Pennsylvanian period of the Paleozoic era, became well established in the Permian period, and underwent an extensive adaptive radiation during the next era, the Mesozoic, sometimes called the Age of Reptiles. At the end of the Mesozoic they rather suddenly became extinct except for modern snakes, lizards, turtles, tortoises, crocodiles, alligators, and one interesting living species, *Sphenodon punctatus,* a rhynchocephalian. The animal classed as a reptile, but most closely resembling the labyrinthodont amphibian from which the reptiles arose, is *Seymouria* (Fig. 36.2). The skeletal features are reptilian but it may have been at least partially aquatic because of the presence of a lateral line system. *Seymouria* lived into the Permian and was quadripedal and lizardlike.

During the early appearance and development of reptiles there was a period of mountain building, and a variety of climatic and geological changes occurred that created new ecological niches. At the height of the Age of Reptiles all sizes of animals developed, ranging from small lizardlike animals to animals extending over 13 m (40 feet) in length. The adaptive radiation led to animals of various food habits, modes of locomotion, semiaquatic and aquatic existences, and even the development of flight.

Anatomy and Physiology

PROTECTION

Living reptiles are protected by several traits. First, their epidermal scales, scutes, and underlying bones of the skin (osteoderms) offer general protection. Second, most reptiles, with a few exceptions such as the Galapagos iguanas (Fig. 36.3) and giant tortoises, lead a fairly secretive existence. Snakes, except when sunning, are usually obscure on the ground, in the trees, or in the water. Lizards often have body postures that conceal them effectively. Alligators and crocodiles on shore often resemble other parts of their environment such as logs, or when in the water they often show only their eyes and nostrils above the surface. Only the tortoises give the impression of lack of correlation between their environment and concealing behavior. This is probably the result of the relative security a tortoise has by the simple presence of its heavy shell and also by the capability of most tortoises to withdraw the head and neck into the shell, even being able to close the front and rear of the plastron against the carapace to leave no unprotected appendages exposed, as in the case of the box tortoise.

Some extinct reptiles also had a variety of dorsal embellishments along their spines. One group had especially elongated neural spines from the vertebrae that supported a saillike membrane vertically over the back. This may have been a device more to radiate (or entrap) heat than to deter aggression, however. Some dinosaurs had large leaflike spines along the

Figure 36.2 *Seymouria,* a primitive reptile. (From the American Museum of Natural History, New York.)

Figure 36.3 Marine iguana, Galapagos Islands. (Photo—E. C. Williams)

back that would have made it difficult for a predator to do damage to the spinal cord. Other reptiles, such as the *Triceratops,* had broad hoodlike expansions of the skull and horns that appeared to guard the throat and neck regions (Fig. 36.4).

Figure 36.4 Triceratops.

LOCOMOTION AND THE SKELETAL SYSTEM

Locomotion in the reptiles has been achieved in more ways than in any other class. Nevertheless, the basic reptilian body plan, as seen in surviving reptiles, is often declared to be primitive. By contrast with their successors, the mammals and the birds, quadriped reptiles do seem to have a simple skeletal orientation insofar as locomotion is concerned. Most chelonians, lizards, rhynchocephalians, and crocodilians have a horizontally oriented axial skeleton. Except in snakes and limbless lizards, the vertebral column shows differentiation into cervical, thoracic, lumbar, sacral, and caudal regions. The thoracic vertebrae bear ribs that meet ventrally at a sternum, except in limbless forms and chelonians. The lumbar vertebrae are variable and provide considerable flexibility. There are only two sacral vertebrae. The number of caudal vertebrae is variable. The appendicular skeleton of living reptiles is relatively simple. A scapula and coracoid plus an intersternum make up most pectoral girdles. There may be additions to this depending on the reptile. The pelvic girdle consists of the three bones, ilium, ischium, and pubis, as seen in amphibians. Both of these girdles articulate the legs (via the humerus or the femur) at right angles to the long axis of the trunk, and the trunk itself may not be supported by the limbs when the animal is at rest. Most living limb-bearing reptiles seem to run by retaining their appendages at this rather inefficient angle. However, some lizards and the crocodilians can lift the body so that the legs orient almost vertically to the ground and can achieve rapid movement.

Bipedalism was an important development, as is evident from such giants as *Tyranosaurus rex* and some of the extinct ostrichlike reptiles (Fig. 36.5), or the modern basilisk lizard. The pelvis played an important role in providing the scaffolding to which strong muscles attached. In the limbless snakes, movement has been achieved in four ways. Some snakes have *horizontal undulatory* motion where in the lateral movements the snake tracks through all the curves of the path of movement. A second form is *rectilinear* where the body remains relatively straight and movement is achieved by elevation of flat horny scutes on the ventral surface, over which the animal shifts its body forward. No lateral movement is apparent. A third movement is *concertina;* this is an alternate curving and straightening of the body. The body forms a series of curves, the tail anchors, and the body is pushed forward until straight. Then the head and neck

Figure 36.5 Ostrich-like reptile (one of the dinosaurs).

anchor while new curves are established and the process repeated. *Sidewinding* is the fourth form of locomotion of snakes. It is difficult to describe. It consists of lateral looping movements where only two parts of the body are in contact with the ground at any one time. The result is a series of J-shaped marks without any connection between them.

An unusual form of locomotion is seen in the so-called flying snake of New Guinea that can glide from tree to tree because of the ability to expand its ribs laterally to produce a planing surface from the belly and flanks.

The skull is the skeletal structure that has undergone four major modifications from the primitive anapsid condition (Fig. 36.6). Fenestration (windowing) of the skull has been the result. The changes that occurred in the evolution of the reptilian skull are the results of reduction in the numbers of bones making up its roof. This is accomplished by the appearance of temporal openings in the sides of the skull. The most primitive form of skull is the *anapsid* form (Fig. 36.6) where there are no openings in the temporal region. In anapsids the muscles that close the jaws lie on the inside edges of the skull; of course, this gives limited oportunity for development. In the other three skull types it is possible for the muscles of the lower jaw to run through the temporal openings and over the exterior of the skull; this is a less restrictive form of anatomy. The condition can be illustrated in humans, whose reptilian ancestors had synapsid skulls. If one places three fingers as follows and tenses the jaw, it can be noted that muscles associated with the jaw run behind a bony bar and extend over the side of the head: Place middle finger on bony bar about 3 cm (1 inch) in front of the ear; place adjacent finger on the muscle overlying the

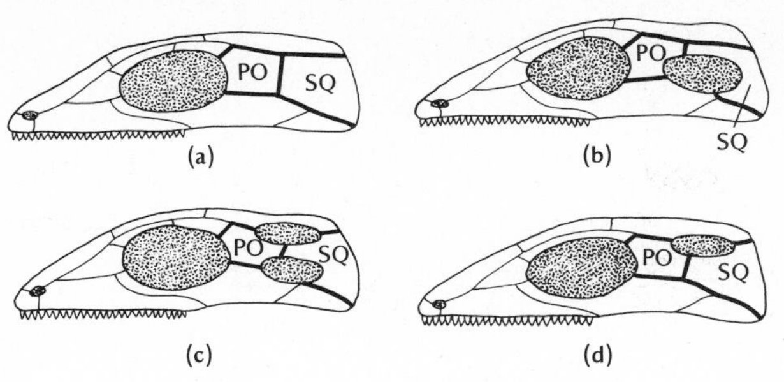

Figure 36.6 Reptilian skull types emphasising differences in the temporal openings. **(a)** Anapsid (no temporal opening, postorbital (po) and squamosal continuous; stem reptiles and in turtles. **(b)** Synapsid: temporal opening at lower edge of postorbital and squamosal; found in mammallike reptiles. **(c)** Diapsid: temporal openings at upper and lower edges of postorbital and squamosal; found in ruling reptiles and reptiles related to birds. **(d)** Parapsid: temporal opening at upper edge of postorbital and squamosal; found in plesiosaurs.

Figure 36.7 Horned lizards. (New York Zoological Society Photo)

lower jaw; place a third finger over the temple. Move the jaw by lowering it or by "gritting" the teeth. It will be noted that the flanking fingers feel muscle movements while the middle finger over the bone does not. This is because muscles operating the jaws run from the jaw, under the bony bar, and over the temporal bone.

One general modification from the amphibian skull is that the roof of the reptilian skull is usually strongly arched by contrast with the broadly flattened skulls of amphibians. A significant feature, retained in the mammals, is the development of the secondary palate. Skulls of some fossil reptiles underwent some unusual changes in proportions of the bones of the face, often displacing the external nares considerably. Also some developed enlarged parts of the skull to form shields and horns (Fig. 36.4). These are rare in living reptiles but can be seen in miniature in the horned lizard (often erroneously called the horned toad) *Phrynosoma* (Fig. 36.7). In some snakes, such as the cobras, the expanded hood is formed by extending cervical ribs laterally to push out the skin.

THE MUSCULAR SYSTEM

The musculature of the reptiles is significantly advanced over the amphibians, even in the living remnants of this class. The flexibility and variability of the skeletal anatomy is accompanied by muscular specialization that allows greater variation in movement than possessed by amphibians.

It was seen that amphibian lateroventral trunk musculature developed three layers, the external and internal obliques and the transversus, plus the longitudinally oriented rectus abdominis. With the evolution of ribs the three layers have been broken up in the thoracic region into three layers of intercostal muscles between the ribs. Another development not seen before is the presence of dermal and skin musculature; this is especially useful in the locomotion of snakes. Limb musculature, as would be expected, especially with the development of bipedalism and the need for good balance and for speed, is highly developed and variable, depending on the functional body plan.

FEEDING AND DIGESTION

Reptiles, extinct or living, are both herbivores and carnivores. The teeth in reptiles relate to the jaws in three ways (Fig. 36.8). Most have teeth attached at the cutting or grasping surface of the jaw and with no depressions or sockets in the jaws; these teeth are called *acrodont.* Some lizards have the teeth adherent to the inner edge of the jaw, either on a ledge or on slight depressions; these are *pleurodont*

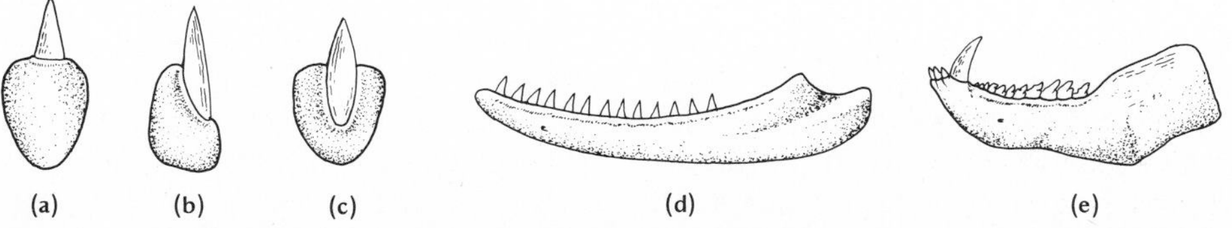

Figure 36.8 Tooth location in reptiles. **(a)** Acrodont. **(b)** Pleurodont. **(c)** Thecodont. **(d)** Diagram of homodont (similar) tooth pattern in jaw. **(e)** Diagram of heterodont (varying) tooth pattern in an advanced mammallike fossil reptile.

teeth. Others, such as crocodiles, have teeth set in pits and are *thecodont.* Most reptiles have little differentiation among the teeth on different parts of the jaws; such similar teeth form *homodont* dentition. In a few living reptiles, and certainly in the extinct therapsid reptiles ancestral to mammals, the *heterodont* condition exists. Incisors, canines, and molars can be distinguished in such animals.

The two venomous lizards of the genus *Heloderma* and the venomous snakes have accompanying tooth modification. Helodermids have pleurodont teeth-bearing grooves. Poison glands in the lower jaw deliver poison to the grooves and into the wounds created when the gila monster holds and twists its prey.

A few snakes may have grooved teeth, but the truly dangerous venomous snakes have fangs derived from modified teeth. The fangs are hollow and the poison glands in the upper jaw deliver poison to the fangs by ducts. Thus they behave something like hypodermic syringes. Some snakes usually do not hold their prey as the Gila monster does. Instead, they strike, deliver the poison via the fangs, and then follow the victim to await its death. This is the behavior of rattlesnakes, for instance. Other venomous snakes, such as cobras, corals, and sea snakes, have relatively shorter fangs and must bite down to inject the venom. Fangs themselves are of different types based on whether they are fixed or hinged. For instance, snakes with grooved teeth are *opisthoglyphs;* those with rigid fangs are *proteroglyphs;* and those with hinged erectile fangs, such as possessed by rattlesnakes, are *solenoglyphs* (Fig. 36.9).

Striking, by snakes, in addition to its usefulness in obtaining food, is also a protective device. Many unwary people are bitten each year as snakes take defensive measures. Most nations attempt to have antisera available for such cases. Antisera are usually produced by collecting the venom manually by forcing the snake to express venom into a container; this is used in small amounts to produce the antisera in mammals.

Food gathering is achieved by snakes through another effective means. The prey is struck and held while it is encircled and squeezed to death; boas and pythons use this procedure.

Another factor useful in obtaining food is that many snakes have a very flexible palate and upper jaw bones that articulate loosely to the skull so that they can be lowered to expand the size of the mouth and pharynx. This permits such snakes to swallow animals larger in girth than the snakes themselves. Accompanying the flexibility of the jaw, of course, is the capability for the gut to stretch.

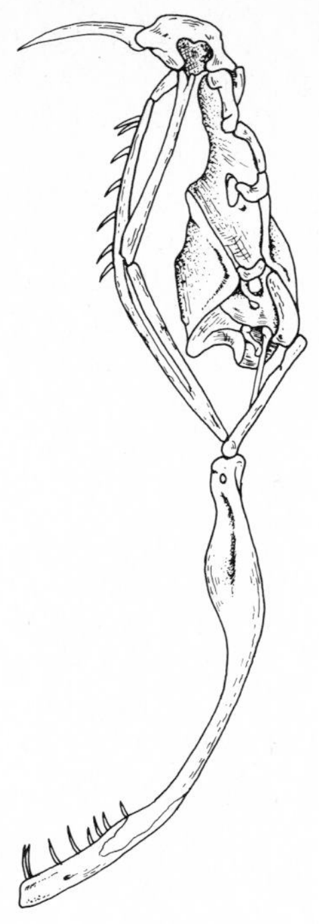

Figure 36.9 Skull of rattlesnake in striking position. In repose the fang folds back in the mouth.

A number of reptiles, both extinct and living, are herbivorous. Many tortoises exhibit this trait and they lack teeth but have horny beaks that help them tear and engulf plant parts.

Terrestrial reptiles have more glands in the mouth than do amphibians. This permits relatively dry food to be moistened. In fact, the poison glands of venomous reptiles are modified oral (salivary) glands.

The tongue may be highly developed. In chameleons, for instance, it is long and can be extended half the length of the animal, where its thickened, sticky tip traps insects. The forked tongue of snakes is very sensitive and functions by converting scents into solution on the tip of the tongue which is then transferred to the vomeronasal (Jacobson's) organs on the roof of the mouth for contact with sensory receptors. When a snake is seen rapidly flicking its tongue it is probing its environment in this way. Some living reptiles such as turtles and alligators have nonprotrusible tongues.

Food is passed to the stomach by a clearly delineated esophagus. It is interesting that the closest living reptilian relative of the birds, the crocodiles, have a modified esophagus or stomach that is gizzardlike and has thick musculature.

The intestines are relatively longer than in amphibians. A new development, probably selected for through the herbivorous traits of evolving reptiles, is the presence in some living reptiles of blind diverticula, caeca, at the junction of the small and large intestines. A caecum usually serves as a bacterial reservoir where plant materials are partially digested by bacterial action for further use by the animal. The straight large intestine opens into the cloaca.

GASEOUS EXCHANGE AND INTERNAL TRANSPORT

Gaseous Exchange There is little or no gaseous exchange through the integument or its modifications by contrast with amphibians. Lungs are the principal places for gas exchange. The reptilian lungs are intermediate between those of amphibians and mammals. In lizards and snakes there is often an asymmetry of lung development. Lizards often have one lung larger than the other, and in snakes only the right lung is fully developed, the left sometimes being fully absent. Some lizards have additional diverticulae from the lungs that extend posteriorly. The presence of this tendency is interesting, since it becomes well developed in the air sacs of birds, and probably some similar relationship developed with the lightening of the skeleton of the pterosaurs. In some sea turtles it is likely that exchange of gases is also accomplished in the cloacal walls.

The lungs are connected to a larynx, made of arytenoid and cricoid cartilages, by a trachea that is reinforced by semicircular cartilaginous rings that may also extend along primary and secondary bronchi (Ch. 10).

Respiration in reptiles may be by the force-pump method used by amphibians. However, some also use the rib and abdominal muscles for sucking in air. Although there is no diaphragm, in some reptiles such as crocodilians the lungs may be partially separated from the abdominal coelom by a transverse septum lying between the lungs and the liver.

Internal Transport Internal transport by the blood is achieved through arterial and venous channels bearing some modifications from those of the amphibians. Of course, there are no gills, and young reptiles in the shell have gaseous exchange through the CAM which is supplied by allantoic blood vessels. The adult reptiles have much more efficient circulatory systems than do the amphibians. The absence of gills and the presence of a metanephric kidney (see below) are associated with this change.

The hearts of various reptiles vary in their efficiencies. There may be as many as six chambers functioning in some, although two of these are not completely separated. On the other hand, the chambers may approach the four characteristic of birds and mammals in the crocodilians. Among the variations that may be seen is the presence of a sinus venosus in some reptiles but its absorption into the right atria of others. The atria are always separated into right and left chambers, but the ventricles vary. There is always an interventricular septum, but it is incomplete in all but the crocodilians, thus permitting exchange of oxygenated and deoxygenated blood between the ventricles of all but the latter.

The conus arteriosus becomes partitioned into three channels rather than the two principal ones seen in amphibians. One channel delivers blood to the lungs from the right ventricle. A second channel delivers blood to the body and head from the left ventricle through the persistent right fourth aortic arch. The third channel exits from the right ventricle to the left fourth aortic arch where it delivers a branch to the viscera before joining the right arch to form the dorsal aorta. This permits some admixture of oxygenated blood from the left ventricle and deoxygenated blood from the right ventricle.

In crocodilians the right and left atria and ventricles are completely separated by septa, creating a heart with two isolated pumping systems. However, it is claimed by some that a small opening (*foramen of Panizzae*) between the left and right vessels derived from the fourth aortic arches permits the admixture of bloods. Some experiments with the caiman suggest, however, that if there is a flow across this opening, it is for the oxygenated blood to enrich the deoxygenated blood, rather than the reverse.

One new development in the venous systems of reptiles is for veins in the trunk to unite to form a distinct posterior vena cava that drains all of the viscera and trunk. Although a renal portal vein exists, it does little filtering at the kidney level as there is a new kind of kidney developed and the renal portal veins travel through this kidney without significantly supplying it.

WATER BALANCE AND EXCRETORY CONTROL

The skins of reptiles have been said to offer little opportunity for water loss due to the presence of a well-developed exoskeleton of epidermally derived keratinized horny scales (Fig. 36.10). However, it has been demonstrated that cutaneous evaporation can be a major avenue of

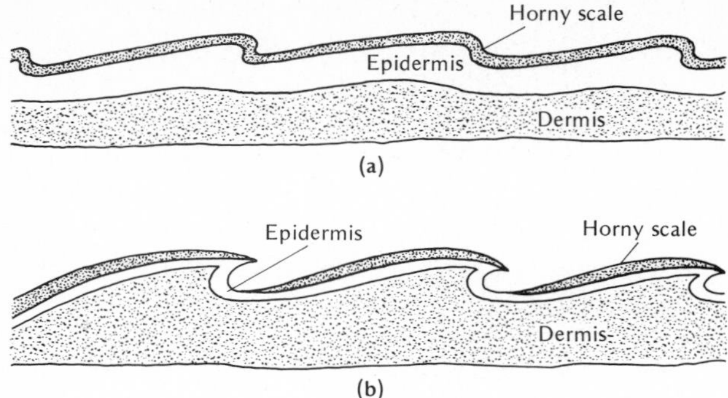

Figure 36.10 Two forms of scales in reptiles. **(a)** Lizard scales. **(b)** Strongly overlapping scales as seen in snakes.

reptilian water loss. The scales are not homologous with fish scales, which are dermal in origin. The reptilian scale develops as a localized thickening of epidermis; these thickenings often overlap. Scale development is assisted by a dermal papilla lying below the scale; this is a well-vascularized area and helps to provide nutrients to the overlying epidermis. In some reptiles, such as crocodilians, dermal plates, *osteoderms,* underlie and reinforce the scales.

In order for growth to occur, the exoskeleton must expand. In most reptiles this is achieved by periodic shedding of the nonliving scales and skin. Shedding may be in pieces or the animal may break out of its intact covering. In snakes, entire empty skins may be found after shedding, including the covering of the eye which is also shed. In fact, some snakes, such as rattlesnakes, are blind during the period just before shedding due to the opacity of the skin over the eye.

Most reptiles need to conserve water. This is achieved in several ways, other than by excluding its loss through the skin. A more efficient kidney, the *metanephros,* develops in reptiles and is retained in all amniotes. The internal anatomy of the metanephros (Ch. 10) resembles that of the mesonephros except that the ducts are longer and assemble into collecting tubules that drain into the ureter, a new duct differing from the archinephric duct of fishes and amphibians. The duct system increases the efficiency in the resorption of water into the blood. The excretory waste of reptiles is in the form of uric acid. This may be delivered in fluid form to the cloaca in aquatic forms such as turtles and crocodiles or in semisolid form in others.

Some desert reptiles do not drink water and obtain it from their food and their metabolism. Some marine reptiles, such as the marine iguanas of the Galapagos Islands, drink sea water or take it in when they eat marine algae. They possess specialized salt-secreting glands located on the head that deliver highly concentrated salt solution to the nasal chamber. Some terrestrial iguanids also use a similar system, and this may assist in water conservation by reducing the salts in the filtered urine, thus allowing for greater absorption of water in the cloaca.

Although a urinary bladder, derived from the floor of the cloaca, is present in some reptiles, it is absent in crocodiles, snakes, and some lizards.

REPRODUCTION

The ovaries and testes are paired. Ovaries may be saclike, resembling those of amphibians, or they may be solid as in other amniotes. The testes retain the opisthonephric ducts as the sperm ducts. Near the testes residual portions of the mesonephric kidneys form epididymises. Lizards and snakes have paired extrusible cloacal structures, the *hemepenes,* that transfer sperm to the cloaca of the female during copulation (Fig. 36.11). Turtles, tortoises, and crocodilians have a penis.

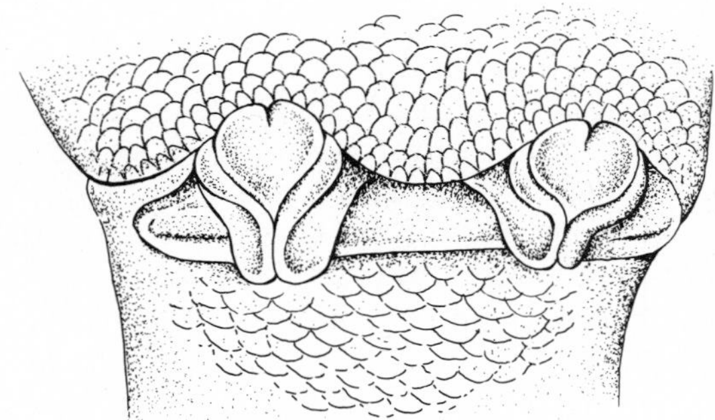

Figure 36.11 Hemipenes of a lizard.

The fate of the egg varies in different reptiles. Large fossilized eggs of dinosaurs have been found, indicating that some of them, at least, were oviparous. On the other hand, skeletons of ichthyosaurs have been found containing small intact ichthyosaur skeletons within, demonstrating the existence of an ovoviviparous or viviparous condition in these extinct forms. Living reptiles demonstrate all three conditions.

Oviparous reptiles lay eggs in numbers varying from one or two to over 1000. Where the eggs are laid varies with the species. Some turtles travel considerable distances to special spawning sites. For instance, female snapping turtles travel overland from the pond or stream to find a suitable spot where a cavity is dug with the hind legs. Marine loggerhead turtles come ashore and dig holes in the sand above the high tide mark with their front flippers, lay their eggs, cover them and return to the sea immediately.

Crocodiles lay their eggs in moist nests or mounds, depositing as many as three dozen eggs in layers. The females may return to the nests.

Some snakes, such as pythons, may lay eggs in pyramids and then wrap themselves around the eggs. Also, the body temperature of the female is said to rise about 3° C (5° F) or more above the normal body temperature of the snake.

Eggs may be round, oval, or elongated. They are usually creamy white. They may be hard and calcified as with some turtles or they may be leathery as with a number of snakes.

Ovoviviparity also exists. For instance, the North American rattlesnakes are of this category. The female retains eggs in her reproductive tract until hatching. Some ovoviviparous reptiles have been noted to break the egg membrane as it protrudes from the cloaca at hatching and, after release of the young, to eat the membrane.

There are a few reptiles that are truly viviparous. This means that either through the yolk sac or the CAM there is an intimate relationship established with the lining of the female's reproductive tract. Viviparousness has been seen in some Australian skinks and snakes as well as in some European and American snakes. For instance, it is said that the garter snake has a placental zone where amino acids are exchanged as well as gases and water. Of course, the presence of viviparity is interesting to students of the evolution of the trait, since those forms showing this trait in living reptiles are diapsids. Birds are evolved from diapsid reptiles, yet they never show viviparity, while the viviparous true mammals arose from synapsid reptiles. This is an excellent example of the situation seen throughout the living world where selection for a particular adaptation has developed more than once. In Chapter 38 we will see that some primitive mammals are oviparous, for instance.

SENSORY RECEPTION AND NEURAL COORDINATION

The central nervous system of reptiles shows a shift of nervous integration from the midbrain as in amphibians to the cerebrum, which is relatively larger than that of amphibians. However, the reptilian central nervous system is still rather small.

Some reptiles have taste buds, and the vomeronasal organs described earlier play an important role. In reptiles with secondary palates, it appears that the sense of smell is also improved due to the greater surface area for olfactory receptors.

The eyes of reptiles resemble those of other amniotes in that accommodation is achieved by changing the curvature of the lens rather than by shifting its position in the eye as with anamniotes. Although movable eyelids are present in some living reptiles, snakes and some burrowing lizards have the eye covered by immovable transparent skin. Other reptiles may have movable and sometimes transparent nictitating membranes.

The pineal or third eye is present in some lizards and *Sphenodon*. It functions through neurosecretion to serve as a biological clock.

Hearing is poorly developed in most reptiles. The lagena is small but longer than in amphibians, and crocodiles have it developed into a cochlea. Snakes and lizards show interesting contrasts. Snakes are deaf in the sense of receiving sound waves through the air. They lack external and middle ears or eustachian tubes. They receive vibrations through the quadrate bone of the jaw, and it transmits the impulses to the columella. Lizards, on the other hand, hear. They often have an external auditory canal (meatus) and an ear drum, middle ear, and eustachian tube.

The interesting and unusual sense organ found in the reptiles is the radiosensitive pits of the pit vipers (Fig. 36.12). These *loreal* pits are located between each nostril and eye. They contain high concentrations of free nerve endings from the ophthalmic and supramaxillary

Figure 36.12 Pit vipers. Pope's pit viper (*top*); Eastern Diamondback Rattlesnake (*bottom*). Note the thermosensitive pit between the nostrils and the eyes. (New York Zoological Society Photos)

branches of the fifth cranial nerve. The opening to the pit is directed forward and the receptors are sensitive to very slight changes in radiant energy, especially long-wave-length infrared. Changes as small as 0.2° C can be distinguished in the surroundings. Pit vipers, even when they are blind during the preshedding period, can accurately locate heated (or chilled) objects such as mice and strike accurately at them.

HORMONAL CONTROL

The reptilian endocrine system is fully developed, functional, and typical of the vertebrates (Ch. 12).

BEHAVIOR

Three forms of behavior in reptiles are of interest. As slow as some reptiles may seem, they have courtship patterns, nevertheless, ranging from head and neck posturings in some turtles to movements designed to display color as in the case of male fence lizards. In the presence of females the lizard raises and lowers the body rhythmically, increasing the frequency the closer he is to the female, to expose blue ventral body patches. If the patches serve adequately as releasers for the females, they accept the males.

Another interesting and similar set of body movements used by lizards relates to their regulation of body temperature. On emerging from burrows or recesses in the morning, some lizards flatten themselves to expose the maximum body surface to the sun. This warms them enough to increase their activity and they proceed with food hunting. As the day passes and the ground becomes warmer, they elevate themselves on their legs to increase circulation of air, minimize contact with the warm ground, and they also alter their body contour to minimize the amount of sun striking them. The elevation process, called stilting, is used at times during the day when the sun is hot and they are not in recesses or burrows. As evening approaches they again spread themselves to receive more warmth. Examination has shown that the lizards retain a relatively narrow range of body temperatures this way, close to that of mammals in fact, even when the surrounding temperatures may be several degrees lower or higher. Regulation of a narrow range is also achieved by the way lizards orient their body to the sun during their basking behavior.

There is an interesting behavioral pattern in some snakes related to hibernation. As winter approaches they may seek sites where they aggregate in large numbers. Interestingly aggregation may involve more than one species. Joint occupancy of one spot, called a *hibernaculum*, has been reported from a variety of places. In one instance over 250 snakes of three species were found together. In another instance 62 snakes of seven species and fifteen amphibians of three species were found together.

Relationship of Reptiles to Other Animals

During the Age of Reptiles, of course, the existing dominant vertebrates bore the same relationship to each other and their environment as the dominant mammals of today do to each other. They filled existing niches and related to each other in food chains. Today the relationship of the reptiles to other animals is slighter. The carnivorous reptiles are predators on other reptiles, amphibians, fishes, and mammals. Of course, due to their size the prey of most reptiles is small, rarely exceeding the size of large rats or rabbits. On the other hand, pythons have been known to ingest adult pigs. Some snakes are predators on the nests of birds, either for their eggs or for the young.

Reptiles also have natural enemies. Some snakes eat other snakes, a noteworthy example being the kingsnake that eats rattlesnakes. There are mammals that are natural enemies of snakes; the skunk and the mongoose are good examples. It has been demonstrated repeatedly that hawks are able to locate and prey on snakes from the air. A point to be made is that the sum of the natural enemies of reptiles serves to keep their numbers in check just as occurs with birds and mammals, for instance.

Relationship of Reptiles to Humans

For as important a class as the reptiles are in terms of their evolutionary significance as predecessors of birds and mammals, probably humans have less relationship with this class than any other, even including the amphibians.

Culturally, snakes have served as gods for some peoples or as symbols of evil for others. Because of their appearance and forms of locomotion, reptiles are usually associated with symbols of ugliness and of cunning, and snakes also serve as phallic symbols. Also the presence of several kinds of venomous reptiles from asps, to rattlesnakes, cobras, and fer-de-lances instills fear in most people. The copperhead undoubtedly bites more people yearly in the western hemisphere than any other snake, although rattlesnakes, in spite of their terrestrial habitat and the presence of horny rattles on the tail, also bite a large number.

Figure 36.13 Tortoises. Galapagos Island tortoise (*left*). (Photo—E. C. Williams) Moorish tortoise (*right*). (New York Zoological Society Photo)

Reptiles play only a minor role in providing food for humanity. Perhaps most known is turtle soup, usually commercially prepared from sea turtles. Some snakes, such as rattlesnakes, are said to be a delicacy when they are skinned and sliced into steaks.

At one time it was highly fashionable to have snakeskin shoes, gloves, and handbags. Especially prized were garments and luggage made from alligator skin. Legislation and public awareness that the alligators are an endangered species may save them from extinction. However, the draining of large swampy areas in the state of Florida may lead to the extinction of these large reptiles, there; however, they are relatively abundant in Louisiana.

Classification (Including Present Status of Subclasses)

Reptiles evolved into six subclasses. It is usually stated that only two of these have living representatives today. However, R. T. Bakker, of Harvard University (see Suggestions for Further Reading), has proposed that birds are living members of a subclass Dinosauria, and that there is no existing full-fledged class Aves or, for that matter, Mammalia. While this is a challenging interpretation, we will retain the more usual classification, not in rejection of Bakker's proposal, but to permit it time better to be understood by teachers and students.

The usual classification of reptiles is based principally on the structure of the skull and the presence or absence of the several kinds of fenestration.

SUBCLASS **Anapsida** Temporal openings in skull absent. Quadrate bone of jaw hinge immovable.

ORDER **Cotylosauria**—Stem reptiles. Extinct. The cotylosaurs had a parietal foramen [central dorsal opening in the roof of the skull associated with parietal (pineal) eye], were quadripedal and lizard-like.

ORDER **Chelonia**—Turtles (aquatic) and tortoises (terrestrial) (Fig. 36.13, Pl. 10). The bodies are enclosed within bony shells consisting of a dorsal carapace and ventral plastron. They lack teeth but have a horny beak. The tongue cannot be extended. They are usually capable of retracting the neck and head into the shell, which may be accomplished by vertical flexure or by a horizontal (side-necked) flexure. There are 19 living families of tortoises and turtles. Among these are the snappers, land tortoises, leatherback sea turtles, true sea turtles, and snake-necked turtles. This is an ancient and stable order. Worldwide except Arctic and Antarctic.

SUBCLASS **Synaptosauria** All members extinct. They possessed a parapsid type of skull. The temporal opening had the postorbital and squamosal bones meet below the temporal opening. Their quadrate was immovable and fused to the skull.

Figure 36.14 Plesiosaur, a sauropterygian.

They existed from the Permian to the Cretaceous, and consist of the plesiosaurs and their relatives. There were two orders.

ORDER **Protosauria** Small and lizardlike.

ORDER **Sauropterygia** This order was amphibious or marine. Some had short necks but others had rather long ones that made it possible for these herbivorous animals to reach below the surface of the water to forage for food. Their limbs (and girdles) reflect their aquatic existence, usually being formed as broad flattened flippers (Fig. 36.14).

SUBCLASS **Ichthyopterygia** All members extinct. They were marine and, in a manner analogous to later and modern dolphins, had acquired a fish-shaped body (Fig. 36.15). The parapsid skull was unusual in that the postorbital and squamosal did not form the border of the temporal fossa. The limbs formed finlike paddles. A dorsal fin was present but lacked skeletal support. The nostrils were set back dorsally on a pointed and fishlike snout. Present between Triassic and Cretaceous.

SUBCLASS **Lepidosauria** The members of this subclass retained quadripedal locomotion. They possess a diapsid skull. Two of the three orders survive today.

Figure 36.16 *Sphenodon* the tuatara, the only living species of rhyncocephalian; from New Zealand. (Steinhart Aquarium, San Francisco)

Figure 36.15 Ichthyosaur.

ORDER **Eosuchia** Extinct. Primitive, slender limbs, lizardlike form. Permian to Eocene.

ORDER **Rhynchocephalia** Represented today by one species, *Sphenodon punctatus,* the tuatara, in New Zealand, where it is protected (Fig. 36.16). Complete diapsid skull.

ORDER **Squamata** This order includes the lizards and snakes, the most abundant reptiles surviving today. The Squamata occupy various niches. The diapsid skull is modified in that the lower temporal opening (infraorbital fossa) lacks a lower bony border. The squamosal bone is movable, a trait of considerable value to some snakes. The body is covered by epidermal scales; sometimes these have bony osteoderms under them. The order arose in the Jurassic. There are two distinctly different suborders, each with a number of families.

SUBORDER **Sauria (Lacertilia)**—Lizards (Fig. 36.17) They range from small to medium large. Traits they usually bear in common are generally quadripedal locomotion, eyelids, and external auditory canal (meatus). They have a worldwide distribution. There are about 3000 species in 20 living families, and there is considerable variation among the families. Some examples are the gekkos, chameleons, iguanas, skinks, lizards, Gila monster, limbless lizards, worm lizards, and flat-footed lizards. Worldwide except Arctic and Antarctic.

SUBORDER **Serpentes (Ophidia)**—Snakes (Fig. 36.18) They range from as small as 8 cm (3 inches) to 9 m (30 feet). In common they are limbless. Their lower jaws are joined by a ligament rather than a suture (as in saurians). Their eyelids are immovable and they lack an external ear opening. The tongue is forked and protrusible. There are about 3000 species in 13 living families. Representatives of this suborder include the boas, pythons, true vipers, pit vipers, sea snakes, blind snakes, and colubrids. Worldwide except Antarctic. Most are temperate or tropical.

SUBCLASS **Archosauria** This group includes the great ruling reptiles of the Mesozoic. While bipedalism was a dominant trait at the height of their reign, the only living members, the Crocodilia, are quadripedal. The archosaurs underwent a great adaptive radiation and demonstrated great diversity. The diapsid skull is characteristic of the group. They arose in the Triassic.

ORDER **Thecodontia** The extinct thecodonts were the primitive members of this subclass. They were carnivores. Their hind limbs were longer than the forelimbs as is characteristic of animals with a bipedal tendency. Their teeth were set in sockets (thecodont) and their long slender skulls were lightened by absence of some bones around the eye and in the palate. Triassic.

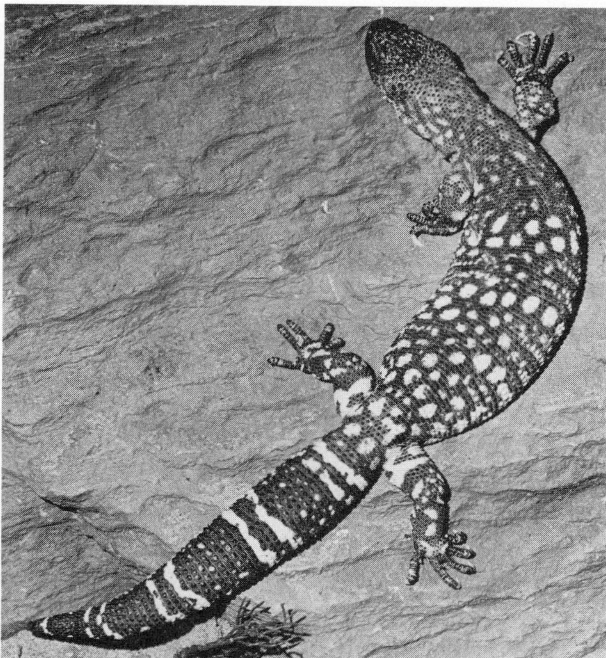

Figure 36.17 Variation among lizards. Sungazer lizard (*top*). Black Tegu (*center*). Gila Monster (*bottom*). (New York Zoological Society Photos)

Figure 36.18 Variation among snakes. Northern Banded Watersnake (*top*). Egyptian Cobra (*center*). African Puff Adder (*bottom*). (New York Zoological Society Photos)

Figure 36.19 Alligator and Crocodile. American Alligator (*top*), immature Mugger Crocodile (*bottom*). (New York Zoological Society Photos)

ORDER **Crocodilia** These are the degenerate survivors of some archosaurs (Fig. 36.19). They have terminal but elevated nostrils and possess a secondary palate, both traits that aid these carnivorous aquatic reptiles. When on land, their locomotion is quadripedal. Swimming is by tail and trunk undulations. There are three living families that include the crocodiles, gavials, alligators, and caimans. North and South America, Asia, East Indies, Australia, and Africa.

ORDER **Pterosauria** The extinct flying reptiles achieved flight or volplaning through the great elongation of the fourth digit, to which a membranous sheet of skin from the length of the trunk was attached (Fig. 36.20). Aiding this trait was the development of light bones that had air chambers in them. Elongated tail present. The teeth were well developed and the pterosaurs probably existed on insects and small animals. Jurassic to Cretaceous.

ORDER **Saurischia** Members of this order expanded into several modes of life (Fig. 36.21). They are known as the reptilelike dinosaurs on the basis of the anatomy of the pelvis. The saurischians were primarily bipedal, and the pelvis retained a three-pronged (triradiate) orientation of the bones. The teeth were present throughout the length of the jaws. Some, with greatly reduced forelimbs, reached awesome heights. Triassic to Cretaceous.

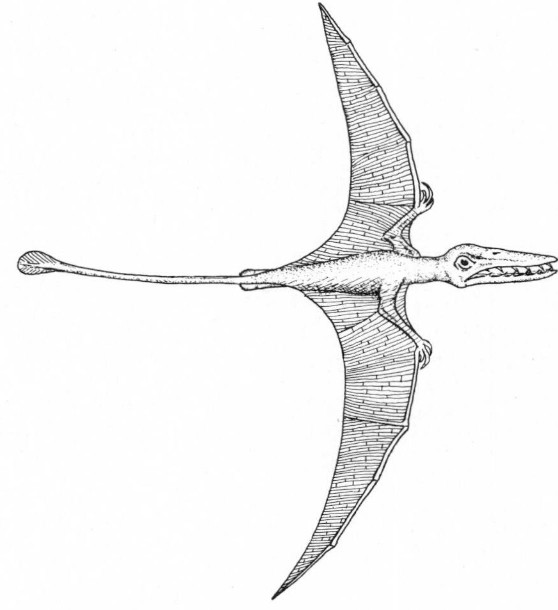

Figure 36.20 Pterosaur.

ORDER **Ornithischia** These were the birdlike dinosaurs (Fig. 36.22). Although they were largely quadripedal, the pelvic girdle was tetraradiate due

Figure 36.21 An herbivorous and a carnivorous saurischian.

to the development of two prongs on the pelvic bones. There was a tendency for the pelvis to orient parallel to the spinal column, a trait retained in birds, and one prong of the pelvis came to lie close to the ischium. The teeth were either totally lacking or absent from the front of the jaws. A number were principally herbivorous. Triassic to Cretaceous.

SUBCLASS **Synapsida** These extinct animals are the mammallike reptiles (Fig. 36.23). The single lateral temporal opening was bounded below by the jugal and the squamosal, a trait retained in part in the mammalian descendents. The skulls of these animals are relatively large.

ORDER **Pelycosauria** The pelycosaurs were primitive synapsids and lacked a secondary palate. They had widely separated external nares. Carboniferous to Permian.

ORDER **Therapsida** The therapsids are the synapsids from which the mammals arose. The external nares were dorsal and near the tip of the snout. The secondary palate was present in advanced forms. Permian to early Jurassic.

Figure 36.22 An herbivorous and a carnivorous ornithischian.

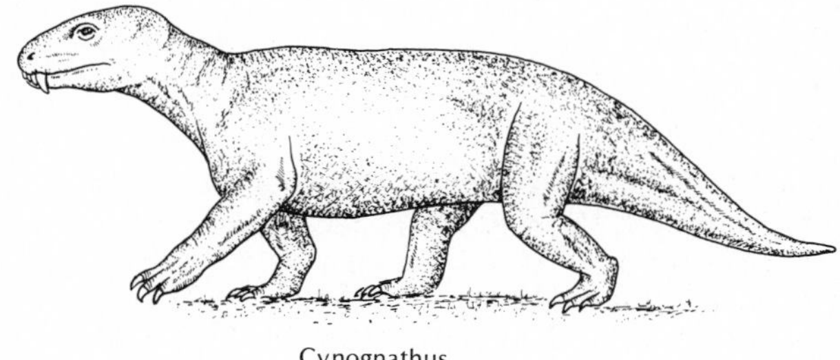

Figure 36.23 A mammallike reptile.

SUGGESTIONS FOR FURTHER READING

Bellairs, A. *The Life of Reptiles,* 2 vols. New York: Universe Books (1970).

Bucherl, W., E. E. Buckley, and V. Deulofeu (Eds.). *Venomous Animals and Their Venoms,* 3 vols. New York: Academic Press (1968).

Carr, A. *Handbook of Turtles of the United States, Canada, and Baja California.* Ithaca, NY: Cornell Univ. Press (1952).

Cochran, D. M., and C. J. Goin. *The New Field Book of Reptiles and Amphibians.* New York: G. P. Putnam's (1970).

Colbert, E. H. *Dinosaurs: Their Discovery and Their World.* New York: E. P. Dutton (1961).

Ditmars, R. L. *The Reptiles of North America.* Garden City, NY: Doubleday (1936).

Goin, C. J., and O. B. Goin, *Introduction to Herpetology,* 2d ed. San Francisco: W. H. Freeman (1971).

Klauber, L. M. *Rattlesnakes: Their Habits, Life Histories, and Influence on Mankind,* 2 vols. Berkeley, CA: Univ. of California Press (1956).

Minton, S. A., Jr., and M. R. Minton. *Venomous Reptiles.* New York: Charles Scribner's (1969).

Pope, C. H. *Snakes Alive and How They Live.* New York: Viking Press (1937).

Pope, C. H. *Turtles of the United States and Canada.* New York: Alfred A. Knopf (1939).

Pope, C. H. *The Reptile World. A Natural History of the Snakes, Lizards, Turtles and Crocodilians.* New York: Alfred A. Knopf (1955).

Schmidt, K. P., and D. D. Davis. *Field Book of Snakes of the United States and Canada.* New York: G. P.

Putnam's (1941).

Schmidt, K. P., and R. F. Inger. *Living Reptiles of the World.* Garden City, NY: Doubleday (1957).

Smith, H. M. *Handbook of Lizards (of the United States and Canada).* Ithaca, NY: Comstock Publ. (1946).

Stebbens, R. C. *A Field Guide to Western Reptiles and Amphibians.* Boston: Houghton Mifflin (1966).

Wright, A. H., and A. A. Wright. *Handbook of Snakes of the United States and Canada,* 2 vols. Ithaca, NY: Comstock Publ. (1957, 1962).

Young, J. Z. *The Life of Vertebrates,* 2d ed. New York: Oxford Univ. Press (1962).

From Scientific American

Bakker, R. T. "Dinosaur Renaissance" (Apr. 1975).

Bogert, C. M. "How Reptiles Regulate Body Temperature" (Apr. 1959).

Carr, A. "The Navigation of the Green Turtle" (May 1965).

Gamow, R. T., and J. E. Harris. "The Infrared Receptors of Snakes" (May 1973).

Gans, C. "How Snakes Move" (June 1970).

Minton, S. A., Jr. "Snakebite" (Jan. 1957).

Riper, W. Van. "How a Rattlesnake Strikes" (Oct. 1953).

37 The Birds

The most colorful class of the vertebrates consists of the birds, class Aves. Coloration is generally achieved through patterns of pigment deposited in the feathers. Feathers are the most distinctive diagnostic feature of these animals inasmuch as only birds have feathers. The coloration of birds leads them to be especially attractive to laymen as is attested by the number of local and national organizations devoted to the study of birds. No other class of vertebrates has such popular appeal.

The development of feathers permitted the invasion of the air in a way only approached by the success of the insects. Bats and the extinct flying reptiles are vastly inferior in their exploitation of the aerial environment by comparison with birds. The feather, which accounts for this success, is a derivative of the skin and grows in several forms. The flight feathers of the wings are built and used in ways that give birds excellence of flight, stability from aerial accidents such as stalls and spins, and efficiency in insulation.

Although there are about 8700 species distributed throughout the world, including the arctic and antarctic areas, with diverse shapes, sizes, and coloration, birds are strikingly uniform as a class. Small but significant variations result in the many orders and families.

In addition to feathers, modern birds lack teeth, but their bills, or beaks, created by horny coverings over the extended bones of the face and jaw, serve some of the grasping, cutting, and tearing achieved by tooth-bearing animals. Also, birds' eyes usually lie in relatively large orbits, often with only a thin bony plate separating one orbit from the other. Birds also all have wings, all are bipedal, all lay eggs, all are homoiotherms, and all have evolved a well-developed nervous system. The dominating part of the avian brain, however, differs from that of mammals.

As we shall see below, loss and fusion of bones has occurred extensively in birds. The most prominent reductions are in the bones of the wing, creating a sparlike structure along which the flight feathers are oriented. The legs also show noteworthy reduction in the number of bones and show several fusions. A third skeletally associated adaptation is the great lightness of bones and the presence of air sacs within them.

Another way in which birds differ from all other classes, except mammals (who independently achieved a similar condition), is that they are *endothermous* and produce their own heat. They do this within relatively narrow temperature limits and, thus, are *homoiotherms.*

Anatomy and Physiology

Birds have evolved with three organ systems highly integrated serving for protection and locomotion. They are the skeletal, muscular, and integumentary systems. No other class so strongly relates the three systems. The significance of this is seen when one considers a bird in flight. The skeleton is beautifully evolved providing strength, lightness of weight, and it is aerodynamically sound. The muscles are arranged in groups that assist in keeping these heavier-than-air animals aloft; their oxygen utilization is also more efficient than in other classes. Over most of the body the feathers make flight possible, provide protection against loss of heat, or are involved in keeping the animal afloat on water. In addition, areas of the integument on the legs retain the capability of producing protective scales and claws.

THE SKELETAL SYSTEM

Several skeletal features are noteworthy. The light weight of the skeleton is achieved in two ways. First, the wings and the legs are reduced from the typical pentadactyl limbs of vertebrates. The principal reduction in the wing is in the wrist, palm, and fingerbones: the carpals and metacarpals and phalanges are reduced in number and fused (Fig. 37.1). The leg has the tibia, fibula, and tarsals reduced and/or fused. The shinbone is a tibiotarsus. The bones of the ankle are fused with the arch bones of the foot to form a tarsometatarsus. The toes may be reduced from the typical four of most birds to as few as two in ostriches. Second, strength in the body (the bird's "fuselage") is obtained in several ways. The bones of the pectoral girdle are anchored to the axial skeleton providing maximum strength. The scapulas are elongated and lie close to the axial skeleton. The clavicles ("wishbone") and the coracoids form a good fulcrum relationship with the broad sternum and contribute to the tripodlike orientation of this girdle with the humerus of the wing. The pelvic girdle especially helps to strengthen the skeleton through a firm union with the pelvic and lumbar areas of the vertebral column to form a rigid support for the legs, the *synsacrum.* In addition, the pelvic girdle is rotated to lie almost parallel to the long axis. The ilium and ischium especially do this, and the pelvic bones do not meet in the ventral midline. Instead, they too grow posterior–laterally. This position of the pelvic bones permits ease of egg-laying in females by creating no skeletal obstruction to the passage of the eggs.

Accompanying the anatomical specializations of the avian skeletal system is an unusual arrangement whereby tubular extensions from the lungs penetrate the long bones.

Another stabilizing influence on the axial skeleton is the presence of *uncinate processes* that bridge from one rib to the adjacent one behind it. These processes provide structural reinforcements for the thoracic cage.

Birds lack well-developed skeletal support in the tail by contrast with their Jurassic ancestor, *Archaeopteryx* (Fig. 37.2). The *pygostyle,* a few fused caudal vertebrae, is all that remains and it is buried in the stubby mound, the *pygidium,* that usually also includes the oil-producing preen gland.

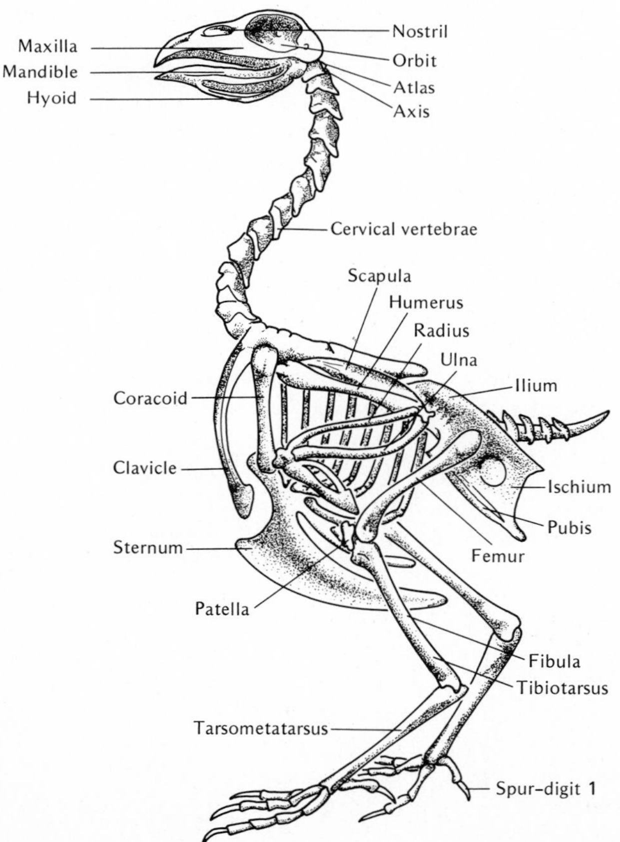

Figure 37.1 Bird skeleton.

Figure 37.2 Archeopteryx, an extinct bird.

Another distinctive skeletal trait of birds is the presence of the highly elongated face bones that make up the beak. The lower jaw of the beak, the mandible, is formed by five bones. The movable bone which articulates with the skull is the articular bone as it is in reptiles and amphibians. Of course, modern birds lack teeth, although the first birds had well-developed conical teeth. The upper and lower members of the beak are overlaid by a horny sheath derived from the integument. It is suggested that the sheath is formed from modified scales, remnants of the reptilian ancestry. The study of beak types is a subject in its own right. The principal point to be made here is that beaks are adaptations for particular feeding habits as D. Lack so clearly showed in a study of Darwin's findings on the Galapagos Islands (Ch. 42). A few feeding types are illustrated by hawk, eagle, or falcon curved beaks used for tearing flesh, scooping beaks as seen in pelicans, flamingos, and spoonbills, crushing as seen in parrots, toucans, and seed-eating birds such as finches, spearing as with the members of wading birds, chiseling seen with woodpeckers, straining beaks of ducks, slender beaks for probing mud by shore birds or for picking up insects as with warblers, or tubular beaks of hummingbirds for sucking nectar. In addition to the functional horny components of beaks, there are additional modifications accompanying them. For instance, pelicans and their relatives have widely expansible sacs under the beak for temporary storage of food, or such structures may be used in sexual display during mating. In some birds, such as hawks and parrots, the proximal part of the upper beak may be soft and fleshy. This structure is called a *cere* (Fig. 37.3).

Figure 37.3 Eastern sparrow hawk. Note cere, a fleshy area around the nostril. (New York Zoological Society Photo)

Another skeletal trait that is noteworthy is the great flexibility of the neck. This is created by the way in which each neck vertebra articulates with the next. The opposing surfaces articulate as curved areas much as a pair of clasped human hands are related to each other. Each surface allows for a small degree of lateral rotation; the collective effect of several cervical vertebrae is to permit great movement of the neck.

THE MUSCULAR SYSTEM

The muscles, tendons, and ligaments of birds also are well-adapted to aerial and perching existences. Whereas in anamniotes and reptiles the prominent musculature is associated with the axial skeleton, in birds axial musculature is greatly reduced, and the bulk of the muscle is associated with the appendicular skeleton. There is a great reduction of the dorsal musculature accompanying the highly fused trunk skeleton. Also, abdominal muscles are poorly developed sheets. The greatest expansion in the musculature is seen in the pectoralis muscles connecting between the sternum (and its keel in carinate birds) and the wing. This muscle in flying birds is relatively enormous and serves to depress the wing in flight to produce the wing stroke. Interestingly, the muscle that is concerned with the upstroke of the wing, the supracoracoideus, is also attached to the sternum and its tendon runs through a pulleylike arrangement to the upper side of the humerus; contraction of this raises the wing.

The muscles of the thigh and shank are usually well developed. The shank muscles are concentrated near the knee and function through long tendons that run past the ankle to the foot. For this reason the visible, and usually scaly, parts of birds' legs are very thin.

Perching birds are able to rest and sleep without falling because the tendons are stretched when the leg is bent as the bird settles on the perch; this flexes the toes to lock the animal into position. A similar process occurs with the grasping of birds of prey.

THE INTEGUMENT

It is the feather, an integumentary derivative, that has made for the distinctive evolution of birds, and an understanding of feather structure, feather types, feather patterns, and use of feathers in flight is in order.

Feathers are of different types (Fig. 37.4). A newly hatched bird principally has *down* feathers that create a fluffy appearance; these conserve heat in the young bird. (Some newly hatched birds are naked, however, or have a scanty covering of down.) Structurally they consist of a *quill* (calamus) partially embedded in the skin and with a cluster of *barbs* emerging from the quill. The barbs are not modified to anchor together and thus produce the fluffy condition with much

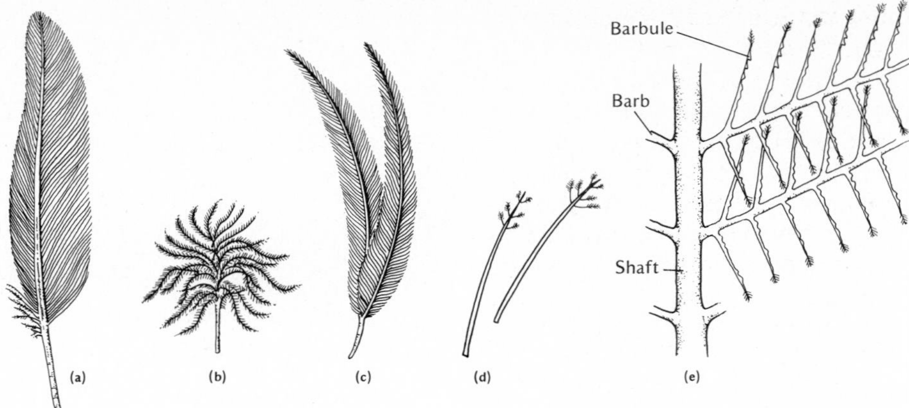

Figure 37.4 Some variations in feathers. **(a)** Wing feather. **(b)** Down feather. **(c)** Feather with accessory feather or aftershaft. **(d)** Filoplume. **(e)** Detail of contour feather. The hooklike irregularities on the barbules interlock to give a smooth, strong, but lightweight surface.

air trapped between the barbs. Adult birds, noteworthily geese and ducks, have down feathers as soft heat-conserving tufts, located under the overlying or contour feathers.

The *contour* feathers conform to the body and help to give general shape and aerodynamic efficiency to the bird. By contrast with the down feathers, contour feathers have definite shapes and are more complicated. In addition to a quill, there is a longitudinal central extension, the *rachis* or shaft, to which the barbs are attached in parallel fashion on each side of the rachis. These closely aligned barbs make up the *vane.* Usually the vane has a smooth appearance. This is achieved by the presence of *barbules,* parallel extensions on each side of a barb; these extensions, related to barbs as barbs are related to the rachis, usually have small hooks. The barbules from one barb overlie the barbules of adjacent barbs. The hooks on the barbules hold the adjoining barbs together with a collective great strength while retaining the lightness of weight so characteristic of feathers. Some contour feathers have vanes with strongly interlocked barbs only distally while the proximal barbs are separate and capable of trapping more air close to the surface of the body. Some birds have a second complete, but small, feather that grows out from the back of the shaft at the junction of the quill and the rachis; such a feather is called an *aftershaft.*

A fourth kind of feather, seemingly a degenerate one, is the *filoplume* or "hair feather," seen remaining on the body of a chicken or duck after it is plucked. This type of feather looks like a hair with a very small tuft of barbs at the tip. A few birds have bristles around the beak and nostrils; these may be modified filoplumes. Also, on some birds, such as herons and parrots, a modified down feather exists. Its terminal barbs break off in very small bits, leading to the name powder-down feathers. The fine powder produced creates the sheen of the feathers and perhaps adds to the waterproofing of the contour feathers or may aid in insulation.

Birds usually are very neat in caring for the feathers through *preening.* Preening does two things. First, by drawing the feathers through the beak, the hooks on the barbules can be interlocked to assure a smooth and aerodynamically efficient surface. Second, by drawing oil from a preen gland in the tail onto the beak, the bird can coat the feathers with oil, contributing to their waterproofing and, in the case of waterfowl such as ducks, assisting in keeping them afloat.

LOCOMOTION

To humanity, from ancient times, as witness the legend of Icarus or the da Vincian interest in human-propelled flight, to modern aviation and hang-gliding, birds have been of interest for their capabilities as airborne animals. What conditions of the anatomy lead to such flying as represented by soaring birds such as eagles, hawks, condors; long-flying birds such as geese; striking such as seen in the birds of prey; or hovering as seen in hummingbirds?

In addition to lightness of weight, muscular specialization, and sleekness of contour, the evolution of the flight feathers of the wing and of the tail feathers provide the specializations that lead to the magnificent flight characteristics seen in most birds. Both flight feathers and most tail feathers have the general anatomy of contour feathers, except that they are heavier and extend beyond the body in use while contour feathers lie close and reduce the resistance.

To appreciate bird flight, consider for a moment a few basic conditions for the flight of a small two- or four-place airplane. Drag (resistance) is normally reduced by the shape (contour) of the fuselage, the tail assembly, and the wing(s). Thrust is produced by the engine. Lift is produced by the wings. Their shape and degree of curvature influence the lift. Lift is achieved through this curvature because, as the cross section of the wing usually shows, the upper surface is more curved than the lower surface. Air traveling past the under surface creates a positive lift, but very important is the negative pressure created by the longer distance the air must travel over the upper surface. If the angle of attack of the wing becomes too great, the air fails to travel smoothly over the upper surface, lift is destroyed, and a stall may occur. This is undesirable in flight, of course, but a good touchdown of an airplane is to land as slowly as possible so that only just at touchdown the plane enters a stall. To do this at safe and slow speeds, planes use flaps that roll out or drop down from the wing either to extend the surface of the wing or to alter its effective angle of attack, or both. While this description is very brief, it identifies physical conditions faced by an airplane. The physical conditions faced by birds are no different. Birds have coped with these physical conditions for 130 million years.

Birds resist drag as do airplanes by a body shape offering little resistance. Birds differ greatly from airplanes by providing thrust by the beat of the wings instead of by engines. The lift is provided by the wing, but since it is in motion, it is constantly accommodating to shifting conditions through the way in which the wing feathers function and in association with the shape of the wing characteristic for the species of bird. There are two ways whereby feathers maintain smooth airflow over the wing. One way is an adaptation seen in slower flying birds where a group of feathers emerging from the first digit forms a leading edge slot to direct air smoothly over the wing. The second way is through the spread of the long primary feathers of the outer area of the wing to create slots for smoothing airflow, to increase the surface for airflow, and to provide appropriate angles of attack.

Wing types reflect the mode of locomotion for which the species is adapted. Some wings are adapted for great maneuverability. Tree dwellers especially show this kind of adaptation. The wing is elliptical, has slotted primary feathers, and can be used for slow speed, sharp turns, and close control of landings and takeoffs. Sparrows and woodthrushes possess wings of this type, for instance.

At the other extreme are wings of birds that remain aloft for long times principally through soaring. The ratio of length to width is high, and the wings are long and narrow. Birds such as gannets and albatrosses have soaring wings. They take advantage of air currents, and flight is usually quite rapid, but there is a lack of maneuverability at slow speeds due to an absence of any slotting in the feather pattern.

Somewhat similar are the high-speed wings; as with faster flying airplanes, the wings are swept back, at least in their outer halves, and usually taper to a point. Falcons, plovers, and sandpipers are representatives.

In contrast to the three types described are the wings of the birds of prey. Owls and eagles are good examples. They have a high lift capacity achieved through breadth of wings, and large amounts of slotting both at the leading edge and through the spread of the primary feathers. This not only makes it easier for them to support the added weight of their prey, but it also permits close control of flight right up to striking the prey. In the case of an owl or hawk, for instance, this can mean fairly rapid descent upon the prey, a spread of the wings and the wing feathers to a near-stall situation, and then a rapid shift to a high-lift situation as the bird lifts its prey into the air.

The descriptions above related to protection and locomotion are only broad examples, and birds vary greatly in the way they employ themselves in locomotion or in flight. For instance, some water birds are remarkably efficient scavengers because their legs are located far to the rear; however, these same birds on land have great difficulty in walking and keeping their balance. The common fowl, however, has a center of gravity closely centered over the legs and can maneuver on land with ease as dogs, foxes, and cats have seen demonstrated. At another extreme of development is that seen by the long-legged wading birds such as ibis, crane, or flamingo. Other birds can spring into flight with ease while some birds with heavier bodies and higher aspect ratio wings, such as geese, must swim for some distance while beating the wings for lift and then literally tread water while partially airborne to gain thrust, and only finally can support themselves fully in the air when adequate airspeed is reached.

At the other extreme of the evolution of birds are those that have had a great reduction in the wings. This has occurred with several orders of birds. On land a noteworthy example is the ostrich. Among the seabirds the penguins are a good example. These birds have a corresponding reduction in the pectoral muscles and in the keel on the sternum; they are acarinate (ratite) birds.

FEEDING AND DIGESTION

Birds can generally be classified as carnivorous, granivorous, insectivorous, and scavengers. The distinction among the birds obtaining food from animal sources, however, is sometimes difficult to make. Some birds prefer to fish actively, for instance, and are carnivorous in that sense, but they may also consume fish that they have not caught alive and might be classified as scavengers under this circumstance.

As has been stated earlier, the beaks of birds correspond in their structure to the food habits of the species. For instance, sharply hooked tips to beaks are most common among flesh-tearing birds, sturdy beaks usually are for crushing, and so forth.

Of course, the feet of birds play important roles in obtaining food also. Direct involvement of the feet in grasping prey, as by an owl, is one way. Indirectly, the anchoring position of the feet can aid in supporting the bird, as exemplified by the support a woodpecker's feet give as it clings to the side of a tree.

Once food is obtained, it passes rather rapidly through the digestive tract, with some exceptions. Food is metabolized rapidly and quite efficiently to provide energy for the high metabolism of most birds.

Food passes through the toothless jaws into the mouth which is poorly supplied with saliva, and although salivary amylase has been found, it is uncertain whether it contributes significantly to digestion. From the pharynx, into which the food passes over the tongue, the food enters the esophagus. The lower end of the esophagus may be expanded with a storage chamber, the *crop* (Fig. 37.5). Temporary storage of food in a crop can occur in both granivorous and carnivorous birds. In pigeons, additionally, the crop has two glandlike structures that secrete a fatty nutrient material, pigeon milk, that is regurgitated by the parents to feed the young. The hormone *prolactin* (Ch. 12) stimulates pigeon milk production.

The stomach of birds is complicated and adapted to achieve crushing action that mammals achieve with their teeth. Food first enters the part of the stomach called the *proventriculus.* This is the glandular enzyme-secreting section. Food then passes into the very muscular gizzard whose inner lining is horny and often corrugated. Birds ingest pebbles and other small hard ob-

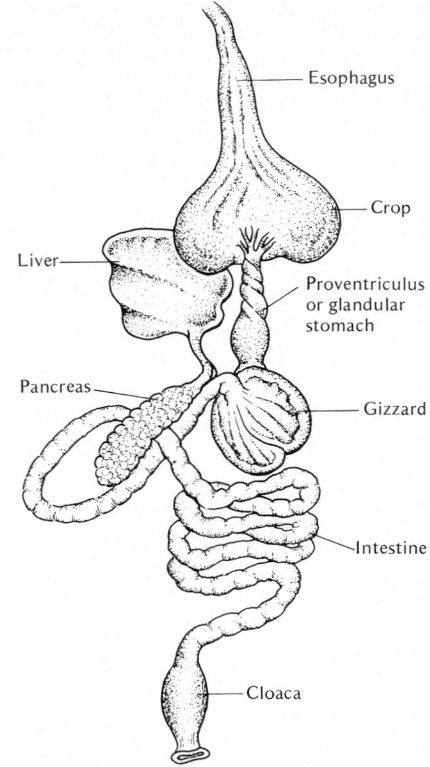

Figure 37.5 Diagram of digestive tract of a grain-eating bird.

jects that lodge in the stomach and assist in grinding up the food. Food passes into the coiled or looped small intestine that receives several ducts from the liver, gall bladder, and pancreas. From the small intestine, residual material passes into a straight short large intestine that empties into the cloaca. At the junction between the small and large intestine there usually are one or two colic *caeca* that probably serve as bacterial reservoirs. If there is indigestible material, such as fur, it is usually ejected through the mouth. Fecal material is not highly dehydrated and is passed from the cloaca in a relatively soft form in many birds.

GASEOUS EXCHANGE AND
INTERNAL TRANSPORT

The Respiratory System Birds possess a very efficient respiratory system that differs in a number of ways from the typical mammalian respiratory system (Ch. 10). Three large differences are the relatively small size and lack of expansion of the paired lungs, the connections of the lungs with nine air sacs located in various parts of the body (Fig. 37.6), and the presence of a voice box (*syrinx*) at the junction of the bronchi with the trachea instead of vocal cords located in the larynx.

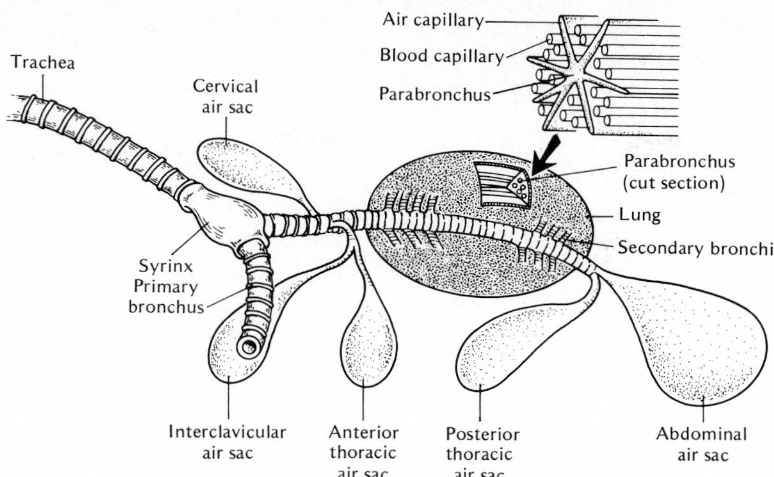

Figure 37.6 Schematic drawing of bird respiratory system. Air entering the trachea passes to each primary bronchus and then along to the lung where some passes to the parabronchi by way of anterior and posterior bronchi, but a large amount of inspired air passes to the posterior air sacs. On expiration most of the air in the posterior air sacs is passed into the lungs, flushing them with another surge of oxygenated air. The anterior air sacs receive the air passed through the lung at inspiration and exhaust it at expiration. As shown in the enlargement, parabronchi expand into air capillaries in close contact with blood capillaries for exchange of gases (only the right side of the system is shown).

The mechanism for moving air into and out of the respiratory system is not fully understood. As birds lack a diaphragm, the movement seems to be performed by the action of the ribs and the sternum. Involvement of the latter has an adaptive advantage as respiratory flow in flight seems to be synchronized with wing movements, and, of course, the wing muscles are strongly attached to the sternum.

Air enters through the nostrils, unless the bird is panting as some do during exposure to high temperatures, and passes through the glottis into the trachea whose expanded anterior end is the voiceless *larynx*. From the larynx air travels through the trachea to and through the syrinx to the paired primary bronchi and into the lungs. The lungs are not made up of blind alveolar sacs as in mammals but instead are filled with numerous air capillaries. At inspiration, air travels through these capillaries that are closely surrounded by blood capillaries. However, most inspired air goes to the posterior thoracic air sacs and the abdominal air sacs; these provide a reservoir of oxygenated air. When the bird exhales, the air from these air sacs is flushed through the air capillaries and into the anterior thoracic air sacs from which it passes to the exterior. The advantage of this system is that oxygen-rich air passes through the air capillaries during expiration as well as at inspiration. This adaptation supports the high metabolic rate characteristic of birds and also serves as an efficient system for cooling the bird. Some of the air sacs have tubular extensions into the pectoral girdle and wings. A number of unproved functions have been attributed to this system. One function that seems valid, however, is to increase the buoyancy of the animal.

The Circulatory System Although evolved from a different group of reptiles than the mammals, the plan of the circulatory system of birds is very similar to that of the mammals. As in mammals (Chs. 11, 38) the heart is a four-chambered double pump. However, the aorta swings to the right from the heart instead of the left, as it is the right, fourth aortic arch that persists in birds. Consistent with the high metabolic rate of birds, the heart rate is high but bears a relationship to the size of the animal. Larger birds have slower rates than small birds, some of whose rates can reach 1000 beats a minute during exercise. The blood vessels supplying the head in some birds are fused. The blood from the limbs travels through the kidneys by renal portal veins, but they do not break into capillaries in the kidney and thus do not function as they do in lower vertebrates. The same kinds of cells are present in the blood as in mammals. In birds, however, it has been possible to demonstrate that lymphocytes principally concerned with producing circulating antibodies (Ch. 11) are derived from a small round structure dorsal to the cloaca, the *bursa of Fabricus,* while other lymphocytes are associated with the thymus.

One function of the blood is to regulate the body temperature. Along with mammals, birds are homoiotherms, but there is greater variation in body temperature in the birds. While the body temperature is high in most birds, about 40–42° C, it may vary as much as 8 degrees in some. Control of the temperature is achieved as it is in mammals through variations in the blood vessels of the skin or by control of the breathing rate. If a bird is exposed to low temperature, it may ruffle its feathers to create an insulating layer of warm air adjacent to the skin.

WATER BALANCE

As with other vertebrates, water balance is controlled mostly by a metanephric kidney that functionally resembles the kidney of reptiles more than that of mammals. The kidneys are lobed and lie in depressions on the internal surface of the synsacrum. They are drained by ureters directly into the cloaca as birds lack a urinary bladder (except ostriches).

Functionally, urine is produced by filtration into renal capsules followed by progressive resorption of water in the tubule of the nephron. Also, on reaching the cloaca, water is absorbed there. The principal metabolic by-products are uric acid and creatine. Uric acid is insoluble and this contributes to the ability of water to be resorbed. Birds lack the counter-current system (Ch. 10) of mammals and thus do not employ sodium ions as they do. However, birds may have to cope with large amounts of salt, especially if they have a marine existence or their habitat is associated with lakes of high alkalinity. Supraorbital glands, called *salt glands,* in these birds function to excrete highly concentrated solutions of sodium chloride. In desert birds, salt glands assist in conserving water because removal of the salt allows for more water to be resorbed, especially by the cloaca. As with reptiles, birds that appear to have "runny noses" are not ill, but are removing salt from the body.

REPRODUCTION

Birds have urogenital systems with a close association between the urinary and reproductive organs. In the male the testes are drained to the cloaca by the vasa deferentia, ducts derived from the embryonic mesonephric kidney. The lower ends of these ducts are expanded as seminal vesicles for storage of sperm. The testes vary in size depending on the breeding season in many birds. During this period there is much spermatogenesis. Between breeding seasons, the testes are very small. While most males copulate with the female by apposing their cloacal surface with that of the female, a few male birds, such as ducks and geese, have a penislike genital tubercle, or copulatory organ, grooved to assist in the transfer of sperm.

Females develop a pair of paramesonephric (Mullerian) ducts (Ch. 15), but early in embryogeny only the left one, and the left ovary, persist. Eggs are ovulated, pass through the ostium of the oviduct, are fertilized in the upper oviduct, and then have a progression of depositions laid down around each one as they pass to the cloaca. These include the depositions of several layers of albumen secreted by the walls of the oviducts, shell membrane secreted next in the progression, and finally the shell and its pigments.

Bird species vary in their egg-laying habits. Of course, domestic fowl have been genetically selected for daily laying capabilities, but in nature birds usually lay seasonally and fall into two groups. Determinate layers lay a fixed number of eggs and removal of any eggs from the nest does not result in additional laying. Indeterminate layers apparently respond to the stimulus of "clutch" size and lay a number of eggs that can comfortably be incubated by the "setting" bird. If eggs are removed, the female will lay additional eggs.

The entire reproductive pattern of birds, including mating, egg-laying, brooding, and so forth, is regulated through extrinsic influences such as light on the nervous and endocrine systems.

Birds often have complicated and ritualistic courting patterns, nesting, and brood-feeding habits that ensure the continuity of the species (see below; Ch. 43).

Nest-building is a common trait among birds, and some nests are remarkably engineered. Many of the songbirds build cupshaped nests of grass, plant fibers, and rootlets, while larger birds may use twigs and eagles use small limbs for their reusable nests. Some birds, such as American robins, line nests with mud. Barn swallows use mud mixed with straw and feathers. Other birds build suspended nests with openings on the sides as, for example, the bushtits or the Baltimore orioles. When swifts build nests, they glue the materials together with a saliva. An extreme of nest-building is illustrated by an Asiatic swift that uses only salivary secretions to build the nest; these nests are collected to make the delicacy, bird's nest soup. Not all birds build nests. Some, such as killdeers, simply lay eggs on a bed of pebbles. Other birds provide scooped-out areas on the ground. A number of marine birds simply lay eggs on rocky ledges of cliffs. A few birds, such as cowbirds, make no nests, but lay eggs in nests of other birds, usually too small to repel the invader, and the young hatches and is raised by the foster parents, usually to the disadvantage of the resident young.

Egg clutch size ranges from one laid in alternate years to 20–24 of quail and domestic fowl. Incubation times range from 9 to 11 days in birds related to sparrows to the rare extreme of 11–12 weeks of one species of albatross; most common birds take from two to three weeks to hatch.

Hatchings fall into two categories, descriptive of mammals as well. *Altricial* species are those whose offspring are helpless at birth and require parental care for some time; *precocial* species are those whose offspring are able to move about and care for themselves to a considerable degree upon hatching or birth. Most birds are altricial and require care for days, weeks, or months. Some birds become independent in one to two weeks, but some large birds, such as the California condor, take up to five months (Fig. 37.7). The nestling period is a trying one for the parent birds, as some young may consume their own weight in food each day. Cooperative action of the parents is often required also as young nestlings are incapable of regulating their body temperatures. This requires brooding by the parents until the nestlings have developed adequate plumage to assist in heat retention.

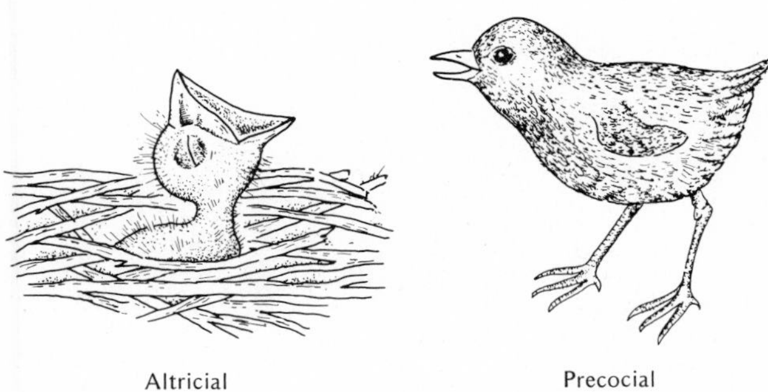

Figure 37.7 Altricial and precocial birds. Helpless, naked "blind" bird in nest represents the altricial type. Newly hatched, feathered, standing bird represents precocial type.

Some birds, storks for instance, that learn to fly and to leave the nest but return to it for some time are termed *fledglings.* They retain a temporary dependence on the parents. Eventually fledglings become completely independent of the parents and leave to establish territories of their own or to migrate.

Precocial birds usually have a strong association with ground dwelling or shore dwelling. Ducks, geese, swans, fowllike birds such as chickens, turkeys, quail, and shorebirds, are good examples. The degree to which precocial birds show independence is variable. Some quail can run about almost as soon as hatched. Gulls often leave the nest after a couple of days but depend on their parents for food until they learn to fly.

Behavioral patterns are established early in birds and are being extensively studied. For instance, specific behavior patterns associated with feeding are triggered in the young upon arrival of the parent at the nest. Also, in precocial birds fixations are developed soon after hatching. In nature these fixations have a value in causing the young to follow or associate with their own parents, a process called imprinting (Ch. 14). Under unnatural and domestic conditions, some young birds become unusually imprinted and may follow their keepers as they might their parents. K. Lorenz, a recent Nobel Prize winner, contributed to our knowledge of imprinting.

SENSORY RECEPTION AND NEURAL COORDINATION

To see and strike prey from an aerial position, or to sweep through a forest and bring flight to a halt at just the right moment to land on a swaying branch, requires a well-coordinated animal readily able to receive stimuli from its environment. The nervous system of birds reflects this as it is very well developed. The brain shows several characteristic advances over the reptilian ancestors and illustrates its independent evolution from the mammalian brain. The cerebral cortex is usually thin and smooth. However, the deep part of the forebrain, the *corpus striatum,* is the integrating center. The olfactory lobes are poorly developed consistent with the reduced sense of smell possessed by most birds, but the optic lobes are greatly enlarged to correspond to the very well-developed vision characteristic of birds. The optic lobes serve as association centers for vision instead of the posterior cerebral cortex as in mammals. The cerebellum is very well developed and is large to correspond to the finely coordinated events that occur between muscles, skeleton, vision, and equilibrium associated with the inner ear. (Fig. 37.8)

Sensory reception is especially developed for vision, hearing, and equilibrium. The eyes of birds usually are relatively large, but they usually lack mobility in their orbits. Birds see objects of attention by turning the head on the flexible neck. While most birds' eyes are located on the sides of the head, some birds do have

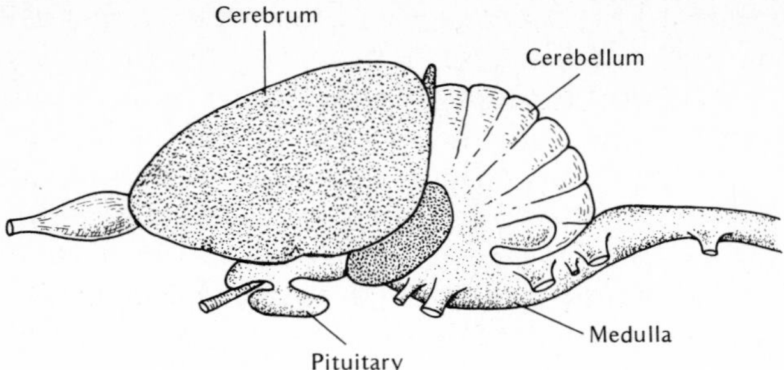

Figure 37.8 Brain of bird. The cerebrum is quite smooth on the surface and the cerebellum is relatively large.

overlapping and binocular vision. The retina of the bird's eyes has both rods and cones, but the proportions vary depending on whether the birds are diurnal or nocturnal. Color vision exists and the visual acuity in dim light is large. Birds have visual acuities ranging from eight to ten times greater than that of humans. Another trait of importance in bird vision is the presence of the *fovea* (Ch. 13) in a pit, requiring sharp focus into this most sensitive area of the retina. Also, a number of birds have two foveas. One, more central, functions for good monocular vision as when the head is cocked at rest or in flight; the other fovea seems to function with binocular vision. A distinctive feature of the avian eye is the presence of a highly vascularized fan-shaped structure, the *pecten,* that extends into the posterior chamber near the optic nerve. Two functions have been attributed to the pecten. One is to provide nourishment to the retina. The other is for the ridges on the pecten to provide an orienting "grid" to assist in the interpretation of the position of the bird.

The auditory apparatus usually lacks external sound-trapping devices except in a few birds. The eardrum lies at the end of the depression, the external auditory canal. The middle ear has only one bone, the columella, homologous with the mammalian stapes (Ch. 13). This transmits vibrations to the cochlea in the inner ear. While the eyes of owls are known to have very acute dim-light vision, in addition, trained barn owls have been used to show that an acute sense of hearing is important to the owl's hunting in the dark. The cochlea is shorter and less coiled than the mammalian cochlea. However, as attested by electronic recordings of songs of birds raised in isolation and with opportunities to learn from other birds, they are able to recognize and distinguish intensities and ranges of sound in a greater degree than can humans.

Of course, the bird's equilibratory apparatus, composed of the semicircular canals, the saccule, and the utricle, is very efficient and closely integrated through the cerebellum.

CHEMICAL COORDINATION AND HORMONAL CONTROL

Endocrine physiology in birds is complicated. For one thing, there is variation among birds in their hormonal actions. But more to the point is the complicated interaction that exists between the birds' internal endocrine activities and external influences such as light, temperature, and magnetic fields, but especially light. The expression of these interactions in birds is best seen in the reproductive patterns and the migratory behavior of birds.

Endocrine structures and secretions as listed in Table 12.1 are also generally applicable to birds. It is noteworthy in birds, though not restricted to them, that the pineal gland, derived from the dorsal area of the forebrain, is influenced by light. The serotonin produced there acts on the hypothalamus through neurosecretion. Probably also involved is melatonin. The stimulus entrains the activities of other endocrine glands. Two kinds of closely related events emerge in birds from this environmental stimulus–endocrine response system. The first general event is the influence on energy storage in the form of fat, usually accompanied by a restlessness. The second event is the actual migration from one area to another, and usually by quite precise and predictable routes and times.

It has been known for about 50 years that lengthened exposure to light usually increases fat storage in birds and triggers their migratory behavior. This is not a simple event, however. Recent evidence suggests that it is not just a seasonally changing set of environmental cues but also includes seasonal changes in the birds' interpretations of the cues. This very complex area of avian physiology is currently under intensive investigation, especially by Donald Farner at the University of Washington and Albert Meier at Louisiana State University. Research in this area can be approached through several methodologies including removal of the pineal gland, blinding, hormonal injections of such substances as prolactin or corticosterones, altering the photoperiods, trapping, holding birds, and so forth.

An outcome of the complex interactions between the environment and the endocrine physiology is the emergence of a restlessness that then leads to migration. Migratory behavior is also yet an open area for research as there remain many unsolved problems.

BEHAVIOR

Migration and Homing While both migration and homing are known in other animals including fish, amphibians, reptiles, and mammals among the vertebrates, and several kinds of arthropods among the invertebrates, the most dramatic and obvious expression of this is seen in the birds. Always mentioned as the most striking example of this capability is that of the arctic tern that migrates between north and south polar areas. Its route is not direct; from North America it travels along the western edge of Europe and Africa to the antarctic regions. A round trip in one year, thus, is about equivalent to circumnavigating the globe at the equator.

Migration has obvious adaptive advantages. In the northern hemisphere, birds during spring and summer can find ready nesting places and food resources. Even those birds sought by predators in northern areas do not persist there long enough to allow the predators to increase their populations as the birds are gone in the fall and winter. The most obvious adaptive advantage is to move to more favorable environments as the ones occupied become less favorable. However, it is still not clear how birds, or other animals, become selected genetically to establish ingrained, regular, and stereotyped migrations that sometimes are so precisely controlled that the birds first appear in local areas within the span of just a few days each year.

At present it is not entirely clear what degree of interaction there is between two events that seem to guide migrations. One of these is guidance by the use of celestial cues. Studies at Cornell University by S. Emlen demonstrated that some birds can be shown to follow celestial cues by training them in a planetarium. Using indigo buntings, he was able to show that birds released at one point in a planetarium would orient relative to the sky to attempt flight in the usual direction independently of whether the skyscape of the planetarium was properly oriented to the actual skyscape or not. Furthermore, the direction, northward or southward, that the buntings would fly could be influenced by the experimentally produced day lengths to which the buntings were exposed before release in the planetarium.

Another influence on migration seems to be magnetic fields of the earth. Also at Cornell University, William Keeton found that homing pigeons bearing magnets that altered the influence of the earth's magnetic field could be disoriented, while pigeons bearing equivalent weights of nonmagnetized metal were capable of homing. Work continues intensively with homing pigeons because homing can be used as a model for understanding migration. Furthermore, homing pigeon studies can subject the birds to even greater problems to solve than that faced by naturally migrating birds because a natural migrator starts out with a precise sense of orientation while experimentally pigeons can be taken in concealment to points unknown to the pigeon and homing orientation examined. In the case of pigeons it seems that they also rely on landmarks that they establish in their familiarization flight patterns after they are released.

A striking example of homing unrelated to the usual flight pattern of the bird was recently demonstrated with a burrow-dwelling British bird, the Manx shearwater. The bird was dug from its burrow, transported by airplane to Boston, and released in Massachusetts. A watch was set up at its burrow and it was seen to arrive there within 12 days of its release in America. This bird was challenged, therefore, with an unusual flight path and one almost totally lacking any surface visual clues.

Mating and Nesting In addition to migration (and navigation) that is so closely tied to the chemical coordination of birds, an especially interesting aspect of bird behavior relates to the mating–nesting patterns. Some of these traits are discussed in more detail in the chapter on behavior. Birds exhibit a number of patterns in pair formation for mating. D. Lack identified five main categories. In some birds the sexes meet only for copulation. Fitting this category are the sage grouses whose males meet on prescribed strutting grounds to display themselves, to fight each other, to "boom" through special movements of the wings over stiff breast feathers, and to attract females to join them on the strutting ground to copulate (Fig. 37.9). Some birds remain together briefly while copulatory stimuli persist. For instance, male humming birds will remain with the female until near nest-building time when they depart or are driven away. Some birds pair quite a while before copulation but once it is achieved, eggs laid, and nests built, the males leave. Ducks, but not all waterfowl, fit this category. Other birds remain together for the breeding season. Here territoriality is very important and the males usually stake out their territories by song and display, the female joins the male, a nest is built, and the two may participate in the incubation of the eggs and the feeding of the young until they are independent. Sparrows and other passerine birds fit this category. Finally,

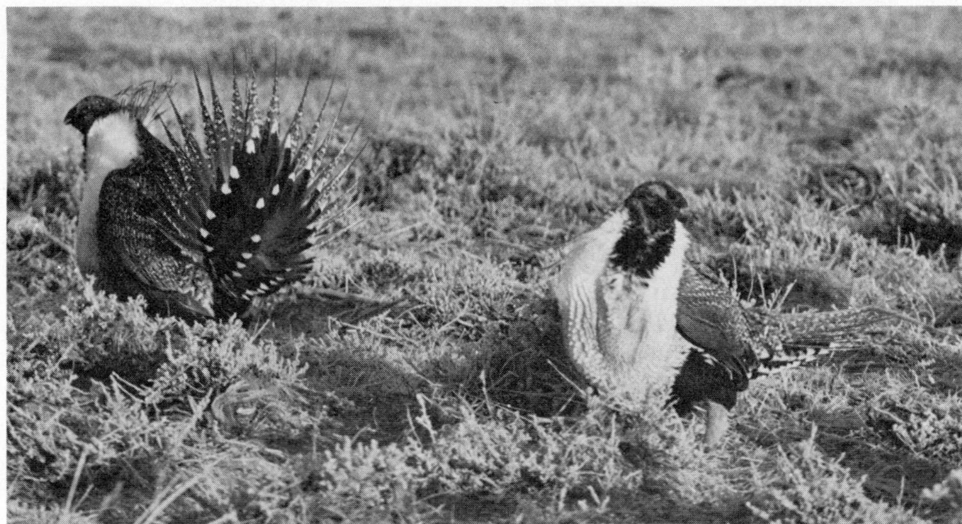

Figure 37.9 Sage Grouse strutting on strutting grounds (New York Zoological Society Photo)

some birds pair for life. Some waterfowl such as swans fit this category.

That one can identify categories and place species of birds in these categories is a reflection of behavior patterns that are strongly determined by the genetics of the birds rather than by the environment. However, birds are not entirely stereotyped in their behavior patterns. Many birds are taught by their parents. This is especially true of some of the storks, eagles, and hawks where learning to fly is under the guidance of the parents.

Importance of Birds to Other Animals

It is probably an accurate generalization to assert that most birds are at the ends of their particular food chains. Even the herbivorous and granivorous birds have few certain predators, although they may from time to time be victims of predation by other birds or by mammals. The predatory birds such as the hawks or the fishing birds undoubtedly serve as controlling forces on the populations that serve as their food. Birds, however, serve as hosts for a variety of kinds of invertebrates, especially arthropods in the form of mites and lice. In fact, often quite specific relationships are established between the ectoparasites and their hosts.

The insectivorous birds display the most common relationship with other animals and play an important role in regulating the populations of some insects.

Importance of Birds to Humans

To the layman birds are important in several ways that can be described as aesthetic. Song birds especially are appealing to humans, not only in nature, but in captivity as witnessed by the numbers of canaries kept in homes. The colorful coat patterns of birds are also attractive to humans. Live birds are kept for the pleasure their color brings as in the cases of canaries, budgerigars (parakeets), cockatoos, and parrots. Also, the feathers of birds have been sought and used by many human societies. Noteworthy was the colorful cape of Hawaiian royalty, made of thousands of bird feathers.

Another source of enjoyment to humans is the sheer pleasure of spotting and identifying species of wild birds either in their natural environment or when they are found out of their habitat. Recently a single Ross' gull was seen on Massachusetts shores. As it is not usually seen south of the Arctic circle, this caused great excitement among ornithologists and bird watchers, and dedicated ornithophils came for hundreds of miles simply to observe and/or photograph this animal.

Birds have played important roles in the folklore and religions of human societies, especially in some Egyptian dynasties. Also, even in primitive societies the gallinaceous birds, represented best by the jungle fowl, have been a source of food. In recent years the common fowl, a descendant of the jungle fowl, has come to play a large role in the nutrition of humans. First, they have been highly bred and selected to lay eggs over a large segment of the year. This in itself is an important industry. Additionally, chickens have come to be more exploited as an economical source of protein in the human diet. Birds are more efficient in converting grain into edible meat than are cattle and do so less expensively. However, there has been a recent change in the

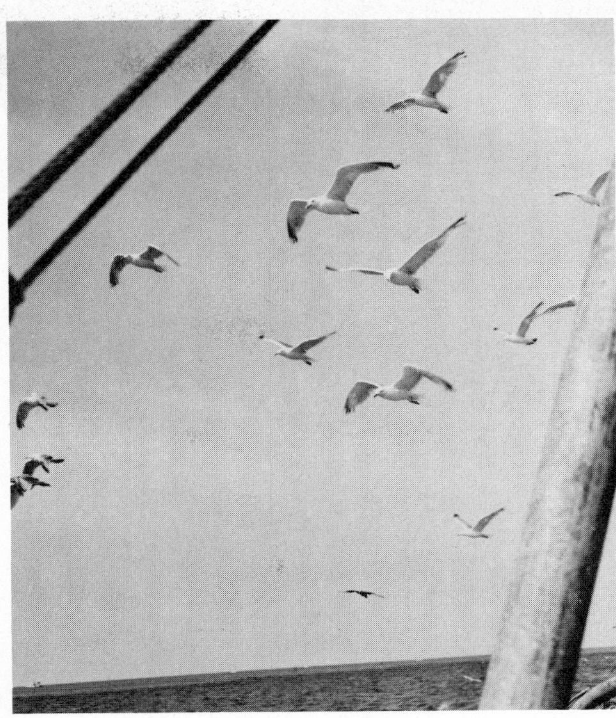

Figure 37.10 Several bird types. Snowy egret (*upper left*). Gulls following fishing boat (*upper right*). Eastern screech owl (*lower left*). Amazon parrot (*lower right*). (New York Zoological Society Photos)

feeding patterns among poultry producers. Today probably 50 percent of the fish proteins harvested from the sea in the form of sardines, herring, and the like, are converted into fish meal that is used as food for poultry. This moves poultry to market faster, but it can have a serious influence on the flow of energy in the biosphere (Chs. 44–47).

The relationship of birds to humans is also seen in two contrasting sets of conditions in which numbers of birds are involved. The first set of conditions relates to the danger of extinction that some birds have suffered from accumulating pollutants produced and used by humans. Two outstanding examples are the American bald eagle and the brown pelican. In both cases DDT accumulated through the food chain and concentrated in the bodies of birds, leading to the production of eggs with shells so thin that they crushed during incubation by the parents. There are now hopeful signs, with restriction on DDT, that young can again survive.

The alternate example is illustrated by starlings and grackles. These birds have multiplied to such numbers in various localities in the United States that they not only have become pests to crops, to peace of mind, and to cleanliness, but they also are threats to humans as vectors for histoplasmosis organisms that cause a debilitating human disease, and as potential viral transmitters. Reducing the populations of these successful birds is a problem of great magnitude. Shooting or poisoning do not achieve the goal without secondary consequences. Recently, in those parts of the country where the climatic conditions favor the procedure, wholesale aerial spraying of roosting birds with detergents has been employed. The detergents remove insulating oils from the feathers, prevent the birds from flying, and with below zero temperatures cooperating, cause the birds to freeze to death. Dead birds on the ground can then be collected and disposed of economically.

The example above is meant to show how some perceive population control in birds. One of the outstanding examples of destruction of birds by humans is the notorious example of the passenger pigeon. In the 1850s these birds had a population of above nine billion (yes, billion). In four decades, in 1895, a packing company would state that they could no longer find them to net for market. The last hope for these birds, that once literally darkened the skies, came in 1909 when an infertile egg was laid by the last living pair. With the death of the male in 1910, it was only a matter of time and on September 1, 1914 the passenger pigeon, as a species, became extinct.

Classification

Birds fall into many orders and families, and it is not possible here to make a detailed study of bird variations through the classification scheme. Also, a single diagnostic trait for determining the larger subdivisions is the variation in the palate. As the palate of the bird is not readily accessible to the beginning zoologist, this trait will not be discussed here; the interested student can pursue the subject from the references.

The birds are usually described as having arisen as a new class from thecodont reptiles during the Jurassic. They have evolved 33 orders; only six have become extinct. The class is divided into two subclasses. Recently, however, it has been proposed that the dinosaurs were warm-blooded, developed feathers, and that the birds are persistent members of an entirely new class than is employed in this book, the class Archosauria, derived from the class Reptilia. While the concept bears careful consideration and may indeed reflect the realities of avian evolution, for the time being the authors will retain the more conventional classification with Aves as a separate class. The interested student is urged to read the article on this subject by R. T. Bakker (see Suggestions for Further Reading).

SUBCLASS **Archaeornithes** These are the ancestral birds. They possessed teeth in both jaws and had three claws on the forelimb (wing). Members of this class did not evolve the strong fusion between the pelvic girdle and the vertebrae that is so characteristic of modern birds. *Archaeopteryx,* the first recognized bird, and a relative, *Archaeornis,* represent the only two known genera.

SUBCLASS **Neornithes** These are the true birds, some groups are extinct, but most survive. These birds either possess or lack a keel on the greatly enlarged sternum. When a keel is absent, the birds are *ratites*, who are flightless; when the keel is present, the birds are *carinate.* These traits, however, are not used as major classification subdivisions. In the descriptions that follow, it should be assumed that a keel is present unless the subject is mentioned.

ORDER **Hesperornithiformes** Extinct ratites with degenerate wings, but possessing teeth.

ORDER **Ichthyornithiformes** These extinct birds lacked teeth, were carinate, and had well-developed wings.

ORDER **Sphenisciformes** Penguins range from medium to large size. These marine ratites have wings modified as flippers. The body is densely covered with feathers and they lack bare patches (apteria) on the skin as is common under the feathers of many birds. The feathers on the flippers are scale-like. Southern coasts of southern continents, Antarctic, and Galapagos Islands.

ORDER **Struthioniformes** The ostriches are very large ratites with soft body plumage but bare thighs; they have but two toes. Africa and Southern Asia.

ORDER **Rheiformes** Though large, the rheas are

smaller than ostriches. These ratites have soft plumage on both trunk and thighs; three toes present. South America.

ORDER **Casuariiformes** The cassowaries and emus are very large ratites with coarse hairlike plumage and with feathered thighs; they have just three toes. Australia, Tasmania, and New Guinea.

ORDER **Aepyornithiformes** These extinct elephant birds achieved heights of 3 m (9 feet). They were ratites and had heavy bodies and legs.

ORDER **Dinornithiformes** The moas are extinct ratites that grew to heights of 4 m (12 feet).

ORDER **Apterygiformes** The kiwis are medium-sized ratites with very reduced wings and with coarse hairlike feathers. New Zealand.

ORDER **Tinamiformes** The tinamous, ranging from small to medium size, are carinates but with palates resembling ratites. Mexico and Central America.

ORDER **Gaviiformes** The loons are diving birds, medium sized, with fully webbed toes.

ORDER **Podicipediformes** The grebes are diving birds, medium sized with lobed, rather than webbed, toes. Nearly worldwide.

ORDER **Procellariiformes** The four families of albatrosses, shearwaters and petrels are oceanic birds especially adept for soaring for long distances. Their nostrils form raised tubes. They range from very small to large. Worldwide.

ORDER **Pelecaniformes** The pelicans and their allies are aquatic or marine birds of medium to large size. Their nostrils are very small or lacking. A pouch is associated with the lower jaw. Members of this order also include boobies and cormorants. All nearly worldwide.

ORDER **Ciconiiformes** The herons, storks, ibises, spoonbills, and flamingos exist in seven families. They are wading birds of small to large size, with relatively long bills, necks, and legs. Worldwide.

ORDER **Anseriformes** The two families of this order are made up of the screamers and the ducks, geese, and swans. They range in size from small to very large. They are aquatic or semiaquatic with webbed or partially webbed feet. Worldwide.

ORDER **Falconiformes** These birds of prey are diurnal and seek food in the daylight. They range from small to very large. As flesheaters they have hooked bills. The nostrils may open through the cere. The feet are very powerful with the hind toe being opposable to the front ones, providing a strong grasp. The five families include the American vultures, Old World vultures, hawks, eagles, ospreys, and falcons. Also, one family (*Sagittariidae*) of secretary birds consists of very large terrestrial birds of prey by contrast with the others that are principally soarers over land. Worldwide.

ORDER **Galliformes** The seven families include the common fowls, sudras, chickens, grouses, pheasants, guinea fowls, and turkeys. Common among them is a tendency to be ground dwelling. Essentially worldwide.

ORDER **Gruiformes** The cranes and rails and their allies are divided into 12 families. These mainly aquatic birds range from very small to very large, have relatively long necks, rounded wings, and short tails. The legs range from medium length to long. Worldwide.

ORDER **Charadriiformes** These small to fairly large birds composing 16 families principally are shorebirds of seacoasts, lakes, and marshes. The members include the gulls, jaegers, skuas, auks, plovers, and sandpipers. The wings are usually long. Worldwide.

ORDER **Columbiformes** Pigeons, doves, and relatives are land birds with loosely attached feathers. The three families include the dodos and the solitaire which became extinct about 150 years ago. Nearly worldwide.

ORDER **Psittaciformes** While the parrots and their allies compose only one family, there are over 300 species that range from very small to fairly large. Characteristic is the short strongly hooked bill which has a fleshy area (cere) over the base of the upper part. The feet are strong. Tropical parts of the world and southern hemisphere.

ORDER **Cuculiformes** The cuckoos and their allies exist as two families. Among these birds are the roadrunners. Nearly worldwide.

ORDER **Strigiformes** Owls are nocturnal birds of prey; small to fairly large. Their eyes are directed forward. The bill is hooked. The wing pattern permits them to shift rapidly from rapid flight to stall-like landings. The feet have talons and one toe is reversible. There are two families, the barn owls and the typical owls. Worldwide.

ORDER **Caprimulgiformes** The goatsuckers are usually nocturnal. The bill is small but the mouth can open widely. There are five families. Worldwide.

ORDER **Apodiformes** The swifts and hummingbirds range from very small to medium small. They have relatively long pointed wings and can fly rapidly. Their legs are very short. Except for extremes of latitudes and some islands, nearly worldwide.

ORDER **Coliiformes** Mousebirds are medium small birds with long tails, short bills, rounded short wings, and red feet and legs. Temperate and tropical North and South America.

ORDER **Trogoniformes** The trogons are medium-sized birds with feathered shanks, colorful loosely attached feathers, long tails. Tropical and subtropical Americas, Africa, Asia, and Philippines to Sumatra.

ORDER **Coraciiformes** The kingfishers and their allies comprise nine families, including such birds as the bee-eaters, rollers, and hornbills. In most families the bill is large proportionate to body size. The plumage is generally brightly-colored. The kingfishers have large heads, strong bills, compact bodies, and short necks and legs. Tropical and subtropical areas.

ORDER **Piciformes** The woodpeckers and allies are small to medium large, with strong feet and legs, and with a reversible outer toe. The six families, besides the woodpeckers themselves (*Picidae*), include the large-billed colorful toucans, the honey-guides, and the puffbirds. The order is worldwide except for New Zealand, Australia, and Madagascar.

ORDER **Passeriformes** This is a very large order of small to medium large perching birds. About 60 percent of the known species are in this order. Representatives among the four suborders of this group are the broadbills, the ovenbirds, lyrebirds, and the large number of songbirds. As is known to even casual observers of the songbirds, they too are variable and range from crows and bowerbirds to small wrens and vireos among the four dozen or so families. Worldwide.

SUGGESTIONS FOR FURTHER READING

American Ornithologists' Union. *Checklist of North American Birds,* 5th ed. Lancaster, PA.: American Ornithologists' Union (1957).

American Ornithologists' Union. *Recent Studies in Avian Biology,* A. Wolfson (Ed.). Urbana, IL: Univ. of Illinois Press (1955).

Audubon, J. J. *The Birds of America.* New York: Macmillan (1937).

Austin, O. L., Jr. *Birds of the World.* New York: Golden (1961).

Fisher, J., and R. T. Peterson. *Birds. An Introduction to General Ornithology.* London: Aldus Books (1971).

Gilliard, E. T. *Living Birds of the World.* New York: Doubleday (1958).

Headstrom, R. *Birds' Nests.* New York: Ives Washburn (1949).

Hellmayr, C. E., C. B. Corg, and B. Conover, *Catalogue of Birds of the Americas and the Adjacent Islands.* Field Museum of Natural History, Zool. Ser. **13,** 15 parts (1918–1949).

Krutch, J. W., and P. S. Eriksson. *A Treasury of Bird Lore.* New York: Paul S. Eriksson (1962).

Marshall, A. J. (Ed.). *Biology and Comparative Physiology of Birds,* 2 vols. New York: Academic Press (1960–1961).

Peterson, R. T. *A Field Guide to the Birds (of Eastern North America),* 2d ed. Boston: Houghton Mifflin (1959).

Peterson, R. T. *A Field Guide to Western Birds (and Hawaii),* 2d ed. Boston: Houghton Mifflin (1961).

Wallace, G. J. *An Introduction to Ornithology,* 2d ed. New York: Macmillan (1963).

Welty, J. C. *The Life of Birds.* Philadelphia: W. B. Saunders (1962).

From Scientific American

Bakker, R. T. "Dinosaur Renaissance" (Apr. 1975).

Cone, C. D. "The Soaring Flight of Birds" (Apr. 1962).

Dilger, W. C. "The Behavior of Lovebirds" (Jan. 1962).

Eklund, C. R. "The Antarctic Skua" (Feb. 1964).

Emlen, J. T., and R. L. Penny. "The Navigation of Penguins" (Oct. 1966).

Frings, H., and M. Frings. "The Language of Crows" (Nov. 1959).

Gilliard, E. T. "The Evolution of Bowerbirds" (Aug. 1968).

Greenwalt, C. H. "How Birds Sing" (Nov. 1969).

Höhn, E. O. "The Phalarope" (June 1969).

Keeton, W. T. "The Mystery of Pigeon Homing" (Dec. 1974).

Lack, D., and E. Lack. "The Home Life of the Swift" (July 1954).

Peakall, D. B. "Pesticides and the Reproduction of Birds" (Apr. 1970).

Sauer, E. G. F. "Celestial Navigation by Birds" (Aug. 1958).

Schmidt-Nielsen, K. "Salt Glands" (Jan. 1959).

Schmidt-Nielsen, K. "How Birds Breathe" (Dec. 1971).

Sladen, W. J. L. "Penguins" (Dec. 1957).

Smith, N. G. "Visual Isolation in Gulls" (Oct. 1967).

Stetner, L. J., and K. A. Matyniak. "The Brain of Birds" (June 1968).

Storer, J. H. "Bird Aerodynamics" (Apr. 1952).

Taylor, T. G. "How an Eggshell is Made" (Mar. 1970).

Thorpe, W. H. "Duet-Singing Birds" (Aug. 1973).

Tickell, W. L. N. "The Great Albatrosses" (Nov. 1970).

Tucker, V. A. "The Energetics of Bird Flight" (May 1969).

Watts, C. R., and A. W. Stokes. "The Social Order of Turkeys" (June 1971).

Welty, C. "Birds as Flying Machines" (Mar. 1955).

38 The Mammals

Mammals, in contrast to birds, show a great variability in their structure, their physiology, and their life styles. Arising from the mammal-like therapsid reptiles in the Jurassic period of the Mesozoic era, the small first mammals have rapidly evolved into nearly every available niche and habitat on the planet. Living mammals range in size from less than 4 cm and weight of less than 3 g to the largest animals on earth, the whales, that can weigh over 100 metric tons and reach 30 meters (100 feet) in length. Mammals are adapted to life in the ocean, in lakes, or streams, to underground dwelling, to life spent mainly in trees, and the bats have taken to the air. Collectively, the mammals show traits of adaptation that rank them with the insects in evolutionary success. The presence of humans among this class of vertebrates has had an added impact on the class's influence on the earth. The nature of this influence, whether collectively constructive or destructive, leads to a value judgment inappropriate to this chapter.

The outstanding diagnostic traits of the living members of the class are the presence of hair at some time in the life of the animal and the presence of mammary glands. Furthermore, in most mammals, development is internal in the uterus for varying degrees of maturation. This, however, is not a diagnostic trait for all mammals since primitive egg-laying mammals (monotremes), such as the echidna and duckbill platypus, exist.

Mammals seem to have evolved gradually from the reptiles and, thus, share a number of traits with them. There are traits that are not shared with reptiles but which require detailed analysis of mammalian structure to observe. One is that the skull is supported on the vertebrae by two processes, called *occipital condyles,* rather than one, as seen in reptiles and birds. Another is that the lower jaw of mammals is composed of just one pair of bones, the dentaries, while reptiles and birds have more. The reduction in the jaw bone between reptiles and mammals is accompanied by the shift of the hinging bones of the reptilian jaws to the malleus (hammer) and the incus (anvil) of the middle ear bones. Also, where reptiles have several sets of teeth replacing lost ones, most mammals have a single deciduous set (milk teeth) that are replaced by a permanent set.

The mammalian brain is noteworthy in that a new evolutionary part of it, the *neocerebrum,* or *neopallium,* that first appeared in reptiles, is highly developed as the cerebral cortex. Associated with this are behavioral traits of inquisitiveness, learning, and initiative not seen in any other class of vertebrates.

As with birds, and quite possibly with the therapsid reptilian ancestors, the mammals are endotherms and possess internal thermoregulating mechanisms that generally maintain the body temperature within narrow limits.

In pursuing the analysis of this class of vertebrates, the reader will frequently be directed to chapters in Part II of this book as that unusual mammal, the human, was used to describe characteristic adaptations to various biological problems faced by animals. Therefore, some details will be assumed to be known to the reader and additions here are meant to emphasize special adaptations seen in the class.

Anatomy and Physiology

PROTECTION AND LOCOMOTION

Some general conclusions can be drawn with respect to adaptations for protection and locomotion. The protective effect of the integumentary system is accomplished with variations in the amount of hair and in the degree to which fat is deposited in the skin. Thus from the standpoint of natural selection, relatively slight variations in the integument provide the animal with ability to survive and cope successfully with the environment. On the other hand, the mammals show great variation in the principal organs of locomotion, the limbs.

THE INTEGUMENT

As described in more detail in Chapter 8, the integument, or skin, is composed of two parts, an outer epithelial layer, the epidermis, and the deeper dermis or corium. The epidermis is highly cellular, has numerous derivatives and variations scattered over the body, while the dermis has a large amount of intercellular material and relatively few specialized derivatives. The principal mammalian epidermal derivatives, other than hair, are nails, claws, hooves, sweat glands, sebaceous glands, mammary glands, and several kinds of specialized glands that are used for identification or to mark territories.

In many mammals hair can be divided into two categories: relatively coarse and long *guard hairs,* and shorter finer insulating *underhair.* The relative amount of underhair is sometimes seasonal, varying through growth or by molting. Usually each species has its own characteristic molting pattern. The *pelage,* as the hairy coat is called, serves as general protection and as an insulator to conserve body heat, in general analogous to the function of feathers in birds. There is a correlation between the amount of coat of mammals in cold regions, where the coat is heavy, and those dwelling in warm environments, where the coat is light.

The degree to which hair covers the surface of the body varies. Humans, of course, have a greatly reduced amount of hair. Aquatic mammals that would be hindered by heavy coats of hair usually have short smooth hair, as in seals, or they may be completely naked except for a few bristles around the mouth, such as the whales and the sea cows (manatees) or dugongs. Two noteworthy modifications of hair are the quills of porcupines and hedgehogs and the horns of rhinoceros.

Where reduction of hair occurs in terrestrial mammals, there are usually accompanying compensatory changes in the integument. For instance, the sparse hair on the skin of elephants is compensated by a greatly thickened skin which is the meaning of the name pachyderm applied to these animals. Also, two unrelated kinds of animals have developed coverings of bony scales covered with a tough keratinized epidermis; these are the armadillos and the pangolins.

By contrast with birds, whose feathers develop similarly, mammals lack a great variety of colors. The generalization about color is that most mammals have coat colors that tend to have them blend with the environment or to break up the outlines of the animal making it difficult to be seen easily. The color traits function both for predators and for prey. Predators of appropriate color or color patterns can approach prey with greater ease. Prey, on the other hand, avoid predation by being protectively colored.

Seasonal variations in hair color may occur in some animals whose summer pelage may be dark and whose winter coats may be white. Weasels, some foxes, and some hares may display this trait.

Patterns of hair length and coarseness also play an adaptive role. In a number of animals the hair around the vulnerable throat may be thicker than at other places on the body. Male lions, for instance, develop manes, and most dogs and their relatives have a thicker pelage around the neck. Another modification of hair in most mammals that serves an important function are the moustachial hairs or *vibrisae.* These whiskerlike hairs are very sensitive to touch as the result of their innervation. These are well developed in burrowing animals, where a sense of touch is especially useful.

Glands of the integument are variable in number and function in mammals. Basically they fall into three types. Directly opening to the surface are the *eccrine* sweat glands, true secretory glands that produce the watery sweat that helps to regulate body heat. While found spread over the body in humans and horses, some mammals lack eccrine sweat glands. Most carnivores, though not all, have greatly reduced concentrations of the glands. These glands are thin tubes whose coiled ends lie in the dermis. Opening into hair follicles is a second type of sweat gland. It produces a denser product than the eccrine sweat glands by a loss of the tip of the secretory cells of the glands; for this reason they are *apocrine* sweat glands. These glands, in humans and some other animals, develop only during sexual maturation. Also, rather than being widespread over the body, they are limited in their distribution to localized areas such as the genitalia, near the mammary glands, axillary areas, the ears, and ear canals. The coiled ends of glands of this kind lie in the fatty subdermis rather than the more reticular dermis.

Specializations of apocrine sweat glands serve two adaptive purposes. On one hand they may be specialized as scent glands that are especially active during mating periods, and thus their secretions may serve as pheromones. On the other hand these glands produce products for marking territories or to serve defensively. Dogs and other carnivores have scent glands useful in attracting mates. Some such as skunks have scent glands that can function in defense.

A third type of skin gland is represented by sebaceous glands which produce a somewhat oily secretion by a breakdown of the entire secretory cell of the gland; these cells are replaced. Most sebaceous glands open into hair follicles and provide an oily protective film to the hair and the integument.

The distribution of the three major kinds of integumentary glands on the body varies with the species. For instance, scent glands vary in their locations. Some, such as canids (dogs, foxes, wolves), have scent glands associated with the base of the tail or the anus; peccaries, on the other hand, have scent glands that produce a strong odor located along the back; some hooved animals have several forms in one species. For example, the pronghorn antelope is said to have 11 scent glands distributed over the body in such places as at the base of the ears, on the rump, or between the digits of the hooves. Other hooved animals, such as some deer, have glands along the tarsal segment of the hind limbs that are smelled by other members of the species and presumably serve for recognition.

Noteworthy among the glands of the class Mammalia, of course, are the mammary glands. These glands show traits in common with both sweat and sebaceous glands. Although mammalian embryos have bilateral ridges, the *milk lines,* running on the ventro–lateral surface between the pectoral and pelvic girdles, the epithelial ingrowths that give rise to the future glands develop (in both sexes) only in localized positions along the line. For instance, in humans and bats one pair develops in the pectoral region, in horses and whales one pair develops in the pelvic region, while cows develop two functional pairs and often a third pair that are usually removed in domesticated stock. Other animals have the nipples of the mammae distributed along the abdomen and thorax, the largest number being 12 pairs plus 1 (25) in opossums where the nipples are located in the pouch or *marsupium*. The average number of young produced at one time is closely correlated with the numbers of mammae that are characteristic of the species.

In humans, and domestic cattle, for instance, the mammae are continuously prominent in the females but are only represented by nipples or localized discolorations in the males. In human females the mammae begin to enlarge with the onset of puberty and usually reach a maximum enlargement at about 20 years of age. In most other animals the mammae enlarge early in association with pregnancy and under the control of the luteinizing and lactogenic hormones of the anterior pituitary.

A common confusion is the distinction between nipples and teats in mammals. Nipples are elevations into which the principal drainage ducts of the mammary glands drain to the exterior. Teats, on the other hand, are elongated extensions at the blind end of which the actual drainage of the mammae occur. Thus when a human infant suckles, it receives milk directly from the milk ducts through a group of openings of the ducts on the surface of the nipple. When a calf suckles, the milk delivered to it is released from a common channel within the teat that in turn receives the milk from the mammary ducts. The primitive egg-laying monotremes do not deliver milk into localized nipples, which are lacking. Instead, the ducts open onto the surface of the skin and onto surrounding hairs where the young lick the milk.

While monotremes acquire milk by licking, most mammals suck on the nipples to draw out milk. However, the cetaceans (whales, porpoises, and dolphins) may have the nipples retracted into slits flanking the vent, and when feeding, the females can forcibly eject milk into the mouths of the young through the use of special muscles, since the young lack lips capable of sucking.

Milk, the sole sustenance of the early young, in addition to water, has fats, proteins, and salts in concentrations characteristic of the species, or, in domesticated strains of cattle, even characteristic of the strains. Thus the butterfat content of Guernsey cows is higher than that of Holsteins. The fat content of marine mammals is especially high, approaching or even equal to 50 percent. Animals with high protein (guinea pigs) or fat (seals) in their milk increase their body weight relatively rapidly. Milk sugar, lactose, is common in many animals, though variable, but is absent in whale milk. The monotremes differ from marsupials and mammals by little free lactose and a high content of fucose, a monosaccharide that is combined with lactose in trisaccharide or tetrasaccharide form.

Teeth The teeth (Chs. 8 and 9), whose enamel is derived from the ectodermal epithelium and whose dentine is derived from mesodermal cells, are integumentary derivatives of especial importance to most mammals but not all because anteaters may completely lack teeth. Mammalian teeth are distinct from the teeth of other classes of vertebrates for two reasons: they have two sets of teeth, making them diphyodont (see below), and the teeth vary both in numbers characteristic of the order to which the animal belongs and in architecture according to the position of the tooth in the series. The teeth in a series beginning at the midline and progressing laterally along each quadrant of the jaw are of four kinds in order: incisors (I), canines (C), premolars (Pm), and molars (M). A dental formula is thus possible for each species. The formula takes into account one side of the jaw, with the upper and lower numbers representing the number of specific teeth in the upper and lower jaws, respectively. For instance, three dental formulas are shown as follows:

	I	C	Pm	M
Coyote	$^3/_3$,	$^1/_1$,	$^4/_4$,	$^2/_3$
Man	$^2/_2$,	$^1/_1$,	$^2/_2$,	$^3/_3$
Pronghorn antelope	$^0/_2$,	$^0/_1$,	$^3/_3$,	$^3/_3$

To determine the total number of teeth in the mouth, the sum of the dental formula is simply doubled. Thus in humans the dental formula sums to 16 and the total number of teeth, including the so-called wisdom teeth, is 32.

The first set of teeth is called the *deciduous* or milk teeth. They may have a different dental formula from the second set or *permanent* teeth. The length of time the deciduous teeth persist varies among animals. In some species they are resorbed prior to birth, while in other species, such as moles, the second set is permanently suppressed and the first set is functional throughout the life of the animal.

Each kind of tooth in the dental formula has its own characteristics in each species (Fig. 38.1). Functionally, incisors serve to cut, gnaw, or grasp. In larger herbivores they may be flat bladelike nippers for grazing, while rodents have specialized incisors with enamel only on the anterior surface and with a continuous growth. As a result, wear must keep abreast of growth in order to prevent malocclusion, a condition in which the cutting edges do not meet, the incisors are overgrown, and the animal is incapable of eating. However, when wear keeps abreast of growth, such incisors wear into efficient chisellike surfaces. It is appropriate here to point out that the common fallacy that rabbits are rodents (they are lagomorphs) is disproved by examining the teeth. Rabbits have an additional small pair of upper incisors located behind the first pair, and the crowns of the incisors are covered with enamel. Some specializations in the incisors are small bladelike structures for bloodletting in vampire bats, tusks of elephants, or small structures of minimal use such as seen in dogs, cats, and other carnivores.

The canines, or dog teeth, are highly developed in carnivores, serving for capture, killing, and tearing. These too can form tusks, as in the case of wild boars, where they serve as defensive or aggressive structures, or as in walruses where they are used to pull themselves onto the ice from the sea or to dislodge shellfish for food. Canines may be entirely lacking, as in rodents, where a gap in the teeth, the diastema,

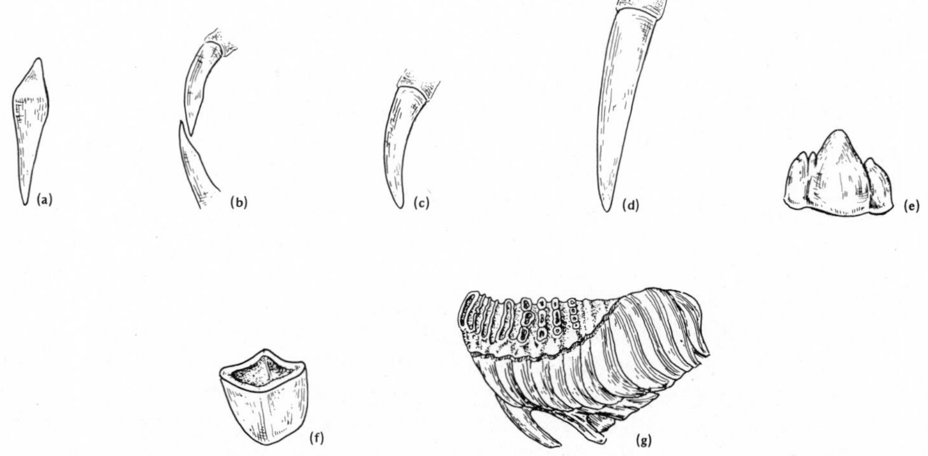

Figure 38.1 Some kinds of teeth. **(a)** Human incisor. **(b)** Rodent incisors. **(c)** Carnivore canine. **(d)** Walrus canine. **(e)** Carnassial (shearing) tooth of a dog. **(f)** Molar of a deer. **(g)** Molar of an elephant.

exists. An interesting condition is that of the narwhal, a cetacean, where only the left upper canine exists in males (rarely in females); there are no other functional teeth and this single canine may grow to as long as nine feet.

The premolars and molars show great diversity, depending on the species. The general pattern of the premolars and molars is similar. They range from broader flat-crowned grinding teeth of herbivorous animals, such as horses and cattle, to slicing bladelike cutting teeth of carnivores. In many herbivores the enamel is folded into the dentine, and wear creates either a series of crescentic ridges or transverse ridges from the enamel becoming less worn than the softer dentine between the enamel folds. These edges assist in the chewing of vegetable matter by such animals as horses, cows, and deer. Carnivorous animals have specializations for cutting through the last premolar of the upper jaw, forming slicing, cutting edges with the first molar of the lower jaw. These sets of cutting teeth in carnivores are called *carnassial* teeth.

An unusual condition exists among the porpoises and dolphins, whose numerous teeth are more or less identical and resemble tapering round-tipped pegs set in each jaw. Animals such as this with uniform teeth are called *homodonts,* while those with classical dental formulas are *heterodonts.*

THE SKELETAL SYSTEM

The skeletal system assists in protection and in locomotion in several ways. By contrast with the skeleton of a bipedal form such as humans, as seen in Chapter 8, most mammals are quadripeds and their skeletons reflect this through several adaptations associated with a four-footed support.

Except that the monotremes retain a pair of coracoid bones characteristic of other classes of vertebrates, and the pouch-bearing mammals (marsupials) have a pair of bones to support the pouch or marsupium, the kinds of bones found in mammals are as described in Chapter 8. For instance, with rare exceptions, and even in giraffes, the number of vertebrae in the neck is seven, as it is in humans. Two noteworthy attributes of the axial skeleton of quadripeds are that the vertebral column usually provides a strong arched support, and with a few exceptions this column extends into the flexible and often useful tail.

The skeletal features that deserve special comment are the limbs (Fig. 38.2) and their girdles and such appendages as horns and antlers.

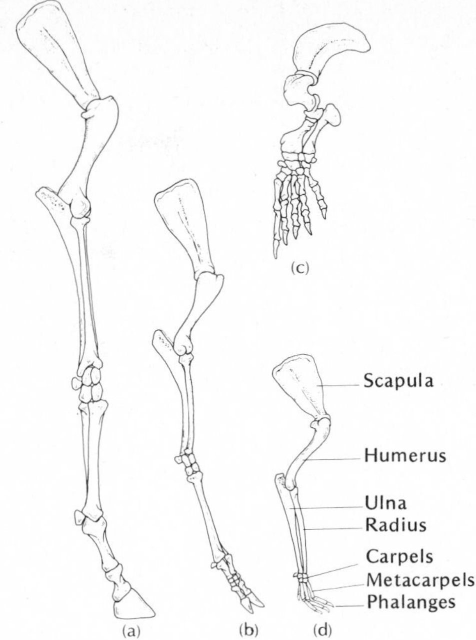

Figure 38.2 Forelimb variations. **(a)** Horse. **(b)** Deer. **(c)** Walrus. **(d)** Cat.

Animals adapted for speed have achieved this with the accompanying loss or reduction of the clavicle of the pectoral girdle. In this way the remaining skeletal components of the forelimb and girdle are freed from articulating directly with the axial skeleton. In horses, for instance, this permits a sweeping stride with the appendicular muscles serving to support the trunk between the forelimbs. By contrast, animals such as the primates that rely on *brachiation,* the use of the forelimbs for climbing or swinging, retain the clavicle as a strong fulcrum articulating with the sternum of the thorax.

It is the appendages of mammals that obviously reflect the mode of life of the particular species; it is possible to group these animals in this respect. Of the terrestrial forms, fast runners are called *cursorial,* jumpers are called *saltatorial,* subterranean forms are *fossorial,* tree dwellers are *arboreal,* gliders are *volant,* and flyers are *aerial* animals. Those animals either totally oriented to the seas or partially associated with seas, lakes, or streams, are, respectively, *marine* or *aquatic* forms.

The cursorial animals usually have long slender limbs. The muscles lie close to the trunk and action is controlled by long tendons. The humerus and femur are usually relatively short and the radius–ulna and tibia–fibula portions elongated. Also, there is little rotation in the joints, and it is usually confined to a forward–backward motion. The cursorial animals fall into

Figure 38.3 Artiodactyls. White-tailed deer (*left*). Thompson's gazelles (*right*). Digits common to all artiodactyls are number three and four. (New York Zoological Society Photos) See also Plate 11 for four-digitted artiodactyl, the hippopotamus.

two major groups, both represented by a reduction in the toes and the presence of hooves. The animals that have even numbers of toes on which they stand are called *artiodactyls* and include cows, pigs, deer, and antelope, for instance (Fig. 38.3). Those that have odd numbers of toes on which they stand are called *perissodactyls* and include fast runners such as horses, with one functional toe, or slower moving tapirs and rhinoceroses, with three functional toes (Fig. 38.4, Pl 2). An animal that shows some cursorial traits in its speed and durability is a member of the cat family, the cheetah. The proportions of the limb bones and the musculature give this handsome predator a capability of running down food from among the rapidly moving hooved animals.

Jumping animals, as might be expected, have the hind limbs disproportionately enlarged and the forelimbs tend to be reduced for grasping. This trait is seen well developed in marsupials such as the kangaroo and wallaby and in placental mammals such as jumping mice and kangaroo rats (Fig. 38.5). All three examples have the tails well developed to assist in balance for these essentially bipedal animals.

Figure 38.4 Perissodactyls. Rhinoceros. The third toe is in common among all perissodactyls. Pair of black rhinoceros (*left*). Head of black rhinoceros showing epidermally derived horn of fused modified hairs (*right*). (New York Zoological Society Photos) See also Plate 2 for single-toed perissodactyls, the zebra.

Figure 38.5 Euro kangaroo female with young in pouch. This marsupial has well-developed hind limbs permitting a strong saltatorial locomotion. (New York Zoological Society Photo)

Figure 38.6 Flying squirrel, a rodent that has volant locomotion. Note the flattened tail and the skin fold visible on the animal's right just above the branch. These stretch widely to permit long gliding movement through the air. (New York Zoological Society Photo)

Subterranean animals have short limbs that often extend laterally and have a strong set of shoulder muscles to provide strength for digging. Also, these forms, such as moles and gophers, have large claws.

Arboreal animals, such as squirrels, not only have well-developed claws and front feet especially capable of grasping, but also the joints rotate freely to allow great variation in motion. Thus a squirrel can be almost equally effective on the branch of a tree whether it is heading away from or toward the ground. The tails of arboreal forms are also effective for balance or for steering. In the arboreal New World monkeys the tail is also prehensile.

The gliding animals, volants, usually achieve this not so much by skeletal changes but by the broad lateral skin folds that extends from forelimbs to hind limbs. The pattern of the hair on the tails of gliders, such as flying squirrels, provides laterally extending flattened surfaces that improve gliding (Fig. 38.6).

The bats are the only truly aerial mammals (Fig. 38.7). They show a combination of noteworthy adaptations of the limbs and of the skin. The hind limbs and tail are small, but the forelimbs are highly modified. The metacarpals and

Figure 38.7 Bats, animals with aerial locomotion. Egyptian fruit bat in flight (*left*). Giant fruit bats (Flying Foxes) suspended at rest (*right*). (New York Zoological Society Photos).

Figure 38.8 Some aquatic and marine mammals. Beaver: Builds homes in ponds but can forage freely on land (*upper left*). Seal: with modified limbs; feeds in the sea but comes onto land awkwardly (*upper right*). Amazon manatee: entirely adapted to a fresh water habitat; herbivorous (*lower left*). Pacific White Sided Dolphins: marine mammals with highly streamlined bodies (*lower right*). (beaver: New York Zoological Society Photo; others: Steinhart Aquarium, San Francisco)

Figure 38.9 Plantigrade (bear) and digitigrade (dog) locomotion.

phalanges are greatly elongated and connected by broad double membranes of skin to provide the flight surfaces. This membrane extends along the trunk to the hind limbs and to the tail. The pectoral and shoulder muscles are powerful and are concentrated on the body; this maintains the light structure of the wings.

Aquatic mammals show varying degrees of adaptation of the limbs or appendages. In beavers, for instance, the principal adaptation is in the tail (Fig. 38.8), while in seals and sea lions the forelimbs and hind limbs are shortened, broadened, and provide the characteristic fore- and hindflippers of these animals.

Marine animals, such as dolphins and whales, show extreme reduction or complete loss of the hind limbs or pelvic girdle, and the front limbs serve principally as stabilizing flippers for the highly streamlined body.

There is another trait that has not been formally covered in the adaptations of the appendages for locomotion. This trait refers to the degree to which the foot contacts the ground in locomotion. Some animals, such as racoons, bears, and humans, walk with the entire foot from the carpals and tarsals to the digits on the ground. This form of architecture is called *plantigrade* (Fig. 38.9). In other animals, such as cats and dogs, there is a complete set of fingers or toes, or at least a total of four, and they walk only on the digits; these animals are *digitigrade*.

Horns and Antlers In addition to epidermally derived horns, as mentioned earlier for the rhinoceros, two limited groups of mammals have developed skeletal specializations on the head in the form of horns and antlers. These two kinds of structures are quite different. Horns are permanent bony outgrowths of the head that are ensheathed in true horn, a keratinized product of the cells surrounding the bony core (Fig. 38.10). In most animals such horns exist on both sexes. The structures are single and tapering but they may take on different shapes, as is obvious by comparing the horns of domestic cattle, longhorn cattle, and the curving massive horns of bighorn sheep. Antelopes provide variations in the usual pattern of horns: male four-

Figure 38.10 Horns and antlers. Indian antelope or blackbuck: male bears spiral horns (*left*). Northern White-tailed deer: buck has annually produced and shed antlers; doe lacks antlers (*right*). (New York Zoological Society Photos)

horned antelopes of India have a small pair in front of the larger ones and they are lacking in the females; pronghorn antelopes have a branch or fork and also there is annual shedding of the outer horny sheath.

Another type of skull outgrowth is found in giraffes and okapis in Africa where bony outgrowths, that originated as cartilage, persist with a covering of skin and hair.

Antlers (Fig. 38.10) are truly amazing adaptations inasmuch as they grow to full size from small outgrowths of the frontal bone, called *pedicels.* They are also shed annually. Thus when one realizes that antlers grow in a few months from the small pedicels to spreads in some moose to as great as six feet from tip to tip, this is obviously a dramatic growth rate. The antlers grow under stimulation of the sex hormones of the pituitary and testosterone from the testis. The hormones stimulate the vascularization of the integument and the pedicels at the site of growth. The integument keeps pace with the bone growth and forms a cover, called *velvet,* due to its texture. After growth is completed, the blood vessels at the base of each antler constrict, the velvet dries, and is sloughed off. After the breeding season ends, the bony base of the antler is weakened by bone resorption, and the antlers are shed during the winter. The process in the males is repeated the next spring.

Antlers and horns serve both for protection and apparently for sexual display. Males employ them in establishing their claim on the females.

FEEDING AND DIGESTION

Types of food and feeding habits are very diverse in the mammals. One immediately thinks of two broad divisions related to food habits, the plant-eating herbivores and the flesh-eating carnivores, whose diet principally consists of herbivores. There are, however, two additional groups, the *omnivores* that eat both plants and animals, and the *insectivores* whose diet is principally insects and other small arthropods.

Of course, the problem of obtaining food in nature is a serious one as the metabolism and size of each species is related to the amount of energy required in the form of food. Interestingly, the smaller mammals face greater challenges to locate food as more food must be eaten relative to total size in smaller animals than in large ones. Thus smaller animals with higher metabolic rates must spend a large amount of time taking in food. Shrews, bats, and mice fall into this category, and small shrews may eat more than their body weight each day. Other animals, especially the carnivores, may eat intermittently, often gorging themselves after a kill and then resting for several days. At the other extreme of size of food acquiring are the whales. Some whales, such as the killer whale, have numerous peglike teeth of similar structure (homodont) while other whales, in spite of their immense size, obtain their food by straining the seas for planktonic organisms. These whales lack functional teeth and achieve the straining through the presence of long horny plates, called *baleen,* suspended in rows from the roof of the mouth. The plates of baleen are fringed to serve as strainers.

The herbivores show interesting adaptations both in the way they obtain food and the way they process it. In the manner in which they take in plant materials they are divisible into four groups: browsers, ungulates that eat leafy food from small plants or shrubs; grazers, ungulates that subsist on grass; gnawers, represented by some rodents such as beavers; and nibblers, such as rabbits. As none of the vertebrates can digest cellulose, yet this is a large part of a herbivore's diet, adaptations that permit bacteria and other microorganisms to break down cellulose exist. The adaptations in herbivores that support such cellulose digestion are of several kinds. First, the intestine is relatively long. In cattle, for instance, the intestine may exceed 50 m in length. Second, there may be saclike outgrowths from the intestine, called *ceca,* that retain the plant food and where microorganisms ferment the food and permit the products to be absorbed. Third, complicated stomachs may be found in the hooved animals called *ruminants.* A ruminant stomach functions in the following way. Food is swallowed into the first of four chambers, the *rumen.* This chamber and the next two are thought to be modifications of the posterior end of the esophagus. In the rumen bacterial action ferments the food, including the cellulose. The food may be returned to the mouth as cud where it is chewed and returned to the rumen for further digestion. In time the partially digested food travels through a honeycombed chamber, the *reticulum,* and into a collecting chamber, the *omasum,* from which it is delivered into the true stomach, called here the *abomasum,* where hydrolytic digestion occurs. As can be illustrated by the more fibrous feces of a horse by contrast with that of a cow, there is greater digestive efficiency in a cow, a ruminant, than in a horse, a nonruminant, whose principal cellulose-digesting structure is a large cecum.

A few animals, such as rabbits, add to the efficiency of cecal digestion by eating their fecal pellets. This recycles the food through the cecum again for additional digestion.

Carnivores and omnivores do not have specializations for digestion that differ from those described in Chapter 9. The carnivores do have skills in hunting, of course, that improve their opportunity to eat. Also, as described earlier, their teeth are adapted for piercing, tearing, and cutting. Omnivores, such as humans and some of the other primates, have a varied diet, but a strong distinction between omnivores and carnivores is sometimes difficult to make because animals normally predaceous and carnivorous will eat fruit, berries, or seeds if conditions become unfavorable.

Insectivores may differ from other mammals in the way they acquire food, but the process of digesting chitin is accomplished by secretion of concentrated gastric juices into the stomach from a specialized pocket formed of the fundus of the stomach, while the remainder of the stomach is lined with highly cornified epithelium to resist the action of the chitin as it travels through the stomach.

GASEOUS EXCHANGE AND INTERNAL TRANSPORT

Mammals have much less flexibility of movement or change of environment than do birds. Thus their adaptations must permit them to cope with the local environmental conditions in which they find themselves. Among the matters of concern under these conditions, other than simple exchange of gases, are problems relating to temperature regulation and water balance. Both the respiratory and circulatory systems play a role here.

The Respiratory System The general architecture of the respiratory system is basically as described in Chapter 10 and is less complicated than that of birds (Ch. 37). However, some modes of life of mammals have resulted in adaptive uses of the basic structure that were not discussed in Chapter 10. The respiratory rate is one trait that is variable among the mammals. For instance, while a human in good physical condition has a respiratory rate of about 12 inspirations per minute, some insectivores will respire as much as 850 times per minute.

The aquatic and marine animals have striking adaptations that permit them to rapidly descend and ascend in the water without suffering from bends or other problems associated with the great pressure changes to which they are exposed. Two animal groups are noteworthy. The seals have flaps or valves in the nostrils. When the seal's head submerges when diving, the pulse will drop dramatically from about 80–85 beats per minute to about 12 beats per minute. This greatly reduces the oxygen requirement as there is a decrease in the metabolism accompanying this drop. Seals can descend 600 m (2000 feet) and remain submerged for about a half hour and return rapidly to the surface without any nitrogen narcosis (bends). The cetaceans are the other group highly adapted for diving to great depths without the great pressures there doing harm. In some dolphins an aid to accomplishing this is through the complete collapse of the alveoli of the lungs by a descent of about 65 m. This prevents gas exchange under these conditions and nitrogen is not forced into the vessels under the high pressures of the depths. This method of alveolar collapse probably operates widely in the cetaceans. Two other conditions in the animals also add to their efficiency. First, the hemoglobin of the blood of whales is more efficient than in terrestrial animals as is the carbon-dioxide-carrying capacity of the blood. Also, there is more myoglobin in the muscle, and this is very efficient in storing oxygen. Muscle and skin also have their blood supply reduced, another mechanism of conserving oxygen; this capability, of course, is related to the ability of muscle to function anaerobically (Ch. 8).

The Circulatory System While the pattern of circulation is the same in mammals, there are some adaptations that relate to circulatory function. Recall (Ch. 11) that one of the major functions of the circulatory system is temperature regulation accomplished principally through the regulation, by vasodilation and vasoconstriction, of the bore of blood vessels at the surface in the skin. Basically, this system prevails in adaptations both for hot environments and for cold ones. The adaptations discussed here will explore more than just the circulatory system, but will always be closely related to it.

First, remember that mammals are homoiotherms and have temperatures usually not varying from 36 to 38° C during the day. Such sustained temperatures permit regulated enzyme action throughout the body, and the enzymes have evolved usually to function most efficiently at these temperatures. To maintain such an essential, the narrow range of temperature requires a delicate equilibrium between heat production by the animal's metabolism (including its food intake) and heat loss. Variations in body temperature do occur in some mammals, however, ranging from significant variations between night and day in insectivores, for instance, or during periods of hibernation as in the case of bears.

In hot environments the problems faced by animals are excessive water loss and excessive heating. Prevention of water loss is accomplished in several ways. The respiratory rate may be slow and deep, thus avoiding excess loss in expired air. Or the number of sweat glands may be reduced. Or the animal may develop efficient means for processing food so that the water produced in the metabolism is conserved; this is called *metabolic* or *oxidation* water. Some desert dwellers, such as kangaroo rats, can subsist entirely on metabolic water if forced to. Accompanying the use of metabolic water usually is a kidney adaptation that results in very concentrated urine.

Some animals inhabiting hot environments use pelage as protection against overheating. One way is to provide a smooth highly reflective coat. Another is to develop a sufficiently thick coat to prevent the penetration of heat.

Modifications in the vascular bed of the surface are among the most obvious adaptations for hot environments. For instance, animals such as rabbits may have large radiating surfaces created by their especially large ears by contrast with rabbits of more temperate or cold climates. Some animals, such as the ungulates, may have areas on the underside of the body that are not so thickly covered with hair and which are highly vascularized. This permits a rather rapid heat loss in these locations. Another adaptation is to permit some body temperature variation. In some larger plains animals the temperature may drop a few degrees at night and rise during the day.

Some mammals simply seek out cooler moister places when it is hot. This may simply be a behavioral trait of seeking shade or a more complex pattern of being fossorial (burrow or subterranean dwelling). The normal size of the animal influences whether it will be fossorial, and most such burrowers are relatively small mammals.

At the other extreme of environment, cold, some reversed adaptations exist. For instance, in cold climates rabbits have much smaller ears than do rabbits occupying hot environments. This relationship of extremities to environmental temperature is called Allen's rule (Ch. 44). However, extremities in cold climates, even though relatively smaller than elsewhere, still present problems. One adaptation of the circulatory system influenced by this problem is both simple on the one hand, and adaptively "clever" on the other hand. Some animals simply let the extremities cool, often close to the freezing point. The obvious problem associated with excessive cooling of the blood flowing into the hypothermic area is met by having the artery and vein serving the extremity lie very close together. Thus when especially cooled blood returns toward the body, there is a counter current heat exchange and the artery warms the blood in the vein before it reaches the trunk. The musculature of mammalian extremities, of course, is not directly affected by such supercooling as can occur at the footpads or hooves, for muscles are located close to the body and the principal structures at the extremities are the tendons that can function at reduced temperatures.

Muscle, however, is very important if other devices for heat conservation in cold environments fail. The reduced temperature at the surface, either directly through cold receptors or through an effect on the blood, can result in generalized stimuli arising from the medulla to cause augmented muscular activity or shivering. Such activity can result in a relatively rather large release of heat. Finally, similar stimuli can result in increased metabolism from stored energy reserves. Therefore we see that there is both shivering and nonshivering thermogenesis at work.

Lastly, just as smaller animals in hot environments behave as fossorial individuals, so too in cold environments there may be a true fossorial pattern or a pattern resembling it simply by burrowing under snow. Snow is an excellent insulator in most cases, having much trapped air within it.

THE UROGENITAL SYSTEM AND REPRODUCTION

The metanephric kidneys of mammals participate in the maintenance of appropriate water balance as well as with metabolites of protein and other metabolism. The efficiency of the countercurrent system (Ch. 10) is great in those animals adapted to environments requiring conservation of water.

All mammals have internal fertilization. However, the development that occurs after fertilization differs in three major ways in mammals. Monotremes lay eggs. Marsupials have the development proceed in the uterus only until the fetus can be born and make its way to the pouch where it completes its development. The rest of the mammals support the development of the young in the uterus until all structures are differentiated. However, the degree to which this is achieved varies. Some are born without hair, with their eyes closed, and generally quite helpless; these are the *altricial* forms. Other mammals produce young with hair, with open eyes, and with an ability to move around; these are the *precocial* forms. Rats, for instance, are altricial, while guinea pigs are precocial.

Mating While some mammalian males are capable of mating at any time, most mating is regulated by the physiological condition of the females, whose mating responses are cyclical, and each cycle is called an *estrous cycle.* By contrast with the human, described in Chapter 15, where there is a menstrual cycle, most mammals have an estrous cycle divisible into four phases: *proestrous, estrous, metestrous,* and *diestrous.* During proestrous, the ovarian follicles grow and the uterine lining thickens. During estrous, when mating occurs, ovulation also occurs. In some mammals the ovulation is spontaneous; in others ovulation is induced in that only after copulation is there release of ova from the ovaries. If mating does not occur or fails, estrous is followed by metestrous during which time the uterine lining regresses and the ovarian follicles form corpora lutea. The fourth period, diestrous, is usually the longest, during which time the corpora lutea regress and the uterus becomes small.

The frequency of estrous cycles varies among mammals. In some rodents the cycle is repeated several times during the breeding season. For instance, unless impregnated, a female rat comes into estrous every four days. Other animals vary. Female dogs are usually "in heat" (estrous) twice a year. Some animals breed but once a year.

If impregnation occurs, the length of the gestation period varies with the different species. Mice and rats have gestation periods of 21 days, dogs take 60 days, moose take about 8 months, and elephants take 20 months.

There is an interesting condition related to gestation that is not fully understood and which occurs in both marsupials and some mammals. This is a condition of delayed implantation. In principle, mating occurs, fertilization is achieved, and the zygote develops into a blastocyst stage when development is arrested until some passage of time when implantation in the uterine wall occurs. For instance, weasels and their relatives mate in the summer, but the resulting blastocysts remain free until midwinter when they implant, and the young are born in the spring.

The numbers of offspring vary among mammals. Large herbivores, cetaceans, members of the seal group, bats, and primates usually have only one young at a time. There is some correlation between the number of young produced and the mortality rate. Thus those animals that occupy low positions in food chains and are preyed upon extensively (Ch. 45) usually have rather large and frequent litters. Voles and mice, for instance, are quite prolific. A very successful animal, the opossum, may raise as many as 12 young to maturity.

SENSORY RECEPTION AND NEURAL COORDINATION

Since animals are adapted to a large variety of environments, it seems natural to assume that there will be a variety of kinds of sensory inputs. One well-developed trait in many mammals is the sense of smell, often accompanied by a greatly increased surface area in the nasal region through scroll-like nasal *conchae.* However, this is not a universal trait of mammals; two outstanding exceptions are humans, who usually have limited olfactory capabilities, and whales, who may have essentially nonfunctional olfactory reception.

Eyes in mammals vary principally in size, nocturnal animals having relatively large eyes.

Generally auditory reception is highly developed throughout the class. The range of receptions varies with the species, however. For instance, dogs can hear far higher pitched sounds (shorter wave lengths) than can humans. Three groups of animals, the cetaceans, the pinnipeds (seals), and the bats, have remarkably well adapted mechanisms for echolocation. In bats and cetaceans the procedure used is for the animal to emit a sound which is reflected from nearby objects, and the information received by the animals is interpreted to guide the movement of the animals. Bats, for instance, can capture insects in the dark by this means, and in experiments to test their capabilities, they can fly through a set of strung wires without striking them. In the case of cetaceans, sound travels very rapidly in water and it also can pass easily through the body of the animal due to the similarity of the density of the body. This ordinarily would create a problem for the whale if both ears received the sound. However, directionality of sound pulses can be maintained because the tympanic bone of the skull that contains the cochlea is suspended free from the skull by ligaments and is insulated in a cavity of air or foamy material. Therefore sound only gets to the cochlea through the ear ossicles. The sounds some cetaceans have emitted have been recorded at as high as 300,000 cycles per second, even higher than those reported for bats.

Recent studies on cetaceans show that they have evolved specializations in the head that assist in beaming sound waves outward as well as for their reception.

The general nervous system (Ch. 13) of mammals is more highly developed than in any other vertebrates. The cerebral hemispheres are dominantly expanded adaptations that account for this. This area of the brain is amazingly complex and ignorance about its complete function remains large in spite of intensive study and research.

BEHAVIOR

Migration and Homing As with the other classes, mammals too have migratory patterns and some can home. The number of land dwellers who do this, however, is few. Noteworthy are the barren-ground caribou, which move out of the Canadian forests in late winter to the tundra where the young are born in June. In July and August they move southward and reach their principal feeding grounds in the forests in September. Elk also migrate and the now nearly extinct plains bison (buffalo) had well-established migrations. There are three species of bats that migrate between northern and southern states in the United States.

Seals and whales probably are the most extensive migrators among mammals. Fur seals are interesting in that there is a sexual difference in their migrations. The breeding area is the Pribilof Islands in the Bering Sea north of the inner Aleutian Islands. The bull seals do not follow the females who migrate with their young to the wintering area off southern California.

Territoriality In many mammals territoriality is established by marking the perimeters of a territory with scent. The canids are noteworthy for urinating to leave their identification. Territory of some of the pinnipeds is established by the bull seal taking a central position in an area and rushing to the perimeter to challenge potential invaders. However, northern elephant seals do not depend so much on a physical territory as on a social hierarchy in which the dominant males stay with the breeding females and the lesser males are prevented from mating with members of the harem. In cetaceans there is growing evidence that the animals may maintain territories by sending out vocalizations; in this way they resemble the birds.

Mammals have developed a number of social patterns among themselves and a number of these are discussed in Chapter 45.

Importance of Mammals to Humans

There are three significant ways in which mammals relate to humans. One is through domestication. There are several expressions of this. Domesticated mammals, such as cats and dogs, may simply bring pleasure as pets; or they may serve as hunters, cats being mousers and dogs participating as pointers, retrievers, and killers. Another is through the products derived from the mammal. Here pelts may serve as clothing or parts of clothing, but more important is the providing of food for humans. It is a rare mammal that cannot serve as food for humans in some part of the world. In the United States the herbivores provide the principal source of food, but elsewhere in the world dogs, for instance, may be a normal part of the diet. The third major relationship of mammals to humans is and has been as a beast of burden. The principal animals that have played this role are horses, cattle, camels, llamas (relatives of camels), elephants, and water buffalo.

In contrast to the constructive relationship with humans, the principal negative one has been the roles of rodents and lagomorphs in the destruction of crops and in the transmission of disease. Another less important one in modern society has been the predation of some mammals, such as wolves, foxes, coyotes, and members of the cat family, both large (lions, tigers) and small (lynx, wildcat), on the domestic animals of humans.

Classification

As mentioned earlier, the mammals arose from the therapsid reptiles. In the process of arriving at the three major subclasses of living mammals, two subclasses, the Multituberculata and the Triconodonta, have become extinct.

Surviving mammals, consisting of approximately 20,000 species and subspecies, fall into two subclasses, the Prototheria, or egg-laying mammals, and the Theria, the viviparous mammals. The Theria are divisible into two living infraclasses collectively with 19 orders.

SUBCLASS **Prototheria** Egg-laying. Mammary secretions but no nipples.

ORDER **Monotremata** Duckbilled platypus, echidna (spiny anteater) (Fig. 38.11). Australia, Tasmania.

SUBCLASS **Theria** Viviparous.

INFRACLASS **Metatheria** Marsupials.

ORDER **Marsupialia** Abdominal pouch or marsupium. Numerous adaptations to fill various niches in Australia, where they are dominant. Opossum in North America. Kangaroos, wallabys, wombats, Tasmanian wolf, and many others.

INFRACLASS **Eutheria** Placental mammals.

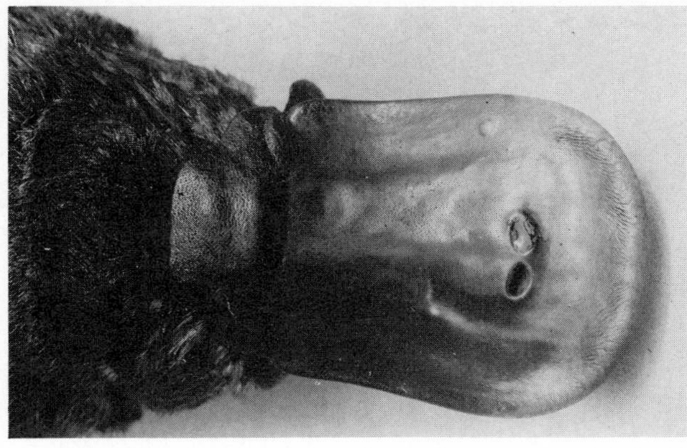

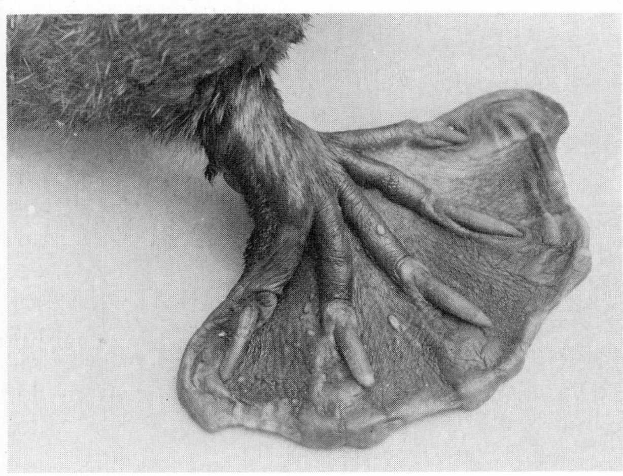

Figure 38.11 Prototherian monotremes. Duckbill platypus (*upper left*). Close-up of platypus' horny bill (*upper right*). Expanded front foot of a platypus to show webbing and claws (*lower left*). Australian echidna or spiny anteater (*lower right*). (New York Zoological Society Photos)

ORDER **Insectivora** Most primitive placental mammals. Pentadactyl limbs. Primitive teeth. Moles and shrews. Worldwide.

ORDER **Dermoptera** Primitive gliding animals. Medium small. Membrane from head to tail. Pentadactyl. Gliding lemurs. Southeast Asia, East Indies, and Philippines.

ORDER **Chiroptera**—Bats. Wings derived from webbed hand. Many species worldwide.

ORDER **Primates**—Lemurs, lorises, tarsiers, marmosets, Old World monkeys, New World monkeys, anthropoid apes, humans. Well-developed cerebral hemispheres. Orbits directed forward, surrounded by bone. Pentadactyl with opposable thumb (and toe). Nails usually flattened. Asia, Africa, North and South America. This order is often subdivided into three suborders: Lemuroidea, Tarsoidea, and Anthropoidea. Humans belong to the Anthropoidea along with monkeys and apes.

ORDER **Edentata**—Anteaters, tree sloths, armadillos (Fig. 39.12). Toothless or with degenerate teeth. Front claws often enlarged for digging. Southern United States, Central and most of South America.

ORDER **Pholidota**—Pangolins (Fig. 39.13). Overlapping scales cover dorsal surface of head and body, outer surface of legs. Scales horny and derived from fused hairs. No teeth. Africa and southeastern Asia.

ORDER **Lagomorpha**—Rabbits, hares, pikas. Worldwide except introduced to Australia.

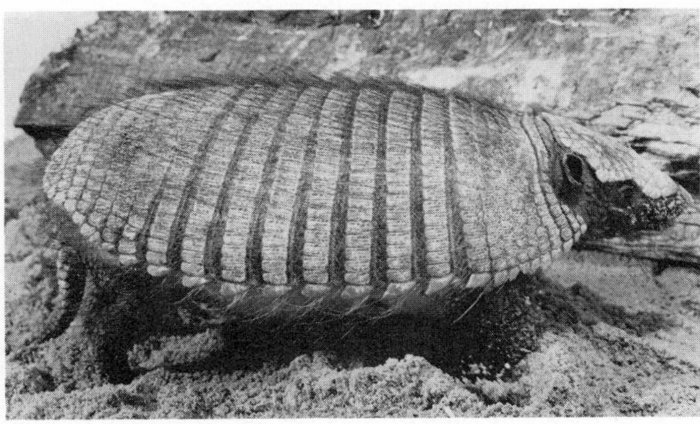

Figure 38.12 Edentates. Nine-banded armadillo (*left*). Hairy armadillo (*right*). New York Zoological Society Photos)

Figure 38.13 Pangolin: has horny scales derived from fused hairs (New York Zoological Society Photo)

ORDER **Rodentia** Large order of many families. Squirrels and squirrellike rodents, gophers, rats, and mice of several kinds, jerboas, porcupines, chinchillas, cavies. Basically gnawing mammals. Worldwide.

ORDER **Cetacea**—Whales, porpoises, dolphins. Marine. Skin lacks hair. Blubber under skin. Nostrils modified as blowhole. Flippers for front limbs. Hind limbs visibly absent. Tail in form of horizontal flukes. Variable number of teeth up to over 200, or may be absent and with baleen as strainers. Worldwide.

ORDER **Carnivora**—Bears and pandas, dogs, wolves and foxes, racoons, mustelids (weasel family), civets, hyenas, cats. 4-clawed toes. Relatively large canines. Worldwide.

ORDER **Pinnipedia**—Sea lions, fur seals, walrus, earless seals. Limbs modified as flippers. Large eyes. Vibrissae (whiskers) well developed. Short or absent tail. Arctic, Antarctic, continental coastlines.

ORDER **Tubulidentata**—Aardvarks. Piglike snout. Long tubular ears. 4 front, 5 rear toes. Muscular tapered tail. No incisors or canines. Teeth lack enamel and tubular. Africa.

ORDER **Proboscidea**—Elephants. Thick skin. Trunk. Large ears. Nearly hairless. Asia, Africa.

ORDER **Hyracoidea**—Hyraxes and Old World conies. Small. Nails hooflike. Reduced or vestigial tail. Africa, southeastern Asia.

ORDER **Sirenia**—Manatees, dugong, sea cow. Front limbs as paddlelike flippers. Hind limbs lacking. Small eyes. Hairless except around mouth. Tail as horizontal flukes. Reduced or absent teeth. Tropical seas, coastal rivers.

ORDER **Perissodactyla**—Odd-toed hoofed mammals, horse family, tapirs, rhinoceroses. Third digit dominates. Elongated head. Africa, Asia, Americas.

ORDER **Artiodactyla**—Even-toed hoofed mammals, swine, peccaries, hippopotamuses, camels, llamas, deer, giraffes, okapis, pronghorns, cattle, antelope, sheep, goats. Toes 3 and 4 dominate. Worldwide.

SUGGESTIONS FOR FURTHER READING

Burt, W. H., and R. P. Grossenheider. *A Field Guide to Mammals (North of Mexico).* Boston: Houghton Mifflin (1952).

Davis, D. E., and F. B. Golley. *Principles of Mammology.* New York: Reinhold (1963).

Eimerl, S., and I. DeVore. Life Nature Library. *The Primates.* New York: Time (1965).

Hall, E. R., and K. R. Kelson. *The Mammals of North America,* 2 vols. New York: Ronald Press (1959).

Lawick-Goodall, J. van, and H. van Lawick. *Innocent Killers.* Boston: Houghton Mifflin (1971).

Mathews, L. H. *The Life of Mammals.* London: Weidenfeld and Nicolson (1969).

Miller, G. S., Jr., and R. Kellogg. *List of North American Recent Mammals.* Washington: U. S. Nat. Museum, Bull. 205 (1955).

Orr, R. T. *Vertebrate Biology.* Philadelphia: W. B. Saunders (1971).

Simpson, G. G. *The Principles of Classification and a Classification of Mammals.* New York: American Museum of Natural History, Bull. 85 (1945).

Walker, E. P. *Mammals of the World,* 3 vols. Baltimore: Johns Hopkins Press (1968).

From Scientific American

Drinker, C. K. "The Physiology of Whales" (July 1949).

Flyger, V., and M. R. Townsend. "The Migration of Polar Bears" (Feb. 1968).

Griffin, D. R. "More About Bat 'Radar' " (July 1958).

King, J. A. "The Social Behavior of Prairie Dogs" (Oct. 1959).

Kooyman, G. L. "The Weddell Seal" (Aug. 1969).

McVay, S. "The Last of the Great Whales" (Aug. 1966).

Modell, W. "Horns and Antlers" (Apr. 1969).

Montagna, W. "The Skin" (Feb. 1965).

Mrosovsky, N. "The Adjustable Brain of Hibernators" (Mar. 1968).

Myers, J. H., and C. J. Krebs. "Population Cycles in Rodents" (June 1974).

Mykytowysz, R. "Territorial Marking by Rabbits" (Mar. 1968).

Pearson, O. P. "Shrews" (Aug. 1954).

Savory, T. H. "The Mule" (Dec. 1970).

Schmidt-Nielsen, K., and B. Schmidt-Nielsen. "The Desert Rat" (July 1953).

Schmidt-Nielsen, K. "The Physiology of the Camel" (Dec. 1959).

Scholander, P. F. "The Master Switch of Life" (Dec. 1963).

Taylor, C. R. "The Eland and the Oryx" (Jan. 1969).

Warren, J. V. "The Physiology of the Giraffe" (Nov. 1974).

39 Phylogeny

Studies of evolution within the subphylum Vertebrata reveal that vertebrates arose from a single ancestral stock and that the trend during the long course of evolution was toward increasing structural complexity and often a more efficient organization. The history can be followed in considerable detail using a number of sources of evidence. Such studies leave little doubt as to the common plan of organization which is basic to all members of the subphylum.

SOURCES OF EVIDENCE

In contrast, finding relationships between the invertebrate phyla presents difficult problems. A common plan of organization is quite readily seen among the members of a given phylum. It is possible to construct reasonable hypotheses which trace the development within a phylum from a single ancestral type, even in the absence of good paleontological evidence. When one attempts to bridge the gap between phyla, however, the relationships are much less clear. These difficulties are largely due to the fact that the evolution of the major invertebrate phyla occurred very early in geologic history. The earliest fossil-bearing rocks of the lower Cambrian contain well-preserved evidence of the presence of many invertebrate phyla—mollusks, annelids, arthropods such as trilobites and crustaceans, and many others. All the major invertebrate groups evolved during the 2 billion or more years of Precambrian times. The rocks of this important period in the history of life on earth have been so changed by crushing, folding, heating, and other means, that little meaningful paleontological evidence can be obtained.

In the absence of a valid fossil record, comparative morphology remains as the main basis for speculation on invertebrate phylogenetic relationships. Such evidence frequently can only be used to arrive at tentative and uncertain conclusions. Present-day forms which are used in such studies are themselves a product of continuing change since the beginning of metazoan evolution.

One of the significant aspects of vertebrate evolution was the appearance of new types of organization and the subsequent radiation of animals embodying these new features into many different environmental situations. Among invertebrates the phenomenon of adpative radiation is seen, perhaps at its best, in the evolution of the insects. There is little evidence to indicate that the various invertebrate phyla developed by a process of adaptive radiation in which a basic type of plan become modified as an adaptation to various environmental situations. It seems more likely that the evolution of the invertebrate phyla occurred by the development of substantially new and different body plans. If relationships exist between two phyla, one must have arisen from the other, or they both evolved from a common ancestral form by the development of new types of organization.

If adult structures of different phyla are used as a basis of comparison, there are commonly so many important differences that meaningful deductions of evolutionary relationships cannot be made. The adult structural features of the members of one phylum seem to be quite distinct from those of all other phyla. In fact, this is the basis for the division of the animal kingdom into phyla. Features which occur in a number of phyla, for example, the triploblastic condition, presence or absence of a coelom, segmentation and other similar features, become of little value because they are not unique.

The most striking resemblances come from a comparison of larval forms, rather than adults. The most pronounced similarities are seen in the very early larval stages. Similarities in larval stages in members of two or more different phyla can mean one of two things. Either the similar features resulted from parallel evolution, or the animals in question evolved from a common ancestor. If the similarities resulted from parallel evolution, they would not serve as evidence for evolutionary relationship. However, the resemblances between larval forms are often too numerous and too detailed to permit acceptance of the parallel evolution hypothesis.

Studies of similarity in developmental stages of mammals, where ample fossil evidence of relationship is also available, indicate the validity of conclusions drawn from comparison of immature forms. Haeckel, with his biogenetic law, stated this concept succinctly in the phrase "ontogeny recapitulates phylogeny." Haeckel suggested that larvae were indicative of the appearance of adult ancestral forms, and that the developmental stages actually told the story of the evolution of the group in question.

Haeckel's concept has been modified in light of more detailed knowledge. It is no longer assumed that evolutionary change occurred only at the end of development. Evolutionary change could have and in fact did take place at any and all stages of the life history. Often the most significant changes did occur in later stages of development, but not exclusively. Therefore when two forms have larval stages which are similar, it indicates that they had a common ancestor with such a larval stage—not that a common ancestor in its adult condition was like the larval stage of its descendants. The adult structure of such a common ancestor, in the absence of fossil evidence, can only be speculated upon.

THE PHYLOGENETIC TREE

The diagram (Fig. 39.1) of the relationships between the phyla which make up the animal kingdom is not by any means the only possible explanation of phylogeny. There are a number of theories, and this diagram embodies several of them. It does give a basis for discussion, and if the student bears in mind the more or less tentative nature of at least certain portions of it, it will be useful.

Protozoa and Lower Metazoa

Some current thinking on the origins of life is presented in Chapter 42. The discussion here will begin with the hypothetical plant–animal ancestor which gave rise to the two great groups of living organisms—the plant kingdom and the animal kingdom.

There is general agreement that the simplest animals are the protozoans and that the most primitive protozans are the flagellates. It is in the flagellate group that organisms with both plant and animal characteristics are found, indicative of common ancestry of the two kingdoms.

The sponges probably were derived separately from the other metazoans from ancestral choanoflagellates. These protozoans have the peculiar protoplasmic collar found also in the collar cells of sponges. The similarity between these specialized cells and their absence in other groups leaves little doubt of the relationship.

As indicated in Chapter 22, the phylum Mesozoa presents problems to the student of phylogeny. The primitive characteristics—internal cell or cells entirely reproductive surrounded by a single layer of cells—may well be the result of degeneration associated with their parasitic existence. In many other groups of internal parasites there are great modifications as compared with free-living relatives. Thus the mesozoans are placed in this position on the basis of their structure, but without any phylogenetic implications.

In Chapter 21 several theories of the origin of the metazoans were discussed. The one accepted for use in this discussion is that of L. Hyman. She proposed that the ciliated planula larva, characteristic of most coelenterates, is representative of a hypothetical ancestor. This *planuloid ancestor* can be readily derived from a colonial flagellate. It could have been either a hollow ball colony like the present-day *Volvox* or, more likely, a type of colony which lacked a central cavity.

The radiate phyla, Coelenterata and Ctenophora, are offshoots of the main line representing an evolutionary dead end. Although their widespread distribution and large numbers are indicative of successful adaptation to their environment, no higher phyla are believed to have been derived from this source.

Origins of the Bilateria

The planuloid ancestor also serves as a possible source of all the rest of the animal kingdom. The present-day acoel flatworms are not greatly different from a simple planula, and such a form could have been the next step in the phylogenetic scale. Although mesoderm appeared in the

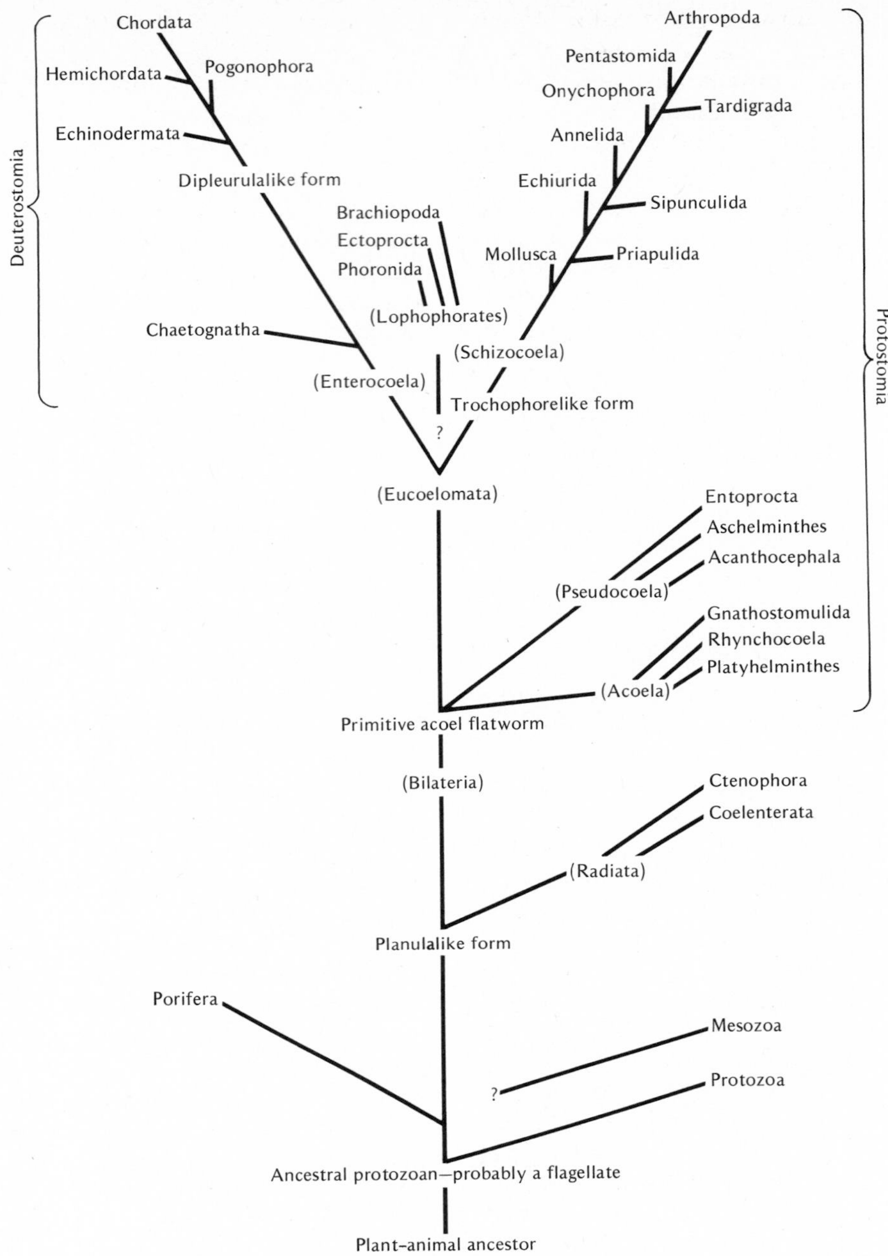

Figure 39.1 Phylogenetic tree.

scyphozoans and became more prominent in the anthozoans, these forms still are basically at the tissue level of organization. The mesoderm did not become a truly significant layer until the platyhelminthes appeared with their well-developed organs and organ systems. All the higher phyla possess the middle germ layer, and it is in this layer that the major advances in the development of complex organ systems occurred.

The bilaterally symmetrical acoel ancestor gave rise to the present day acoel flatworms which in turn were the source of the acoelomate phyla, the Platyhelminthes and the Rhynchocoela. The rhynchocoels have clearly defined affinities with the platyhelminthes as indicated by many common structural features, but are more advanced with the appearance of the second opening into the digestive tract, the anus. The newly established phylum Gnathostomulida also is placed with the acoels, even though it has some features which indicate affinities with the pseudocoelomate phyla.

The pseudocoelomate phyla could also have arisen from the primitive acoel flatworms rather than from any of the more advanced flatworms

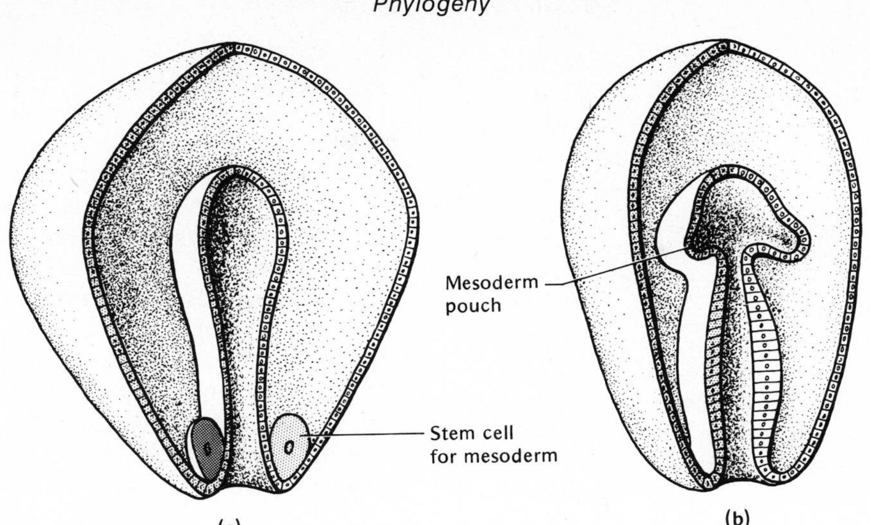

Figure 39.2 Comparison of the method of mesoderm formation in protostomes and deuterostomes. **(a)** Protostome method. **(b)** Deuterostome method.

or the rhynchocoels. Further studies may point to the gnathostomulids as the link between the acoels and the pseudocoels. The acanthocephalans with their highly specialized parasitic modifications are separated from the aschelminthes, but have close affinities with that group. Similarly, the pseudocoelous entoprocts, probably most closely related to the rotifers, are placed in a phylum by themselves.

The third offshoot from the acoel ancestor produced the eucoelomate animals with their completely lined body cavity. It seems likely that large complex animals could not have developed without the appearance of the coelom. Such a fluid-filled cavity was necessary for the development of more complex musculature which could use such a cavity as a hydrostatic skeleton.

The method of coelom formation is the basis for the split in the phylogenetic tree at this point into two main lines of development, one, the schizocoels, culminating in the arthropods, and the other, the enterocoels, culminating in the chordates.

The Protostomia

The acoela, pseudocoela, and schizocoela have a number of features in common which indicate a close phylogenetic relationship. They are called the *Protostomia* and have in common the following features: spiral, determinate cleavage; mesoderm derived in large part from teloblast cells set aside early in development [Fig. 39.2 (a)]; the coelom arising as a split in a solid mass of mesodermal cells; the embryonic blastopore or the surrounding region becoming the mouth of the adult; the trochophore larva (Fig. 39.3) is characteristic of the schizocoels and it is foreshadowed by a number of similar larval types in the acoels and the pseudocoels.

The trochophore larva and other protostome features serve to link the several phyla in the line leading to the arthropods. The appearance of segmentation in the annelids is foreshadowed by the segmented mollusk, *Neopilina*. The relationship between the annelids and arthropods is clearly seen in the onychophorans, which possess some characteristics of each of these two phyla (Ch. 28).

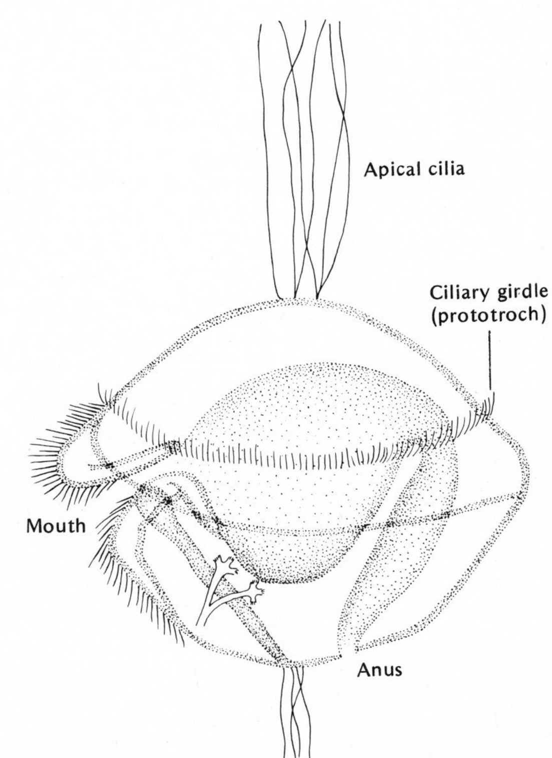

Figure 39.3 A trochophore larva. Characteristic of many schizocoelous protostomes.

The Deuterostomia

Members of the other main line of development, culminating in the phylum Chordata, are called the *Deuterostomia.* Deuterostomes have in common the following characteristics: radial, indeterminate cleavage; mesoderm and coelom arising as outpocketings of the primitive gut [Fig. 39.2 (b)]; the embryonic blastopore or the surrounding region becoming the anus of the adult; trochophore larvae are absent but other larval similarities are found in some of the phyla.

The lophophore phyla present a number of problems since among them are found certain deuterostome characteristics and also certain protostome characteristics. This may indicate their origin from an ancestral type which gave rise to both the deuterostomes and the coelomate protostomes. It may even be that they are the link between the two groups with the deuterostomes evolving from a lophophorate ancestor. The protostome characteristics of the lophophorates include the fate of the blastopore which becomes the mouth in phoronids and branchiopods, and the presence of a trochophorelike larva in all three. Coelom formation is enterocoelous in some brachiopods, clearly a deuterostome characteristic. Only the phoronids have spiral cleavage, and the others are more nearly radial. An interesting fact pointed out by Steiner is that the lophophorates differ from other protostomes in their complete loss of cephalization. It is tempting to suggest that this lack of cephalization was passed on to the other deuterostomes and that the development of a head in the vertebrates constitutes a new feature, independent of the head development seen in the protostomes.

Looking now at the deuterostome side of the tree, the chaetognaths appear first in the line of development. Examination of adult features of these organisms reveals characteristics mostly allied to the pseudocoelomate aschelminths. The body wall has only longitudinal muscles, and the intestinal wall lacks muscles entirely as in the nematodes. The intestine is not covered with the regular coelomic epithelium. The nervous system is quite similar to that seen in the nematodes. The only structures in adult chaetognaths which are typical of coelomate forms are the ciliated coelomate funnels of the sperm ducts. It is the developmental history of the chaetognaths which reveals the reason for placing them with the deuterostomes. Cleavage is holoblastic, radial, and indeterminate. Coelom and mesoderm are formed by a modified enterocoelous method. There is no larval form, making comparison of this feature impossible. Taking all these facts into account, it seems clear that the chaetognaths evolved from a primitive deuterostome ancestor very shortly after it diverged from the protostomes. This would account for the retention of certain primitive characteristics.

It is postulated that the rest of the deuterostome line evolved from a primitive ancestral form called a dipleurula, similar to the larva of some echinoderms. In addition to possession of the deuterostome characteristics mentioned above, the echinoderms, although radially symmetrical as adults, are bilaterally symmetrical in their larval stages. The echinoderm bipinnaria larva is remarkably similar to the tornaria larva of the hemichordates (Fig. 39.4).

The hemichordata have all of the deuterostome characteristics and they are placed between the echinoderms and the chordates. As mentioned in the discussion of the phylum, the position of the pogonophora, formerly thought to be related to the hemichordates, is now in question. Until further evidence is available, they are left in this position.

The larval resemblances mentioned above link the echinoderms with the hemichordates. The gill slits present in some members of the phylum as well as a dorsal hollow nerve cord link the hemichordates with the phylum Chordata. This is not to say that the echinoderms or the hemichordates are ancestral to the chordates, rather that a common ancestral type gave rise to all three.

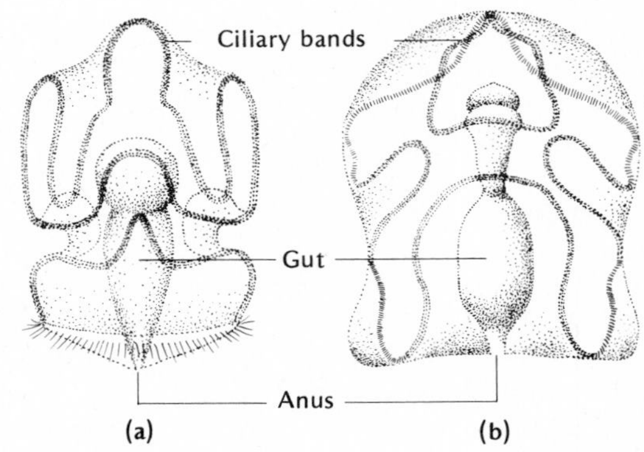

Figure 39.4 Comparison of larvae. **(a)** Tornaria larva of a hemichordate. **(b)** Echinoderm larva.

SUGGESTIONS FOR FURTHER READING

Barnes, R. D. *Invertebrate Zoology,* 3rd ed. Philadelphia: W. B. Saunders (1974).

Clark, R. B. *Dynamics in Metazoan Evolution.* Oxford: Clarendon Press (1964).

Dougherty, E. C. (Ed.). *The Lower Metazoa.* Berkeley, CA: Univ. of California Press (1963).

Hadzi, J. *The Evolution of the Metazoa.* New York: Macmillan (1963).

Hyman, L. H. *The Invertebrata,* vols. 1–6. New York: McGraw-Hill (1940–1967).

Parker, T. J., and W. A. Haswell. In *Textbook of Zoology, Invertebrates,* 7th ed., A. J. Marshall and W. D. Williams (Eds.). New York: American Elsevier (1972).

The Dynamics of Species

In Part V the focus is on evolution, behavior (primarily innate), and ecology. It will be obvious in the discussions that genetic information plays an important role in our understanding of evolution, behavior, and ecology.

Evolution is a central thread that ties all areas of zoology together. Throughout the book many references are made to evidences of evolution of animals and humans, utilizing a wide variety of fields: comparative morphology, comparative embryology, comparative biochemistry, paleontology, taxonomy, and geographical distribution. Many of the problems in geographical distribution in the past have been explained in recent years by the acceptance of the concept of continental drift which is discussed. Recent discoveries have added to our understanding of human evolution.

In considering the mechanisms that operate in the process of evolution, the contributions of Lamarck and Darwin are outlined first. At the base of evolutionary thought is the concept of population (the members of a species) as the unit of evolution. The modern concept of the mechanisms of the process focuses on the following: the nature and stability of the genetic material, mutations, sexual reproduction with its genetic recombinations, natural selection, reproductive isolation, and several other factors of lesser importance. Most discussions of evolution focus exclusively on the origin of new species. Here, time and extinctions are considered as playing roles in the formation of the higher taxonomic categories.

The human, although a part of the whole process of biological evolution, is unique in that a second evolution, cultural evolution, has been superimposed on his biological evolution.

Behavior, both innate and learned, is the product of a long evolution. Learned behavior was discussed in Part II. In Part V the focus is on innate behavior. Some of the genetic, evolutionary, and environmental aspects of such behavior are pointed out.

Ecology deals with the study of relationships between organisms and their environment. While important in itself, this study of ecology is also fundamental to the understanding of evolution. For a long time ecology focused on plant and animal communities without considering humans as a part of the system. But in recent years we have come to realize that humans are very important parts of such systems. Only recently have we begun to realize the need for greater conservation of our energy sources.

Part V closes with a discussion of the world and the future of humanity.

A portion of the coral reef community, the Great Barrier Reef, Queensland, Australia. (Photo – E. C. Williams)

40 Evidences of Evolution

According to the concept of organic evolution, the various kinds of living things in the world today have been (and are being) derived from previously existing ancestral forms by descent with modification. Although the studies thus far made in this field do not reveal the exact nature of the origin of life, they do indicate that once simple forms of life appeared on the earth, there followed, during millions and millions of years, a gradual development of many kinds of organisms with increasing complexity of structure. Numerous references have been made to this process in other sections of the text. It is now time to examine the concept in some detail. Although the idea of evolution is a very old one, it has had widespread acceptance only in the past hundred or so years. All aspects of biological study contribute to an understanding of evolution, and the concept is certainly one of the most important unifying principles in the whole field.

In a study of organic evolution it is necessary to distinguish between two aspects of the study: *fact* and *explanation.* The many evidences of evolution constitute the fact; ideas about the mechanisms involved in the process, which include the various theories of evolution, constitute the explanation, or attempted explanation. After considering some of the more important sources of evidence of evolution, some of the interpretations of the causal mechanisms will be discussed.

EVIDENCE FROM COMPARATIVE ANATOMY

Studies in comparative anatomy provide many evidences of evolution. In such a study we are concerned with *homologous structures*—structures that have the same general arrangement of parts and which arise in a similar way from similar embryonic structures. Homologous structures may be quite diverse in function. The skeletal parts of vertebrate appendages may be used to illustrate homologous structures.

The forelimbs of vertebrates will be used to illustrate the point, but the hind limbs would serve equally well. An examination of Fig. 40.1 shows that all the different forelimbs have the same general arrangement of parts. The primitive vertebrate forelimb is a five-fingered (pentadactyl) appendage having a *humerus, radius* and *ulna, carpals, metacarpals,* and *phalanges.* All vertebrate forelimbs are modifications of this primitive type. In the bat the metacarpals and phalanges have been greatly elongated for the support of the membranous wing. In the whale, the whole structure is greatly shortened and thickened to serve as a flipper. The great modifications in the bird are found in the metacarpals and phalanges. The only digit left in the forelimb of the adult horse is the third. The enlarged nail forms the hoof, the radius and ulna are fused together, and the metacarpal of the single digit is enlarged and elongated to form the *cannon bone.* Closely attached to this bone are the two *splint bones,* which are the reduced metacarpals of the second and fourth digits.

In the course of vertebrate evolution, many modifications have occurred which represent adaptations for special modes of existence. Some vertebrate forelimbs are adapted for grasping, some for running, some for flying, and some for swimming, but all are constructed upon the same basic pattern. To the biologist these homologous structures indicate relationship. All vertebrates have this basic limb pattern because they inherited it from an ancestor that had pentadactyl (five-digit) limbs.

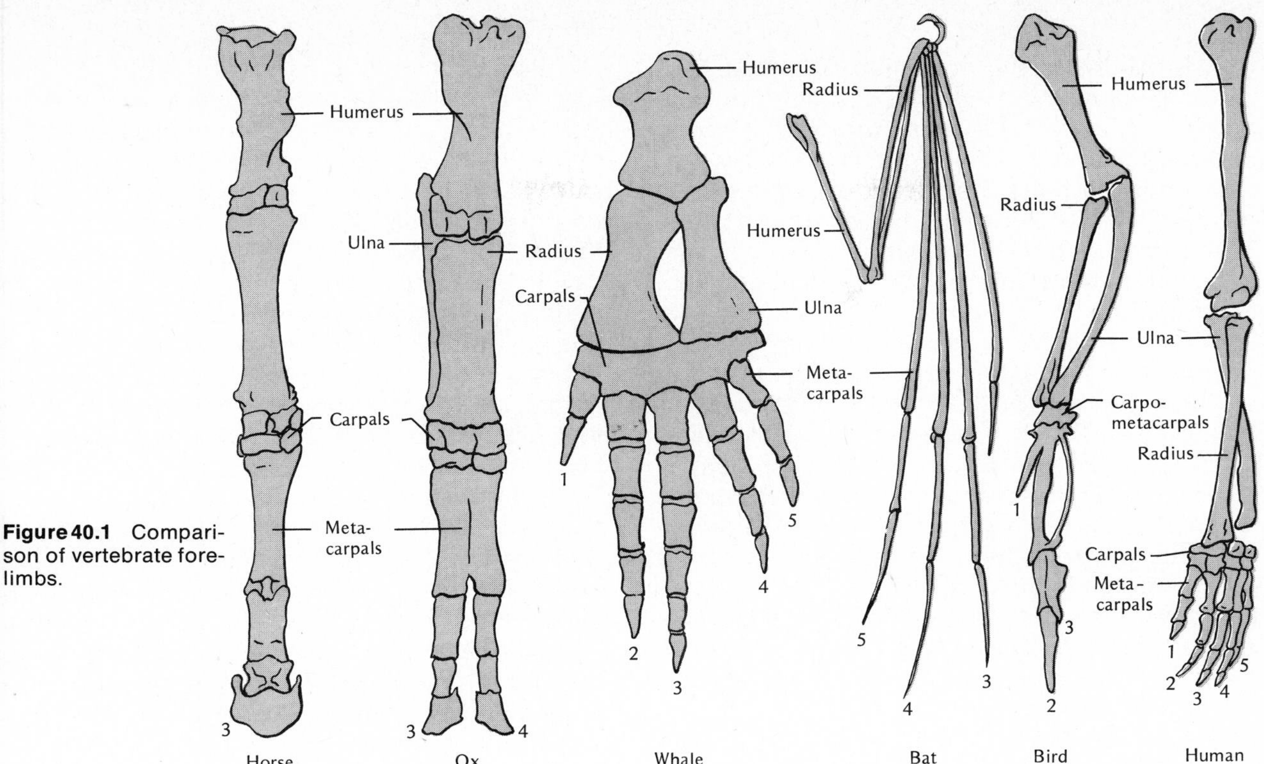

Figure 40.1 Comparison of vertebrate forelimbs.

All insect legs are composed of five parts that always occur in the same order. Starting at their attachment on the thorax, these parts are coxa, trochanter, femur, tibia, and tarsus. These parts are single segments except the tarsus, which varies from one to five segments. There have been many adaptive changes in insect legs during the evolution of the group. Many insects have a simple type of walking legs, such as the first and second legs of the grasshopper. Figure 40.2 shows the middle leg of a grasshopper. The hind leg of the grasshopper, however, is modified for jumping. The forelimbs of a mole cricket are modified for digging; the legs of a hog louse are adapted for clinging to hairs; the legs of a diving beetle are constructed for swimming; and the claws, pulvilli, and hairs of the tarsi of the common housefly make it possible for the fly to walk upside down.

There are many other modifications of the primitive-type insect leg, but perhaps the most interesting series of special adaptations is that found on the legs of the worker honey bee. The series of special adaptations found on these

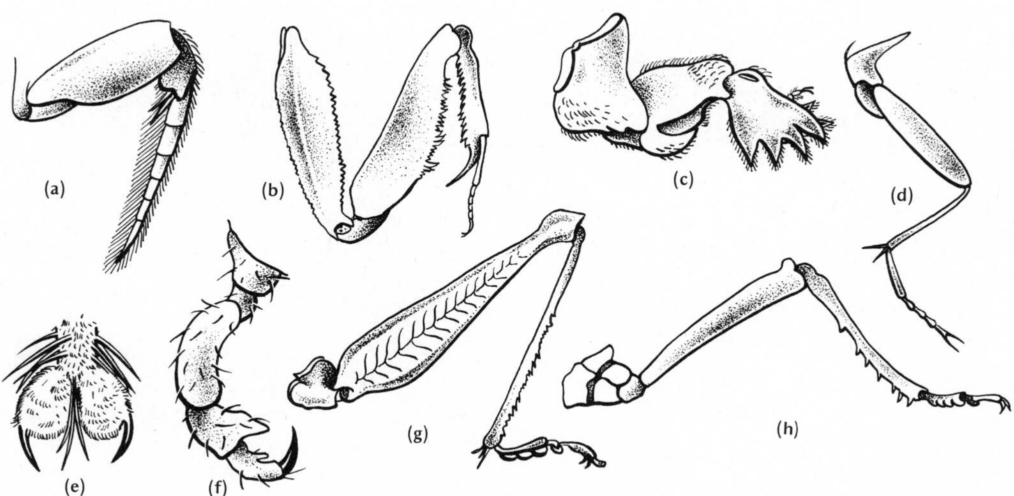

Figure 40.2 Modifications of insect legs. **(a)** Paddlelike hindleg of a water beetle. **(b)** Grasping organ, the foreleg of a praying mantis. **(c)** Digging foreleg of a mole cricket. **(d)** Hind leg of a ground beetle, for running. **(e)** The foot of a house fly has claws and sticky climbing pads; **(f)** The foreleg of a head louse is adapted for grasping hair. **(g)** Hind leg of a grasshopper, for jumping. **(h)** Relatively unspecialized middle leg of a grasshopper.

legs functions in the collection of pollen. This material serves as a source of protein for both adults and larvae. The undersurface of the body and the basal segments of the legs are covered with plumose hairs that become covered with pollen as the bee collects nectar from flowers. The special modifications of the legs serve to concentrate the scattered pollen into a compact mass (Fig. 40.3).

Each *foreleg* has a fringe of short stiff hairs on one edge of the inner surface of the tibia that serves as an *eye brush* for cleaning the compound eyes. The first segment of the tarsus of all of the legs is enlarged, and this segment of the forelegs is covered with long hairs that serve as *pollen brushes* for the removal of the pollen from the plumose hairs on the forepart of the body. A notch in this first segment of the tarsus, together with a spur on the end of the tibia, constitutes the *antenna cleaner.* The bee removes the pollen that adheres to the antennae by pulling them through these notches.

The *middle leg* is less specialized than the foreleg. The first tarsal segment is covered with long hairs that serve as a *pollen brush* for removing pollen from the forelegs and the thorax. The *spine* on the end of the tibia functions in the removal of wax from the wax glands on the ventral region of the abdomen.

The most specialized legs in the series are the *hind legs.* The inner surface of the first tarsal segment bears several rows of *pollen combs* that are used to comb out the pollen from the middle legs and the posterior part of the abdomen. A single stout comb, the *pecten,* is located at the end of the tibia. This is used to remove the pollen from the combs of the opposite leg. The pollen so removed falls on the flattened upper end of the first tarsal segment. This flattened region is called the *auricle.* When the leg is flexed, the auricle is pressed against the end of the tibia; this action compresses the pollen and pushes it into the *pollen basket* on the outer surface of the tibia. A concavity in the outer surface of the tibia forms the bottom of the basket and long curved hairs form its sides.

Any interested person, in a place where honey bees are active, can easily find a worker bee with a large lump of pollen in each of its baskets. When the baskets are filled, the bee returns to the hive and places the pollen in special cells. This complex series of special adaptations is one of the most remarkable to be found anywhere in the animal kingdom.

Not only have the legs of insects undergone many modifications in their evolution, but so also have the mouth parts. During arthropod evolution the mouth parts have been derived from the primitive type of appendage. The more primitive insects, such as the grasshopper, have *chewing* mouth parts. Three pairs of head appendages form the mouth parts (Fig. 30.53): the *mandibles,* the *maxillae,* and the second maxillae fused to form the *labium.* Many variations of the

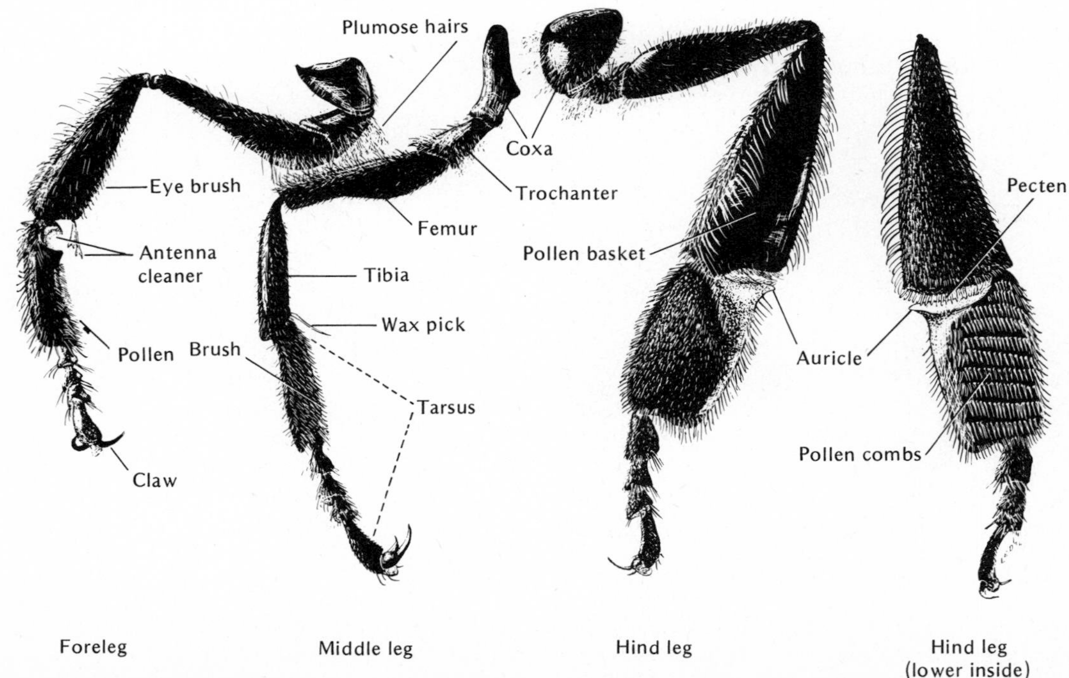

Figure 40.3 Legs of the honey bee.

primitive chewing mouth parts are found in the different insect groups. The honey bee has a chewing-lapping type of mouth parts. The mandibles are similar to the chewing type, but the maxillae and labium are elongated to form a sort of lapping tongue. A variety of forms, including lice, mosquitoes, and aphids, have *piercing-sucking* mouth parts. There are several structural variations of this type, but they all follow the same general pattern. The labium forms a long protective tube for the other parts. The mandibles and maxillae are long slender structures that serve in piercing the skin of an animal or the epidermis of a plant, and also serve as the food channel up which the liquid is drawn. In the strictly *sucking* type of mouth part, found in butterflies and moths, the mandibles are usually absent and the labium is represented only by the labial palps. The functional structure is formed by the two elongated maxillae. Each maxilla forms a half tube, and when the two are together, they form a long sucking proboscis. This proboscis is coiled under the head except when feeding.

There are other modifications of insect mouth parts, but these will suffice to indicate that many adaptive changes have occurred. Throughout the evolution of the arthropods, many modifications of the basic biramous appendage have occurred. Each of these modifications which persisted enabled its possessor to inhabit a different habitat and to lead a different type of existence, that is, the modification was adaptive. Within a given group, such as the insects, the many different modifications of legs and mouth parts resulted in a great variety of forms adapted to life in different habitats. Such a process is called *adaptive radiation* or *divergent evolution* and is found many times in the course of evolution in many different groups.

Another source of evidence of evolution from comparative anatomy is found in *vestigial,* or *rudimentary,* structures. These are reduced and generally useless structures that are found in many plants and animals, relatives of which often have the structures in a fully developed and functional condition. The snakes as a group are limbless vertebrates, but in the group to which the pythons and boas belong, the members have vestiges of a pelvic girdle and hind limbs. Externally there is no sign of hind limbs in the whale, but embedded in the blubber are the much-reduced bones of the pelvic girdle and hind limbs. There are more than 100 vestigial structures in humans. Some of our vestigial structures are ear muscles, tail (coccyx), tail muscles, and possibly the appendix.

In the study of comparative anatomy, structures are often found that have the same function and are superficially alike, such as the wing of a bird and the wing of a butterfly, yet are quite different in origin and in structural design. Such structures are said to be *analogous.* They have arisen in the evolutionary process through adaptations of quite different organisms to similar modes of life.

The flying adaptation was developed independently in three different groups of vertebrates: reptiles, birds, and bats. The pterodactyl, an extinct flying reptile, had a wing formed by a fold of skin stretched between the body and the posterior surface of the forelimb and a greatly elongated outer digit. In birds the feathers are inserted on all three major segments of the forelimb. The wing of the bat is composed of a membrane extending from the body to the forelimb, where its main support is formed by the four very elongated digits. When a comparison of the bones of the forelimbs of these three flying vertebrates is made, they are seen to be homologous. However, as organs of flight, they are *only* analogous; different structures have been involved in adaptations for flight.

There are many instances where different organisms that are not very closely related have developed similar adaptations for life in similar habitats. This phenomenon is often referred to as *convergent evolution.* The development of the same fusiform type of body for life in the water by a shark (fish), the extinct *Ichthyosaurus* (reptile), and a dolphin (mammal) illustrates this very well.

EVIDENCE FROM DEVELOPMENT

The embryos of higher animals repeat many of the stages through which embryos of lower animals have passed. This has been referred to as *recapitulation.* This concept, as originally used by von Baer, indicated that some of the developmental stages of an organism are similar to some of the developmental stages of its ancestors. Unfortunately, however, Haeckel modified the concept to mean that in its development, the individual passes through stages like the adult stages of its ancestors. Modern students of development insist that Haeckel's version is wrong. As a matter of fact, our present knowledge of the heriditary mechanisms tends to support the views of von Baer. Several examples of recapitulation have been described in a previous section. In the development of any mammalian embryo, the heart is a four-chambered in-series structure, as it is in fish embryos; then it develops partitions of the auricles (atria) similar to those of amphibian embryos, followed by ventricular division that is incomplete for a period, as it is in the embryos of reptiles. The existence of a common trochophore larval stage in the development of many mollusks and many annelids is also typical of this process.

There are many examples of recapitulation during embryonic development, but one of the best is afforded by a comparison of different vertebrate embryos at comparable stages in development (Fig. 40.4). In the first stage in the illustration all the embryos look very much alike. All have similar segmentally arranged *somites* and similar *pharyngeal arches* and *pharyngeal clefts.* In the second row the *limbbud* primordia are forming on all embryos in a similar way, and all of them have embryonic tails. The embryos of the lizard, chick, pig, and human have strong resemblances, yet those of the fish and the salamander are beginning to assume recognizable forms. At this stage, gills have formed from the tissue lining the gill clefts of both the fish and the salamander. Later each embryo has developed features that indicate fairly clearly its definitive nature.

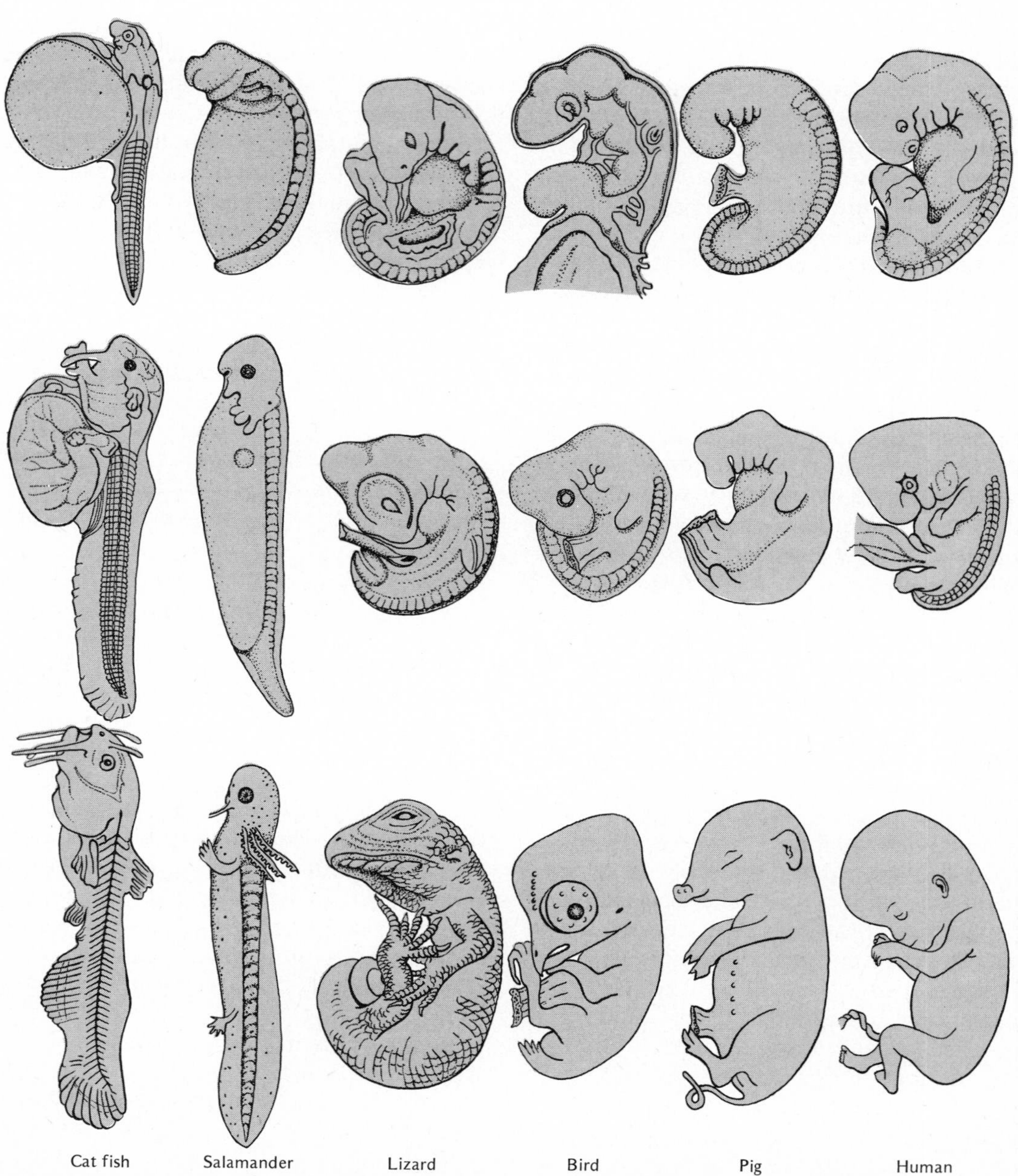

Figure 40.4 Comparison of vertebrate embryos at early, middle, and late developmental stages.

Why, it may be asked, do the embryos of reptiles, birds, and mammals develop similar pharyngeal modifications without ever having functional gills? The best answer to this seems to be the same one that was given as an explanation of the presence of homologous bones in the appendages of adult vertebrates. They have developed from a common ancestor and they possess a common heredity for the manifestation of these characteristics, and their persistence usually means they have adaptive value. For instance, although gills are not formed, the pharynx is a site of similar derivatives in all vertebrates.

EVIDENCE FROM PALEONTOLOGY

Paleontology is the study of the life of past geologic ages, and it is based on the fossil remains of ancient plants and animals. The evidence of evolution provided by fossils is the most forceful and direct evidence of evolution which we have. A *fossil* is some evidence of an animal or plant that lived a long time ago.

Kinds of Fossils

In the main, the hard parts of organisms (teeth, skeletal parts, and shells) are preserved as fossils. Occasionally animals are preserved with little or no change from the time of death. In 1929 a Russian worker described the finding of a frozen mammoth (an extinct form related to elephants) in Siberia. It was estimated that the animal had been preserved in the frozen state for approximately 25,000 years, yet the flesh was so well preserved that it could be, and was, eaten by dogs. Numerous specimens of insects, spiders, and mites have been found preserved in amber in the Baltic region of Europe. During the Oligocene period, about 38 million years ago, northern Europe was covered with coniferous forests. The trees in these forests exuded a sticky resin that trapped the spiders, mites, and insects. The resin ultimately hardened into amber, with the arthropods embedded in it. In some cases the preservation is so good that the colors do not seem to have changed. Land animals occasionally have been covered with windblown sand or volcanic ash, or have been trapped in bogs, quicksand, or asphalt pits, and their hard parts have been preserved.

Still other fossils are in the form of *molds,* or *casts.* The shells, bones, and other parts of organisms may have become embedded in mud or silt that later became rock. In time the entire original specimen may have dissolved, leaving a cavity. The wall of this cavity contains an impression of the exterior of the original structure, and this constitutes a natural mold. If some plastic substance is pressed into such a mold, the external replica of the original object can be obtained. Natural casts are formed in a slightly different way. If a clamshell, with its two valves closed, is buried in mud or sand, a natural mold may be formed, as has just been described, or a natural cast may be formed. If, in the burial process, mineral matter fills the cavity and then the shell is completely dissolved, later leaving the mineral matter in the cavity in a hardened condition, a natural cast will be formed.

The two most important conditions favoring fossilization are the possession of hard parts and immediate burial. It should be clear that when a dead animal or plant is left exposed, it soon disappears as the result of the activity of scavengers and the bacteria and fungi of decay. Even the bones of horses and cattle that die on the plains soon disappear. The chance of any exposed organism becoming fossilized is very slight indeed. Immediate burial is a prerequisite to the process; by far the most common kind of burial favorable for fossilization is that provided by waterborne sediments.

Sedimentary Rocks

To understand the nature of the formation of *sedimentary rocks,* one must take into account certain natural processes that are occurring in the world today and that have been occurring since the crust of the earth was formed. By the process of *erosion,* through the action of wind and rain, and of freezing and thawing, rocks are gradually broken into the small particles that form soil. Through the action of rain the particles of the soil are carried into streams and rivers and ultimately into lakes or oceans. This sedimentary material carries with it the bodies of many aquatic organisms and also the bodies of terrestrial forms that happen to be swept along by the streams and rivers. The hard parts of some of these organisms may be preserved, and in the course of time the sedimentary deposit, owing to the pressure of the water above it and also to chemical reactions, is converted into sedimentary rock. The nature of the sedimentary material determines the kind of sedimentary rock formed; limestone, sandstone, and shale are familiar kinds. The method of formation of sedimentary rocks clearly distinguishes them from *igneous rocks,* which were formed by the solidification of molten material when the earth cooled, and are being formed by cooling of magma expelled from active volcanoes.

An understanding of the layers of sedimentary rock found in many places on the earth must take into account another natural process, the changes that have occurred in the past and that are still occurring between land and sea. Slow and gradual changes in level between the land and the sea are now in process. For instance, it is now recognized that the land on the eastern coast of the United States is gradually sinking into the sea, whereas the land on the Pacific coast is gradually rising. The sinking of land below the sea is called *submergence*, and the rising of land above the sea is called *emergence.* In the geologic past, many regions of the earth have undergone a series of submergences and emergences. As a consequence, in many places a whole series of different layers of sedimentary rocks is found. A good place to see such a series of layers, or strata, is the Grand Canyon of the Colorado (Fig. 40.5).

Just as a person looks at a brick wall and realizes that the bottommost tier of brick was the first laid by the brick mason and the upper tiers were last in order, so must a person who views the mile-deep series of strata in the Grand Canyon realize that the deepest layers were deposited first and the other layers in succession at later periods of time. This simple concept in geology is called the *law of superposition.* If each formed layer of rock were left undisturbed until another layer was deposited on top of it, one would have a perfect series for the study of fossil forms. The rock record of the earth, however, is not so complete. Layers deposited under water can emerge as land and be partially or completely eroded away. If a new submergence then occurs and a new layer of sedimentary rock is formed, there will be an unconformity between the two strata. This lack of sequence between layers makes the study and identification of the strata difficult, but over 150 years ago in England, William Smith determined that each stratum is characterized by certain index fossils. Thus it is possible to identify similar strata in different parts of the world.

Another factor complicating the interpretation of the geologic record involves the numerous foldings and splittings that have occurred in the earth's crust. Mountain ranges are formed by such foldings, and quite often a split in the fold occurs with a thrust of part of the fold over the rest. As a result of this, older strata are found to lie over younger ones. Also, fossils that were once formed may be destroyed in the formation of *metamorphic rock.* As a result of great pressure and heat, deep layers may melt. When this material later solidifies again, the fossils originally present will usually be lost. Limestone is a form of sedimentary rock rich in fossils. When limestone melts and crystallizes into metamorphic rock the result is marble.

In spite of the paucity of fossil formation, the unconformities in the rocks, the overthrusts,

Figure 40.5 Grand Canyon of Colorado River showing many layers of stratified rock. (From Union Pacific Railroad, with permission.)

the destruction of fossils by erosion and by the formation of metamorphic rocks, and other factors not mentioned here, the story of the rocks is a very convincing one with reference to evolution.

Determining the Age of Rocks

In the past, geologists and paleontologists were able to make fairly accurate estimations of the age of different rock strata by using the known rate of the accumulation of salt in the oceans. More recently the use of radioactive elements has provided a better method. The element uranium changes into an isotope of lead and helium through a long series of transformations. The rate of this change is known, and the rate of change is independent of the conditions under which it occurs. This rate is such that 7,600,000 g of uranium yield 1 g of lead per year. Thus when a piece of igneous rock is found that contains both uranium and lead, its age can be determined. If sedimentary rock containing fossils is associated with this igneous rock, its age is assumed to be the same. Only a few rocks have been dated by this method, but fortunately they are widely scattered in geologic time. The oldest rocks that have been dated by radioactive measurements are over 3000 million years old. Others are Cambrian deposits over 500 million years old, Permian deposits about 270 million years old, and Eocene deposits about 55 million years old (Table 40.1).

Recently the transformation of radioactive potassium to argon and rubidium to strontium has been used in a similar way for dating fossil-bearing rocks. By the mid-1960s, rocks from a number of localities were dated at approximately 3500 million years. There are indications that still older rocks exist. The exact age of the earth is still undetermined, but evidence from several sources suggests an age in the neighborhood of 4500 million years.

For very recent fossils, 20,000 years old or less, the decay of carbon-14 has been very useful. All organisms contain a relatively constant proportion of this isotope, and upon death no additional isotope is incorporated. All organisms are constantly incorporating ^{14}C in life in a balance equal to that in the atmosphere and at death ^{14}C starts to decay to ^{12}C with no new ^{14}C added. Measurement of the remaining isotope gives an estimate of the time since death.

The Geological Timetable

Geologists, as a result of their studies of the strata of sedimentary rocks in the different regions of the world, have classified geologic history into six *eras.* The oldest era with fossils is the *Archeozoic* (era of primitive life) and this is followed in turn by the *Proterozoic* (era of early life), the *Paleozoic* (era of ancient life), the *Mesozoic* (era of medieval life), and the *Cenozoic* (era of modern life). The Paleozoic, Mesozoic, and Cenozoic eras are divided into *periods,* and the periods of the Cenozoic into epochs. There is evidence that between the different eras there were widespread geologic disturbances called *revolutions.* In some of these revolutions a large part of the existing forms of life was destroyed. Table 40.1 shows the eras and some of their subdivisions, the approximate duration of each era, some of the important geological features, and the characteristic animals and plants. Plants are added to make the period complete.

It is not possible to describe the many interesting aspects of the fossil record here. However, a few of the outstanding things should be mentioned. The Mesozoic era, often called the "Age of Reptiles," is interesting for a number of reasons. The largest animals that have lived on land, the dinosaurs (Fig. 40.6), were dominant during this era. Many different orders of reptiles evolved and flourished, but only four kinds—lizards, snakes, crocodiles, and turtles—persisted into the Cenozoic. It is clear that both the birds and mammals evolved from reptilian ancestors during this era. The earliest fossil bird, called *Archaeopteryx,* was found in the rocks of the Jurassic period (Fig. 40.7). It was about the size of a crow and in certain respects quite like a reptile. The tail, quite unlike that of modern birds, was long with a row of feathers on each side. The jaws were equipped with conical teeth. The wings were small relative to body size, and three of the digits on each forelimb persisted, armed with claws. It is probable that the forelimbs were used for climbing as well as for flying.

John Ostrom of Yale University has concluded that every bird in the world today is a lineal descendant of the coelurosaurs, an infraorder of small bipedal carnivorous saurischian dinosaurs that flourished from late Triassic times until the end of the Cretaceous. Ostrom has examined the four existing fossils of *Archaeopteryx* and has concluded that were it not for the fact that clear imprints of feathers accompany the four skeletons, all four would probably have been identified as coelurosaurs. It is his opinion that those forms which possessed feathers at that time in geologic history survived, while those that were bare-skinned did not, because the feathers gave them insulation from extreme heat and cold.

The mammals of the Mesozoic were generally small and inconspicuous when compared with

TABLE 40-1 GEOLOGICAL TIMETABLE

Eras (Millions of Years Ago)*	Periods	Epochs	Geological Features	Plants	Invertebrates	Vertebrates
Cenozoic (70)	Quaternary (2)	Recent (0.01) Pleistocene (2)	Periodic glaciations			
	Tertiary (68)	Pliocene (10) Miocene (25) Oligocene (35) Eocene (55) Paleocene (70)	Climate warm in the beginning but gradually cooling. Formation of Alps and Himalayas.	Development and spread of modern flowering plants. Rise of grasses. Rise of herbs.	Arthropods and mollusks most abundant. Appearance of modern invertebrate types.	Recent epoch marked by man. Archaic mammals declined after Eocene. The order of modern mammals evolved in the later epochs. Rise of anthropoids.
Mesozoic (230)	Cretaceous (135)		Great swamps in early part. In late part, Rocky Mountains and Andes formed.	Rapid development of angiosperms	Extinction of ammonites. Spread of insects.	Extinction of dinosaurs. Spread of birds. Rise of primitive mammals.
	Jurassic (180)		Great continental seas in western U.S. and parts of Europe.	Conifers and cycads dominant. First angiosperms.	Maximum of ammonites. Insects abundant, including social insects.	Dominance of dinosaurs. First birds. Early mammals.
	Triassic (230)		Climate warm. Great desert areas.	Spread of cycads and conifers. Seed ferns disappear.	Limulus present. Marine invertebrates decline in numbers	First dinosaurs. Mammal-like reptiles.
Paleozoic (600)	Permian (280)		Appalachians and Urals formed. Glaciation and aridity.	First cycads and conifers.	Last of trilobites. Expansion of ammonites.	Expansion of reptiles.
	Pennsylvanian (320)		Mountain building. Great coal swamps.	Extensive coal formations in swamp forests.	First insect fossils.	First reptiles.
	Mississippian (345)		Warm humid climate. Shallow inland seas.	Lycopsids, horsetails, and seed ferns dominant. First coal deposits.	Culmination of crinoids.	Spread of sharks. Rise of amphibians.
	Devonian (405)		Emergence of land. Some arid regions.	Lycopsids, horsetails, ferns, seed ferns. First forests.	Brachiopods flourishing. Decline of trilobites.	First amphibians. Rise of fishes.
	Silurian (425)		Mild climate. Great inland seas. Taconic Mountains.	First known land plants, psilopsids.	Corals, brachiopods, eurypterids. First land invertebrates, arachnids.	Rise of ostracoderms (primitive fishes).
	Ordovician (500)		Great submergence of land. Mild in arctic regions.	Marine algae.	Climax of trilobites; cephalopods, corals, starfishes.	First vertebrates, armored fishes.
	Cambrian (600)		Mild climate. Lowlands and inland seas.	Algae, especially marine forms.	Trilobites dominant. All phyla represented.	
Proterozoic (1200)			Rocks chiefly sedimentary. Glaciation. Grand Canyon.	Few fossils [sponges, protozoa (Radiolaria), worm burrows, algae]. Thallophytes and most invertebrate phyla probably evolved.		
Archeozoic (3500)			Few sedimentary rocks. Rocks mostly igneous or metamorphosed.	Indirect evidence of life from graphite and limestone, but no recognizable fossils except possibly bacteria.		
Azoic (4500)			Igneous rocks.	No life present.		

* Numbers given in parentheses indicate approximate time since beginning of era, period, or epoch.

the large reptiles. However, with the extinction of all the very large reptiles at the end of the Mesozoic, the Cenozoic era is marked by the great adaptive radiation of mammals, and it is often called the "Age of Mammals." The rocks of the Cenozoic era furnish, in many instances, very detailed records of the evolution of different mammalian lines, including those leading to our modern horses, elephants, and camels.

Evolution of the Horse

Figure 40.8 illustrates some of the stages in the evolution of the modern horse. The earliest known horse, *Eohippus,* lived in the Eocene epoch. It was about the size of a fox terrier dog, but with a longer head. Its legs were short, with four toes on each front foot and three on each hind foot. The third digit was somewhat longer than the others. All the toes were placed on the ground and used in walking.

Eohippus was a forest dweller, subsisting on soft vegetation. The molar teeth were much like human molars. This animal lived in North America during the Eocene and migrated to Europe during the same epoch. A number of different lines evolved from *Eohippus,* but only some of

Figure 40.6 Skeleton of dinosaur from Cretaceous rocks. (From American Museum of Natural History, with permission.)

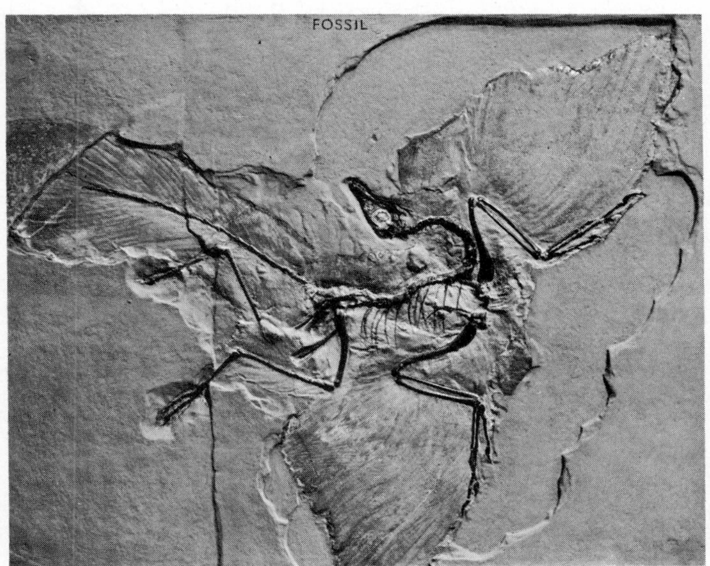

Figure 40.7 (*left*) Fossil of *Archaeopteryx*. (*right*) Restoration of *Archaeopteryx*. (From American Museum of Natural History, with permission.)

the stages in the direct line of evolution to the modern horse are shown in the figure.

A number of changes or adaptations are seen in the evolution of the modern horse from its ancestor in the Eocene. The most important of these changes are the following. The enlargement and elongation of the third digit, with a loss of the other digits; the elongation of the forepart of the skull; the development of the premolars and molars into high-crown, continuously growing grinders; and a general increase in body size.

All these changes were adaptations for life in open plains country where the primary source of food was grass. The enlargement of the third digit and the loss of the others, together with the general increase in body size, made possible

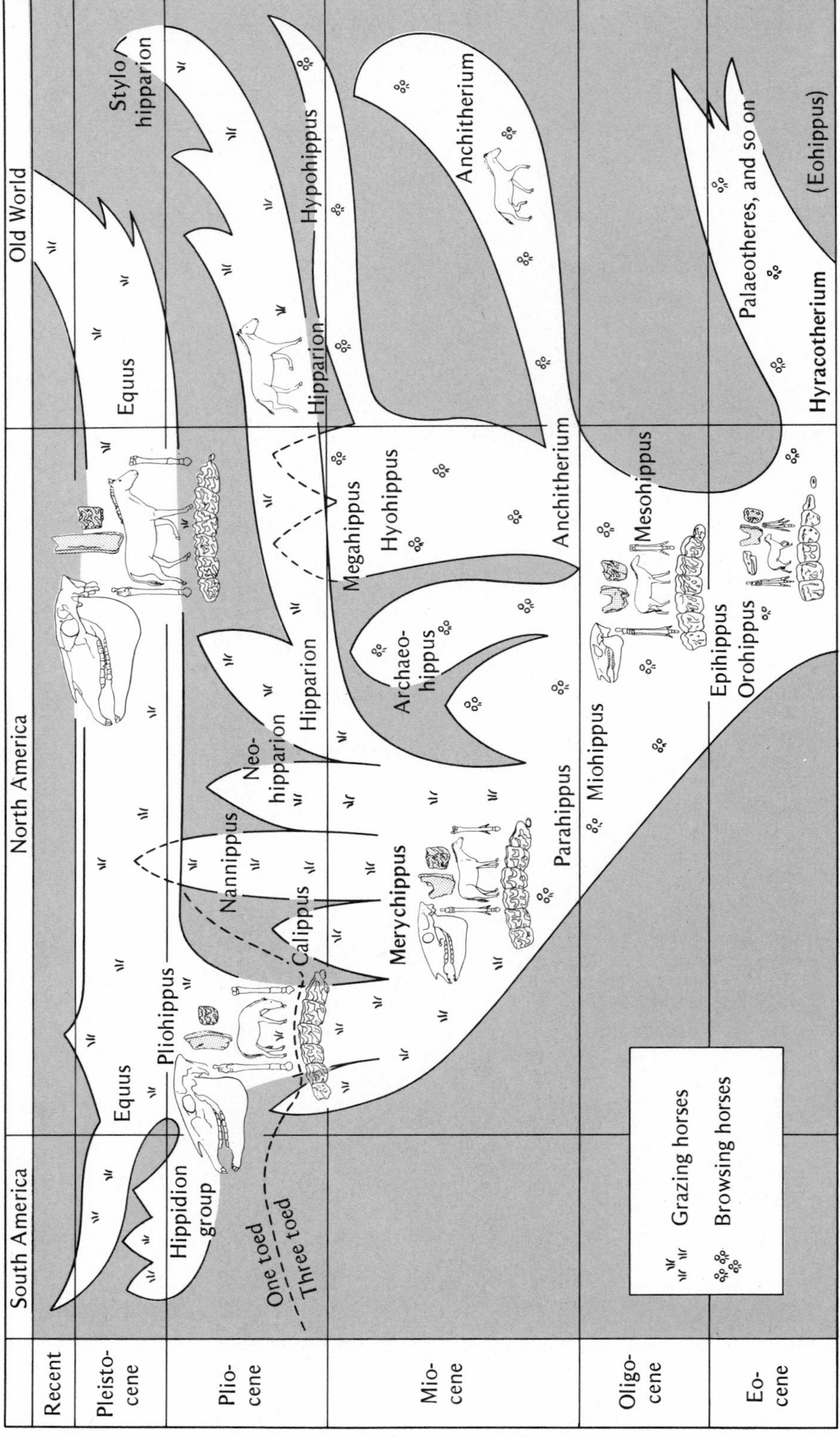

Figure 40.8 Evolution of horse showing radiation and phylogenetic relationships. Around the drawings of certain horses are characteristic diagnostic features, each drawn to comparative scale and arranged with skull and hind foot to the left, forefoot to the right, upper tooth above, set of premolars and molars below. Small cloverlike design indicates browsers; small tufts indicate grazers. (Modified from G. G. Simpson, *Horses,* © 1951 by Oxford University Press, with permission.)

greater speed in escaping from enemies on the grassy plains. The elongation of the forepart of the skull and the elongation of the neck are adaptations for grazing. The changes in the teeth are associated with the grinding of the coarse siliceous grasses. The evolution of these grasses occurred simultaneously with the evolution of the horse and probably served as a selective force acting upon the horses.

Merychippus, a horse of the Miocene, was about the size of a small pony. It carried all its weight on the third digit of each foot with the other digits dangling at the sides. The first representative of the modern horse, *Equus*, appeared in late Pliocene times. In the modern horse the side toes are reduced to the vestigial splint bones that are often fused to the metacarpal or metatarsal of the enlarged third digit. During the Pleistocene, the genus *Equus* achieved nearly worldwide distribution.

As mentioned before, many different kinds of genera of horses evolved during the Cenozoic, and only a part of them were in the direct line of evolution of *Equus*. *Eohippus* gave rise to several different lines, *Merychippus* gave rise to others, and so on. Some of them remained as browsers, others became grass feeders, but by the time of the appearance of the modern horse, all the other lines had become extinct (Fig. 40.8).

It is interesting to note that although the continent of North America provided the place for the greater part of horse evolution, horses became extinct on this continent by the close of the Pleistocene. The causes of this extinction are not fully understood. At any rate there were no horses in North America when the continent was discovered by the white man. The so-called "wild horses" in both North and South America are descendants of horses, brought over by the early explorers and colonists, that escaped or were allowed to run wild.

General Nature of the Fossil Record

The most important features of the fossil record can be summarized as follows.

1. The most primitive forms of life are found in the oldest rocks.
2. Moving up through the various strata, from older to more recent formations, there is a succession of higher and more complex forms of life.
3. In many instances a group arose in one period or era and remained scarce, but in the next period or era it became dominant after undergoing adaptive radiation.
4. There have been many extinctions of large groups, but after the establishment of all major phyla, some species of each phylum have persisted down to the present.
5. None of the past forms of life are exactly like any of those now living.

EVIDENCE FROM GEOGRAPHICAL DISTRIBUTION

Other interesting evidences of evolution are found in studies of the present distribution of different kinds of animals and plants on the earth, especially when this distribution is considered in connection with the fossil record.

One might assume, on the theory of special creation, that the regions of the world with similar climates and similar environmental conditions would be inhabited by the same kinds of animals and plants. But this is not so generally. The conditions for life in the African and American deserts are much the same, yet the flora and fauna of the two regions are quite different. Tropical rainforests in different parts of the world afford the same conditions for existence, but the animals and plants in an African rainforest are not the same as those found in a South American rainforest.

A very interesting aspect of the study of geographical distribution is the phenomenon of *discontinuous distribution*. A few examples will illustrate. The only places where marsupials are native in the world today are Australia, the eastern United States, and South America. There are many kinds of marsupials found in Australia: kangaroos, koalas, wombats, and bandicoots, to mention a few. The opossum is the only marsupial in the United States, and there are a few marsupials in South America. When the white man first appeared in Australia, there were no placental mammals on the entire continent except for some bats, the native human inhabitants, and some dogs and mice that were probably introduced when the first humans arrived there. That Australia is suited for placental mammals has been demonstrated; all that have been introduced there have thrived.

This discontinuous distribution of marsupials can be accounted for by the geological record. There is evidence that Australia was connected with the other land masses during the Mesozoic, and at that time the marsupials spread from South America to Antarctica and then to Australia. After the land connections were broken, placental mammals, except for flying forms and humans, were unable to reach Australia. In addition, the fossil record shows that marsupials were at one time widely distributed over Europe, Asia, and North America. In competition with placental mammals, the marsupials became extinct in most regions of the earth other than Australia.

The distribution of the members of the camel family is another case. Members of this family are found today in parts of Africa, Asia, and South America. The llama of the Andes is a member of the camel family. The fossil record shows that members of the camel group once existed over Asia, Europe, and North America. During a part of the Cenozoic era, Asia and North America were connected in the Bering Strait region, and this served as a bridge for the spread of many different forms. But after the widespread distribution of the camel group, extinction occurred in most of the regions, leaving the present discontinuous distribution.

The flora and fauna of oceanic islands also provide evidence of evolution. Oceanic islands are of recent geological origin. In the main, the species of animals and plants found on such islands are peculiar to them but are most like those forms found on the nearest mainland. As a young man, Charles Darwin became interested in the animals on the Galapagos Islands, 500 miles off the west coast of South America. He found that 23 of the 26 species of land birds on the islands were peculiar to the archipelago, yet they were all obviously related to the birds of South America. He found 11 kinds of giant tortoise, each kind inhabiting a different island. They were all related, but those farthest apart were most unlike. There were no land mammals on the islands and no amphibians, yet many habitats on the island were suitable for their existence.

In considering the evidence of evolution from geographical distribution, one must keep in mind these things. Each type of organism had a place of *origin* in the beginning. As the type increased in numbers, the members would tend to spread slowly or *disperse* from this region of origin into all suitable territory, and during this process evolutionary changes might occur. Often *barriers* of some kind, such as oceans or high mountains, stopped the dispersal process. In the course of time, because of geological changes or action of enemies, *extinctions* over large areas might occur, and this would result in a discontinuous distribution.

Continental Drift

In the above discussion reference was made to a former land bridge in the Bering Strait region. Other former land bridges are known. The Isthmus of Panama now constitutes a land bridge between North and South America. However, this bridge has not always existed. Through much of the geologic past the two Americas were not connected. That land bridges have played important roles in the past movements of plant and animals is a certainty. In the past it has been assumed that the continents in general have been fixed masses, that there has been no lateral movement of continents. In explaining the distribution of fossils, where very similar forms are found in widely separated regions, such as the finding of very similar fossils of snails and earthworms in Africa and South America, hypothetical former land bridges between the two continents have been proposed. The idea of continental drift was originally proposed by A. Wegener early in this century. According to his thinking, at one time the continents of Africa and South America were connected. (Anyone looking at the western coastline of Africa and the eastern coastline of South America on a map will see that they fit together like two pieces of a jig-saw puzzle.) Wegener presented many evidences supporting his concept of continental drift, but geologists and geophysicists were not attracted ot it. However, in the past decade evidence from a variety of measurements and observations by geophysicists has convinced most workers in this field that continental drift has occurred, and, for that matter, is still occurring. This concept has been heralded as a revolution in the earth sciences that will rank in its effects on the earth sciences with organic evolution and its effect on biology.

According to this concept, about 200 million years ago during the Mesozoic, the various land masses of the earth had moved together and formed one continuous land mass, called *Pangaea* (Fig. 40.9). Twenty million years later this single mass had separated into two masses —*Laurasia* and *Gondwana.* Laurasia contained North America, Greenland, and Eurasia. Gondwana contained South America, Africa, India, Australia, and Antarctica. These two land masses broke up and the present continents moved apart through the rest of the Mesozoic and through the Cenozoic to give the positions found today. By the end of the Mesozoic the South Atlantic had widened into a major ocean. Widening of the North Atlantic came later.

With this concept of the history of the earth, it is no longer necessary to postulate hypothetical land bridges to explain the finding of very similar fossils on widely separated continents. Colbert has recently described how the distribu-

Figure 40.9 Schematic reconstruction of Pangaea. On the east the northern land masses were separated from the southern land masses by Tethys Sea. With the separation of Laurasia and Gondwana this sea extended between them from east to west. In this period there was no Atlantic Ocean. (Modified from E. H. Colbert, *Wandering Lands and Animals,* © 1973 by E. P. Dutton and Company, by permission).

tion of many land vertebrates and the fossils of many vertebrates can best be explained on the basis of continental drift. For example, the fossil remains of the reptile, *Lystrosaurus,* have been found in South Africa, Antarctica, and India. This distribution can be easily understood on the assumption that continental drift has occurred (see Suggestions for Further Reading).

The idea of continental drift is now associated with the modern geological concept of plate tectonics. According to this the surface of the earth is divided into a number of huge plates, each including not only a continental mass but some of the surrounding ocean basin as well, and these plates are and have been constantly moving in relation to each other. The interested student should consult *Continents Adrift, The Dynamic Earth,* and *The Restless Earth,* listed at the end of the chapter.

Associated with the movement and collision of these plates are earthquakes, volcanic action, and mountain formation. One of the most interesting movements, to this writer, was the movement of the Indian plate to join with Laurasia in the formation of the present Indian peninsula. This was a long journey across the Tethys Sea, and it opened the Indian Ocean behind it. The collision occurred about 50 million years ago and resulted in the formation of the world's greatest mountains, the Himalayas.

You may ask, "What was the situation with respect to land masses before Pangaea?" Much remains to be learned, but Palmer (1974), in his article "Search for the Cambrian World," indicates that there were probably four major continents in the north and Gondwanaland, as described above, in the south at that time in the earth's history.

EVIDENCE FROM BIOCHEMISTRY

Probably the first use of biochemical techniques in evolutionary studies was the use of the precipitation test developed by Nuttall early in this century. This test is based on the antigen–antibody reaction already discussed. Nuttall prepared antihuman serum by injecting a rabbit with small amounts of human blood serum over a period of time. When blood was then drawn from this rabbit and the serum was allowed to separate, this serum was used in making tests. By mixing a small amount of this rabbit serum with human blood, a heavy precipitate was formed. This indicated that the rabbit serum contained antibodies that precipitated certain proteins in human blood. This was a good test for human blood, but it was not entirely species specific. When some of the serum from the immunized rabbit was added, in equal amounts, to five test tubes, and to these were then added sera from a human, an ape, an Old World monkey, a New World monkey, and a lemur, a precipitate formed in each tube, in decreasing amounts, however, from human to lemur. This serological test then supported the theories of primate relationship that had been arrived at earlier from studies in comparative anatomy.

Since the work of Nuttall, the precipitin tests have been used many times to verify or clarify other relationships. For instance, it has been possible to show that the horseshoe crab is related to the Arachnida, that whales are most closely related to the even-toed ungulates like the hog, and that the rabbit, long placed with the rodents, is not very closely related to them.

Evolution of Hemoglobins

From what has been said in earlier discussions, it is fair to say that the DNA and protein molecules in every living organism are living documents of evolutionary history. A new discipline, *paleogenetics,* has emerged with the goal of determining how evolution proceeds at the molecular level. Although only a little is known, as yet, about the linear sequence of nucleotides that carry the code for a single gene in a molecule of DNA, the sequence of amino acid residues in several polypeptide chains has been determined. Each amino acid in such chains was specified by a three-letter code in the DNA. Emile Zuckerkandl states that enough is now known about the amino acid sequence in several polypeptides to enable a paleogeneticist to begin the study. Zuckerkandl selected the hemoglobin molecule for his study, which involved the testing of three basic postulates. (1) Polypeptide chains present in living organisms today have arisen by evolutionary divergence from similar polypeptide chains that existed in the past. The present and past chains would have many of the amino acids in the same places in the chains and would be homologous. (2) A gene existing in some organism of the past might occasionally be duplicated, with the result that it is present at two or more sites in the genome of present-day organisms. Such an organism with two or more homologous genes would have two or more homologous polypeptide chains. Since these homologous genes could undergo mutations independently, their derived polypeptide chains would not be identical in all their amino acids. (3) The mutations most commonly selected in the evolutionary process are those that result in the replacement of a single amino acid in a polypeptide chain.

The study involved the protein *myoglobin* as well as various hemoglobins. Myoglobin serves as an oxygen repository in muscle, and an ancestral myoglobin apparently was the base molecule from which the hemoglobins were derived. John Kendrew, in 1958, determined the three-dimensional structure and the amino acid sequence in sperm whale myoglobin. This was the first complete determination of the structure of any protein molecule. Since that time the structure of several hemoglobins has been determined; these include hemoglobin chains from humans, gorilla, pig, horse, cow, and rabbit. The three-dimensional structure of myoglobin and the various hemoglobin chains are very similar. While the myoglobin molecule is a single chain associated with an iron-containing heme group, a hemoglobin molecule consists of four polypeptide chains, with each chain enfolding a heme group.

In adult humans the principal kind of hemoglobin is composed of two α chains and two β chains, and it is assumed that they have a common ancestry. In people the β chains are sometimes replaced with other chains -γ, δ, and ϵ chains. The ϵ chain is found for only a brief period in early development. The γ chain replaces the β chain during most of the embryonic development, and in adult life a small amount of the hemoglobin contains δ chains instead of β chains.

The amino acid sequence is known for all the chains except the ϵ chain. The α chain contains 141 amino acid residues, whereas the β, γ, and δ chains all contain 146. The differences could be explained by deletions or addition in the genetic material. When the α and β chains are compared, there are 77 sites different and 64 the same. There are only 39 differences in amino acids between the β and γ chains, and only 10 between the β and δ chains. There are much greater differences between the whale myoglobin and the hemoglobin chains of humans. There are 37 sites alike in the myoglobin and human α chains, and 35 alike in the myoglobin and human β chains.

How can it be argued that all these chains are homologous when there are so many differences between some of them? It seems quite improbable that different and unrelated polypeptide chains could evolve in such a way as to have the same function, the same three-dimensional configuration, and a substantial number of amino acids at corresponding sites. The marked difference in amino acid sequence would seem to be an indication that a long time has elapsed since they diverged from a common ancestor.

The number of differences in amino acid sequence between the α and β chains of the horse, the pig, the cow, and the rabbit and the corresponding human chains were determined. The mean difference is 22 changes in the two chains. Assuming that the rate of mutations is the same in both chains, this would be an average of 11 changes per chain. From other evidence it is assumed that the common ancestor of present-day mammals lived about 80 million years ago. This means that the average time to establish an amino acid substitution is about 7 million years (Fig. 40.10).

This information has been used to estimate roughly the time of the origin of the different human chains and the whale myoglobin chain from an ancestral chain through the process of gene duplications. The rough estimates of the time of these divergences are as follows: the β and δ chains, 35 million years ago; the γ chain, 150 million years ago; the α chain, 380 million years ago; and the myoglobin chain, 650 million years ago.

Zuckerkandl points out that techniques are now available for the determination of ancestral residues and for the construction of molecular phylogenetic trees. Such studies in the future,

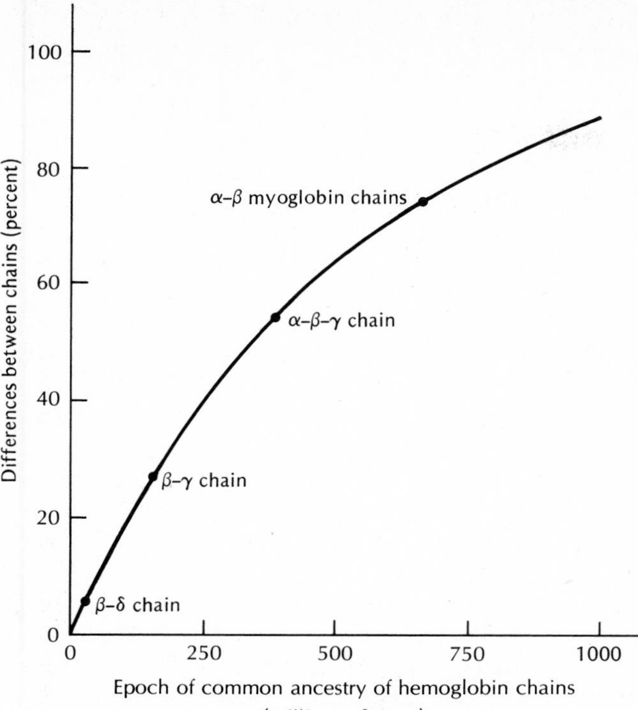

Figure 40.10 Age of ancestral hemoglobin–myoglobin chains. (From E. Zuckerkandl, "The Evolution of Hemoglobin", © 1965 by Scientific American, with permission.)

when many more polypeptide chains have been analyzed for their amino acid sequences, should provide additional evidence of evolution.

EVIDENCE FROM DOMESTICATION

Every state fair, every dog or cat show, and every horse show furnishes numerous examples of evolution through domestication. In a comparatively short period of time, with reference to the whole process of evolution, humans through their practices of selection and breeding, have developed a tremendous variety of domesticated animals and plants. The different varieties of dogs, chickens, cattle, horses, and so forth, have been derived from original wild types. Darwin recognized this and made use of it in formulating his theory, to be discussed later. He realized that, in domestication, humans do the selecting in the evolutionary process, whereas in nature all the factors of the environment act in the process.

EVIDENCE FROM TAXONOMY AND GENETICS

The fact that living organisms can be fitted into a scheme of classification (as discussed in Chapter 20), involving species, genera, families, orders, classes, and phyla, is best interpreted as indicating evolutionary relationship. Also the fact that the genetic material in nearly all species is found as variations of DNA makes evolution in all of its ramifications more readily comprehensible.

SUGGESTIONS FOR FURTHER READING

Brosseau, G. E. *Evolution.* Dubuque, IA: William C. Brown (1967).

Calder, N. *The Restless Earth.* New York: Viking Press (1972).

Colbert, E. H. *Wandering Lands and Animals.* New York: E. P. Dutton (1973).

Dodson, E. D. *Evolution.* New York: Reinhold (1960).

Hochachka, P. W., and G. N. Somero. *Strategies of Biochemical Adaptation.* Philadelphia: W. B. Saunders (1973).

McAlester, A. L. *The History of Life.* Englewood Cliffs, NJ: Prentice Hall (1968).

Merrell, D. J. *Evolution and Genetics.* New York: Holt, Rinehart and Winston (1962).

Moody, P. A. *Introduction to Evolution.* New York: Harper and Row (1968).

Palmer, A. R. "Search for the Cambrian World," *Amer. Scientist,* **62,** 216–224 (1974).

Romer, A. S. *The Vertebrate Story.* Chicago: Univ. of Chicago Press (1959).

Simpson, G. G. *Horses.* New York: Oxford Univ. Press (1951).

Simpson, G. G. *This View of Life; The World of an Evolutionist.* New York: Harcourt, Brace and World (1964).

Tax, S. (Ed.). *Evolution After Darwin,* 3 vols. Chicago: Univ. of Chicago Press (1960).

From Scientific American

Barghoorn, E. S. "The Oldest Fossils" (May 1971).

Hallam, A. "Continental Drift and the Fossil Record" (Nov. 1972).

Wilson, J. Tuzo (Ed.). *Continents Adrift* (Readings from *Scientific American*). San Francisco: W. H. Freeman (1971).

41 Human Evolution

The preceding chapter presented abundant evidence that animals have evolved. What about humans? This chapter considers this question. Although there are, at the present time, a few gaps in the fossil record and a few places where the relationships of some of the fossils are not clear, the general picture is clear—humans have evolved like all other living things. In earlier discussions the relation of humans to other vertebrates has been indicated. A comparison of the different organ systems in humans and any other vertebrate shows a similarity of structure throughout. Humans, like all other mammals, possess hair and mammary glands. As a primate, the human being shares with the other members of the order such characteristics as one pair of mammary glands, nails instead of claws, hands with opposable thumbs for grasping and handling objects, and eyes directed anteriorly instead of laterally. Another characteristic of primates, when compared with other mammals, is the proportionately greater brain development that seems to have originated as an adaptation or specialization for life in trees.

ORIGIN OF PRIMATES

In the late Mesozoic the two main groups of mammals, the marsupials and placentals, were in existence. Placental mammals of this period were small insect eaters belonging to the order Insectivora and, at present, are known only from fossils collected in Mongolia. They were generalized in nature, and it is believed that they gave rise to the other orders of placental mammals, including the primates. The tree shrews (Fig. 41.1) of the early Eocene, descendants of the insectivores, were probably the first primates and the ancestors of higher primates.

Figure 41.1 Tree shrew. (From San Diego Zoo, with permission.)

EVOLUTION OF PRIMATES

During the Eocene several lines developed from this ancestral stock of primates. One line gave rise to the *lemurs.* They are small tree-living animals and at present are found mainly on the Island of Madagascar. They resemble monkeys in many ways, having nails instead of claws and a long tail. However, their faces are characterized by a prominent snout much like that of a dog. Another line developed into the tarsiers (Fig. 41.2) that are found today on some of the islands of the East Indies. They were also small with long tails, but the snout was greatly reduced and they had very large eyes and stereoscopic vision. The lemurs and tarsiers and a few other forms are placed in the suborder Prosimii. Some of the Eocene prosimians were the ancestors of the higher primates. There is some

Figure 41.2 Tarsier. This little animal is about the size of a kitten and is the most advanced of the prosimians. Its upper lip is free from the gums below it, and its hands and feet are adapted for grasping. (From San Diego Zoo, with permission.)

evidence that the line that gave rise to the living tarsiers might have been ancestral to the higher groups.

The higher primates—monkeys, apes, and humans—are placed in the suborder Anthropoidea. The New World monkeys (family Cebidae) diverged from the ancestral stock late in the Eocene or in early Oligocene. These monkeys are flat-nosed with widely spaced nostrils in contrast to the other higher primates in which the nostrils are close together and point downward. A striking trait of the New World monkeys is their prehensile tail by which most of them can hang or swing from branches or which they can use like a hand. The New World monkeys have no direct bearing on the ancestry of the other primates.

Figure 41.3 shows suggested relationships among the other primates. The Old World monkeys (family Cercopithecidae) probably arose from the ancestral line in the early Oligocene. They lack prehensile tails and some, like the baboon, have become terrestrial, living in rocky open country. The other higher primates are placed in the superfamily Hominoidea—the apes in family Pongidae and humans in the family Hominidae. There are four groups of living apes—the gibbons, the orangutans, the gorillas, and the chimpanzees (Figs. 41.4–41.7).

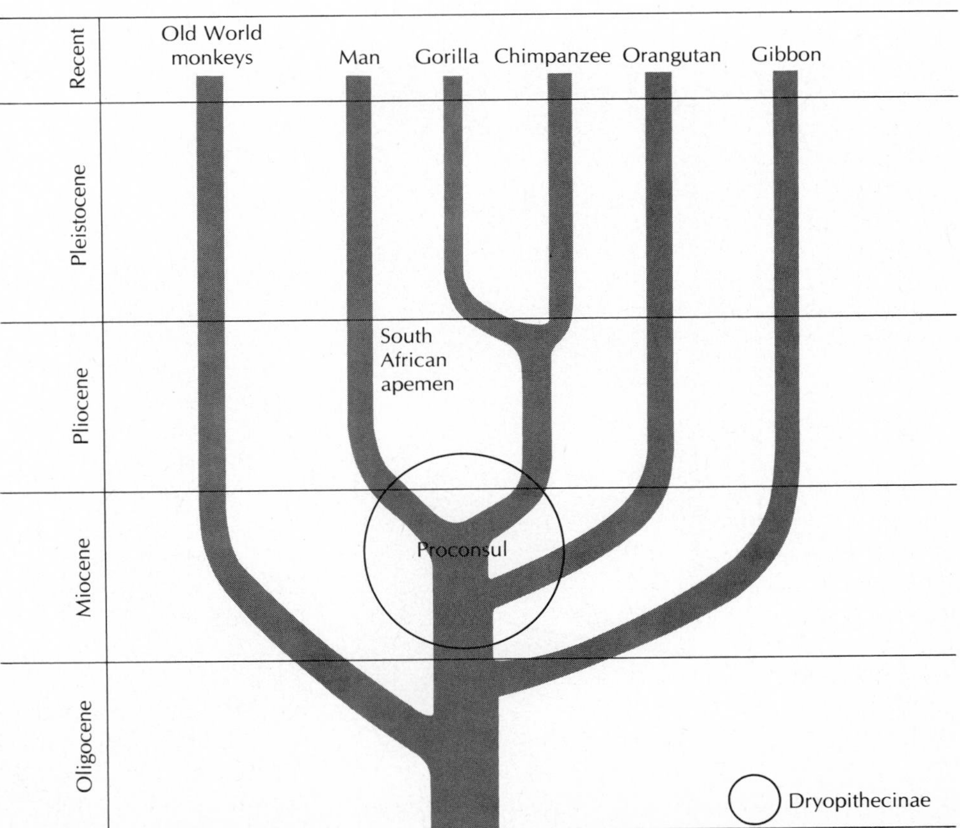

Figure 41.3 Family tree of humans and higher primates. (Modified from Moody, *Introduction to Evolution,* © 1953 by Harper and Row, with permission.)

Figure 41.4 Gibbon. (From New York Zoological Society, with permission.)

Figure 41.6 Chimpanzee, mother and baby. (From Diego Zoo by Ron Garrison, with permission.)

Figure 41.5 Orangutans. The larger individual is a male, the smaller one, a female. (From New York Zoological Society, with permission.)

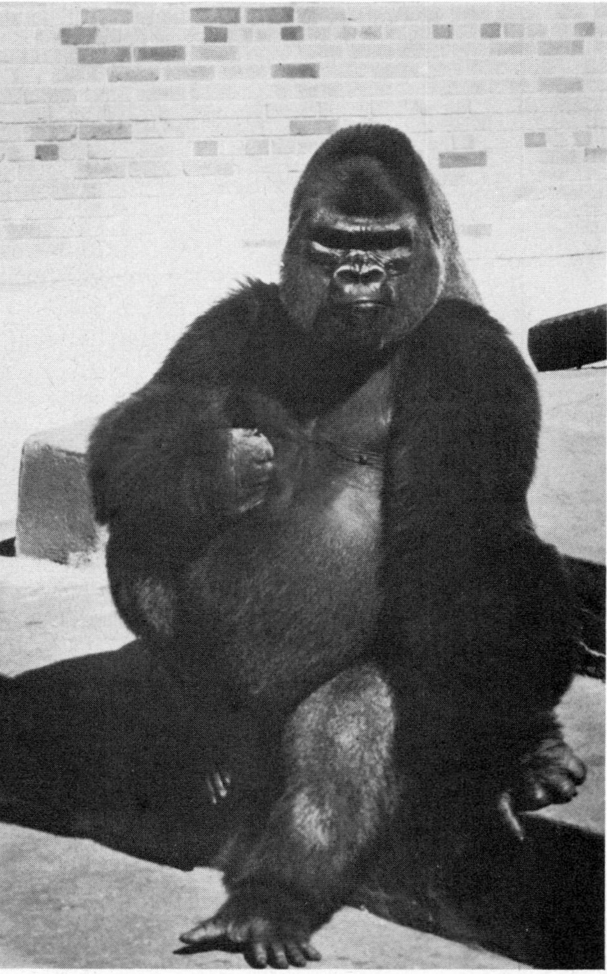

The gibbon (Fig. 41.4) is indicated as having originated from the ancestral line late in the Oligocene. These forms, which are found today in Southeastern Asia and the East Indies, are most unlike the other three types of great apes. They have slender bodies and are highly specialized for swinging through the trees by their long arms. The encircled region on Figure 41.3, labeled Proconsul, represents a group of fossil apes of the Miocene from Africa, Europe, and India. One genus of these, *Dryopithecus* (formerly called *Proconsul*), is among the less specialized and seems to be close to the ancestral forms of the other higher apes and of humans. They did not have the very long arms which modern brachiators have by which they move hand over hand from one tree branch to another. Also they did not have the "simian shelf" on their jaws, which supports the strong muscles used in tearing and chewing the bark of trees; nor did they have the very large canine teeth characteristic of the higher apes. The orangutan of Sumatra and Borneo is considered to have been derived from this ancestral group, whereas the chimpanzee and gorilla of Africa are believed to have diverged from a common ancestor later. With the exception of the gorilla, all the living great apes spend most of their lives in trees. Although a few of the monkeys lack tails, all of the apes do. In general the apes have relatively larger brains than the monkeys and are more intelligent.

FOSSIL HOMINIDS

Although the number of human and subhuman fossils is not large, the record is extensive enough to reveal the evolutionary trends. It would take us too far afield to describe all the hominid fossils that have been found; only a few of the well authenticated ones are considered.

One of the problems in this area of study is to determine what distinguishes a true human from a manlike ape. Several criteria are used and the following are among the important ones.

1. A true human has upright posture, that is, he is bipedal. An expert anthropologist can tell by examining the femur and certain other bones whether the specimen walked upright or on all fours. Upright posture is an adaptation associated with life on the ground in contrast to life in trees.
2. A true human will have a brain distinctly larger than the brain of apes. It is customary to give the capacity of the skull in cubic centimeters in referring to brain size.
3. The teeth of humans generally are smaller than the teeth of apes and the canine teeth are also small. The changes in the teeth were associated with a change in diet from a vegetarian to one of at least some meat.
4. The teeth in the jaws of apes are arranged in a U shape, while in humans they form a curved row (Fig. 41.8).

Figure 41.7 (*opposite*) Gorilla. (From New York Zoological Society, with permission.)

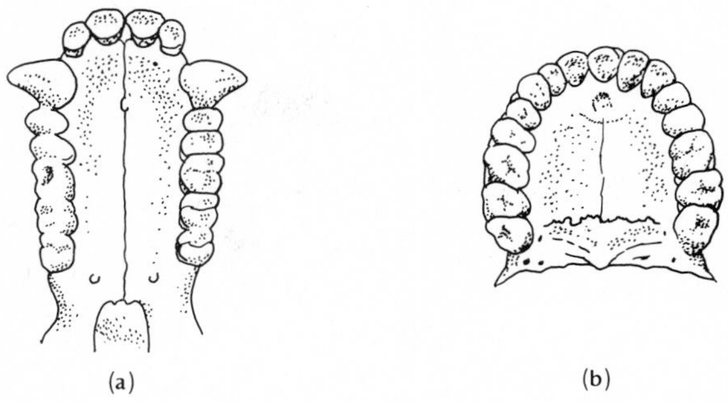

Figure 41.8 Dentition pattern in upper jaws. **(a)** ape. **(b)** Human.

5. A true human is a tool maker. In studying fossil remains tools must be present with the fossil remains, or be found nearby, if toolmaking is to be associated with the specimen.

One of the most distinguishing characteristics of humans is the use of language. Ordinarily it is not possible to tell whether a fossil specimen used language, although Alan Walker believes that he has detected in the underside of skull 1470 (to be discussed later) an area similar to Broca's area in modern man, the part of the brain that makes speech possible.

Three genera of the Hominidae are recognized: *Ramapithecus, Australopithecus,* and *Homo.*

Ramapithecus

Two species of *Ramapithecus* are recognized; one from East Africa and the other from the India–Pakistan border. They were found in late Miocene deposits and are dated from 9 to 14 million years ago. Parts of jaws and several teeth were found in each case. At the present these fossils are considered to be the oldest known fossils in the family Hominidae, and they were probably derived from the dryopithecines. Their teeth (Fig. 41.9) were more humanlike than apelike. The molar teeth were broad, shortened, flat, and compressed and of about equal size to those in humans, and the canine teeth were apelike in shape but about the size of human canines.

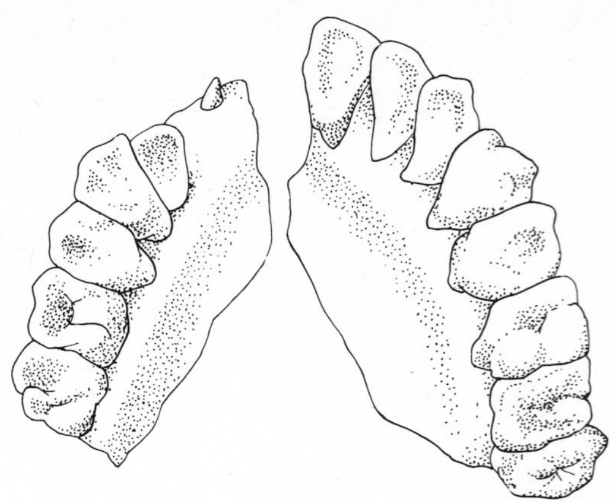

Figure 41.9 Part of the upper jaw of *Ramapithecus*. Compare with Figure 41.8. The shape is more humanlike than apelike.

Australopithecus

There is a long gap in the human fossil record from 14 to 9 million years ago (the period of *Ramapithecus*) to the time of the oldest known Australopithecine, 5.5 million years ago. However, an ancestral strain or strains leading from *Ramapithecus* to *Australopithecus* must have existed during the Pliocene.

Australopithecus africanus was first found by R. Dart in 1924 in South Africa and was recognized as a possible missing link in the ancestry of humans. Since then parts of over 300 specimens have been unearthed. They have been found not only in South Africa but also in East Africa (Tanzania, Kenya, Ethiopia) and Java. Fossils of *A. africanus* are dated from 5.5 to 1.5 millions years ago, while those of the related species, *A. boisei,* are dated from 3 to 1.2 million years ago.

The evidence indicates that *A. africanus* (Fig. 41.10) was adapted for bipedal locomotion with essentially human dentition. Estimates of height based on single leg bones range from 1.30 m to 1.68 m (4 feet, 3 inches to 5 feet, 6 inches). They were taller generally than *Ramapithecus,* but shorter generally than their successor *Homo.* The cranial capacity ranges from 435 to 724 cubic centimeters (cm^3). The face was rather massive, the jaws were large, the brow ridges were well developed, and there was no chin. In general it can be stated that the earlier specimens of *Australopithecus* were more like *Ramapithecus,* while the later specimens were more like *Homo.* There was great variation in their structure during the $4\frac{1}{2}$ million years of their existence.

In recent years many of the fossil specimens of this genus were discovered by a man and wife team, Louis and Mary Leakey, working in Olduvai Gorge, located in Tanzania, East Africa. In 1959 Mary Leakey unearthed some primitive

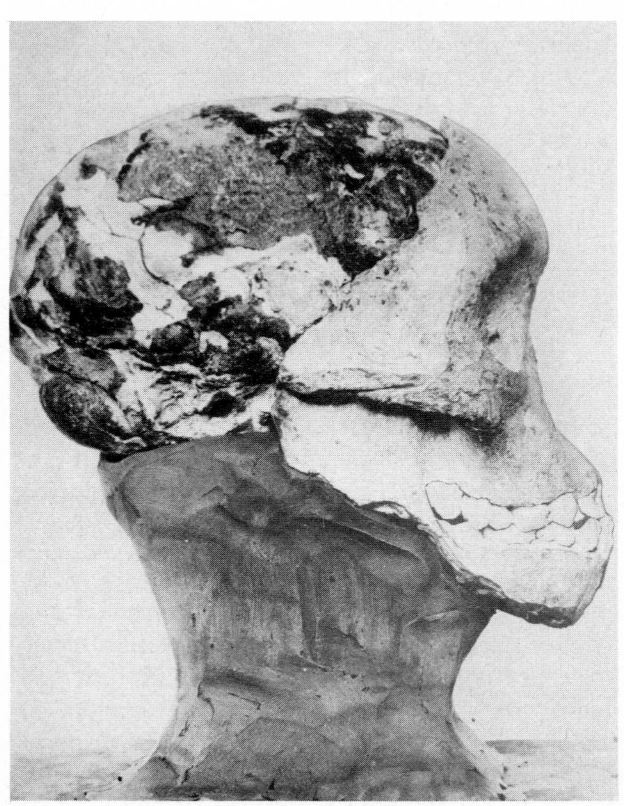

(a)

(b)

Figure 41.10 **(a)** Skull and **(b)** reconstruction of *Australopithecus africanus.* (From American Museum of Natural History, with permission.)

stone tools and an almost complete skull, which, to their amazement, turned out to be 1¾ million years old. This was almost a million years older than the oldest specimen of *Homo* known at that time. L. Leakey created a new genus for this specimen, *Zinjanthropus,* and stated that this represented the connecting link between the australopithicines and modern humans.

However, in 1961 the Leakeys made an even more striking find in the same deposits. They found another skull that looked much more similar to modern humans than *Zinjanthropus.* The teeth were more like those of modern humans, the cranium was much larger than that of *Zinjanthropus,* and it lacked the gorillalike crest at the top of the head. Leakey called this creature *Homo habilis* and said that he was the true ancestor of humans. Leakey decided that *Zinjanthropus* was an australopithecine that lived alongside of *Homo habilis* for a long period before becoming extinct. There is now fairly general agreement on this, and *Zinjanthropus boisei* is now called *Australopithecus boisei.* This species was more massive than *A. africanus,* was a herbivore with large grinding molars and reduced front teeth, and with a cranial capacity of about 530 cm^3.

By 1972 many workers were of the opinion that what Leakey has described as *Homo habilis* was just an advanced *Australopithecus africanus* and that *Homo erectus* evolved from this strain.

RECENT FINDS

Two discoveries, one in 1972 and one in 1974, have cast some doubt on the correctness of the human lineage as expressed in the preceding paragraph.

In 1972 Richard Leakey, son of Louis and Mary Leakey, and his party, working at Lake Rudolf in Kenya, found a human skull with a cranial capacity of about 800 cm^3. This is about one third larger than the *Homo habilis* skull. And even more important is the fact that it was dated as almost 3 million years old, at least a million years older than *Homo habilis.* If the dating is correct, it means that a member of the genus *Homo* existed alongside the australopithecines for a long period of time. Richard Leakey has not named this specimen but refers to it as Skull no. 1470. If other studies bear out the authenticity of this dating, this would be further evidence for discarding *Homo habilis* as a direct ancestor of modern humans. According to Richard Leakey, this may mean that there was a common ancestor of both the *Homo* line and the *Australopithecus* line in the period between 4 and 5 million years ago. The *Australopithecus* line divided into the two species (already described) and the *Homo* line, of which skull no. 1470 is a part, gave rise to *Homo erectus* about a million years ago.

The latest finds, those of 1974, were made in Hadar river in north-central Ethiopia by Alemeyu Asfew of the four-nation Afar research expedition. He was working with D. Carl Johanson of Case Western Reserve University. The finds here included a complete upper jaw, half of another upper jaw, and half of a lower jaw, all with teeth. The dating of these finds places them back as far as 4 million years ago. According to Johansen the small size of the teeth in these jawbones may well mean that genus *Homo* was walking, eating meat, and using tools as much as 4 million years ago. This latest find could fit right into the concept of Richard Leakey as given in the preceding paragraph.

Homo erectus

Fossils of this species have been found in Africa (Olduvai), Java, Algeria, Germany, Hungary, and China. The oldest specimens were found in Africa and Java and are about 1 million years old. Outside of their skulls, the skeletons were much like that of modern humans. The face was larger and the jaws were more powerful than those of modern humans. The skull was thick and heavily built. There was no chin. The cranial capacity was larger than any australopithecine, ranging from about 900 to almost 1100 cm^3. Compared to modern humans it was a muscular, stocky, thickset creature with perhaps slightly longer arms and shorter legs. This human mastered fire, often lived in caves, and survived the northern winters.

The earliest specimen of this species was unearthed at Trinil in Java in 1891 by Eugene Dubois. He called it *Pithecanthropus erectus,* and it is often referred to as the "Java ape man." The first findings there consisted of a skull cap, a jaw fragment, and a femur bone. Later a num- of other fossils were discovered there.

Davidson Black discovered a number of fossils in a cave near Peking, China. These humans lived at about the same time as the Java men and had the same characteristics. They left their hearths, their tools, their food remains, and many of their own bones in the cave. Black named this human *Sinanthropus pekinensis,* but it is now placed in the species *Homo erectus,* along with the Java man and others. *Homo erectus* existed in the period from about 1 million years ago to about 300,000 years ago. After that time our ancestors become almost indistinguishable from modern humans.

Homo sapiens

The fossil record is extremely sparse in the period between 300,000 and 100,000 years ago. There have been only eight finds in this period, four substantial ones and four fragmentary ones. There is enough evidence, however, to say the following about the humans of this period:

1. Their brain size was well within the range of modern humans.
2. Their bodies were indistinguishable from modern humans.
3. They had long skulls and heavily built faces and jaws.

NEANDERTHAL MAN (HOMO SAPIENS NEANDERTHALENSIS[1])

This man, originally called *Homo neanderthalensis,* lived in Europe from over 100,000 to about 35,000 years ago. The skeletal remains, some of them quite complete, of over 100 individuals have been studied. They have been found in deposits in Asia Minor, Africa, and Siberia, as well as in widely separated parts of Europe (Fig. 41.11).

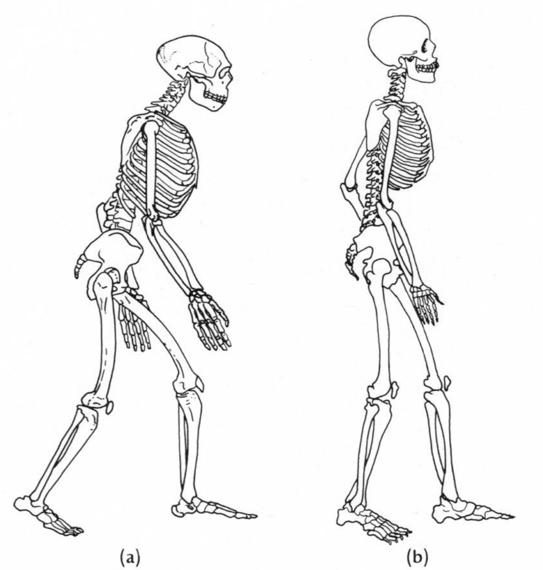

Figure 41.11 **(a)** Skeleton of Neanderthal man. **(b)** Skeleton of modern human.

Neanderthal man was about 1.50 m (5 feet) tall and powerfully built. He walked with a slight stoop. His hands and feet were large. The cranial capacity was about 1400 cm^3, about the same as modern humans. However, the shape of the skull was different; the forehead was low and slanting, there were heavy ridges projecting over the eyes, and there was no chin.

This man lived primarily in caves, used fire, and made flint tools. There is evidence that the Neanderthal man buried his dead with ornaments and flint tools, which has been interpreted as an indication that he had some form of belief in immortality of the spirit.

Between 50,000 and 35,000 years ago the Neanderthal people were replaced by people like ourselves. At this time it is not possible to say just what happened. In part of its range (western Europe) Neanderthal man showed little variability, and his replacement was rather sudden. In another part of his range (southern Africa) the two forms overlapped in time. In still another part of his range (western Asia) intermediate specimens have been found. A number of skeletons of Upper Pleistocene age were found at Mount Carmel in Israel that show an odd mixture of Neanderthal and Cro-Magnon characteristics. Some were tall, some had chins, and others did not. Some had high foreheads, some had low foreheads, and some had an intermediate condition. Because of his rather abrupt disappearance in western Europe, some students have proposed the extermination of the Neanderthal type there by the more modern form who came in from the east. Today more and more workers are suggesting that the Neanderthal race in western Europe just evolved into their successors or that they were genetically swamped by the invaders in crossbreeding.

CRO-MAGNON MAN (*HOMO SAPIENS SAPIENS*)

This is one of the first modern humans. Fossils of this type have been found that date back at least 40,000 years. Although specimens have been found in several parts of Europe, western Asia and Africa, Cro-Magnon man is primarily known from numerous specimens found in a French cave. The males averaged over 1.80 m (6 feet) in height, whereas the females were about 1.65 m (5 feet, 5 inches). The skull was of the modern type with a high forehead, no heavy ridges over the eyes, and a distinct chin.

In addition to using stone for tools and weapons, this man also used bone. Numerous bone needles have been found which they apparently used in making clothing. The Cro-Magnons are best known for their excellent drawings of contemporary animals on the walls of caves in France and Spain.

The origin of the Cro-Magnons is not known, but one theory holds that they developed in Asia and then invaded Europe. They were con-

[1] Two subspecies of *Homo sapiens* are now recognized: *H sapiens neanderthalensis* and *H. sapiens sapiens.*

temporaries of the Neanderthal man for many years and ultimately replaced him. (Fig. 41.12) The age of the Cro-Magnons was from about 40,000 to 10,000 years ago. They were expert hunters but did not develop agriculture or domesticate animals. These cultural activities came later, as did the use of metals in tool making.

Possible Relationships of Prehistoric Men

In the past it was customary to describe each new fossil and then represent it as a separate branch of the evolving human line. As a consequence none of the fossil men were ever indicated as being ancestral to any other type of man. The present tendency among students of the field is to move away from this view and to consider that all the prehistoric men, beginning at least with *Homo erectus,* belonged to the same genus, *Homo*. This concept assumes that throughout human evolution hominids living at any one time, although members of different races, have been capable of exchanging genes through interbreeding.

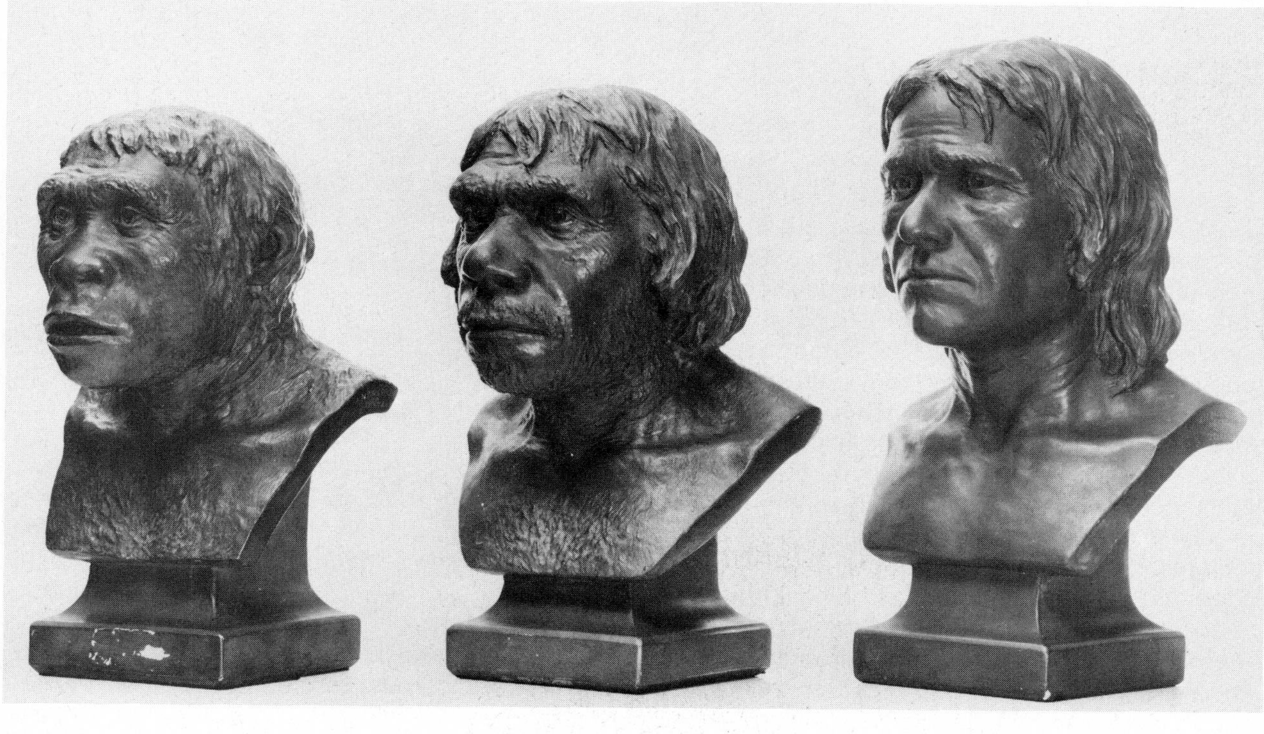

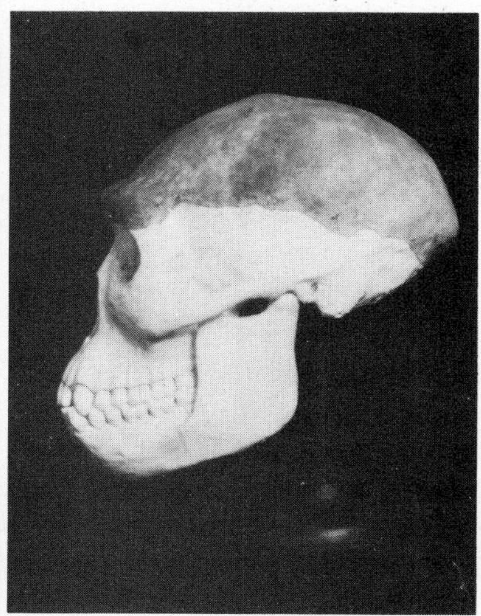

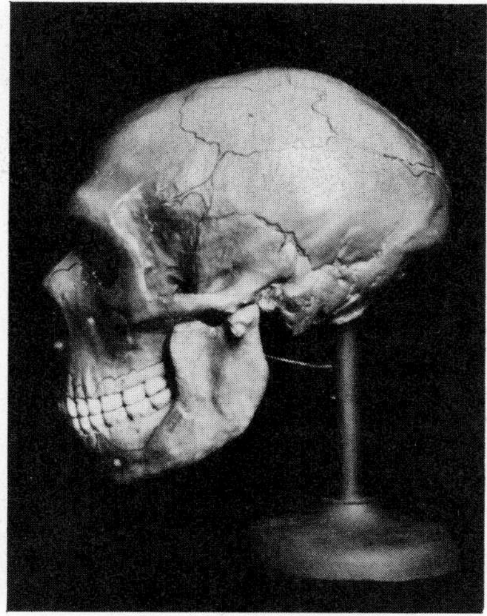

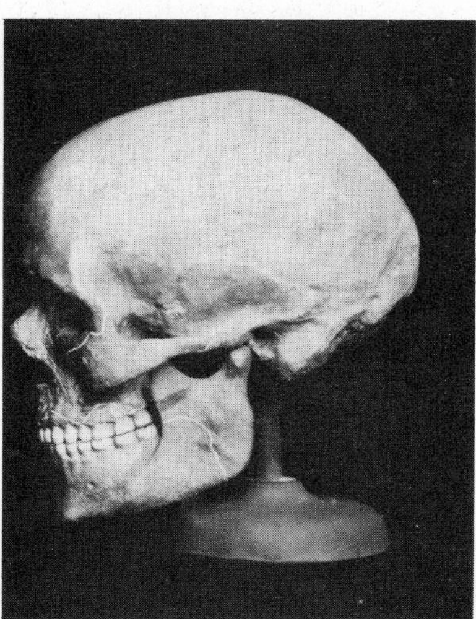

Figure 41.12 Prehistoric men. From left to right the Java ape man (*Homo erectus*). Neanderthal man (*Homo sapiens neandertalensis*). Cro-Magnon man (*Homo sapiens sapiens*). Restorations *above,* skulls *below.* (From American Museum of Natural History, with permission.)

If the recent discoveries of fossil men dating back 3–4 million years turn out to be authentic, we have the following line of descent in human evolution starting in the Miocene: dryopithicines, *Ramapithecus,* unknown direct ancestor of both the *Homo* and the *Australopithecus* lines, *Homo* sp., *Homo erectus,* and *Homo sapiens.* If these recent discoveries turn out to be other australopithicines, then the line of descent is as follows: dryopithicines, *Ramapithecus, Australopithecus, Homo erectus,* and *Homo sapiens.* It appears that the transition from an apelike form to modern humans was a gradual process.

Figure 41.13 Some Stone Age tools. **(a)** Some of the earliest humanmade tools, known as choppers. **(b)** Neanderthal tools. **(c)** Ground stone ax (left) and a Cro-Magnon blade tool (right). (From American Museum of Natural History, with permission.)

CULTURAL EVOLUTION

Most of the above discussion has been concerned with the biological evolution of humans —changes in posture, brain size, dentition, and so forth—changes that involve gene changes and the selection processes. With respect to their biological evolution humans have changed little in the past 35 million years. But with human biological evolution there has come into existence another kind of evolution—*cultural evolution.* This new evolution is a product of biological evolution, but different from it.

Our ancient ancestors lived in the tropics; later ancestors became adapted to live in temperate climates, and still later ones to cold climates. Some of our apelike ancestors used tools, but only humans and our humanlike ancestors made tools. There has been an evolution of stone tools (Fig. 41.13). Early humans began hunting in groups, then they developed fire and started living in caves. By Cro-Magnon times they were making clothing and practicing art. All these activities are a part of human culture and its evolution. Culture involves learning and instruction, and for instruction to occur there must be means of communication, that is, language. So although the fossil record does not tell us when language began, it can be assumed that, when our ancestors began making tools and hunting in groups and sharing food, some kind of language existed.

Culture is the learned part of behavior. It involves ideas, laws, customs, traditions, inventions, and all the other learned responses that regulate human societies. Culture is transmitted, not by specific genes, but by conditioning, training, and learning. The study of culture and cultural evolution unites the fields of biology, anthropology, psychology, and sociology. Humans, through their language, are able to transmit the knowledge learned in each generation to the next. Humans, of all the products of organic evolution, are the one kind of organism able to exercise control over their environment. They have already done much in this direction, and the possibilities for the future are great. This means that in this new cultural evolution humans can slowly determine the future course of their evolution, both biological and cultural, by instituting human purpose and will in its direction. But humans also have the power for self-destruction. The responsibility for what happens belongs to everyone.

RACES OF MODERN HUMANS

If the same taxonomic criteria are applied to humans as are applied to other species, there is but one human species living at the present time. The entire human population shows great variability, and it is divided into a number of races, but it is one species because members of the different races can and do interbreed. David Merrell states that probably the only racial cross that has not occurred is between Eskimos and African bushmen. It is not possible to draw sharp lines of distinction between racial groups because one race blends with another in the region of contact. Humans have always been wanderers, and the present rapid movement of people,

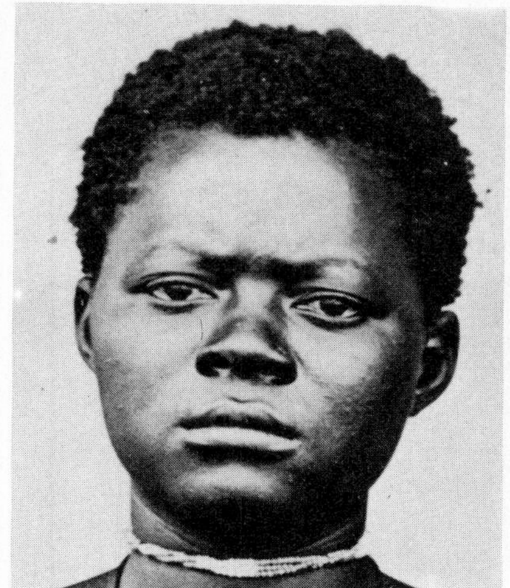

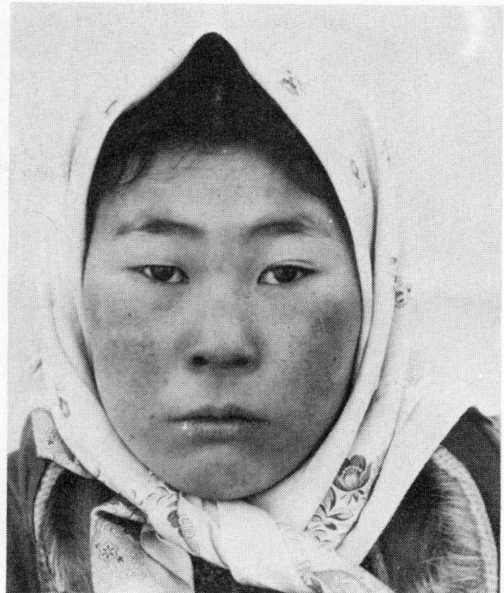

Figure 41.14 Major human races. **(a)** Negroid. **(b)** Caucasoid. **(c)** Mongoloid. (From American Museum of Natural History, with permission.)

throughout the world is tending to break down further the distinction between the races. There is no such thing as a pure race of humans. The human races differ from each other in the incidence of certain of their genes; this constitutes the basic distinction between races.

There is no general agreement on the number of distinct human races; more than 30 have been recognized. Three rather distinct racial groups are described below: the Negroid, the Caucasoid, and the Mongoloid (Fig. 41.14). In terms of numbers and widespread distribution, these three are major human races.

Prior to the settling of the Americas, members of the Negroid race were found in tropical Africa and in the islands of the western Pacific from New Guinea to Fiji. The Negroids are usually dark-skinned with black woolly hair, sparse body and facial hair, broad flat noses, and thick lips. The Caucasoid race was native to Europe and western Asia, southeast into the tropics of India and to North Africa. Caucasoids generally have rather light skin, long narrow noses, relatively straight hair, fairly abundant facial and body hair, and medium to thin lips. However this so-called "white" race varies all the way from the blond blue-eyed Scandinavian to the dark-eyed dark-skinned native of India. The Caucasoid race is quite variable, and a number of subgroups are recognized, such as the Mediterranean, the Nordic, and the Alpine. The Mongoloid race is native to North and East Asia. Members of the Mongoloid race generally have straight black hair, sparse facial and body hair, yellow to brown skin, brown eyes, an eye fold, and a flattened face with high cheekbones. This race includes not only the yellow-skinned Asians and Eskimos, but also, in many classifications, the American Indian. Certainly the American Indians had Mongoloid ancestors.

The characteristics of each race stated above are only general ones. Individuals within each race exhibit great variation. In recent years a more precise method of differentiating between human races has been undertaken using the differences in the frequencies of the different blood group genes. William Boyd has thus proposed a fivefold race classification: Caucasoid, Negroid, Mongoloid, American Indian, and Australoid (the aboriginal population of Australia).

There is no evidence that any race is inherently better than another. So far all attempts to demonstrate hereditary differences in intelligence between races have failed. Of course, what is measured as intelligence is strongly influenced by learning opportunities.

SUGGESTIONS FOR FURTHER READING

Boyd, W. *Genetics and the Races of Man.* Boston: Little, Brown (1950).

Campbell, B. G. *Human Evolution.* Chicago: Aldine Publ. (1974).

Dobzhansky, T. *Mankind Evolving.* New Haven, CT: Yale Univ. Press (1962).

Eaton, T. H. *Evolution.* New York: W. W. Norton (1970).

Ehrlich, P. R., and R. W. Holm. *The Process of Evolution.* New York: McGraw-Hill (1963).

Goldsby, R. A. *Race and Races.* New York: Macmillan (1971).

Howells, W. *Mankind in the Making,* rev. ed. New York: Doubleday (1967).

Leakey, R. E. "Skull 1470." *National Geographic,* **143:** 6 (1973).

Life Nature Library. *Early Man.* New York: Time-Life Books (1970).

Merrell, D. J. *Evolution and Genetics.* New York: Holt Rinehart and Winston (1962).

Stebbins, C. L. *Processes of Organic Evolution.* Englewood Cliffs, NJ: Prentice-Hall (1971).

Tobias, P. V. "Early Man in East Africa." **Science, 149:** 3679 (1965).

Washburn, S. L., and P. C. Jay (Eds.). *Perspectives on Human Evolution,* vol. 1. New York: Holt, Rinehart and Winston (1968).

From Scientific American

Howells, W. W. "The Distribution of Man" (Sept. 1960).

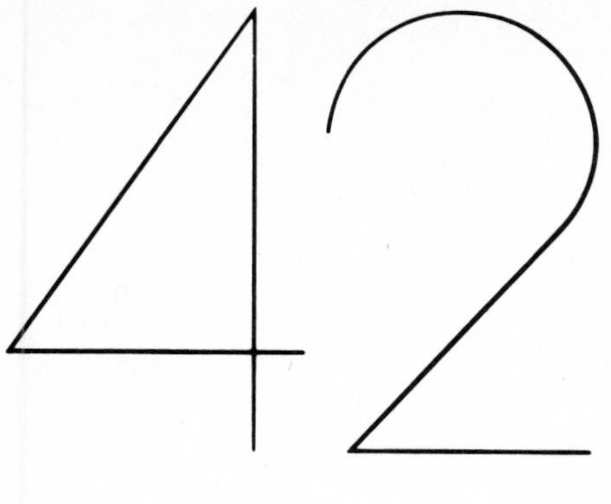

The Mechanism of Evolution and the Origin of Life

Two explanations have been offered for the origin of the different forms of life—special creation and evolution. Many evidences have been presented in previous chapters to support the concept of evolution, and almost without exception modern biologists are convinced of evolution. But now we must ask the question: What are the factors responsible for the process?

Although a vague idea of evolution was held by some of the early Greek philosophers, no important theory to explain it appeared until the nineteenth century. One of the first of these theories, which is now only of historical importance, was that of Jean Baptiste de Lamarck.

LAMARCK

Lamarck spent the early part of his life as a botanist. Then, at the age of 50, he turned his attention to zoology, particularly to the study of invertebrates. (We are indebted to Lamarck for the terms "invertebrate" and "biology.") As a result of his systematic studies he became convinced that species were not constant but rather were derived from preexisting species. This idea was in conflict with the view of the period—that of fixity of species. As a result Lamarck's views were challenged by most of the biologists of that time, particularly by Georges Cuvier, and Lamarck's influence was not great.

In 1809 Lamarck published *Philosophie Zoologique,* which included his theory explaining the changes that occur in the formation of new types. Lamarck believed that environmental influences are the chief causes of evolutionary change. According to him, when an animal's environment changes, its needs change, and this leads to special demands on certain organs. Organs used more extensively would enlarge and become more efficient. Conversely, an organ, or organs, no longer used would degenerate and atrophy. He postulated that such changed characteristics would be transmitted to the offspring.

Lamarck's theory, then, was based on the ideas of *use and disuse* and the *inheritance of acquired traits.* He visualized the evolution of the giraffe as follows. An original deerlike animal, finding the supply of grass and herbs inadequate, started to feed on the leaves of trees. It needed greater height to reach the higher leaves, and in the process of reaching, its neck became longer and longer. In the course of generations the long neck became a more accentuated feature, and our modern giraffe was the result. His theory was simple, and it had some appeal. Everyone knows that exercise results in larger muscles. The theory also afforded a means of explaining reduced or vestigial structures. Through disuse the eyes of a cave animal, for example, might become functionless and might even disappear.

The great deficiency in the Lamarckian theory was the assumption of the hereditary transmission of acquired traits. It must be recalled that at this time in history all notions of inheritance were vague and fanciful. No one tried to test this part of Lamarck's thesis until late in the century, when the German worker August Weismann tried cutting the tails off mice for many successive generations. At the end of his experiment the mice in the last generation grew tails as long as their ancestors. Since Weismann's time, many workers have devised experiments to test this hypothesis, and the net result has been the same, acquired traits or modifications are not inherited. In this connection it can be pointed out that mutilation experiments, such as Weismann performed, were not necessary. The Chinese bound the feet of their women for many generations, yet this has not resulted in any modification of the feet of present-day Chinese women. Also, Jewish boys have been circumcised for thousands of years, but this has not resulted in a tendency toward reduction of the prepuce in this group.

DARWIN

The most impressive study of evolution made by any single man was that of Charles Darwin. His principal publication, *The Origin of Species by Means of Natural Selection or the Preservation of Favored Races in the Struggle for Life,* which appeared in 1859, contained not only a masterful assemblage of evidence for evolution, but also his theory which, with certain modifications, is still accepted today.

Charles Darwin (Fig. 42.1) was born in Shrewsbury, England, in 1809, the son of a physician. He spent three years at Edinburgh studying medicine and another three years at Cambridge studying theology. Although his university record was creditable, his heart was not in either medicine or the ministry. As a boy he became interested in natural history and collecting natural objects of all kinds. He spent much of his time at the universities finding out more about botany, zoology, and geology. When, following his graduation from Cambridge, he was offered the opportunity to travel around the world as a naturalist on H.M.S. *Beagle,* his father agreed to let him go.

So at the age of 22 Darwin set out on a voyage that lasted five years and affected profoundly his own life and ultimately the thinking of much of mankind. When he embarked on this voyage, he believed, as did most people at the time, in the fixity of species—of special creation. When he returned, the concept of evolution was firmly fixed in his mind. He made numerous observations during the five-year period that led him to this view. He was especially impressed with the striking adaptations found in the animals and plants that he collected and with the progressive changes in the different species of a given genus as the expedition moved along the South American coast. In the Galapagos Islands he found many species of a given genus confined to a single island, yet resembling other species on other islands and the mainland. Such facts, Darwin reasoned, could be explained only by the supposition that species may become modified.

After his return home he began to gather all the information he could on this idea of the modification of species. For over 20 years he collected data, performed experiments, read extensively, and thought about a possible mechanism that would explain the modification of species. During this long period in which he amassed his evidence and developed his theory, many things affected his thinking, of which three in particular deserve mention here.

Figure 42.1 Charles Darwin. (From British Information Service, with permission.)

He had taken with him on the voyage of the *Beagle* a copy of Charles Lyell's *Principles of Geology,* which had been published only a short time before. This book attempted to explain the past changes in the earth's surface on the basis of processes now in operation. It extended greatly the older ideas about the age of the earth, and it gave Darwin a basis for appreciating some of the fossil material that he found. In addition, Lyell's insistence on explaining the past history of the earth in terms of processes still operating affected Darwin's thinking about the causes of evolution.

Darwin was aware of the similarities of many domesticated animals and plants to wild forms and realized that members of our society, for centuries, selected individual plants and animals with desirable traits for breeding. In this way the various lines of domesticated forms have been developed. He found himself asking the question: Might not some sort of selection operate in nature to produce the various highly adapted forms found on the earth? While Darwin was thinking about this possibility, he happened to read the famous *Essay on Population* by T. Malthus. From this he obtained a suggestion that was to serve as the basis for his explanation of evolution. Malthus, who was a mathematician and an economist as well as a minister, was concerned with problems of the increasing human population. He pointed out that human population, as well as plant and animal populations, tends to increase at such rates that its numbers outstrip its resources, and because there is not food or space enough for all, there is a struggle for existence. Malthus argued that in time the human population would outstrip the possible food supply and therefore that steps should be taken to control human reproduction. He proposed late marriage and sexual restraint as means of doing this and stated that unless steps were taken by humans to limit their population, checks apparently beyond their control—in the form of famine, disease, and war—would limit it instead.

Darwin was familiar with the struggle for existence that occurs everywhere in nature. He reasoned that in such situations, favorable variations would be preserved and unfavorable ones destroyed, and the result would be the formation of new species.

Darwin conceived this theory of evolution when he was 29 years old, but he did not publish his ideas until many years later. Meanwhile he continued to collect data bearing on the question of evolution. It was not until 1842 that he wrote out a first brief abstract of his theory. By 1844 he had expanded this to a document of 230 pages. In 1858 before the final manuscript of *The Origin of Species* was completed, Darwin received a manuscript from Alfred Wallace, who had been studying the natural history of the Malay Archipelago, with the request that Darwin read it, and if he found it worthy, to forward it to Lyell for publication. When Darwin read the manuscript, he found that Wallace had proposed a theory of natural selection similar to his own. He forwarded the manuscript to Lyell for publication, but Lyell, who had become acquainted with Darwin's work over the years, arranged to have Darwin submit an abstract of his own work, to be presented simultaneously with Wallace's paper. In the following year Darwin's complete work was published. It is interesting that two great naturalists developed the idea of natural selection independently. But Darwin, who spent much of his life in the study of this theory and who organized such an overwhelming mass of evidence in support of the concept of natural selection, made the greater contribution.

Darwin's Theory of Natural Selection

Darwin's theory is based on three observable facts and two deductions drawn from these facts.

FIRST OBSERVABLE FACT

Every species, in the absence of environmental checks, tends to increase in a geometric manner. If a population of a given species doubles in one year and if there are no checks on its increase, it will quadruple the next year, and so on.

The great reproductive potential of different species may be easily observed in nature. It has been estimated that a common Atlantic coast oyster may shed as many as 80 million eggs in one season. A single pair of English sparrows would be the ancestors of over 275 billion individuals in 10 years if they and their descendants could reproduce at their natural rate without any checks. Darwin calculated that even a pair of elephants which are about the slowest breeding animals known, could, in the absence of any checks, have 29 million descendants at the end of 800 years.

SECOND OBSERVABLE FACT

This tremendous reproductive potential is not realized in nature; the size of the population of each species remains relatively constant over long periods of time. This is not to say that given species will not vary some from year to year. This relative constancy in the numbers of each species is discussed again in Chapter 45 in connection with the concept of a balance in nature.

First Deduction Because the great reproductive potential of different organisms is not realized in nature (that is, because not all individuals that are produced do survive), *there must be a struggle for existence.*

THIRD OBSERVABLE FACT

Within every species there is individual variation. For instance, no two human beings are exactly alike, and so on.

Second Deduction In the struggle for existence those individuals that have favorable variations will survive in proportionately greater numbers and will produce a proportionately greater percent of the next generation. This is *natural selection.*

Great Controversy

With the appearance of Darwin's book in 1859 a veritable storm of controversy broke out. Many people became interested in his presentation;

the arguments for and against that developed were often heated. The concept was opposed primarily on ethical and religious grounds. But the sheer weight of evidence and the logic of Darwin's presentation finally convinced a majority of the educated people. The concept of evolution affected many aspects of our thinking and it will continue to do so. The impact of Darwinism on social and economic thinking has been unfortunate in certain respects. The idea of "the survival of the fittest" has been erroneously interpreted by many to mean survival only by "tooth and claw," an interpretation that has led in many cases to rationalizations for the attitude of "every man for himself" in social and economic affairs. But this view of Darwin's theory and of its present-day modification is a misinterpretation.

The process of evolution is not all "tooth and claw." Plants have evolved, too, and there is no bloody competition between them. In all forms, plants and animals alike, a part of the selective process involves the action of inanimate nature, factors such as drought, storm, moisture, temperature, and others that are considered in more detail in Chapter 44. Active *competition* between organisms is a part of the process, but as Warder Allee and others have pointed out, *cooperation* has in many cases been involved in survival. Natural selection results in the survival of those forms that are best integrated with the various factors of the environment in which they live. Survival of the fittest merely indicates those whose offspring survive to reproduce.

Darwin's theory had one great weakness; it did not explain the origin and transmission of variations. Although Darwin criticized many aspects of Lamarck's theory, he did not deny that acquired characteristics can be transmitted. He realized that the nature of inheritance was unknown, but he devised a working hypothesis to explain how acquired variations may be transmitted. According to his hypothesis the parts of the body give off tiny particles, which he called *pangenes*, into the body fluids, and these are collected in the eggs or sperms. Later, when a fertilized egg undergoes development, according to the concept, the pangenes present are responsible for the particular features of the new individual. This hypothesis, of course, has been proved untenable. However, it should be remembered that Darwin (and Lamarck, too) did as well as they could with the information they had. Mendel published his paper on the mechanism of inheritance seven years after Darwin published the theory of evolution, but it was 1900 before these laws of inheritance became generally known to the scientific world and helped in bringing rationality to the theory of evolution.

MODERN VIEWS OF EVOLUTION

Many studies have been made during the past 75 years to explain further the mechanism of evolution. As a result it can be said that the general thesis of Darwin, except his ideas about the origin and transmission of variations, has been upheld. There are many different aspects of the mechanism of evolution, and no one is willing to say that we understand the entire process by any means. What is presented here is only an outline of some of the major features.

Mutations

Darwin made no distinction between environmentally induced variations (which we now know are not inherited) and inheritable variations. It is now recognized that mutations (see Ch. 18) are the raw materials for evolution by natural selection. These changes in genes and in chromosome configurations, together with the recombinations that result in sexual reproduction, are the only known inheritable changes that occur in organisms. Natural selection does not cause these changes, but it plays an important role in determining which of them survive.

Hugo de Vries, a Dutch botanist and one of the discoverers of Mendel's paper, first focused attention on the importance of mutations. In his studies on the evening primrose he found several individual plants that differed markedly from the rest, and these new types bred true. He regarded some of these new types as elementary species, and as a result of his studies he announced his "mutation theory" of evolution. He considered that new species arise by sudden changes or steps rather than by gradual processes. According to de Vries it was mutations and not selection that should be considered as the primary factor in evolution. Later studies on the same material showed that some of the new types described by de Vries were not true mutations but were recombinations. Though some mutations may cause major changes, most of the ones that have been studied produce relatively slight changes. The great importance of de Vries' contribution was in directing attention to the study of mutations.

SELECTION OF MUTATIONS

When a mutation occurs that is beneficial (that gives some survival advantages to its possessor), it will probably appear with increasing frequency in subsequent generations. The rate at which the new gene replaces the original one in the population will be determined by a number of factors: whether the new mutation is dominant or recessive, the intensity of the selection, the rate at which the mutation occurs in the population, the rate of reverse mutation, the breeding structure of the population, and the size of the population. On the basis of the studies that have been made it must be assumed that many, many generations are needed for the incorporation of an ordinary mutation into the majority of the individuals of a large population.

In the earlier discussion of mutations it was pointed out that most mutations are deleterious in their effects and thus are not of survival value. How does this fit into our present concept? At first glance it might seem to mean that mutations cannot be the source of new materials for the evolutionary process. However, a little thought will show the following fact. All organisms now living in the world are the products of hundreds of thousands of years of this process of selecting favorable mutations, and they are well adapted for the particular environment in which they live. During this process of evolving, when the environmental conditions have remained relatively constant for a long period of time, most of the mutations that could give survival value have occurred and have been selected. Thus it would be surprising if beneficial mutations occurred with very great frequency.

Also, in the discussion of mutations it was pointed out that the rate of mutations is low and that most mutations produce only slight effects. Is it possible, then, to explain the results of the process of evolution on the basis of mutations as a part of the process? George Simpson, in his study of horse evolution, has estimated that something over 5 million years was required for the formation of each of the genera in the evolution of the horse. Assuming that about 45 million years were involved, and making certain assumptions about the size of the population involved, it has been possible to show that if a very conservative estimate of the rate of mutation is assumed, mutations producing very slight effects could have produced the changes in the size of the teeth of the horse, the modifications in the legs, and other characteristics.

The discussion so far has implied that selection deals with single mutations, without reference to the whole gene complex. In general this is not so. Modern genetic studies have shown that many, if not most, genes produce manifold effects in the developmental process. Thus the selective forces will act on different gene combinations or complexes. In the discussion on heredity we have already considered how, as a result of sexual reproduction, many, many *recombinations* of genes are produced through independent assortment and crossing over. It is from this great range of variation that the best combinations are selected.

One of the weak points of the original theory of natural selection—one recognized by Darwin as such—was that it did not adequately explain the existence of the many apparently nonadaptive characteristics found in organisms. Every student of taxonomy and natural history knows that quite often some of the more obvious characteristics that distinguish two related species apparently have no survival value. This is now understandable on the basis of our knowledge of heredity. Some genes that give no survival value may become fixed in a population by chance (see the subsequent discussion of genetic drift); others may be carried in the gene pool of the population for generations in the heterozygous state. Thus it is not necessary to assume that all genes of an organism are beneficial or that all characters of an organism have survival value. As one worker has expressed it, all that is necessary for survival is that the gene pool of the species provide in excess of 50 percent overall adaptation for the members of the species.

This reservoir of nonadaptive genes in the gene complex of each species appears to be an important feature in the process of evolution. The size of this reservoir, or the degree of adaptation, will vary from species to species. It is at the time of great environmental changes that this reservoir of genes becomes important. A species that is very highly specialized for a particular environment may be unable to survive a radical change in the environment. On the other hand, a species that is not so specialized and that has a larger reservoir of nonadaptive genes (for the original environment) may survive in a very different environment because some of the originally nonadaptive genes now have survival value. Many workers call this preadaptation.

The Population as the Unit in Evolution

An individual organism does not evolve; only a population does. Great strides have been made in recent years in the study of population genetics as related to the process of evolution. A population, as discussed here, is a group of organisms of the same species that interbreed with one another in a particular geographical area. As a result of the interbreeding in the population, there is a free flow of genes within it. All genes within the population are often referred to as the *gene pool* of the population.

What happens within a population when random mating occurs in successive generations? This was first answered in 1908, independently, by G. H. Hardy and by Wilhelm Weinberg, and we now refer to their findings as the Hardy–Weinberg law.

Before discussing this law, a review of simple genetics is in order. Suppose a cross of two rabbits, homozygous black with homozygous brown, is made. In this case the gene for black (*B*) is dominant over the gene for brown (*b*). Such a cross would result in all heterozygous (*Bb*) black offspring in the F_1 generation. Now if one or several matings of F_1 individuals are made to obtain many F_2 offspring, the theoretical results can be represented on a Punnett square as follows:

		Eggs	
		B	*b*
Sperm	*B*	*BB*	*Bb*
	b	*Bb*	*bb*

According to the Punnett square, one would expect that three fourths of the F_2 generation would be black ($\frac{1}{4}BB$, $\frac{1}{2}Bb$) and one fourth would be brown (*bb*). If large numbers were obtained, this ratio would be approximated in the actual results.

Now assume that 1000 of these F_2 rabbits are distributed as follows: 250 *BB*, 500 *Bb*, and 250 *bb*. These are placed on an island where the conditions for their existence are quite favorable. What would the next generation be like? Assume the following conditions: (1) mating is purely random; (2) there is no selection acting, that is, all the rabbits are equally viable and all are capable of producing the same number of offspring; (3) no mutation occurs; and (4) no rabbits are added from the outside and none are lost. It would be a long and tedious process to set up all possible matings within the population of 1000 rabbits. This, however, is not necessary. A look at the original F_1 cross that created the F_2 generation in the first example above shows $Bb \times Bb$. Here one has the same number of *B* genes and *b* genes. One can represent the frequency of *B* as 0.5 and the frequency of *b* as 0.5. Now if the F_2 generation is examined, it becomes obvious that the same frequency of these alleles (or 0.5*B* and 0.5*b*) exists.

To determine the frequency of the genes, of the genotypes, and of the phenotypes of the next generation that the 1000 rabbits on the island would produce, one can thus represent not the cross of two individuals, as was done above, but what would happen when the whole population reproduces, assuming the conditions outlined above. Because there are as many *B* genes as *b* genes in both the males and females of the population, the gametes can be shown as follows:

		Eggs	
		0.5*B*	0.5*b*
Sperm	0.5*B*	0.25*BB*	0.25*Bb*
	0.5*b*	0.25*Bb*	0.25*bb*

This generation of rabbits on the island, which would total several thousand, will have the same frequency of genes (0.5*B*, 0.5*b*), of genotypes (0.25*BB*, 0.5*Bb*, 0.25*bb*), and of phenotypes ($\frac{3}{4}$ black, $\frac{1}{4}$ brown) that the simple F_2 laboratory cross had. This population is in equilibrium, and this is the type of relationship discovered by Hardy and Weinberg. According to their law, in the absence of forces that change gene frequencies, the relative frequencies of alleles in a population tend to remain constant from generation to generation.

This relationship can be expressed in the form of an expanded binomial. The proportion of *B* alleles in a population can be represented by p, and the proportion of *b* alleles by q, and $p + q = 1$. In the expanded binomial $(p + q)^2$, p^2 will represent the proportion of *BB* individuals in the population; $2pq$, the proportion of *Bb* individuals; and q^2, the proportion of *bb* individuals.

A very simple situation has been used to illustrate this, a situation in which the frequency of each of the alleles was the same, 0.5. However, this is not the type of situation usually found in nature. When two alleles are present, their frequencies may vary from almost 0 to almost 1.

Another hypothetical population where the alleles are not equal in frequency is now examined. Consider a population of rabbits on an isolated island in which the brown rabbits constitute 64 percent of the population and the black rabbits, 36 percent. What is the gene frequency here, and what percent of the black rabbits are homozygous and what percent are heterozygous? One of the handy things about the Hardy–Weinberg formula is that if the percentage of the recessive genotype (the brown rabbits in this case) is known, one can determine the answers to these questions. Consider the expression, $p^2 + 2pq + q^2$. In this case q^2 stands for the proportion of brown rabbits, and the frequency of the *b* gene can be determined by taking the square root of 0.64. This is 0.8, or 80 percent. The frequency of *p* (black gene) must be 0.2, or 20 percent, because $p + q = 1$. The term p^2, which represents the proportion of the homozygous (*BB*) rabbits, would be $0.2 \times 0.2 = 0.04$, or 4 percent. The proportion of the heterozygous (*Bb*) black rabbits is represented by $2pq$, and this would be $2 \times 0.8 \times 0.2 = 0.32$, or 32 percent. So the black rabbits in the population, which constitute 36 percent, would be composed of 4 percent homozygous individuals and 32 percent heterozygous individuals. What would be the situation in the next generation?

Again this can be represented as before:

		Eggs	
		0.2*B*	0.8*b*
Sperm	0.2*B*	0.04*BB*	0.16*Bb*
	0.8*b*	0.16*Bb*	0.64*bb*

The results are the same, and this would then continue generation after generation in the absence of forces that change gene frequencies. This type of situation—in which the alleles of a given pair are not present with equal frequencies—is the most common type found in natural populations. People who have an acquaintance with only the most elementary concepts of genetics sometimes assume that all dominant genes are expressed in three fourths of the population and that all recessive genes are expressed in one fourth of the population, but this is not the case.

The populations that have been discussed so far were in equilibrium. What happens in a population not in equilibrium? One can illustrate this with another hypothetical case. Assume that 1000 black rabbits are placed on an island free of other rabbits and that half of them have the *BB* genotype and half the *Bb* genotype. What will their offspring be like? Assume the same conditions as before. All the genes contributed by the *BB* individuals will be *B*(0.5) and, in addition, half the genes furnished by the *Bb* individuals will be *B*(0.25). Hence the frequency of the *B* genes is 0.75. The frequency of the *b* genes is 0.25. Then $p = 0.75$ and $q = 0.25$.

$$\begin{aligned}(p + q)^2 &= p^2 + 2pq + q^2 \\ &= (0.75)^2 + 2 \times (0.75) \times (0.25) + (0.25)^2 \\ &= 0.5625 + 0.375 + 0.0625\end{aligned}$$

Thus there will be 56.25 percent *BB,* 37.5 percent *Bb,* and 6.25 percent *bb,* or a total of 93.75 percent black and 6.25 percent brown individuals. The parental population was composed of only black rabbits, but this one has some brown rabbits. Is this population now in equilibrium? This can be answered by determining the frequencies of genes *B* and *b*. The percent of individuals having only *B* genes is 56.25, so they will contribute that percentage to the gene pool; 37.5 percent are heterozygous, so they will contribute 18.75 percent *B* genes to the pool and 18.75 percent *b* genes to the pool; 6.25 percent of the individuals have only *b* genes and they will contribute that percentage to the pool. Thus

$$p = 56.25 + 18.75 = 75 \text{ percent}$$
$$q = 6.25 + 18.75 = 25 \text{ percent}$$

This result is the same as that at the start, and so the population is now in equilibrium and the same frequencies of genes, of genotypes, and of phenotypes will recur in generation after generation under the conditions outlined. This example demonstrates that when a population is not in equilibrium, it tends to attain equlibrium in one generation.

In this discussion very simple hypothetical populations were used in order to keep the mathematical analysis simple. However, it can be stated that the Hardy–Weinberg principle applies to natural populations wherever it has been tested. Snyder has studied the frequencies of the genes that determine the *M, MN,* and *N* blood groups in several different human populations. In this case gene *M* in the homozygous state determines the *M* group, gene *N* in the homozygous state determines the *N* group, and genes *M* and *N* together determine the *MN* group. All three of these blood groups can be detected serologically. In a sample of 1200 Swedes the distribution was as follows: 0.361 *MM,* 0.470 *MN,* and 0.169 *NN.* Using the Hardy–Weinberg formula, the frequency of *M* $\sqrt{0.361}$ is approximately 0.60, while the frequency of *N* $\sqrt{0.169}$ is approximately 0.40. The estimated frequency of *MN* $[2 \times (0.60) \times (0.40)]$ would be approximately 0.480. This corresponds very closely to the observed 0.470. The same type of agreement was found in the samples of other human populations studied.

The Hardy–Weinberg law states that the relative frequency of alleles in a population tends to remain constant from generation to generation when (1) the population is large enough so that accidents of sampling may be ignored; (2) mating occurs at random; (3) mutation does not occur or, if it does, the rate is the same in both directions; (4) all members of the population are viable and have equal reproductive rates; and (5) there is no emigration or immigration involved.

This is a very important principle in population genetics, but it describes a situation in equilibrium. This is a static condition, and evolution does not occur under these conditions. For evolution to occur there must be some disturbance in the gene frequencies of the population. The four primary factors that disturb genetic equilibrium are mutation, selection, genetic drift, and differential migration. The first three of these are discussed below.

MUTATIONS IN POPULATIONS

Different genes mutate at different rates. Assume gene *A*, which mutates at such a rate that one gene *a* is found in every 100,000 gametes formed. Assume also that the organisms that have gene *a* are as fit to survive and reproduce their kind as those with gene *A*. In this situation the frequency of gene *A* will decrease and that of gene *a* will increase in each succeeding generation. Gene *A* might mutate itself out of existence, but this is not likely because of the process of backmutation. Gene *a* will probably mutate to *A* with its own rate of mutation. In many cases studied the rate of backmutation (from the recessive to the dominant gene) is usually lower. At any rate, under these conditions an equilibrium will be established in the population when the number of changes of *A* to *a* is the same as the number of changes of *a* to *A* in each generation. It is possible that such a tendency to reach an equilibrium of opposing mutation rates may be the basis for the persistence of many alternative traits found in human and other populations.

The situation described above is not the usual one. As mentioned before, most mutations have a harmful effect, which varies from lethal, as when the particular mutation is present in the homozygous recessive state, to very slight. Suppose that we have a recessive mutation *a* which, when present in the homozygous state (*aa*), results in the production of only 99 offspring that are viable and capable of reproduction, in comparison with every 100 offspring that the *AA* and *Aa* individuals produce. In this situation the frequency of gene *A* increases with each generation. Although the *selection pressure* in this case is low, the results over a long period of time would be definite; gene *a* would be eliminated from the population unless other factors intervened.

In most situations one cannot assume that only mutation or only selection is operating. Selection works against the spread of deleterious mutations in a population, but a certain number of such mutations of various kinds are introduced into the gene pool of a population in each generation. What is the general nature of the results of the action of these two opposing forces? When more mutations are produced than eliminated, the frequency of the mutant allele will increase; when more mutations are eliminated than formed, the mutant allele will become less frequent; and when the number of mutant genes produced equals the number of such genes eliminated, an equilibrium will be established. Theodosius Dobzhansky (see Suggestions for Further Reading) has described an interesting situation in fruit flies, where the equilibrium reached is affected by the survival qualities of the heterozygote. In this situation the heterozygotes have greater survival value than either homozygous type under certain conditions. The sickle-cell condition in humans is another example.

SELECTION

One of the difficult concepts in population genetics involves the rates of selection. In most cases complete selection against a dominant (lethal) mutation will occur in one generation; complete selection against the homozygous recessive is slower; and partial selection against the homozygous recessive is much slower. Assume a population in which the mutation *a* represents almost 100 percent of the genes at that locus in the population. It can be shown, by methods developed by population geneticists, that with complete selection operating on the homozygous recessives, their frequency will be reduced from almost 100 percent to 10 percent in about two generations. However, it will take 7 generations to reduce it from 10 percent to 1 percent, 22 generations to reduce it from 1 percent to 0.1 percent, and 68 generations to reduce it from 0.1 percent to 0.01 percent. Thus the rarer any trait becomes, the less effective is any further selection against it. This is something that should be kept in mind when eugenic sterilization programs are considered. In such programs the traits involved are usually rare ones, and there is no possibility at present of approaching complete selection against them.

GENETIC DRIFT

The concept of *genetic drift,* developed by S. Wright, is concerned with changes in gene frequencies in small populations. A gene may be fixed purely by chance in a small population in the course of a few generations. In the discussion on genetics it was emphasized that whether one obtains actual genetic ratios that correspond with the theoretically expected ratios is dependent upon having large numbers of offspring. This is because the assortment of genes into gametes and the combination of gametes to form zygotes are random processes. In small breeding populations where only a few offspring are produced, great fluctuations from expected ratios may occur by chance. Thus in a very small population consisting of two mating pairs, if one individual contains a new mutation *a*, its genotype would be *Aa* and that of the other three would be *AA*. It is entirely possible, purely on the basis of chance, that if this individual with the new mutation and its mate produce six offspring that live to maturity, and the other pair produces four that live to maturity, five of the offspring of the first pair will have the mutant gene *a*. Thus out of 10 individuals to produce the next generation half of them would contain the mutant gene. Assuming that there is no selection against this gene, it would have very good possibilities of spreading further in subsequent generations.

On the other hand, the opposite might occur in the same situation. None of the offspring of the first pair, purely by chance, might receive the gene *a*, and it would be lost immediately. This chance fixation of genes is called *genetic drift,* and there seems to be little doubt that it plays a role in the evolutionary process. Conditions that favor genetic drift occur in nature. One such situation is that of a very small group completely isolated on an island. Another situation favoring the operation of this chance fixation of genes occurs at times when large populations are reduced to very small size as the result of epidemics or weather. It may be that many of the so-called nonadaptive characters that distinguish one species from related ones may have arisen as mutations that eventually became established by genetic drift.

The ideal situations for rapid evolution are found in populations of medium size that are broken up into a number of smaller populations, each of which is almost completely isolated from the others. In these relatively small populations, mutant genes may become established either through selection or through drift. After a gene that has survival value has become established in the small population, if the isolation is only partial and if individuals in this one small subpopulation migrate and mate with individuals in other subpopulations, the stage is set for the modification of the whole species. It should be pointed out also that species that are divided into a number of small subpopulations, because they have several partly differentiated groups, are better able to survive major environmental changes than are single large species lacking this differentiation.

SURVIVAL OF THE FITTEST

From the studies on population genetics a more plausible interpretation of "the survival of the fittest" has emerged. In Darwin's day in the minds of many the "fittest" meant literally the strongest in a physical sense. This was the origin of the "tooth and claw" concept, a concept that unfortunately was taken over in other areas and was used to justify exploitation of the weak. On the basis of the knowledge today, it is apparent that *the fittest are merely those in a population that produce the most offspring in the next generation.* This is the modern concept of the survival of the fittest, and it is more in line with what is seen in nature. As has been mentioned, the old idea of tooth and claw never did fit the evolution of plants, and they are certainly an integral part of the whole process. From this modern point of view natural selection, then, is the differential reproduction of certain genotypes. The population geneticist describes the evolutionary process in terms of changes in the frequency of genes in a population. Indeed, the expression "survival of the fittest" can be applied with more accuracy to the behavior of genes over many generations than to that of individual organisms in one generation.

Role of Isolation in Speciation

The process of evolution involves more than just change in the characteristics of a species; it involves the formation of new species, or *speciation.* Beyond this it involves the differentiations that are of genus significance, family significance, order significance, and so on. Speciation may occur in two ways: by replacement and by branching. Speciation by replacement may occur in the following way. A freely interbreeding population exists in a given area. Over a long period of time, the descendants of this population remain as one freely interbreeding group. Mutation and selection operate over the years, and after many generations the population has so changed in some of its characteristics that a taxonomist would call the later group a new species. There is, of course, no way of testing whether the later group would be able to interbreed with the original group. However, it has been assumed by many biologists that this may have occurred in the past, and if so, it would be speciation by replacement.

How can one species give rise to another or to several other species? According to one definition, a species is a group of similar organisms that freely interbreed and produce viable and fertile offspring. In such a group, when a beneficial mutation arises, it will ultimately be incorporated in the gene pool of the species. What causes a division?

There is general agreement that speciation, in the sense under discussion now, involves some kind or kinds of *reproductive isolation*. Mechanisms that prevent successful reproduction between members of two or more populations that have descended from the same original population are called *isolating mechanisms*. A number of such mechanisms have been identified, and several of them are discussed later.

It is further recognized that for these reproductive isolating mechanisms to evolve, the separated populations of an original single group must be separated spatially or geographically for a long period of time. The only clearcut exception to this, speciation by polyploidy, is considered in a later section. Geographical isolation for long periods of time, however, does not necessarily produce reproductive isolation and thus speciation. The catalpa trees of eastern North America and eastern Asia are fully fertile, yet they have been separated for a few million years. In many instances, however, such spatial separation does lead to reproductive isolation and speciation.

Darwin was impressed with his findings of distinct, but closely related, species on many of the different islands of the Galapagos archipelago, and it is this kind of phenomenon that many modern biologists have studied. When an original population is divided into two or more groups by geographical barriers that prevent interbreeding between them, in the course of time different mutations may become incorporated in the gene pools of the different groups. Often these differences are of such a nature that the separated groups, when they come in contact again, do not interbreed; thus species have been formed. Over the long haul, geographic isolation is seldom permanent. Changes in geography, migrations resulting from great population pressure, or chance dispersal during storms may bring separated groups in contact again. If they then interbreed and have fertile offspring, speciation has not occurred; but if they do not interbreed, or if they interbreed and have sterile offspring, new species have been produced; reproductive isolation has occurred. This does not take into account speciation in asexually reproducing organisms.

Oceans and mountain ranges with deep valleys between provide this type of spatial isolating mechanism. It is well established that there are more different species of the same genus in mountainous country than in plains regions. For instance in the eastern part of this country, there are eight species of cottontail rabbits, whereas in the mountainous regions of the west there are 23 species of these rabbits. Often in mountainous country many of the plants and animals found in deep valleys, which are separated by high peaks but which may be only a few miles apart, are peculiar to those valleys. Darwin found similar situations in the Galapagos Islands.

David Lack, in his recent studies of "Darwin's finches" in the Galapagos Islands, believes that only through geographical isolation combined with ecological specialization could the great variety of finches have been produced. Finches belong to one of the largest families of birds, which include sparrows, cardinals, goldfinches, and canaries, among others. The more typical

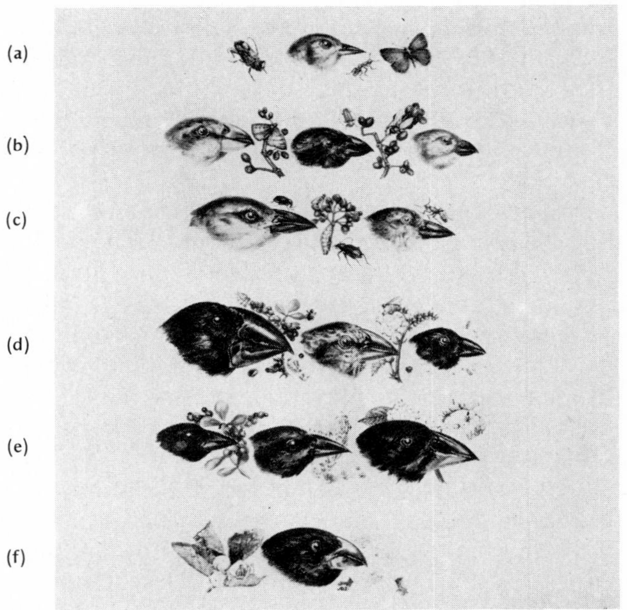

Figure 42.2 Darwin's finches. **(a)** Warbler type—insect eater. **(b)** Large, medium, and small tree finches—primarily insect eaters. **(c)** Tool-using finch and mangrove finch—the tool user is something like a woodpecker and uses a cactus spine as a tool to pry out insects. **(d)** Large, medium, and small ground finches—primarily seed eaters. **(e)** Sharp-beaked ground finch, cactus ground finch, and large cactus ground finch—primarily seed eaters. **(f)** Vegetarian tree finch. (From Jean Zallinger, as reproduced in *Life,* September 8, 1958, with permission.)

forms possess stout conical bills adapted to crush seeds. In Darwin's finches (Fig. 42.2) the different species are differentiated morphologically, primarily by the nature of their bills. Some feed on seeds, some on leaves and fruit, some on insects in the manner of the woodpecker, and others in the manner of the warbler, also on insects.

These finches are all clearly related, and they are quite different from any of those on the mainland of South America; yet everything points to their origin from the South American group. Several of these different types are now found inhabiting the same island. Lack's explanation of this evolution is, briefly, as follows. Some of the South American finches originally reached one of the Galapagos Islands. They were seed eaters and in time population pressure caused some to migrate to another island. The conditions there were such that this group, through mutations and selection, gradually became modified morphologically for food-getting and life in a different habitat. In time, some of this group migrated back to the original island. Now the members of the two groups could inhabit the same island because they had different ecological requirements and they were reproductively isolated. As time went on, the other differences were added to the original ones, while the groups were occupying the same general region. Lack believes that several such separations, leading to specialization along different lines in the different islands and subsequent remeetings of the groups, would explain the adaptive radiation that has occurred in this group of birds.

Lack supports his argument by pointing out that on Cocos Island, a single island located north of the Galapagos Islands in the Pacific and hundreds of miles from any other land mass, where the conditions for life are quite similar to those on the Galapagos Islands, only one kind of finch is found, and it is placed in a different genus from any of the finches on the Galapagos. Apparently such adaptation leading to speciation can occur on an archipelago but not on a single small island.

A number of different kinds of isolating mechanisms have been identified, and they can be placed in two categories: those that prevent fertilization and zygote formation, and those that prevent development after fertilization occurs or that cause weak or sterile hybrids.

In the first category difference in habitat may be the effective mechanism. Two populations may live in the same general area but because of differences in specific habitats they do not meet to cross-fertilize. This situation is found between many closely related species of trees. In other cases mating between two closely related species does not occur because their gametes are produced at different seasons of the year. In many instances in animals small differences in behavior patterns in courtship, such as calls and songs, between two closely related species prevent mating. An interesting example of this type in *Drosophila* is described in Chapter 43. In many of the flowering plants differences in the structure of the flower are a very effective way of preventing cross-pollination between two closely related species because of the close adaptation of each insect pollinator.

In the second category all the effects are due to genic disharmony, that is, the genes brought in from the two lines do not work in harmony in the cells of the hybrid, or of its progeny. Death may occur at any stage in development or a very weak hybrid may result. A vigorous hybrid may be formed, but it is sterile. This may be due to the abnormal development of the gonads or it may be due to the failure to produce normal gametes in the process of meiosis. In some cases viable hybrids are formed that are at least partly fertile, but the F_2 generation is weak, abnormal, or sterile.

There is much interest in the genetic basis of these isolating mechanisms. Several experimental studies indicate that all isolating mechanisms are controlled by several genes in multiple-factor inheritance.

An illustration of this is found in Dobzhansky's studies of backcross progeny between two species of *Drosophila, D. pseudoobscura* and *D. persimilis.* He found the male F_1 hybrids to be completely sterile because of much reduced abnormal testes, while the females were partly fertile. When the females were backcrossed to either parent, the males of such crosses showed a wide range of testis sizes. All of the chromosomes in the cultures used of the parental species could be identified by means of recessive genes which they bore, and Dobzhansky was able to determine that the degree of abnormality of the testes of a fly was directly proportional to the number of his autosomes which were derived from a different species from that which contributed his single X chromosome. This means that genes tending to reduce testis size in the hybrids are present on every autosome of the parental species.

Also, related species are usually separated from each other, not by one, but by several different isolating mechanisms. All this means that the process of speciation usually involves the incorporation of many mutations over a long period of time.

EVOLUTION IN PROGRESS

In the preceding discussion evolution has been described as a process that has been going on for many millions of years and is still going on. How much of the process, if any, are we able to observe? The importance of the time factor in the process has been emphasized, and with this in mind perhaps one should not expect to be able to observe many significant changes in the course of a lifetime. In the course of several lifetimes one could expect a little more—and still more in the course of many generations.

It has been pointed out that there is a difference between just the modification of species and the formation of new ones. Several things can be pointed out about the observed modification of species. Reference has already been made to the great number of different varieties of domesticated plants and animals that were developed in historical time. In these cases we have done the selecting, and although this differs from what happens in nature, it shows the modifiability of species under selection and thus has a bearing on the problem. There are, in addition, a number of cases where species have been observed to change in nature, and a few of these cases are mentioned.

Records show that populations of certain species of insects have, in recent times, changed from predominantly light-colored forms to dark-colored forms. These changes have occurred in industrial regions characterized by large amounts of smoke and soot. Originally the light-colored forms blended well with their background. When black mutants appeared, they were easy prey for birds. But now, with a darker background in many regions the dark-colored forms have become the prevailing types. Kettlewell, who furnished the photographs for Figure 42.3, has studied this phenomenon (*industrial melanism*) in England.

His chief study was on the light and dark forms of *Biston betularia.* The black form was first reported from Manchester, England, in 1848. The black form is caused by a dominant mutation. Collections made in 1958 showed that 90 percent were black and only 10 percent light. Kettlewell released marked moths (both the dark and the light varieties) in a polluted woods near Manchester and also in an unpolluted woods. He recorded their captures on film. The results were quite definite: more of the light moths were captured in the polluted woods and more of the black moths were captured in the unpolluted woods. In recent years there has been a drive on in England to cut down on the smoke in industrial regions. It is interesting that a count in 1974 on the frequency of the two forms of *Biston betularia* in the Manchester region showed an increase in the light forms over the 1958 count.

Figure 42.3 *Biston betularia*, the peppered moth, and its black form, *carbonaria.* **(a)** Light and dark forms on a soot-covered oak trunk near Birmingham, England. **(b)** Two forms on a lichen-covered tree trunk in a soot-free region. (From experiments of H. B. D. Kettlewell, with permission.)

In society's attempts to control various insects that are injurious to crops or that are household pests, a number of new chemicals have been developed. In many cases a specific chemical has been found that would effectively control a given pest for a few seasons, but which seemed later to lose its effectiveness. A good example is resistance to DDT. A study of some of these cases has revealed that resistant strains have developed. There seems to be little doubt that in some cases the resistance was a product of mutated genes and was inherited. A similar situation has been observed in the fungus that causes wheat rust. As mentioned earlier, plant breeders have been able to develop rust-resistant strains of wheat. Such a strain may work well in a given locality for a few years, after which it may become susceptible to rust. It now appears that this happens because a new strain of fungus has developed. In this situation the fungus must either become adapted to the new strain of wheat or die out. So just as the plant breeder practices artificial selection on the wheat, natural selection operates on mutations in the fungus and new strains appear.

With regard to the observation of the formation of new species, there is less to say. By and large, the first steps in speciation seem to require spatial isolation. An exception to this, however, is found in the phenomenon of polyploidy—a phenomenon rather common in the plant kingdom. As a result of irregularities during cell division, a plant may come to have more than two haploid sets of chromosomes in each of its cells. The increase in number of chromosomes by multiples of haploid sets is called *polyploidy*. Such a condition may arise in various ways from diploid ancestors. One way this can happen is by the formation of diploid germ cells. If a diploid egg is fertilized by a diploid sperm nucleus, the result is a zygote with the $4n$ condition, called a *tetraploid*. Many tetraploids are entirely fertile among themselves or are self-fertile. But there is a high degree of sterility between the tetraploid and its ancestral diploid type. In such cases reproductive isolation has occurred without spatial isolation. Studies of the chromosomes of plants indicate that polyploidy has been a rather common method of speciation in plants. This has not been true, however, with animals.

It was stated in an earlier chapter that in the past most organisms have been classified primarily on the basis of morphological features. In the discussion in this chapter the statement was made that separated populations of a species must remain apart spatially a long time in order for isolating mechanisms to become established, if they are to develop into separate species. If this is so, then it should be possible to find related separated groups that are in the process of becoming completely reproductively isolated.

In Chapter 20 reference was made to John Moore's studies on the frog, *Rana pipiens*, which has a distribution from Minnesota to Texas. When specimens from Minnesota and Texas were crossed no viable offspring were produced, yet matings from closely adjacent regions between Minnesota and Texas always gave viable offspring. Jens Clausen has described another interesting situation in a group of plants of the genus *Layia* of the daisy family (often called "tidy tips") in California. These species differ in morphology, geographical distribution, and habitat preferences.

Six species have been recognized. Figure 42.4 shows their distribution. Geographically they fall into three groups. One consists of two species: (C) in the region around San Francisco Bay and (F) directly east of the first in the foothills of the Sierras. The second group consists of three species. These are located about 250 miles south of San Francisco: one (J) on the coast, another (M) about 50 miles inland, and the third (L) about 100 miles inland. The third group consists of one species (P), and is located along the coast from southern California to north of San Francisco Bay. For part of its distribution it overlaps the distribution of members of the first two groups.

Clausen has made matings between all the species, and in all cases vigorous hybrids re-

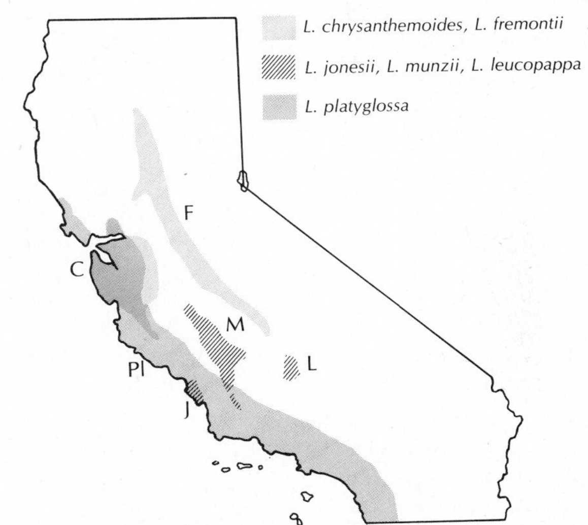

Figure 42.4 Distribution of six species of genus *Layia*, or "tidy tips" (Compositae), found in California. Letters on map are the first letters of the species. (After Stebbins, *Processes of Organic Evolution*, © 1966 by Prentice-Hall, with permission.)

sulted. Hybrids between the two species in the first group showed 25 to 30 percent of normal fertility. The same was true for the hybrids between the three species in the second group. This might be taken to indicate that these species are really on the borderline between subspecies and species. Hybrids between populations of the first group and those in the second group were much more often sterile. Their chromosomes do not pair normally in meiosis. This would indicate that these *two groups* have certainly reached the stage of separate species. Hybrids involving the third group are almost completely sterile. Furthermore, where members of this species overlap other species in their distribution, they occupy quite different habitats and there are no signs of gene exchanges between them and the members of the other groups.

This study provides an example of evolution in progress; for example, not all these species have developed complete isolating mechanisms.

ORIGIN OF HIGHER TAXONOMIC CATEGORIES

In our taxonomic system species, genera, families, orders, classes, and phyla are recognized. An account has been given above of the processes involved in the formation of new species. Have these same processes been involved in the formation of the higher categories, or have other factors been involved?

G. Ledyard Stebbins has given an interesting discussion of this problem. He says that the answer to the above question depends on the answer to the following questions: "Are the categories of the systematic hierarchy intrinsic entities which the naturalist merely discovers, or are they groups which naturalists themselves have established in order to understand better the complex patterns of living beings in nature?" Stebbins goes on to say that if the categories are intrinsic entities, it would be predicted that the more intensively a group was studied by different workers, the more easily could they agree on the limits of the categories. On the other hand, if the categories are created by investigators, it might be predicted that they would have difficulty in agreement on the limits of the categories because each worker would have somewhat different ideas as to which characteristics are more important.

When the different groups of organisms are considered it is found that the first prediction holds for many of them. For example, the orders of mammals have been recognized as such for over a hundred years, and modern knowledge has, with only few exceptions, confirmed the system. Within many of the orders of mammals modern families are also widely agreed upon. A similar situation exists in higher plants. In the pines and other conifers, genera are well defined and agreed upon. However in some of the flowering plants, such as grasses, which are flourishing and evolving, genera are often hard to define. In a similar way among the mammals, the rodents, which are still flourishing, are often difficult to delineate.

This seems to indicate that groups that originated a very long time ago show clearcut characteristics that can be agreed upon easily, whereas groups that originated late in the evolutionary process may provide difficulties for the student of classification. What does this mean? According to Stebbins we can only understand the origin of higher categories by taking into account *extinctions*—extinctions of intermediate forms. When many extinctions have occurred, gaps between groups are formed. This applies to the gaps between genera, families, orders, classes, and even phyla. When intermediate groups are still living, there are no gaps and the student of classification has trouble.

The factors or processes described above (mutation, recombination, natural selection, chromosomal change, and reproductive isolation) operate at all times in the process of evolution and result in the production of new species. But to understand the origin of the higher taxonomic categories, the factors of extinction and time expanses of millions and millions of years must also be taken into account.

SUMMARY

Much remains to be discovered about the mechanisms of evolution. No attempt has been made here to discuss all details of the mechanisms that have been studied because some of them are too involved for an introductory presentation. Rather, an attempt has been made to outline broadly the major features of the mechanisms that are recognizable today. The following have been considered as essential: (1) mutations, which provide the really new material for the process; (2) sexual reproduction, which provides a means for increasing the diversity in a population through the recombinations that result from independent assortment and crossing over; (3) natural selection, which is involved in the perpetuation of some of the mutations that arise; (4) population, which is the unit of evolution; and (5) reproductive isolation, which is essential for species formation.

The amount of evidence for organic evolution is tremendous, and those who have studied the evidence agree that all organisms, past and present, are products of this process. Some individuals say that scientists do not agree about evolution. An individual who makes such a statement does not take into account the two aspects of evolution: the *fact*, as seen in the evidence, and the *explanation*. There is agreement about the fact, but sometimes scientists disagree about the details of the mechanism. However, more and more agreement is being reached in this area.

The process, as described, involves much randomness. It has been a much-branching process with many blind alleys, not a straight-line process leading to a single goal. There has been much of what Simpson calls "opportunism" in the process. In view of these aspects, some are inclined to ask whether this does not disprove a Creator. It is not possible to go into all ramifications of that question here, but it can be said that science does not disprove a Creator. Perhaps a fair statement of the situation is this. Some *believe* that organic evolution is a process entirely materialistic in its origin and operation; others, partly because science does not tell us about first causes—about the origin of the laws and properties of matter in general under which organic evolution operates—and partly because they see some idea of design in the whole process, *believe* in a Creator.

THE ORIGIN OF LIFE—A HYPOTHESIS

The discussion of evolution has been based on the assumption that a very simple form of life arose in the dim and distant past and that from this simple beginning, with the operation of the various causal factors of evolution that have been described, has resulted the great variety of living forms that have lived in the past and those that are now living. In this study quite a bit has been said about the results of evolution and about the mechanisms operating, but practically nothing has been said about the origin of life.

Louis Pasteur provided the final proof that spontaneous generation of life does not occur under present conditions. Darwin and his successors have discredited the old idea of the special creation of living things just as we find them in the world today. As a result of these two findings, biologists for many years had no real anchor as far as the origin of life was concerned. It was not until 1936, with the publication of Alexander Oparin's *The Origin of Life* that a reasonable working hypothesis appeared. Although we may never be able to prove that life first arose in a specific way, it is now possible, with the information available, to formulate a working hypothesis concerning the conditions and circumstances under which life might have arisen. This working hypothesis or modifications of it may make it ultimately possible for biochemists to simulate in the test tube the conditions under which the first living entities arose on the earth many millions of years ago and thus demonstrate how the first living things might have started. The hypothesis involves spontaneous generation of a sort, but not the appearance of life all in one stroke and not under the same conditions that are found on the earth at present. Rather what one may think of as the first living thing arose gradually, through many steps and in a changing environment quite different from that of today.

Actually, to envisage the origin of life one must broaden the meaning of the term "evolution." Students must consider an evolution of chemical elements, an evolution of inorganic chemical compounds, and an evolution of organic compounds before they consider living matter first arising. From the data of astronomy and chemistry there is evidence that our universe began between five and six billion years ago, possibly in an explosion. Matter was in the form of elementary particles, such as electrons, protons, and neutrons. Soon these particles combined to form the elements helium and hydrogen, and later the other elements were formed. In time compounds formed, first inorganic and later organic. It is estimated that our sun and its planets were formed a billion years later, which would make our earth about four and a half billion years old.

Proof that spontaneous generation does not occur now is not proof that it could not have occurred under other environmental conditions. What were the conditions of the earth and its atmosphere during the first two billion years of its existence? As the mass that made our earth cooled, atoms of the various elements combined to form simple compounds. In the surface gas of the earth mass were atoms of hydrogen, nitrogen, oxygen, and carbon, and these combined to form water vapor (H_2O), ammonia (NH_3), and methane (CH_4). According to Harold Urey the primitive atmosphere of the earth contained water vapor, ammonia, methane, and hydrogen gas; there was no free oxygen and little, if any, carbon dioxide. Upon further cooling, a crust formed on the earth, vapors and gases condensed, and rain started to fall. In time the seas were formed, in which were dissolved ammonia and methane, in addition to salts and minerals washed in by the rains. Much high-energy radiation in the form of ultraviolet light came in from the sun. The atmosphere was characterized by violent thunderstorms and much lightning. Such was the stage, according to the hypothesis upon which the first signs of life were ultimately to appear.

It was stated in Chapter 3 that, in addition to much water and minerals, living material is composed of the four basic kinds of organic compounds: carbohydrates, fats, proteins, and nucleic acids. We also know that even though many different organic compounds have been synthesized by chemists in the laboratory, organic compounds in nature are only formed by living organisms in cellular syntheses. Is it possible that organic compounds were formed in nature before the origin of life? In the present working hypothesis it is assumed that formation of organic compounds did indeed precede the first living things and that during the first two billion years of the earth's history many kinds of organic molecules were formed and accumulated. Some writers have referred to the contents of the seas of this period as "organic soup." There was energy available for such syntheses in the form of ultraviolet light and of lightning; raw materials were present; although no enzymes were present, there were millions and millions of years available in which these possible reactions could occur; and the compounds could have accumulated, once formed, because there were no decay bacteria around and there was no free oxygen to attack them.

Is there any evidence that organic compounds will form under conditions that simulate those assumed to exist in the early history of the earth? The now famous experiments of Stanley Miller in 1953 provide such evidence. Miller's experiments involved circulating a mixture of water vapor, methane, ammonia, and hydrogen gas in a closed system continuously for a week and over an electric spark (see Fig. 42.5). The circulation was maintained by boiling water in one limb of the apparatus and condensing it in the other. When the liquid was analyzed, using paper chromatography, sizable amounts of the amino acids glycine and alanine were detected along with traces of a few other amino acids and some unidentified organic compounds. In a later experiment Miller was able to identify several other amino acids.

The source of the energy for the syntheses in Miller's experiment was an electric spark, which simulated the lightning in the primeval skies. Other investigators have used ultraviolet and high temperatures as energy sources for similar syntheses. In 1970 a husband and wife team, Akibs and Nurit Bar-Nun, working at Cornell University, were able to show that shock waves, such as might have been produced by thunderclaps or by meteors plunging into the atmo-

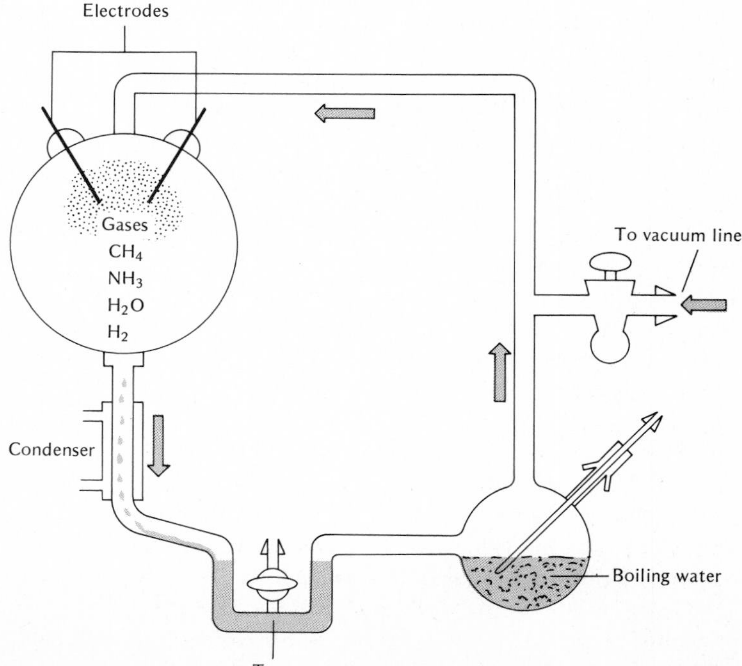

Figure 42.5 Spark discharge apparatus used in Miller's experiment simulating conditions assumed to exist early in earth's history.

sphere, could have been involved in such syntheses.

They filled one end of a tube with a mixture of ammonia, methane, ethane, and water vapor. This mixture was separated by a thin plastic membrane from the other end of the tube containing the relatively chemically inert gas, helium. These experimenters increased the helium pressure until the membrane broke. This produced a shock wave, and the temperature increased momentarily several thousand degrees. In several different experiments they found that at least four different amino acids had been formed. They also found a relatively high yield, with 36 percent of the ammonia present converted into amino acids. Thus another possible mechanism for such syntheses has been demonstrated.

It has been shown that many kinds of organic molecules have a natural tendency to form large aggregates, and that in some of these the molecules orient in a specific way to provide a certain organization or structure. The organization of collagen is a good example. Aggregates of various kinds may interact, forming larger and more complex entities. It has been suggested that growing aggregates, when they reach a certain size, may break up, with each particle able to grow. It may have been, as Oparin suggests, that selection started to operate in the ancient seas, favoring those aggregates that could most readily capture from the environments the molecules necessary for their growth. The question of the nature of the first particles that exhibited precise replication is still to be solved, but from what we know about genes and viruses, the assumption is made that this must have started when something like DNA was present. Since the original work of Miller, many other experiments have been performed. It has been shown that a number of important compounds, including organic acids, purines, pyrimidines, and many of the amino acids, may be synthesized from methane, ammonia, hydrogen, and water under prebiological conditions.

It is assumed that, in time, the changing conditions on the earth prevented the further formation of organic compounds. The original replicating particles, which may have been viruslike, presumably gave rise to cellular forms. But just how the first cells came into existence we do not know. For a long period these simple forms lived upon the organic compounds in their environment. It should be noted here that these first forms of life were *heterotrophs,* not *autotrophs,* as the older theories assumed. According to the present hypothesis, these first living organisms utilized the organic material around them, in a form of fermentation, for their energy sources. As has been seen in Chapter 5, fermentation or anaerobic respiration is the first stage in the respiration of most living forms today and is the only form of respiration used by some organisms. It is not a very efficient process, but it does suffice for simple organisms. In this next long period of the evolution these primitive anaerobic forms were adding carbon dioxide to the atmosphere, but they were also depleting the supply of organic compounds that had formed and accumulated. The production of carbon dioxide was important because this, as we know now, had to be present before any simple form of photosynthesis could appear. It is assumed that in this period, and before the store of organic compounds had become exhausted, one series of mutations (of the many that must have occurred in these simple forms) provided the mechanism for photosynthesis. This made possible the great diversity of life found today.

The first simple photosynthetic organisms were the first autotrophic organisms in the process of evolution. They, of course, began adding oxygen to the atmosphere. The addition of free oxygen to the atmosphere was important for two reasons: (1) with free oxygen present, organisms could evolve the mechanisms for aerobic respiration, a much more efficient form of respiration; and (2) with free oxygen present, a layer of ozone (O_3) formed high in the atmosphere and filtered out ultraviolet radiations so that life was possible on land and in the air below the protective layer. We assume that from these first autotrophs evolved all other living things—the long succession of photosynthetic plants and the many kinds of heterotrophs, both plant and animal.

It was stated above that we do not know just how the first cells came into existence. However, it must be inferred that they were of the procaryote type. The oldest known fossils (3.1 billion years old) from South Africa were bacterialike. Furthermore many of the fossils from the Gunflint formation in Canada (1.9 billion years old) resemble blue-green algae. The oldest known eucaryotic fossils are of green algae in the Bitter Springs deposits in Australia, which are 1 billion years old. How did eucaryotic cells develop from procaryotic cells? Again we really do not know. However, some biologists have suggested that many of the organelles that characterize eucaryotic cells were once independent organisms that somehow came to live symbiotically in larger host cells. In a recent article Margulis describes how such a transformation might have taken place.

Not long ago there was no basis for a reasonable working hypothesis about the origin of life on this earth. Now that we have one, and with astronomers telling us that there are other planets like the earth scattered throughout the vast reaches of the universe, many predict that life has probably evolved on some of those other planets.

SUGGESTIONS FOR FURTHER READING

Darwin, C. *On the Origin of Species by Means of Natural Selection.* (A reprint of the second edition.) New York: Oxford Univ. Press (1951).

Dobzhansky, T. *Genetics and the Origin of Species.* New York: Columbia Univ. Press (1956).

Mayr, E. *Animal Species and Evolution.* Cambridge: Harvard Univ. Press (1963).

Miller, S. L. "A Production of Amino Acids under Possible Primitive Earth Conditions." *Science,* **117,** 528 (1953).

Miller, S. L., and L. E. Orgel. *The Origins of Life on the Earth.* Englewood Cliffs, NJ: Prentice-Hall (1973).

Mueller, G. "Organic Microspheres from the Pre-Cambrian of South-West Africa." *Nature,* **235,** 90 (1972).

Oparin, A. I. *The Origin of Life,* 3d ed. New York: Academic Press (1961).

Simpson, G. G. *The Major Features of Evolution.* New York: Columbia Univ. Press (1953).

Stebbins, G. L. *Processes of Organic Evolution,* 2d ed. Englewood Cliffs, NJ: Prentice-Hall (1971).

From Scientific American

Dobzhansky, T. "The Genetic Basis of Evolution" (Jan. 1950).

Ehrlich, P. R., and P. Raven. "Butterflies and Plants" (June 1967).

Eiseley, L. C. "Charles Darwin" (Feb. 1956).

Glaessner, M. F. "Precambrian Animals" (Mar. 1961).

Kettlewell, H. B. D. "Darwin's Missing Evidence" (Mar. 1959).

Lack, D. "Darwin's Finches" (Apr. 1953).

Margulis, L. "Symbiosis and Evolution" (Aug. 1971).

Wald, G. "The Origin of Life" (Aug. 1954).

43 Behavior: Social and Other Aspects

In Chapter 14 the subject of behavior was introduced; there the focus was on learned behavior. Some of the basic types of innate behavior are examined in this chapter, and some of the techniques that modern workers have used to gain further understanding of these processes are described. In addition brief discussions of the modifications of innate behaviors by learning, social behavior, the genetics of behavior, the evolution of behavior, the role of the hypothalamus in behavior, and rhythmic behavior are included.

The simplest of the innate behaviors in animals is the reflex. Reflexes were discussed in Chapters 13 and 14. Also in Chapter 14 some examples of the complicated activities of animals that are primarily innate and that are often called *instincts,* or fixed action patterns, were presented. More of these are considered here.

TAXES

Before we deal further with instincts, we present a brief consideration of a simple type of response, known as taxis. A *taxis* is the orientation of an organism to some aspect of its environment. A taxis differs from a reflex in that the whole organism is involved in a taxis, whereas only a part may be involved in a reflex. At the beginning of this century Jacques Loeb, a pioneer in the study of behavior, used the term *tropism* for this type of response. Loeb had been impressed with the studies on plant tropisms in which it had been demonstrated that the growth movements of plants are due to the differences in stimulation of light, gravity, and other factors on the two sides of a stem or root. Loeb developed a theory of *animal tropisms,* or *forced movements,* and his aim was to show that the pattern of animal movement was caused by differences in stimulation on its two sides. Most workers today reserve the term "tropism" for the orientation of plants by growth and use the term "taxis" for orientation movements of animals that are continuously and specifically guided by an external stimulus.

Most of the orientation movements of animals that clearly fit into this category are those involving light (phototaxes). A good example of this is the grayling butterfly, which flies toward the sun in its escape from predators. As shown in Figure 43.1, the butterfly orients to the sun by turning to that position in which both eyes are equally stimulated. If one eye of the butterfly is experimentally blinded, it flies in circles. This demonstrates its dependence on bilateral optical stimulation for the normal escape reaction.

INSTINCTIVE BEHAVIOR

In recent years some psychologists have objected to the term "instinct." There are several reasons for this. The word has been used very loosely to describe many habitual or unconscious acts. It has also been used to imply a mysterious or vitalistic force. Some behaviors have been described as instincts when obviously they can be modified by experience or learning. Those who object to the term argue that since all behavior is a product of both genetic factors and environmental factors, a category that implies that a behavior is entirely innate should not be used. The authors feel, however, that many complex behaviors are primarily innate and that the use of the term is justified.

Figure 43.1 Grayling butterfly flying toward sun. After release it orients itself so that both eyes are equally illuminated by the sun.

The types of behavior discussed here as instincts have a few identifying characteristics: they are species-specific; they can be performed without an opportunity for learning, although they may be modified with experience; they are more or less stereotyped; and they are immediately adaptive. An examination of a few examples illustrates these points.

William Van der Kloot (see Suggestions for Further Reading) has described the spinning of a cocoon by a caterpillar of the cecropia moth and has reported some interesting experiments in his attempts to analyze the process. In the life cycle of this moth, the larvae undergo five molts. At the time of each of the first four molts, the larva stops feeding, finds a twig, and spins a thin sheet of silk on the twig. The silk pad serves as an anchor for the old skin as it is pulled away. Just before the fifth molt, the larva again stops feeding, finds a twig, and then begins to spin out silk, but this time it spins out a thread about a mile long and weaves it into a complicated cocoon. Figure 43.2 shows such a cocoon, which is composed of three distinct layers. The outer layer is dense, blunt at one end, and pointed at the other. The middle layer is a spongy meshwork. The inner layer is dense, like the outer layer and with a similar shape. The cocoon is

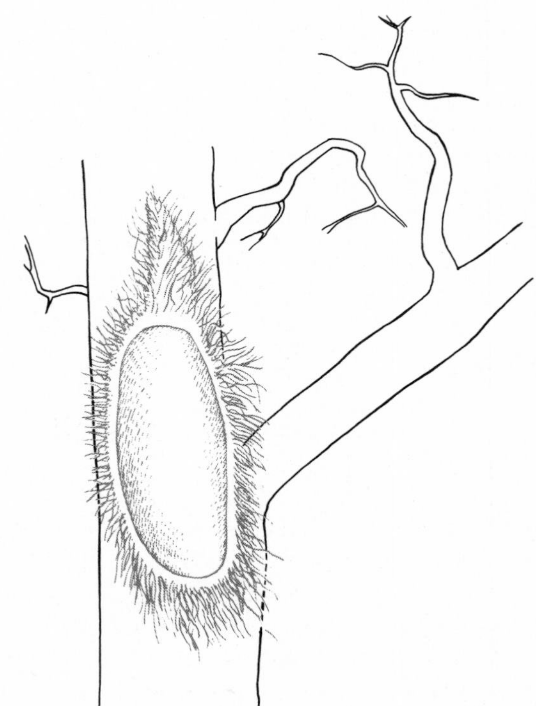

Figure 43.2 Structure of cocoon of giant American silkworm. There are three layers: a dense outer layer, a middle spongy layer, and a dense inner layer which contains the pupa. The upper end of the outer and inner envelope is loosely spun. (After Van der Kloot, *Behavior,* © 1968 by Holt, Rinehart and Winston, with permission.)

woven from the outside in. The structure of the cocoon is so characteristic of the species that it is an excellent taxonomic criterion. Within the cocoon the caterpillar molts into a pupa. The silk is spun loosely at the pointed ends of the cocoon, and the silk here can be pushed aside by the emerging moth in the springtime. It has been demonstrated that hormones are involved in much of this behavior (see Ch. 30).

Van der Kloot set out to determine just how this caterpillar forms its cocoon. In its movements the caterpillar stretches, bends, and sways, with the forepart of its body moving through sweeping areas as the silk thread is pulled from the spinneret. It was difficult to see any pattern in the movements. To determine the pattern he watched the process carefully in the laboratory, using the setup shown in Figure 43.3.

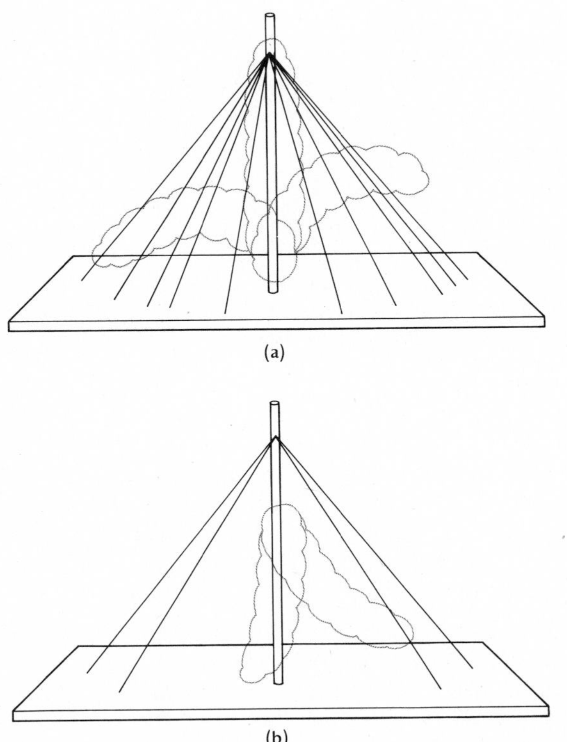

Figure 43.3 **(a)** Movements of a spinning caterpillar when facing upward. **(b)** Movements of a spinning caterpillar when facing downward. (After Van der Kloot, *Behavior,* © 1968 by Holt, Rinehart and Winston, with permission.)

A rod fixed vertically to a small board was used. By replacing the caterpillars on the board each time they wandered off, they eventually began to spin. The first steps in the formation of the outer layer are shown in Figure 43.3(**a**). With its head upward, the caterpillar makes the tentlike supports, producing a cone-shaped structure. Then in the lower figure, with its head downward, it makes the floor. When the caterpillar touches a network of the silk already laid down, it makes figure-eight movements. These movements lay the silk as a meshwork instead of a straight thread. At the apex of the cone there is not enough room for the caterpillar to make the figure-eight movements, and therefore in this region the cocoon consists only of vertical strands of silk. This region serves as the escape hatch for the emerging moth later on. The other layers are laid down inside this outer layer. Since the larva changes its position regularly, it must have some kind of built-in clock. Van der Kloot performed many experiments in his attempts to understand this behavior. One experiment was directed to the question, "Why are the escape valves or hatches always at the pointed ends of both the inner and outer membrane?" It was observed that the valves of both the outer and inner membranes are spun with the caterpillar facing upward. He allowed a caterpillar to spin a normal outer envelope and then turned the outer envelope upside down. In this situation the valve of the inner envelope is spun with the larva facing upward as usual, but the inner valve is at the opposite end of the cocoon and the moth will be trapped. Gravity, then, seems to play an important role in orienting the behavior of a spinning caterpillar. The formation of a cocoon with the inner and outer valves at opposite ends, however, does not occur in natural situations.

Another series of observations and experiments was designed to determine which factors are involved in changing the behavior of the caterpillar, so that the cocoon is divided into outer and inner envelopes. Larvae taken from completed outer envelopes and allowed to begin again, spin only an inner envelope. A caterpillar taken at the onset of spinning and sewed in an outer envelope taken from another larva, spins a complete cocoon of its own. When a caterpillar was removed from a half-finished outer envelope and was allowed to begin again, it formed an outer envelope of half-normal thickness and then a normal inner envelope. It appeared that the caterpillar was in some way able to measure the amount of silk spun in the outer envelope before starting on the inner envelope.

If a larva is kept on a flat surface, it cannot form a cocoon, but it will spin out the silk in a flat sheet. When a caterpillar is placed inside an inflated balloon, it spins out the silk in a thin sheet on the inside of the balloon. If a larva that has spun about 60 percent of its silk on a flat sheet is then placed in a proper situation, it will form only the inner envelope of a cocoon. It must be able to measure the amount of silk spun. That this is so was demonstrated by the removal of one silk gland and part of the other. These caterpillars always began by spinning an outer envelope. If silk remained after the outer envelope was completed, the inner envelope was started. They never spun less than the normal amount of silk in the outer envelope.

The regions of the caterpillar's brain involved in the control of spinning behavior were located by cutting and electrical cautery. About half of the brain can be removed without affecting spinning behavior. The regions essential for normal cocoon spinning involve the part of the brain called the "mushroom bodies." In the cecropia caterpillar these bodies make up a relatively small part of each brain hemisphere, but in insects such as bees and ants, which have a more highly developed behavior, these bodies are much larger.

Here in the spinning behavior of the cecropia caterpillar we find a complex series of actions that are quite stereotyped and innate. The behavior is species-specific and is performed without an opportunity for learning. Such behaviors can be thought of as arising in evolution through mutations and selection, just as structural adaptations arise in the process.

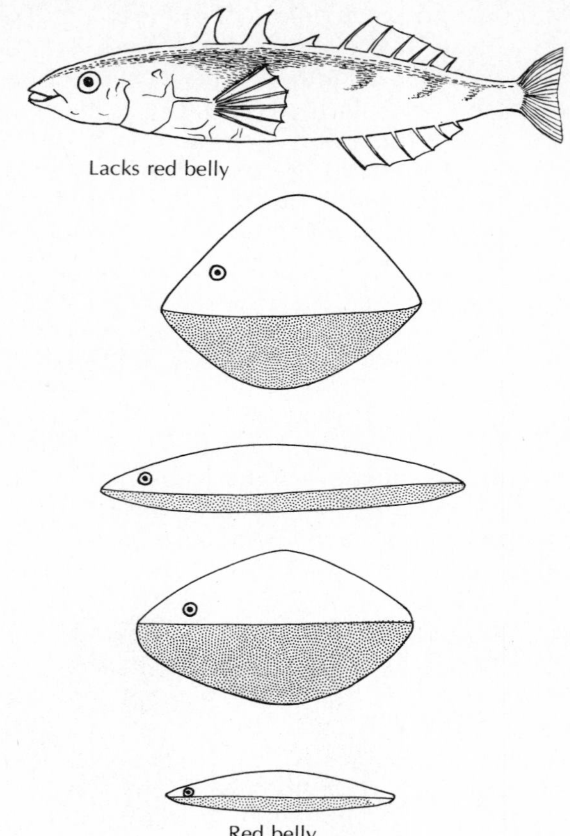

Figure 43.4 Fighting in the stickleback was elicited by the four models with unusual shapes but with red bellies. The true model, lacking the red belly, usually did not elicit fighting. (After Tinbergen, *The Study of Instinct,* © 1951 by Oxford University Press, with permission.)

Stimuli for Instinctive Behavior

The ethologists have contributed much to our understanding of instinctive behavior and they stress how instinctive behavior often depends on some special condition within the organism. This is often referred to as *motivation* and is determined by hormones, chemical states, and influences from various neural centers. For example, in numerous animals many aspects of reproductive behavior depend on the presence of sex hormones; in the absence of the hormones there is no response to strong sexual stimulation, while only minimal stimulation is required when the hormone concentration is high. Another point made by the ethologists is that a particular stimulus triggers a given instinctive behavior. Such stimuli have been called *releasing stimuli, sign stimuli,* and *signaling devices.* It is the ethologist's conception that each instinctive act is held in check by a neural-inhibiting mechanism that is released by a specific releasing stimulus when the organism is in the proper internal state. What are some examples of releasing stimuli?

In the spring of the year the male three-spined stickleback (fish) establishes a territory, which he defends against all other adult males of the species. Niko Tinbergen (see Suggestions for Further Reading) was able to show that the defense reaction is triggered by the red belly of another male. Figure 43.4 shows five models that were used to demonstrate this. One is an exact replica of a male stickleback, except that it lacks the red belly. The other four look little like a stickleback, but all have red bellies and all served as releasers of the defense reaction. David Lack observed a similar situation in the European robin. He found that a male robin would attack a tuft of red feathers more readily than it would a mounted young robin lacking a red breast (Fig. 43.5). But all releasers are not red colors! Male mosquitoes of the genus *Aedes* are attracted to females by a particular sound made by the female in flight. When a sound similar in pitch to that made by the female mosquito is made on a tuning fork, the male mosquitoes fly to the tuning fork. There is nothing about the tuning fork that resembles a female mosquito in sight and smell, yet male mosquitoes will react to this particular sound time after time. Males of some species of insects will react to the scent

Figure 43.5 The mounted young robin **(a)** with a dull brown breast was rarely attacked while the tuft of red feathers **(b)** was attacked often. (After Tinbergen, *The Study of Instinct,* © 1951 by Oxford University Press, with permission.)

of a female of its species. Such males will attempt to copulate with pieces of paper on which the female odor has been placed. Many organisms secrete substances (pheromones, Chs. 12 and 30) that affect the behavior of others.

Edward Wilson has recently described the activity of three insect sex attractants, taken from the silkworm moth, the gypsy moth, and the cockroach. The substance given off by the gypsy moth, *disparlure,* is so potent that 0.01 microgram (μg) (which is the average content of one female moth) would, if properly distributed, be adequate to excite more than a billion male moths! Ants secrete one pheromone to make a trail to food and another that triggers alarm reactions. It is now believed that much of the behavior of social insects is set off by different pheromones.

Many behavior patterns are complex, involving several separate activities occurring in sequence. The courtship behavior in the three-spined stickleback is a good example. Tinbergen has studied this behavior and has found that each activity serves as the initiator or the sign stimulus for the next. As mentioned above, each male stickleback leaves the school in the early spring and stakes out a territory of its own. It then builds a nest, and at this time its pink belly changes to the characteristic red color. When a female whose abdomen is enlarged with eggs enters the territory of the male, the male is stimulated to swim toward the female in a series of zigzag movements (Fig. 43.6). This continues until the female notices the male; then she in turn is stimulated to swim toward the male in a head-up posture. The male turns and swims toward the nest; the female follows. At the nest the male makes a series of rapid thrusts with his snout in the entrance and then turns on his side and raises his dorsal spines toward the female; the female enters the nest and rests there. The male then prods the tail of the female, which induces her to lay eggs. The female then leaves the nest, and the male enters and releases sperm to fertilize the eggs. Following this, the male chases the female away and starts to look for another female.

A given male may escort up to five females to a single nest, and the behavior is the same in each case. Each movement is the releaser for the next activity. It has been shown, for instance, that the prodding of the female's tail by the male is necessary to induce egg-laying in the female. However, this stimulus can be effectively duplicated by prodding the female's tail with a glass rod. After a few trips to the nest with different females, the male's mating impulses subside. His color begins to change, and he becomes hostile to females. The male guards the nest and fans water over the eggs to increase the oxygen supply until the eggs hatch.

Releasing stimuli produced by different members of the same species actually constitute means of communication. One of the classical studies in animal communication is that of Karl von Frisch on the language of honeybees. When a scout bee locates the source of nectar, it returns to the hive and communicates this to other worker bees in the hive. Von Frisch identified the bees with dots of paint of different colors and used an observation hive made with a glass wall so that he could observe what was happening inside. He also used marked feeding places. He found that two types of dance are involved in this communication.

When a source of food is located near the hive [within 84 m (275 feet)], the scout bee performs a *round dance* (Fig. 43.7) on its return. The dance consists of circling first to the right, then to the left, and repeating this pattern over and over

Figure 43.6 Courtship behavior in the stickleback. The male (shaded) performs a zigzag dance direct to the female (not shaded). The female turns toward the male and assumes an upright posture; then the male swims toward the nest and the female follows. The male shows the female the nest by placing his snount in the entrance and rolling on his side, the female then enters the nest. The male prods the female at the base of the tail which stimulates her to lay eggs. After the female leaves the nest, the male enters and fertilizes the eggs (After Tinbergen, The Study of Instinct, © 1951 by Oxford University Press, with permission.)

with great vigor. The dance excites the workers in the vicinity of the dancer, and they follow her with their antennae held close to her. Soon the worker bees begin to leave the hive, and in a short time they appear at the feeding place. The round dance conveys information that there is food nearby. The worker bees detect the particular food odor present on the scout bee with their antennae.

For greater distances, which may involve a few miles, another dance is used. Von Frisch called this the *wagging dance* (Fig. 43.7). In this dance, according to von Frisch, the bee communicates both the direction and the distance to the source of food. At the start of this dance the bee runs along a straight line, wagging her abdomen. Then she turns in a semicircle back to the starting point and dances the straight line again. This time she turns in a semicircle in the opposite direction and goes back to the starting point, ending one cycle. The pattern is repeated over and over. The speed of the dance indicates the distance away; the farther away the food, the slower the dance.

Recently William Wenner has made observations and experiments that suggest that the bee uses sound to communicate the distance during the wagging dance. He found that the bee emitted a train of sound during each straight run of its wagging dance and that the average length of the sound trains during a given dance was directly proportional to the distance that the bee had traveled to the source of food.

This dance is usually performed inside the hive on the vertical surface of the honeycomb. Von Frisch found that the scout bee indicates the direction of the food source in relation to the direction of the sun. This must be done on the vertical surface of the comb inside the hive. But how? If the food source is in the direction of the sun, the straight-line part of the dance is vertically performed, with the bee's head pointing upward. If the food source is away from the sun, in this part of the dance the bee is vertical, with

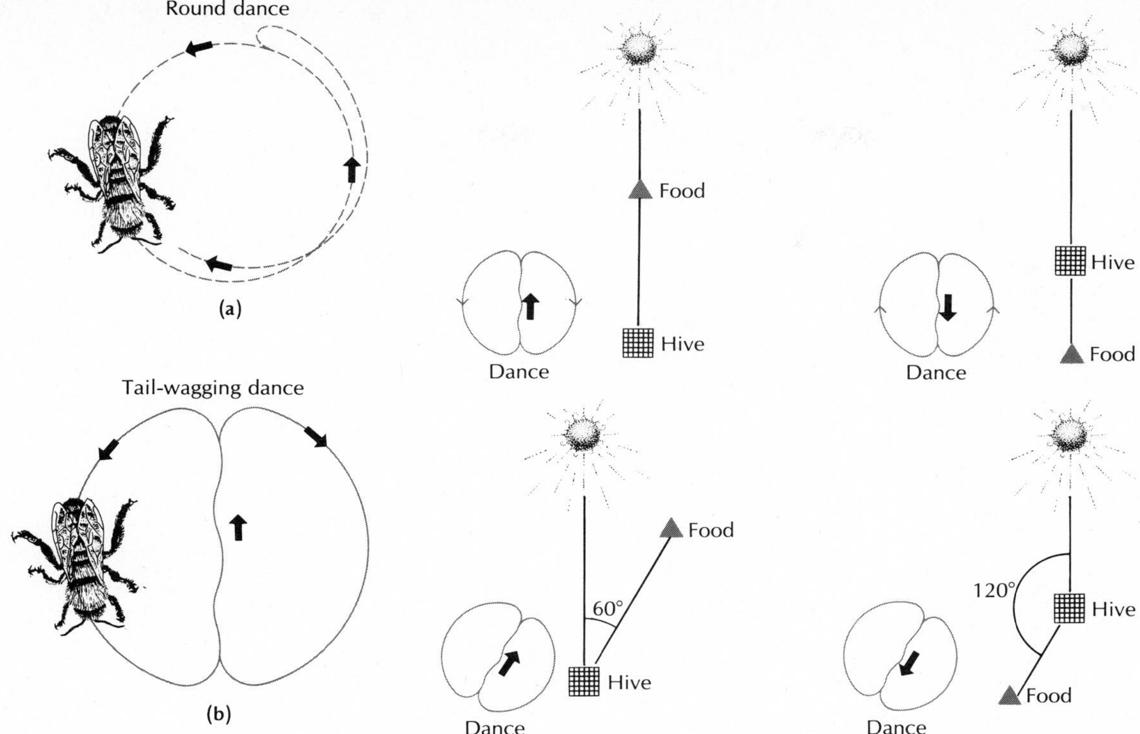

Figure 43.7 Dances of honeybees. **(a)** Round dance performed when food is less than 90 m from the hive. **(b)** Wagging dance. The other drawings show the relation between the angle of the dance and the vertical with the angle between the sun and the food. (From Karl von Frisch, *Bees, Their Vision, Chemical Senses, and Language,* © 1950 by Cornell University Press, with permission.)

the head pointing down. If the food is at some angle to the right or to the left of the sun, the straight-line dance is performed at the appropriate angle to the right or to the left of the vertical (Fig. 43.7).

Bees can use the sun as a guide for their foraging and communication activities, even on very cloudy days, if there is only a small area of blue sky. This is possible because their compound eyes are able to analyze the polarization of the light reflected from the particles in the atmosphere. Since the plane of polarization of light from any point in the sky is related in a fixed manner to the position of the sun and to the position of the observer, detection of this plane indicates to the bee where the sun is.

This system of communication is effective because worker bees, after watching the scout bee, leave the hive and find the source of food while the scout bee remains behind. This behavior of the scout bee in the dance and of the other bees in finding the source of food is innate. Different strains of bees show some variations in the dance.

To say that the responses of the bee just described are innate does not imply that the bee is incapable of learning. Von Frisch demonstrated that honeybees could be trained to associate certain background colors with sugar solutions and to distinguish between certain background patterns, such as a solid triangle versus an open triangle or a single bar from a double bar. William Van der Kloot (see Suggestions for Further Reading) states that it is necessary for bees to see the sun moving across the sky for the development of the behavior described above. He cites this observation: "A colony of bees was reared in a cellar, illuminated by a stationary lamp. The hive was then moved out-of-doors. At first the bees were unable to use the sun as a compass for locating food or dancing. When retested five days later, the sun was used as a compass in the usual way."

Modifications of Innate Behavior through Learning

There seems to be little doubt that the complex activities of the larva of the cecropia moth, described earlier, are unlearned and innate. However there are many other examples of innate behaviors that are modified through learning. The example of the bee using the sun as a compass, just described, is a case in point.

Some birds, such as thrushes, sing the song typical of the species even though they have been completely isolated from other members of the species at hatching. With the chaffinch, however, the situation is different. William Thorpe has studied the singing of chaffinches that were separated at hatching from their parents and reared in isolation. Such birds do not hear any song except the singing of the other hand-reared chaffinches. They all sing a similar song, but it is not the normal chaffinch song. Their song has a narrower range of frequencies, lacks the trills of the ordinary song, and does not have the final flourish. Thorpe isolated another group of birds that had been reared by their parents for the first six months of life. At this stage the birds had never sung a note. They were kept in a soundproof room with birds of other species, each singing its own song. Some months later the chaffinches began to sing. Their song was very similar to the normal chaffinch song except for the terminal part. Apparently the young birds during the six months with their parents had learned the greater part of the chaffinch song, even though they did not actually sing until several months later. Experiments show that when such experience is necessary, birds are able to pick out the song of their own species from the background of other bird songs —they are particularly sensitive to the song of their own species.

The ordinary observer will be convinced that many complicated activities must be learned. In Chapter 14 the reader was warned that to reach such a conclusion may often be a mistake. Each case must be investigated. The flying movements of a pigeon are very complex. To test whether these movements are innate or learned, young pigeons were confined in narrow earthenware tubes until other young pigeons of the same age were flying. When the confined birds were released, they flew just as well as the other young birds. These birds are able to fly as soon as maturation has occurred. As Tinbergen has pointed out, the clumsy nature of a young pigeon's first attempts to fly is probably due to the fact that the urge to fly was present before maturation was completed. However, studies on gulls and many other birds indicate that they must learn to land against the wind; this is not an innate behavior.

Old, experienced squirrels can quickly crack a hazel nut. This kind of nut has a groove in the shell, and an experienced squirrel will gnaw at the groove and quickly deepen it. It then turns the nut around and with a quick hard bite opens the nut. A young squirrel that has never cracked a hazel nut will, at first, gnaw vigorously all over the surface of the nut, unable to open it. The young squirrel must learn by experience to gnaw at the groove in the nut.

GROUP BEHAVIOR

Individual members of all sexually reproducing species of animals must associate with members of the opposite sex at least for reproduction. In some instances this is the only close association shown, that is, such organisms spend the rest of their time as solitary individuals. But in most cases members of the same species have associations with others of that species in varying degrees. The types of communications described above (visual, sound, chemical) are utilized not only at times of reproduction but within the various intraspecific groupings that exist. In general (there are some exceptions, such as the chaffinch song mentioned above) the signaling devices found in group communications are innate, although the responses to the signals may be improved by learning in many animals.

Animal Aggregations

In the study of animal communities, one often finds, in addition to isolated individuals, fairly closely aggregated groups of animals of the same species. Some of them are temporary aggregations, such as the breeding aggregations of frogs in the spring, the dense aggregations of box-elder bugs on a tree trunk during the summer, and the roosting aggregations of certain birds. Some of the aggregations involve physical contact, as in paramecia, planarian worms, sow bugs, and many others. Other aggregations occur in which physical contact is not the rule. Such aggregations are found in swarms of gnats, colonies of ants and bees, schools of fishes, flocks of birds, herds of various ungulates (deer, horses, and so forth), packs of wolves, and groups of various other mammals, including humans.

It has long been established that crowding, or high population density, has harmful effects. How then can one explain this tendency in many species to form aggregations?

Numerous experimental studies have demonstrated that a group of organisms is often able to survive where an isolated individual cannot. This has been shown in a variety of aquatic organisms, including protozoans, worms, and fishes, when they are exposed to toxic substances in the water. In one experiment a group of 10 goldfish and an isolated goldfish were exposed to the same dose of colloidal silver. The group survived but the isolated fish died. In this case it was shown that the slime secreted by the grouped fish changed much of the colloid into a less toxic form, and they were then not exposed to a lethal dose. A group of water fleas, *Daphnia,* will survive longer in the same volume of highly alkaline water than will a single one. In this case the grouped animals give off more carbon dioxide that neutralizes the alkali. Almost everyone who has turned over a log in the woods has found dense clusters of sow bugs, or pill bugs. This aggregation is one that conserves moisture. If 100 sow bugs are placed in a dish in dry air and a single one is placed in another dish alone, the single one soon dies from desiccation, but the group survives. There is greater survival in a group of planarians exposed to ultraviolet light than there is with the same number exposed individually. In the densely packed group many of the individuals are protected by shading. The heat-conserving value of a hibernation aggregation seems obvious. It can be demonstrated in the laboratory with mice. A group exposed to extreme cold will survive, whereas isolated individuals will die.

Some of the mechanisms that give survival value to some of the aggregations of higher animals are obvious. The many pairs of eyes in a herd of wild horses or in a flock of birds afford greater protection from enemies than does a single pair. Many sets of teeth in a pack of wolves are better insurance against hunger than is one set. Also the phenomenon of leadership is manifested. This brings organization and survival value to the group.

Thus survival value has been demonstrated for a number of different aggregations, and it may be assumed that there is survival value in all such groups that are not simply chance aggregations. Warder Allee has referred to this greater group survival as "unconscious cooperation." A genetic basis for behavioral patterns that have adaptive value for group survival seems implicit in the evolution of animals that form aggregations. This tendency to form aggregations, this unconscious cooperation, is found widespread in the animal kingdom, and is one of the basic principles of biology.

Animal Societies

Societies are highly integrated aggregations of individuals of the same species with a division of labor. The evolution of truly social animals has occurred independently in termites, bees, ants, and primates. Although the different societies of insects have evolved separately, they have much in common.

Termite colonies have complex social organizations. Termites are rather primitive insects related to cockroaches. A colony is composed of individuals belonging to three primary castes, the *reproductives,* the *workers,* and the *soldiers* (Fig. 43.8). The typical reproductive forms have wings, eyes, and functional reproductive organs. Periodic swarms of reproductives occur, usually in the spring, when the insects leave the parent colony, fly forth in search of mates, shed their wings, and settle down to start a new colony. Male and female termites live together throughout their lives (the ant and bee societies are primarily female societies, with the males functioning only to fertilize the female reproduc-

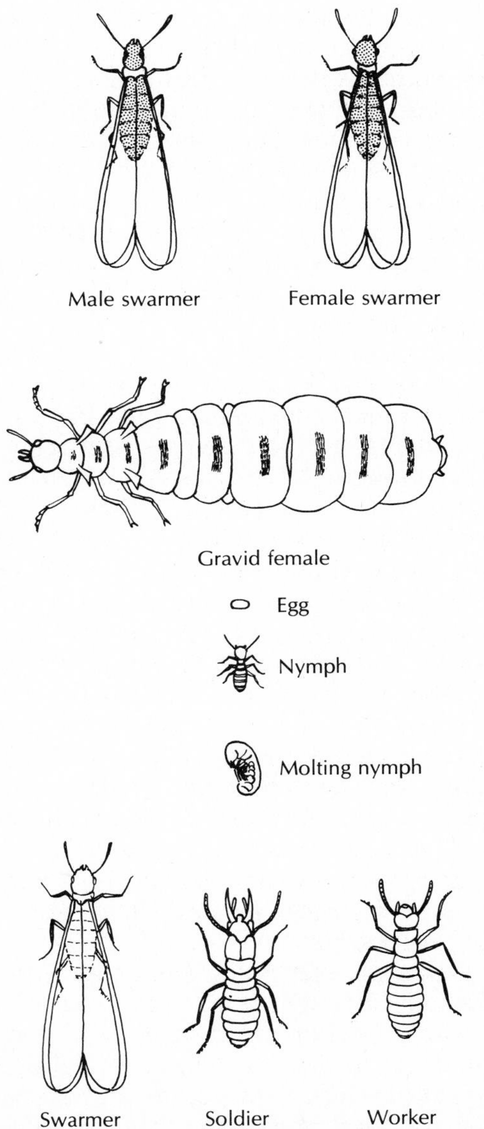

Figure 43.8 Different castes in the termite *Reticulitermes flavipes.*

tives). After mating, the termite queen starts to lay eggs. In many species, the abdomen of the queen expands until it is many times its original size. It is said that a mature queen termite produces more offspring than any other individual animal in the world. It is estimated that the queens in some species produce as many as 50 million offspring during a lifetime of twenty years or more. The eggs do not all develop into eyeless, wingless, sterile workers that eat wood, build the nests and tunnels, and feed the other members of the colony. Some develop into soldiers, and a variety of types are found in the different species. They are all specialized for protection. Some have very heavy mandibles for fighting; some have their heads elongated to serve as squirt guns for the ejection of a defensive fluid; and some have curiously crossed mandibles that are used for making warning signals on the substrate.

In an ordinary colony most of the individuals are workers. The problem of how, out of eggs produced by the same female, some of the nymphs develop into reproductives, some into soldiers, and some into workers, is a fascinating one. As a result of the experiments of Sol Light it has been established that the presence of males, females, or soldiers, respectively, inhibits the development of the same caste from nymphs. Young nymphs obtain their fauna of wood-digesting flagellates by licking the anal openings of adult members of the colony. Both reproductives and soldiers exude pheromones that are passed to the nymphs when they lick other members of the colony. This inhibits the development of either reproductives or soldiers until the population increases beyond a certain threshold. The worker does not inhibit the development of other workers, but inhibition by reproductives and soldiers results in the development of workers.

The mechanism for differentiation of the castes is different in the honeybee. The males are haploid. The worker (sterile female) and drone (male) larvae are fed royal jelly (a very nutritious secretion of the esophageal glands of worker bees) for the first two or three days, and then bee bread (a mixture of nectar and pollen) for another three days. The queen larvae are fed entirely with royal jelly, and this is the determining factor.

The high integration of insect societies with their striking division of labor and their general working together for the common good is a fascinating product of evolution. When compared to human societies, the differences are great. In the insect society each individual is in a fixed caste. Although insects exhibit some capacity for learning, their behavior is primarily instinctive. On the other hand, humans are characterized by great learning ability, by intelligence, and by reasoning ability. Psychological division of labor characterizes human society. Of all the products of evolution, human society is the only one that has the possibility of determining its future evolution.

It is often stated that with a few exceptions the behavior of humans is so affected by culture (learning) that innate patterns are hard to discern. However, Irenäus Eibl-Eibesfeldt (see Suggestions for Further Reading) discusses a number of instances of human innate behavior including the nursing behavior, the grasping behavior, and crying and smiling in human infants. He points out that children who are born blind or deaf–blind smile, laugh, and cry, and show expressions of anger, pouting, fear, and sadness —just like seeing children—although they could not have imitated anyone.

Eibl-Eibesfeldt has reported a number of observations, using the fast-motion film technique, that indicate other innate human behaviors. In an analysis of fast-motion films of persons eating it was found that individuals eating alone look up into the distance quite frequently with the gaze often sweeping automatically to the sides as if scanning the horizon. Eibl-Eibesfeldt postulates this to be an innate alert behavior against enemies, although there is very little danger for humans eating alone today. Another interesting case, using the film technique, had to do with some complex human expressions that do not seem to be culturally determined. Agreement in the smallest detail was found in the flirting behavior of girls from Samoa, Papua, France, Japan, certain tribes in Africa, and certain tribes of South American Indians. Eibl-Eibesfeldt discuss other instances of fixed human action patterns as well as internal motivating mechanisms, innate releasing mechanisms, releasers, and innate learning dispositions.

One of the great dangers in human societies today is aggressiveness as seen in wars and in the riots and mob violence in many of the large cities in this country and throughout the world. One group of experimenters holds that aggressive behavior is learned. John Scott was able to produce very aggressive male mice by enabling them to win fights repeatedly, and he produced peaceful male mice by raising them with females and by daily handling them and stroking them gently. Other workers have shown experimentally that success in fighting makes mice more aggressive and that aggression is dampened by repeated losses in fights.

Konrad Lorenz and several of his followers consider that aggression is a true instinct. They find also that there are strong indications that the dynamic instinct concept of aggression holds for humans, yet they admit that no clear-cut proof for an innate aggressive drive has yet been presented.

A number of experiments suggest that aggressive behavior is innate. The male hormone is known to stimulate aggressiveness. In human adolescents with feelings of inferiority the male hormone increases aggressiveness. On the other hand, *stilbestrol* provides striking control of irritable aggression. In some experiments at Princeton, investigators tested the effects of *carbachol,* a drug that mimics the effects of acetylcholine, which plays a role in the transmission of nerve impulses. A dozen rats that normally never killed mice were injected with carbachol in a specific site of the lateral hypothalamus, which is considered to be a center of emotion. Subsequently each of the treated rats killed mice that were placed in their cages. These killings were very precise, with no incomplete or ineffective attacks. The kill was made with a very precise bite through the cervical region of the spinal cord. Carbachol treatment was sufficient to trigger the complete behavior pattern, even though the rats had never before performed or witnessed the response.

It is possible to produce fighting behavior in chickens by means of electrical brain stimulation. This technique has been applied to many animal types. Electrical stimulation of the lateral hypothalamus of a cat resulted in aggressive displays with subordinate members of its group, but a careful avoidance of the dominant cat in his group. The phenomenon of rank order (peck order) is found in many animal groups, including many species of birds, wolves, wild dogs, prairie dogs, horses, monkeys, chimpanzees, and many others. In such a social hierarchy some members of the group will fight for authority, and the lower ranking animals accept this authority. This capacity to submit makes stable groups possible. The rank order is changed from time to time. A monkey colony is ruled by a dominant male. When such a dominant male monkey was stimulated in the thalamus he showed increased aggressiveness to another male who represented a challenge to his authority, but he did not show this aggressiveness to the female who was his favorite partner. According to José Delgado, stimulation of the brain determined the emotional state of hostility, but behavioral performance depended on the individual characteristics of the stimulated animal, including learned skills and previous experience.

This area certainly needs more study. The human environment is constantly changing in this age of great technological developments, population pressures, and environmental pollution. More effort should be placed on a better understanding of the bases of human aggressiveness if human societies are to be preserved. Eibl-Eibesfeldt closes his book, *Ethology—The Biology of Behavior,* with this statement: "Our knowledge is still filled with gaps, but we are encouraged to continue our research in line with biological viewpoints, especially as it has been shown that, especially with respect to our social behavior, our actions are to an important extent preprogrammed by phylogenetic adaptations. Our social behavior, particularly, is clearly disrupted today by certain changes in the environment, and only insight into the casual relationships can lead to successful therapy. Only the exact knowledge of the determinants of our behavior will lead to its eventual mastery. If we know, for example, which releasing stimulus situations arouse certain impulses to act, then we can either seek them out or avoid them. Insight into the workings of our innate motivating mechanisms will make their control easier, as is true for all insights into relationships which allow us to extricate ourselves from a right stimulus–response chain by virtue of our mental capacities. We have already discussed the capacity to detach ourselves from a problem and view it from a distance, an ability that is highly developed in humans. This ability to gain perspective allows us to contemplate the consequences of our actions and to choose among several alternative courses of action. Yes, up to a certain point humans can, with the help of this capacity, act against their drives. It is in a way the basis of our specifically human freedom. The prerequisite for a responsible decision is, however, the causal understanding of those behavior mechanisms which underlie our behavior. The less we know about them, the more blindly we will follow their dictates."

SOME GENETIC ASPECTS OF BEHAVIOR

What is known about the inheritance of behavior patterns? Geneticists have traditionally studied morphological characters more than other types of characters. Behavior patterns that involve many structures, muscles, and nerve paths are more difficult to analyze. However, a start has been made in this area, as a few examples will show.

Variations in given behavior patterns are found in different strains of the same species. One good example of this is the difference in the mating behavior of a yellow male fruit fly as compared with a normal wild type of fly (the yellow mutation causes a yellow body color instead of the gray body color of the wild type). In the mating behavior of *Drosophila melanogaster* the male approaches the female, orients his body at right angles to the female, and taps her on the abdomen. At this stage the female may move away or she may remain on the same spot. If she remains, the male than raises the wing nearest the female and rapidly vibrates it. After the male has vibrated his wing, he will lick the genitalia of the female, jump on her back, and attempt to copulate.

The yellow males are less successful than normal males. Margaret Bastock has studied carefully the differences in the behavior of the yellow and wild-type males. The differences are slight. The yellow males spend slightly less time vibrating and slightly more time orienting; the wild male spends 22 percent of its time vibrating, whereas the yellow male spends 18 percent (the whole process up to copulation may occur in less than a minute). Bastock found that there is one chance in 20 that the difference in vibrating between two types comes by chance alone. This difference in behavior is so slight that it is difficult to think of it as the basis for the great difference in success in mating. However, the discovery that different species of *Drosophila* are kept from interbreeding because of such differences (to be discussed later) strengthens this conclusion.

Another interesting example is found in a study of dogs. Some breeds, such as bloodhounds, foxhounds, springer spaniels, and beagles, are known as "open trailers." When they follow the fresh trail of a rabbit or other game, they make a loud baying sound. Other breeds, such as the collie, airedale, English setter, and fox terrier, are mute trailers; they will follow and overtake their prey without making any sounds. Eight different crosses have been made between these two types of trailers. The results were always the same, regardless of the type of the mother. The first generation always barked when following a fresh trail, but the bark was the yapping of the mute trailer rather than the baying of an open tracker. One collie–hound cross was followed to the second generation. Here independent assortment of the genes for hair length and type of bark produced all four possible combinations: short hair and baying, short hair and yapping, long hair and yapping, and the double recessive of long hair and baying.

Lorenz and his colleagues have been studying the evolution of behavior in several species of surface-feeding ducks. In closely related species the reproductive isolating mechanisms are often not strong enough to prevent the mating of such species under experimental conditions. Workers have been able to induce mating between some of these species, with the production of fertile hybrids. They have studied a number of elementary innate motor patterns exhibited by these ducks in their courtship behavior. Figure 43.9 illustrates 10 of these patterns, which are common to three species. The different species display these patterns welded into different combinations. First-generation hybrids show new combinations of motor patterns in their crosses. For example, in crosses between chiloe teals and Bahama pintails, the hybrid progeny regularly perform the head-up-tail-up, although neither parent is capable of this.

William Dilger has studied the behavior of lovebirds, small African parrots in the genus *Agapornis*. Unlike most of the other members of the parrot family, all lovebirds make nests. They use leaves, bark, paper, or other pliable material for nest construction. Using their beaks they cut the nest material into strips or small pieces. The method of carrying the material to the nest varies in the different species. Fisher's lovebird carries a strip of material in its bill, whereas a peachfaced lovebird carries it tucked among the feathers on the lower part of its back. Dilger was able to mate these two species and obtain viable hybrids. The hybrids show neither of these two patterns in clearcut form. At first the hybrids seemed completely confused in the nest-building attempts. They could cut the strips but had difficulty in determining how to carry them. They were able only to get the material to the nesting site when they carried it in their bills. Materials tucked into the feathers always dropped out on the flight to the nesting place. In the first attempt to build a nest they carried material in their beaks 6 percent of the time. After two months they carried material in their beaks 41 percent of the time, but they continued to make movements associated with tucking. Dilger reports that after three years these hybrids behave like the Fisher's lovebirds—they carry nesting material in their bills with only

Figure 43.9 Ten courtship poses found in the Mallard, a surface-feeding duck. (1) Initial bill-shake; (2) head-flick; (3) tail shake; (4) gruntwhistle; (5) head-up-tail-up; (6) turn toward the female; (7) nodswimming; (8) turning the back of the head; (9) bridling; and (10) down–up. (From Konard Z. Lorenz, with permission.)

occasional attempts to tuck it in their feathers. This experiment shows the great difficulty that hybrid lovebirds have in learning to use one innate pattern over another, even when the other is never successful. Since the behavior of the hybrids is so complex in character, it suggests that the carrying behavior is determined by many genes.

EVOLUTION OF BEHAVIOR

In the evolution of species, different mechanisms may be involved in the production of reproductive isolation. In a number of instances a difference in the courting or mating behavior is involved. This is at least part of the basis for reproductive isolation among the different species of *Drosophila.* The mating behavior of 11 species of *Drosophila* found in North America has been carefully studied. The movements of courtship are similar from species to species; they all exhibit the orientation, vibration, and jumping phases described above. But there are subtle differences, and as a result, interbreeding rarely occurs. One comparison illustrates this.

The chief difference between the mating behavior of males of *D. pseudoobscura* and those of *D. persimilis* is the frequency of wing vibration. The *D. persimilis* males vibrate at a lower frequency, but this small difference almost invariably keeps females of the other species from accepting them. The *D. persimilis* males also beat their wings at a lower frequency during flight. It has been suggested that selection was for two different frequencies of wing beating and that when the two populations came together they were reproductively isolated because of the difference in courtship.

In recent years Lorenz (see Suggestions for Further Reading) has done much to elucidate the evolution of behavior patterns. He has indicated that phylogeny can be studied through comparative behavior as well as through comparative morphology, and that there are homologous behavior traits as well as homologous structures. Sometimes these traits of behavior are common to the members of a genus, a family, an order, or a class. Sometimes they are common to an even larger taxonomic group. Oskar Henroth first described such a widespread motor pattern, one that is shared by the amniotes—the reptiles, the birds, and the mammals. This involves scratching movements. If you have watched a dog scratch its head or a bird preen its head feathers, you realize that both carry out these movements in the same way. The dog props itself on its two front legs and one hind leg, and then moves the other hind leg forward in front of its shoulder to scratch its head. A bird, in preening its head feathers, lowers its wing and extends the claw of its limb forward in front of its shoulder. This seems like a clumsy way to scratch, for it would appear that it would be easier for the bird to move its claw directly to its head without moving its wing, since the wing is folded on its back, out of the way (Fig. 43.10).

Most birds use the over-the-shoulder method of scratching, but a few, such as the larger par-

Figure 43.10 Scratching behavior of dog and European bullfinch. (From Konrad Z. Lorenz, "The Evolution of Behavior," © 1958, by *Scientific American*, with permission.)

rots, have lost this trait. They use their claws for feeding, and they use the same direct motion for scratching. Most parakeets use the typical bird motion for scratching, and they do not use their claws in feeding. However, the Australian broadtailed parakeet has learned to eat with its claw, which it raises directly to the bill. But in scratching, it reaches its claw around its lowered wing. No one has been able to teach this species of parakeet to scratch without lowering its wing. According to Lorenz, such deep-seated behavior traits tend to be masked by learned behavior in higher animals, but they are revealed clearly in lower forms. Such traits change slowly in evolution, but they do change. As a result, these traits and their changes can be studied in tracing phylogenies, just as the slowly evolving skeletal structures are used in tracing phylogenies.

Tinbergen and his associates at Oxford University have made a very detailed study of the innate behavior of gulls and terns. Most gulls are beachcombers, and they nest on the ground. One gull, the kittiwake, is quite different. It lives over the open sea except when it is breeding, and it nests on tiny ledges on the steepest of cliffs. The male of most species of gulls stakes its claim to a nesting territory by making the *long call* with its body in the *oblique* posture, its head up and its tail down. To advertise its nesting site it performs the *choking* movements, a series of down-and-up movements of the head. These behaviors have been modified in the kittiwake. On its tiny ledge, territory and nest sites are identical. It has lost the oblique posture and the long call, and it uses choking alone for display purposes.

A number of other evolutionary relationships are known. Since there is much interest in this area of study at the present, much more will be discovered in the coming years.

THE HYPOTHALAMUS AND BEHAVIOR

Many American psychologists and neurophysiologists have attacked the problems of instinctive behavior in a different way. They use laboratory studies in contrast to the field studies of the ethologists. They emphasize the motivational aspects of such behavior and start with the conception that in the case of many behavior patterns the pattern can be analyzed into a *drive* directed to a *goal,* and with the attainment of the goal there is a reduction of the drive or *satiation.* As an example, a hungry animal seeks food, finds it, eats it, and then stops eating. The drive is the striving for food; the goal is the food.

Since the work of Walter Hess in the 1930s (for which he shared a Nobel Prize in 1949), it has been known that the region of the brain primarily involved in the regulation of drives is the hypothalamus, a part of the diencephalon. Several techniques have been involved in determining the roles of different regions of the hypothalamus: actual destruction, stimulation of specific regions by electrodes, and stimulation of specific regions by the injection of chemicals.

If a certain region on both sides of the hypothalamus of a rat is destroyed with an electric needle, the animal becomes a ravenous eater and will soon double its weight if food is provided. On the other hand, if different regions adjacent to these are destroyed, the rat will cease to eat. The first regions mentioned have been called the *satiety center;* the second, the *eating center.* Similar results are obtained when these regions are stimulated through electrodes. The question arises as to how the satiety center is activated or stimulated in a normal animal. There is evidence that glucose in the blood acts on this center. When experimental animals are fed gold thioglucose, the cells in the satiety center are selectively killed. The cells apparently have a high affinity for glucose, and the accumulated gold is toxic. Such animals will keep on eating. However, this may not be the complete story. Sebastian Grossman at Yale has found that the injection of noradrenalin into a site in the brain just above the hypothalamus would cause a well fed rat to start eating again.

In another experiment the injection of a 5-percent solution of common salt into a particular region in the middle of the hypothalamus of goats resulted in the goats immediately beginning to drink large quantities of water.

Alan Fisher found that when he injected testosterone into a region of the brain of a rat just in front of and a little to one side of the hypothalamus, the rat exhibited male sexual activity. Male sexual activity occurred in both male and female rats after the injection.

A number of workers have used chickens and turkeys in experiments involving local stimulations of different regions of the diencephalon. In this work it has been possible to elicit both simple movements, such as sitting, standing, preening, grooming, and neck stretching, as well as more complex patterns, such as seeking and eating food, seeking and drinking water, escaping from predators, and a variety of defense movements. This is an area of active study today, and therefore many more discoveries can be expected.

RHYTHMIC BEHAVIORS

A number of cyclic phenomena involving behavior are found in a great variety of organisms. Some recur each year, each lunar month, each day, or with the changes in the tides. There is much interest today in those cycles that recur approximately every 24 hours, and they have been called *circadian rhythms.* Such rhythms have been described for plants. They are also found in a great variety of animals, but only a few examples are mentioned here.

Many animals are active by day and many others are active by night. The cockroach is active by night. The surprising thing about its rhythmic behavior, as well as that of other forms that have been studied, is that the rhythm persists whether the animals are kept in constant light or in constant darkness. Cockroaches will maintain the same rhythm for several days, whether they are in constant light or in constant darkness. A cockroach will show a peak of activity during the first half of the night. As the name "circadian" suggests, most of these rhythms are not exactly 24-hour rhythms; some are a little more or a little less. If the animal is kept in either complete light or complete darkness, the peak of activity will come later and later on successive days if the period is longer than 24 hours, and earlier and earlier if the period is shorter.

A number of properties of biological rhythms have been demonstrated. The rhythms are independent of temperature within the range normally encountered. They are insensitive to a great variety of chemical inhibitors. With the use of sodium cyanide one might expect that the rhythmic period would become appreciably lengthened because sodium cyanide depresses respiration. However, this does not happen; the amplitude of the rhythm may be diminished, but the period remains unaltered. The phase of the rhythm is not necessarily restricted to a particular time of day. This has been demonstrated in both plants and animals. When plants that exhibit the sleep-movement rhythm are moved into the laboratory and are subjected to a reversal of the normal day and night light conditions, they quickly adjust to the new light conditions with the leaves in the raised condition during the changed period of illumination. When such a plant is placed in constant conditions (dark or light), the new phase relationship may persist for several days. Thus the phase can be changed. To many the evidence seems clear that these biological rhythms are innate—they are not learned. This has been demonstrated to the satisfaction of many raising animals from birth in static laboratory conditions. Such developing organisms either develop a rhythm *de novo*, or they lack a rhythm until they are subjected to a single nonperiodic stimulus. In nature, fruit flies emerge from the pupal case only at dawn. If fruit fly eggs are laid and allowed to develop in constant conditions, the adults eventually emerge at all times of the day and the population is without this rhythm. When, however, such arrhythmic flies, after living for 25 generations in constant conditions, have their larvae and pupae treated with a single light flash, the rhythm is again initiated in the population.

At the present the fundamental nature of these rhythms is unknown, and there are two schools of thought with respect to the operating mechanisms. One school holds that the mechanism involved is entirely innate, and the term *biological clock* is used for this innate mechanism. In nature organisms show rhythms that are of the same lengths as the occurrence of tides, days, months, and years. When organisms are placed in experimental situations where there are no obvious clues to these natural environmental cycles and where the rhythms still persist, many workers conclude that the organisms must have some internal innate timer that measures these periods. The other school of thought questions whether the so-called controlled and constant conditions of the laboratory experiment are truly constant for the organism. They propose that such factors as gravity, geomagnetism, the electrostatic field, and background radiation may be affecting the rhythmic behavior, even under the controlled conditions of the laboratory. The advocates of this hypothesis postulate that the organism's biological clock has the capacity to receive timing information from the environment and to transduce it into the observable biological rhythms. Both internal and external timing mechanisms may be involved.

There are many problems to be solved in connection with biological clocks, and more and more workers are becoming interested in these problems. Humans are also subject to such rhythms. Body temperature, heart beat, blood pressure, and hormone secretion are a few of the basic physiological processes that are known to vary according to a circadian rhythm. Anyone who has taken long jet flights from one time zone to another has experienced the problem of having to adjust to the new situation. This period of adjustment of basic processes does not occur after long flights north or south in the same time zone.

SUGGESTIONS FOR FURTHER READING

Brown, F. A., J. W. Hastings, and J. D. Palmer. *The Biological Clock.* New York: Academic Press (1970).

Davis, D. E. *Integral Animal Behavior.* New York: Macmillan (1966).

Dethier, V. G., and E. Stellar. *Animal Behavior,* 2d ed. Englewood Cliffs, NJ: Prentice-Hall (1964).

Eibl-Eibesfeldt, I. *Ethology—The Biology of Behavior.* New York: Holt, Rinehart and Winston (1970).

McGill, T. E. (Ed.). *Readings in Animal Behavior.* New York: Holt, Rinehart and Winston (1965).

Tinbergen, N. *Social Behavior in Animals,* 2d ed. New York: Barnes and Noble (1967).

Van der Kloot, W. G. *Behavior.* New York: Holt, Rinehart and Winston (1968).

From Scientific American

Benzer, S. "Genetic Dissection of Behavior" (Dec. 1973).

Esch, H. "The Evolution of Bee Language" (Apr. 1967).

Fisher, A. E. "Chemical Stimulation of the Brain" (June 1964).

Frisch, K. von. "Dialects in the Language of the Bees" (Aug. 1962).

Hess, E. H. "Imprinting in a Natural Laboratory" (Aug. 1972).

Lorenz, K. Z. "The Evolution of Behavior" (Dec. 1958).

Tinbergen, N. "The Courtship of Animals" (Nov. 1954).

Tinbergen, N. "The Evolution of Behavior in Gulls" (Dec. 1960).

Wenner, A. M. "Sound Communication in Honeybees" (Apr. 1964).

Wilson, E. O. "Animal Communication" (Sept. 1972).

44 Species Ecology

INTRODUCTION

The term *oekologie* was defined by Ernst Haeckel in 1866 as "the body of knowledge concerning the economy of nature—the investigation of the total relations of the animal to its inorganic and organic environment." The term is derived from the Greek *oikos,* meaning house, and *logos*, a discourse on. A commonly used definition of the word ecology is "a study of the relationship between an organism and its environment." Environment is used in this sense to include the living (*biotic*) as well as the nonliving (*abiotic*) environment.

While humans have been interested since the dawn of history in many aspects of what we now include in the general field of ecology, it was not until the time of Linnaeus and those who took up his great work after him that scientists began to look at the living world from the viewpoint of the ecologist. In the middle of the last century they began to measure environmental factors and to correlate these measurements with phenomena which they observed in nature. Both Darwin and Wallace made many significant ecological observations, and their concept of natural selection is strongly based on ecological principles, since either the biotic or the abiotic environment is more often than not the instrument of the selection process. The concept of the web of life introduced by Darwin is the forerunner of modern sophisticated studies of energy flow in nature.

In the present day of an exploding human population, with the multitude of problems which this explosion entails, there is a great interest in ecology. One might say that ecology as a discipline has at last "come into its own." Some of today's most pressing problems—a need for increased productivity; dwindling energy supplies; pollution of air, soil, and water; protection of plant and animal species from extinction; the cycling of potent poisons and radioactive isotopes in our food chains—are directly or indirectly ecological in nature, and they call for people trained in, or at least acquainted with, the general field of ecology. Conservation is in large part the application of ecological principles to the problems of maintaining natural balance in the face of the pressure of civilization. These problems will be discussed more fully later.

Since ecology is concerned with the complete environment, animal ecologists must have some familiarity with plants as well. Plants form a significant and often major portion of the environment for many animals. Therefore, although the emphasis will be upon the animal component of the ecological systems, there will be some discussion of the interactions between plants and animals in these systems.

For convenience, ecology is often said to consist of two aspects, the ecology of the individual —*autecology* (Gk. *autos,* self) and the ecology of groups or communities—*synecology* (Gk. *syn,* with). The remainder of this chapter will deal with the ecological aspects of the life of the individual and of groups of the same species. In Chapter 45 the emphasis will be on the biotic community as a whole.

THE ABIOTIC ENVIRONMENT

Water

There is no single most important environmental factor in the surroundings of organisms, but because water is so important to any kind of life, it is an appropriate factor with which to open this discussion. As mentioned earlier, living matter has evolved as it has, in part, because of the unique properties of water. Variations in the amounts and in the state of water have profound effects on organisms.

The first forms of life originated in water, and land organisms have been derived, through evolution, from aquatic organisms. Evolution of both aquatic and land organisms has involved many adaptations. The problem of locomotion in water has been solved by a variety of adaptations in the different animal groups. Many of these, such as cilia and flagella in protozoans, jet propulsion in the squid, and the fins of fishes, have been discussed. The streamlined bodies of fishes and whales are adaptations permitting rapid movement through the water. Other swimming adaptations include the flattened tail found in the whale and the alligator, the flippers of the seal, and the modified wings of the penguin.

Many aquatic animals attain large size with little supporting skeleton. The largest of all animals, the whale, a mammal which became secondarily adapted for aquatic life, could not support its weight on land. A whale stranded on the beach soon dies because it cannot breathe. Its great weight, unsupported by water, collapses the lungs. The transition of animals from water to land has therefore necessitated the development of supporting structures too.

Land animals vary in their degree of adaptation for life out of the water. Animals with thin soft skins such as frogs, salamanders, and worms are limited to a moist environment. Only animals with a relatively impermeable body covering can live in the drier habitats. Reptiles, birds, mammals, and many insects have such an integument. The development of lungs and tracheal systems are also adaptations for life out of the water.

The methods evolved for the elimination of nitrogenous wastes and fecal material are water-saving devices in many land animals. Insects, reptiles, and most birds conserve water by disposing of their nitrogenous wastes as solid uric acid. The deposition of dry feces is found in most insects, reptiles, birds, and some mammals. The dry feces of rodents and antelopes contrast with the more liquid feces of cattle.

Some animals are adapted to live virtually without water intake. Drywood termites and powder-post beetles are good examples. Perhaps an even more striking example is the clothes moth. Such forms subsist entirely on dry food and are able to satisfy their water requirements from the water of metabolism. The camel is able to go long periods without drinking because it uses the water obtained from the oxidation of the fat in its hump.

Light

Light furnishes the energy for the photosynthesis of basic carbohydrates by green plants. Light is necessary not only for photosynthesis, but also for the synthesis of plant pigments.

In the life cycles of many plants and animals there are seasonal periodicities associated with the light factor. This effect is due not so much to intensities of light as to differences in total amount of light and in the relative lengths of day and night. The response of organisms to day length is known as *photoperiodism.*

Photoperiodism plays an important role in the life cycles of many animals, particularly in the control of the reproductive cycle. The most extensive studies have been made on birds in connection with migration. It had not been possible to correlate the arrival of birds in the spring with temperature or other common aspects of the weather, nor was it possible to correlate the southward migration of birds with cold weather or lack of food. It was finally recognized that the length of day was the only environmental factor that could account for the exact timing of bird migrations. It has been shown that the greater amount of light received during the long days of spring causes an increase in the size of the gonads, and this, in some unknown way, sets off the northward migration. After the breeding season the diminishing day lengths of late summer and autumn cause a decrease in gonad size, and this provides the stimulus for the southward migration. Although photoperiodism seems to be basically involved in setting off bird migrations, it does not explain the evolutionary origin of the migratory habit nor how the birds find their way.

In mammals there are also marked effects of day length on breeding cycles and on coat color. Deer come into heat with the shortening days of fall, while ferrets breed in the spring when the days are lengthening. Artificial manipulation of day length under laboratory conditions indicates that it is in fact the day length which induces these phenomena. The snowshoe hare develops a white coat in response to the shortening days of autumn regardless of other factors such as temperature or the presence or absence of snow.

Perhaps the most obvious effect of light on animals is their utilization of it to become aware of the beneficial and harmful aspects of the environment. In this connection a variety of photoreceptors have evolved in the various animal groups. Pigment spots capable of a differential reaction to light are found in many phyla, including protozoans, coelenterates, platyhelminthes, mollusks, annelids, echinoderms, and lower chordates. Some have a lens that concentrates the light. Actual image perceiving has evolved separately in the arthropods, mollusks, and vertebrates. With the development of image-perceiving eyes, adaptations in animal coloration have evolved. Some provide concealing coloration, others conspicuous coloration. One very common type of camouflage or protective coloration is a simple matching of the background with respect to color and pattern. A katydid among green leaves or a walking stick on a twig is quite difficult to distinguish from its surroundings. Desert animals, in the main, display a pale coloration in contrast to the darker hues of the inhabitants of humid regions. There is evidence that indicates that pale or dark coloration has evolved in many species as a result of selective survival as influenced by their conspicuousness against the background. A good example is the peppered moth mentioned in Chapter 42 (Fig. 42.3).

Intense light may be lethal to certain animals just as it is to certain plants. An earthworm, for instance, is soon killed in intense sunlight. Pigment formation in the outer coverings of animals is common in those animals that are able to withstand medium to high intensities of light.

The animals with which we are most familiar are active during daylight hours. But there is a large number of animals whose activity is confined chiefly to periods of darkness. Many of the relatively defenseless forms forage mostly at night when they are less likely to be detected by their enemies. Park has pointed out that a patch of woods may be inhabited by two sets of animals that practically never meet because one set is active only during the night and the other is active only during the day. That such a diurnal rhythm of activity may be deeply ingrained in given organisms has been demonstrated in laboratory experiments. In one such experiment the deer mouse, *Peromyscus,* which is normally active at night and quiescent during the day, continued to display a diurnal rhythm in its behavior after seven months in continuous darkness. Other diurnal or circadian rhythms were discussed in Chapter 43.

The adaptations of animals that live in dim light are interesting. In the "twilight zone" of the oceans, many of the fishes have very large eyes and some of them have light-producing organs (Fig. 45.21). The luminescence of the firefly is a mating signal. The adaptive significance of bioluminescence is not understood for many organisms, but it seems to be clear that it aids in vision in some cases.

Cave-dwelling animals have become adapted for existence in an environment completely devoid of light. Extensive development of tactile, auditory, and chemical senses enables them to carry out their normal functions. In many of these *cavernicoles* there is also a lack of pigmentation and degeneration or complete loss of eyes. Figure 44.1 is a photograph of the blind white crayfish, *Orconectes pellucidus pellucidus,* of southern Indiana caves. The loss of pigment and of eyes seems to be due to a lack of selection pressure for their maintenance. Thus the limited energy sources can be used for other purposes.

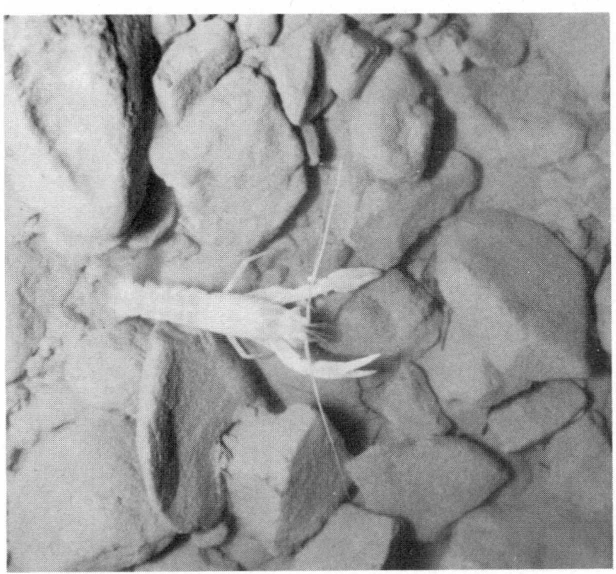

Figure 44.1 Blind and unpigmented cave crayfish. (Photo—E. C. Williams)

Temperature

Temperature is an important factor in determining the distribution of both plants and animals. Both the range of temperature in a region and the rate at which temperature changes diurnally and seasonally are important in determining whether a given species can survive. The temperature range for the life processes in most organisms is between 0 and 50° C; there is great variation in the temperature tolerances among individual species. A species which can tolerate a wide range of temperature is said to be *eurythermal,* while one with a narrow range is *stenothermal.* The prefixes *eury* (Gk. *eurys,* wide) and *steno* (Gk. *stenos,* narrow) are used in a similar manner for other environmental factors such as salinity (*euryhaline* and *stenohaline*). Some animals are killed at temperatures well above zero. This may be accounted for on the basis that all vital processes are slowed down by lowered temperature, and in some cases a particular vital function is stopped (by enzyme inactivation) before the freezing point is reached. The Alaskan black fish is reported to be able to live after having been frozen solid. Most Arctic fish do not venture south into waters as warm as 10° C, yet a temperature of 10° C is far too cold for survival of most tropical fish.

Hot springs furnish the warmest environment known to be inhabited by active organisms. Two kinds of algae have been found living in the hot springs of Yellowstone National Park at a temperature of 75° C. There are reports that the larvae of brine flies may live at a temperature of 45° C and that amoeboid protozoans have been taken from water at 51° C. Most heat-tolerant animals live in water below 40° C. However, heat deaths may occur at very low temperature. Certain snow-dwelling algae are killed at temperatures higher than 4° C, and many animals are killed by heat before the temperature reaches 20° C. In general, fishes and marine invertebrates are less resistant to increased temperature than are terrestrial insects and mammals.

There are a number of ways, many of them physiological, in which organisms may be adapted for survival in extreme temperatures. For example, the soft, fleshy parts of most plants are killed by freezing temperatures. However, by losing water, increasing salt concentration, and binding water in colloidal form, all of which tend to lower the freezing point, the leaves of winter rye and evergreen trees can survive extreme cold. Dry seeds, spores, and cysts do not freeze because there is no free water in them. Seeds, cysts, tardigrades, and nematodes have survived fairly long experimental exposures to liquid air at about −193° C.

The removal of water also provides resistance to high temperatures. Cysts of the protozoan *Colpoda* have survived exposure to dry heat of 100° C for three days. There is some evidence that certain organisms are adapted to withstand high temperatures because they are able to synthesize lipids with a higher melting point.

One of the most interesting adaptations for partially solving the problem of changing environmental temperature is that found in birds and mammals in which a relatively constant body temperature is maintained, as discussed in Chapters 37 and 38.

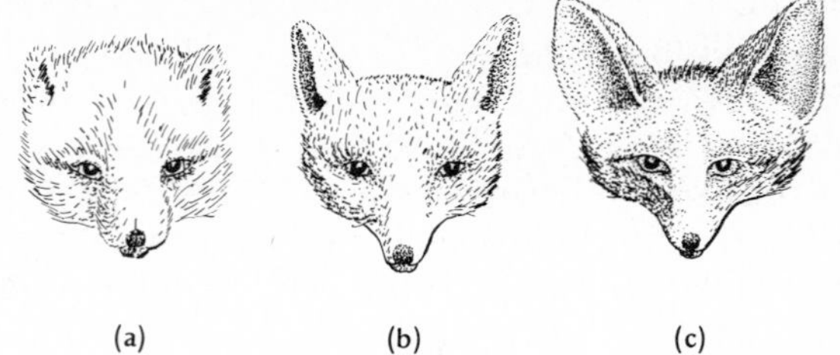

Figure 44.2 Illustration of Allen's rule. **(a)** Arctic fox, *Alopex lagopus.* **(b)** Red fox, *Vulpes vulpes.* **(c)** Desert fox, *Fennecus zerda.* (From Hesse, Allee, and Schmidt, *Ecological Animal Geography,* © 1937 by John Wiley and Sons, with permission.)

Structural adaptations which help many animals adapt to cold include the development of fat, fur, or feathers. There are two other morphological adaptations of birds and mammals which are less well known but easily demonstrated. In general, extremities such as tails, ears, and beaks are shorter in inhabitants of colder regions than in those which live in warmer regions. This is sometimes called *Allen's rule.* The classic example is seen in Figure 44.2 which shows the relative size of ears in three species of fox (Pl. 4). The adaptive significance of this feature is obvious. The other is *Bergmann's rule,* which points out that birds and mammals in colder regions are larger than related species which live in warmer areas. The significance of this adaptation is the fact that a larger mammal or bird has less surface area per unit of weight than a smaller one, and thus proportionately less heat is lost by radiation than in a smaller animal. The advantages are reversed in warmer climates where loss of heat by radiation is a beneficial feature. The largest bears, polar bears and Kodiak bears, are found in the far north; the smaller black bear is found in more temperate regions. Penguins range in size from a species which has a body length of 1 m or more in Antarctica, to one with a length of 0.49 m in the equatorial Galapagos Islands.

Some organisms which do not have physiological or structural adaptations for surviving extremes of temperature have evolved behavioral adaptations. Migration, as described for birds, is a behavioral adaptation that enables some kinds of animals to avoid cold temperatures. In some cases, animals *hibernate* and thus escape the cold. Hibernation occurs in a variety of animals, including certain mammals such as the woodchuck, and certain reptiles, amphibians, insects, and mollusks. Hibernating animals may be found in caves, burrows, in crevices, under rocks, and in the mud. During hibernation, metabolism is greatly reduced; there is a drop in body temperature, even among the so-called warm-blooded forms, and a slowing down of the heartbeat and respiratory rate. The black bear and the skunk are not true hibernators. Their body temperature, pulse, and respiratory rate are not appreciably lowered. During warm spells both leave their dens and walk abroad. Other animals go into a state of dormancy and escape the heat of the summer. Summer dormancy is called *estivation.* Many insects and certain spiders, mollusks, fish, snakes, and mammals estivate.

Substratum

The substratum is the surface upon which an organism rests or moves, or the solid material within which it lives part or all of the time. In aquatic environments many plankton organisms and many fish have no substratum at any time, but the great majority of animals and plants have a substratum for at least part of their existence. Flying birds and insects all come to rest on the substratum. A water strider uses the surface of the water as a substratum. Many organisms make use of living substrata. Examples of this are numerous—scale insects growing on plants, barnacles growing on horseshoe crabs, and Spanish moss growing upon the branches of trees. But the great majority of animals and plants utilize the crust of the earth, whether it be dry land or the bottoms of aquatic habitats, as a substratum.

Animals have a great variety of adaptations for their relations with the substratum. Among these are a number of types of walking appendages, claws, tails, burrowing mechanisms, suckers, and adhesive disks. Such adaptations are found in both water and land animals. Vertebrate legs are modifications of fish fins, and insect legs are modifications of the biramous appendages of the ancestral trilobites.

There are great physical differences in the land surface. A great variety of special adaptations in animals is correlated with these differences in the land surface. In rocky regions and in plains regions with hard soil, the running efficiency of animals is improved by the possession of small resistant feet with a reduced number of toes, as found in the horse, deer, and ostrich. The pentadactyl foot of the ancestral horse was adequate on the soft forest floor, but the one-toed foot of its descendants was much better for the open plains. Animals that live on a sandy substratum, like the camel, have spread-out feet, and wading birds that live in marshes are characterized by webbed feet that present a large surface.

Many animals burrow into the substratum. Such forms include certain rodents, birds, reptiles, and numerous insects. Forms such as the mole, the earthworm, and many insects spend much of their lives underground. All of these burrowing forms are limited to regions where the soil is suitable for burrowing.

The supply of certain chemicals in the soil substratum may be a limiting factor for certain animals. Land snails with calcareous shells are restricted to regions where calcium is readily available. Many herbivorous mammals require an abundant supply of salt for the maintenance of a proper ionic balance with the large amount of potassium contained in their plant food. Thus they are restricted to regions where salt is available.

Of the different physical factors that have been considered, the soil substratum is the only one that can be modified greatly by biological activity. This is important both from the standpoint of evolution and for our future economy.

SOIL STRUCTURE AND FORMATION

The soil in a broad sense consists of organic components, alive and dead, and inorganic materials. Included among the inorganic materials are air, water, and minerals. The mineral content of a particular soil is determined by the kind of underlying parent rock from which the soil was formed. According to their size, mineral particles can be of three general types: sand (2–0.02 mm), silt (0.02–0.002 mm), and clay (less than 0.002 mm). The proportions in which these components are combined regulate soil texture. A good loam, for example, is about one part clay, two parts sand, and two parts silt. Soil structure refers to the way that sand, silt, and clay particles are aggregated to form larger units.

Soil formation is initiated by the physical and chemical breakdown of the parent rock. The physical fragmentation occurs by freezing and thawing, the action of plant roots, and other abrasive action. Chemical breakdown involves oxidation, reduction, hydration, and other related reactions, including the action of microorganisms. Solution by ground water is aided by the presence of carbonic acid and humic acids produced by lichens and other plants. Until organic matter—plant parts, animal wastes and remains, acted upon by bacteria—becomes thoroughly mixed with the mineral matter, there is no soil. This mixing of the organic matter (commonly called *humus*) with the minerals is accomplished by the action of burrowing animals, such as rodents, insects, and, above all, earthworms. As mentioned in Chapter 27, Darwin determined that earthworms may bring to the surface as much as 40 metric tons of castings per hectare (18 tons per acre) in a single year. Soil is in fact a very complex material which is organized in a very specific way. The nature of the soil is determined by a number of factors including the parent material, amount of rainfall, and the presence of suitable living organisms. In turn, the very nature of the environment is determined by the nature of the soil; without suitable soil there can be no significant plant growth, and without the plants there can be little or no animal life.

It is apparent even to the casual observer of a basement excavation or a road cut that the soil is not homogeneous throughout; rather there seem to be more or less distinct layers. All of these layers, or *horizons* as they are called, constitute the soil profile. The uppermost layer is the A horizon or the topsoil (Fig. 44.3). Its thickness varies depending on the climate and the vegetation of a particular region. In a temperate deciduous forest it is often about 30 cm (1 foot) in thickness. The soil of the A horizon is darker and of a looser texture than the underlying B horizon.

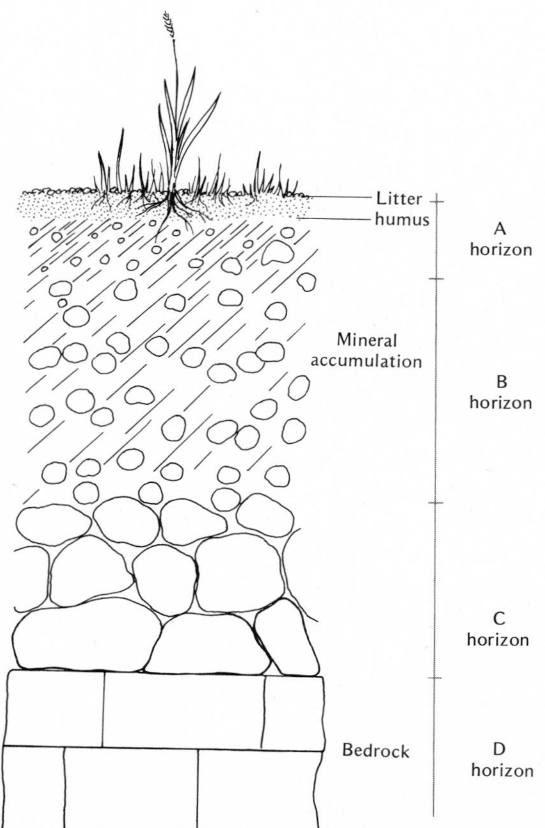

Figure 44.3 Soil profile.

Plant and animal debris collect on the surface of the A horizon forming the litter. Just beneath the litter is the humus where microbial action brings about a decay of the organic material. The remainder of the A horizon is rich in minerals and is of high organic content. As rain water percolates through this layer, minerals are leached out and they move into the B horizon.

The B horizon is the region of mineral accumulation. Soil particles of this zone are smaller and usually more compacted than in the A horizon. Unlike the A layer this layer has little biological activity. It may be more than 1 m thick in a deciduous forest.

Beneath the B horizon is the weathered parent material, the C horizon, and then the D horizon, which is the unweathered parent material or bed rock.

A classic example of the importance of the soil substratum can be seen east of Chattanooga, Tennessee, in a region called Copperhill (Fig. 44.4). A copper smelting operation released toxic fumes which killed the vegetation in the area. In this denuded area the unprotected topsoil was soon washed away and today there is no growth in the area. It is a sterile desert in the midst of luxuriant forests. It will be many years, if ever, before life returns to this region since significant amounts of toxic materials still remain.

Other Abiotic Factors

It is not possible in an introductory text to discuss all the abiotic factors of the environment. Other abiotic factors that affect organisms and that have played roles in their evolution are atmospheric gases, chemicals in solution, air currents, water currents, gravity, pressure, radiations other than light, and sound. The interested reader will find these discussed in any modern ecology text.

The Law of the Minimum

In 1840 Justus Liebig, who was studying the effects of various chemicals on agricultural crops, proposed that in the growth of plants the nutrient material which is least plentiful in proportion to the plant's requirements will be the factor which limits its growth. This concept is often called *Leibig's law of the minimum.* Later workers expanded the concept to include animals and all the abiotic environmental factors. In 1911 Shelford expanded this basic concept still further to include maximal tolerances as

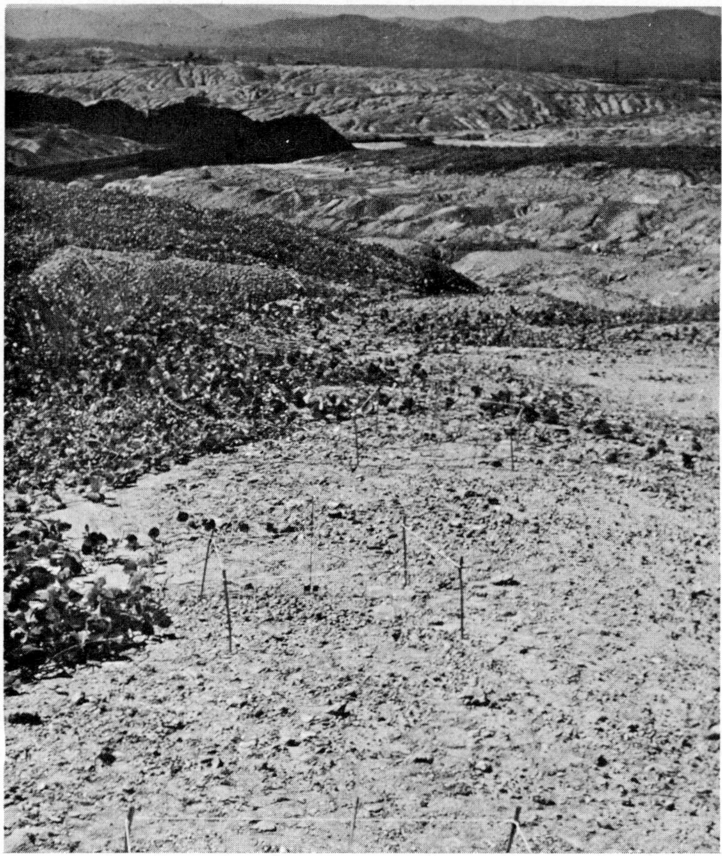

Figure 44.4 Copperhill, Tennessee. A region completely denuded of soil as a result of toxic fumes from a copper smelter which killed the vegetation. (From Martin Witkamp, with permission.)

well as minimal requirements and proposed what has come to be called *Shelford's law of tolerance*. In addition to minimal and maximal tolerances, Shelford includes the concept of the *optimum*—that point at which a given factor is acting at a level best suited to the requirements of a given organism. Figure 44.5 shows how the principle of tolerance operates.

Microclimate

Such abiotic factors as temperature and relative humidity must be measured near an organism if the effects of these environmental factors are to be correlated in any meaningful way with its activities. There are great differences in such factors at various points in the environment. For example, the temperature differences at 1 m above the ground, at ground level, and 15 cm below the soil may be quite large. It is not very meaningful to measure temperature at 1 m above the ground and attempt to determine temperature effects on an organism which lives in the surface litter. The term *microclimate* is used for the specific conditions to which an organism is exposed in its particular habitat as opposed to the general climate in the region.

THE BIOTIC ENVIRONMENT

The biotic factors in the environment include all the other organisms—both plant and animal

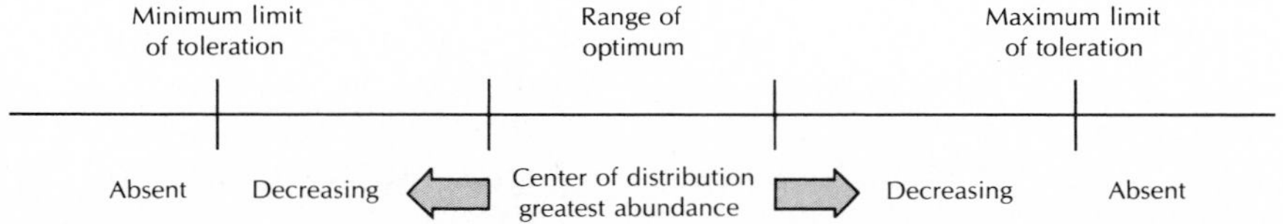

Figure 44.5 Shelford's law of tolerance.

—with which an animal interacts either directly or indirectly.

Biogeochemical Cycles

There are finite amounts of all the elements required by living organisms. Were it not for the continual recycling of these elements, permitting the use of atoms of carbon, nitrogen, phosphorus, and so forth, over and over again, life on the planet Earth would have ceased long ago. The cycles which allow for the use and reuse of essential elements are called *biogeochemical cycles.*

In some cases materials are returned to the available pool as fast as they are removed; in others, material may be tied up in biological systems for brief periods; while in others these necessary elements may be buried deep in the earth for long periods before being released in a form available for use by living organisms. The biogeochemical cycles are either *gaseous* with the atmosphere or the hydrosphere as a reservoir, or *sedimentary* in which the earth's crust is the reservoir. As examples of biogeochemical cycles we will briefly look at the cycles of nitrogen, carbon, phosphorus, and water.

THE NITROGEN CYCLE

Although molecular nitrogen is a major constituent of the earth's atmosphere, the nitrogen molecule is relatively inert and is not a suitable source of the element for most living organisms. Green plants assimilate nitrogen mainly in the form of nitrate. This nitrate then contributes nitrogen to the proteins, nucleic acids, and other important biological compounds. These organic nitrogenous compounds produced by plants, in turn, serve as the nitrogen source for the entire animal kingdom. Animals excrete a significant quantity of nitrogenous waste, though the end product that is excreted varies from one group of animals to another. The nitrogen of these waste products and of dead plant and animal bodies is converted by microorganisms to ammonia. All the reactions involved in the conversion of nitrogenous organic compounds to ammonia constitute *ammonification.* The ammonia is then converted to nitrate; this change is called *nitrification* and is carried out by two specialized groups of *aerobic* bacteria. Aerobic bacteria are those bacteria which must live in an atmosphere containing oxygen. Nitrification occurs in two steps: ammonia is oxidized to nitrite by bacteria of the genus *Nitrosomonas;* nitrite is oxidized to nitrate by *Nitrobacter.*

Some types of bacteria, the denitrifying bacteria, use nitrate as a final hydrogen acceptor; as a result, molecular nitrogen is formed in a process called *denitrification* and is therefore removed from the cycle. If molecular nitrogen were biologically inert, denitrification would deplete the supply of this element available for growth. However, there are two groups of bacteria which can convert molecular nitrogen into usable compounds, that is, they may carry on the process of *nitrogen fixation.* One type of nitrogen fixation results from a partnership between a green plant and a bacterium, neither of which can fix nitrogen alone. The bacteria are members of the *Rhizobium* group, they enter the root hairs of leguminous plants (clover, alfalfa, peas, beans, and so on) and form nodules on the roots. Within the nodule cells, molecular nitrogen is fixed by the bacteria into nitrate which is usable by the plant. This kind of relationship, in which two organisms live together, both benefiting by the association, is called *mutualism.* (This will be discussed more fully in Chapter 45.) Nitrogen fixation is also carried out by several free-living microorganisms. The principal ones are certain blue-green algae and bacteria belonging to the *Azotobacter* and *Clostridium* groups. The interactions of the various organisms involved in the nitrogen cycle are illustrated in Figure 44.6.

THE CARBON CYCLE

This cycle, although important, is not as complex as the nitrogen cycle (Fig. 44.7). *Carbon dioxide* is the key compound in the entire cycle. It is used by green plants in the process of photosynthesis to form simple carbohydrates. These carbohydrates, in turn, are utilized in the synthesis of other organic material by both plants and animals. Ultimately, through the various kinds of respiration—including putrefactions and fermentations—carbon emerges once more in the form of carbon dioxide. (Both of the latter processes are the result of bacterial activity. *Putrefaction* refers to the anaerobic breakdown of proteins, and *fermentation* refers to the anaerobic breakdown of carbohydrates. *Anaerobic* means that the organisms involved live in an atmosphere which lacks oxygen. *Decay* generally includes the breakdown of all kinds of material under either aerobic or anaerobic conditions.) In addition to these processes, the burning of fuels and other oxidations add constantly to the supply of carbon dioxide.

Thus with the contributions of carbon dioxide from various sources and the use of carbon di-

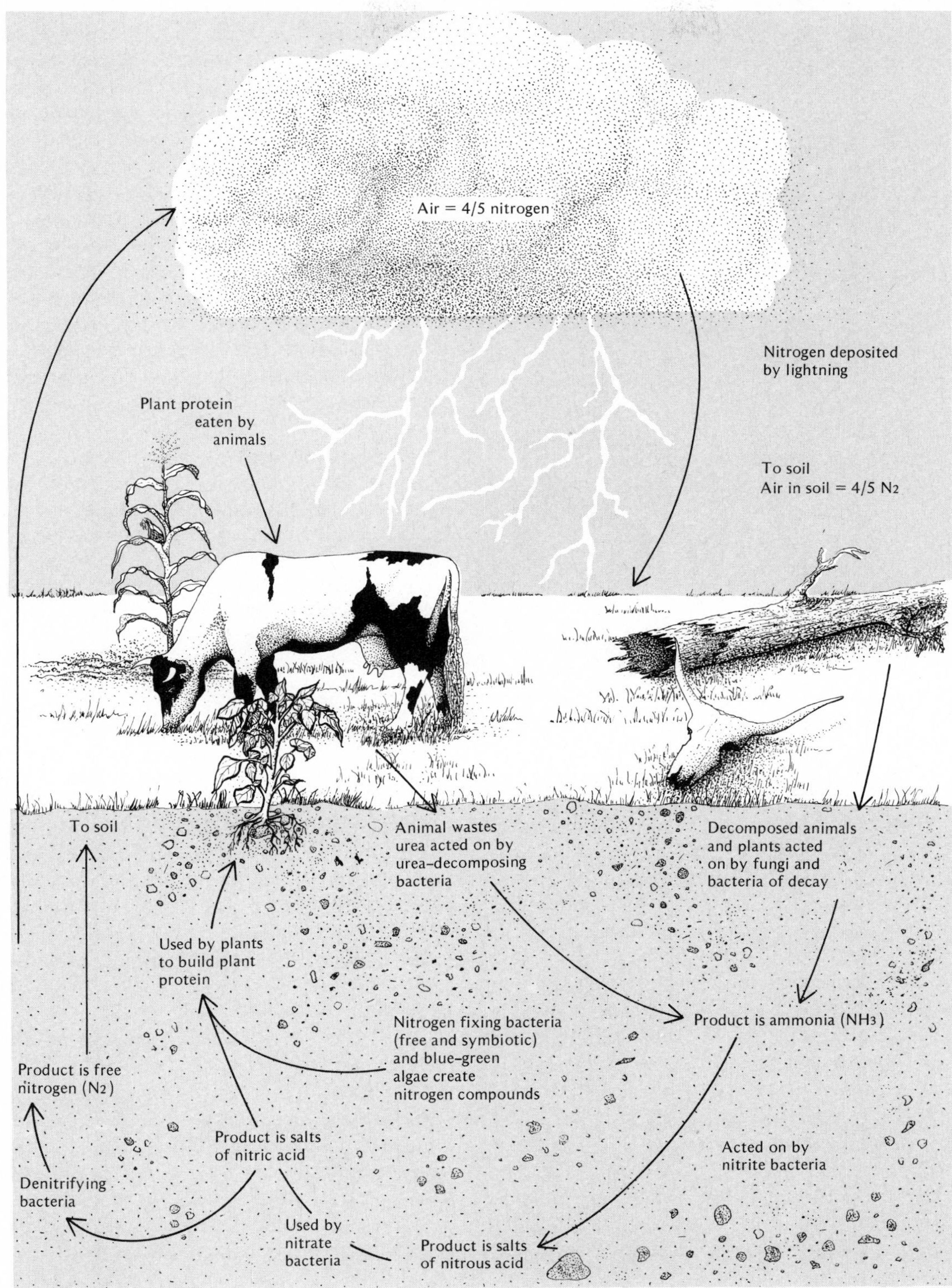

Figure 44.6 The nitrogen cycle.

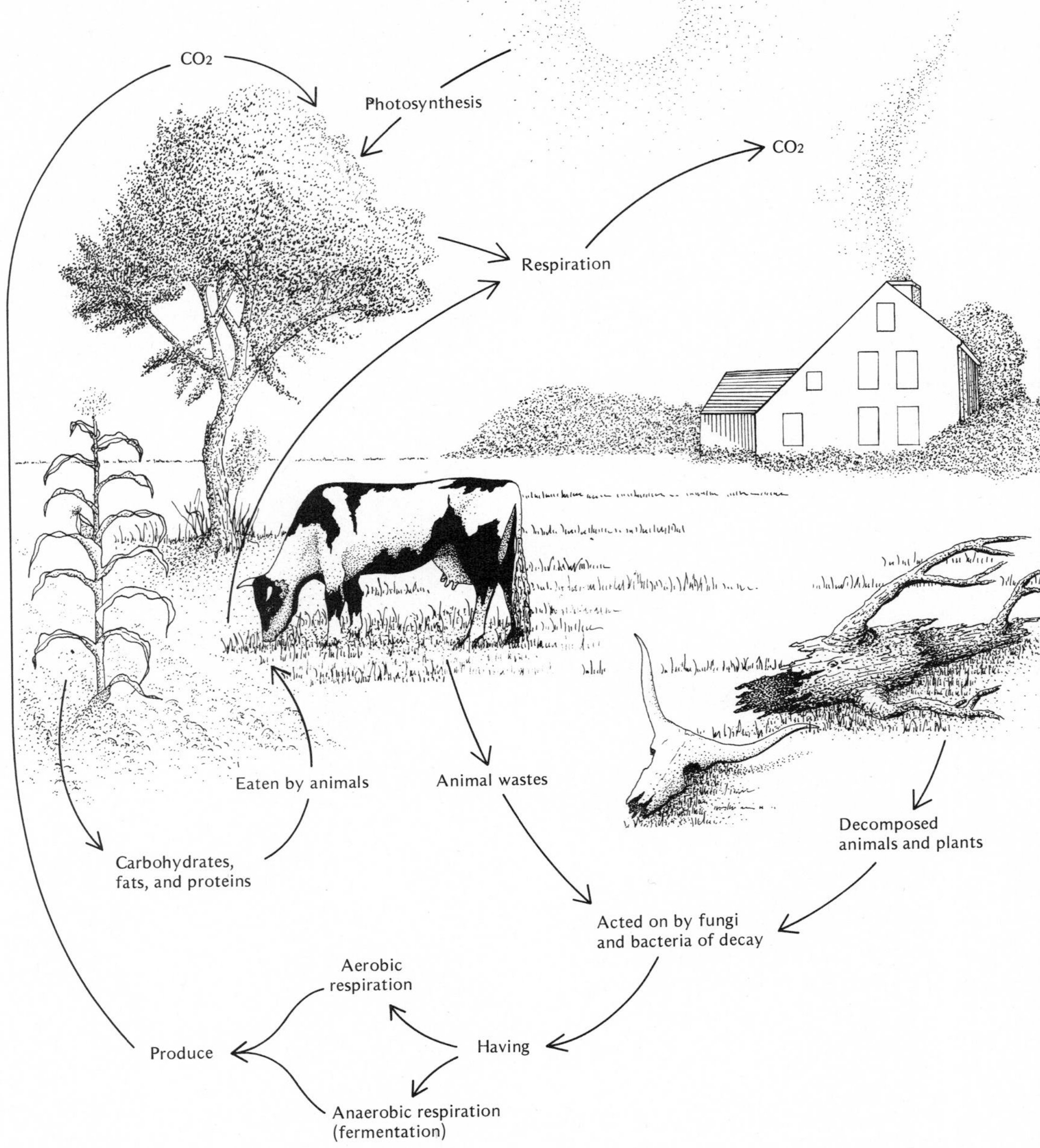

Figure 44.7 The carbon cycle.

oxide by plants in photosynthesis, an active carbon cycle is kept in operation.

THE PHOSPHORUS CYCLE

Phosphorus, involved in the major mechanisms of energy storage and utilization in living organisms (ATP) is perhaps the most critical element in terms of long-range requirements (Fig. 44.8). It seems probable that we are losing phosphorus from the available usable pool at a greater rate than its return. The main reservoir for phosphorus is in deposits laid down in past geologic ages. Natural erosion and exploitation by humans make it available. It is lost to the sea where a significant amount ends up in deep sediments where it is effectively removed from the available reservoir. It will only become available when great uplift movements bring the deep strata to the surface, or dwindling supplies make deep sea mining operations necessary.

Sea birds and fish return a certain amount of phosphorus to the pool, but not enough to offset the continual loss to deep sediments. Between 60,000 and 70,000 tons of phosphorus are re-

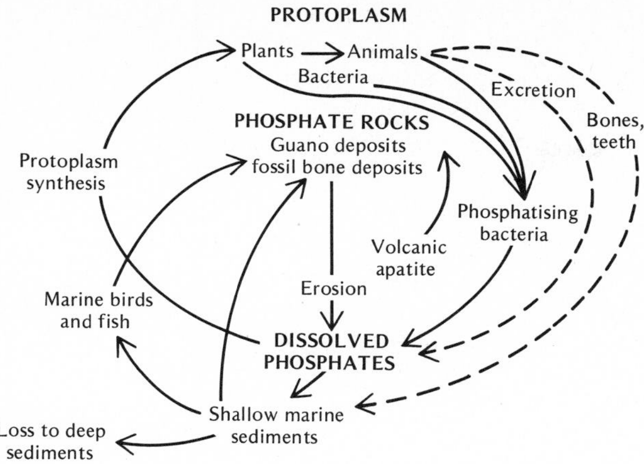

Figure 44.8 The phosphorus cycle. The ratio of phosphorus to nitrogen in aquatic environments is approximately 1:23. Loss of phosphorus by leaching is on the order of 34 metric tons per square kilometer per year.

turned to the land each year through fishing operations and in some areas, such as the islands off the coast of Peru, large deposits of guano are built up by the great bird populations living there. Over two million tons of phosphate are mined annually and used mainly as fertilizer. Much of this phosphorus washes out of the soil, finds its way to the sea, and is lost for the foreseeable future.

THE HYDROLOGIC CYCLE

Water, the most abundant inorganic compound in cells and necessary for all life, also has its cycle—the *hydrologic cycle.* Water evaporates from rivers, oceans, lakes, and from the surface of the land as well as its inhabitants, both plant and animal, and forms clouds. Upon cooling, the water vapor of clouds condenses to form rain or snow. This cycle is repeated over and over again.

Habitat and Ecological Niche

The term *habitat* is used by the ecologist, as it is by the layperson, to indicate a specific place, that is, the place where an organism lives. Thus we speak of a marine habitat, desert habitat, freshwater pond habitat, and so on.

But the term *ecological niche* means a great deal more than that. Although the word niche is normally used to indicate a specific place, it is used by the ecologist to mean the total interaction of a given organism with its environment, including its part in energy acquisition and transfer, nesting and other spatial requirements, harmful or beneficial effects on other components of the community, and so on. (Marston Bates has used the term *role* in the same sense.) The concept was originally introduced by Charles Elton, the eminent British animal ecologist. One of Elton's classic examples may serve to clarify its meaning. The arctic fox in summer feeds upon the eggs of a bird, the guillemot, and in winter an important portion of its diet comes from the remains of seals killed by polar bears. In tropical Africa the spotted hyena depends upon the remains of zebras killed by lions and upon ostrich eggs. Thus these two mammals, half a world apart, occupy essentially the same ecological niche (Pls. 3, 4). At least it is the same as far as their feeding requirements are concerned. Many other examples can be found among insects, birds, fishes—indeed, there is an organism to fill every conceivable niche in any natural environment.

Conversely, no two species can occupy exactly the same niche in any given habitat. This will be discussed more fully in the section on competition in Chapter 45.

Population Ecology

Animals interact with members of the same species, as well as with other species of animals and with plants. Some of the interactions between members of the same species, intraspecific interactions, will be discussed below. Since a group of interacting individuals of the same species constitutes a population, this area of ecological study is often called *population ecology.*

MATING

In dioecious, sexually reproducing animals, the minimum interaction that can occur is mating between male and female. There is a great variation in the permanence of association of male and female, from the black widow spider, which may devour her mate soon after mating, to the monogamous union of a coyote pair which each year breed and rear a new brood of young that leave the parental influence only when they are mature.

PARENTAL CARE

The number of offspring usually reflects the amount of parental care given to the young. In those invertebrates in which parental care is lacking, a large number of eggs are produced—up to several million in many cases. With those forms provided with brood pouches the number ranges from 100 to 1000, while great protection is correlated with a production an order of magnitude less, from 1 to 10 eggs. In those species which produce large numbers of eggs, the mortality is very high and the production of many eggs is the mechanism which evolved, sustaining the species in spite of the great initial loss. Birds and mammals which care for their young during the early days of life at least, reproduce in the lowest range, often in the case of mammals, as few as one.

Population Growth Patterns

Animal populations in nature may fluctuate in numbers from year to year, but in the absence of great physical change in the environment or of human interference, this fluctuation is around a certain norm. If a population is greatly reduced by an epidemic or some other agency, it will increase to its former size.

The pattern for the growth of many populations is the same. The growth curve for populations of bacteria, yeast, protozoans, fruit flies, rabbits, and humans is of the same form, an S-shaped *logistic* curve (Fig. 44.9). The increase in a population is proportional to the number of individuals in the population. There is a period of time when the population increases at its maximum rate. (This maximum possible rate of reproduction is a species characteristic, and it is called the *biotic potential* of the species.) This is followed by a period of decreasing multiplication and ends with a stationary population where the number of deaths equals the number of births. In some organisms the biotic potential is measured in terms of hours or days, and in others in terms of years; yet the nature of the growth curve is the same. What causes the decline in the growth that ends in the stationary phase? The causes vary with different species,

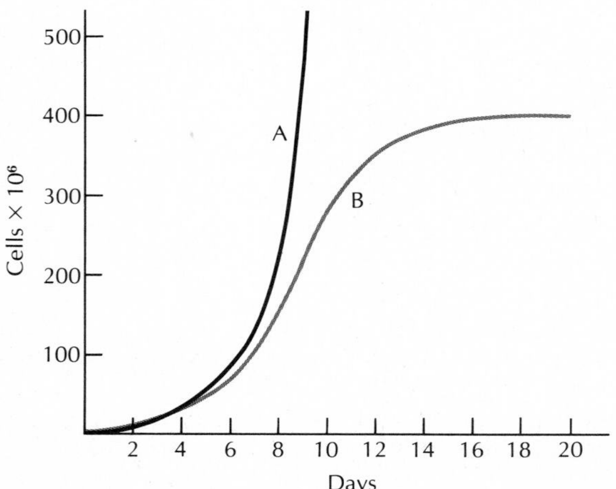

Figure 44.9 Multiplication of yeast cells. Abcissa—time in days; ordinate—number of cells (in millions). A—Theoretical rate of multiplication with unlimited space and food; B—actual rate of multiplication in a definite limited amount of nutrient solution.

and the term applied to the sum total of the limiting influences is *environmental resistance.* A primary cause in most populations is lack of food. In addition, metabolic wastes, disease, and conflicts within the population may act as checks on further growth. However, if none of these were operative, limitation of space, or standing room, would limit further increase.

With this knowledge of the growth of populations, it is now possible to devise ways to apply it in harvesting the maximum yield of lumber trees, fish, deer, and other natural resources. The harvesting rate must not exceed the maximum rate of increase in the population. When the population is very small, at the bottom of the curve, the removal of a few individuals will have a marked depressing effect. Because the maximum rate of increase occurs just before the population growth slows down to level off, the harvesting should be done at this stage. If the number removed is equal to the depressing effect on the rate of increase by the environment, the population will continue to increase at its maximum rate.

There is now considerable interest in applying this knowledge about the growth of populations to human populations. This will be discussed in Chapter 46.

Factors Affecting Population Size

Many factors combine to keep animal populations at a certain level. Those factors which constitute a type of self-limiting mechanism are said to be *density dependent.* These are the factors which have an increasingly greater impact on population growth as the size of the population increases, and conversely, they have less and less effect as the population size decreases. Density-dependent factors are always biotic in nature. *Density-independent* factors are environmental factors such as weather, overall food supplies, cover, and so forth. They operate independent of population density. They may cause very severe changes in population size—even to the point of extinction. Our concern here will be with density-dependent factors.

REDUCTION IN LITTER SIZE

In a variety of animals the number of offspring produced per female is related to the density of the population. In laboratory cultures of *Daphnia,* the common water flea, the number of offspring per female per day ranged from 5 with a density of one adult per cubic centimeter to about one when the density was 10 adults per cubic centimeter.

INTRASPECIFIC COMPETITION

The most vigorous competition in nature occurs among organisms with the most similar requirements. Darwin said "the struggle will almost invariably be most severe between individuals of the same species, for they frequent the same district, require the same food, and are exposed to the same dangers."

Competition takes a variety of forms. Among those requirements for which there may be competition are food, water, mates, nesting sites, territories, shelter, and even a place to "sit down" in the case of sessile animals and most plants.

TERRITORIALITY

Many vertebrates confine their day-to-day movements within certain particular boundaries. Normally the nesting site will be at or near the center of this area which has come to be called the *home range* of the animal. In some cases, for example many species of nesting birds, there is an inner smaller area called the *territory* which will be actively defended against all members of the same species except for mates. A male bird "sets up shop" in a given area, often commencing to build a nest. He will actively drive away all other males of his own species but pay little attention to birds of another species. He will sing and go through various courtship activities when a female of his species is near, eventually winning a mate and building a nest. The vigorous defense of the territory usually continues until the young fledgling birds are able to fend for themselves. There is seldom, if ever, an overlap of territory for members of the same species. The approximate size of the territory seems also to be a species-specific factor and it is related to the overall food requirements of the nesting pair and their young. After the young have left the nest, the territoriality disappears and the birds of a given species often join in larger flocks feeding at will in all portions of the community.

Since each breeding pair requires a specific territory, some males and females do not find suitable areas and hence do not mate. This is a type of check on population growth and constitutes one of the stabilizing influences on the numbers of the species involved. The advantages of a territory are largely related to food requirements. The parents can forage for necessary food for their young without severe competition from other mating pairs. As a result, all successful mating pairs in the region benefit. When the young are able to seek their own food, the pressure is removed and all birds can roam over the entire region.

The phenomenon of territoriality is found in many other groups besides the birds. In most cases the adaptive significance is probably much the same. In a study of a box turtle population (*Terrapene carolina carolina*) in a deciduous forest in west central Indiana, it was found that each turtle seems to have a more or less definite home range averaging between 100 and 135 m in diameter. Though there was no evidence for a defended territory, since many home ranges overlap, it nevertheless seems probable that territoriality has a correlation with the size of the turtle population. This holds fairly steady at about 10 or 12 adult turtles per hectare in the 40 hectare tract (100 acres). Figure 44.10 shows the home ranges of three turtles. The observations were made over a nine-year period, so the home ranges for these long-lived vertebrates seem to be quite stable.

MIGRATION

Some groups of animals travel from one habitat to another, usually on a seasonal basis. Although birds offer the most familiar example, such *migrations* away from one habitat—with a subsequent return to the same habitat by at least some portion of the original population—are found among many other groups of animals. Animals other than birds which migrate each year include fish of many kinds, caribou, whales, and butterflies. The migrating populations of these animals may encounter adverse conditions during their travels and thus the population size in the following year may be affected. Mechanisms whereby migrating animals, particularly birds, can travel thousands of miles and return to the same nesting site are called *homing* instincts. The workings of these homing instincts are not clearly understood; much work remains to be done on them (Ch. 37).

EMIGRATION

Some animals will periodically move out of their home range and never return. Such movements are called *emigration.* This has an obvious effect on the total population size. The classic example of emigration is in the European lemming (*Lemmus lemmus*), in which species the periodic population increase at approximately four-year intervals seems to set off an instinctive behavior pattern whereby the animals move out from their overcrowded habitat in the highland areas of the Scandinavian peninsula, often ending up at the edge of the sea where many perish. All the lemmings do not emigrate, so this mech-

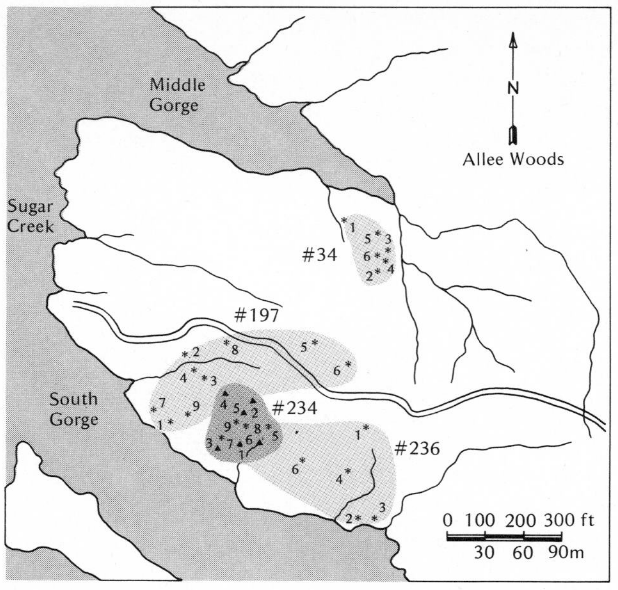

TURTLE 34
1. 7 August 1958
2. 6 July 1959
3. 20 July 1964
4. 13 July 1965
5. 28 July 1965
6. 26 June 1967

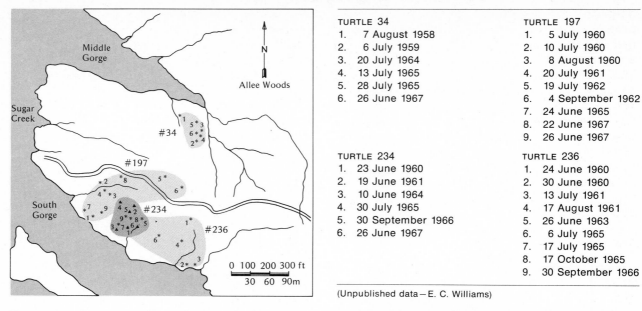

TURTLE 197
1. 5 July 1960
2. 10 July 1960
3. 8 August 1960
4. 20 July 1961
5. 19 July 1962
6. 4 September 1962
7. 24 June 1965
8. 22 June 1967
9. 26 June 1967

TURTLE 234
1. 23 June 1960
2. 19 June 1961
3. 10 June 1964
4. 30 July 1965
5. 30 September 1966
6. 26 June 1967

TURTLE 236
1. 24 June 1960
2. 30 June 1960
3. 13 July 1961
4. 17 August 1961
5. 26 June 1963
6. 6 July 1965
7. 17 July 1965
8. 17 October 1965
9. 30 September 1966

(Unpublished data—E. C. Williams)

Figure 44.10 Map showing home ranges of four box turtles, *Terrapene carolina,* in Allee Memorial Woods, Parke County, Indiana. Numbered circles indicate location of captures, and dates for each capture are given at right.

anism of regulation of population size has important survival value to the species. Emigration of one species may also bring about movement of its predators. The snowy owl of northern Canada depends largely on the lemming for its food. When the lemming populations decline, the snowy owl often moves southward and appears in the northern portions of the United States, where it feeds on other small rodents. (It does not breed at these latitudes, however, and either remains until it perishes or moves northward the following year.)

Population Cycles

The periodic increase in lemming populations mentioned above is an example of a wide-spread biological phenomenon. Many of these outbreaks are cyclic with 3–4-year cycles occurring in lemmings, voles, arctic foxes, snowy owls, ptarmigan, and sockeye salmon as examples. Another type of cycle has intervals of 9–10 years, and among the animals which display it are ruffed grouse, snowshoe hares, muskrats, Canada lynx, mink, and Atlantic salmon.

The causes of these cycles are as yet not well understood. Often there is a remarkable synchrony in widely separated geographic areas. This suggests a broad climatic control. However, no single climatic factor has been correlated with the cycles. Sun spots are often mentioned, but their cycles are of the order of 7–17 years with an average of 11.2 years and no match with any of the well-known cycles.

Predator–prey relationships are reflected in many cycles. The snowy owl and the lemming are mentioned above. A similar relationship occurs in the case of the Canada lynx and the snowshoe hare. They have a 9–10 year cycle, with the lynx cycle lagging about a year behind that of the hare.

Animal Societies

While the various degrees of social interaction between members of the same species could well be discussed here, it seemed more appropriate to include that discussion in Chapter 43.

ECOLOGY AND EVOLUTION

In a number of important areas the study of evolution leans heavily on ecology. The ecological nature of the work of Darwin and Wallace has already been mentioned. Evolution can take place only through mutations which occur in isolated species and prove successful in the process of adaptation and natural selection.

Mutations

Many external influences have been shown to increase the mutation rate in laboratory populations. In nature the causes of mutation are not known. However, cosmic radiation, ultraviolet radiation, and radiation from naturally occurring radioactive materials in the environment may all contribute. Under certain circumstances, temperature may also be mutagenic. These environmental influences are of ecological importance and thus bring the ecologist into the picture. The establishment of a mutation, whatever its cause, involves interactions within populations. The modern concept is that it is the entire population rather than individual organisms upon which natural selection operates. Many ecologists are directly concerned with studies of populations and they have made numerous contributions to the knowledge of the establishment, maintenance, and interaction of the natural populations.

Isolation

The necessity of isolation in the process of speciation has already been pointed out. Many of the factors involved in isolation are environmental in nature and are therefore ecological. For example, a slight difference in habitat selection may effectively isolate two populations. Organisms may lack adaptations which would enable them to pass certain topographic barriers, and hence they may become isolated. Differences in behavior patterns may result in sexual isolation when complex courtship patterns are involved.

Adaptation and Natural Selection

The prime basis of natural selection is the appearance of features which adapt an organism to a particular set of environmental conditions. The sorting out of the better adapted organisms at the expense of those which are less well adapted—with the result that the better-adapted organisms contribute most to the succeeding generation—is natural selection. In most cases the agent of the selective process is basically ecological. It may be competition, predation, ability (or inability) to attract a mate, operation of Shelford's law of tolerance (as explained earlier), or any one of many other ecological factors.

SUGGESTIONS FOR FURTHER READING

See list of references at the end of Chapter 45.

45 Interspecies and Ecosystem Ecology

TYPES OF INTERACTION AMONG SPECIES

Interactions among species may involve just two species or may be among several species. Our focus first will be on specific interactions involving only two species—or at most a small number of them. Some of the interactions are *antagonistic.* Included in this category are competition, predation, parasitism, and allelopathy. Other interactions are *beneficial,* at least to one species, while being harmful to neither. These interactions include: commensalism and mutualism. The word *symbiosis* (Gk. *sym,* with; *bios,* life) has been used in various ways in the classification of interactions between species. For many years it was used for a single kind of interaction, one in which both of the interacting species benefited by the relationship. More recently the tendency for ecologists has been to take the term in its literal sense and include all interactions in which two species actually "live together," without regard to benefit or harm to the species concerned. Thus the term includes parasitism, commensalism, and mutualism.

Antagonistic Interactions

COMPETITION

It was pointed out in Chapter 44 that the most rigorous competition occurs between members of the same species. However, there is also a very real competition between members of different species. Sessile marine animals of all kinds compete with each other and with plants for the limited surface area in the intertidal and subtidal zones. Among plants there is continual competition for light as well as for surface area. Herbivores with rather general requirements are all competitors for the same food; for example, grasshoppers, mice, sheep, and bison all compete for grasses. Predators are competitors for the same prey in many instances.

Laboratory studies have revealed that species with similar requirements cannot live permanently in the same environment. One or the other will eventually become extinct. Gause's pioneer work on competition between protozoan species demonstrated this (Fig. 45.1). The fact that a given ecological niche can be occupied by only one species is often called *Gause's principle.* In nature the best examples of the operation

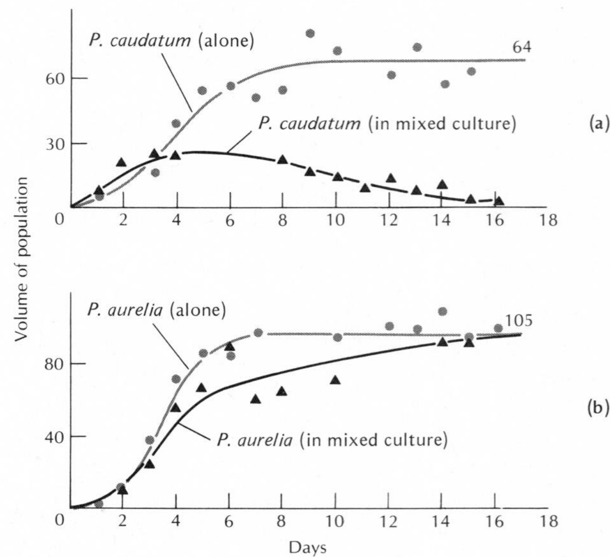

Figure 45.1 The effect of competition on populations of protozoans. **(a)** Growth of population volume of *Paramecium caudatum* alone in a controlled environment with a fixed density of bacterial food at the beginning of the experiment compared with the population volume of *P. caudatum* in competition with *P. aurelia* under the same conditions. **(b)** Growth of population volume of *P. aurelia* alone and in competition with *P. caudatum.* (Redrawn from G. C. Gause, *Science* **79,** 16-17, 1934, with permission.)

of Gause's principle can be seen in cases where exotic species have been either accidentally or purposely introduced into a new situation. The American gray squirrel, *Sciurus carolinensis,* was introduced into the British Isles, and in several areas it has completely replaced the native red squirrel, *Sciurus vulgaris.*

Introduced species often present great problems not only of competition but of damage through predation and other causes. Such problems will be discussed later in this chapter in the section entitled "Balance of Nature."

PREDATION

Predation is a mode of existence whereby an organism captures and devours another organism. With few exceptions, predators are animals. The insectivorous plants, such as the pitcher plant or the Venus fly trap are exceptions. Normally predators are relatively large in size as compared with their prey, and later discussions will point out that the number of predators must be significantly smaller than the number of prey. In some cases predators act in concert and thus are able to attack animals larger than themselves. A pack of wolves versus a musk ox would be an example. Other predators may successfully cope with larger prey by means of poisonous venoms, as is the case of many snakes. In addition to the obvious characteristics of speed, claws, and teeth, numerous special adaptations related to predation have evolved. A few of these are webs of spiders for entrapment of prey, nematocysts of coelenterates, and the eversible stomach of a starfish.

In the normal functioning of an ecosystem the predators play a significant role in maintaining a population which is in balance with the food produced.

PARASITISM

A *parasite* is an organism which lives in or on and at the expense of another organism, its *host.* In many ways parasitism is just a special form of predation. It is often said that the predator lives on capital while the parasite lives on interest. Parasitic species are found among almost all groups of plants and animals. In many cases the parasite is highly modified for its special mode of life. Often there are complex life cycles involving one or more hosts. The role of parasites in balancing populations is probably equally important to that of the predators.

One interesting type of parasitism which was mentioned in Chapter 37 is *social parasitism* seen in some birds. The European cuckoo and the American cowbird are examples. These birds do not build nests of their own. They lay their eggs in the nests of other birds and leave them to be incubated and the young cared for by the unwitting foster parents. In the case of the cuckoo, the young intruder when it hatches pushes its foster nest mates out of the nest.

ALLELOPATHY

Some organisms produce substances which are highly toxic to others. This relationship is called *allelopathy* (Gk. *allelon,* one another; *pathos,* suffering). Certain fungi and bacteria produce toxic materials which inhibit the growth of other species. The reader has undoubtedly had personal experience with the efficiency of such substances in the form of penicillin, streptomycin, aureomycin, and so forth.

Pioneer species becoming established in a disturbed area such as an abandoned field, produce substances which inhibit the bacteria associated with the nitrogen cycle. This slows down the rate of invasion by other species which have high nitrogen requirements. Some plants produce allelopathic substances which inhibit seedling growth, thus conserving scarce water and mineral supplies. The common walnut tree produces a compound which inhibits many types of plants including its own seedlings. The compound, called *juglone,* is a naphthoquinone.

Surprisingly enough, allelopathy even occurs in the sea. A small red dinoflagellate of the genus *Gonyaulax* produces a highly toxic metabolite which is lethal to fish and many other marine animals. A great increase in numbers of these organisms causes a phenomenon known as the "red tide." Humans can be secondarily affected by eating marine organisms which have ingested the dinoflagellate. Because of this danger, there is a permanent quarantine on clams in Alaska and on mussels from May through October in California. The active ingredient of the poison, isolated from clams, is saxitoxin. Its effect on humans is through the nervous system where it causes paralysis by blocking the nerve impulse, apparently through inhibition of the increase in sodium conductance of the nerve membrane which normally accompanies the action potential.

Beneficial Interactions

COMMENSALISM

The word commensal (L. *com,* with; *mensa,* table) is taken to mean "at the same table"—in short, messmates. In this type of interaction there is a definite benefit for only one of the two species involved; however, the second one is not harmed. Although most of the commensal relationships are related to feeding, there are other cases in which the benefit is in terms of shelter or protection rather than one of food acquisition. Such associations probably started with one species tolerating the presence of the other near, on, or in its body. If the tolerated "guest" derived some benefit without interfering with the host, the relationship would tend to persist. Many workers think that parasitism probably started in such associations. It is also considered that mutualism might have started in commensalism. If the host species became adapted to gain some advantage because of the presence of the guest species, mutualism would be established.

A striking example of commensalism is that of the remora fishes and the sharks to which they attach. This group of fishes have their dorsal fins modified into suction disks by means of which they attach to the underside of a shark. They ride along attached to the shark, occasionally detaching themselves long enough to collect fragments of food from the shark's meal. Remoras also attach themselves to large sea turtles. Other examples are certain beetles that live in the nests of meadow mice, the owl that nests in the burrows of prairie dogs, the crab that lives within the oyster's shell, and the tropical fish that lives within the cloaca of the sea cucumber.

MUTUALISM

When both members of a close association benefit from living together, the relationship is called *mutualism.* Often such associations are obligatory. A good example of this is found in the relationship of termites and their intestinal protozoans. Termites feed on wood, but have no enzymes to digest it. The protozoans in the termite intestines are able to break down the cellulose into sugar, using some of it in their own metabolism and leaving some for the termite. Cleveland has shown that when all the protozoans in a termite are killed, the termite soon dies. No one, yet, has been able to culture these protozoans outside of the termite on artificial media.

Muscatine and his coworkers demonstrated that intracellular algae, zoochlorellae, living in a mutualistic relationship with *Paramecium bursaria* liberate significant amounts of soluble glucose and maltose, while related free-living strains of algae do not. From 5.4 to 86.7 percent of the total photosynthate was released from the intracellular species, while the free-living forms released only 0.4–7.6 percent of their total, principally as glycolic acid. These results clearly indicate that the relationship between alga and protozoan is more than just a chance union. Actually, metabolic modifications have occurred in the alga, making possible the production of significant amounts of usable food materials for its protozoan host. This kind of relationship between algae and animals is found in a variety of forms, including sponges, coelenterates, flatworms, annelids, and mollusks. The giant clam, *Tridacna*, of South Pacific coral reefs has vast numbers of intracellular algae in the edge of its mantle. It even has small lenslike structures in the mantle which are believed to function in focusing light on the deeper-lying algal cells.

Ecological Succession

A very special interaction between numbers of organisms, both plants and animals, and their environment is seen in the phenomenon of ecological succession. In nature, stable communities may sometimes be disrupted and the conditions for life may be changed entirely. For example, such a change is brought about when a forest is completely burned over. As time goes on, different forms appear, and these gradually replace the first forms. Such a replacement is known as ecological succession.

One of the best places to study such succession in this country is along the southern shores of Lake Michigan (Fig. 45.2). It was once a much larger lake, and as its shores receded they have left successively younger sand dunes. There is a series of communities beginning at the shores of the lake and continuing inland several kilometers. Starting on the open beach there is no permanent plant life. Here one finds a community of organisms, mostly insects, that feed on the drift material, and in turn are fed on by others in the community. The initial plant community is found in the foredune region and is composed of sandbinding grasses and the sand cherry. Back of the foredune region will be found a community characterized by cottonwoods; behind this is a community dominated by pine trees; beyond this is a series of oak communities; and finally the beech–maple forest community is reached. The beech–maple forest is a stable community and will undergo no further change unless disturbed by major climatic changes. The beech–maple forest is called the *climax* stage of this succession. The entire series of communities leading up to the climax is called a *sere*. Figures 45.3–45.6 are a series of photographs taken near the eastern border of the Indiana Dunes State Park. They show the first four stages of the successional sere.

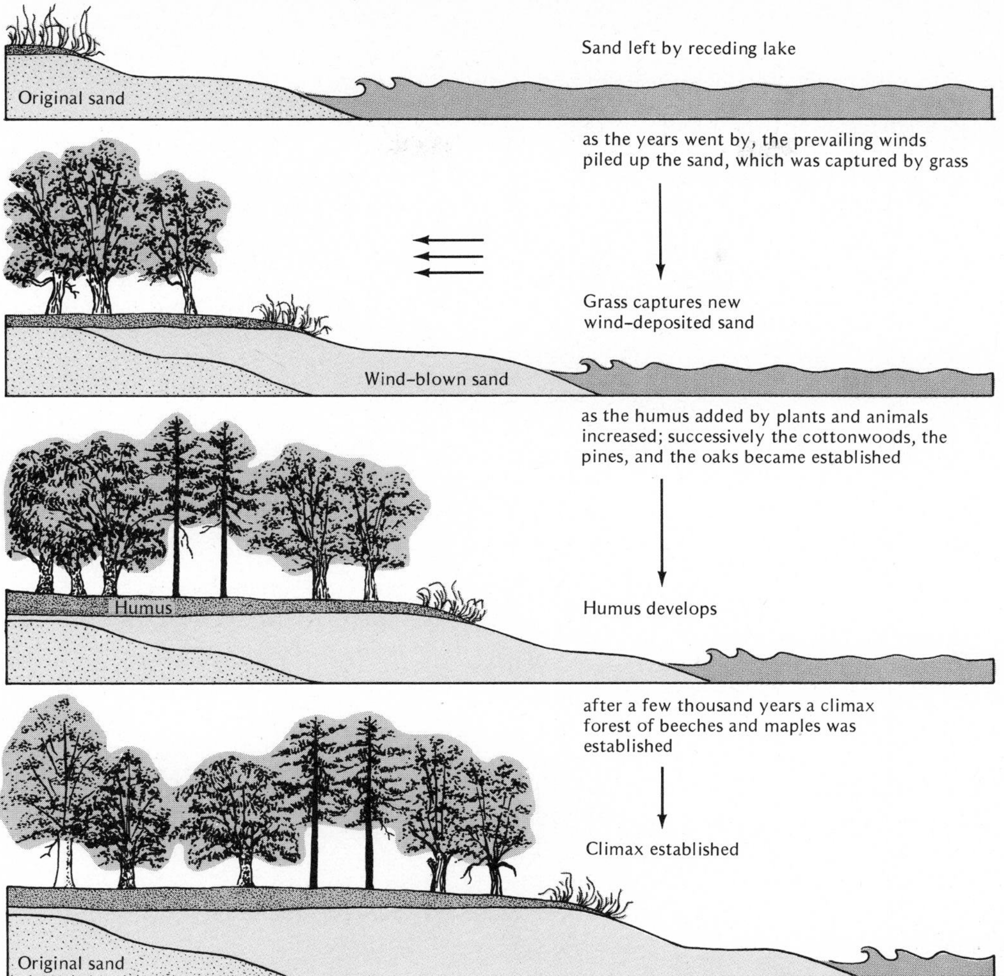
Sand left by receding lake
Original sand
as the years went by, the prevailing winds piled up the sand, which was captured by grass
Grass captures new wind-deposited sand
Wind-blown sand
as the humus added by plants and animals increased; successively the cottonwoods, the pines, and the oaks became established
Humus
Humus develops
after a few thousand years a climax forest of beeches and maples was established
Climax established
Original sand

Figure 45.2 Diagram of succession in the dunes of northern Indiana. (Modified after Allee, Emerson, Park, Park, and Schmidt, *Principles of Animal Ecology,* © 1949 by W. B. Saunders and Company, with permission.)

Each of the stages in the succession was preceded in past time by an earlier, less moist community. Thus the beech–maple forest was preceded by oak forests and these by a pine forest, and so on.

The causes of such a succession are complex and not completely known. Both physical and biotic factors are involved. With the establishment of grasses and the sand cherry, the sand is held in place. As time goes by, these plants die and humus is added to the soil. Ultimately, conditions develop in which cottonwoods, then pines and oaks may grow. In each case the particular kind of plant modifies the soil by adding humus, making it possible for the next type in the series to grow. The trees that appear in the earlier stages of the succession can grow in full sunlight. With the development of forests comes shade and each type of tree seedling varies in its tolerance to shade. In dense shade the seedlings of pines and oaks do not thrive but those of beech and maple grow successfully. Thus each community or stage modifies the conditions in such a way that it becomes more suitable for the next stage. *In effect, each community drives itself out.* In moving from the lakeshore back through the stages of succession, one finds a number of differences in the environment that the different kinds of plants have produced. There is a gradual increase in humus from stage to stage, in the amount of soil moisture, and in the relative humidity; there is a progressive decrease in total light intensity, in wind velocity, and in the rate of evaporation.

Associated with the plant succession is a succession of animal life. Each plant community in the sere is inhabited by a characteristic assemblage of animals. Some animals may be restricted to a single stage while others may be found in several. Shelford has shown an interesting sequence of species of tiger beetles (family Cicindelidae). On the beach there are two species, *Cicindela hirticollis* and *C. cuprescens.* The upper beach and grassy zone are characterized by *C. lepida.* The ecotone or transition zone between the cottonwood and

Figure 45.3 Ecological succession in the sand dunes of northern Indiana I. Beach and foredune stages. (Photo – E. C. Williams)

Figure 45.4 Ecological succession in the sand dunes of northern Indiana II. Transition from foredune to cottonwood stage. (Photo – E. C. Williams)

Figure 45.5 Ecological succession in the sand dunes of northern Indiana III. Pine stage. (Photo – E. C. Williams)

Figure 45.6 Ecological succession in the sand dunes of northern Indiana IV. Oak stage. The area shown here is only about 131 m (430 ft.) from the water's edge. (Photo—E. C. Williams)

pine is occupied by *C. formosa,* while *C. scutellaris* is most common in the pines. Finally in the oaks and beech–maple climax the green tiger beetle, *Cicindela sexguttata,* is the only common form. Although the plants play a primary role in the development of the various stages, the whole sere involves changes in both the plant life and the animal life, and therefore this is truly a *biotic succession.*

Not all climax communities are beech and maple forests. There are numerous kinds of climax communities in different regions with different climates. Even within a given region the climax may vary from place to place due to differences in soil, drainage, or other modifying factors.

Anyone interested in following through the stages in a succession in a fairly short period of time can do it easily in the laboratory. When a hay infusion is placed in a jar and allowed to stand, protozoans of a variety of kinds appear and in a definite sequence, as was first reported by Woodruff. The changes in the culture are primarily due to biotic influences. The first forms to appear are bacteria. These are followed in turn by small monad flagellates, the ciliate *Colpoda,* hypotrich ciliates, paramecia, stalked ciliates like *Vorticella,* and finally amoebae.

The Ecotone

The transition zone between two communities such as grassland and forest is called the *ecotone.* This is a very special type of situation since it has certain attributes of each of the community types and in addition has certain attributes all its own. The animals living in the ecotone will be partly those of one community and partly those of the other. In addition there will be certain animals present which are normally found *only* in the ecotone. Thus the diversity of species in the ecotone is often greater than that in either of the neighboring areas. The phenomenon of increased diversity in the zone between two community types has been called the *edge effect.*

THE ECOSYSTEM, BASIC FUNCTIONAL ECOLOGICAL UNIT

All the various animal populations living in a given habitat may be spoken of as an *animal community,* while the plants can be called collectively a *plant community.* The term *biotic community* is used to include both the plants and the animals in a particular habitat.

Living organisms cannot exist divorced from a nonliving environment. Therefore ecologists use a broader, more inclusive term, the *ecosystem,* which is defined as an interacting community of organisms together with their abiotic environment. Ecosystem ecology, as opposed to community ecology, is the study of the dynamics and function of the community.

Ecosystem Components

There are four components to a complete ecosystem.

1. *Abiotic substances,* the nonliving components of the environment.
2. *Producers.* These are *autotrophic* organisms (Gk. *auto,* self; *trophos,* feeder). Producers are usually green plants which utilize light energy and simple inorganic material to synthesize carbohydrates and in turn use this basic material to produce more complex organic compounds.
3. *Consumers.* These as well as the next com-

ponent are *heterotrophic* (Gk. *heteros,* other; *trophos,* feeder). These are mainly animals which consume other organisms or parts of organisms.

4. *Decomposers.* These are mainly bacteria and fungi which break down organic material into simpler compounds, utilizing some of the material themselves and releasing the rest into the environment in a form which can be reused by producers.

The term *macroconsumer* is sometimes used for the consumers and the term *microconsumer* for the decomposers.

The actual organisms which make up the three living components in an ecosystem are quite different when one compares a balanced aquarium with a pond or a pond with a deciduous forest community—yet in each case there are producers, consumers, and decomposers. The general ecological niche in each case is occupied.

A small freshwater pond is an excellent example of an ecosystem. Figure 45.7 shows the components of this system graphically.

Abiotic Substances This component includes the water with all the dissolved minerals, gases, and suspended organic matter which it contains; the soil of the bottom of the pond with its mineral and organic sediments; and the air above the surface of the pond.

Producers The most significant producers are the microscopic floating plants, the *phytoplankton.* Most of them are algae, the simplest type of green plant. Along the sides of the pond will be found the second type of producer, the rooted plants. Although these plants, because of their size, are more obvious, they are usually much less important producers than the microscopic algae.

Consumers These are the animal inhabitants of the pond. There are several levels in the consumer compartment. The *primary consumer* is the *herbivore,* feeding directly upon the plant materials. Elton has used the term "key industry animals" for these organisms which are first in the chain to convert plant material into animal material. Some of these herbivorous animals are microscopic free-swimming animals which are at the mercy of the currents, and they constitute the *zooplankton.* Together with the microscopic phytoplankton they make up the *plankton.* Zooplankton include the protozoans, rotifers, and

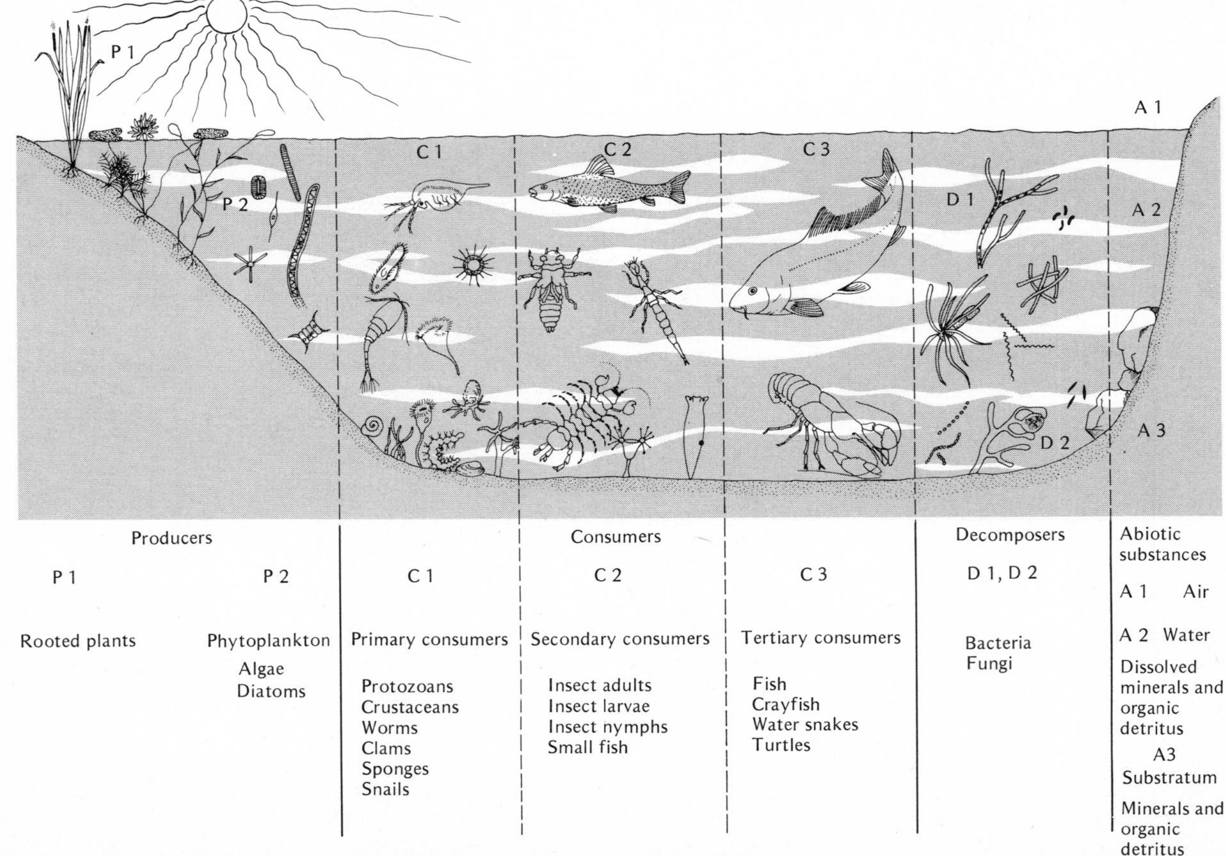

Figure 45.7 A freshwater pond as an example of an ecosystem. All major components are represented, namely, the producers, consumers at three levels, decomposers, and abiotic substances. (The organisms are not drawn to scale.)

small crustaceans. Other primary consumers live in or on the substratum—worms and burrowing insect larvae, for example. Feeding upon the herbivores are several levels of *carnivores,* called progressively *secondary* and *tertiary consumers,* depending upon the steps removed from the primary consumers. The carnivorous forms include larger insect larvae, adult insects, crayfish, and fish.

Decomposers The animals are continually releasing organic waste materials, and all of the biotic components will eventually die. The wastes and dead remains sink to the bottom where they are acted upon by the bacteria and fungi of decay. Some decomposition will occur in the water, but most of it occurs on the bottom where the bulk of the organic debris comes to lie.

Energy Flow in the Ecosystem

FOOD CHAINS

The nutritive interactions of the biotic components of the ecosystem are often called *food chains.* The concept of a lake or pond as a self-sufficient system is not a new one. Stephen Forbes, in 1887, published a paper entitled "The Lake as a Microcosm." He described the relations of the black bass with the other inhabitants of the pond approximately as follows.

The food of the black bass consists of other fishes, insects, crayfish, and small crustaceans. The other fishes feed on algae and small crustaceans; the insects feed on small crustaceans; the crayfish is omnivorous; and the small crustaceans feed on algae. Competitors include other fishes, certain mollusks and insects, and the carnivorous plant, the bladderwort. Predators on the bass are other fishes, turtles, water snakes, birds, beetles, water bugs, and dragonfly naiads.

The organisms in every habitat show similar interrelations. Shelford has determined them in a variety of habitats. Figure 45.8 shows his results for an Illinois deciduous forest. He also described several food chains in a prairie where the prairie grasses are the major producers. The buffalo (now supplanted by cattle) feeds on the grass, and people feed on the buffalo. In another food chain, field mice feed on the grasses and they, in turn, are eaten by horned owls, coyotes, skunks, and weasels.

Darwin described a variation of this type of relationship. He said that the amount of clover seed produced in a given region varies with the number of cats present. His explanation was as follows. The more cats in a region, the fewer mice; the fewer field mice, the more bumble bees; the more bumble bees, the better the pol-

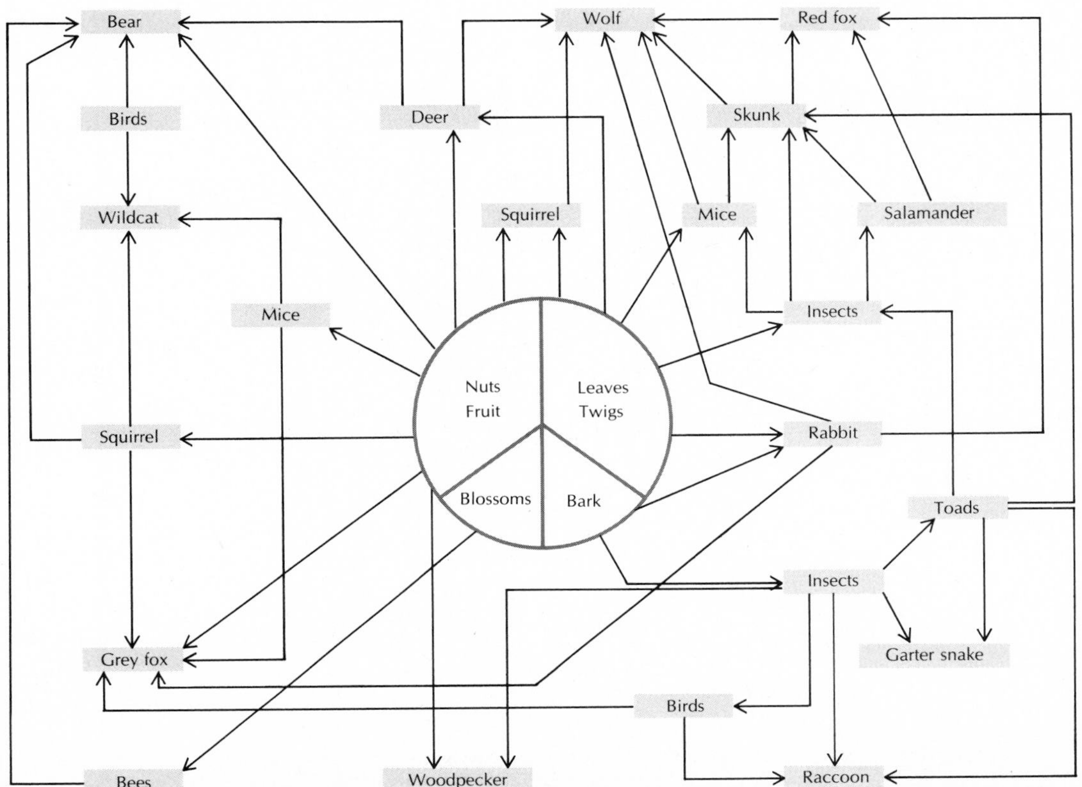

Figure 45.8 Food chains in an Illinois deciduous forest. (After Shelford, with permission.)

lination of clover and, thus, the more clover seed. Huxley, in telling Darwin's story, added one more link to the chain. He said that the amount of clover seed is related to the number of spinsters in the community; the more spinsters, the more cats, and so on.

ECOLOGICAL PYRAMIDS

The relationships in a food chain have been described by Elton as a *pyramid of numbers.* At the bottom of each pyramid is the plant life that first synthesizes food, utilizing the energy from the sun. Just above are the key-industry animals. Above these are other animal types, the ones at each higher level feeding on the ones below. The number of organisms in the lowest level of the pyramid is very great, greater than in any other level, and their size is small. In the successively higher levels, the number of individuals declines and their size increases until, at the top of the pyramid, the number of individuals is relatively small and they are large in size.

In general, the smallest organisms are found at the bottom level and the largest at the top, with gradations in size in between. In the case of parasites the pyramid is inverted as far as numbers are concerned. Large numbers of very small parasites are collectively equivalent to a small number of large carnivores. Figure 45.9 shows an actual pyramid of numbers based on animals living in the surface litter of the Panama rain forest. Figure 45.10 illustrates the same principle at the level of vertebrate animals.

In addition to the pyramid of numbers, there is also the *pyramid of biomass,* the amount of living matter in a particular environment. The total biomass of the basic plant life is greater than that of the key-industry animals. The total biomass of the key-industry animals is greater than that of the forms that prey on them, and so on.

THERMODYNAMIC LAWS AND THE ECOSYSTEM

According to the first law of thermodynamics (Ch. 3), one form of energy may be converted into other forms, but in any such conversions there is neither loss of nor creation of energy. But according to the second law of thermodynamics, in any conversion of energy from one form to another, there is always a decrease in the amount of useful energy; some energy in the form of heat is dissipated into the environment. Every time organic compounds are passed from one organism to another, they are decreased in amount, and some of their energy is lost as heat. Because of this loss of energy from one level of consumption to another, there can be no cycle for the reuse of energy similar to the cycles of carbon, nitrogen, and other elements described previously. (Figure 45.11 illustrates the flow of energy through an ecosystem.) In general the energy is reduced in magnitude by 100 from primary producers to the initial herbivore consumer and by 10 for each step thereafter in succeeding carnivorous members of the food chain. If 1000 kcal of light energy per square meter per day are fixed by green plants, then 10 kcal would be converted to herbivore tissue, 1.0 kcal to the first carnivore, 0.1 to the second, and so forth.

The number of links in food chains varies from two to five or six. Rarely are there more. The most efficient food chains, from the standpoint of energy relations, are those with the fewest links. It is no accident that people in densely populated countries, such as China and

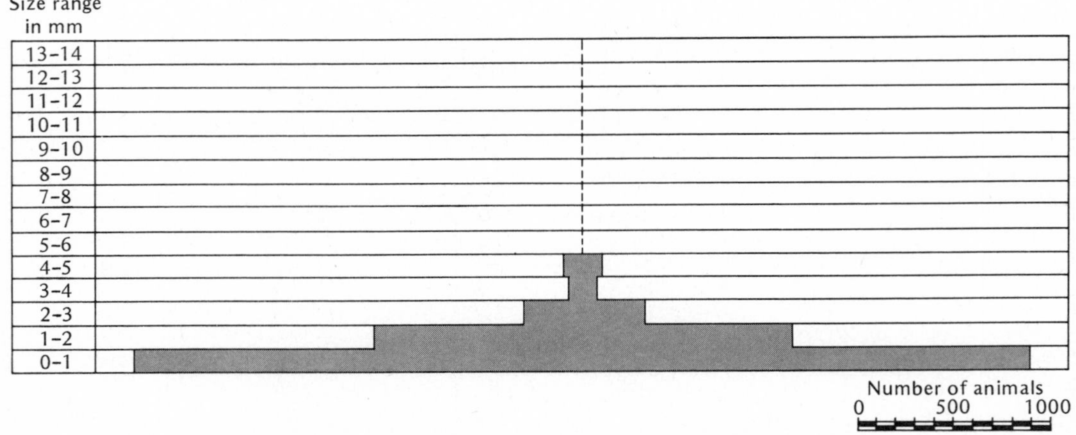

Figure 45.9 Eltonian pyramid of numbers, based on quantitative sampling of litter-dwelling organisms of the Panama rain forest floor. Vertical divisions are size ranges and the horizontal scale is in terms of numbers of animals. (From Williams, *Bulletin of the Chicago Academy of Sciences,* **6,** 4, 1940, with permission.)

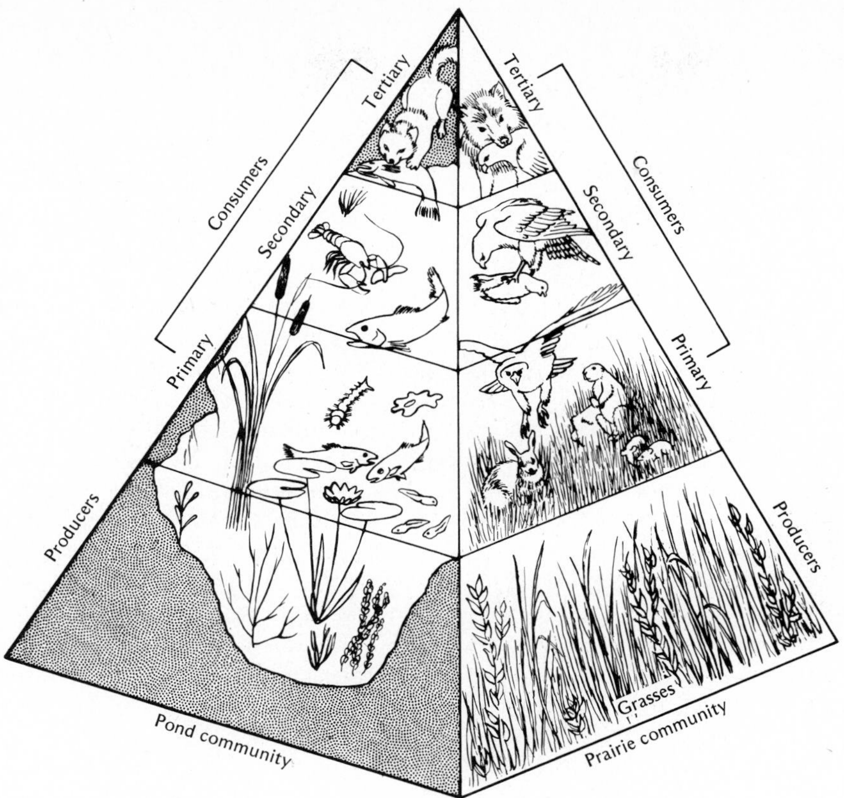

Figure 45.10 Pyramid of numbers. Producers, usually green plants, exist in greatest numbers. Animals which consume producers, herbivores or primary consumers, exist in greater numbers than successive series of secondary and tertiary consumers. On the right face of the pyramid this relationship of numbers is suggested for a prairie community and on the left for a pond community.

India, depend primarily on plant foods. It can be said, then, that steak is a luxury ecologically as well as economically.

TECHNIQUES FOR THE STUDY OF ENERGY FLOW

The availability of radioisotopes opens up entirely new avenues of attack on the problem of energy flow in the ecosystem. Detailed analysis of food chains or webs can now be made. For example, one can spray a plot of ground with a solution containing a suitable radioisotope—phosphorus^{32} is quite useful because of its short half-life—and follow the isotope by sampling plants and animals on subsequent days. Initially the isotope will be found in the plants. Shortly thereafter it will appear in the herbivores, and later in the carnivores. More detailed information can be obtained on specific food chains by limiting the isotope to a single kind of plant (Fig. 45.12). The isotope can be injected into the plant in a suitable carrier, or it can be topically applied to the leaves, which will readily absorb it. In such experiments any herbivore in which the radioisotope can be detected is known to be a primary consumer of the plant which was inoculated. Secondary consumers are also identifiable in the same manner. The amount of isotope and the time after inoculation in which it is detected can be used to estimate whether predators are secondary, tertiary, or even further removed in the chain from the producer. Those predators in which the isotope first appears and in which the levels are greatest are probably the secondary consumers, while those in which the isotope does not appear for some days and in which levels are somewhat lower are probably tertiary.

More detailed studies of energy turnover are feasible by use of these techniques. If an herbivore feeds upon plant material uniformly saturated with a radioisotope, it will reach a level of equilibrium whereby the intake of new isotope balances that which is lost through normal excretion or defecation. By determining the normal rate of loss of isotope it is possible to estimate the consumption necessary to maintain a constant level. Further, it is possible to treat the plant material in a calorimeter and thus determine the actual intake in terms of calories. The advantages of these techniques are that insects can be left on their normal food plants in

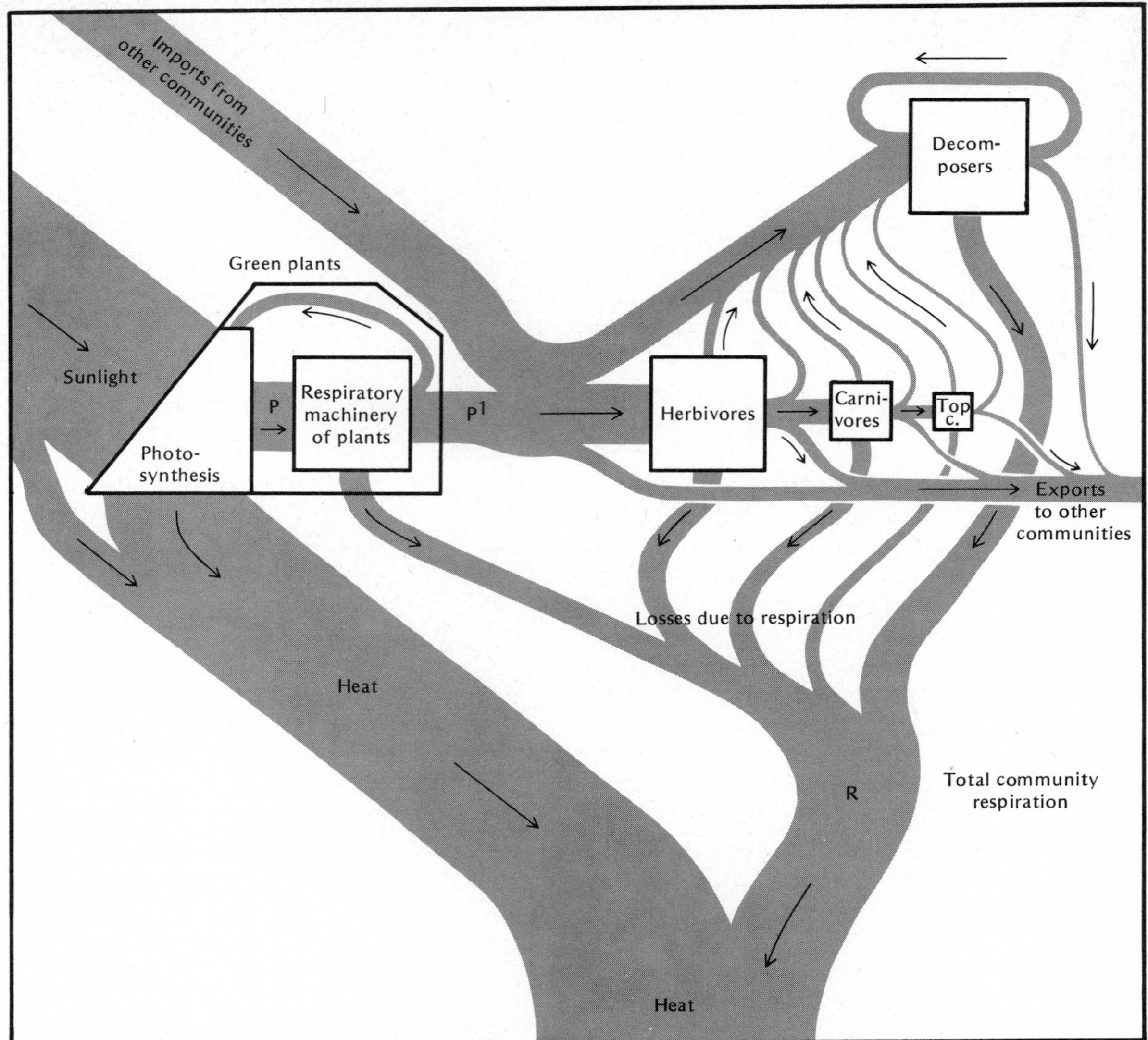

Figure 45.11 A generalized diagram of energy flow in an ecosystem. Energy from the sun reaching the ecosystem is largely lost as heat. A small fraction is utilized by the green plants—the *producers* of the ecosystem—in photosynthesis. The *gross primary production* (P) is in part used by the green plants in their respiration, and the rest is available to *consumers* as *net primary production* (P^1). Some materials may enter the ecosystem as imports, the amount varying greatly depending upon the ecosystem. The *herbivores* in turn use up a large proportion of the available food in their own respiration with some net production available for the *carnivores.* The amount remaining for the top carnivore is very small, and beyond this little surplus remains. Each level of the ecosystem—producers, herbivores, and carnivores—loses a great deal of energy as heat due to respiration (R). Unassimilated materials, as well as wastes and remains, move to the *decomposers* which in turn lose much via respiration and have almost no surplus available. Little is left for export to other systems. As in the case of imports, this factor will vary a great deal with different ecosystems. This system illustrates the operation of the laws of thermodynamics, since inflows balance outflows, and each level loses a great deal of energy in the form of heat. (From H. T. Odum, *Limnology and Oceanography,* vol. 1, p. 113, 1956, with permission.)

the field and the values obtained will approximate those under normal conditions. Such information is invaluable in estimating the total energy requirements of a given ecosystem.

Systems Ecology—Mathematical Modeling of Ecosystems

The application of systems analysis to ecology is *systems ecology.* Systems analysis consists of converting the physical and biological components of an ecosystem into mathematical terms and using these terms in mathematical constructs, called *models.* The model is an abstract representation of the real-world ecosystem. Of necessity, it is greatly simplified because only the most significant variables out of a very large number can be used. Figure 45.11 is an example of an ecological model.

There are usually four parts to a mathematical model. *System* or *state variables* are the numeri-

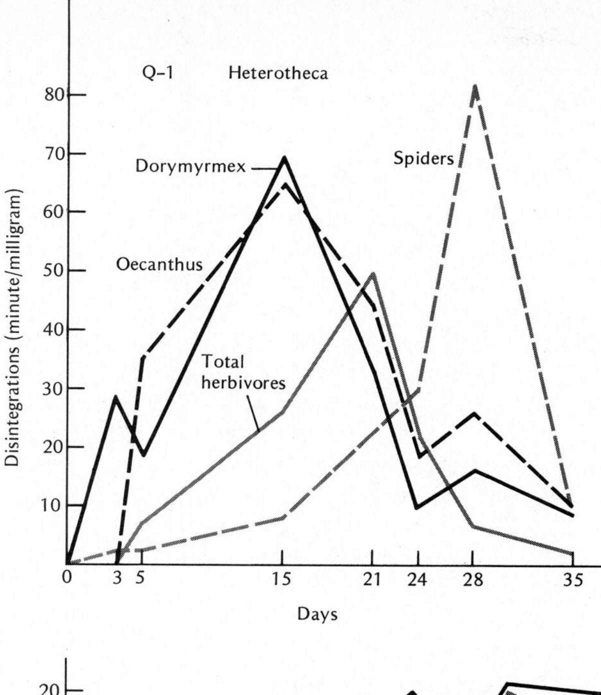

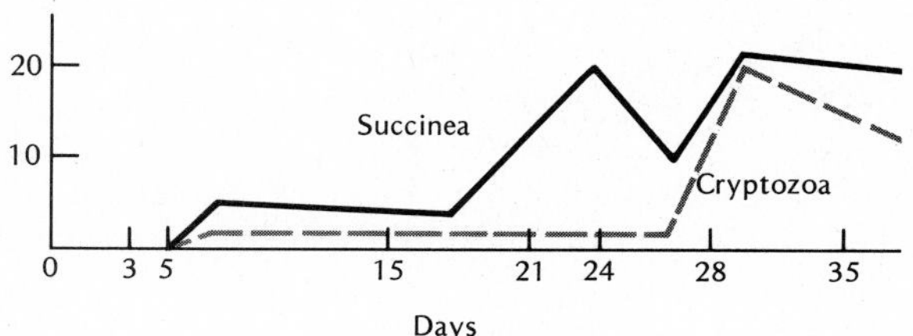

Figure 45.12 Movement of phosphorus-32 through a food chain. The radioisotope was applied on day 0 to mature plants, *Heterotheca subaxillaris,* in an 81-m^2 old field plot; 575 plants were tagged by application to the leaves of a solution containing the ^{32}P. *Oecanthus* is a tree cricket; *Dorymyrmex* is a small ant; "total herbivores" includes *Oecanthus,* grasshopper nymphs, and plant-feeding Hemiptera, Homoptera, and Coleoptera; *Succinea* is a snail; and cryptozoa (soil dwellers) includes ground beetles, crickets, darkling beetles, and roaches. (From Schultz and Klement, *Radioecology,* © 1963 by Reinhold Book Corporation, a subsidiary of Chapman-Reinhold, New York, with permission.)

cal representations of the state or condition of the system at any one time. Information on production, consumption, population size, and other such features are involved. *Transfer functions* are equations which represent interactions between various compartments of the system. The arrows between the parts of the energy flow diagram in Figure 45.11 represent such transfer functions. *Forcing functions* are equations relating to inputs into the system which are not affected by the system but which do have important effects on it. In the energy flow diagram the sunlight and the imports from other communities are forcing functions. The fourth part of the model are mathematical constants, called *parameters.* The basic mathematical tools of matrix algebra and differential and linear equations are used in the construction of mathematical models.

Models have been most useful in the analysis of various material resources where yields under varying conditions can be determined and optimum conditions for efficient resource utilization can be established.

Evolution of the Ecosystem

In the chapters on evolution, the primary emphasis was upon the evolution of individual species and on the mechanisms involved. It was shown that individual species have come to be what they are today through adaptation and complex interactions. In the same way, whole groups of species—both plants and animals—have interacted with each other and with their environment, and have concurrently become adapted to each other.

Natural selection does not act only on structure and function. It also acts on behavior and social patterns as well. In Chapter 44 the concept of ecological niche was presented. Much of the evolutionary sorting out that has gone on in the evolution of the ecosystem has resulted in the selection of those organisms best adapted for a specific niche—a process called *niche segregation.* Among the advantages of such segregation is a reduction in competition. It also avoids interference between members of the community and allows for a more smoothly operating ecosystem. Finally the segregation of each species into its own niche allows for greater diversity of species since the available resources are more efficiently utilized.

The end result of these evolutionary phenomena is the ecosystem which we can study today. Thus we say that ecosystems have evolved just as individual species have evolved. The fact that one can recognize the various biomes even though they are composed of different species on the different continents is adequate proof that a biome is not a haphazard group of plants and animals; it is a complex unit which has come to be what it is today through the evolutionary process.

Ecosystems and Society

In Chapter 46 we will deal specifically with human ecology. At this point let us briefly consider the impact of society on natural ecosystems and the ecological basis of sound conservation practices.

THE BALANCE OF NATURE

The relationship of population densities of the different species that make up a biotic community is often spoken of as the *balance of nature.* The interaction of the different species in the food chains is such that the numbers of each species are kept relatively constant. Even those organisms that are not eaten by other organisms have parasites that tend to keep them in check. Although the population densities of every member species of a community fluctuate from season to season and from year to year, these fluctuations usually have fairly definite limits. Barring major physical changes in the environment, it seems unlikely that in recent times many species have been completely eliminated from a stable community except through human intervention.

Society is only gradually realizing the importance of not disturbing this balance of nature. The idea is basic in all conservation work. We cannot introduce new species in a region or completely eliminate a species without the danger of completely disrupting the balance. The introduction of rabbits into Australia, which had none of their natural enemies to keep them in check, resulted in a rabbit plague in that country. Large sums have been spent in trying to control them. The introduction of an infectious disease, *Myxomytosis,* in 1950 has gone a long way toward solving the problem. Examples of other introductions are numerous. English sparrows—and more recently, starlings—became pests after they were introduced into this country. The European cornborer, once introduced here, soon spread through the corn belt and caused great economic losses. In this day of rapid travel from one continent to another, the problem of accidental introduction of exotic species is a serious one. Even where careful immigration restrictions are enforced, as in the state of California, new species can be accidentally introduced and sometimes become established. Intensive international cooperation is needed to reduce this danger.

But we are continually disturbing the balance of nature in more obvious ways than by the introduction of new species. We cut down entire forests, plow up large areas of grassland, and drain swamps and marshes, to mention only a few. All of these things disrupt the balance of nature, often with drastic consequences. The "dust bowls" in the Southwest, and the treeless, eroded countryside in China are examples. We now realize that much of our plains country should never have been plowed for crops. Increasingly, one is being brought to the realization that water is one of the critical resources. The removal of the forests and the drainage of swamps and lakes has hastened the return of the water to the sea. With all the modern developments, more and more water is needed. As a result, in many regions, areas that had been drained are being returned to the natural state and many new lakes are being constructed.

CONSERVATION

Fortunately, recognition of the need for forest and wildlife conservation also is becoming more widespread. With modern forest management, as explained in the section on population growth, it is now possible to have a sustained yield from our forests. A sustained yield is obtained when new timber is being produced as rapidly as it is being removed by cutting, fire, and all other causes.

With forest conservation, the conservation of many forms of wildlife becomes easier. But it is one thing to believe in conservation and quite another to practice conservation effectively. In this connection, conservationists have learned from experience that predation is not necessarily a bad thing. Many now recognize that some kind of balance between the predators and the protected species is, in the end, better for the protected form. Thus we see that in wildlife management, as in all forms of conservation, it is necessary to make use of all the available ecological information.

THE BIOMES OF THE WORLD

While the ecosystem is the basic functional unit in ecology, there is yet another unit based on the interaction of regional climates with the biotic and abiotic components which results in a large, easily recognizable community unit, called the *biome.* The biome is the largest land unit which is convenient to recognize. It is usually characterized by a particular general type of vegetation, not by any particular species. For example, the *deciduous forest* biome is characterized by deciduous trees, but they are of many different species. Similarly, the *grassland* biome is characterized by grasses of a number of kinds. In each case, however, the biome is not a region which is completely uniform. Local differences in soil or topography result, for example, in grassy areas in the deciduous forest biome or trees in the grassland biome, and so forth. In any of these areas, bodies of water such as lakes, rivers, and ponds may occur. Thus a biome may be considered as a composite of many ecosystems, or as an ecosystem itself.

A look at a map of the world showing the major biomes (Fig. 45.13) will point up some significant features of the biome concept. Only two of the biomes are even remotely continuous around the globe. The tundra and the northern coniferous forest stretch across the northern hemisphere. The flora and fauna of these two biomes in North America are much more similar to those of their counterparts in Eurasia in terms of actual species makeup than any other biome counterparts. Although temperate deciduous forest biomes are found in eastern North America, Europe, central China, and southeastern Australia, and there are some similarities in plant species, the animal components are in most cases quite dissimilar.

The plants and animals of widely separated biomes of the same type are usually distinctive in terms of the actual species, yet ecologically the similarities are quite striking. The deserts of the western hemisphere are characterized by the presence of cacti, plants whose characteristics are well known to all. In the deserts of Africa one finds members of a completely different family, the *Euphorbiaceae,* which are amazingly like our own cacti in appearance and growth form.

Plants and animals which perform the same role in different ecosystems are said to fill the same ecological niche. Often the organisms filling the same niche are called *ecological equivalants.* In essence, these ecological equivalents make it possible to divide the earth's surface into the recognizable units which we call biomes.

Tundra Biome

The vast portions of North America and Eurasia north of the timberline make up the tundra biome (Pl. 47). The tundra landscape is gently rolling, with lakes and bogs in the lowlands. The climate is rigorous, with but a sixty-day growing season and with mean monthly temperatures varying between –35° C (–30° F) and 13° C (55° F). Much of the tundra is above the Arctic Circle and it receives almost continuous sunlight for approximately half of the year. Precipitation is sparse; between 30 and 50 cm (12 and 30 inches) per year, decreasing northward.

During the summer months the soil thaws only to a depth of a few centimeters in most places. Below this is the permafrost—permanently frozen soil. When the upper part does thaw, the water cannot sink into the permafrost

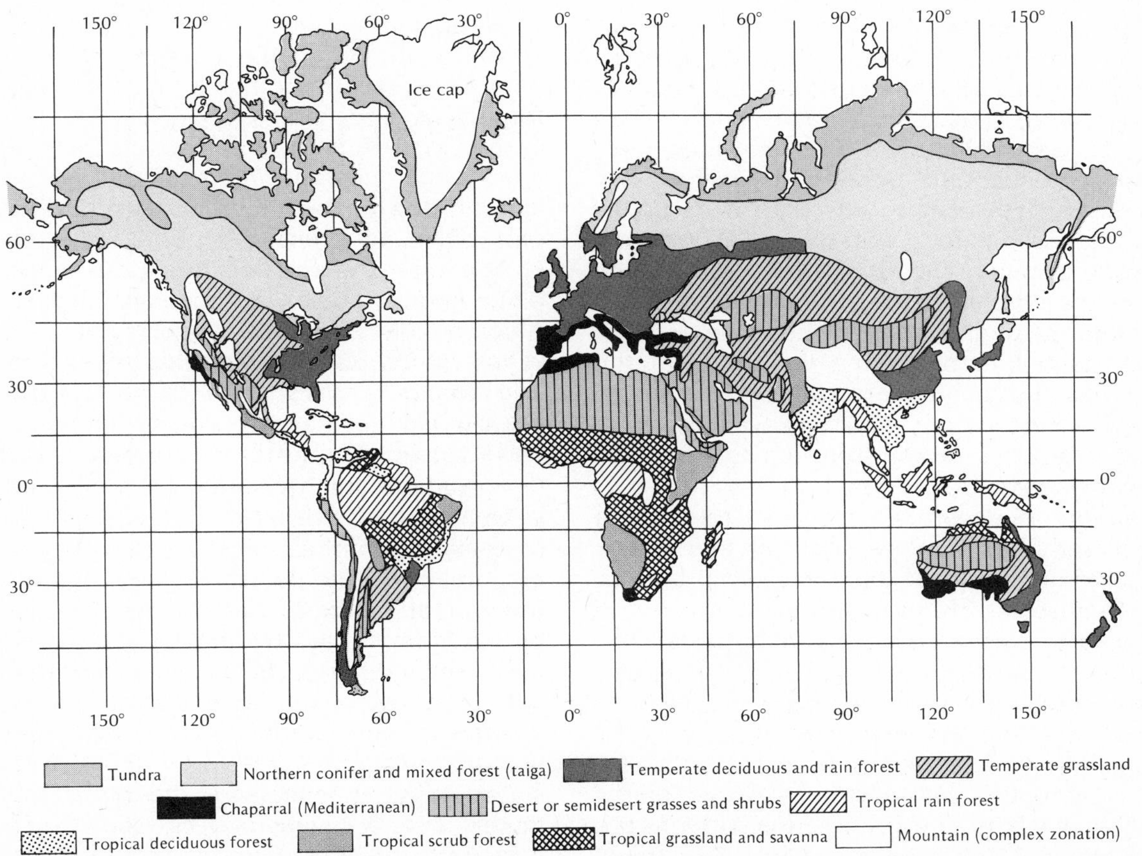

Figure 45.13 Map of the major biomes of the world.

and thus the standing water produces great expanses of shallow ponds and bogs. Mosquitoes and other insects which undergo development in water are often unbelievably abundant during this time.

The vegetation of the tundra gives the impression of a short grass prairie, but closer examination reveals sphagnum mosses, lichens (reindeer moss), sedges, rushes, grasses, low shrubs such as bilberry, and some flowering herbaceous plants. Most plants are stunted as compared with related forms in warmer climates, an adaptation which affords them some protection against the drying and abrasion to which they are subjected by blowing sand, sleet, and snow. In this stern environment seeds do not readily germinate and most plants are perennials with a good capacity for vegetative reproduction.

A surprising number of birds and mammals are found in the tundra. The caribou of North America and its close relative, the reindeer of northern Europe, are the most abundant large mammals. The caribou migrates each year to southern portions of the tundra, often moving into the forested regions to the south. Carnivores include the gray wolf, arctic fox, wolverine, and polar bear.

The musk ox, formerly found throughout much of the arctic tundra, is now restricted to North America and Greenland, and its numbers are dwindling.

Lemmings are the most abundant mammals, and their numbers rise and fall in a very regular three-to-four-year cycle, with enormous populations developing in the peak years. Other small mammals include the arctic hare, the suslik (which is a relative of the ground squirrel), and several kinds of shrews and voles. The suslik is the only truly hibernating mammal of the arctic.

Most tundra animals are white in the winter months, many remaining so all year around. Others acquire a darker color during the short summer period.

Although birds are abundant in the tundra during the short summer, most of them leave before winter arrives. They take advantage of the abundance of food in the form of fish and insects, but most species are ill-adapted to withstand the rigors of an arctic winter. The ptarmigan and the snowy owl are among the few permanent residents of the tundra.

There is little if any tundra in the antarctic, most of the land surface being permanently covered by ice and snow. On the other hand, the alpine vegetation found at high elevations in many other parts of the world is quite similar to that of the arctic tundra.

Northern Coniferous Forest or Taiga

Like the tundra, the northern coniferous forest extends around the northern portion of the globe (Pl. 41). It is just south of the tundra and is continuous except for the extreme western portion of Alaska and the extreme eastern part of Siberia. In addition, it extends as far south as Costa Rica at higher elevations of the great mountain chain which runs along the western edge of the continent. The following discussion will deal specifically with the coniferous forest biome of Canada and the northern United States, but the general nature of the flora and fauna of the Eurasian taiga and of the forest found at high elevation in the western mountain chain is much the same (Fig. 45.14).

Rainfall in the eastern portion of the biome ranges from 50 to 125 cm (25–50 inches) a year and from 25 to 50 cm (10 to 20 inches) in the west. Most of the precipitation comes in the summer months but there is some in all seasons. Mean monthly temperatures vary from a low of –34° C (–30° F) to a high of 23° C (73° F). There is permafrost under the forest in its northernmost portion. Because of the width of the biome, there is a great spread in the length of the growing period: from 60 days in the north to 150 days in the south.

There are few mountains and they are not of high elevation. The forest is found at elevations ranging from sea level to 360 m (1200 feet) except at the northern edge of the Appalachians where it is found up to about 760 m (2500 feet). As a result of the glaciation of much of the biome, many lakes are present.

Dominant trees of the forest are balsam fir, white spruce, black spruce, and red spruce. Along the banks of streams and rivers, tamarack, willow, paper birch, and poplar trees are found. The dominant evergreens are well adapted for the rigorous climate in which they live. Their leaves (needles) prevent excessive loss of water during winter and dry periods, and they can also withstand freezing temperatures. Their branches are flexible and thus can hold heavy snow without breaking. The dead needles remain on the tree for some time with the result that huge crown fires, started by lightning, are common. Over fifty species of insects, including the spruce budworm, the pine sawfly, and the tent caterpillar, are a serious problem, often destroying large areas of forest.

The large mammals of the biome are the moose and the woodland caribou, which was formerly more abundant than at present. Carnivores include the gray wolf, lynx, wolverine, ermine, fisher, and marten. Black bear and

Figure 45.14 Tuolumne meadow, Yosemite National Park at an altitude of 2621 m (8,600 ft.) is the largest subalpine meadow in the high Sierra. (Photo—E. C. Williams)

porcupine are common, the former being an omnivore and the latter, feeding on bark, often doing serious damage by girdling trees. The snowshoe rabbit also does considerable damage to young trees. Like the lemming of the tundra, the snowshoe rabbit has a regular population cycle, this one of nine to ten years. Other small mammals found throughout the biome are flying squirrels and red squirrels, chipmunks, shrews, and mice.

Birds are abundant, with the spruce grouse, great horned owl, pine siskin, red crossbill, and black-backed three-toed woodpecker being characteristic. Many warblers breed in the forest but migrate south for the winter.

Temperate Deciduous Forest

Temperate deciduous forests originally covered eastern North America, Europe, southern Japan, and eastern China. Many of the same genera (though not the same species) of both plants and animals are found in all these widely separated regions, indicating a former geographic connection. The temperate deciduous forest biome of the eastern United States will be used as an example (Pl. 46).

Since the biome extends from the center of the Great Lakes region south to the Gulf of Mexico, there is considerable spread in the climatic features. Precipitation ranges from 75 to 150 cm (30 to 60 inches) a year, being highest in the Gulf States. It is rather evenly distributed over the year. Mean monthly temperatures from north to south vary from lows in January of −12° to 15° C (10° to 60° F) to highs of 21° to 27° C (70° to 80° F). The growing season varies from 150 days in the north to 300 days in the south.

The major characteristic of the biome is its climax forest dominated by broad-leaved trees which shed their leaves each fall. Below the dense forest canopy are several understories of smaller trees and shrubs which are also deciduous. The shrubs are often sparse in the dense shade of the deep forest, but are quite evident at the forest edge and in clearings within the forest resulting from windfallen trees. In the spring, before the trees leaf out fully and cut off the light, a wide variety of flowering herbaceous plants are characteristic.

In the far south the deciduous trees are supplanted by broad-leafed evergreens such as magnolia and live oak. Along the southern coastal plain, the southern pine forest may seem like an anomaly within the eastern deciduous forest biome. However, this forest type would be supplanted by deciduous trees if it were not for the periodic fires which prevent normal succession from occurring. Much of the burning is purposeful in order to maintain the commercially valuable stands of pine.

Because of the great area covered by the biome, there are important regional differences in climate as well as differences in soil and topography. As a result, there are many subdivisions of the biome, each with a distinctive climax forest type. Among the recognized types are the beech–maple forest of the north–central region; the maple–basswood forest of Minnesota and Wisconsin; the oak–hickory forest of Missouri, Arkansas, and southern Illinois, as well as parts of the Gulf and South Atlantic states and the rich dense, mixed-mesophytic forest of the unglaciated Appalachian plateau, with a large number of dominant tree species including white basswood, yellow buckeye, and tulip poplar.

When Europeans arrived on the scene, the eastern deciduous forest was essentially intact, covering most of the area. In the ensuing years the original forest has been almost completely destroyed for agricultural purposes or seriously modified by logging. It is estimated that not over 0.1 percent of the original forested area has been immune from at least partial cutting.

Many of the original native animals are now either extinct or essentially absent from the biome. Those which are extinct include the passenger pigeon, which was present by the hundreds of millions, the colorful Carolina parakeet, and the eastern bison. Formerly present, but now existing in other areas, were the wolf, the mountain lion, and the wapiti or American elk.

The white-tailed deer and wild turkey were originally very characteristic animals throughout the biome and they, along with the black bear, are commonly present in many areas which have been allowed to return to their original forested state. Other important mammals of the biome are the gray fox, striped skunk, raccoon, bobcat, opossum, gray squirrel, southern flying squirrel, mouse, mole, and shrew. Many of the smaller birds such as cuckoos, flycatchers, and vireos are not permanent residents. Among those which are year-round residents we find the tufted titmouse, several species of woodpeckers, white-breasted nuthatch, two kinds of owls, and several hawks.

Reptiles of the biome include the box turtle, garter snake, timber rattlesnake, and copperhead. Amphibians are represented by several kinds of salamanders and frogs. Invertebrates are abundant, including land snails and slugs, countless species of insects and spiders, and a variety of millipedes.

Temperate Grassland or Prairie

Temperate grasslands cover vast areas of North America (prairie and plains), Eurasia (steppe), and South America (pampas). In almost all these areas the original vegetation has been replaced with cultivated grasses such as wheat and corn, or introduced domestic cattle have greatly modified the plant cover. Until late in the last century, most of these areas still retained their native vegetation, which supported great herds of large herbivorous mammals such as the bison of North America.

All the temperate grasslands are remarkably similar in climate, topography, and plant and animal types. In the following discussion the North American temperate grasslands biome will be used as the example (Pl. 40). The descriptions are largely based on the conditions which existed prior to the invasion of the area by white settlers.

The North American temperate grassland biome reaches from western Indiana to California and from Texas to Alberta and Saskatchewan. In the east and central part of the grassland, this vast area is continuous—with the exception of strips of forested land in the river valleys—and the terrain is either flat or gently rolling. In the west the great mountain ranges interrupt the continuity of the grassland, but large expanses still occur in lower elevations between the ranges.

The wide spread from north to south provides a tremendous variation in temperature. In the north the low mean monthly temperature is 10° C (50° F), and in the south the summer temperatures often exceed 32° C (90° F). The growing season in the north may be as short as 100 days, while in the south, frost rarely if ever occurs.

The controlling climatic feature for the temperate grasslands, however, is precipitation. Rainfall in the eastern part of the grassland biome may be as much as 100 cm (40 inches) a year and decreases in a rather regular gradient moving westward, to a low of about 25 cm (10 inches). In general the high rainfall in the east would be sufficient to support forests, but the spread of trees into the grassland is prevented by high evaporation rates, periodic severe and prolonged droughts, and formerly by intermittent fires. The latter kill any tree seedlings but do little harm to the grasses. In fact, it is generally believed that periodic burning is beneficial to the maintenance of a healthy grassland.

The moisture gradient from east to west was responsible for a variance in the type of native grassland that existed prior to cultivation. In the eastern part the grasses were tall (1.5–3 m), grading to mid-height grasses (0.5–1.5 m), and finally to short grasses (less than 0.5 m) in the western part. The tall and mid-height grasses (the latter usually called mixed grasses) made up the prairie while the region of short grasses was called the plains.

The most characteristic animals were the bison and the pronghorn antelope, populations of the former are estimated to have been about 45,000,000 in 1600 and those of the latter almost as large. Although the bison was found throughout the biome, the pronghorn occurred only as far east as western Missouri. These two grazing animals were important factors in determining the nature of the vegetation and in the actual maintenance of the grassland system. Jack rabbits were also important in the mixed- and shortgrass regions, as were the prairie dogs, while the cottontail rabbit was found in the long-grass portions. Pocket gophers, ground squirrels, several kinds of rats, and a number of species of mice rounded out the mammalian herbivores. They were all preyed upon by the gray or buffalo wolf, coyote, swift fox, weasel, badger, and ferret. Several kinds of hawks were important predators also, including the rough-legged hawk and the ferruginous hawk.

The prairie chicken, a grassland relative of the ruffed grouse, was a very characteristic bird. Others were the meadowlark, horned lark, and dickcissel. Reptiles were represented by the prairie rattlesnake, blue racer, bull snake, and garter snake. Insects were quite abundant, with many species of grasshopper, many true bugs, leaf beetles, and ants, all of which fed upon the grasses. In the winter animals were almost completely absent above ground. The bison moved south while the pronghorn moved to the west and south, mainly to the pine forests of the mountain foothills. Most of the birds also moved south, and the reptiles and small mammals hibernated in dens or in burrows beneath the surface of the ground.

Tropical Rain Forest

Tropical rain forests are found at low elevations in equatorial regions with an annual rainfall greater than 200 cm (80 inches). The three regions in which extensive rain forests occur are Central and South America, where the vast dense forests of the Amazon and Orinoco basins form the world's largest continuous rain forest; Central and West Africa, in the basins of the Zambesi, Congo, and Niger, as well as on the island of Madagascar to the east; and discontinuously on the Indian subcontinent, on the Malay Peninsula, south and east into the East Indies, the Philippines, New Guinea, and northeastern Australia (Fig. 45.15). Although the plants and animals differ, these biomes are all quite similar (Pl. 45).

There is little if any variation in mean temperature from month to month; daily high temperatures are in the 30°–35° C range (85°–95° F), and daily lows are from 18°–27° C (65°–80° F). Actually, the dense forest moderates the

Figure 45.15 Rain forest, Queensland, Australia. Note the zonation of the vegetation (see diagram, Fig. 45.16). The epiphytic ferns occupy a niche which is filled by bromeliads, members of the pineapple family, in the tropical rain forests of the western hemisphere. (Photo—E. C. Williams)

temperatures so that within the forest itself the highs are not much above 29° C (82° F) and lows are about 21° C (70° F). Rainfall is over 200 cm (80 inches) a year, in some places reaching well over 1000 cm (400 inches). There is always some rainfall each month, but usually there will be one or two so-called dry seasons in which less than 12 cm (5 inches) of rain fall each month. Many plants grow continuously, and there is no time when there are not some trees in flower, some in fruit, and others in the process of shedding their leaves.

Although stratification is found in temperate deciduous forests and other forest types, it reaches a peak of development in the tropical rain forest (Fig. 45.16). On Barro Colorado Island, Panama Canal Zone, there are eight recognizable strata: (1) the air above the forest; (2) tall trees which emerge from the canopy, over 40 m tall; (3) the upper forest canopy, a continuous green layer, 25–30 m above the ground; (4) the second-story trees, 10–20 m tall; (5) small trees, 5–10 m tall; (6) shrubs to 3 m; (7) the forest floor; and (8) the subterranean stratum.

Only minimal amounts of light penetrate the dense canopy, and as a consequence there is little or no herbaceous growth on the forest floor, except in the clearings due to fallen trees and at the forest edge. Great vines or lianas twine around the trunks of the forest giants, growing upward into the light. The strangler fig starts out as such a vine, but it eventually surrounds its supporting tree, develops sturdy woody roots, stems, and branches, and finally replaces the tree on which it had started to grow.

Large numbers of other kinds of plants, called *epiphytes,* are found on the trunks and branches of the great trees. The lovely orchid is such a plant, found in both the old- and new-world tropics. In the new world many of the epiphytes are members of the pineapple family, the *Bromeliaceae.* Significant amounts of water are normally present in the spaces at the bases of their leaves, and many small aquatic organisms live in these tiny isolated "ponds" high in the air. In at least one species of tree frog, eggs are laid and early development takes place in this unusual environment.

Probably the most unique feature of the tropical rain forest is the great diversity of species of both plants and animals. In the most diverse types of temperate deciduous forest one finds a maximum of 10 to 15 species of trees, and extensive natural stands of a single species are not uncommon. In contrast, over 500 species of trees and 800 species of smaller woody plants have been described in the rain forest of equatorial Africa. In the rain forest there may be miles between the closest specimens of the same species of the larger trees. One can find more species of trees in a square mile than in all of Europe.

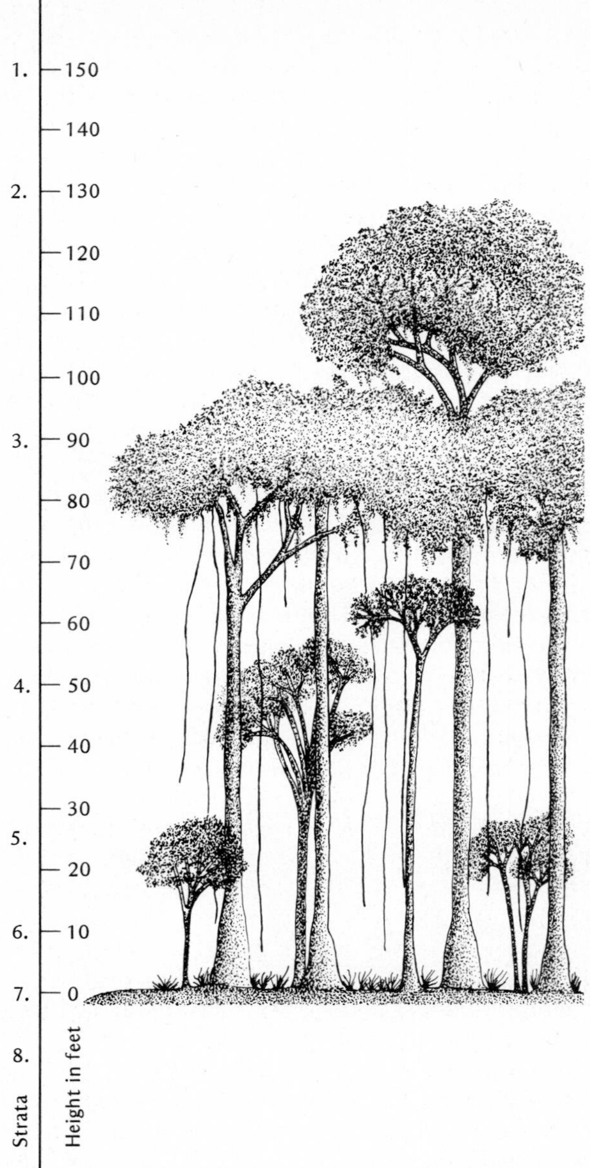

Figure 45.16 Stratification in the Panamanian rain forest.

Diversity of species is also found in the fauna of the rain forest. A 16.8-square kilometer (km^2) (6.5 squares miles) island in the Panama Canal Zone, Barro Colorado Island, is the site of a tropical research station, and its fauna is better known than that of most rain forests. Over 20,000 different species of insects have been found there, many times the number of species found in temperate forests of much larger size.

A visitor to the rain forest is usually impressed by a seeming absence of animal life. In large part this is due to the fact that the majority of animals are arboreal, spending much if not all of their time high in the canopy, feeding on flowers, fruits, leaves, and insects. Among the important arboreal mammals in the American rain forest are the howler monkeys, spider monkeys, capuchin monkeys, and the squirrel marmoset. Prehensile-tailed porcupines and a relative of the raccoon, the kinkajou, which also has a prehensile tail, are common. The sloth is so well adapted for arboreal life that it is quite helpless on the ground. It spends its life hanging upside down from the branches, lazily feeding on the abundant supply of leaves. Bats abound, with some fruit eaters, some insect eaters, and even one species which feeds on fish. Over 100 species of birds inhabit the canopy of Barro Colorado Island, including the fruit-eating toucan with its outlandish beak, parrots, parakeets, and trogons. Insect feeders such as woodpeckers find plenty of food in the tree tops and on the upper trunks. A number of arboreal or tree-dwelling geckos and lizards are also present, including the giant iguana, whose appearance reminds one of its long-dead dinosaur ancestors. Some animals are equally at home either on the ground or in the trees. In this category are the coati, ocelot, opossum, some species of squirrels, many lizards, and snakes.

The forest floor has its own assemblage of inhabitants as well. Mammals include the puma, the tapir (which also spends some of its time in the water), peccaries, agoutis, rats, and mice. Ground-dwelling birds include the greater tinamou and the great curassow. Numerous snakes and lizards are also found in the American rain forest, including such poisonous species as the bushmaster, *fer-de-lance,* and coral snake.

Insects are abundant at all levels of the forest, with ants and termites probably the most numerous. The army ants are very characteristic, living in enormous colonies which construct a temporary nest or bivouac of their own bodies like a swarm of bees. These large clusters, suspended under a log, consist of chains or nets in which workers and soldiers form living links or meshes. Raiding parties move out from the nest over the forest floor in search of insects and other soft-bodied invertebrates upon which these carnivores feed. The leaf-cutting ants are also plentiful. Long columns of ants in single file, each carrying a freshly-cut piece of leaf many times larger than itself, are a familiar sight along the forest floor. The head of the column suddenly disappears as the ants move into their undergound nest. The leaves are used as a mulch in the cultivation of fungus gardens which provide the colony with its food.

As a result of the constant high humidity, there are species of leeches and planarians, normally aquatic organisms, which lead a terrestrial existence in the rain forest. Peripatus, a wormlike animal with features of both annelid worms and arthropods, is found in the warm humid environment of the rain forest. The relative silence within the rain forest during the day is usually noticed by those who visit it for the first time. If they remain until nightfall, which occurs quite abruptly in the tropics, there is a sudden burst of sound from howler monkeys, many birds, and above all, insects and tree frogs. The insects and frogs keep up their loud chorus for much of the night, while the birds and monkeys soon settle down to rest until just before dawn, when they join in the finale.

Tropical Grassland or Savanna

Extensive savanna biomes are found to the north and south of the tropical rain forest biomes of South America and Africa, and there is a strip of savanna across the northern part of Australia. The most extensive savanna biome is in Africa, and the following account will deal largely with that, more specifically with the savanna which straddles the equator in East Africa (Pl. 43).

The African savanna extends north to the Sahara Desert and south almost to the southeastern tip of the continent. It is continuous from north to south by means of a strip between the great equatorial rain forest in the west and the semiarid scrub or thorn forest which spans the equator in the east. Rainfall is from 75 to 125 cm (30–50 inches) a year, but there are prolonged dry seasons between the rainy ones. Temperatures range from highs in mid afternoon of 30°–35° C (85°–95° F) to lows at night of 18°–24° C (65°–75° F). At higher elevations these figures are of course lower. As in the tropical rain forest biome, there is more variation from day to night than from month to month.

The savanna differs from the temperate grassland in the presence of scattered trees, either singly or in clumps, along with the characteristic grasses and sedges. All these plants have in common a great resistance to fire and drought. As in the temperate grasslands, periodic fires have been important factors in determining the nature of the vegetation. In the absence of fire, other types of trees and shrubs would have invaded and greatly modified the system. The grasses, some of which belong to the same genera that make up the American prairie and plains, provide the dominant vegetation of the savanna.

The most characteristic tree is the flat-topped acacia, but almost as prominent are the large euphorbia trees, the ecological equivalent of our American cacti. Other trees include the squat, massive-trunked baobab tree (representatives of which are among the oldest known living trees on earth today), and the sausage tree whose great fruits, 1 m long and 15 cm in diameter, hanging by long slender stalks, prove the common name of the tree to be quite appropriate. Several species of palm, including the stately borassus palm, also occur.

In many cases the clumps of trees grow around old termite mounds. These mounds, often 1.5 m high, are scattered in a fairly regular fashion over the savanna. It seems probable that the termites, which are fungus growers and carry large amounts of dead vegetation into their undergound galleries for mulch, greatly enrich the soil in the area near the nest and thus provide a favorable place for the trees to grow.

Nowhere in the world is there anything to compare with the varied population of large herbivorous mammals which evolved in the favorable environment of the African savanna. The more common large animals which feed chiefly or entirely on grasses are the wildebeeste (or gnu), zebra (Pl. 2), topi, African buffalo, impala, Grant's gazelle, Thompson's gazelle, hartebeeste, eland, wart hog, elephant (Pl. 5), rhinoceros, and hippopotamus (Pl. 11). The latter spends its days in the water, but comes out on land at night to feed on grasses. Browsers (animals which feed on leaves and twigs) include the giraffe and the gerenuk.

Just as the American bison and the pronghorn antelope migrated south and west to escape the rigors of winter, so do the great herds of African herbivores migrate from place to place with changes in rainfall and resultant changes in the amount of grass available.

Associated with the great herbivore populations we find a rather large number of carnivores and scavengers in the mammal population, including the lion (Pl. 8), leopard, cheetah, hyena (Pl. 3), jackal, and Cape hunting dog.

In Tanzania, East Africa, most of which is savanna, there are over 280 species of land mammals, probably more than in any other area of comparable size in the world. Included in this number are 89 small herbivores which feed on grasses and their seeds. Fifty-seven of these are rats and mice and 10 species are squirrels. Forty-five large herbivores, 37 of which are antelopes, are found. Ninety-one mammals are insect eaters, including 56 species of bats and 19 shrews. Many of them feed on a variety of insects, but the peculiar armor-plated pangolin feeds only on termites. Forty-one species are carnivorous, including many which scavenge upon the leavings of those which track and kill their own prey. Eleven species of bats live on fruits as do some of the ten species of monkeys. The abundant baboons are primarily flesh eaters.

Birds are very abundant in the savanna, feeding on the plentiful seeds and insects as well as preying on other inhabitants or living as scavengers. Several species of storks, ostriches, bustards of several kinds, guinea-fowl, many species of weaver birds (which live in intricate nests of woven grasses), and the colorful sunbirds are a few examples of the nonpredatory birds (Pl. 7). Predators include several eagles, the African kite, and many kinds of hawks. The white-backed vulture and a number of other species are to be found wherever there is carrion.

Reptiles are well represented in the savanna. Many, like the crocodile and monitor lizard, are found in and near the rivers which flow through the region. Among terrestrial reptiles, the African chameleon is one of the best-adapted of animals, with several striking characteristics in addition to its well-known ability to change color. The digits of its feet enable it to grasp a branch or twig quite firmly because two of them are directly opposite the other three. With its long sticky tongue, the chameleon can capture an insect at a distance almost as great as its entire body length. The most remarkable adaptation is in the eyes, which are located within conical, moveable, turretlike structures. These eyes can be pointed in any direction in a full 360-degree arc, and they operate independently, making it possible for one eye to look forward while the other keeps tabs on the situation to the rear.

Other reptiles include numerous other kinds of lizards, such as the colorful agama lizard, also skinks, geckos, and numerous snakes. The large reticulated python often inhabits abandoned termite nests. Poisonous snakes include the puff adder, gaboon viper, and a number of kinds of cobras.

Chaparral or Mediterranean Biome

Chaparral biomes on the North American continent are found inland from the coastal plain of central and southern California extend-

Figure 45.17 Mellee scrub forest of South Australia. (Photo – E. C. Williams)

ing into Baja California, and in scattered areas on north to the Oregon border. The shores of the Mediterranean (except for the eastern part of the north coast of Africa), including much of Spain and Italy, are also chaparral, the local name for which is maqui. The third major chaparral biome is along the south coast of Australia, where it is called the mellee scrub (Fig. 45.17). The following discussion will deal with the southern California chaparral biome (Pl. 42).

As in the grassland, the rainfall – particularly the pattern of its fall during the year – is the most important climatic factor. Rainfall ranges from 50–75 cm (20–30 inches) a year, usually closer to 50 cm. Most of the rain falls between November and May. For example, in one area the annual rainfall is 55 cm (21.7 inches) but only 1.25–4 cm (0.5–1.5 inches) fall in June, July, August, and September, with most of that falling in September. The average frostfree period is over 240 days, with many parts almost never experiencing frost. High temperatures [over 33° C (90° F)] are common in the summer time. Along with the rainfall pattern, fires are important in determining the nature of the chaparral biome. The plants are resistant to fire damage and many are actually aided by periodic burning, through stimulation of seed germination by the fire's heat and consequent production of many new sprouts.

Plants of the chaparral include chamiso, Christmasberry, scrub oak, leather oak, interior live oak, manzanita, buckbush, mountain mahogany, and redberry. The trees and shrubs are characterized by having hard, thick, evergreen leaves. Many of the trees are quite stunted and have a shrublike growth form. Seldom do they reach a height of over 2.50 m (8 feet).

Mammals of the chaparral formerly included the grizzly bear and the puma. Today these are absent, but many other mammals still abound, including the mule deer, bobcat, gray fox, coyote, wood rats, and several species of skunks. Brush rabbits and chipmunks, as well as several kinds of mice, are also found. The most common birds of the dry season are the common bushtit, the rufous-sided towhee, and the wren tit. The rainy season sees the appearance of several sparrows, hermit thrushes, ruby-crowned kinglets, and many warblers.

A number of lizards and snakes, including several rattlesnake species, are found in the chaparral, but none are limited to this biome. Most of them are found in adjacent biomes, particularly the more arid ones. The dry nature of the climate precludes the presence of many amphibians, but there is a tree frog which breeds in temporary ponds resulting from heavy rains in early winter.

Insects are common, including ants, leafhoppers, grasshoppers, beetles, and flies. Millipeds and centipedes are sparse, living under rocks. Many of these invertebrates undergo a period of dormancy comparable with hibernation during the dry season—a condition termed estivation.

The Desert Biome

The desert biome is characterized by sparse vegetation, less than 25 cm (10 inches) of rainfall per year, and often extremely high temperatures (Pl. 44). The plants and animals of desert communities are highly specialized for life under these conditions. The plant life of deserts includes annuals, which grow only for short periods when moisture is present; the succulents, such as cacti and yuccas, which store water; the desert shrubs, such as the creosote bush and sagebrush; and occasionally mosses, algae, and lichens. Desert insects and reptiles are well adapted to this habitat because of their impervious integuments and their dry excretions. A number of desert animals are able to survive in the absence of water because they make use of metabolic water. The pocket mouse and the kangaroo rat are able to live indefinitely on dry seeds without any drinking water. They remain in burrows by day and conserve water by excreting very concentrated urine. The wood rat, on the other hand, eats some succulent plants to obtain water. Birds are most abundant in those deserts where some water or succulent foods are available.

The Sea

Some authors consider the sea as an additional biome, since it is a large, distinct, and easily recognizable community unit. However, biomes are (strictly speaking) terrestrial. Further, although the sea is characterized by many different types of habitats, they are all interconnected and some organisms move freely from one to the other. It seems appropriate, therefore, to speak of the sea as the marine ecosystem.

It is a very complex ecosystem, to be sure, with many different kinds of communities. Temperature, salinity, and depth all affect the nature of these communities and present major barriers to the free movement of organisms which have specific adaptations to certain environmental conditions. The major currents, such as the Gulf Stream, which keep the waters of the ocean in continuous circulation, also affect the nature of marine communities, as do the waves and tides. Another important phenomenon, called *upwelling,* occurs where winds move surface waters away from steep coastal slopes, bringing cold water to the surface from the depths rich in such nutrients as phosphates and nitrates; and the regions of such upwelling are often among the most productive regions of the ocean.

The major marine habitats are shown in Figure 45.18. The offshore area of the ocean over the sloping continental shelf is known as the *neritic region.* This is divided into the intertidal, subtidal, and lower neritic regions. The area of the ocean beyond the continental shelf is the *oceanic region.* The upper portion of the open ocean that is penetrated by light (the upper 100 m, on the average) is called the *euphotic zone.* The deeper parts of the ocean are divided into the *bathyal zone,* down to about 2000 m, and below that the *abyssal zone.* The great "deeps," below 6000 m, are often called the *hadal zone.*

Marine organisms in general are divided into two groups depending upon whether they live in the open water or upon the bottom. Those organisms which live in the open water are said to be *pelagic* (Gk. *pelagos,* the sea) as opposed to those which live on the bottom and are called collectively the *benthos* (Gk. *benthos,* depth of the sea). The benthos includes both attached organisms and those organisms which move in or on the substratum. Bottom dwellers are generally distinct for each of the neritic regions, and the type found in each region is largely dependent on the type of bottom, whether sand, rock, or mud.

Pelagic organisms are of two types. The microscopic plants and animals which are unable to move against the currents are collectively called the *plankton* (Gk. *planktos,* wanderer), and the larger free-swimming animals which are able to move at will either with or against the current are called the *nekton* (Gk. *nektos,* swimming). The planktonic plants constitute the *phytoplankton*, while the animals constitute *zooplankton* (Pls. 35–38). The planktonic diatoms and photosynthetic flagellates are the most important producers throughout the ocean, although the brown and red algae are important in limited areas. Primary consumers are princi-

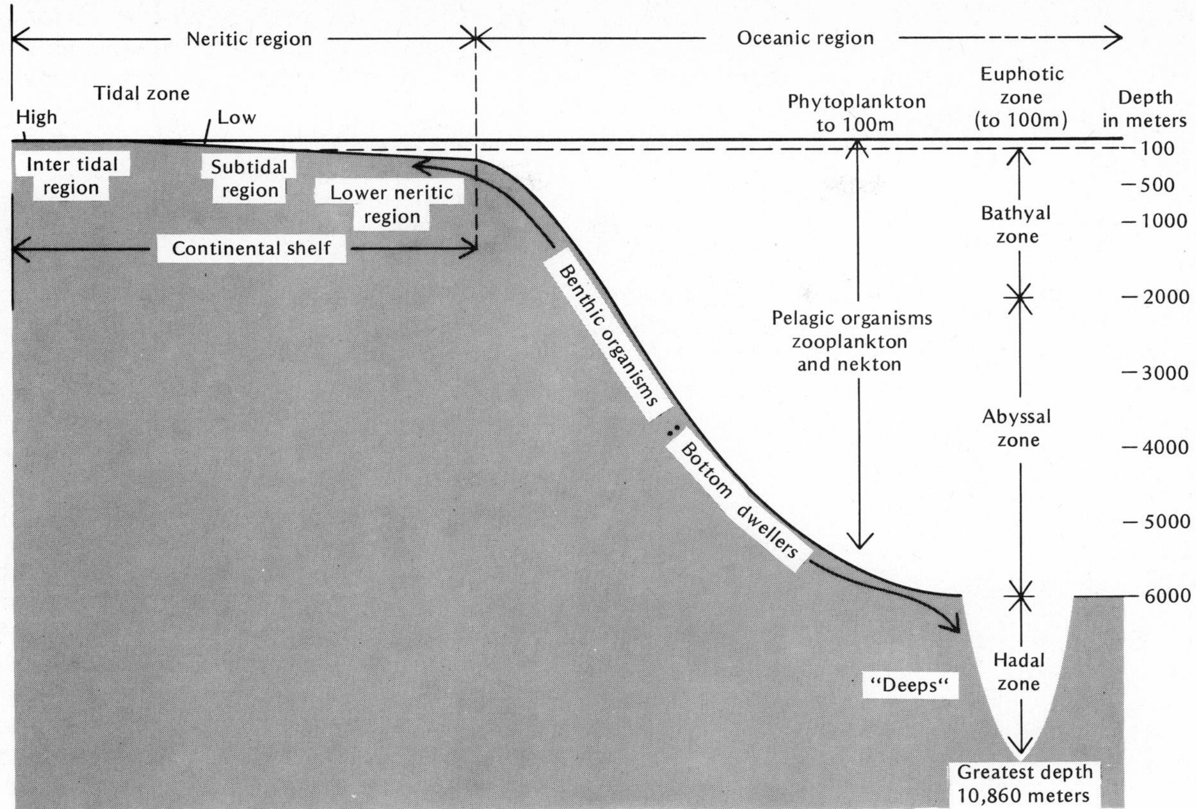

Figure 45.18 Marine habitats.

pally zooplankton. Larger animals either feed on plankton, living or dead, or they are carnivorous.

More is known about the communities of the intertidal and subtidal regions than about those of the lower neritic region. On a typical sandy shore the dominant animals in the intertidal region might include ghost shrimps, burrowing isopods, burrowing amphipods, mole crabs, polychaete worms, and beach clams; and the dominant forms in the subtidal region might include hermit crabs, sand dollars, burrowing shrimps, ascidians, and copepods.

In contrast, on a rocky shore the dominant animal forms in the intertidal region might include barnacles, oysters, and mussels, with such forms as sea anemones, sea urchins, and corals in the subtidal region (Pl. 39). On mud bottoms various kinds of burrowing clams and snails are found. Many of the bottom dwellers are adapted to feed on plankton and detritus (Fig. 45.19).

The organisms of the intertidal region are subjected to alternating periods of exposure to the air and to the strong action of waves. They are adapted to these environmental factors in a number of ways. The sea weeds, such as the common *Fucus,* have tough outer coverings and pliable bodies. Such animals as crabs, barnacles, mollusks, and starfishes have hard calcareous shells, whereas others like the sea anemones have a leathery outer covering. Many animals are able to burrow in the sand to escape exposure to the air.

Neritic pelagic communities include a great variety of fish species, larger crustaceans, turtles, seals, whales, and others. Most of the fish are plankton feeders. The adult fishes feed mostly on zooplankton and are secondary consumers. Some fish, such as sharks, are predators. The great commercial fisheries of the world are located primarily in neritic regions, or not far from the continental shelf, in regions of cold water upwelling, for example, the great fishing banks off Newfoundland.

The euphotic zone of the oceanic region is inhabited by plankton organisms and a variety of nektonic forms, including many kinds of fishes. Mackerel and herring are among the commercial fishes found in the oceanic region. Whales may be present, as well as many kinds of birds such as petrels, frigate birds (Fig. 45.20), and terns.

The bathyal and abyssal zones are incomplete

Figure 45.19 Rocky intertidal zone near Pacific Grove, California. (Photo – E. C. Williams)

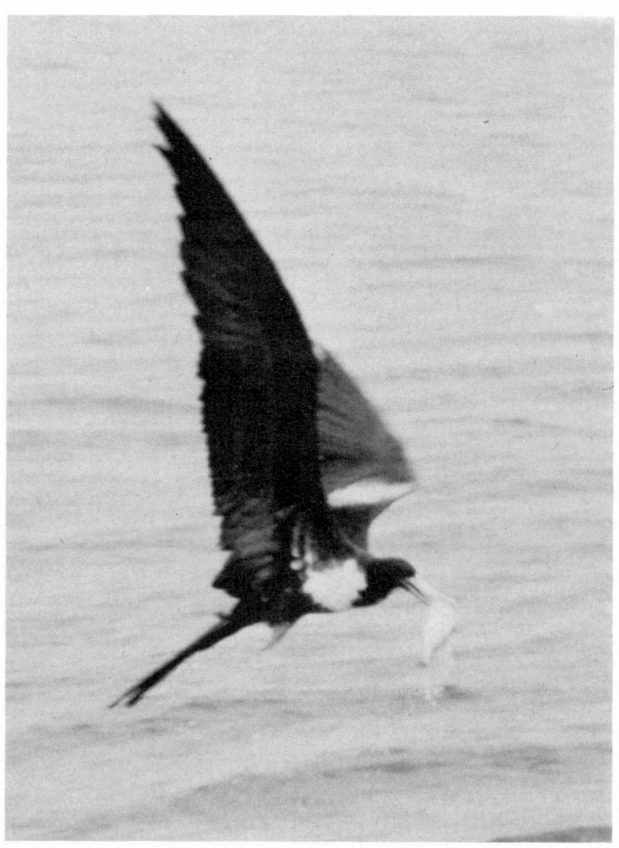

Figure 45.20 A frigate-bird, *Fregata magnificens,* scooping a fish from the surface of the sea near Charles Island, Galápagos Islands. With a wing span of over seven feet and a skeletal weight of about 114 grams (.25 lb) it has the greatest wing area relative to its weight of any other bird. It also pirates food from other sea birds such as boobies, forcing them to regurgitate recently caught fish in mid-air. (Photo – E. C. Williams)

ecosystems since there can be no producers where light does not penetrate. The food of the inhabitants consists partly of the dead bodies of plants and animals that rain down from above, and partly of each other. A variety of species of fishes, crustaceans, echinoderms, worms, and mollusks is found in these zones and on the ocean floor. Many of the inhabitants possess light-producing structures (Fig. 45.21). This is a significant adaptation in this permanently darkened habitat.

The deep-sea forms are adapted to withstand great pressure. In the hadal zone at the bottom of the great deeps in the Pacific, organisms are subjected to pressures greater than 1000 atmospheres. This is over 1 metric tone pressing down upon every square centimeter. The organisms can withstand these tremendous pressures because the pressure is the same inside their bodies as it is on the outside. For many years it was believed that living organisms could not exist under such conditions of external pressure. However, in 1951 scientists on the Danish Galathea expedition collected organisms near the Philippine Islands at a depth of 10,500 m. At this depth they were able to collect 61 sea cucumbers, 17 anemones, 2 bivalve mollusks, and 1 crustacean.

In general the population density of living organisms in the open ocean is much less than that found in the neritic region, but because of its large area and volume the total biomass in the open ocean is greater. The oceans as a whole support an astonishing amount of life.

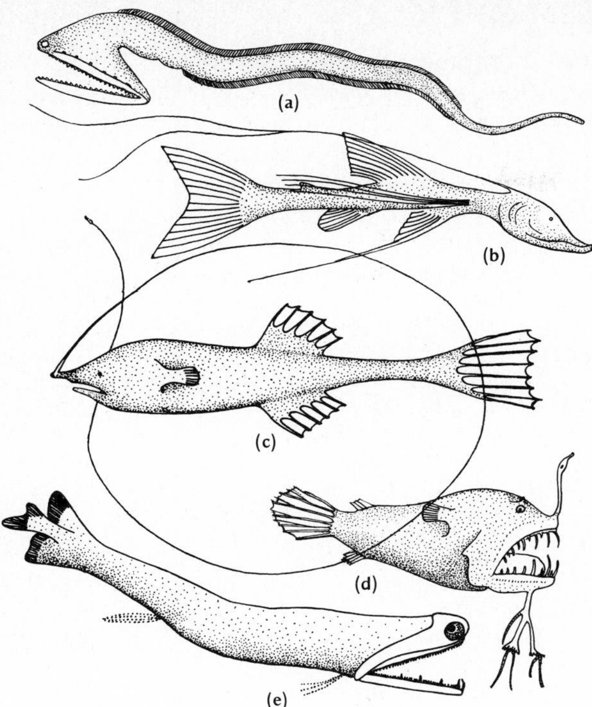

Figure 45.21 Some deep sea fishes. Note the enormous mouths and the peculiar appendages, some of which bear light-producing structures. The species and the depths where they are found are as follows. **(a)** *Macropharynx longicaudus,* 3500 m. **(b)** *Bathypteroid longicauda,* 550 m. **(c)** *Gigantactis macronema,* 2500 m. **(d)** *Linophryne macrodon,* 1500 m. **(e)** *Malacostus indicus,* 2500 m. (From Sverdrup, Johnson, and Fleming, *The Oceans, Their Physics, Chemistry, and General Biology,* © 1942. by Prentice-Hall, with permission.)

However, even though 70 percent of the earth is covered by oceans, the total net annual production on land is about twice that in the oceans. The contrast between biomass of land and sea is even greater with total terrestrial plant biomass being 500 times that in the sea. Animal production in the sea is about three times that of the land, primarily because of the rapid turnover in plankton populations. Actual animal biomass of the two is about equal.

ZOOGEOGRAPHIC REGIONS

There is an even larger unit which includes both biomes and ecosystems, which is important both to ecology and to studies of speciation and evolution. This unit is the *zoogeographic region.* Alfred Russel Wallace, the British scientist mentioned in the chapter on evolution, was a student of animal distribution. His studies played a significant role in giving him a clue to the concept of natural selection, just as did Darwin's studies of a similar nature. In 1877 Wallace published a two-volume treatise on animal distribution. In this work he divided the world into six regions which have come to be called *Wallace's realms.* This division is essentially similar to the breakdown proposed by a British ornithologist named Sclater in a paper which he presented before the Linnaean Society in 1857, and published in 1858. The basis of this breakdown was a distinctive and relatively uniform flora and fauna. The realms, shown in Figure 45.22, are large continental land masses more or less isolated from other land masses by

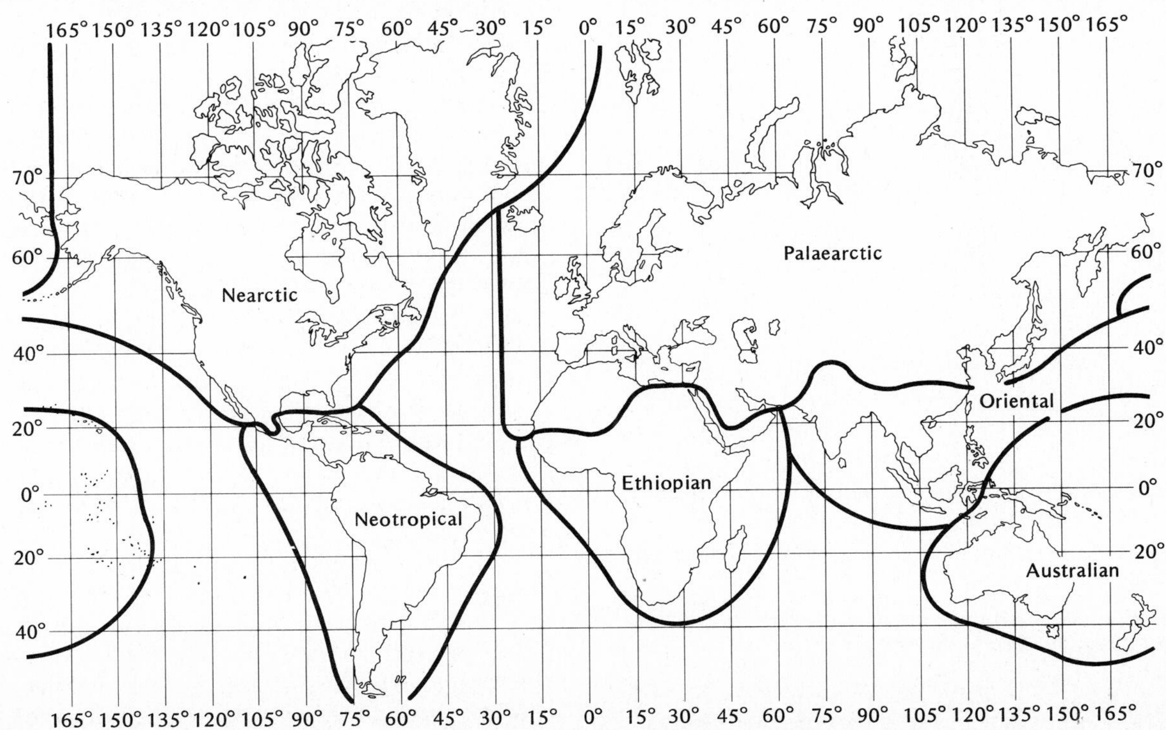

Figure 45.22 Wallace's realms—the zoogeographic regions of the world.

significant ecological barriers. In the past there have been changes in the degree of isolation of these regions, so that from time to time animals could pass from one region to another. However, there have been long periods of isolation during which the distinctive fauna of each region had an opportunity to evolve with little or no significant introduction of genes from related forms in neighboring regions. The role of isolation is most clearly seen in the Australian region with its unique marsupial fauna, which includes ecologically equivalent forms in most of the niches occupied by placental mammals in other parts of the world. Table 45.1 lists the zoogeographic regions and indicates a number of representative mammals for each region.

Table 45.1 Representative Mammals of the Zoogeographic Regions

Ethiopian Region
Hippopotamus
Giraffe
Elephant, also in oriental region
Old World monkey, also in oriental region
Ape, also in oriental region
Oriental Region
Tarsier
Tree shrew
Flying lemur
Tapir, also in neotropical region
Palaearctic Region
Panda
Camel
Horse and relatives, also in Ethiopian region
Beaver, also in nearctic region
Bear, also in oriental, nearctic, and neotropical regions
Mole, also in nearctic region
Nearctic Region
Pronghorn antelope
Pocket gopher, also in neotropical region
Porcupine, also in neotropical region
Neotropical Region
New World monkey
Anteater
Sloth
Vampire bat
Opossum, also in nearctic region
Armadillo, also in nearctic region
Australian Region
Six families of marsupials
Monotreme

SUGGESTIONS FOR FURTHER READING
(Chapters 44 and 45)

Allee, W. C., A. E. Emerson, O. Park, T. Park, and K. P. Schmidt. *Principles of Animal Ecology.* Philadelphia: W. B. Saunders (1959).

Bates, M. *The Forest and the Sea.* New York: Random House (1960).

Boughey, A. S. *Ecology of Populations.* New York: Macmillan (1968).

Carson, R. L. *The Sea around Us.* New York: Oxford Univ. Press (1950).

Clarke, G. L. *Elements of Ecology,* rev. ed. New York: Wiley (1965).

Darlington, P. J. *Zoogeography.* New York: Wiley (1957).

Elton, C. *Animal Ecology.* London: Sidgwick and Jackson (1947).

Elton, C. *The Ecology of Invasions.* New York: Wiley (1958).

Emlen, J. M. *Ecology: An Evolutionary Approach.* Reading, MA: Addison-Wesley (1973).

Hazen, W. E. *Readings in Population and Community Ecology.* Philadelphia: W. B. Saunders (1964).

Kendeigh, S. C. *Ecology.* Englewood Cliffs, NJ: Prentice-Hall (1974).

Kormondy, E. J. *Readings in Ecology.* Englewood Cliffs, NJ: Prentice-Hall (1965).

Krebs, C. J. *Ecology.* New York: Harper and Row (1972).

Odum, E. P. *Fundamentals of Ecology,* 3d ed. Philadelphia: W. B. Saunders (1971).

Odum, E. P. *Ecology.* New York: Holt, Rinehart and Winston (1963).

Ricklefs, R. E. *Ecology.* Newton, MA: Chiron Press (1973).

Shelford, V. E. *The Ecology of North America.* Urbana, IL: Univ. of Illinois Press (1963).

Slobodkin, L. B. *Growth and Regulation of Animal Populations.* New York: Holt, Rinehart and Winston (1961).

Smith, R. L. *Ecology and Field Biology.* New York: Harper and Row (1966).

White, W., Jr., and F. J. Little (Eds.). *North American Encyclopedia of Ecology and Pollution.* Philadelphia: North American Publ. (1972).

From Scientific American

Bell, R. H. V. "A Grazing System in the Serengeti" (July 1971).

Bolin, B. "The Carbon Cycle" (Sept. 1970).

Bormann, F. H., and G. E. Likens. "The Nutrient Cycles of an Ecosystem" (Oct. 1970).

Brower, L. P. "Ecological Chemistry" (Feb. 1969).

Cloud, P., and A. Gibor. "The Oxygen Cycle" (Sept. 1970).

Cole, L. C., "The Ecosphere" (Apr. 1958).

Deevey, E. S., Jr. "Mineral Cycles" (Sept. 1970).

Delwiche, C. C. "The Nitrogen Cycle" (Sept. 1970)

Dietz, R. S. "The Sea's Deep Scattering Layers" (Aug. 1962).

Fairbridge, R. W. "The Changing Level of the Sea" (May 1960).

Gates, D. M. "The Flow of Energy in the Biosphere" (Sept. 1971).

Gregg, M. C. "The Microstructure of the Ocean" (Feb. 1973).

Legett, W. C. "The Migration of the Shad" (Mar. 1973).

Penman, H. L. "The Water Cycle" (Sept. 1970).

Savory, T. H. "The Hidden Lives" (Oct. 1968).

Wecker, S. C. "Habitat Selection" (Oct. 1964).

Wynne-Edwards, V. C. "Population Control in Animals" (Aug. 1964).

46 Human Ecology–Human Population and the Environment

The ecosphere, described in the preceding chapter, with its great variety of organisms and their complex interrelationships and interdependencies, is the product of a very long evolution. Humans are a part of this total interrelationship, but they are newcomers. In terms of geologic time, they arrived only yesterday. Their brain power has enabled them to alter many aspects of nature, and much of the alteration has been done without realizing the ultimate consequences. Alterations of the environment, and in a sense the desecration of the environment, have created problems that no longer can be ignored. Many ecologists have been talking about the damages wrought on the environment for a long time, but by 1960 a more general recognition of this started. By 1970, through newspapers, magazines, television, radio, seminars, conferences, special lectures, and special courses, more and more people were being made aware of the many environmental problems. It is significant that the annual American Association for the Advancement of Science (AAAS) meetings between 1969 and 1975 devoted half of the time of each meeting to conferences and discussions on environmental problems. In 1971 the Second National Biological Congress had this for its theme. So although the proper study of ecology includes humans and all other organisms, as well as the physical aspects of the environment where they live, a separate chapter dealing with our changes of the environment and the possible consequences of these changes seems in order.

Since we are relative newcomers to this planet (our genus has been around for only about two million years), why is there so much concern now about the changes we have made and are making in the environment? Two interrelated phenomena are involved: the rapidly growing world population and the many technological advances that have resulted from scientific discoveries. Every reader should have some familiarity with both of these phenomena. With so many people on the earth and with so many technological developments, the environment is changing so rapidly that there is great concern as to whether this is now leading to a situation where we will no longer be able to survive on this planet. Among the many problems which we face are providing enough food for all mankind, finding sufficient energy sources for our civilization, and preventing intolerable pollution of our environment.

HUMAN POPULATION

Chapter 42 considered Darwin's reasoning that led to his concept of natural selection, and it was pointed out that the population of each species, although it may vary some from year to year, remains relatively constant over long periods of time. Also in Chapter 44 the discussions of population growth pointed out that the growth of any population of organisms in a finite environment may be expressed by a sigmoid or logistic curve. Such a population, after a period of exponential growth, reaches an *asymptote,* where it may remain relatively constant if the food supply remains constant and if other environmental factors do not intervene. Figure 46.1 shows the growth of the human population of the world over a very long period of time. An examination of this graph shows that for a long time in human history, growth was either absent or very slow. Then there was a very long period with a very slow increase. By the year 1 A.D. there were 0.25 billion people in the world. When the pilgrims landed on Plymouth Rock in 1620 there

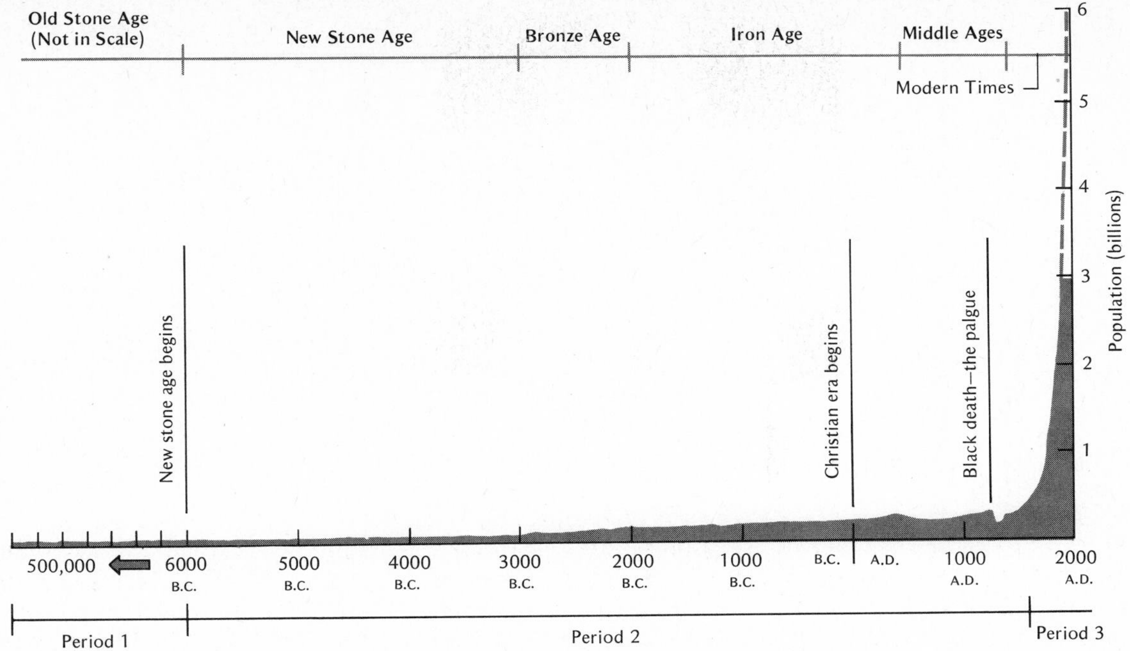

Figure 46.1 Growth of human numbers. (From *Population Bulletin,* February 1962, © by Population Reference Bureau, with permission.)

were about 0.5 billion. Figure 46.2 shows how the doubling time has shortened since 1750. The present rate of world population increase is 2.0 percent a year. At that rate numbers double every 35 years. With a 1-percent increase a year, the doubling time is 70 years, while with a 3-percent annual increase, the doubling time is about 23 years.

The primary factors responsible for the rising growth of human population over the years include better tools for hunting, the development of agriculture, the Industrial Revolution, and the development of modern medicine and public health measures. These factors have so far allowed human population to keep growing while populations of other organisms have remained relatively constant. The striking growth in recent years is primarily the result of better food production and distribution, and better medical care and public health measures. As long as we were just hunters and as long as nothing was known about diseases and their prevention or about what kinds of food to eat for a proper diet, the human population grew very little.

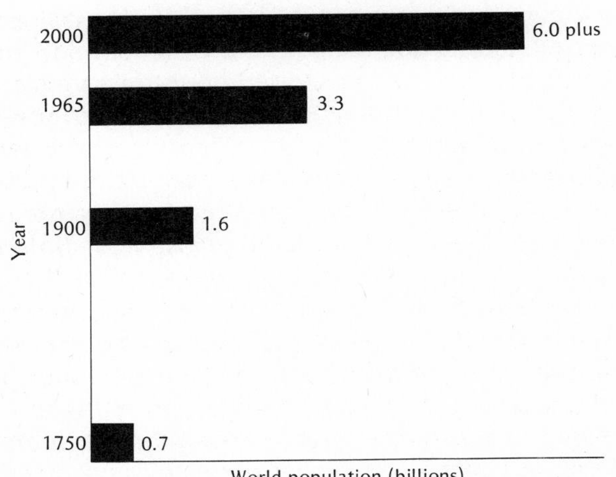

Figure 46.2 Shrinking span of time needed to approximately double the world's population. (From *Population Bulletin,* October 1965, © by Population Reference Bureau, with permission.)

The growth of human population differs in various parts of the world. Nations are usually divided into two groups in this connection: developed or industrial nations and developing or underdeveloped nations. Included in the first category are the United States, Canada, the European nations, the Soviet Union, Australia, New Zealand, and Japan. Included in the second group are the nations of Asia, Africa, and Latin America. Figure 46.3 shows at a glance some of the differences between some of the nations in these two categories. Nations in the first category have a high per-capita income and a relatively low rate of population increase, whereas the opposite is true in the nations in the second category. Besides these two characteristic differences, a number of others stand out. In the developed countries the people have better diets, better medical attention, better education, better roads and transportation facilities, better communication facilities, and more capital for further development than do the people of de-

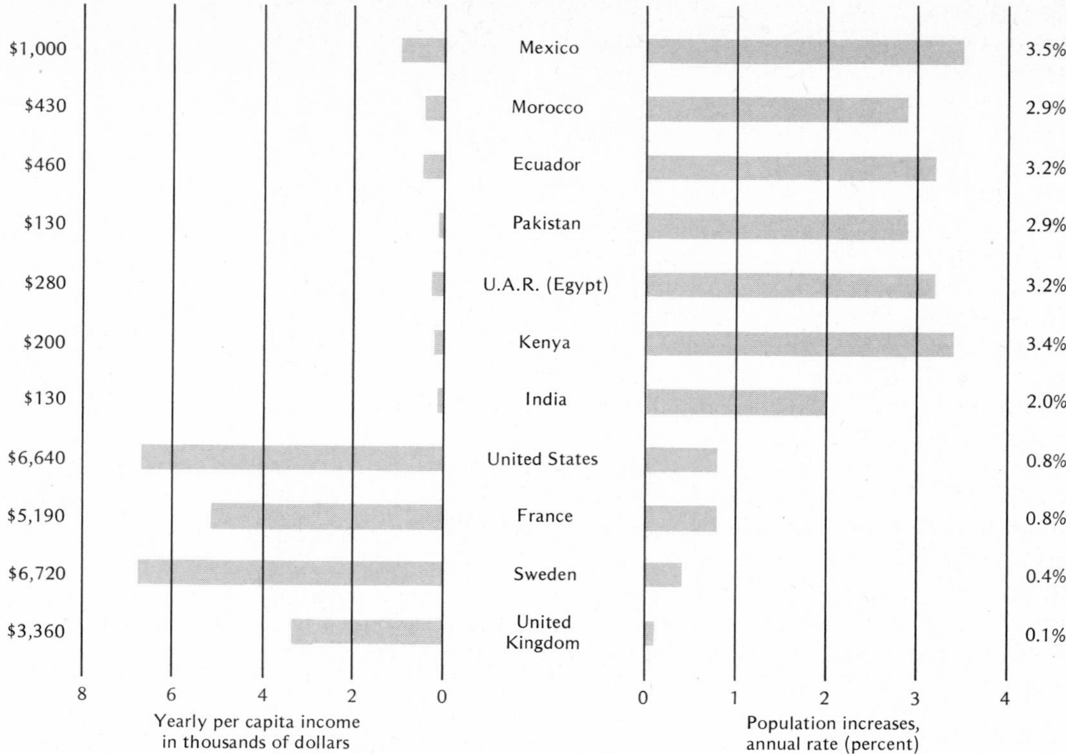

Figure 46.3 Per capita income and population increase of selected countries about 1976. (Based on 1976 World Population Data Sheet, © by Population Reference Bureau, with permission.)

veloping nations. Today approximately one third of the people in the world live in the developed regions and two thirds live in the developing or underdeveloped regions.

In the developed nations a shift from a high birth rate and high death rate to a lower birth rate and a much lowered death rate came gradually with the industrial revolution and the accompanying developments in medicine and public health. On the other hand, during the past 50 years the death rates in the developing countries have been lowered drastically. In this period health teams from the United States and other developed countries have carried information about modern medicine to all parts of the world. As a result malaria and many other diseases that killed so many in the underdeveloped countries were brought under partial control. In Ceylon in one year (1948–1949) the death rate dropped from 20 to 14 per 1000. When birth rates remain high and the death rates suddenly drop, the rate of population growth zooms. India, for instance, had a population of about 520 million in 1968. It is projected that it will have a billion by 1994. This is indeed a gloomy projection, when India was unable to feed adequately even half of her population in 1968.

It was mentioned above that one of the products of the developed nations, modern medicine and public health techniques, has been carried to nations all over the world. However, many of the other technological developments, including modern agricultural practices and machinery, have not been utilized on a broad scale in these nations. The reason is largely economic. When most of the energy of a nation is used simply to provide the necessary food to remain alive, capital needed to purchase farm and other machinery and to educate people to use such machinery on farms and in factories is not accumulated. Also, it should be mentioned that some of the underdeveloped countries are poor in fossil fuels and other essential material resources. Can an underdeveloped country approach the status of the developed nations? The United States and other countries have provided aid in the form of food during periods of starvation, of medical help, of education for some of the people, of technical specialists, of educators, and of outright cash gifts. These have been great humanitarian efforts, and they are to be applauded, but these alone cannot solve the problems. The nations themselves must do more. How and what? These are very difficult problems, but one thing seems clear: the problems of the underdeveloped nations cannot be solved until they greatly reduce their rates of population growth. Only then will they be able to accumulate the capital necessary for better farm machinery, better factories, better education, better roads, and so forth. Only when a family has money left over after all the necessities of living are paid for can capital be built up. The same concept applies to nations also.

How can population growth be brought under control? For the past 20 years India has been operating a fairly extensive family planning service in its efforts to slow population growth. A United Nations team that recently visited India reported that of the 100 million couples in the fertile ages, only a little over 7 million have been provided with contraceptive services by the government. Also, between 1965 and 1968, well over half of the IUD (intrauterine device) insertions and sterilizations occurred after the subjects had already had four children. During the 19 years that the program has operated, India's birth rate has dropped slightly, from 41.7 to 38.4 per 10,000 population. This is taken by many to mean that voluntary family planning is not a method for reducing population growth in any major sense.

Food

In 1798 Thomas Malthus published his *Essay on Population.* In it he pointed out that population, when unchecked, increases in a geometrical ratio, whereas subsistence (food) increases only in an arithmetical ratio. He concluded that on a finite earth such a situation would result in human misery through starvation and pestilence, unless there was willfull control of reproduction. Malthus had no way of knowing about the technological developments that would come in the form of better agricultural techniques, better industry and transportation, and better medical and health practices. Since 1798 the world population has grown from a little under 1 billion to about 4 billion at the present time. In his own day Malthus had many critics and he has had many more over the years. But new techniques in agriculture were developed, more food was produced, all regions of the world were explored, more land was brought under production, new strains of food crops and food animals were developed, and extensive irrigation was augmented with fertilizers and pesticides. This made possible food for many more millions of human mouths. Was Malthus wrong? For a long period of time most people have behaved as though he were. But in the opinion of many people today, a Malthusian specter seems to be hanging over the heads of mankind. It must be recognized that the earth is finite, and there must be a limit to the number of human beings who can exist here. Despite the wishful thinking of some, there is no possibility that the excess of mankind on the earth can be transported to some other planet. Garrett Hardin (see Suggestions for Further Reading) has clearly documented this. Despite the fact that more food continues to be produced, it is becoming apparent that these increases are not keeping up with population growth. A recent United States report indicates that despite the increases in world food production, the per-capita food available has decreased. The United States Department of Agriculture has estimated that this country will not be in a position to supply food to underdeveloped nations after 1984. In 1974 a committee of the National Research Council indicated that the United States could have food problems of its own by the turn of the century.

In recent years the development of new higher yielding strains of rice and wheat (the "green revolution") has caused some to be optimists about the food problem. The development of such new strains is important for hungry millions, yet the analysis of the situation by many students in the field does not give cause for much long-term optimism. It seems that very favorable weather was, in part, responsible for the high yields. It also appears that to succeed with these strains the farmers must be able to irrigate in a certain way, must have fertilizers to apply, must have the proper equipment, and, when a crop is produced, must have roads and transportation facilities for moving the grain to market. This all takes capital, which most of the farmers in India and similar countries do not have. It is significant that Norman Borlaug, when he received the Nobel Prize in 1970 for his research in developing new high-yielding varieties of wheat, commented to the effect that unless the frightening power of human reproduction is curbed, the success of the green revolution will be ephemeral only.

In 1974 Borlaug, when testifying before a Senate Select Committee on Food and Nutrition, again expressed his concern about the growing world population in relation to food supply. He said, "If there should be a major drought anywhere in the world at the present time, I hate to think what would happen. We would have disasters in which tens of millions of people would die, and we could do nothing about it."

The outlook is gloomy. The only way that human suffering can be lessened, in addition to a great reduction in population growth, is for all of the food-producing regions of the world to produce as much food as possible.

Does this country have a population problem? The United States reached a population of 100 million in 1915 and 200 million in 1968. It is projected that we may have 300 million by 2000. In spite of the fact that a congressional committee has shown that we do have pockets of starvation in the United States, it is well known that there is, at present, enough food for all if it is properly distributed. As a consequence there is in the United States perhaps a majority at this time who would say that we have no population problem. But is lack of food the only index or consequence of overpopulation? Are not crime and riots in the large cities and the great problems of environmental pollution of many kinds the products of large numbers of people?

Urban Areas

In 1790 most Americans lived in rural areas. By 1968 seven out of ten Americans lived in or near a city. Now three out of four live in urban areas. This movement to the cities has been going on for years, due to the great mechanization of farming and to increased industrialization in the cities, and the trend is continuing. When the United States passed the 200 million mark in 1968, it was the fourth nation in history to reach that size. The population of mainland China is estimated at 800 million, India at 500 million, and the Soviet Union at 240 million.

The rapid growth of the United States population is a matter of deep concern, not only because of the numbers but because of the concentration of the people in urban areas. Detroit has 4448 people per square mile; Chicago, 5671; Washington, 13,148; and Manhattan Island, 67,870.

With this trend to move to the cities, the Population Reference Bureau envisions the majority of Americans, by the year 2000, massed in three *megalopolises.* Boswash, with an estimated 80 million people, would stretch along the East Coast from Boston to Washington. Chippitts, with 40 million, would follow the Great Lakes from Chicago to Pittsburgh. Sansan, with 45 million, would extend from San Francisco through Los Angeles to San Diego. These great urban areas are a part of our ecology, a part of our total environment. Although United States birth rates, as well as the annual percentage of population increase, have been declining, there is a possibility for change in this. More and more young women who were born during the historic baby boom of 1946 to 1955 are entering their peak child-bearing years. In 1960 there were 11 million women aged 20 to 29 in this country. By 1980 there will be 20.1 million in that age bracket. Whether or not they bring on another baby boom will depend on their family-size goals and their ability to put these goals into practice. At any rate the Population Reference Bureau is projecting that we will have 400 million by 2030 —double that in 1968. Imagine what conditions will be like with twice as many people around.

Perhaps you live in a large city; if not, you have probably visited one recently. Evidences of overcrowding and overpopulation are seen on every hand. Attention was called to the crowded slums by the riots in the summer of 1967. There are problems of traffic congestion, crowded subways, garbage collection, pollution of many kinds, crowded schools, inadequate playgrounds, to mention but a few. Are not these aspects of urban existence evidence that a population problem exists purely apart from the question of food?

The question is sometimes raised, what is the maximum population that the earth can support, or, what is the maximum population that this country can support? Such questions do not seem very relevant if it means that humans would have to live like maggots in a sore, piled on one another. More and more people are beginning to ask, what is the maximum population that we can have and still allow each individual to enjoy life? To many, the quality of life seems more important that just the numbers that might exist here. In this connection, a recent National Academy of Sciences report contained this statement: "Indeed a human population less than the present one would offer the best hope of comfortable living for our descendants, long duration of the species, and the preservation of environmental quality."

POLLUTION

Unhappily, our culture sees its proper role as dominating nature, rather than living in harmony with it. It is a culture that equates *growth* with *progress,* and considers both as self-evidently desirable. It is a culture that often considers undeveloped land to be wasted land.

Through our technological developments we have been polluting the environment with disastrous effects. In this connection one thing seems obvious: we have put many things into use without any knowledge of their ultimate effects on the environment. The United States Food and Drug Administration has estimated that we are now exposing ourselves and our environment to over 0.5 million manufactured substances, and the number is growing by 400 to 500 new chemicals each year. Only a small fraction of these substances has been tested for toxicity to marine algae that produce some 70 percent of the earth's annual supply of oxygen. Nor have most of them been tested for toxicity to the equally important organisms involved in the cycles of nitrogen and other essential elements.

For about two million years human beings have survived and multiplied on the earth. They have become adapted to the environment which includes animals, plants, microorganisms, soil, water, and air. All of these, ourselves included, are tied together in an elaborate network of mutual relationships. In the preindustrial world the environment appeared to have an unlimited store of clean water and air. It seemed reasonable, as the need arose, to vent smoke into the sky and sewage into rivers in the expectation that the huge reserves of uncontaminated air and water would effectively dilute and degrade the pollutants. Now there is simply not enough air and water on earth to absorb current wastes without effect. Although some attempts have been made to cut down on the wastes, both human and industrial, dumped into our streams and lakes, not nearly enough has been done.

Water

Lake Erie is a classic example of extreme water pollution. This large lake has been overwhelmed by pollutants. In its natural state Lake Erie was a balanced system in which water plants, microorganisms, and many kinds of animals lived together in an intricate harmony. By 1968 a large portion of Lake Erie was dead. Sewage, industrial wastes, and the runoff from heavily fertilized farm lands had loaded the water of the lake with so much phosphate and nitrate that the natural biology of the lake had been destroyed. With the excess phosphate and nitrate in the lake, algae grew to tremendous densities and then died. Much of the dissolved oxygen in the lake was used up by bacteria when algae were decomposed. This process is called *eutrophication.* In the absence of oxygen, fish and other aquatic animals die. A start has been made on the restoration of Lake Erie. The same fate may be in store for Lake Michigan unless great changes are made rapidly. According to a recent report by a committee of the National Academy of Sciences, within 20 years city wastes and agricultural runoff are expected to overwhelm the biology of most of the nation's waterways unless the present direction is changed. In 1974 a higher incidence of certain types of cancer in the city of New Orleans was associated with certain carcinogens present in the drinking water which is taken from the Mississippi River.

The problem of having enough fresh usable water is becoming acute. Any assessment of the magnitude of water pollution must consider the relationship between the total available supply of fresh water and the quantity of waste-carrying water. The total available supply is the average annual stream flow that discharges into the oceans from continental United States. This amounts to about 41.6 billion hectoliters (hl) (1100 billion gallons) a day. The waste water is the quantity returned to stream flow after use, with its quality altered one way or another. In 1954 some 11 billion hl (300 billion gallons) of the total were withdrawn daily, of which 3.8 billion hl (100 billion gallons) were consumed and about 7.5 billion hl (200 billion gallons) were returned to the streams. For the year 2000, with the same water usage habits, it is projected that 33.7 billion hl (889 billion gallons) will be returned. Thus in 1954 withdrawals amounted to less than one third of the total and waste-ridden returns less than one fifth. However, in the year 2000 withdrawals will be a little over four fifths, and polluted returns will be a little over two thirds of the stream flow.

Pollutants entering water sources are classified broadly into the following eight categories: domestic sewage and other oxygen-demanding wastes; infectious agents; plant nutrients; chemicals such as insecticides, herbicides, and detergents; other minerals and chemicals; sediment from land erosion; radioactive substances; and heat from power and industrial plants.

It is obvious that our fresh water will have to be purified and recycled for use. Each of the above categories represents a different problem. The last category (heat from power and industrial plants) constitutes a new form of pollution, *thermal pollution.*

The principal source of thermal pollution is the electric power industry. In 1968 generating plants accounted for about three fourths of the 2271 billion hl (60,000 billion gallons) of water used in the United States for industrial cooling. By the year 2000 it is estimated that the electric power industry, using more atomic power, will require the disposal of 20 million billion British thermal units (Btu) of waste heat per day. To carry off this heat by natural waters would require a flow through power plants of about one third of the average daily fresh water runoff, or approximately 13.9 billion hl (366 billion gallons).

The Federal Water Pollution Control Administration (now a part of the Environmental Protection Agency) has stated that waters above 33° C (93° F) are essentially uninhabitable for all fishes. Many rivers reach a temperature of 32° C (90° F) or more in the summer through natural heating alone. Also the effluents of many of the power plants of the future can be expected to elevate the temperature of a receiving river by 6° C (10° F). It is for this reason that many people are now objecting to some of the power plants under construction until it is established that they will have devices (condensers, wet towers, dry towers) that will return effluent water at a temperature that is compatible with the temperature requirements of the river or lake.

The oceans are also coming in for a share of the present concern about pollution. Every reader probably knows about the oil spillages off the California, Florida, and Louisiana coasts. It is not possible at present to estimate the damage to marine life, let alone the damage of these disasters to beaches. On July 12, 1970, Thor Heyerdahl ended his voyage from Morocco to Barbados in a papyrus boat. He had stories to tell about regions of the Atlantic Ocean through which they had passed where there was floating refuse and where the water was so visibly polluted with globules of unknown nature that the men did not want to dip their toothbrushes in the water. In March of 1970 a report was made on a study of the dumping area in the Atlantic for New York City. This area is located about 19 km (12 miles) south of Long Island and 19 km (12 miles) east of the New Jersey coast. As viewed from a helicopter, this area, about 32 km (20 miles) across, appeared like black ooze in a sea of blue. For the past 40 years sewage sludge, dredge spoils, and industrial acids have been dumped into this area from barges. In addition the Hudson River carries much raw sewage and industrial wastes into the area. At the center of the dumping area biologists found no life. The report stated that fish and vegetation in the dump area have been poisoned, and that some forms such as horseshoe crabs, snails, and sea worms no longer exist there. The report concluded that if no more waste was dumped in this area, it would take from 10 to 100 years for the ocean currents to clear up this huge "dead sea" of pollution.

Heavy metals are known to be toxic in relatively low concentrations to most organisms. Since the introduction of tetraethyl lead into gasoline in the last four decades, the volume of lead has increased many times that of earlier levels in ocean surface waters. The same thing can be said about mercury, which is used as a catalyst in industrial chemical processes, as electrodes in the chemical industry, and in pesticides. It is estimated that between 4500 and 9000 metric tons (5000–10,000 tons) of this element are lost each year, and that much of it reaches the oceans. A few years ago in a coastal town in Japan a number of people became ill with a strange disease that was first called *Minamata Bay disease.* Those afflicted became seriously ill, palsied, blind, and bald. About 50 people died. Later it was determined that these people had been poisoned by methyl mercury chloride, a waste product from the manufacture of certain plastics. The Japanese people obtained the poison in fish and shellfish which they had eaten. The poison was concentrated in these food organisms in the food chains found in the ocean. The government there soon regulated the use and disposal of mercury contaminants.

In addition to lead and mercury, other heavy metals such as arsenic, selenium, iron, copper, cadmium, and zinc are also being discharged and are building up in the oceans. Some of these are also air pollutants. The Environmental Protection Agency has classified mercury, beryllium, and asbestos as hazardous air pollutants.

Another source of environmental pollution is the use of pesticides. Readers of newspapers and magazines during the period since 1970–1971 were certainly made aware of the controversies over the use of DDT and related compounds. DDT is a chlorinated hydrocarbon, and it is extremely stable. Instead of breaking down rather rapidly, it accumulates in the soil and water and remains intact for years. It has been used for many years in the control of insects, and this has made possible much greater food production in many regions of the world. Its harmful effects on the environment, however, have become apparent in the last few years. Recent studies show that penguins in Antarctica have high concentrations of DDT in their tissues —and Antarctica is hundreds of miles away from any region where DDT is used. There is good evidence now that much of its spread is by the air. DDT and many other pesticides have been used widely without the users' really knowing the full and ultimate effects on the total environment.

It has been assumed by the pesticide industry that in the amounts used DDT could not be harmful to humans and other vertebrates. But it is now clear that one process that occurs in nature was overlooked—the process of a substance becoming more and more *concentrated* as it passes up through a food chain. Now there are several well studied food chains showing how the concentration of a pesticide is increased in a food chain. Here is one for DDT in

Lake Michigan: bottom mud, 0.04 parts per million (ppm); shrimp, 0.44 ppm; whitefish, 5.6 ppm; herring gull, 98.8 ppm. We can be the last link in such a food chain. The situation involving mercury poisoning in a Japanese village was referred to earlier. In 1969 a large catch of Coho salmon caught in Lake Michigan was declared inedible because of its high content of DDT and other pesticides. This food chain effect is taking its toll in many predatory birds (hawks, owls, ospreys, and eagles). The peregrine falcon no longer breeds in the northeastern United States. Recent experimental studies indicate that the effect that prevents reproduction in this case has to do with the formation of strong egg shells. With high concentrations of DDT the shells are very thin, and they are either broken or the developing embryo undergoes desiccation and dies.

A campaign is now underway to develop other effective pesticides that will breakdown in short periods of time and also to develop more forms of biological control over pests. A number of such controls are now is use. Recently it has become possible to control leaf hoppers in California by using a tiny wasp. The leaf hoppers destroy grapes. For a number of years they were kept in control with DDT, but then they developed an immunity to DDT. The use of the wasp has proven effective, and it has also reduced the cost of leaf hopper control by 87 percent.

Air

We have been polluting the air ever since we started using fire. The Industrial Revolution with its smoke, soot, and ash speeded this pollution. Now in the United States over half the air pollution comes from the exhausts of motor vehicles. The exhaust from automobiles contains carbon monoxide and nitrogen oxides, among other things. The nitrogen oxides, in the presence of sunlight, react with waste hydrocarbons from gasoline to form peroxyacyl nitrate (PAN) which, along with ozone, is one of the most toxic substances in smog. Los Angeles, because of its location near the ocean and with mountains all around suffers more from smog than perhaps any other American city. During certain periods the air does not circulate freely over the city, and a layer of smog builds up (Fig. 46.4). The effects of the smog on certain plant crops in the Los Angeles area have been known for some

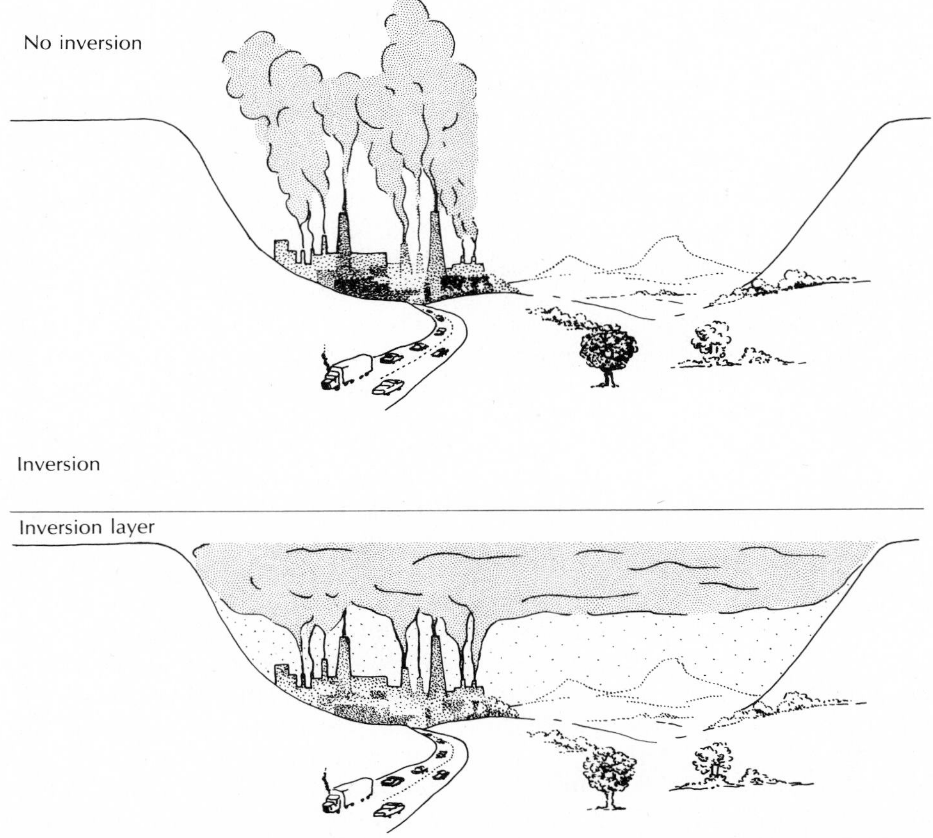

Figure 46.4 Temperature inversion, in which a layer of warm air overlays a layer of cooler air, trapping air pollution close to the ground. (From P. R. Ehrlich and A. H. Ehrlich, *Population, Resources, Environment:* Issues in Human Ecology, 2d ed., © by W. H. Freeman and Company, with permission.)

time. In one experimental study three varieties of spinach were grown side by side under the same conditions. One variety was completely damaged, one was damaged somewhat, and one produced a salable crop. Other large metropolitan areas are also plagued with smog. Not only does it affect plants, but there is good evidence that, in addition to the irritation of eyes and throats, it causes respiratory troubles (emphysema, bronchitis, and also lung cancer) in humans. The incidence of emphysema increased 400 percent in California in a decade. So serious is this problem that Congress passed the Air Quality Act of 1967. This provided, among other things, that beginning with 1968 all new cars must pass Federal emission standards. The Air Quality Act of 1970 provides for a progressive tightening of standards on auto emissions to 1975. Most 1975 model cars were equipped with catalytic converters which should result in a 90-percent reduction in hydrocarbon and carbon monoxide emissions below the allowable emission level of 1970 cars. At this time (1975) there is some fear that the catalytic converters will add a new problem to air pollution control because they add sulfates (which are converted to sulfuric acid) to the air. Because of the energy crisis (discussed below) and to allow automobile makers more opportunity to concentrate on improving fuel economy, the emission standards for hydrocarbons and carbon monoxide are not to be increased for 1976 models. Also the emission standard for nitrogen oxides of 3.1 g per mile 2.4/km) was extended through 1976.

Something less than half the air pollution comes from industries, power plants, the heating of buildings, and the burning of solid wastes. Sulfur oxide and particulate matter from these sources are especially damaging to structures, buildings, and clothing, as well as to living things. The social costs of the air pollutants are measured in billions of dollars each year.

Solid Wastes

In the Solid Waste Disposal Act of 1965, Congress defined solid waste as "garbage, refuse, and other discarded materials, including solid waste materials from industrial, commercial, and agricultural operations, and from community activities." This certainly includes a wide range of materials and objects.

In this connection the testimony of Jean Mayr of Harvard University before one of the congressional committees is interesting. He said, "The need to control population growth in the rich countries is no less urgent than in the poor countries. The environmental degradation that has accompanied population and economic growth in the rich nations is more of a problem to the world than the pressure put on food supplies by the poor nations' burgeoning populations."

Mayr continued, "The rich occupy more space, consume more of each natural resource, disturb the ecology more, and create more land, air, water, chemical, thermal, and radioactive pollution than the poor."

In 1966 the United States, with only 6 percent of the world's population, consumed 34 percent of the world's energy production, 29 percent of all steel production, and 17 percent of all timber cut. Mayr further pointed out, using solid wastes as an example, how the United States demonstrates strikingly that the superrich tend to be superpolluters. He said, "We spread 48 billion rust proof cans and 26 billion nondegradable bottles over our landscape every year. We produce 362,000 metric tons (800 million pounds) of trash a day, a great deal of which ends up in our fields, our parks, and our forests. Only one third of the 453,000 metric tons (1 billion pounds) of paper we use every year is reclaimed. Nine million cars, trucks, and buses are abandoned every year, and while some of them are used as scrap, a large though undetermined number are left to disintegrate slowly in backyards, in fields and woods, and on the sides of highways." The greatly increased use of nonreturnable glass containers, aluminum cans, and plastic containers has greatly aggravated the problems of solid waste disposal. These substances do not decompose in landfill operations, and the plastic containers give off noxious fumes when burned.

It is becoming clear that much of the material that is now being discarded as solid waste must be recycled. Our earth's metal resources are not unlimited. Figure 46.5 shows the situation that the United States faces with respect to some of the commonly used metals.

In addition to the pollution of the land by solid wastes, more and more land each year is despoiled through various mining activities. Mounds of waste, often tens of meters high and covering extensive land areas, accumulate adjacent to many mining operations. Prior to 1965, over 1.5 million hectares (4 million acres) were covered with unsightly piles of waste. In 1980 it is expected that more than 2 million hectares (5 million acres) will have been defaced by these operations.

Noise

There are other types of pollutants, but only one more is briefly described. This is the pollu-

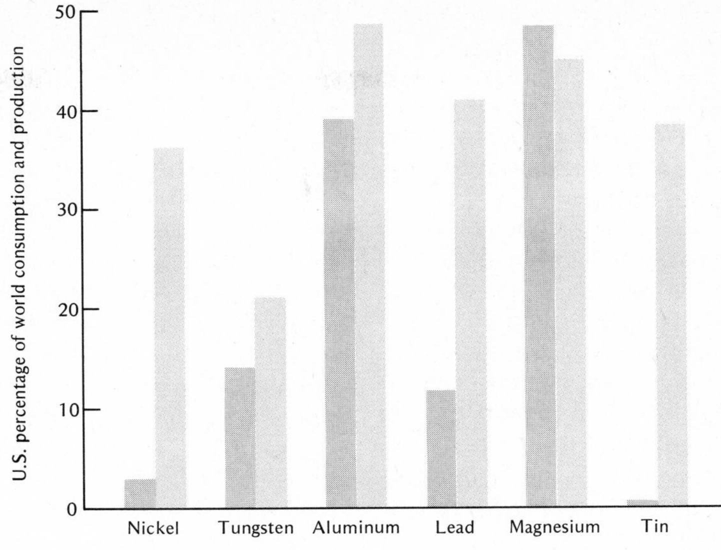

Figure 46.5 U. S. share of world consumption and production of selected minerals, 1967. (From *Population Bulletin,* February 1970, © Population Reference Bureau, with permission.)

tion of the environment by noises produced in our present-day industrial society.

In Chapter 13 the delicate nature of the end organs of hearing, the hair cells in the organ of Corti in the inner ear, were described. There is clear evidence now that these hair cells can be permanently damaged if they are subjected to repeated sounds of high intensity before they have an opportunity to recover. Most of us are subjected to many annoying and often damaging sounds much of the time. These come from radios, television sets, stereo systems, power tools, vacuum cleaners, automobiles, motorcycles, trucks, aircraft, power lawn mowers, outboard motors, and factory noises, to name a few of the sources.

The problem of noise has been defined in terms of the basic unit used to measure the intensity of sound. This unit is the decibel (dB), which is a tenth of the larger unit, the bel, named in honor of Alexander Graham Bell. One decibel is equivalent to the faintest sound that can be heard by the human ear. The decibel scale is a logarithmic one. Some people feel discomfort with sounds of 85 dB, whereas most do not feel discomfort until the loudness reaches 115dB. Pain is usually felt at 145 dB. Figure 46.6 shows the levels in decibels of some common noises.

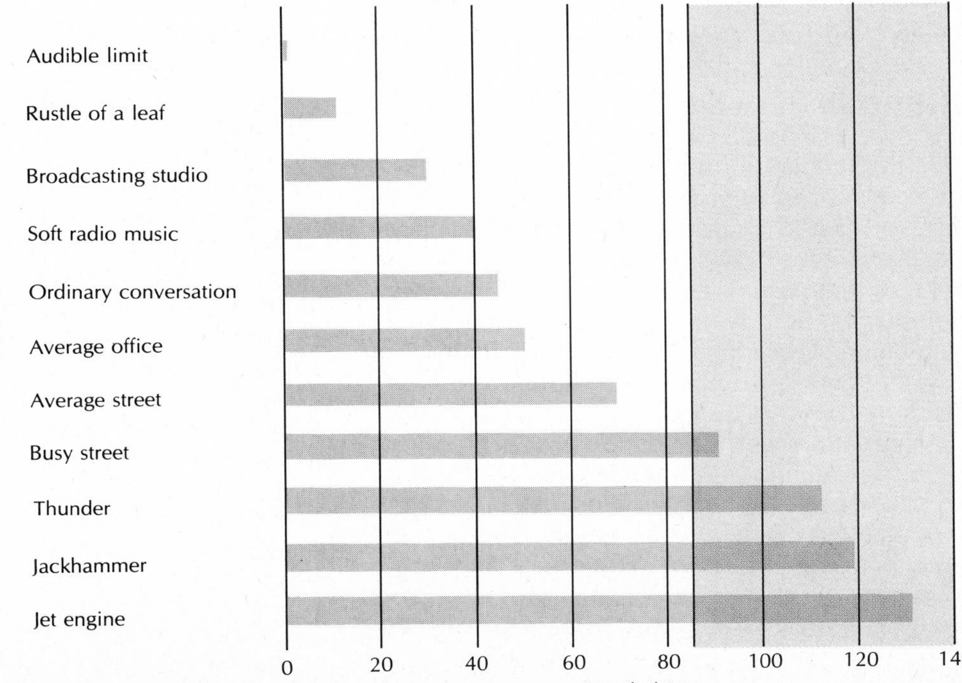

Figure 46.6 Levels of common noises. (From *The World Book Year Book,* © 1970 by Field Enterprises Educational Corporation, with permission.)

Physicians first became aware of this problem about a century ago when they first examined boilermakers suffering from noise deafness. Since then workers from a variety of industries have developed noise deafness, and millions of dollars in compensation have been claimed for these injuries.

Rock music, which so many young people enjoy, is usually electronically amplified to 100 dB and over. There is good evidence now that such music produces temporary deafness in certain ranges in the listeners and also that members of rock bands, who are subjected to such sounds for a period of a few years, suffer permanent hearing damage.

In addition to the impairment of hearing, there is good medical evidence now that noise affects more than the ears. There is evidence from research in both the United States and Europe that indicates that noise increases the level of cholesterol in the blood and that it raises blood pressure; that it causes small blood vessels to constrict; that it not only causes the constriction of blood vessels in fingers and eyes, but that the opposite effect, dilation of blood vessels in the brain, is produced (which could be an explanation of the headaches caused by noises); and that it causes the pupils of the eyes to dilate. This last effect helps to explain why surgeons, watchmakers, and others who do close work are bothered by noise—their eyes are forced to change focus and this causes eyestrain and headache.

The Noise Control Act of 1972 establishes authority for EPA to issue standards for identified sources. Its first report on federal noise control programs was published in June 1975.[1]

ENERGY

At first glance it may seem strange to include a discussion of energy in a chapter on human ecology. But it properly belongs here because the primary sources of energy in the development of this country even now, are the fossil fuels—coal, petroleum, and gas. These are all of biological origin and their combustion results in some of the main pollutants of our environment.

As a nation we have been slow in recognizing our problems with energy, even slower than we have been in recognizing our problems with population growth and pollution. This is certainly not because information about the energy situation has not been available. In 1952 President Truman's Materials Policy Commission took stock of our resources, and warned of coming shortages if the nation failed to act to assure supplies. Using material from the report of this Commission, Edward R. Murrow gave a dramatic TV report in 1954. His report started out with these statements: "There never was a nation that consumed so much coal and steel and oil and copper and lumber and water and strange minerals and everything that comes out of the earth, and at the same time gave so little thought to where it comes from. You may get a jolt this evening to realize that although America is the land of plenty, the plenty is giving out." The Commission concluded its report with words to this effect. The country is no longer self-sufficient in the raw materials needed for life and growth. Moreover, a developing shortage of materials, including energy, might over the next quarter-century bring about an economic crisis and a dangerous dependency on foreign sources of supply.

In 1954, Harrison Brown, in his "The Challenge of Man's Future," called attention in a forceful way to the rapid depletion of our energy sources. This point was emphasized again in 1969 in a publication "Resources and Man" by the Committee on Resources and Man of the National Academy of Sciences—National Research Council. This publication pointed out that since the earth's deposits of fossil fuels are finite in amount and nonrenewable, energy from this source can be obtained for only a limited period of time. It estimated that the earth's coal supplies are sufficient to serve as a major source of industrial energy for two or three centuries. And it estimated that the corresponding period for petroleum, both because of its smaller initial supply and because of its more rapid rate of consumption, is only about 70–80 years. The report indicated that in the United States the peak of crude-oil production was already past and that the peak of natural gas production would occur about 1980.

According to Dixie Lee Ray, former Chairman of the Atomic Energy Commission, each person in the United States now uses 8100 kW of electric power each year. We use 16 times as much electricity as we did in 1920. The total gross energy consumption doubled between 1950 and 1970, and it is scheduled to double again by 1990. About 25 percent of this represents our electrical demands with another 25 percent going for transportation. Industrial production uses 30 percent, and 20 percent is required for our commercial and household needs. Oil and gas supply more than 75 percent of these energy needs. *In 1973, 35 percent of our total petroleum*

[1] See *Environmental Quality*, **6**.

supply was from imports. At the present time the United States consumes about 35 percent of the world's energy.

It took the Arab oil embargo in 1973 to make the American people aware of our energy situation. All readers will remember this. Gasoline was soon in short supply. Stations soon ran out of gasoline each day, and it was customary to see long lines of customers waiting to buy gasoline. The price soon increased from 35 cents per gallon to 55 cents. By 1974 the supply was ample again, but the price remained very high, and this has seriously affected our balance of payments with foreign countries. As a result President Nixon started Project Independence. The project would aim to make the United States self-sufficient energywise by 1985. At the present time (1976) a complete program has not been developed.

However, in early 1975 a National Research Council study was released which stated that the government's goal of energy self-sufficiency by 1985 is "essentially impossible." Their conclusion was based on their findings that the nation's remaining resources of oil and gas are considerably smaller than previously estimated by the Interior Department and other government agencies. The report states that there is little that the United States can do in the near future to increase oil and gas production by significant amounts. It goes on to recommend that the country should turn its attention to stronger conservation efforts.

For the short haul attempts will certainly be made to increase our production of oil by drilling off shore on the continental shelf. But it appears that we will have to depend more and more on coal for increased energy production. There are real environmental problems associated with increased coal production. Much of the coal that is readily available has a high sulfur content. To make this coal suitable for use in many places, it will be necessary to improve the present techniques for removing sulfur from the effluents before they leave the stacks. There are large coal reserves in the West that are low in sulfur content. Much of this could be made available by strip mining. However, we now have a law which provides that any area that is strip-mined must be restored. In many parts of the West where this coal is located there is not enough water available to get plant life started in the reclaiming process. The Government now has an 11.3 billion dollar five-year program for energy research and development. A good part of this will be used in trying to solve the problems mentioned above, and, in addition, for finding feasible methods for the gasification and liquefaction of coal.

Tremendous amounts of oil are present in Colorado, Utah, and Wyoming in association with shale. The technology for recovering this oil so far has been very expensive and further research is needed. This operation also has environmental problems. According to one report, water to support a 1-million barrel per day industry may be available, but a 3–5 million barrel per day industry using current technology might require all the surface water in the oil shale regions of Colorado, Utah, and Wyoming. Furthermore, a 1 million barrel per day industry using current technology would require disposal of 1.36 million metric tons (1.5 million tons) of waste material per day.

In addition to the fossil fuels there are other sources that should help in making us self-sufficient. These include water power, geothermal power, use of methanol, nuclear fission, nuclear fusion, solar power, and the use of hydrogen as an energy source. Increased research on several of these is now underway.

FACING THE PROBLEMS OF POPULATION AND POLLUTION

Population

Although at the time of writing this not much can be said about how we go about solving our energy problems, some definite things can be said about what we are doing to solve the problems of population and pollution.

So far as the position of our government on population is concerned we have come a long way in the last few years. The attitude of President Eisenhower, until after his retirement, was that birth control is not the business of government. Both Presidents Kennedy and Johnson officially supported programs of family planning services throughout the world. The first presidential message to Congress dealing only with population problems was delivered by President Nixon on July 18, 1969. In it he asked this question: "Finally we must ask: how can we better assist American families so that they will have no more children than they want to have?" This was an important message. Certainly the prevention of the births of unwanted children is a desirable goal. However, this goal, if achieved, will not in itself cut down materially on population growth. In this message the President recommended the establishment of a Commission on Population Growth and the American Future. This Commission was appointed and it consisted of 24 members representing most sectors of the society. The chairman was John D. Rockefeller, III.

On May 5, 1972, the Commission, after working two years, delivered the report to the President (see Suggestions for Further Reading). The report definitely indicated that this country does have population problems. The President rejected two of the Commission's recommendations: one on abortion[1] and the other on providing contraceptives for teenagers, but he did indicate that the recommendations would be taken into account in policy and budget formulations.

The Commission's report is very broad in scope, and it is not possible to give a digest of it here. Among the many things considered are the effects of population growth on the economy, use of minerals, use of energy, use of water, use of land, food production, and pollution. On the economy the report states, "We have looked for, and have not found, any convincing economic argument for continued national population growth. The health of our economy does not depend on it. The vitality of business does not depend on it. The welfare of the average person certainly does not depend on it." In the matter of water supply the Commission's research found that growing population and economic activity will cause the area of water shortage to spread eastward and northward across the country in the decades ahead. The report says, "Few will like the austerity created by the need to conserve on something as fundamental as water. The rate of national population growth will largely determine how rapidly we must accomplish these changes."

A birth rate slightly below replacement level existed in 1974. This might mean that the people in this country are getting the message about the need to reduce reproduction. This does not mean, however, that the population did not grow. It grew because of the much greater number of women in the child-bearing age as compared with the number of women in the age group that produced the deaths. It is to be hoped that this lowered trend will continue but, as pointed out above, we have the potential for another "baby boom" in the great number of women in the child-bearing age, and this will continue for a number of years.

On the international scene the picture is not so bright. As of the end of 1974 most of the developing nations were not willing to recognize the need for bringing down their runaway birth rates. These nations had an opportunity to make a commitment both at the World Population Conference in Bucharest in August, 1974, and at the World Food Conference in Rome in November, 1974. Instead they banded together to pass resolutions suggesting that the real problem was that wealthy nations consume too much and give away too little.

[1] It should be noted that the Supreme Court has now ruled that state laws which prohibit a pregnant woman from deciding herself to have an abortion during the first three months of pregnancy are unconstitutional.

Pollution

THE GOVERNMENT

Our government has been concerned with problems of pollution of various kinds for a long time. But in facing these problems there was no overall policy. Some aspects of pollution have been handled by the Department of Interior, some by the Department of Agriculture, some by the Department of Health, Education and Welfare, some by the Department of Commerce, and so on. In both the House and the Senate many different committees have been concerned with different aspects of the same problem. What is needed now is an overall policy and then some new programs. In consultation with his Council on Environmental Quality, President Nixon in mid-1970 proposed executive reorganization plans for the establishment of the two new agencies—an independent Environmental Protection Agency (EPA) and a National Oceanic and Atmospheric Administration (NOAA). Establishment of the two agencies soon followed. EPA has responsibility for air and water pollution and solid wastes. It is made up of units transferred from other federal departments, primarily from HEW and the Department of Interior. NOAA is charged with exploring and predicting what is happening to our oceans and climate, and with taking appropriate remedial steps. It is housed in the Department of Commerce. This brings together the Environmental Science and Service Administration, already in the Department of Commerce and involves the Weather Bureau and the Coast and Geodetic Survey, Interior's Bureau of Commercial Fisheries, the Lake Survey Office of the Army Corps of Engineers, the Sea Grant Program of the National Science Foundation, and the Marine Technology Program of the Bureau of Mines.

Since the Environmental Protection Agency was formed on December 2, 1970, a good start has been made in attacking various problems of pollution. Progress has been made in the control of both air and water pollution, working through State agencies. A ban was finally imposed on the general use of DDT and certain related insecticides. The Resource Recovery Act of 1970 puts an important new emphasis on recycling. This act authorizes funds for demonstration grants for recycling systems and for studies of methods to encourage resource recovery. The Office of Solid Waste Management Programs in EPA has a program to replace 5000 of the 15,000 open dumps in this country with sanitary land fills. EPA is engaged in many many activities. Detailed reports of this work may be found in the "Environmental Quality" reports for 1970, 1971, 1972, 1973, 1974, and 1975 prepared by the Council of Environmental Quality (see Suggestions for Further Reading).

INDUSTRY

Industry has a real responsibility in helping solve the environmental problems. For years industry has considered our air and water as commons, and the resultant pollution has been great. However, a perusal of the recent reports to stockholders from many corporations in the country indicates that some of them, at least, are now quite aware of what they have been doing and that they are now spending large sums of money to bring their pollution under control.

EDUCATION

The newspapers, magazines, and television stations during the past several years have performed a great role in educating the general public to the problems of environmental pollution. We can expect this to continue, but more is needed. The facts about both population and pollution must be presented to the young throughout their educational experience. Already many college courses are involved in this. Also there is evidence that this information is becoming part of the programs in grade and high schools.

One other important area in education should be mentioned. It should be obvious by now that the problems of population and pollution are not only ecological in the strictly biological sense, they also involve the physical sciences, engineering, medicine, economics, political science, psychology, religion, and many other areas. To solve the urgent problems facing mankind in the next few decades many people will be needed who are generalists in the whole area but who have special training in one or a few of the areas. This will necessitate interdisciplinary programs of training in the colleges and universities. It is heartening to note that the institutions of higher education in the United States are responding. In 1958 there were only 20 institutions with interdisciplinary programs dealing with ecological problems. Now there are over 200 pertinent and viable programs with others in the developmental stages.

THE INDIVIDUAL

Whatever is done in the solution of these great problems depends ultimately on what individual citizens believe and do. Are you willing to pay your share of the cost of an environmental cleanup in the form of higher taxes and higher priced products? Are you willing to cease being a polluter? Will you work with others to attain a goal of a stabilized population and a cleaner and safer environment? Will you use our food and energy resources carefully? Many more questions could be asked, but you can ask them yourself. If we can set our house in order in the United States, there may be some hope for the rest of the world. If we do not, the prospects seem dim.

The interested reader should consult the articles by Davis, Ehrlich, Hardin, Miles, Ogburn, and White listed in Suggestions for Further Reading.

SUGGESTIONS FOR FURTHER READING

Benarde, M. A. *Our Precarious Habitat.* New York: W. W. Norton (1970).

Brewer, M. F. "Population: The Future Is Now." *Population Bull.,* **28** (Apr. 1972).

Brown, H. *The Challenge of Man's Future.* New York: Viking Press (1954).

Cloud, P. E., Jr. (Ed.). *Resources and Man.* San Francisco: W. H. Freeman (1969).

Commoner, B. *The Closing Circle.* New York: Alfred A. Knopf (1971).

Davis, K. "Population Policy: Will Current Programs Succeed?" *Science,* **158:** 3801, 730–739 (Nov. 10, 1967).

Ehrenfeld, D. N. *Biological Conservation.* New York: Holt, Rinehart and Winston (1971).

Ehrlich, P. R., and A. H. Ehrlich. *Population, Resources, Environment,* 2d ed. San Francisco: W. H. Freeman (1972).

Environmental Quality (Annual Reports of the Council on Environmental Quality). Washington: U.S. Government Printing Office, vol. 1 (1970); vol. II (1971); vol. III (1972); vol. IV (1973); vol. V (1974); vol. VI (1975).

Hardin G. (Ed.). *Population, Evolution and Birth Control.* San Francisco: W. H. Freeman (1969).

Hardin, G. "The Tragedy of the Commons." *Science,* **162:** 3859, 1243-1248 (Dec. 13, 1968).

Johnson, W. H., and W. C. Steere (Eds.). *The Environmental Challenge.* New York: Holt, Rinehart and Winston (1974).

Meadows, D. H., *et al. The Limits to Growth.* New York: Signet (1972).

Miles, R. E., Jr. "Whose Baby is the Population Problem?" *Population Bull.,* **26** (June 1970).

Ogburn, C., Jr. "Why the Global Income Gap Grows Wider." *Population Bull.,* **26** (June 1970).

Population and the American Future (Report of the Commission on Population Growth and the American Future). New York: Signet (1972).

Shepard, P., and D. McKinley (Ed.). *The Subversive Science.* Boston: Houghton Mifflin (1969).

Udall, S. *1976, Agenda for Tomorrow.* New York: Harcourt, Brace, Jovanovich (1968).

Waldron, I., and R. E. Ricklefs. *Environment and Population.* New York: Holt, Rinehart and Winston (1973).

White, L. "The Historical Roots of Our Ecologic Crisis." *Science,* **155:** 1203-1207 (1967).

From Scientific American

Brown, L. R. "Human Food Production as a Process of the Biosphere" (Sept. 1970).

"Energy and Power" (Sept. 1971). (Entire issue devoted to this topic.)

Frejka, T. "The Prospects for a Stationary World Population" (Mar. 1973).

Goldwater, L. T. "Mercury in the Environment" (May 1971).

Haagen-Smit, A. J. "The Control of Air Pollution" (Jan. 1964).

Huxley, J. "World Population" (Mar. 1956).

Nevers, N. de. "Enforcing the Clean Air Act of 1970" (June 1973).

Newell, R. E. "The Global Circulation of Atmospheric Pollutants" (Jan. 1971).

47 The Biotic World and the Future of Humanity

Many examples of organisms meeting the problems of life were detailed in the previous chapters. Just as the problems are diverse, the solutions to these problems are diverse. In general, the structures of all organisms at all levels of the biological hierarchy are designed for function—for meeting the problems that organisms face. The problem approach of studying organisms in relation to their physical and biotic conditions provides a basis for understanding the vast complexities of the biotic world.

Those organisms that are more efficient than others in solving their problems will contribute greater numbers to the next generation. This is the basic statement of Darwinian evolutionary theory (Ch. 42). Those organisms that have changed by random mutation will produce offspring that will have these mutations tested by natural selection. Thus "the ecological theater and the evolutionary play" have produced a great variety of autotrophic and heterotrophic organisms adapted for life in many specific environments. Evolution focuses the disciplines of biology on the problems that organisms face. Biochemistry, physiology, genetics, anatomy, ecology, and behavioral biology contribute to the understanding of the basic evolutionary mechanisms.

As environmental conditions change, organisms that survive these changes must have adaptations that offer survival advantages in the new environment's conditions. This story has been repeated millions of times during the period of life on earth. Dinosaurs have become extinct, millions of humans have died from malaria, some insects have become resistant to pesticides, and the peppered moth has developed a dark form.

What are the implications of these observations to the human population? Humans are faced with the basic problems of all organisms. Also, the ability to manipulate the environment expands the number of potential problems that humanity will face sooner or later.

A listing of society's problems in the biotic world is as difficult as the listing of the concepts of biology in Chapter 1. Everyone will find omissions from the list, but five problems that are likely to appear on most lists are

1. Food production
2. Population control
3. Disease
4. Conservation of natural resources
5. Endangered species

[The obvious human problems of war, crime, aggression (see Berkowitz and Moyer, Suggestions for Further Reading), poverty, the energy crisis, racial prejudice, drug abuse, and political immaturity are not listed although biology does have a role in most of them.[1]] Does the study of biology influence the formulation of this list of problems and the proposals for solutions?

FOOD PRODUCTION

Many of the world's people suffer from malnutrition. These nutritional problems are twofold: the amount of food and the right kinds of food. The roles of carbohydrates, lipids, and proteins as sources of calories are discussed in Chapter 9. The importance of vitamins, essential amino acids, essential fatty acids, and minerals in nutrition is outlined in Part II. Will humanity be able to produce, gather, and distribute enough of the right kinds of food to support the world's population?

[1] All of these topics would serve as points for extra research and reading.

The purpose of agriculture is to provide a regular source of a variety of foods without the uncertainties associated with those cultures dependent upon fishing, hunting, and collecting. One of the most important changes in human history came with the domestication of plants and animals and the rise of the agricultural way of life. This probably began about 7000 B.C. along with the use of new and improved stone tools. Domestication of plants was probably a gradual development, but eventually the ideas of carrying and planting seeds emerged. As a result, people learned to channel the nutrients and energy of the ecosystem in the directions of their own needs and choosing. The dog, pig, and chickens were probably the first animals kept around villages, perhaps as much for pets as for food.

Agriculture developed on a trial-and-error rather than on a scientific basis. The deep rich soils of river basins produced the best crop yields, and the early centers of civilization developed in river valleys of the Mediterranean and western Asia. The agricultural industry, as it is known today in technologically advanced countries, developed with the Industrial Revolution. Scientific agriculture has produced enormous bounties by the application of chemical fertilizers, the use of herbicides and pesticides, the development of irrigation systems, and the application of genetics to agricultural practices. There is no basis for complacency since the peak of food surplus in the United States around 1966 would have fed the hungry in the world for only a few months.

In addition to agricultural genetics, other possible ways to increase food production in the world include

1. Extensive use of marine organisms for fertilizer, livestock feed, and human food
2. Conversion of the fishing industry from a hunting mode of operation to a farming mode
3. Use of the microorganisms grown on coal or oil as a source of protein
4. Use of desalinized water for irrigation of farm lands; since many of the world's deserts are close to oceans, cheap desalinization may be extremely useful
5. More effective means for combating plant and animal diseases, including elimination of pests
6. More effective weed control
7. Better harvesting, processing, and preserving techniques
8. Better understanding of diverse soils, climates, and food preferences in each region of the world may be useful

Ultimately, all these approaches depend on advancements in understanding basic sciences and development of technologies. Biology textbooks often discuss these possibilities in some detail. Here we consider three further basic biological points which may affect increased food production.

POPULATION CONTROL

The dynamics of the human population as a basic biological phenomenon and as applied ecology were discussed in detail in Chapter 46. This section will discuss one way of attacking the problem of relentless human population growth: birth control. This discussion necessarily relies on many facts and concepts discussed in earlier chapters, particularly Chapters 12 and 15.

The traditional methods of rhythm, withdrawal before ejaculation (*coitus interruptus*), and mechanical devices, such as the diaphragm or condom, have an element of unpredictability, lack of sexual satisfaction, and/or inconvenience. Spermicidal jellies, foams, and creams and the use of the vaginal douche after intercourse have a degree of ineffectiveness also. Birth control pills for women became available in 1953 when the ovulation-inhibiting properties of progesterone and some of its derivatives as administered orally, intravaginally, and subcutaneously were reported. Research on chemical control of ovulation produced a combination of steroids (estrogen, progesterone, and progesterone derivatives) that had maximum contraceptive efficiency and optimum control of menstrual periodicity. The antifertility properties of these steroid combinations are extraordinary. Oral contraceptives have changed population patterns and birth rates in most of the world's industrialized nations almost single-handedly. The birth control pill has many advantages over other forms of control, but recently it has come in for much criticism, too. The pill has been linked to blood clots, migraine headaches, elevated blood pressure, decreased mental acuity, adverse effects in diabetes, liver malfunction, and cervical cancer. A relationship between oral contraceptives and high blood pressure was first noted in young women in 1967. In 1974 a research report indicated that women on the pill are nine times more vulnerable to heart disease than are other women. Also in 1974, preliminary data were published linking the pill to birth defects, particularly limb-reduction malformations. Although limb-reduction malformations are rare (one in 4000 births), the incidence rose between

1968 and 1973, and the use of oral contraceptives rose in the same period. In 1975 about 10 million women in the United States were taking oral contraceptives. Of course, birth control pills are prescribed only by physicians. For most women the benefits of oral contraception outweigh adverse effects. Whether birth control pills do have undesirable side effects should be determined by a physician.

Physicians are being cautioned to avoid using pregnancy tests in which hormones are administered. The physician sometimes gives a large dose of estrogen to a woman who has missed a menstrual period. If the period begins soon after, the woman is not pregnant, but if the period does not begin, she often is pregnant and the fetus is exposed to the hormones during early crucial stages of development. Further, a physician should not prescribe birth control pills until he is certain that the woman is not already pregnant. Even though the birth control pill is the most effective contraceptive, it definitely places the burden of contraception on the woman.

A birth control pill without side effects is being pursued vigorously in several research laboratories. Work at the Baylor College of Medicine focuses on the progesterone receptor in the uterus. This protein receptor molecule binds both natural and synthetic progestins (a group of substances which includes progesterone) and prepares the uterine lining for implantation. If more were known about the biochemistry of the receptor protein, perhaps an agent could be found which blocks the receptor site, and the uterine lining would not be prepared for implantation. Since the normal action of progesterone would be blocked without changing hormone levels or adding synthetic hormones, such an antifertility agent would not cause the adverse side effects that are being correlated with the pills currently in use.

Other than the pill, the most effective means of contraception is the intrauterine device (IUD). This method also places the burden on the female and the mechanism of action of the intrauterine device is not understood in detail. Recently complications in the use of IUDs have been recognized. Two of them, the Dalkon Shield and the Majzlin Spring (Fig. 47.1) have been removed from the market because of adverse effects such as bacterial infection or the inability to be removed. The oral pill and the IUD have the lowest failure rates of the chemical and physical contraceptive methods. In general they have been improved greatly in the past ten years and are now the most popular methods in the United States.[2]

Surgical sterilization is a method of birth control that will be considered by many persons in the future. In the female, sterilization involves tying the Fallopian tubes. In the male, surgery, called vasectomy, involves cutting and tying the vas deferens from each testis before it passes from the scrotum through the inguinal canals, and not more than half an hour is required for the complete procedure. After a vasectomy there is about a two-month period when sperm may persist in the ejaculate. Birth control procedures are necessary during this time.

A male birth control pill is also being researched. A combination of a synthetic androgen and estrogen has been shown to stop sperm production in 12 to 18 weeks, and normal sperm production was recovered in 35 to 40 weeks after stopping treatment.

Thus at the present time for a couple that has completed its family and wants to be absolutely certain that they will have no more children, vasectomy for the husband should be an attractive solution. Indeed, if vasectomy is such an attractive approach, why is it not used extensively? At Cornell University some biology professors conducted a survey of students and faculty with respect to the desirability of various forms of birth control. This survey showed that there is a rather weak enthusiasm for permanent methods of birth control. There is a rather widespread, and incorrect, belief that a vasectomy makes a man impotent, decreases sex drive, and leads to an inability to ejaculate. Sterilization by a vasectomy is confused with castration. The secretions added to the semen by the three sets of glands (see Ch. 15) are added normally in a vasectomized male, and the testicular hormones leave the testes via the blood. Thus there is no medical evidence that vasectomized males possess reduced sex drives. In a study of 68 men who had a vasectomy 44 percent felt their enjoyment of sex had increased and 55 percent thought that their wives were more sexually responsive. Only three percent said that their enjoyment of sex had decreased.

Another method of birth control that needs to be mentioned is *abortion.* Abortion has received much attention in the press since some states have modified greatly their abortion laws and the Supreme Court has ruled that abortion during the first three months of pregnancy is a matter of concern only to a woman and her physician. Probably more states will lower the requirements under which an abortion may be

[2] The interested reader is directed to the articles by Jaffe and Parr (Suggestions for Further Reading) which are available in most college and university libraries.

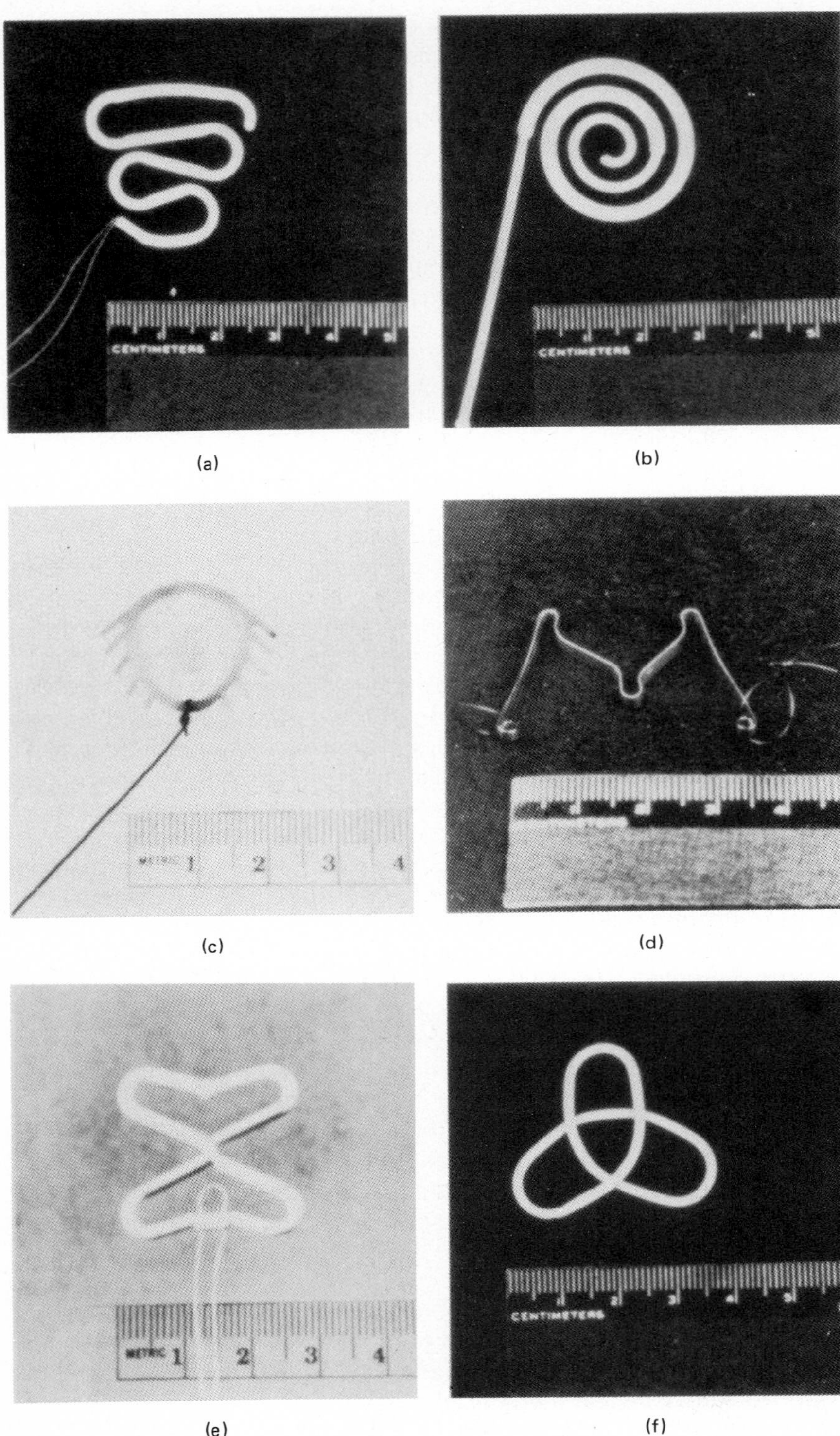

Figure 47.1 Intrauterine devices. **(a)** Lippes loop. **(b)** Gynekoil. **(c)** Dalkon shield. **(d)** M device. **(e)** Birnberg bow. **(f)** Shamrock device.

performed. That abortion is an important birth control method is evidenced by the fact that it played a major role in the stabilization of the population of Japan after World War II.

The attitude of many people toward abortion is one of strong disapproval. But as the population problem of the world continues to increase, abortion will have to be considered as a meaningful approach to birth control. Other people think that the laws concerning abortion are too restrictive still and the legislatures are moving too slowly. As some small solace, we can refer to 1916 when Margaret Sanger opened her first family planning clinic in Brooklyn. She was promptly arrested by the vice squad. Attitudes toward abortion are changing. The word "abortion" has ceased to be a hush-hush word and the pros and cons of abortion are debated openly and freely.

Brain hormone manipulation is a potentially powerful contraceptive technique. Figure 15.5 shows the interaction of the hypothalamus of the brain and the anterior lobe of the pituitary gland: the luteinizing hormone-releasing factor (LRF) travels from the hypothalamus to the pituitary via a portal system where, in proper concentration, it causes the release of luteinizing hormone. This hormone, in turn, stimulates the release of estrogen and progesterone from the ovarian follicle and the corpus luteum, respectively. LRF, a peptide, has been isolated, its primary structure determined, and has been synthesized. One goal of this research is to make LRF analogs that differ only slightly from the normal sequence. Perhaps an analog can be found which competes with the normal LRF at its site of action. If the action of estrogen and luteinizing hormone is thereby prevented, ovulation does not occur. Eight LRF analogs have been found which inhibit LRF action in test animals.

Injections of synthetic LRF can be used to induce ovulation. Such induction could permit intercourse without fear of pregnancy for the remainder of the menstrual cycle. Ovulation induction could provide the basis for a reliable rhythm method which might have significance on population control among Roman Catholics. The rhythm method is the only birth control method now sanctioned officially by the Roman Catholic Church since it does not suppress ovulation.

Both LRF inhibitors and induction ovulation by synthetic LRF may be safer than present contraceptive pills because they are more specific in their action. One major hurdle that needs to be overcome in both forms of brain-hormone manipulation is the need to find an oral form and avoid the need for injection.

Selective immunization against certain substances in the reproductive cycle may be a contraceptive possibility. For example, both male and female experimental animals have been injected with sperm or testes extracts; subsequent sperm production was depleted in the male and fertility was impaired in the female. Several laboratories are studying other kinds of immunological possibilities. For example, hamster ovaries were ground up and certain antigens from the ovaries injected into rabbits. The antihamster ovary antibodies have now been injected into female hamsters to determine if the antibodies reach the reproductive tract in high enough concentration to keep the sperm from penetrating the eggs.

Contraception and other birth control research requires much more work. The prostaglandins (Ch. 12) may be developed into important contraceptives. The birth control topic is an excellent interface of several disciplines: biology, chemistry, law, ethics, psychology, sociology, religion, and economics.

DISEASE

To a certain degree, the history of civilization converges around the attempt of the human species to cope with disease. The various diseases which have plagued humans may be classified simply for our purposes:

1. Infectious: diseases caused by bacteria, viruses, fungi, and rickettsias
2. Parasitic: diseases caused by protozoans, flatworms, and round worms
3. Genetic: diseases affecting metabolic pathways and development

There is some overlap of the fundamental aspects of infectious and parasitic diseases, but these pose no problems in our discussions here.

Infectious Diseases

Prior to 1940 many infectious diseases stalked the world: syphilis, tuberculosis, pneumonia, dysentery, meningitis, influenza, childbed fever, and others. The development of antibiotics and other drugs has been most significant in decreasing death by infectious diseases. Recent evidence, however, shows that syphilis and gonorrhea are increasing in association with the current acceptance of permissiveness among teenagers and young adults and, in the case of gonorrhea, the appearance of strains resistant to antibiotics. Often there is failure to recognize the disease or to seek readily available treatment.

For twelve years, A. Fleming's 1928 discovery of penicillin was unregarded and unused. Other British workers, stimulated by medical problems arising in World War II, showed in 1940 that penicillin was stable after it was purified and dried. With purified samples, penicillin was shown to cure infections in animals including humans. Because of war pressure on British laboratories, industry, and government, the United States initiated programs of mass production of this drug which proved to be remarkably nontoxic to humans and particularly effective toward many susceptible bacteria.

Modern chemotherapy against bacteria actually began in Germany in 1935 with the production of sulfanilamide and a group of related compounds known as sulfonamides, or "sulfa drugs." Sulfonamides inhibit bacterial growth by preventing synthesis of folic acid. Folic acid is a vitamin required by humans. (A serious deficiency of folic acid results in sprue, a disease characterized by intestinal disorders, anemia, and loss of hair.) Sulfanilamide, like other sulfonamides, is similar in structure to *p*-aminobenzoic acid, which is a normal component of folic acid. Sulfanilamide blocks the synthesis of folic acid by competing with *p*-aminobenzoic acid for the active site of the enzyme that catalyzes the incorporation of *p*-aminobenzoic into folic acid. The sulfonamides do not interfere with their metabolism. The sulfonamides led to the concept of antimetabolic drugs, and the penicillins gave great impetus to the search for new antibiotics. Probably many drugs affect the structure of membranes as the basis of their action. As the structure of membranes becomes better understood, it may be possible to elucidate in great detail how some drugs work.

Relatively little success has been seen in conquering viral diseases as compared with bacterial diseases. Since viruses are composed of a nucleic acid (DNA or RNA) and protein, any drug that is applied to interfere with the nucleic acids and proteins of the virus will interfere also with those of cells. Although certain tissues of the body can tolerate such drug stresses, tissues with a rapid rate of division or metabolism cannot. Probably a good place for inhibiting viral diseases is at the stage of viral assembly—wrapping up the nucleic acids in the protein coat. No counterpart of this process is known in cells, so they should not be affected by such a drug. One compound, rifampicin, has been shown to have a strong potency against viruses; it seems to inhibit attachment of the protein coat subunits to the nucleic acid core. On the other hand, many viral diseases are treated by immunological means. For smallpox, measles, and mumps, this means contracting the disease and forming antibodies against it. Of course, injection of a weakened or killed virus strain will bring forth the immunological response also. Further, an extremely detailed analysis of viral life cycles may reveal good places to attack viral disease.

Are new infectious diseases likely to arise? This question is answered by the recurrent epidemics of influenza. Presumably mutation of the virus, which causes the flu, produces new strains that sweep through human populations. Influenza, "the last of the great plagues," killed at least 2200 people during the winter of 1973 (the London flu). The battle against disease-causing agents will be a continual one, but with proper sanitation practices and continued research we have the tools to conquer many epidemic diseases. One serious potential threat is the transference of drug resistance between strains of bacteria. One strain of bacteria that is resistant to a drug may conjugate with another strain that is resistant to another drug with the result that the new recombinant strain is resistant to both drugs. By such repeated exchanges of genetic material, strains may develop that are resistant to several drugs. Such organisms are potentially dangerous; epidemics of diseases caused by such organisms may be extremely difficult to control. Many bio-medical scientists are concerned about the possibility of laboratory strains of such recombinant bacteria escaping into the environment. Since both the gypsy moth and African killer bee, problems much in the news, are results of such escapes, this is a serious possibility. In addition to epidemic diseases caused by bacteria, viruses, fungi, rickettsias, and other organisms, other diseases such as parasitic diseases, cancer, and genetic diseases present serious problems for humanity.

Parasitic Diseases

Malaria is the number one killer in all human history. It remains a major disease problem in the warm regions of South America, Africa, and Asia.[3] A major accomplishment in the United States has been the elimination of malaria as an endemic disease. Increased knowledge of the life cycle of *Plasmodium* (Ch. 21), the protozoan which causes malaria, improved drugs, the use of pesticides in killing the mosquito carrier, and the elimination of potential breeding spaces for the vector species all have contributed to this success. However, a significant number of United States servicemen have contracted malaria in Southeast Asia. (There has been some transmittal of malaria through common use of hypodermic syringes at illicit drug parties.) In addition to malaria, the prevalent diseases caused by a vector-transmitted protozoan are Chaga's disease, kala azar, and African sleeping sickness.

Kala azar is due to a protozoan that has some resemblance to the trypanosomes, but is placed in a distinct group (Leishmania) by some workers. The protozoans multiply in the stomach of the sand fly. Rodents, dogs, and humans can be

[3] Currently there are 100 million cases of malaria per year and 1 million malaria deaths per year in the world.

infected when the fly takes a blood meal. Around the turn of the century (1890–1905) kala azar was one of the major plagues of Southeast Asia. The general disease is caused by multiplication of the parasites in the spleen and liver.

African sleeping sickness is the result of infection of the blood and then the brain with the trypanosome transmitted by the tsetse fly. There are millions of square kilometers in tropical Africa where trypanosomes carried by the fly make it impossible to keep cattle. Table 47.1 summarizes some human infections derived from animal vectors.

Table 47.1 Important Human Diseases Caused by Vector-Borne Agents

DISEASE	VECTOR	MICROORGANISM OR VIRUS
Yellow fever	Mosquito	Virus
Typhus	Louse or flea	Rickettsia
Plague	Flea	Bacterium
Malaria	Mosquito	Protozoan
Sleeping sickness	Tsetse fly	Protozoan
Kala azar	Sand fly	Protozoan
Chaga's disease	True bug	Protozoan

Flatworms which infect humans include the flukes (class Trematoda) and tapeworms (class Cestoda, Ch. 24). The Chinese liver fluke (*Clonorchis sinensis*), sheep liver fluke (*Fasciola hepatica*), and blood flukes (*Schistosoma*) are the trematodes that have most seriously affected human health. The most serious fluke affecting humans is the blood fluke. The secondary host is the snail and, humans are infected by direct entry of the cercariae through the skin while bathing in or drinking water which contains cercariae. The adults inhabit the blood vessels. The eggs possess sharp spines and gradually pass through the capillary walls, the tissues, and finally to the gut or bladder, from which they are voided with the wastes. The great damage is caused by eggs which fail to make their way into the bladder or gut. The resulting reaction of the host tissue produces the disease called *schistosomiasis.* As a result of the control of the waters of the Nile River by the High Aswan dam, the snail population and the incidence of schistosomiasis has increased dramatically in the Nile valley in recent years.

Genetic Diseases

Infectious, epidemic, viral diseases, and some forms of cancer may be easy to cure compared to genetic diseases. The better understood inherited diseases were discussed in Chapters 6 and 19. What are the prospects for curing human diseases? This genetic engineering would take two approaches:

1. Altering the development of the individual so that he or she may lead a normal life
2. Altering the cells of the individual so that the genetic defect is not expressed

Many genetic diseases are controlled already by alteration of development. Phenylketonuric infants may be placed on diets low in phenylalanine and grow to maturity with normal characteristics. The detection and alteration of development are relatively simple in this case. Galactosemic infants can be detected relatively easily also. Feeding them a galactose-free diet prevents the bone malformations and cataracts associated with the disease. Injection of insulin into the diabetic is an alteration of the development of the individual also. In all these cases the alteration is a relatively simple procedure, but at maturity these individuals can pass their defective genes on to their offspring. Certainly in the case of diabetes many persons have lived to produce offspring only with the help of insulin injections.

The second approach, altering cells or cell populations, is a difficult task. One method is to replace defective cells with normal cells from a donor. This has been tried in a few cases, and the results are encouraging. In the case of the diabetic, some islets of Langerhans (or at least some β cells) would be transplanted into the pancreas of the affected individual. In order to be able to do this, the immune system of rejection would have to be controlled. The successful transplantation of cells has been done, interestingly enough, with individuals having diseases of the immune system. These patients could not produce antibodies (totally or partially), and the cells they received were bone marrow cells. In 1973 a report of 29 children suffering from diseases which involved immune incompetence revealed that 25 recovered after receiving bone cell transplants.

Many biologists have felt that the introduction of normal genes into the cells of individuals affected with genetic diseases would be a logical approach. Others have suspected that this type of genetic engineering would be years or decades away. In late 1971 such a transfer of genes was accomplished. Cells were taken from an individual with galactosemia (see Chs. 6 and 19) and grown in vitro. A bacterial virus, one capable of transduction and carrying the normal "galactosemic" gene from *Escherichia coli,* was added to the cell culture. After incubation of the human cells and the bacterial viruses, tests showed that the cells had been "cured" of the galactosemic deficiency. The actual test involved hybridization of messenger RNA from the incubated cells to DNA of the virus. Messenger RNA from control cells did not hybridize to the viral DNA. Additionally, the normal enzyme was found in the cells and the ability to make this normal enzyme was passed on to the daughter cells at mitosis. The next step will be to try to establish such transformed cells in an organism; this would be a true genetic transplant. This type of experiment is now underway with laboratory animals.

The first successful enzyme therapy in humans was reported in 1973. Fabry's disease is due to a faulty fat-metabolizing enzyme. Because the abnormal enzyme does not break down lipids in the blood as it should, they accumulate in the body. Enough of the enzyme was isolated from human placental tissue for injection into two individuals with Fabry's disease. The results look promising in that they show a reduction of the fats transported from the bloodstream to the kidneys. Although the amino acid sequence of the enzyme has not been determined, workers do know that it contains about 200 amino acids. After the amino acid sequence and the details of the three-dimensional structure are known, it should be possible to synthesize the enzyme. If a synthetic enzyme can be made, therapy would become available to many patients with Fabry's disease. The same line of research is being used for Gaucher's disease, which is due also to a defective enzyme for lipid metabolism.

Amniocentesis, the technique of examining cells from the amniotic fluid, has been developed to a fine state of art (Chs. 7 and 19). The loose cells floating in the amniotic fluid are sloughed from the lower digestive tract, skin, and respiratory tract of the fetus. These cells can be removed and examined. Karyotype preparation will show whether a chromosomal imbalance exists or not. Furthermore, a number of metabolic diseases can be detected during the first three months of pregnancy. The success of these tests for enzymes is directly related to the basic research on enzymes that has gone on for years. Recently the detection of Tay-Sachs disease has become possible. This disease (Ch. 19) is an accumulation of lipids in neurological tissues and always leads to death, usually in the third or fourth year. With these techniques of testing fetal cells during the early stages of pregnancy, the parents have the option of terminating pregnancy rather than having a child with a birth defect born to them. In 1975 prenatal diagnosis of thalassemia by amniocentesis was accomplished.

CONSERVATION OF NATURAL RESOURCES

The natural resources of the developed countries of the world have been subjected to severe demands during the past 200 years. Rather suddenly, the problems of depletion of land and mineral resources as well as contamination of air and water resources are national concerns. Chapter 46, "Human Ecology," specified many details of human ecology: population, food, and pollution. Facing the problems of population and pollution was discussed also.

The quest for security is an almost universal phenomenon among individuals, families, organizations, and nations. The conservation philosophy on national levels has sprung from this basic need for security-engaging activity. Land as the basic natural resource has been involved in the dualism of the conservationist philosophy: one group of conservationists has promoted land use and the other group has promoted land nonuse (preservation). By the close of the War Between the States, the land constituting the contiguous 48 states had been acquired. The Homestead Act of 1862 accelerated the settlement (use) of the newly acquired land. Society's agricultural use of land has been primarily to produce food, feed, fiber, and other useful plant crops, and when America was first settled, the soil was generally considered to be inexhaustible. The agricultural economy is mainly dependent on the topsoil layer. Whenever the topsoil is damaged by loss of organic matter, erosion, salinity, or alkalinity, reclamation of ruined land is expensive. The infamous Dust Bowl of the Great Plains during the 1930s was due to excessive cultivation of land that should never have been cleared of the native grass cover. The dry conditions of the 1930s accelerated the process. In addition to land damage from agricultural mismanagement, overgrazing, deforestation, strip mining, spoil banks, roadside erosion, and flooding have contributed to land misuse. The Soil Conservation Service was established early in 1935 and the Soil Conservation Act became effective on April 27, 1935, "for development and prosecution of a long-time program of soil and water conservation." Since then the Soil Erosion Service and the Soil Conservation Service, two existing federal agencies, have operated under the name of the latter agency. Today the Soil Conservation Service provides aid in land management by recognizing 124 different conservation practices, but 25 leading practices cover more than 90 percent of all applied practices. The most common are contouring, crop rotation, strip cropping, stubble mulching, terracing, cover crops, turning under green crops, irrigation, drainage, fertilizing, mulching, windbreaks, pasture development, grassed waterways, and woodland harvest.

Many other nations are actively involved in soil conservation, but most of this work has been started as a result of practices developed in the United States, beginning 40 years ago. At least 40 nations have sent technical conservationists or trainees to acquire better understanding of the worldwide problems of land management. Land deterioration has been almost universal under agricultural land use. Since land use is increasing in area and intensity, the seriousness of mismanagement is becoming critical to the future of our current civilization and its practices.

Conservation of land as the time-honored basic natural resource has been emphasized in this section on the conservation of natural resources. However, in view of the future of civilization, other natural resources must be remembered: air, water, mineral, grasslands, forests, recreational areas, wilderness areas, and wildlife. Each of these resources has its unique characteristics and biological components.

The natural resource, wildlife, is singled out for special attention since obliteration of a species is forever, whereas natural feedback signals tell us where other natural resources are in danger of severe damage. The Survival Service Commission (of the International Union for Conservation of Nature and Natural Resources) is the guiding organization for data on endangered species. This organization publishes two data books: Red and Black. The Red Data Book lists those species in *danger* and the Black Data Book lists extinct species. The year 1600 is an important date for wildlife conservationists since all the higher vertebrates (birds and mammals) that have become extinct since 1600 are known. In 1600 there were 4226 living species of mammals; since then 36 have become extinct and about 120 are endangered. In 1600 there were about 8684 living species of birds; since then 94 have become extinct and about 187 are endangered. Thus about one percent of the higher animals have become extinct since 1600 and about 2.5 percent are endangered.

There are five main factors that lead to rarity or extinction of a species: natural causes, hunting, introduced predators, introduced competitors, and habitat alteration or destruction. About one fourth of the birds and mammals that have become extinct since 1600 probably did so naturally. Human activity, directly or indirectly, is probably responsible for the rest. The Labrador duck, heath hen, and passenger pigeon are the most popularly known examples of extinction under the heel of civilization.

CONCLUSION

This chapter is rather diverse in its purpose of relating zoology to the future of humanity. Certainly most biological problems and applications of zoological knowledge to problems do not relate in a simple cause and effect relationship. That is because the constructs of zoology are in nets or webs (*The Web of Life*) rather than in straight lines. The concept of the web of life has been expressed in many ways in this text. In this chapter on biology and the future of humanity the idea can be conveyed simply: nature and human nature are intertwined in the web of life. Every reasonable person in an industrialized society understands that environmental problems exist. The problems of the environment will not be solved with straight-line thinking but with multiple approaches. The educated citizen must have an acquaintance with the basic sciences, biological and physical, which relate to solving the problems. Whether one is concerned about legislation that affects the conservation programs, nuclear wastes and contamination, birth-control clinics, or medical practices, the better decisions will be made by the better informed individuals. The educated citizen must be able to participate in the decision making, whether it be as a voter, as a legislator, as a governmental employee, or as a representative of business. Lacking this participation, decisions affecting the majority of the population may be made by *ad hoc* groups or committees of specialists.

Chapters 46 and 47 have explored some applications of science and technology to human problems. The future is neither completely optimistic nor completely dismal. The authors hope that the principles of zoology presented in this text will be enduring stimuli for the readers of today to become the leaders of tomorrow in the solutions to problems related to all levels of the biological hierarchy.

SUGGESTIONS FOR FURTHER READING

Many books, particularly in paperback, have been published recently on human and environmental problems. The following list is the nucleus for a comprehensive, independent study.

Abortion: The Continuing Controversy. Population Bull., **28** (1972). (Population Reference Bureau, Inc., 1755 Massachusetts Ave., N.W., Washington, DC 20036.)

Anderson, P. K. *Omega, Murder of the Ecosystem and Suicide of Man.* Dubuque, IA: Wm. C. Brown (1971).

Bausam, H. T. *Science for Society: A Bibliography,* 3d ed. Washington, DC: American Assoc. for the Advancement of Science (1972).[4]

Benarde, M. A. *Our Precarious Habitat,* rev. ed. New York: W. W. Norton (1973).

Berkowitz, L. "The Case for Bottling up Rage." *Psychology Today,* **7:** 24-31 (1973).

Boughey, A. S. *Man and the Environment: An Introduction to Human Ecology and Evolution,* 2d ed. New York: Macmillan (1975).

Boughey, A. S. *Ecology Populations,* 2d ed. New York: Macmillan (1973).

Boughey, A. S. (Ed.). *Readings in Man, the Environment, and Human Ecology.* New York: Macmillan (1973).

Bresler, J. B. *Environments of Man.* Reading, MA: Addison Wesley (1968).

Burnet, M., and D. O. White. *Natural History of Infectious Disease,* 4th ed. New York: Cambridge Univ. Press (1972).

Cartwright, F. F. *Disease and History.* New York: Thomas Y. Crowell (1972).

Clapham, W. B., Jr. *Natural Ecosystems.* New York: Macmillan (1973).

Detwyler, T. R. *Man's Impact on the Environment.* New York: McGraw-Hill (1971).

Esser, A. H. *Behavior and Environment.* New York: Plenum Press (1971).

Fisher, J., N. Simon, and J. Vincent. *Wildlife in Danger.* New York: Viking Press (1969).

Gillette, R. "Endangered Species: Diplomacy Tries Building an Ark." *Science,* **179:** 777-780 (1973).

[4] An important bibliography for a person starting a detailed study of biology and human affairs. The sections are (1) Reference, (2) Science, Technology, and Society, (3) Resources and Environment, (4) Education, (5) Health, and (6) Conflict and Population. A general index is included.

Hardin, G. *Stalking the Wild Taboo.* Los Altos, CA: William Kaufmann (1973).

Heiser, C. B., Jr. *Seeds to Civilization.* San Francisco: W. H. Freeman (1973).

Idyll, C. P. *The Sea Against Hunger.* New York: Thomas Y. Crowell (1971).

Leisner, R. S., and E. J. Kormandy. *Population and Food.* Dubuque, IA: Wm. C. Brown (1971).

Lennette, E. H., and J. F. A. McManus. "Contributions of the Biological Sciences to Human Welfare." *Federation Proceedings,* **31** (1972).

Marx, J. L. "Birth Control: Current Technology, Future Prospects." *Science,* **179:** 1222-1223 (1973).

Maugh, T. H., III. "Influenza: The Last of the Great Plagues." *Science,* **180:** 1042-1044 (1973).

Miller, M. W., and G. G. Berg (Eds.). *Chemical Fallout.* Springfield, IL: Charles C Thomas (1969).

Moyer, K. E. "The Physiology of Violence." *Psychology Today,* **7:** 35-38 (1973).

Murdoch, W. W. (Ed.). *Environment.* Stanford, CT: Sinauer Assoc. (1971).

Odum, H. T. *Environment, Power and Society.* New York: Wiley (1971).

Parr, E. L. "Contraception with Intrauterine Devices." *BioScience,* **23:** 281-286 (1973).

Phillips, J., Jr. *Environmental Health.* Dubuque, IA: Wm. C. Brown (1971).

Ravetz, J. R. *Scientific Knowledge and Its Social Problems.* New York: Oxford Univ. Press (1971).

Smith, G. H. (Ed.). *Conservation of Natural Resources,* 4th ed. New York: Wiley (1971).

Strobbe, M. A. *Understanding Environmental Pollution.* St. Louis: C. V. Mosby (1971).

Wagner, K. A., P. C. Bailey, and G. H. Campbell. *Under Siege: Man, Men, and Earth.* New York: Intext Educational Publ. (1973).

Waldron, I., and R. E. Ricklefs. *Environment and Population: Problems and Solutions.* New York: Holt, Rinehart and Winston (1973).

Zinberg, N. E., and J. A. Robertson. *Drugs and the Public.* New York: Simon and Schuster (1972).

From Scientific American

The entire issue of September, 1976, is devoted to *Food and Agriculture.*

Jaffe, F. S. "Public Policy on Fertility Control." *Scientific American,* **229:** 17-23 (1973).

Glossary

Abdomen Belly region of vertebrates that contains the viscera; in arthropods, the posterior region of the body.

Abomasum True stomach of the cud-chewing animals.

Aboral Side opposite the mouth.

Abortion The expulsion of the human fetus prematurely. May be caused by the removal of the ovaries of pregnant females.

Absorption Taking up of water and dissolved substances by cells or tissues.

Absorption spectrum Plot of absorbed light versus wavelength. Absorption spectra are related to the structure of the absorbing molecule.

Acarinate Birds whose sternum lacks a keel; flightless birds.

Accommodation Automatic adjustment of the eye to near and far vision.

Accretion Growth by addition to the surface.

Acetogenin Class of biomolecules that includes lipids and steroids; these compounds are synthesized from acetic acid.

Acetylcholine Chemical substance liberated at or near nerve endings and thought to be involved in conduction across synapses.

Aciculum Stiff bristle in a polychaete parapodium.

Acid Group of molecules that donates protons, for example, hydrochloric acid.

Acini Clusters of cells at the ends of ducts, as in the pancreas.

Acromegaly Condition that develops if an oversecretion of the growth-promoting hormone starts after an individual reaches maturity.

Acrosome Cap on the tip of most sperm involved in establishing a direct contact of the spermatozoon with the egg and the penetration of the egg.

Actin Protein involved in the contraction of muscle cells.

Action current Electric current which flows between a region of excitation and neighboring unexcited regions.

Active transport Movement of materials across membranes by means of a carrier (perhaps an enzyme).

Adaptation Any characteristic of an organism which fits it to live and reproduce in a particular environment.

Adaptive radiation Formation in evolution of a number of species adapted to different ways of life from a single ancestral species.

Adenine Purine found in nucleic acids.

Adenosine diphosphate (ADP) Molecule composed of adenosine (adenine plus ribose) and two phosphate groups. ADP can be transformed into ATP by the addition of a third phosphate group, a process which requires energy.

Adenosine triphosphate (ATP) Primary energy-carrying molecule in the cell. *See* ADP.

Adipose Pertaining to fat.

Adrenergic Nerve fibers that secrete noradrenaline.

Adsorption Attachment of molecules of one substance to the surface of another substance.

Afferent Leading to or toward a certain region. Opposite of efferent.

Afterbirth Placenta and fetal membranes discharged after the birth of the young.

Agglutinins Antibodies that cause antigens or whole cells to stick together in clumps.

Agranular reticulum Pairs of parallel membranes with smooth outer surfaces due to the lack of ribosomes enclosing narrow cavities of varying shapes; smooth endoplasmic reticulum.

Albinism Metabolic disease characterized by reduced amounts of melanin in the skin.

Alcaptonuria Metabolic disease characterized by urine which turns dark on exposure to the air.

Alcohol Any compound which contains the $-\overset{|}{\underset{|}{C}}-OH$ group.

Aldehyde Compound containing the group: $-\overset{O}{\overset{\|}{C}}-H$

Allantois Extraembryonic membrane which arises as an outgrowth of the cloaca in reptiles, birds, and mammals.

Alleles Alternative forms of a gene which occupy similar loci in homologous chromosomes.

Allelopathy Production of substances by one organism which are highly toxic to other organisms.

Allergy Hypersensitive state acquired through exposure to many common substances such as pollen, certain foods, hair, dust, and so forth. The sensitivity is manifested in many ways including hives, asthmatic attacks, headaches, and nasal congestion.

Allotopic Refers to proteins which have different conformations in different locations, for example, when localized in a membane versus free in solution.

Alpha helix Helix found in certain structural proteins such as alpha keratin and in sections of globular proteins.

Altricial Birds and mammals who because of their restricted development are confined to the nest for a period after birth.

Alveolus Small terminal cavity in a lung.

Ambulacrum Echinoderm radius in which its water vascular system lies.

Amine Any of a class of compounds derived by replacement of the hydrogen atoms of ammonia by organic groups.

Amino acid Organic acid containing an amino group (NH_2) and a carboxyl group (COOH); a building block of proteins.

Ammonification All the reactions involved in the conversion of the nitrogenous organic compounds to ammonia.

Amnion Extraembryonic membrane that immediately surrounds the embryos of reptiles, birds, and mammals.

Amniocentesis Removal of cells from the amniotic fluid for testing for chromosomal abnormalities or biochemical defects.

Amniote Vertebrates whose embryonic development takes place within the amniotic cavity as in the case of reptiles, birds, and mammals.

Amoeboid Form of movement in cells resembling that of an amoeba. Pseudopods are formed.

Amphiblastula Type of larva found in certain sponges.

Amphoteric compounds Compounds that can react with an acid or a base.

Amplexus Act of clasping of the female by a male amphibian and deposition of sperm over the eggs she lays.

Ampulla Swelling at a point of attachment of the semicircular canal and utriculus containing a cluster of sensory hair cells but lacking otoliths; also a bulblike swelling at the base of some types of echinoderm tube feet.

Anabolism Constructive phase of metabolism in which complex substances are synthesized.

Anaerobic In the absence of free oxygen.

Analogous structures Structures similar in function but different in evolutionary origin.

Anamnestic response Early onset of antibody production with the secondary introduction of the antigen.

Anaphase Third phase of mitosis and meiosis where the chromosomes migrate from the equator of a spindle toward its poles.

Anatomy Branch of morphology that deals with the gross structure of organisms.

Anemia Condition resulting from a deficiency of hemoglobin in the blood.

Aneuploid Variation of numbers of chromosomes in a haploid set.

Animal pole That pole of the egg (opposite the vegetal pole) where most of the cytoplasm is and where the polar bodies are formed.

Anions Negatively charged ions.

Antenna Jointed moveable sense organ found on the heads of many arthropods.

Anterior chamber Fluid-filled chamber lying in the eye between cornea and iris.

Antibiotic Chemical produced by microorganisms which has the ability to inhibit the growth of other microorganisms. Many synthetic antibiotics have been produced by chemical means.

Antibody Specific protein produced by an organism which can combine specifically with its antigen.

Anticodon Three nucleotides in tRNA which are complementary to a codon of mRNA. The complementary association of the codon and anticodon is fundamental to the genetic code.

Antigen Substance which, when introduced into the body of an organism, stimulates the production of antibodies; usually protein or carbohydrate.

Aortic arch Pair of blood vessels leading around the walls of the pharynx from the ventral aorta to the dorsal aorta.

Aphasia Result of the destruction by disease or accident of one or more of the association areas of the human cortex.

Apoenzyme Protein part of an enzyme.

Apopyle Excurrent opening of a flagellated chamber in sponges.

Archenteron Gastrula cavity; the primitive gut.

Area opaca Region in the blastodisk around the periphery where the cells have not completely separated from the yolk.

Area pellucida Region in the blastodisk over the subgerminal space.

Arteriosclerosis Disorder of arteries in which there are abnormal deposits of cholesterol and other lipids in the walls; hardening of the arteries.

Artery Blood vessel that carries blood away from the heart.

Artificial parthenogenesis Artificially stimulated development of an egg without penetration of sperm into the egg. Can be achieved by various methods such as pricking, ultraviolet light.

Ascending tracts Those tracts in the white matter of the cord of the central nervous system that conduct impulses to higher levels of the cord and to the brain.

Ascon Body form in sponges of intermediate complexity.

Asexual reproduction Reproduction of a new individual from a part of the old one; there is no fusion of parts.

Assimilation Processes by which an organism converts the end products of digestion and other materials into cellular material.

Association areas Large areas in the human cerebral cortex, particularly in the frontal lobes and in the parietal lobes, in which neurons make connections with all the other areas or centers.

Aster The structure of the centriole and astral rays formed following the separation of the centrioles in the early prophase of mitosis in animal cells.

Astigmatism Common eye defect in which either the cornea or the lens of the eye is irregular in its curvature.

Astral ray Strands or microtubules radiating from the poles of a mitotic spindle.

Asymmetry Absence of symmetry.

Atomic orbital Space in which an electron might be found; an atomic orbital encompasses one nucleus.

ATP *See* Adenosine triphosphate.

Atrium Chamber of the heart that delivers blood to the ventricles.

Auditory canal Canal leading from the exterior to the tympanic membrane.

Autogamy Self-fertilization.

Autonomic nerves Nerves that carry involuntary impulses from the central nervous system; control functions such as secretion of glands, contraction and expansion of blood vessels, peristalsis, heart action.

Autoradiograph Radiograph of an object made by the object's own radioactivity, such as a radiograph produced by a radioactive substance in cells or tissues.

Autosomes Chromosomes other than sex chromosomes.

Autotrophic Refers to organisms that can manufacture their own food.

Avicularium Zooid of an ectoproct colony, shaped like a bird's beak and functioning in protection.

Avoiding reaction Refers to action of a paramecium when it strikes an obstacle, stops, reverses its ciliary stroke, swims backward for a short distance, then pivots and swims forward again. It is the basis for the trial and error behavior seen in this organism.

Axon Process of a neuron that conducts impulses away from the cell body.

Axonal flow Movement of substances synthesized in an axon through the cytoplasm to other parts of the axon.

Back cross Mating of an individual with an unknown genotype with a parent that is homozygous recessive for all loci in question.

Bacteriophage Virus that infects bacterial cells.

Barr body Inactive X chromosome that is located on the inside of the nuclear membrane in certain cells. Generally there are n-1 Barr bodies where n is the number of X chromosomes in a cell. Also called sex chromatin.

Base Chemical compound or ion (such as hydroxide ion, OH^-) that can combine with a proton.

Behavior Response of an organism to a stimulus or to stimuli. The stimuli may originate outside the organism or within the organism (secretions, changes in chemical states, and the like).

Benthos Organisms which live on the bottom of lakes or seas.

Beta-oxidation Oxidation of fatty acids by removal of two carbons at a time.

Bilateral symmetry Arrangement in which the right and left halves of the body parts are mirror images of each other.

Binary fission Type of asexual reproduction in which the organism divides into two approximately equal parts.

Biological clock Biological mechanism of an endogenous rhythm.

Biomass Weight of the living organisms per unit area.

Biome Large, easily recognizable communities, for example, grasslands, deciduous forests, deserts, tundras, and so forth.

Biomolecule Any molecule with a specific role and found in living organisms.

Biosphere Area of the earth that is inhabited by living organisms.

Biotic potential Highest possible rate of population increase, resulting from maximum birth rate and minimum death rate.

Bipinnaria Larval stage of a starfish.

Biramous With two branches, as in the basic arthropod appendage.

Blastocoel Fairly large fluid-filled cavity in the center of the mass of blastomeres composing the blastula.

Blastocyst Hollow vesicle or blastula in primate development containing an inner cell mass.

Blastodisk Disk of tissue in early embryogenesis from which the embryo develops.

Blastopore Opening where cells move from the surface to the interior to form the primitive gut during embryogenesis.

Blastula Early stage of development in which the cells are typically arranged in the form of a hollow ball.

Blepharoplast Basal granule of a flagellum or cilium = kinetosome.

Blind spot Region of the retina in which the axons of the neurons converge to form the optic nerve.

Body stalk Mass of tissue which connects the embryo (developing) with the chorion.

Bolus Spherical mass of food shaped by the tongue in preparation for swallowing.

Botany Study of plants.

Brachial plexus Network of fibers from the first and third spinal nerves to the second to form the brachial nerve.

Brain Enlarged anterior end of the central nervous system.

B-type lymphocyte Small white blood cell derived from the bone marrow in mammals that can become a plasma cell.

Budding Reproduction by formation of a new individual from a small process of the parent organism; may involve one cell or a cluster of cells.

Buffer Substance or mixture of compounds that can neutralize both acids and bases without appreciably changing the original pH.

Byssus Tough leathery threads secreted by mussels as attachment to their substratum.

Calorie Unit used in studies on nutrition and metabolism. The amount of heat to raise the temperature of 1 kg of water 1° C. This is a large calorie. The small calorie is one thousandth of this measure.

Capillaries Very small thin-walled blood vessels that connect arteries and veins, and through whose walls exchanges occur between the blood and the cells of the body.

Captaculum Knobbed tentacles located around the mouth of scaphopod mollusks.

Carbohydrate Class of organic compounds consisting of carbon, hydrogen, and oxygen; hydrogen and oxygen usually being in the proportion found in water (H_2O).

Carboxylic acid Organic compound containing the group

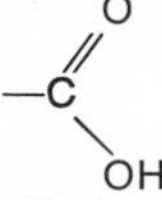

Carinate Birds with a prominent keel on the sternum to which the pectoral muscles of flight are anchored.

Castration Removal of testes or ovaries, emasculation.

Catabolism Destructive metabolism in which larger molecules are broken down into smaller ones.

Catalyst Substance that usually accelerates the rate of chemical change without being used up in the process.

Cataract Condition of the lens of the eye causing it to become opaque.

Cations Positively charged ions.

Cell Structural unit of organisms, consisting of cytoplasm and nucleus and surrounded by the cell membrane; in plant cells there is a cell wall outside the membrane.

Cell cycle Sequence of events in one interphase through the next interphase of the division cycle of a cell.

Cell-mediated immunity Immunity in which cells actually participate in hypersensitivity such as with tissue transplantation and with allergies.

Cellulose Complex carbohydrate (polysaccharide) which is the chief constituent of the plant cell wall.

Central nervous system Brain and spinal cord of vertebrates.

Centrifugation Separation of materials according to their densities by rotation in a centrifugal field.

Centriole Structure between which the spindle forms during mitosis in animal cells.

Centromere Point of spindle fiber attachment in the prophase of cell mitosis; also called kinetochore.

Centrosome Small body near the nucleus which gives rise to the centrioles during prophase of mitosis.

Cephalothorax Fused head and thorax in some arthropods.

Cercaria Free-swimming trematode larva which leaves snail and enters next host either directly of after encystment.

Cerebellum Part of the brain just anterior to the medulla which functions in modulating complex muscular movements and equilibrium.

Cerebral hemisphere One half of the cerebrum.

Cerebrum Large-lobed anterior part of the brain which functions in conscious sensations, learning, and memory.

Chelate Claw-bearing, as an appendage.

Chelating agent Compound which binds metal ions.

Chelicera Pincerlike anterior appendages of horseshoe crabs, spiders, scorpions, and other members of the subphylum Chelicerata.

Chitin Tough horny organic material which forms the basis of the arthropod exoskeleton.

Chlorocruorin Respiratory pigment in certain annelids which appears red or green, depending on its concentration.

Chlorophyll Green pigment of plants; involved in photosynthesis.

Cholecystokinin Hormone secreted by the duodenum; acts to release bile.

Cholinergic Refers to nerve fibers that secrete acteylcholine.

Cholinesterase Enzyme that breaks down acetylcholine. It is important at synapses between neurons.

Chordamesoderm Area of the roof of the archenteron in the frog embryo.

Chorioallantoic membrane Refers to the close union between the allantois and the chorion.

Chorion Outer extraembryonic membrane of reptiles, birds, and mammals. In mammals it contributes to the formation of the placenta.

Chorionic villi Fingerlike outgrowths of the chorion which invade the uterine wall and provide sites of exchange between mother and embryo.

Choroid coat Middle of the three layers of the human eye. It is a black pigmented layer that absorbs imperfectly focused light rays.

Chromatids Duplicated strands of chromosomes held together by the kinetochore. After division of the kinetochore, the chromatids are called daughter chromosomes.

Chromatin Diffuse chromosomal material in the nucleus. It is readily stainable and contains DNA, RNA, and proteins. The term *chromatin* is also used for the prokaryotes.

Chromatography Method for separation of compounds. Paper chromatography is one of the most common forms.

Chromatophore Cell specialized for the production of pigment.

Chromosome complement Quantity of chromosomes characteristic of a cell or organism.

Chromosome constrictions Narrow areas on chromosomes believed to represent sites where DNA is extended and active in transcription.

Chromosomes Deeply staining bodies composed of chromatin and located in the nucleus; they contain genes.

Cilia Fine hairlike protoplasmic processes that beat in unison to move the cell or to move particles over its surface.

Ciliary body Structure on the inner surface of the retina where it joins the choroid coat and whose smooth muscles control the curvature of the lens by altering the tension on the suspensory ligament.

Circadian Rhythmic biological cycles of about 24 hours in duration.

Cirrus Movable fingerlike extension from the body surface; in ciliates a leglike organelle formed by the fusion of many cilia.

Cistron Equivalent to a structural gene; the unit of genetic material that specifies one polypeptide.

Citric acid cycle *See* Krebs cycle.

Class Taxonomic category; a subdivision of a phylum; a group of next higher rank than an order.

Cleavage Divisions of the fertilized egg to form the multicellular blastula.

Climax Terminal community of a succession, which maintains itself relatively unchanged unless the environment changes.

Clitellum Thickened glandular region of the body wall of the earthworm that secretes the cocoon.

Clitoris Erectile structure of the female; homologous with the terminal part of the penis.

Cloaca Common chamber into which the intestine, the kidney ducts, and the genital ducts empty.

Cnidoblast Stinging cell of coelenterates which contains a nematocyst.

Cochlea Coiled auditory part of the inner ear.

Cocoon Specialized capsule in which many invertebrates deposit their eggs; the structure constructed by many insect larvae as a covering for the subsequent pupa.

Codon Three nucleotides which specify an amino acid in protein synthesis. There are DNA codons and mRNA codons. Also called a code word.

Coelom General body cavity lined entirely with a membrane of mesodermal origin.

Coenzyme Nonprotein part of an enzyme.

Colinearity Point-for-point correspondence between linear structures, for example, DNA and its polypeptide product are colinear.

Colloblast Specialized adhesive cell peculiar to the tentacles of ctenophores.

Colloidal system The cytoplasm of a cell is, in part, a colloidal system; the particles (macromolecules or aggregates of smaller molecules) in suspension range in size from 1 to 100 mm.

Colony Group of individuals, unicellular or multicellular, living together in close association.

Commensalism The living together of two species in which one is benefited and the other is neither benefited nor harmed.

Community Organisms of a habitat.

Complementation Functional nonoverlapping of genetic mutants.

Compound Substance consisting of two or more different atoms in chemical union.

Compound eye Complex eye found in many arthropods consisting of many separate visual elements (ommatidia).

Cone Receptor cell in the retina for responding to bright light and color.

Conjugation In protozoans the temporary union of two organisms for the exchange of micronuclei.

Constitutive heterochromatin Permanent or long-lasting heterochromatin.

Consumers Those organisms in an ecosystem which feed upon other organisms; often divided into primary consumers, secondary consumers, and so on.

Continental drift Movement of the different continental masses throughout geologic history; for example, the separation and movement apart of South America and Africa.

Control Part of an experiment which is a reference or standard for the experimental part.

Conus arteriosus Anteriormost part of the embryonic heart in all vertebrates that persists in the adult to some degee in all but mammals.

Convergence Superficial similarities between unrelated organisms that are adapted to live in similar environments.

Copulation Act of introduction of sperm cells into the body of a female by a male animal.

Corepressor Substance which, when present, inhibits the formation of a specific enzyme.

Cornea Thin transparent layer of the eyeball through which light enters.

Corpora allata Endocrine glands behind the brain in an insect's head which secrete the juvenile hormone.

Corpus luteum Structure in the ovary that produces the hormone progesterone.

Cortex Relating to the external part; for example, the outer part of an organ such as the adrenal cortex, the cerebral cortex.

Countercurrent exchange Process which, for example, forms the basis for the establishment of the osmotic gradient in the medulla of the kidney in the formation of urine.

Cowper's gland Accessory male sex gland that delivers a fluid vehicle to discharging sperm.

Coxal gland Excretory structure found in the basal segment of the leg in some arthropods.

Crenation Shrinking of red blood cells in hypertonic solution.

Cretinism Condition resulting from a very low output of thyroxin in a newborn child; growth is stunted and mental development is greatly retarded.

Cristae Folds of the inner membrane of a mitochondrion.

Crop Pouchlike enlargement of the esophagus in birds; also the thin-walled portion of earthworm gut anterior to the gizzard.

Crossing over Exchange of corresponding segments of homologous chromatids during synapsis.

Cryptorchidism Condition whereby the testes do not descend into the scrotum; causes sterility.

Ctene Paddlelike locomotor structure of ctenophores, formed by fusion of many cilia.

Cydippid Larval stage of a ctenophore.

Cyphonautes Larval form of the phylum Ectoprocta.

Cyst Inactive stage of certain organisms when enclosed in a resistant wall or sac.

Cytochrome oxidase Last enzyme in the electron transport system; it transfers electrons to oxygen.

Cytochrome system Group of iron-containing enzymes that transmit hydrogen to free oxygen.

Cytokinesis Changes taking place in the cytoplasm of a cell during cell division.

Cytology Study of cells.

Cytoplasm Material of the cell exclusive of the nucleus.

Dalton Unit of atomic and molecular weight. The hydrogen atom has an atomic weight of 1 dalton.

Deamination Removal of an amino (NH_2) group from a compound.

Decay General breakdown of organic materials involving microorganisms.

Decibel Unit used in measuring loudness of sound.

Decomposers Organisms, such as bacteria and fungi, in an ecosystem that convert dead organisms into plant nutrients.

Deletion Loss of a chromosome part.

Dendrites Processes of a neuron that carry impulses to the cell body.

Denitrification Formation of molecular nitrogen by denitrifying bacteria which use nitrate as a final hydrogen acceptor.

Density-dependent factors Factors which vary in the intensity of their action with the size of the population.

Density-gradient centrifugation Centrifugation in a gradient of viscous solution such as sucrose or dense salt such as cesium chloride. Particles are separated on the basis of their molecular weight.

Density-independent factors Factors whose effect on population size is independent of the population density.

Dermis Inner layer of the skin, under the epidermis.

Desmosomes Modification of plasma membranes that binds cells together in some tissues.

Determinate cleavage The fate of the blastomeres is determined very early in development.

Deuterostomia Coelomate animals in which the blastopore becomes the anus and a mouth opening forms at the opposite end.

Diabetes mellitus Disease caused by inadequate supply of insulin; results in elevated blood sugar.

Diaphragm Muscular partition separating the thoracic cavities from the abdominal cavity in mammals.

Diastole Relaxation of the heart chambers.

Diencephalon One of the regions in the brain during the development of the nervous system; the site of the future thalamus.

Differentiation Specialization of cells and tissues for particular functions during development.

Diffusion Movement of particles (ions and molecules) as a consequence of their kinetic energy, the movement tending toward equal distribution.

Digestion Hydrolysis of complex insoluble food substances into soluble substances that may be absorbed.

Dihybrid Cross involving two pairs of contrasting characteristics.

Dioecious Having the male and female sex organs in separate individuals.

Dipeptide Two amino acids linked by a peptide bond.

Dipleurula Hypothetical ancestor of the deuterostomia.

Diploblastic organisms Having only two embryonic germ layers.

Diploid Possessing chromosomes in homologous pairs.

Disaccharide Formed by combination of two simple sugars.

Diurnal Active in the day time.

DNA or deoxyribonucleic acid Type of nucleic acid found mostly in chromosomes; carries the hereditary codescript.

DNA polymerase Enzyme which catalyzes DNA replication. DNA polymerases I, II, and III are known; DNA polymerase III is probably the true synthetic enzyme.

Doliolaria Barrel-shaped larva of crinoids and holothuroids.

Dominant Gene whose phenotypic effect masks the effect of its recessive allele.

Dorsal Pertaining to the upper side, as opposed to ventral.

Dorsal root ganglion Aggregation of sensory neuron cell bodies located dorsal and lateral to the central nervous system. The ganglia are associated with each spinal nerve.

Dosage compensation Theory that one X chromosome in each cell of a female (XX) is selectively inactivated.

Double bond Linkage between two atoms that is represented by two lines, for example, $CH_2{=}CH_2$.

Down Feathers associated with newly hatched birds.

Drosophila melanogaster Common fruit fly or vinegar fly. An important organism for genetic, developmental, and biochemical studies.

Dyad One half of each tetrad.

Ecdysone Insect hormone which stimulates molting between instars.

Echinoderm Member of phylum Echinodermata; deuterostomatous enterocoels with an adult body plan which is superficially radially symmetrical on a plan of five parts.

Ecological niche Role of a plant or animal in the community in which it lives.

Ecology Study of the relations of organisms to their environment, both animate and inanimate.

Ecosystem Community and its abiotic environment; the functional unit of ecology.

Ecotone Transition zone between two ecosystems.

Ectoderm Outer layer of cells in the gastrula.

Ectoplasm Outer layer of cytoplasm in a cell.

Ectotherm Vertebrate whose body temperature is obtained from external sources such as the sun.

Edema A condition, also called dropsy, resulting from the accumulation of lymph in the tissue spaces, causing swelling.

Effector Muscle or gland.

Efferent Carrying away from, as a nerve fiber that carries impulses from the central nervous system to an effector.

Electrolytes Substances, like table salt (NaCl), which dissociate into ions. Most of the inorganic compounds in cells are electrolytes.

Electromagnetic radiation Any radiation in the electromagnetic spectrum from gamma rays to long radiowaves.

Electron transport system *See* Hydrogen transport system.

Electroplax Catalyzed electric organs, derived from modified muscles and found in elasmobranch and bony fishes.

Element Substance containing only one kind of atom.

Elytra Thick anterior wings of beetles which cover the membranous posterior pair of wings.

Embolus Clot that breaks loose from its origin and lodges elsewhere, as in the brain.

Embryo Organism in an early stage of development.

Embryology That branch of biology which deals with the developmental stages of organisms.

Embryonic induction Process whereby one embryonic tissue stimulates adjacent tissues to develop in a way they would not develop independently.

Endemic Occurring in a particular geographic region only.

Endergonic A reaction that requires energy.

Endocrine Pertaining to the glands that produce hormones, ductless glands.

Endoderm Innermost layer of the gastrula destined to form the gut and associated structures.

Endometrium Lining of the uterus.

Endoplasm Inner region of the cytoplasm, surrounded by the ectoplasm.

Endoplasmic reticulum System of vesicles or canaliculi in the cytoplasm of cells as revealed by the electron microscope.

Endothelium Single layer of thin flat cells forming the lining of blood vessels and lymphatics in vertebrates.

Endotherm Vertebrate whose body temperature is obtained primarily from its own metabolism.

Energy Capacity to do work.

Energy level One of a series of states in which matter may exist. Each energy level is separated from the others by finite energy differences.

Energy of activation Energy barrier to a reaction, the amount of energy which must be possessed by molecules before they can react.

Engram Definite and permanent trace left by a stimulus in the nervous system, a unit of memory.

Enterocoel Type of coelom formed by the outpouching of the primitive gut.

Entomology Branch of biology that studies insects.

Entropy Form of energy which cannot be used to perform work. A measure of randomness of a system.

Environment Surroundings, biological and physical, of an organism.

Environmental resistance Sum total of external influences which act to limit population growth.

Enzyme Catalyst produced by a living organism, protein or partly protein in nature.

Ephyra Free-swimming larva of a scyphozoan coelenterate.

Epiblast Surface of the area pellucida in a developing chick.

Epiboly Gastrulation by the overgrowth of animal pole cells over those of the vegetal pole.

Epidermis Outer layer of the skin.

Epididymis Structure which passes sperm to the vas deferens; it is derived from the embryonic mesonephros.

Epithelium General type of tissue that covers a body or structure or lines a cavity.

Equilibrium Condition of forward and reverse reactions occurring at the same rate.

Erythroblastosis fetalis Severe anemia in infants which is due to Rh incompatibility between mother and fetus.

Estivation State of inactivity induced by the heat and dryness of summer.

Estradiol One of the steroid hormones produced by the ovarian follicle.

Estrogens Hormones produced by the ovary that initiate female secondary sex characteristics and act on the uterine mucosa.

Estrus Mating period in female mammals (other than primates), characterized by intensified sexual urge.

Euchromatin Lightly staining diffuse or "relaxed" chromatin.

Eugenics Science of improving the human race through the application of genetic information.

Eukaryote Cells with membrane-organized organelles. All cells and organisms except bacteria and blue-green algae. All animal cells are eukaryotic.

Euphenics Experimental modification of human development by physiological and embryological methods.

Euploid Variation of the number of haploid sets of chromosomes in an organism.

Eustachian tube Canal connecting the middle ear with the pharyngeal region.

Eutrophication Aging of lakes and other bodies of water through the increase in concentration of nitrogen, phosphorus, and other plant nutrients. The pollution of waters with sewage and fertilizer hastens the process and causes great changes in plant and animal life.

Evolution Process in which the different kinds of living organisms have developed from a few simple forms of life, or from a single form. Descent with modification.

Excretion Elimination of the waste products of metabolism.

Exergonic Reaction that liberates energy.

Exocrine Glands that secrete into ducts or spaces, as contrasted with endocrine; the salivary glands are examples.

Exophthalmia Condition of protruding eyeballs associated with hyperthyroidism.

Exponential growth rate Doubling of the population with each generation time.

Exteroceptors Those receptors that receive stimuli from outside the body.

Extraembryonic membranes Four membranes, chorion, amnion, yolk sac, and allantois, which lie external to the embryo in development and mostly are lost at hatching or birth.

F_1 First filial generation.

F_2 Second filial generation.

Facilitation Passage of a series of impulses over a given synapse enabling subsequent ones to pass more readily.

Facultative anaerobes Organisms that can grow either in the presence or in the absence of oxygen.

Facultative heterochromatin Heterochromatin which changes to euchromatin (sometimes inactive and sometimes active).

Feces Waste material discharged from the alimentary tract.

Feedback Return of information derived in a chain of events to or near the beginning of the chain and where the information acts as a positive or negative regulator of the chain of events.

Fermentation Decomposition of carbohydrates or derivatives by living microorganisms under anaerobic conditions.

Fertilization Union of the egg nucleus and sperm nucleus to form a zygote.
Fertilizin Sperm-attracting substance in the outer jelly coat of sea urchin eggs.
Fetus Young mammal during the later stages of development in the uterus.
Fibrinogen Soluble blood protein which is converted to fibrin in the formation of a clot.
Filopodia Long thin pseudopods which may branch, but do not anastomose.
Fission Asexual reproduction by division into two or more parts.
Flagellum Long whiplike protoplasmic process, longer than a cilium, used in locomotion.
Flexible-site hypothesis Theory that an enzyme moves during its activity.
Follicle-stimulating hormone (FSH) Hormone of the anterior lobe of the pituitary that stimulates ovarian follicle development.
Food chain Chain of organisms in a natural community with each link in the chain feeding on the one below it and being eaten by the one above.
Forcing functions Equations that relate to important inputs into an ecological model which are not affected by the system.
Fossil Remains of an organism living at some former time or evidence of its presence; for example, mold, cast, petrifaction, and so forth.
Fovea centralis Small spot on the retina containing only cone cells; the principal point of focus of the light.
Functional group Configuration of atoms which react with other such configurations or functional groups.

G_1 (Gap 1) Stage Period in the interphase before DNA synthesis and chromosome replication begins. This stage varies considerably in different kinds of cells.
G_2 (Gap 2) Stage Period in the interphase succeeding the synthesis stage and before prophase of mitosis.
Gamete Mature germ cell.
Gametogenesis Production of sex cells, eggs, and sperm.
Ganglion Group of nerve cell bodies outside the central nervous system.
Gastrovascular cavity Body cavity used for both digestion and circulation and having only a single opening to the outside.
Gastrula Stage of development in which cell movements have resulted in the formation of a cavity, an archenteron.
Gastrulation Process during embryogenesis whereby surface materials move to the interior; the process also identifies the future anterior or posterior end of the mature organism.
Gel Semisolid phase of a colloidal system in which water is the discontinuous phase.
Gemmule Overwintering body of sponges (mostly freshwater).
Gene Unit of heredity, carried in a chromosome, capable of self-reproduction and mutation.
Genetic code Mechanism whereby information written in DNA is read and proteins are synthesized in accordance with that information.
Genetic drift Chance fixation or fluctuation of a gene or a group of genes in a small population.
Genetics Study of inheritance in living things.
Genome Total amount of genetic information present in a cell or organism.
Genotype Genetic constitution of an organism.
Genus Taxonomic subdivision between species and family, a group of closely related species.
Germ layer Embryonic layer of tissue; ectoderm, mesoderm, or endoderm.
Germinal epithelium Epithelial area of ovary and testis in which primordial germ cells reside.
Gestation Period of development of a fertilized egg in the uterus.
Glochidium Small bivalve larva of freshwater clams which is parasitic on fish.
Glucagon Polypeptide hormone of the pancreas that acts in the liver to elevate the blood glucose.
Glucose $C_6H_{12}O_6$, the most common simple sugar; also called dextrose.
Glycogen Complex polysaccharide, often called animal starch.
Glycolysis Fermentation; the anaerobic breakdown of foodstuffs; the steps in sugar breakdown up to the formation of pyruvic acid.
Glycoprotein Protein combined with a carbohydrate.
Glycosidic bond Common type of bond between two sugars in disaccharides, oligosaccharides, and polysaccharides.
Goiter Enlargement of the thyroid gland.
Golgi complex System of membrane-bounded cavities in the cytoplasm of cells, usually of the smooth type; also called dictyosomes; involved in secretory activity.
Gonad Reproductive organ, ovary or testis.
Gondwana Pangaea first broke up into two land masses—Gondwana and Laurasia. Gondwana contained the future continents of South America, Africa, Antarctica, Australia, and the region now known as India.
Granular reticulum Pairs of parallel membranes enclosing narrow cavities of varying shapes with rough outer surfaces because of the presence of ribosomes or small dense bodies.
Gray crescent Slightly lighter area in the fertilized amphibian egg which lies approximately opposite the entrance point of the sperm and which is caused by a shift of the surface pigments and cytoplasm.
Gray matter Parts of the central nervous system so named because of tone or color, but usually areas of large numbers of nerve cell bodies and little or no myelination of the neurons. Examples are the center of the spinal cord, the cerebral cortex, and the cerebellar cortex.
Green revolution Term applied to agricultural advances due primarily to improved genetic strains of rice, wheat, and corn.
Growth Increase in the mass of living material.
Growth rate Increase in the number of individuals per unit time.
Guanine Purine found in nucleic acids.
Gynandromorph Sexual mosaic having cells which differ in their sex chromosomes.

Habit Acquired or conditioned reflex response.
Habitat Location where an organism is commonly found.
Habituation Process in which an organism becomes accustomed to and ceases to respond to a stimulus.
Haploid Having only a single set of chromosomes, as in gametes.
Hardy–Weinberg law In the absence of forces that change gene frequencies, the relative frequencies of alleles in a population tend to remain constant from generation to generation.
Helix Spiral; found in DNA and many proteins.
Hemipenes Penislike double structure employed in copulation by some male reptiles such as snakes.
Hemoglobin Iron-containing protein, capable of carrying oxygen, which gives the blood its red color.

Hemophilia Inherited abnormality in humans in which there is a delay in the clotting of the blood.

Hermaphroditism Condition in which one individual possesses both male and female reproductive organs; monoecious.

Heterochromatin Densely staining or condensed chromatin.

Heterodonts Vertebrates whose teeth vary in shape and function.

Heterosis Hybrid vigor.

Heterotrophic Refers to organisms that cannot manufacture their own food and must rely on some outside source.

Heterozygous Having two different alleles at similar loci on homologous chromosomes.

Hexose sugars Six carbon sugars such as glucose, galactose, and fructose.

Histology Microscopic study of tissues.

Histone Any of a class of basic proteins. Often found in chromosomes and may be involved in genetic regulation.

Holoblastic Cleavage of eggs whereby the cleavage furrows pass entirely through the eggs.

Holozoic Type of nutrition found in most animals which involves the ingestion and digestion of solid food.

Homeostasis Maintenance of the constancy of the internal environment.

Home range That area regularly traversed by an animal in search of food and mates, and caring for young.

Homing Trait of many vertebrates of returning to a site of origin.

Homoiothermal Having constant self-regulated body temperature; "warm-blooded," as in birds and mammals.

Homologous chromosomes Chromosomes that bear genes which control the same characteristics.

Homology Fundamental similarity of structures due to genetic relationship.

Homozygous Having two alike alleles at similar loci on homologous chromosomes.

Hormone Chemical substance produced in one part of an organism and transported in the blood to another part of the organism where it produces a specific effect.

Humus Organic material in the soil; derived from decaying plant and animal material.

Hybrid Organism resulting from the crossing of two species; also used for a heterozygote.

Hydrogen bond Weak interaction between hydrogen that is covalently bonded to an atom and which interacts with nonbonded electrons on other atoms. Important in the three-dimensional structure of proteins and nucleic acids.

Hydrogen transport system Structurally and functionally organized system which generates ATP in cells.

Hydrolase Enzyme that adds water to a molecule and splits it in two.

Hydrolysis Decomposition of a compound in which water is taken up.

Hypertonic Solution surrounding a cell the osmotic concentration of which is greater than that of the cell fluids.

Hypothalamus Lower part of the diencephalon; a center for numerous functions such as sleep, blood pressure, body temperature; secretes some hormones.

Hypotonic Solution surrounding a cell whose osmotic concentration is less than that of the cell fluids.

Identical twins Twins arising from the same zygote and thus having the same genetic constitution.

Imago Adult insect.

Immune response Reaction of the immune system to foreign material.

Immunity Ability to react against disease-producing organisms and certain other foreign substances.

Implantation Entrance of the developing ovum into the wall of the uterus in placental animals.

Impulse Stimulus conveyed by the nervous system and associated with changes in permeability on the surfaces of neurons.

Incus Middle ear bone between the malleus and stapes.

Independent assortment Independent distribution of genes of different allelic pairs located in different pairs of homologous chromosomes, during meiosis.

Inducer Substance which when present causes the formation of a specific enzyme.

Inducible enzyme Produced by a cell or organism in response to an inducer molecule; for example, in response to a change in the chemical environment.

Informational macromolecule Large polymeric molecule which contains bioinformation in the sequence of its basic building block biomolecules; proteins and nucleic acids.

Ingestion Process of taking solid food into the body.

Inguinal canal Connection between the body cavity and the cavity of the scrotum; ducts from the testes and blood vessels pass through it.

Initiator codon Code word which begins translation; recognized by a tRNA bearing *N*-formylmethionine: AUG.

Inner cell mass Suspended mass of cells in the blastocyst of mammals and which will be the site of development of the embryo.

Inner ear Complex structure that receives impulses associated with hearing and equilibrium.

Insight Ability to perceive relations between simple situations.

Instar Stage between molts in an insect.

Instinct Complex unconditioned reflex behavior.

Insulin Protein hormone produced in the pancreas; stimulates the conversion of blood sugar into glycogen.

Integument Refers to the skin and its derivatives.

Interoceptors Those sensory receptors that are affected by changes that take place within the body.

Interphase Stage between cell divisions.

Interstitial cells Cells lying between germ cells in the testes and which produce male hormones.

Intrauterine device One of a variety of materials which, when inserted into the uterus, inhibits implantation of the developing blastocyst; a contraceptive device.

Intussusception Growth from within; biological growth.

In utero Refers to developmental conditions in the uterus, usually in relationship to the development of an embryo or fetus.

Invertebrate Animal without a vertebral column.

Ion Electrically charged atom or molecule.

Iris Pigmented part of the eye containing involuntary (smooth) muscles that control the diameter of the pupil.

Irritability Response of living organisms to stimuli or changes in the environment.

Islets of Langerhans Clusters of cells that produce the hormones of the pancreas.

Isomer Compound with the same empirical formula but a different structure as another compound.

Isosteric inhibition Inhibition of an enzyme by a molecule that binds to the active site, preventing the substrate molecule from being bound, but it is not changed by the enzyme.

Isotonic Solution surrounding a cell whose osmotic concentration is the same as that of the cell fluids.

Isotope Chemical element having the same properties as another element, but differing slightly in atomic weight.

Isozymes Enzymes that catalyze specific reactions but exist in several molecular forms.

Juvenile hormone Hormone produced in insects by the corpora allata. It stimulates larval or nymphal growth and inhibits metamorphosis into the pupa or development of adult characteristics.

Karyotype Photographic record of the chromosome morphology of an organism made by arranging the chromosomes according to their morphologies.

Ketone Compound containing $\overset{O}{\overset{\|}{-C-}}$ attached to two carbon atoms, for example, $CH_3-\overset{O}{\overset{\|}{C}}-CH_3$

Kinetic energy Energy of motion.

Kinetochore Specific point or region of a chromosome where a spindle fiber becomes attached; sometimes called a centromere.

Kinety system Coordinating system of fibrils and granules found in ciliated protozoans.

Krebs cycle Also called citric acid cycle; a part of the aerobic phase of cellular respiration in which citric acid is continuously reformed.

Labium Lower lip of insect.

Labrum Upper lip of insect.

Lactate dehydrogenase (LDH) Enzyme that exists in several forms as isozymes; acts on lactic acid.

Lacuna Cavity in cartilage or bone containing living cells.

Larva Immature stage of an animal that must undergo metamorphosis before becoming an adult.

Lateral plate mesoderm Embryonic mesoderm lying lateral to somites and in which the coelom will develop in the amphibian.

Laurasia Pangaea first broke up into two masses—Laurasia and Gondwana. Laurasia contained the future continents of North America and Eurasia (minus India).

Learning Processes that produce adaptive change in individual behavior as a result of experience.

Lens Crystalline structure that lies behind the pupil in the eye and whose curvature can be changed to focus light on the retina.

Leukocyte White blood cell.

Linkage In heredity the tendency for genes located in the same chromosome to stay together.

Lipids One of the most common kinds of organic compounds found in cells. They do not dissolve in water, but in solvents like ether, chloroform, or alcohol. As a group, they function in living organisms to form structural components, fuels, and storage materials.

Locus Specific place; a point on a chromosome.

Lophophore Tentaculated extension of the anterior portion of the body which embraces the mouth and not the anus and contains an extension of the coelom.

Lumen Internal cavity of a structure.

Luteinizing hormone (LH) Hormone of the anterior lobe of the pituitary that stimulates the conversion of the ovarian follicle to corpus luteum.

Lymph Fluid derived from the blood plasma which bathes the cells of the body.

Lymphocyte Type of nongranular nonphagocytic white blood cell.

Lysins Those antibodies that lead to disruption (lysis) of cells.

Lysosomes Structures in cells that are bounded by a membrane but have no inner membrane and contain large quantities of hydrolyzing enzymes.

Macronucleus Large type of nucleus found in ciliate protozoans associated with metabolic functions.

Macrophages Tissue or blood phagocytes that not only digest debris in their cytoplasm but also participate in immune reactions.

Madreporite Sievelike external opening of the echinoderm water-vascular system.

Malleus Middle ear bone in contact with the tympanic membrane.

Malpighian body Structure in the vertebrate kidney composed of a glomerulus and Bowman's capsule.

Mandible Jaw.

Mantle Extension of the body wall in mollusks and brachiopods which secretes the shell and is important in respiration.

Mass Fundamental property of an atom, molecule, or body which is usually considered to be a measure of its matter, although mass varies with the velocity of a body.

Mastax Muscular jaw-containing pharynx of rotifers.

Maturation divisions Two meiotic divisions in which the chromosome number is reduced from the diploid to the haploid.

Maxillae Pair of head appendages on many arthropods; upper jawbones in vertebrates.

Medulla Posteriormost region of the vertebrate brain; also the inner part of an organ as opposed to the outer part.

Medusa Jellyfish.

Megalopolis Name given to an area where cities have grown together; for example, Boswash—the area extending from Boston south through Washington.

Meiosis Two cell divisions found in the formation of gametes in animals and of spores in plants, in which the chromosome number is reduced from the diploid to the haploid.

Meiotic drive Probability that nondisjuncted chromosomes will remain in the cytoplasm rather than go into a polar body during meiosis.

Meroblastic Cleavage in eggs with large amount of yolk where cell divisions initially are restricted to a localized area of the surface of the egg.

Mesencephalon Middle of the three embryonic first divisions of the brain; the midbrain develops from it.

Mesenchyme Mesoderm whose cells are not in solid sheets or in solid groups, but are relatively loosely associated with each other. Produces such tissues or structures as connecting tissues and cartilage.

Mesentery Thin sheet of tissue that supports the organs in the body cavity.

Mesoderm Middle layer of embryonic cells between ectoderm and endoderm.

Mesoglea Jellylike noncellular layer between epidermis and gastrodermis of coelenterates.

Mesonephric ducts Ducts delivering products from the mesonephric kidneys.

Mesorchium Mesentery supporting the testis.

Mesovarium Mesentery that supports the ovaries.

Messenger RNA (mRNA) A ribonucleic acid which specifies the synthesis of proteins by virtue of the "code" contained in it.

Metabolism Sum total of the chemical reactions that occur in living cells and organisms.

Metachronous Coordinated beat of successive cilia resulting in waves of movement passing down the organism.

Metamorphosis Transformation of a larval form into an adult, such as a tadpole to a frog or a caterpillar to a moth.

Metaphase Second phase of mitosis and meiosis where chromosomes align at the equator of a spindle.

Metencephalon Anterior subdivision of the rhombencephalon from which the cerebellum will develop.

Micrometer Unit of microscopic measurement; the one thousandth part of a millimeter.

Micronucleus Small type of nucleus found in ciliate protozoa associated with reproductive functions.

Microvilli Minute cylindrical processes which line the intestine and kidney tubules.

Middle ear The part of the ear containing the middle ear bones for transmitting vibrations from the tympanic membrane to the inner ear. It is connected to the mouth by the eustachian tube.

Migration Periodic departure from and return to its habitat by an animal; often seasonal in nature.

Miracidium Ciliated larval stage which develops in the trematode egg and enters a snail where it forms a sporocyst.

Mitochondria Small granular or rod-shaped bodies found in the cytoplasm and involved in cellular respiration.

Mitosis Ordinary cell division. In a restricted sense, only the nuclear division in which the daughter chromosomes move to opposite poles of the spindle.

Model Ecologically, a mathematical construct which is an abstract representation of a real-world ecosystem.

Mole Gram molecule, the molecular weight of a substance expressed in grams.

Molecular genetics Study of the molecular nature of the genetic material.

Molt To shed the outer covering.

Monoecious Having male and female gonads in the same individual; hermaphroditic.

Monohybrid Genetic cross involving one pair of contrasting characteristics.

Monosaccharide Simple sugar.

Morgan Unit of recombination or crossing over, equal to one percent crossing over.

Morphology Study of form and structure.

Morula Tight cluster of cells found in early stages of development.

Mosaic Organism that is made up of patches of genetically diverse cells.

Motor nerve Nerve that conducts impulses away from the central nervous system.

Mucosa Mucous membrane lining the alimentary tract.

Mutation Change in the gene that breeds true; also addition of extra chromosomes or parts.

Muton Smallest unit of genetic material that undergoes mutation.

Mutualism Organisms of two different species living in close association whereby both benefit by the relationship.

Myelencephalon Posterior subdivision of the rhombencephalon from which the medulla will develop.

Myelin sheath Fatty ensheathing of some nerve fibers; assists the rate of conduction of nerve impulses.

Myoblast Embryonic muscle cell that develops into a myotube by fusion with other myoblasts.

Myoglobin Oxygen-binding protein found in muscle.

Myosin Protein found in muscle cells and involved in contraction.

Myxedema Condition resulting from a decrease in the secretion of thyroxin; associated with a basal metabolic rate much lower than normal.

Nanometer 10^{-9} meters.

Nares External and internal openings of the air passages in vertebrates.

Natural selection Differential survival of certain genotypes in a population through successive generations.

Nekton Aquatic organisms which can actively swim and move against currents, if necessary.

Nematocyst Stinging structure of coelenterates composed of a flask and thread.

Neotenic Capacity to become sexually mature in the larval state, as with some amphibians.

Nephridium Type of excretory organ found in many invertebrates.

Nephritis Most common type of kidney disease caused by a bacterial infection of the capillaries of the glomeruli.

Nephron Unit of structure of vertebrate kidneys consisting of a renal corpuscle and uriniferous tubule.

Neritic Region of the ocean from the shore to the edge of the continental shelf.

Nerve Bundle or cable of nerve fibers.

Nerve fiber Filamentous process of a neuron or nerve cell.

Nerve impulse Physicochemical excitation that passes along a nerve fiber.

Nerve net Continuous network of nerve cells throughout the body without synapses.

Neural crest Thin strip of cells (source of sensory neurons) formed by the closure of the neural folds.

Neural fold Rim of the neural plate formed by the development of the edges of the plate; the rim tapers off at the blastopore.

Neural plate Thickened layer of overlying ectoderm that has formed into columnar cells.

Neural tube In embryogenesis, the tubular structure formed by the closure of the neural folds that will form the central nervous system.

Neurilemma Outer sheath that surrounds the axons of the motor neurons and dendrites and the axons of sensory nerves that make up the peripheral nerves.

Neuromuscular junction Point of close contact between neurons and muscle fibers where the impulse is transmitted from neuron to muscle.

Neuron Nerve cell.

Neurosecretion Substance, often a hormone, secreted by cells of the nervous system.

Neurulation Formation of the nervous system in animals.

Nicotinamide adenine dinucleotide Oxidation-reduction cofactor found extensively in cells.

Node of Ranvier Position along a neuron where one Schwann cell contacts an adjacent Schwann cell.

Nondisjunction Failure of daughter chromosomes to separate in meiosis.

Nonelectrolyte Substance which does not dissociate into ions.

Norepinephrine Hormone secreted by the adrenal medulla which in general causes effects similar to sympathetic stimulation; also secreted at ends of sympathetic nerve fibers.

Notochord Rod-shaped structure that forms a skeletal axis in chordates. Although present in the embryonic stages of all chordates, it is replaced by the vertebrae in the vertebrates.

Nuclear envelope Double membrane that surrounds the nucleus of a cell.

Nuclear sap More fluid material contained within the membrane of a cell nucleus.

Nucleic acid Organic acid composed of joined nucleotide complexes, the principal types are deoxyribosenucleic acid (DNA) and ribosenucleic acid (RNA).

Nucleolar organizer Position on a chromosome controlling the formation of a nucleolus.

Nucleolus Spherical body found within the nucleus of most cells, containing RNA.

Nucleotide Molecule composed of joined phosphate, pentose sugar, and a purine or pyrimidine base.

Nucleus Central part as in nucleus of a cell or atom.

Nymph One of a series of immature stages in certain insects.

Ocellus Simple eye of many types of invertebrates.

Olfactory fatigue Phenomenon by which one becomes unaware of a particular odor shortly after noticing it (that is, walk into a room where an odor is obvious at first, then it seems to disappear).

Olfactory nerve Cranial nerve that transmits impulses for smell to the brain.

Ommatidium One unit of the arthropod compound eye.

Omnivore Animals that eat both plant and animal food.

Ontogeny Developmental history of an organism.

Oogenesis Process in the ovary whereby oogonia divide in meiosis and differentiate into eggs.

Operator gene Region of DNA at which transcription is initiated. Controls structural genes of its operon.

Operon Functional unit of transcription. The operator gene and the structural genes under the control of the operator gene.

Opisthonephros Type of kidney characteristic of fish, amphibians, and reptiles. Its internal organization is similar to an embryonic mesonephros.

Opsonins Antibodies in contact with some bacteria that make the latter especially susceptible to phagocytosis.

Order One of the taxonomic categories; two or more families placed together form an order, and two or more orders form a class.

Organ Structurally integrated group of tissues which performs a specific function, as the heart, liver, and so forth.

Organelle Organlike structures in single-celled animals.

Organic Refers to compounds of carbon or to organisms.

Organic evolution *See* Evolution.

Organogenesis Coming into being and differentiation of organs.

Osmosis Diffusion of water through a semipermeable membrane.

Ostium Literally mouth; anterior opening into an oviduct; openings into the heart in animals with open circulatory systems; and in the body wall of sponges.

Otoliths Calcareous structures lying on specialized patches (maculae) in the sacculus and utriculus. Shifts in their positions stimulate neurons to send impulses to the brain and are interpreted for equilibrium.

Oval window Window against whose membrane the stapes relays vibrations to the inner ear.

Ovary Female reproductive organ in animals that produces eggs.

Oviducts Tubes through which eggs pass in going to the uterus, or to the outside.

Ovoviviparous Embryonic development to various stages of the egg in the oviduct, but without the formation of a placenta.

Ovulation Discharge of an egg from the ovary.

Oxidation Combination with oxygen or a nonmetal, removal of hydrogen, removal of electrons.

Oxytocin Hormone produced in the hypothalamus and stored in the posterior lobe of the pituitary; stimulates the muscles of the uterus.

Palp Feelerlike appendage.

Pangea Name given to the single land mass that existed in Permean times.

Parameters Mathematical constants used in ecological models.

Parapodia Flat lateral appendages in polychaete worms used in locomotion and respiration.

Parasite Organism that lives on or in another and at its expense.

Parasympathetic Subdivision of the autonomic division of the nervous system whose nerve fibers emerge from the cranial and sacral parts of the central nervous system. Its functions are antagonistic to those of the sympathetic subdivision.

Parthenogenesis Development of an egg without fertilization.

Pectines Comblike sensory structures on the ventral surface of a scorpion's abdomen.

Pedigree Genealogical record, usually in chart form.

Pelage Covering of a mammal, referring to its hair, fur, or wool.

Pelagic Oceanic habitat zone, the open waters of the ocean.

Pellicle Thin membrane found outside the plasma membrane in many animal cells.

Penicillin Antibiotic produced by *Penicillium* fungi; acts on growing bacterial cells by interfering with cell-wall synthesis.

Pentadactyl Having all four limbs normally terminating in five digits.

Peptide Compound containing two or more amino acids in which the carboxyl group of one is linked to the amino group of the other.

Peptide bond Amide bond between two amino acids in peptides and proteins.

Pericardium Membranous sac that surrounds the heart.

Perinuclear cisterna Cavity between the two membanes which form the nuclear envelope.

Perissodactyl Having an uneven number of toes as in such forms as horses and rhinoceroses.

Peristalsis Wavelike muscular contractions of the wall of a tube, as in the alimentary tract.

Peritoneum Thin membrane that lines the body cavity and forms the external coverings of the organs of the viscera.

Permafrost Permanently frozen deeper soil layers of the tundra.

Phagocytosis Ingestion of solid particles by phagocytes, for example, white blood cells.

Pharyngeal clefts Ingrowths of the embryonic ectoderm which join and sometimes perforate with the pharyngeal pouches.

Pharyngeal pouches Lateral outpouchings of the wall of the embryonic pharynx; may open to the exterior as in fish. Several structures are derived from the pouches, for instance, the thymus, the parathyroids.

Pharynx Region between the mouth cavity and the esophagus.

Phenocopy Organism which exhibits different phenotypes under different environmental conditions.

Phenotype Appearance of an individual in contrast to its genetic constitution.

Phenylketonuria Metabolic disease characterized by faulty phenylalanine metabolism and mental retardation.

Pheromone Substances secreted by one individual that elicit a physiological or behavioral response from an individual of the same species. An insect sex attractant.

Phosphoarginine Compound containing high-energy phosphate found mainly in the muscles of animals in the annelid–arthropod series.

Phosphocreatine Compound containing high-energy phosphate found mainly in the muscles of animals of the echinoderm–chordate series.

Phosphorylation ATP production from ADP and P, or, more generally, addition of phosphate to a molecule.

Photoperiodism Response of organisms to differing day lengths.

Photosynthesis Manufacture of a simple carbohydrate from CO_2 and water in the presence of chlorophyll with light as a source of energy and with electrons released. Oxygen is generally the electron acceptor.

Phylogeny Evolutionary history of a group of organisms.

Physiology Study of function, whether in cells, organs, or entire organisms.

Pilidium Helmet-shaped larval form characteristic of nemertean worms.

Pinna Fleshy external extension of the ear.

Pinocytosis Intake of fluid droplets by a cell.

Pinosomes Vacuoles formed by the plasma membrane, involved in pinocytosis.

Pitch Tone or frequency of sound.

Pituitary gland Gland located below the brain and releasing numerous hormones of a variety of functions.

Placenta Combination of extraembryonic membranes of the embryo and wall of the uterus of the female mammal which is a center for exchanges of nutrients, gases, and wastes, between the mother and the embryo.

Planck's constant Fundamental constant of quantum mechanics; the ratio of the energy to the frequency of the quantum, 6.62×10^{-27} erg•seconds.

Plankton Free-floating microscopic aquatic organisms, both plant and animal.

Planula Ciliated larval form characteristic of coelenterates.

Plasma Fluid fraction of blood in which corpuscles are suspended.

Plasma cell Cell derived from a B-type lymphocyte actively synthesizes immunoglobulins.

Plasmagel Outer region of the endoplasm; stiff and jellylike.

Plasma membrane Very thin membrane that surrounds and differentiates from the cytoplasm of the cell.

Plasmasol Inner part of the endoplasm; fluid in nature.

Plasmolysis Shrinkage of a cell when placed in hypertonic solution due to loss of water.

Platelets Minute granular bodies in mammalian blood involved in blood clotting.

Pleated sheet Secondary structure of certain structural proteins in which individual polypeptide chains are in an extended nonhelical arrangement, and in which the adjacent chains run in a parallel or antiparallel direction.

Plexus Network, usually of nerves or blood vessels.

-ploid Combining form used in cytology and genetics to indicate the number of chromosomes.

Ploidy Phenomenon of changes in the chromosome number.

Poikilothermal Animals without internal temperature controls; "cold-blooded."

Polar bodies Very small nonfunctioning cells that are produced in the maturation of an egg cell.

Polymer Large molecule composed of many similar units, for example, polysaccharides, proteins, and nucleic acids.

Polymorphism Occurrence of more than one body form in the life cycle of an animal.

Polynucleotide Polymer molecule of nucleotides.

Polyp Attached stage of a coelenterate, cylindrical in shape, with mouth and tentacles at the free end.

Polypeptide Many amino acids linked by peptide bonds.

Polyploid Containing one or more extra sets of chromosomes.

Polysaccharides Complex carbohydrates composed of many simple sugars linked together.

Polysome Two or more ribosomes attached to an mRNA.

Polytene Duplication of chromosomes in the absence of cell division. Usually there are many duplications, thereby giving rise to large chromosomes.

Population Group of interacting individuals of the same species.

Porocyte Specialized cell on the surface of some sponges containing an intracellular pore.

Porphyrin Complex ring structure found in heme and chlorophyll molecules.

Position effect Phenomenon of varying phenotype with the different arrangement of the same number of genes on the chromosomes; for example, bar eye in the fruit fly.

Potential energy Energy of position.

Precipitins Antibodies that cause precipitation from the soluble to the insoluble state.

Precocial Birds and mammals who are capable of activity and feeding shortly after being hatched or being born.

Primary structure Sequence of monomer units in a polymer.

Primer Initiator molecule.

Primitive streak Region in early chick embryo where cells move from outside to inside; equivalent to blastopore in other forms.

Primordia Beginnings of embryonic structures, for example, limb primordia.

Prochromosomes "Naked" nucleic acid of viruses and prokaryotes.

Proctodeum Ectodermally lined terminal portion of the gut in many animals.

Producers Organisms, green plants, able to synthesize organic compounds from inorganic substances.

Progesterone Hormone produced by the corpus luteum that acts in association with estradiol and regulates the condition of the uterine mucosa.

Proglottid One of the segments of a tapeworm.

Prokaryote Cells without membrane-organized organelles; bacteria and blue-green algae.

Prolactin Hormone of the anterior lobe of the pituitary and which regulates milk production after pregnancy.

Pronephros One type of vertebrate kidney; it forms in the anteriormost part of the strip of tissue known as nephrogenous tissue, and is the first type of kidney to appear embryologically.

Pronuclei Nucleus of the sperm and egg which unite to form the nucleus of the zygote.

Prophase First phase of mitosis and meiosis when chromosomes become distinct structures.

Prosencephalon Most anterior of the three embryonic first divisions of the brain; the cerebrum and thalamus develop from it.

Prosopyle Incurrent opening into the flagellated chamber of a sponge.

Prostaglandins Group of cyclic and oxygenated fatty acids that have potential as birth control agents; resemble hormones in action, but effects are localized and compounds are not transported over long distances.

Prostate Accessory male sex gland that delivers a fluid vehicle to discharging sperm.

Protandrous Condition in which an animal first shows characteristics of a male and then changes to characteristics of a female.

Protein Complex organic compounds composed of numerous amino acid molecules.

Protogynous Condition in which an animal first shows the characteristics of a female and then change to characteristics of a male.

Protoplasm Name sometimes used for the living substance.

Protoplast All the cell structure which is inside and including the cell or plasma membrane.

Pseudocoelom False body cavity; one that is not a cavity in the mesoderm as a true coelom; found in certain lower invertebrate phyla.

Pseudopodium Temporary projection of a cell such as found in sarcodina or certain white blood cells which functions in locomotion and feeding.

Pupa Superficially quiescent stage in the life cycle of insects, in which reorganization of larval tissues into adult tissues occurs.

Pupil Opening in the iris whose size can be changed to control the amount of light passing to the interior of the eye.

Purine Type of nucleobase found in nucleic acids. The other type is the pyrimidine.

Putrefaction Decomposition of proteins by living microorganisms, usually under anaerobic conditions.

Pygidium Oil gland containing area supported by caudal vertebra in birds.

Pyramid of numbers Numerical relationships between the various organisms involved in a particular food chain; producers are most abundant, primary consumers next, and so forth. Similar concepts are involved in *pyramid of biomass* and *pyramid of energy.*

Pyrimidine Type of nucleobase found in nucleic acids. The other type is the purine.

Pyruvic acid CH_3-CO-COOH, an important metabolite in cellular respiration.

Quadrate bone Upper bone that forms the hinge of the jaw in all vertebrates but mammals; in mammals it becomes the middle ear bone, the incus.

Quaternary structure Two or more subunits in a molecule, for example, hemoglobin.

Radial symmetry Type of symmetry in which the parts are regularly arranged about a central axis and where the organism may be divided into two equal halves by more than one cut through the central axis.

Radical Group of atoms that act as a unit.

Radula Rasping organ in the anterior part of the alimentary tract of all mollusks except the Pelecypoda.

Reading frame Group of three nucleotides in mRNA.

Recapitulation Occurrence during development of stages thought to have occurred in the embryonic development of its ancestors.

Receptor Sense organ.

Recessive Gene whose phenotypic expression is masked by its dominant allele.

Recon Smallest unit of genetic material that can undergo recombination.

Rectum Terminal portion of the digestive tract.

Redia Trematode larval stage, found in the snail host; gives rise to more redia or to cercaria.

Reduction Combination with hydrogen, lowering of oxidation state or valence, gain of electrons.

Redundancy Repetition of DNA sequences linearly; currently applied to repeating sequences of cistrons.

Reflex Automatic response to a stimulus which may or may not elicit a conscious sensation.

Reflex arc Functional unit of the nervous system and consisting of a receptor, sensory, association, and motor neurons, and ending with an effector, either a muscle or gland; produces an involuntary response.

Regeneration Replacement of lost or injured parts.

Regulator gene Produces the repressor substance, a protein, which turns structural genes on and off.

Relaxation period Period following the contraction period during which the muscle returns to its original length.

Relaxin Hormone which makes the pubic symphysis more flexible and relaxes the birth canal to assist delivery of the young.

Renal corpuscle Structure in the vertebrate kidney composed of a glomerulus and Bowman's capsule.

Replication Process of exact duplication. Each original item gives rise to two identical items which are also identical to the original item.

Repressible enzyme Enzyme that is no longer synthesized by a cell when a certain small molecule (corepressor) is added to the culture medium.

Repressor Protein which recognizes operator genes. Association of the repressor with the operator gene inhibits messenger synthesis.

Reproductive isolation Absence of gene exchange between different species due to a variety of mechanisms —morphological, behavioral, physiological, and so forth.

Residual air Air which remains in the lungs after a forced expiration and which cannot be removed.

Respiration Oxidative release of energy from molecules; may occur aerobically or anaerobically.

Rete mirabile Network of blood vessels associated with gas bladders in fish.

Retina Inner layer of the eye which contains light receptors.

Retinene Carotenoid pigment associated with rhodopsin in vision.

Rhabdite Rodlike mucus-producing structure in the ectoderm of many turbellarians.

Rhodopsin Compound of a protein and a pigment that is involved in the reception of dim light by rods.

Rhombencephalon Posteriormost of the three embryonic first divisions of the brain; the cerebellum and medulla develop from it.

Riboflavin Member of the vitamin B complex; vitamin B_2. Found in some flavoproteins.

Ribosomal RNA (rRNA) RNA found in ribosomes. Prokaryotic ribosomes contain 5 S, 16 S, and 23 S rRNA; eukaryotic ribosomes contain 5 S, 7 S, 18 S, and 28 S rRNA.

Ribosomes Small ribonucleoprotein particles free in the cytoplasm or on the surface of the vesicles of the endoplasmic reticulum; the sites of protein synthesis.

RNA or ribonucleic acid Found in nuclei and in cytoplasm; involved in the synthesis of proteins.

Rod Receptor cell in the retina for responding to dim light but not to color.

Round window Membrane-covered opening between the middle ear chamber and the tympanic canal of the cochlea.

Sacculus Part of the inner ear that receives stimuli associated with equilibrium as it relates to orientation in space (static equilibrium).

Sagittal Pertaining to the median vertical longitudinal plane.

Salt Compound produced by the reaction of a base with an acid.

Salt glands Specialized glands associated with the eyes or nasal areas of marine birds and reptiles.

Saprozoic Type of nutrition in which organisms absorb predigested nutrients from decaying organic matter.

Sarcolemma Thin membrane that bounds each fiber of muscle tissue.

Sarcomere Unit, in muscle fibers, extending from one Z membrane to the next.

Scale Horny or bony plates on the body surface, as in fish and reptiles.

Schwann cell Cells ensheathing nerve fibers. The way in which they are wrapped around neurons determines whether the neurons are myelinated or nonmyelinated.

Sclera Outer coat of the human eye. A tough opaque layer of connective tissue that protects the inner structures and helps to maintain the shape of the eyeball.

Scolex Head of a tapeworm.

Scrotum Saclike extension from the body cavity which houses the testes in mammals.

Secretion Material synthesized and discharged by a cell or group of cells.

Segregation Separation of homologous chromosomes in meiosis.

Semen Ejaculate of sperm and fluids from accessory male sex glands.

Semicircular canal One of three thin canals in the inner ear that receives stimuli associated with motion (dynamic equilibrium).

Seminal vesicle Accessory male sex gland that delivers a fluid vehicle to discharging sperm.

Seminiferous tubules In the testis, tubules along whose walls spermatogenesis occurs.

Semipermeable membrane One that can be penetrated by different substances to different degrees.

Sense organ Multicellular receptor including sensory cells and associated structures.

Sensory nerve Nerve that conducts an impulse toward the central nervous system.

Septum Membrane separating two cavities.

Sere Series of communities which follow one another in a definite sequence in ecological succession, ending in a climax that is typical for the particular climate and geographical region.

Serosa Membrane that covers coelomic organs in vertebrates.

Serum Clear fluid portion of the blood which separates on clotting.

Serum albumin Chief protein of human blood plasma.

Serum globulin One type of protein present in blood plasma; may function as antibody.

Sessile Nonmotile and usually attached to the substratum.

Sex chromatin Inactivated X chromosome in cells with two or more X chromosomes; Barr body.

Sex chromosomes Name given to the odd pair of chromosomes, often called X and Y, which are involved in determining the sex of the individual.

Sinus gland Storage-release center for the hormones produced by the X organ in crustaceans. Both the sinus gland and the X organ are located in the eye stalk.

Siphonoglyph Ciliated groove in pharynx of sea anemones which passes materials down into the gastrovascular cavity.

Smog Fog over a city containing industrial smoke, the exhausts of automobiles, and substances produced by photochemical reactions.

Society Integrated organized group of individuals, such as a colony of termites, in which there is considerable specialization and division of labor.

Sol Fluid phase of a colloidal system.

Solenocyte Type of flame cell with very long flagella.

Somatic cell Cells other than germ cells.

Somite One of the segments into which the body of many animals is divided.

Species The subdivision of a genus. The members of a species are capable of freely interbreeding and producing fertile offspring.

Spectrophotometer Device that measures the amount of light absorbed by a solution.

Sperm Male sex cell or gamete.

Spermatid Cells produced by the second meiotic division in the testis which metamorphose into sperm cells.

Spermatogenesis Process in the testis whereby spermatogonia undergo meiosis and differentiate into spermatozoa.

Spermatophore Specialized structure deposited by a male and containing in or on it a package of sperm that will be picked up by the female.

Sphincter Circular muscle which closes a tubular opening upon contraction.

Spicules Small bodies of calcium carbonate or silicon dioxide which form the supporting skeleton of certain sponges and other organisms.

Spinal cord Part of the central nervous system posterior to the brain.

Spindle fiber Microtubules running from the centromeres of chromosomes to the poles of mitotic spindles or from pole to pole.

Spiracle One of the external openings of the respiratory system of many terrestrial arthropods.

Spiral cleavage Cleavage in which the mitotic spindles tip to produce daughter cells oriented in a spiral pattern.

Spongin Fibrous material which makes up the skeleton of some sponges.

Stapes Innermost of the three middle ear bones.

State variables Numerical representations of the condition of an ecological model at any particular time.

Static equilibrium Interpretation of orientation in space and not involving movement.

Stimulus Environmental change which elicits responses in living organisms.

Stock (genetic) Line of organisms that are maintained for genetic purposes.

Stomodeum Ectodermally lined portion of the mouth, and often the pharynx, in many animals.

Strobilization Process of forming a series of transverse constrictions in the life cycle of scyphozoan jellyfishes with each body so formed becoming an immature jellyfish.

Structural gene Gene which specifies a polypeptide by the process of transcription; also called a cistron.

Substrate Molecule upon which an enzyme acts.

Succession Occupation of an area by an orderly sequence of plant and animal communities.

Supplemental air Air which may be forcefully expired after a normal expiration (about 1000 cm^3).

Symbiosis Organisms of two different species living in close association with one another; the relationship may be mutualism, commensalism, or parasitism.

Sympathetic Subdivision of the autonomic division of the nervous system and whose nerve fibers emerge from the thoracic and lumbar parts of the spinal cord. Its functions are usually antagonistic to those of the parasympathetic subdivision.

Symphysis pubis Where the two pubic bones unite.

Synapse Region of nerve impulse transfer between two neurons.

Synapsis Close association of homologous chromosomes in the early stages of meiosis.

Synaptic fatigue Cessation of impulse transfer at a synapse after a constantly repeated stimulus.

Synaptic summation Transmission at a synapse following a series of impulses.

Synaptic vesicles Small vesicles located in the knoblike endings of axon processes, involved in the transmittal of impulses.

Synaptonemal complex Synapsed chromosomes with a medial element as seen with the electron microscope.

Syncytium Multinucleate animal tissue lacking cell boundaries.

Syngamy Reproduction by the union of gametes.

Synthesis Formation of a more complex substance from simpler ones.

Syrinx Voice box of birds located at the junction of the primary bronchi.

Systems ecology Application of systems analysis to ecological problems.

Systole Contractions of the heart chambers.

Taxis Movement of an organism toward or away from a stimulus.

Taxonomy Study of the classification of organisms.

Telencephalon Anterior subdivision of the prosencephalon from which the cerebrum will develop.

Telophase Fourth phase of mitosis and meiosis where the chromosomes reorganize new nuclei at the two poles of a spindle.

Template Specific surface, pattern, or mold.

Tentacles Long flexible processes, usually tactile or prehensile in nature, located about the mouth in many invertebrates.

Termination codon Code word which ends translation, a nonsense codon, is not recognized by a tRNA: UAA, UGA, UAG.

Territory All or part of an animal's home range which is actively defended against intruders.

Tertiary structure Superfolding of the secondary structure of a polymer.

Test cross Mating of an individual with an unknown genotype with one that is homozygous recessive for all loci in question.

Testes Male organs which produce sperm.

Testosterone Male sex hormone produced in the testes.

Tetrad Group of four chromatids in a pair of homologous chromosomes after synapsis.

Tetraploid Twice the usual diploid number of chromosomes.

Thalamus Middle part of the diencephalon, a relay and reflex center.

Thalassemia Inborn genetic defect prevalent in Mediterranean peoples; characterized by lack or reduced amounts of alpha or beta chains of hemoglobin.

Thorax Chest region, or part of the body between head and abdomen.

Threshold Amount of stimulus that must be reached to cause a nerve impulse.

Thrombin Enzyme of the blood derived from its inactive state prothrombin and which participates in blood clotting.

Thrombocytes Small colorless structures that function in the formation of blood clots.

Thrombus Clot in a blood vessel, heart, or lymph vessel.

Thymine Pyrimidine found in nucleic acids.

Tidal air The 500 cm^3 of air that is taken in and out of the lungs with each inspiration and expiration in normal breathing.

Tissue Group or aggregation of similar cells specialized to perform a similar function.

Tolerance Refers to the acceptance of foreign materials such as tissue grafts that ordinarily would be rejected immunologically.

Tornaria Ciliated free-swimming larva of some hemichordates, similar to larva of many echinoderms.

Trace element Mineral required in extremely small amounts.

Trachea Windpipe of vertebrates; the air tubes of certain arthropods.

Transcription Synthesis of RNA on a DNA template.

Transduction Transferral of genetic material from one cell to another by means of a bacterial virus.

Transfer functions Equations which represent interactions between various parts of an ecological model.

Transfer RNA (tRNA) A small ribonucleic acid which carries amino acids to the sites of protein synthesis.

Transformation Release of hereditary material from one bacterial cell into the culture medium where it is taken up by another cell.

Translation Synthesis of protein under the direction of mRNA.

Transmitter substance Substance that crosses between an axon and a dendrite at the synapse and relays the nerve impulse.

Triose Three-carbon compounds formed in cellular respiration and in photosynthesis (commonly used for 3-phosphoglyceraldehyde and dihydroxyacetone phosphate).

Tripeptide Three amino acids linked by peptide bonds.

Triploblastic Having three germ layers.

Triploidy Condition in which all chromosomes of a cell exist in sets of three.

Trisomy Condition in which one or more kinds of chromosomes in cells will have three homologous members, while other kinds of chromosomes in the cell exist in normal pairs.

Tritium Radioactive isotope of hydrogen, ^3H. Useful in labeling proteins and nucleic acids and in autoradiography.

Trochophore Ciliated top-shaped free-swimming larva characteristic of some marine annelids, mollusks, and certain other organisms.

Tropism Involuntary response or movement of an organism, the direction of which is determined by the source of the stimulus.

T-type lymphocyte Lymphocyte that is responsible for cell-mediated immunity.

Tympanic canal Fluid-filled canal in the cochlea that carries vibrations.

Tympanic membrane Ear drum.

Typhlosole Dorsal fold of intestinal wall of the earthworm and certain other invertebrates which projects into the gut cavity and increases the absorptive surface.

Ultrastructure Structure of cells and tissues as revealed by the electron microscope.

Ultraviolet light Light between 174 and 350 nm in wavelength.

Umbilical cord Cord that connects the fetus of mammals and the placenta.

Unit membrane Trilaminar membrane composed of an outer and inner layer of protein molecules with layers of lipid material between. Most of the membranous structures of cells have this basic pattern of construction.

Uremia Kidney disease caused by the gradual accumulation of urea and other wastes in the blood and which produces a toxicity that may lead to death.

Ureter Duct that leads from a kidney to the urinary bladder.

Urethra Duct that leads from the urinary bladder to the exterior.

Urogenital Pertaining to both the excretory and the reproductive systems.

Uterus Expanded terminal part of the oviduct modified for the storage of eggs, or for housing and nourishing the developing embryo.

Utriculus Inner ear structure associated with static equilibrium; it shares its function with the sacculus.

Vacuole Cavity within the cell filled with water and materials in solution.

Vagina Part of the female genital tract that usually receives the intromittant organ.

Vasectomy Removal of a part (usually) of the vas deferens (sperm duct) as a contraceptive practice.

Vasopressin Hormone produced in the hypothalamus and stored in the posterior lobe of the pituitary. It stimulates water resorption in the kidney.

Vegetal pole That pole (opposite the animal pole) in the egg at which food material is concentrated.

Veins Blood vessels conducting blood toward the heart.

Veliger Larval stage in many mollusks following the trochophore stage.

Velum Flat band of tissue extending inward from the edge of the bell of hydrozoan medusae.

Ventral Pertaining to the lower side, as opposed to dorsal.

Ventral horn Ventrolateral part of the gray matter of the spinal cord. Cell bodies of motor neurons are numerous here.

Ventricle A chamber of the heart that pumps blood into arteries; also a cavity in the brain.

Vertebral column Series of bones running in the mid-dorsal line of a vertebrate and surrounding the spinal cord.

Vestibular canal Fluid-filled canal in the cochlea that transmits vibrations from the stapes to the organ of Corti.

Vestigial structure Rudimentary or degenerate structure.

Villi Small fingerlike projections of the lining of the small intestine.

Virus Acellular particle which contains a nucleic acid core and a protein or protein-lipid (or membrane) coat. Cause of certain diseases such as polio and smallpox.

Viscera Organs in the body cavity.

Visceral peritoneum Surface of the coelom that covers the derivatives of the gut such as the duodenum.

Visceral reflexes Reflexes over which we have no conscious control.

Vitamin Substance that, in very small amounts, is necessary for normal metabolism and that cannot be synthesized by a particular organism.

Vitreous humor Jellylike material filling the large, or vitreous, chamber of the eye and through which light passes to the retina.

Viviparous Bringing forth living young.

Wobble hypothesis Hypothesis put forth by F. H. C. Crick; the third letter in an mRNA codon may vary with no change in the specificity of the codon.

Yolk Stored food materials in egg cells.

Yolk sac Extra embryonic membrane covering the yolk, when present, or as a yolkless sac suspended from the gut of the embryo.

Zoogeography Study of the distribution of animals.

Zygote Cell formed by the union of male and female gametes; a fertilized egg.

Index

Index